The Mediterranean Sea

Map of Europe and the Mediterranean from the copy to nineteenth century of Catalan Atlas of 1375

The Catalan Atlas of 1375

The Catalan Atlas is a large scale map, dated 1375, that is made up of six leaves of vellum. The first two leaves reveal various astrological and cosmological data of the period, highlighting the known qualities of the world and illustrated with various charts and diagrams. The last four vellum leaves contain the Atlas itself. It has been argued that the Catalan Atlas was the product of the Majorcan Jewish mapmaker Abraham Cresques (d.1387) and his son Jehuda, who were patronized by the king of Aragon Pedro IV (1336–1387) in the latter half of the fourteenth century. As a royally commissioned piece, the Catalan Atlas is thought to have been a gift from the royal house of Aragon to the newly crowned king of France, Charles VI (1368–1422). The Catalan Atlas illustrates, however, a much more complex – or global – view of the world than just the interaction between a French and a Spanish monarch and their patronage of a Jewish mapmaker. At least three cultural groups generally distinguished by their religion into Islamic, Jewish and Christian peoples often coexisted side by side, creating societies that demonstrated remarkable integration, especially in their cultural output. The Catalan Atlas created in 1375, at the end of the medieval period in Europe, can be seen as a product of this fluorescence of interaction. This multi-cultural interaction, facilitated by the communication between these social groups, led to a different, more global way of viewing the world. It is in a cosmological, geographic, cultural and performative sense that the Catalan Atlas should be seen as a document illustrating this more global view. It is an attempt to use the accumulated information taken from communication between cultures to portray a more complete idea of the world while illustrating how cosmology, trade, religion and politics coexist. It represents a more complete and complex vision of communication between medieval cultures emerges. The world it describes is not one of cultural isolation but of a steady and growing knowledge, not just of the geographic contours of the world, but of the diverse people who inhabit it. (This text originates from the paper "Looking Beyond: Globalization in the Catalan Atlas of the Fourteenth Century", by Kathleen Holland, University of North Texas, 2010.)

Stefano Goffredo • Zvy Dubinsky
Editors

The Mediterranean Sea

Its history and present challenges

Springer

Editors
Stefano Goffredo
Department of Biological
Geological and Environmental Sciences
Alma Mater Studiorum – University of Bologna
Bologna, Italy

Zvy Dubinsky
Faculty of Life Sciences
Bar Ilan University
Ramat Gan, Israel

ISBN 978-94-007-6703-4 ISBN 978-94-007-6704-1 (eBook)
DOI 10.1007/978-94-007-6704-1
Springer Dordrecht Heidelberg New York London

Library of Congress Control Number: 2013946128

Printed on acid-free paper

Springer is part of Springer Science+Business Media (www.springer.com)

Preface

The present volume brings together a multidisciplinary, up-to-date survey of the current status of our knowledge regarding the Mediterranean. It begins with the very birth of Mare Nostrum along with all of the Earth's oceans, "Let the waters under the heaven be gathered together unto one place, and let the dry land appear: and it was so … and the gathering together of the waters called the Seas … and it was good".

The geology chapters relating past events that shaped the Mediterranean, as we know it, are followed by discourses of current regimes and nutrient fluxes, and paleoclimatology. These abiotic aspects set the stage for a detailed discussion of the main components of the Mediterranean ecosystems and their trophic levels: chemosynthesis, primary productivity, divided into phytoplankton, seaweeds and seagrasses, and moving through the various heterotrophic taxa, bacteria, corals, mollusks, bryozoans, and crustaceans, through fishes. All discussed taxa are viewed from their earliest Mediterranean evidence, present status and future threats.

The past, present and likely future features of the Mediterranean ecosystems are discussed from structural through functional aspects, in relation to paleoclimatic evidence, current processes and future scenarios. Ongoing changes in biodiversity are presented as driven by the increasing influx of Lessepsian Indo-Pacific invaders, facilitated by the exponential growth of cargo shipping and the "tropicalization" of the Mediterranean.

A major focus of the volume is the discourse of Global Climate Change processes and their main marine expressions, seawater warming and acidification, focusing on their likely future effects on sensitive taxa, on the balance between bioconstruction and bioerosion processes, ecosystem level changes and their socioeconomic consequences and opportunities. Effects of the steeply accelerating changes in the water level, temperature and acidity of the Mediterranean are covered in the book's chapters, describing how these are likely to force adjustments of maricultural practices and health safeguards, as they increase security instability likelihood of tensions and conflicts. All of these require policy and legislation responses.

Human relations with the Sea are examined, from our ancestors timidly skirting its shores, while mixing genes and metes among surrounding tribes, all the way through the intense marine traffic of our times, from exploiting the "Royal purple" to current mariculture, overfishing and oil drilling and exploitation. Mankind's responsibilities towards safeguarding the Mediterranean biodiversity and its cultural heritage summarize the volume's message: "And the LORD God took the man, and put him into the Garden of Eden to dress it and to keep it".

The volume is written by leading authorities in all relevant fields and will be an indispensable addition to the bookshelves of specialists and advanced students, alerting them to the value of a broad view of the Mediterranean and its abiotic features shaping life in its waters and along the long shorelines, the cradle of Western Civilization. It is unique in the breath of its scope, covering geology, climatology, oceanography, ecology, botany, zoology, science of humanity, history, culture, through socioeconomic analysis and forecasting.

Bologna, Italy Stefano Goffredo
Ramat Gan, Israel Zvy Dubinsky

The idea leading to the publication of this book was born during the implementation of the EU project "CoralWarm – Corals and Global Warming: The Mediterranean *versus* the Red Sea", funded by the European Research Council under the European Union's Seventh Framework Programme (*FP7/2007-2013)/*ERC *grant agreement* n° [*249930;* www.CoralWarm.eu]

Contents

Part IV Mediterranean Man and Sea: Myths, Origins, Challenges and Opportunities

Contributors

Paolo G. Albano Department of Experimental Evolutionary Biology, University of Bologna, Bologna, Italy

Alberto Barausse Laboratorio di Analisi dei Sistemi Ambientali, Università di Padova, Padova, Italy

Ilana Berman-Frank Mina & Everard Goodman Faculty of Life Sciences, Bar Ilan University, Ramat Gan, Israel

Francesca R. Bosellini Department of Chemical and Geological Sciences, University of Modena and Reggio Emilia, Modena, Italy

Hans Günter Brauch Faculty of Political and Social Sciences, Free University of Berlin, Berlin, Germany

Peace Research and European Security Studies (AFES-PRESS), Mosbach, Germany

M. Namik Çağatay Geology Department, School of Mines, Istanbul Technical University, Istanbul, Turkey

Bruno Capaccioni Department of Earth and Environmental-Geological Sciences, University of Bologna, Bologna, Italy

Erik Caroselli Marine Science Group, Department of Biological, Geological and Environmental Sciences, Alma Mater Studiorum – University of Bologna, Bologna, Italy

Silvia Cocito Marine Environment and Sustainable Development Unit, ENEA, La Spezia, Italy

Rafael Conde-Álvarez Estudios de Flora y Vegetación (EFYVE), Málaga, Spain

Francesca de' Donato Department of Epidemiology, Lazio Regional Health Service, Rome, Italy

Enrico Dinelli Department of Biological, Geological, and Environmental Sciences, University of Bologna, Bologna, Italy

Dafna M. DiSegni Berglas School of Economics, Tel Aviv University, Tel Aviv, Israel

Awantha Dissanayake Institute for East China Sea Research, Nagasaki University, Nagasaki, Japan

Marta Estrada Institut de Ciències del Mar, Consejo Superior de Investigaciones Científicas, Barcelona, Spain

Elena Fabbri Department of Biological, Geological, and Environmental Sciences, University of Bologna, Bologna, Italy

Félix L. Figueroa Departamento de Ecología, Facultad de Ciencias, Universidad de Málaga, Málaga, Spain

Antonio Flores-Moya Departamento de Ecología, Facultad de Ciencias, Universidad de Málaga, Málaga, Spain

B.S. Galil National Institute of Oceanography, Israel Oceanographic and Limnological Research, Haifa, Israel

Josep-Maria Gili Institut de Ciències del Mar, Consejo Superior de Investigaciones Científicas, Barcelona, Spain

Stefano Goffredo Department of Biological, Geological and Environmental Sciences, Alma Mater Studiorum – University of Bologna, Bologna, Italy

Department of Biological, Geological and Environmental Sciences, Bologna, Italy

Menachem Goren Department of Zoology, George S. Wise Faculty of Life Sciences, Tel Aviv University, Tel Aviv, Israel

Ignacio Hernández Departamento de Biología, Facultad de Ciencias del Mar y Ambientales, Universidad de Cádiz, Cádiz, Spain

Razy Hoffman The Mina and Everard Goodman Faculty of Life Sciences, Bar-Ilan University, Ramat Gan, Israel

David Iluz The Mina and Everard Goodman Faculty of Life Sciences, Bar-Ilan University, Ramat Gan, Israel

Arie W. Janssen Naturalis Biodiversity Center (NCB), RA Leiden, The Netherlands

Nathalie Korbee Departamento de Ecología, Facultad de Ciencias, Universidad de Málaga, Málaga, Spain

Alexandra C. Kraberg Biologische Anstalt Helgoland, Alfred-Wegener Institute for Polar and Marine Research, Helgoland, Germany

Nurit Kress Israel Oceanographic & Limnological Res, National Institute of Oceanography, Haifa, Israel

Michael Krom Earth and Biosphere Institute, School of Earth and Environment, University of Leeds, Leeds, UK

Charney School of Marine Sciences, University of Haifa, Haifa, Israel

Petar Kružić Laboratory for Marine Biology, Department of Zoology, Faculty of Science, University of Zagreb, Zagreb, Croatia

Chiara Lombardi Marine Environment and Sustainable Development Unit, ENEA, La Spezia, Italy

Maria Helena Trindade Lopes F.C.S.H. – Universidade Nova de Lisboa, Lisbon, Portugal

Donata Luiselli Department of Biological, Geological, and Environmental Sciences, University of Bologna, Bologna, Italy

Teresa Madurell Institut de Ciències del Mar, Consejo Superior de Investigaciones Científicas, Barcelona, Spain

António Marques Unidade de Investigação em Valorização dos Produtos da Pesca e Aquacultura, INRB, I.P./L-IPIMAR, Lisbon, Portugal

Yossi Mart Recanati Institute for Marine Studies, University of Haifa, Haifa, Israel Michmoret School of Marine Sciences, Ruppin Academic Center, Michmoret, Israel

Cecilia McHugh Lamont-Doherty Earth Observatory, Columbia University, Palisades, NY, USA

Queens College, City University of New York, Flushing, NY, USA

Jesús M. Mercado Centro Oceanográfico de Málaga, Instituto Español de Oceanografía, Málaga, Spain

Anna Maria Mercuri Dipartimento di Scienze della Vita, Università di Modena e Reggio Emilia, Modena, Italy

Basile Michaelidis Laboratory of Animal Physiology, Department of Zoology, School of Biology, Faculty of Science, Aristotle University of Thessaloniki, Thessaloniki, Greece

Paola Michelozzi Department of Epidemiology, Lazio Regional Health Service, Rome, Italy

Caterina Morigi Stratigraphy Department, Geological Survey of Denmark and Greenland (GEUS), Copenhagen K, Denmark

Maria Pia Nardelli Dipartimento di Scienze della Vita e dell'Ambiente, Università Politecnica delle Marche, Ancona, Italy

Laboratoire des Bio-Indicateurs Actuels et Fossiles (BIAF), CNRS, UMR6112, LPGN, LUNAM Université, Université d'Angers, Angers CEDEX 01, France

Alessandra Negri Dipartimento di Scienze della Vita e dell'Ambiente, Università Politecnica delle Marche, Ancona, Italy

Maria Leonor Nunes Unidade de Investigação em Valorização dos Produtos da Pesca e Aquacultura, INRB, I.P./L-IPIMAR, Lisbon, Portugal

Irene Olivé CCMar – Center of Marine Sciences, Universidade do Algarve, Faro, Portugal

Thierry Pérez CNRS UMR 7263 Institut Méditerranéen de Biodiversité et d'Ecologie marine et continentale (IMBE), Aix-Marseille Université, Marseille, France

J. Lucas Pérez-Lloréns Departamento de Biología, Facultad de Ciencias del Mar y Ambientales, Universidad de Cádiz, Campus de Excelencia Internacional del Mar (CEIMAR), Cádiz, Spain

Ángel Pérez-Ruzafa Departamento de Ecología e Hidrología, Facultad de Biología, Campus de Espinardo, Universidad de Murcia, Murcia, Spain

Vincenzo Picotti Department of Earth and Environmental-Geological Sciences, University of Bologna, Bologna, Italy

Hans-O. Pörtner Alfred-Wegener-Institut für Polar- und Meeresforschung, Integrative Ökophysiologie, Bologna, Germany

Luca Palmeri Laboratorio di Analisi dei Sistemi Ambientali, Università di Padova, Padova, Italy

Katja T.C.A. Peijnenburg Netherlands Centre for Biodiversity Naturalis, RA Leiden, The Netherlands

Institute for Biodiversity and Ecosystem Dynamics, University of Amsterdam, Amsterdam, The Netherlands

Michelle E. Portman Faculty of Architecture and Town Planning, Technion–Israel Institute of Technology, Haifa, Israel

Eyal Rahav Mina & Everard Goodman Faculty of Life Sciences, Bar Ilan University, Ramat Gan, Israel

Rui Rosa Laboratório Marítimo da Guia, Centro de Oceanografia, Faculdade de Ciências da, Universidade de Lisboa, Cascais, Portugal

Sergi Rossi Institut de Ciència i Technologia Ambientals, Barcelona, Spain

William B.F. Ryan Lamont-Doherty Earth Observatory, Columbia University, Palisades, NY, USA

Ana Sabatés Institut de Ciències del Mar, Consejo Superior de Investigaciones Científicas, Barcelona, Spain

Anna Sabbatini Dipartimento di Scienze della Vita e dell'Ambiente, Università Politecnica delle Marche, Ancona, Italy

Bruno Sabelli Department of Biological, Geological and Environmental Sciences, University of Bologna, Bologna, Italy

Laura Sadori Dipartimento di Biologia Ambientale, Università "La Sapienza", Rome, Italy

Enric Saiz Institut de Ciències del Mar, Consejo Superior de Investigaciones Científicas, Barcelona, Spain

Rafael Sardá Institut d'Estudis Avançats de Blanes, Consejo Superior de Investigaciones Científicas, Blanes, Spain

Stefania Sarno Department of Biological, Geological, and Environmental Sciences, University of Bologna, Bologna, Italy

Marco Sazzini Department of Biological, Geological, and Environmental Sciences, University of Bologna, Bologna, Italy

Christine H.L. Schönberg Australian Institute of Marine Science (AIMS), Oceans Institute at the University of Western Australia, Crawley, WA, Australia

Jürgen Scheffran Institute of Geography, Research Group Climate Change and Security (CLISEC), KlimaCampus Excellence Initiative, University of Hamburg, Hamburg, Germany

Mordechai Shechter Natural Resource and Environmental Research Center, University of Haifa, and School of Sustainability, Interdisciplinary Center (IDC), Herzlyia, Israel

Nikolaos Skliris National Oceanography Center, University of Southampton, Southampton, UK

Inna Sokolova Department of Biology, University of North Carolina at Charlotte, Charlotte, NC, USA

Noga Stambler The Mina & Everard Goodman Faculty of Life Sciences, Bar-Ilan University, Ramat-Gan, Israel

Jarosław Stolarski Institute of Paleobiology, Polish Academy of Sciences, Warsaw, Poland

Marco Taviani ISMAR-CNR, Bologna, Italy

Biology Department, Woods Hole Oceanographic Institution, Woods Hole, MA, USA

Paul D. Taylor Department of Earth Sciences, Natural History Museum, London, UK

José Templado Department of Biodiversity and Evolutionary Biology, Museo Nacional de Ciencias Naturales, Consejo Superior de Investigaciones Científicas, Madrid, Spain

Lars Tomanek Department of Biological Sciences, Center for Coastal Marine Sciences and Environmental Proteomics Laboratory, California Polytechnic State University, San Luis Obispo, CA, USA

Jean Vacelet CNRS UMR 7263 Institut Méditerranéen de Biodiversité et d'Ecologie marine et continentale (IMBE), Aix-Marseille Université, Marseille, France

Dina Vachtman Department of Geography and Environmental Studies, Ben Gurion University, Be'er Sheva, Israel

Department of Earth Sciences, University of Bergen, Bergen, Norway

Dolors Vaqué Institut de Ciències del Mar, Consejo Superior de Investigaciones Científicas, Barcelona, Spain

Juan J. Vergara Departamento de Biología, Facultad de Ciencias del Mar y Ambientales, Universidad de Cádiz, Campus de Excelencia Internacional del Mar (CEIMAR), Cádiz, Spain

Agostina Vertino University Milano-Bicocca, Milan, Italy

Karen H. Wiltshire Biologische Anstalt Helgoland, Alfred-Wegener Institute for Polar and Marine Research, Helgoland, Germany

Max Wisshak Marine Research Division, Senckenberg am Meer, Wilhelmshaven, Germany

Part I

Geology, Chemical and Physical Oceanography

The Geological Origins and Paleoceanographic History of the Mediterranean Region: Tethys to Present

1

Vincenzo Picotti, Alessandra Negri, and Bruno Capaccioni

Abstract

In this chapter we briefly discussed the main steps of the evolution of the Mediterranean sea, from the Tethys to its closure, until the present days. It is worth highlighting the importance of the physical processes controlling the final shape and bathymetry of the basin, namely the subduction/collision of Africa and Eurasia and associated microplates, and the erosion and sediment dispersal. Both carbonate platforms and organic rich layers (ORL or sapropels) represent the sink of inorganic and organic Carbon, respectively. Therefore, their occurrence and evolution is critical for the carbon cycle, with feedbacks on the global biogeochemical cycle. This chapter highlights how the most interesting geological objects are the result of the interaction between the biological and physical processes.

Keywords

Mediterranean region • Evaporites • Sapropels • Tethys • Paleoceanography • Paleoecology

Introduction

The Mediterranean we presently see is the result of a long geological history of rifting, spreading, subduction and collision of plates and microplates since the Mesozoic (Fig. 1.1). At the beginning of that time, a wide ocean, the Tethys, was encased between the continental masses of the Pangea, forming a large gulf, wedging to the west into this supercontinent. The subsequent geological history brings to the modern Mediterranean through the collision between African and Eurasian continents almost consuming all the former oceanic crust of the Tethys.

Nowadays, it is possible to distinguish the Western Mediterranean from the Eastern, separated by the Sicily channel and characterized by different age and setting.

The Eastern Mediterranean includes the oldest portion of oceanic lithosphere preserved from subduction in two remnants: the Ionian and Herodotus abyssal plains, at present among the deepest portion of the Mediterranean. They consists of an old oceanic lithosphere, possibly of pre-Jurassic to Cretaceous age that is subducting under the Eurasian plate. This subduction forms the Mediterranean Ridge, an accretionary prism of deformed rocks, scraped off from the subducting plate. Associated with this subduction are some features that characterize the entire Eastern Mediterranean: the opening of the Aegean sea, a back-arc basin created by stretching of the upper (Eurasian) plate, due to the retreat of the subducting plate. The Anatolian microplate started drifting to the west at the same time of the Aegean spreading, i.e. around 23 My ago, basically driven by the collision of Arabia against Eurasia, and the occurrence of a free boundary to the west, i.e. the Aegean opening. Toward the northwest, the Ionian oceanic lithosphere subducts again part of the Eurasian plate, with the accretionary prism affecting a wide area known as Calabrian arc (see Fig. 1.1). The Adriatic Sea

V. Picotti • B. Capaccioni (✉)
Department of Earth and Environmental-Geological Sciences, University of Bologna, Bologna, Italy
e-mail: vincenzo.picotti@unibo.it; bruno.capaccioni@unibo.it

A. Negri
Department of Marine Sciences, Polytechnic University of Marche, Ancona, Italy
e-mail: a.negri@univpm.it

S. Goffredo and Z. Dubinsky (eds.), *The Mediterranean Sea: Its history and present challenges*,
DOI 10.1007/978-94-007-6704-1_1, © Springer Science+Business Media Dordrecht 2014

Fig. 1.1 Main geostructure of the Mediterranean region

is a relict epicontinental sea formed by patches of carbonate platforms and intervening basins, whose most prominent are the Apulian platform and the Umbro-Marchean-Adriatic basin. The sea is bordered by active margins, documenting the subduction of the Adriatic plate to the northeast under the Dinarides, as well as to the southwest under the Apennines.

The Western Mediterranean is much younger with respect to the Eastern (Fig. 1.1). It developed as the now consumed oceanic crust of the African plate (the so-called Neo-Tethys, or Alpine Tethys) was subducted under Eurasia. The retreat of this oceanic crust forced the upper (Eurasian) plate to stretch and open back-arc basins, that developed in relatively short time periods, as is the case of the Ligurian-Provençal basin, started around 20 My and stopped at 15, the Alboran and Algerian basins (between 16 and 8 My) and the Tyrrhenian (between 7 and 2 My). The Corsica – Sardinia block was rifted off from Eurasia and drifted to the present-day setting by the opening of the Ligurian-Provençal basin. Also the Apenninic chain was rifted and rotated by the opening of the Tyrrhenian Sea.

A Brief Synthesis of the Geological Evolution of the Mediterranean Region

To describe the evolution of the marine environments along time, it is necessary to recall the main steps that brought from the Tethys to the present Mediterranean. The starting point is around the beginning of the Mesozoic, the time of the supercontinent Pangea, when the marine basins formed a gulf within a large continental mass: the Tehtyan gulf. The breakup and dispersal of the African and Eurasian plates, interspersed with microplates, such as Adria and Anatolia created a series of sub-basins collectively called Tethys, that became a proto-Mediterranean after the Eocene, a time of collision of continental masses against Eurasia. This process brought about the progressive separation of the eastern basins, also known as Paratethys, presently represented by the residual Black Sea and Caspian Sea. The spreading of the Western Mediterranean basins and the consumption of the Tethyan oceanic crust dominated the Miocene and Pliocene Mediterranean (23–3 My), until it reached a final configuration at the end of the Miocene. The last 0.6 My of the Miocene are very important for the evolution of the Mediterranean, because of the severe drawdown that affected the whole basin, and the associated dramatic changes in the sedimentation dynamics. In fact huge amounts of evaporites (gypsum, anhydrite and halite) deposited, especially in the deepest portion of the basin, where they reach hundreds of meters in thickness. After the opening of the Gibraltar strait, the normal oceanic conditions re-established, and the Plio-Pleistocene Mediterranean developed in conditions gradually similar to the present-day.

In the following chapters, the history of opening of the Tethyan and closure and formation of the Mediterranean is provided with some more detail. The first chapter will present the evolution of the Tethys from a paleogeographic and paleoecologic point of view, whereas the most important events characterizing the evolution of the Mediterranean after the collision will be focused on the events of the last 6 My, i.e. between the end Miocene and the Holocene.

Paleoecology of Tethyan Carbonate Platforms as Trackers of Its Geological Evolution and Associated Paleoceanographic Changes

We briefly presented the rich and complex geological history of the Mediterranean. In order to highlight the main evolutionary steps of the basin from the Mesozoic on, we will deal with one of the geological objects more sensitive to the paleoceanographic changes: the carbonate platforms.

Geologists call carbonate platforms the accumulations of carbonate sediments directly deriving from biotic activity in the shallow sea, i.e. formed by skeletal rests of organisms or mud crystalized after biotic activity (e.g. Tucker et al. 1990). The present day global distribution of the carbonate platforms allows associating paleoceanographic significance to the their main ecologic characters. Warm, oligotrophic and transparent shallow water favors photosynthetic or photosymbiontic organisms. Carbonate platforms developed in such intertropical settings includes skeletal remains of green calcareous algae, corals, calcareous sponges and endosymbiontic thick-shelled bivalves, as well as non-skeletal mud or ooids crystalized thanks to photosynthetic organisms, such as bacteria and cyanophyceae among others. To recall the main ecological relationships, these platforms have been called Photozoan (James 1997). Cool, mesotrophic and turbid water favors suspension feeders, such as mollusks, foraminifers, bryozoans, and, especially in the past, brachiopods and crinoids. Water turbidity tends to select red calcareous algae with respect to green algae, basically absent. The non-skeletal fraction is virtually lacking, due to the strongly reduced photosynthesis, hampering the productivity of bacteria and cyanophyceae. Due to their living strategy, these platforms, typical of temperate latitude, or even to areas of upwelling and river mouth vicinity in tropical regions, are known as Heterozoan (James 1997).

In Fig. 1.2 we compiled an evolutionary diagram of the occurrence of carbonate platforms in the Tethyan and Mediterranean basins from the Late Triassic on, based on the ecological subdivision previously mentioned. The diagram shows in a qualitative way the relative abundance and the main ecological characters of the platforms, based on published reviews (e.g. Bosellini 1989; Kiessling et al. 2002) and our own experience. The productivity of the platforms varied through time, due to several reasons, related to local and global variations/perturbations of the climate, such as plate configurations changes, carbon cycle perturbations, or the relative sea-level. In Fig. 1.2 we add three curves, to track local and global variations: the dashed line marks the paleolatitudinal evolution of a point presently at around 40° on the Apulia platform (see Fig. 1.1), according to Stampfli and Borel (2002). The paleolatitude do track the changes of the whole basin to match the variations of local plate configurations that can affect sea-surface temperature or the current patterns, therefore directly impacting the carbonate platform communities (see Carannante et al. 1988). The continuous curve, taken from Bown et al. (2004) shows the evolutionary trend of calcareous nannoplankton, worldwide, and is intended to track the global evolution of planktonic primary producers. The closeness of this trend with the curve of relative sea-level speaks about its control on the speciation of the calcareous nannoplankton. Of course, the misfits document the presence of other processes, controlling the evolution of species in general, such as the K/T boundary (bolide impacts) or the so-called Oceanic Anoxic Events of the Cretaceous (global perturbations of the Carbon cycle).

Throughout the Late Triassic and the Early Jurassic until the end Pliensbachian the Tethyan gulf was at low latitude protected from the zonal flux by the occurrence of the Pangea Supercontinent to the west, therefore the scarce terrestrial runoff and the wide shelves favoured the abundance of Photozoan platform, locally interrupted by climatic perturbations, such as the end-Norian humid period (Iannace and Frisia 1994), or the early Sinemurian platform crisis (Picotti and Cobianchi 1996). At the end of Pliensbachian, a perturbation of the Carbon cycle occur, whose effects lasted till the Early Toarcian (Cobianchi and Picotti 2001).

After the Pliensbachian, the carbonate productivity shifted dramatically toward the Heterozoan, with abundance of crinoids, brachiopods and minor non skeletal sediments, such as ooids. At the same time, the paleolatitude was moving toward the tropic, reaching 28° at the Aalenian, around 180 Ma, due to the opening of the proto-Atlantic and its seaways with the Tethys. This opening of a new corridor possibly changed the trophic status of oceanic waters in the intertropical belt. This perturbation was globally recorded by the first decrease of calcareous nannoplankton species richness (see Fig. 1.2), and ended around the Bathonian, a critical period for platforms, with reworking of ooids toward the basin. The Callovian and Oxfordian are periods of renewed Photozoan platform, spreading all over the Tethyan margins with abundant corals, calcareous sponge and green algae. The Photozoan carbonate productivity was high, but slowly decreasing up to the Valanginian, a time of crisis possibly associated to a sea-level fall (Fig. 1.2). The paleolatitude was well within the intertropical belt for most of the Cretaceous, whereas the productivity was shifting, for the near absence of corals and calcareous sponge and the presence of large foraminifers, red algae and bivalves. Among these latter, the prevailing were the Rudists, a group of thick-shelled pelecypods, whose occurrence of endosymbionts is dubious, locally

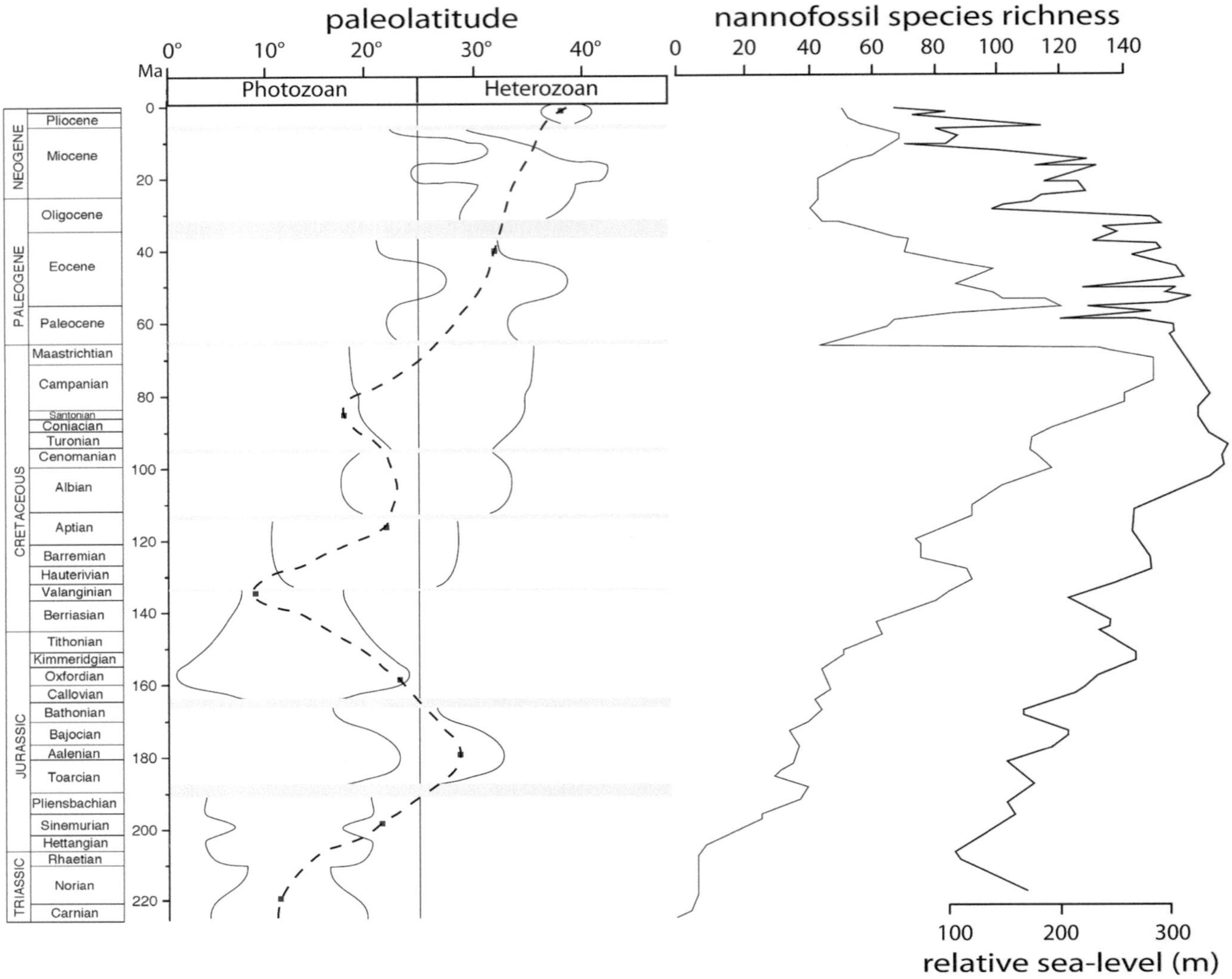

Fig. 1.2 History of the Tethyan carbonate platform compared with local and global events. *Width of brackets* represents the distribution of Photozoan and Heterozoan platforms around the present-day Mediterranean. *Grey bars* are periods of crisis and non – deposition of platform carbonates. *Dashed line* represents the paleolatitude of a point presently placed at 40° at the Apulian platform, *dots* represent the data of Stampfli and Borel (2002). Nannofossil species richness and sea-level curve after Bown et al. (2004)

forming patches also in deep-water setting. For these reasons, we showed in Fig. 1.2 an important component of Heterozoan producers. Two oceanic events, widely recorded on the Tethyan realm, but also worldwide, the so-called Oceanic Anoxic Events of Early Aptian and Late Cenomanian have been indicated. These events represents major perturbation of the Carbon cycle, with the worldwide expansion of the Oxygen Minimum Zone and the deposition of thick and widespread organic rich deposits (black shales, see the section "The Anoxic Events (Sapropels)").

Their impact on the shallow water carbonate platforms is difficult to ascertain, given the relatively short duration and the difficult platform-basin stratigraphic correlation. It is widely recognized, however, that the shallow water carbonate factory cannot stand strong perturbations of the Carbon cycle, such as those described, therefore serious stops of the carbonate factories (drowning unconformities of Schlager 1993) are expected in coeval platform intervals. However, these unconformities on the geological record of the shallow water platform are not evident or well explored in the literature.

Throughout the Late Cretaceous, the species richness of the nannoplankton increased almost gradually, due to the dispersal of the continents and the increase of the sea-level, opening new ecological niches. The productivity of the Tethyan platform was relatively high, but it collapsed at the Cretaceous – Tertiary (K-T) boundary, one of the most important extinction in the Earth history, together with the richness of calcareous nannoplankton species. The recovery after the K-T extinction was relatively rapid for the carbonate platforms, however the main producers were mostly red algae, large forams and bryozoans, with minor contribution of corals. Paleogene platforms, however, are less abundant than the Cretaceous, for the double cause of the continuous northward drift of the Tethyan realm, passing the 30° of paleolatitude within the Eocene, and the increasing elevation of the peri- Tethyan belts, shedding more and more clastics to the sea, therefore diminishing the niches for carbonate productivity.

Late Eocene to Oligocene was a time of minimum richness of the calcareous nannofossil species and of pronounced low global sea-level. The Tethyan ocean at that time was severely reduced, due to collision of the Indian continent and other microplates to the west that closed most of the seaways and separated a large internal basin, the Paratethys, mostly filled now (Pannonian plains), but still represented by the Caspian and Black seas. The Mediterranean seaway to the east was interrupted and the local currents dramatically changed: this brought about the near absence of carbonate platforms.

The recovery of the carbonate productivity occurred during the Late Oligocene, but the Photozoan organisms never recovered, due to the increasingly high latitude. Only few periods of time were warm enough to allow patch corals to spread into the shelves: the late Burdigalian (Miocene Climatic Optimum, around 18 Ma) and the Tortonian to Early Messinian (Late Miocene, around 7 Ma), the latter dominated by Porites corals. The carbonate platforms of other times, less and less abundant through time are dominated by mollusks, bryozoans, forams and red algae, locally showing hyper-calcified skeletons during particularly warm periods.

After the late Messinian salinity crisis (around 5.6 Ma, see next chapter) and especially with the onset of Northern Hemishpere glaciations (around 3 Ma), the occurrence of carbonate platforms are more and more patchy, mostly as isolated banks of mollusks, documenting the increasing climate variability and associated variable paleoceanographic conditions.

After the Mediterranean Closure

In the early 1960s, seismic soundings in the Mediterranean found ubiquitous evidence of an acoustic reflector a few hundred meters below the sea floor and closely simulating its topography – named the M layer. In August of 1970 the DSDP ship Challenger was positioned in the western Mediterranean, south of the Balearic Islands, in almost 3,000 m of water. The geologists on board looking for the source of that prominent sub-sea-floor seismic feature drilled into a thick layer of anhydrite (calcium sulfate, precipitated only from saline ground-water originally as gypsum) – and stromatolites (organic fossils of algal mats likely formed in an intertidal environment). These sediments indicated that 6 million years ago the Mediterranean, which had likely similar bathymetry as today and had been deep sea for many millions of years, had been a series of brackish lakes and desiccated tidal flats, (Hsu 1983). Research since then has built up the picture of one of the greatest evaporitic basins in Earth history, not only in shallow marine environments, but also widespread in the deeper Mediterranean.

The chronology of the event, although difficult because of lacking fossil remains, is based on 87Sr, ∂18O, and ∂13C isotopes, magnetostratigraphy, astrochronology, and stratigraphical studies where uplift and exhumation have exposed sequences, mainly in Italy and Spain. Although still matter of debate the age of the progression of the event are:

7.24/6.88 Ma – alternations of marine marls and sapropels (organic rich sediments, see later) in Sicilian and Northern Apennine strata, reflecting gradual modification of water exchange with the Atlantic ocean;

5.96 ± 0.02 Ma – synchronous transition to regressive evaporite deposition over the entire Mediterranean. Deposition of Lower Evaporite unit (Fig. 1.3), possibly indicating a relative sea-level fall of 200 m or even more;
5.8–5.5 Ma – complete isolation established. In basins erosional surface/M-layer (lowstand in sequence stratigraphy terms), on land deeply incised fluvial channels cut.
Diachronous onset of transgressive Upper Evaporites (Sicily and deep basins) and 'Lago Mare' deposits with anomalous marine water masses, from brackish to fresh-water, fed by (warmer?) wetter climate in river drainage areas (Willett et al. 2006).

5.33 Ma – abrupt synchronous return to full marine conditions. Miocene/Pliocene boundary. (Krijgsman et al. 1999).

It is now widely accepted that the closure of the marine passages between the Atlantic and the Mediterranean, and the consequent negative hydraulic budget, is responsible of the large volumes of evaporites spread all over the Mediterranean Sea. The important question, which is still debated, is the mechanism which caused the isolation of the Mediterranean and the major sea-level drawdawn – glacioeustatic climate change, tectonic movements, or a combination?

However, a fact is that the present-day precipitation-evaporation budget in the Mediterranean basin is negative. Without a significant inflow of ocean water from Atlantic, the Mediterranean sea level cannot be sustained. An important feature pointing to the dramatic drawdown of the Mediterranean is that, during the lowstand, rivers incised deep canyons. The building of the Aswan Dam in the 1950s, 1,200 km from the sea, was complicated by the discovery of

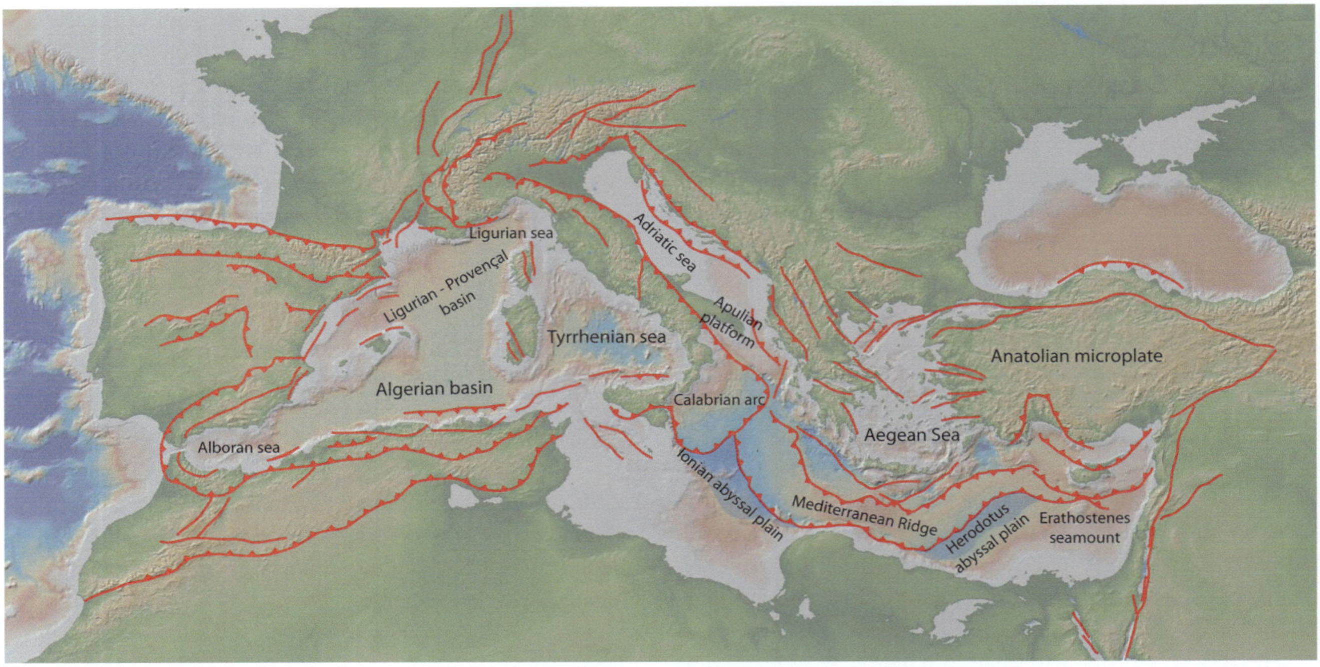

Fig. 1.3 The Lower Evaporite unit in the Northern Apennines consists in well bedded gypsum deposited in tidal flats (Photo courtesy: G.B. Vai)

a deep narrow gorge in granite hundreds of metres below sea level, and found to be 2,500 m deep below the Nile at Cairo. The Rhone and Ebro also have gorges 1,500 m below sea level, filled with late Miocene gravels and early Pliocene marine fossils. Other buried gorges and channels have been found in Israel and Syria and linked to the mouths of modern rivers all round the Mediterranean. Atlantic and Pacific canyons are linked to turbidity canyons on the continental shelf, whereas the Mediterranean ones all appear to be drowned river valleys (Hsu 1983).

Another important question is how did the Messinian crisis end? The beginning of the Pliocene was warmer, and rainfall increased in drainage areas, but the sudden return to full marine conditions with deep- sea fossils, cannot be explained by a climatic change. The maximum sea-level risewas found to have been reached 170 Ka before oceanic re- flooding. Faulting or tectonic slumping opening the Gibraltar gateway in the Betic-Rif mountain arc have been suggested, but have not been supported by any study. A recent paper (Loget and Van Der Driesche 2006) proposed the backward retreat of the river incision to explain the opening of the Gibraltar V-shaped strait.

The Anoxic Events (Sapropels)

An important feature already appearing at the end of the Miocene is the deposition of sediments rich in organic matter that, according to Hilgen (1991), occur cyclically (approximately every 22 kyr), this cyclicity being related to the Milankovič theory. Milankovič, in fact, mathematically theorized that variations in eccentricity, axial tilt, and precession of the Earth's orbit determined climatic patterns on Earth through orbital forcing. This cyclical deposition is interrupted by the evaporites sedimentation, but occurs once again in the Early Pliocene. In fact, after the return to normal marine condition, the Mediterranean sea experienced dramatic anoxic events that swiped out almost completely the life on the sea bottom. These events are materialized as dark sediments called sapropels.

Sapropel (a contraction of ancient Greek words sapros and pelos, meaning putrefaction and mud), is a term used in marine geology to describe dark-coloured sediments, rich in organic matter, which occur, alternate to normal sediment, in the Mediterranean. Those sediments were studied in great detail after two Scientific cruises (DSDP Leg 13 and Leg 42). Since the very beginning, the researchers inferred that, since those levels are interbedded as dark layers within lighter "open marine" sediments, the formation of these layers should refer to short-lived, but dramatic, alterations in oceanographic conditions probably linked to climatic cyclical changes. Since then, many models dealing with the origin of those layers have been proposed. Among the various processes that have been invoked to explain sapropel formation (see Rohling 1994 and Rohling et al. 2009 for a review), the "stagnation/anoxia" and the "increased productivity" models are the most discussed. According to the stagnation/anoxia model, anoxic bottom conditions are caused by a

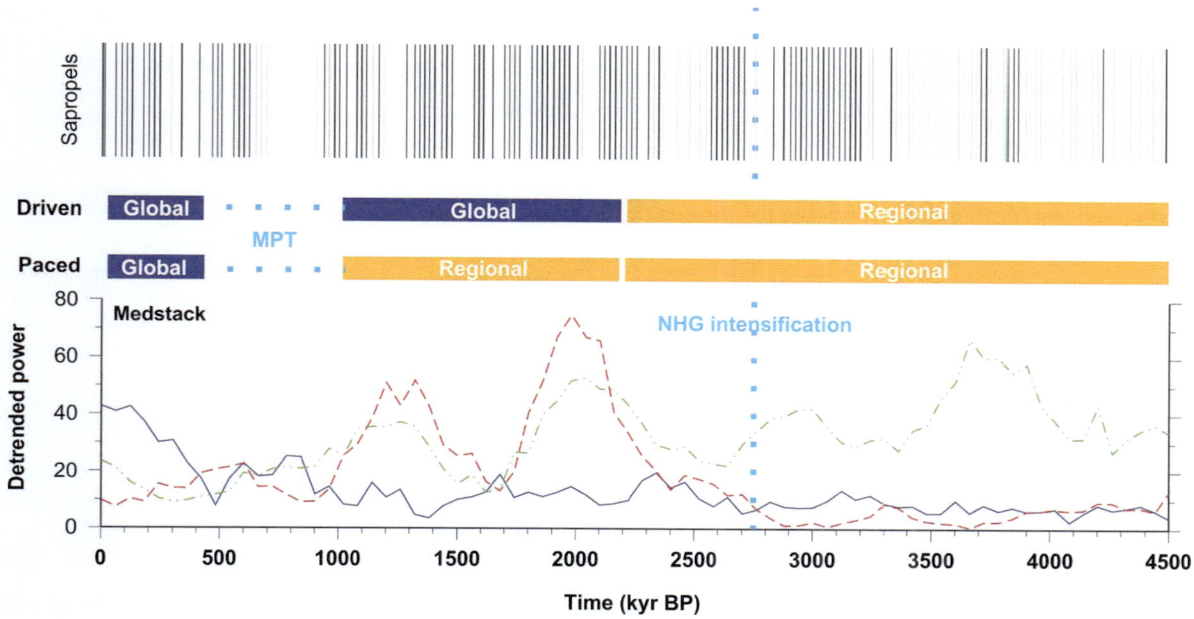

Fig. 1.4 From *bottom* to *top*: Medstack detrended power in the 23, 41 and 100 k band (Colleoni et al. 2012); Sapropels (*black*, organic content>2 %) and *grey layers* succession from Emeis et al. (2000b) and Lourens et al. (1996). The *vertical dotted line* corresponds to the intensification of the Northern Hemisphere Glaciation (~2.75 Ma). *MPT* Mid Pleistocene Transition

strong stratification of the water column that prevents vertical mixing and oxygen supply to the bottom waters. The origin of this stratification was explained as due to increased Nile river runoff linked to the periodic enhancement of the African-Asian monsoons (Rossignol-Strick 1983, 1985) and by increased rainfalls and river discharge along the northern part of the Eastern Mediterranean Sea (Cramp et al. 1988; Rohling and Hilgen 1991). Instead, the "increased productivity" model, sapropel deposition is linked to enhanced organic matter flux (Calvert 1983; Calvert et al. 1992), since the present production of organic matter in the eastern Mediterranean cannot account for the high values of organic carbon (TOC) characterizing these layers (Calvert 1983). However, sapropels are probably the result of a combined interplay of the two causes that are not mutually exclusive and some authors (Rohling and Gieskes 1989; Castradori 1993; Rohling 1994; Emeis et al. 1998, 2000a, b), proposed a mechanism in which stratification and productivity could have been combined.

Sapropels occur in the whole Mediterranean basin, but their origin has been and still is matter of debate. Those layers represent a response of the Mediterranean system to the climatic perturbation and preferentially occur during time intervals of high CO2 in the atmosphere (Negri et al. 2012), when the system reacts sinking Organic Carbon to the sea bottom. Then, sapropels or ORL must be seen as a precious source of information, useful for the understanding of the evolution of our changing planet. In this frame, the recently published paper by Colleoni et al. (2012), in fact show that the sapropel deposition is related to monsoon intensification, and the response of the Mediterranean is linked to the precession signal but also to the influence of the high latitude processes (e.g. glaciations). This paper examines the behavior of the planktonic foraminifera signal in a composite stack encompassing the last 4.5 Ma. They show (Fig. 1.4) that the interruption of the cyclical deposition of sapropels corresponds the so-called Middle Pleistocene Transition (MPT), when the glacial intervals alternate to the interglacials forced by the 100 kyr cyclicity (eccentricity), instead of the 41 kyr (obliquity). This means that the signal of ice formation at the high latitudes penetrates to the low latitude and perturbs the processes, causing the deposition of sapropels. In addition, the Authors evidence that in the first part of the Pliocene before the NHG (Northern Hemisphere Glaciation) intensification the occurrence of gray intervals (roughly: not well developed sapropels) suggests that the monsoon intensity was reduced during the Early Pliocene compared to the Late Pleistocene and was able to expand further to North Africa only during exceptional precession times (Larrasoaña et al. 2003). Indeed, less and less gray layers were deposited and more and more black sapropels were formed but only during the largest summer insolation peaks, with increasing correlation between the African monsoon index and the occurrence of sapropels.

Finally, the $\delta^{18}O$ signal evidence that the Mediterranean Sea (Fig. 1.4), before the intensification of the NHG, shows a regional signal (related to the monsoons) precession. Afterwards, but before the MPT, the signal is mixed, dominated by the polar obliquity signal. After the MPT, also the response of the Mediterranean sea is synchronized with the Oceans showing fluctuations related to the 100 kyr periodicity.

This strengthens the suggestion that sapropels must be considered not only a geological accident, but instead they must be explored at a global scale to decipher the response of a system like the Mediterranean to extreme climatic events.

Acknowledgments The authors warmly thank for their thorough review G.B. Vai (University of Bologna) and M. Cobianchi (University of Pavia). G.B. Vai kindly provided the picture of Fig. 1.3.

References

Bosellini A (1989) Dynamics of Tethyan carbonate platform. In: Crevello PD, James LW, Sarg JF, Read JF (eds) Controls on carbonate platform and basin evolution, Soc Econ Paleont Miner Spec Publ. University of California, Berkeley

Bown PR, Lees JA, Young JR (2004) Calcareous nannoplankton evolution and diversity through time. In: Thiestein HR, Young JR (eds) Coccolithophores, from molecular processes to global impact. Springer, London

Calvert SE (1983) Geochemistry of Pleistocene sapropel and associated sediments from the Eastern Mediterranean. Oceanol Acta 6:255–267

Calvert SE, Nielsen B, Fontugne MR (1992) Evidence from nitrogen isotope ratios for enhanced productivity during formation of eastern Mediterranean sapropels. Nature 359:223–225

Carannante G, Esteban M, Milliman JD, Simone L (1988) Carbonate lithofacies as paleolatitude indicators: problems and limitations. Sediment Geol 60:333–346

Castradori D (1993) Calcareous nannofossils and the origin of eastern Mediterranean sapropels. Paleoceanography 8:459–471

Cobianchi M, Picotti V (2001) Sedimentary and biological response to sea-level and paleoceanographic changes of a Lower-Middle Jurassic Tethyan platform margin (Southern Alps, Italy). Palaeogeogr Palaeoclim Palaeoecol 169:219–244

Colleoni F, Masina S, Negri A, Marzocchi A (2012) Plio–Pleistocene high–low latitude climate interplay: a Mediterranean point of view. Earth Planet Sc Lett 319–320:35–44

Cramp A, Collins M, West R (1988) Late Pleistocene-Holocene sedimentation in the NW aegean sea: a paleoclimatic-paleoceanographic reconstruction. Palaeogeogr Palaeoclim Palaeoecol 68:61–77

Emeis KC, Sakamoto T (1998) The sapropel theme of Leg 160. Proc. Ocean Drill Prog Sci Res 160:29–36

Emeis KC, Struck U, Schulz HM, Rosenberg R, Bernasconi S, Erlenkeuser H, Sakamoto T, Martinez-Ruiz F (2000a) Temperature and salinity variations of Mediterranean sea surface waters over the last 16,000 years from records of planktonic stable oxygen isotopes and alkenone unsaturation ratios. Palaeogeogr Palaeoclim Palaeoecol 158:259–280

Emeis KC, Sakamoto T, Wehausen R, Brumsack HJ (2000b) The sapropel record of the eastern Mediterranean sea – results of ocean drilling program leg 160. Palaeogeogr Palaeoclim Palaeoecol 158:371–395

Hilgen FJ (1991) Astronomical calibration of Gauss to Matuyama sapropels in the Mediterranean and implication for the geomagnetic polarity time scale. Earth Planet Sci Lett 104:226–244

Hsu KJ (1983) When the Mediterranean was a desert: a voyage of the Glomar Challenger. Princeton University Press, Princeton

Iannace S, Frisia S (1994) Changing dolomitization styles from Norian to Rhaetian in the southern Tethys realm. IAS Spec Publ 21:75–90

James NP (1997) The cool-water carbonate depositional realm. In: James NP, Clarke JAD (eds) Cool-water carbonates. SEPM, Society for Sedimentary Geology, Special Publication, Tulsa

Kiessling W, Flugen E, Golonka J (2002) From patterns to process: the future of reef research. In: Kiessling W, Flugen E, Golonka J (eds) Phanerozoic reef patterns, vol 72, SEPM (Socity of Sedimentary Geology) Special Publication. Society for Sedimentary Geology, Tulsa, pp 735–743

Krijgsman W, Hilgen FJ, Raffi I, Sierro FJ, Wilson DS (1999) Chronology, causes and progression of the Messinian salinity crisis. Nature 400:652–655

Larrasoaña JC, Roberts AP, Rohling EJ, Winklhofer M, Wehausen R (2003) Three million years of monsoon variability over the northern Sahara. Climate Dyn 21:689–698

Loget N, Van Den Driessche J (2006) On the origin of the strait of Gibraltar. Sediment Geol 188–189:341–356

Lourens LJ, Hilgen FJ, Zachariasse WJ, Van Hoof AAM, Antonara- kou A, Vergnaud-Grazzini C (1996) Evaluation of the Plio-Pleistocene astronomical time scale. Paleoceanography 11:391–413

Negri A, Colleoni F, Masina S (2012) Mediterranean sapropels: a mere geological problem or a resource for the study of a changing planet? Alp Mediterr Quat 25:81–89

Picotti V, Cobianchi M (1996) Jurassic periplatform sequences of the eastern Lombardian Basin (Southern Alps). The deep-sea record of the tectonic evolution, growth and demise history of a carbonate platform. Mem Sci Geol 48:171–219

Rohling EJ (1994) Review and new aspects concerning the formation of eastern Mediterranean sapropels. Mar Geol 122:1–28

Rohling EJ, Gieskes WWC (1989) Late quaternary changes in Mediterranean intermediate water density and formation rate. Paleoceanography 4:531–545

Rohling EJ, Hilgen FJ (1991) The eastern Mediterranean climate at times of sapropel formation: a review. Geol Mijnb 70:253–264

Rohling EJ, Abu-Zied R, Casford JSL, Hayes A, Hoogakker BAA (2009) The marine environment: present and past. In: Woodward JC (ed) The physical geography of the Mediterranean. Oxford University Press, Oxford

Rossignol-Strick M (1983) African monsoon an immediate climate response to orbital insolation. Nature 304:46–49

Rossignol-Strick M (1985) Mediterranean quaternary sapropels, an immediate response of the African monsoon to variations of insolation. Palaeogeogr Palaeoclimatol Palaeoecol 49:237–263

Schlager W (1993) Accommodation and supply – a dual control on stratigraphic sequences. Sediment Geol 86:111–136

Stampfli GM, Borel GD (2002) A plate tectonic model for the Paleozoic and Mesozoic constrained by dynamic plate boundaries and restored synthetic oceanic isochrons. Earth Planet Sci Lett 1–2:17–33

Tucker ME, Wilson JL, Crevello PD, Sarg JR, Read JF (1990) Carbonate platforms. Facies, sequences and evolution. Wiley, New York

Willett SD, Schlunegger F, Picotti V (2006) Messinian climate change and erosional destruction of the central European Alps. Geology 34:613–616

A Channeled Shelf Fan Initiated by Flooding of the Black Sea

2

William B.F. Ryan, Dina Vachtman, Cecilia McHugh, M. Namik Çağatay, and Yossi Mart

Abstract

High-resolution mapping and reflection profiling reveal a depositional fan on the Black Sea SW shelf fed from the Strait of Istanbul (Bosporus). The fan is constructed with an initial deposit of pebbles mixed with glacial and post-glacial shell debris. The pebbles are identical in their composition to quartzite and gabbro recovered in drill cores from the Bosporus Strait. Directly above the pebble layer are mollusks and foraminifera of Mediterranean provenance dated at 6.9 ka bp (uncorrected). Synchronicity between the onset of fan construction and arrival of Mediterranean fauna suggests an origin linked to the connection of the Black Sea's lake with the global ocean. The volume of the chaotic interior of the fan is comparable in magnitude to the volume excavated from the floor of the Bosporus Strait. We propose that when the exterior ocean breached the sill of this inlet, it transformed into an outburst of saltwater that gained energy as it enlarged the inlet. Torrents stripped the glacial and post-glacial covering from its pathway and scattered entrained debris in sheets and mounds as far away as the edge of the shelf. Even in areas where the pre-existing cover survived, its eroded surface attests to the passage of flooding water.

Keywords

Black Sea SW shelf • Black Sea s lake • Bosporus Strait • Depositional fan • Flooding water • Foraminiferans • Gabbro • Glacial and post-glacial shell debris • Global ocean • High-resolution mapping • Mediterranean fauna • Mollusks • Pebbles • Quartzite • Reflection profiling • Strait of Istanbul (Bosporus)

W.B.F. Ryan
Lamont-Doherty Earth Observatory, Columbia University, Palisades, NY 10964, USA
e-mail: billr@ldeo.columbia.edu

D. Vachtman
Department of Earth Sciences, University of Bergen, Bergen N-5007, Norway

Department of Geography and Environmental Studies, Ben Gurion University, Be'er Sheva, Israel

C. McHugh
Queens College, City University of New York, Flushing. NY 11367, USA

Lamont-Doherty Earth Observatory, Columbia University, Palisades, NY 10964, USA

M.N. Çağatay
Geology Department, School of Mines, Istanbul Technical University, Ayazaga, 80626 Istanbul, Turkey

Y. Mart (✉)
Recanati Institute for Marine Studies, University of Haifa. Haifa 31905, Israel

Michmoret School of Marine Sciences, Ruppin Academic Center, Michmoret, Israel
e-mail: yossimart@gmail.com

S. Goffredo and Z. Dubinsky (eds.), *The Mediterranean Sea: Its history and present challenges*, DOI 10.1007/978-94-007-6704-1_2, © Springer Science+Business Media Dordrecht 2014

Introduction

The Black Sea is the world's largest (534 ×10^3 km^3) and deepest (2.2 km) body of water whose physical, chemical and biological properties are regulated by the nature of its connection to the external ocean as well as to its extensive fluvial system (Strakhov 1971; Sorokin 1983; Murray et al. 1991). Today this connection is via the Strait of Istanbul (Bosporus) through which excess fresh water is expelled to the Sea of Marmara and saltwater from the Mediterranean is supplied (Oğuz et al. 1990; Özsoy et al. 1993). During this two-way exchange, the saltwater passes through the Bosporus as an undercurrent. Upon entering the Black Sea, it flows across the continental shelf. The plume remains mostly confined within a network of channels and experiences minimal turbulent entrainment with overlying water (Di Iorio and Yüce 1999; Gregg and Öszoy 1999; Özsoy et al. 2001, 2002). It is only at the shelf break and beyond where the major mixing with the Black Sea cold intermediate layer takes place (Bogdanova 1963; Latif et al. 1991; Özsoy et al. 1993). The Black Sea deep-water is the product of this mixing between the two-end member compositions (Murray et al. 2007).

At times of past low global sea level this inflow did not exist. The Black Sea transformed into a giant inland lake (Arkhangel'skiy and Strakhov 1938; Chepalyga 1984). Assemblages of diatoms in deep-sea drill cores (Schrader et al. 1979) and mollusks from coastal outcrops (Svitoch et al. 2000; Sorokin 2011) indicate that during the past five million years the Black Sea has been predominately a fresh to brackish lake with only 8–10 marine incursions in late Pleistocene interglacial stages coinciding with brief episodes of high global sea level (Zubakov 1988). This paper addresses the last reconnection of the Mediterranean with the Black Sea. Most researchers consider this reconnection and subsequent mixing of waters to have been a gradual process with the surface of the two seas at the same level when the post-glacial global ocean surface reached the elevation of Bosporus inlet (Ross et al. 1970; Kuprin et al. 1974; Fedorov 1988; Lane-Serff et al. 1997; Aksu et al. 2002; Hiscott et al. 2002; 2007a, b). However, an alternate and now controversial hypothesis has proposed an abrupt flooding of saltwater into the lake whose surface lay below the inlet (Ryan et al. 1997a, b; 2003).

North of Bosporus exit, the continental shelf of the Black Sea has a conspicuous delta-shaped apron of sediment crossed by a splay of channels leading to the heads of submarine canyons (Fig. 2.1). The surface of this deposit has been mapped with multibeam echo-sounding methods (Di Iorio et al. 1999; Flood et al. 2009; Okay et al. 2011). The interior has been imaged by seismic reflection profiles (Aksu et al. 2002; Algan et al. 2002, 2007; Gökasan et al. 1997, 2005; Okay et al. 2011). The channel network is clearly sourced from the Bosporus. We adopt the term 'shelf fan' from Flood et al. (2009) to describe this feature.

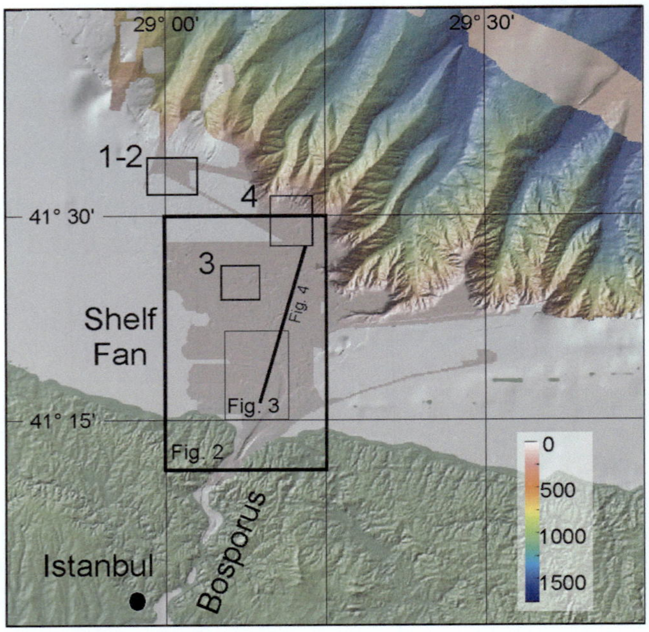

Fig. 2.1 Shaded relief bathymetry of the margin of the Black Sea north of the Bosporus Strait. Bathymetry is from digital elevation models obtained with multibeam swath mapping (Lericolais et al. 2003; Di Iorio and Yüce 1999). Shown are the locations of the three surveys reported here and Figs. 2.3, 2.4 and 2.5

Prior Observations

The northward-flowing undercurrent in the Bosporus has been known and exploited by mariners since antiquity (Scholten 1974). The floor of the 22-km long Strait has now been mapped in remarkable detail by the Department of Navigation, Hydrography and Oceanography of the Turkish Navy (Gökasan et al. 2005). On its winding passage from the Sea of Marmara the Bosporus deepens northwards from a −33 m sill near Istanbul to its exit into the Black Sea at −75 m. Although containing some substantial sediment fill in a bedrock gorge, bathymetry and seismic reflection profiling show that below a depth of around −50 m the valley has an inner, more sinuous channel that has been eroded into earlier fill. This entrenched inner channel continues out onto the Black Sea shelf as a straight chute. After 8 km it turns in a broad curve to the NW (Fig. 2.1).

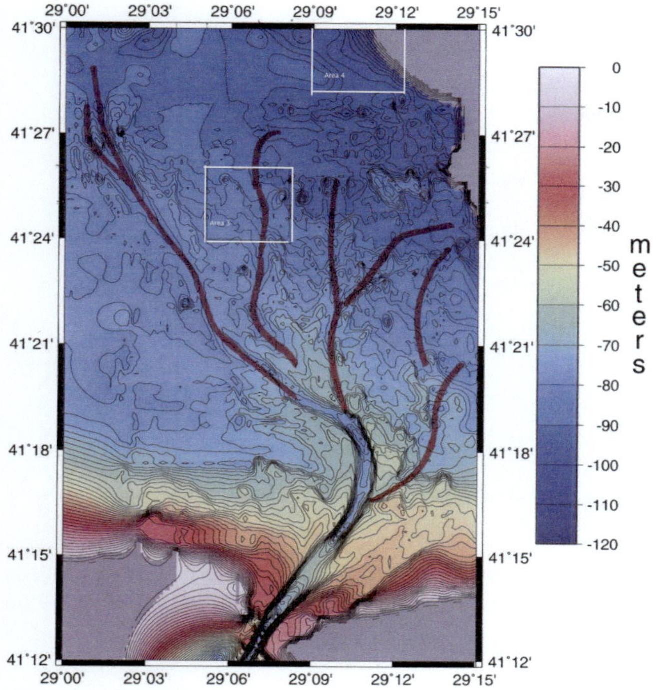

Fig. 2.2 Contours at a 2-m interval of the shelf fan showing in red the paths of the channels and their bifurcations. (elevation model from; Di Iorio and Yüce 1999). Boxes with *white* outline demark surveys of areas 3 and 4

Fig. 2.3 The main channel trunk curving to the NW and branching around 'tear-shaped 'islands' (*upper left*) produced by the changing paths of the channels crossing through the shelf fan and tear-shaped elevations (*lower left*) caused by upstream accumulation of sediment against mounds protruding through the fan surface. These especially large protrusions are abundant on the inner shelf near the exit of the Bosporus Strait. Illumination is from the *upper right* (NE)

Shelf Fan Morphology

The shelf fan displays an active main trunk with numerous secondary channels as well as abandoned trunks and their channels (Di Iorio et al. 1999) (Fig. 2.2). Although the lateral levees of the main active channel now block older pathways, there are some escape routes through low regions of the levees. Flow pathways have shifted during the construction of the fan, leaving inactive distributaries (Fig. 2.3). This evolution has constructed a fan deposit that is thick under levees and thin within the interfluves and with many bumps, mounds, pinnacles and tear shaped islands.

The levee tops and backsides are crenulated by sub-parallel asymmetric bedforms of 1–4 m in height and spaced 100–500 m apart. These bedforms are particularly evident on channel floors as well as on the flanks of levees, where the active channel exiting the Bosporus turns to the northwest. In the bend, the steepest sides of the asymmetric bedforms face away from the channel and the ridge crests are oriented more or less parallel to the levees. Outside of the levees, the asymmetric bedforms are nearly perpendicular to the levees. Even where aligned in fields and parallel to each other, the crests of bedforms are not straight, but somewhat sinuous. The levees and interfluves of the distal channels are thoroughly crenulated. Overall, there seems to be a somewhat systematic SE-NW orientation to the bedform crests.

Many protrusions dot the surface of the shelf fan both in the distal portions of the shelf as well as in the environs of the channel levees (Fig. 2.3). Sedimentation on the lee side of protrusions results in tear-shaped knolls.

Seismic Stratigraphy

A shelf-wide unconformity appears in reflection profiles crossing all Black Sea margins (Ryan et al. 1997a,b, 2003; Aksu et al. 2002; Genov 2004; Algan et al. 2007; Lericolais et al. 2007a,b; Glebov and Shel'ting 2007). Where sampled, this unconformity has been identified as an erosion surface developed during a lowstand of the lake (Fig. 2.4). On the SW Black Sea shelf the reflector that corresponds to this unconformity has been labeled α (Aksu et al. 2002). A paleo-shoreline of the lake has been documented on the Ukraine (Evvsylekov and Shimkus 1995; Ryan, et al. 1997a), Russian (Ivanova et al. 2007), Romanian (Lericolais et al. 2007a, b) and Bulgarian (Dimitrov et al. 1979) margins at a depth circa −90 to −95 m. The shoreline appears in reflection profiles as a prominent wave-cut terrace, steep shore-face with berm and an adjacent belt of relict coastal dunes (Ryan et al. 1997b; Naudts et al. 2006; Lericolais et al. 2009).

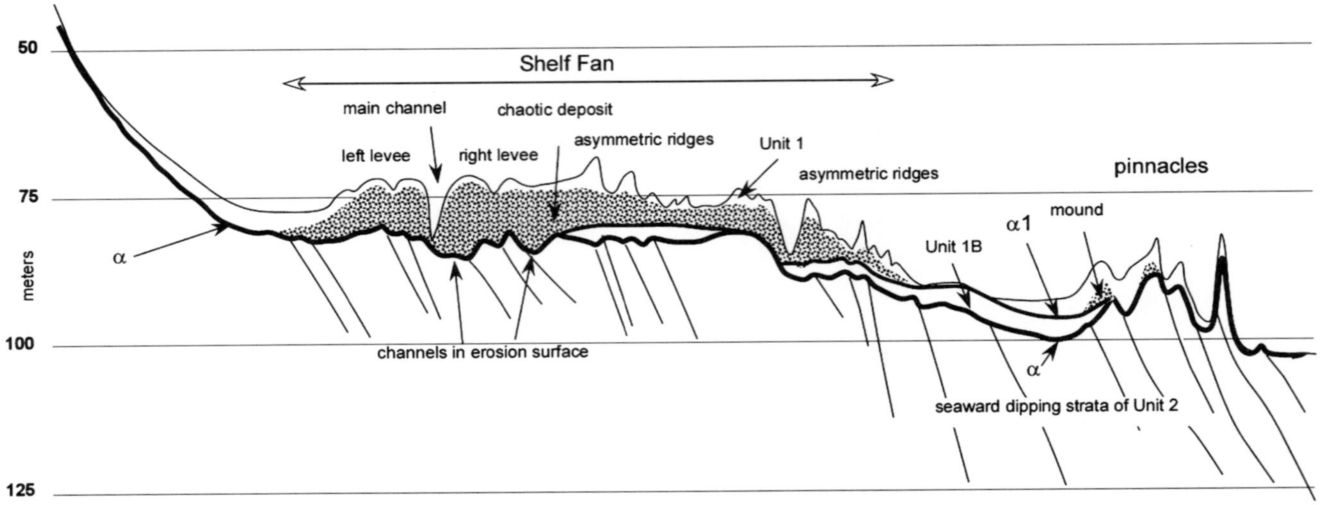

Fig. 2.4 Reflection profile extending from the mouth of Bosporus Strait, crossing the Shelf Fan and two of its channels and reaching Area 4. The interior "chaotic deposit)" of the fan consists of highly-reverberant strata. This deposit is draped by a more acoustically transparent cover. The Shelf Fan has been built upon reflector α1. Unit 1B is present only in the mid region of the Shelf Fan at depths between −80 and −95 m

Litho- and Bio-Stratigraphy

Where sampled by cores landward of the paleo-shoreline, the erosion surface is covered with a relatively thin (5–20 cm) deposit composed mostly of sand- to gravel-size fragments of freshwater mollusks (called a shell hash) with rare intact specimens of *Dreissena sp.* dated between 8.8 and 8.2 ka bp[1] (Major et al. 2002, 2006; Lericolais et al. 2007a, b). Some of the individually-dated *Dreissena* specimens co-occur in the same sample with *Monodacna caspia* and *Turricaspia caspia lincta* dated somewhat younger and between 7.9 and 7.2 ka bp and occasionally with *Cardium exiggum* (also called *Cardium edule* or *Parvicardium exiguum*) dated at 7.15 ka bp. All of the intact specimens (i.e., still articulated when sampled) of fresh to brackish affinity from this shell hash on these three margins have an $^{87}Sr/^{86}Sr$ isotopic composition (0.70897 – 0.70912) indicative of an on-going mixing of Mediterranean water having a value of 0.709155 with Black Sea water having a distinctly different isotopic composition of 0.70885 (Major et al. 2006). The shell hash often contains stained, abraded, polished and fragmented specimens. When such specimens are individually dated, they produce scattered ages. These shells, the shell debris (all bleached white) and detrital sand with occasional gravel that makes up the bulk of the shell hash, have been interpreted to be a lag deposit derived from the substrate while the surface of the continental shelf (i.e., its wave ravinement surface) was transgressed by the rising level of the lake (Lericolais et al. 2007a, b). Directly above the shell hash, one finds *Cardium exiggum* with an age no earlier than 7.15 ka bp (Ryan et al. 1997a, b, 2003; Major 2006). The fully euryhaline species *Mytilus galloprovincialis* appears with an age of 6.9 ka bp.

It is only at locations beyond the paleo-shoreline that one finds sandy mud with mollusks living during the period preceding Younger Dryas stadial and Bølling-Allerød interstadials of the terminal Pleistocene (Major et al. 2002; Ryan et al. 2003).

Methods

In July-Aug 2005, we investigated the internal structure of the shelf fan with reflection profiles calibrated by coring. Our surveys aboard the R/V *Mediterranean Explorer* deployed an Edgetech SB424 chirp reflection profiler navigated by GPS. N-S and E-W track lines were spaced 185–150 m apart. Each of the three survey regions was approximately 4 by 6 km on a side. The digital field data were sampled at a 0.125 ms interval across a 0.2 s window and recorded in jsf format. The files were converted to SEGY format for shipboard analysis with the TKS Kingdom Suite® software package. We digitized seafloor and subsurface reflectors to generate bathymetric and sediment thickness maps to assist the targeting of sites for sampling. Gravity coring was accomplished with a 10-cm diameter and 3-m long barrel propelled by a 400 kg core head. Cores were split, photographed, described and sampled at sea. Radiocarbon dates were obtained on ultrasonically cleaned single shells at the NOSAMS facility in Woods Hole, MA and at the Archaeometric Laboratory of the Weizmann

[1] Lower case bp indicates uncorrected AMS measurements. Calibration to calendar years is not provided due to inadequate knowledge of radiocarbon reservoir ages, as will be discussed. All dating is with single specimens.

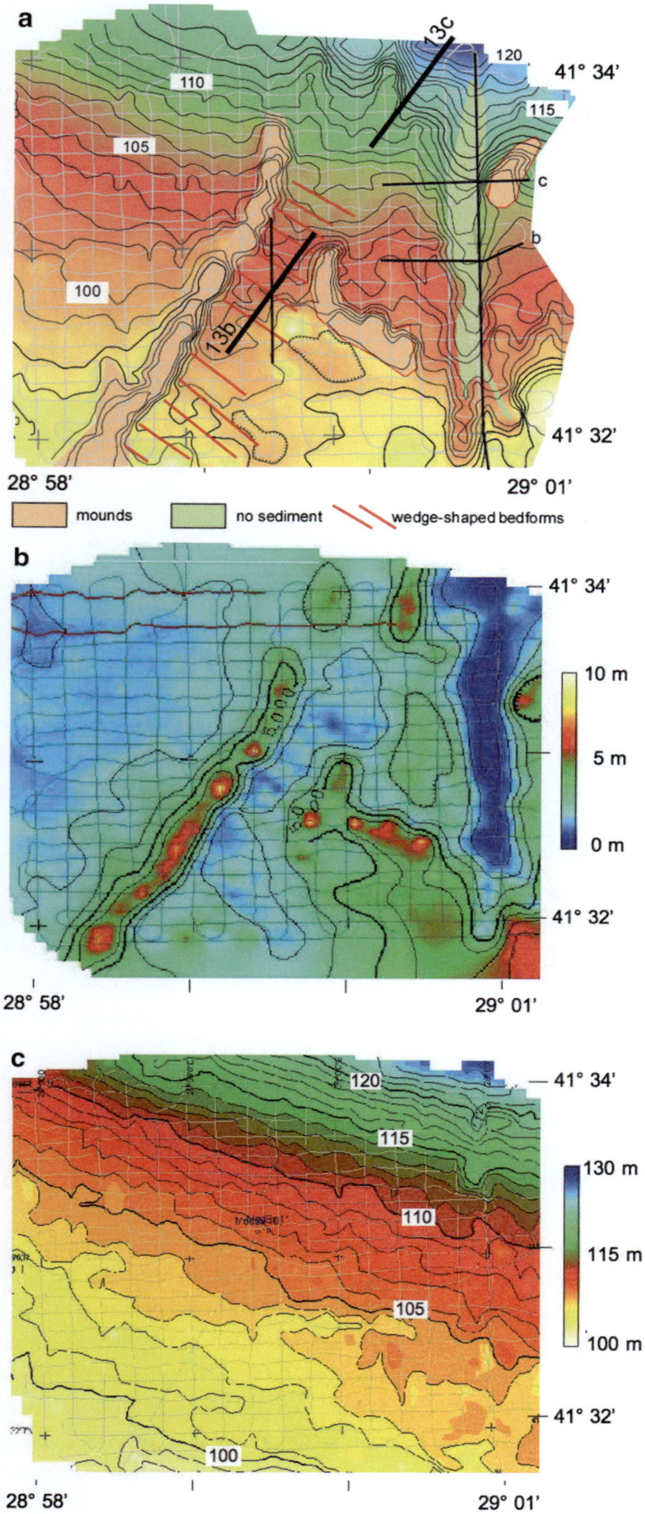

Fig. 2.5 Area 1–2. (**a**): Bathymetry contours showing a chain of mounds trending down slope towards the NE. Also depicted are the crests of asymmetric bedforms, and a channel where the sediment veneer above unconformity α is swept clean. Lines 13a and 13b are profiles of Aksu et al. (2002) and are labeled according to their figure number. The other lines (a, b, c) show the locations of chirp profiles illustrated in Fig. 2.6. (**b**): Thickness in meters of the Unit 1 sediments reflector α. The mounds (up to 9 m in relief) belong to Unit 1 and have formed on the α surface. Sediment is very thin to absent on the floor of the channel on the eastern side of the survey area. (**c**): Contours in meters of the α reflector revealing a regionally-smooth surface without channels and dipping to the NE

Institute of Science, Rehovot, Israel. Cores are archived (one half each) at the Istanbul Technical University and the University of Haifa.

We conducted our exploration in three rectangular regions and one transit line (Fig. 2.1). Two surveys are located near the shelf edge where the sediment cover above reflector α is sufficiently thin to penetrate by gravity coring. The third survey is on the eastern right-hand side (looking down stream) of the main channel at a location midway across the shelf and displaying two abandoned channels that are still evident in the bathymetry. A single chirp line crosses the channeled fan from shelf edge to inner shelf. Cores were obtained in areas 1–2 and 4.

New Results

Survey Area 1–2

Survey Area 1–2 is located between 40 and 50 km downstream from the Bosporus exit and therefore in the distal channel-levee complex close to the shelf edge at a depth of −95 to −125 m (Fig. 2.5a). Two channels cross this region. One channel extends south to north along the east side of the survey. For the most part, its floor is clean of the sediment cover that elsewhere overlies reflectors α and α1 of Aksu et al. (2002). The other channel is partly filled. It runs SW to NE and is bounded on its left bank by a linear chain of knolls. These mostly-conical hills have a relief of up to 12 m high above the underlying unconformity upon which they have formed (Fig. 2.5b). The authors adopt the term 'mounds' as the descriptive term for these knolls so as to be consistent with Flood et al. (2009) and Okay et al. (2011). However, in order to distinguish them from other mound-like features, these particular ones are called tall mounds.

The tall mounds are composed of an upper layer of wavy reflectors overlying an acoustically amorphous (e.g., lacking stratification) interior. Cores MedEx05-10 and MedEx05-11 were extracted, respectively, from the base and from the top of the same mound. Both cores recovered soft mud. However the core from the summit penetrated deeper and reached repetitive beds of large (5 cm) shells of *Mytilus galloprovincialis*. The valves of these shells were still articulated and were hosted in a matrix of stiff mud with voids produced by gas expansion after recovery. The chirp profiles display a drape of 2–5 m thickness for the wavy reflectors above the

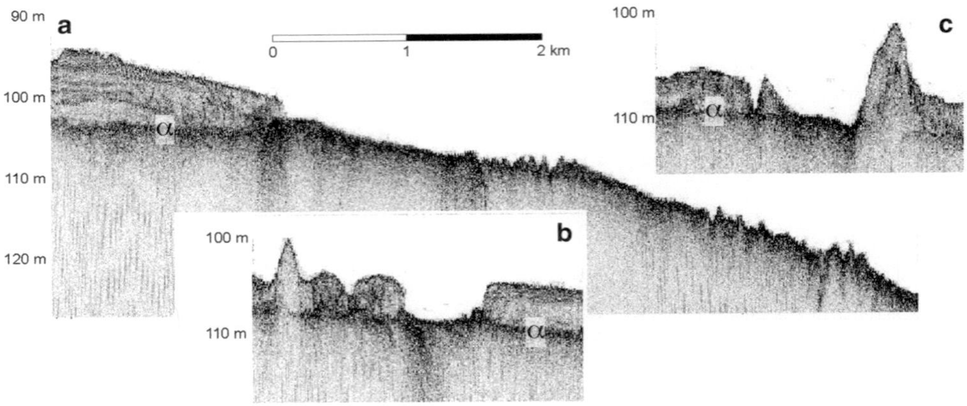

Fig. 2.6 Chirp lines *a*, *b*, and *c* (located in Fig. 2.5a), showing the absence of the Unit 1 sediment cover along the axis of the eastern channel. The saw-tooth nature of the reverberant bottom is the result of truncated underlying seaward-dipping beds. The reflective sub-surface horizon corresponds to reflector α

non-stratified interior that forms the primary relief and foundation of the tall mounds. Other numerous, but much lower relief (<1 m high), bumps populate the erosion surface. A few orient in lines, most are scattered. We also use the nomenclature of Aksu et al. (2002) to refer to these features as 'mounds', but call them small mounds. The latter are distributed everywhere across reflector α1. Their interiors contain reflectors that tend to be inclined in the upslope direction.

The floor of the abandon channel is populated by asymmetric, wedge-shaped bedforms spaced 150 – 500 m apart and built over the small mounds. The crests of these bedforms orient perpendicular to the channel axis; their steeper flanks commonly face downstream (Fig. 2.5a). These wedge-shaped features have evolved from amorphous (without stratification) cores that appear to have nucleated over the small mounds. The amorphous cores are followed by the progressive growth of inclined clinoforms advancing in the downdip direction and then capped by an upper layer of generally wavy reflectors that show progressive growth in the updip direction. Where sampled by coring the upper layer consists of mud with marine mollusks, foraminifera and ostracods.

The surface of reflector α is regionally smooth, but locally jagged, and deepens seaward towards the shelf edge at progressively steeper inclines (Fig. 2.5c). When the reflector is viewed in dip lines (Fig. 2.6), its small-scale roughness results from differential erosion of the underlying and more-steeply dipping strata. The toplap terminations of these underlying strata have been attributed to subaerial erosion (Aksu et al. 2002; Algan et al. 2007). The floor of the channel without observable sediment fill displays a high acoustic backscatter with amplitude equivalent to the strong backscatter from the floor of the main trunk (Flood et al. 2009).

Survey Area 4

Area 4 is also located on the outer shelf, but extends beyond the shelf break and onto the upper slope (Fig. 2.7a). Asymmetric bedforms are also present in this region of the shelf fan. However, these features are not necessarily confined to channel floors as in Area 1–2, but are widespread with lengths of crests reaching several kilometers. The narrow and steep flanks face down the regional dip and the broad and smooth flanks with their cover of wavy reflectors occur on the updip side. At the base of the steep sides, there are depressions where the substrate is exposed. Several tall mounds with heights up to 9 m occupy the SW region of the survey area. Except for the tall mounds the sediment thickness above reflector α is less than 2.5 m (Fig. 2.7b). However, just beyond the shelf break the sediment cover increases substantially. When stripped of the cover above reflector α, the surface of the underlying substrate is regionally smooth with the exception of a few gullies on the uppermost slope. Some of these gullies have cut into the shelf break (Fig. 2.7c).

Core MedEx05-13, at a depth of −118 m, penetrated a 53 cm-thick layer of soft mud, passed through reflectors α1 and α where they are indistinguishable from each other and bottomed in the substrate of steeply-dipping and truncated strata belonging to Unit 2 of Aksu et al. (2002). The sediment corresponding to reflector α is a 17 cm-thick bed of sub-rounded gravel and pebbles. The pebbles range from 0.2 to 4 cm in size. One of the larger pebbles is a metamorphic-grade quartzite (Fig. 2.8a). Another is a mildly-metamorphosed gabbro.

The pebble deposit is clast-supported with a matrix of firm sand containing disarticulated shells of *Dreissena sp.* and small gastropods of mixed ages. One *Dreissena* specimen was radiocarbon dated at 10.75 ± 0.07 ka bp and another abraded and eroded specimen at 28.2 ± 0.3 ka bp (Fig. 2.8b). A sharp angular contact separates the pebble bed from stiff

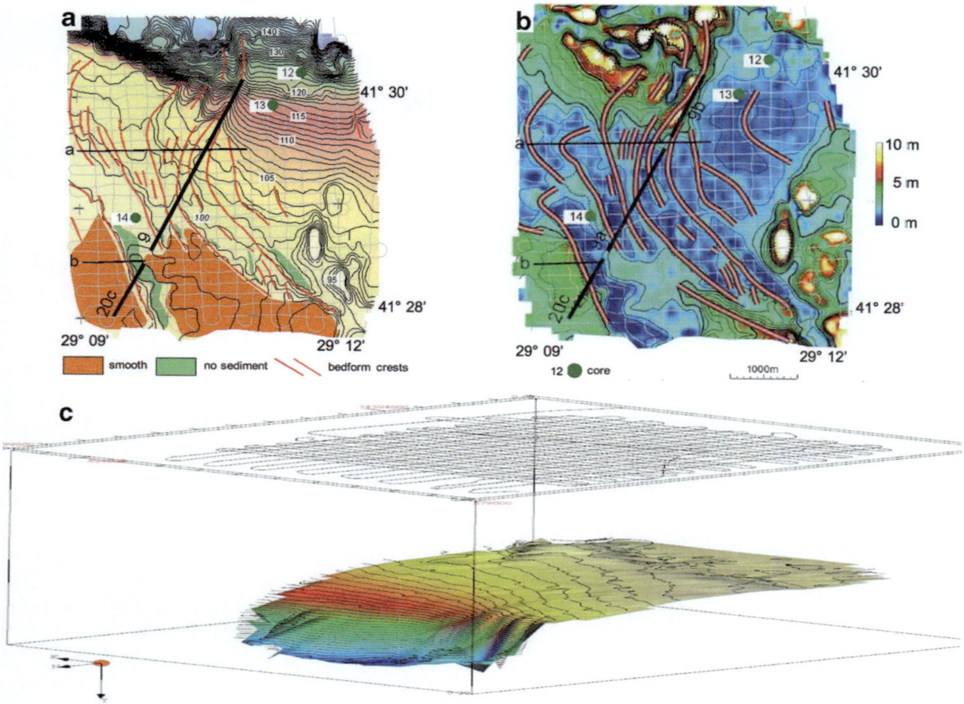

Fig. 2.7 Area 4. (**a**): Bathymetric contours showing the crests of asymmetric bedforms that curve and cross isobaths. Green dots are MedEX05 cores 12, 13 and 14. Lines 9 and 20c are profiles of Aksu et al. (2002) and are labeled according to their figure number. (**b**): Thickness in meters of Unit 1 above reflector α. All of the mounds have formed on the α surface. (**c**): Perspective view (from the NW) of the regionally-smooth α surface without channeling prior to the development of the overlying Shelf Fan

gray clay below containing specimens of Mytilus sp. both scattered and in densely packed beds. The Mytilus sp. are articulated and very well preserved with lustrous nacre (Fig. 2.8c). However, these Mytilus specimens from below reflector α are radiocarbon inactive and thus older than 50 ka bp.

The sediment above the pebble layer is soft olive gray mud with large specimens of *Mytilus galloprovincialis*. A specimen from the base of this layer provides an age of 6.86±0.06 ka bp. Cores MedEx05-12 and MedEx05-14 only recovered the upper wavy sediments above reflector α1, consisting of similar olive gray mud with abundant *Mytilus galloprovincialis*.

Survey Area 3

Area 3 is located in the mid shelf between −80 and −95 m where asymmetric bedforms on the shelf fan are oriented approximately NW-SE. There are fewer tall mounds here, but many small ones. (Fig. 2.9a). The tallest large mound stands 8 m above the surrounding seafloor. All mounds, both small and large, have accreted on reflector α1. The tall mounds have acoustically amorphous interiors that absorb sound energy. The attenuation of sound energy within the interiors partly to completely masks reflector α1 and the strata below this reflector.

The asymmetric bedforms are a lens-shaped deposit that grew from an initial core situated beneath the crest. When viewed in cross-section the cores appear to have acted as the site of nucleation above small reflective mounds. However, when mapped in plan view the small mounds and their overlying cores are elongate deposits whose orientation determined the strike for the subsequent growth of the bedform. At the base of the steeper down-dip flanks there are, once again, narrow depressions where the cover above reflector α1 is either extremely thin or absent. The distinguishing feature of this survey region is the greater thickness of the layer separating reflectors α1 and α and comprising subunit 1B of Aksu et al. (2002). In Area 3, this layer thins seaward and is not prevalent in the survey region beyond the −88 m isobath.

The sediment thickness map (Fig. 2.9b) reveals thin regions that may serve as passages for the plume of Mediterranean water that has kept them more or less sediment free. The floor of the western gap displays high acoustic backscatter. Along portions of both pathways, there are local exposures of reflector α1 and even reflector α.

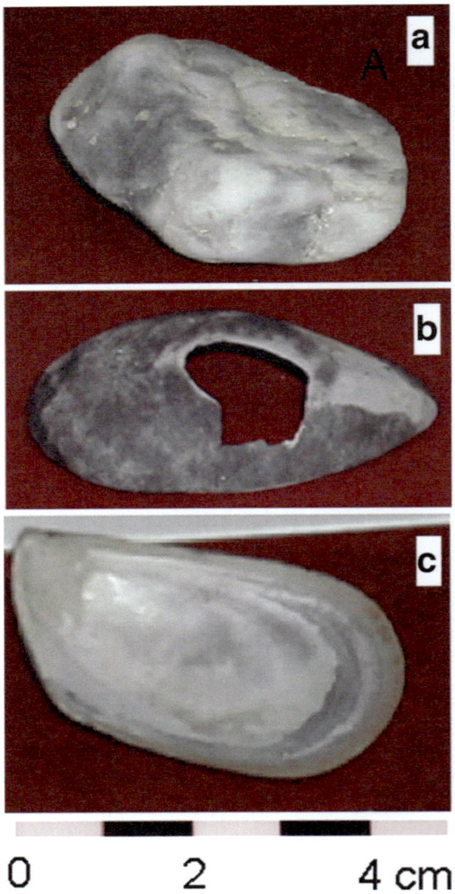

Fig. 2.8 (**a**) Sub-rounded pebble of quartzite from core MedEX05-13 recovered from the α reflector surface.; (**b**) abraded specimen of *Dreissena* dated at 22.6 ka bp from the same level as the pebbles; (**c**) intact specimen of *Mytilus galloprovincialia* with an age>50 ka

Transit Line

A chirp line extends from Survey Area 4 southward across the shelf fan toward the Bosporus Strait (Fig. 2.4). Additional lines to the west have been added in Fig 2.10. As the shelf fan is crossed, the layer between reflectors α1 and α first thickens and then thins to eventually disappear altogether at depths shallower than 75 – 70 m. It is here and shoreward where the shelf fan has been constructed directly on a jagged surface corresponding to reflector α. The initial deposit of the middle and inner shelf fan that provides the foundation for the bedforms, levees, and tall mounds is strongly reverberant, suggesting internal scattering from point sources. Its lack of stratification gives it a "chaotic" appearance in chirp profiles (Okay et al. 2011). The reverberant layer forms much of the underlying relief of the shelf fan. An upper wavy and more acoustically transparent layer of more uniform and reduced thickness drapes the "chaotic" foundation.

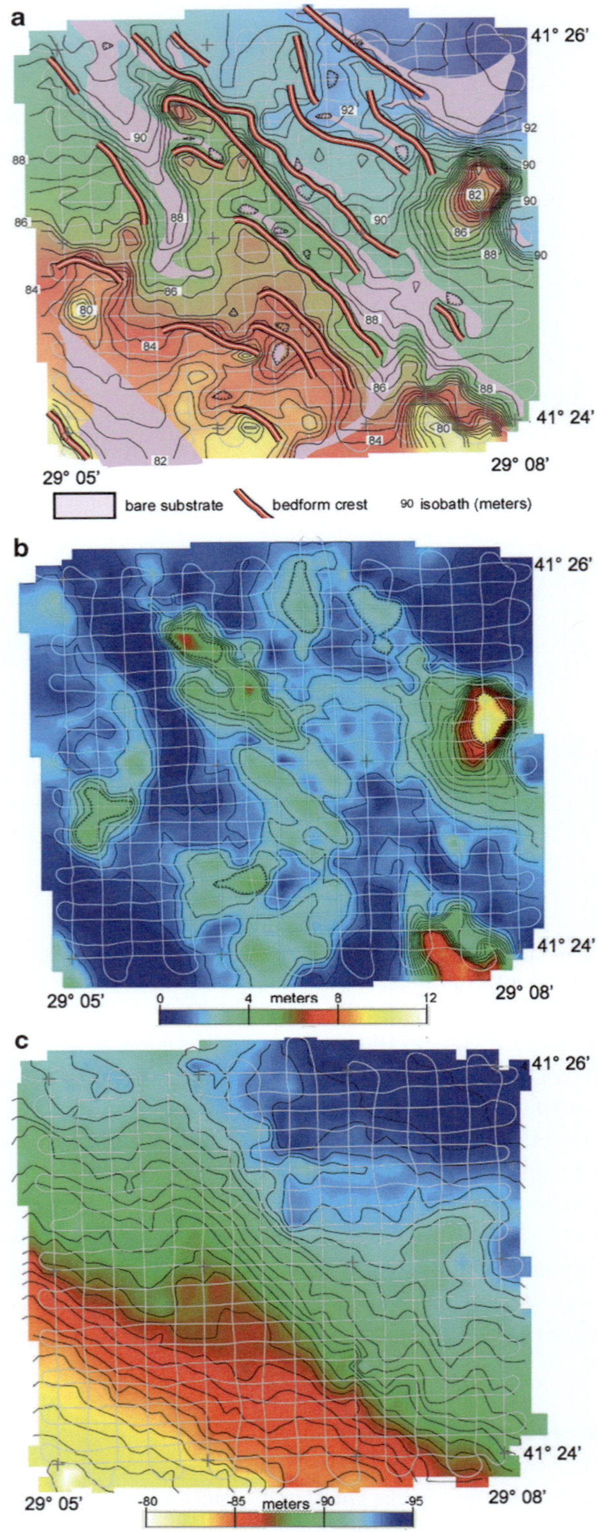

Fig. 2.9 Area 3. (**a**): Bathymetric contours showing crests of asymmetric bedforms. (**b**): Thickness in meters of Unit 1 sediments in meters above reflector α. Sediment is thin to absent in the channels that flow around the mounds and asymmetric bedforms. (**c**): Contours for the reflector α surface, again revealing this surface to be regionally smooth and dipping in the NE directions towards the shelf break

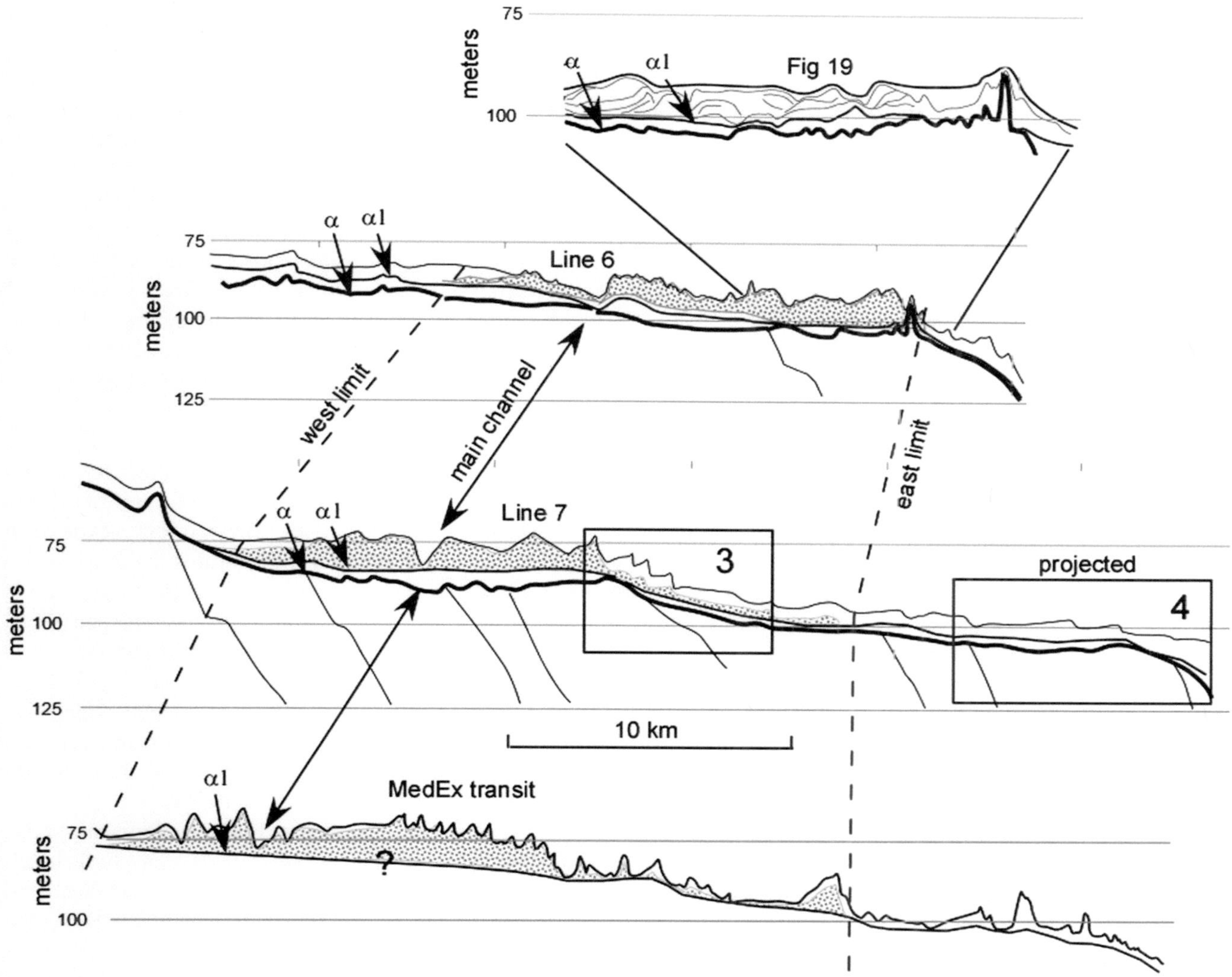

Fig. 2.10 Tracing and interpretation of profile 19 from Aksu et al. (2002), lines 6 and 7 from Algan et al. (2007) and the MedEx transit chirp line across the shelf-fan. These lines show the limits of the chaotic, reverberant deposit (stippled) and the position of this deposit directly on reflector αl and, in places, on reflector α. Line 7 passes through survey area 3. The bounds of survey area 4 are projected onto Line 7

Discussion

Hypothesis to Be Evaluated for the Origin of the Shelf Fan

The shelf fan is a unique feature on the margin of the Black Sea. The northward splay of the channels and branching pattern of the fan are clear evidence of sediment delivery from the Bosporus (Flood et al. 2009). In contrast, no similar sediment deposit is present where the ancient Danube, Dniepr and Don Rivers once extended onto the shelf during glacial times (Popescu et al. 2004; Ryan et al. 1997b; Konikov 2007; Glebov and Shel'tring 2007; Lericolais et al. 2009, 2011). Three explanations for sediment delivery from the Bosporus are discussed.

First, the fan may have been initiated as a sub-aerial delta sourced by a river flowing out of the Bosporus valley during a lowstand of the Black Sea lake, an option discussed by Özsoy et al. (2001).

Second, the fan was built more or less continuously over thousands of years by the inflowing Mediterranean undercurrent beginning ~8.0 ka and well after reconnection of the Black Sea with the Mediterranean at 10.5 ka (Flood et al. 2009).

Third, the deposit originated with the opening of the Bosporus at a time when the shore of the Black Sea lake was located on the outer shelf (Lericolais et al. 2003).

Lowstand River Delta

If the deposit was a river delta, one might ask why is such a delta so extraordinarily large and well-preserved here and not elsewhere? The drainage area feeding a delta via a Bosporus river is confined to the narrow strip of land between the Marmara and Black Seas. This surface area is miniscule compared to drainage basins of the major rivers emptying into the Black Sea from the interior of Asia and Europe. Yet, none of the rivers such as the Danube, Dnieper and Don have shelf fans on their offshore shelves.

Construction Delayed After the Initial Mediterranean Connection

Construction of the shelf fan from a subaqueous plume of Mediterranean origin is supported by many observations. The sediment cover is almost always thin to absent along the floors of active channels crossing through the survey areas. The local exposure of the underlying substrate suggests ongoing scour and/or non-deposition in the path of the saline inflow, especially where it is confined within channels or has broken through gaps in the levees (Flood et al. 2009). Continuous inflow with occasional over-bank spillage is also consistent with the progressive lateral growth of the mud-rich asymmetric bedforms populating the channel floors and levee backsides. Transport of Mediterranean water carrying suspended sediment across the fan is a compelling explanation for the observed accumulation on the upstream flanks of these bedforms. The large 200–1,000 m wavelengths and 1–5 m heights of the asymmetric bedforms are similar in many respects to abyssal mudwaves formed beneath thick nepheloid layers under conditions of supercritical flow (Flood 1988).

In respect to the timing of the birth of the shelf fan, the *Mytilus galloprovincialis* sampled directly above the $\alpha 1$ reflector in core MedEx05-13 provides a date of 6.8 ka bp. This age is practically identical to the observed appearance of this species elsewhere on the Romanian, Ukraine and Russian shelves (Major et al. 2006) and hardly distinguishable from the 6.7 ka bp age for the first appearance of benthic foraminifera overlying reflector $\alpha 1$ elsewhere on the fan in core B2KS02 (Okay et al. 2011). The dating links fan deposition quite firmly to a time when or shortly after Black Sea salinity was elevated sufficiently high to both enable colonization by euryhaline fauna and completely exclude the prior fresh to brackish fauna.

But how long did it take between the entry of Mediterranean water and the rise of salinity to the threshold indicated by the mollusks and foraminifera? Hiscott et al. (2007a, b) elaborate on this timing. They propose an unabated Black Sea outflow from ~10.5 ky bp to the present. Accordingly, this outflow formed a cascading river in the Bosporus Strait until 9 ka bp when the level of the surface of the external ocean and the Marmara Sea, in particular, reached the elevation of the Bosporus sill. Hiscott et al. (2007a, p. 112) write "from 9 to 8.4 ka bp, the outflow was sufficiently strong to prevent any measurable amount of saline water from entering the Black Sea, and so during this time, there was only one-way (outward) flow through the strait." This deduction was predicated on the observation of a delta-like deposit on the Marmara shelf south of Bosporus that allegedly formed between 10.5 and 9 ka bp and had been sourced by sediments swept southward out of the strait by the outflow. However, this delta-like deposit has since been cored and firmly dated as originating after 7 ka bp (Eris et al. 2007). Furthermore, detailed mapping shows that this deposit was delivered from a nearby river and not from the strait (Gökasan et al. 2005).

The age of 8.4 ka bp for the start of the inflow comes from the strontium isotopic signature of Black Sea mollusks (Ryan et al. 2003; Major et al. 2006). The $^{87}Sr/^{86}Sr$ measured in *Dreissena sp.* shells shows a shift towards the value of the global ocean beginning around that time. However, there is an apparent gap between 8.4 ka bp for *Dreissena* and 7.2 ka bp for the first euryhaline mollusks. Is this gap real or an artifact? If an artifact, this gap may be caused by unaccounted for radiocarbon reservoir ages for the incoming seawater and its influence on the ages of the marine mollusks and the prior lake water and its contribution to the measured ages of the brackish to freshwater mollusks?

As first proposed by Strakhov (1971), the inflowing saline water descended into the bottom of the lake due to its initial large excess density. In response to the growing supply of salty bottom water, the reserve of nutrients (N, P, SiO_2) and dissolved inorganic carbon that had been stored in the lake's freshwater interior would be "uplifted… into… the zone of photosynthesis." When ingested by mollusks living during the period of salinity change, this old carbon (with its reduced ^{14}C activity) would start the radiocarbon clock within the shell beginning with the lake's ^{14}C reservoir age.

Accurate chronologies require knowledge of the lake's reservoir age. Ryan (2007) wiggle-matched variations in stable isotopes and other proxies in Black Sea cores with the Greenland GISP ice-core measurements and derived an approximate 1,600 year reservoir age for the lake at the time of the post-glacial meltwater pulse. Kwiecein et al. (2008) compared radiocarbon ages from ostracod and mollusk shells with the known ages of volcanic ash layers (tephrachronology) to achieve a similar estimate of 1,400 years for the slightly older glacial period. Soulet et al. (2011) correlated variations in Ca and Ti/Ca signals in Black Sea core MD04-2790 to isotopic excursions in the Hulu Cave stalagmites and calculated a $1{,}570 \pm 380$ year reservoir age for the same early post-glacial period. These three independent investigations document that not only the lake's interior, but also its coastal waters had been able to acquire a large reservoir age.

The ^{14}C reservoir estimates ages for the Black Sea s water after the Mediterranean connection only come from modern shells in museum collections that were collected alive at a known place and year. The calculated reservoir ages for these shells range from 460 years (Jones and Gagnon 1994) to 415 years (Siani et al. 2000). Using an alternate method of directly measuring the dissolved inorganic carbon ^{14}C activity in the Black Sea water column, Fontugne et al. (2009) found that modern surface waters are generally in equilibrium with the ^{14}C atmospheric concentration. However, carbon ^{14}C activity decreases with water depth and offshore distance resulting in reservoir ages that may reach or exceed 200 years at the shelf edge and reach more than 1,000 years in the deep interior.

The marine mollusks discussed previously and with age spans of 6.9–7.2 ka bp were recovered at sites on today's shelf. Although the reservoir age of seawater 7,000 years ago is likely to have been influenced by depth and distance, the particular carbon ^{14}C activity incorporated into the shell might have also been further controlled by whether the mollusk lived on the seabed or in the mud. For example mussels, such as *Mytilus galloprovincialis*, attach to the substrate and filter the water in which they are bathed. In contrast, clams, such as *Cardium sp.*, live in the substrate of previously deposited sediment where they come into contact with carbonate minerals, decaying organic matter and dissolved methane. River water also contributes a significant reservoir age due to dissolution of bedrock and input of old groundwater in the watershed (e.g., 1,420 years for the Colorado River (Goodfriend and Flessa 1997)). Mollusks on and in the beds of rivers, estuaries, lagoons, coastal sea and the open sea might live simultaneously in waters possessing diverse reservoir ages. As a result their shells would display a sizeable spread of measured radiocarbon ages.

One way to examine the abruptness of the initial salinization of the Black Sea's lake is to investigate cores recovered at sites that were submerged before and after the connection at Bosporus. The presumption is that in quiet, deep-water environments sediment accumulation is likely to have been continuous throughout the period of salinity change. There are two cores of particular relevance (BLKS9811 at –500 m and GGC18 at 971 m depth). In both cores radiocarbon measurements were made on bulk organic carbon sampled from the uppermost part of the freshwater Unit 3 sediment and just a few cm from the base of the Unit 2 sapropel (Ross and Degens 1974). The measured dates are 8.59 ± 0.045 (Popescu 2002) and 8.65 ± 0.04 ka bp, (Filipova-Marinova et al. 2013), respectively. An age of 7.35 ± 0.08 ka bp for the base of the sapropel has been determined from many cores (Jones and Gagnon). Thus in cores BLKS9811 ands GGC18 one finds an apparently age offset across the Unit 2/3 boundary exceeding 1,000 years. To account for this offset, there could have been either a depositional hiatus of this duration at the same stratigraphic position in each core, or a substantial reduction in reservoir age during the transition from lake to sea.

A sample with shells of ostracods located 1 cm below the base of the Unit 2 in core BLKS9810 at water depth of −378 m has an $^{87}Sr/^{86}Sr$ value of 0.709086. This measurement indicates that some Mediterranean water had already been introduced prior to the deposition of the Unit 2 sapropel (Major et al. 2006). Exactly when the intrusion occurred is as yet unknown. The next sample of ostracods 10 cm further down in this core is distinctly lacustrine with a composition of 0.708908.

Initiated with the Entry of Mediterranean Water Coincident with a Lowstand Lake

There are two cores from the shelf fan with pebble- and gravel-size rock fragments mixed with shells of freshwater *Dreissen sp.*: MedEx05-13 from the distal fan (already discussed), and M00-12 (Aksu et al. 2002) from beneath the left levee crest of the main trunk in the mid fan (looking downstream). Core 5 (Algan et al. 2007), located less than 1 km from core MAR00-12 and on the same left levee, sampled fragmented shells of *Dreissena sp.* in a coarse sand layer that stopped the penetration of the core barrel. Flood et al. (2009) attribute the shells and pebbles in M00-12 to be material reworked from the underlying unconformity α.

However, based on their depths in the cores in comparison to the thickness of the shelf fan deposits at the respective core sites, the pebbly sand in core M00-12 and the fragmented shells of *Dreissena sp.* in Core 5 can not be lag deposits associated with unconformity α but, instead, is material from the interior of the shelf fan itself. In the case of core MedEx05-13, the pebble deposit was sampled from the level of reflector α1 where it merged with reflector α. These subsurface coarse deposits indicate that earlier in the fan's history, the energy of the Mediterranean inflow was greater than at present.

Figures 13, 20 and 22 of Aksu et al. (2002) illustrate that reflector α1 serves as the foundation for the numerous small and "irregular mounds" that are present and widely-scattered throughout each survey area. In turn, these mounds serve as the underpinning for the initial acoustically amorphous (e.g., lacking internal reflectors) "core" of the asymmetric bedforms.[2] The mounds often align in straight or curving pathways that resemble trajectories. Their dense arrangement in rows beneath the lateral levees to each side of the currently active and abandoned main trunks suggests that the shelf fan is a deposit inherited from precursor mounds.

Aksu et al. (2002) suggested that the tall mounds might be mud volcanoes. These authors were the first to note that some mounds line up in straight paths as documented in

[2] Although these asymmetric bedforms were previously interpreted as "barrier island/beaches on the basis their geometry" (Aksu et al. 2002), this interpretation is no longer advocated (Flood et al. 2009).

Survey region 1–2. The presence of expansion pockets in the cores from the tall mounds is supportive of gas venting from below. However coring through an entire mound, as was accomplished with B2KS02, did not reveal an internal body of extruded or intruded mud originating from the underlying substrate (Okay et al. 2011). What the coring did show is a rapid growth rate for the reverberant interior of the mounds and a slowdown for the accumulation of the upper drape of wavy reflectors. The reflection profiles reveal that the mounds were preceded by erosion capable of washing away some and in various regions all of the preceding Unit 1B sediments and leaving a jagged and channeled surface on the top of the older inclined strata belonging to Unit 2.

The velocity of ~20–40 cm s^{-1} of the present saline plume is driven primarily by the salinity difference of ~12–16 psu between Black Sea surface water and the inflowing Mediterranean water with some reduction due to their differing temperatures (Özsoy et al. 2001). The calculated density contrasts are ~0.009–0.012 g cm^{-3} (Flood et al. 2009). Based on the analysis of pore water extracted from a long core on the Black Sea continental slope and accounting for advection and diffusion, Soulet et al. (2010) conclude that the salinity of the Black Sea lake before its Mediterranean connection was no more than 1 psu. Thus the density contrast between the initially inflowing saltwater and the fresh lake might have been two to three times greater at the time of the connection than today.

When confined to channels the modeled peak mean velocity of the modern density-driven current is fully capable of suspending fine sand and transporting coarse sand in bedload (Hiscott 1994). Due to the earlier higher density contrast, one might expect flow-velocities competent enough to transport the gravel and even small pebble-size material recovered in cores M0012 (Aksu et al. 2002). However, concurrent with the addition of more and more Mediterranean water, the density contrast would drop and the velocity of the plume would lessen. The stepwise construction of the asymmetric bedforms beginning with mounds overlain by a core of acoustically amorphous material, followed in turn by highly-reflective clinoforms and then passage to the wavy cover might all have been the consequence of a diminishing carrying capacity for the inflow.

Çağatay et al. (2000) and Kerey et al. (2004) report on a 40 m long borehole, BPMP-14, in the north-central part of the Bosporus Strait that penetrated down through the entire fill and into the Paleozoic basement. Starting at the bedrock contact, the succession of sediment begins with sand containing *Dreissena* shells and ostracods. This lower unit has a radiocarbon age of 26.1 ka bp measured on *Dreissena* near its bottom and 16.6 ka bp on *Dreissena* from its mid-section (Çağatay et al. 2000). The lower unit is abruptly capped by a fining-upward sequence of large pebbles, gravel and sand with shells of *Ostrea*, *Mytilus* and *Cardium*, followed by clays with *Ostrea* fragments and lastly with a coarsening-upward deposit with sand at the top. The pebbles at the base of this upper unit consist of gabbro and vein quartz with the same description as those recovered in core MedEX05-13 from the outer Black Sea shelf fan. The fining-upwards texture indicates an initial high-energy environment, followed by decreasing strength. The mollusks and assemblage of foraminifera in the upper unit signal sediment accumulation during the passage of Mediterranean water through the strait.

A date of 5.3 ka bp on an *Ostrea* shell from the base of the upper unit makes this specimen the oldest known euryhaline species of Mediterranean provenance sampled from the subsurface of the Bosporus Strait (Çağatay et al. 2000). Its age, delayed 2,000 years after the first appearance of *Cardium* elsewhere on the Black Sea shelf, attests that the erosive and non-depositional power of the bottom current of saline water passing through the strait. The inflow may have been vigorous for quite some time before its diminishing velocity allowed the accumulation of a permanent deposit at the location of the borehole.

If the entry of Mediterranean water occurred at a time when the lake's shoreline resided on the outer shelf, the incoming water would begin with a hydraulic head tied to the level of the external Marmara Sea. As a first trickle evolved into a stream and the stream grew in energy to the threshold where it could then enlarge the entry portal, the positive feedback would quickly rupture the Bosporus divide. However, unlike the overtopping and subsequent washing away of a natural or man-made dam, the supply of water from the exterior ocean would have been more or less infinite and not just the volume of the supplying reservoir.

We consider it likely that the outburst phase was rich in sediment scoured from the entry portal and delivered to an apron on the shelf in sheets of chaotic debris. The outburst phase would then be followed by a water-flood phase. The density contrast between the incoming saltwater and the air it would replace would be 30 times greater than the aforementioned contrast concerning saltwater and freshwater. The cascade of saltwater would travel across the debris apron in jets. According to the shock-capturing numerical model of Siddall et al. (2004) there would have been a hydraulic jump where the cascade of water met the shore of the lake. Downstream of the jump the cascade would turn into a jet capable of penetrating into the interior of the lake. As the lake filled and the shelf submerged, the jet would no longer have been guided by pre-existing channels and thus may have been capable of a wide scouring of the shelf. Although the numerical model of Siddall et al. (2004) did not include a sediment dynamics module, the authors argue that erosion by the jet would activate a positive feedback between the flow and the shelf bed, leading to entrenchment and flow channeling through the debris apron. The water-flood phase shaped the outburst debris into the small mounds scattered widely as the initial foundation of the shelf fan. The early tendency towards confinement might have provided a certain stability to the flow, distributing the mounds in straight and curvilinear paths. The updip stratification in the mounds is indicative of supercritical flow. Asymmetric bedform wavelengths of 150–500 m indicate flow velocities of 15–25 m s^{-1} (Kennedy 1963). The energy of turbulence is high enough to make channel avulsions and bifurcations likely, resulting in the observed splay of channels.

Where the jets traveled in streams across the shelf, velocities would certainly have been capable of transporting cobbles and pebbles in the size range recovered in core MedEX05-13. The scattered small mounds on the outer shelf are possibly deposits of a rapidly shifting braid plain as the flume plucked material from the debris apron and scoured material from Unit 1B. The evolution from multiple distributaries to a single active channel could represent the evolution from a terrestrial fan to a submerged fan as the lake expanded and deepened. Such progression accounts for the two-part structure of the fan deposit as well as the fining-upward character of the sediments. Once the rising shoreline entered the Bosporus, the jet would transition to a stable undercurrent (Siddall et al. 2004).

The coincidence in time between the appearance of euryhaline fauna (mollusks and foraminifera) of Mediterranean provenance in the Black Sea as recorded in their presence in the sediment recovered in cores from immediately above reflector α1 and the onset of the construction of the shelf fan directly over reflector α1 is considered an important constraint for hypotheses related to fan origin. A gradual rise in salinity over an interval of thousands of years after the connection had been made does not account for the synchronicity. No *insitu* lacustrine fauna (e.g., s specimens with articulated valves) are found above reflector α1 in the cores from the shelf fan, only shell debris. Conversely, no euryhaline fauna are found below reflector α1, except in the much older Unit 2 strata below reflector α.

If there was such a sudden flooding of the Black Sea's lake, how profound was the inrush of water? From their synthesis using seismic reflection and bathymetric surveys in the Bosporus Strait and on the Black Sea shelf just beyond the exit from the strait Gökasan et al. (2005) have recognized an inner channel carved into prior fill and displaying an increased depth of erosion from south to north towards the Black Sea. Their estimate of the volume of sediment eroded is 2×10^8 m^3. This volume is comparable in magnitude to the chaotic deposit in the subsurface of the shelf fan.

How long might it have taken for the lake surface to rise to the level of the exterior ocean, once the inflow started? The volume of water to fill the lake from its –90 m isobath to its –20 m isobath is approximately 3×10^4 km^3. Using the inflow rate calculated by Siddall et al. (2004), the resulting duration would have been on the order of 15–20 years and perhaps more if it took awhile for the trickle to evolve into an stream with eroding power. This duration is only a little longer than the 5-year duration first estimated by Ryan et al. (1997a) or the 8-year duration of Myers et al. (2003).

Examples of Other Flooding Events

With the break-up of ice-dams, some pro-glacial lakes in North America emptied abruptly to produce megafloods (Baker et al. 1981; Clarke et al. 1984). Downstream, torrents of lake water carved outflow channels into the terrestrial landscape before reaching the sea. The outburst floods from Lake Missoula in Montana produced a channeled erosion surface called 'Scabland' with tear-shaped islands within the channels and chaotic coarse deposits scattered across the flood path (Bretz 1923; 1927; 1969). Estimates of maximum discharge reach 10 million m^3/s or nearly 900 km^3 per day (Baker 1981). According to Shaw et al. (1999), the total reservoir volume that drained through the Scabland may have been on the order of 10^5 km^3. This stored volume, about three times more than the amount of water required to fill the Black Sea's lake from its –90 m isobath to its –20 m isobath, would have sustained a continuous discharge for a period of about 100 days.

The chaotic outwash deposits in the Channeled Scabland in southern Washington State, United States are exemplified in the Starbuck Section upstream of the Wallula Gap. The outwash contains rhythmic sequences with depositional bedforms hosting boulders, cobbles and gravel along with ripple drift cross-laminated sand indicative of transport primarily by bedload (Baker 1978). It is from a comparison of depositional bedforms in southern Washington to those observed in the Black Sea's shelf fan and from the similarity in composition of the pebbles sampled near the shelf edge and those recovered by drilling in the Bosporus Strait that we attribute the pebble deposit in the fan to a outwash origin.

Another catastrophic meltwater flood event has been linked to the formation of the Hudson Shelf Valley (Donnelly et al. 2005; Thieler et al. 2007). This flood drained postglacial lakes in the Hudson and Ontario basins through a breach in the terminal moraine dam located at the Narrows between Staten Island and Long Island, New York. The initial outburst phase delivered an acoustically amorphous high-volume deposit as thick as 5 m, interpreted as a non-cohesive sediment debris flow (Uchupi et al. 2001). The amorphous deposit rests unconformably on eroded channel fill. The massive internal character alludes to rapid emplacement (Thieler et al. 2007). A prominent feature of the valley floor is an elongate field of bedforms confined within channel levees. The bedforms have heights of 3–6 m and wavelengths of 150–300 m. Crests are oriented perpendicular to the valley axis. The Hudson Shelf Valley bedforms are remarkably similar to the bedforms observed on the channel floors of the Black Sea shelf fan, especially in respect to their height, wavelength, cross-channel orientation and updip internal cross-stratification.

Thieler at al. (2007) attribute the shaping of the bedforms to the succeeding water-flood phase while the glacial lakes continued to empty. The volume of the total discharge has been estimated at 3,200±480 km^3 (Rayburn et al. 2005) or about 10 % the magnitude of water required to fill the Black Sea's lake from its pre-flood shoreline to the level of the external ocean.

Combination of Processes

The catastrophic flood explanation does not rule out the role of the Mediterranean undercurrent in subsequently shaping the bulk of the fan in a sustained subaqueous environment. Nor does it exclude the presence of a terrestrial delta at the edge of the lake before the inflow began. The flood hypotheses is present to account primarily for the scattering of mounds and chaotic deposits on surfaces sculpted by erosion and secondarily to link the introduction of fauna of Mediterranean provenance with the initiation of fan growth above the erosion surfaces. The flood hypothesis predicts that when more of the deep interior of the fan is sampled by coring or drilling, the inventory of a pebbly outwash will enlarge.

Conclusions

Surveys in a few key locales on the continental shelf of the Black Sea north of the Bosporus confirm that a channel-levee complex in the shape of a shelf fan consists of two depositional sequences. The earliest deposit is acoustically amorphous, internally-chaotic and coarse. The chaotic deposit is coeval with a series of small mounds that are scattered across the outer shelf and aligned along the principal pathways of the channel network. Both the mounds and the fan above nucleate right on reflector $\alpha 1$ where it is present or reflector α where Unit 1B is absent. The upper sequence of the shelf fan is finer-grained and wavy in reflection profiles. This cover has accumulated from sediment delivered by channeled flow as well as by over-bank spillage.

The shelf fan owes its origin to multiple stages and processes. We propose that it began as a channeled scabland scoured by an outburst of water and sediment rushing into the Black Sea's lake and triggered when the exterior ocean breached the Bosporus divide. As the surface of the lake rose and its shorelines retreated landward, the mostly-eroded and current-sculpted landscape transformed into an apron of deposition. The energy of the Bosporus jet diminished and the flow became confined to channels. The sediment that is observed between reflectors $\alpha 1$ and α on the outer shelf is absent landward of the −75 m isobath. If this sediment had once been there, it was removed simultaneous with the initiation of the fan. Here on the inner shelf, the base of the fan rests directly on a jagged erosion surface corresponding to reflector α.

A first the shelf fan was supplied mostly by material eroded from the inner channel of the Bosporus Strait. The volume of this material is similar to the volume of the chaotic deposit beneath the levees, but this does not mean that all of the fan deposit was supplied from the Bosporus. Mediterranean water is still shaping the fan to the present time. However, coincident with the rise of salinity of the Black Sea, the density contrast between its surface water and the inflowing plume has diminished. As a result, the velocity of plume has slowed along with its ability to transport coarse material. The dating of shells in the mud from the wavy layer of the fan suggests that this slowing became significant 2,000 years after the initial flood, more or less coincident with the fining-upward sediment layer sampled in borehole BPMP-14 in the north-central part of the Bosporus.

Acknowledgements We thank Danny Schaffer, Master of the *R/V Mediterranean Explorer* and his crew for their expertise in obtaining the nautical and scientific objectives of the expedition. The project was made possible with the financial support of the Eco-Ocean organization of Herzliya, Israel. Daniela Di Iorio and Seda Okay generously contributed their digital elevation models of multibeam bathymetry. Liviu Giosan provided the Edgetech Chirp equipment and participated in the expedition. Discussions with him at sea and during earlier stages of the manuscript preparation are gratefully acknowledged. We show our appreciation to Candace Major, Seda Okay, Gilles Lericolais. Oya Algan, Erkan Gökasan, and Kadir Eris for their contributions of data and ideas.

References

Aksu AE, Hiscott RN, Mudie PJ, Gillespie H, Yasar D (2002) Last glacial–Holocene sea surface temperature and salinity variations in the Black and Marmara sea: stable isotopic, planktonic foraminiferal and coccolith evidence. Mar Geol 190:119–149

Algan O, Gokasan E, Gazioglu C, Yucel Z, Alpar B, Guneysu C, Kırcı E, Demirel S, Sarı E, Ongan D (2002) A high-resolution seismic study in Sakarya Delta and Submarine Canyon, southern Black Sea shelf. Cont Shelf Res 22:1511–1527

Algan O, Ergin M, Keskin S, Gökaşan E, Alpar B, Ongan D, Kırcı-Elmas E (2007) Sea-level changes during the late Pleistocene-Holocene on the southern shelves of the Black Sea. In: Yanko-Hombach V, Gilbert AS, Panin N, Dolukhanov P (eds) The Black Sea flood question: changes in coastline, climate, and human settlement. Springer, New York

Arkhangel'skiy AD, Strakhov NM (1938) Geological structure and history of the evolution of the Black Sea. Izvestiya Akademii Nauk SSSR 46:103–104

Baker VR (1978) Large-scale erosional and depositional features of the channeled scabland. In: Baker VR, Nummedal D (eds) The

channeled scabland. Planetary Geology Program, Office of Space Research, NASA, Washington, DC
Baker VR (1981) Catastrophic flooding, the origin of the channeled scabland. Dowden/Hutchinson & Ross, Philadelphia
Berman-Frank I, Rahav E (2012) Ch. 8: Nitrogen fixation as a source for new production in the Mediterranean Sea: a review. In: Stambler NE (ed) Life in the Mediterranean Sea: a look at habitat changes. Nova Science Publishers, New York, pp 199–226
Bethoux JP, Copin-Montegut G (1986) Biological fixation of atmospheric nitrogen in the Mediterranean Sea. Limnol Oceanogr 31:1353–1358
Bethoux JP, Morin P, Madec C, Gentili B (1992) Phosphorus and nitrogen behavior in the Mediterranean Sea. Deep-Sea Res 39: 1641–1654
Bogdanova AK (1963) The distribution of Mediterranean waters in the Black Sea. Deep-Sea Res 10:665–672
Bonnet S, Grosso O, Moutin T (2011) Planktonic dinitrogen fixation along a longitudinal gradient across the Mediterranean Sea during the stratified period (BOUM cruise). Biogeosciences 8:2257–2267
Bretz JH (1923) The Channeled Scabland of the Columbia Plateau. J Geol 31:617–649
Bretz JH (1927) Channeled Scabland and the Spokane flood. Wash Acad Sci J 17:200–211
Bretz JH (1969) The lake Missoula floods and the channeled Scabland. J Geol 77:505–543
Çağatay MN, Görür N, Algan O, Eastoe C, Tchepalyga A, Ongan D, Kuhn T, Kusçu Ì (2000) Late glacial- Holocene palaeoceonography of the sea of Marmara: timing of the last connections with the Mediterranean and the Black Seas. Mar Geol 167:191–206
Capone DG, Zehr JP, Paerl HW, Bergman B, Carpenter EJ (1997) *Trichodesmium*, a globally significant marine cyanobacterium. Science 276:1221–1229
Chepalyga AL (1984) Inland sea basins. In: Velichko AA (ed) Late Quaternary environments of the Soviet Union. University of Minnesota Press, Minneapolis
Clarke GKC, Mathews WH, Pack RT (1984) Outburst floods from glacial lake Missoula. Quat Res 22:289–299
Di Iorio D, Yüce H (1999) Observations of Mediterranean flow into the Black Sea. J Geophys Res 104:3091–3108
Di Iorio D, Akal T, Guerrini P, Yüce H, Gezgin E, Öszoy E (1999) Oceanographic measurements of the West Black Sea: June 15 to July 1996. Saclant Undersea Research Centre Report SR-305, La Spezia
Dimitrov PS, Govberg LI, Kuneva-Abadzhieva VI (1979) Marine Quaternary deposits of the periferal region of the shelf from the western part of the Black Sea. Oceanol Bulgarian Acad Sci 5:67–78
Donnelly JP, Driscoll NW, Uchupi E, Keigwin LD, Schwab WC, Thieler ER, Swift SA (2005) Catastrophic meltwater discharge down the Hudson River valley: a potential trigger for the Intra-Allerød Cold Period. Geology 33:89–92
Eris KK, Ryan WB, Çağatay MN, Sancar U, Lericolais G, Menot G, Bard E (2007) The timing and evolution of the post-glacial transgression across the sea of Marmara shelf south of Istanbul. Mar Geol 243:57–76
Evsylekov YD, Shimkus KM (1995) Geomorphological and neotectonic development of outer part of continental margin to the south of Kerch Strait. Oceanology 35:623–628
Fedorov PV (1988) The problem of changes in the level of the Black Sea during the Pleistocene. Int Geol Rev 30:635–641
Filipova-Marinova M, Pavalov D, Coolen M, Giosan L (2013) First high resolution marinopalynological stratigraphy of Late Quaternary sediments from the central part of the Bulgarian Black Sea area. Quat Int 293:170–183
Flood RD (1988) A lee wave model for deep-sea mudwave activity. Deep-Sea Res A35:973–983
Flood RD, Hiscott RN, Aksu AE (2009) Morphology and evolution of an anastomosed channel network where saline underflow enters the Black Sea. Sedimentology 56:807–839
Fontugne M, Guichard F, Strechie C, Lericolais G (2009) Reservoir age of the Black Sea waters during anoxic periods. Radiocarbon 51:969–976
Garcia N, Raimbault P, Gouze E, Sandroni V (2006) Nitrogen fixation and primary production in western Mediterranean. C R Biol 329:742–750
Genov I (2004) Way and time of swell formation near the Bulgarian Black Sea shelf. C R Acad Bulgare Sci 57:95–98
Glebov AY, Shel'ting SK (2007) Sea-level changes and coastline migrations in the Russian sector of the Black Sea: application to the Noah's Flood hypothesis. In: Yanko- Hombach V, Gilbert AS, Panin N, Dolukhanov PM (eds) The Black Sea flood question: changes in coastline, climate, and human settlement. Springer, New York
Gökasan E, Toroğlu H, Ecevitoğlu B, Gorum T, Türker A, Tok B, Çağlak F, Birkan H, Şimşek M (2005) Evidence and implications of massive erosion along the Strait of Istanbul (Bosphorus). Geo-Mar Lett 25:324–342
Gökasan E, Demirbağ E, Oktay FY, Ecevitoğlu B, Simsek M, Yüce H (1997) On the origin of the Bosphorous. Mar Geol 140:183–197
Goodfriend GA, Flessa KW (1997) Radiocarbon reservoir ages in the Gulf of California: roles of upwelling and flow from the Colorado River. Radiocarbon 39:139–148
Gregg M, Öszoy E (1999) Mixing on the Black Sea shelf north of the Bosphorus. Geophys Res Lett 26:1869–1872
Gruber N, Sarmiento JL (1997) Global patterns of marine nitrogen fixation and denitrification. Global Biogeochm Cycle 11:235–266
Hiscott RN (1994) Loss of capacity, not competence, as the fundamental process governing deposition from turbidity currents. J Sediment Res A64:209–214
Hiscott RN, Aksu AE, Yasar D, Kaminski MA, Mudie PJ, Kostylev VE, McDonald JC, Isler FI, Lord AR (2002) Deltas south of Bosphorus Strait record persistent Black Sea outflow to the Marmara sea since 10 ka. Mar Geol 190:95–118
Hiscott RN, Aksu AE, Mudie PJ, Marret F, Abrajano T, Kaminski MA, Evans J, Cakıroglu AI, Yasar D (2007a) A gradual drowning of the southwestern Black Sea shelf: evidence for a progressive rather than abrupt Holocene reconnection with the eastern Mediterranean Sea through the Marmara Sea Gateway. Quat Int 167–168:19–34
Hiscott RN, Aksu AE, Mudie PJ, Kaminski MA, Abrajano T, Yasar D, Rochon A (2007b) The Marmara Sea Gateway since 16 ka: non-catastrophic causes of paleoceanographic events in the Black Sea at 8.4 and 7.15 ka. In: Yanko-Hombach V, Gilbert AS, Panin N, Dolukhanov PM (eds) The Black Sea flood question: changes in coastline, climate, and human settlement. Springer, New York
Ibello V, Cantoni C, Cozzi S, Civitarese G (2010) First basin-wide experimental results on N_2 fixation in the open Mediterranean Sea. Geophys Res Lett 37
Ivanova EV, Murdmaa IO, Chepalyga AL, Cronin TM, Pasechnik IV, Levchenko OV, Howe SS, Manushkina AV, Platonova EA (2007) Holocene sea-level oscillations and environmental changes on the eastern Black Sea shelf. Palaeogeogr Palaeoclimatol Palaeoecol 246:228–259
Jones GA, Gagnon AR (1994) Radiocarbon chronology of Black Sea sediments. Deep-Sea Res 41:531–557
Kennedy JF (1963) The mechanics of dunes and antidunes in erodible-bed channels. J Fluid Mech 16:521–544
Kerey E, Meriç E, Tunğöu C, Kelling G, Brenner RL, Doğan AU (2004) Black Sea-Marmara Sea Quaternary connections: new data from the Bosphorus, Istanbul, Turkey. Palaeo 204:277–295
Konikov EG (2007) Sea-level fluctuations and coastline migration in the northwestern Black Sea area over the last 18ky based on high-resolution lithological-genetic analysis of sediment architecture.

In: Yanko-Hombach V, Gilbert AS, Panin N, Dolukhanov PM (eds) The Black Sea flood question: changes in coastline, climate, and human settlement. Springer, New York

Kress N, Thingstad TF, Pitta P, Psarra S, Tanaka T, Zohary T, Groom S, Herut B, Mantoura RFC, Polychronaki T, Rassoulzadegan F, Spyres G (2005) Effect of P and N addition to oligotrophic Eastern Mediterranean waters influenced by near-shore waters: a microcosm experiment. Deep-Sea Res II 52:3054–3073

Kuprin PN, Scherbakov FA, Morgunov II (1974) Correlation, age, and distribution of the postglacial continental terrace sediments of the Black Sea. Baltica 5:241–249

Kwiecien O, Arz H, Lamy F, Wulf S, Bahr A, Röhl U, Haug GH (2008) Estimated reservoir ages of the Black Sea since the last glacial. Radiocarbon 50:99–118

Lane-Serff GF, Rohling EJ, Bryden HL, Charnock H (1997) Postglacial connection of the Black Sea to the Mediterranean and its relation to the timing of sapropel formation. Paleoceanography 12:169–174

Latif MA, Öszoy E, Oğuz T, Ünlüata Ü (1991) Observations of the Mediterranean inflow into the Black Sea. Deep-Sea Res 38: S711–S723

Lericolais G, Popescu I, Panin N, Ryan WBF Guichard F (2003) Last rapid flooding in the Black-sea. In: Uscinowicz S, Zachowicz J (eds) Rapid transgressions into semi- enclosed basins. IGCP 464 Polish Geological Institute, Gdansk

Lericolais G, Popescu I, Guichard F Popescu SM (2007a) A Black Sea lowstand at 8500 yr B.P. indicated by a relict coastal dune system at a depth of 90 m below sea level. In: Harff J, Hay WW, Tetzlaff DM (eds) Coastline changes: interrelation of climate and geological processes. GSA Sp Pap 426:171–188

Lericolais G, Popescu I, Guichard F, Popescu SM, Manolakakis I (2007b) Water-level fluctuations in the Black Sea since the Last Glacial Maximum. In: Yanko-Hombach V, Gilbert AS, Panin N, Dolukhanov PM (eds) The Black Sea flood question: changes in coastline, climate, and human settlement. Springer, New York

Lericolais G, Bulois C, Gillet H, Guichard F (2009) High frequency sea level fluctuations recorded in the Black Sea since the LGM. Glob Planet Change 66:65–75

Lericolais G, Guichard F, Morigi C, Popescu I, Bulois C, Gillet H, Ryan WBF (2011) Assessment of Black Sea water-level fluctuations since the last glacial maximum. Geol Soc Am Sp Pap 473:1–18

Loh AN, Bauer JE (2000) Distribution, partitioning and fluxes of dissolved and particulate organic C, N and P in the eastern North Pacific and Southern Oceans. Deep-Sea Res 47:2287–2316

Man-Aharonovich D, Kress N, Bar Zeev E, Berman-Frank I, Beja O (2007) Molecular ecology of *nifH* genes and transcripts in the eastern Mediterranean Sea. Environ Microbiol 9:2354–2363

Major CO (2006) The co-evolution of Black Sea level and composition through the deglaciation and its paleoclimatic significance. Quart Sci Rev 25:2031–2047

Major C, Ryan W, Lericolas G, Hajdas I (2002) Constraints on Black Sea outflow to the Sea of Marmara during the last glacial- interglacial transition. Mar Geol 190:19–34

Marty JC, Garcia N, Rairnbault P (2008) Phytoplankton dynamics and primary production under late summer conditions in the NW Mediterranean Sea. Deep-Sea Res I 55:1131–1149

Montoya JP, Voss M, Kahler P, Capone DG (1996) A simple, high-precision, high-sensitivity tracer assay for N_2 fixation. Appl Environ Microbiol 62:986–993

Murray JW, Top Z, Öszoy E (1991) Hydrographic properties and ventilation of the Black Sea. Deep-Sea Res 38:S663–S689

Murray JW, Stewart K, Kassakian S, Krynytzky M, DiJulio D (2007) Oxic, suboxic, and anoxic conditions in the Black Sea. In: Yanko-Hombach V, Gilbert AS, Panin N, Dolukhaov PM (eds) The Black Sea flood question: changes in coastline, climate, and human settlement. Springer, New York

Myers PG, Wielki C, Goldstein SB, Rohling EJ (2003) Hydraulic calculations of postglacial connections between the Mediterranean and the Black Sea. Mar Geol 201:253–267

Naudts L, Greinert J, Artemov Y, Staelens P, Poort J, Van Rensbergen P, De Batist M (2006) Geological and morphological setting of 2778 methane seeps in the Dnepr paleo-delta, northwestern Black Sea. Mar Geol 227:177–199

Oğuz T, Öszoy E, Latif MA, Ünlüata Ü (1990) Modeling of hydraulically controlled exchange flow in the Bosphorus Strait. J Phys Ocean 20:945–965

Okay S, Jupinet B, Lericolais G, Çifçi G, Morigi C (2011) Morphological and stratigraphic investigation of a Holocene subaqueous shelf fan, north of the Istanbul Strait in the Black Sea. Turkish J Earth Sci 20:287–305

Özsoy E, Ünlüata Ü, Top Z (1993) The evolution of Mediterranean water in the Black Sea: interior mixing and material transport by double diffusive intrusions. Prog Ocean 31:275–320

Özsoy E, Di Iorio D, Gregg M, Backhaus J (2001) Mixing in the Bosphorus Strait and the Black Sea continental shelf: observations and a model of the dense water outflow. J Mar Syst 31:99–135

Özsoy E, Rank D, Salihoğlu I (2002) Pycnocline and deep mixing in the Black Sea: stable isotope and transient tracer measurements. Estuar Coast Shelf Sci 54:621–629

Pantoja S, Repeta DJ, Sachs JP, Sigman DM (2002) Stable isotope constraints on the nitrogen cycle of the Mediterranean Sea water column. Deep-Sea Res I 49:1609–1621

Popescu I (2002) Analyse des processus sedimentaires recents dans l'eventail profond du Danube (mer Noire). Dissertation, Université de Bucarest – Université de Bretagne occidentale

Popescu I, Lericolais G, Panin N, Normand A, Dinu C, Le Drezen E (2004) The Danube submarine canyon (Black Sea): morphology and sedimentary processes. Mar Geol 206:249–265

Postgate J (1998) The origins of the unit of nitrogen fixation at the University of Sussex. Notes Rec R Soc 52:355–362

Rahav E, Herut B, Stambler N, Bar-Zeev E, Mulholland RM, Berman-Frank I (2013a) Uncoupling between dinitrogen fixation and primary productivity in the Eastern Mediterranean Sea. JGR Biogeosci 18:195–202

Rahav E, Herut B, Levi A, Mulholland RM, Berman-Frank I (2013b) Springtime contribution of dinitrogen fixation to primary production across the Mediterranean. Ocean Sci 9:1–10

Rayburn JA, Knuepfer PLK, Franzi DA (2005) A series of large, late Wisconsinan meltwater floods through the Champlain and Hudson Valleys, New York State, USA. Quat Sci Rev 24:2410–2419

Rees AP, Law CS, Woodward EMS (2006) High rates of nitrogen fixation during an *in-situ* phosphate release experiment in the Eastern Mediterranean Sea. Geophys Res Lett 33

Ridame C, Le Moal M, Guieu C, Ternon E, Biegala IC, L'Helguen S, Pujo-Pay M (2011) Nutrient control of N_2 fixation in the oligotrophic Mediterranean Sea and the impact of Saharan dust events. Biogeosciences 8:2773–2783

Ross DA, Degens ET, MacIlvaine J (1970) Black Sea: recent sediment history. Science 170:163–165

Ross DA, Degens ET (1974) Resent sediments of the Black sea. The Black Sea: geology, chemistry and biology. American Association of Petroleum Geologists Sp Pub, Tulsa

Ryan WBF (2007) Status of the Black Sea flood hypothesis. In: Yanko-Hombach V, Gilbert AS, Panin N (eds) The Black Sea flood question: changes in coastline, climate and human settlement. Springer, New York

Ryan WBF, Pitman WC III, Major CO, Shimkus K, Moskalenko V, Jones GA, Dimitrov P, Gorur N, Yuce H (1997a) Abrupt drowning of the Black Sea shelf. Mar Geol 190:119–126

Ryan WBF, Pitman WC III, Major CO, Shimkus K, Moskalenko V et al (1997b) An abrupt drowning of the Black Sea shelf at 7.5 kyr BP. Geo-Eco Mar 2:115–125

Ryan WBF, Major CO, Lericolais G, Goldstein SL (2003) Catastrophic flooding of the Black Sea. Annu Rev Earth Planet Sci 31:525–554

Sachs JP, Repeta DJ (1999) Oligotrophy and nitrogen fixation during eastern Mediterranean sapropel events. Science 286:2485–2488

Sandroni V, Raimbault P, Migon C, Garcia N, Gouze E (2007) Dry atmospheric deposition and diazotrophy as sources of new nitrogen to northwestern Mediterranean oligotrophic surface waters. Deep-Sea Res I 54:1859–1870

Scholten R (1974) Role of the Bosporus in Black Sea chemistry and sedimentation. In: Degens ET, Ross DA (eds) The Black Sea: geology, chemistry, and biology. American Association of Petroleum Geologists Special Publication, Tulsa

Schrader HJ, Ross DA, Gvirtzman G (1979) Quaternary paleoclimatology of the Black Sea basin. Sediment Geol 23:165–180

Shaw J, Munro-Stasluk M, Sawyer B, Beaney C, Lesemann J-E, Musacchio A, Rains B, Young RR (1999) The Channeled Scabland: back to Bertz? Geology 27:605–608

Siani G, Paterne M, Arnold M, Bard E, Métivier B, Tis-nerat N, Bassinot F (2000) Radiocarbon reservoir ages in the Mediterranean Sea and Black Sea. Radiocarbon 42:271–280

Siddall M, Pratt LJ, Helfrich KR, Giosan L (2004) Testing the physical oceanographic implications of the suggested sudden Black Sea infill 8400 years ago. Paleoceanography 19, PA1024. doi:10.1029/2003PA000903

Somot S, Sevault F, Deque M (2006) Transient climate change scenario simulation of the Mediterranean Sea for the twenty-first century using a high-resolution ocean circulation model. Climate Dyn 27:851–879

Sorokin Yu I (1983) The Black Sea. In: Miller CB, Ketchum BH (eds) Estuaries and enclosed seas. Elsevier, Amsterdam

Sorokin VM (2011) Correlation of upper Quaternary deposits and Paleogeography of the Black and Caspian Seas. Stratigr Geol Correl 19:563–578

Spatharis S, Skliris N, Meziti A, Kormas KA (2012) First record of a Trichodesmium erythraeum bloom in the Mediterranean Sea. Can J Fish Aquat Sci 69:1444–1455

Stihl A, Sommer U, Post AF (2001) Alkaline phosphatase activities among populations of the colony-forming diazotrophic cyanobacterium *Trichodesmium* spp. (cyanobacteria) in the Red Sea. J Phycol 37:310–317

Soulet G, Delaygue G, Vallet-Coulomb C, Böttcher ME, Sonzogni C, Lericolais G, Bard E (2010) Glacial hydrologic conditions in the Black Sea reconstructed using geochemical pore water profiles. Earth Planet Sci Lett 296:57–66

Soulet G, Ménot G, Garreta V, Rostek F, Lericolais G, Bard E (2011) Black Sea "Lake" reservoir age evolution since the last Glacial – Hydrologic and climatic implications. Earth Planet Sci Lett 308:245–258

Strakhov NM (1971) Geochemical evolution of the Black Sea in the Holocene. Lithol Miner Resour 3:263–274

Svitoch AA, Selivanov AO, Yanina TA (2000) Paleohydrology of the Black Sea Pleistocene Basins. Water Res 27:594–603

Thieler ER, Butman B, Schwab WC, Allison MA, Driscoll NW, Donnelly JP, Uchupi E (2007) A catastrophic meltwater flood event and the formation of the Hudson Shelf Valley. Palaeogeogr Palaeoclimatol Palaeoecol 246:120–136

Uchupi E, Driscoll N, Ballard RD, Bolmer ST (2001) Drainage of late Wisconsin glacial lakes and the morphology and late Quaternary stratigraphy of the New Jersey–southern New England continental shelf and slope. Mar Geol 172:117–145

Zubakov VA (1988) Climatostratigraphic scheme of the Black Sea Pleistocene and its correlation with the oxygen isotope scale and glacial events. Quat Res 29:1–24

Past, Present and Future Patterns of the Thermohaline Circulation and Characteristic Water Masses of the Mediterranean Sea

3

Nikolaos Skliris

Abstract

The aim of this section is to provide a comprehensive overview of the research findings concerning the thermohaline circulation of the Mediterranean Sea. The decadal/inter-decadal variability and long-term evolution of the thermohaline properties of the Mediterranean Basin are extensively discussed. We focus on the major climate transient thermohaline events and their links to atmospheric variability and anthropogenic/environmental changes that dramatically changed the deep hydrology and strongly affected the marine ecosystems of the Mediterranean basin during the last 20 years. This section also includes a synthesis of the results of future projections of the thermohaline circulation derived by climate model simulations of the Mediterranean region covering the twenty-first century. The expected effects of the projected thermohaline circulation changes on the marine ecology are also discussed.

Keywords

Mediterranean Sea • Thermohaline circulation • Salinity • Temperature • Trends • Climate change

Abbreviations

ADW	Adriatic Deep Water
AW	Atlantic Water
BSW	Black Sea Water
CDW	Cretan Deep Water
CIW	Cretan Intermediate Water
DWF	Deep Water Formation
EMDW	Eastern Mediterranean Deep Water
EMED	Eastern Mediterranean
EMT	Eastern Mediterranean Transient
LIW	Levantine Intermediate Water
LSW	Levantine Surface Water
MOW	Mediterranean Overflow Water
NACW	North Atlantic Central Water
NADW	North Atlantic Deep Water
TDW	Tyrrhenian Deep Water
TMW	Transitional Mediterranean Water
WIW	Western Intermediate Water
WMDW	Western Mediterranean Deep Water
WMED	Western Mediterranean
WMT	Western Mediterranean Transition

N. Skliris (✉)
National Oceanography Center, University of Southampton, Southampton, UK
e-mail: N.Skliris@noc.soton.ac.uk

Introduction

The Mediterranean is a semi-enclosed basin connected to the Atlantic Ocean by the narrow strait of Gibraltar (width~13 km, sill depth~300 m) and to the Black Sea by the Dardanelles/Marmara Sea/Bosphorus system (Fig. 3.1). It is composed by two major sub-basins the eastern (EMED) and western (WMED) Mediterranean, connected by the Sicily strait (width~35 km, sill depth~300 m). The Mediterranean Sea is characterized by negative surface freshwater and heat budgets. The whole functioning of the Mediterranean as a

S. Goffredo and Z. Dubinsky (eds.), *The Mediterranean Sea: Its history and present challenges*,
DOI 10.1007/978-94-007-6704-1_3, © Springer Science+Business Media Dordrecht 2014

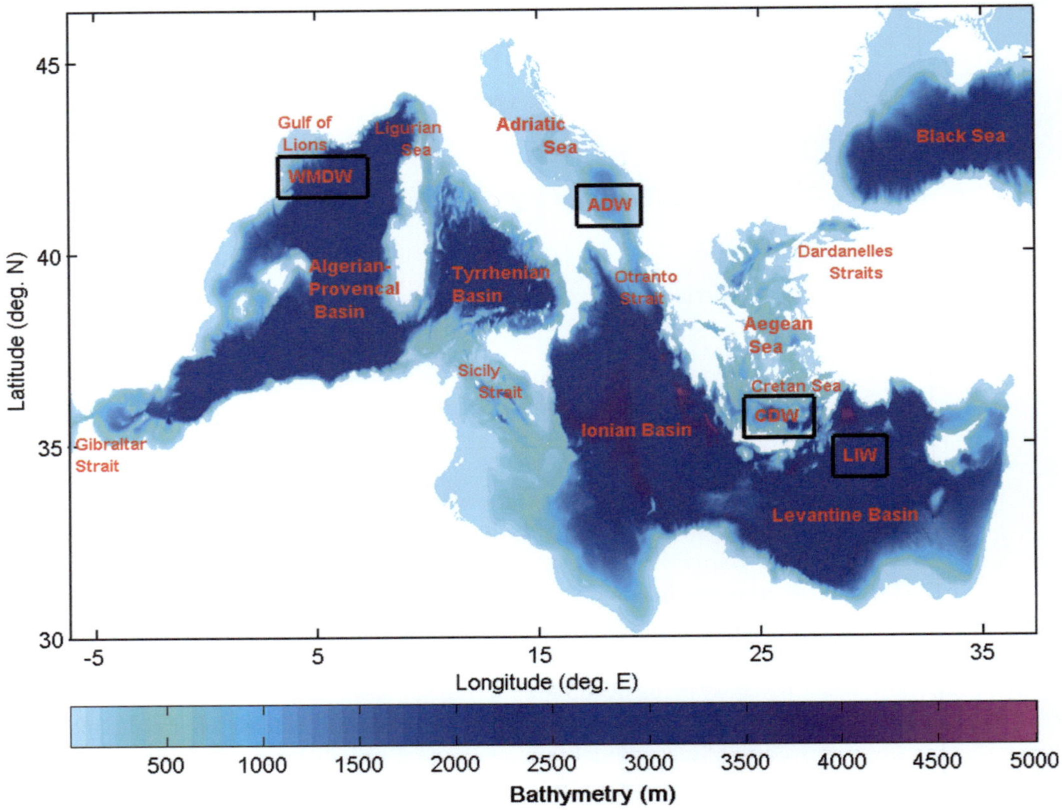

Fig. 3.1 Geography and bathymetry map of the Mediterranean Sea. Main areas of dense water formation (*black boxes*) and dense water masses discussed in the text are also depicted

concentration basin (i.e. evaporation exceeds precipitation and river runoff) is evident from the transformation of the inflowing low salinity Atlantic surface waters (i.e. about 36.2 at the Strait of Gibraltar) into very saline Mediterranean intermediate and deep waters (with a deep water salinity of about 38.4 and 38.7 in WMED and EMED, respectively). The density contrast between the highly saline Mediterranean Water and the relatively fresh Atlantic Water (AW) drives an inverse estuarine circulation in the Strait of Gibraltar. Mediterranean water of high salinity exits into the Atlantic to balance the freshwater loss through the surface.

The Mediterranean thermohaline circulation is typically defined by an open vertical cell comprising the whole basin and two closed vertical cells bounded within the western and eastern sub-basins, respectively. The open thermohaline cell evolves the Atlantic Water (AW) spreading in the surface layer from the Gibraltar strait to the eastern Mediterranean, transformed into Levantine Intermediate Water (LIW) through excessive heat loss and evaporation in the northeastern Levantine basin which then spread in the opposite direction at intermediate depths throughout the whole basin to finally exit through the bottom layer of the Gibraltar strait into the Atlantic Ocean. The two closed thermohaline cells resembling the global conveyor belt are driven by Deep Water Formation (DWF) in the two sub-basins. Traditionally DWF is proposed to occur during winter mainly in the cyclonic gyre of the Gulf of Lions where the Western Mediterranean Deep Water (WMDW) is formed and in the Southern Adriatic cyclonic gyre where the Eastern Mediterranean Deep Water (EMDW) is formed. Until the mid-1980s the thermohaline circulation of the Mediterranean Sea was considered to be in a quasi-steady state with the properties of the intermediate/deep waters presenting very low spatiotemporal variability. However this picture of a linear, stationary thermohaline circulation definitively changed in the last two decades. The discovery of more localized circulation features in the Eastern Mediterranean coincided with the most dramatic climatic shift of the twentieth century. During this climatic event the main source of deep water formation driving the eastern conveyor belt changed from the Adriatic to the Aegean Sea strongly affecting the characteristic intermediate and deep water masses of the whole basin (e.g. Roether et al. 1996; Klein et al. 1999). Recent observational studies indicate that a new large climate transient also occurred in the western basin after the mid-2000s significantly affecting WMDW properties (e.g. Schroeder et al. 2010). A secondary mechanism of dense water formation in the Mediterranean is that associated with cascading of dense water formed over the continental shelf. Interestingly, both large open sea convection and cascading of dense shelf water generally occur at the same locations, i.e. the Gulf of Lions, the Adriatic and Aegean seas (Durrieu de Madron et al. 2005), all these regions being characterised by strong northerly continental winds, implying large ocean heat loss and evaporation during winter.

Due to its size and limited exchange at the Gibraltar Strait, Mediterranean Sea dynamics is particularly sensitive to climatic/anthropogenic perturbations. Variations in the climatic forcing and environmental changes such as river freshwater discharge reduction due to dam constructions cannot be compensated for by the Atlantic inflow but mainly by variations in the water mass characteristics of the basin (Bethoux and Gentili 1999). Recent studies show a rapid sea surface warming throughout the basin with the Mediterranean warming trend being several times larger than the global ocean average over the last two decades (Belkin 2009; Skliris et al. 2012). Observations also indicate strong warming and salinification signals over the second half of the twentieth century in the intermediate/deep Mediterranean waters of both sub-basins (Rohling and Bryden 1992; Bethoux and Gentili 1999; Rixen et al. 2005) as well as in the Mediterranean outflow in the sub-tropical North Atlantic (Curry et al. 2003; Lozier and Stewart 2008; Durack and Wijffels 2010). Modification of hydrological properties of the deep layers are generally considered as signatures of changes which took place in the surface layer induced by anomalous air-sea fluxes, which then propagate in the abyssal waters via dense water formation processes. The presence of several sites of dense (intermediate and deep) water formation along with the very active thermohaline circulation result in a much faster propagation of surface signatures toward the deeper layers of the basin. The Intergovernmental Panel on Climate Change (IPCC) 4th Assessment Report (IPCC 2007) highlights the Mediterranean region as a "hot spot", one of the most vulnerable regions to climate change. Climate model projections indicate an accelerating warming and drying of the Mediterranean region during the twentieth century with strong implications on the Mediterranean thermohaline circulation and ocean climate evolution (Somot et al. 2006, 2008).

The thermohaline circulation and dense water formation processes are of great importance for the biogeochemical cycles and marine ecosystems of the Mediterranean Sea. DWF and the associated intense mixing of the water column induce an increase of the surface layer nutrients content enhancing primary production. DWF is also a key process for the deep ocean ventilation and the sediment transport and coastal carbon export to the deeper layers with strong implications for the deep sea ecosystems and the seabed morphology of the Mediterranean basin. Environmental/climatic-induced changes in the Mediterranean characteristic water masses are expected to rapidly spread throughout the basin by the thermohaline open and closed cells affecting both surface and deep marine ecosystems. Moreover, the study of the Mediterranean thermohaline circulation and its long-term variability is also important for the assessment of the changing global climate system. Due to its intense thermohaline circulation strongly resembling the global conveyor belt and its easier accessibility the Mediterranean is often regarded as a natural laboratory basin to study global ocean dynamic processes such as DWF and the overturning thermohaline circulation and to anticipate the global ocean response to the effects of climate/environmental changes. Furthermore, the Mediterranean outflow in the North Atlantic Ocean is shown to provide significant amounts of salty waters to the Nordic and Greenland Seas contributing to the preconditioning of Northern Atlantic Deep Water (NADW) formation (e.g. Reid 1979), a key component of the global thermohaline circulation regulating the global climate system.

Past and Present Status of the Thermohaline Circulation

The Open Thermohaline Cell

Atlantic Water

AW enters the Gibraltar Strait at a rate of about 1 Sv (1 Sv = 10^6 m^3/s) and it may be identified as a low salinity vein in the surface layer of the major part of the Mediterranean Sea. Along its eastward route from the Gibraltar strait into the western basin and through the Sicily Strait to the eastern basin AW is progressively modified particularly due to the strong evaporation and mixing with surrounding waters and looses much of its freshwater signal. Its salinity gradually increases from about 36.2 at the Gibraltar Strait to about 37.5 at the Sicily Strait and exceeds 38.6 at the eastern part of the Levantine basin. At basin scale the AW circulation is defined by an along-slope cyclonic pattern as already described in the early twentieth century in the pioneer work of Nielsen (1912) and much later by Ovchinnikov (1966). In the 1980s and 1990s observational studies of the sub-basin/mesoscale circulation in both the WMED (e.g. La Violette et al. 1990; Millot 1991) and the EMED (e.g. Robinson et al. 1991; Malanotte-Rizzoli et al. 1997) and later during the 2000s the development of high-resolution operational forecast systems at regional/sub-regional and coastal shelf scales (e.g. Pinardi et al. 2003) demonstrated a much more complex picture of the Mediterranean Sea hydrodynamics. The broad-scale surface cyclonic circulation pattern was shown to be largely modified locally by intense mesoscale activity, gyre

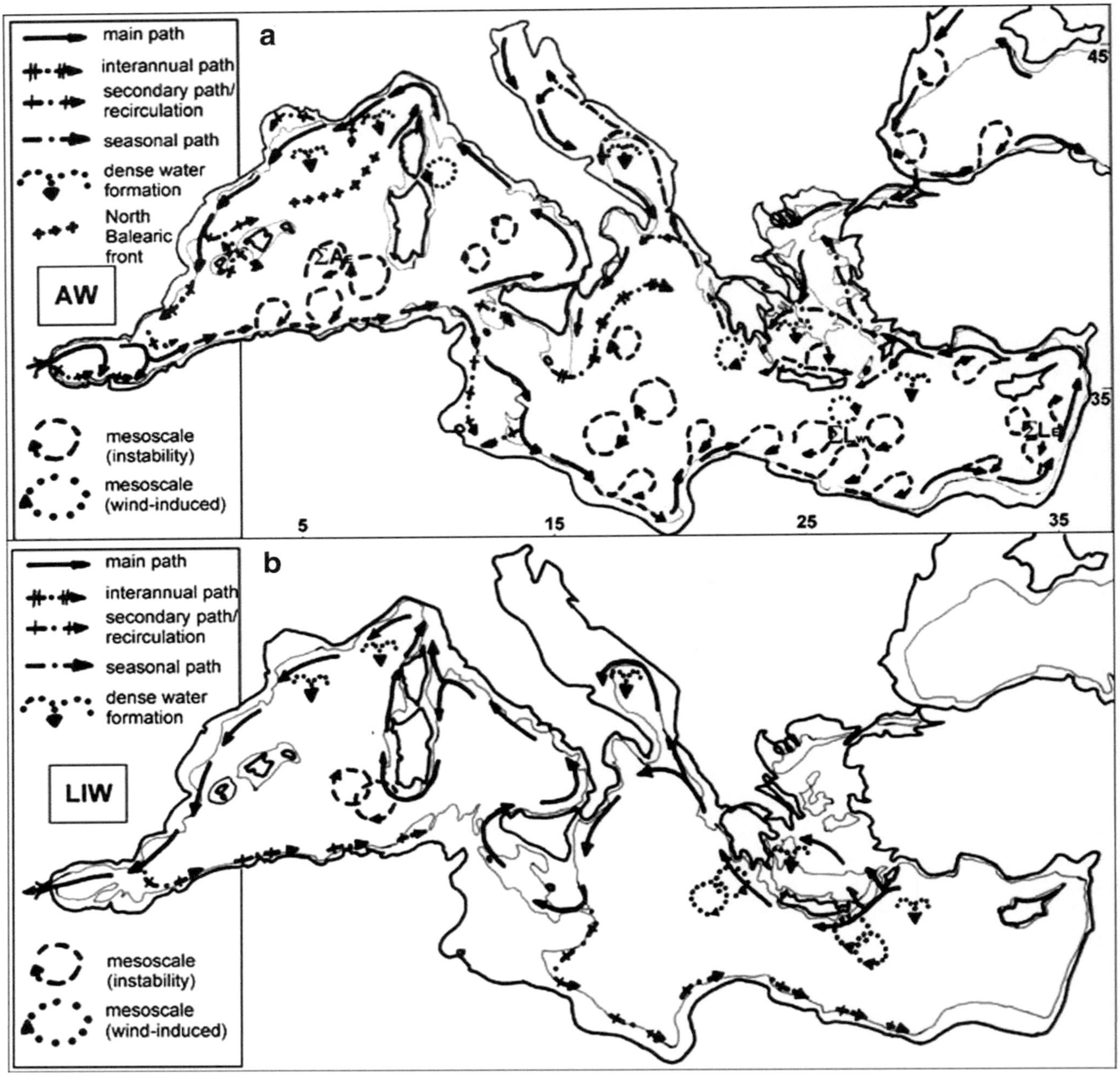

Fig. 3.2 Mediterranean circulation patterns for (**a**) the Atlantic Water and (**b**) the Levantine Intermediate Water (From Millot and Taupage 2005)

formation and highly energetic currents (Fig. 3.2a). In the southern parts of both sub-basins the strong along-slope currents, namely the Algerian Current in WMED and the Lybio-Egyptian current in EMED, become often unstable and lead to the formation of strong mesoscale anticyclonic eddies which then are detached from the slope intruding several hundreds of kilometres in the interior of the basin (Millot and Taupage 2005). These eddies reach horizontal scales of 10–100ths of kms and in some cases may penetrate into intermediate and even the bottom layers of the two sub-basins.

Levantine Intermediate Water

The vertical "steady state" thermohaline circulation of the Mediterranean is first described by Wust (1961). Along its cyclonic route around the basin AW is transformed into dense (intermediate and deep) waters through open sea convection in areas characterised by strong surface heat and freshwater loss during winter. LIW is mainly formed in the area of the cyclonic Rhodes Gyre in late February/early March (Ovchinnikov et al. 1987; Lascaratos et al. 1993). LIW is the water mass formed with the largest volume in the

Mediterranean basin with its formation rate being comparable to that of the Atlantic inflow at the Gibraltar Strait, i.e. of the order of 1 Sv (Lascaratos 1993). The primary source for LIW is the Levantine Surface Water (LSW), a very saline and relatively warm water mass that is formed by intensive heating and evaporation in the Eastern Levantine Basin during summer and is advected to the Rhodes Gyre area by the general cyclonic circulation (Hecht et al. 1988). The winter cooling increases the density of the saline and cold center of the gyre and the newly formed water mass sinks isopycnally to the periphery of the gyre. During the maximum formation rate convection often exceeds 300 m depth at the center of the Rhodes gyre. LIW is formed at a maximum σ_θ of about 29.10, a salinity around 39 psu, and a temperature around 15.5 °C (Lascaratos 1993). Small-scale baroclinic eddies develop around the convection area facilitating the spreading of the newly formed water (Wu and Haines 1996; Nittis and Lascaratos 1998). This very saline and relatively warm water mass spreads at intermediate depths over almost the entire Mediterranean and its westward pathway until reaching the Gibraltar Strait may be easily detected as a subsurface salinity maximum at a depth range between 200 and 600 m. A detailed circulation schematic of LIW in the Mediterranean basin is presented by Millot and Taupage (2005) (Fig. 3.2b). Sur et al. (1992) reported LIW formation in the major part of the northern Levantine Sea during the severe winter of 1992 while during the same period they found deep water formation in the center of the Rhodes Gyre. Apart from LIW, dense water formation at intermediate depths is also observed in other areas such as the Cretan Sea (Southern Aegean) where the Cretan Intermediate Water (CIW) is formed and in the western basin where the Western Intermediate Water (WIW) is formed.

The Closed Thermohaline Cells

DWF mainly occurs through open sea convection in northern parts of the basin characterized by a strong cyclonic circulation and intense continental (cold and dry) northerly winds. The doming of isopycnal surfaces in the center of cyclonic gyres provides a reduced stability allowing open sea convection during winter extreme cooling episodes. Apart from the surface hydrological properties mainly controlled by the characteristics of the inflowing Atlantic waters and local air-sea heat and freshwater fluxes another important factor controlling deep ocean convection in the Mediterranean basin is the salt preconditioning of the intermediate layers. Several studies have demonstrated the crucial role of LIW in Mediterranean deep water formation in both the eastern and the western sub-basins (e.g. Wu and Haines 1996; Skliris and Lascaratos 2004). The entrainment of this saline water mass within the formation sites induces a density increase in the upper layers that allows deeper convection.

Eastern Mediterranean Deep Water

Until the late 1980s observational studies had shown that EMDW was mainly formed within the South Adriatic cyclonic gyre (e.g. Pollack 1951; Ovchinnikov et al. 1987). Very dense water, the so-called Adriatic Deep Water (ADW), is formed there and spreads through the bottom layer of the Otranto Strait to fill the deepest parts of the Ionian and Levantine basins. Other water masses contributing to the DWF process in the Southern Adriatic is the LIW and a dense shelf water which forms in the shallow northern part of the Adriatic Sea during winter by intense cooling and evaporation driven by the local northerly winds (Bora). This very dense shelf water cascades southward along the western Adriatic coast and mixes with the highly saline LIW within the Southern Adriatic cyclonic gyre strongly preconditioning ADW formation. ADW exits through the Otranto Strait at a rate of about 0.3 Sv (Roether and Schlitzer 1991) with a typical sigma-theta around 29.2, a temperature around 13 °C and a salinity around 38.65 (Gacic et al. 1996; Lascaratos et al. 1999). Before the 1990s EMDW was characterised by very low spatiotemporal variability and the bottom layers of the Ionian and Levantine basins had quite homogeneous hydrological properties (T~13.3 °C, S~38.67) (Schlitzer et al. 1991).

However, the Aegean Sea was also reported as a possible secondary source of EMDW since the early studies of the Mediterranean circulation (Nielsen 1912; Pollak 1951). Before the 1990s the Cretan Deep Water (CDW), formed in the deep Cretan basin, was suggested to produce small amounts of warmer (T~14 °C) and saltier water (S~38.9) of relatively lower density (σ_θ~29.15) than the EMDW of Adriatic origin mainly affecting intermediate depths of the Northern Levantine basin adjacent to the Eastern Cretan Arc straits (Theocharis et al. 1999). CIW and CDW were generally found in isolated lenses at intermediate depths below the LIW and above the EMDW of Adriatic origin (Schlitzer et al. 1991).

Western Mediterranean Deep Water

WMDW is mainly formed within the cyclonic gyre of the Gulf of Lions through open sea convection. DWF processes in the Gulf of Lions were extensively investigated for the first time by the MEDOC group during several cruises conducted in the area in the late 1960s (MEDOC Group 1970) while later numerous observational and modelling studies followed (e.g. Gascard 1978; Madec et al. 1991; Leaman and Schott 1991; Schott et al. 1996; Testor and Gascard 2006; Smith et al. 2008). DWF occurs during strong persistent northerly winds (Mistral) in winter which induce excessive heat loss reaching 1,000 W/m^2. The convection area has typically a diameter ranging from of a few 10ths of kms to 100 km while it can extend vertically down to the sea bottom (~2,400 m depth). The newly formed water has a typical σ_θ of about 29.10–29.11 a temperature of about 12.7–12.8 and a salinity of about 38.40–38.45. WMDW formation rate is estimated to be about 0.3 Sv (Lascaratos et al. 1993). Sporadic and less intense winter convection events were observed in the Balearic Sea (Salat and Font 1987) and in the Ligurian Sea (Sparnocchia et al. 1994). Apart from the air-sea heat/freshwater fluxes the formation rate and properties of WMDW are shown to be highly regulated by the characteristics and particularly the salinity of the LIW and AW reaching the formation sites (e.g. Krahman and Schott 1998). The AW salinity is affected by local atmospheric conditions during its residence time in the Western Mediterranean while the LIW salinity is directly linked to the surface salinity in its formation site in the northeastern Levantine basin and therefore to the Eastern Mediterranean freshwater budget. The high submarine canyon density and the strong meteorological forcing over the costal shelf of the Gulf of Lions render it also a preferential site of dense shelf water cascading which has been shown to induce dense water formation within the north-western Mediterranean basin (Durrieu de Madron et al. 2005; Canals et al. 2006, 2009). In this case dense water is formed over the continental shelf due to the intense wind-driven surface heat and freshwater loss and then flows as a strong bottom gravity current along the continental slope until reaching the same density as the ambient waters. While typically dense water formation reaches intermediate depths (~500 m) during extreme surface cooling episodes large density plumes were evidenced through the canyons of the Gulf of Lions which act as very efficient conduits facilitating the transport of coastal shelf dense and sediment/carbon-rich water into the deepest part of the north-western Mediterranean basin (>2,000 m) (Canals et al. 2006; Pusceddu et al. 2010).

Climate Transient Events

Eastern Mediterranean Transient

During the late 1980s/mid-1990s a large transition occurred in the Eastern Mediterranean thermohaline closed cell with the main source of EMDW shifting from the Adriatic to the Aegean (Roether et al. 1996, 2007; Klein et al. 1999; Theoharis et al. 1999; Lascaratos et al. 1999; Malanotte-Rizzoli et al. 1999; Tsimplis et al. 2006) (Fig. 3.3). The densities in the Cretan Sea at outflow depths started to increase since 1987 and from the early 1990s unusually large amounts of CDW were produced and spread out through the Cretan Arc Straits into the deep Levantine basin and simultaneously lifted up the old EMDW of Adriatic origin by about 500 m (Roether et al. 2007). The newly formed CDW was more saline (~39), warmer (~14 °C) and much denser than the old EMDW, resulting in large increase of the salt and heat contents of the deep layer. This dramatic shift in the Eastern Mediterranean conveyor belt was defined of transient nature by Klein et al. (1999), and definitively named the Eastern Mediterranean Transient (EMT). The density of the newly formed CDW progressively increased from typical σ_θ values of 29.15–29.2 prior to the event to an extreme value of about 29.4 in 1994 (Theocharis et al. 1999). Roether et al. (2007) estimated that during the peak EMT period between mid-1992 and late 1994 the mean outflow of CDW from the Cretan Arc Straits amounted to about 2.8 Sv which is an order of magnitude larger than the typical ADW formation rate. By 1995 about 20 % of the bottom layer of the eastern Mediterranean was replaced by this new denser water mass (Lascaratos et al. 1999). Although the Mediterranean Sea experiences intense variability on different time scales, the EMT is the strongest signal of climatic variability ever observed, and seems to be associated with many different dynamical aspects of the circulation, water mass formation, and air-sea interaction. Large deep water formation in the Aegean Sea is linked with anomalous air-sea heat and freshwater fluxes which, in turn, are associated with large-scale atmospheric teleconnection patterns. A strong impact on the winter net air-sea heat flux of the Aegean Sea is observed for the East Atlantic/West Russian pattern (Josey et al. 2011) whereas freshwater flux and particularly precipitation variations in the Mediterranean are highly correlated with the North Atlantic Oscillation (NAO) index (Jones et al. 1997; Tsimplis and Josey 2001; Mariotti and Dell'Alquila 2011). Some investigators of the EMT argued that there were two distinct phases, one preconditioning phase where the salinity of the Aegean Sea was considerably increased over the period 1987–1991 followed by an extreme surface cooling phase related to the exceptionally cold winters of 1992 and 1993 (e.g. Lascaratos et al. 1999; Theocharis et al. 1999; Roether et al. 2007). In a recent modeling study of the EMT Beuvier et al. (2010) highlighted the intense surface heat/water losses and wind stress during the winters of 1992 and 1993, as major triggering elements considerably lowering the stratification of the Aegean Sea leading to exceptionally large deep water formation. The enhancement of salt preconditioning for DWF in the Aegean Sea was attributed to various mechanisms. Tsimplis and Josey (2001) proposed as the key process driving the salt content increase a drastic increase in the freshwater loss due to both an evaporation increase and a precipitation reduction linked to NAO. Samuel et al. (1999) suggested that the intensification of northerly winds over that period enhanced the water mass exchange at the Cretan Arc Straits, leading to an increased inflow of the highly saline LIW into the Aegean Sea. Another proposed mechanism for salt preconditioning in the South Aegean Sea was

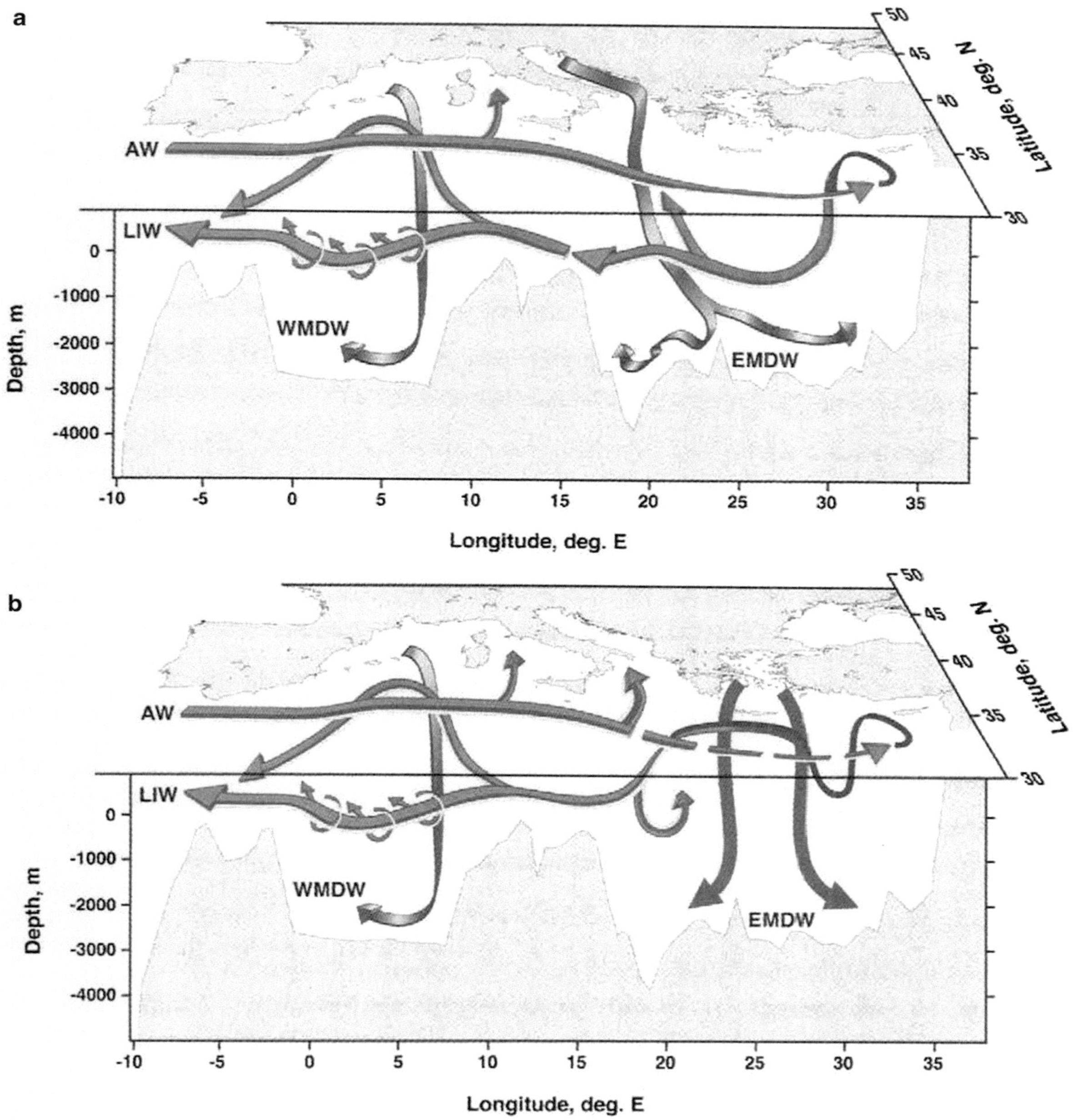

Fig. 3.3 Mediterranean thermohaline circulation schematic prior to the EMT (**a**) and during the EMT (**b**) (From Tsimplis et al. 2006)

associated with wind-induced circulation changes in the Eastern Mediterranean basin, namely a deep three-lobe anticyclonic structure in the southeastern Levantine deflecting the LIW path into the Cretan Sea through the eastern Cretan Arc Straits and, at the same time, a strong gyre formed in the Ionian Sea blocking the eastward route of low salinity MAW into the Levantine basin (Malanotte-Rizzoli et al. 1999). However, this short-term salt preconditioning period seems to be superimposed on a long-term period of salt increase due to long-term changes in the salinity of the LSW and LIW entering the South Aegean (Skliris and Lascaratos 2004; Skliris et al. 2007). The comparison between data from October 1961 Chain survey and those from September 1987 Meteor survey (Roether and Schlitzer 1991) revealed a near disappearance of the relatively low salinity intermediate layer between 500 and 1,000 m in the Cretan Basin (Boscolo and Bryden 2002). Just after the Nile damming and over the 1960s and early 1970s a large increase in the surface salinity of the eastern Levantine Basin is observed in historical data (Gertman and Hecht 2002; Skliris and Lascaratos 2004). Over the same period no significant trend was observed in the Evaporation-Precipitation field over the

Levantine basin, suggesting that the large surface salinity increase could be triggered by the drastic Nile outflow reduction (Skliris and Lascaratos 2004). The LSW salinity increase was also proposed to induce an enhanced preconditioning for LIW formation and a large increase in the salinity of the newly formed intermediate waters which, in turn, spread this salinification signal westward via the LIW circulation throughout the whole basin (Skliris and Lascaratos 2004, Skliris et al. 2007). Rohling and Bryden (1992) found long-term increasing trends in the LIW core salinity in the Levantine Basin prior to the EMT which they attributed to the damming of major rivers such as the Nile since the early 1960s. The Black Sea Water (BSW) outflow through the Dardanelles Strait was also reduced since the 1950s mainly due to the diversion of Russian rivers for irrigation (Rohling and Bryden 1992; Bethoux and Gentili 1999). Zervakis et al. (2000) suggested that the EMT was initiated by DWF in the northern part of the Aegean Sea, mainly triggered by the BSW freshwater input reduction and the large winter heat loss, and then this newly formed dense water overflowed into the deeper Cretan basin.

Observations indicate that after 1995 the EMT is relaxed and the main source of EMDW progressively shifted again back to the South Adriatic (Klein et al. 2000; Manca et al. 2002). According to Roether et al. (2007) during the peak EMT period between 1992 and 1994 about the 75 % of the total CDW outflow was delivered while later outflow up to 2002 only added another 20 % of the total. The shift to the traditional Adriatic source has been linked to a reversal of the upper circulation in the Ionian Sea from anticyclone to cyclonic (Borzelli et al. 2009). Moreover an increase in the salinity of the southern Adriatic Sea is evidenced since 1997 attributed to the advection of high salinity intermediate waters of Aegean origin produced during the EMT whilst the properties of the new EMDW of Adriatic origin have become saltier and warmer than those prior to the EMT (Rubino and Hainbucher 2007). However, there is clear evidence that, although considerably slowed down, the EMT is an ongoing process which will probably continue to affect the thermohaline circulation and hydrological properties of the Mediterranean Sea in the coming decades (Roether et al. 2007).

Western Mediterranean Transition

Recent observational studies show an extensive WMDW renewal over 2004–2006 evidenced by an abrupt and exceptionally large warming and salinification of the bottom layer of WMED (López-Jurado et al. 2005; Schroeder et al. 2006, 2008; Smith et al. 2008). Observations indicate a broad-scale event with the newly formed denser (saltier and relatively warmer) water rapidly spreading into the bottom layer of the Ligurian Sea, the Balearic Sea and the Algero-Provencal basin (Schroeder et al. 2006). This event is considered to be of comparable importance with the EMT in terms of the induced broad-scale thermohaline changes in WMED, and recently named as the Western Mediterranean Transition (WMT) (CIESM 2009). Furthermore, similarly to some EMT analyses, WMT seems to be triggered by a progressive salt content increase in the upper/intermediate layers prior to the event followed by an anomalous strong winter surface cooling period leading to extensive deep water formation. The enhanced salt preconditioning was suggested to be partially attributed to the EMT signal that propagated from the Levantine to the western basin intermediate waters through the Sicily Strait over the previous years (Schroeder et al. 2006, 2010; Font et al. 2007). Gasparini et al. (2005) demonstrated the arrival of the EMT signal to the Sicily Strait by the end of 1992 and Schröeder et al. (2006) showed that progressively saltier and warmer intermediate waters reached first the Tyrrhenian Sea and then the Ligurian Sea until 2004. During the severe winter of 2004–2005 exceptionally large heat loss led to excessive DWF extending from the Gulf of Lions to the Catalan sub-basin inducing large increases in the heat and salt contents of the deep layers (López-Jurado et al. 2005; Schroeder et al. 2006, 2010; Font et al. 2007). In a recent modelling study Herrmann et al. (2010) investigated the factors regulating the exceptional convection event of the winter 2004–2005 in the NW Mediterranean. The authors proposed the strong buoyancy loss over that winter as the key factor responsible for the intensity of the convection in terms of depth and volume of newly formed WMDW. The authors also suggested that the propagation of the EMT signal in the NW Mediterranean potentially doubled the volume of this new deep water by inducing a deepening of the heat and salt maximum that weakened the stratification prior to the convection event. In February 2006 an exceptionally strong open sea convection event was also monitored in the DYFAMED station within the Ligurian Sea which led to the mixing of almost the whole water column (>2,000 m) and abruptly increased the salt and heat content of WMDW by mixing with the saltier and warmer LIW (e.g. Marty and Chiaverini 2010). The latter authors proposed as one of the triggering factors of this convection event a drastic precipitation reduction inducing increased surface salinities in the Ligurian Sea during the previous 3 years (2003–2005). Moreover, during the cold winters of 2004–2005 and 2005–2006, along with strong open sea convection, dense shelf water cascading through the canyons of the Gulf of Lions, typically resulting in intermediate water formation (~500 m), reached much deeper layers (i.e. >2,000 m) significantly contributing to the deep water formation of the western basin (Canals et al. 2006; Pusceddu et al. 2010). Apart from the anomalous winter atmospheric forcing another proposed cause for the production of the large shelf density plumes during these winters was the pronounced decrease of river freshwater inputs within the Gulf of Lions.

The Mediterranean Outflow

Until the early eighties observational studies indicated that the Mediterranean outflow at the bottom layer of the Gibraltar strait was composed predominantly by LIW (occupying intermediate depths of ~100–200 m) with a smaller contribution from WMDW (occupying the deepest layer) (e.g. Baringer and Price 1997). However more recent observational studies demonstrated the presence of two other water masses contributing to the outflow namely the Tyrrhenian Dense Water (TDW), which basically is a mixture of dense waters originated from the eastern basin with WMDW, and Western Intermediate Water (WIW) produced locally in the western basin (Millot et al. 2006, Millot 2009). These studies also showed important changes in the composition of the outflow during the early 2000s, when a new, more saline and warmer form of TDW was evidenced as the major contributor of the deep outflowing waters at the Gibraltar Strait. This water mass was produced in the Tyrrhenian Sea probably by mixing of newly formed dense waters overflowing the Sicily Channel originated from the eastern basin during the EMT (Millot 2006). As the Mediterranean water exits the Gibraltar strait it mixes with the surrounding North Atlantic Central Water (NACW) in the Gulf of Gadiz to produce a new water mass known as Mediterranean Overflow Water (MOW) (Potter and Lozier 2004; Lozier and Sindlinger 2009). The pathway of MOW is clearly evidenced as a warm and salty tongue spreading westwards at intermediate depths (~1,000–1,500 m) across the subtropical North Atlantic (Curry et al. 2003; Potter and Lozier 2004). A part of MOW turns to flow northwards at mid-depth along the European continental slope, eventually reaching the Nordic Seas. It has been suggested that by providing high salt content waters MOW contributes to the preconditioning of North Atlantic Deep Water formation and thus plays a significant role in the global meridional overturning circulation (Reid 1979; McCartney and Mauritzen 2001; Candela 2001). However, the northward penetration of MOW was recently shown to be an intermittent feature depending upon the expansion/contraction of the North Atlantic subpolar gyre and the location of the subpolar front, which, in turn, is shown to vary with NAO (Lozier and Stewart 2008).

Long-Term Hydrological Changes

Various observational/modelling studies have shown the high sensitivity of the Mediterranean "climate" to anthropogenic/climatic perturbations (e.g. Rohling and Bryden 1992; Wu and Haines 1998; Krahman and Schott 1998; Bethoux and Gentili 1999; Tsimplis and Josey 2001; Skliris et al. 2007). Small changes in the surface freshwater input accounting for 5–10 % of the mean freshwater budget induced by anthropogenic/climatic perturbations such as damming of rivers and changes in the precipitation field are shown to have large and long-term impacts on the thermohaline circulation and the hydrological properties of the basin (Wu and Haines 1998; Tsimplis and Josey 2001; Skliris et al. 2007). As mentioned before the observed large changes in the Mediterranean intermediate and deep waters are generally considered as signatures of changes which took place in the surface layer and propagated in the deep layers by the thermohaline circulation cells within the two sub-basins. The signatures of evolutions due to climatic/environmental changes may remain for a long period in the system with the time-scale response of the Mediterranean deep waters to reach a new thermohaline equilibrium state being about 60–70 years (Skliris and Lascaratos 2004). One may distinguish between short-term abrupt transient events mainly driven by winter meteorological forcing extremes in the two sub-basins (i.e. such as the EMT and WMT), atmospheric decadal/inter-decadal natural variability signals induced by large-scale teleconnection patterns and long-term anthropogenic global climate change patterns on the regional scale of the Mediterranean Sea. All these signals acting on different time-scales are superimposed making difficult the interpretation of the observed changes in the hydrological properties of the Mediterranean Sea.

The air-temperature over-land anomalies in the Mediterranean region during the last five centuries indicate an unprecedented strong warming from the mid-1970s onwards, featuring the hottest summer decade 1994–2003 in the entire record (1500–2003) (Luterbacher et al. 2004; Xoplaki et al. 2006). Recent studies using satellite AVHRR-derived Sea Surface Temperature (SST) data have demonstrated an accelerated surface warming of the Mediterranean Sea during the last 25 years (Criado-Aldeanueva et al. 2008; Nykjaer 2009; Skliris et al. 2012). Skliris et al. (2012) showed a strong eastward increasing warming pattern with the mean warming rate being about 0.26 °C/decade in WMED and about 0.42 °C/decade in EMED over 1985–2008 (Fig. 3.4). Mediterranean SSTs showed an exceptionally high warming during the 1994–2003 period reaching about 1 °C/decade which is one of the highest warming rates recorded in the global ocean. The Mediterranean Sea is one of the first oceanic regions where the temperature increase was linked to greenhouse effects and global warming (Béthoux et al. 1990). Belkin et al. (2009) argued that the observed rapid surface warming in the enclosed and semi-enclosed European Seas such as the Mediterranean, surrounded by major industrial/population agglomerations, may have resulted from the observed large terrestrial warming directly affecting the adjacent coastal seas, whilst regions of freshwater influence seem to play a special role in modulating and exacerbating global warming effects on the regional scale. However, it is difficult to discriminate between the regional patterns of global climate change and

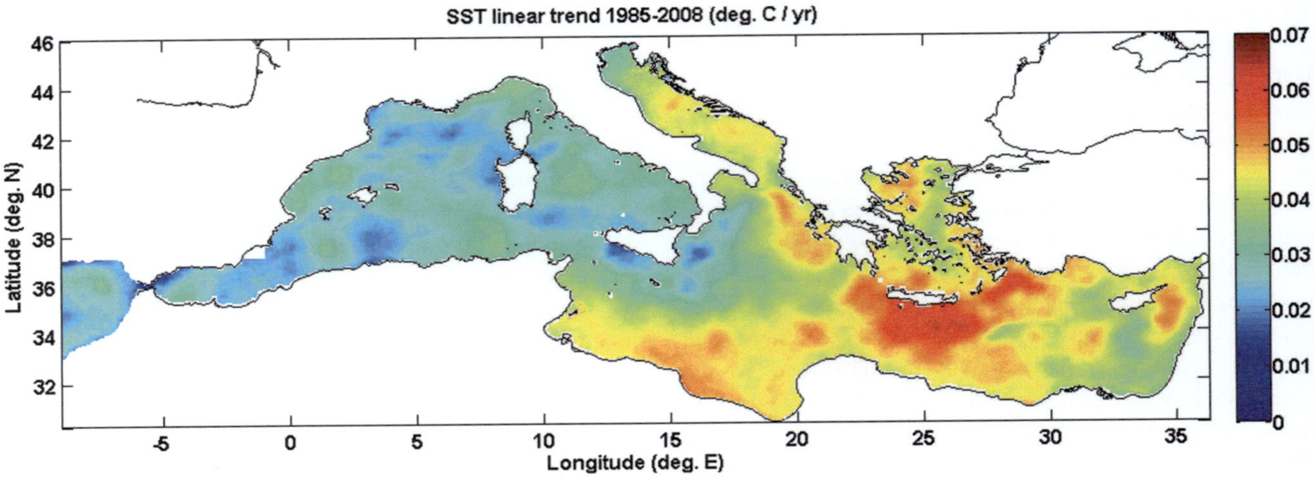

Fig. 3.4 Satellite-derived SST linear trend over 1985–2008 (From Skliris et al. 2012)

the anomalies induced by natural modes of large-scale atmospheric variability generally known to strongly affect the Mediterranean climate from multi-decadal to centennial timescales (Lionello et al. 2006). On the other hand, it is difficult to assess which part of the large-scale atmospheric variability is natural and which part is also induced by global climate change. The Mediterranean SST multi-decadal variations are shown to be highly correlated with the variations of teleconnection patterns of the North Atlantic such as the East Atlantic Pattern (EAP) and the Atlantic Multidecadal Oscillation (AMO) (Mariotti et al. 2010; Skliris et al 2012; Mariotti and Dell'Alquila 2011). Rixen et al. (2005) provided evidence for consistent temperature changes in the WMED and the North Atlantic, explained by similarities in the atmospheric heat fluxes anomalies strongly correlated to NAO. Moreover, Vargas-Yanez et al. (2010) showed that the long-term temperature variability in the upper 200 m layer of the western Mediterranean highly correlates with the heat absorbed by the upper North Atlantic Ocean. Based on both satellite and in-situ observations Skliris et al. (2012) suggested that a part of the strong surface warming signal during the last two decades was advected in the Mediterranean by the Atlantic inflow.

Long-term salinity increases in the Mediterranean basin have been linked with changes in the surface freshwater budget mainly due to large scale precipitation negative anomalies (Krahman and Schott 1998; Bethoux and Gentili 1999) and to reduced river freshwater inputs (Rohling and Bryden 1992, Skliris et al. 2007). The precipitation rate over the Mediterranean Sea is controlled by the northern hemisphere atmospheric circulation mainly expressed by NAO (Hurrell 1995). The long-term increase of the NAO index from the late 1960s until the mid-1990s was found to be strongly correlated with a reduction of the precipitation in the Mediterranean area (Krahmann and Schott 1998; Tsimplis and Josey 2001). Another long-term decrease of freshwater input in the Mediterranean Sea is due to decreasing river freshwater discharge reflecting the impact of recent climate change and river damming. A decrease larger than 50 % in the freshwater discharge of rivers feeding directly or indirectly (i.e. i.e. the Black Sea rivers impact on the) the Mediterranean Sea was estimated since the second half of the nineteenth century due to dam constructions (Poulos and Drakopoulos 2001) with the major damming occurring over the last 50 years. Recently Ludwig et al. (2010) estimated a reduction in the river freshwater discharge in the Mediterranean basin of at least 20 % between 1960 and 2000. After the construction of the Aswan Dam in 1964, the runoff of the Nile, the river with the largest water load and drainage basin in the Mediterranean until the early 1960s, was drastically reduced (by more than 90 %) affecting the salt budget and the thermohaline circulation of the whole basin (Rohling and Bryden 1992; Skliris and Lascaratos 2004; Skliris et al. 2007). The runoff of the Ebro River, one of the major rivers draining into the Western Mediterranean Basin, was also abruptly reduced (by more than 60 %) due to damming in the early 1960s. Salinity increases in the eastern basin are also associated with a reduction of the Black Sea Water freshwater input through the Dardanelles Straits due to the control of Russian rivers draining into the Black Sea since the 1950s (Rohling and Bryden 1992). Recent observational studies also indicate a long-term evaporation increase driven by the rapid sea surface warming during the last decades implying an increase of latent heat loss from the sea surface (Mariotti et al. 2010; Romanou et al. 2010; Skliris et al. 2012). Finally the AW freshwater input was also proposed to be a contributor to the interannual/decadal scale variability of the Mediterranean salt content (Millot 2007). The latter author reported a continuous large salinification of the AW inflow between 2003 and 2007, with the salinity increasing trend during this short period being much larger than the decadal-scale trends observed in the Mediterranean characteristic water masses.

Long-term warming and salinification trends have been observed in the intermediate and the deep layers of WMED during the second half of the twentieth century. Rohling and Bryden (1992) reported increasing trends in the salinity and the temperature of the WMDW of about 0.9 10^{-3} psu/year and 1.6 10^{-3} °C/year, respectively, over the 1955–1989 period. They also found an increasing trend in the LIW core salinity (of about 2.5 10^{-3} psu/year) in the north-eastern part of the Balearic basin, near the formation site of WMDW. The authors argued that the salinity changes in the intermediate/deep waters of WMED were mainly linked to the Nile damming and they were imported from the eastern basin by the LIW inflow through the Sicily strait. Leaman and Schott (1991) and Rohling and Bryden (1992) postulated that the warming trend in the WMDW could have resulted from a continuous increase in the average salinity of LIW reaching the Gulf of Lions. Since LIW is mixed throughout the water column during winter convection events in this area, the upper layers would not have to be cooled as much to reach density values characteristic of the WMDW (σ_{θ}.~29.11). Bethoux and Gentili (1999) reported temperature and salinity increasing trends over the 1959–1997 period in WIW within the Ligurian Sea of about 6.8 10^{-3} °C/year and 1.8 10^{-3} psu/year, respectively. Over the same period, observed salinity and temperature increasing trends in the deep water of the Algerian-Provençal basin were estimated to be about 1.1 10^{-3} psu/year and 3.5 10^{-3} °C/year (Bethoux and Gentili 1999). The authors used a simplified 20-box model to explain the deep water signatures of the western basin over the 1959–1997 period in terms of heat and freshwater flux changes across the sea surface. The temperature increase was attributed to the decrease of long-wave radiation (i.e. due to the increase of the greenhouse effect) while the major contributor to the salinity increase was shown to be the Nile outflow reduction. Continuous trends in the properties of the intermediate/deep layers were also observed throughout the western basin. Sparnocchia et al. (1994) reported salinity and temperature increasing trends over the 1950–1987 period in the LIW layer of the Sicily Straits (1.6 10^{-3} psu/year and 6.5 10^{-3} °C/year) and of the Ligurian Sea (1.9 10^{-3} psu/year and 9.1 10^{-3} °C/year). Fuda and Millot (2002) reported increasing trends in the salinity and temperature of the Tyrrhenian Deep Water (TDW) at depths higher than 3,000 m of about 0.6 10^{-3} psu/year and 1. 10^{-3} °C/year, respectively, over the 1960–1990 period. Krahman and Schott (1998) reported, in addition to the WMDW salinity and temperature trends, a large salinity increase in the upper 0–70 m layer of the northwestern Mediterranean from 1960 to 1990. Over the same period, they found that the LIW salinity increase in that region was not strong enough to be solely responsible for the deep water trends. They authors concluded that WMDW trends were mainly caused by local variations in the surface freshwater fluxes of the north-western Mediterranean, mainly decreasing precipitation and Ebro River runoff since the early 1960s. In a numerical modelling study Skliris et al. (2007) showed that about 50 % of the WMDW salinity increase during the second half of the twentieth century could be explained by the changes in the LIW originated from the eastern basin (mainly due to the Nile damming) reaching the WMDW formation sites while the remaining part could be explained by local freshwater budget changes in the western basin (mainly due to a precipitation reduction). During the WMT the long-term WMDW salinity/temperature slowly increasing trends since the late 1950s are disrupted when strong deep water formation and rapid WMDW renewal by much warmer and saltier waters, partially originated from the eastern basin during the EMT, induced abrupt and excessive increases in the heat and salt contents of the deep layer. Schroeder et al. (2010) estimated a total salinity and temperature increase at the bottom of the new WMDW layer of 0.024 psu and 0.042 °C, respectively, between 2004 and 2008.

In contrast with WMED where salinity and temperature variations in the deep layer during the second half of the twentieth century are characterised by long-term quasi-monotonic increasing trends, the evolution of hydrological properties in the bottom layer of the Levantine basin shows no clear trends and much larger decadal-scale variability. A large increase in the salinity of both the surface (~0.15 psu) and the deep (~0.1 psu at 2,000 m depth) layers of the Levantine Basin is observed in historical data between the early 1960s and the early 1970s (Skliris and Lascaratos 2004). This event mainly affecting the salinity of the EMDW within the Levantine basin was suggested to be induced by the dramatic reduction of the Nile outflow in the early 1960s and a prolonged period of heat loss increase resulting in deep water formation in the Rhodes Gyre area and propagation of the surface salinification signal towards the deep layers (Skliris et al. 2007). However observations during the mid-1980s indicate that there was no any imprint left of this event within the Levantine basin and that the deep layer salinity was restored again to the reduced levels of the early 1960s. During the EMT large salinification and warming trends were evidenced in the deep Levantine basin as a consequence of the large amounts of newly formed denser, much saltier and warmer waters of Aegean origin that filled the bottom layer of this basin. Between the late 1980s and the mid-1990s the Levantine basin average salinity and temperature at 2,000 m depth increased by about 0.1 psu and 0.3 °C, respectively (Skliris and Lascaratos 2004). Manca et al. (2004) reported increasing trends in the EMDW salinity (~10^{-3} psu/year) and temperature (~3.10^{-3} °C/year) within the bottom layer of the Ionian basin (>1,200 m) from the mid-1950s to the end of 1990s. Salinity increasing trends were also observed at intermediate depths throughout the Levantine and Ionian Basins as well as in the Adriatic Sea. Rohling and Bryden (1992) reported increasing trends in the LIW core salinity in both the western Levantine Basin (~2.10^{-3} psu/year) and the eastern Ionian Basin (~4.5 10^{-3} psu/year) between the mid-1950s and the early 1970s. Vilibic (2002) reported a salinity increasing trend of about 3.5 10^{-3} psu/year in the bottom water of the Jabuka Pit (Middle Adriatic) over the 1951–1989 period. Vilibic and Orlic (2001) compared salinity measurements in the 100–500 m interval of the South Adriatic Pit, conducted at the beginning of the twentieth century and during the 1967–1990 period. They found a large increase (>0.05) in the average salinity, which they mainly attributed to the reduction of the Nile river discharge.

A more integrated picture of the temperature and salinity decadal variability and long-term trends in the two Mediterranean sub-basins is provided by Rixen et al. (2005) who exploited the MEDAR group database (MEDAR Group 2002). This is a yearly three-dimensional dataset consisting of quality checked temperature and salinity profiles in the Mediterranean from 1945 to 2002 interpolated onto a 0.2×0.2° horizontal grid by a Variational Inverse Model (VIM) (Brankart and Brasseur 1998). Rixen et al. (2005) investigated the basin-average temperature/salinity variations for the upper 150 m layer as well as for the intermediate (150–600) and deep (600-bottom) layers of EMED and WMED over 1950–2000. The authors showed a significant warming trend in the upper 150 m of both sub-basins from the mid-1980s onwards, except from a brief cooling period in the early 1990s during the EMT peak phase. For the bottom layer of both sub-basins small increasing trends are obtained until the mid-1980s followed by an accelerated warming. Bottom layer temperature increase was larger in EMED, suggesting that the surface warming signal was rapidly propagated in the deep part of the basin during the EMT. On the other hand, the authors indicated no long-term trends but strong decadal-scale variability in the upper/intermediate layers with a small decreasing temperature trend in WMED and a marked cooling in EMED from the mid-1970s until the mid-1980s. Concerning salinity variations the authors found a significant increasing trend in the upper layer salinity in WMED over 1960–1990. As mentioned before this period was characterised by decreasing precipitation over the western basin strongly correlated with an increasing positive NAO index. Salinity variations show quasi-continuous increasing trends in both the intermediate and bottom layers of WMED over 1950–2000. In EMED there is a small increasing salinity long-term trend in the intermediate layer whilst the evolution of bottom layer salinity shows strong decadal variability driven by the abrupt transient events with large increasing trends during the 1960s/early 1970s and then during the EMT in the 1990s.

Although there is large interannual/decadal scale variability in the thermohaline properties of the Mediterranean Sea, a coherent and robust trend of increased salinity and warmer temperature is evidenced in key Mediterranean water masses over the last 50 years and this "Mediterranean" signal of change is also clearly depicted in the subtropical North Atlantic intermediate waters (Curry et al. 2003; Potter and Lozier 2004; Durack and Wijffels 2010). A significant warming and salinification of MOW in the eastern North Atlantic was reported by Potter and Lozier (2004). The authors found increasing trends of about 0.1 °C/decade and 0.03 psu/decade in the MOW core temperature and salinity, respectively, over 1950–2000.

Future Projections of the Thermohaline Circulation

The Mediterranean Sea surface warming as observed during the last 20 years is projected to be accelerated in the context of global warming affecting the DWF processes and the thermohaline circulation. This rapid surface warming may significantly diminish density and thus may increase stratification in the upper layer in the near future. This in turn may lead to a reduction of dense water formation rates in the various deep/intermediate water formation sites of the basin slowing down the thermohaline circulation. However, on the other hand, the sea surface warming is also shown to drive a long-term latent heat loss/evaporation increase in the basin during the last few decades (Mariotti 2010; Mariotti and Dell' Alquila 2011; Romanou et al. 2010; Skliris et al. 2012). The evaporation increase combined with a precipitation and river discharge reduction, as already observed over the second half of the twentieth century, would induce a salinity increase thus favouring a density increase and a reduced stratification in the upper layer. As a consequence the warming and salinification signals could be, at least partially, density compensated moderating the regional impact of global climate change on the Mediterranean stratification. The future status of stratification which in turn determines the DWF rates and thus the whole overturning thermohaline circulation of the Mediterranean depends on the relative contribution of the salinity and temperature trends in defining the surface density evolution. The long-term salinity/temperature increasing trends observed in the WMDW over the second half of the twentieth century resulted in a moderate increasing trend in the deep layer density (e.g. Rohling and Bryden 1992; Bethoux and Gentili 1999). On the other hand, Potter and Lozier (2004) found that the warming and salinification changes in the MOW within the eastern North Atlantic over 1950–2000 were density compensated.

The future climate change was first investigated by means of global-scale Atmosphere or coupled Atmosphere/Ocean/

land General Circulation Model (AGCM and AOGCM, respectively) projections based on different anthropogenic greenhouse gases and aerosols emission scenarios covering the twenty-first century. These emission scenarios are defined by the IPCC working group (WP III) in charge of the assessment of options for mitigating climate change (IPCC 2000). According to the latest IPCC assessment report (IPCC-AR4 2007), the climate over the Mediterranean basin is expected to become both warmer and drier during the twenty-first Century. Global climate model projections indicate that annual mean temperatures of the Mediterranean region are likely to increase more than the global mean (Christensen et al. 2007). The Mediterranean warming trend is expected to be much stronger during summer with the air temperature increase over the sea reaching on average 3–4 °C at the end of the twenty-first century (Giorgi and Lionello 2008). Annual precipitation and number of precipitation days are likely to decrease in most of the Mediterranean area while the risk of summer draughts is likely to increase. Mariotti et al. (2008) used Coupled Model Intercomparison Project phase 3 (CMIP3) multi-model simulations based on the SRES A1B emissions scenario to predict Mediterranean water cycle changes during the twenty-first century. Based on the multi-model ensemble average the authors estimated a 24 % increase in the freshwater loss over the Mediterranean Sea by 2070–2099 due to precipitation reduction and enhanced evaporation driven by the sea surface warming. River freshwater inputs were also projected to be significantly reduced further exacerbating draught conditions in the Mediterranean Sea. These model projections for the twenty-first century are consistent with the current trends observed during the second half of the twentieth century for all the surface freshwater budget components (i.e. precipitation, evaporation and river runoff), suggesting that the hydrological cycle has already started to change significantly in the Mediterranean region.

In recent years the analysis of the climate change at the regional scale is performed by dynamically downscaling global-scale AGCMs or AOGCMs using higher resolution limited area Atmosphere or coupled Atmosphere/Ocean Regional Climate Models (ARCMs and AORCMs, respectively). By resolving smaller-scale topographically controlled features and taking into account the specificity/complexity of regional dynamics such dynamical downscaling provides a more adequate representation of the regional climate changes. Although the majority of both regional and global climate prediction models are qualitatively consistent in producing the future warming and drying trends in the Mediterranean area there are still large uncertainties concerning both the magnitude and spatial patterns of change. Apart from the different representation of physical processes among the models uncertainties also arise from the intense natural variability of the Mediterranean climate. The Mediterranean region is a mid-latitude transitional zone between the warm and dry climate of the North Africa and the cold and wet climate of the Central Europe whilst at the same time is under the influence of the atmospheric teleconnection systems of the North Hemisphere, from the North Atlantic to the African and Indian Monsoon patterns, strongly modulating its climatic conditions at decadal/inter-decadal timescales. Whilst there have been several multi-model prediction studies concerning the "atmospheric" climate of the Mediterranean region (e.g. Giorgi and Lionello 2008; Mariotti et al. 2008) there are only a few published works concerning climate projections of the Mediterranean Sea thermohaline circulation and hydrological properties. A few EU funded projects such as SESAME and CIRCE which ended in 2011 addressed the issue of the Mediterranean future climate projections focusing on the ocean component and their results are highly expected by the oceanographic community of the Mediterranean Sea. In an early study Thorpe and Big (2000) used a high resolution (1/4° horizontal resolution) ocean model forced by air-sea fluxes from a global IPCC coarse resolution model based on a doubling CO2 emissions scenario over the twenty-first century. A very strong sea surface warming is simulated with the SST increase reaching on average 4 °C at the end of the twenty-first century. Increasing temperature and salinity trends were found in the deep layers while the stratification was strongly enhanced leading to a considerable weakening of the thermohaline circulation. Somot et al. (2006) used a higher resolution ocean model (1/8° horizontal resolution) to predict thermohaline changes of the Mediterranean Sea forced by the air-sea fluxes of an ARCM derived from a dynamical downscaling (at 50 km resolution) from a global AOGCM based on the IPCC A2 emission scenario covering the twenty-first century. Results showed that the basin averaged surface net heat loss decreased from 6.2 to 2.4 W m^{-2} and the freshwater loss increased from 0.72 to 1.01 m $year^{-1}$ by the end of the century. The Mediterranean Sea response to the air-sea flux changes resulted in a significant warming and salinification for both the surface (on average +3.1 °C and +0.48 psu) and deep layer (on average +1.5 °C and +0.23 psu) of the basin. These hydrological changes induced a strong weakening of the thermohaline circulation mainly driven by the surface density decrease and a subsequent excessive decrease in both the intermediate water formation rate (by about 40 %) and in the DWF rate (by about 80 %) with respect to present climate conditions (Fig. 3.5). Results also indicated that the volume exchange through the Strait of Gibraltar was only slightly reduced (from 1.17 to 1.12 Sv) while both the Mediterranean outflow and Atlantic inflow became significantly warmer and saltier at the end of the scenario simulation. Overall, the model predicted a reduction of total heat gain by 13 % and an increase of salt loss by 5 % through the Gibraltar Strait by

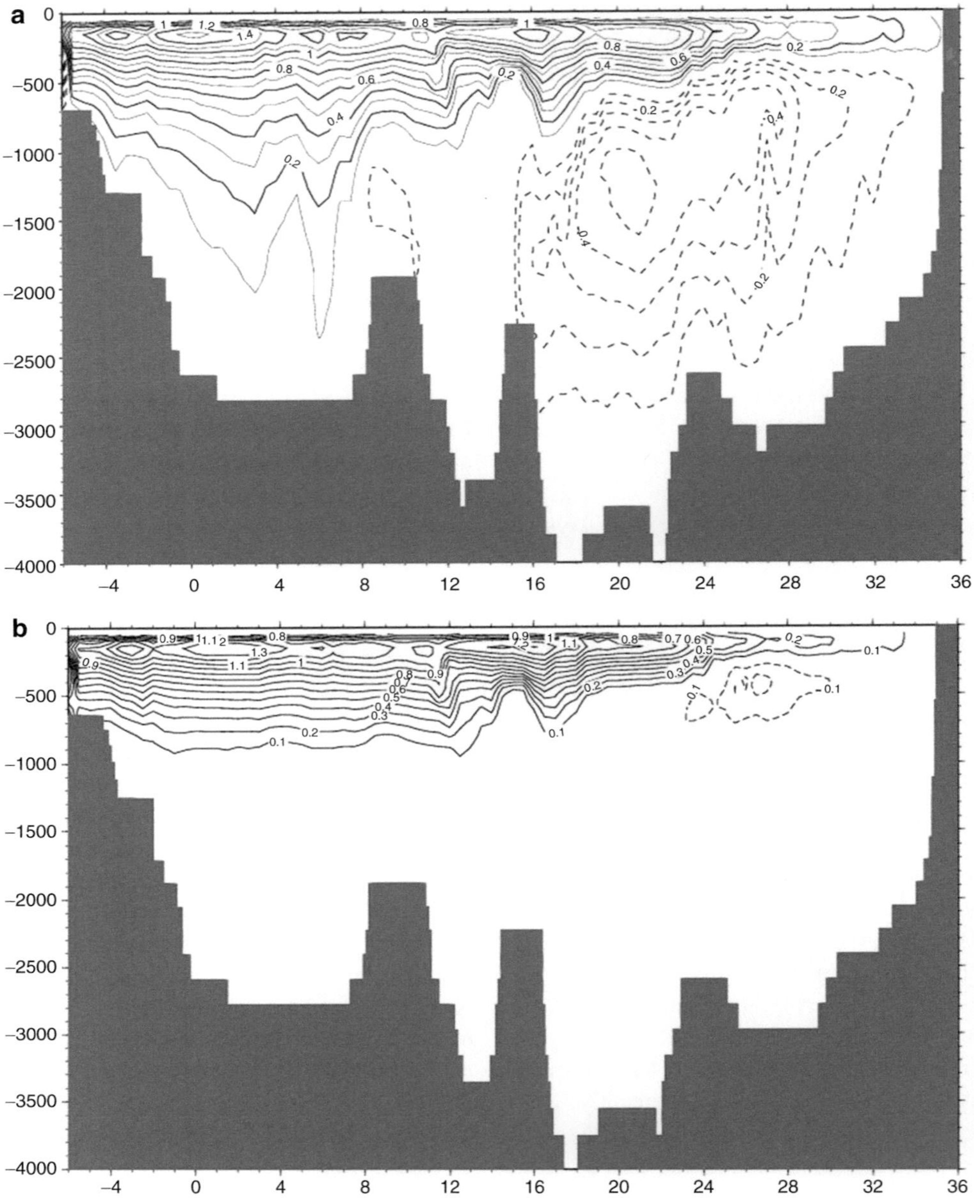

Fig. 3.5 Vertical sections averaged over the 2070–2099 period for the global Mediterranean Zonal Overturning stream Function (**a**) for a control run simulating the present climate and (**b**) for an IPCC-A2 scenario run (From Somot et al. 2006)

the end of the twenty-first century. The authors also indicated that the future predictions of the Mediterranean outflow in the North Atlantic were consistent with the current observed changes over the second half of the twentieth century. Somot et al. (2008) performed the same projection based again on the A2 emissions scenario and using the same ocean model component but this time they applied a fully coupled Atmosphere/Ocean Regional Climate Model (AORCM) to predict ocean/atmosphere climate changes in the Europe/Mediterranean region. They also compared the AORCM results with those of a non-coupled atmospheric regional model (ARCM) under the same emissions scenario. Results from the AORCM simulation showed that the yearly mean basin-average SST increased by +2.6 °C with a maximum in summer and autumn (+2.9 °C) in good agreement with the results of the previous one-way coupled ocean simulation by Somot et al. (2006) but also with the results of the corresponding ARCM simulation. However the authors found that the proper representation of Mediterranean SSTs in the higher resolution AORCM resulted in the amplification of the climate change signal over large parts of Europe with respect to the corresponding ARCM. In particular the warming was found to be significantly higher throughout the year in many areas of Europe, winters were found to be wetter over northern Europe and summers drier over southern and eastern Europe. Herrmann et al. (2008) used the Somot et al. (2006) Mediterranean ocean climate simulation outputs to force a higher resolution model of the NW Mediterranean in order to investigate the impact of climate change on dense water formation over the shelf of the Gulf of Lions. Based on the model results the authors estimated that between the twentieth century and the end of the twenty-first century the volume of dense water annually formed on the shelf was reduced by about 50 % while that cascading from the shelf edge was reduced by 90 %.

Implications of Thermohaline Circulation and Hydrological Changes on Mediterranean Marine Ecosystems

The global sea surface warming during the last decades has already a significant impact on marine ecosystems (e.g. Behrenfeld et al. 2006; Polovina et al. 2008). According to Belkin et al. (2009) the Mediterranean is one of the Large Marine Ecosystems (LMEs) which is strongly affected by the very rapid surface warming i.e. the observed warming rate is four times the global rate over 1978–2003. This rapid warming was shown to have large effects on marine biodiversity disrupting functions in various marine ecosystems and shifting the biogeochemical equilibrium of the Mediterranean Sea. As previously discussed a potential consequence of the rapid warming is the weakening of the thermohaline circulation and the increased stratification of the upper layer. The subsequent reduced mixing of the water column may result in lower upward turbulent nutrient transports, thus inhibiting new primary production in the Mediterranean Sea. Barale et al. (2008) analysing the Seawifs-derived surface chlorophyll variability in the Mediterranean Sea from 1998 to 2003 showed a general decrease of surface chlorophyll monthly mean values over the basin interior. The authors have interpreted the decreasing chlorophyll trend in the open sea in terms of increased nutrient-limitation, resulting from reduced vertical mixing and a more stable stratification associated with the rapid surface warming of the Mediterranean Sea. This pattern is consistent with the decreased primary productivity levels already evidenced at global scale as a response to the warming climate (e.g. Boyd and Doney 2002; Sarmiento et al. 2004; Doney 2006; Behrenfeld et al. 2006; Polovina et al. 2008).

However, at short-term scales enhanced primary productivity levels were observed in particular locations of the Mediterranean over the last two decades associated with the strong climatic transient events. In the highly productive NW Mediterranean a small increase of phytoplankton biomass has been observed over 1991–1999 (Marty et al. 2002) while an exceptionally high chlorophyll a integrated biomass was observed over 2003–2006 (Marty and Chiaverini 2010). The latter period was characterised by intense winter cooling which resulted in the dramatic convection events of the WMT (see section "Climate transient events"). Marty and Chiaverini (2010) evidenced a very intense diatom bloom in the DYFAMED site within the Ligurian Sea following the excessive convection event of February 2006. During this event the strong mixing affected almost the whole water column (>2,000 m) leading to a very efficient nutrient replenishment of the surface layer strongly stimulating new primary production. The above authors concluded that the NW Mediterranean Sea primary productivity was increasing over the last two decades. Moreover during the convection events of 2004–2006 large amounts of sedimentary organic matter originated from the continental shelf were rapidly transported by dense shelf water cascading through the canyons of the Gulf of Lions and accumulated into the deepest part of the area resulting in a large drop in benthic abundance and biomass (e.g. Pusceddu et al. 2010).

Touratier and Goyet (2009) studied the evolution of anthropogenic CO_2 in the DYFAMED station from the early-1990s to the mid-2000s. They found much higher concentrations of anthropogenic CO_2 than those in the Atlantic Ocean but the temporal trend was decreasing in the intermediate and deep layers of the water column and this decrease was highly correlated with a decrease in the dissolved oxygen and an increase in both temperature and salinity. The authors explained these trends by an invasion of older water masses occurred during this period partially originated from the eastern basin as a consequence of the EMT. During WMT the strong convection events produced large volumes of deep waters as a mixture of warm and very saline intermediate waters from the eastern basin and local waters from the DWF sites. These newly formed deep waters were sufficiently dense to uplift the existing deep waters, which resulted in an apparent increase in the age of the overlying waters and consequently in a decrease in the anthropogenic CO_2 and oxygen contents (Touratier and Goyet 2009).

Increased primary productivity levels were also observed in the highly oligotrophic Cretan Sea during the EMT climatic event. Over the peak EMT period (1992–1994) huge amounts of the dense CDW were produced and outflowed through the Cretan Arc Straits into the deep Levantine Basin uplifting several hundred meters the bottom nutrient-rich layer occupied by older EMDW of Adriatic origin. As a consequence a new nutrient-enriched intermediate layer was formed in EMED, the so-called Transitional Mediterranean Water (TMW). At the same time nutrient concentrations decreased and oxygen concentrations increased in the bottom layer of the Levantine basin due to the replacement of the old

EMDW by the newly formed oxygen-rich/nutrient-poor CDW. Following the general cyclonic circulation the nutrient-rich TMW progressively intruded mainly through the Eastern Cretan Arc Straits into the Cretan Sea compensating for the massive CDW outflow (Souvermezoglou et al. 1999). By 1994 a well defined 'minimum temperature/salinity/oxygen and maximum nutrient' intermediate layer was established in the Cretan Sea. Nutrient concentrations at intermediate depths in the Cretan Sea nearly doubled in 1994–1995 as compared with those observed during previous years until 1992 (Souvermezoglou et al. 1999; Souvermezoglou and Krasakopoulou 2000). Intense winter mixing and cyclonic gyre formation in the area resulted in a nutrient replenishment of the upper layer (Psarra et al. 2000). An intense diatom bloom then occurred which led to exceptionally high seasonal mean primary productivity levels (i.e. about 220 and 160 mg C m^2 day^{-1} at the continental shelf and slope areas, respectively) observed for the first time in this oligotrophic area (Psarra et al. 2000).

Taken into account the projected accelerated warming in the Mediterranean region during the twenty-first century (e.g. Somot et al. 2006; Somot et al. 2008) the impact on Mediterranean marine ecosystems is expected to be amplified over the coming decades. As mentioned in the previous section current regional climate model projections indicate that the thermohaline circulation will significantly slow down and DWF both through open ocean convection and through dense shelf water cascading will be considerably reduced during this century (Somot et al. 2006; Herrmann et al. 2008) suggesting that new primary production could be strongly inhibited in the Mediterranean Sea. Decreasing productivity levels are particularly expected to be evidenced in the highly productive DWF areas such as the Gulf of Lions where deep convection events typically result in the nutrient replenishment of the surface layer through extensive mixing of the water column strongly stimulating primary production.

On the other hand, in contrast with the oligotrophic open sea, several coastal areas particularly along the highly populated and industrialised northern Mediterranean coastline are progressively characterised by higher eutrophication levels due to increasing anthropogenic nitrogen/phosphorous inputs mainly derived by the run-off from agricultural land as well as by raw or poorly treated urban and industrial wastewaters. The projected enhanced stratification/reduced mixing as well as the decreased shelf-slope exchange/dense shelf water cascading over the twenty-first century (e.g. Herrmann et al. 2008) could further exacerbate eutrophication phenomena in the Mediterranean coastal zones by inducing an accumulation of the increasing anthropogenic nutrient loads over the coastal shelf. The weakening of the thermohaline circulation and subsequent decrease in shelf-slope exchange is also likely to affect coastal carbon exports to the deep ocean reducing the sequestration rate of anthropogenic carbon.

Another factor contributing to the sensitivity of the Mediterranean marine ecosystems is the large number of invasive Red Sea tropical species observed since the opening of the Suez Canal in 1869 (Streftaris et al. 2005), a situation which could be further aggravated by the observed accelerated sea surface warming and increased stratification in EMED. Moreover, some Mediterranean species inhabiting intermediate/deep waters appear particularly susceptible to hydrological changes. Cartes et al. (2011) suggested that long-term changes in the LIW properties at its formation site within the northeastern Levantine basin mainly attributed to the Nile damming was probably a triggering factor for the extinction/drop in abundance of deep-sea shrimp species typically inhabiting the LIW layer off Catalonian coasts within the northwestern Mediterranean (i.e. located several thousand kilometres away from the Nile mouth).

Summary

The Mediterranean Sea is characterised by a very active thermohaline circulation strongly resembling the global conveyor belt and for this reason is often considered as an appropriate scaled-down model to study the global overturning circulation and to anticipate the ocean response to environmental/climate changes. Recent observational studies demonstrated that far from being stationary the thermohaline circulation presents a very high spatiotemporal variability driven by anthropogenic environmental changes and the strong interannual and decadal scale natural variability of the air-sea heat/freshwater fluxes. The intense climatic variability of the basin is particularly highlighted during two extreme climate transient events, the EMT during the 1990s and the WMT during the 2000s which resulted in dramatic changes in the thermohaline circulation and hydrological properties of its two sub-basins strongly affecting marine ecosystems. Both climate transients were triggered by an enhanced salt preconditioning in the upper/intermediate layers followed by extreme winter surface cooling events that led to massive DWF and an abrupt and drastic increase of the heat and salt contents of the bottom layer. Intense mixing of the water column and changes in the spatial distribution of key water masses during these transient events induced a strong nutrient enrichment of the euphotic zone stimulating primary production. Moreover DWF processes during these climatic transients (i.e. both open sea convection and dense shelf water cascading through canyon areas) resulted in efficient exports of surface carbon-rich water and coastal shelf sedimentary organic matter into the deepest part of the basin strongly affecting the benthic communities.

Although there is a large interannual/decadal variability observations show spatially coherent multi-decadal warming and salinification trends in key intermediate and deep water masses of the basin. These long-term trends are linked with changes in the surface freshwater/heat budgets reflecting both natural and anthropogenic impacts. The signals of surface changes seem to be rapidly propagated in the intermediate/deep layers via the very active open and closed vertical thermohaline circulation cells of the basin. Important salinification and warming signals over the second half of the twentieth century are also evidenced in the Mediterranean outflow at the bottom layer of the Gibraltar Strait as well in the MOW within the North Atlantic. The extent to which the long-term salinification and warming of the Mediterranean outflow will affect the global overturning circulation remains an open question (Lozier and Stewart 2008; Lozier and Sindlinger 2009). Observational studies also indicate a rapid surface warming of the Mediterranean Sea during the last two decades related with the global-scale warming pattern. However, the Mediterranean Sea surface warming rate is several times higher with respect to the global mean. A few available climate regional model projections for the Mediterranean Sea demonstrate that the current sea surface warming will accelerate over the twenty-first century. Although the surface freshwater loss over the Mediterranean is also projected to further increase over the same period, the simulated surface warming and salinification trends are not density compensated but induce instead a significant surface density decrease. As a consequence, stratification is enhanced in the upper layer resulting in a considerable weakening of the thermohaline circulation and a dramatic reduction of dense water formation rates throughout the basin (e.g. Somot et al. 2006). Such drastic hydrodynamic changes are expected to have a large impact on the Mediterranean marine ecosystems by inhibiting both primary productivity levels and surface organic carbon exports to the deep parts of the basin.

References

Barale V, Jaquet JM, Ndiaye M (2008) Algal blooming patterns and anomalies in the Mediterranean sea as derived from the SeaWiFS data set (1998–2003). Rem Sens Environ 112:3300–3313

Baringer MO, Price JF (1997) Mixing and spreading of the Mediterranean outflow. J Phys Oceanogr 27:1654–167

Behrenfeld M, O'Malley R, Siegel DA, McLain C, Sarmiento J, Feldman G, Falkowski P, Boss E, Milligan A (2006) Climate-driven trends in contemporary ocean productivity. Nature 444:752–755

Belkin M (2009) Rapid warming of large Marine ecosystems. Progr Oceanogr 81:207–213

Bethoux JP, Gentili B (1999) Functioning of the Mediterranean sea: past and present changes related to freshwater input and climate changes. J Mar Syst 20:33–47

Béthoux JP, Gentili B, Raunet J, Tailliez D (1990) Warming trend in the Western Mediterranean deep water. Nature 347:660–662

Beuvier J, Sevault F, Herrmann M, Kontoyiannis H, Ludwig W, Rixen M, Stanev E, Béranger K, Somot S (2010) Modeling the Mediterranean sea interannual variability during 1961–2000: focus on the Eastern Mediterranean Transient. J Geophys Res 115, C08017. doi:10.1029/2009JC005950

Borzelli GLE, Gacic M, Cardin V, Civitarese G (2009) Eastern Mediterranean Transient and reversal of the Ionian sea circulation. Geophys Res Lett 36, L15108. doi:10.1029/2009GL039261

Boscolo R, Bryden H (2002) Causes of long-term changes in Aegean sea deep water. Oceanol Acta 24:19–527

Boyd W, Doney SC (2002) Modelling regional responses by marine pelagic ecosystems to global climate change. Geophys Res Lett 29:1806. doi:10.1029/2001GL014130

Brankart JM, Brasseur P (1998) The general circulation in the Mediterranean sea: a climatological approach. J Mar Syst 18:41–70

Canals M, Puig P, Durrieu de Madron X, Heussner S, Palanques A, Fabres J (2006) Flushing submarine canyons. Nature 444: 354–357

Canals M, Danovaro R, Heussner S, Lykousis V, Puig P, Trincardi F, Calafat AM, Durrieu de Madron X, Palanques A, Sanchez-Vidal A (2009) Cascades in Mediterranean submarine grand canyons. Oceanography 22:26–43

Candela J (2001) Mediterranean water and the global circulation. In: Siedler G, Church J, Gould J (eds) Ocean circulation and climate: observing and modelling the global ocean. Academic, New York

Cartes JE, Maynou F, Fanelli E (2011) Nile damming as plausible cause of extinction and drop in abundance of deep-sea shrimp in the western Mediterranean over broad spatial scales. Progr Oceanogr 91:286–294

Christensen JH, Hewitson BA, Busuioc A et al (2007) Regional climate projections. In: Solomon S, Qin D, Manning M, Chen Z, Marquis M, Averyt KB, Tignor M, Miller HL (eds) Climate change 2007: the physical science basis contribution of working group I to the fourth assessment report of the intergovernmental panel on climate change. Cambridge University Press, Cambridge

CIESM (2009) Dynamics of Mediterranean deep waters. CIESM Workshop Monographs 38. CIESM, Monaco

Criado-Aldeanueva F, Del Río J, García-Lafuente J (2008) Steric and mass-induced Mediterranean sea level trends from 14 years of altimetry data. Glob Planet Chang 60:563–575

Curry R, Dickson B, Yashayaev I (2003) A change in the freshwater balance of the Atlantic Ocean over the past four decades. Nature 426:826–828

Doney SC (2006) Plankton in a warmer world. Nature 444:695–696

Durack PJ, Wijffels SE (2010) Fifty-year trends in global ocean salinities and their relationship to broad-scale warming. J Climate 23:4342–4362

Durrieu de Madron X, Zervakis V, Theocharis A, Georgopoulos D (2005) Comments to 'Cascades of dense water around the world ocean'. Progr Oceanogr 64:83–89

Font J, Puig P, Salat J, Palanques A, Emelianov M (2007) Sequence of hydrographic changes in NW Mediterranean deep water due to the exceptional winter of 2005. Sci Mar 71:339–346

Fuda J-L, Millot C (2002) Warming and salting of the Tyrrhenian Deep Water. CIESM Workshop Ser 16:51–54

Gacic M, Kovacevic V, Manca B, Papageorgiou E, Poulain PM, Scarazzato P, Vetrano A (1996) Thermohaline properties and circulation in the Strait of Otranto. CIESM Sci Ser 2:117–145

Gascard JC (1978) Mediterranean deep water formation, baroclinic instability and oceanic eddies. Oceanol Acta 1:315–330

Gasparini GP, Ortona A, Budillon G, Astraldi M, Sansone E (2005) The effect of the Eastern Mediterranean Transient on the hydrographic characteristics in the Strait of Sicily and in the Tyrrhenian Sea. Deep-Sea Res I 52:915–935

Gertman I, Hecht A (2002) Annual and long term changes in the salinity and the temperature of the waters of the South-eastern Levantine Basin. CIESM Workshop Ser 16:75–78

Giorgi F, Lionello P (2008) Climate change projections for the Mediterranean region. Glob Planet Chang 63:90–104

Hecht A, Pinardi N, Robinson A (1988) Currents, water masses, eddies and jets in the Mediterranean Levantine Basin. J Phys Oceanogr 18:1320–1353

Herrmann M, Estournel C, Déqué M, Marsaleix P, Sevault F, Somot S (2008) Dense water formation in the Gulf of Lion shelf: impact of atmospheric interannual variability and climate change. Cont Shelf Res 28:2092–2112

Herrmann M, Sevault F, Beuvier J, Somot S (2010) What induced the exceptional 2005 convection event in the northwestern Mediterranean basin? Answers from a modeling study. J Geophys Res 115, C12051. doi:10.1029/2010JC006162

Hurrell JW (1995) Decadal trends in the North Atlantic Oscillation, regional temperatures and precipitation. Science 269:676–679

IPCC (2000) Emission scenarios. A special report of the Intergovernmental Panel on Climate Change. Working Group III. Cambridge University Press , Cambridge

IPCC (2007) Climate change 2007: the physical science basis. Fourth assessment report of the Intergovernmental Panel on Climate Change. Cambridge University Press, Cambridge

Jones PD, Jonsson T, Wheeler D (1997) Extension to the North Atlantic Oscillation using early instrumental pressure observations from Gibraltar and south-west Iceland. Int J Climatol 17:1433–1450

Josey SA, Somot S, Tsimplis M (2011) Impacts of atmospheric modes of variability on Mediterranean Sea surface heat exchange. J Geophys Res 116, C02032. doi:10.1029/2010JC006685

Klein B, Roether W, Manca BB, Bregant D, Beitzel V, Kovacevic V, Luchetta A (1999) The large deep water transient in the Eastern Mediterranean. Deep-Sea Res I 46:371–414

Klein B, Roether W, Civitarese G, Gačić M, Manca BB, Ribera d' AM (2000) Is the Adriatic returning to dominate production of Eastern Mediterranean deep water? Geophys Res Lett 27:3377–3380

Krahman G, Schott F (1998) Longterm increases in western Mediterranean salinities and temperatures: anthropogenic and climatic sources. Geophys Res Lett 25:4209–4212

La Violette P (1990) The Western Mediterranean Circulation Experiment (WMCE): introduction. J Geophys Res 95:1511–1514

Lascaratos A (1993) Estimation of deep and intermediate water mass formation rates in the Mediterranean Sea. Deep-Sea Res II 40:1327–1332

Lascaratos A, Nittis K (1998) A high-resolution model three-imensional numerical study of Intermediate Water formation in the Levantine Sea. J Geophys Res 103:18497–18511

Lascaratos A, Roether W, Nittis K, Klein B (1999) Recent changes in deep water formation and spreading in the eastern Mediterranean Sea: a review. Progr Oceanogr 44:5–36

Leaman KD, Schott FA (1991) Hydrographic structure of the convection regime in the Gulf of Lions in winter 1987. J Phys Oceanogr 21:575–598

Lionello P, Malanotte-Rizzoli P, Boscolo R, Alpert P, Artale V, Li L, Luterbacher J, May W, Trigo R, Tsimplis M, Ulbrich U, Xoplaki E (2006) The Mediterranean climate: an overview of the main characteristics and issues. In: Lionello P, Malanotte-Rizzoli R, Boscolo P (eds) Mediterranean climate variability. Elsevier, Amsterdam

Lopez-Jurado J, Gonzalez-Pola C, Velez-Belchi P (2005) Observation of an abrupt disruption of the longterm warming trend at the Balearic Sea, western Mediterranean sea, in summer 2005. Geophys Res Lett 32, L24606. doi:10.1029/2005GL024430

Lozier S, Sidlinger L (2009) On the source of Mediterranean Overflow Water property changes. J Phys Oceanogr 39:1800–1819

Lozier MS, Stewart NM (2008) On the temporally varying northward penetration of Mediterranean overflow water and eastward penetration of Labrador Sea Water. J Phys Oceanogr 38:2097–2103

Luterbacher J, Dietrich D, Xoplaki E, Grosjean M, Wanner H (2004) European seasonal and annual temperature variability, trends, and extremes since 1500. Science 303:1499–1503

Madec G, Chartier M, Delecluse P, Crepon M (1991) A three-dimensional numerical study of deep-water formation in the northwestern Mediterranean Sea. J Phys Oceanogr 21:1349–1371.

Malanotte-Rizzoli P, Manca BB, Ribera d' AM, Theocharis A, Bergamasco A, Bregant D, Budillon G, Civitarese G, Georgopoulos D, Michelato A, Sansone E, Scarazzato P, Souverrmezoglou E (1997) A synthesis of the Ionian hydrography, circulation an water mass pathways during POEM-Phase I. Progr Oceanogr 39:153–204

Malanotte-Rizzoli P, Manca BB, d'Alcala MR, Theocharis A, Brenner S, Budillon G, Ozsoy E (1999) The eastern Mediterranean in the 80s and in the 90s: the big transition in the intermediate and deep circulations. Dyn At Oceans 29:365–395

Manca BB, Kovacevic V, Gacic M, Viezzoli D (2002) Dense water formation in the Southern Adriatic Sea and spreading into the Ionian Sea in the period 1997–1999. J Mar Syst 33:133–154

Manca BB, Burca M, Giorgetti A, Coatanoan C, Garcia M-J, lona A (2004) Physical and biochemical averaged vertical profiles in the Mediterranean regions: an important tool to trace the climatology of water masses and to validate incoming data from operational oceanography. J Mar Syst 48:83–116

Mariotti A (2010) Recent changes in the Mediterranean water cycle: A pathway toward long-term regional hydroclimatic change? J Climate 23:1513–1525

Mariotti A, Dell'Aquila A (2011) Decadal climate variability in the Mediterranean region: roles of large-scale forcings and regional processes. Clim Dyn 38:1129. doi:10.1007/s00382-011-1056-7

Mariotti A, Zeng N, Yoon J, Artale V, Navarra A, Alpert P, Li L (2008) Mediterranean water cycle changes: transition to drier 21st century conditions in observations and CMIP3 simulations. Environ Res Lett 3:044001. doi:10.1088/1748-9326/3/044001

Marty JC, Chiaverini J (2010) Hydrological changes in the Ligurian Sea (NW Mediterranean, DYFAMED site) during 1995–2007 and biogeochemical consequences. Biogeosciences 7:2117–2128

Marty JC, Chiaverini J, Pizay MD, Avril B (2002) Seasonal and internannual dynamics of nutrients and phytoplankton pigments in the western Mediterranean Sea at the DYFAMED timeseries station (1991–1999). Deep-Sea Res II 49:1965–1986

McCartney MS, Mauritzen C (2001) On the origin of the warm inflow to the Nordic Seas. Prog Oceanogr 51:125–214

MEDAR Group (2002) Mediterranean and Black Sea database of temperature, salinity and biogeochemical parameters climatological atlas, 4 CDROM. IFREMER, Brest

MEDOC Group (1970) Observation of formation of deep water in the Mediterranean Sea, 1969. Nature 227:1037–1040

Millot C (1991) Mesoscale and seasonal variabilities of the circulation in the Western Mediterranean. Dyn At Oceans 15:179–215

Millot C (2007) Interannual salinification of the Mediterranean inflow. Geophys Res Lett 34, L21609. doi:10.1029/2007GL031179

Millot C (2009) Another description of the Mediterranean Sea outflow. Progr Oceanogr 82:101–124

Millot C, Taupier-Letage I (2005) Circulation in the Mediterranean Sea. In: Hutzinger O (ed) The Mediterranean Sea, the handbook of environmental chemistry 5. Springer, Berlin

Millot C, Candela J, Fuda JL, Tber Y (2006) Large warming and salinification of the Mediterranean outflow due to changes in its composition. Deep-Sea Res I 53:656–666

Nielsen JN (1912) Hydrography of the Mediterranean and Adjacent Waters. Rep Dan Oceanogr Exp 1:72–191

Nykjaer L (2009) Mediterranean Sea surface warming 1985–2006. Clim Res 39:11–17

Ovchinnikov LM (1966) Circulation in the surface and intermediate layers of the Mediterranean. Oceanology 6:48–59

Ovchinnikov LM, Zats VI, Krivosheya VG, Nemirovsky MS, Udodov AI (1987) Winter convection in the Adriatic and formation of deep eastern Mediterranean waters. Ann Geophys 5:89–92

Pinardi N, Allen I, Demirov E, De Mey P, Korres G, Lascaratos A, Le Traon P-Y, Maillard C, Manzella G, Tziavos C (2003) The Mediterranean ocean forecasting system: first phase of implementation (1998–2001). Ann Geophys 21:3–20

Pollak MI (1951) The sources of deep water in the eastern Mediterranean Sea. J Mar Res 10:128–152

Polovina JJ, Howell EA, Abecassis M (2008) The ocean's least productive waters are expanding. Geophys Res Lett 35, L03618. doi:10.10 29/2007GL031745

Potter R, Lozier S (2004) On the warming and salinification of the Mediterranean outflow waters in the North Atlantic. Geophys Res Lett 31, L01202. doi:10.1029/2003GL018161

Poulos SE, Drakopoulos PG (2001) A reassessment of the Mediterranean river runoff. Rapp Comm Int Mer Medit 36:76

Psarra S, Tselepides A, Ignatiades L (2000) Primary productivity in the oligotrophic Cretan Sea (NE Mediterranean): seasonal and interannual variability. Progr Oceanogr 46:187–204

Pusceddu A, Mea M, Gambi C, Bianchelli S, Canals M, Sanchez-Vidal A, Calafat A, Heussner S, Durrieu De Madron X, Avril J, Thomsen L, Garcìa R, Danovaro R (2010) Ecosystem effects of dense water formation on deep Mediterranean Sea ecosystems: an overview. Adv Oceanogr Limn 1:51–62

Reid JL (1979) On the contribution of the Mediterranean Sea outflow to the Norwegian-Greenland Sea. Deep Sea Res I 26:1199–1223

Rixen M et al (2005) The Western Mediterranean Deep Water: a proxy for climate change. Geophys Res Lett 32, L12608. doi:10.1029/200 5GL022702

Robinson AR, Golnaraghi M, Leslie N, Artegiani A, Hecht A, Lazzone E, Michelato A, Sansone E, Theocharis A, Unluata U (1991) Structure and variability of the Eastern Mediterranean general circulation. Dyn At Oceans 15:215–240

Roether W, Schlitzer R (1991) Eastern Mediterranean deep water renewal on the basis of chlorofluoromethane and tritium. Dyn At Oceans 15:333–354

Roether W, Manca BB, Klein B, Bregant B, Georgopoulos D, Beitzel V, Kovacevic V, Luchetta A (1996) Recent changes in Eastern Mediterranean deep waters. Science 271:333–335

Roether W, Klein B, Manca BB, Theocharis A, Kioroglou S (2007) Transient Eastern Mediterranean deep waters in response to the massive dense-water output of the Aegean Sea in 1990s. Prog Oceanogr 74:540–571

Rohling EJ, Bryden HL (1992) Man-induced salinity and temperature increases in western Mediterranean water. J Geophys Res 97:11191–11198

Romanou A, Tselioudis G, Zerefos CS, Clayson C, Curry JA, Andersson A (2010) Evaporation-precipitation variability over the Mediterranean and the Black Seas from satellite and reanalysis estimates. J Climate 23:5268–5287

Rubino A, Hainbucher D (2007) A large abrupt change in the abyssal water masses of the eastern Mediterranean. Geophys Res Lett 34, L23607. doi:10.1029/2007GL031737

Salat J, Font J (1987). Water mass structure near and offshore the Catalan coast during the winter of 1982 and 1983. Ann Geophys 5:49–54.

Samuel S, Haines K, Josey S, Myers PG (1999) Response of the Mediterranean Sea thermohaline circulation to observed changes in the winter wind stress field in the period 1980–1993. J Geophys Res 104:7771–7784

Sarmiento JL, Slater R, Barber R, Bopp L, Doney SC, Hirst AC, Kleypas J, Matear R, Mikolajewicz U, Monfray P, Soldatov V, Spall SA, Stouffer R (2004) Response of ocean ecosystems to climate warming. Global Biogeochem Cycles 18:GB3003. doi:1029/2003GB002134

Schlitzer R, Roether W, Oster H, Junghans H, Hausmann M, Johannsen H, Michelato A (1991) Chlorofluoromethane and oxygen in the Eastern Mediterranean. Deep-Sea Res I 38:1531–1551

Schott F, Visbeck M, Send U, Fisher J, Stramma L, Desaubies Y (1996) Observations of deep convection in the Gulf of Lions, northern Mediterranean, during the winter of 1991/1992. J Phys Oceanogr 26:505–524

Schroeder K, Gasparini GP, Tangherlini M, Astraldi M (2006) Deep and intermediate water in the western Mediterranean under the influence of the Eastern Mediterranean Transient. Geophys Res Lett 33, L21607. doi:10.1029/2006GL027121

Schroeder K, Ribotti A, Borghini M, Sorgente R, Perilli A, Gasparini GP (2008) An extensive Western Mediterranean Deep Water renewal between 2004 and 2006. Geophys Res Lett 35, L18605. doi:10.102 9/2008GL035146

Schroeder K, Josey SA, Herrmann M, Grignon L, Gasparini GP, Bryden HL (2010) Abrupt warming and salting of the Western Mediterranean Deep Water after 2005: Atmospheric forcings and lateral advection. J Geophys Res 115, C08029. doi:10.1029/2009JC005749

Skliris N, Lascaratos A (2004) Impacts of the Nile River damming on the thermohaline circulation and water mass characteristics of the Mediterranean Sea. J Mar Syst 52:121–143

Skliris N, Sofianos S, Lascaratos A (2007) Hydrological changes in the Mediterranean Sea in relation to changes in the freshwater budget: a numerical modelling study. J Mar Syst 65:400–416

Skliris N, Sofianos S, Gkanasos A, Mantziafou A, Vervatis V, Axaopoulos P, Lascaratos A (2012) Decadal scale variability of sea surface temperature in the Mediterranean Sea in relation to atmospheric variability. Ocean Dyn 62:13–30

Smith RO, Bryden HL, Stansfield K (2008) Observations of new western Mediterranean deep water formation using Argo floats 2004–2006. Ocean Sci 4:133–149

Somot S, Sevault F, Déqué M (2006) Transient climate change scenario simulation of the Mediterranean Sea for the twenty-first century using a high-resolution ocean circulation model. Clim Dyn 27:851–879

Somot S, Sevault F, Déqué M, Crépon M (2008) 21st century climate change scenario for the Mediterranean using a coupled atmosphere–ocean regional climate model. Glob Planet Chang 63:112–126

Souvermezoglou E, Krasakopoulou E (2000) Chemical oceanography in the Cretan Sea: changes associated to the transient. Medit Mar Sci 1:91–103

Souvermezoglou E, Krasakopoulou E, Pavlidou A (1999) Temporal variability in oxygen and nutrients concentrations in the southern Aegean Sea and the straits of the Cretan Arc. Progr Oceanogr 44:573–600

Sparnocchia S, Manzella GMR, La Violette P (1994) The interannual and seasonal variability of the MAW and LIW core properties in the Western Mediterranean Sea. In: La Violette P (ed) Seasonal and interannual variability of the Western Mediterranean sea. AGU, Washington, DC

Streftaris N, Zenetos A, Papathanassiou E (2005) Globalisation in marine ecosystems: the story of non-indigenous marine species across European seas. Oceanogr Mar Biol 43:419–453

Sur HI, Ozsoy E, Unluata U (1992) Simultaneous deep and intermediate convection in the Northern Levantine Sea, winter 1992. Oceanol Acta 16:33–43

Testor P, Gascard JC (2006) Post convection spreading phase in the northwestern Mediterranean Sea. Deep Sea Res I 53: 869–893. doi:10.1016/j.dsr.2006.02.004

Theocharis A, Nittis K, Kontoyiannis H, Papageorgiou E, Balopoulos E (1999) Climatic changes in the Aegean Sea influence the Eastern Mediterranean thermohaline circulation (1986–1997). Geophys Res Lett 26:1617–1620

Thorpe RB, Bigg GR (2000) Modelling the sensitivity of Mediterranean outflow to anthropogenically forced climate change. Clim Dyn 16:355–368

Touratier F, Goyet C (2009) Decadal evolution of anthropogenic CO_2 in the north western Mediterranean sea from the mid-1990s to the mid-2000s. Deep-Sea Res I 56:1708–1716

Tsimplis MN, Josey SA (2001) Forcing of the Mediterranean Sea by atmospheric oscillations over the North Atlantic. Geophys Res Lett 28:803–806

Tsimplis MN, Zervakis V, Josey SA, Peneva EL, Struglia MV, Stanev EV, Theocharis A, Lionello P, Malanotte-Rizzoli P, Artale V, Tragou E, Oguz T (2006) Changes in the oceanography of the Mediterranean sea and their link to climate variability. In: Lionello P, Malanotte-Rizzoli P, Boscolo R (eds) Mediterranean climate variability. Elsevier, Amsterdam

Vargas-Yáñez M, Moya F, García-Martínez MC, Tel E, Zunino P, Plaza F, Salat J, López-Jurado JL, Serra M (2010) Climate change in the Western Mediterranean Sea 1900–2008. J Mar Syst 82:171–176

Vilibic I (2002) Long term changes in Adriatic thermohaline properties. CIESM Workshop Ser 16:56–59

Vilibic I, Orlic M (2001) Least-squares tracer analysis of water masses in the South Adriatic (1967–1990). Deep-Sea Res I 48:2297–2330

Wu P, Haines K (1996) Modeling the dispersal of Levantine Intermediate Water and its role in Mediterranean deep water formation. J Geophys Res 101:6591–6607

Wu P, Haines K (1998) The general circulation of the Mediterranean Sea from a 100-year simulation. J Geophys Res 103:1121–1135

Wust G (1961) On the vertical circulation of the Mediterranean Sea. J Geophys Res 66:3261–3271

Xoplaki E, Luterbacher J, Gonzalez-Rouco JF (2006) Mediterranean summer temperature and winter precipitation, large-scale dynamics, trends. Il Nuovo Cimento 29:45–54

Zervakis V, Georgopoulos D, Drakopoulos PG (2000) The role of the North Aegean in triggering the recent Eastern Mediterranean climatic changes. J Geophys Res 105:26103–26116

4 Past, Present and Future Patterns in the Nutrient Chemistry of the Eastern Mediterranean

Michael Krom, Nurit Kress, Ilana Berman-Frank, and Eyal Rahav

Abstract

This review summarises the distribution of dissolved nutrients (nitrate, phosphate and silicate) over the period of modern measurements (~30 years) and aspects of the biogeochemical processes which control their distribution in the Eastern Mediterranean Sea (EMS). The levels of nitrate, phosphate and silicic acid (~6 μ mol N kg^{-1}, 0.25 μ mol P kg^{-1} and 11 μ mol Si kg^{-1} in deep water) in the EMS are much lower than all other parts of the ocean due to the unusual anti-estuarine circulation which exports nutrient replete intermediate water at the Straits of Sicily and to the relatively young age of the deep water ($\tau \approx 120$ years). Dissolved oxygen decrease and nutrients increase eastwards, in the direction of the deep-water circulation with nitrate increasing by ~0.5 μ mol kg^{-1} from the Western Ionian to the Eastern Levantine, phosphate by 0.05 μ mol kg^{-1} and silicic acid by ~4.5 μ mol kg^{-1} representing the amount (rate) of organic matter and biogenic silica (BSi) breakdown over the residence time of Deep Water in the basin. The East Mediterranean Transient interrupted this simple pattern causing lower nutrients in the deep water and upwelled nutrients into the upper layers. The high nitrate/phosphate found in the deep water (25–28:1) is due to a combination of high Nitrate:Phosphate waters advected from the surface of the Adriatic during deep water formation during the P-limited winter bloom and P recycling more efficiently than N from the descending particulate organic matter (POM). High N:P ratios in POM and dissolved organic matter (DOM) show that the entire EMS is P starved. There are seasonal changes in nutrient limitation with conventional P limitation during the winter phytoplankton bloom which becomes N&P co-limitation in early summer and can, under some circumstances, become even N limited in mid-summer in the nutrient starved upper waters. N_2 fixation rates are constantly low in the EMS compared to somewhat higher levels in the western MS. Nutrient budgets have been used to explain that the reason for the unusual N:P ratio in the basin is high N:P ratio in the external inputs combined with low denitrification rates caused by the ultra-oligotrophic status of the basin. However the external nutrient

M. Krom (✉)
Earth and Biosphere Institute, School of Earth and Environment, University of Leeds, Woodhouse Lane, Leeds LS2 9JT, UK

Charney School of Marine Sciences, University of Haifa, Haifa, Israel
e-mail: M.D.Krom@leeds.ac.uk

N. Kress
Israel Oceanographic & Limnological Res, National Institute of Oceanography, P.O.Box 8030, Tel Shikmona, Haifa 31080, Israel
e-mail: nurit@ocean.org.il

I. Berman-Frank • E. Rahav
Mina & Everard Goodman Faculty of Life Sciences, Bar Ilan University, Ramat Gan 52900, Israel
e-mail: ilana.berman-frank@biu.ac.il

S. Goffredo and Z. Dubinsky (eds.), *The Mediterranean Sea: Its history and present challenges*,
DOI 10.1007/978-94-007-6704-1_4,

(both riverine and atmospheric) inputs have increased dramatically between 1950 and 2000. Estimates of the pristine flux of nutrients by riverine sources to the EMS are 2.0×10^9 moles N/y and 0.095×10^9 moles P/y while the atmospheric flux has increased by 85 % for NOx and 65 % for NH3 between 1910 and the end of the twentieth century. It is suggested that climate change may result in dramatic threshold changes in trophic status of the EMS if surface circulation rates decrease as has been predicted.

Keywords
Nutrient cycling • Anti-estuarine circulation • Seasonal P limitation • Anthropogenic inputs • Climate change

Introduction

The Eastern Mediterranean Sea (EMS) is an unusual part of the global ocean. It is ultra-oligotrophic with primary productivity approximately half that of the Sargasso Sea (Krom et al. 2003). The nutrient concentrations in the deep water are amongst the lowest measured anywhere in the oceans despite the fact that the EMS is almost land-locked and has considerable nutrient inputs (both atmospheric and riverine), much of it anthropogenic, particularly from its northern shores. Its low nutrient concentration is due to a combination of the relatively young age of the water masses and to the unusual anti-estuarine circulation in which nutrient depleted waters flow in through the Straits of Sicily while nutrient enriched intermediate waters flow out. The nitrate: phosphate (N:P) ratio in the deep water is highly unusual being almost twice the Redfield ratio (Krom et al. 1991) while the N: P ratios of both dissolved and particulate organic matter is also far in excess of 16:1 making the whole system P starved. Recent work has explained this unusual N: P ratio as being due to high N: P ratios in the external supply combined with very low denitrification rates due to the ultra-oligotrophic nature of the basin (Krom et al. 2010). Together these unusual properties make the EMS an ideal natural laboratory to examine the nature of the controls on nutrient cycling and hence important processes related to carbon uptake in the ocean.

In this chapter we review the distribution of inorganic nutrients in the water column of the EMS over the past 30 years, the period over which modern nutrient measurements have been made. This time period includes the changes in the physical circulation caused by the Eastern Mediterranean Transient (EMT) event. The vertical distribution of dissolved and particulate organic matter is presented together with changes in chlorophyll-a and Deep Chlorophyll Maximum (DCM) across the basin. In examining the biogeochemical and nutrient cycling processes in the EMS, we describe why the deep waters of the EMS have such an unusual N:P ratio even though the uptake by phytoplankton in surface waters is approximately Redfieldian. Although primary productivity in the EMS is P limited during the main winter phytoplankton bloom, later in the year the nature of the nutrient limitation changes for different parts of the ecosystem including the phytoplankton becoming N and P co-limited. We also review the most recent measurements of N_2 fixation rates across the basin which display very low rates throughout the year. Thus, in the EMS new N sources from current N_2 fixation in the photic layers are negligible and contribute only a small percentage to new and primary production in the basin.

The review continues by examining the use of nutrient budgets to define biogeochemical processes within the basin. We show how the input terms in those budgets have changed with time and discuss the implications as regards the proportion of anthropogenic nutrients in the system. We finish by describing how changes in the physical circulation have caused significant modifications in biogeochemical processes in the past and speculate, how environmental and climate change might cause major changes in the future and what further studies are needed to understand the workings of this semi-enclosed sea.

General and Mesoscale Circulation and Changes in the Circulation

The general circulation of the Mediterranean Sea is anti-estuarine. Surface water of Atlantic origin (AW), low in nutrients, flows through Gibraltar and the Sicily Straits and spreads to the Ionian and Levantine, being modified along the way . This Modified Atlantic water (MAW) circulation in the EMS displays a number of mesoscale features with large variability in extent, shape and time scales (Malanotte-Rizzoli et al. 1999; POEM-Group 1992; Robinson et al. 1991). Levantine Intermediate Water (LIW), formed in the Levantine Basin, with higher nutrient concentrations, flows at intermediate depths (200–500 m) westwards from the Levantine Basin, through the Cretan Passage with the major flow directed towards the Sicily Straits, and a minor part introduced to the Adriatic Sea. The deep waters in the EMS are isolated from the Western Basin by the shallow sill depth of the Sicily Straits. Until the early 1990s the deep layer was

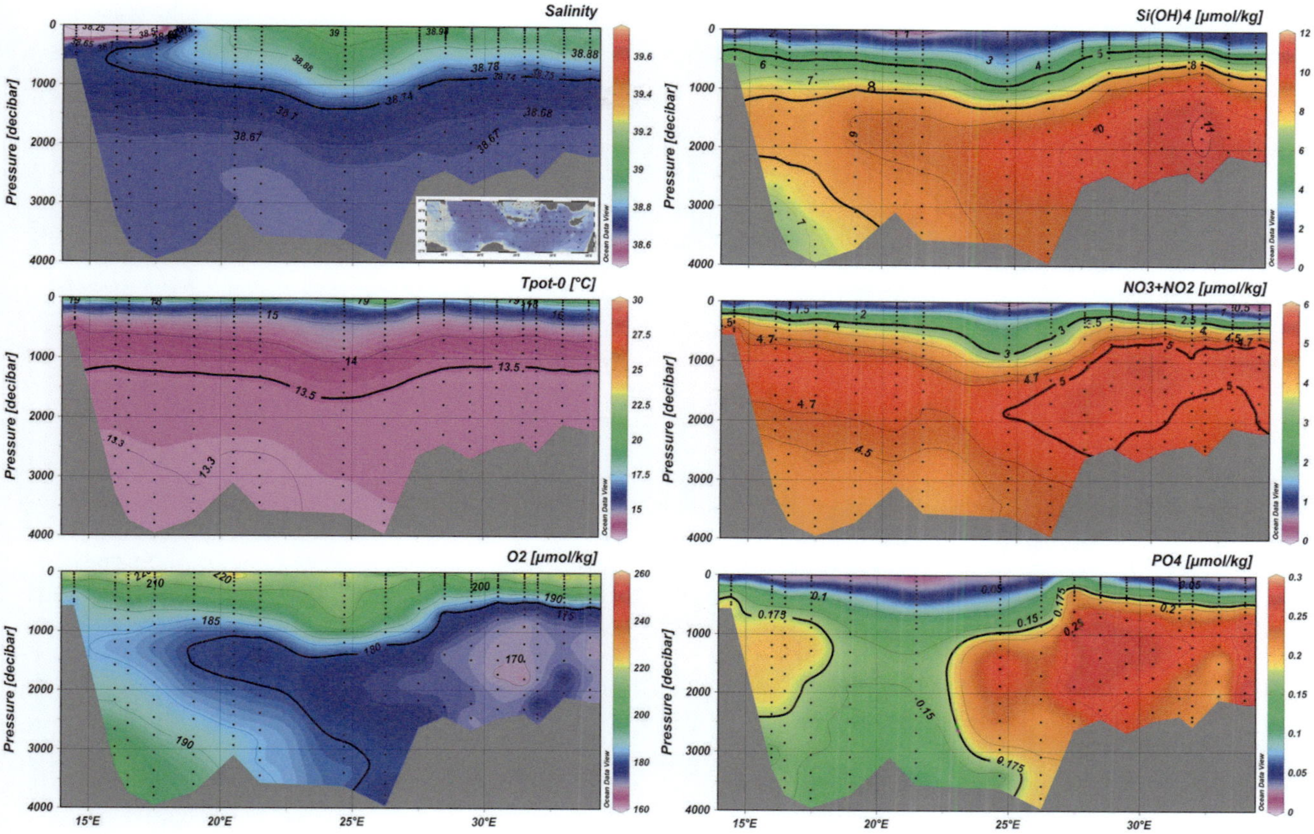

Fig. 4.1 Vertical sections of salinity, potential temperature, dissolved oxygen, silicic acid, nitrate and phosphate along the west–east cross section of the Eastern Mediterranean in 1987. Stations and data points are indicated by *filled dots* (see *inset map*) (Adapted from Schlitzer et al. 1991; Klein et al. 1999)

occupied exclusively by dense water formed in the Southern Adriatic, known as Adriatic Deep Water (ADW) (Fig. 4.1). A combination of meteorological and hydrological factors led to the establishment of the Aegean as a new source of deep water in the area at the beginning of the 1990s, changing the circulation in the basin (Roether et al. 2007 and references therein). The event was termed the Eastern Mediterranean Transient (EMT). The EMT formed a new water mass, the Cretan Sea Outflow water (CSOW) (Klein et al. 1999). This new, younger, dense water was warmer and more saline than the older ADW and could be recognised by an inversion in the TS diagrams. The CSOW was also characterized by higher dissolved oxygen and lower nutrient concentrations than the ADW (Roether et al. 1996). The main outflow of CSOW (ca. 75 %) was delivered between mid 1992 and late 1994. It propagated preferentially westwards and more slowly towards the east. In 1991 there were strong signals of the CSOW at stations south of Crete located only to the north of the Eastern Mediterranean Ridge (EMRidge). By 1995 all the stations in the Levantine showed strong TS inversions while by 1999, the EMT was well established in the whole EMS (Roether et al. 1996; 2007; Klein et al. 2003; Kress et al. 2003) (Fig. 4.2). A snapshot of the evolution of the CSOW in the Levantine Basin taken in September 2008 (Fig. 4.3) showed the continuing change and erosion of the deep water structure. From 27 °E eastwards, potential temperature and salinity increased from the original Adriatic Deep water (ADWo) towards the bottom due to the CSOW, while close to Crete, a new deep water was present (ADWn) which was slightly less saline and colder compared to the CSOW (Manca et al. 2002, 2006; Rubino and Hainbucher 2007).

Description of the Present Distribution of Nutrients and Dissolved Oxygen and Changes Over the Past 25 years

Description of Typical Vertical Profiles of Nutrients and Dissolved Oxygen

The upper layer (0–150 m) across the EMS is quite homogeneous in dissolved oxygen and nutrient concentrations, influenced only by the strong mesoscale field (Figs. 4.1, 4.2, and 4.3). The surface waters are generally saturated with dissolved oxygen; indeed in summer a slight oversaturation

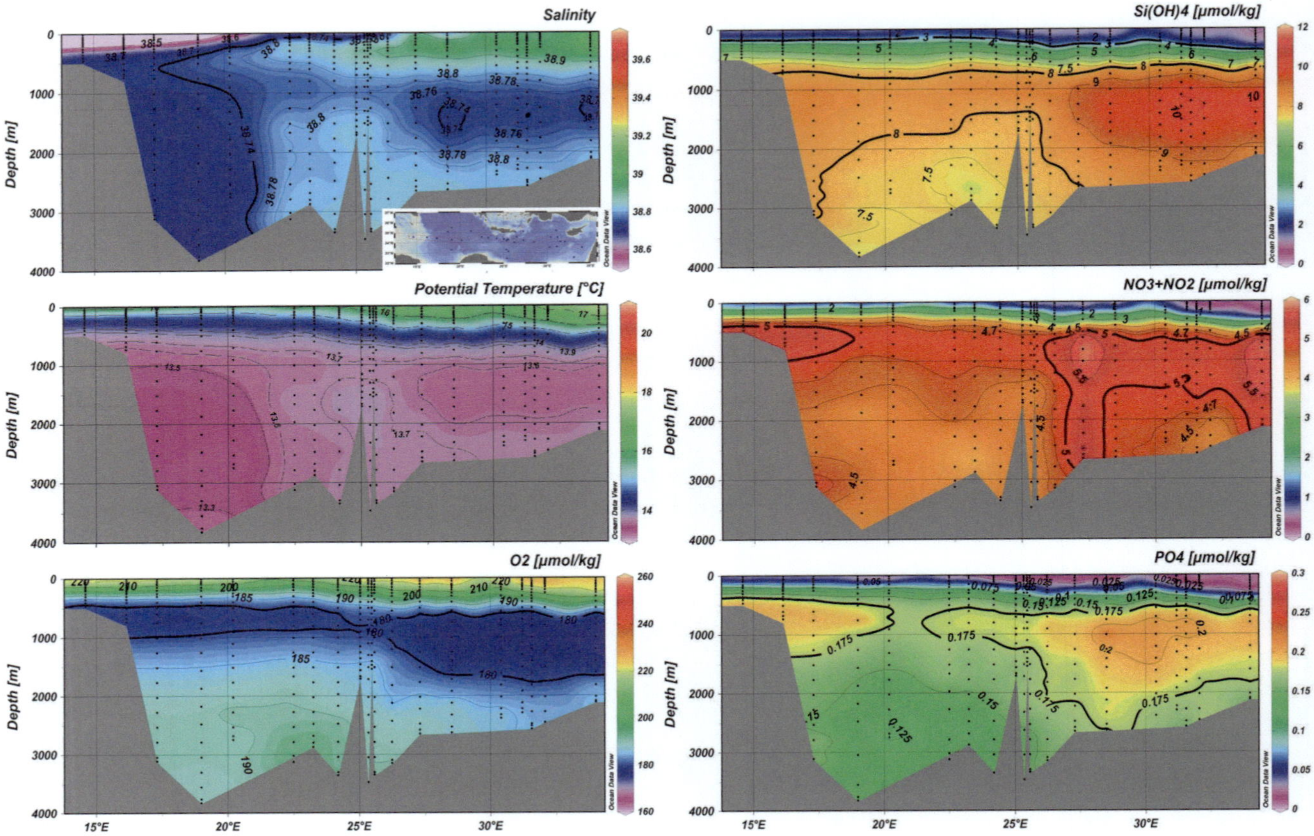

Fig. 4.2 Vertical sections of salinity, dissolved oxygen, silicic acid, nitrate and phosphate along the west–east cross section of the Eastern Mediterranean in 1999. Stations and data points are indicated by *filled dots* (see *inset map*) (Adapted from Kress et al. 2003)

often appears in the MAW (Kress and Herut 2001). The nutrient concentrations are very low, with phosphate in the nanomolar level throughout the year, while nitrate is at the nanomolar level in the summer and slightly higher in winter (~0.5–1 µmole kg^{-1}) and silicic acid close to 1 µmole kg^{-1} (e.g. Krom et al. 2005; Kress et al. 2003; Pujo-Pay et al. 2011)

Typical composite depth profiles of dissolved oxygen, nitrate and silicic acid in the Levantine and central Ionian Basin after the EMT event are presented in Fig. 4.4 (Kress et al. 2003). Dissolved oxygen concentrations are high at the surface, decrease to minimal values at the 500–1,500 m layer and increase towards the bottom (Levantine) or down to 2,500 m, the concentrations remaining then essentially constant (Ionian). The nitrate depth profile is a mirror image of the dissolved oxygen profile: low concentrations at the surface, an increase to maximal values at the 500–1,500 m layer and a decrease towards the bottom or down to 2,500 m. The silicic acid depth profile is similar to that of nitrate, but the maximal layer is located deeper, due to the different remineralization processes, biological for nitrate and chemical dissolution for silicic acid.

Cross Basin Distribution and How They Have Changed with Time

In comparison to the physical evolution, the changes in the distributions of dissolved oxygen and dissolved nutrients across the Eastern Mediterranean in the mid and deep waters (i.e. >300 m) have been less described in the literature (Schlitzer et al. 1991; Roether et al. 1996; Klein et al. 1999; 2003; Kress et al. 2011; 2003). In 1987, prior to the EMT (Fig. 4.1), the cross-basin distribution of dissolved oxygen and nutrients in the deep water was essentially uniform from 1,000 m and below, except for the Ionian that exhibited a mid-depth layer at ca. 1,200 m with minimum oxygen and maximum nutrients concentrations (Min_{Ox}/Max_{Nut}) (with 178, 9.3, 5.0 and 0.16 µmole kg^{-1} for oxygen, silicic acid, nitrate and phosphate, respectively in the Mid Ionian) due to the intrusion of newly added ADW at depth toward the base of the western continental slope (Schlitzer et al. 1991). Dissolved oxygen decreased and nutrients increased eastwards, in agreement with the direction of the deep-water circulation. No mid depth Min_{Ox}/Max_{Nut} layer was present in the Levantine Basin, and the waters were essentially

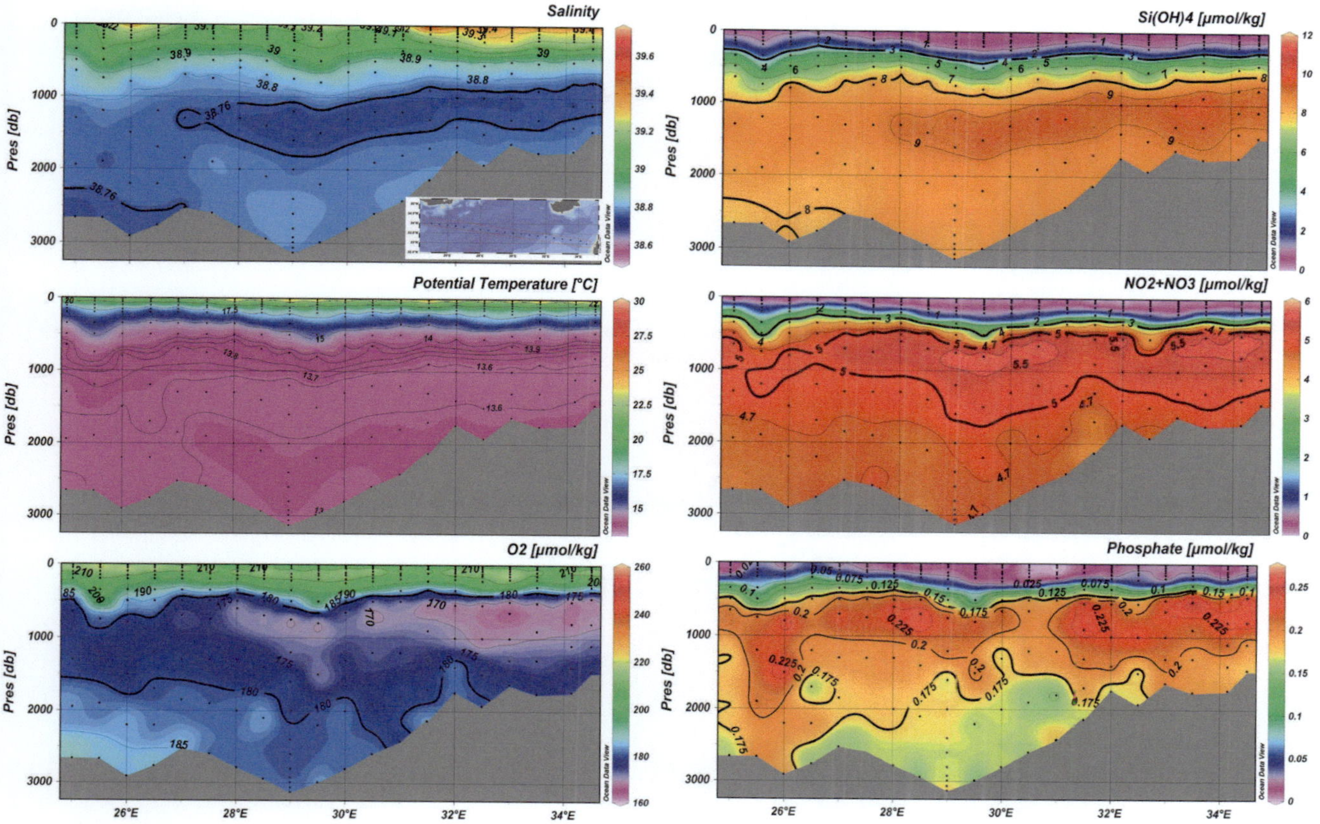

Fig. 4.3 Vertical sections of salinity, dissolved oxygen, silicic acid, nitrate and phosphate along the west–east cross section of the Levantine Basin in 2008. Stations and data points are indicated by *filled dots* (see *inset map*) (Adapted from Kress et al. 2011)

homogeneous below 1,200 m depth with 170 µmole kg^{-1} dissolved oxygen and 10.5, 4.9 and 0.21 µmole kg^{-1} silicic acid, nitrate and phosphate, respectively. The nitrate increased by ~0.5 µmole kg^{-1} from the Western Ionian to the Eastern Levantine, while the phosphate increased by only 0.05 µmole kg^{-1} . Silicic acid increased by ~4.5 µ mol kg^{-1} while dissolved oxygen decreased by 20 µmole kg^{-1} . These nutrient changes represent the amount (rate) of organic matter and biogenic silica (BSi) breakdown at depth in the system over the residence time of ADW in the basin.

This simple pattern changed during the 1990s as a result of the EMT event. In 1991, the CSOW was confined to the vicinity of the Cretan Arc, but by 1995, ca. 6 years after the start of the EMT event its influence was present in most of the EMS. The deep water was no longer uniform and a pronounced Min_{Ox}/Max_{Nut} layer was detected at the 500–1,500 m depth range depending on the basin sampled (Klein et al. 1999; Lascaratos et al. 1999). This layer corresponded to the older ADWo that was pushed up by the younger CSOW. The Min_{Ox}/Max_{Nut} layer continued to change as the EMT evolved (Klein et al. 2003; Kress et al. 2003).

Nine years from its onset, in 1999, the CSOW was found in the whole basin and clearly discernible in the dissolved oxygen and nutrient vertical distributions (Fig. 4.2) (Klein et al. 2003; Kress et al. 2003). The Min_{Ox}/Max_{Nut} layer, the old ADW, was found at the 300–1,500 m depth interval, thickening and deepening eastwards. The Min_{ox} concentrations were similar in the Levantine and Western Ionian (175 µmole kg^{-1}) but located at different depth layers, 600–1,500 m and 300–800 m, respectively (Fig. 4.4). The Min_{Ox} at the Cretan Passage was slightly higher (175–180 µmole kg^{-1}) indicating the presence of younger water. The maximal concentrations of phosphate and nitrate (> 0.175 and > 5.5 µmol kg^{-1}, respectively) were similar in the Levantine and the Western Ionian with lower concentrations at the Cretan Passage and Eastern Ionian. The maximal concentration of silicic acid increased eastwards, from ca. 8.5 to >9.5 µmol kg^{-1} due to increasing age of the water mass. Maximal values were centered at 1,200–1,400 m, deeper than the maxima for nitrate and phosphate due to the different remineralisation pathways (slower chemical dissolution for silicic acid as opposed to biological remineralisation for N and P). The CSOW was very noticeable in the

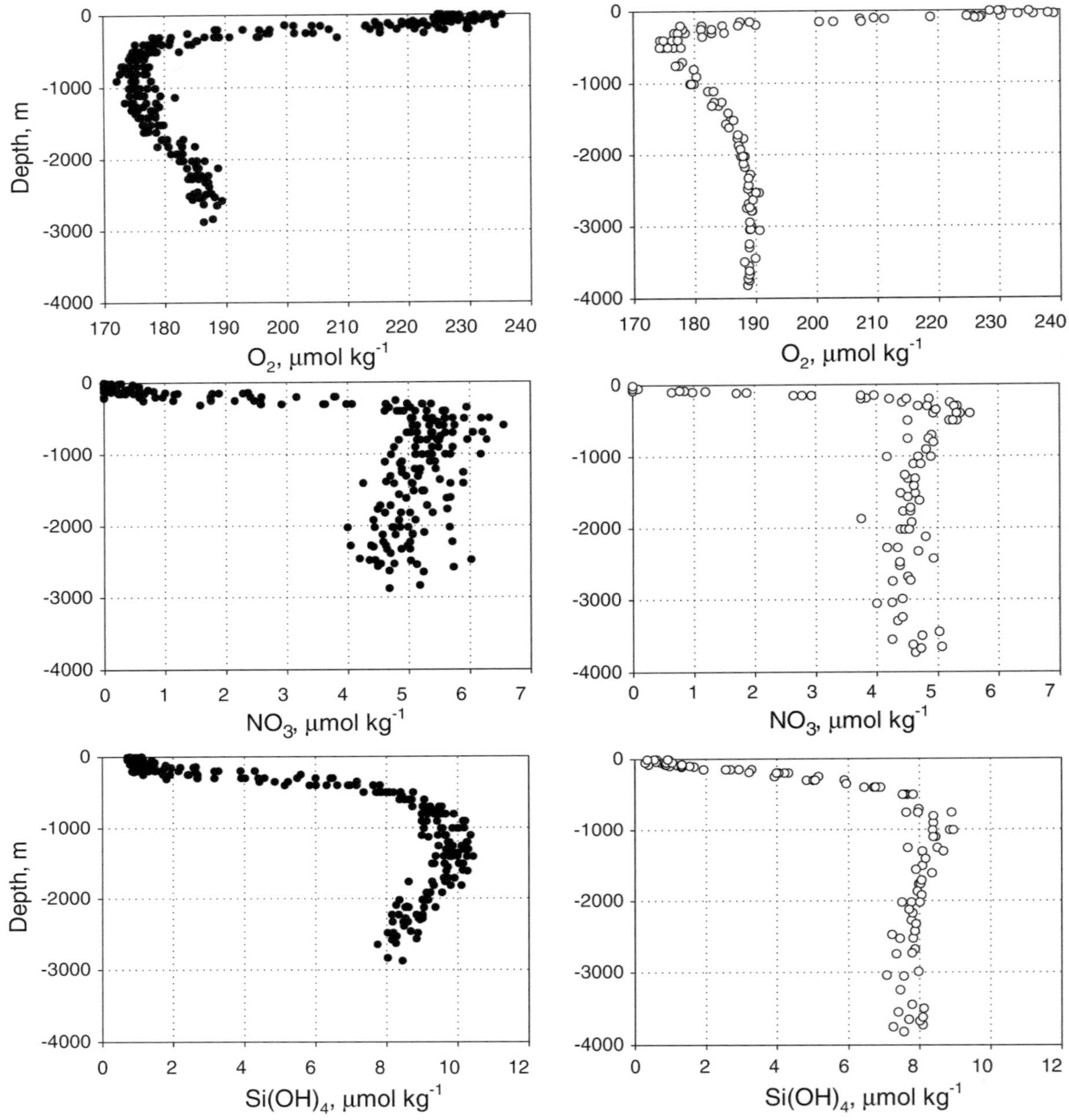

Fig. 4.4 Composite depth profiles of dissolved oxygen, nitrate and silicic acid in the Levantine (*left panel, filled circles*) and the central Ionian basin (*right panel, open circles*) in April–May 1999, after the EMT event (Adapted from Kress et al. 2003)

Levantine, the concentrations of nutrients decreasing from the maxima in the ADW to minima of 4.7, 0.17 and 8.2 µmol kg^{-1} for nitrate, phosphate and silicic acid, respectively in the CSOW. Before the EMT event, nitrate and phosphate concentrations were constant below 1,200 m while silicic acid concentrations continued to increase with depth (Kress and Herut 2001). In the Eastern Ionian and Western Crete (19–25.5 °E), below 2,000 m, the concentrations of nutrients were lower (<4.4, <0.15 and < 7.8 µmol kg^{-1} for nitrate, phosphate and silicic acid, respectively) and dissolved oxygen higher than those found in the Levantine and in the westernmost Ionian, indicating a source of younger water.

By 2008, ca. 12 years after the maximum CSOW outflow, the effect of the EMT could still be identified in the Levantine Basin (Fig. 4.3). The Min_{Ox} layer was centred at ca. 900 m, more emphasized in the eastern part of the transect (<175 µmol kg^{-1} up to 27 °E) and eroded towards the west (minimal concentrations of 180–175 µmol kg^{-1}). The concentrations increased towards the bottom, in particular westwards of 27 °E, indicating mixing and erosion of the CSOW

Table 4.1 Concentration of DOC measured by high temperature oxidation in the Eastern Mediterranean compared to values determined in other subtropical regions of the world's oceans

Location	Depth (m)	DOC (μmol kg^{-1})	Depth (m)	DOC (μmol kg^{-1})	Reference
SE Levantine	Photic zone	65–100	500–1,200	40–60	Krom et al. (2005)
Eastern Mediterranean	Biological layer	62 ± 5.9	Deep layer	41 ± 1.4	Pujo-Pay et al. (2011)
Ionian Sea	Surface	50–73	400	31–62	Seritti et al. (2003)
Ionian Sea – April 02	0–100	66 ± 6	200–500	53 ± 4	Santinelli et al. (2010)
Ionian Sea – March 08	0–100	58 ± 6	200–500	45 ± 4	Santinelli et al. (2010)
Eastern Med June 07	0–100	64 ± 5	200–500	46 ± 4	Santinelli et al. (2010)
Western Med May 05	0–100	62 ± 9	200–500	48 ± 5	Santinelli et al. (2010)
Western Med March 08	0–100	57 ± 7	200–500	44 ± 3	Santinelli et al. (2010)
Average Whole Med	0–100	57–68	200–500	44–53	Santinelli et al. (2010)
N.Pacific central Gyre (ALOHA)	0–50 m	100–120	400	40	Hansell (2002)
Sargasso sea (BATS)	0–50 m	60–70	400	40–50	Hansell (2002)
North-West Mediterranean	Surface	100	400	60	Raimbault et al. (1999b)

Table 4.2 Concentration of DON measured by High Temperature Combustion (HTC) in the Eastern Mediterranean compared to values determined by HTC, Persulphate oxidation (PO) or UV oxidation (UV) in other subtropical regions of the world's oceans

Location	Depth (m)	DON (μmol kg^{-1})	Depth (m)	DON (μmol kg^{-1})	Reference
SE. Levantine	Photic zone	3–11	500–1,200	1–2	Krom et al. (2005) – HTC
Eastern Mediterranean	Biologic layer	4.7 ± 0.6	Deep layer	3.1 ± 0.5	Pujo-Pay et al. (2011) – PO
Northern N.Pacific	Surface	8–10	200–4,000	6–8	Koike and Tupas (1993) – HTC
Equatorial Pacific	Upper 200 m	3–7			Raimbault et al. (1999b) – PO
Sargasso sea	Surface	4–5.5	250–1,000	2.1–5	Hansell and Carlson (2001) – UV
NW. Mediterranean	Surface	5	400	3	Raimbault et al. (1999a) – PO

in the east and penetration of younger waters in the west. The $Max_{NO3/PO4}$ was located at 400–1,500 m depth interval with maximal concentrations of > 5 and > 0.2 μmol kg^{-1}, respectively, the upper isoline (i.e. lines of constant concentration) shoaling eastwards up to 400 m depth. The concentrations decreased towards the bottom, except for phosphate at 25.5–26.5 °E. The Max_{Si} was positioned at the 700–1,750 m depth interval, centered at 1,200 m, similar to the position of the $Min_{Sal/T}$. This layer shoaled eastwards with the upper 9 μmol kg^{-1} isoline reaching 700 m depth. There were indications of deep intrusion of seawater with lower silicic acid concentration in the vicinity of Crete.

Dissolved and Particulate Organic Carbon and Nutrients

The concentration of dissolved organic carbon (DOC) measured in the SE. Levantine Basin (Krom et al. 2005) is similar to that measured in the Ionian Sea (Seritti et al. 2003) and across the entire EMS in both the surface layers and deep waters (Santinelli et al. 2010; Pujo-Pay et al. 2011) (Table 4.1). These DOC concentrations are only slightly lower than values measured in the subtropical gyres of the N.Atlantic and the N.Pacific (Hansell 2002) which are similar to the concentrations measured in the N.W. Mediterranean (Raimbault et al. 1999a). Likewise the dissolved organic nitrogen (DON) values are relatively constant across the EMS and similar to values in other oligotrophic regions of the world's ocean (Table 4.2). It is somewhat surprising that the DOC and DON contents are as high as those determined in other sub-tropical areas since the annual primary production rates in the EMS are much lower (Psarra et al. 2000; Siokou-Frangou et al. 2010) . A likely contributory factor to this is that while DOC and DON produced during respiration and grazing of phytoplankton is probably low in the EMS, their consumption by heterotrophic bacteria is likely also to be low because the bacteria are strongly P limited (Thingstad et al. 2005). Sufficient data are not available yet to determine whether there is a measurable temporal trend in DOC and/or DON in surface waters as shown in the Sargasso Sea (BATS station, http://bats.bios.edu/) and as predicted by a I-D nutrient flux-reservoir model by Powley et al. (2012)

By contrast, the dissolved organic phosphorus (DOP) concentration in surface waters of the EMS (40–60 nmol kg^{-1}) is considerably lower than values obtained in both the Sargasso Sea (20–500 nmol kg^{-1}) and the N.Pacific Subtropical Gyre (150–270 nmol kg^{-1}; Table 4.3) . Despite this, DOP (dissolved organic phosphorus) is the largest reservoir of phosphorus in the surface water column of the EMS compared to a maximum of a few nanomoles of DIP and < 18 nmol kg^{-1} of particulate P (Table 4.4). This residual

Table 4.3 Concentration of DOP in the Eastern Mediterranean compared to concentrations in other subtropical regions of the world's oceans

Location	Depth (m)	DOP (nmol kg^{-1})	Depth (m)	DOP (nmol kg^{-1})	Reference
S.E. Levantine	Photic zone	50–60	500–1,200	40	Krom et al. (2005) – UV
Eastern Mediterranean	Biologic layer	40 ± 20	Deep layer	20 ± 20	Pujo-Pay et al. (2011) – PO
N.Pacific Subtropical Gyre	Surface	270	900	120	Williams et al. (1980) – UV
N.Pacific Subtropical Gyre	0–100 m	150–200	900	30	Smith et al. (1986) – UV
Sargasso sea	Surface	100–500			Cavender-Bares et al. (2001) – UV
Sargasso sea	Surface	74 ± 42			Wu et al. (2000) – UV
Sargasso sea (BATS)	0–200 m	20–120			Lomas et al. (2010)
North-West Mediterranean	Surface	130	400	Bdl	Raimbault et al. (1999a) – PO

UV Ultra Violet oxidation, *PO* Persulfate oxidation

Table 4.4 Concentration of POC, PON and POP in the Eastern Mediterranean together with calculated C:N:P molar ratios, assuming C = 106. Values for C:N:P ratio for DOM in deep water are also given

Location	Depth	POC (μM)	PON (μM)	POP (nM)	Molar ratio (C = 106)		Reference
					N	P	
SE. Levantine	0-50 m (*)	3	0.32	n.m.	11.34		Krom et al. (2005)
SE. Levantine	0-50 m (**)	3.7	0.39	9.1	11.31	0.28	Krom et al. (2005)
SE. Levantine 30 km west of Haifa	Surface			18			Kress et al. (2010)
Eastern Mediterranean	Biogenic layer	3.1	0.3	14	10.3	0.47	Pujo-Pay et al. (2011)
SE. Levantine	DCM (*)	2.6	0.3	n.m.	12.29		Krom et al. (2005)
SE. Levantine	DCM (**)	2.3	0.29	7.6	12.92	0.4	Krom et al. (2005)
SE. Levantine	350-2,500 m (*)	1.5	0.12	n.m.	9.77		Krom et al. (2005)
SE. Levantine	350–2,500 m (**)	1.2	0.11	3.8	9.61	0.36	Krom et al. (2005)
Eastern Med.	Deep layer	1.12	0.07	5	6.6	0.47	Pujo-Pay et al. (2011)
Measured DOM							
SE. Levantine	Photic zone			9	0.06		Krom et al. (2005)
Eastern Med.	Biological layer			8	0.07		Pujo-Pay et al. (2011)
SE. Levantine	Deep layer			8–13	0.05–0.06		Krom et al. (2005)
Eastern Med.	Deep layer			8	0.05		Pujo-Pay et al. (2011)

Data for the Eastern Mediterranean are from Pujo-Pay et al. (2011) sampled in June/July 2008
Locations marked with (*) are from stations outside the Cyprus Eddy while stations marked as (**) are from stations within the core of the Cyprus warm-core eddy sampled in May 2002 (Krom et al. 2005)

DOP was considered biologically unreactive during the CYCLOPS experiment (Thingstad et al. 2005; Zohary et al. 2005) because there was only limited primary production even though N was present as ammonium (~80 nmol kg^{-1}).

However, it is becoming increasingly evident that inorganic and organic P species other than the traditional orthophosphate can be utilized by phytoplankton and bacteria to sustain productivity, in particular in oligotrophic seas (Mather et al. 2008; Ranhofer et al. 2009; Diaz et al. 2008; Dyhrman et al. 2006). A measure of the degree to which DOP is utilized by micro-organisms is the activity of the alkaline phosphatase (APA), an enzyme that can hydrolyze P-esters to phosphate (Thingstad and Mantoura 2005; Ranhofer et al. 2009). The measured activity of APA in the Levantine Basin in May 2002 was 2.2 nM P h^{-1} (Thingstad and Mantoura 2005), similar to the APA activity measured at BATS, in summer and fall (Lomas et al. 2010) but more than 10 times the values observed in the North and South Pacific Gyres (Duhamel et al. 2010) which, though oligotrophic, are not particularly P limited. Further studies are needed to understand the nature and bioavailability of DOP in the EMS in comparison with other less P starved low productivity waters.

Biogeochemical and Nutrient Cycling Processes

Processes Related to the Unusually High N:P Ratio in the EMS

Nitrate: Phosphate Ratio in the Deep Water

One of the first unusual features to be recognised in the EMS was the high nitrate to phosphate ratio in deep waters of the Eastern Mediterranean, much higher than the Redfield ratio of 16 (Redfield et al. 1963). High N: P molar ratios in the deep waters were measured by Krom et al. (1991) – N: P = 28 ± 3:1; Civitarese et al. (1998) – N: P = 25–27:1; Yilmaz and Tugrul (1998) – N: P = 27–28.5:1; Krasakopoulou et al.

Table 4.5 Calculated nutrient inputs and outputs from the Eastern Mediterranean basin

Source	N input/output	P input/output	Molar N:P ratio	Reference
Atmospheric input[a]	147	0.9	77	Kocak et al. (2010)
Riverine input (1993)[b]	44.8	1.4	16	Ludwig et al. (2009)
Black Sea	8	0		Krom et al. (2004)
Total Inputs to basin	**200**	**2.3**	**43**	
Straits of Sicily	142	4.4	32	Krom et al. (2004)
Sediment deposition	27	1.0	27	Krom et al. (2004)
Sediment denitrification	10	0		Krom et al. (2004)
Total output from basin	**170**	**5.4**	**33**	

All values are given in 10^9 moles/year. The data table is updated from that given in Krom et al. (2004, 2010). The values in bold are the summed values of input and output

[a]Based on a compilation of data published between 1999 and 2010 and thus represents an average value for nitrate and ammonia depositional fluxes over a period of approximately 20 years of measurements. The total flux of nitrate (100) is approximately twice that of the total ammonia flux (47)

[b]The estimate of total N input from rivers is taken from Ludwig et al. (2009) using the estimate made for 1993. All other values in the table are those calculated in Krom et al. (2004)

(1999) – N: P=24–36:1; and Kress and Herut (2001) – N: P=25±3:1, Kress et al. (2003) – 26–31:1, among others. These values are higher than those in the Western Mediterranean (23:1) and almost double the values found in the Eastern North Atlantic (16:1) and almost everywhere else in the global ocean.

Various hypotheses had been put forward to explain the unusually high ratio including high rates of nitrogen fixation, adsorption of phosphate onto Saharan dust, preferential remineralisation of P over N, nitrate enriched waters forming the ADW. The high N:P ratio in the EMS has now been explained as being due to high N:P ratio in the external sources of nutrients (atmospheric and riverine) (Table 4.5) combined with low rates of denitrification in the EMS (Krom et al. 2004, 2010). This is caused by the ultraoligotrophic status of the basin which results in few areas having enough residual OM to cause the sediment to become anoxic and hence to enable microbial denitrification to take place.

The Eastern Mediterranean as a P Starved System

Recent data has shown that it is not only the major dissolved inorganic nutrients which have a ratio of > 16:1 but all nutrient species (particulate and organic) have N: P ratios far in excess of Redfield (Table 4.4). The DOC:DON:DOP ratios in the photic zone is 106: 8–9: 0.06–0.07, which is depleted in N and highly depleted in P relative to the Redfield ratio (Krom et al. 2005; Pujo-Pay et al. 2011). The POC:PON:POP ratio of particulate matter measured within the photic zone, a combination of live and dead cells and faecal pellets, is 106:10.3–11.3:0.28–0.47, much greater than Redfield's, indicating nutrient depletion in particulate matter as well. High carbon to nutrient ratios are not uncommon in ultra-oligotrophic systems where carbon fixation by primary producers continues after N and P become totally depleted (e.g. Berman-Frank and Dubinsky 1999) and rapid recycling occurs. However the very high N:P ratios observed in the photic zone of the EMS combined with higher release of N than P below the photic zone was interpreted by Krom et al. (2005) as evidence of the rapid and very efficient recycling of P within the photic zone compared to that of N as seen in other areas such as the N. Pacific (Loh and Bauer 2000). They concluded that DOP is preferentially mineralised compared to DOC and DON resulting in increased C:P and N:P ratios with depth. In the EMS, by contrast, there is a much smaller decrease in DOP with depth than for either DON or DOC . This implies that in the EMS, DOP is recycled very efficiently in the photic zone and at lower depths the DOP is too refractory to break down much.

Seasonal Changes in Primary Production and Nutrient Limitation

The high N:P ratio was the basis of the hypothesis that the productivity of the Eastern Mediterranean was simply limited by the availability of phosphorus. However, recent results have shown that the nutrient limitation in the Eastern Mediterranean is more complex than it at first appeared and that there are seasonal differences in the ultimate nutrient limiting specific phytoplankton groups as well as different members of the surface microbial community (Thingstad et al. 2005; Zohary et al. 2005; Tanaka et al. 2011).

The annual major phytoplankton bloom in the EMS starts as soon as deep winter mixing occurs in early winter (e.g. December – Fig. 4.5) and nutrients are supplied into the photic zone. The bloom continues and increases in intensity throughout the winter months as the depth of mixing increases adding more nutrients until early spring, typically March in the SE. Levantine when the seasonal thermocline begins to form (Hecht et al. 1988). This unusual pattern is

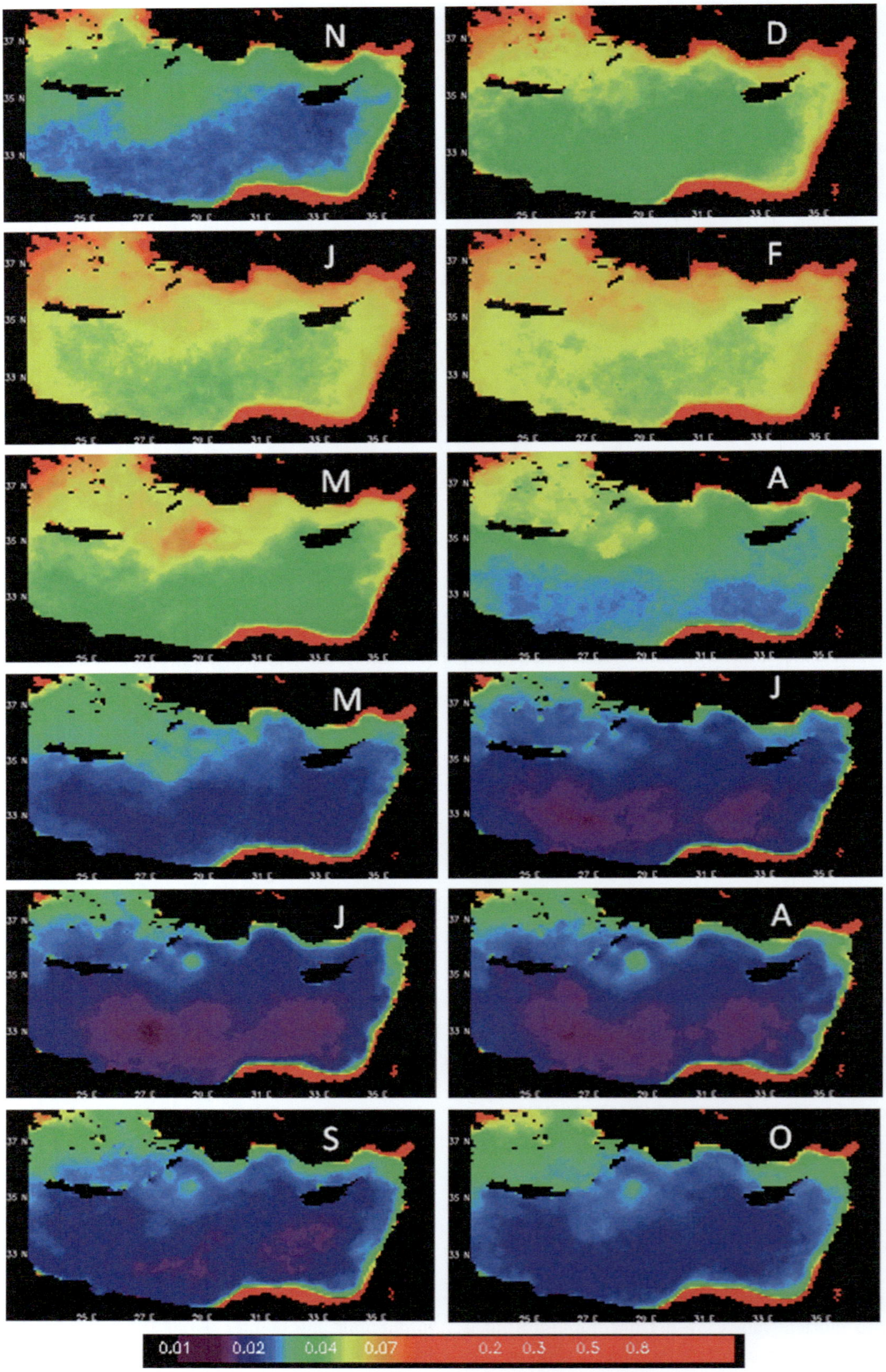

Fig. 4.5 Monthly SeaWiFS "climatological" images of chlorophyll over the seasonal cycle. The images are based on input data from Sept 1997 to August 2004 and chl-a was computed using the Bricaud et al. (2002) algorithm for the Mediterranean (Groom et al. 2005). The images were provided by Steve Groom (NERC Earth Observation Data Acquisition and Analysis Service, Plymouth)

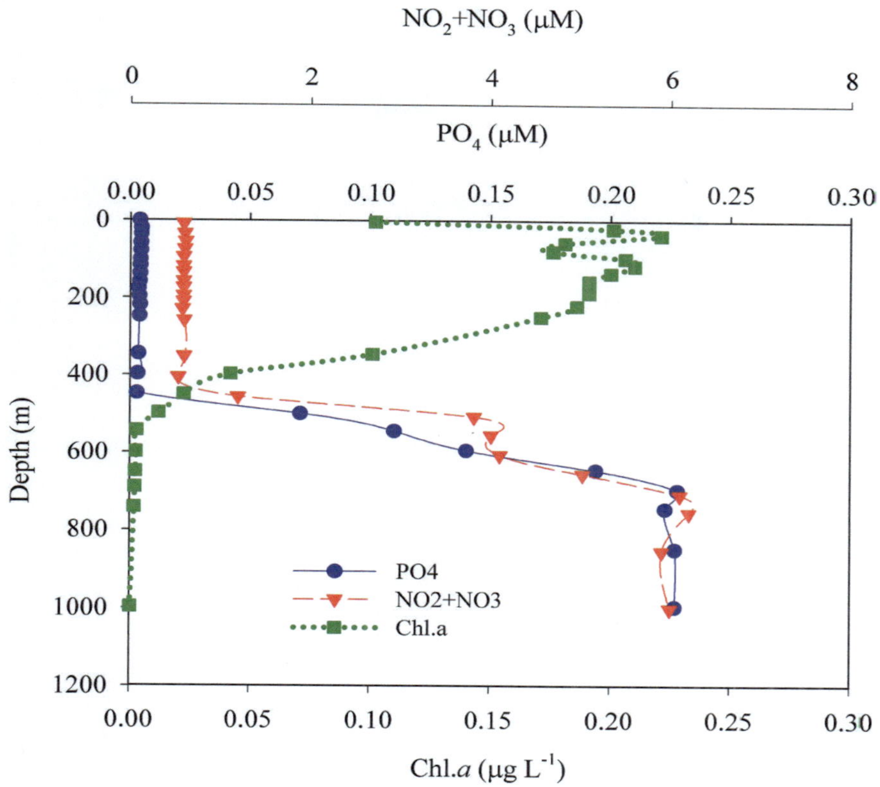

Fig. 4.6 Nutrient distribution during simple P-limited phytoplankton bloom in winter (February 1989) in the warm core Cyprus eddy showing that all the phosphate has been consumed leaving residual nitrate in the water column (Modified from Krom et al. 1992)

probably controlled by the typical weather pattern in the region which consists of a few days of cold often wet weather which causes the upper layers of the water column to mix, followed by several days of relatively warmer clear weather which causes a short term water column stability and allows the phytoplankton to bloom thereby exploiting the nutrients mixed into the surface waters.

This winter phytoplankton bloom is conventionally P limited. Deep winter mixing advects water with an N:P ratio >16:1 to the photic zone (e.g. Krom et al. 1992) containing nutrient concentrations typical of LIW (Levantine intermediate water) that is on the order of 0.03 μmolP kg^{-1} (Kress and Herut 2001) and between 1.3 and 2.0 μmolN kg^{-1} (Schlitzer et al. 1991) i.e. with an N:P ratio of 40–60:1. The bloom ceases as soon as the waters in the photic zone run out of the phosphate (Fig. 4.6). Characteristically this occurs when there is 0.3–1 μmolesN kg^{-1} of nitrate remaining in the surface waters (Krom et al. 1992; Kress and Herut 2001). This residual nitrate has been shown to be isotopically heavy (16–40 ‰) compared with 3–12 ‰ at depth (Emeis et al. 2010; Struck et al. 2001). This pattern of heavier N isotopes is characteristic of water in which the phytoplankton runs out of the limiting factor, in this case phosphate, before all the nitrate has been taken up by biological productivity. Additional biological evidence for P-limitation of the surface waters of the Eastern Mediterranean has been obtained from a series of observations on phytoplankton activity and specific microcosm experiments (Vukadin and Stojanovski 1976; Becacos-Kontos 1977; Pojed and Kveder 1977; Berland et al. 1980; Bonin et al. 1989). Zohary and Robarts (1998) also showed that in winter, bacteria from various sites in the EMS are also P-Limited.

Starting generally in March/April a seasonal thermocline develops firstly in the SE. Levantine which then expands to the north and west (Fig. 4.5). As this happens, the primary productivity and plankton biomass is reduced in the uppermost layers and becomes concentrated at the top of the nutricline in a deep chlorophyll maximum (DCM). The DCM, which is characteristic of the summer in the EMS (April to November) is approximately the same depth (100–120 m) across the basin with decreasing concentrations of chlorophyll (a proxy to total phytoplankton biomass) and concurrent decrease in nutrients (Fig. 4.7). However this pattern is not simple because of the complex mesoscale circulation in the EMS.

By late spring and early summer (May-June), the phytoplankton production has become N & P co-limited (Tanaka et al. 2011) although heterotrophic bacterial growth is still P limited (Thingstad et al. 2005). At that time, the seasonal thermocline is present, a deep chlorophyll maximum has

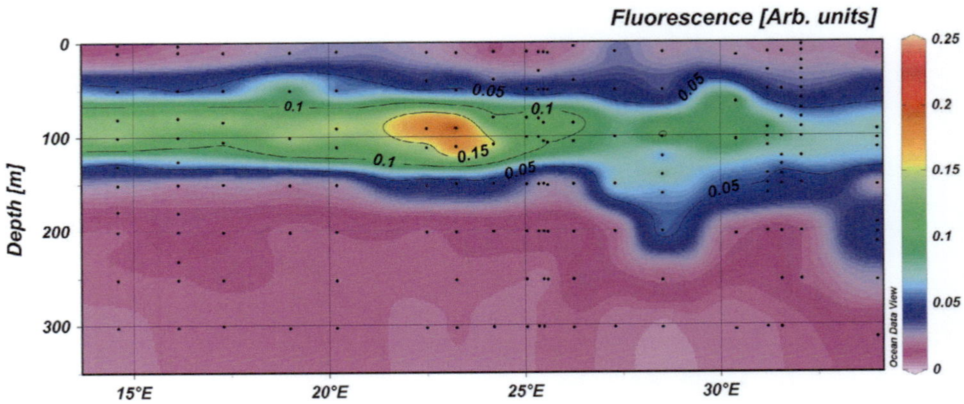

Fig. 4.7 Vertical section of in situ chl-a fluorescence (arbitrary units) across the Eastern Mediterranean sampled in April-May 1999. Stations and data points are indicated by *filled dots* (Adapted from Kress et al. 2011)

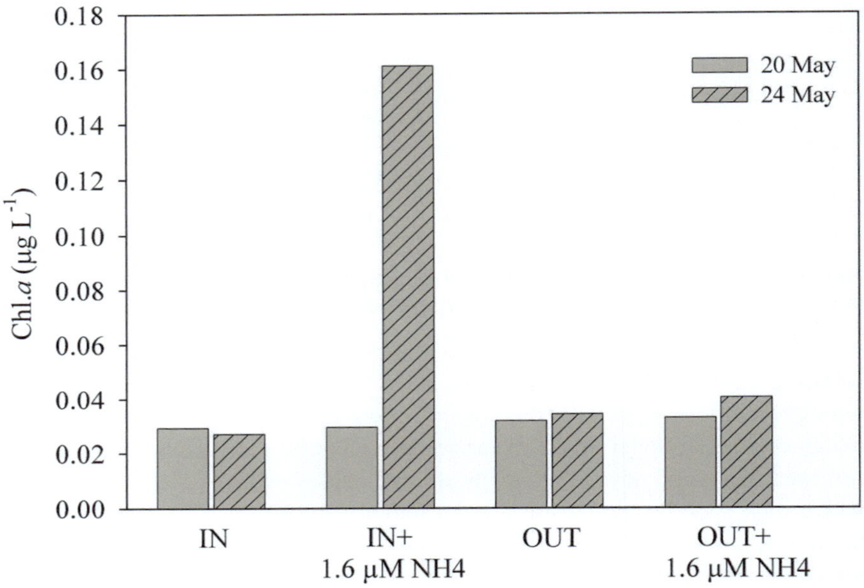

Fig. 4.8 A shipboard microcosm experiment in which the effect of ammonium addition on phytoplankton growth as determined by changes in Chlorophyll-a concentration was followed using IN patch (phosphate enriched waters) and OUT patch (control waters). For more details see Zohary et al. (2005)

developed, and the surface waters have become depleted in nitrate as well as phosphate (e.g. Kress and Herut 2001; Krom et al. 2005). This change in phytoplankton nutrient limitation in May was observed and identified by microcosm experiments carried out in the framework of the CYCLOPS P addition Lagrangian study (Krom et al. 2005; Zohary et al. 2005). In the microcosm experiment (Fig. 4.8) surface waters preconditioned with phosphate added to a patch of water (IN) and water from outside the patch (control) were incubated with 1.6 µmol kg^{-1} of ammonium. The results showed the classic behaviour of an N & P co-limited system with the only large increase in Chl-a concentrations in the treatment where ammonium was added to seawater preconditioned with phosphate. This change in nutrient limitation was explained by Thingstad et al. (2005). At the end of winter mixing, the surface water has residual nitrate and no free phosphate (see Fig. 4.6). After stratification develops, the phytoplankton in the water column above the nutricline receives very limited amounts of new nutrients and most of the production is sustained by recycled nutrients produced by respiration, grazing and biological oxidation of organic matter (PON – particulate organic nitrogen -, and POP – particulate organic phosphorus). The recycled nutrients include ammonium, nitrate, phosphate, that are directly available for phytoplankton uptake, and DON and DOP. While some of the DON compounds can be made available to phytoplankton, most of it cannot. Therefore the N available for phytoplankton growth decreases with recycling. The situation for P is different. When the particulate P is recycled there is no significant pool of non-bioavailable P formed i.e. if significant amounts of DOP are formed then it is either directly bioavailable to phytoplankton or after the action of enzymes such as APA. The surface waters have high activity of APA which is excreted by both bacteria and microphytoplankton to enable them to access part of the DOP pool (e.g. Thingstad and Mantoura 2005; Tanaka et al. 2011). As a result, more of the recycled P is available for phytoplankton uptake than N, eventually removing all of the 'excess' dissolved N (mainly nitrate) from the 'winter excess' and causing productivity to become N & P co-limited. Heterotrophic bacteria that are able to access the DON pool were P limited during the same time (Thingstad et al. 2005).

In the summer (June/July), similar nutrient limitation experiments were performed in the framework of the BOUM cruise (BOUM: Biogeochemistry from the Oligotrophic to the Ultra-oligotrophic Mediterranean) using phytoplankton collected from 20 m i.e. phytoplankton from the highly nutrient depleted waters well above the DCM (Tanaka et al. 2011). These results showed that although the organisms and system were P starved with DON:DOP values of 114–190:1 and PON:POP values of 24–28:1, there was only an increase in growth of osmotrophs when N+P or N only were added i.e. the system was no longer P limited. They suggested that as the stratified period progresses the pool of bioavailable N and P becomes very small in the surface layers and the most limiting nutrient for osmotrophic growth shifts sporadically among N, P and N&P (c.f. Hecky and Kilham 1988)

Is the Nutrient Uptake in the Surface Waters Redfieldian?

While the data is somewhat equivocal it seems that nutrient uptake in the EMS is often close to Redfieldian (i.e. 16:1). In a series of microcosm experiments carried out as part of the CYCLOPS programme, Kress et al. (2005) added ammonium and phosphate to EMS surface waters and obtained a nutrient uptake ratio of 18±3.1. Zohary et al. (2005) subsequently carried out similar microcosm experiments in the core of the Cyprus Eddy and found an uptake ratio of 25:1. However these experiments were carried out by adding ammonium to surface water collected within artificially phosphate-enriched patch from the CYCLOPS addition experiment. Prior exposure to excess dissolved phosphate may have preconditioned the phytoplankton to take up nutrients in an N:P ratio exceeding 16:1. The nutrient budget calculations of Krom et al. (1992) yielded an N:P ratio for new production of phytoplankton in the core Cyprus eddy of 20:1. In another indirect estimate of the N:P ratio of nutrient uptake, assuming the nutrient supply to the photic zone contains concentrations typical of LIW (Kress and Herut 2001; Schlitzer et al. 1991) and that the phytoplankton uptake occurs in a Redfieldian ratio of 16;1, then the residual nitrate which would be left in the surface water would be 0.8–1 μmol kg^{-1}, which agrees with the typical range of observed values in winter in the EMS (Kress and Herut 2001; Thingstad and Mantoura 2005). Thus phytoplankton growth in the surface water is approximately Redfieldian, certainly much lower than the 28:1 molar ratio measured in the Deep Water. However it has been suggested that P is more efficiently recycled from POM than N in surface (see section "Seasonal changes in primary production and nutrient limitation") and intermediate water. This results in the POM which reaches Deep Water having a N:P ratio higher than 16:1. Indeed the measurements made of POM in deep water by Krom et al. (2005) found N:P of 26:1 though Pujo-Pay et al. (2011) determined values close to 16:1 (Table 4.4). When this POM is broken down it releases more N than P.

In addition, there is an unusual (high) nutrient ratio of preformed nutrients in the EMS (Civitarese et al. 1998). It is known that ADW is formed in the S.Adriatic in winter simultaneously with the P limited major phytoplankton bloom i.e. when phosphate is entirely depleted and residual nitrate remains (see section "Seasonal changes in primary production and nutrient limitation"). The measured nitrate:phosphate ratio in the middle of the S.Adriatic in winter is ~33:1 (Zavatarelli et al. 1998) and this is the water advected into the deep Ionian Basin. The results of the 1-D flux reservoir model (Powley et al., 2012) confirm this explanation.

Nitrogen Fixation

Biological Dinitrogen (N_2) fixation can contribute new N sources to N starved systems and significantly induce further nutrient uptake by subsequent blooms of primary producers. Early geochemical and isotopic models suggested that N_2 fixation rates were high in the Mediterranean Sea (Bethoux and Copin-Montegut 1986, Bonin et al. 1989, Bethoux et al. 1992, Gruber and Sarmiento 1997, Pantoja et al. 2002, Sachs and Repeta 1999). Yet, actual measured rates utilizing the ^{15}N-uptake method (Montoya et al. 1996) are now accumulating from throughout the Mediterranean Basin bridging spatial and temporal scales (Garcia et al. 2006, Rees et al. 2006, Sandroni et al. 2007, Bar Zeev et al. 2008, Marty et al. 2008, Ibello et al. 2010, Yogev et al. 2011, Bonnet et al. 2011, Rahav et al. 2013a, b). The published N_2 fixation rates from the Western Basin are typically higher both volumetrically and areally than those measured from the Eastern Basin during all measured seasons (Berman-Frank and Rahav 2012, Rahav et al. 2013a, b). Volumetric rates ranged from non-detected to 0.3 nmol N L^{-1} day^{-1} in the Eastern Basin, compared to non-detected to 17 nmol N L^{-1} day^{-1} in the Western Basin (Fig 4.9a). Areal integrated rates changed spatially and temporally with rates in the Western Basin ranging from>3 to 50 μmol N m^{-2} day^{-1} (Sandroni et al. 2007, Ibello et al. 2010), compared with usually less than 10 μmol

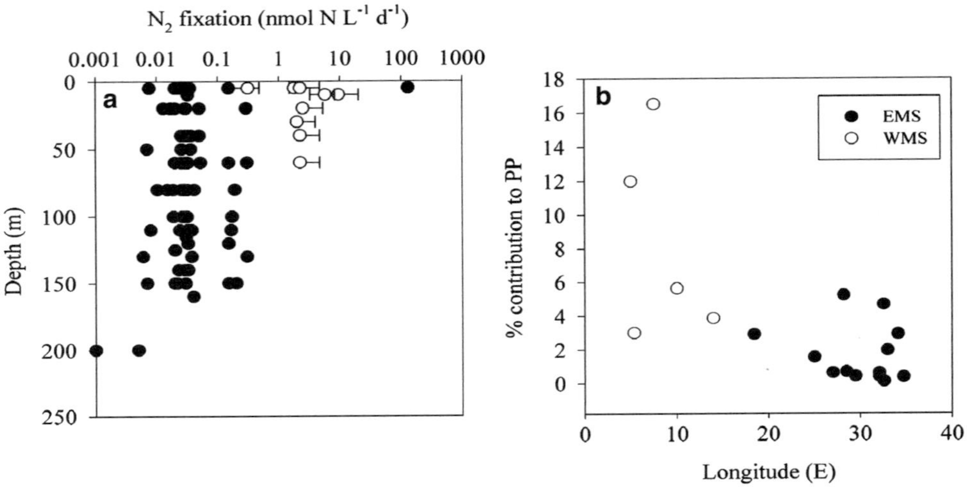

Fig. 4.9 Summary depth distribution of measured volumetric rates of N_2 fixation measured across the Mediterranean Sea (**a**) and the percentage contribution of the fixed N to primary productivity (**b**). Data were compiled from Yogev et al. 2011, Rahav et al. (2013a, b) (EMS), Ibello et al. (2010), Bonnet et al. (2011) (EMS+WMS), Rees et al. (2006), Garcia et al. (2006), Sandroni et al. (2007), Marty et al. (2008) (WMS)

N m^{-2} day^{-1} for the Eastern Basin (Berman-Frank and Rahav 2012; Yogev et al. 2011; including a coastal Turkish station with rates of 4 μmol N m^{-2} day^{-1} (Ibello et al. 2010) and 0–0.4 μmol N m^{-2} day^{-1} during summertime in an anticyclonic eddy in the Ionian Sea (Bonnet et al. 2011)). Interestingly, low rates of two studies from the EMS are accompanied by the observations that ~35 % of all samples taken (accounting for spatial and temporal samplings) were below the detection limit (Ibello et al. 2010, Yogev et al. 2011). The contribution of N_2 fixation to primary productivity in the EMS is extremely low with a maximum of 5 % (Fig. 4.9b) compared to the WMS, where N_2 fixation comprised up to 16.5 % of primary productivity (Fig. 4.9b).

The low measured rates of N_2 fixation correspond with a taxonomically diverse yet limited abundance of diazotrophic organisms found throughout the EMS including unicellular representatives, heterotrophic diazotrophs, filamentous, and symbiotic associations (Man-Aharonovich et al. 2007; Bar-Zeev et al. 2008; Yogev et al. 2011; Bonnet et al. 2011). Of the few conspicuous diazotrophs in the Eastern Mediterranean Sea, the diatom-cyanobacterial associations (*Rhizosolenia-Richelia, Hemiaulus-Richelia*) were consistently present (10–55 heterocysts L^{-1} maximal abundance) in two routinely monitored stations off the Israeli coast (Bar Zeev et al. 2008). Despite the microscopic and molecular evidence no blooms of *Hemiaulus-R. intracellularis* were observed in over 8 years of monitoring and sampling in the EMS and in the above stations (personal observation; Bar Zeev et al. 2008).

Conspicuously rare is the appearance of the globally important filamentous non-heterocystous *Trichodesmium* in the Mediterranean Sea. In the EMS two species of *Trichodesmium* (*T. erythraeum* and *T. thiebautii*) have been sporadically observed in extremely low abundance (<1 trichome per 1,000 L) and only one bloom (from the Aegean Sea) has been reported (Spatharis et al. 2012). The rarity of *Trichodesmium* in the Mediterranean is surprising since this organism is routinely observed in both coastal and open waters of the adjacent Red Sea (Stihl et al. 2001) and in the Atlantic Ocean (Capone et al. 1997). Limitation by bioavailable inorganic P may control *Trichodesmium* populations in the Mediterranean although *Trichodesmium* can utilize other forms of P such as phosphonates (Dyhrman et al. 2006).

The limited contributions of N_2 fixation in the Mediterranean Sea and particularly in the EMS have been routinely attributed to P limitation (Bonnet et al. 2011, Ridame et al. 2011) as dinitrogen fixers require P for the energetically expensive process of N_2 fixation (16 ATP per mole of N_2 fixed) (Postgate 1998). However, this explanation is not always consistent with experimental results. While some P additions increased N_2 fixation rates in an anticyclonic eddy of the EMS (BOUM Station. C) during the stratified summer (Ridame et al. 2011), at other times and stations P enrichments did not enhance N_2 fixation (e.g. BOUM, Station B (Ridame et al. 2011); pelagic EMS (Yogev et al. 2011); and cyclonic and anticyclonic eddies (Rahav et al. 2013a). Furthermore, dust enrichment in microcosm experiments (high N and P in dust) did not show consistent trends with regards to N_2 fixation in the EMS. A significant increase was observed at the anticyclonic Station C (Ridame et al. 2011), while N_2 fixation was not elevated at the Ionian Sea (Station B, Ridame et al. 2011), and at either the Cyprus Eddy or the Rhodes Gyre during July 2009 (Rahav et al. 2013a). Moreover, the published studies do not report any significant correlations between N_2 fixation rates and any of the physio-chemical parameters such as temperature, nutrient availability, or biological competition. The limited samplings, methodological underestimation of ^{15}N uptake

(Mohr et al. 2010), or the generally low N_2 fixation rates measured in the EMS that were near or below detection limits at times, constrain our current understanding of what parameters limit N_2 fixation in the EMS.

Physics Induced Bio-chemical Processes

The EMT event provided the opportunity to investigate the effect of physical changes in circulation on chemical and biological depth distribution and processes. As mentioned in section "Cross basin distribution and how they have changed with time", the younger CSOW was introduced into the deep waters of the EMS and caused the older nutrient rich ADW to be uplifted to mid depths. In the Eastern Levantine Basin, from 2002 to 2010, this uplift increased the nitrate and phosphate concentrations at the base of the nutricline by 0.5 and 0.04 μmole kg^{-1}, respectively (400–500 m). Concurrently, the DCM depth range widened and became shallower, with a slight increase in chl-a concentrations, most evident at the continental slope. The concentrations at the surface waters increased as well since March 2010 in the open sea and earlier (Sep 2008) in the surface waters at the continental slope. There was no concurrent trend of increase in nitrate and phosphate at the DCM, but silicic acid concentrations decreased at the upper layers. Therefore, it is reasonable to assume that the increase in concentration at the base of the nutricline increased the supply of nutrients to the ultra-oligotrophic photic zone, where they were rapidly utilized by the phytoplankton. It is possible that the increased supply of nitrate and phosphate from below changed the phytoplankton community structure in the upper layers, increasing the relative contribution of diatoms and hence depleting silicic acid. However, there are no data on the phytoplankton community from this time series that can directly support this hypothesis.

Basin-wide oceanographic transects conducted in 1991 (Yacobi et al. 1995), 2001 (Kress et al. 2011), and 2008 did not show a clear trend in Chl-a concentrations, nor did satellite imagery analysis up to 2002 (D'Ortenzio et al. 2003). In October 1991, Chl-a ranged from 0.01 to 0.25 μg l^{-1} (depth integrated chlorophyll ranged from 17 to 35 mg m^{-2} with higher values in the Western Levantine). In October 2001, concentrations ranged from 0.01 to 0.28 μg l^{-1} (integrated values of 16.5–23.1 mg m^{-2}) while in September 2008, from 0.01 to 0.33 μg l^{-1} (depth integrated 10–27.1 mg m^{-2}). It is evident that the occasional basin-wide snapshots are not enough to indicate trends that may be cyclic and short lived. High frequency surveys are needed if we are to identify long term trends in this important parameter. However it is clear that despite large changes in external nutrient supply (see section "Changes in the nutrient supply from 1950 to 2000"), there have been no equivalent large changes in chlorophyll content or any other regional measurement of phytoplankton productivity.

Use of Nutrient Budgets to Define Biogeochemical Processes

Oceanic, Terrestrial and Atmospheric Nutrient Inputs

The Mediterranean in general and the Eastern Basin in particular represent almost closed systems in which total nutrient budgets have been used successfully to explain and quantify the relative importance of various biogeochemical processes within the system (e.g. Ribera d'Alcalà et al. 2003; Krom et al. 2004; Béthoux et al. 2002). The most recent total nutrient budget of the Eastern Basin first described by Krom et al. (2004, 2010) has been updated here (Table 4.5) to include more recent atmospheric inputs to the basin. The model output is balanced (~10 % error) for total N with 200 $\times 10^9$ molesN/year imported to the basin and 179 $\times 10^9$ molesN/year exported mainly through flux of nutrients with intermediate water through the Straits of Sicily. The major conclusion drawn from this budget was that N_2 fixation is not significant in the EMS contrary to previous suggestions (Bethoux et al. 1986, 2002; Pantoja et al. 2002; Rees et al. 2006 and others) . This conclusion was confirmed subsequently by *in-situ* measurements (Bar-Zeev et al. 2008; Ibello et al. 2010; Yogev et al. 2011; Bonnet et al. 2011; Rahav et al. 2013a, b) (see section "Nitrogen fixation")

There are currently two problems with this budget : the phosphate budget does not balance and the external input of nutrients does not change with time. Considerably more P is exported from the basin (5.4 × 10^9 molesP/year) than is input (2.3 × 10^9 molesP/year). The export values are rather well constrained since they were calculated from measured values for dissolved phosphate at the Straits of Sicily and total P deposition rates in the sediment. By contrast the values for the input of bioavailable P is far less known. It is possible that both atmospheric P and riverine P fluxes do not take into account chemical species which become bioavailable within the basin. Thus the atmospheric input measured by Kocak et al. (2010) is only the leachable P and does not include any non-leachable P such as iron-bound P and apatite which are known to be present in Saharan dust (Nenes et al. 2011) and which might be bioavailable under the P starved conditions of the EMS water column. Likewise, the riverine input was calculated from dissolved inorganic P alone and does not include any particulate P or organic P which are brought down the rivers and might subsequently be converted to bioavailable P in the adjacent water column and sediment. DOP comprises an additional possible source of bioavailable P not included in the total budget. DOP is somewhat higher in surface waters (i.e. incoming waters) than in the outflowing LIW. However there are no data available for DOP fluxes through the Straits of Sicily nor is there any information of the relative bioavailability of surface vs intermediate water

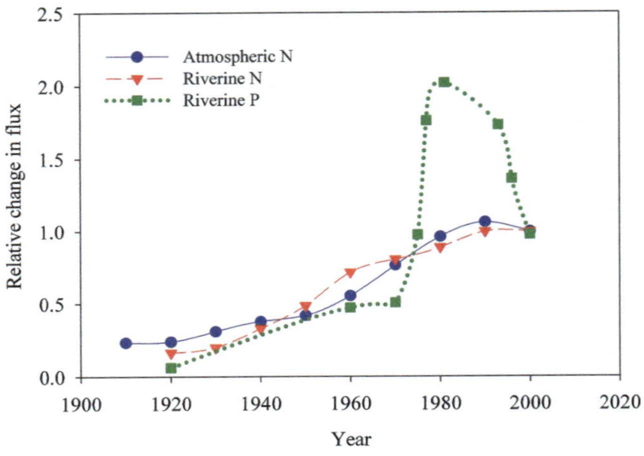

Fig. 4.10 Figure showing the relative changes in external nutrient input to the EMS from the beginning of the twentieth century to 2000. The riverine data is from Ludwig et al. (2009) extrapolated back to 1920 assuming the major rivers behave in a similar manner to the Rhine (Zobrist and Stumm 1981). The Atmospheric flux uses data from an Alpine ice core modified by modelling from Preunkert et al. (2003) and Fagereli et al. (2007). In order to calculate a total atmospheric N input it was assumed that ⅔ of the N flux to EMS is NO_X and ⅓ is NH_4 based on measurements made in Crete (Kocak et al. 2010)

DOP. In a flux-reservoir model designed to quantitatively model the biogeochemical cycling in the EMS, van Cappellen et al. (2012) increased the P inputs by adding DOP which come close to closing the P budget.

Changes in External Nutrient Fluxes into the EMS with Time

Changes in the Nutrient Supply from 1950 to 2000

The other problem with the total nutrient budgets calculated thus far is that they do not take into account the changing fluxes of important parameters with time. It is known that the external supply of nutrients (N & P) from both rivers and atmospheric inputs to the EMS has increased significantly over the past century, particularly since 1960 as a result of anthropogenic inputs.

Thus the total N and P inputs from riverine sources to the EMS have been calculated by Ludwig et al. (2009) from 1963 to 1998 (Fig. 4.10). These data show a systematic increase in total N to the basin with the flux increasing by a factor of 5. The changes in riverine P are more complex. There was an increase in P through 1965 followed by a sharp decrease caused by the effect of the completion of the Aswan dam stopping the flow of dissolved and particle borne P into the S.E. Levantine Basin (Ludwig et al. 2009). Over the next 15 years, up to 1980, P increased as a result of anthropogenic discharges into rivers from sewage and other sources. However from the late 1980s to the present day, P fluxes into the basin have decreased as a result of abatement procedures designed to reduce the effect of P pollution within the rivers.

The changes in atmospheric N (NO_x and NH_4) have been estimated using measured data from alpine ice cores modelled to calculate atmospheric fluxes for NO_x (Preunkert et al. 2003) and for NH_4 (Fagerli et al. 2007). Pro-rating these data for the EMS using measured atmospheric flux data from Crete (Kocak et al. 2010) and Israel (Herut et al. 1999; Carbo et al. 2005), it has been possible to determine the changing flux of N since 1950. The changing flux of P is more difficult to estimate since no long term record has been developed. It is complicated by the fact that there is a large 'natural' source of P to the EMS in the form of Saharan dust which contains an unknown proportion of bioavailable P. Recent data have shown that the fraction of bioavailable P in this dust has been modified by the effect of acidic atmospheric processes caused mainly by anthropogenic inputs from Southern Europe (Nenes et al. 2011). Further work is needed to understand the nature and magnitude of these atmospheric processes and their effect on the supply of this key nutrient to the EMS.

What Was the Level of Nutrient Supply Prior to Anthropogenic Inputs?

There are no data available to directly determine the anthropogenic increases in nutrients in rivers flowing into the EMS prior to 1950. However calculations to determine the nutrient content of the 'pristine' Rhine suggests that there was significant nutrient pollution of the river already by 1950 and it is likely that there would be similar changes in rivers such as the Nile and Po flowing into the EMS. Powley et al. (2012) have used these values and estimated that the riverine inputs to the EMS were $N = 2.0 \times 10^9$ MolesN/year and $P = 0.095 \times 10^9$ MolesP/year (i.e. ~5 % and 10 % of the 2,000 values). The relative increase in external supply is shown in Fig. 4.10. The input flux of atmospheric supply for both N species has increased by 85 % for NO_x and by 65 % for NH_4 between 1910 and 20 and the end of the century.

Possible Factors Which Will Influence Future Changes in Nutrient Distribution

Future Circulation Changes That Might Affect Nutrient Distribution

The most dramatic change in circulation in the recent past has been the EMT event which has caused changes in nutrient distribution in the EMS. It is still not determined whether this is part of an on-going cyclic pattern of changes in the

EMS which may be natural or as a result of anthropogenically caused environmental and/or climate change. At present the most striking temporal change observed in physical parameters has been the increase in salinity (from 39.3 to 39.6) at the surface of the Eastern Levantine that started by mid-2005, with no concurrent changes in temperature. Salinity has continued to increase within the upper layers (down to 300 m) until the end of 2010. Historical data has shown a cyclic pattern in salinity: lower salinities at the beginning of the 1980s and 2000s and higher salinities at the beginning of the 1990s (onset of the EMT) and since 2006. This salinity increase, identified at the easternmost part of the basin, should be followed and analyzed not only in the local context but basinwide, for it is assumed that a similar increase in salinity in the area may have conditioned the EMT event in the 1990s. Cyclic pattern changes have also been observed in the upper layer circulation in the Ionian Basin that changed from cyclonic to anti-cyclonic circulation and back (Civitarese et al. 2010; Gacic et al. 2010). Both these changes in circulation resulted in observed changes in nutrient distribution and it is likely that future changes will also modify nutrients and hence primary productivity.

Is the EMS Particularly Vulnerable to Climate and Environmental Change?

In a more general way, it is known that the nutrient status of the EMS has changed dramatically over the past 10 k years as a result of natural climate change. The evidence for this is found in sapropels deposited in the deeper waters (>500 m) of the EMS basin. The present ultra-oligotrophic conditions in the EMS result are characterised by calcareous marl sediments with very low concentrations of organic C (0.1–0.2 %) . However interbedded with these sediments are a series of organic C rich sapropel layers. The most recent of these sapropel layers (S-1) began at ~9.8 ky ^{14}C BP and finished at ~5.7 ky ^{14}C BP (de Lange et al. 2008). During the S-1 period there was both higher primary productivity and restricted circulation within the deeper water which resulted in sub-oxic to anoxic conditions. While the precise controls which gave rise to this dramatic change in trophic status in the EMS is still being debated, it is generally agreed that the climate during that period was more humid and river flow into the basin (e.g. the river Nile) was higher. These changes in climate resulted in both reduced water exchange through the Straits of Sicily and more restricted formation of deep water. The changes in circulation predicted as a result of modern climate change (e.g. Somot et al. 2006) may also have profound effects on the export rate of nutrients in the EMS and thus have the potential to result in drastic changes in the trophic status of the basin

Acknowledgements The ideas which are contained in this review have developed over many years as a result of conversations with many colleagues. While many of those interactions ended up as papers which have been referenced here, others were more informal. Our first scientific interaction with the Eastern Mediterranean was inspired by Hava Hornung, Tom Berman, Arthur Hecht and the POEM group. Since then MK has particularly appreciated the interactions with Frede Thingstad, Fauzi Mantoura, Barak Herut and the other members of the CYCLOPS team, and with Kay Emeis, Philippe van Cappellen and Helen Powley who together have developed a quantitative description of processes in the EMS.

References

Bar Zeev E, Yogev T, Man-Aharonovich D, Kress N, Herut B, Beja O, Berman-Frank I (2008) Seasonal dynamics of the endosymbiotic, nitrogen-fixing cyanobacterium *Richelia intracellularis* in the eastern Mediterranean sea. ISME J 2:911–923

Becacos-Kontos T (1977) Primary production and environmental factors in an oligotrophic biome in the Aegean sea. Mar Biol 42:93–98

Berland BR, Bonin DJ, Maestrini SY (1980) Azote ou phosphore? Considérations sur le "paradoxe nutritionnel" de la Mer Méditerranée. Oceanol Acta 3:135–141

Berman-Frank I, Dubinsky Z (1999) Balanced growth of photosynthesizing aquatic organisms: myth and reality. Bioscience 49:29–37

Berman-Frank I, Rahav E (2012) Ch. 8: Nitrogen fixation as a source for new production in the Mediterranean Sea: a review. In: Stambler NE (ed) Life in the Mediterranean Sea: a look at habitat changes. Nova Science Publishers, New York, pp 199–226

Bethoux JP, Copin-Montegut G (1986) Biological fixation of atmospheric nitrogen in the Mediterranean Sea. Limnol Oceanogr 31:1353–1358

Bethoux JP, Morin P, Madec C, Gentili B (1992) Phosphorus and nitrogen behavior in the Mediterranean Sea. Deep-Sea Res 39:1641–1654

Béthoux JP, Morin P, Ruiz-Pino DP (2002) Temporal trends in nutrient ratios: chemical evidence of Mediterranean ecosystem changes driven by human activity. Deep-Sea Res II 49:2007–2016

Bonin DJ, Bonin MC, Berman T (1989) Mise en evidence experimentale des facteurs nutritifs limitants de la production du micronanoplancton et de l'ultraplancton dans une eau cotiere de la Mediterranee orientale (Haifa Israel). Aquat Sci 51:129–159

Bonnet S, Grosso O, Moutin T (2011) Planktonic dinitrogen fixation along a longitudinal gradient across the Mediterranean Sea during the stratified period (BOUM cruise). Biogeosciences 8:2257–2267

Bricaud A, Bosc E, Antoine D (2002) Algal biomass and sea surface temperature in the Mediterranean Basin: intercomparison of data from various satellite sensors, and implications for primary production estimates. Remote Sens Environ 81:163–178

Capone DG, Zehr JP, Paerl HW, Bergman B, Carpenter EJ (1997) *Trichodesmium*, a globally significant marine cyanobacterium. Science 276:1221–1229

Carbo P, Krom MD, Homoky WB, Benning LG, Herut B (2005) Impact of atmospheric deposition on N and P geochemistry in the southeastern Levantine basin. Deep-Sea Res II 52:3041–3053

Cavender-Bares KK, Karl DM, Chisholm SW (2001) Nutrient gradients in the Western North Atlantic Ocean: relationship to microbial community structure and comparison to patterns in the Pacific Ocean. Deep-Sea Res I 48:2373–2395

Civitarese G, Gacic M, Vetrano A, Boldrin A, Bregant D, Rabitti S, Souvermezoglou E (1998) Biogeochemical fluxes through the Strait of Otranto (Eastern Mediterranean). Cont Shelf Res 18:773–789

Civitarese G, Gacic M, Lipizer M, Borzelli GLE (2010) On the impact of the Bimodal Oscillating System (BiOS) on the biogeochemistry

and biology of the Adriatic and Ionian seas (Eastern Mediterranean). Biogeosciences 7:3987–3997
D'Ortenzio F, Ragni M, Marullo S, Ribera d'Alcalà M (2003) Did biological activity in the Ionian sea change after the eastern Mediterranean Transient? Results from the analysis of remote sensing observations. J Geophys Res-Oceans 108:8113
de Lange GJ, Thomson J, Reitz A, Slomp CP, Principato MS, Erba E, Corselli C (2008) Synchronous basin-wide formation and redox-controlled preservation of a Mediterranean sapropel. Nat Geosci 1:606–610
Diaz J, Ingall E, Benitez-Nelson C, Paterson D, de Jonge MD, McNulty I, Brandes JA (2008) Marine polyphosphate: a key player in geologic phosphorus sequestration. Science 320:652–655
Duhamel S, Dyhrman ST, Karl DM (2010) Alkaline phosphatase activity and regulation in the North Pacific Subtropical Gyre. Limnol Oceanogr 55:1414–1425
Dyhrman ST, Chappell PD, Haley ST, Moffett JW, Orchard ED, Waterbury JB, Webb EA (2006) Phosphonate utilization by the globally important marine diazotroph *Trichodesmium*. Nature 439:68–71
Emeis KC, Mara P, Schlarbaum T, Möbius J, Dähnke K, Struck U, Mihalopoulos N, Krom M (2010) External N inputs and internal N cycling traced by isotope ratios of nitrate, dissolved reduced nitrogen, and particulate nitrogen in the eastern Mediterranean sea. J Geophys Res 115, G04041
Fagerli H, Legrand M, Preunkert S, Vestreng V, Simpson D, Cerqueira M (2007) Modeling historical long-term trends of sulfate, ammonium, and elemental carbon over Europe: a comparison with ice core records in the Alps. J Geophys Res 112:D23S13
Gacic M, Borzelli GLE, Civitarese G, Cardin V, Yari S (2010) Can internal processes sustain reversals of the ocean upper circulation? The Ionian sea example. Geophys Res Lett 37, L09608
Garcia N, Raimbault P, Gouze E, Sandroni V (2006) Nitrogen fixation and primary production in western Mediterranean. C R Biol 329:742–750
Groom S, Herut B, Brenner S, Zodiatis G, Psarra S, Kress N, Krom MD, Law CS, Drakopoulos P (2005) Satellite-derived spatial and temporal biological variability in the Cyprus Eddy. Deep-Sea Res II 52:2990–3010
Gruber N, Sarmiento JL (1997) Global patterns of marine nitrogen fixation and denitrification. Global Biogeochm Cycle 11:235–266
Hansell DA (2002) DOC in the global ocean carbon cycle. In: Hansell DA, Carlson CA (eds) Biogeochemistry of marine dissolved organic matter. Academic, San Diego
Hansell DA, Carlson CA (2001) Biogeochemistry of total organic carbon and nitrogen in the Sargasso sea: control by convective overturn. Deep-Sea Res II 48:1649–1667
Hecht A, Pinardi N, Robinson AR (1988) Currents, water masses, eddies and jets in the Mediterranean Levantine Basin. J Phys Oceanogr 18:1320–1353
Hecky RE, Kilham P (1988) Nutrient limitation of phytoplankton in freshwater and marine environments: a review of recent evidence on the effects of enrichment. Limnol Oceanogr 33:796–822
Herut B, Zohary T, Robarts RD, Kress N (1999) Adsorption of dissolved phosphate onto loess particles in surface and deep eastern Mediterranean water. Mar Chem 64:253–265
Ibello V, Cantoni C, Cozzi S, Civitarese G (2010) First basin-wide experimental results on N_2 fixation in the open Mediterranean Sea. Geophys Res Lett 37, L03608, doi:10.1029/2009GL041635
Klein B, Roether W, Manca BB, Bregant D, Beitzel V, Kovacevic V, Luchetta A (1999) The large deep water transient in the Eastern Mediterranean. Deep-Sea Res I 46:371–414
Klein B, Roether W, Kress N, Manca BB, d'Alcala MR, Souvermezoglou E, Theocharis A, Civitarese G, Luchetta A (2003) Accelerated oxygen consumption in eastern Mediterranean deep waters following the recent changes in thermohaline circulation. J Geophys Res Oceans 108:8107
Kocak M, Kubilay N, Tugrul S, Mihalopoulos N (2010) Long-term atmospheric nutrient inputs to the eastern Mediterranean: sources, solubility and comparison with riverine inputs. Biogeosciences 7:4037–4050
Koike I, Tupas L (1993) Total dissolved nitrogen in the northern North Pacific assessed by a high-temperature combustion method. Mar Chem 41:209–214
Krasakopoulou E, Souvermezoglou E, Pavlidou A, Kontoyannis H (1999) Oxygen and nutrient fluxes through the Straits of the Cretan Arc (March 1994–January 1995). Prog Oceanogr 44:601–624
Kress N, Herut B (2001) Spatial and seasonal evolution of dissolved oxygen and nutrients in the southern Levantine Basin (eastern Mediterranean sea): chemical characterization of the water masses and inferences on the N : P ratios. Deep-Sea Res I 48:2347–2372
Kress N, Manca BB, Klein B, Deponte D (2003) Continuing influence of the changed thermohaline circulation in the eastern Mediterranean on the distribution of dissolved oxygen and nutrients: physical and chemical characterization of the water masses. J Geophys Res Oceans 108:8109
Kress N, Herut B, Gertman I (2011) Nutrient distribution in the Eastern Mediterranean before and after the transient event. In: Stambler N (ed) Life in the Mediterranean Sea: a look at habitat changes. Nova, New York
Kress N, Thingstad TF, Pitta P, Psarra S, Tanaka T, Zohary T, Groom S, Herut B, Mantoura RFC, Polychronaki T, Rassoulzadegan F, Spyres G (2005) Effect of P and N addition to oligotrophic Eastern Mediterranean waters influenced by near-shore waters: a microcosm experiment. Deep-Sea Res II 52:3054–3073
Krom MD, Kress N, Brenner S, Gordon LI (1991) Phosphorus limitation of primary productivity in the eastern Mediterranean sea. Limnol Oceanogr 36:424–432
Krom MD, Brenner S, Kress N, Neori A, Gordon LI (1992) Nutrient dynamics and new production in a warm-core eddy from the eastern Mediterranean sea. Deep-Sea Res A 39:467–480
Krom MD, Groom S, Zohary T (2003) The Eastern Mediterranean. In: Black KD, Shimmield GB (eds) The biogeochemistry of marine systems. Blackwell Publishing, Oxford
Krom MD, Herut B, Mantoura RFC (2004) Nutrient budget for the eastern Mediterranean: implications for phosphorus limitation. Limnol Oceanogr 49:1582–1592
Krom MD, Woodward EMS, Herut B, Kress N, Carbo P, Mantoura RFC, Spyres G, Thingstad TF, Wassmann P, Wexels-Riser C, Kitidis V, Law CS, Zodiatis G (2005) Nutrient cycling in the south east Levantine basin of the eastern Mediterranean: results from a phosphorus starved system. Deep-Sea Res II 52:2879–2896
Krom MD, Emeis KC, Van Cappellen P (2010) Why is the eastern Mediterranean phosphorus limited? Prog Oceanogr 85:236–244
Lascaratos A, Roether W, Nittis K, Klein B (1999) Recent changes in deep water formation and spreading in the eastern Mediterranean sea: a review. Prog Oceanogr 44:5–36
Lomas MW, Burke AL, Lomas DA, Bell DW, Shen C, Dyhrman ST, Ammerman JW (2010) Sargasso sea phosphorus biogeochemistry: an important role for dissolved organic phosphorus (DOP). Biogeosciences 7:695–710
Loh AN, Bauer JE (2000) Distribution, partitioning and fluxes of dissolved and particulate organic C, N and P in the eastern North Pacific and Southern Oceans. Deep-Sea Res 47:2287–2316
Ludwig W, Dumont E, Meybeck M, Heussner S (2009) River discharges of water and nutrients to the Mediterranean and Black sea: major drivers for ecosystem changes during past and future decades? Prog Oceanogr 80:199–217
Man-Aharonovich D, Kress N, Bar Zeev E, Berman-Frank I, Beja O (2007) Molecular ecology of *nifH* genes and transcripts in the eastern Mediterranean Sea. Environ Microbiol 9:2354–2363.
Malanotte-Rizzoli P, Manca BB, d'Alcala MR, Theocharis A, Brenner S, Budillon G, Ozsoy E (1999) The eastern Mediterranean in the 80s

and in the 90s: the big transition in the intermediate and deep circulations. Dynam Atmos Oceans 29:365–395

Manca BB, Kovacevic V, Gacic M, Viezzoli D (2002) Dense water formation in the southern Adriatic sea and spreading into the Ionian sea in the period 1997–1999. J Marine Syst 33–34:133–154

Manca BB, Ibello V, Pacciaroni M, Scarazzato P, Giorgetti A (2006) Ventilation of deep waters in the Adriatic and Ionian seas following changes in thermohaline circulation of the eastern Mediterranean. Climate Res 31:239–256

Marty JC, Garcia N, Rairnbault P (2008) Phytoplankton dynamics and primary production under late summer conditions in the NW Mediterranean Sea. Deep-Sea Res I 55:1131–1149

Mather RL, Reynolds SE, Wolff GA, Williams RG, Torres-Valdes S, Woodward EMS, Landolfi A, Pan X, Sanders R, Achterberg EP (2008) Phosphorus cycling in the north and south Atlantic ocean subtropical gyres. Nat Geosci 1:439–443

Montoya JP, Voss M, Kahler P, Capone DG (1996) A simple, high-precision, high-sensitivity tracer assay for N_2 fixation. Appl Environ Microbiol 62:986–993

Nenes A, Krom MD, Mihalopoulos N, Van Cappellen P, Shi Z, Bougiatioti A, Zarmpas P, Herut B (2011) Atmospheric acidification of mineral aerosols: a source of bioavailable phosphorus for the oceans. Atmos Chem Phys 11:6265–6272

Pantoja S, Repeta DJ, Sachs JP, Sigman DM (2002) Stable isotope constraints on the nitrogen cycle of the Mediterranean Sea water column. Deep-Sea Res I 49:1609–1621

POEM-Group (1992) General circulation of the Eastern Mediterranean. Earth Sci Rev 32:285–309

Pojed I, Kveder S (1977) Investigation of nutrient limitation of phytoplankton production in northern Adriatic. Thalass Jugosl 13:13–27

Postgate J (1998) The origins of the unit of nitrogen fixation at the University of Sussex. Notes Rec R Soc 52:355–362

Preunkert S, Wagenbach D, Legrand M (2003) A seasonally resolved alpine ice core record of nitrate: comparison with anthropogenic inventories and estimation of preindustrial emissions of NO in Europe. J Geophys Res 108:4681

Psarra S, Tselepides A, Ignatiades L (2000) Primary productivity in the oligotrophic Cretan sea (NE Mediterranean): seasonal and interannual variability. Prog Oceanogr 46:187–204

Pujo-Pay M, Conan P, Oriol L, Cornet-Barthaux V, Falco C, Ghiglione JF, Goyet C, Moutin T, Prieur L (2011) Integrated survey of elemental stoichiometry (C, N, P) from the Western to Eastern Mediterranean Sea. Biogeosciences 8:883–899

Rahav E, Herut B, Stambler N, Bar-Zeev E, Mulholland MR, Berman-Frank I (2013a) Uncoupling between dinitrogen fixation and primary productivity in the Eastern Mediterranean Sea. JGR Biogeosci 18:195–202

Rahav E, Herut B, Levi A, Mulholland MR, Berman-Frank I (2013b) Springtime contribution of dinitrogen fixation to primary production across the Mediterranean. Ocean Sci 9:1–10

Raimbault P, Diaz F, Pouvesle W, Boudjellal B (1999a) Simultaneous determination of particulate organic carbon, nitrogen and phosphorus collected on filters, using a semi-automatic wet-oxidation method. Marine Ecol-Progr Ser 180:289–295

Raimbault P, Slawyk G, Boudjellal B, Coatanoan C, Conan P, Coste B, Garcia N, Moutin T, Pujo-Pay M (1999b) Carbon and nitrogen uptake and export in the equatorial Pacific at 150°W: evidence of an efficient regenerated production cycle. J Geophys Res 104:3341–3356

Ranhofer M, Lawrenz E, Pinckney J, Benitez-Nelson C, Richardson T (2009) Cell-specific alkaline phosphatase expression by phytoplankton from Winyah Bay, South Carolina, USA. Estuar Coasts 32:943–957

Redfield AC, Ketchum BH, Richards FA (1963) The influence of organisms on the composition of sea-water. In: Hill MN (ed) The sea. Interscience, New York

Rees AP, Law CS, Woodward EMS (2006) High rates of nitrogen fixation during an *in-situ* phosphate release experiment in the Eastern Mediterranean Sea. Geophys Res Lett 33, L10607, doi:10.1029/2006GL025791

Ribera d'Alcalà M, Civitarese G, Conversano F, Lavezza R (2003) Nutrient ratios and fluxes hint at overlooked processes in the Mediterranean sea. J Geophys Res 108:8106

Ridame C, Le Moal M, Guieu C, Ternon E, Biegala IC, L'Helguen S, Pujo-Pay M (2011) Nutrient control of N_2 fixation in the oligotrophic Mediterranean Sea and the impact of Saharan dust events. Biogeosciences 8:2773–2783

Robinson AR, Golnaraghi M, Leslie WG, Artegiani A, Hecht A, Lazzoni E, Michelato A, Sansone E, Theocharis A, Ünlüata Ü (1991) The eastern Mediterranean general circulation: features, structure and variability. Dynam Atmos Oceans 15:215–240

Roether W, Manca B, Klein B, Bregant D, Georgopoulos D, Beitzel V, Kovacevic V, Luchetta A (1996) Recent changes in eastern Mediterranean deep water. Science 271:333–335

Roether W, Klein B, Manca BB, Theocharis A, Kioroglou S (2007) Transient eastern Mediterranean deep waters in response to the massive dense-water output of the Aegean sea in the 1990s. Prog Oceanogr 74:540–571

Rubino A, Hainbucher D (2007) A large abrupt change in the abyssal water masses of the eastern Mediterranean. Geophys Res Lett 34, L23607

Sachs JP, Repeta DJ (1999) Oligotrophy and nitrogen fixation during eastern Mediterranean sapropel events. Science 286:2485–2488

Sandroni V, Raimbault P, Migon C, Garcia N, Gouze E (2007) Dry atmospheric deposition and diazotrophy as sources of new nitrogen to northwestern Mediterranean oligotrophic surface waters. Deep-Sea Res I 54:1859–1870

Santinelli C, Nannicini L, Seritti A (2010) DOC dynamics in the meso and bathypelagic layers of the Mediterranean sea. Deep-Sea Res II 57:1446–1459

Schlitzer R, Roether W, Oster H, Junghans HG, Hausmann M, Johannsen H, Michelato A (1991) Chlorofluoromethane and oxygen in the eastern Mediterranean. Deep-Sea Res A 38:1531–1551

Seritti A, Manca BB, Santinelli C, Murru E, Boldrin A, Nannicini L (2003) Relationships between dissolved organic carbon (DOC) and water mass structures in the Ionian sea (winter 1999). J Geophys Res 108:8112

Siokou-Frangou I, Christaki U, Mazzocchi MG, Montresor M, Ribera d'Alcalá M, Vaqué D, Zingone A (2010) Plankton in the open Mediterranean sea: a review. Biogeosciences 7:1543–1586

Smith SV, Kimmerer WJ, Walsh TW (1986) Vertical flux and biogeochemical turnover regulate nutrient limitation of net organic production in the north Pacific Gyre. Limnol Oceanogr 31:161–167

Somot S, Sevault F, Deque M (2006) Transient climate change scenario simulation of the Mediterranean Sea for the twenty-first century using a high-resolution ocean circulation model. Climate Dyn 27:851–879

Spatharis S, Skliris N, Meziti A, Kormas KA (2012) First record of a Trichodesmium erythraeum bloom in the Mediterranean Sea. Can J Fish Aquat Sci 69:1444–1455

Stihl A, Sommer U, Post AF (2001) Alkaline phosphatase activities among populations of the colony-forming diazotrophic cyanobacterium *Trichodesmium* spp. (cyanobacteria) in the Red Sea. J Phycol 37:310–317

Struck U, Emeis K-C, Voß M, Krom MD, Rau GH (2001) Biological productivity during sapropel S5 formation in the eastern Mediterranean sea: evidence from stable isotopes of nitrogen and carbon. Geochim Cosmochim Acta 65:3249–3266

Tanaka T, Thingstad TF, Christaki U, Colombet J, Cornet-Barthaux V, Courties C, Grattepanche JD, Lagaria A, Nedoma J, Oriol L, Psarra S, Pujo-Pay M, Van Wambeke F (2011) Lack of P-limitation of phytoplankton and heterotrophic prokaryotes in surface waters of three anticyclonic eddies in the stratified Mediterranean sea. Biogeosciences 8:525–538

Thingstad TF, Mantoura RFC (2005) Titrating excess nitrogen content of phosphorous-deficient eastern Mediterranean surface water using alkaline phosphatase activity as a bio-indicator. Limnol Oceanogr Methods 3:94–100

Thingstad TF, Krom MD, Mantoura RFC, Flaten GAF, Groom S, Herut B, Kress N, Law CS, Pasternak A, Pitta P, Psarra S, Rassoulzadegan F, Tanaka T, Tselepides A, Wassmann P, Woodward EMS, Riser CW, Zodiatis G, Zohary T (2005) Nature of phosphorus limitation in the ultraoligotrophic eastern Mediterranean. Science 309:1068–1071

Vukadin I, Stojanovski L (1976) C:N:Si:P ratio in the waters of the middle and south Adriatic. Commun Int ExplSci Mer Medit Rapp PV Reun 23:41–43

Williams PM, Carlucci AF, Olson R (1980) Deep profile of some biologically important properties in the Central North Pacific Gyre. Oceanol Acta 3:471–476

Wu J, Sunda W, Boyle EA, Karl DM (2000) Phosphate depletion in the western north Atlantic ocean. Science 289:759–762

Yacobi YZ, Zohary T, Kress N, Hecht A, Robarts RD, Waiser M, Wood AM (1995) Chlorophyll distribution throughout the southeastern Mediterranean in relation to the physical structure of the water mass. J Marine Syst 6:179–190

Yilmaz A, Tugrul S (1998) The effect of cold- and warm-core eddies on the distribution and stoichiometry of dissolved nutrients in the Northeastern Mediterranean. J Marine Syst 16:253–268

Yogev T, Rahav E, Bar-Zeev E, Man-Aharonovich D, Stambler N, Kress N, Béjà O, Mulholland MR, Herut B, Berman-Frank I (2011) Is dinitrogen fixation significant in the Levantine Basin, east Mediterranean sea? Environ Microbiol 13:854–871

Zavatarelli M, Raicich F, Artegiani A, Bregant D, Russo A (1998) Climatological biogeochemical characteristics of the Adriatic sea. J Marine Syst 18:227–263

Zobrist J, Stumm W (1981) Chemical dynamics of the Rhine catchment area in Switzerland, extrapolation to the 'pristine' Rhine river input to the ocean. In: Martin JM, Burton JD, Eisma D (eds) River inputs to ocean systems. United Nations, New York

Zohary T, Robarts RD (1998) Experimental study of microbial P limitation in the eastern Mediterranean. Limnol Oceanogr 43:387–395

Zohary T, Herut B, Krom MD, Fauzi C, Mantoura R, Pitta P, Psarra S, Rassoulzadegan F, Stambler N, Tanaka T, Frede Thingstad T, Malcolm S, Woodward E (2005) P-limited bacteria but N and P co-limited phytoplankton in the eastern Mediterranean – a microcosm experiment. Deep-Sea Res II 52:3011–3023

Marine Chemosynthesis in the Mediterranean Sea

5

Marco Taviani

Abstract

The Mediterranean Sea contains a vast spectrum of chemosynthetic habitats from shallow marine to bathyal depths. These habitats (hydrothermal vents, cold seeps, reducing sediment) are home to bacteria and archaea acting as primary producers using the energy obtained by oxidizing reducing compounds in fluids (e.g., H_2S and hydrocarbons such as CH_4) to synthesize organic matter. Such sites may make a large microbial biomass available to consumers and promote the development of complex symbiotic relationships between prokaryotes and hosting eukaryotes. Shallow water (<200 m) chemosynthetic niches are pervasive in the Mediterranean where reduced sediment are present (lagoons, seagrass beds, prodeltaic settings etc.) being exploited primarily by sulphur-oxidizing bacteria, some of which are symbiotic with metazoans. Particular cases of shallow chemosynthetic habitats are hydrothermal vents, submarine caves with sulphur springs, cold seeps including active pockmarks and reducing sediments in areas of high organic deposition. Deep-water chemosynthetic habitats (>200 m) with also metazoans with chemoautotrophic symbionts are present at various sites in the Mediterranean, encompassing both cold seeps and hydrothermal vents. Cold seeps are diffuse and geologically diverse, and comprise mud volcanoes, brine pools and active pockmark fields, whose reducing environments are often exploited by endemic chemosymbiotic metazoans, including bivalves and siboglinid polychaetes. Such environments are clustered in the Eastern Mediterranean (Nile Deep Sea Fan, Anaximander mud volcano field, Olimpi mud volcano field, Eratosthenes Seamount, Calabrian Arc, Marmara deep fault systems) but equally occur in the Strait of Sicily, Adriatic, Tyrrhenian and Alboran Seas. Deep hyperhaline anoxic basins occur in the Eastern Mediterranean and even host metazoans that live in anoxic conditions. Deep-water hydrothermal vents characterized by microbial communities and metazoans (siboglinid polychaetes) are documented so far only from some Tyrrhenian seamounts (Marsili, Palinuro) and Aegean submerged volcanic craters (Santorini, Kolumbo). The geo-biological relevance of such complex habitats often situated in the high seas, calls for an international effort towards the implementation of proper protection and sustainable management.

Keywords

Mediterranean basin • Chemosynthesis • Chemosymbiotic metazoans • Reducing bottoms • Hydrothermal vents • Cold seeps • Recent • Fossil • Deep sea

M. Taviani (✉)
ISMAR-CNR, Bologna, via Gobetti 101, 40129 Bologna, Italy

Biology Department, Woods Hole Oceanographic Institution, 266 Woods Hole Road, Woods Hole, MA 02543, USA
e-mail: marco.taviani@bo.ismar.cnr.it

S. Goffredo and Z. Dubinsky (eds.), *The Mediterranean Sea: Its history and present challenges*,
DOI 10.1007/978-94-007-6704-1_5,

Introduction

Chemosynthesis can provide the energetic requirements for life in specialised marine ecosystems. In contrast with photosynthesis, chemosynthesis does not require sunlight for reactions and may therefore take place in darkness. It works through microbial-mediated processes that can take place in the presence or absence of free oxygen, transforming energy stored in reduced chemicals (hydrogen sulphide, methane, molecular hydrogen, ammonia, insoluble iron sulphides and elemental sulphur) into organics, mainly through sulphur oxidation or reduction, and methane oxidation. Reactions led by chemotrophic microbial consortia as primary producers may result in producing considerable prokaryotic biomass available to heterotrophs. Chemosynthesis in the ocean can support endemic metazoans hosting specific epi- or endosymbionts for their nutritional requirements (e.g., Felbeck et al. 1981; Fisher 1990; McMullin et al. 2000; Van Dover 2000; Ott et al. 2004; Cavanaugh et al. 2006; Dubilier et al. 2006, 2008; Stewart and Cavanaugh 2006; Hourdez and Lallier 2007; Speight and Henderson 2010; Duperron 2010; Vrijenhoek 2010; Duperron et al. 2011; German et al. 2011; Roeselers and Newton 2012). Chemosynthesis related metazoans are present in the geologic record since Proterozoic times (Little 2002; Campbell 2006; Kiel and Little 2006; Kiel 2010; Taviani 2011, with references therein).

Virtually unappreciated until a little more than two decades ago, the relevance of chemosynthetic processes in the deep marine realm of the Mediterranean basin is by now fully recognized (Foucher et al. 2009; Taviani 2011). In fact, the Mediterranean contains a large variety of environments structured through the interaction of fluid expulsion and microbial consortia conducive to spectacular habitats (Olu-Le Roy et al. 2004; Werne et al. 2004; Foucher et al. 2009; Dando 2010; Danovaro et al. 2010a). Furthermore, not only the Mediterranean houses at present cold and hot vents, but it also boasts one of the best world's fossil legacy of such environments (Taviani 2001, 2011).

Modern Mediterranean Chemosynthetic Habitats

At shallow depth, reducing sulphide-rich environments, exploitable by chemosynthetic organisms, occur commonly in brackish lagoons, seagrass meadows, prodeltaic organic-rich sediments among others, always intermixed with photosynthetic communities. In fact, organic-rich reducing sites are the dominant chemosynthetic habitat of the Mediterranean Sea but will not be treated in great detail in this chapter, which is largely focused upon chemosynthetic deep sea environments and their associated metazoans (Fig. 5.1).

These sulphidic habitats require physiological flexibility in inhabiting eukaryotic organisms (Hagerman 1998; Thiermann et al. 2000), some of which show complex interactions with microbes, such as for example colonial ciliates (*Zoothamnium niveum*: Rinke et al. 2007), or solemyid (*Solemya togata*), lucinid (e.g., *Loripes lucinalis, Anodontia fragilis, Myrtea spinifera*) and thyasirid (e.g., *Thyasira flexuosa*) bivalves, all known to host sulphur-oxidising symbiotic bacteria (e.g., Rinke et al. 2006; Taylor and Glover 2010; Dreier et al. 2012). An even more extreme case of chemosynthetic relation between bacterial symbionts and metazoans is offered by the gutless marine oligochaete *Olavius algarvensis* inhabiting seagrass sediment, whose highly complex metabolic pathways seem to also include the use of hydrogen and CO as an energy source (Kleiner et al. 2012).

Submarine caves within limestone bedrock are known to occur along the Tyrrhenian coast, some of which were formed by and contain active sulphide-rich springs that sustain a trophic chain dominated by prokaryotes. The best examples are found at Capo Palinuro in southern Italy (Fig. 5.2), where sulphide-containing warm springs support dense *Beggiatoa*-like bacterial mats and other chemoautotrophic prokaryotes (Maugeri et al. 2010) that in turn sustain some metazoans including sponges, cnidarians and polychaetes, often of abnormally large size (Alvisi et al. 1994; Morri et al. 1994; Mattison et al. 1998; Canganella et al. 2007; Southward et al. 2009; Dando 2010). Many metazoans sharing this habitat are covered by bacterial filaments (Southward et al. 2009). However, besides consumption of bacterial-originated and photosynthetically-derived organic matter, no conclusive evidence has been found so far that some such metazoans also use microbial symbiosis (P. Dando pers. com. 2012).

Some shallow-water 'caves' on volcanic coasts contain specialized endemic chemosymbiotic macrofauna. This is the case on the western side of Sicily near Taormina where a relatively large lucinid bivalve (*Lucinoma speleum* Palazzi and Villari 2001) has been described from soft bottom sediment. This taxon is co-generic to species known to host sulphur-oxidising bacteria in their tissues, such as *L. borealis* (shallow marine) and *L. kazani* (deep marine). *L.* aff. *speleum* is recorded also from volcanic islands surrounding Sicily, such as Salina island in the Aeolian archipelago (M. Oliverio pers. com.). Nothing is known about its ecology and there is a need for an investigation of its habitat.

Hydrothermal Vents

The vast majority of charted hydrothermal vents in the Mediterranean are found at shallow depth (Dando et al. 1999; Dando 2010 for a review). Although influencing the structure of the local living communities and sustaining a

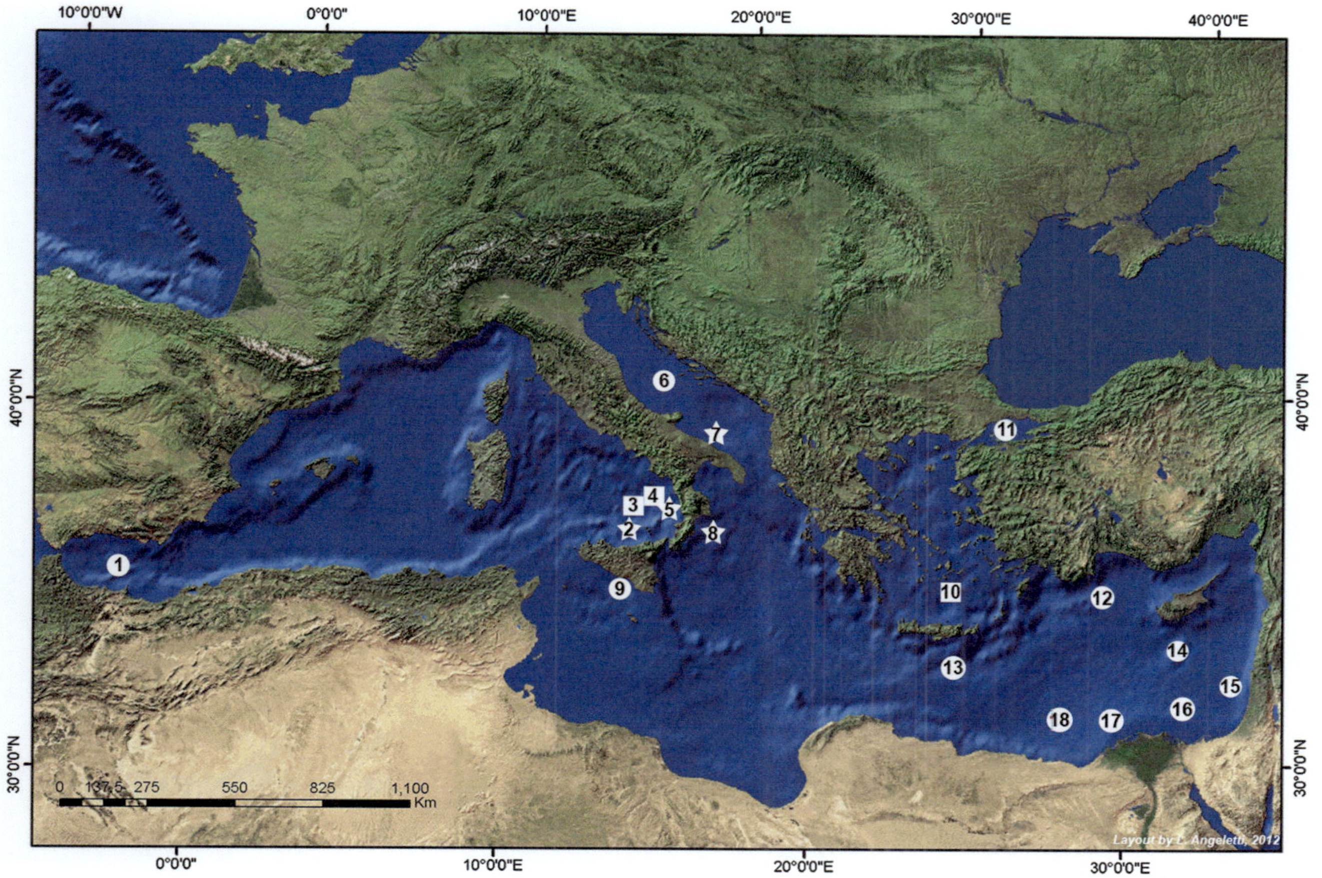

Fig. 5.1 Map of the Mediterranean sea showing location of modern deep-sea hydrothermal vents (*squares*) and cold seeps (*circles*) discussed in this article containing specialized chemosymbiotic metazoans; *stars* indicate deep seep sites likely hosting similar communities. From West to East: *1* Alboran mud volcanoes, *2* Enarete Seamount vents, *3* Marsili Seamount vents, *4* Palinuro Seamount, *5* Paola Basin pockmarks, *6* Horseshoe hill pockmarks at Pomo/Jabuka, *7* Bari Canyon area, *8* Calabrian Arc mud volcanoes and pockmarks, *9* Gela Basin pockmark field, *10* Kolumbo and Santorini craters, *11* Marmara submarine deep fault system, *12* Anaximander mud volcano field, *13* Olimpi mud volcano field, *14* Eratosthenes Seamount, *15* Israel continental margin, *16–18* Nile Deep Sea Fan cluster (eastern, central and western provinces)

variety of microorganisms (Manini et al. 2008; Maugeri et al. 2010), these shallow water hydrothermal vents (Fig. 5.2) host few, if any, vent-specific eukaryotic organisms (e.g., Thiermann et al. 1997; Gamenick et al. 1998; Morri et al. 1999; Cocito et al. 2000; De Biasi and Aliani 2003; Panieri et al. 2005; Dando 2010; Bianchi et al. 2011).

Well-developed prokaryotic mats have been recently imaged from a few hundreds to 500 m in the Aegean Kolumbo and Santorini underwater craters, at places associated with vigorous hot venting (Carey and Sigurdsson 2007; Carey et al. 2011; Bell et al. 2012). First metagenomic exploration of microbial communities from such submarine vents reveals that they present a high microbial diversity. Various seamounts in the Tyrrhenian Sea have been or are hydrothermally active (Dando et al. 1999). Mounds of microbial origin have been detected on the submerged portions of the Aeolian arc, including the Enarete Seamount (Carey et al. 2012). An ROV survey has documented the presence of living sibloglinid polychaetes on the Marsili Seamount (Lott and Zimmerman 2012). The species is similar, or identical to *Lamellibrachia anaximandri* Southward et al. 2011, a cold seep species (Olu-Le Roy et al. 2004; Werne et al. 2004; Hughes and Crawford 2006; Southward et al. 2011). The exploration of the Tyrrhenian Sea submarine volcanoes by E/V *Nautilus* using the ROV *Hercules* also led to the first discovery of Siboglinidae tubeworm clusters and bacterial mats on the Palinuro Seamount surrounding hot vents at 600 m (Carey et al. 2012). These recent discoveries paralleling in a way what happened with Mediterranean cold vent communities, is a further proof of our deficiency in the full knowledge of deep-sea ecosystems (Danovaro et al. 2010a). It will hopefully inspire a more detailed investigation of the many submarine quiescent and active volcanic settings in the deep Mediterranean, whose best candidates are located in the Tyrrhenian Sea, the Strait of Sicily, and the Aegean Sea.

Fig. 5.2 Shallow water hydrothermal vents and underwater caves with sulfidic springs, Aegean and Tyrrhenian Seas: (**a**) Emission of fluids and gas bubbles from the shallow seabed along the Hellenic Arc; Kos island, Greece, Aegean Sea (© C.N. Bianchi, photo taken in 1981); (**b**) Gardening of bacterial-filament by the tubicolous polychaete *Phyllochetopterus socialis* at the interface between sulphur-rich reducing water and oxygenated seawater; Grotta Sulfurea, Palinuro, Italy (© C.N. Bianchi, photo taken in 1992); (**c**) Thick mat of *Beggiatoa* filaments developing on a rocky wall in sulphidic water; Grotta Azzurra, Palinuro, Italy (© C.N. Bianchi, photo taken in 1992)

Cold Seeps

Cold seeps are characterized by the release of sulphide, sometimes accompanied by hydrocarbons and can support, in deeper water, diverse benthic communities (Paull et al. 1984; Sibuet and Olu-Le Roy 1998, 2002; Judd and Hovland 2007; Vanreusel et al. 2009; Cordes et al. 2010). Mediterranean cold-seep environments include seeping pockmarks, mud volcanoes, and brine pools. Gas hydrates which are sites often colonized by chemosynthetic communities, occur at present in this basin (De Lange and Brumsack 1998; Woodside et al. 1998; Perissoratis et al. 2011), but uncommonly due to the Mediterranean 13 °C homothermal situation limiting their stability. Other reducing environments include anoxic sediments and organic falls, including sunken wood, mammal falls, and shipwrecks (see Taviani 2011 for a summary).

In the Mediterranean Sea, hydrocarbon cold seeps are documented from relatively shallow to bathyal depths. Seeps at shallow depth are typically deprived of seep-specific chemosynthetic eukaryotes (Dando 2010), thus differing from their deep-water counterparts. As for hydrothermal vents, shallow hydrocarbon seep habitats like those described from the Adriatic shelf (Capozzi et al. 2012; Curzi 2012), exert some influence on local benthic biota as seen, for example, in augmented benthic foraminifera density next to *Beggiatoa* mats (Panieri 2003, 2006). Seeps at bathyal depths often sustain specialized chemosynthetic communities. Many such sites are by now known in the deep Mediterranean Sea (Charlou et al. 2003; Lykousis et al. 2004; Olu-Le Roy et al. 2004; Armijo et al. 2005; Dupré et al. 2007; Hilário et al. 2011; Coleman et al. 2012), and many more will surely be discovered in the future. The first record of any Mediterranean deep chemosymbiotic organism dates back to the nineteenth century when the Austro-Hungarian ship *Pola* took bottom samples from what it is known today as the Nile Deep Sea Fan. These were bivalves in the families Mytilidae, Vesicomyidae, Lucinidae and Thyasiridae, described by Sturany (1896), obviously unaware of their peculiar niche. These were only found again about 15 years ago when

Corselli and Basso (1996) recorded chemosymbiotic communities from the Napoli Dome. Since then, such discoveries have multiplied (Cosel and Salas 2001; Salas and Woodside 2002; Olu-Le Roy et al. 2004; Gofas 2004; Werne et al. 2004; Hughes and Crawford 2006; Smriglio and Mariottini 2002, 2006; Ritt et al. 2010; Hilário et al. 2011; Mayer et al. 2011; Shank et al. 2011; Southward et al. 2011). A preliminary and very conservative list of metazoans associated with deep-water cold seepage in the Mediterranean Sea accounts for at least a dozen species (6–8 molluscs, 1–2 decapods, 4 siboglinids), but this number will certainly increase in the next years. Besides triggering highly-specialized chemosynthetic biota, cold seeps enhance meiofauna biodiversity (Zeppilli et al. 2011). Finally, they act as deep-sea nurseries for elasmobranchs that are known to attach their egg cases to *Lamellibrachia* tubes (Treude et al. 2011).

To date, the Eastern Mediterranean, characterized by a number of mud volcanoes and other reducing habitats, hosts most of the known situations of chemosynthetic communities linked to methane escape and sulphidic bottoms (Olu-Le Roy et al. 2004; Foucher et al. 2009; Mayer et al. 2011). These include the Anaximander mud volcano field south of Turkey, Olimpi mud volcano south of Crete, the Eratosthenes Seamount south-west of Cyprus (Fig. 5.3), the Calabrian Arc, the Nile Deep Sea Fan, the Marmara deep fault systems, and the continental margin of Israel. Remarkably, the Eastern Mediterranean fluid seepages take place under contrasting scenarios linked to both active and passive margin kinematics (Perissoratis et al. 2011). The resulting high heterogeneity in bottom types, nature of seeping fluids types and rates of flux provide habitats for many bacterial niches (Grünke et al. 2011) which produce authigenic mineral precipitates and sustain chemosynthetic macrobenthos (Aloisi et al. 2002; Olu-Le Roy et al. 2004; Gontharet et al. 2007; Foucher et al. 2009; Himmler et al. 2011). This environmental heterogeneity is equally reflected in the trophic ecology of communities directly associated with these habitats, as shown by the study of the Amsterdam and Napoli mud volcanoes There the main food webs are thiotrophic through free-living and symbiotic sulphur-oxidizing bacteria that sustain *Lamellibrachia anaximandri*, lucinids (*Lucinoma kazani, Myrtea amorpha*), thyasiridids (*Thyasira striata*), and vesicomyids (*Isorropodon perplexum*). Methanotrophy is an important source of carbon for some organisms in these communities (i.a. the mytilid *Idas* sp, ampharetid, capitellid, and spionid polychaetes). Based upon stable isotope signatures Carlier et al. (2010) demonstrate that chemosymbiotic macrobenthos gains dissolved organic carbon from an array of sources, again reflecting the high environmental heterogeneity of these habitats also at small scale. Regarding the typology of bacteria associated with the macrobenthos, various studies are targeting the Eastern Mediterranean deep reducing habitats, resulting in the identification of various 16S rRNA gene phylotypes. The dominant symbiont in *Lucinoma kazani* is a sulphide-oxidizing symbiont similar to that found in *L. aequizonata* and putatively performing the sulphide oxidation and chemoautotrophy (Duperron et al. 2007). On the same line, Brissac et al. (2011) provide evidence that endosymbiotic bacteria contained in *Myrtea* sp. and *Thyasira* sp. bivalves from the Nile Deep Sea Fan at ca. 1,700 m are close to their respective counterparts living in shallow waters, suggesting taxonomic identity. The small vesicomyid *Isorropodon perplexum* harbours a single chemoautotrophic symbiont related to other symbionts reported from this family (Rodrigues et al. 2012). The siboglinid vestimentiferan *Lamellibrachia* contains a gamma-proteobacterium in its trophosome performing thiotrophism (Duperron et al. 2009). A thyasirid from this area is also documented to host a putative sulphur-oxidizing gamma-proteobacterium (Rodrigues and Duperron 2011). The bathymodioline mytilid *Idas modiolaeformis* was found to host as many as six different endosymbiotic bacteria capable of likely performing sulphide and methane oxidation (Duperron et al. 2008). Bacterial symbionts of this hermaphroditic species only occur in the gills and are absent in gametes (Gaudron et al. 2012).

Recent experiments using standardized colonization devices deployed next to reducing habitats in the Eastern Mediterranean (CHEMECOLI) have documented active larval settlement by chemosymbiotic bivalves on various organic substrates (Gaudron et al. 2010). However, very little is known at present about the dispersal mechanisms and connectivity of Mediterranean and adjacent eastern Atlantic deep-sea chemosynthetic sites. Remarkably, vestimentiferan DNA was identified in Mediterranean picoplankton from shallow water samples (Marie et al. 2006).

A somewhat intermediate bathymetric situation between shallow and deep cold seeps is represented by pockmarks located in the central Adriatic immediately east of the Jabuka/Pomo pit (Figs 5.1 and 5.5). Here, dead but still articulated and fresh shells of the 'bathyal' lucinid *L. kazani* (C. Salas pers. com.) occur as shallow as 200 m, its range being previously known to be between ca. 500 and 1,700 m (Duperron et al. 2007).

Finally, anoxic deep-sea brine lakes and pools thrive of microbial life (Borin et al. 2009; La Cono et al. 2011; Stock et al. 2012) but are seemingly inhospitable to chemosymbiotic macrofauna (Olu-Le Roy et al. 2004). However, three new-for-science metazoan species belonging to Loricifera, (the last phylum discovered so far), possibly associated with endosymbiotic prokarya have been recently reported from L'Atalante basin, one of the deep hyperhaline anoxic basins (DHABs: L'Atalante, Urania, Bannock, Thetis) in the Eastern Mediterranean under permanent anoxic conditions (Danovaro et al. 2010b). Such discovery has stimulated a great debate about the phylogenetic origin of these organisms (Mentel and Martin 2010) and promoted a renewed

Fig. 5.3 Deep water hydrothermal vents on Palinuro Seamount, Tyrrhenian Sea; pictures taken by the ROV *Hercules* during the 2011 Field Season of E/V *Nautilus* (© Ocean Exploration Trust/Institute for Exploration): (**a**) The robotic arm of *Hercules* probing an area of low temperature venting with bacterial mats; (**b**) Fe-oxide columnar vents with microbial communities; (**c**) Dense aggregate of live siboglinids (possibly *Lamellibrachia anaximandri*) on rocky volcanic substrate affected by active high-temperature hydrothermalism; note also an octopus leaving the scene. Deep water cold seep environments with chemosymbiotic metazoans at ca. 950 m water depth, Eratosthenes Seamount, Levantine Basin; pictures taken by the ROV *Hercules* during the 2010 Field Season of E/V *Nautilus* (© Ocean Exploration Trust/Institute for Exploration): (**d**) Large field of chemosymbiotic lucinid bivalves next to black stained limestone substrate; (**e**) Concentration of dead lucinid clams (likely *Lucinoma kazani*) both articulated and loose with interspersed siboglinid bushes (*Lamellibrachia*) and an occasional non-seep sea-urchin; (**f**) Live siboglinids (possibly *Lamellibrachia anaximandri*) in fissures of limestone bedrock

interest in the systematic exploration of the anoxic regions of the deep sea (Levin 2010).

Historical Legacy of Chemosynthesis

In 1988, Italian geologists suggested that a limestone of Miocene age, occurring as sparse individual blocks, resulted from diagenetic processes in the presence of methane-enriched fluids as revealed by its strongly depleted carbon stable isotope composition (Clari et al. 1988). Such limestones, known in the literature as the collective name of “Calcare a lucina” (Fig. 5.4) are scattered along peninsular Italy and Sicily (Conti et al. 1993; Terzi 1993; Ricci Lucchi and Vai 1994; Conti and Fontana 1999; Taviani 2001), and embed very large lucinid bivalves, therefore their name. These methanogenetic carbonates display a wide spectrum of sizes from small concretions up to metric to decametric chemoherms (sensu Aharon 1994).

Soon after this pioneer study, the occurrence of fossil large vesicomyid clams was reported, the first of their kind in the Mediterranean. These fossils (figures in Taviani 1996, 2001)

Fig. 5.4 Fossil legacy of deep-sea cold seepage in the Mediterranean basin: (**a**) A Pliocene-age carbonate chemoherm protruding off the deep-sea clay sediments pertaining to the Argille Azzurre Fm, Torrente Stirone, Parma, Italy (photo taken in 2010); (**b**) A series of carbonate chimneys presumably linked to hydrocarbon defluidization within hemipelagic sediments, same site as above (photo taken in 2010); (**c**) Cluster of large bathymodioline mussels, among the most emblematic deep-water cold seep chemosymbionts to be found in upper Miocene (Tortonian) methanogenetic limestone, Romagna Apennine (photo taken by L. Angeletti); (**d**) Example of a canonical "Calcare a lucina" s.s.: sliced rock surface revealing densely packed lucinid clams within Miocene-age methanogenetic limestone, Romagna Apennines, Italy

tentatively attributed to *Calyptogena*, were associated with turbidites within an olistostrome of the Miocene–age Marnoso-Arenacea Fm (Berti et al. 1994). Within a few years the relevance of past and recent deep sea cold seepage in the Mediterranean basin became apparent. To date, imprints of past geofluid expulsion from Cenozoic (mostly Miocene and subordinately Pliocene) onto deep-water sea-bottoms of the Mediterranean basin and its precursors are not uncommon (Fig. 5.4), largely but not exclusively linked to compressional tectonic geologic scenarios (Aharon 1994; Vai et al. 1997; Conti and Fontana 1999, 2005; Taviani 2001, 2004, 2011; Clari et al. 2004; Conti et al. 2004, 2008). Fossil occurrences of past hydrocarbon/sulphide seepage encompasses a variety of situations including mud volcanoes (Clari et al. 2004), submarine slides (Lucente and Taviani 2005), and putative gas hydrates (Barbieri et al. 2000; Conti and Fontana 2002; Pierre et al. 2002; Pierre and Rouchy 2004; Martire et al. 2010). Many these paleoseeps contain specialized invertebrates (Taviani 2011). At times also traces of chemosynthetic microbial life, in the form of fabrics, bio-signatures or body fossils, can be observed (Cavagna et al. 1999; Peckmann et al. 1999, 2004; Barbieri and Cavalazzi 2004, 2005). The chemosymbiotic associated fauna is rather diverse and overall resembles modern analogues in the ocean, although sensibly differing from the ones inhabiting the deep Mediterranean basin at present (Taviani 1994, 2001, 2011). The most frequent specialized animal life found in association with imprints of past cold seeps and reducing habitats are bivalves, followed by gastropods, polychaetes and decapods. The seepage setting also seems to influence the quality of the benthic non-seep fauna (Panieri et al. 2009).

Besides old records that pre-date the discovery of modern deep-water chemosynthetic communities and therefore in need of reinterpretations, a limited taxonomic information about Mediterranean fossil seeps, organic falls and reducing sediments is sparse in the scientific (Taviani 1994, 2011; Conti et al. 1996; Conti and Fontana 1998; Venturini et al. 1998; Clari et al. 2004; Pierre and Rouchy 2004; Lucente and Taviani 2005; Dominici et al. 2009; Higgs et al. 2011; Janssen and Krylova 2012; Vinn et al. 2012a, b), and vernacular (Taviani 1996; Terzi and Sami 2007; Taviani and Ceregato 2009) literature. Based upon literature records and unpublished data, a preliminary and very conservative estimate of the diversity of truly chemosymbiotic macrobenthos associated with deep-sea hydrocarbon and reducing sediment provides the following list: Bivalvia: Solemyidae (2 species), Mytilidae (3 species), Lucinidae (6–8 species), Thyasiridae (1–2 species), Vesicomyidae (8 species); Gastropoda: Colloniidae (1 species), Phenacolepadidae (1 species), plus vestimentiferans (1 species), and serpulids (1 species). To this, one should add various other organisms occurring as occasional visitors (e.g. buccinid gastropods) or settling on the authigenic carbonate bedrock (e.g. scleractinian corals, serpulid polychaetes). Hydrocarbon seepage was also locally influencing, sometimes detrimentally by lowering diversity or abundance, the non-chemosymbiotic faunal assemblages as shown by data on benthic foraminifera and ostracodes from Miocene-age deep water sites (Barbieri and Panieri 2004; Panieri et al. 2009; Russo et al. 2012). A throughout taxonomic evaluation of the macrobenthos associated with deep-water Neogene paleoseeps is in progress (M.T, in association with S. Kiel and M. Sami).

Finally, concerning deep-sea hydrothermal vents, the only known fossil example to date is the Cretaceous occurrences associated with the Troodos ophiolitic complex in Cyprus (Little et al. 1999). The identified macrobenthos comprise gastropods, vestimentiferans, and serpulids. It is more than likely that the Tethyan Mediterranean had multiple vent situations but their heritage has been either consumed within the related oceanic crust or metamorphosed.

Threats to Mediterranean Chemosynthetic Habitats

The concept that recent chemosynthetic habitats (namely hydrothermal vents, cold seeps, gas hydrates) require protection has been at the beginning and for a while somewhat neglected by the scientific community. In the last few years, however, a mounting awareness is arising about the ecological, biological and geological interest of such habitats, and of their vulnerability to anthropogenic stressors, with special emphasis on deep-sea settings (UNEP 2007; Van Dover et al. 2012). At a global scale, chemosynthetic habitats are exposed to various anthropogenic stressors, such as trawling, hydrocarbon extraction, and ore mining calling for guidance in policing their correct management and protection (Van Dover et al. 2011, with references therein).

Regarding Mediterranean chemosynthetic habitats, shallow and deep, the question is whether they are in principle exposed to some or all of such threats, and there is some consensus that they are. The entire Mediterranean is obviously also exposed to the problem of both solid waste dumping and oil spills (WWF/IUCN 2004; UNEP 2010) that may impact chemosynthetic sites. Fishing in the high seas has a strong potential of impacting these habitats either through bottom trawling, especially on pockmark-peppered muddy bottoms, or through abandoned fishing gear, as for the '*cannizzi*' dumped in the pockmark area in the Gela basin. Hydrocarbon extraction already well established in the Mediterranean especially on continental shelves, will likely turn into offshore exploration and drilling in the near future impacting local ecosystems. At present, one important methanogenetic site located in the central Adriatic and focus of attention since the 1990s of the past century (Capozzi et al. 2012), became almost inaccessible due to the establishment of oil industry infrastructures. Ore mining for massive sulphides or manganese is at present still a distant menace although it could become reality in the future (Gemmell et al. 2009). Volcanic seamounts, on the other hand, could be targeted for geothermal energy, as for example the Marsili Seamount in the Tyrrhenian basin. Furthermore deep-sea hydrothermal vents were not even contemplated for protection in the Mediterranean based on the presumed absence of conspicuous chemosynthetic communities (WWF/IUCN 2004:44). The very recent documentation of *Lamellibrachia* vestimentiferans thriving on the Marsili Seamount (Lott and Zimmerman 2012), and possibly on the Palinuro Seamount (Carey et al. 2012) is necessarily calling for their urgent inclusion among the sites to be protected. It is also an evident clue that other seamounts in the Mediterranean may equally host still uncharted complex chemosymbiotic metazoan-hosting communities.

The recent compilation by Salomidi et al. (2012) of the biotopes characterizing the European seas in front of goods and services does not include a mention to cold seep chemosynthetic habitats in the Atlantic ocean, Mediterranean, and Black seas. In fact, Salomidi et al. (2012) identify solely deep-sea hydrothermal vents, as well as anoxic sulphide mud and anaerobic microbial reefs in the Black Sea (Michaelis et al. 2002; Treude et al. 2005; Reitner et al. 2005). Following Salomidi et al. (2012)'s proposed scheme, goods and services offered by hydrocarbon cold seeps met partly or totally all categories in their table 1: especially (i) raw materials (i.a., bioprospecting), (ii) climate regulation (i.a., methane release or trapping), (iii) chemosynthesis and primary production, (iv) nutrient cycling, and (v) maintenance of biodiversity.

A Call for Protecting Chemosynthetic Habitats in the Mediterranean Past and Present

MPAs for Extant Habitats

The careful evaluation of potential anthropogenic threats to these habitats is now resulting in recommendations for the establishment of Marine Protected Areas (UNEP 2010,

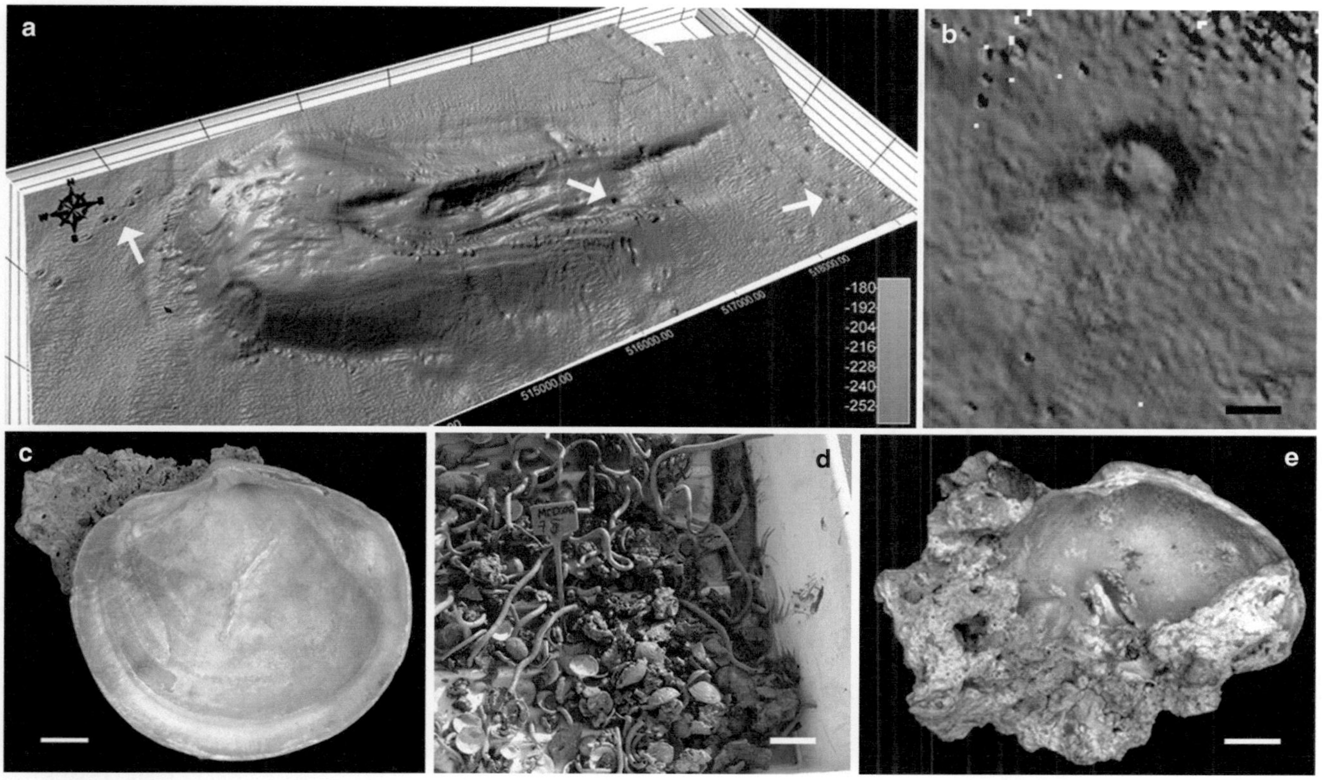

Fig. 5.5 Examples of hydrocarbon pockmarks: Adriatic Sea: (**a**) multiswath bathymetric map showing the "Horseshoe" structure near Jabuka/Pomo, note the sediment surface pitted by numerous pockmarks (*arrows*) in 200 m water depth; Gela basin pockmark field in the Strait of Sicily, ca. 800 m water depth: (**b**) individual pockmark containing chemosymbiotic metazoans and authigenic carbonates, bar = 100 m, (**c**) shell of *Lucinoma kazani* showing external encrustation by authigenically-cemented carbonate mudstone; (**d**) sample showing the benthic assemblage containing many dead shells of lucinid bivalves (*Lucinoma kazani*), and vestimentiferan empty tubes (*Lamelli-brachia* sp.); note also the presence of numerous authigenic carbonate mudstone concretions; (**e**) valve of the chemosymbiotic vesicomyid *Isorropodon perplexum* embedded in authigenically-cemented carbonate mudstone

OCEANA MedNet 2011). At present, candidates for which a protection is sought are the deep sea mud volcanoes cold seeps such as the Olympic and Anaximander fields, the Deep Nile Fan in the Eastern Mediterranean (WWF/IUCN 2004; OCEANA MedNet 2011), and the Jabuka/Pomo Pit (UNEP 2010).

To those, a few more sites are already by now recommendable for protection, i.e.:

1. the Eratosthenes Seamount, site of abundant cold seep macrofauna with chemosymbiotic tubeworms and bivalves, discovered by the E/V *Nautilus* Exploration (Mayer et al. 2011; Fig. 5.4d–f)
2. the recently discovered pockmark field in the Gela Basin (Taviani et al. 2013), containing a unique subfossil to present record of chemosymbiotic macrofauna (Fig. 5.5b–e)
3. the "Horseshoe hill" (Fig. 5.5a) in the central Adriatic next to Jabuka/Pomo Pit (work in progress), to date the shallowest occurrence of *Lucinoma kazani*
4. the Alboran mud volcano field (Hilário et al. 2011; Somoza et al. 2012)
5. the Sicilian coastal volcanic "caves" near Taormina (Palazzi and Villari 2001), site of *Lucinoma speleum*
6. The Marmara submarine deep fault systems inhabited by sibloginids and chemosymbiontic bivalves (Armijo et al. 2005; Zitter et al. 2008; Ritt et al. 2010; Crémière et al. 2012; Tryon et al. 2012)
7. The continental margin of Israel (Akko pockmarks and Palmachin hydrocarbon cold seeps) hosting tubeworms (likely *Lamellibrachia*) and bivalves (Coleman et al. 2012)
8. The Tyrrhenian seamounts (Marsili and Palinuro) hosting *Lamellibrachia* vestimentiferans (Carey et al. 2012; Lott and Zimmerman 2012).

This list will likely increase in the near future as a result of further exploration of the deeper parts of the Mediterranean. Additional candidates for hosting significant chemosynthetic products, including specialized organisms, are the mud volcanoes identified in the Paola Basin, south-eastern Tyrrhenian Sea (Gamberi and Rovere 2010), and the Bari Canyon area, where sand and mud volcanoes have been noticed and exhumed carbonate chimneys are known to occur at depth, although no definitive documentation of chemosynthetic communities do exist at present (Trincardi et al. 2007).

Geosites

Besides the active recent marine chemosynthetic sites, the unique fossil heritage of these habitats strongly deserves protection and proper management. The Mediterranean fossil chemosynthetic world described in a previous chapter is a legacy of universal value, rivalling other sites, such as the West American Pacific margin, New Zealand and Japan (Campbell 2006). The quasi-totality of known past chemosynthesis is related to cold seepage s.l. and outcrops along the Apennine chain and in Sicily. Only a fraction has at present the privilege to be protected by national or regional laws, and rather typically only because they chancy occur in a general protected area. Case-in-points are the Stirone river Pliocene chemoherm in Northern Italy (Stirone River Regional Park, www.parcostirone.it), and some Miocene 'Calcare a lucina' occurrences in the Romagna Apennine (The Karst Park Carnè www.venadelgesso.org/itinerari/brisighella/parco-carne. htm), under the jurisdiction of local park authorities. Clearly, many more fossil cold vent sites should be considered for becoming protected geosites. Their general punctiform occurrence, small volume and poor appreciation of their natural value by the civil community, are progressively causing a quick and irreparable loss of substantial parts of this heritage, already including sites discussed few years ago. Regarding hydrothermal (hot) vents, the unique Cretaceous fossiliferous sites described by Little et al. (1999) from the Troodos ophiolitic complex in Cyprus are of uttermost scientific importance calling for their strict protection.

Acknowledgments This paper is dedicated to the memory of Michael Ivanov, a gentle man, a keen scientist and a friend who immensely contributed to the appreciation of deep-water chemosynthetic situations in the Atlantic Ocean, Mediterranean, and Black seas. This article used images provided by the E/V *Nautilus* Exploration Program, Expeditions NA015, NA010, NA023. Katherine Croff Bell (Chief Scientist, E/V *Nautilus* Exploration Program, Ocean Science and Exploration Center, Narraganset) and Larry Mayer (University of New Hampshire, Durham) generously gave access to data and images from their explorations on deep sea chemosynthetic habitats in the Mediterranean Sea. Carlo Nike Bianchi kindly supplied underwater pictures from shallow water hydrothermal vents and caves with sulphur springs. Captains, Officers, Crew and Scientific Staff of cruises R/V *Urania* Arcadia, Medcor and Decors are thanked for their cooperation. Lorenzo Angeletti (Ismar-CNR, Bologna) helped with figure preparation. Thanks are due to Marina Cunha, Katherine Croff Bell, Bella Galil, Michael Marani, Catherine Pierre and Ann Vanreusel for sharing views and useful information on aspects of deep sea chemosynthetic habitats in the Mediterranean. Paul Dando (Maine Biological Association of the UK, Plymouth), Katherine Croff Bell (Ocean Science and Exploration Center, Narraganset), Stefano Aliani (Ismar-CNR, La Spezia), Carlo Nike Bianchi (DiSTAV University of Genova, Genova), Roberto Danovaro (Polytechnic University of Marche, Ancona), Wolfang-Christian Dullo (Geomar, Kiel), Gregor Eberli (Rosenstiel School of Marine & Atmospheric Science, Miami), Stéphane Hourdez (Station Biologique, Roscoff), Steffen Kiel (Georg-August University, Göttingen), Giuliana Panieri (Ismar-CNR, Bologna), and Antonio Pusceddu (Polytechnic University of Marche, Ancona) kindly revised the ms providing many constructive suggestions. This article is a contribution to the EU Hermione (contract n° 226354) and Coconet (contract n° 287844) programmes, to the PRIN 2009 project "*Condotti carbonatici legati all'espulsione di fluidi ricchi in idrocarburi* (Carbonate conduits linked to hydrocarbons enriched seepages)", and to the RITMARE project. ISMAR-CNR Bologna scientific contribution n. 1763.

References

Aharon P (1994) Geology and biology of modern and ancient submarine hydrocarbon seeps and vents. Geo Mar Lett 14:69–230

Aloisi G, Bouloubassi L, Heijs SK, Pancost R, Pierre C, Sinninghe Damsté JS, Gottschal JC, Forney LJ, Rouchy J-M (2002) CH_4-consuming microorganisms and the formation of carbonate crusts at cold seeps. Earth Planet Sci Lett 203:195–203

Alvisi M, Barbieri F, Colantoni P (1994) Le grotte marine di Capo Palinuro. In: Alvisi M, Colantoni P, Forti P (eds) Grotte Marine d'Italia. Memoire dell'Istituto Italiano di Speleologia, Bologna

Armijo R, Pondard N, Meyer B, Uçarkus G, Mercier de Lépinay B, Malavieille J, Dominguez S, Gustcher M-A, Schmidt S, Beck C, Çagatay N, Çakir Z, Imren C, Eris K, Natalin B, Özalaybey S, Tolun L, Léfevre I, Seeber L, Gasperini L, Rangin C, Emre O, Sarikavak K (2005) Submarine fault scarps in the Sea of Marmara pull-apart (North Anatolian Fault): implications for seismic hazard in Istanbul. Geochem Geophys Geosyst 6, Q06009

Barbieri R, Aharon P, Taviani M, Ricci Lucchi F, Vai GB (2000) Were gas hydrates associated with Miocene-age cold seeps of the Apennines? In: EAGE conference on geology and petroleum geology, St Julians Malta, 1–4 October 2000, Extended Abstracts Book

Barbieri R, Cavalazzi B (2004) Astrobiological implications of microbial morphologies in cold fluid-generated carbonates. Adv Space Res 33:1262–1267

Barbieri R, Cavalazzi B (2005) Microbial fabrics from Neogene cold seep carbonates, Northern Apennine, Italy. Palaeogeogr Palaeoclimatol Palaeoecol 227:143–155

Barbieri R, Panieri G (2004) How are benthic foraminiferal faunas influenced by cold seeps? Evidence from the Miocene of Italy. Palaeogeogr Palaeoclimatol Palaeoecol 204:255–275

Bell KLC, Nomikou P, Carey SN, Stathopoulou E, Polymenakou P, Godelitsas A, Roman C, Parks M (2012) Continued exploration of the Santorini volcanic field and Cretan Basin, Aegean Sea. Oceanography 25(Suppl):30–31

Berti M, Cuzzani MG, Landuzzi A, Taviani M, Aharon P, Vai GB (1994) Hydrocarbon-derived imprints in olistostromes of the Early Serravallian Marnoso-arenacea Formation, Romagna Apennines (Northern Italy). Geo Mar Lett 14:192–200

Bianchi CN, Dando PR, Morri C (2011) Increased diversity of sessile epibenthos at subtidal hydrothermal vents: seven hypotheses based on observations at Milos Island, Aegean Sea. Adv Oceanogr Limnol 2:1–31

Borin S, Brusetti L, Mapelli L, D'Auria G, Brusa T, Marzorati M, Rizzi A, Yakimov M, Marty D, de Lange GJ, Van der Wielen P, Bolhuis H, McGenity TJ, Polymenakou PN, Malinverno E, Giuliano L, Corselli C, Daffonchio D (2009) Sulfur cycling and methanogenesis primarily drive microbial colonization of the highly sulfidic Urania deep hypersaline basin. Proc Natl Acad Sci USA 106:9151–9156

Brissac T, Rodrigues CF, Gros O, Duperron S (2011) Characterization of bacterial symbioses in *Myrtea* sp. (Bivalvia: Lucinidae) and *Thyasira* sp. (Bivalvia: Thyasiridae) from a cold seep in the Eastern Mediterranean. Mar Ecol 32:198–210

Campbell KA (2006) Hydrocarbon seep and hydrothermal vent paleoenvironments and paleontology: past developments and future research directions. Palaeogeogr Palaeoclimatol Palaeoecol 232:362–407

Canganella F, Bianconi G, Kato C, Gonzalez J (2007) Microbial ecology of submerged marine caves and holes characterized by high levels of hydrogen sulphide. Rev Environ Sci Biotechnol 6:61–70

Capozzi R, Guido FL, Oppo D, Gabbianelli G (2012) Methane-derived authigenic carbonates (MDAC) in northern-central Adriatic Sea: relationships between reservoir and methane seepages. Mar Geol 332–334:174–188

Carey S, Sigurdsson H (2007) Exploring submarine Arc volcanoes. Oceanography 20:80–89

Carey S, Bell KLC, Nornikou P, Vougioukalakis G, Roman CN, Cantner K, Bejelou K, Bourbouli M, Fero Martin J (2011) Exploration of the Kolumbo volcanic rift zone. Oceanography 24(Suppl):24–25

Carey S, Bell KLC, Rosi M, Marani M, Nornikou P, Walker SL, Faure K, Kelly J (2012) Submarine volcanoes of the Aeolian Arc, Tyrrhenian Sea. Oceanography 25(Suppl):32–33

Carlier A, Ritt B, Rodrigues CF, Sarrazin J, Olu K, Grall J, Clavier J (2010) Heterogeneous energetic pathways and carbon sources on deep eastern Mediterranean cold seep communities. Mar Biol 157:2545–2565

Cavagna S, Clari P, Martire L (1999) The role of bacteria in the formation of cold seep carbonates: geological evidence from Monferrato (Tertiary, NW Italy). Sediment Geol 126:253–270

Cavanaugh CM, McKiness ZP, Newton ILG (2006) Marine chemosynthetic symbioses. In: Dworkin M, Falkow S, Rosenberg E, Schleifer KH, Stackebrandt E (eds) The prokaryotes, 3rd edn. A handbook on the biology of bacteria: symbiotic associations, biotechnology, applied microbiology. Springer, New York

Charlou JL, Donval JP, Zitter T, Roy N, Foucher JP, Woodside J (2003) Evidence of methane venting and geochemistry of brines in Mud Volcanoes of the Eastern Mediterranean Sea. Deep Sea Res I 50:941–958

Clari P, Gagliardi C, Governa ME, Ricci B, Zuppi GM (1988) I Calcari di Marmorito: una testimonianza di processi diagenetici in presenza di metano. Boll Mus Reg Sc Nat Torino 6:197–216

Clari P, Cavagna S, Martire L, Hunziker J (2004) A Miocene mud volcano and its plumbing system: a chaotic complex revisited (Monferrato, NW Italy). J Sediment Res 74:662–676

Cocito S, Bianchi CN, Morri C, Peirano A (2000) First survey of sessile communities on subtidal rocks in an area with hydrothermal vents: Milos Island, Aegean Sea. Hydrobiologia 426:113–121

Coleman DF, Austin JA, Ben-Avraham Z, Makowsky Y, Tchernov D (2012) Serafloor pockmarks, deepwater corals, and cold seeps along the continental margin of Israel. Oceanography 25(Suppl):40–41

Conti S, Fontana D (1998) Recognition of primary and secondary Miocene lucinid deposits in the Apennine chain. Mem Soc Geol 50:101–131

Conti S, Fontana D (1999) Miocene chemoherms of the northern Apennines Italy. Geology 7:927–930

Conti S, Fontana D (2002) Sediment instability related to fluid venting in Miocene authigenic carbonate deposits of the northern Apennines (Italy). Int J Earth Sci 91:1030–1040

Conti S, Fontana D (2005) Anatomy of seep-carbonates: ancient examples from the Miocene of the northern Apennines (Italy). Palaeogeogr Palaeoclimatol Palaeoecol 227:156–175

Conti S, Gelmini R, Ponzana L (1993) Osservazioni preliminari sui calcari a Lucine dell'Appennino Settentrionale. Atti Soc Nat Mat Modena 124:35–56

Conti S, Gelmini R, Ponzana L, Sighinolfi GP (1996) "Il Calcare a *Lucina pomum*" della successione epiligure dell'Appennino modenese: stratigrafia, sedimentologia e dati geochimici. Miscell Geol Accad Naz Sci Lett Arti Modena Coll Studi 15:105–139

Conti S, Fontana D, Gubertini A, Sighinolfi G, Tateo F, Fioroni C, Fregni P (2004) A multidisciplinary study of middle Miocene seep-carbonates from the northern Apennine foredeep (Italy). Sediment Geol 169:1–19

Conti S, Fontana D, Lucente CC (2008) Authigenic seep-carbonates cementing coarse-grained deposits in a fan-delta depositional system (middle Miocene, Marnoso-arenacea Formation, central Italy). Sedimentology 55:471–486

Cordes EE, Cunha M, Galéron J, Mora C, Olu-Le Roy K, Sibuet M, Van Gaever S, Vanreusel A, Levin LA (2010) The influence of geological, geochemical, and biogenic habitat heterogeneity on seep biodiversity. Mar Ecol 31:51–65

Corselli C, Basso D (1996) First evidence of benthic communities based on chemosynthesis on the Napoli mud volcano (Eastern Mediterranean). Mar Geol 132:227–239

Cosel R, Salas C (2001) Vesicomyidae (Mollusca: Bivalvia) of the genera *Vesicomya, Waisiuconcha, Isorropodon* and *Callogonia* in the eastern Atlantic and the Mediterranean. Sarsia 86:333–366

Crémière A, Pierre C, Blanc-Valleron M-M, Zitter T, Çağatay N, Henry P (2012) Methane-derived authigenic carbonates along the North Anatolian fault system in the Sea of Marmara (Turkey). Deep Sea Res I 66:114–130

Curzi PV (2012) Central Adriatic geology and fishing. Aracne Editrice, Bologna

Dando PR (2010) Biological communities at marine shallow-water vent and seep sites. In: Kiel S (ed) The vent and seep biota, Topics in geobiology 33. Springer Science-Business Media, Dordrecht

Dando PR, Stüben D, Varnavas SP (1999) Hydrothermalism in the Mediterranean Sea. Prog Oceanogr 44:333–367

Danovaro R, Company JB, Corinaldesi C, D'Onghia G, Galil B, Gambi C, Gooday AJ, Lampadariou N, Luna GM, Morigi C, Olu K, Polymenakou P, Ramirez-Llodra E, Sabbatini A, Sardà F, Sibuet M, Tselepides A (2010a) Deep-Sea biodiversity in the Mediterranean Sea: the known, the unknown, and the unknowable. PLoS One 5:e11832

Danovaro R, Dell'Anno A, Pusceddu A, Gambi C, Heiner I, Kristensen RM (2010b) The first metazoa living in permanently anoxic conditions. BMC Biol 8:30

De Biasi AM, Aliani S (2003) Shallow-water hydrothermal vents in the Mediterranean sea: stepping stones for Lessepsian migration? Hydrobiologia 503:37–44

De Lange G, Brumsack HJ (1998) The occurrence of gas hydrates in eastern Mediterranean mud dome structures as indicated by porewater composition. In: Henriet JP, Mienert J (eds) Gas hydrates: relevance to world marine stability and climate change. The Geological Society of London, London

Dominici S, Cioppi E, Danise S, Betocchi U, Gallai G, Tangocci F, Valleri G, Monechi S (2009) Mediterranean fossil whale falls and the adaptation of molluscs to extreme habitats. Geology 37:815–818

Dreier A, Stannek L, Blumenberg M, Taviani M, Sigovini M, Wrede C, Thiel V, Hoppert M (2012) The fingerprint of chemosymbiosis: origin and preservation of isotopic biosignatures in the non-seep bivalve *Loripes lacteus* compared to *Venerupis aurea*. FEMS Microbiol Ecol 81:480–493

Dubilier N, Blazejak A, Ruhland C (2006) Symbioses between bacteria and gutless marine oligochaetes. Prog Mol Subcell Biol 41:251–275

Dubilier N, Bergin C, Lott C (2008) Symbiotic diversity in marine animals: the art of harnessing chemosynthesis. Nat Rev Microbiol 6:725–740

Duperron S (2010) The diversity of deep-sea mussels and their bacterial symbioses. In: Kiel S (ed) The vent and seep biota, Topics in geobiology 33. Springer Science-Business Media, Dordrecht

Duperron S, Fiala-Médioni A, Caprais J-C, Olu K, Sibuet M (2007) Evidence for chemoautotrophic symbiosis in a Mediterranean cold seep clam (Bivalvia: Lucinidae): comparative sequence analysis of bacterial 16S rRNA, APS, reductase and rubisCO genes. FEMS Microbiol Ecol 59:64–70

Duperron S, Halary S, Lorion J, Sibuet M, Gaill F (2008) Unexpected co-occurrence of six bacterial symbionts in the gills of the cold seep mussel *Idas* sp. (Bivalvia: Mytilidae). Environ Microbiol 10:433–445

Duperron S, de Beer D, Zbinden M, Boetius A, Schipani V, Kahil N, Gaill F (2009) Molecular characterization of bacteria associated with the trophosome and the tube of *Lamellibrachia* sp., a siboglinid annelid from cold seeps in the eastern Mediterranean. FEMS Microbiol Ecol 69:395–409

Duperron D, Guezi H, Gaudron SM, Popristova P, Wenzhöfer F, Boetius A (2011) Relative abundances of methane- and sulphur-oxidizing symbionts in the gills of a cold seep mussel and link to their potential energy source. Geobiology 9:481–491

Dupré S, Woodside J, Foucher J-P, de Lange G, Mascle J, Boetius A, Mastalerz V, Stadnitskaia A, Ondréas H, Huguen C, Harmeniégnies F, Gontharet S, Loncke L, Deville E, Niemann H, Omoregie E, Olu-Le Roy K, Fiala-Médioni A, Dählmann A, Caprais J-C, Prinzhofer A, Sibuet M, Pierre C, Damsté JS, the NAUTINIL Scientific Party (2007) Seafloor geological studies above active gas chimneys off Egypt (Central Nile Deep Sea Fan). Deep Sea Res I 54:1146–1172

Felbeck H, Childress JJ, Somero GN (1981) Calvin-Benson cycle and sulphide oxidation enzymes in animals from sulphide-rich habitats. Nature 293:291–293

Fisher CR (1990) Chemoautotrophic and methanotrophic symbiosis in marine invertebrates. CRC Crit Rev Aquat Sci 2:399–436

Foucher J-P, Westbrook GK, Boetius A, Ceramicola S, Dupré S, Mascle J, Mienert J, Pfannkuche O, Pierre C, Praeg D (2009) Structures and drivers of cold seep ecosystems. Oceanography 22:58–74

Gamberi F, Rovere M (2010) Mud diapirs, mud volcanoes and fluid flow in the rear of the Calabrian Arc Orogenic Wedge (southeastern Tyrrhenian sea). Basin Res 22:452–464

Gamenick L, Abbiati M, Giere O (1998) Field distribution and sulphide tolerance of *Capitella capitata* (Annelida: Polychaeta) around shallow water hydrothermal vents off Milos (Aegean Sea). A new sibling species? Mar Biol 130:447–453

Gaudron SM, Pradillon F, Pailleret M, Duperron S, Le Bris N, Galil F (2010) Colonization of organic substrates deployed in deep-sea reducing habitats by symbiotic species and associate fauna. Mar Environ Res 70:1–12

Gaudron SM, Demoyencourt E, Duperron S (2012) Reproductive traits of the cold-seep symbiotic mussel *Idas modiolaeformis*: gametogenesis and larval biology. Biol Bull 222:6–16

Gemmell JB, Petersen S, Monecke T, Hannington MD, Lackschewitz K, Augustin N, Gibson H, Perrin K, Sharpe R, Simpson K (2009) High to intermediate sufidation, shallow marine, sulfide mineralisation in the South-eastern Tyrrhenian Sea, Italy. In: Williams PJ et al (eds) Smart science for exploration and mining. Proceedings of the 10th biennial SGA meet, Townsville Australi, 17–20 August 2009, pp 438–440

German CR, Ramirez-Llodra E, Baker MC, Tyler PA, the ChEss Scientific Steering Committee (2011) Deep-water chemosynthetic ecosystem research during the census of marine life decade and beyond: a proposed deep-ocean road map. PLoS One 6:e23259

Gofas S (2004) Geographical differentiation in *Clelandella* (Gastropoda: Trochidae) in the northeastern Atlantic. J Mollusc Stud 71:133–144

Gontharet S, Pierre C, Blanc-Valleron M-M, Rouchy JM, Fouquet Y, Bayon G, Foucher JP, Woodside J, Mascle J, The Nautinil Scientific Party (2007) Nature and origin of diagenetic carbonate crusts and concretions from mud volcanoes and pockmarks of the Nile deep-sea fan (eastern Mediterranean Sea). Deep Sea Res II 54:292–1311

Grünke S, Felden J, Lichtschlag A, Girnth A-C, De Beer D, Wenzhöfer F, Boetius A (2011) Niche differentation among mat-forming, sulfide-oxidizing bacteria at cold seeps of the Nile Deep Sea Fan (Eastern Mediterranean Sea). Geobiology 9:330–348

Hagerman L (1998) Physiological flexibility; a necessity for life in anoxic and sulphidic habitats. Hydrobiologia 375(376):241–254

Higgs ND, Little CTS, Glover AG, Dahlgren TG, Smith CR, Dominici S (2011) Evidence of Osedax worm borings in Pliocene (3Ma) whale bone from the Mediterranean. Hist Biol 24:269–277

Hilário A, Comas MC, Azevedo L, Pinheiro L, Ivanov MK, Cunha MR (2011) First record of a Vestimentifera (Polychaeta: Siboglinidae) from chemosynthetic habitats in the western Mediterranean Sea-Biogeographical implications and future exploration. Deep Sea Res I 58:200–207

Himmler T, Brinkmann F, Bohrmann G, Peckmann J (2011) Corrosion patterns of seep-carbonates from the eastern Mediterranean Sea. Terra Nova 23:206–212

Hourdez S, Lallier FH (2007) Adaptations to hypoxia in hydrothermal-vent and cold-seep invertebrates. Rev Environ Sci Biotechnol 6:143–159

Hughes DJ, Crawford M (2006) A new record of the vestimentiferan *Lamellibrachia* sp. (Polychaeta: Siboglinidae) from a deep shipwreck in the eastern Mediterranean. Mar Biodivers Rec 1:1–3

Janssen R, Krylova EM (2012) Bivalves of the family Vesicomyidae from the Neogene Mediterranean basin (Bivalvia: Vesicomyidae). Arch Mollusk 141:87–113

Judd AG, Hovland M (2007) Seabed fluid flow. Impact on geology, biology, and the marine environment. Cambridge University Press, Cambridge

Kiel S (2010) The fossil record of vent and seep molluscs. In: Kiel S (ed) The vent and seep biota, Topics in geobiology 33. Springer Science-Business Media, Dordrecht

Kiel S, Little CST (2006) Cold-Seep molluscs are older than the general marine mollusc fauna. Science 313:1429–1431

Kleiner M, Wentrup C, Lott C, Teeling H, Wetzel S, Young J, Chang Y-J, Shah S, VerBerkmoes NC, Zarzycki J, Fuchs G, Marker S, Hempe K, Voigt B, Becher D, Liebeke M, Lalk M, Albrecht D, Hecker M, Schweder T, Dubilier N (2012) Metaproteomics of a gutless marine worm and its symbiotic microbial community reveal unusual pathways for carbon and energy use. Proc Natl Acad Sci USA 109:E1173–E1182

La Cono V, Smedile F, Bortoluzzi G, Arcadi E, Maimone G, Messina E, Borghini M, Oliveri E, Mazzola S, L'Haridon S, Toffin L, Genovese L, Ferrer M, Giuliano L, Golyshin PN, Yakimov MM (2011) Unveiling microbial life in new deep-sea hypersaline Lake Thetis. Part I: Prokaryotes and environmental settings. Environ Microbiol 13:2250–2268

Levin LA (2010) Anaerobic metazoans: no longer an oxymoron. BMC Biol 8:31

Little CTS (2002) The fossil record of hydrothermal vent communities. Cah Biol Mar 43:313–316

Little CTS, Cann JR, Herrington RJ, Morisseau M (1999) Late Cretaceous hydrothermal vent communities from the Troodos ophiolite, Cyprus. Geology 27:1027–1030

Lott C, Zimmerman J (2012) Do Mediterranean tubeworms like it hot? First discovery of tubeworms at hydrothermal vents outside of the Pacific Ocean. Max-Plank-Institut für Mikrobiologie. http://www.mpi-bremen.de/en/Recent_News.html. Accessed 28 Sept 2012

Lucente C, Taviani M (2005) Chemosynthetic communities as fingerprints of submarine sliding linked fluid seepage, Miocene deep-sea strata of the Romagna Apennines, Italy. Palaeogeogr Palaeoclimatol Palaeoecol 227:176–190

Lykousis V, Alexandri S, Woodside J, Nomikou P, Perissoratis C, Sakellariou D, de Lange G, Dählmann A, Casas D, Rousakis G, Ballas D, Ioakim C (2004) New evidence of extensive active mud volcanism in the Anaximander mountains (Eastern Mediterranean): the "ATHINA" mud volcano. Environ Geol 46:1030–1037

Manini E, Luna GM, Corinaldesi C, Zeppilli D, Bortoluzzi G, Caramanna G, Raffa F, Danovaro R (2008) Prokaryote diversity and virus abundance in shallow hydrothermal vents of the Mediterranean Sea (Panarea Island) and the Pacific Ocean (north Sulawesi-Indonesia). Microbiol Ecol 55:626–639

Marie D, Zhu F, Balagué V, Ras J, Vaulot D (2006) Eukaryotic picoplankton communities of the Mediterranean Sea in summer assessed by molecular approaches (DGGE, TTGE, QPCR). FEMS Microbiol Ecol 55:403–415

Martire L, Natalicchio M, Petrea CC, Cavagna S, Clari P, Dela Pierre F (2010) Petrographic evidence of the past occurrence of gas hydrates in the Tertiary Piedmont Basin (NW Italy). Geo Mar Lett 30:461–476

Mattison RG, Abbiati M, Dando PR, Fitzsimons MF, Pratt SM, Southward AJ, Southward EC (1998) Chemoautotrophic microbial mats in submarine caves with hydrothermal sulphidic springs at Cape Palinuro, Italy. Microb Ecol 35:58–71

Maugeri TL, Bianconi G, Canganella F, Danovaro R, Gugliandolo C, Italiano F, Lentini V, Manini E, Nicolaus B (2010) Shallow hydrothermal vents in the southern Tyrrhenian Sea. Chem Ecol 26:285–298

Mayer L, Croff Bell KL, Ballard R, Nicolaides S, Konnaris K, Hall J, Tibor G, Austin JA jr, Shank T (2011) Discovery of sinkholes and seeps on Eratosthenes Seamount. Oceanography 24(Suppl):28–29

McMullin ER, Bergquist DC, Fisher CR (2000) Metazoans in extreme environments: adaptations of hydrothermal vent and hydrocarbon seep Fauna. Gravit Space Biol Bull 13:13–24

Mentel M, Martin W (2010) Anaerobic animals from an ancient, anoxic ecological niche. BMC Biol 8:32

Michaelis W, Seifert R, Nauhaus K, Treude T, Thiel V, Blumenberg M, Knittel K, Gieske A, Peternecht K, Pape T, Boetius A, Amann R, Jorgensen BB, Widdel F, Peckmann JR, Pimenov NV, Gulin MB (2002) Microbial reefs in the Black Sea fuelled by anaerobic oxidation of methane. Science 297:1013–1015

Morri C, Cinelli F, Bianchi CN (1994) Sessile fauna gigantism in a submarine cave with sulphur springs. Cave Diving 6:4–9

Morri C, Bianchi CN, Cocito S, Peirano A, De Biase AM, Aliani S, Pansini M, Boyer M, Ferdeghini F, Pestarino M, Dando P (1999) Biodiversity of marine sessile epifauna at an Aegean island subject to hydrothermal activity: Milos, eastern Mediterranean Sea. Mar Biol 135:729–739

OCEANA MedNet (2011) MPA network proposal for the Mediterranean Sea OCEANA. http://oceana.org/sites/default/files/reports/OCEANA_MEDNet_ING_16012012_0.pdf. Accessed 11 Nov 2012

Olu-Le Roy K, Sibuet M, Fiala-Médioni A, Gofas S, Salas C, Mariotti A, Foucher JP, Woodside J (2004) Cold seep communities in the deep eastern Mediterranean Sea: composition, symbiosis and spatial distribution on mud volcanoes. Deep Sea Res I 51:1915–1936

Ott JA, Bright M, Bulgheresi S (2004) Symbiosis between marine nematodes and sulphur-oxidizing chemoautotrophic bacteria. Symbiosis 36:103–126

Palazzi S, Villari A (2001) Molluschi e brachiopodi delle grotte sottomarine del Taorminese. La Conchiglia 32(Suppl a 297):1–56

Panieri G (2003) Benthic foraminifera response to methane release in an Adriatic Sea pockmark. Riv It Pal Strat 109:549–562

Panieri G (2006) Foraminiferal response to an active methane seep environment: a case study from the Adriatic Sea. Mar Micropaleontol 61:116–130

Panieri G, Gamberi F, Marani M, Barbieri R (2005) Benthic foraminifera from a recent, shallow-water hydrothermal environment in the Aeolian Arc (Tyrrhenian Sea). Mar Geol 218:207–229

Panieri G, Camerlenghi A, Conti S, Pini G-A, Cacho I (2009) Methane seepages recorded in benthic foraminifera from Miocene seep carbonates, Northern Apennines (Italy). Palaeogeogr Palaeoclimatol Palaeoecol 284:271–282

Paull CK, Hecker B, Commeau R, Freemanlynde RP, Neumann C, Corso WP, Golubic S, Hook JE, Sikes E, Curray J (1984) Biological communities at the Florida Escarpment resemble hydrothermal vent taxa. Science 226:965–967

Peckmann J, Thiel V, Michaelis W, Clari P, Gaillard C, Martire L, Reitner J (1999) Cold seep deposits of Beauvoisin (Oxfordian; southeastern France) and Marmorito (Miocene; northern Italy): microbially induced authigenic carbonates. Int J Earth Sci 88:60–75

Peckmann J, Thiel V, Reitner J, Taviani M, Aharon P, Michaelis W (2004) A microbial mat of a large sulphur bacterium preserved in a Miocene methane-seep. Geomicrob J 21:247–255

Perissoratis C, Ioakim C, Alexandri S, Woodside J, Nomikou P, Dahlmann A, Casas S, Heeschen K, Amman H, Rousakis G, Lykousis V (2011) Thessaloniki mud volcano, the shallowest gas hydrate-bearing mud volcano in the Anaximander Mountains, Eastern Mediterranean. J Geol Res 2011:1–11

Pierre C, Rouchy J-M (2004) Isotopic compositions of diagenetic dolomites in the Tortonian marls of the western Mediterranean margins: evidence of past gas hydrate formation and dissociation. Chem Ecol 205:469–484

Pierre C, Rouchy J-M, Blanc-Valleron M-M (2002) Gas hydrate dissociation in the Lorca Basin (SE Spain) during the Mediterranean Messinian salinity crisis. Sediment Geol 147:247–252

Reitner J, Peckmann J, Reimer A, Schumann G, Thiel V (2005) Methane-derived carbonate buildups and associated microbial communities at cold seeps on the lower Crimean shelf (Black Sea). Facies 51:66–79

Ricci Lucchi F, Vai GB (1994) A stratigraphic and tectonofacies framework of the "calcari a Lucina" in the Apennine Chain, Italy. Geo Mar Lett 14:210–218

Rinke C, Schmitz-Esser S, Stoeker K, Nussbaumer AD, Molnár DA, Vanura K, Wagner M, Horn M, Ott JA, Bright M (2006) "*Candidatus* Thiobios zoothamnicoli", an ectosymbiotic bacterium covering the giant marine ciliate *Zoothamnium niveum*. Appl Environ Microbiol 72:2014–2021

Rinke C, Lee R, Katz S, Bright M (2007) The effects of sulphide on growth and behaviour of the thotrophic *Zoothamnium niveum* symbiosis. Proc R Soc B 274:2259–2269

Ritt B, Sarrazin J, Caprais J-C, Noël P, Gauthier O, Pierre C, Henry P, Desbruyères D (2010) First insights into the structure and environmental setting of cold-seep communities in the Marmara Sea. Deep Sea Res I 57:1120–1136

Rodrigues CF, Duperron S (2011) Distinct symbiont lineages in three thyasirid species (Bivalvia:Thyasiridae) from the eastern Atlantic and Mediterranean Sea. Naturwissen 98:281–287

Rodrigues C, Cunha M, Olu K, Duperron S (2012) The smaller vesicomyid bivalves in the genus *Isorropodon* (Bivalvia, Vesicomyidae, Pliocardiinae) also harbour chemoautotrophic symbionts. Symbiosis 56:129–137

Roeselers G, Newton ILG (2012) On the evolutionary ecology of symbioses between chemosynthetic bacteria and bivalves. Appl Microbiol Biotechnol 94:1–10

Russo A, Pugliese N, Serventi P (2012) Miocene ostracodes of cold seep settings from northern Apennines (Italy). Rev Micropal 55:29–38

Salas C, Woodside J (2002) *Lucinoma kazani* n.sp. (Mollusca: Bivalvia): evidence of a living benthic community associated with a cold seep in the eastern Mediterranean Sea. Deep Sea Res I Oceanogr Res Pap 50:1391–1409

Salomidi M, Katsanevakis S, Borja A, Braeckman U, Damalas D, Galparsoro I, Mifsud R, Mirto S, Pascual M, Pipitone C, Rabaut M, Todorova V, Vassilopoulou V, Fernández V (2012) Assessment of goods and services, vulnerability, and conservation status of European seabed biotopes: a stepping stone towards ecosystem-based marine spatial management. Mediterr Mar Sci 13(1):49–88

Shank TM, Herrera S, Cho W, Roman CN, Croff Bell KL (2011) Exploration of the Anaximander mud volcanoes. Oceanography 24(Suppl):22–23

Sibuet M, Olu-Le Roy K (1998) Biogeography, biodiversity and fluid dependence of deep-sea cold-seep communities at active and passive margins. Deep Sea Res II 45:517–567

Sibuet M, Olu-Le Roy K (2002) Cold seep communities on continental margins: structure and quantitative distribution relative to geological and fluid venting patterns. In: Wefer G, Billett D, Hebbeln D, Jørgensen BB, Schlüter M, Van Weering T (eds) Ocean margin systems. Springer, Berlin

Smriglio C, Mariottini P (2002) *Lurifax vitreus* Warèn & Bouchet, 2001 (Gastropoda, Orbitestellidae), first report from Mediterranean Sea. Boll Malacol 38:45–47

Smriglio C, Mariottini P (2006) Molluschi di singolari ecosistemi. Scienza online. http://www.scienzaonline.com/scienze-naturali/-279-malacologia/2917-molluschi-di-singolari-ecosistemi.html. Accessed 28 Sept 2012

Somoza L, Medialdea T, León R, Ercilla G, Vázquez JT, Farran M, Hernández-Molina J, González J, Juan C, Fernández-Puga MC (2012) Structure of mud volcano systems and pockmarks in the region of the Ceuta Contourite Depositional System (Western Alborán Sea). Mar Geol 332–334:4–26

Southward AJ, Kennicutt MC, Herrea-Alcalà J, Abbiati M, Airoldi L, Cinelli F, Bianchi CN, Morri C, Southward EC (2009) On the biology of submarine caves with sulphur springs: appraisal of $^{13}C/^{12}C$ ratios as a guide to trophic relations. J Mar Biol Assoc UK 76:265–285

Southward EC, Andersen AC, Hourdez A (2011) *Lamellibrachia anaximandri* n.sp., a new vestimentiferan tubeworm (Annelida) from the Mediterranean, with notes on frenulate tubeworms from the same habitat. Zoosystema 33:245–279

Speight M, Henderson P (2010) Marine ecology: concepts and applications. Wiley-Blackwell, Chichester

Stewart FJ, Cavanaugh CM (2006) Bacterial endosymbioses in *Solemya* (Mollusca: Bivalvia)-model systems for studies of symbiont-host adaptation. Antonie Van Leeuwenhoek 90:343–360

Stock A, Breiner H-W, Pachiadaki M, Edgcomb V, Filker S, La Cono V, Yakimov MM, Stoeck T (2012) Microbial eukaryote life in the new hypersaline deep-sea basin Thetis. Extremophiles 16:21–34

Sturany R (1896) Zoologische Ergebnisse VII Mollusken I (Prosobranchier und Opistobranchier: Scaphopoden; Lamellibranchier) gesammelt von SM Schiff "Pola" 1890–1894. Denkschr Akad Wissensch Wien Math-Naturwiss K 63:1–6

Taviani M (1994) The "calcari a Lucina" macrofauna reconsidered: deep-sea faunal oases from Miocene-age cold vents in the Romagna Apennine, Italy. Geo Mar Lett 14:185–191

Taviani M (1996) La scoperta delle oasi di mare profondo nel Miocene italiano. Paleocronache 1:7–14

Taviani M (2001) Fluid venting and associated processes. In: Vai GB, Martini PI (eds) Anatomy of an orogen: the Apennines and adjacent Mediterranean basins. Kluwer Academic Publishers, Dordrecht

Taviani M (2004) Shaping the biogeography of the Mediterranean basin: one geologist's perspective. Biogeography 24:15–22

Taviani M (2011) The deep-sea chemoautotroph microbial world as experienced by the Mediterranean metazoans through time. Lect Notes Earth Sci 131:277–295

Taviani M, Angeletti L, Ceregato A, Foglini F, Froglia C, Trincardi F (2013) The Gela Basin pockmark field in the strait of Sicily (Mediterranean Sea): chemosymbiotic faunal and carbonate signatures of postglacial to modern cold seepage. Biogeosciences 10:4653–4671. doi:10.5194/bg-10-4653-2013

Taviani M, Ceregato A (2009) Vite primordiali negli abissi. Darwin 32:82–87

Taylor JD, Glover EA (2010) Chemosymbiotic bivalves. In: Kiel S (ed) The vent and seep biota, Topics in geobiology 33. Springer Science-Business Media, Dordrecht

Terzi C (1993) The "Calcari a Lucina" (Lucina limestone) of the Tuscan-Romagna Apennines as indicators of Miocene cold seep activity (northern Apennines, Italy). G Geol 55:71–81

Terzi C, Sami M (2007) Antiche "oasi di mare profondo": l'ex cava di Pietralunga e i calcari a Lucina del Brisighellese. In: Contarini E, Sami M (eds) Da un mare di pietra, le pietre per il mare: l'ex cava di Pietralunga. Carta Bianca Editore, Ravenna

Thiermann F, Akoumianaki I, Hughes JA, Giere O (1997) Benthic fauna of a shallow-water gaseohydrothermal in the Aegean Sea (Milos, Greece). Mar Biol 128:149–159

Thiermann F, Vismann B, Giere O (2000) Sulphide tolerance of the marine nematode *Oncholaimus campylocercoides* – a result of internal sulphur formation? Mar Ecol Prog Ser 193:251–259

Treude T, Knittel K, Blumemberg M, Seifert R, Boetius A (2005) Subsurface microbial methanotrophic mats in the Black Sea. Appl Environ Microbiol 71:6375–6378

Treude T, Kiel S, Linke P, Peckmann J, Goedert JL (2011) Elasmobranch egg capsules associated with modern and ancient cold seeps: a nursery for marine deep-water predators. Elasmobranch egg capsules associated with modern and ancient cold seeps: a nursery for marine deep-water predators. Mar Ecol Prog Ser 437:175–181

Trincardi F, Foglini F, Verdicchio G, Asioli A, Correggiari A, Minisini D, Piva A, Remia A, Ridente D, Taviani M (2007) The impact of cascading currents on the Bari Canyon System, SW-Adriatic Margin (Central Mediterranean). Mar Geol 246:208–230

Tryon MD, Henry P, Hilton DR (2012) Quantifying submarine fluid seep activity along the North Anatolian Fault Zone in the Sea of Marmara. Mar Geol 315–318:15–28

UNEP (2007) Report on the expert workshop on ecological criteria and biogeographic classification system for marine areas in need of protection. UNEP, Nairobi

UNEP (2010) The Mediterranean Sea Biodiversity: state of the ecosystems, pressures, impacts and future priorities. UNEP, Nairobi

Vai GB, Taviani M, Conti S, Aharon P (eds) (1997) Cold-E-Vent hydrocarbon seepage and chemosynthesis in tethyan relic basins: products, processes and causes. In: An international field workshop to be held in Bologna and nearby Apennines, 23–26 June 1997, Abstracts with Program, Bologna

Van Dover CL (2000) The ecology of deep-sea hydrothermal vents. Princeton University Press, Princeton

Van Dover CL, Smith CR, Ardron J, Arnaud S, Beaudoin Y, Bezaury J et al (2011) Environmental management of deep-sea chemosynthetic ecosystems: justification of and considerations for a spatially based approach. International Seabed Authority. http://hal.archives-ouvertes.fr/docs/00/66/78/38/PDF/ISA_Tech_Report_9.pdf. Accessed 28 Sept 2012

Van Dover CL, Smith CR, Ardron J, Dunn D, Gjerde K, Levin L, Smith S, the Dinard Workshop Contributors (2012) Designating networks of chemosynthetic ecosystem reserves in the deep sea. Mar Policy 36:378–381

Vanreusel A, Andersen AC, Boetius A, Connelly D, Cunha MR, Decker C, Hilario A, Kormas KA, Maignien L, Olu K, Pachiadaki M, Ritt B, Rodrigues C, Sarrazin J, Tyler P, Van Gaever S, Vanneste H (2009) Biodiversity of cold seep ecosystems along the European margins. Oceanography 22:110–127

Venturini S, Selmo E, Tarlao A, Tunis G (1998) Fossiliferous methanogenetic limestones in the Eocene flysch of Istria (Croatia). G Geol 60:219–234

Vinn O, Kupriyanova EK, Kiel S (2012a) Systematics of serpulid tubeworms (Annelida, Polychaeta) from Cretaceous and Cenozoic

hydrocarbon-seep deposits in North America and Europe. N Jb Geol Paläont Abh 266:315–325
Vinn O, Kupriyanova EK, Kiel S (2012b) Serpulids (Annelida, Polychaeta) at Cretaceous to modern hydrocarbon seeps: ecologic and evolutionary patterns. Palaeoeogr Palaeoclimatol Palaeoecol. doi:org/10.1016/j.palaeo.2012.08.003
Vrijenhoek RC (2010) Genetics and evolution of deep-sea chemosynthetic bacteria and their invertebrate hosts. In: Kiel S (ed) The vent and seep biota, Topics in geobiology 33. Springer Science-Business Media, Dordrecht
Werne JP, Haese RR, Zitter T, Aloisi G, Bouloubassi I, Heijfs S, Fiala-Médioni A, Pancost RD, Sinninghe Damsté JS, de Lange G, Forney LJ, Gottschal JC, Foucher JP, Mascle J, Woodside J (2004) Life at cold seeps: a synthesis of biogeochemical and ecological data from Kazan mud volcano, eastern Mediterranean Sea. Chem Geol 205:367–390
Woodside J, Ivanov MK, Limonov AF (1998) Shallow gas and gas hydrates in the Anaximander Mountains region, Eastern Mediterranean Sea. In: Henriet JP, Mienert J (eds) Gas hydrates: relevance to world margin stability and climate change. The Geological Society of London, London
WWF/IUCN (2004) The Mediterranean deep-sea ecosystems: an overview of their diversity, structure, functioning and anthropogenic impacts, with a proposal for conservation. IUCN, Malaga
Zeppilli D, Canals M, Danovaro R (2011) Pockmarks enhance deep-sea benthic biodiversity: a case study in the western Mediterranean Sea. Divers Distrib 18:832–846
Zitter TAC, Henry P, Aloisi G, Delaygue G, Çagatay MN, Mercier de Lepinay B, Al-Samir M, Fornacciari F, Tesmer M, Pekdeger A, Wallmann K, Lericolais G (2008) Cold seeps along the main Marmara Fault in the Sea of Marmara (Turkey). Deep Sea Res I 55:552–557

Part II

Ecology: Taxa and Trophic Levels

6 Microbial Components

Marta Estrada and Dolors Vaqué

Abstract

This chapter presents an overview of the diversity, distribution and ecology of major groups of microbial plankton in the Mediterranean Sea, including phytoplankton, viruses, heterotrophic prokaryotes and flagellates, and ciliates. Some protists with hard structures like diatoms, thecate dinoflagellates, coccolithophorids and tintinnids have been relatively well studied from a morphological point of view, but in general microbial diversity is poorly known, in particular with respect to prokaryotes and the smallest eukaryotes. This situation is rapidly changing, in a large part due to the incorporation of molecular techniques. The general oligotrophy of the Mediterranean, which increases from west to east, is reflected in a strong contribution of the picoplankton and the microbial food web. However, a variety of nutrient-enrichment mechanisms, including winter mixing, mesoscale hydrographic structures and land runoff, which operate at various spatio-temporal scales, may enhance primary production and result in the intermittent dominance of diatoms and the herbivorous food web. During the stratification period, a deep chlorophyll maximum is a general feature throughout the basin and plays a substantial role in the fertility of the Mediterranean.

Keywords

"Phytoplankton mandala" • Bacteriophages • Blooming taxa • Chemotaxonomic and molecular techniques • Chlorophyll maximum, Colonial diatoms • Cryptophytes • Cyanobacterial genera • Dinophyceae • Eutrophication • Free-living pelagic bacterial cells • Global change • Heterotrophic bacteria • Large-scale phytoplankton community • Long-term trends • Mediterranean photic zone • Microbial diversity • Microbial DNA • Microbial eukaryotes • Microbial food web • Microbial plankton • Nano-and picoplanktonic flagellates • Pelagophyceae • Phytoplankton biomass • Phytoplankton blooms • Pigment chemotaxonomic approach • Plankton ecosystem dynamics • Plankton • Planktonic ciliates • Planktonic community • Prasinophyceae • Primary production • Prymnesiophytes • Seasonal succession of phytoplankton • Silicoflagellates • The mediterranean • Toxic benthic dinoflagellate • Viral community • Viral effects on bacterial biomass and production • Viral infection • Viruses • Winter phytoplankton bloom

M. Estrada (✉) • D. Vaqué
Institut de Ciències del Mar, Consejo Superior de Investigaciones Científicas, Barcelona, Spain
e-mail: marta@icm.csic.es

S. Goffredo and Z. Dubinsky (eds.), *The Mediterranean Sea: Its history and present challenges*, DOI 10.1007/978-94-007-6704-1_6, © Springer Science+Business Media Dordrecht 2014

Introduction

In the Mediterranean Sea, as in other marine ecosystems, microbial components of the plankton are major agents in biogeochemical cycles and account for most of the biodiversity. In spite of their importance, many groups of marine microbes are still poorly known, although the situation is rapidly changing thanks to the introduction of molecular techniques. The structure and dynamics of Mediterranean plankton has been considered in a number of collective volumes such as Margalef (1985), Moraitou-Apostolopoulou and Kiortsis (1985) and Minas and Nival (1988). Recently, Siokou-Frangou et al. (2010) published an updated review of studies conducted during the last 25 years in offshore epipelagic waters and Coll et al. (2010) presented a synthetic account of Mediterranean marine biodiversity. This chapter addresses the biodiversity and ecology of the major microbial groups in the Mediterranean, as part of the ecological background for a book on history and present challenges. In the following section ("Main groups and trophic roles of microbial plankton"), we present a brief account of the taxonomic affiliation and functional roles of microbial groups. The main fertilization mechanisms in the Mediterranean and the general patterns of temporal distribution and vertical organization of phytoplankton biomass are considered in sections "Fertilization mechanisms in the Mediterranean" and "Overview of diversity of the main microbial groups". The next section ("Phytoplankton"), devoted to phytoplankton, presents an overview of the taxonomic diversity of the main groups, a brief review of harmful algal events in the Mediterranean and some insights obtained from the study of long-term series. Section "Viruses and heterotrophic microbes" deals with the distribution and diversity of viruses, heterotrophic prokaryotes and flagellates, and ciliates. Finally, section "Microbial and classical food webs. The example of the Catalan Sea" deals with the structure of the planktonic food webs and describes a case study in the NW Mediterranean.

Main Groups and Trophic Roles of Microbial Plankton

Until about three decades ago, the classical view of the planktonic food web featured phytoplankton of sizes above a few micrometers as primary producers, microzooplankton such as ciliates, as herbivores, and zooplankton (mainly copepods), as a bridge between the former trophic levels and fishes. Phytoplankton, roughly defined as the autotrophic component of the planktonic food web, was considered as mostly composed of eukaryotic microalgae. However, many exceptions had long been recognized, as some groups classically considered as phytoplankton, like the dinoflagellates, comprise heterotrophic taxa that prey on other organisms, and many autotrophic forms are also capable of phagotrophy and osmotrophy. Phototrophic organisms of sizes below a few micrometers had been described, but the available techniques did not allow their reliable enumeration. The occurrence of bacteria, considered mainly as decomposers, was known, but the methods used at the time, mostly based on culture techniques derived from medical microbiology, allowed only the detection of a small proportion of them. This general picture changed around the mid 1970s with the incorporation of DNA staining and epifluorescence microscopy, which allowed a reliable observation and counting of prokaryotic and eukaryotic cells <2–3 μm in size, known collectively as "picoplankton", an operational category that complements those of "nanoplankton" (>2 μm and >20 μm) and "microplankton" (>20 μm and <200 μm). These techniques and the later addition of automatic counting by means of flow cytometry demonstrated the abundance and ubiquity not only of heterotrophic bacteria, but also of autotrophic cyanobacteria of the genera *Synechococcus* and *Prochlorococcus*, and of auto- and heterotrophic picoeukaryotes. The use of radioactive markers helped to point out the magnitude of the bacterial assimilation of organic matter originated from phytoplankton or from allochthonous sources, and the role of picoplankton as food of small heterotrophic flagellates that could be in turn consumed by other microheterotrophs and by zooplankton (Ducklow et al. 1986). The pathway in which dissolved organic matter was assimilated by bacteria and reintroduced into the marine food web via heterotrophic pico- or nanoflagellates and microzooplankton received the name of "microbial loop" and has been shown to be particularly important in the functioning of oligotrophic aquatic systems such as the Mediterranean. More recently, the application of molecular techniques to the study of microbial DNA has led to the detection of an immense amount of previously unknown biodiversity. Among other surprises, came the realization that Archaea, a group of prokaryotes previously known from extreme environments, and now considered together with bacteria and eukaryotes as one of the three main branches of life, were widespread in the pelagic marine domain (Karner et al. 2001). Given the rapid changes in taxonomic knowledge, there is not yet a generally accepted classification system for the different taxa of planktonic microbes within the so-called tree of life of living organisms (Pennisi 2003), although there is increasing agreement on the basic features. Work carried out in the last decades has uncovered links between certain groups and particular metabolic activities. For example, 16S rDNA and nitrogenase genes have been targeted to show the presence of several lineages of unicellular diazotrophic cyanobacteria (Zehr et al. 2003) and other genetic procedures have allowed the detection of heterotrophic bacteria that can enhance their energy sources by means of aerobic anoxygenic photosynthesis (Jiao et al. 2007) or light-driven proton pumps based on proteorhodopsin (DeLong and Béjà 2010). A new phycobilin-containing group (the picobiliphytes) was described by Not et al. (2007); they were initially described as algae, but could be heterotrophs (Massana 2011). Another recent development has been the proof that viruses are the most abundant and ubiquitous entities of aquatic communities (Bergh et al. 1989; Weinbauer 2004; Suttle 2007, and references therein). The net effect of viruses in the pelagic food web, described as 'the viral shunt' by Fuhrman (1999), consists of transforming the particulate organic matter of the host into more viruses, and returning biomass in the form of dissolved and colloidal organic matter.

Fertilization Mechanisms in the Mediterranean

The Mediterranean is the largest quasi-enclosed sea on Earth (Siokou-Frangou et al. 2010) and functions as a negative estuary because of the excess of evaporation over precipitation plus runoff. However, the Mediterranean has also been considered as one of the most complex marine environments of the planet (d'Ortenzio and Ribera d'Alcalà 2009) due to the variety of physical processes that operate in it, including deep water formation, thermohaline circulation, sub-basin gyres and mesoscale activity. The exchange of water at Gibraltar, where denser and relatively nutrient-rich Mediterranean water flows out at depth and fresher Atlantic water enters at surface has contributed to the low nutrient content of deep mediterranean waters as compared to other world oceans (McGill 1965). In turn, the eastern basin presents an anti-estuarine exchange with respect to the western basin. These relationships originate an eastward transport of surface waters and a westward transport of intermediate waters (Pinardi and Masetti 2000), and underlie the general Mediterranean gradient of increased oligotrophy towards the east. The wind distribution, and the morphology and characteristics of the basin contribute to a water circulation pattern with a dominance of cyclonic gyres in the north and anticyclonic gyres in the south of the basin. Deep convection can occur in winter in the Gulf of Lions and parts of the Eastern Mediterranean (EM), such as the Aegean Sea (MEDOC-Group 1970; Zervakis et al. 2004).

The main mechanisms of nutrient input in the Mediterranean photic zone include winter mixing, coastal upwelling and other mesoscale features, and land runoff. Eolic inputs are also significant for certain elements. However, the efficacy of some fertilization mechanisms such as mixing or upwelling is limited by the relatively low nutrient content of the deep Mediterranean waters. Winter mixing occurs throughout the Mediterranean, but the mixing depth varies and with it the degree of nutrient enrichment of the surface layers (Estrada et al. 1985). Wind-driven upwelling occurs in some areas with favourable coastal profiles, such as the coast of the Gulf of Lions. In the Alboran sea, another important upwelling region is influenced by the system of anticyclonic eddies associated to the jet of Atlantic water entering the basin (Tintoré et al. 1991). In addition to upwelling, the Atlantic water inflow contributes to nutrient enrichment by causing turbulent mixing in the Straits of Gibraltar, which entrains nutrients from the deeper Mediterranean waters into the euphotic zone, and by producing fertilization spots associated with eddies of the Atlantic Current along the Algerian coast (Taupier-Letage and Millot 1988; Morán et al. 2001). Other important mesoscale and submesoscale features in the Mediterranean include gyres, frontal regions and filaments (Wang et al. 1988; Iermano et al. 2009). The cyclonic circulation in the northern part of the western basin is bound by a series of shelf-slope fronts and leaves a central divergence that may be a site of deep convection in winter and, because of the shallower depth of the pycnocline, of enhanced nutrient supply to the upper water layers in summer (Estrada 1996). The shelf-slope fronts, including the Ligurian, Catalan and Balearic Fronts, are affected by meanders, eddies and filaments, that together with the ageostrophic coastal circulation could originate enrichment events (Estrada and Margalef 1988; Font et al. 1988). Both the central divergence and these fronts appear to be important in the fertility of the NW Mediterranean (Marty et al. 2002; Estrada 1996). Other permanent or quasi-permanent sub-basin gyres with winter convective events are the Rhodos Gyre (NW Levantine sea) and the South Adriatic Gyre (Siokou-Frangou et al. 2010). Coastal fertilization due to land runoff from rivers is particularly important in the areas of influence of large rivers like the Rhône and the Ebre, in the NW Mediterranean, the Po in the North Adriatic Sea and the Nile, but input from land due to smaller rivers, storms and wastewater discharge may be locally significant (see Stambler, this volume, for primary production figures).

Overview of Diversity of the Main Microbial Groups

The earlier studies of microbial plankton in the Mediterranean and other marine regions of the world centered on taxa with cellular characteristics that could be described under optical microscopy, like diatoms, thecate dinoflagellates, coccolithophorids and silicoflagellates among the phytoplankton, and Tintinnids, Foraminifera or Radiolaria among the microheterotrophs. In fact, many early illustrated books and checklists were based on Mediterranean samples (e. g. Jörgensen 1920; Pavillard 1925; Schiller 1928). Later on, the introduction of electron microscopy allowed a better description of morphological species diversity and helped to establish the biogeography of some taxa. In the last decades, the implementation of molecular genetics has brought a rapid change of ideas and concepts to the field of both prokaryotic (Bacteria and Archaea) and eukaryotic (Protists) microbial diversity. Techniques now in use may provide a measure of the diversity of organisms (for example, by determining 16SrRNA sequences) or of the number of taxonomic units in an environmental sample (by means of metagenomic approaches). The new findings highlight the problem posed by the inconsistencies in the morphological, biological or phylogenetic species concept when applied to marine microbial communities (Coll et al. 2010). The picture that has emerged is that of a high microdiversity, with accepted morphospecies encompassing a number of cryptic or pseudocryptic variants.

The rapidly changing state of taxonomical knowledge makes it difficult to assess the number of microbial taxa in the Mediterranean and to establish comparisons with other areas. Hofrichter et al. (2002) estimated that about 4,400 species of protists had been described in the region, although this figure must be taken with caution, given the problems with cryptic and pseudocryptic taxa mentioned above. The compilation of Velasquez (1997) for the Western Mediterranean (WM) recorded 96 genera and 736 diatom species and Gómez (2006) listed 104 genera and 673 dinoflagellate species, none of them endemic, in contrast with the situation for many macroscopic organisms. The biodiversity of coccolithophores in the WM was revised by Cros (2001), who compiled 166 taxa. The recent recognition that holococcolithophores, previously considered as a separate family, are part of a life cycle that includes holo- and heterococcolith-bearing stages was based, mainly, on combination coccospheres sampled in the WM (Cros et al. 2000) and has resulted in an overhaul of coccolithophore taxonomy. Much less known is the biodiversity of naked flagellates and small-sized picoplankton, although recent work combining optical and electron microscopy with pigment chemotaxonomy, molecular techniques and experimental cultures is clarifying phylogenetic relationships and uncovering much unknown diversity.

Assessing bacterial biodiversity is hindered not only by the lack of knowledge of the existing organisms, but also by the difficulties in reaching a consensus about bacterial "species". Bacterial richness has been shown to peak in the tropics (Pommier et al. 2007; Fuhrman et al. 2008) and, for the latitude of the Mediterranean (30–45°N), a number of detectable "operational taxonomic units" between 100 and 150 would seem logical. In fact, Zaballos et al. (2006) reported a similar value that, once extrapolated using statistical techniques, suggested a number of ca. 360 for surface waters (Coll et al. 2010). A slightly smaller value was found for the coastal Blanes Bay Microbial Observatory (Alonso-Saez et al. 2007) based on a different approach. However these techniques usually refer only to the dominant organisms. Recent application of new methodologies (such as metagenomics and 454- tag sequencing) will in the near future provide more accurate estimates (Coll et al. 2010).

Phytoplankton

Temporal Distribution and Vertical Organization of Phytoplankton Biomass

Measurements of phytoplankton biomass covering at least a year cycle are numerous in coastal areas but rare in the open sea. However, satellite imagery has shown the main features of the distribution of phytoplankton biomass in open waters of the Mediterranean (Morel and André 1991; Antoine et al. 1995; Bosc et al. 2004). Using SeaWIFS data, D'Ortenzio and Ribera d'Alcalá (2009) defined seven open sea bioprovinces with distinct seasonal patterns and concluded that a temperate regime, with a marked late winter-early spring bloom was only observed in the northern part of the NW Mediterranean (their "blooming" province) and, intermittently, in a few other areas, while a subtropical mode, with seasonal biomass enhancement centered in January but lasting for 2–3 months, occurred in most of the basin. A recurrent autumn peak attributed to breakup of the thermocline has been detected in the "blooming" province (D'Ortenzio and Ribera d'Alcalá 2009) and in some coastal regions (Margalef 1969). In a study of monthly composites of SeaWIFS images (from 1997 to 2004) for squares of 0.25° (longitude) × 0.25° (latitude), representative of 11 subregions of the NW Mediterranean, Morales (2006) found a generally unimodal seasonal variability, with a chlorophyll *a* (chl *a*) concentration maximum occurring in January off the Iberian Coast (Valencia) around 40°N, in February in the Catalan Sea, near the sill between Majorca and Minorca, in March in two coastal subregions near Barcelona and Fréjus (Southern France), around 41° and 43°N, respectively, and in April in the center of the Gulf of Lions. Interannual variability can also be very high, depending on factors such as climate variations that affect the intensity of winter mixing or the amount of precipitation (Siokou-Frangou et al. 2010). Superimposed to these general patterns, there is a strong smaller-scale spatio-temporal variability, in particular in areas of intense mesoscale activity such as the Liguro-Provençal Catalan front and the Alboran Sea (Morán et al. 2001; Estrada 1996; Siokou-Frangou et al. 2010). In coastal waters and estuarine areas of the Mediterranean, the interactions among topography, wind patterns and highly variable continental water inputs originate a large interannual, seasonal and short-term diversity in the distribution patterns of phytoplankton biomass (see section "The seasonal succession of phytoplankton" and Stambler, this volume). During blooms, chl *a* concentrations in offshore surface waters of the NW

Mediterranean reach often around 2–3 mg m^{-3} (Vidussi et al. 2001; Marty et al. 2002; Estrada et al. 1993, 1999), but there are point records of values up to 6 mg m^{-3} in February 2009, during the FAMOSO cruise in the Gulf of Lions (M. Estrada, unpublished data). High chl *a* concentrations have also been measured in the Alboran Sea (e. g., 7.9 mg m^{-3}, Arin et al. 2002).

After the relatively short period winter mixing, there is a progressive stabilization of the water column until the pycnocline starts to breakup in the fall. As the winter-spring phytoplankton bloom consumes the nutrients of the upper layers, the peak of phytoplankton biomass migrates downwards, following the nutricline. During most of the year, the vertical distribution of phytoplankton in the open Mediterranean is characterized by the presence of a deep chlorophyll maximum (DCM), consisting in general of a combination of higher cell numbers and enhanced chlorophyll content per cell, with the second factor becoming more important later in the stratification period. The DCM is closely associated with the nutricline and its depth, which ranges from around 50 m in the NW Mediterranean (Estrada 1985) to 120 m in the Levantine Basin (Dolan et al. 2002), is determined by the availability of both sufficient nutrients and light to carry out photosynthesis. As happens with deep chl *a* maxima of tropical and subtropical regions (Herbland and Voituriez 1979), the Mediterranean DCM is accompanied by a slightly shallower oxygen maximum, indicative of a relative primary production maximum, and by a nitrite maximum (Estrada et al. 1993). The chl *a* concentration within the Mediterranean DCM may reach values around 1.5 mg m^{-3} and can be very patchy, indicating an important spatio-temporal dynamism (Estrada 1985) related to processes such as mesoscale variability or internal waves.

Phytoplankton Composition

Cyanobacteria

Well represented cyanobacterial genera in the Mediterranean include *Synechococcus* and *Prochlorococcus,* which are the main components of the prokaryotic picophytoplankton, and two filamentous nitrogen-fixers, the free-living *Trichodesmium* and *Richelia intracellularis*, which is an endosymbiont of diatom genera such as *Rhizosolenia* and *Hemiaulus*. Peaks of cyanobacterial picoplankton tend to occur from summer to autumn (Bernardi Aubry et al. 2006; Charles et al. 2005) but have been recorded also in spring (Cerino et al. 2011). Latasa et al. (2010), who compared the distribution of various phytoplankton taxonomic groups in bloom, postbloom and stratification conditions in the NW Mediterranean, classified *Synechococcus* as mesotrophic and *Prochlorococcus* as oligotrophic. *Synechococcus* may be an important component of offshore (Denis et al. 2010) and coastal picophytoplankton (Agawin et al. 1998; Bernardi Aubry et al. 2006; Cerino et al. 2011), while *Prochlorococcus* was first reported offshore, but may be also important nearshore (Charles et al. 2005). In the survey that Denis et al. (2010), carried out in winter 1995 across the whole Mediterranean basin, *Synechococcus* represented a 65 % of the overall ultraphytoplanktonic carbon biomass while the contribution of *Prochlorococcus* did not exceed 6.5 %. The authors did not detect clear large-scale patterns, but found a significant relationship between *Synechococcus* abundance and mesoscale structures in the Adriatic and Ionian sub-basins. *Prochlorococcus* was represented by two ecotypes, one occupying the upper 75 m and the other adapted to low light and growing mainly below 75 m. Garczarek et al. (2007) used amplification and sequencing of the pcb gene (encoding the major light-harvesting proteins) of *Prochlorococcus* in a late summer-early autumn Mediterranean-wide survey, to find at least four different ecotypes in the photic zone, large microdiversity, and strong vertical but low horizontal heterogeneity. A vertical stratification with *Synechococcus* in the upper 40 m and prochlorophytes at the base of the chl *a* maximum was found by Ghiglione et al. (2008) during early autumn 2004 in the DYFAMED site of the Gulf of Lions. The depth relationships of *Synechococcus* and *Prochlorococcus* are in a large part related to the presence of ecotypes adapted to factors such as different light levels and nutrient availability, although temperature can also be a factor. For example, Mella-Flores et al. (2011) compared the ecotype diversity of *Synechococcus* and *Prochlorococcus* during cruises carried out 9 years apart and reported that the dominant clades of the two genera in both cruises were typical of temperate waters, while groups typical of (sub)tropical and warm waters were only present in low concentrations. Based on these results, the authors concluded that there had not yet been a substitution of clades related to climate change.

The occurrence of the nitrogen-fixing *Trichodesmium* and diatom – *Richelia intracellularis* consortia is frequent during the stratification season, albeit in low concentrations (Béthoux and Copin-Montégut 1986). Krom et al. (2010) concluded that their impact on the global nitrogen budget in the Mediterranean is insignificant, in part due to strong P limitation, but the sporadic finding of *Trichodesmium* sp. filaments and the consistent presence of *Hemiaulus hauckii* and *Rhizosolenia styliformis* hosting *Richelia intracellularis* in the DCM of the whole Mediterranean basin during the stratification period led Crombet et al. (2011) to suggest that their role could be more important than previously accepted. Recent studies have pointed out the presence of diazotrophic unicellular cyanobacteria in coastal areas of the Eastern (Man-Aharonovich et al. 2007) and NW Mediterranean Sea (Le Moal and Biegala 2009) and in open waters across the entire basin, where the dominant unicellular diazotrophs were α-proteobacteria (Le Moal et al. 2011). However, the role of these organisms in the nitrogen budget of the Mediterranean Sea is still unclear.

Diatoms (Bacillariophyceae)

Large colonial diatoms are the principal phytoplankters in most seasonal blooms and represent the typical primary producers of the so-called classical food web, leading from phytoplankton to zooplankton and fish. However, diatoms can be also important in oligotrophic situations (Scharek et al. 1999) and some taxa have sizes close to the picoplankton range. In the Mediterranean, diatoms dominate the winter-spring bloom in many areas, although their proliferation may be of short duration and easily missed by field measurements. Chemotaxonomic determinations in the DYFAMED area have shown that the contribution of diatoms to phytoplankton biomass is maximal in January or February (Marty et al. 2002). In March 1995, the phytoplankton of a station occupied by Latasa et al. (2010) in the open NW Mediterranean (41.45°N, 05.10°W) was dominated by *Pseudo-nitzschia* spp., *Chaetoceros* subgenus *Hyalochaete* spp., *Thalassiosira* spp., *Guinardia delicatula*, *Bacteriastrum delicatulum* and *Dytilum brightwellii*. In the Catalan Sea, the main nano- and microplankton taxa inshore of the Catalan Front in winter and early spring were the diatoms *Chaetoceros* spp, *Thalassiosira* spp. and *Pseudo-nitzschia* spp. (Estrada 1991; Estrada et al. 1999). The genera *Chaetoceros* and *Pseudo-nitzschia*, which dominate winter –spring proliferations, accompanied by others like *Thalassiosira*, *Bacteriastrum*, *Rhizosolenia* and *Leptocylindrus*, account also for phytoplankton biomass maxima in mesoscale features such as the Catalan Front (Estrada 1991) and the upwelling of the Alboran Sea (Arin et al. 2002) and for high chl *a* patches at the DCM (Estrada 1985; Estrada et al. 1993; Crombet et al. 2011), sustaining the view that intermittent pulses of growth can occur at the DCM level, originated, for example by hydrographic perturbations that expose bodies of relatively nutrient- rich waters to enhanced irradiance levels (Estrada 1991). In coastal and estuarine areas, the seasonal sequence of diatom assemblages tends to be locally persistent. For example, off the coasts of Barcelona and Castelló, in the NW Mediterranean, Margalef (1969) described the consistent appearance of a winter-spring maximum, dominated by *Chaetoceros* spp., together with cryptomonads and flagellates, a late spring peak, with *Rhisosolenia* and *Pseudo-nitzschia*, and an autumn peak with *Thalassionema nitzschioides*, *Asterionella japonica* and *Thalassiothrix mediterranea*. An autumn diatom assemblage with *Thalassionema nitzschioides* is a recurrent feature in the seasonal cycle in the bay of Els Alfacs, a part of the Ebre Delta (Llebot et al. 2011). In the Bay of Naples, after a winter-spring bloom with *Chaetoceros* spp., *Pseudo-nitzschia delicatissima*, *Thalassionema bacillaris* (in addition to autotrophic flagellates), there was a substitution of the dominant diatom species by smaller-sized ones like *Skeletonema pseudocostatum*, *Chaetoceros tenuissimus* and *Chaetoceros socialis*; other species, including *Leptocylindrus danicus*, were generally recorded in autumn. The most abundant taxa succeeded and overlapped each other in a fairly regular pattern, although there was strong interannual variability (Ribera d'Alcalà et al. 2004). Although genera such as *Chaetoceros* and *Thalassiosira* are commonly found in all diatom-dominated blooms, the particular species present of these genera and the co-occurring diatom taxa can vary at different places and times (Estrada 1991; Siokou-Frangou et al. 2010).

In addition to the above-mentioned blooming taxa, diatoms like *Hemidiscus cuneiformis* and *Coscinodiscus radiatus*, have been found consistently within the DCM of the WM, but only in low concentrations. Together with representatives from other groups (*Dictyocha fibula* and *Oxytoxum margalefi*), these species appeared to be part of a "shade flora" assemblage (Estrada 1991). Another group of large-sized diatom species of genera like *Rhizosolenia*, *Proboscia* and *Hemiaulus* tend to occur in low numbers in the euphotic zone of stratified waters. As noted above, some of these *Rhizosolenia* and *Hemialus* species host the diazotrophic cyanobacterial symbiont *Richelia intracellularis*; their present contribution to nitrogen fixation seems to be small (Krom et al. 2010), but it has been suggested that extensive Rhizosolenid mats and their symbionts may have supported nitrogen fixing during sapropel episodes (Sachs and Repeta 1999; Stambler, this volume). Another diatom typical of relatively oligotrophic conditions is *Leptocylindrus mediterraneus*, which occurs in association with the heterotrophic flagellate *Solenicola setigera* and with potentially diazotrophic unicellular cyanobacteria (Buck and Bentham 1998). More studies are needed also to assess the importance in offshore waters of small-sized species like *Minidiscus trioculatus* and *Arcocellulus* sp., which were abundant in winter samples from the open NW Mediterranean (Delgado et al. 1992; Percopo et al. 2011; Cros and Estrada, unpublished data).

Dinoflagellates (Dinophyceae)

As a group, the Dinophyceae include a bewildering variety of ecophysiological niches. Dinoflagellates are often very abundant in marine samples but most of them are naked nanoplankton-sized forms that cannot be properly identified in optical microscopy examinations and are often overlooked or lumped in categories such as "small dinoflagellates" or "small gymnodinioids". Chemotaxonomic analysis may fail to give an accurate measure of dinoflagellate abundance because peridinin, the pigment used as a dinoflagellate biomarker, is not present in many of these organisms. Widespread nano- and microplankton dinoflagellate genera like *Noctiluca*, *Protoperidinium* and *Warnowia*, and species of *Gyrodinium* and *Gymnodinium*, among others, are heterotrophic and play a functional microzooplankton role. Many genera, like *Ceratium* (*Neoceratium*, a new name recently proposed for the genus by Gómez et al. 2010, is still under discussion) and *Alexandrium* can be mixotrophic (Jacobson and Anderson 1996) while others like *Dinophysis* comprise species with a varying degree of autotrophic capabilities. *Ornithocercus*, *Histioneis* and *Cytharistes*, open water taxa typically present at low concentrations, host endosymbiotic cyanobacteria that could help with nitrogen fixation (Gordon et al. 1994). The variety of trophic strategies and the ability to migrate vertically between the shallow well-illuminated layers and the nutricline underlie the prevalence of dinoflagellates during the stratification period. In general, they are more abundant within the upper layers of the euphotic zone, but species like *Oxytoxum margalefi*, are part of the "shade flora" of the DCM level (Estrada 1991). Some taxa, referred loosely as "red tide dinoflagellates" in Margalef's Mandala (see section "Generalities on phytoplankton community structure and function"), are frequent producers of blooms when relatively high water column stability, that allows biomass accumulation, is accompanied by nutrient availability. These forms include genera like *Prorocentrum*, *Scrippsiella* and *Alexandrium*, frequently found in coastal waters. Some species or strains of *Alexandrium* and other genera produce toxins that can accumulate in members of higher levels of the trophic web and their proliferations may have noxious effects on human health or on other ecosystem organisms (see section "Harmful algal blooms in the Mediterranean"). In some cases, as happens with several species of *Dinophysis*, a low concentration of individuals is sufficient to cause toxic effects.

Prymnesiophyceae

Many prymnesiophytes belong to the nanophytoplankton size category, although pigment-based and molecular techniques have recently pointed out their important contribution to the picophytoplankton (Liu et al. 2009). The best studied group is that of the coccolithophores, which possess a cell cover of calcareous platelets. Coccolithophores are well represented in the Mediterranean (Ignatiades et al. 2002; Malinverno et al. 2003; Siokou-Frangou et al. 2010). As mentioned above, they tend to be important in mesotrophic situations, as in the open waters of the NW Mediterranean in winter-spring (Estrada et al. 1999; Latasa et al. 2010) and autumn (Cros 2001). The most widespread species is *Emiliania huxleyi*, which can be dominant both in coastal (Bernardi Aubry et al. 2004; Cerino et al. 2011) and open sea regions (Estrada et al. 1999), a finding perhaps related to the high intraspecific variability of this taxon. There seems to be some ecological differentiation between hetero- and holococcolithophores, with a higher proportion of heterococcolithophores at the DCM level and of holococcolithophores in the more oligotrophic upper part of the euphotic zone (Cros 2001). Curiously, species like *Calyptrosphaera sphaeroidea*, generally reported in low concentrations, have produced high-density proliferations in zones like the Tarragona Harbour and the Algiers coast (Cros et al. 2002; Illoul et al. 2008). The distributions of non-calcifying prymnesiophytes, which include widespread genera like *Chrysochromulina*, are poorly known with the exception of the colonial forms of species of *Phaeocystis*, which can be very abundant in open waters of the Catalan Sea, in winter (Estrada 1991).

Other Groups

Cryptophytes are often an important component of the nanophytoplankton (Margalef 1969), but species lists derived from light microscopy should be reconsidered (Cerino and Zingone 2006). Based on microscopic observations or on the detection of alloxanthin, their pigment marker, a substantial contribution of cryptophytes has been found, for example, in offshore waters of the NW Mediterranean in winter-spring (Marty et al. 2002; Latasa et al. 2010) and in relationship with mesoscale fertilization areas (Margalef 1971). Novarino (2005) found 11 morphospecies in samples from Barcelona coastal waters and the Alboran Sea, and noted the abundance of *Plagioselmis prolonga*. In the Gulf of Naples, Cerino and Zingone (2006) used cultures and electron microscopy to identify 16 morphospecies, and reported that the maximum cryptomonad concentration occurred in spring-summer and autumn. The most abundant species was *Hemiselmis* sp., while *Plagioselmis prolonga* was the most frequently found, although at lower concentrations. The regular presence of *Plagioselmis prolonga* was confirmed by McDonald et al. (2007).

The silicoflagellates, a group of Chrysophyceae recognizable by their tubular skeleton of silica, are mainly represented by the genera *Dictyocha* and *Distephanus*. They tend to show population maxima in winter and to be present in subsurface levels during the stratification period (Estrada et al. 1985). Another morphologically recognizable chrysophyte is *Meringosphaera mediterranea* (Cerino et al. 2011).

Other important phytoplankton groups in the Mediterranean Sea are the Prasinophyceae and the Pelagophyceae. Both groups include widespread picoplankton-sized representatives, but the Prasinophyceae comprise also frequent nanoplankton genera, like *Tetraselmis* and *Pyramimonas*, and the striking microplanktonic species *Halosphaera viridis*. Latasa et al. (2010) found that picoplanktonic Prasinophyceae were the most abundant group, after diatoms, during the bloom period in the open NW Mediterranean. Marie et al. (2006) used flow citometry, gradient gel electrophoresis and quantitative PCR to assess the abundance and composition of picoplankton (cells <3 μm) and the distribution of the prasinophyte genera *Micromonas*, *Bathycoccus* and *Ostreococcus* across a transect between the Morocco upwelling and the EM. The quasi -absence of *Micromonas*, which was replaced mainly by *Bathycoccus* and sporadically by *Ostreococcus*, agreed with the general oligotrophy of the Mediterranean. In contrast,

Micromonas was abundant in the coastal waters of the Gulf of Naples (Zingone et al. 1999) and in Blanes Bay, where together with *Bathycoccus* and *Ostreococcus* dominated the sequences obtained in winter and fall (Massana et al. 2004). In fact, *Micromonas pusilla*, the most commonly cited species of the genus *Micromonas*, appears to comprise several independent clades occupying specific niches, as shown by Foulon et al. (2008) for Blanes Bay.

The pigment signature of Pelagophyceae (which they share with chrysophytes) has been found at a number of locations in the Mediterranean (Claustre et al. 1994; Marty et al. 2002). Latasa et al. (2010) reported that pelagophytes were abundant during the winter-spring bloom, but presented also significant concentrations in more oligotrophic periods. In general, Pelagophyceae tended to be more important at depth, a feature that has been related to nutrient availability (Claustre et al. 1994; Barlow et al. 1997; Marty et al. 2002) and low light adaptation (Latasa et al. 2010). Pelagophytes have been investigated by means of molecular methods in coastal areas such as the Gulf of Naples, where they occurred between December and March (McDonald et al. 2007) and Blanes Bay (Massana et al. 2004).

Generalities on Phytoplankton Community Structure and Function

The available information on large-scale phytoplankton community distribution patterns in the Mediterranean basin is scarce (Siokou-Frangou et al. 2010) and limited to particular times of the year. Based on samples from a cruise that visited nine stations along a longitudinal transect from 4° 95′ E to 32° 67′ E, Ignatiades et al. (2009) reported that diatoms dominated over the water column (1–120 m depth) in the three westernmost stations (between the Strait of Sicily and the western basin), while dinoflagellates and coccolithophores were more abundant in the eastern stations. These trends coincided with a decrease in water column chl *a* and primary production and an increase of phytoplankton diversity towards the east, in agreement with the more oligotrophic character of the Eastern Mediterranean (EM). In consonance with the generally oligotrophic character of the Mediterranean, the proportion of prokaryotic and eukaryotic picoplankton with respect to total phytoplankton biomass exceeds often 50 % (Siokou-Frangou et al. 2010). Picoplankton maxima tend to occur in summer, when temperature is highest and nutrient availability tends to be lowest, although there is substantial spatio-temporal variability (Agawin et al. 1998; Bernardi Aubry et al. 2006; Arin et al. 2002). Superimposed on the picoplankton component, micro and nanoplankton may dominate at particular times of the year or in areas of strong hydrodynamic activity. Margalef (1978) considered the input of external energy, in the form of water convection and turbulence, and nutrient availability were the main selective factors of the dominant life-forms of phytoplankton, as represented by major taxonomic groups. He proposed a conceptual model, later called the "phytoplankton mandala", which ordained diatoms, coccolithophores and dinoflagellates along a sequence reflecting the ecological succession of phytoplankton following a fertilization event such as the seasonal mixing of the water column. Diatoms, with grow rapidly in nutrient-rich, turbulent conditions, are the typical dominants in the winter-spring bloom in coastal areas and in deep mixing zones such as the central area of the Ligurian Sea (Estrada et al. 1985; Marty et al. 2002; Siokou-Frangou et al. 2010), but they may also proliferate and form high chl *a* patches within the DCM (Estrada 1985, 1999; Crombet et al. 2011). Dinoflagellates, which may show migratory behaviour and encompass a wide variety of physiological strategies, become relatively more abundant in the nutrient-depleted upper water layers during the stratification period, while coccolithophorids tend to occupy an intermediate position with respect to trophic conditions. Margalef (1978) distinguished a subgroup of red tide dinoflagellates, which would be favoured by situations of high stability and nutrient availability. Later works have proposed modifications or expansions of the mandala (Smayda and Reynolds 2001; Cullen et al. 2002). The relationship of other taxonomic groups with environmental conditions is less known, particularly for nano-and picoplanktonic flagellates, although the situation is rapidly changing with the expansion of the use of chemotaxonomic and molecular techniques (Latasa et al. 2010; Charles et al. 2005).

The Seasonal Succession of Phytoplankton

The seasonal succession of phytoplankton has been investigated in numerous coastal locations of the Mediterranean, but just a small number of them has been followed regularly during more than a few years. In offshore waters, only the DYFAMED station in the Ligurian Sea (Marty et al. 2002) has been studied for more than a decade (Siokou-Frangou et al. 2010), although in some open water regions, the repetition of cruises in different seasons over different years has provided a general picture of the seasonal phytoplankton variability. A case-study for the Catalan Sea (Estrada 1999) will be described below.

Using a pigment chemotaxonomic approach, Marty et al. (2002) reported that the eukaryotic biomass at the DYFAMED site was dominated by prymnesiophytes (containing 19′hexanoyloxyfuxcoxanthin) except during a short period of diatom dominance (inferred from fucoxanthin), from January to March, generally associated with the period of maximal chl *a* except on some occasions in which it was related to nanoflagellates. The diatom peak

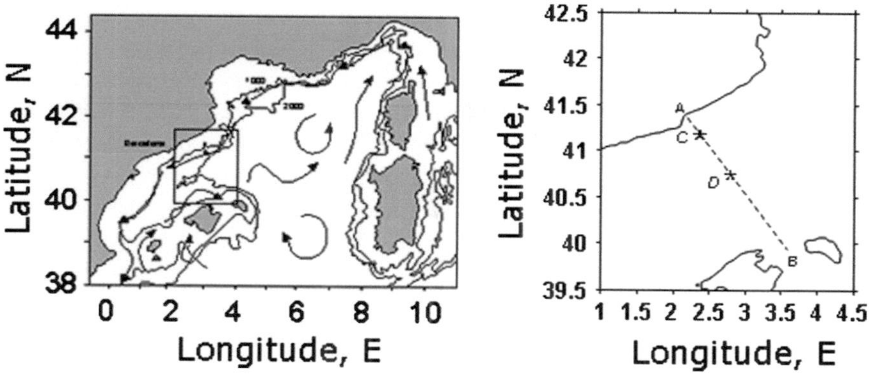

Fig. 6.1 (*Left*) Scheme of the circulation in the NW Mediterranean. (*Right*) Position of the shelf (*C*) and central divergence (*D*) stations mentioned in section "Microbial and classical food webs. The example of the Catalan Sea". *The line* indicates the situation of the "main transect"

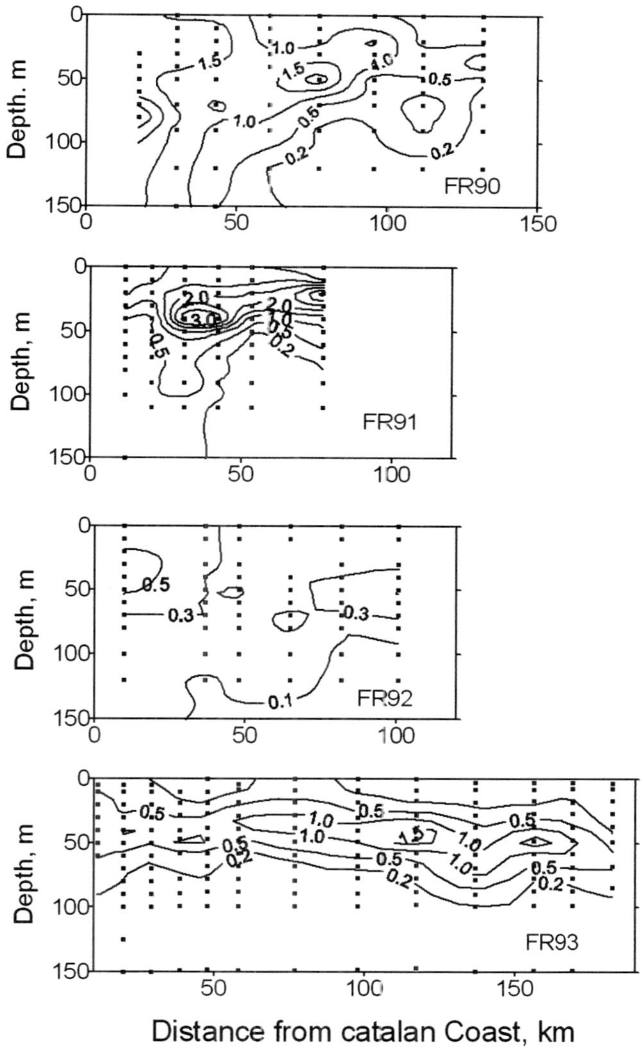

Fig. 6.2 Cruises FR90, FR91, FR92 and FR93. Distribution of chlorophyll *a* (mg C m^{-3}) along the main transect of Fig. 6.1

was followed about 1–2 month later by prymnesiophytes and chrysophytes/pelagophytes (these latter, indicated by 19′butanoyloxyfuxcoxanthin). The maximum contribution of prokaryotes, mainly represented by *Synechococcus*, occurred between August and November; prochlorophytes reached their peak abundance at the end of the stratification period, but persisted during the winter. The presence of cryptophytes (alloxanthin) coincided with that of diatoms, and the peridinin (a dinoflagellate marker) maximum occurred in parallel or slightly later than that of prymnesiophytes and chrysophytes/pelagophytes.

The seasonal variability of phytoplankton composition in the Catalan Sea was described by Estrada (1999), based on data from the cruises FRONTS 90 (FR90, 9–16 February 1990), FRONTS 91 (FR91, 10–21 April 1991), FRONTS 92 (FR92, 15 October-4 November 1992) and VARIMED 93 Phase 2 (FR93, 1–8 June 1993). Studies concerning biological distributions in these cruises can be found in Alcaraz et al. (1994), Margalef (1995), Calbet et al. (1996), Estrada (1996) and Estrada et al. (1999). The surveys visited a variable number of stations within the Catalan Sea, but the phytoplankton data considered here will be those of the transect between Barcelona and the Balearic Islands ("main transect"), which was occupied in all the cruises (Fig. 6.1). The phytoplankton was examined by R. Margalef using the inverted microscope technique, after sedimentation of 100 ml of water in composite chambers. The distribution of physico-chemical variables (data not shown) and chl *a* (Fig. 6.2) along the principal transect featured the characteristic central zone of elevation of the isopycnes, limited in the continental side by the Catalan Front and by the Balearic front in the Balearic Islands side. During FR90, the water column inshore of the Catalan Front was well mixed, but there was already appreciable stratification offshore, with a DCM at 50 m depth. The Catalan Front was also well marked during April 1991 (FR91), but in late spring (FR93) the whole transect was covered by a layer of warm water and featured a well-developed DCM (Fig. 6.2). In the fall (FR92), there was still an appreciable stratification of the upper layers, but the surface waters had been cooling and the pycnocline had started to erode. A principal component analysis was carried out with the log-transformed abundance data of 49 phytoplankton taxa that occurred in all the cruises and that were present, at least, in 15 % of the samples. The three first principal components explained 36.1 % of the variance, within the range usually found in this kind of studies. As can be seen in Fig. 6.3, which presents the distribution of the species in the statistical space of the components, the taxa tended to group together according to taxonomic affinities, reflecting a similarity in the ecological response of phylogenetically-related organisms.

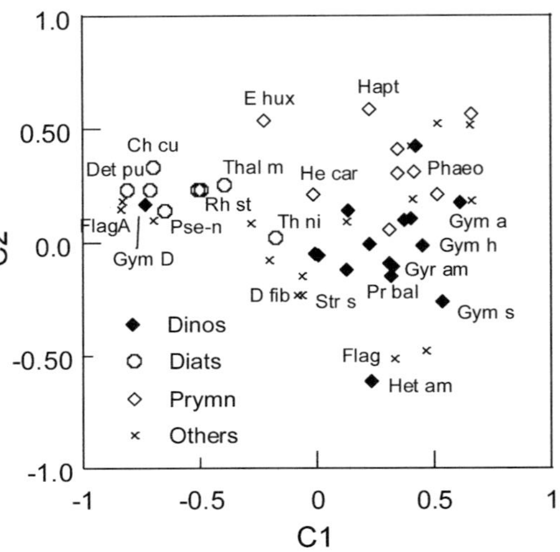

Fig. 6.3 Position of the extremes of the taxa vectors in the space of the first (C1) and second (C2) principal components. Legend: Dinos, Dinoflagellates; Gym a, *Gymnodinium acutissimum*; Gym h, *Gymnodinium hamulus*; Gym s, *Gymnodinium* sp.; Gym D, *Gymnodinium* sp D; Gyr am, *Gyrodinium acutum* "small"; Het am, *Heterocapsa* sp.; Pr balt, *Prorocentrum balticum*; Diats, Diatoms; Ch cu, *Chaetoceros curvisetus*; Det pu, *Detonula pumila*; Pse-n, *Pseudo-nitzschia* spp.; Th ni, *Thalassionema nitzschioides*; Thal m, *Thalassiosira* spp.; Prymn, Prymnesiophytes; E hux, *Emiliania huxleyi*; He car, *Helicosphaera carteri*; Hapt., Unidentified haptophytes; Phaeo, *Phaeocystis* sp.; Others, Other groups; D fib, *Dictyocha fibula*; Flag, Unidentified flagellates; FlagA, Unidentified flagellate A; Str s, *Strombidium* sp. (ciliate)

Most diatoms, including *Detonula pumila* and species of *Chaetoceros*, *Pseudo-nitzschia* and *Thalassiosira*, showed negative correlations (or loadings) with principal component 1 (PC1) while dinoflagellates and coccolithophores presented positive correlations. In turn, PC2 was positively correlated with the diatoms, the cryptophyte *Rhodomonas* sp., the coccolithophores, *Phaeocystis* sp. and other non-calcifying prymnesiophytes, and negatively correlated with the silicoflagellate *Dictyocha fibula* and with most dinoflagellates, represented mainly by species of *Heterocapsa*, *Amphidinium*, *Gymnodinium* and *Prorocentrum*. PC3, which was positively correlated with most taxa, reflected the distribution of total cell abundance (data not shown). The position of the sample points in the space of PC1 and PC2 showed a marked division between the samples of FR90 in the negative side of PC1, reflecting the dominance of winter bloom diatoms inshore of the Catalan Front during this cruise, and the samples from the other seasons, grouped in the positive side of PC1 (Fig. 6.4a). With respect to PC2, the sequence FR91-FR92-FR93 (spring, fall, late spring) from the positive to the negative side, can be related to the increasing stratification from FR91 to FR92 and FR93 (Fig. 6.4b). The shift of prymnesiophytes to dinoflagellates from the positive to the negative side of the correlations with PC2 can be interpreted in terms of Margalef's mandala, as reflecting a successional sequence of better adaptation to decreased turbulence.

In coastal and estuarine areas, seasonal cycles vary depending on local factors, as found by Cloern and Jassby (2010) for data sets from around the world. A winter phytoplankton bloom or at least an increase of biomass, allowed by stable meteorological conditions or stratification of the water column is a common feature in Mediterranean phytoplankton communities (Margalef 1969; Duarte et al. 1999; Bernardi Aubry et al. 2004; Ninčević Gladan et al. 2010).

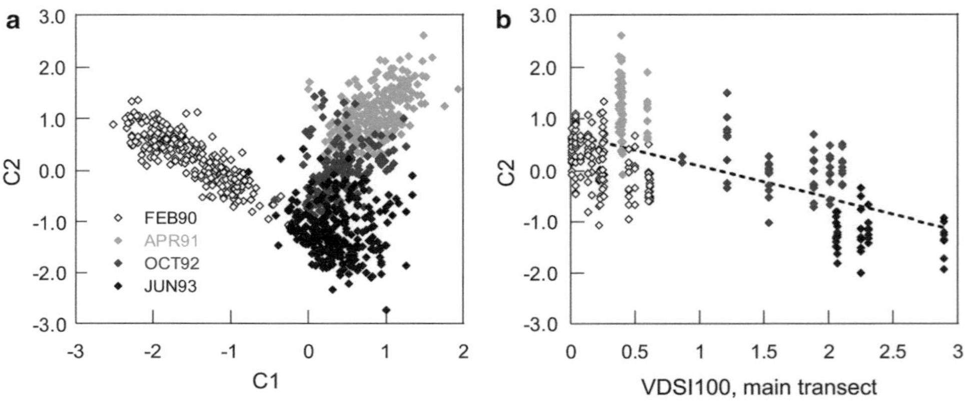

Fig. 6.4 (**a**) Distribution of the points corresponding to the samples of the four cruises in the space of the first (C1) and second (C2) principal components. Cruises, FEB90 = FR90 (February 1990), APR91 = FR91 (April 1991), OCT92 = FR92 (October–November 1992), JUN93 = FR93 (June 1993). (**b**) Relationship between the values of the C2 and a vertical stratification index, VSI100, calculated as the sum of the absolute values of the differences in water density every 10 m, for a water column between 0 and 100 m depth:

$$VSI100 = \sum_{i=1}^{10} [\,|\,(\sigma_t(i*10) - \sigma_t((i-1)*10))\,|\,]\text{, where } i = 1, 2, \ldots 10$$

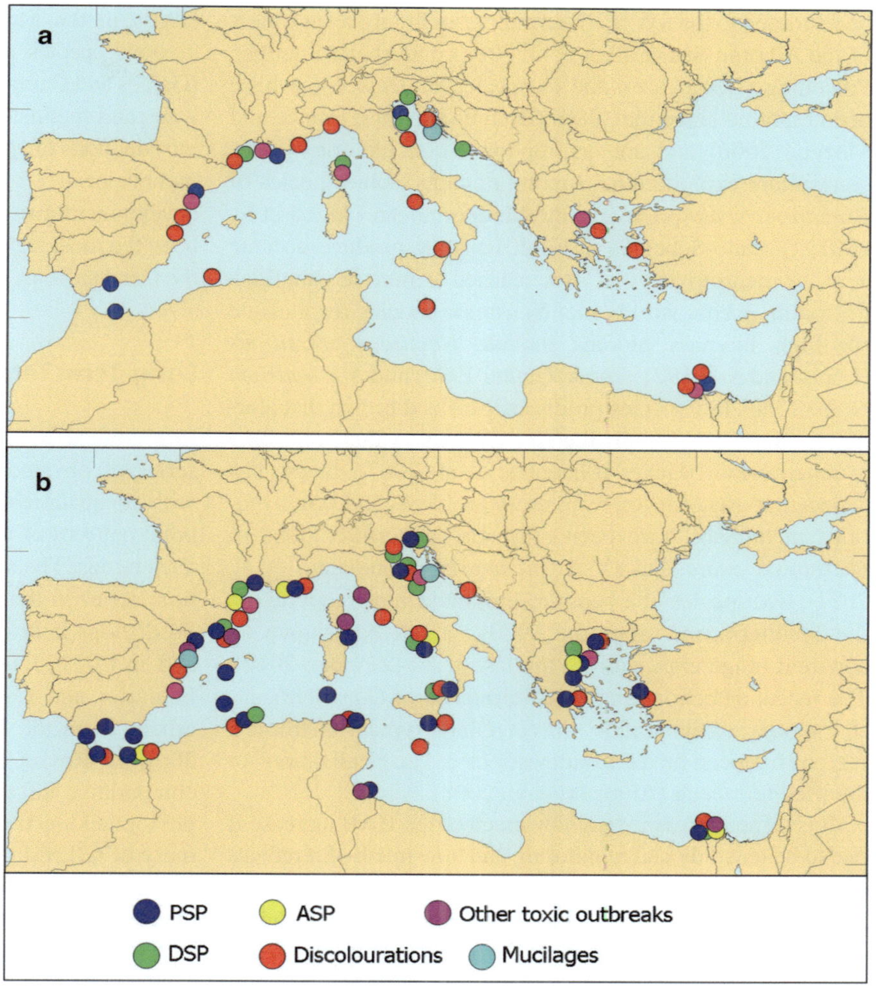

Fig. 6.5 Harmful events along Mediterranean coasts (excluding those attributed to *Ostreopsis*). Based on (**a**) reports up to 1995 (Jacques and Sournia 1978–1979; Honsell et al. 1995) and (**b**) updated findings (Reproduced from Zingone 2010)

Often, there is also a fall peak, that has been related to higher nutrient availability due to erosion of the thermocline (Margalef 1969) or to stable weather conditions, which allow the exploitation of nutrients of coastal origin (Zingone et al. 2003; Mura et al. 1996). Several zones present peaks driven by nutrient runoff at times like late spring or summer (Bernardi Aubry et al. 2004; Ribera d'Alcalà et al. 2004). In the Gulf of Trieste (Northern Adriatic), during the period 1989–2002, the autumn peak of phytoplankton biomass was the highest one on the annual scale and was related to enhanced freshwater inputs (Mozetič et al. 2012)

Harmful Algal Blooms in the Mediterranean

Phytoplankton blooms are part of plankton ecosystem dynamics. However, on occasions, microalgal proliferations may be perceived as detrimental due to the production of high accumulations of biomass or of toxicity that can affect other ecosystem organisms or human health. In this case, proliferations are referred to as "harmful Algal Blooms" (HABs), although it must be taken into account that this is a purely operational designation without any objective ecological meaning. As can be gleaned from Margalef's mandala, the occurrence of phytoplankton blooms, whether harmful or not, will be favoured by nutrient enrichment, which may enhance phytoplankton growth and low turbulence (or high water residence times), which reduce cell dispersion; another factor may be a reduction of grazing pressure. Until the mid 1980s, HABs, then generally referred to as "red tides", had been rarely recorded in the Mediterranean and no significant toxic episodes were known (Jacques and Sournia 1978–1979). The first contamination of shellfish by Paralytic Shellfish Poisoning (PSP) toxins, due to *Gymnodinium catenatum*, occurred in 1987 in Southern Spain (Bravo et al. 1990). Since then, HABs have been detected throughout the Mediterranean (Fig. 6.5). The most widespread PSP-producers in the region are *Alexandrium minutum* (Garcés and Camp 2012), and *Alexandrium catenella*, which seems to have increased its geographic distribution on recent years (Vila et al. 2001b; Collos et al. 2009); *G. catenatum* continues to be present, but appears to be restricted to the SW Mediterranean, including open waters of the Alboran Sea (Calbet et al. 2002). Diarrhetic Shellfish Poisoning (DSP) cases due to species of *Dinophysis* have been reported since the 1980s (Boni et al. 1993; France and Mozetič 2006) and are an important concern for bivalve aquaculture exploitations. Among diatoms, some species of *Pseudo-nitzschia* present in the Mediterranean (Amzil et al. 2001; Quijano-Scheggia et al. 2010) may produce domoic acid, a neurotoxin that has caused Amnesic Shellfish Poisoning events worldwide. Non-toxic species responsible for high biomass blooms include *Noctiluca scintillans* (López and Arté 1971; Fonda Umani 1996) and *Alexandrium taylori*. This last species produces localized brown discolorations that deteriorate the recreational value of some Mediterranean beaches (Giacobbe et al. 2007). In the last decade, reports of fever, respiratory problems and skin irritations in humans have been attributed to proliferations of *Ostreopsis ovata* and *O.* cf. *siamensis* (Mangialajo et al. 2011), two species of a benthic –epiphytic genus reported in the Mediterranean since the 1970s, which has shown an apparent range expansion in the last years (Zingone 2010). The recent detection in the Mediterranean of *Gambierdiscus* sp., another toxic benthic dinoflagellate, typical of tropical regions has been linked to suggestions of "tropicalization" of the Mediterranean (Aligizaki et al. 2008).

It is difficult to ascertain how much of the HAB increase is due to better study and monitoring and how much of it reflects a real trend (Garcés and Camp 2012). Eutrophication could be a potential factor in some areas. During part of the twentieth century, the increment of nitrate and phosphate discharges in coastal waters may have driven increases of chl *a* concentrations (Šolić et al. 1997) and high biomass blooms in some coastal zones but, after about 1990, the reduction of nutrient sources (for example, of phosphate from detergents), the expansion of wastewater treatment and, in same cases, the decrease of river discharges, reversed the eutrophication trend in many locations (Garcés and Camp 2012; Mozetič et al. 2010) but not in others (Wyatt 2010; Polat 2010). Often, however, there is no clear relationship between high biomass or toxic blooms and eutrophication (Zingone 2010). In the Northern Adriatic, for example, the absence of red tides since the late 1980s has been associated to a decrease in nutrient availability, but other changes such as the return of mucilage or "mare sporco" events, apparently not related to eutrophication (see section "Long-term temporal variability"), point out to hydrological and ecological changes that could be related to large-scale climate drivers (Conversi et al. 2010). In the Thau Lagoon, the increment of temperature and the reduction of phosphate loads resulted in a decrease of phytoplankton biomass but increased the abundance of *Synechococcus* and promoted the development of the toxic *Alexandrium minutum*, which apparently preyed on the cyanobacteria (Collos et al. 2009). Other anthropogenic factors affecting the occurrence of HABs in the Mediterranean are the human spreading of noxious species and the building of coastal infrastructures (Garcés and Camp 2012). While increasing transfers of ballast water and shellfish stocks may have been responsible for some introductions of species into new regions, the increment of the number or size of confined water bodies suitable for HAB development due to constructions in the coastline may have been the most significant cause of HAB increase in areas like the Catalan Coast (Garcés et al. 2000).

Long-Term Temporal Variability

In a context of global change, the collection of long time series of physico-chemical and biological ecosystem data with adequate resolution is a necessary tool to uncover temporal patterns of variability and to detect signatures of long-term trends. The availability and characteristics of biological time series in the Mediterranean were examined in CIESM Workshops held in 2003 (CIESM 2003) in Split (Croatia) and in 2009 (CIESM 2010) in Tunis. A special issue of Estuaries and Coasts (Zingone et al. 2010a) with papers arising from the AGU-Chapman Conference carried out in 2007, in Rovinj (Croatia), contained research results and comparative information concerning several Mediterranean phytoplankton time series. As noted in the executive summary of CIESM (2003), there are more non-biological than biological Mediterrranean time series and, among the later, the frequency decreased from higher to lower trophic levels. A list with information on some ongoing Mediterranean time series sites measuring microbial plankton variables is presented in Table 6.1. In addition, water quality parameters and plankton composition are being sampled at many littoral locations (not included in the table) as part of monitoring requirements (Le Bec et al. 1996; Goberville et al. 2010; Vila et al. 2001a).

Corresponding to the interannual and seasonal patterns in forcing variables, the available Mediterranean time series exhibit strong interannual and within-year variability. Ascertaining long-term signals that could be caused by climate change is difficult due to strong interannual fluctuations and the relatively short length of the available series. However, some consistent trends have been reported. For example, Marty et al. (2002) attributed an increase of pico- and nanoplankton biomass during the period 1990–1999, at the DYFAMED station, to a lengthening of the stratification period. A latter work (Marty and Chiavérini 2010) confirmed an increment of chl *a* biomass between 1995 and 2006, but the responsible organisms for the most recent years were mainly diatoms, rather than flagellates. Marty and Chiavérini (2010) related the strong winter-spring bloom of 2006 to the enhanced introduction of nutrients to the surface waters caused by the intense deep water convection of the winter

Table 6.1 Information on some ongoing long time series in estuarine, coastal and open waters of the Mediterranean sea

Location	Variables sampled (non-exhaustive list)	Approximate starting date	Representative publications
NW Mediterranean, DYFAMED site (43° 25′ N, 7° 52′ E)	Hydrography, nutrients, phytoplankton pigments	1991	Marty (2002), Marty et al. (2002), and Marty and Chiavérini (2010)
Northern Adriatic Sea, Italy/Slovenia/Croatia (several sites)	Hydrography, nutrients, chlorophyll *a*, phytoplankton	1970–1999, (gaps)	Bernardi Aubry et al. (2004), Viličić et al. (2009), and Mozetič et al. (2010)
Mid-Adriatic Sea, Croatia Kaštela Bay (43° 31′ N, 16° 22′ E), Offshore station (43° 00′ N, 16° 20′ E)	Hydrography, nutrients, bacterioplankton, chlorophyll *a*, primary production (PP), phytoplankton (Phyto), zooplankton	1950s (Phyto) 1962 (PP) 1960s–1980s (other variables)	Šolić et al. (1997), Marasović et al. (2005), and Ninčević Gladan et al. (2010)
Tyrrhenian, Gulf of Naples, Italy (40 ° 48.5′ N 14° 15′ E)	Hydrography, nutrients, phytoplankton, zooplankton	1984	Ribera d'Alcalà et al. (2004) and Zingone et al. (2010a, b)
NW Mediterranean, Blanes Bay (41.67 °N, 2.80 °E)	Hydrography, nutrients, Bacterioplankton (B), chlorophyll *a* (Chl *a*), primary production (PP), phytoplankton (Phyto),	1992 (Chl *a*) 1998 (B, PP, Phyto)	Duarte et al. (1999), Galand et al. (2010), and Gasol (2007)
NW Mediterranean, Thau Lagoon, France (43 ° 24′ N – 3 ° 36′ E)	Hydrography, nutrients, phytoplankton >5 μm (Phyto), picophytoplankton (Pico)	1971 1987 (Phyto) 1991 (Pico)	Collos et al. (2009)
NW Mediterranean, Point B, Villefrance-sur-mer, France (43 ° 41 N, 07 °19 E)	Hydrography, nutrients, zooplankton	1974 (with gaps) 1995	García-Comas et al. (2011)
NW Mediterranean, els Alfacs Bay (Ebre Delta), Spain	Hydrography, chlorophyll *a*, phytoplankton	1990	Llebot et al. (2011) and Fernández-Tejedor et al. (2010)
Eastern Mediterranean, Aegean Sea, Saronikos Gulf, Greece	Hydrography, nutrients, chlorophyll *a*, zooplankton	1989	Christou 1998
NW Mediterranean, Baleares transect, Mallorca, Spain (39.48 °N, 39.40 °N, 39.34 °N, all at 2.43 °E)	Hydrography, nutrients, chlorophyll *a*, zooplankton	1993	Fernández de Puelles et al. (2007) and Fernández de Puelles and Molinero (2008)

2005/2006, and associated the greater intensity of recent convection events to drought episodes decreasing the input of freshwater to the NW Mediterranean. In the Gulf of Naples, Zingone et al. 2010b related a negative trend in phytoplankton biomass with meteorological changes causing a decrease in water column stability, while Tunin-Ley et al. (2009) reported that some likely stenothermic *Ceratium* species had disappeared from the surface layers in response to water warming. As described in section "Harmful algal blooms in the Mediterranean", the decrease in phytoplankton abundance and biomass found in the Adriatic Sea over the last 20 years (Mozetič et al. 2010) has been related to a reduction of nutrient loads and to a lower river outflow, reflected in positive anomalies of the surface salinity. The changes in the Northern Adriatic encompassed also higher trophic levels (mesozooplankton and gelatinous plankton). The proposed scenario (shift in trophic control and identification of drivers) is presented in Mozetič et al. (2012). Variations in river discharge have also produced shifts in the appearance of the phytoplankton maximum (Viličić et al. 2009; Mozetič et al. 2010). In contrast to the Northern Adriatic, Ninčević Gladan et al. (2010) did not find modifications in the timing of the winter and spring phytoplankton maxima of a 30 year phytoplankton series in Kaštela Bay (Middle Adriatic), but reported an increase of phytoplankton abundance, in particular of dinoflagellates, from the mid-1980s to the 1990s, coinciding with years of high North Atlantic Oscillation (NAO) index and with a large scale change in the Northern Hemisphere (Conversi et al. 2010). These authors found that the relationship between chl *a* and primary production with the NAO index varied in different seasons, reflecting the NAO influence on local weather factors (such as temperature, winds and precipitation).

Viruses and Heterotrophic Microbes

Viruses

Studies on viruses in open and coastal waters of the Mediterranean Sea are still scarce. To date most of Mediterranean works have addressed the viral effects on bacterial biomass and production rather than the characterization of the viral community. Several publications (Table 6.2) reveal that viral abundances in the surface of coastal waters vary from 0.3×10^7 to 3.9×10^7 viruses mL^{-1}, while lower values occur in off-shore surface waters (0.02×10^7 to 2.5×10^7 viruses mL^{-1}) and still lower ones in

Table 6.2 Site, sampling data depths and variables measured and source of the studies considered in this review

Location	Date	Depth (m)	N	Variables	References
NW Mediterranean	September 2001	4 (coastal)	25	BA, VA, VBM, VBR	Bettarel et al. (2002)
	May 2005–April 2007	5 (coastal)	24	Chl *a* BA, VA, VBM, VBR	Boras et al. (2009)
	June 1995	5–200	42	Chl *a* BA, VA, VBM, VBR	Guixa -Boixereu et al. (1999a, b)
	June 1999	5–200	10	BA, VA, VBM, VBR	Weinbauer et al. (2003)
Alboran Sea	October and November 2004	1–200	6	BA, VA, BP, VBR	Magagnini et al. (2007)
W Mediterranean	October and November 2004	1–200	16	BA, VA, BP, VBR	Magagnini et al. (2007)
	June–July 2008	1–200	16	BA, VA, BP, VBR	Christaki et al. (2011)
Thyrrenean Sea	October and November 2004	1–200	11	BA, VA, BP, VBR	Magagnini et al. (2007)
Sicily chanel	October and November 2004	1–200	15	BA, VA, BP, VBR	Magagnini et al. (2007)
Adriatic Sea	May 1991–November 1992	0.5 (coastal)	5	Chl *a* BA, VA, VBR	Weinbauer et al. (1993)
	January–February 2001	1–200	2	BA, VA, BP, VBR	Corinaldesi et al. (2003)
	April–May 2003	0.5 (coastal)	1	BA, VA, BP, VBR	Bongiorni et al. (2005)
Ionian Sea	October and November 2004	1–200	19	BA, VA, BP, VBR	Magagnini et al. (2007)
	June–July 2008	1–200	17	BA, VA, BP, VBR	Christaki et al. (2011)
Levantine basin	June–July 2008	1–200	17	BA, VA, BP, VBR	Christaki et al. (2011)

Chl a chlorophyll *a* concentration, *BA* bacterial abundance, *VA* Viral abundance, *BP* bacterial production, *VBR* ratio of viral abundance respect to bacterial abundance, *VBM* viral mortality on bacteria, *N* number of data corresponding to viral abundance

deep waters (Magagnini et al. 2007). In the Mediterranean, as elsewhere, viral abundances increase from oligotrophic to more eutrophic areas. This means from coastal to open sea waters, and from the western to the eastern basin (Siokou-Frangou et al. 2010; Christaki et al. 2011).

Existing data (Table 6.2) show that bacterial and viral abundance are more coupled with chl *a* in open than in coastal waters (Fig. 6.6a, b). Furthermore, correlations between bacterial and viral abundance are higher ($n=218$, $r=0.677$, $p<0.001$) than between virus abundance and chl *a* concentration ($n=67$, $r=0.425$, $p<0.01$). This suggests that bacteria are more probable virus hosts than phytoplankton. The virus to bacteria ratio (VBR) varied between 7 and 69 (20 ± 11) inshore and between 1 and 93 (19 ± 20) in the upper 200 m of the open sea. Similar values of virus abundances and VBR could be found in other marine systems with comparable levels of fertility (Weinbauer 2004). The strong variability of the VBR may be due to factors such as (1) viruses may belong to different types of hosts, and or (2) viruses are in different phases of infection, increasing or decreasing in the water column as a consequence of their release from or their entrance into host cells at a particular sampling time. For instance, in the North Adriatic Sea an increase of VBR covariated with an increase of released viruses per cell (burst size) (Weinbauer et al. 1993), while the opposite relationship was observed by Bongiorni et al. (2005). Furthermore, based on data from Table 6.2, bacterial production (BP) increased from oligotrophic to eutrophic waters, and was positively and significantly correlated with viral ($n=97$, $r=0.556$, $p<0.05$) and bacterial abundances ($n=97$, $r=0.698$, $p<0.01$). Weinbauer et al. (1993) found a similar trend in the Adriatic Sea.

Along a gradient from eutrophy to oligotrophy in the Adriatic Sea, viral infection, estimated from viral lysis production, affected from 40 to 1.3 % of bacteria (Bongiorni et al. 2005). In the same study, burst size increased with the productivity of the system. The few available data indicate that lysogenic infection is lower than could be expected for the oligotrophic Mediterranean conditions. Reported values oscillate between 2 and 14 % of the total active bacteria (Bongiorni et al. 2005). Viral infection in open and coastal waters of the NW Mediterranean accounted for less than 20 % of the total bacterial mortality, suggesting that grazing by protists was the dominant pathway of bacterial losses (Guixa-Boixereu et al. 1999a, b; Bettarel et al. 2002). However, viral activity can occasionally prevail over grazing by HNF, for example in coastal waters with high bacterial abundances (Weinbauer and Peduzzi 1995; Boras et al. 2009).

Using enriched cultures, Alonso et al. (2002) isolated and characterized 26 bacteriophages from the W Mediterranean Sea. Morphological studies revealed that most of them were included in two of the three tailed families (Siphonviridae, Myoviridae) of the Caudovirales order, while viruses belonging to the Podoviridae (the third family) were not observed. All bacteriophages had icosahedral heads. Based on protein patterns, phages were grouped in 11 classes. Sizes oscillated between 30 and >100 nm (Weinbauer and Peduzzi 1994; Alonso et al. 2002) and it was observed that bacteria with different morphotypes hosted viruses with different sizes. Thus, viruses between 30 and 60 nm were mainly infecting rods (74 %) and spirillae bacteria (100 %), while viruses sized between 60 and 110 nm were found inside of cocci (65.5 %).

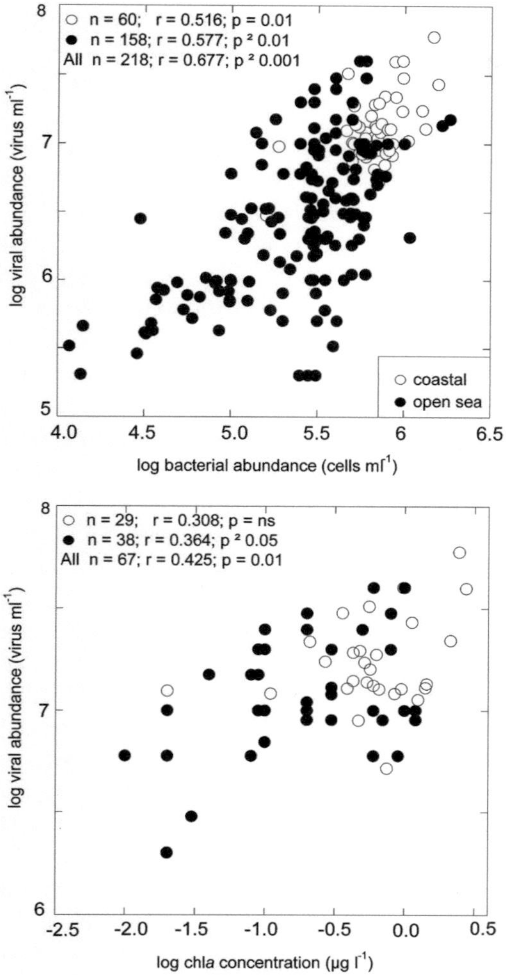

Fig. 6.6 Relationships between (**a**) bacterial and viral abundances, and (**b**) chlorophyll *a* concentration and viral abundance for coastal and open sea Mediterranean waters

In studies carried out in the Gulf of Naples, Zingone et al. (1999, 2006) found a highly dynamic seasonal fluctuation between the abundance of the picoeukaryote *Micromonas pusillla* and that of the viruses infecting it. Although changes of *M. pusilla* concentrations could be influenced by viral infection, there was no evidence that viruses could terminate host blooms, and the decline of *M. pusilla* populations did not appear to be related to viral impact. *M. pusilla* virus strains sampled on different dates showed distinct molecular characteristics, indicating that viral diversity for a single host can be very high. Zingone et al. (2006) isolated two different viruses (MpVN1 and MpVN2) infecting *M. pusilla.* Both viruses had similar sizes (100–130 nm) and icosahedral shape, but MpVN1 attached to the host with a tail of about 0.2 µm that was absent in MpVN2. The infectivity of these two viruses, tested in 11 *M. pusilla* isolates, showed no relationships with the geographic origin or the phylogenetic diversity of the host strains. A host strain cultured after recovery became resistant to infection from the same virus. This observation indicates that acquired immunity could have important implications concerning susceptibility of host strains to infection and host-virus dynamics in the natural environment. A potential consequence is that viral infection could change host diversity, although there are not yet studies of this effect in the Mediterranean Sea.

Heterotrophic Prokaryotes

Due to space limitations, this subsection will focus mainly on heterotrophic bacteria. However, it must be remembered that marine prokaryotes, including both bacteria and archaea, may present a rich variety of trophic strategies, including photoheterotrophy and chemolithotrophy. One of the firsts studies of bacterial distribution in the open Mediterranean, carried out in ultra-oligotrophic waters of the Levantine Sea (Zohary and Robarts 1992), showed that bacterial abundance at (3×10^8 cells L^{-1}) was around the lower threshold of the world ocean (Cho and Azam 1990). It was suggested that some key factor like the availability of inorganic nutrients, in particular phosphorus, could limit bacterial production in the Mediterranean Sea, and several experimental approaches were implemented (revision of Siokou-Frangou et al. 2010, and references therein) to test the nutrient control of bacterial production. In a Lagrangian experiment in surface waters of the Levantine Sea, the addition of phosphate produced a surprising response: a decline in chl *a* concentration and a rise in bacterial production. It was hypothesized that while phytoplankton was concurrently nitrogen- and phosphorus- limited, bacterial growth was mainly phosphorus- limited (Thingstad et al. 2005; Pitta et al. 2005; Zohary et al. 2005). Nevertheless, while phosphorus is usually the limiting nutrient, nitrogen and carbon limitation or co-limitation also occurs, and the type of limitation can vary with slight changes in competition, nutrient concentrations and grazing pressure (Sala et al. 2002; Van Wambeke et al. 2002, 2009). As a result, large changes in abundance, production and proportions of particle-attached to free-living bacteria have been documented over time scales of only a few hours (Mével et al. 2008).

The identification of the existing bacterial phylotypes and their temporal succession is a necessary first step to understand their role in the functioning of the system. Fingerprinting techniques such as DGGE, TRFLP, SSCP or ARISA (see, Dorigo et al. 2005 for an overview of methodologies) have been used to compare bacterial communities and establish their scales of variability. However, although, we have advanced in the current knowledge of prokaryotic diversity, these molecular studies are still far from answering specific questions on the ecosystem functioning worldwide (see review, Höfle et al. 2008). Since the late 1990s, a number of molecular studies carried out in the Mediterranean have improved our understanding of different aspects of bacterial community structure (Ghiglione et al. 2005; Coll et al. 2010, and references therein). The first microbial diversity studies from Mediterranean Sea microcosm experiments revealed an important variety of unknown microorganisms (Pukall et al. 1999) and allowed the unexpected detection of a high number of isolates that were phylogenetically close to *Sulfitobacter pontiacus*, a species only represented until then by two strains from the Black Sea (Sorokin 1995). Later, Schauer et al. (2000) were the first to provide a seasonal study of the bacterioplankton composition coastal sites of the NW Mediterranean, using the DGGE technique (denaturing gradient gel electrophoresis). They found marked seasonal changes of the bacterial assemblages and reported that spatial variability among some of the stations was consistent with different hydrographic conditions. In a new study determining the time-scale of changes in the dominant bacterial populations of Blanes Bay, Schauer et al. (2003) reported the taxonomic composition of the main bacterioplankton groups over time. Alonso-Saez et al. (2007) continued the work of Schauer et al. (2003) in the same station, and presented a comprehensive assessment of seasonal changes in bacterial diversity by means of different approaches (PCR-based and the direct CARD-FISH method).

In investigations realized by Acinas et al. (1997), community fingerprinting by 16S rDNA restriction analysis applied to Western Mediterranean offshore waters showed that the free-living pelagic bacterial cells were very different from those aggregated or attached to particles of more than about 8 μm. Acinas et al. (1999) studied also free and attached assemblages from three depths (5, 50, and 400 m) by cloning and sequencing the 16S rDNA obtained from the same samples, and used scanning electron microscopy to detect morphological patterns. As expected, the sequences retrieved from free or attached assemblages were very different. The subsample of attached bacteria contained low diversity, with close relatives of a well-known marine species, *Alteromonas macleodii*, representing the vast majority of the clones at every depth. On the other hand, the free-living assemblage was highly diverse and varied with depth. At 400 m, close relatives of cultivated Proteobacteria predominated, but near the surface most clones were related to phylotypes of the SAR11 cluster of the Proteobacteria. Moesender et al. (2001) found that the composition of surface and deep-water free bacterial communities was highly complex and reported pronounced differences among the attached and free-living bacterial communities throughout the water column (10–1,200 m) of the open EM (Aegean Sea). All studies, whether based on PCR-based clone libraries, metagenomic studies or fluorescence *in situ* hybridization, concur to identify members of the SAR11 group as some of the most abundant Mediterranean bacteria, with contributions of 25–45 % of the sequences (Zaballos et al. 2006; Alonso-Saez et al. 2007; Feingersch et al. 2010). These are commonly followed by other Alphaproteobacteria, such as the members of the Rhodobacteriaceae, particularly *Roseobacter*, which tend to be more common in coastal regions and during algal blooms. Apart of phototrophic Cyanobacteria (*Prochlorococcus* and *Synechococcus*), diverse culturable (Alteromonadales) and unculturable Gammaproteobacteria and Bacteroidetes represent the remaining bacterial diversity, with some differences with depth and distance to land. Rather similar community structures have been described at the Aloha station in the Pacific (Feingersch et al. 2010) and in surface waters of the Greenland Sea (Zaballos et al. 2006), although these latter authors found a larger share of Alteromonadales in the Ionian Sea, an observation that was not corroborated by Feingersch et al. (2010) in waters further East. Blümel et al. (2007) focused on the depth-specific distribution of Bacteroidetes in the EM, while García-Martínez and Rodríguez-Valera (2000) examined the microdiversity of the SAR11 and the Crenarchaeota group I in the WM. The number of studies addressing the distribution of Archaea in Mediterranean waters has increased substantially in the last decade, but the abundance patterns, diversity and metabolic activity of these prokaryotes are still poorly known. For example, Yakimov et al. (2009) found Mediterranean-specific archaeal ecotypes in bathypelagic waters, and Tamburini et al. (2009) in the Tyrrhenian, and De Corte et al. (2009) in the EM examined the depth distribution of bacteria and of different archaeal groups. Galand et al. (2010) found a strong seasonality of archaeal assemblages in Blanes Bay (NW Mediterranean), matching the variability of the environment and the rest of the planktonic community.

Heterotrophic Pico- and Nanoflagellates

Colourless and pigmented eukaryotic pico- and nanoflagellates in open and coastal marine systems are mainly dominated by small cells (70–80 % <5 μm) with total abundances between 10^5 and 10^6 cells L^{-1} (Zohary and Robarts 1992; Christaki et al. 1999, 2001) and 10^5–10^7 cells^{-1} L^{-1} (Massana 2011), respectively. For instance, in the Blanes Bay Microbial Observatory, along a time-series of 9 years, abundances of pigmented and colorless flagellates ≤3 μm (called picoeukaryotes) averaged 4.9×10^6 cells L^{-1} and 9.4×10^5 cells L^{-1}, respectively. Both groups showed a clear seasonality, with phototrophic picoeukaryotes (PPE) being higher in winter, and heterotrophic picoeukaryotes (HPE) peaking in summer (Fig. 2 of Massana 2011). Picoeukaryotes accounted for the largest fraction of eukaryotes year-round: PPE explained on average 82 % of total pigmented cells and heterotrophic picoeukaryotes represented 83 % of total colourless cells. In contrast, smaller eukaryotic cells (≤2 μm) constituted a more variable and minor fraction of pigmented (PPE) and colourless (HPE) cells, respectively (Massana 2011). Similar picoeukaryote abundances were found in other sites (Jürgens and Massana 2008). The ubiquity and relative stability of the picoeukaryotes appears to be related to bottom-up forcing, including environmental constraints and resource availability, and top-down effects such as predation and viral infection, which may control the realized abundances. In general, it seems that microbial eukaryotes in oligotrophic systems are controlled by resources, while predation control is more common in productive systems (Gasol 1994). Both, bottom up and top-down control could operate on individuals and populations, so that intrinsic specific differences in resource acquisition and predation-viral susceptibility may affect the final observed trends (Saura et al. 2011).

Heterotrophic nanoflagellates ≤5 μm are important bacterivores. For example, bacterivory consumed from 45 to 87 % of bacterial production in an east–west Mediterranean transect (Christaki et al. 2001). Large variability of bacterivory (<10–100 % consumption of bacterial production) was reported by Vaqué et al. (2001) for two different transects in the NW-Mediterranean Sea. These authors also showed that bacterivory was positively correlated with the abundance of bacteria with a high DNA content. A study in the Aegean Sea showed that mixotrophic nanoflagellates were relatively abundant (10–15 % of the stock). However, BP consumption by these protists was of the order of only 5 % (Christaki et al. 1999). A similar percentage of mixotrophic nanoflagellates was found in coastal waters (11–18 %) of the NW Mediterranean Sea, but with a contribution of 50 % to the total grazing by flagellates (Unrein et al. 2007). This activity was relatively high throughout the year (35–65 %) and was negatively correlated with soluble reactive phosphorus concentrations, suggesting that mixotrophic flagellates would be using their phagotrophic capability when nutrients such as P were limiting (Unrein et al. 2007).

Reports on heterotrophic pico- and nanoflagellate diversity are scarce in the Mediterranean Sea (Review of Massana 2011 and references therein). For instance, Massana et al. (2004) constructed genetic libraries of picoeukaryotes from surface coastal picoplankton of Blanes Bay (NW Mediterranean Sea, 41°40′N, 2°48′E) and found that the best-represented phylogenetic groups were novel alveolates-I (36 % of clones), dinoflagellates (17 %), novel marine stramenopiles or MAST (10 %), prasinophytes (5 %) novel alveolates-II (5 %), and cryptophytes (4 %). In the Mediterranean-wide study of Marie et al. (2006) cited in section "The seasonal succession of phytoplankton", the picoplankton displayed characteristics typical of oligotrophic oceanic areas with concentrations down to 1,000 cells mL^{-1} in the Eastern Basin. The most abundant eukaryotic sequences recovered by gradient gel electrophoresis represented uncultivated heterotrophic marine groups: alveolates I (16 %) and II (26 %) and a newly discovered group (env Nansha, 17 %) that could be related to Acantharians. Prasinophyceae (photosynthetic green algae) accounted for 10 % of the sequences, whereas Cercozoa, Stramenopiles, Polycystinea, dinoflagellates and ciliates provided minor contributions.

Ciliates

In the Mediterranean Sea, as anywhere in coastal and open marine environments, planktonic ciliates are dominated by the order Oligotrichida (Lynn and Small 2000). Within that order, the naked aloricate forms are the most important group (Margalef 1963; Travers 1973; Rassoulzadegan 1977, 1979; Vaqué et al. 1997). In one of the few studies dealing with the Mediterranean open waters, Dolan and Marrasé (1995) found that ciliate concentrations in the Catalan Sea averaged 230 cells L^{-1} at surface and about 850 cells L^{-1} at the DCM, with a total ciliate biomass of about 125 mg C m^{-2} for the 0–80 m layer, and reported a significant correlation between heterotrophic ciliate abundance and the chl *a* concentration. Pérez et al. (2000) studied the daily vertical distribution of planktonic ciliates in open waters of the Ligurian Sea and found average abundances ~3.3×10^3 cells L^{-1}, with a maximum in May of ~10^4 cells L^{-1}. These authors showed also that a portion of the heterotrophic ciliate population was able to migrate from 20 to 30 m depth during the day to the surface at night or in early morning and that ciliate abundance was positively correlated to zeaxanthin concentration, implying the possibility that ciliates concentrated around *Synechococcus* maxima and fed on them. Also in the open Ligurian Sea, Tanaka and Rassoulzadegan (2002) studied water column profiles and found that the abundance of ciliates, heterotrophic nanoflagellates and bacteria decreased by three, two and one order of magnitude over depth (5–2,000 m), respectively.

Since most of the primary production in the Mediterranean is due to nano- and picophytoplankton, one can expect that ciliates feed mainly on these size groups, as found in coastal and laboratory studies (i.e. Pierce and Turner 1992; Rassoulzadegan 1978; Sherr et al. 1986; Ferrier and Rassoulzadegan 1991; Christaki et al. 1999). Along a Mediterranean transect, Dolan et al. (1999) found changes in both ciliates and chl *a* concentration with a sevenfold W-to-E decline (2.8–0.4 and 0.48–0.07 mg C m^{-3}, for ciliates and chl *a*, respectively), although the lower chl *a* concentrations (0.1 mg chl *a* m^{-3}) of the eastern basin stations corresponded to a relatively high stock of ciliates (0.5 mg C m^{-3}, approx. 40 mg C m^{-2}). Pitta et al. (2001) reported a twofold decrease in ciliate concentration from west to east ($26.9–11.2 \times 10^6$ cells m^{-2} and

84.8–41.5 mg C m^{-2}) in abundance and biomass, respectively. The relationship between ciliate abundance and chl *a* concentration (Dolan and Marrasé 1995; Pitta et al. 2001) was stronger in the west than in the east (Siokou-Frangou et al. 2010), suggesting a better coupling with the phytoplankton stock in the west. Dolan and Marrasé (1995) reported that ciliates could have a grazing impact reaching 50 % of the primary production in the Catalan Sea and Pitta et al. (2001) found similar values in stations of the WM. Pérez et al. (2000) showed that ciliates were able to consume between 8 and 40 % of a primary production varying between 11 and 5.5 mg C m^{-3} day^{-1} in the Ligurian Sea. The estimates of ciliate consumption of primary production reported in Dolan et al. (1999) were lower than those measured by Pitta et al. (2001), especially in the EM; however, both studies seemed to be in agreement regarding the increasing importance of ciliates as primary production consumers towards the east.

The taxonomic diversity patterns of ciliates have been investigated with regard to tintinnids. Although they represent only a small fraction of the ciliate community, generally <10 % of total cell numbers or biomass (Dolan and Marrasé 1995; Vaqué et al. 1997; Dolan 2000; Dolan et al. 2002), their distinctive morphology makes them ideal organisms for the study of changes in the structure of microzooplankton communities (Thompson et al. 1999; Dolan 2000). A west-to-east increasing gradient of tintinnid diversity in the Mediterranean Sea was positively related to the depth of the DCM and inversely related to the chl *a* concentration (Dolan 2000; Dolan et al. 2002). In contrast, Pitta et al. (2001) did not observe any obvious west–east trend in tintinnid diversity but noted a peak in species richness in central stations. The size-class diversity of potential phytoplankton prey was identified by Dolan et al. (2002) as a major mechanism influencing tintinnid richness, suggesting that tintinnid diversity reflects that of the resources rather than competitive interaction or predation. No clear relationship between tintinnid diversity, chl *a* concentration and water column stability was found by Cariou et al. (1999) in the NW Mediterranean, and recent studies confirmed that water column structure alone appeared to be a poor predictor of temporal changes in tintinnid diversity, supporting the view that it could be directly linked to characteristics of food resources (Dolan et al. 2006).

Microbial and Classical Food Webs. The Example of the Catalan Sea

The general oligotrophy of the Mediterrranean has been associated with a widespread dominance of small phytoplankton and the channelling of an important proportion of the carbon fluxes through the microbial food web (Siokou-Frangou et al. 2010). However, there is a large variability associated with the west to east gradient of oligotrophy and with the intermittent occurrence of fertilization events linked to the seasonal cycle, mesoscale features or other processes. In agreement with the increasing oligotrophy towards the east, heterotrophic to autotrophic biomass ratios ranged in general from 0.5 to 3.0 in the WM (Christaki et al. 1996; Pedrós-Alió et al. 1999) and from 0.9 to 3.9 in the Aegean Sea (Siokou-Frangou et al. 2002, 2010). Phytoplankton production tends to be several times higher than bacterial production (Pedrós et al. 1999; Christaki et al. 2011), although the strong variability typical of metabolic rate estimates (Christaki et al. 2011) and differences in methodology make it difficult to compare values from different studies. The contribution of bacterioplankton in the open sea tends to be more important than near the coast, but there are exceptions (see below). For example Krstulović et al. (1995) report that BP amounted 9–28 % of PP in eutrophic Kaštela Bay and 10–40 % in an open sea station in the middle Adriatic. The dominance of picoplankton may be interrupted by proliferations of nano and microplankton, and activation of the classical or herbivorous food web, after nutrient enrichment events. These pulses of growth tend to be short-lived and difficult to sample, leading to a probable underestimation of their contribution. The intermittent dominance of the herbivorous pathway may occur also within the DCM layers, in particular where the chlorophyll maximum is relatively shallow, as happens in the central divergence of the NW Mediterranean (Estrada 1996). Some of these characteristics can be appreciated in Fig. 6.7, based in part on data from Pedrós-Alió et al. 1999, which represents major biomasses and fluxes of the microbial food web in two stations located on the shelf and in the central divergence of the NW Mediterranean (marked C and D in Fig. 6.1). As can be expected, phytoplankton biomass (PB) and production (PP) were higher during the winter phytoplankton bloom than during the summer stratification period. At the same time, bacterial biomass (BB) and production (BP) were lower in winter, specially on the shelf, contributing to the higher PB/BB and PP/BP ratios in winter. The flux of carbon to metazooplankton through the herbivorous food web was also greater in winter, although the metazooplankton biomass was similar during both seasons, presumably because metazooplankton had not yet experienced a substantial increase in response to the phytoplankton bloom. The standing stocks and C fluxes of the shelf and the divergence stations were fairly similar in winter, while a seasonal phytoplankton bloom was happening at both stations. In summer, BB did not change much from the coast to the open sea, while BP was higher on the shelf, probably in response to dissolved organic carbon inputs from land, while in contrast what could be expected for an open sea location. PB and PP were also similar in both stations, although diatom abundance and mesozooplankton ingestion rates were higher on the shelf. The summer enhancement of the classical food web activity in the central divergence can

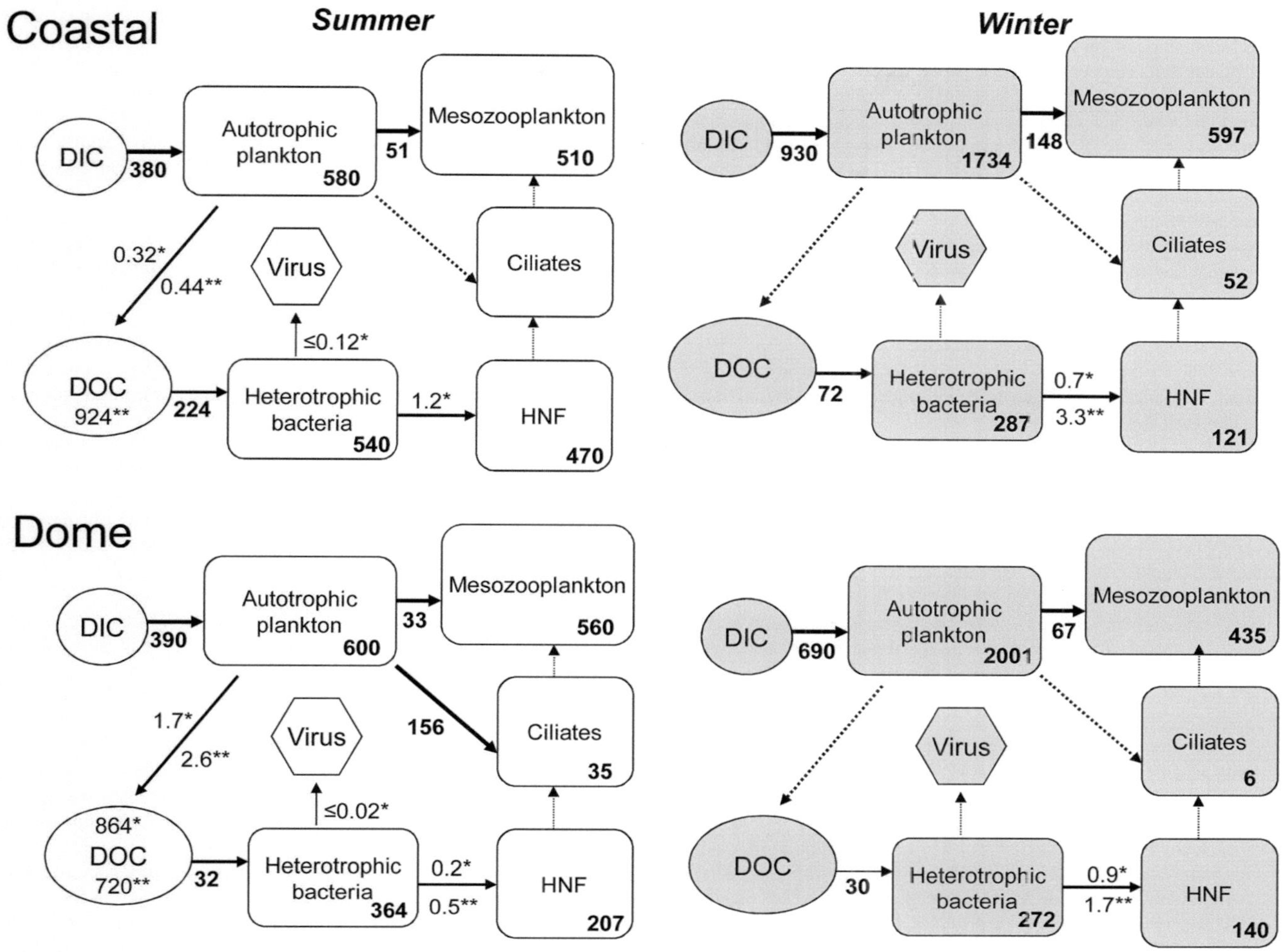

Fig. 6.7 Summary of microbial biomasses and fluxes in the planktonic ecosystem of the Catalan Sea, for two reference stations, one located on the shelf (*C* in Fig. 6.1) and another above the central divergence (*D* in Fig. 6.1). *Boxes* represent depth-integrated biomasses over 0 to 60–80 m (mg C m^{-2}) and *arrows* indicate fluxes (* and **, surface and DCM values, respectively, in mg C m^{-3} d^{-1}; *bold numbers* refer to depth-integrated values in mg C m^{-2} day^{-1}). The summer figures are based on pooled data from cruises carried out in late spring of 1993, 1995 and 1996, and the winter ones are taken from a cruise in March 1999. DOC concentrations are from Doval et al. (1995); phytoplankton and bacterial data from Pedrós-Alió et al. (1999), Morán and Estrada (2005), Vaqué et al. (2001) and Vaqué (unpublished data); ciliate predation rates were derived from Dolan and Marrasé (1995); mesozooplankton biomass was taken from Calbet et al. (1996, 2002), Saiz et al. (1999) and Alcaraz (unpublished data); predation rates were calculated from the gut content or from egg production rates assuming a gross growth efficiency of 40 %

be explained, as noted in section "Fertilization mechanisms in the Mediterranean", because the shallow level of the nutricline in this zone facilitates the introduction of nutrients into relatively well-illuminated water layers (Estrada 1996). Siokou-Frangou et al. (2010) have highlighted the potential flexibility and efficiency of the planktonic food web in the Mediterranean Sea, characterized by a high diversity of trophic roles and with connections spanning a continuum between the microbial loop and the classical food web. These features may contribute to the maintenance of a higher heterotrophic biomass per unit of primary production and a more efficient energy transfer to the upper trophic levels in the Mediterranean and other oligotrophic seas, than in more eutrophic marine systems (Margalef 1986; Alcaraz et al. 1985).

Acknowledgments We acknowledge the support of grant 2009 SGR 588 of the Generalitat de Catalunya, and projects FAMOSO (CTM2008-06261-C03-01) and FLAME (CGL2010-16304) of the Spanish Ministry of Science and Innovation. We are very grateful to Drs. M. Alcaraz, A. Calbet and E. Saiz for information on zooplankton biomass and activity.

References

Acinas SG, Rodríguez-Valera F, Pedrós-Alió C (1997) Spatial and temporal variation in marine bacterioplankton diversity as shown by RFLP fingerprinting of PCR amplified 16S rDNA. FEMS Microbiol Ecol 24:27–40

Acinas SG, Anton J, Rodriguez-Valera F (1999) Diversity of free-living and attached bacteria in offshore Western Mediterranean waters as depicted by analysis of genes encoding 16S rRNA. Appl Environ Microbiol 65:514–522

Agawin NSR, Duarte CM, Agusti S (1998) Growth and abundance of Synechococcus sp. in a Mediterranean Bay: seasonality and relationship with temperature. Mar Ecol Prog Ser 170:45–53

Alcaraz M, Estrada M, Flos J, Fraga F (1985) Particulate organic carbon and nitrogen and plankton biomass in oligotrophic and upwelling systems. In: Bas C, Margalef R, Rubiés P (eds) Simposio Internacional sobre las áreas de afloramiento más importantes del Oeste Africano. CSIC, Barcelona, pp 435–438

Alcaraz M, Saiz E, Estrada M (1994) Excretion of ammonia by zooplankton and its potential contribution to nitrogen requirements for primary production in the Catalan Sea (NW Mediterranean). Mar Biol 119:69–76

Aligizaki K, Nikolaidis G, Fraga S (2008) Is *Gambierdiscus* expanding to new areas? Harmful Algal News 36:6–7

Alonso MC, Rodriguez J, Borrego JJ (2002) Characterization of marine bacteriophages isolated from the Alboran Sea (Western Mediterranean). J Plankton Res 24:1079–1087

Alonso-Saez L, Balagué V, Sa EL, Sánchez O, González JM, Pinhassi J, Massana R, Pernthaler J, Pedrós-Alió C, Gasol JM (2007) Seasonality in bacterial diversity in north-west Mediterranean coastal waters: assessment through clone libraries, fingerprinting and FISH. FEMS Microbiol Ecol 60:98–112

Amzil Z, Fresnel J, Le Gal D, Billard C (2001) Domoic acid accumulation in French shellfish in relation to toxic species of *Pseudo-nitzschia multiseries* and *P. pseudodelicatissima*. Toxicon 39:77–82

Antoine D, Morel A, André J (1995) Algal pigment distribution and primary production in the eastern Mediterranean as derived from Coastal Zone Color Scanner observations. J Geophys Res 100: 16193–16209

Arin L, Morán XAG, Estrada M (2002) Phytoplankton size distribution and growth rates in the Alboran Sea (SW Mediterranean): short term variability related to mesoscale hydrodynamics. J Plankton Res 24:1019–1033

Barlow RG, Mantoura RFC, Cummings DG, Fileman TW (1997) Pigment chemotaxonomic distributions of phytoplankton during summer in the western Mediterranean. Deep Sea Res II 44:833–850. doi:10.1016/S0967-0645(96)00089-6

Bergh Ø, Børsheim KY, Bratbak G, Hedal M (1989) High abundance of viruses found in aquatic environments. Nature 340:467–468

Bernardi Aubry B, Berton A, Bastianini M, Socal G, Acri F (2004) Phytoplankton succession in a coastal area of the NW Adriatic, over a 10-year sampling period (1990–1999). Cont Shelf Res 24:97–115

Bernardi Aubry F, Acri F, Bastianini M, Pugnetti A, Socal G (2006) Picophytoplankton contribution to phytoplankton community structure in the Gulf of Venice (NW Adriatic Sea). Int Rev Hydrobiol 91(1):51–70

Béthoux JP, Copin-Montégut G (1986) Biological fixation of atmospheric nitrogen in the Mediterranean Sea. Limnol Oceanogr 31:1353–1358

Bettarel Y, Dolan JR, Hornak K, Lemée R, Masin M, Pedrotti ML, Rochelle-Newall E, Simek K, Sime-Ngando T (2002) Strong, weak, and missing links in a microbial community of the NW Mediterranean Sea. FEMS Microbiol Ecol 42:451–462

Blümel M, Suling J, Imhoff JF (2007) Depth-specific distribution of bacteroidetes in the oligotrophic eastern Mediterranean Sea. Aquat Microb Ecol 46:209–224

Bongiorni L, Magagnini M, Armeni M, Noble R, Danovaro R (2005) Viral production decay rates and life strategies along a trophic gradient in the North Adriatic Sea. Appl Environ Microbiol 71:6644–6650

Boni L, Milandri A, Poletti R, Pompei M (1993) DSP cases along the coast of Emilia-Romagna (northwestern Adriatic Sea). In: Smayda TJ, Shimizu Y (eds) Toxic phytoplankton blooms in the sea. Elsevier, Amsterdam, pp 475–481

Boras JA, Sala MM, Vázquez-Dominguez E, Weinbauer MG, Vaqué D (2009) Annual changes of bacterial mortality due to viruses and protists in an oligotrophic coastal environment (NW Mediterranean). Environ Microbiol 11:1181–1193

Bosc E, Bricaud A, Antoine D (2004) Seasonal and interannual variability in algal biomass and primary production in the Mediterranean Sea, as derived from 4 years of SeaWiFS observations. Glob Biogeochem Cycle 18:GB1005. doi:10.1029/2003GB002034

Bravo I, Reguera B, Martínez A, Fraga S (1990) First report of *Gymnodinium catenatum* Graham on the Spanish Mediterranean coast. In: Granéli A, Sundström B, Edler L, Anderson DM (eds) Toxic marine phytoplankton. Elsevier, New York, pp 449–452

Buck K, Bentham WN (1998) A novel symbiosis between a cyanobacterium, *Synechococcus* sp., an aplastidic protist, *Solenicola setigera*, and a diatom, *Leptocylindrus mediterraneus*, in the open ocean. Mar Biol 132:349–355

Calbet A, Alcaraz M, Saiz E, Estrada M, Trepat I (1996) Planktonic herbivorous food webs in the Catalan Sea (NW Mediterranean): temporal variability and comparison of indices of phytozooplankton coupling based on state variables and rate processes. J Plankton Res 18:2329–2347

Calbet A, Saiz E, Alcaraz M (2002) Copepod egg production in the NW Mediterranean: effects of winter environmental conditions. Mar Ecol Prog Ser 237:173–184

Cariou JB, Dolan JR, Dallot S (1999) A preliminary study of tintinnid diversity in the NW Mediterranean Sea. J Plankton Res 21:1065–1075

Cerino F, Zingone A (2006) A survey of cryptomonad diversity and seasonality at a coastal Mediterranean site. Eur J Phycol 4:363–378

Cerino F, Bernardi Aubry F, Coppola J, La Ferla R, Maimone G, Socal G, Totti C (2011) Spatial and temporal variability of pico-, nano- and microphytoplankton in the offshore waters of the southern Adriatic Sea (Mediterranean Sea). Cont Shelf Res 44:94–105. doi:10.1016/j.csr.2011.06.006

Charles F, Lantoine F, Brugel S, Chrétiennot-Dinet MJ, Quiroga I, Rivière B (2005) Seasonal survey of the phytoplankton biomass, composition and production in a littoral NW Mediterranean site, with special emphasis on the picoplanktonic contribution. Estuar Coast Shelf Res 65:199–212

Cho BC, Azam F (1990) Biogeochemical significance of bacterial biomass in the ocean's euphotic zone. Mar Ecol Prog Ser 63:253–259

Christaki U, Van Wambeke F, Christou ED, Conan P, Gaudy R (1996) Food web structure variability in the surface layer, at a fixed station influenced by the North Western Mediterranean Current. Hydrobiologia 32:145–153

Christaki U, Van-Wambeke F, Dolan JR (1999) Nanoflagellates (mixotrophs, heterotrophs, and autotrophs) in the oligotrophic eastern Mediterranean: standing stocks, bacterivory and relationships with bacterial production. Mar Ecol Prog Ser 181:297–307

Christaki U, Giannakourou A, Van Wambeke F, Gregorí G (2001) Nanoflagellate predation on auto-and heterotrophic picoplankton in the oligotrophic Mediterranean Sea. J Plankton Res 23:1297–1310

Christaki U, Van Wambeke F, Lefevre D, Lagaria A, Prieur L, Pujo-Pay M, Grattepanche JD, Colombet J, Psarra S, Dolan JR, Simé-Ngando T, Conan P, Weinbauer MG, Moutin T (2011) The impact of anticyclonic mesoscale structures on microbial food webs in the Mediterranean Sea. Biogeosciences 8:1839–1852

Christou E (1998) Interannual variability of copepods in a Mediterranean coastal area (Saronikos Gulf, Aegean Sea). J Mar Syst 15:523–532

CIESM (2003) Mediterranean biological time series. In: Bryand F (ed) CIESM workshop monographs 22, Monaco

CIESM (2010) Phytoplankton responses to Mediterranean environmental changes. In: Bryand F (ed) CIESM workshop monographs 40, Monaco

Claustre H, Kerherve P, Marty JC, Prieur L, Videau C, Hecq JH (1994) Phytoplankton dynamics associated with a geostrophic front: ecological and biogeochemical implications. J Mar Res 52:711–742

Cloern JE, Jassby AD (2010) Patterns and scales of phytoplankton variability in estuarine–coastal ecosystems. Estuaries Coast 33:230–241

Coll M, Piroddi C, Steenbeek J, Kaschner K, Rais Lasram FB, Aguzzi J, Ballesteros E, Nike Bianchi C, Corbera J, Dailianis T, Danovaro R,

Estrada M, Froglia C, Galil BS, Gasol JM, Gertwagen R, Gil J, Guilhaumon F, Kesner-Reyes K, Kitsos MS, Koukouras A, Lampadariou N, Laxamana E, López-Fé de la Cuadra CM, Lotze HK, Martin D, Mouillot D, Oro D, Raicevich S, Rius-Barile J, Saiz-Salinas JI, San Vicente C, Somot S, Templado J, Turon X, Vafidis D, Villanueva R, Voultsiadou E (2010) The biodiversity of the Mediterranean Sea: estimates, patterns, and threats. PLoS One 5:11842–11843

Collos Y, Bec B, Jauzein C, Abadie E, Laugier T, Lautier J, Pastoureaud A, Souchou P, Vaquer A (2009) Oligotrophication and emergence of picocyanobacteria and a toxic dinoflagellate in Thau lagoon, southern France. J Sea Res 61:68–75

Conversi A, Fonda Umani S, Peluso T, Molinero JC, Santojanni A, Edwards M (2010) The Mediterranean Sea regime shift at the end of the 1980s, and intriguing parallelisms with other European basins. PLoS One 5:e10633. doi:10.1371/journal.pone.0010633

Corinaldesi R, Crevatin E, del Negro P, Marini M, Russo A, Fonda-Umani S, Danovaro R (2003) Large-scale spatial distribution of virioplankton in the Adriatic Sea: testing the trophic state control hypothesis. Appl Environ Microbiol 69:2664–2673

Crombet Y, Leblanc K, Quéguiner B, Moutin T, Rimmelin P, Ras J, Claustre H, Leblond N, Oriol L, Pujo-Pay M (2011) Deep silicon maxima in the stratified oligotrophic Mediterranean Sea. Biogeosciences 8:459–475. doi:10.5194/bg-8-459-2011

Cros L (2001) Planktonic coccolithophores of the NW Mediterranean. PhD thesis, Universitat de Barcelona, Publicacions Universitat de Barcelona, Barcelona

Cros L, Kleijne A, Zeltner A, Billard C, Young JR (2000) New examples of holococcolith-heterococcolith combination coccospheres and their implications for cocolithophorid biology. Mar Micropaleontol 39:1–34

Cros L, Arbós D, Arin L, Camp J, Estrada M, Fortuño JM, Morales-Blake A, Sampedro N, Riba J, Vila M (2002) A *Calyptrosphaera sphaeroidea* bloom in waters of the Tarragona harbour (NW Mediterranean). J Nannoplankton Res 24:87

Cullen JJ, Franks PJS, Karl DM, Longhurst A (2002) Physical influences on marine ecosystem dynamics. In: Robinson AR, Mc Carthy JJ, Rothschild BJ (eds) Biological-physical interactions in the sea, The sea, 12. Wiley, New York, pp 297–336

D'Ortenzio F, Ribera d'Alcalá M (2009) On the trophic regimes of the Mediterranean Sea: a satellite analysis. Biogeosciences 6:139–148

De Corte D, Yokokawa T, Varela MM, Agogué H, Herndl GH (2009) Spatial distribution of Bacteria and Archaea and amoA gene copy numbers throughout the water column of the Eastern Mediterranean Sea. ISME J 3:147–158

Delgado M, Latasa M, Estrada M (1992) Variability in size distribution of the phytoplankton across the Catalan front of the north-west Mediterranean. J Plankton Res 14:753–771

DeLong EF, Béjà O (2010) The light-driven proton pump proteorhodopsin enhances bacterial survival during tough times. PLoS Biol 8(4):e1000359. doi:10.1371/journal.pbio.1000359

Denis M, Thyssen M, Martin V, Manca B, Vidussi F (2010) Ultraphytoplankton basin-scale distribution in the Eastern Mediterranean Sea in winter: link to hydrodynamism and nutrients. Biogeosciences 7:2227–2244. doi:10.5194/bg-7-2227-2010

Dolan J (2000) Tintinnid ciliate diversity in the Mediterranean Sea: longitudinal patterns related to water column structure in late spring—early summer. Aquat Microb Ecol 22:69–78

Dolan J, Marrasé C (1995) Planktonic ciliate distribution relative to a deep chlorophyll maximum: Catalan Sea, N.W. Mediterranean, June 1993. Deep Sea Res I 42:1965–1987

Dolan JR, Vidussi F, Claustre H (1999) Planktonic ciliates in the Mediterranean Sea: longitudinal trends. Deep Sea Res I 46:2025–2039

Dolan JR, Claustre H, Carlotti F, Plounevez S, Moutin T (2002) Microzooplankton diversity: relationships of tintinnid ciliates with resources, competitors and predators from the Atlantic coast of Morocco to the Eastern Mediterranean. Deep Sea Res I 49:1217–1232

Dolan JR, Lemée R, Gasparini S, Mousseau L, Heyndrickx C (2006) Probing diversity in the plankton: using patterns in tintinnids (planktonic marine ciliates) to identify mechanisms. Hydrobiologia 555:143–157

Dorigo U, Volatier L, Humbert JF (2005) Molecular approaches to the assessment of biodiversity in aquatic microbial communities. Water Res 39:2207–2218

Doval MD, Pérez FF, Berdalet E (1995) Dissolved and particulate organic carbon and nitrogen in the Northwestern Mediterranean. Deep Sea Res I 46:511–527

Duarte CM, Agustí S, Kennedy H, Vaqué D (1999) The Mediterranean climate as a template for Mediterranean marine ecosystems: the example of the northeast Spanish littoral. Prog Oceanogr 44:245–270

Ducklow HW, Purdie DA, Williams PJ, Davies JM (1986) Bacterioplankton: a sink for carbon in a coastal marine plankton community. Science 232:865–867

Estrada M (1985) Deep phytoplankton and chlorophyll maxima in the Western Mediterranean. In: Moraitou-Apostolopoulou M, Kiortsis V (eds) Mediterranean marine ecosystems. Plenum Press, New York, pp 247–277

Estrada M (1991) Phytoplankton assemblages across a NW Mediterranean front: changes from winter mixing to spring stratification. In: Ros J, Prat N (eds) Homage to Ramon Margalef; or, Why there such pleasure in studying nature, Oecologia aquatica 10. Universitat de Barcelona, Barcelona, pp 157–185

Estrada M (1996) Primary production in the northwestern Mediterranean. Sci Mar 60(suppl 2):55–64

Estrada M (1999) Hidrodinàmica i fitoplàncton en el mar Català. Mem Real Acad Ci Barcelona 58:187–247

Estrada M, Margalef R (1988) Supply of nutrients to the Mediterranean photic zone across a persistent front. Oceanol Acta 9 (Spec. issue):133–142

Estrada M, Vives F, Alcaraz M (1985) Life and the productivity of the open sea. In: Margalef R (ed) Western Mediterranean, Key environments. Pergamon Press, Oxford, pp 148–197

Estrada M, Marrasé C, Latasa M, Berdalet E, Delgado M, Riera T (1993) Variability of deep chlorophyll maximum characteristics in the Northwestern Mediterranean. Mar Ecol Prog Ser 92:289–300

Estrada M, Varela R, Salat J, Cruzado A, Arias E (1999) Spatio-temporal variability of the winter phytoplankton distribution across the Catalan and North Balearic fronts (NW Mediterranean). J Plankton Res 21:1–20

Feingersch R, Suzuki MT, Shmoish M, Sharon I, Sabehi G, Partensky F, Béjà O (2010) Microbial community genomics in Eastern Mediterranean sea surface waters. ISME J 4(1):78–80

Fernández de Puelles ML, Molinero JC (2008) Decadal changes in hydrographic and ecological time-series in the Balearic Sea (western Mediterranean), identifying links between climate and zooplankton. ICES J Mar Sci 65:311–317

Fernández de Puelles ML, Alemany F, Jansà J (2007) Zooplankton time-series in the Balearic Sea (Western Mediterranean): variability during the decade 1994–2003. Prog Oceanogr 74:329–354

Fernández-Tejedor M, Delgado M, Garcés E, Camp J, Diogène J (2010) Toxic phytoplankton response to warming in two Mediterranean bays of the Ebro Delta. In: Bryand F (ed) Phytoplankton responses to Mediterranean environmental changes, CIESM workshop monographs 40. CIESM, Monaco, pp 83–88

Ferrier C, Rassoulzadegan F (1991) Density-dependent effects of protozoans on specific growth rates in pico- and nanoplanktonic assemblages. Limnol Oceanogr 36:657–669

Fonda Umani S (1996) Pelagic production and biomass in the Adriatic Sea. Sci Mar 60(suppl 2):65–77

Font J, Salat J, Tintoré J (1988) Permanent features in the circulation of the Catalan sea. Oceanol Acta 9(Spec issue):51–57

Foulon E, Not F, Jalabert F, Cariou T, Massana R, Simon N (2008) Ecological niche partitioning in the picoplanktonic green alga

Micromonas pusilla: evidence from environmental surveys using phylogenetic probes. Environ Microbiol 10:2433–2443

France J, Mozetič P (2006) Ecological characterization of toxic phytoplankton species (*Dinophysis* spp., Dinophyceae) in Slovenian mariculture areas (Gulf of Trieste, Adriatic Sea) and the implications for monitoring. Mar Pollut Bull 52:1504–1516

Fuhrman JA (1999) Marine viruses and their biogeochemical and ecological effects. Nature 399:541–548

Fuhrman JA, Steele JA, Hewson I, Schwalbach MS, Brown MV, Green JL, Brown JH (2008) A latitudinal diversity gradient in planktonic marine bacteria. Proc Natl Acad Sci USA 105(22):7774–7778

Galand P, Gutiérrez-Provecho C, Massana R, Gasol JM, Casamayor EO (2010) Inter-annual recurrence of archaeal assemblages in the coastal NW Mediterranean Sea (Blanes Bay Microbial Observatory). Limnol Oceanogr 55:2117–2125

Garcés E, Camp J (2012) Habitat changes in the Mediterranean sea and the consequences for harmful algal blooms formation. In: Stambler N (ed) Life in the Mediterranean Sea: a look at habitat changes. Nova Science Publishers Inc, Hauppauge

Garcés E, Masó M, Vila M, Camp J (2000) Harmful algae events in the Mediterranean: are they increasing? Harmful Algal News 20:1 and 10

García-Comas C, Stemmann L, Ibanez F, Berline L, Mazzocchi MG, Gasparini S, Picheral M, Gorsky G (2011) Zooplankton long-term changes in the NW Mediterranean Sea: decadal periodicity forced by winter hydrographic conditions related to large-scale atmospheric changes? J Mar Syst 87:216–2226

García-Martínez J, Rodríguez-Valera F (2000) Microdiversity of uncultured marine prokaryotes: The SAR11 cluster and the marine Archaea of Group I. Mol Ecol 9:935–948

Garczarek L, Dufresne A, Rousvoal S, West NJ, Mazard S, Marie D, Claustre H, Raimbault P, Post AF, Scanlan DJ, Partensky F (2007) High vertical and low horizontal diversity of *Prochlorococcus* ecotypes in the mediterranean sea in summer. FEMS Microbiol Ecol 60:189–206

Gasol JM (1994) A framework for the assessment of top-down vs. bottom-up control of heterotrophic nanoflagellates abundance. Mar Ecol Prog Ser 113:291–300

Gasol JM (2007) Microbial observatories as a tool to detect and describe changes in marine (microbial) diversity and ecosystem functioning: lessons learnt from the Blanes Bay Microbial Observatory. In: O'Sullivan G, McDonough N (eds) Anthropogenic and climate change impacts on marine biodiversity and ecosystem function. MarinEra report 5, OOstende, Belgium, pp 21–22

Ghiglione J, Larcher M, Lebaron P (2005) Spatial and temporal scales of variation in bacterioplankton community structure in the NW Mediterranean Sea. Aquat Microb Ecol 40:229–240

Ghiglione JF, Palacios C, Marty JC, Mével G, Labrune C, Conan P, Pujo-Pay M, Garcia N, Goutx M (2008) Role of environmental factors for the vertical distribution (0–1000 m) of marine bacterial communities in the NW Mediterranean Sea. Biogeosciences 5:1751–1764

Giacobbe MG, Penna A, Gangemi E, Masó M, Garcés E, Fraga S, Bravo I, Azzaro S, Penna A (2007) Recurrent high-biomass blooms of *Alexandrium taylori* (Dinophyceae), a HAB species expanding in the Mediterranean. Hydrobiologia 580:125–133

Goberville E, Beaugrand G, Sautour B, Tréguer P, Somlit-Team (2010) Climate-driven changes in coastal marine systems of Western Europe. Mar Ecol Prog Ser 408:129–148

Gómez F (2006) Endemic and Indo-Pacific plankton in the Mediterranean Sea: a study based on dinoflagellate records. J Biogeogr 33:261–270

Gómez F, Moreira D, López-García P (2010) *Neoceratium* gen. nov., a new genus for all marine species currently assigned to *Ceratium* (Dinophyceae). Protist 161:35–54

Gordon N, Angel D, Neori A, Kress N, Kimor B (1994) Heterotrophic dinoflagellates with symbiotic cyanobacteria and nitrogen limitation in the Gulf of Aqaba. Mar Ecol Prog Ser 107:83–88

Guixa-Boixereu N, Vaqué D, Gasol JM, Pedrós-Alió C (1999a) Distribution of viruses and their potential effect on bacterioplankton in an oligotrophic marine system. Aquat Microb Ecol 19:205–213

Guixa-Boixereu N, Lysnes K, Pedrós-Alió C (1999b) Viral lysis and bacterivory during a phytoplankton bloom in a coastal water microcosm. Appl Environ Microbiol 65:1949–1958

Herbland A, Voituriez B (1979) Hydrological structure analysis for estimating the primary production in the tropical Atlantic Ocean. J Mar Res 3:87–101

Höfle MG, Kirchman DL, Christen R, Brettar I (2008) Molecular diversity of bacterioplankton: link to predictive biogeochemistry of pelagic ecosystems. Aquat Microb Ecol 53:39–58

Hofrichter R, Petz W, Kremer B, Schmidt-Rhaesa A, Schaumann K (2002) Biodiversität und Probleme der Systematik: ein Überblick. In: Hofrichter R (ed) Das Mittelmeer – Fauna, Flora, Ökologie, vol II/1. Spektrum Akademischer Verlag, Heidelberg/Berlin, pp 24–50, Kap. 1

Honsell G, Nichetto P, Sidari L, Tubaru A (1995) Toxic dinoflagellates in the Mediterranean sea. Giornale Botanico Italiano 129:321–403

Iermano I, Liguori G, Iudicone D, Buongiorno Nardelli B, Colella S, Zingone A, Saggiomo V, Ribera d'Alcalá M (2009) Dynamics of short-living filaments and their relationship with intense rainfall events and river flows. In: European Geosciences Union, General Assembly 2009, Wien, Abstract: EGU2009-8508

Ignatiades L, Psarra S, Zervakis V, Pagou K, Souvermezoglou E, Assimakopoulou G, Gotsis-Skretas O (2002) Phytoplankton size-based dynamics in the Aegean Sea (Eastern Mediterranean). J Mar Syst 36:11–28

Ignatiades L, Gotsis-Skretas O, Pagou K, Krasakopoulou E (2009) Diversification of phytoplankton community structure and related parameters along a large-scale longitudinal east–west transect of the Mediterranean Sea. J Plankton Res 31:411–428

Illoul H, Masó M, Fortuño JM, Morales-Blake A, Cros L, Seridji R (2008) Potentially harmful microalgae in coastal waters of the Algiers area (Southern Mediterranean Sea). Cryptogam Algol 29:261–278

Jacobson DM, Anderson DM (1996) Widespread phagocytosis of ciliates and other protists by marine mixotrophic and heterotrophic thecate dinoflagellates. J Phycol 32:279–285

Jacques G, Sournia A (1978–1979) Les «eaux rouges» dues au phytoplancton en Méditerranée. Vie Milieu 28–29:175–187

Jiao N, Zhang Y, Zeng Y, Hong N, Liu R, Chen F, Wang P (2007) Distinct distribution pattern of abundance and diversity of aerobic anoxygenic phototrophic bacteria in the global ocean. Environ Microbiol 9:3091–3099

Jörgensen E (1920) Mediterranean Ceratia. Report on the Danish oceanographical expeditions 1908–1910 to the Mediterranean and adjacent Seas, Copenhagen, Denmark, vol II. J1, pp 1–110

Jürgens K, Massana R (2008) Protistan grazing on marine bacterioplankton. In: Kirchman DL (ed) Microbial ecology of the oceans, 2nd edn. Wiley, New York, pp 383–441

Karner MB, DeLong EF, Karl DM (2001) Archaeal dominance in the mesopelagic zone of the Pacific Ocean. Nature 409:507–510

Krom MD, Emeis KC, Van Cappellen P (2010) Why is the eastern Mediterranean phosphorus limited? Prog Oceanogr 85:236–244

Krstulović N, Pucher-Petković T, Šolić M (1995) The relation between bacterioplankton and phytoplankton production in the mid Adriatic Sea. Aquat Microb Ecol 9:41–45

Latasa M, Scharek R, Vidal M, Vila-Reixach G, Emelianov M, Gutiérrez-Rodríguez A, Gasol JM (2010) Preferences of phytoplankton groups for waters of different trophic status in the northwestern Mediterranean Sea. Mar Ecol Prog Ser 407:27–42

Le Bec C, Belin C, Gaertner JC, Beliaeff B, Raffin B, Ibanez F (1996) Séries temporelles du réseau de surveillance de phytoplancton (REPHY). Étude de deux zones de la côte Ouest Méditerranée. Oceanol Acta 20:102–108

Le Moal M, Biegala IC (2009) Diazotrophic unicellular cyanobacteria in the northwestern Mediterranean Sea: a seasonal cycle. Limnol Oceanogr 54:845–855

Le Moal M, Collin M, Biegala IC (2011) Intriguing diversity among diazotrophic picoplankton along a Mediterranean transect: a dominance of rhizobia. Biogeosciences 8:827–840. doi:10.5194/bg-8-827-2011

Liu H, Probert I, Uitz J, Claustre H, Aris-Brosou S, Frada M, Not F, de Vargas C (2009) Extreme diversity in noncalcifying haptophytes explains a major pigment paradox in open oceans. Proc Natl Acad Sci USA 106:12803–12808

Llebot C, Solé J, Delgado M, Fernàndez–Tejedor M, Camp J, Estrada M (2011) Hydrographical forcing and phytoplankton variability in two semi-enclosed estuarine bays. J Mar Syst 86:69–86

López J, Arté P (1971) Aguas rojas en las costas catalanas. Investig Pesq 35:699–708

Lynn DH, Small EB (2000) Phylum Ciliophora. In: Lee JJ, Leedale GF, Bradbury P (eds) An illustrated guide to the protozoa. Society of Protozoologists/Allen Press inc, Kansas, pp 371–656

Magagnini M, Corinaldesi C, Monticelli LS, De Domenico E, Danovaro R (2007) Viral abundance and distribution in mesopelagic and bathypelagic waters of the Mediterranean Sea. Deep Sea Res I 54:1209–1220

Malinverno E, Ziveri P, Corselli C (2003) Coccolithophorid distribution in the Ionian Sea and its relationship to eastern Mediterranean circulation during late fall to early winter 1997. J Geophys Res Oceans 108:8115. doi:10.1029/2002JC001346

Man-Aharonovich D, Kress N, Bar Zeev E, Berman-Frank I, Béjà O (2007) Molecular ecology of *nifH* genes and transcripts in the Eastern Mediterranean Sea. Environ Microbiol 9:2354–2363

Mangialajo L, Ganzin N, Accoroni S, Asnaghi V, Blanfun A, Cabrini M, Cattaneo-Vietti R, Chavanon F, Chiantore M, Cohu S, Costa E, Fornasaro D, Grossel H, Marco-Miralles F, Mas M, Re A, Rossi AM, Sala MM, Thibaut T, Totti C, Vila M, Lemée R (2011) Trends in *Ostreopsis* proliferation along the Northern Mediterranean coasts. Toxicon 57:408–420

Marasović I, Ninčević Ž, Kušpilić G, Marinović S, Marinov S (2005) Long-term changes of basic biological and chemical parameters at two stations in the middle Adriatic. J Sea Res 54:3–14

Margalef R (1963) Role des ciliés dans le cycle de la vie pélagique en Méditerranée, Rapports et Procés-Verbaux des Réunions – Commission Internationale pour l'Exploration Scientifique de la Mer Méditerranée, Monaco, vol 17, pp 511–512

Margalef R (1969) Composición específica del fitoplancton de la costa catalano-levantina (Mediterráneo occidental) en 1962–1967. Investig Pesq 33:345–380

Margalef R (1971) Distribución del fitoplancton entre Córcega y Barcelona, en relación con la mezcla vertical del agua, en marzo de 1970. Investig Pesq 35:687–698

Margalef R (1978) Life-forms of phytoplankton as survival alternatives in an unstable environment. Oceanol Acta 4:493–509

Margalef R (1985) Western Mediterranean, Key environments. Pergamon Press, Oxford, 363 pp

Margalef R (1986) Ecologia. Ediciones Omega, Barcelona, 951 pp

Margalef R (1995) Tenim idea de com funciona la Mediterrània? Arxius de les seccions de ciéncies Institut d'Estudis Catalans 100:127–136

Marie D, Zhu F, Balague VJR, Vaulot D (2006) Eukaryotic picoplankton communities of the Mediterranean Sea in summer assessed by molecular approaches (DGGE, TTGE, QPCR). FEMS Microbiol Ecol 55:403–415

Marty JC (2002) The DYFAMED time-series program (French-JGOFS). Deep Sea Res II 49:1963–1964

Marty JC, Chiavérini J (2010) Hydrological changes in the Ligurian Sea (NW Mediterranean, DYFAMED site) during 1995–2007 and biogeochemical consequences. Biogeosciences 7:2117–2128. doi:10.5194/bg-7-2117-2010

Marty JC, Chiavérini J, Pizay MD, Avril B (2002) Seasonal and internannual dynamics of nutrients and phytoplankton pigments in the western Mediterranean Sea at the DYFAMED time-series station (1991–1999). Deep Sea Res II 49:1965–1986

Massana R (2011) Eukaryotic picoplankton in surface oceans. Annu Rev Microbiol 65:91–110

Massana R, Balagué V, Guillou L, Pedrós-Alió C (2004) Picoeukaryotic diversity in an oligotrophic coastal site studied by molecular and culturing approaches. FEMS Microbiol Ecol 50:231–243

McDonald SM, Sarno D, Scanlan D, Zingone A (2007) Genetic diversity of eukaryotic ultraphytoplankton in the Gulf of Naples during an annual cycle. Aquat Microb Ecol 50:75–89

McGill DA (1965) The relative supplies of nitrate, phosphate and silicate in the Mediterranean Sea. Rapp Proc Ver CIESMM 18, pp 737–744

MEDOC-Group (1970) Observation of formation of deep water in the Mediterranean Sea, 1969. Nature 227:1037–1040

Mella-Flores D, Mazard S, Humily F, Partensky F, Mahé F, Bariat L, Courties C, Marie D, Ras J, Mauriac R, Jeanthon C, Mahdi Bendif E, Ostrowski M, Scanlan DJ, Garczarek L (2011) Is the distribution of *Prochlorococcus* and *Synechococcus* ecotypes in the Mediterranean Sea affected by global warming? Biogeosciences 8:2785–2804. doi:10.5194/bg-8-2785-2011

Mével G, Vernet M, Goutx M, Ghiglione JF (2008) Seasonal to hour variation scales in abundance and production of total and particle-attached bacteria in the open NW Mediterranean Sea (0–1000 m). Biogeosciences 5:1573–1586

Minas HJ, Nival, P (1988) Océanographie pélagique Méditerranéenne. Oceanol Acta 9 (Spec. issue):155–162

Moesender MM, Winter C, Herndl GJ (2001) Horizontal and vertical complexity of attached and free-living bacteria of the eastern Mediterranean Sea, determined by 16S rDNA and 16S rRNA fingerprints. Limnol Oceanogr 46:95–107

Moraitou-Apostolopoulou M, Kiortsis V (eds) (1985) Mediterranean marine ecosystems, vol 8, NATO conference series. I, ecology. Plenum Press, New York

Morales A (2006) Estudio multitemporal de la clorofila superficial en el Mar Mediterráneo Noproccidental, evaluada a partir de datos SeaWIFS: septiembre de 1997 a agosto de 2004. PhD thesis, Universitat de Barcelona, CSIC, Universitat Politècnica de Catalunya

Morán XAG, Estrada M (2005) Winter pelagic photosynthesis in the NW Mediterranean. Deep Sea Res I 52:1806–1822

Morán XAG, Taupier-Lepage I, Vazquez-Dominguez E, Ruiz S, Arin L, Raimbault P, Estrada M (2001) Physical-biological coupling in the Algerian Basin (SW Mediterranean): influence of mesoscale instabilities on the biomass and production of phytoplankton and bacterioplankton. Deep Sea Res I 48:405–437

Morel A, André J (1991) Pigment distribution and primary production in the western Mediterranean as derived and modeled from Coastal Zone Color Scanner observations. J Geophys Res 96:12685–12698

Mozetič P, Solidoro C, Cossarini G, Socal G, Precali R, Francé J, Bianchi F, De Vittor C, Smodlaka N, Fonda Umani S (2010) Recent trends towards oligotrophication of the Northern Adriatic: evidence from chlorophyll *a* time series. Estuar Coast 33:362–375

Mozetič P, Francé J, Kogovšek T, Talaber I, Malej A (2012) Plankton trends and community changes in a coastal sea (northern Adriatic): bottom-up vs. top-down control in relation to environmental drivers. Estuar Coast Shelf Sci 115:138–148. doi:10.1016/j.ecss.2012.02.009

Mura MP, Agustí S, del Giorgio PA, Gasol JM, Vaqué D, Duarte CM (1996) Loss-controlled phytoplankton production in nutrient-poor littoral waters of the NW Mediterranean: in situ experimental evidence. Mar Ecol Prog Ser 130:213–219

Ninčević Gladan Ž, Marasović I, Grbec B, Skejić S, Bužančić M, Kušpilić G, Matijević S, Matić F (2010) Inter-decadal variability in phytoplankton community in the middle Adriatic (Kaštela Bay) in relation to the North Atlantic Oscillation. Estuar Coast 33:376–383. doi:10.1007/s12237-009-9223-3

Not F, Valentin K, Romari K, Lovejoy C, Massana R et al (2007) Picobiliphytes: a marine picoplanktonic algal group with unknown affinities to other eukaryotes. Science 315:252–254

Novarino G (2005) Nanoplankton protists from the western Mediterranean Sea. II. Cryptomonads (Cryptophyceae = Cryptomonadea). Sci Mar 69:47–74

Pavillard J (1925) Bacillariales. In: Report of the Danish oceanographical expeditions 1908–1910 to the Mediterranean and adjacent Seas. Biology, J4, vol II. A. F. Host, Copenhagen, pp 1–72

Pedrós-Alió C, Calderon-Paz JI, Guixa-Boixereu N, Estrada M, Gasol JM (1999) Bacterioplankton and phytoplankton biomass and production during summer stratification in the northwestern Mediterranean Sea. Deep Sea Res I 46:985–1019

Pennisi E (2003) Drafting a tree. Science 300:1694. doi:10.1126/science.300.5626.1694

Percopo I, Siano R, Cerino F, Sarno D, Zingone A (2011) Phytoplankton diversity during the spring bloom in the northwestern Mediterranean Sea. Bot Mar 54:243–267

Pérez MT, Dolan JR, Vidussi F, Fukai E (2000) Diel vertical distribution of planktonic ciliates within the surface layer of the NW Mediterranean (May 1995). Deep Sea Res I 47:479–503

Pierce RW, Turner JT (1992) Ecology of planktonic ciliates in marine food webs. Rev Aquat Sci 6:139–181

Pinardi N, Masetti E (2000) Variability of the large scale general circulation of the Mediterranean Sea from observations and modelling: a review. Palaeogeogr Palaeoclimatol Palaeoecol 158:153–173

Pitta P, Giannakourou A, Christaki U (2001) Planktonic ciliates in the oligotrophic Mediterranean Sea: longitudinal trends of standing stocks, distributions and analysis of food vacuole contents. Aquat Microb Ecol 24:297–311

Pitta P, Stambler N, Tanaka T, Zohary T, Tselepides A, Rassoulzadegan F (2005) Biological response to P addition in the Eastern Mediterranean Sea: the microbial race against time. Deep Sea Res II 52:2961–2974. doi:10.1016/j.dsr2.2005.08.012

Polat S (2010) Abundance and distribution patterns of phytoplankton in the coastal waters of northeastern Mediterranean. In: Bryand F (ed) CIESM workshop monographs 40, Monaco, pp 29–32

Pommier T, Canbäck B, Riemann L, Boström KH, Simu K, Lundberg P, Tunlid A, Hagström A (2007) Global patterns of diversity and community structure in marine bacterioplankton. Mol Ecol 16:867–880

Pukall R, Butefuß D, Frühling A, Rohde M, Kroppenstedt RM, Burghardt J, Lebaron P, Bernard L, Stackebrandt E (1999) *Sulfitobacter mediterraneus* sp. nov., a new sulfite-oxidizing member of the alpha-Proteobacteria. Int J Syst Bacteriol 49:513–519

Quijano-Scheggia S, Garcés E, Andrée KB, de la Iglesia P, Diogène J, Fortuño JM, Camp J (2010) *Pseudo-nitzschia* species on the Catalan coast: characterization and contribution to the current knowledge of the distribution of this genus in the Mediterranean Sea. Sci Mar 74:395–410

Rassoulzadegan F (1977) Evolution annuelle des ciliés pélagiques en Méditerranée Nord-Occidentale. Ciliés oligotriches 'non tintinnidés' (Oligotrichina). Ann Inst Océanogr Paris 53:125–134

Rassoulzadegan F (1978) Dimensions et taux d'ingestion des particules consommées par un Tintinnidé: *Favella ehrenbergii* (Clap. et Lachm.) Jörg Cilié pélagique. Ann Inst Oceanogr Paris 54:17–24

Rassoulzadegan F (1979) Évolution annuelle des ciliés pélagiques en Méditerranée nord-occidentale. II. Ciliés oligotriches. Tintinnidés (Tintinnina). Investig Pesq 43:417–448

Ribera d'Alcalà M, Conversano F, Corato F, Licandro P, Mangoni O, Marino D, Mazzocchi MG, Modigh M, Montresor M, Nardella M, Saggiomo M, Sarno D, Zingone A (2004) Seasonal patterns in plankton communities in a pluriannual time series at a coastal Mediterranean site (Gulf of Naples): an attempt to discern recurrences and trends. Sci Mar 68(suppl 1):65–83

Sachs JP, Repeta DJ (1999) Oligotrophy and nitrogen fixation during Eastern Mediterranean sapropel events. Science 286:2485–2488

Saiz E, Calbet A, Irigoien X, Alcaraz M (1999) Copepod egg production in the western Mediterranean: response to food availability in oligotrophic environments. Mar Ecol Prog Ser 187:179–189

Sala MM, Peters F, Gasol JM, Pedrós-Alió C, Marrasé C, Vaqué D (2002) Seasonal and spatial variations in the nutrient limitation of bacterioplankton growth in the northwestern Mediterranean. Aquat Microb Ecol 27:47–56

Saura A, Massana R, Boras JA, Forn I, Vila-Reixach G, Vaqué D (2011) Effect of viruses on marine stramenopile (MAST) communities in an oligotrophic coastal marine system. J Plankton Res 33:1709–1718

Scharek R, Tupas LM, Karl DM (1999) Diatom fluxes to the deep sea in the oligotrophic North Pacific gyre at Station ALOHA. Mar Ecol Prog Ser 182:55–67

Schauer M, Massana R, Pedrós-Alió C (2000) Spatial differences in bacterioplankton composition along the Catalan coast (NW Mediterranean) assessed by molecular fingerprinting. FEMS Microbiol Ecol 33:51–59

Schauer M, Balagué V, Pedrós-Alió C, Massana R (2003) Seasonal changes in the taxonomic composition of bacterioplankton in a coastal oligotrophic system. Aquat Microb Ecol 31:163–174

Schiller J (1928) Die planktischen vegetationen des Adriatischen Meeres, C. Dinoflagellata, I, II. Arch Protistenkd 61:45, 62:119–166

Sherr EB, Sherr BF, Paffenhöfer GA (1986) Phagotrophic protozoa as food for metazoans: a "missing" trophic link in marine pelagic food webs? Mar Microb Food Webs 1:61–80

Siokou-Frangou I, Bianchi M, Christaki U, Christou ED, Giannakorou A, Gotsis O, Ignatiades L, Pagou K, Pitta P, Psarra S, Souvermezoglou E, Van Wambeke F (2002) Carbon flow in the planktonic food web along a gradient of oligotrophy in the Aegean Sea (Mediterranean Sea). J Mar Syst 33–34:335–353

Siokou-Frangou I, Christaki U, Mazzocchi MG, Montresor M, Ribera d'Alcalá M, Vaqué D, Zingone A (2010) Plankton in the open Mediterranean Sea: a review. Biogeosciences 7:1543–1586

Smayda TJ, Reynolds CS (2001) Community assembly in marine phytoplankton: application of recent models to harmful dinoflagellate blooms. J Plankton Res 23:447–461

Šolić M, Krstulović N, Marasović I, Baranović A, Pucher-Petković T Vučetić (1997) Analysis of time series of planktonic communities in the Adriatic Sea: distinguishing between natural and man-induced changes. Oceanol Acta 20:131–143

Sorokin DY (1995) *Suljtobacter pontiacus* gen. nov., sp. nov. - a new heterotrophic bacterium from the Black Sea, specialized on sulfite oxidation. Mikrobiologiya 64:295–305

Suttle CA (2007) Marine viruses — major players in the global ecosystem. Nat Rev Microbiol 5:801–812

Tamburini C, Garel M, Al Ali B, Mérigot B, Kriwy K, Charrière B, Budillon B (2009) Distribution and activity of Bacteria and Archaea in the different water masses of the Tyrrhenian Sea. Deep Sea Res II 56:700–712

Tanaka T, Rassoulzadegan F (2002) Full-depth profile (0–2000 m) of bacteria, heterotrophic nanoflagellates and ciliates in the NW Mediterranean Sea: vertical partitioning of microbial trophic structures. Deep Sea ResII 49:2093–2107

Taupier-Letage I, Millot C (1988) Surface circulation in the Algerian basin during 1984. Oceanol Acta 9(Spec issue):79–86

Thingstad TF, Krom MD, Mantoura RFC, Flaten GAF, Groom S, Herut B, Kress N, Law CS, Pasternak A, Pitta P, Psarra S, Rassoulzadegan F, Tanaka T, Tselepides A, Wassmann P, Woodward EMS, Wexels Riser C, Zodiatis G, Zohary T (2005) Nature of phosphorus limitation in the ultraoligotrophic Eastern Mediterranean. Science 309:1068–1071

Thompson GA, Alder VA, Boltovskoy D, Brandini F (1999) Abundance and biogeography of tintinnids (Ciliophora) and associated microzooplankton in the Southwestern Atlantic Ocean. J Plankton Res 21:1265–1298

Tintoré J, Gomis D, Alonso S, Parrilla G (1991) Mesoscale dynamics and vertical motion in the Alboran sea. J Phys Oceanogr 21:811–823

Travers M (1973) Le microplancton du golfe de Marseille: variations de la composition systématique et de la densité des populations. Téthys 5:31–53

Tunin-Ley A, Ibanez F, Labat J-P, Zingone A, Lemée R (2009) Phytoplankton biodiversity and NW Mediterranean Sea warming: changes in the dinoflagellate genus *Ceratium* in the 20th century. Mar Ecol Prog Ser 375:85–99

Unrein F, Massana R, Alonso-Sáez L, Gasol JM (2007) Significant year-round effect of small mixotrophic flagellates on bacterioplankton in an oligotrophic coastal system. Limnol Oceanogr 52:456–459

Van Wambeke F, Christaki U, Giannakourou A, Moutin T, Souvemerzoglou K (2002) Longitudinal and vertical trends of bacterial limitation by phosphorus and carbon in the Mediterranean Sea. Microb Ecol 43:119–133

Van Wambeke F, Ghiglione J-F, Nedoma J, Mével G, Raimbault P (2009) Bottom up effects on bacterioplankton growth and composition during summer-autumn transition in the open NW Mediterranean Sea. Biogeosciences 6:705–720

Vaqué D, Blough HA, Duarte CM (1997) Dynamics of ciliate abundance, biomass and community composition in an oligotrophic coastal environment (NW Mediterranean). Aquat Microb Ecol 12:71–83

Vaqué D, Casamayor EO, Gasol JM (2001) Dynamics of whole community bacterial production and grazing losses in seawater incubations as related to the changes in the proportions of bacteria with different DNA content. Aquat Microb Ecol 25:163–177

Velasquez ZR (1997) Fitoplancton en el Mediterráneo Noroccidental. PhD thesis, Universitat Politècnica de Catalunya

Vidussi F, Claustre H, Manca BB, Luchetta A, Marty JC (2001) Phytoplankton pigment distribution in relation to upper thermocline circulation in the eastern Mediterranean Sea during winter. J Geophys Res 106:19939–19956

Vila M, Camp J, Garcés E, Masó M, Delgado M (2001a) High resolution spatio-temporal detection of potentially harmful dinoflagellates in confined waters of the NW Mediterranean. J Plankton Res 23:497–514

Vila M, Garcés E, Masó M, Camp J (2001b) Is the distribution of the toxic dinoflagellate *Alexandrium catenella* expanding along the NW Mediterranean coast? Mar Ecol Prog Ser 222:73–83

Viličić D, Djakovac T, Burić Bosak S (2009) Composition and annual cycle of phytoplankton assemblages in the northeastern Adriatic Sea. Bot Mar 52:291–305

Wang DP, Vieira MEC, Salat J, Tintoré J, La Violette PE (1988) A shelf/slope frontal filament off the northeast Spanish coast. J Mar Res 46:321–332

Weinbauer MG (2004) Ecology of prokaryotic viruses. FEMS Microbiol Rev 28:127–181

Weinbauer MG, Peduzzi P (1994) Frequency, size and distribution of bacteriophages in different marine bacterial morphotypes. Aquat Microb Ecol 108:11–20

Weinbauer MG, Peduzzi P (1995) Significance of viruses versus heterotrophic nanoflagellates for controlling bacterial abundance in the Northern Adriatic Sea. J Plankton Res 17:1851–1856

Weinbauer MG, Fucks D, Peduzzi P (1993) Distribution of viruses and dissolved DNA along a coastal gradient in the Northern Adriatic Sea. Appl Environ Microbiol 59:4074–4082

Weinbauer MG, Brettar I, Höfle MG (2003) Lysogeny and virus-induced mortality of bacterioplankton in surface, deep, and anoxic marine waters. Limnol Oceanogr 48:1457–1465

Wyatt T (2010) Can we detect meaningful changes in mediterranean phytoplankton? In: Bryand F (ed) CIESM workshop monographs 40, Monaco, pp 15–18

Yakimov MM, La Cono V, Denaro R (2009) A first insight into the occurrence and expression of functional amoA and accA genes of autotrophic and ammonia-oxidizing bathypelagic Crenarchaeota of TyrrhenianSea. Deep Sea Res II 56:748–754

Zaballos M, López-López A, Ovreas L, Bartual SG, D'Auria G, Alba JC, Legault B, Pushker R, Daae FL, Rodríguez-Valera F (2006) Comparison of prokaryotic diversity at offshore oceanic locations reveals a different microbiota in the Mediterranean Sea. FEMS Microbiol Ecol 56(3):389–405

Zehr JP, Jenkins BD, Short SM, Steward GF (2003) Nitrogenase gene diversity and microbial community structure: a cross-system comparison. Environ Microbiol 5:539–554

Zervakis V, Georgopoulos D, Karageorgis AP, Theocharis A (2004) On the response of the Aegean Sea to climate variability: a review. Int J Climatol 24:1845–1858

Zingone A (2010) Harmful algal blooms in the Mediterrranean: an historical overview. In: Bryand F (ed) CIESM workshop monographs 40, Monaco, pp 19–24

Zingone A, Sarno D, Forlani G (1999) Seasonal dynamics in the abundance of *Micromonas pusilla* (Prasinophyceae) and its viruses in the Gulf of Naples (Mediterranean Sea). J Plankton Res 21:2143–2159

Zingone A, Licandro P, Sarno D (2003) Revising paradigms and myths of phytoplankton ecology using biological time series. In: Bryand F (ed) CIESM workshop monographs 22, Monaco, pp 109–114

Zingone A, Natale F, Biffali E, Borra M, Forlani G, Sarno D (2006) Diversity in morphology, infectivity, molecular characteristics and induced host resistance between two viruses infecting *Micromonas pusilla*. Aquat Microb Ecol 45:1–14

Zingone A, Phlips EJ, Harrison PJ (2010a) Multiscale variability of twenty-two coastal phytoplankton time series: a global scale comparison. Estuar Coasts 33:224–229. doi:10.1007/s12237-009-9261-x

Zingone A, Dubroca L, Iudicone D, Margiotta F, Corato F, Ribera d'Alcalà M, Saggiomo V, Sarno D (2010b) Coastal phytoplankton do not rest in winter. Estuar Coasts 33:342–361. doi:10.1007/s12237-009-9157-9

Zohary T, Robarts R (1992) Bacterial numbers, bacterial production, and heterotrophic nanoplankton abundance in a warm core eddy in the Eastern Mediterranean. Mar Ecol Prog Ser 84:133–137

Zohary T, Herut B, Krom MD, Mantoura RFC, Pitta P, Psarra S, Rassoulzadegan F, Stambler N, Tanaka T, Thingstad TF, Woodward EMS (2005) P-limited bacteria but N&P co-limited phytoplankton in the Eastern Mediterranean – a microcosm experiment. Deep Sea Res II 52:3011–3023. doi:10.1016/j.dsr2.2005.08.011

The Mediterranean Sea – Primary Productivity

7

Noga Stambler

Abstract

Primary production, the production of organic carbon molecules from carbon dioxide and water by converting sunlight energy to chemical energy through the process of photosynthesis, is the basis of our biosphere. The total Mediterranean Sea primary productivity is 1 % of the global primary productivity (Uitz et al. 2010). The entire Mediterranean Sea as a large marine ecosystem (LME) is characterized as a low-productivity ecosystem, <150 gC m^{-2} $year^{-1}$ and, as such, is considered a Class III ecosystem (Aquarone et al. 2009). Based on Nixon's (1995) definition, it is a mesotrophic sea, yet many parts of this LME are oligotrophic, and the Levantine Basin (eastern Mediterranean) is even ultraoligotrophic. Global and local abiotic fluctuations cause variation in the primary-producer communities and their ability to photosynthesize, thus leading to modification of the food web.

Keywords

Primary productivity • Phytoplankton • Sapropels • Light • Global change

The Past – Paleoecology: Fossil Records

Primary productivity of the Mediterranean Sea basin changed drastically during geologic time. Photosynthetic eukaryotes evolved in the Proterozoic oceans about 1.5 billion years ago, yet only in the past 70 million years primary producers, such as diatoms, dinoflagellates, and coccolithophores, became dominant in the modern seas (Falkowski et al. 2004).

An increase in marine productivity exported to the deep sea is considered to be related to deposition of organic carbon-rich sediments named sapropels (Fig. 7.1) [Gallego-Torres et al. (2011)]. Sapropels from the late Pleistocene, 6,000–9,500 years ago, were found in both the eastern and western basins of the Mediterranean (Sachs and Repeta 1999; Arnaboldi and Meyers 2006; Mobius et al. 2010). During the sapropel episodes, primary-producer populations included picocyanobacteria, *Synechococcus* (cyanobacterium), and *Posidonia* (sea grass) as well as rhizosolenid diatom mats (In Sachs and Repeta 1999). The increase in primary production during this time was associated with the bloom of nitrogen-fixing bacteria (review in Gallego-Torres et al. 2011). It should be noticed that the nutrient-regime conditions in the Pleistocene or Holocene were different from those of modern times. During sapropel events, there was probably increased riverine P fluxes to surface waters, leading to enhanced diazotrophy, while during nonsapropel sediment deposition, the nitrogen was supplied from the North Atlantic (Higgins et al. 2010). Sapropel formation was controlled by hydrographic, e.g., salinity-enhanced stratification and redox conditions in the Mediterranean (Higgins et al. 2010).

N. Stambler (✉)
The Mina & Everard Goodman Faculty of Life Sciences,
Bar-Ilan University, Ramat-Gan 52900, Israel
e-mail: drnogas@gmail.com

S. Goffredo and Z. Dubinsky (eds.), *The Mediterranean Sea: Its history and present challenges*,
DOI 10.1007/978-94-007-6704-1_7,

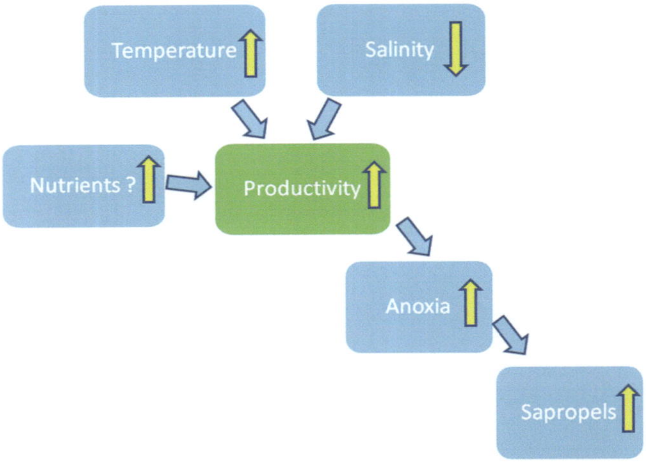

Fig. 7.1 The hypothesis connecting sapropel formation and high productivity (Based on Emeis and Sakamoto 1998)

The Mediterranean Sea Today: Ecology

The Primary Producers in the Mediterranean Sea

On the one hand, there are many primary producers in the Mediterranean Sea, including (number of species in parentheses): Bacillariophyceae (736), Dinoflagellata (673), Rhodophyta (657), Chlorophyta (190), Coccolithophores (166), (Coll et al. 2010). On the other hand, only a few of these species are dominant; most of them are rare and can be found only in specific locations. Picophytoplankton (<2 μM), mainly *Prochlorococcus* and *Synechococcus* (Cyanophyceae), are the dominant species in the entire Mediterranean Sea, especially during stratification, whereas *Prochlorococcus* makes up to 100 % of the community in the deep chlorophyll maximum (DCM). However, some species of pennate diatoms and small dinoflagellates are also found at the DCM (Kimor et al. 1987; Nincevic et al. 2002; Crombet et al. 2011). From the microphytoplankton, dinoflagellates and coccolithophores dominate at the eastern and diatoms at the western Mediterranean Sea (Ignatiades et al. 2009). The pelagic phytoplankton species composition differs from that of the population on the coast, where dinoflagellates and diatoms are dominant (e.g., Kimor et al. 1987; Gómez and Gorsky 2003; Charles et al. 2005; Psarra et al. 2005; Estrada and Vaqué 2013). The abundance and composition of microplankton are a function of location and time of year (e.g., Gómez and Gorsky 2003; Siokou-Frangou et al. 2010; Estrada and Vaqué 2013).

However, in some cases, the maximal contribution of cells <2 μM to total primary production can reach 71 % in pelagic compared to 44 % in neritic water (Magazzu and Decembrini 1995). It should be noted that for the entire Mediterranean Sea, in spite of the dominance of the picophytoplankton in terms of cellular abundance, they do not play a dominant role in primary production because of their low carbon-uptake rates compared to larger cells (Uitz et al. 2010).

Spacial Aspect of Primary Production in the Mediterranean

The last 25 years of plankton studies (including those on primary production) of the epipelagic offshore waters of the Mediterranean Sea were reviewed in the impressive paper of Siokou-Frangou et al. (2010). Primary production exhibits a west-east decreasing trend and ranges between 20 and 230 g Cm^{-2} year^{-1} (e.g., Antoine et al. 1995; Ignatiades et al. 2009; Siokou-Frangou et al. 2010; Yogev et al. 2011). In the eastern Mediterranean Sea, annual primary productivity is low, for example, at the Cretan Sea, the average annual gross primary productivity is 59–80 gC m^{-2} year^{-1}, and only ~ 29 gC m^{-2} year^{-1} at the Levantine Basin (e.g., Psarra et al. 2000; Siokou-Frangou et al. 2010; Yogev et al. 2011). The highest primary production was observed in the western Mediterranean (DYFAMED) and it varied from 86 to 232 gC m^{-2} year^{-1} (Marty and Chiaverini 2002). Haptophytes dominate the phytoplankton biomass all year long, except during the stratified summer period, which is characterized by prochlorophytes (Marty et al. 2002).

The total primary productivity of the Mediterranean Sea [excluding coastal areas, i.e., bathymetry <200 m, based on series data of surface chlorophyll from satellite observations with Sea viewing Wide Field of view Sensor (SeaWiFS)], is ~0.5 (Gt C year^{-1}). From the total primary productivity, 0.2 Gt C year^{-1} (36 %) is due to productivity of microphytoplankton cells (size >20 mm), including diatoms and dinoflagellates, 0.2 Gt C year^{-1} (42 %) is due to nanophytoplankton cells (size 2–20 mm), e.g., prymnesiophytes, pelagophytes, and cryptophytes, and 0.1 Gt C year^{-1} (22 %) is due to picophytoplankton cells (size of <2 mm), e.g., prokaryotes, cyanobacteria, and prochlorophytes (Uitz et al. 2010).

Seasonal Hydrodynamics of the Water Column and Nutrient Control

Plankton-community abundance and diversity are a dynamic system that continuously changes according to the specific conditions, including temperature, salinity, nutrients, vertical mixing and turbidity by wind. These changes occur all the time, with notable changes between days, months, and years (e.g. Marty and Chiaverini 2002; Gernez et al. 2011; Lasternas et al. 2011). Fluctuations in phytoplankton standing crop and primary productivity in response to weather phenomena such as heavy rains or storms, are more pronounced in the neritic station compared to the pelagic station (Azov 1986). In most parts of the Mediterranean Sea, the seasonal dynamics of phytoplankton concentration follows a typical temperate cycle, with very low biomass in summer and higher biomass in winter (2–3 months),

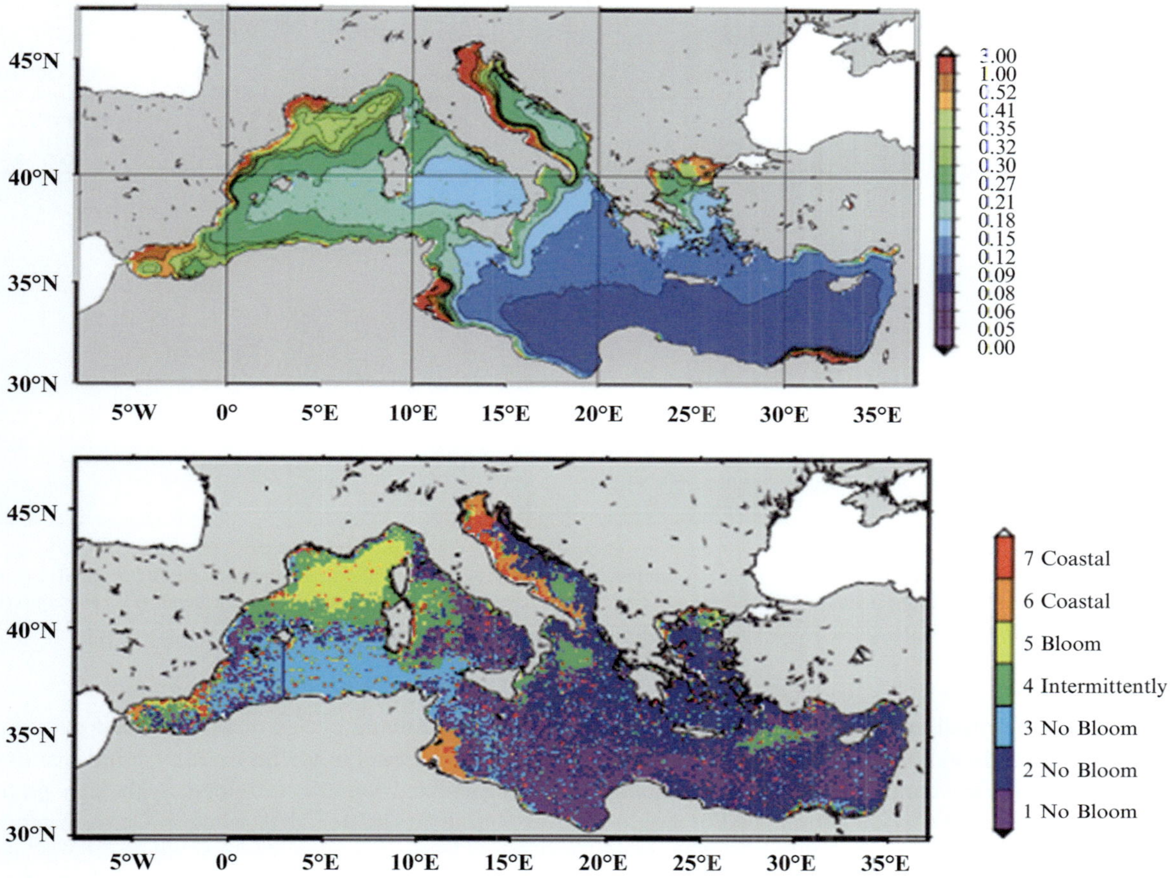

Fig. 7.2 Spatial distribution of (**a**) satellite-derived chlorophyll *a*; and (**b**) the clusters obtained from k-means analysis, as reported by D'Ortenzio and Ribera d'Alcal´a (2009)

increasing for a short time in late winter/early spring (Marty et al. 2002; D'Ortenzio and d'Alcala 2009). Spring bloom occurs regularly in the northwestern Mediterranean and other parts of the sea, while there is less pronounced bloom in the Levant area (D'Ortenzio and d'Alcala 2009).

As stated before, the trophic state of the Mediterranean Sea ranges from oligotrophic at the western Mediterranean Sea to ultraoligotrophic at the eastern Mediterranean Sea (Fig. 7.2). The sea oligotrophy is due to low nutrient concentrations, especially of inorganic phosphorus and nitrogen. The nutrient concentration is the major limitation of primary production in the Mediterranean Sea throughout the year, especially during summer stratification (Berland et al. 1980; Owens et al. 1989; Krom et al. 1991; Thingstad et al. 2005; Zohary et al. 2005; Krom et al. 2010). Within the Mediterranean Sea, the eastern Mediterranean Sea is a relatively small ocean basin with an unusually high nitrate to phosphate ratio (28:1) in the deep waters (Krom et al. 1991). As such, the typical winter phytoplankton bloom is considered as P- rather than N-limited (Krom et al. 1991, 2010) even though major *in situ* microcosm experiments show that N and P co-limited the phytoplankton in the eastern Mediterranean (Zohary et al. 2005; Tanaka et al. 2011). Silica (Si) can limit the growth of diatoms only in some parts of the sea (Crombet et al. 2011).

In the Mediterranean Sea, temperature stratification usually develops during high air temperatures. During stratification, the warmer, less saline, surface water is separated from the deeper, colder, and more saline water (Fig. 7.3). Salinity decreases from the surface downwards; however, the decline is not smooth (e.g., Yacobi et al. 1995). This stratification inhibits the flux of nutrients towards the surface layer, preventing autotrophic biomass accumulation in the surface layer. Transfer from mixing to stratification and vice versa often results in autumnal algal blooms. The temporal changes in phytoplankton communities lead to primary-production seasonal cycles, with irregular productivity maximum at the end of winter (Psarra et al. 2000; Krom et al. 2003). In the Levantine Basin, the difference between the summer stratification season (10 gC m^{-2} $period^{-1}$) and the winter mixing season (19 gC m^{-2} $period^{-1}$) is almost double (Yogev et al. 2011).

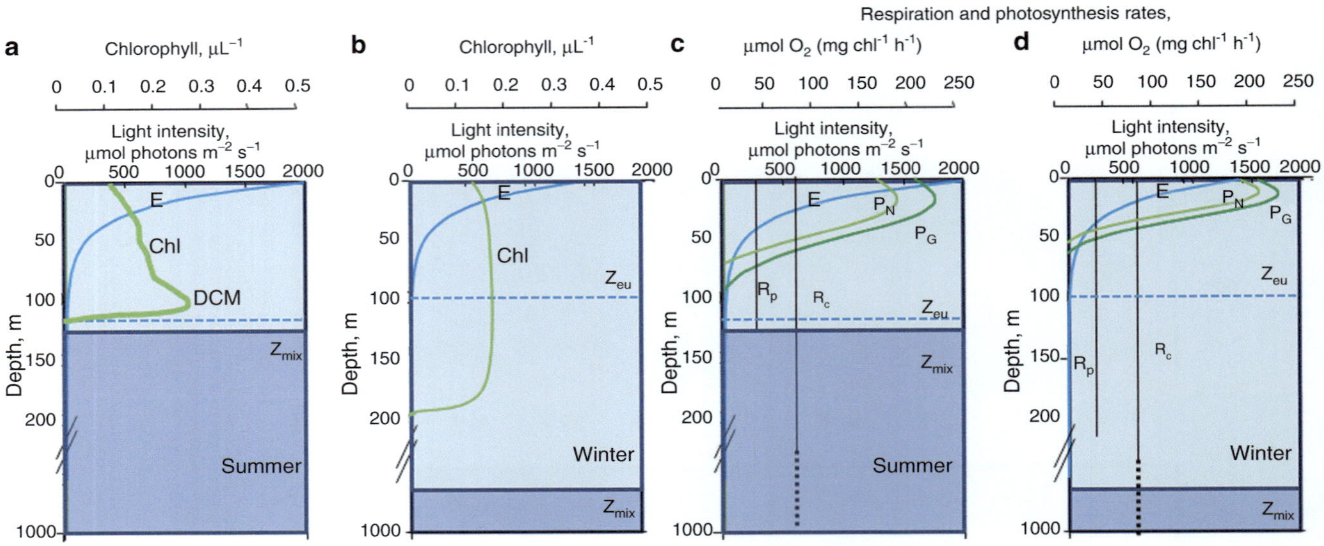

Fig. 7.3 Mediterranean Sea water-column characterization in summer (**a**, **c**) and winter (**b**, **d**). (**a** and **b**) Light intensity (*E*), chlorophyll concentration (*Chl*) as function of depth [deep chlorophyll maximum (*DCM*), Z_{mix} depth of mixing, Z_{eu} euphotic depth]. (**c** and **d**) Respiration, photosynthesis, and light as function of depth (P_N net photosynthesis, P_G gross photosynthesis, *R* respiration)

High nutrient concentrations were observed along the coasts and lagoons near major cities as well as at river estuaries and deltas. Nutrient enrichment, i.e., eutrophication, is a result of the mixing down to the bottom, and/or anthropogenic pollutions (e.g. Skejic et al. 2011). These parts of the sea are considered mesotrophic, with larger producers (microplankton >20 μM) and high levels of primary productivity (Vidussi et al. 2001; Puigserver et al. 2002; Mangoni et al. 2008; D'Ortenzio and d'Alcala 2009).

Mesoscale hydrodynamic structures, such as eddies and gyres, control biomass and primary production. High productivity was observed at gyre and upwelling areas (Vidussi et al. 2001; D'Ortenzio and d'Alcala 2009; Rahav et al. 2013). The Rhodes Basin has cyclonic circulation providing a strong exchange between the surface and intermediate-deep waters through upwelling and vertical convective overturning processes. The annual primary production in the Rhodes Basin is estimated as 97 gC m^{-2} year^{-1}, which is comparable with that found in the northwestern Mediterranean (Napolitano et al. 2000). In winter, the Rhodes Gyre was estimated to have 1.5 times higher primary production compared to the Cyprus Eddy (Vidussi et al. 2001). In the anticylconic circulation eddy, organic particles sink and dissolved organic matter is transported out of the euphotic zone. Eddy surface waters are relatively depleted of nutrients compared to the surrounding water and, as such, have lower primary production. In the Cyprus (Shikmona) Eddy, which has a low chlorophyll core, the yearly, average integrated production is low – about 69 gC m^{-2} year^{-1} (Groom et al. 2005). This is lower than that of the surrounding water of the south Levantine Basin (102–109 gC m^{-2} year^{-1}, Bosc et al. 2004). Annual cycles of biomass and daily primary production observed in the Ligurian cyclonic area in spring (April) exhibited widespread bloom with high production levels (up to 0.9 gC m^{-2} d^{-1}), while in October, production was much lower, reaching maximal values (0.2 gC m^{-2} d^{-1}) in the southern Tyrrhenian waters. During spring, there was a peak both in production (up to 0.55 gC m^{-2} d^{-1}) and biomass (0.47 mgChl m^{-3}), while in summer there was a peak in production, with no increase in biomass. This leads to an increased productivity-to-biomass (P/B) ratio, which is related to the increased daily PAR irradiance (Lazzara et al. 2010). Mesoscale hydrodynamic-structure primary production also depends on location. Surface primary production was significantly different at three anticyclonic eddies: in the Western Basin, the Ionian Basin, and the Levantine Basin (24, 18, and 7 nM C h^{-1}, respectively), although all were sampled within a few days of each other during summer (July 2008). The phytoplankton communities were not P-limited at these three sites during summer, but were consistently limited by N (Tanaka et al. 2011).

While, in the northwestern Mediterranean, the diazotrophy N_2 fixation process supplies up to 55 % of new production during summer, when primary productivity is very low (Garcia et al. 2006; Sandroni et al. 2007), in the Levantine Basin area, N_2 fixation contributes only ~1 and 2 % of new production and ~0.25 and 0.5 % of primary production for the mixed (winter) and stratified (spring–fall) periods, respectively (Yogev et al. 2011). Dust events can play an important role in supplying nitrogen and phosphorus to the southeast Mediterranean, mainly during stratification. Dust enrichment was estimated to correspond to 10–20 or 50–80 %

of new production in the Levant Basin (Krom et al. 1992; Kress and Herut 2001; Herut et al. 2002).

Light Control

The mid-latitude location in the Mediterranean Sea exposes the area to relatively high sunlight intensity during the year, reaching a peak in summer. There is an exponential decrease of light intensity and change in its spectrum in the water column (review in Stambler 2012, Fig. 7.3). The phytoplankton productivity is limited by light availability associated with deep mixed-layer depths (MLDs). The oligotrophic conditions, characterized by low chlorophyll concentrations and low yellow substance, lead to very clear water that allows light penetration to deep depths. For example, in the Levantine Basin, the euphotic depth, calculated as the depth reached by 1 % of surface light, is 123 m. The deep chlorophyll maximum (DCM) develops following emergence of the permanent seasonal thermocline and the limitation of the nutrients in the upper layers of the column water. In the Almeria-Oran frontal zone (southwest Mediterranean), the DCM develops down to 65 m (L'Helguen et al. 2002), while in the eastern Mediterranean Sea, it is usually at 100–140 m (Yacobi et al. 1995). Most of the biomass during stratification is at the DCM. The DCM depth at 80–150 m is closer to the depth reached by 0.1 % of the surface light. This depth should be defined in these cases, with very deep DCM as the euphotic depth (see in Stambler 2012). Growth at the DCM in deep water between 0.1 and 1 % of surface light (review in Stambler 2012) leads to photoadaptation and photoacclimation of the phytoplankton. Phytoplankton pigment content per cell is higher at the deeper depth, while the chlorophyll-to-biomass ratio is low at the surface and high at the DCM (Li et al. 1993). During stratification, when the water column is stable, there is sufficient time for picophytoplankton cells to photoacclimate. These are expressed in the increase of chlorophyll per cell with depth. The opposite is true during the mixing period (winter time), when the mixing depth can reach hundreds of meters, much below the euphotic zone, but the time the algal cell travels through the light gradient (within the mixed layer) is shorter than the time required for photoacclimation (Dubinsky and Stambler 2009). In addition to the higher chlorophyll concentration per biomass of *Prochlorococcus* cells at the DCM compared to surface cells, their small cell size (~0.5 µm diameter) enhances their ability to absorb light by increasing the surface-to-volume ratio (Chisholm 1992). This ratio is also an advantage in increasing nutrient diffusion.

Two main *Prochlorococcus* ecotypes are well recognized around the world: low light-adapted ecotype (LL), which is mainly found at the DCM depth; and high light-adapted ecotype (HL), which is mainly distributed in the upper layer of the water column (Partensky et al. 1999; Garczarek et al. 2007). In the Mediterranean Sea in summer, there are four genetically different ecotypes occupying distinct but overlapping light niches in the photic zone. LL ecotypes grow below the thermocline (Garczarek et al. 2007). In spite of the adaptation of the *Prochlorococcus* high light-adapted ecotype, it was shown that it is much more sensitive to surface light (high light with UV) compared to *Synechococcus*. The surface light decreases in *Prochlorococcus* cell-specific chlorophyll and causes cell death (Sommaruga et al. 2005). On sunny days, UV radiation has a lethal effect on *Prochlorococcus* down to 26 m, while such an effect was limited to the upper surface for *Synechococcus* (Llabres et al. 2010). Two distinct populations of *Synechococcus* also occur in the Mediterranean Sea. They differ in their fluorescence signatures and are found at different locations: one is dominant on the surface of the West Basin water while the other occupies deeper layers, especially at the Ionian Sea stations (Lasternas et al. 2010).

Photoacclimation occurs not only with phytoplankton distributed at different depths but also during day cycle, and is even more pronounced between seasons. Winter phytoplankton from the northwest Alboran Sea have a low C:Chl-a ratio and a higher chlorophyll-to-cell^{-1} ratio, attributing photoacclimation to either lower-incident irradiance in the upwelling area or longer residence time of the cells in deeper water due to increasing mixing depth in winter (Reul et al. 2005). The C:Chl-a ratio was higher at surface waters than at the subsurface chlorophyll maximum (~30 m) (Reul et al. 2005). The increase in chlorophyll concentration per phytoplankton cell in response to low light results in a decrease in the chlorophyll-specific absorption coefficient [$a^*(\lambda)$], e.g., in the Alboran Sea; there is a decrease of 30 % in $a^*(\lambda)$ at the DCM depth with respect to the surface (Mercado et al. 2006). This is a consequence of a 'packaging effect'– self-shading of one chlorophyll molecule one on the other, reducing the efficiency of chlorophyll molecular-to-absorbed light (review in Dubinsky and Stambler 2009).

The vertical gross primary production pattern depends on the primary producer's biomass and species diversity as well as its photoadaptation and photoacclimation (Fig. 7.3). Throughout the Mediterranean Sea, gross primary production showed a diversity of vertical profiles, ranging from lack of vertical structure to strong vertical heterogeneity with surface or deep maxima. This variability did not depend on the pattern of increasing gross primary production from east to west (Regaudie-de-Gioux et al. 2009). Primary production is a function of the ability of the community to photosynthesize. In offshore Mediterranean waters in the Catalano-Balearic Sea (northwest Mediterranean) during winters 1999 and 2000, when the water column was mixed below the euphotic zone, primary productivity reached up to 2 gC m^{-2} d^{-1}. The photosynthetic parameters displayed vertical differences within the mixed layer, mainly by increasing the light intensity at an incipient saturation of photosynthesis (Ek) with depth, indicating that photoacclimation rates were faster than the mixing rate (Moran and Estrada 2005). The phytoplankton community exhibits seasonal changes in the photosynthetic parameters: the saturation irradiance (E_k) follows the pattern of the incident irradiance (E_0), e.g., it is highest during summer and lower the rest of the year. The initial slope of the photosynthesis-versus-irradiance (P vs. E) curve was highest in autumn and winter and lowest in summer. Light variation between summer and winter, in combination with nutrients, leads to a change in chlorophyll concentration and in maximum rate of photosynthesis (Pmax) per biomass (Moran and Estrada 2005). Exposure to higher light will yield higher primary production unless light intensity is already at a saturation level for photosynthesis or so high that it causes photoinhibition (the latter can only occur at the surface) (Fig. 7.3). In the Bay of Banyuls-sur-Mer (northwest Mediterranean), it was clearly shown that primary production per chlorophyll is higher at 3 m than at 24 m, and the major variation between the depths was for light intensity (Charles et al. 2005). As the study was done at a bay, the primary-production values were high, reaching up to 67 mg Cm^{-3} d^{-1} (Charles et al. 2005).

The results of environmental conditions including light, nutrients, temperature, salinity, and many other factors, lead to low primary production in the Mediterranean Sea (annual values during 1998–2001) for the Western Basin were 163 gC m^{-2} $year^{-1}$ and for the Eastern Basin – 121 gC m^{-2} $year^{-1}$ (Bosc et al. 2004).

Future Scenarios – Ecological Effects

On a local scale at the Mediterranean Sea, we hope that pollution from toxic material, such as spilled oil, will not reduce phytoplankton production in the future.

According to some models, on a global scale, chlorophyll concentration has already been declining since the late 1,800 s. In low-latitude oceans, there is a reduction in the net primary-production (NPP) phenomenon, which is related to climatic changes leading to an increase in sea-surface temperature (SST) (Behrenfeld et al. 2006; Boyce et al. 2010). However, based on other models, chlorophyll and global marine primary production have increased over the past two decades (Chavez et al. 2011).

The response to climate change will be an individual response, and different species and ecotypes will have a different susceptibility to altered climatic conditions: some will bloom, while others will become extinct. Although increasing CO_2 (atm) and temperature may increase phytoplankton growth rates, the changes in temperature will probably not have a direct effect on diatoms or *Synechococcus*, but will influence *Prochlorococcus*. Changes in CO_2 will probably not have a significant direct effect on diatoms or *Synechococcus* (Boyd et al. 2010).

The recent 33-year (1974–2006) increase of sea-surface temperature of the Mediterranean caused a 40 % lengthening of summer conditions, which produced enhanced stratification (Coma et al. 2009). However, so far, no significant variations were found in the *Synechococcus* and the *Prochlorococcus* populations (Mella-Flores et al. 2011).

In the oligotrophic Mediterranean Sea, an increase in SST will increase stratification and cause the mixed layer to be shallower, which will further limit nutrient supply to phytoplankton and, thus, reduce primary production (Fig. 7.4). In this scenario, light will not be a limiting factor. However, the shallow-mixed layers can result in exposure of the phytoplankton population to high light and UV radiation, which can lead to cellular damage and inhibited rates of photosynthesis and growth. In this case, it could be that picophytoplankton, mainly *Synechococcus*, will have an advantage over the diatoms as they have effective cross-sections for photochemistry and fast metabolic repair of Photosystem II after photoinactivation (Finkel et al. 2010).

The Intergovernmental Panel on Climate Change (IPCC, Solomon et al. 2007) models predicted a major increase in CO_2 concentration (from ~400 to 600–900 ppm) and, as a result, a decrease in seawater pH (from ~8.1 to ~7.8) (in 2100). In the Mediterranean Sea, there has been a significant reduction (0.05–0.14) in pH since the pre-industrial era (Touratier and Goyet 2011). Continuous seawater acidification would affects calcifying phytoplankton, such as coccolithophores, which play a major role in the primary productivity of the oligotrophic Mediterranean Sea (Coll et al. 2010). Reducing coccolithophore concentration will increase the penetration of light to deeper layers (Balch and Utgoff 2009) and might slightly enhance primary production.

As stratified water coccolithophores might bloom since they are more efficient than diatoms in nutrient acquisition at low concentrations, this bloom will occur in spite of the pH drop (Cermeno et al. 2008). The alternative scenario is that they will block the penetration of light and, by that, will decrease the euphotic zone and decrease even more chlorophyll concentration and primary production in the Mediterranean Sea (Fig. 7.4).

A general decrease in the yearly and monthly means of chlorophyll values has been observed in the Mediterranean Sea over a 25-year period (1979–2003). This decrease can be interpreted as a symptom of increased nutrient limitation resulting from reduced vertical mixing due to a more stable stratification of the basin, in line with the general warming trend (Behrenfeld et al. 2006; Barale et al. 2008).

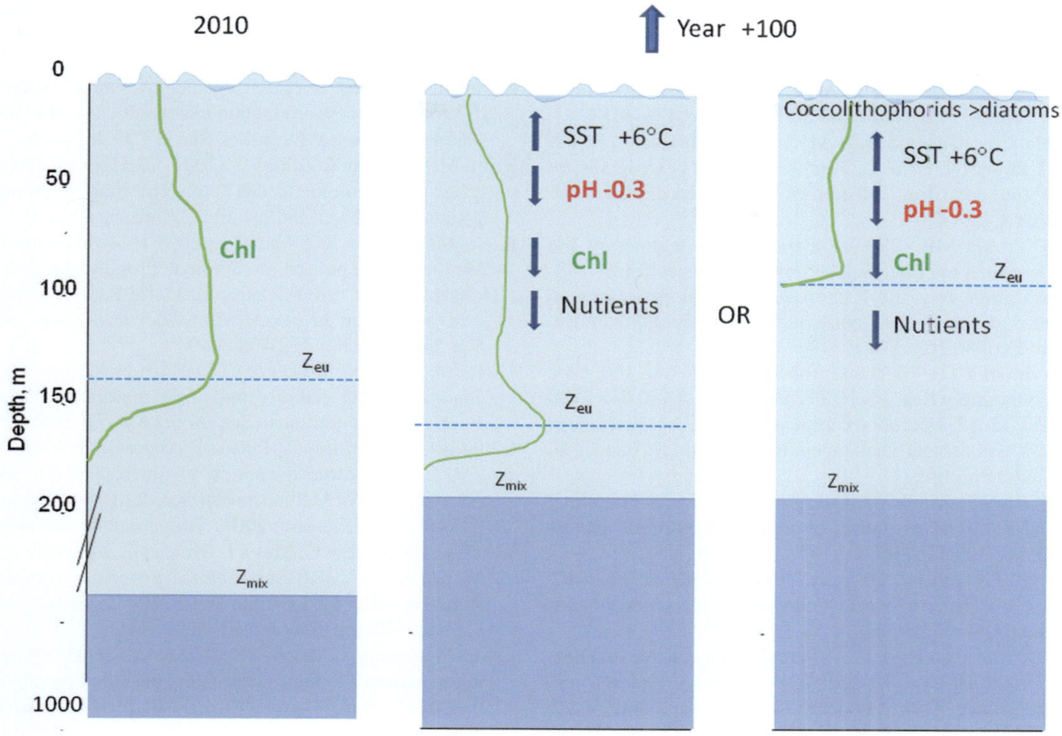

Fig. 7.4 The present and future state of the Mediterranean Sea. Optional hypothetical changes in physical, chemical, and biological parameters as a result of global climate change (*Chl* chlorophyll, *SST* sea surface temperature, Z_{eu} euphotic depth, Z_{mix} mixing depth)

References

Antoine D, Morel A, Andre JM (1995) Algal pigment distribution and primary production in the eastern Mediterranean as derived from coastal zone color scanner observations. J Geophys Res-Oceans 100:16193–16209

Aquarone MC, Adams S, Mifsu P (2009) LMEs and regional seas iv Mediterranean Sea www.lme.noaa.gov/LMEWeb/LME_Report/lme_26.pdf, 187–200. Accessed 10 Mar 2012

Arnaboldi M, Meyers PA (2006) Patterns of organic carbon and nitrogen isotopic compositions of latest Pliocene sapropels from six locations across the Mediterranean Sea. Palaeogeogr Palaeoclimatol Palaeoecol 235:149–167

Azov Y (1986) Seasonal patterns of phytoplankton productivity and abundance in nearshore oligotrophic waters of the Levant Basin (Mediterranean). J Plankton Res 8:41–53

Balch WM, Utgoff PE (2009) Potential interactions among ocean acidification, coccolithophores, and the optical properties of seawater. Oceanography 22:146–159

Barale V, Jaquet J-M, Ndiaye M (2008) Algal blooming patterns and anomalies in the Mediterranean Sea as derived from the Sea WiFS data set (1998–2003). Remote Sens Environ 112:3300–3313

Behrenfeld MJ, O'Malley RT, Siegel DA, McClain CR, Sarmiento JL, Feldman GC, Milligan AJ, Falkowski PG, Letelier RM, Boss ES (2006) Climate-driven trends in contemporary ocean productivity. Nature 444:752–755

Berland BR, Bonin DJ, Maestrini SY (1980) Azote ou phosphore? Considerations sur le "paradoxe nutrionnel" de la mer Mediteranee. Oceanol Acta 3:135–142

Bosc E, Bricaud A, Antoine D (2004) Seasonal and interannual variability in algal biomass and primary production in the Mediterranean Sea, as derived from 4 years of SeaWiFS observations. Global Biogeochem Cy GB1005. doi: 10.1029/2003GB002034

Boyce DG, Lewis MR, Worm B (2010) Global phytoplankton decline over the past century. Nature 466:591–596

Boyd PW, Strzepek R, Fu FX, Hutchins DA (2010) Environmental control of open-ocean phytoplankton groups: now and in the future. Limnol Oceanogr 55:1353–1376

Cermeno P, Dutkiewicz S, Harris RP, Follows M, Schofield O, Falkowski PG (2008) The role of nutricline depth in regulating the ocean carbon cycle. Proc Natl Acad Sci USA 105:20344–20349

Charles F, Lantoine F, Brugel S, Chretiennot-Dinet MJ, Quiroga I, Riviere A (2005) Seasonal survey of the phytoplankton biomass, composition and production in a littoral NW Mediterranean site, with special emphasis on the picoplanktonic contribution. Estuar Coast Shelf Sci 65:199–212

Chavez FP, Messie M, Pennington JT (2011) Marine primary production in relation to climate variability and change. Annu Rev Mar Sci 3:227–260

Chisholm S (1992) Phytoplankton size. In: Falkowski P, Woodhead A (eds) Primary productivity and biogeochemical cycles in the sea. Plenum Press, New York

Coll M, Piroddi C, Steenbeek J, Kaschner K, Lasram FBR, Aguzzi J, Ballesteros E, Bianchi CN, Corbera J, Dailianis T, Danovaro R, Estrada M, Froglia C, Galil BS, Gasol JM, Gertwagen R, Gil J, Guilhaumon F, Kesner-Reyes K, Kitsos M-S, Koukouras A, Lampadariou N, Laxamana E, Lopez-Fe de la Cuadra CM, Lotze HK, Martin D, Mouillot D, Oro D, Raicevich S, Rius-Barile J, Ignacio Saiz-Salinas J, San Vicente C, Somot S, Templado J, Turon X, Vafidis D, Villanueva R, Voultsiadou E (2010) The biodiversity of the Mediterranean Sea: estimates, patterns, and threats. PLoS One 5:1–36

Coma R, Ribes M, Serrano E, Jimenez E, Salat J, Pascual J (2009) Global warming-enhanced stratification and mass mortality events in the Mediterranean. Proc Natl Acad Sci USA 106:6176–6181

Crombet Y, Leblanc K, Queguiner B, Moutin T, Rimmelin P, Ras J, Claustre H, Leblond N, Oriol L, Pujo-Pay M (2011) Deep silicon maxima in the stratified oligotrophic Mediterranean Sea. Biogeosciences 8:459–475

D'Ortenzio F, D'Alcala MR (2009) On the trophic regimes of the Mediterranean Sea: a satellite analysis. Biogeosciences 6:139–148

Dubinsky Z, Stambler N (2009) Photoacclimation processes in phytoplankton: mechanisms, consequences, and applications. Aquat Microb Ecol 56:163–176

Emeis K-C, Sakamoto T (1998) The sapropel theme of Leg 160. Proc Ocean Drill Program Sci Res. doi:10.2973/odp.proc.sr.160.058.1998

Estrada M, Vaqué D (2013) Microbial components in The Mediterranean Sea: its history and present challenges. In: Goffredo S, Baader H, Dubinsky Z)eds) Springer

Falkowski PG, Katz ME, Knoll AH, Quigg A, Raven JA, Schofield O, Taylor FJR (2004) The evolution of modern eukaryotic phytoplankton. Science 305:354–360

Finkel ZV, Beardall J, Flynn KJ, Quigg A, Rees TAV, Raven JA (2010) Phytoplankton in a changing world: cell size and elemental stoichiometry. J Plankton Res 32:119–137

Gallego-Torres D, Martinez-Ruiz F, Meyers PA, Paytan A, Jimenez-Espejo FJ, Ortega-Huertas M (2011) Productivity patterns and N-fixation associated with Pliocene-Holocene sapropels: paleoceanographic and paleoecological significance. Biogeosciences 8:415–431

Garcia N, Raimbault P, Gouze E, Sandroni V (2006) Nitrogen fixation and primary production in western Mediterranean. C R Biol 329:742–750

Garczarek L, Dufresne A, Rousvoal S, West NJ, Mazard S, Marie D, Claustre H, Raimbault P, Post AF, Scanlan DJ, Partensky F (2007) High vertical and low horizontal diversity of *Prochlorococcus* ecotypes in the Mediterranean Sea in summer. FEMS Microbiol Ecol 60:189–206

Gernez P, Antoine D, Huot Y (2011) Diel cycles of the particulate beam attenuation coefficient under varying trophic conditions in the northwestern Mediterranean Sea: observations and modeling. Limnol Oceanogr 56:17–36

Gomez F, Gorsky G (2003) Annual microplankton cycles in Villefranche Bay, Ligurian Sea, NW Mediterranean. J Plankton Res 25:323–339

Groom S, Herut B, Brenner S, Zodiatis G, Psarra S, Kress N, Krom MD, Law CS, Drakopoulos P (2005) Satellite-derived spatial and temporal biological variability in the Cyprus Eddy. Deep Sea Res II 52:2990–3010

Herut B, Collier R, Krom MD (2002) The role of dust in supplying nitrogen and phosphorus to the Southeast Mediterranean. Limnol Oceanogr 47:870–878

Higgins MB, Robinson RS, Carter SJ, Pearson A (2010) Evidence from chlorin nitrogen isotopes for alternating nutrient regimes in the Eastern Mediterranean Sea. Earth Planet Sci Lett 290:102–107

Ignatiades L, Gotsis-Skretas O, Pagou K, Krasakopoulou E (2009) Diversification of phytoplankton community structure and related parameters along a large-scale longitudinal east–west transect of the Mediterranean Sea. J Plankton Res 31:411–428

Kimor B, Berman T, Schneller A (1987) Phytoplankton assemblages in the deep chlorophyll maximum layers off the Mediterranean coast of Israel. J Plankton Res 9:433–443

Kress N, Herut B (2001) Spatial and seasonal evolution of dissolved oxygen and nutrients in the Southern Levantine Basin (Eastern Mediterranean Sea): chemical characterization of the water masses and inferences on the N:P ratios. Deep-Sea Res Pt I 48:2347–2372

Krom MD, Kress N, Brenner S, Gordon LI (1991) Phosphorus limitation of primary productivity in the Eastern Mediterranean Sea. Limnol Oceanogr 36:424–432

Krom MD, Brenner S, Kress N, Neori A, Gordon LI (1992) Nutrient dynamics and new production in a warm-core eddy from the eastern Mediterranean-sea. Deep-Sea Res Pt I 39:467–480

Krom MD, Groom S, Zohary T (2003) The Eastern Mediterranean. In: Black KD, Shimmield GB (eds) The biogeochemistry of marine systems. Blackwell Publishing, Oxford, pp 91–122

Krom MD, Emeis KC, Van Cappellen P (2010) Why is the Eastern Mediterranean phosphorus limited? Prog Oceanogr 85:236–244

L'Helguen S, Le Corre P, Madec C, Morin P (2002) New and regenerated production in the Almeria-Oran front area, eastern Alboran Sea. Deep Sea Res Pt I 49:83–99

Lasternas S, Agusti S, Duarte CM (2010) Phyto- and bacterioplankton abundance and viability and their relationship with phosphorus across the Mediterranean Sea. Aquat Microb Ecol 60:175–191

Lasternas S, Tunin-Ley A, Ibanez F, Andersen V, Pizay MD, Lemee R (2011) Short-term dynamics of microplankton abundance and diversity in NW Mediterranean Sea during late summer conditions (DYNAPROC 2 cruise; 2004). Biogeosciences 8:743–761

Lazzara L, Marchese C, Massi L, Nuccio C, Maselli F, Santini C, Pieri M, Sorani V (2010) Sub-regional patterns of primary production annual cycle in the Ligurian and North Tyrrhenian seas, from satellite data. Riv Ital Telerilevam 42:87–102

Li WKW, Zohary T, Yacobi YZ, Wood AM (1993) Ultraphytoplankton in the eastern Mediterranean-Sea – towards deriving phytoplankton biomass from flow cytometric measurements of abundance, fluorescence and light scatter. Mar Ecol Prog Ser 102:79–87

Llabres M, Agusti S, Alonso-Laita P, Herndl GJ (2010) *Synechococcus* and *Prochlorococcus* cell death induced by UV radiation and the penetration of lethal UVR in the Mediterranean Sea. Mar Ecol Prog Ser 399:27–37

Magazzu G, Decembrini F (1995) Primary production, biomass and abundance of phototrophic picoplankton in the Mediterranean Sea – a review. Aquat Microb Ecol 9:97–104

Mangoni O, Modigh M, Mozetic P, Bergamasco A, Rivaro P, Saggiomo V (2008) Structure and photosynthetic properties of phytoplankton assemblages in a highly dynamic system, the Northern Adriatic Sea. Estuar Coast Shelf S 77:633–644

Marty JC, Chiaverini J (2002) Seasonal and interannual variations in phytoplankton production at DYFAMED time-series station, northwestern Mediterranean Sea. Deep Sea Res II 49:2017–2030

Marty JC, Chiaverini J, Pizay MD, Avril B (2002) Seasonal and interannual dynamics of nutrients and phytoplankton pigments in the western Mediterranean Sea at the DYFAMED time-series station (1991–1999). Deep Sea Res II 49:1965–1985

Mella-Flores D, Mazard S, Humily F, Partensky F, Mahé F, Bariat L, Courties C, Marie D, Ras J, Mauriac R, Jeanthon C, Bendif EM, Ostrowski M, Scanlan DJ, Garczarek L (2011) Is the distribution of *Prochlorococcus* and *Synechococcus* ecotypes in the Mediterranean Sea affected by global warming? Biogeosciences 8:4281–4330

Mercado JM, Ramirez T, Cortes D, Sebastian M, Reul A, Bautista B (2006) Diurnal changes in the bio-optical properties of the phytoplankton in the Alboran Sea (Mediterranean Sea). Estuar Coast Shelf S 69:459–470

Mobius J, Lahajnar N, Emeis KC (2010) Diagenetic control of nitrogen isotope ratios in Holocene sapropels and recent sediments from the Eastern Mediterranean Sea. Biogeosciences 7:3901–3914

Moran XAG, Estrada M (2005) Winter pelagic photosynthesis in the NW Mediterranean. Deep Sea Res I 52:1806–1822

Napolitano E, Oguz T, Malanotte-Rizzoli P, Yilmaz A, Sansone E (2000) Simulations of biological production in the Rhodes and Ionian basins of the eastern Mediterranean. J Marine Syst 24:277–298

Nincevic Z, Marasovic I, Kuspilic G (2002) Deep chlorophyll-a maximum at one station in the middle Adriatic Sea. J Mar Biol Assoc UK 82:9–19

Nixon SW (1995) Coastal marine eutrophication – a definition, social causes, and future concerns. Ophelia 41:199–219

Owens NJP, Rees AP, Woodward EMS, Mantoura RFC (1989) Size-fractionnated primary production and nitrogen assimilation in the northwest Mediterranean Sea during January 1989. Water Pollut Res Bull 13:126–135

Partensky F, Hess WR, Vaulot D (1999) *Prochlorococcus*, a marine photosynthetic prokaryote of global significance. Microbiol Mol Biol Rev 63:106–127

Psarra S, Tselepides A, Ignatiades L (2000) Primary productivity in the oligotrophic Cretan Sea (NE Mediterranean): seasonal and interannual variability. Prog Oceanogr 46:187–204

Psarra S, Zohary T, Krom MD, Mantoura RFC, Polychronaki T, Stambler N, Tanaka T, Tselepides A, Thingstad TF (2005) Phytoplankton response to a Lagrangian phosphate addition in the Levantine Sea (Eastern Mediterranean). Deep Sea Res II 52:2944–2960

Puigserver M, Ramon G, Moya G (2002) Spatial and temporal distribution of phytoplankton in a Mediterranean estuarine canal system. J Coastal Res 18:39–51

Rahav E, Herut B, Stambler N, Bar-Zeev E, Mulholland RM, Berman-Frank I (2013) uncoupling between dinitrogen fixation and primary productivity in the eastern Mediterranean Sea. J Geophys Res Biogeosci 118:195–202

Regaudie-De-Gioux A, Vaquer-Sunyer R, Duarte CM (2009) Patterns in planktonic metabolism in the Mediterranean Sea. Biogeosciences 6:3081–3089

Reul A, Rodriguez V, Jimenez-Gomez F, Blanco JM, Bautista B, Sarhan T, Guerrero F, Ruiz J, Garcia-Lafuente J (2005) Variability in the spatio-temporal distribution and size-structure of phytoplankton across an upwelling area in the NW-Alboran Sea, (W-Mediterranean). Cont Shelf Res 25:589–608

Sachs JP, Repeta DJ (1999) Oligotrophy and nitrogen fixation during eastern Mediterranean sapropel events. Science 286:2485–2488

Sandroni V, Raimbault P, Migon C, Garcia N, Gouze E (2007) Dry atmospheric deposition and diazotrophy as sources of new nitrogen to northwestern Mediterranean oligotrophic surface waters. Deep Sea Res I 54:1859–1870

Siokou-Frangou I, Christaki U, Mazzocchi MG, Montresor M, D'Alcala MR, Vaque D, Zingone A (2010) Plankton in the open Mediterranean Sea: a review. Biogeosciences 7:1543–1586

Skejic S, Marasovic I, Vidjak O, Kuspilic G, Gladan ZN, Sestanovic S, Bojanic N (2011) Effects of cage fish farming on phytoplankton community structure, biomass and primary production in an aquaculture area in the middle Adriatic Sea. Aquac Res 42:1393–1405

Solomon S, Qin D, Manning M, Alley RB, Berntsen T, Bindoff NL, Chen Z, Chidthaisong A, Gregory JM, Hegerl GC, Heimann M, Hewitson B, Hoskins BJ, Joos F, Jouzel J, Kattsov V, Lohmann U, Matsuno T, Molina M, Nicholls N, Overpeck J, Raga G, Ramaswamy V, Ren J, Rusticucci M, Somerville R, Stocker TF, Whetton P, Wood RA, Wratt D (2007) In: Solomon S, Qin D, Manning M, Chen Z, Marquis M, Averyt KB, Tignor M, Miller HL (eds) Technical summary. Climate change 2007: the physical science basis. Contribution of working group I to the fourth assessment report of the intergovernmental panel on climate change. Cambridge University Press, Cambridge, p 74

Sommaruga R, Hofer JS, Alonso-Saez L, Gasol JA (2005) Differential sunlight sensitivity of picophytoplankton from surface Mediterranean coastal waters. Appl Environ Microbiol 71:2154–2157

Stambler N (2012) Underwater light field of the Mediterranean Sea. In: Stambler N (ed) Life in the Mediterranean Sea: a look at habitat changes. Nova Science Publishers, New York

Tanaka T, Thingstad TF, Christaki U, Colombet J, Cornet-Barthaux V, Courties C, Grattepanche JD, Lagaria A, Nedoma J, Oriol L, Psarra S, Pujo-Pay M, Van Wambeke F (2011) Lack of P-limitation of phytoplankton and heterotrophic prokaryotes in surface waters of three anticyclonic eddies in the stratified Mediterranean Sea. Biogeosciences 8:525–538

Thingstad TF, Krom MD, Mantoura RFC, Flaten GAF, Groom S, Herut B, Kress N, Law CS, Pasternak A, Pitta P, Psarra S, Rassoulzadegan F, Tanaka T, Tselepides A, Wassmann P, Woodward EMS, Riser CW, Zodiatis G, Zohary T (2005) Nature of phosphorus limitation in the ultraoligotrophic eastern Mediterranean. Science 309:1068–1071

Touratier F, Goyet C (2011) Impact of the Eastern Mediterranean transient on the distribution of anthropogenic CO_2 and first estimate of acidification for the Mediterranean Sea. Deep Sea Res I 58:1–15

Uitz J, Claustre H, Gentili B, Stramski D (2010) Phytoplankton class-specific primary production in the world's oceans: seasonal and interannual variability from satellite observations. Global Biogeochem Cy 24, Article no. GB3016. doi:10.1029/2009GB003680

Vidussi F, Claustre H, Manca BB, Luchetta A, Marty JC (2001) Phytoplankton pigment distribution in relation to upper thermocline circulation in the eastern Mediterranean Sea during winter. J Geophys Res-Oceans 106:19939–19956

Yacobi YZ, Zohary T, Kress N, Hecht A, Robarts RD, Waiser M, Wood AM (1995) Chlorophyll distribution throughout the Southeastern Mediterranean in relation to the physical structure of the water mass. J Marine Syst 6:179–190

Yogev T, Rahav E, Bar-Zeev E, Man-Aharonovich D, Stambler N, Kress N, Beja O, Mulholland RM, Herut B, Berman-Frank I (2011) Is dinitrogen fixation significant in the Levantine Basin, East Mediterranean Sea? Environ Microbiol 13:854–871. doi:10.1111/j.1462-2920.2010.02402.x

Zohary T, Herut B, Krom MD, Mantoura RFC, Pitta P, Psarra S, Rassoulzadegan F, Stambler N, Tanaka T, Thingstad TF, Woodward EMS (2005) P-limited bacteria but N and P co-limited phytoplankton in the Eastern Mediterranean – a microcosm experiment. Deep Sea Res II 52:3011–3023

Autochthonous Seaweeds

8

Félix L. Figueroa, Antonio Flores-Moya, Juan J. Vergara, Nathalie Korbee, and Ignacio Hernández

Abstract

The Mediterranean Sea is regarded as the largest and deepest enclosed sea in Earth and represents a marine biodiversity hot spot. The current taxonomic classification of macroalgae in the Mediterranean is well known, with 1,124 species and at least 20 % of endemic species. The origin of these seaweeds is diverse and is a consequence of the geological history of the Mediterranean basin. The present seaweed flora is composed of tropical elements, species coming from the boreal Atlantic Ocean, subtropical Atlantic species, cosmopolitan/panoceanic species, and endemic elements. Analytical biogeography studies based on the deductive-hypothetical approach has allowed exploration of the links between ecological variables and the geographical distribution of selected taxa of Mediterranean seaweeds. This miniature ocean has various sources of disturbances interacting synergistically and therefore, providing an insight into a major question: how resilient are marine ecosystems, and how will their current functioning be modified in the future? Mediterranean is a priority area for studies on climate change, as in this area is documented range shifts and recent appearance of new warm-water species, included tropical macroalgae. The effect of climatic change is difficult to untangle from direct anthropogenic activities. Morphology plasticity and physiological mechanisms determine the capacity to acclimate to global climate change. The conservation efforts made in this ecosystem include the design of a network of Marine Protected Areas that, by definition, higher level of protection than its surrounding areas. Recent national legislation, as the Spanish list of wild species under special protection and the list of threatened species (2011) includes also for the first time 12 species of macroalgae. Additionally, the protection of macroalgae is also enhanced by the protection of particular marine habitats under the Habitat Directive 92/43/EEC. The use of macroalgae for the estimation of the ecological status of coastal and transitional waters of the Mediterranean, in the frame of the Water Framework Directive (2000/60/EC) is a challenging applied subject. It is expected that the monitoring and further management plans for the coast will decrease the probable impact amplification in the coming years and will maintain the ability of the Mediterranean Sea to continue the provision of essential ecological services.

F.L. Figueroa (✉) • N. Korbee
Departamento de Ecología, Facultad de Ciencias, Universidad de Málaga, Campus Univ. de Teatinos s/n 29071, Málaga, Spain
e-mail: felix_lopez@uma.es; nkorbee@uma.es

A. Flores-Moya
Departamento de Biología Vegetal, Facultad de Ciencias, Universidad de Málaga, Campus Univ. de Teatinos s/n 29071, Málaga, Spain
e-mail: floresa@uma.es

J.J. Vergara • I. Hernández
Departamento de Biología, Facultad de Ciencias, del Mar y Ambientales, Universidad de Cádiz, 11510. Puerto Real, Cádiz,, Spain
e-mail: juanjose.vergara@uca.es; ignacio.hernandez@uca.es

S. Goffredo and Z. Dubinsky (eds.), *The Mediterranean Sea: Its history and present challenges*,
DOI 10.1007/978-94-007-6704-1_8,

Keywords
Acidification • Biological indicators • Climate change • Ecophysiology • Macroalgae • Mediterranean Sea • Macroalgal protection • Temperature • UV radiation

Introduction: What It Is Seaweed?

"Seaweed" or "marine macroalgae" is a term to name macroscopic, multi-celled and complex plant-like organisms, generally benthic. The term includes species classified in two different kingdoms (Plantae and Chromista) and distributed among three different phyla: Rhodophyta (red algae), Chloro-phyta (green algae), and Heterokontophyta (brown algae). Green algae are included in Plantae while red and brown algae are Chromista. They are all photosynthetic organisms with chlorophyll *a* as the main pigment. As these three groups have not a common ancestor, seaweeds are considered as a polyphyletic group. In addition, some tuft-forming blue-green algae (Cyanobacteria) are sometimes regarded as seaweeds by some authors. For these reasons, seaweed is considered a colloquial term and lacks a formal definition. However, it is a relevant term from an ecological point of view because it pools the macroscopic, primary producers from the benthos on hard substrata.

How Many Different Seaweeds Are There in the Mediterranean Sea?

The Mediterranean Sea is regarded as the largest and deepest enclosed sea in Earth and represents a marine biodiversity hot spot. The number of seaweed species occurring in the Mediterranean was estimated by Giaconne (1974) to be 1,000. The current taxonomic classification of macroalgae in the Mediterranean is well known, with 1,124 species described (a number slightly higher if infraspecific taxa are considered) and at least 20 % of endemic species. This percentage of endemisms is particularly high in certain zones, as the Messina Strait (Cinelli 1981). Estimations have been made of 277 species in Heterokontophyta (Ribera et al. 1992; Coll et al. 2010) and 190 in Chlorophyta (Gallardo et al. 1993; Coll et al. 2010). The most abundant taxonomic group are the rhodophyceans (Gómez-Garreta et al. 2001). An estimation of the number of red algae, based on analytical biogeography hypothesis, proposed that around 700 species could occur in the Mediterranean (Báez et al. 2004), and 657 species have been compiled until now (Coll et al. 2010). Although a significant increase in the rate of description of new seaweed species is not expected (Collins et al. 2010), the description of new taxa continues (Alongi et al. 2008; Rodríguez-Prieto and De Clerck 2009; Serio et al. 2011). Nevertheless, it has been noted that floristic surveys, using modern taxonomical criteria, are needed in the Adriatic Sea and the southern and eastern shores of the Mediterranean Sea (Coll et al. 2010). The total estimates of Mediterranean species of macroalgae represented 10.9 % of their counterparts worldwide. This percentage of shared species with global estimations ranged between 17.3 % for Phaeophyceans (the highest for macrophytes and metazoans) and 7.6 % for Chlorophyta (Coll et al. 2010).

The Origin of Seaweeds in the Mediterranean Sea

The origin of Mediterranean seaweeds is diverse and is a consequence of the geological history of the Mediterranean basin (reviewed by Lüning 1990; Bianchi and Morri 2000).

The Mediterranean Sea is a remnant of the once extensive Tethys Ocean, an eastward-open equatorial water-body that indented Pangea during the Triasic (220 my BP). During the Cretaceous, the movement separating the European and African plates caused the connection of the newly-formed ocean to the older Indopacific Ocean (120 my BP). It is thought that, at that time, the Tethys harboured a highly diverse warm-water biota and this could be the origin of the tropical floral element represented today by, among others, *Halimeda*, *Flabellia*, *Valonia* and *Acetabularia* in Chlorophyta; *Hypnea musciformis*, *Amphiroa rigida*, *Liagora viscida* and *Digenea simplex* in Rhodophyta; and *Sargassum* in Heterokothophyta.

In the Oligocene (30 my BP), a shrinkage of the Tethys Ocean and the diminution of the warming influence of the Tethys on the world oceans produced cold water conditions elsewhere. This may have resulted in the extinction of the greater part of the tropical elements in the Mediterranean area. During the Miocene (around 10 my BP), the Isthmus of Suez was formed, separating the Mediterranean from the Indopacific. At the Late Miocene (Messinian stage, 6 my BP), the connection with the Atlantic also closed for 500,000 years. Europe and Africa collided at Gibraltar and it is thought that the global sea level fell by as much as 50 m (Adams 1981; Stanley and Wezel 1985). The consequence of the closing at both Suez and in Gibraltar, together with the fact that the Mediterranean basin had, as at present, a negative hydrological balance (water losses by evaporation are higher than the inflow by rivers and rain), caused the Mediterranean to dry repeatedly between 6 and 5 my BP (the Messinian salinity crisis). An alternative scenario proposes a Mediterranean transformed into a series of large evaporitic lakes during the Messinian salinity crisis. Although shallow-water biota may have survived through the Neogene (Stanley 1990; Myers 1996), the biodiversity was drastically reduced and only with the re-opening of the Strait of Gibraltar at the dawn of the Pliocene (5 my BP) was the Mediterranean Sea repopulated by species of boreal Atlantic origin.

During the Quaternary, the alternation between ice ages and warm interglacial periods resulted in repeated immigration waves of species of boreal Atlantic or subtropical origins, respectively. In particular, during the glacial periods the seawater temperatures at the surface were 2–6 °C lower during the summer than they are nowadays during the same period (Thunell 1979). The main consequences were that the remaining tropical species were eradicated and that some Atlantic floral elements remained in refugia in the coolest places of the Mediterranean Sea or in deep waters. These species are recognized today as glacial relicts, i.e., *Fucus virsoides*, deep-water Desmarestiales in the Adriatic Sea, deep-water Laminariales and Tilopteridales from the Sicily Channel, the Strait of Messina and the Alboran Sea, and the red alga *Plocamium cartilagineum* (Cinelli 1981, 1985; Flores-Moya 2012). In postglacial times, warm-temperate species probably arrived in the Mediterranean from the tropical West African coast, as is suggested by a continuous distribution of many Lusitanic-Mediterranean species.

The cosmopolitan floral elements are represented by *Ulva* spp. in Chlorophyta and *Scytosiphon lomentaria* in Heterokontophyta. The cosmopolitan species are often found in degraded areas. Meanwhile, the Mediterranean endemic seaweeds account for around 20 % of the flora (Giaconne 1974; Coll et al. 2010); for example, several species of the brown algal genus *Cystoseira* (Oliveras Plá and Gómez-Garreta 1989).

To summarize, the present seaweed flora is composed of tropical elements, species coming from the boreal Atlantic Ocean, subtropical Atlantic species, cosmopolitan/ panoceanic species, and endemic elements. In historic times, the flora has been enriched with Lessepsian immigrants, species coming from the Red Sea to the Eastern Mediterranean Sea through the Suez Canal after the opening in 1869 (e.g. the green alga *Caulerpa racemosa*).

Special mention must be made of introduced and invasive species. It is thought that since the beginning of the twentieth century, the number of species introduced to the Mediterranean has nearly doubled every 20 years, accounting for about 84 species at present (Boudouresque and Verlaque 2002). Eight of these species have been considered as invasive, i.e., playing a conspicuous role in the recipient ecosystems because they take the place of keystone species and/or are economically harmful (the red algae *Acrothamnion preissii*, *Asparagopsis armata*, *Lophocladia lallemandii*, *Womersleyella setacea*; the brown algae *Sargassum muticum*, and *Stypopodium schimperi*; and the green algae *Caulerpa racemosa* and *C.taxifolia*) (Boudouresque and Verlaque 2002).

Spatial Patterns of Seaweed Richness in the Mediterranean Sea

Much effort has been expended to describe the bathymetric patterns of seaweed communities (zonation patterns) in the Mediterranean Sea (reviewed by Ros et al. 1985; Lüning 1990). It must be stressed that, because seaweeds are photosynthetic organisms, they are restricted to the photic zone (from the mediolittoral zone to the deepest limit in the circalittoral zone, where around 1–0.1 % of surface irradiance is available). In the clearest waters of the western Mediterranean the deepest limit of the photic zone is reached at around 110 m (Ballesteros 1994), whereas a somewhat deeper limit occurs in the oligotrophic waters of the eastern area (Ballesteros 2006). It has been proposed that species richness increases from the highest levels of the mediolittoral down to the lower infralittoral and upper circalittoral communities, and then decreases along the circalittoral (Coll et al. 2010).

The spatial distribution patterns of Mediterranean seaweeds and their determining factors have been less studied than their zonation patterns. A usual approach has been based on the description of geographical distributions of seaweed taxa and, by using inductive hypotheses, to infer the historical causes explaining the distributions (van den Hoek 1975, 1982; Giaconne and Geraci 1989; Verlaque 1994). However, analytical biogeography studies based on the deductive-hypothetical approach has allowed exploration of the links between ecological variables and the geographical distribution of selected taxa of Mediterranean seaweeds. For instance, a relevant model is the geographical distribution of the genus *Cystoseira*, which has several endemic species in the Mediterranean Sea (Oliveras Plá and Gómez-Garreta 1989; see section "The origin of seaweeds in the Mediterranean Sea"). It is supposed that all these endemic taxa are neoendemics (which originated after re-opening of the Strait of Gibraltar at the dawn of the Pliocene 5 my BP), with species being replaced by others of similar morphology (vicariant species) from the west to the east. A significant U-shaped relationship was found between species richness in *Cystoseira* and current flown the western Mediterranean. It was hypothesized that species richness in *Cystoseira* depends on single-and few-celled dispersion stages at low water velocities, but in higher-velocity water currents the presence of pneumatocysts in several species of *Cystoseira* could facilitate long-range dispersal of thalli (Báez et al. 2005). Representative species of macroalgae in the Mediterranean

Fig. 8.1 Species of macroalgae in the Mediterranean in the eulittoral area (Rocky shores and ponds) , red algae as *Corallina elongata* (**a**), *Risoella verruculosa* (**b**) and *Porphyra rosengurttii* (**c**), green algae as *Ulva rigida* (**c**) and brown algae as *Padina pavonica* (**d**)

eulittoral and subtidal areas are presented (Figs. 8.1, 8.2 and 8.3). Algae from rocky shores and ponds (eulittoral) are presented in Fig. 8.1: red algae as *Corallina elongata* (1a), *Risoella verruculosa* (1b) and *Porphyra rosengurttii* (1c), green algae as *Ulva rigida* (1c) and brown algae as *Padina pavonica* (1d). Species from subtidal area are presented in Fig. 8.2 as the green algae *Acetabularia mediterranea* (2a) and *Halimeda tuna* (2b) and red algae *Digenea simplex* (2c). Species of macroalgae in the Mediterranean sea in subtidal area as the brow algae *Cystoseira* sp. (3a) and *Laminaria ochroleuca* (3b) growing in deep water.

Potential Effects of Climatic and Non-climatic Drivers on Mediterranean Seaweeds

The Mediterranean is now considered a 'biodiversity hotspot' (Bianchi and Morri 2000) and a 'miniature ocean' by physical oceanographers (Béthoux and Gentili 1999). This miniature ocean can serve as a giant mesocosm of the world's oceans, with various sources of disturbances interacting synergistically and therefore, providing an insight into a major question: how resilient are marine ecosystems, and how will their current functioning be modified in the future? (Lejeusne et al. 2009).

The coastal waters of Mediterranean Sea, as in other parts of the world, are continuously exposed to increasing human pressure through activities such fisheries, energy production, trade and tourism. Thus, the effect of climatic change is difficult to untangle from direct anthropogenic activities. The latter often reduces the resilience of the marine and coastal ecosystems, which then become more vulnerable to stresses due to climate forcing.

Macroalgae show a high diversity and they are complex in functional morphology and ecological role. They range from small, structurally simple, filamentous turfs a few millimetre high, or heavily calcified crustose forms to large leathery macrophytes. Thus, morphology plasticity and physiological mechanisms determine the capacity to acclimate to global climate change.

Fig. 8.2 Species of macroalgae in the Mediterranean sea in the subtidal area as the green algae *Acetabularia mediterranea* (**a**) and *Halimeda tuna* (**b**) and red algae *Digenea simplex* (**c**) and *Asparagopsis taxiformis* (**d**) (Photographs (**a**, **b** and **d**) by permission of J. Carlos Moreno)

Fig. 8.3 Species of macroalgae in the Mediterranean sea in subtidal area as the brow algae *Laminaria ochroleuca* (**a**) and *Cystoseira* sp. (**b**) growing in deep water (Photographs by permission of J. Carlos Moreno)

Temperature and Carbon Sink

Recently Louanchi et al. (2009) reported a review on datasets of temperature, salinity, oxygen, nutrients, chlorophyll and a surface layer box model to reconstruct dissolved inorganic carbon (DIC), total alkalinity and carbon dioxide fugacity (f_{CO_2}), mixed layers filed in the Mediterranean Sea from 1960 to 1990. According to the model, an atmospheric CO_2 increase of 40 μatm over this period induced an increase of 30 μmol l^{-1}DIC surface waters. The Mediterranean Sea has transformed from a source of 0.62 Tg C $year^{-1}$ for atmospheric CO_2 in 1960 to a net sink of −1.98 Tg C $year^{-1}$ in 1990. The pH change in this period was not significant, but according to future projections of increasing CO_2 sink, a substantial acidification is expected during twenty-first century in the Mediterranean Sea. Simultaneously other important impact is expected, the seawater saturation state with respect to a variety of carbonate minerals will decrease and seawater could become unsaturated in many environments with respect to algae with Mg-calcite mineral (as in the coralline macroalgae or *Padina* spp., the unique calcified brown macroalgae) and even aragonite (a less costly calcification present in Rhodophyta such as Nemalionales, Cryptonemiales and Peyssoneliales, or Chlorophyta as Dasycladales and Caulerpales).

Mediterranean is a priority area for studies on climate change (Mieszkowska et al. 2008), as in this area is documented range shifts and recent appearance of new warm-water species, included tropical macroalgae (Walther et al. 2002). It is known that the sea surface temperature has increased 0.13 °C per decade over the last 50 year (IPCC 2007). In fact, it has been reported a simultaneous increase in the abundance of thermotolerant species and the decrease or disappearance of cold waters stenothermic species since the 1980s in the Mediterranean Sea (Francour et al. 1994; Lejeusne et al. 2009).

Global climate change will likely result in higher seawater temperatures, stronger stratification and increased inflows of freshwater and nutrients to coastal waters in many areas (Rabalais et al. 2009). Thus, it is generally expected enhanced primary production with increased macroalgal standing stocks and more frequent or severe hypoxia events. In temperate coasts, perennial canopies of Fucales such as *Cystoseira* spp. and their associated understory could be replaced by mats of turf-forming algae near points of discharge of nutrients (Thibaut et al. 2005). Only five of fourteen species of Fucales reported at the end of the nineteenth century are currently present in the Albères Coast (France, NW Mediterranean). According to historical data there has been a steady decrease of all populations since the 1940s (Thibaut et al. 2005). Seven taxa now extinct were considered frequent and some of them were the dominant species in several phytobenthic assemblages. Moreover, only one of the five species left, shows no signs of regression (*Cystoseira compressa*), two are considered as rare (*Cystoseira caespitosa, Cystoseira zosteroides*), and one is very rare (*Cystoseira elegans*) (Thibaut et al. 2005). *Cystoseira mediterranea*, a species that was reported to make a continuous belt along the shores of the Albères coast, has almost disappeared from some areas. Deep-water species have been affected by an increase in water turbidity and, probably, chemical pollution and direct alga destruction attributed to net fishing. If degradation of the environmental conditions continues, the remaining *Cystoseira* species will face a most unwelcome prospect (Thibaut et al. 2005).

It is also interesting to evaluate the changes of macroalgae assemblages in Mediterranean lagoons. Inside the Mar Menor lagoon (SE Spain), changes in hydrodynamics due to the enlargement of the Estacio inlet in 1972 produced a decrease in salinity and lower extreme temperature, this conditions allowed the colonization by the algae *Caulerpa prolifera,* which substituted the previous meadows of the seagrass *Cymodocea nodosa,* now restricted to shallow sandy bottoms (Pérez-Ruzafa et al. 2012). The high biomass of the main primary producer, *C. prolifera*, covers most of the bottom, and probably plays a role in the resistance of the lagoon to eutrophication processes through the high uptake of nutrients from the water column and their retention in the sediments, avoiding high phytoplankton densities. Some authors suggest that, if climate change predictions become true, the current status of the lagoon is likely to collapse, since future environmental conditions could make *C. prolifera* unable to reach positive values of net photosynthesis, and eutrophication processes are expected to appear (Lloret et al. 2008).

Ocean Acidification and Photosynthetic Metabolism

There is high level of uncertainty about how coastal ecosystems will be affected by rapid ocean acidification caused by anthropogenic CO_2, due to a lack of data. The few experiments to date have been performed at short-term (<1 year) and reveal mixed responses depending on the species examined and the culture conditions applied. A doubling CO_2 produced an increase of photosynthetic activity of 52 % in the temperate red algae *Lobentaria articulate* (Kübler et al. 1999) and up to 130 % in other species (Gao et al. 1993); consequently, these species enhanced their growth (Gao et al. 1993; Kübler et al. 1999; Riebesell et al. 2007).

The sensitivity to the acidification of all algal groups is expected to be complex, due to the interactions between the effects of pH and CO_2 in the enhancement of photosynthesis. There are wide mechanisms of use of inorganic carbon ranged from algae that rely on diffusive CO_2 entry that are limited by actual CO_2 concentration, to high efficient use of

CO_2 in algae with carbon concentration mechanisms by the action of carbonic anhydrase (Mercado et al. 1998). Calcified algae are particularly sensitive to ocean acidification, i.e., a decrease in pH from 8 to 7.5 dramatically reduced calcification of the aragonite forming green macroalga *Halimeda tuna* (Borowitzka and Larkum 1976). Calcareous crustose algae (CCA) are the group most likely to be affected by ocean acidification, as they are highly sensitive to reductions of calcification saturation state. The impacts of increased CO_2 on CCA may include not only reduced calcification, but may ultimately include dissolution of calcified skeletons (Kleypas et al. 2006; Orr et al. 2005). While high CO_2 levels may enhance photosynthesis in calcifying algae such as *Halimeda* spp. (De Beer and Larkum 2001) and *Corallina pilulifera* (Gao et al. 1993), this stimulation will be offset by a decreased calcification as a result of decreased carbonate saturation state; overall outcome is difficult to predict. However, although increased ocean CO_2 concentration may enhance rates of photosynthesis and growth (particularly in species without carbon concentrating mechanisms), such increases may be limited by the availability of other limiting nutrients (Raven et al. 2005). If terrestrial nutrient inputs increase in coastal waters, a synergetic effect of enhanced nutrient and CO_2 levels on growth of turf algae could occur, in contrast to calcifying ones. Acidification may also increase the susceptibility of calcified algae to grazing and erosion, and may lead a reduction in sand production, with a significant loss of habitats (i.e. *Halimeda* banks) and a shift from calcifying to non-calcifying algae (Andersson et al. 2011). Mercado et al. (1998) reported a relationship between carbon concentration mechanisms (CCM) and light energy availability in intertidal macroalgae, but not with inorganic carbon availability (Mercado et al. 1998). Intertidal algae with emersion periods presented higher photosynthetic rates (Mercado et al. 1998) and carbon uptake (Flores-Moya et al. 1998) due to the higher availability of CO_2 than in submerged algae.

There is no information on the potential for adaptation of algal turfs, upright macroalgae or CCA to ocean acidification. In particular, the adaptive capacity of CCA is critical for reef structures, but this capacity is likely to be low, given that calcification is purely a physico-chemically mediated process. There may be some potential for adaptation by CCA and calcified upright algae by secreting less soluble skeletons (i.e. lower content of magnesium calcite in calcite skeletons) as found in the articulated algae *Amphiroa* sp. (Corallinales) from other areas (Caribbean sea, Andersson et al. 2011).

Organisms may be unable to build their skeletons as oceans acidify over the next 100 years (Orr et al. 2005). This may have a combined effect with other stress sources, such as global warming (Hoegh-Guldberg et al. 2007). However, attempts to determine whether expectations on the basis of laboratory experiments and modeled predictions translate to field conditions, they have been hindered by the difficulty of imitating ocean acidification conditions *in situ* for a sufficient time period to affect macroalgal communities. Natural CO_2 fluxes from volcanic vents and high heat flow areas amounts to less than 0.5 % of anthropogenic emissions to the global carbon budget, but can alter local ocean chemistry (Williams et al. 1992), can be excellent macrocosms for studying climate change effects. Marine CO_2 vents are abundant in the Mediterranean, especially around Italy and Greece where they typically eject volcanic fluids containing up to 1–2 % hydrogen. Hall-Spencer et al. (2008) showed that along gradients of normal pH (8.1–8.2) to lowered pH (mean 7.8–7.9, minimum 7.4–7.5) in a volcanic vent area in Ischia (Italy), typical rocky shore communities with abundant calcareous organisms shifted to communities lacking scleractinian corals, with significant reductions in sea urchins and coralline algal abundance. The species populating the vent sites comprise a suite of organisms that are resilient to naturally high concentrations of CO_2 and indicate that ocean acidification may benefit invasive non-native (non-calcified) algal species (Hall-Spencer et al. 2008). Additionally, Porzio et al. (2011) describe the effects of increasing CO_2 levels on macroalgal communities along a pH gradient caused by volcanic vents: macroalgal habitat differed at taxonomic and morphological group levels along a pH gradient. The vast majority of the 101 macroalgal species studied were able to grow, with only a slight 5 % decrease in species richness as the mean pH felt from 8.1 to 7.8. However, this small fall in species richness was associated with shifts in community structure as the cover of turf algae decreased disproportionately. These authors showed that many macroalgal species are tolerant of long-term elevations in CO_2 levels but that macroalgal habitats are altered significantly as pH drops, contributing to a scant but growing body of evidence concerning the long-term effects of CO_2 emissions in vegetated marine ecosystems.

UV Radiation

Although increase in the chlorine concentration in the stratosphere has slowed down, reflecting the execution of the Montreal Protocol, the estimated time for recovery of the ozone layer is unconvincing and will rely on the impacts of climate change on the stratosphere (Weatherhead and Andersen 2006; Mckenzie et al. 2007; Seckmeyer et al. 2008). At some sites of Northern Hemisphere, UVB (280–315 nm) irradiance may continue to increase because of continuing reduction in aerosol extinction since 1990. The ozone depletion is affecting the UVB/UVA ratios since only UVB is increased. The increasing in this ratio can have important effects on repair capacity and biochemical cycles (Zepp et al. 2007). The recovery of the ozone layer is expected to delay to 2070 due to the decrease of the temperature in the stratosphere as influence of climate change (Seckmeyer et al. 2008). Hegglin and Sheperd (2009) using physical models, predicted that the stratosphere to troposphere ozone flux will increase by 23 % between 1965 and 2095 due to climate change, a much larger effect than the recovery of stratospheric ozone, and consequently, the UV index will increase 4 % and up to 20 % in southern high latitudes in late spring and summer; in the Mediterranean area UV index will decrease i.e. less amount of UVB compared to 1980 values. Thus, the analysis of the effects of increased or decreased UVB and climate changes (acidification and warming) on aquatic ecosystems have to be monitored during this century.

The individual effect of increased UV radiation, temperature or CO_2 has been studied on photosynthesis, nutrient incorporation and growth in algae from different latitudes, included Mediterranean (reviewed by Häder and Figueroa 1997). However, short-term experiments are unable to observe the effects of chronic exposure to stressors on macroalgal species and their acclimation (see Hurd et al. 2009). The mechanisms for acclimation to global climate change include photoinhibition, photoprotection, nutrient uptake systems and patterns of growth, reproduction and morphogenesis (Häder and Figueroa 1997; Figueroa et al. 2002; Gómez et al. 2004; Korbee et al. 2005).

In aquatic ecosystems, the increase of UVB by ozone depletion have been related to damage of DNA, RNA, proteins and lipids (Buma et al. 1997; Helbling et al. 2001), decrease of enzyme activities (Viñegla et al. 2006), photoinhibition of photosynthesis (Figueroa et al. 1997, 2002), inhibition of growth (Altamirano et al. 2000) and inhibition of different reproductive stages (Wiencke et al. 2000; Altamirano et al. 2003). Macroalgae present different sensitivity to UVB according to species, morphology and life cycle (Dring et al. 1996; Altamirano et al. 2003). However, aquatic organisms can counteract the deleterious effects of UV radiation by photoprotective mechanisms as the DNA photorepair mediated by PAR and UV radiation (Mitchell and Karentz 1993), accumulation of lipidic and water soluble antioxidants and the activation of antioxidant enzymes (Cockell and Knowland 1999) and finally the accumulation of UV-screen photo protectors such as mycosporine-like aminoacids (MAAs) in red macroalgae (Korbee et al. 2005, 2006; Korbee-Peinado et al. 2004), phenolic compounds in brown algae (Connan et al. 2004; Abdala et al. 2006) and trihydroxicoumarins in the green algae *Dasycladus vermicular is* (Pérez-Rodríguez et al. 1998).

Interactive Effect of Climate Change Variables: Temperature, UV Radiation and Acidification

Temperature influences all aspects of algal growth and physiological rates. Taking all these characteristics into account it is crucial to evaluate the interactive effects of light, temperature, ocean acidification and nutrient status to estimate the vulnerability and the capacity of adaptation of Mediterranean coastal ecosystems to climate change. The environmental variables can interact changing the physiological and ecological responses with antagonist or synergistic effects. However, most of the studies conducted on climate change have evaluated the individual effects, being the studies focussed on interaction of factors very scarce (Bischof et al. 2006; Häder et al. 2007; Gao and Zheng 2010; Porzio et al. 2011). There are potentially complex interactions between calcification, rising temperatures and increasing nutrients, and there is strong evidence in articulated calcareous algae that calcification are enhanced three times the photosynthesis: respiration rate (Gattuso et al. 1999; Kleypas et al. 2006).

Only few studies report the combined effect of UV radiation and acidification (Gao and Zheng 2010; Porzio et al. 2011). A recent study on *Corallina sessilis* suggests that UV radiation and ocean acidification act synergistically to reduce photosynthesis, calcification, light-capturing pigments and growth, and to increase the absorptivity of UV-protective compounds in this species (Gao and Zheng 2010). Additionally, in phenolic producers' species, as *Padina pavonica* and other green algae, increased UV radiation can increase the excretion of phenols causing inhibition of calcification (Porzio et al. 2011). Additionally, a reduced calcification implies a decreasing of the reflectance of solar radiation (including UV radiation), and consequently, making these species more vulnerable to this radiation (Beach et al. 2006). However, the interactive effects of decreased pH and UV radiation on macroalgae are still unclear. Thus, in view of the continuous increase in atmospheric CO_2 concentration and progressive ocean acidification, coralline algae may calcify and photosynthesize less and less, especially when the harm caused by solar UV radiation is considered.

Finally, the combined effect of acidification/CO_2, temperature and UV radiation can be also modulated or affected by nutrient availability. Nutrients availability affects a variety of processes in macroalgae, such as defence mechanisms and repair the damage caused by UV radiation. However, the relation between effects of nitrogen and phosphorus limitation and the effects of UV radiation on primary production and community structure of marine macrophytes are still unclear (Bischof et al. 2006). It is known that nitrogen limitation affects a lot of processes in macrophytes not only the photosynthetic capacity (Pérez-Loréns et al. 1996), but also the content of proteins (Vergara et al. 1995), and photoprotection mechanisms against UV radiation (Korbee et al. 2005). The most common indicator of nutrient limitation in marine macroalgae is the carbon to nitrogen ratio (C:N); low values indicate nitrogen replete growth conditions. This ratio has been used also to assess the sensitivity of species exposed to UVB radiation (Sterner and Elser 2002). The fall in the C:N ratio means an increase in the nutritional quality of the organisms to be transmitted to the rest of the food web (Hessen et al. 2008).

Sea Level Rise

Another expected impact of climate change is the sea level rise that could have a great potential to alter the structure of assemblages of rocky shores. This can affect the biodiversity of coastal areas through alterations of features of the substratum, including the geometry of shorelines. Field measurements on the rocky shore of Calafuria (NW Mediterranean Sea) showed that a rise in sea level in the range of 5–50 cm would increase the availability of steep substrata (>40 °) for assemblages of algae (Vaselli et al. 2008). This authors suggested that sea level rise, by increasing the proportion of vertical substrata at the expense of horizontal surfaces, will lead to the expansion of assemblages dominated by encrusting coralline algae (Vaselli et al. 2008).

Future Research to Evaluate the Vulnerability of the Macrophytes to Climate Change

The studies on the ecological status and vulnerability to climate change of aquatic ecosystems need integration and coordination. There are a good number of reports on ecophysiology of macrophytes focused on functional approaches such as photosynthesis or nutrient assimilation (Häder and Figueroa 1997; Figueroa et al. 2002; Bischof et al. 2006) including also morpho-functional aspects and growth–reproductive studies (Altamirano et al. 2003; Gómez et al. 2004). However the research on the effect of stress factors related to climate change and the adaptation strategies on the structure-function of macroalgal communities and ecosystems is still very scarce (Bischof et al. 2006). Previous work has demonstrated that temperature and UV radiation increases, applied separately, can alter aquatic food –web structure and function (Mostajir et al. 1999).

Only experimental studies under controlled conditions in large mesocosms will allow researchers to (1) control forcing factors to mimic model predictions, (2) evaluate the response of quasi entire food webs to climate change (e.g. microbial food webs) and (3) acquire pertinent data toward the development of predictive global change models integrating both physical and biological components (Nouguier et al. 2007). In fact, several researches using mesocosms have demonstrated the effect of UV radiation on the structure and function of aquatic food webs (Wangberg et al. 2001; Hernando et al. 2006).

Multifactorial experiments conducted with artificial lamps or under solar radiation, the results obtained under natural conditions along several years, latitudinal gradients and seasonality; as well, the comparison of these results with historical data and the correlation with abiotic factors (temperature, radiation, nutrients, pCO2, pH) will help to evaluate the vulnerability and the capacity of adaptation of macroalgae communities to new scenarios of climate change.

Present Status of Mediterranean Seaweeds Communities and Its Relation to the Ecological Status of Coastal Waters

In the *Mare Nostrum*, near 17,000 marine species has been described so far (Coll et al. 2010), an estimated 7 % of the world's marine biodiversity. Despite the past and current threads to diversity in coastal areas and continental shelves of this sea caused by human activities (e.g. benthic habitat loss and degradation, pollution, eutrophication) and the combined pressure of climate change (e.g. tropicalization, acidification) there are still coastal fringes with a remarkable excellent ecological status, not only regarding macroalgae but many other taxonomic groups (Bianchi and Morri 2000). Examples of these areas are the Gibraltar Strait, the Gata Cape in the southeastern Iberian Peninsula, Capo Carbonara in Sardinia, the coastal front of the Regional Natural Park of Corsica or the northern Aegean Sea.

Overall, the current taxonomic classification of macroalgae in the Mediterranean is well known. However, there are still several genera (e.g. *Ulva*, *Porphyra*, *Ectocarpus*, *Peyssonelia*) that claims for a revision of their taxonomy based on molecular tools. Currently, the macroalgal biodiversity is being influenced by the introduction of alien species (e.g. *Caulerpa racemosa*, *Asparagopsis taxiformis*), especially from the Red Sea through the Suez Canal (Erythrean aliens or Lessepsian migrants). As an example, 11 % of the present Rhodophyta are considered alien species in the Mediterranean. These introductions are generally related to human activities as large-scale mariculture or vessel-related operations (anchorage, maritime transport). The ecological impact of invasive species has clearly increased in the last decades but, with some exceptions (e.g. Molnar et al. 2008), it has not been addressed properly.

Despite the acceptable taxonomical classification of macroalgae, the present knowledge of the macroalgal flora is biased towards the western Mediterranean and here, the northern European shores cope the majority of studies. There are regions whose benthic vegetation has been intensively and comprehensively studied, as the northern Adriatic Sea (Rindi and Battelli 2005) or the Catalan Coast (Ballesteros et al. 2007) but there are still shores not extensively studied in the African coast (e.g. Benhissoune et al. 2001), eastern regions of the Levantine Sea or the Aegean Sea, so any generalization about macroalgae distribution must be taken with caution. This disequilibrium is associated in part to the higher number of Universities and research centres in the western European countries, but still, the inventory of the phycological marine flora in some western countries may be improved.

The distribution of marine diversity in the Mediterranean (and hence surely macroalgae, as diversity distribution is generally associated with a productivity gradient) is rather heterogeneous. Overall, there seems to be a decreasing gradient of species richness from northwest to southeast species. In a bathymetric pattern, a key feature of spatial variability in marine habitats, species richness generally increases from the highest levels of mediolittoral rocks down to the lower sublittoral zones. However, the highest species richness can be found in the upper circalittoral zone, with high algal biodiversity at depths greater than 50 m (Ballesteros 1990; Ballesteros et al. 2009). At greater depths, the species richness tends to be nil when irradiance approaches to intensities below compensation points for photosynthesis due to the necessity of light for seaweed survival. However, the present knowledge of the macroalgal community decreases also with depth, so that the composition of deep-water algal assemblages is little known (Ballesteros et al. 2009).

Increasing efforts for environmental protection stem from the Mediterranean Action Plan (1975) and the following Barcelona Convention and continue through the recent Marine Strategy Framework Directive (MSFD, 2008/56/EC). The conservation efforts include the design of a network of Marine Protected Areas (MPAs) that, by definition, enjoy a higher level of protection than its surrounding areas. The MPAs constitute an effective management tool and are designed for the purpose of protection and conservation of species, ecosystems or ecological functions and services of the marine environment. Most of these areas are in the northern European shores and the Anatolian Peninsula and currently covers less than 5 % of the Mediterranean surface (if Pelagos Sanctuary is included). However, the current Mediterranean MPA system is not representative or coherent (Abdulla et al. 2009) so that a future-challenging topic is the effective design of a coherent network of MPAs under an ecosystemic approach.

The Barcelona Convention, renamed the Convention for the Protection of the Marine Environment and the Coastal Region of the Mediterranean (1995), has addressed the preservation of the biodiversity of the sea. In particular, the Protocol concerning special protected areas and biological diversity included a list on endangered or threatened species (annex II; recently amended by the Decision IG.19/12) that, for the first time incorporated 16 macroalgal species plus all the species of the phaeophycean genus *Cystoseira* (except *C. compressa*). Interestingly, recent national legislation, as the Spanish list of wild species under special protection and the list of threatened species (2011) includes also for the first time 12 species of macroalgae.

Finally, the protection of macroalgae is also enhanced by the protection of particular marine habitats under the Habitat Directive 92/43/EEC (annex I), particularly coastal habitats, and other species requiring particular management measures by reason of the potential impact of their exploitation in the wild (annex V), specifically the rodophyceans *Lithothamnium coralloides* and *Phymatholithon calcareum.*

The use of macroalgae for the estimation of the ecological status of coastal and transitional waters of the Mediterranean, in the frame of the Water Framework Directive (2000/60/EC) is a challenging applied subject. This will require the necessity of a new generation of phycologists, augmenting hopefully the shortage of these scientists in some Mediterranean countries (e.g. Spain). Macroalgae as a key biological quality element has been the object of indices based on information from the littoral and upper-sublittoral communities. In particular, two indices have been developed to estimate the ecological status of coastal waters in the Mediterranean ecoregion: the cartography of littoral (CARLIT; Ballesteros et al. 2007), and the ecological evaluation index (EEI; Orfanidis et al. 2003).

Briefly, the CARLIT index is based on the cartography of the algal communities of the rocky shoreline in different sectors, each one characterized by a community category with a sensitivity level. As the presence and abundance of macroalgal communities respond in part to the natural geomorphological variability of the coastal environment, different geomorphological variables and categories are described for every sector of coast. The environmental quality assessment of a stretch of coast is estimated as a ratio between the environmental quality in a particular site and the environmental quality in a reference site with similar geomorphological characteristics, those considered geomorphological relevant situations. These reference sites have been selected in undisturbed or scarcely disturbed rocky shores from three sites representing the Western Mediterranean coasts (excluding Alboran Sea), one in the Regional Natural Park of Corsica and the other in two MPA in the Balearic Island. The CARLIT index was firstly applied to the Catalan coast (Ballesteros et al. 2007) and since then it has been applied to other Mediterranean coasts from Italy (Asnaghi et al. 2009), France, Malta or Slovenia (Blanfuné et al. 2011), Croatia (Nikolic et al. 2011) and more recently to coastal waters in the Alboran Sea, including reference sites for this zone (Bermejo et al. 2013).

The EEI method has been applied to coastal and transitional waters and is based in the coverage (%) of intertidal macrolagae as classified in two ecological state groups (ESG) representing alternative ecological states: pristine (ESG I) or degraded (ESGII). The mean absolute abundance of these groups is non-linearly corresponded to five ecological status classes that are related linearly with the value of the index (Orfanidis et al. 2003). This index was firstly estimated in the coasts of Greece (Orfanidis et al. 2001) but also it has been successfully applied to the Slovenian (Orlando-Bonaca et al. 2008) and Croatian coasts (Nikolic et al. 2011).

These two indices and other developed for macroalgae in transitional waters (e.g. MaQI; Sfriso et al. 2009) are set up to assess the ecological status of the Mediterranean waters according to the requirements of the DMA and then the identification of signals indicating human impact and the effect of anthropogenic pressures on the littoral zone. It is expected that the monitoring and further management plans for the coast will lessen the probable impact amplification in the coming years and will maintain the ability of the Mediterranean sea to continue the provision of essential ecological services.

Acknowledgements The financial support by the Projects CGL08-05407 C03-01 and RNM-5750, and Research groups RNM 295 and RNM 115, as well as the contract OT2010/102 between Consejería de Medio Ambiente (Junta de Andalucía) and University of Cádiz are thankful.

References

Abdala-Díaz RT, Cabello-Pasini A, Pérez-Rodríguez E, Conde Álvarez RM, Figueroa FL (2006) Daily and seasonal variations of optimum quantum yield and phenolic compounds in *Cystoseira tamariscifolia* (Phaeophyta). Mar Biol 148:459–465

Abdulla A, Gomei M, Hyrenbach D, Notarbartolo-di-Sciara G, Agardy T (2009) Challenges facing a network of representative marine protected areas in the Mediterranean: prioritizing the protection of underrepresented habitats. ICES J Mar Sci 66:22–28

Adams CG (1981) An outline of the tertiary paleogeography. In: Cock LRM (ed) The evolving earth. British Museum (Natural History), Cambridge University Press, London

Alongi G, Cormaci M, Furnari G (2008) *Chylocladia wynnei* sp. nov. (Champiaceae, Rhodophyta) from the Mediterranean Sea. Bot Mar 51:145–150

Altamirano M, Flores-Moya A, Conde F, Figueroa FL (2000) Growth seasonality, photosynthetic pigments and C and N content in relation to environmental factors: a field study on *Ulva olivascens* (Ulvales, Chlorophyta). Phycologia 39:50–58

Altamirano M, Flores-Moya A, Figueroa FL (2003) Effects of UV radiation and temperature on growth germlings of three species of *Fucus* (Phaeophyceae). Aquat Bot 75:9–20

Andersson AJ, Mackenzie FT, Gattuso JP (2011) Effects of ocean acidification on benthic processes, organisms, and ecosystems. In: Gattuso JP, Hansson L (eds) Ocean acidification. Oxford University Press, Oxford

Asnaghi V, Chiantore M, Bertolotto RM, Parravicini V, Cattaneo-Vietti R, Gaino F, Moretto P, Privitera D, Mangialajo L (2009) Implementation of the European Water Framework Directive: natural variability associated with the CARLIT method on the rocky shores of the Ligurian Sea (Italy). Mar Ecol 30:505–513

Báez JC, Real R, Vargas JM, Flores-Moya A (2004) Revisión crítica de los estudios sobre biogeografía de macroalgas marinas del Mediterráneo. Acta Bot Malacitana 29:5–11

Báez JC, Olivero J, Real R, Vargas JM, Flores-Moya A (2005) Analysis of geographical variation in species richness withinthe genera *Audouinella* (Rhodophyta), *Cystoseira* (Phaeophyceae) and *Cladophora* (Chlorophyta) in the western Mediterranean Sea. Bot Mar 48:30–37

Ballesteros E (1990) Structure and dynamics of the assemblage of *Cystoseira zosteroides* (Turner) C. Agardh (Fucales, Phaeophyceae) in the Northwestern Mediterranean. Sci Mar 54:217–229

Ballesteros E (1994) The deep-water *Peyssonnelia* beds from the Balearic Islands (Western Mediterranean). PSZNI Mar Ecol 15:233–253

Ballesteros E (2006) Mediterranean coralligenous assemblages: a synthesis of the present knowledge. Oceanogr Mar Biol 44:123–195

Ballesteros E, Torras X, Pinedo S, García M, Mangialajo L, de Torres M (2007) A new methodology base on littoral community cartography by macroalgae for the implementation of the European Water Framework Directive. Mar Pollut Bull 55:172–180

Ballesteros E, Garrabou J, Hereu B, Zabala M, Cebrián E, Sala E (2009) Deep-water stands of *Cystoseira zosteroides* C. Agard (Fucales, Ochrophyta) in the Northwestern Mediterranean: insights into assemblage structure and population dynamics. Estuar Coast Shelf S 82:477–484

Beach KS, Borgeas HB, Smith CM (2006) Ecophysiological implications of the measurement of transmittance and reflectance of tropical macroalgae. Phycologia 45:450–457

Benhissoune S, Boudoresque CF, Verlaque M (2001) A check-list of marine seaweeds of the Mediterranean and Atlantic coasts of Marocco. I. *Chlorophyceae willes*. l. Bot Mar 44:171–182

Bermejo R, de la Fuente G, Vergara JJ, Hernández I (2013) Application of the CARLIT index along a biogeographical gradient in the Alboran Sea (European Coast). Mar Pollut Bull. Available online 10.1016/j.marpolbul.2013.04.011

Béthoux JP, Gentili B (1999) Functioning of the Mediterranean Sea: past and present changes related to freshwater input and climate changes. J Marine Syst 20:33–47

Bianchi CN, Morri C (2000) Marine biodiversity of the Mediterranean Sea: situation, problems and prospects for future research. Mar Pollut Bull 40:367–376

Bischof K, Gómez I, Molis M, Hanelt D, Karsten U, Lüder U, Roleda MY, Zacher K, Wiencke C (2006) Ultraviolet radiation shapes seaweed communities. Rev Evironm Sci Biotechnol 51:141–166

Blanfuné A, Markovic L, Thibaut T (2011) Assessment of the CARLIT methodology in the Mediterranean rocky water bodies. In: 5th European Phycological Congress, Rhodes

Borowitzka MA, Larkum AWD (1976) Calcification in the green alga *Halimeda* III. The sources of inorganic carbon for photosynthesis and calcification and a model of mechanism of calcification. J Exp Bot 27:879–893

Boudouresque CF, Verlaque M (2002) Biological pollution in the Mediterranean Sea: invasive versus introduced macrophytes. Mar Pollut Bull 44:32–38

Buma AGJ, Engelen AH, Gieskes WWC (1997) Wavelength-dependent induction of thymine dimers and growth rate reduction in the marine diatom *Cyclotella* sp exposed to ultraviolet radiation. Mar Ecol Prog Ser 153:91–97

Cinelli F (1981) Biogeography and ecology of the Sicily channel. Proc Int Seaweed Symp 10:235–240

Cinelli F (1985) On the biogeography of the benthic algae of the Mediterranean. In: Moraitou-Apostolopoulou M, Kiortsis V (eds) Mediterranean marine ecosystems. Plenum Press, New York

Cockell CS, Knowland J (1999) Ultraviolet radiation screening compounds. Biol Rev 74:311–345

Coll M, Piroddi C, Steenbeek J et al (2010) The biodiversity of the Mediterranean Sea: estimates, patterns, and threats. PLoS ONE 5:1–36

Connan S, Goulard F, Stiger V, Deslandes E, Ar Gall E (2004) Interspecific and temporal variation in phlorotanin levels in an assemblage of brown algae. Bot Mar 47:410–416

De Beer D, Larkum AWD (2001) Photosynthesis and calcification in the calcifying algae *Halimeda discoidea* studied with microsensors. Plant Cell Environ 24:1209–1217

Dring MJ, Wagner A, Boesko J, Lüning K (1996) Sensitivity of intertidal and subtidal red algae to UVA and UVB radiation, as monitored by chlorophyll fluorescence measurements: influence of collection depth, season and length of radiation. Eur J Phycol 31:293–302

Figueroa FL, Mercado J, Jiménez C, Salles S, Aguilera J, Sánchez-Saavedra MP, Lebert M, Häder DP, Montero O, Lubián L (1997) Relationship between bio-optical characteristics and photoinhibition of phytoplankton. Aquat Bot 59:137–251

Figueroa FL, Jiménez C, Viñegla B, Pérez-Rodríguez E, Aguilera J, Flores-Moya A, Altamirano M, Lebert M, Häder DP (2002) Effects of solar UV radiation on photosynthesis of the marine angiosperm *Posidonia oceanica* from southern Spain. Mar Ecol Prog Ser 230:59–70

Flores-Moya A (2012) Warm temperate seaweed communities: a case study of deep water kelp forests from the Alborán Sea (SW Mediterranean Sea) and the Strait of Gibraltar. In: Wiencke C, Bischof K (eds) Seaweed ecophysiology and ecology. Springer, Berlin

Flores-Moya A, Gómez I, Viñegla B, Altamirano M, Pérez-Rodríguez E, Maestre C, Caballero RM, Figueroa FL (1998) Effects of solar radiation on photosynthetic performance, pigment content and enzymatic activities related to nutrient uptake, of the endemic Mediterranean red alga *Rissoella verruculosa*. New Phytol 139:673–683

Francour P, Boudouresque CF, Harmelin JG, Harmelin-Vivien M, Quignard JP (1994) Are the Mediterranean waters becoming warmer? Information from biological indicators. Mar Pollut Bull 28:523–526

Gallardo T, Gómez-Garreta A, Ribera MA, Cormaci M, Furnari G et al (1993) Check-list of Mediterranean seaweeds. II: *Chlorophyceae wille* s. l. Bot Mar 36:399–421

Gao K, Zheng Y (2010) Combined effects of ocean acidification and solar UV radiation on photosynthesis, growth, pigmentation and calcification of the coralline alga *Corallina sessilis* (Rhodophyta). Glob Change Biol 16:2388–2398

Gao K, Aruga Y, Asada K, Ishihara T, Akano T, Kiyohara M (1993) Calcification in the articulated corallina alga *Corallina pilulifera* with special reference to the effect of elevated CO_2 concentration. Mar Biol 117:129–132

Gattuso JP, Frankignoulle M, Smith SV (1999) Measurement of community metabolism and significance in the coral reef CO_2 source-sink debate. Proc Natl Acad Sci 96:13017–13022

Giaconne G (1974) Tipologia delle comunita fitobentoniche del Mediterraneo. Mem Biol Mar Oceanogr 4:149–168

Giaconne G, Geraci RM (1989) Biogeografia delle alghe del Mediterraneo. Anales Jard Bot Madrid 46:27–33

Gómez I, Figueroa FL, Ulloa N, Morales V, Lovengreen C, Huovinen P, Hess S (2004) Patterns of photosynthetic performance in intertidal macroalgae from southern Chile in relation to solar radiation. Mar Ecol Prog Ser 270:103–116

Gómez-Garreta A, Gallardo T, Ribera MA, Cormaci M, Furnari G, Giaccone G, Boudouresque CF (2001) Checklist of Mediterranean seaweeds. III. *Rhodophyceae rabenh*. 1. *Ceramiales oltm*. Bot Mar 44:425–460

Häder DP, Figueroa FL (1997) Annual review: photoecophysiology of marine macroalgae. Photochem Photobiol 66:1–14

Häder DP, Helbling EW, Williamson CE, Worrest RC (2007) Effect of UV radiation on aquatic ecosystems and interactions with climate change. Photochem Photobiol Sci 10:242–260

Hall-Spencer JM, Rodolfo-Metalpa R, Martin S, Ransome E, Fine M, Turner SM, Rowley SJ, Tedesco D, Buia MC (2008) Volcanic carbon dioxide vents show ecosystem effects of ocean acidification. Nature 454:96–99

Hegglin MI, Sheperd T (2009) Large climate-induced changes in ultraviolet index and stratosphere-to-troposphere ozone flux. Nat Geosci 2:687–691

Helbling EW, Villafañe VE, Buma AGJ, Andrade M, Zaratti F (2001) DNA damage and photosynthetic inhibition induced by solar ultraviolet radiation in tropical phytoplankton (Lake Titicaca, Bolivia). Eur J Phycol 36:157–166

Hernando M, Schloss I, Roy S, Ferreyra G (2006) Photoacclimation to long-term UVR exposure of natural Sub-Antarctic phytoplankton communities: fixed surface incubations versus mixed mesocosms. Photochem Photobiol 82:923–935

Hessen DO, Leu E, Færøvig PJ, Petersen SF (2008) Light and spectral properties as determinants of C:N:P-ratios in phytoplankton. Deep-Sea Res PT II 55:2169–2175

Hoegh-Guldberg O, Mumby PJ, Hooten AJ et al (2007) Coral reefs under rapid climate change and ocean acidification. Science 318:1737–1742

Hurd CL, Hepburn CD, Currie KI, Raven JA, Hunter KA (2009) Testing the effects of ocean acidification on algal metabolism: considerations for experimental designs. J Phycol 45:1236–1251

IPCC (2007) Climate change 2007, the physical science basis. Summary for policy makers. Intergovernmental panel on climate change. Cambridge University Press, Cambridge

Korbee-Peinado N, Abdala-Díaz R, Figueroa FL, Helbling EW (2004) Ammonium and UVR stimulate the accumulation of mycosporine-like aminoacids (MAAs) in Porphyra columbina (Rhodophyta) from Patagonia, Argentina. J Phycol 40:248–259

Korbee N, Huovinen P, Figueroa FL, Aguilera J, Karsten U (2005) Availability of ammonium influences photosynthesis and the accumulation of mycosporine-like amino acids in two *Porphyra* species (Bangiales, Rhodophyta). Mar Biol 146:645–654

Korbee N, Figueroa FL, Aguilera J (2006) Accumulation of mycosporine-like amino acids (MAAs): biosynthesis, photocontrol and ecophysiological functions. Rev Chil Hist Nat 79:119–132

Kleypas JA, Feely RA, Fabry VJ, Langdon C, Sabine CL, Robbins LL (2006) Impacts of ocean acidification on coral reefs and other marine calcifiers: a guide for future research. NSF/NOAA/The U.S. Geological Survey, St. Petersburg

Kübler JE, Johnston AM, Raven JA (1999) The effects of reduced and elevated CO_2 and O_2 on the seaweed *Lomentaria articulata*. Plant Cell Environ 22:1303–1310

Lejeusne C, Chevaldonné P, Pergent-Martini C, Boudouresque CF, Pérez T (2009) Climate change effect on a miniature ocean: the highly diverse, highly impacted Mediterranean sea. Trends Ecol Evol 25:250–260

Lloret J, Marín A, Marín-Guirao L (2008) Is coastal lagoon eutrophication likely to be aggravated by global climate change? Estuar Coast Shelf S 78:403–412

Louanchi F, Boudjakdji M, Nacef N (2009) Decadal changes in surface carbon dioxide and related variables in the Mediterranean Sea as inferred from a coupled data diagnostic model approach. ICES J Mar Sci 66:1538–1546

Lüning K (1990) Seaweeds. Their environment, biogeography, and ecophysiology. Wiley, New York

McKenzie RL, Aucamp PJ, Bais AF, Björn LO, Ilyas M (2007) Changes in biologically active ultraviolet radiation reaching the Earth's surface. Photochem Photobiol Sci 6:218–231

Mercado JM, Gordillo FJ, Figueroa FL, Niell FX (1998) External carbonic anhydrase and affinity to inorganic carbon in intertidal macroalgae. J Exp Mar Biol Ecol 221(2):209–220

Mieszkowska N, Moore P, Hawkins S, Burrows M (2008) Intertidal species in marine climate change impacts. Annual Report Card 2007–2008

Mitchell DL, Karentz D (1993) The induction and repair of DNA photodamage in the environment. In: Young AR, Björn LO, Moan J, Nultsch W (eds) Environmental UV photobiology. Plenum Press, New York

Molnar JL, Gamboa RL, Revenga C, Spalding MD (2008) Assessing the global threat of invasive species to marine biodiversity. Front Ecol Environ 6:485–492

Mostajir B et al (1999) Experimental test of the effect of ultraviolet-B radiation in a planktonic community. Limnol Oceanogr 44:586–596

Myers AA (1996) Species and generic gamma-scale diversity in shallow-water marine Amphipoda with particular reference to the Mediterranean. J Mar Biol Assoc UK 76:195–202

Nikolic V, Zuljevic A, Antolic B (2011). Macroalgae as bioindicators in the Adriatic sea: the application of CARLIT and EEI methods. In: 5th European Phycological Congress. Abstract 9D.20. Rhodes

Nouguier JB, Mostajir B, Le Floch E, Vidussi F (2007) An automatically operated system for simulating global change temperature and ultraviolet B radiation increases: application to the study of aquatic ecosystem responses in mesocosm experiments. Limnol Oceanogr: Methods 5:269–279

Oliveras Plá M, Gómez-Garreta A (1989) Corología del género *Cystoseira* C. Agardh (Phaeophyceae, Fucales). Anales Jard Bot Madrid 46:89–97

Orfanidis S, Panayotidis P, Stamatis N (2001) Ecological evaluation of transitional and coastal waters: a marine benthic macrophytes-based model. Medit Mar Sci 2:45–65

Orfanidis S, Panayotidis P, Stamatis N (2003) An insight to the ecological evaluation index (EEI). Ecol Indic 3:27–33

Orlando-Bonaca M, Lipej L, Orfanidis S (2008) Benthic macrophytes as a tool for delineating, monitoring and assessing ecological status: the case of Slovenian coastal waters. Mar Pollut Bull 56:666–676

Orr JA, Fabry VJ, Aumont O (2005) Anthropogenic ocean acidification over twenty-first century and its impact on calcifying organisms. Nature 437:681–686

Pérez-Lloréns JL, Vergara JJ, Pino RR, Hernández I, Peralta G, Niell FX (1996) The effect of photoacclimation on the photosynthetic physiology of *Ulva curvata* and *Ulva rotundata* (Ulvales, Chlorophyta). Eur J Phycol 331:349–359

Pérez-Rodríguez E, Gómez I, Figueroa FL (1998) Effects of UV radiation on photosynthesis and excretion of UV-absorbing pigments of *Dasycladus vermicularis* (Chlorophyta, Dasycladales) from southern Spain. Phycologia 37:379–387

Pérez-Ruzafa A, Marcos C, Bernal CM, Quintino V, Rodrigues AM, García-Sánchez M, Pérez-Ruzafa IM (2012) *Cymodocea nodosa vs Caulerpa prolifera*, causes and consequences of a long term history of interaction in macrophyte meadows in the Mar Menor coastal lagoon. Estuar Coast Shelf S 110:101–115

Porzio L, Buia MC, Hall-Spencer JM (2011) Effects of ocean acidification on macroalgal communities. J Exp Mar Biol Ecol 400: 278–287

Rabalais NN, Turner RE, Díaz RJ, Justic D (2009) Global change and eutrophication of coastal waters. ICES J Mar Sci 66:1528–1537

Raven JA, Ball LA, Beardall J, Giordano M, Maberley SC (2005) Algae lacking carbon-concentrating mechanisms. Can J Bot 83:879–890

Ribera MA, Gómez A, Gallardo T, Cormaci M, Furnari G et al (1992) Check-list of Mediterranean seaweeds. I: fucophyceae (Warming, 1884). Bot Mar 35:109–130

Riebesell U, Schulz KG, Bellerby RGJ, Botros M, Fritsche P, Meyerhofer M, Neill C, Nondal G, Oschlies A, Wohlers J, Zollner E (2007) Enhanced biological carbon consumption in a high CO2 ocean. Nature 450:545–548

Rindi F, Battelli C (2005) Spatio-temporal variability of intertidal algal assemblages of the Slovenian coast (Gulf of Trieste, northern Adriatic Sea). Bot Mar 48:96–105

Rodríguez-Prieto C, De Clerck O (2009) *Leptofauchea coralligena* (Faucheaceae, Rhodophyta), a new species from the Mediterranean Sea. Eur J Phycol 44:107–121

Ros JD, Romero J, Ballesteros E, Gili JM (1985) Diving in a deep blue sea. In: Margalef R (ed) Western Mediterranean. Pergamon Press, London

Seckmeyer G, Pissulla D, Glandori M, Henriques D, Johnson B, Webb A et al (2008) Variability of UV irradiance in Europe. Photochem Photobiol 84:172–179

Serio D, Catra M, Collodoro D, Nisi A (2011) *Ceramium cormacii* sp. nov. (Ceramiaceae, Rhodophyta), a new Mediterranean species epizoic on loggerhead sea turtles (*Caretta caretta*). Bot Mar 54:545–550

Sfriso A, Facca C, Ghetti PF (2009) Validation of the Macrophyte Quality Index (MaQI) set up to assess the ecological status of Italian marine transitional environments. Hydrobiologia 617: 117–141

Stanley DJ (1990) Med desert theory is drying up. Oceanus 33:14–23

Stanley DJ, Wezel FC (1985) Geological evolution of the Mediterranean basin. Springer, New York

Sterner RW, Elser JJ (2002) Ecological stoichiometry. Princeton University Press, Princeton

Thibaut T, Pinedo S, Torras X, Ballesteros E (2005) Long-term decline of the populations of Fucales (*Cystoseira* spp. and *Sargassum* spp.) in the Albéres coast (France), north-western Mediterranean). Mar Pollut Bull 50:1472–1489

Thunell RC (1979) Pliocece-pleistocene paleotemperature and paleosalinity history of the Mediterranean Sea: results from DSDP sites 125 and 132. Mar Micropaleontol 4:173–187

Van den Hoek C (1975) Phytogeographic provinces along the coasts of the northern Atlantic Ocean. Phycologia 14:317–330

Van den Hoek C (1982) Phytogeographic distribution groups of benthic marine algae in the North Atlantic Ocean. Helgoländer Meersun 35:153–214

Vaselli S, Bertocci I, Maggi E, Benedetti-Cecchi L (2008) Assessing the consequences of sea level rise: effects of changes in the slope of the substratum sessile assemblages of rocky seashores. Mar Ecol Prog Ser 368:9–22

Vergara JJ, Bird KT, Niell FX (1995) Nitrogen assimilation following NH_4^+ pulses in the red alga *Gracilariopsis lemaneiformis*: effect on C metabolism. Mar Ecol Prog Ser 122:253–263

Verlaque A (1994) Inventaire des plantes introduites en Méditérranée: origines et repercusión sur les environnement et les activités humaines. Oceanol Acta 17:1–23

Viñegla B, Segovia M, Figueroa FL (2006) Effect of artificial UV radiation on carbon and nitrogen metabolism in the macroalgae *Fucus spiralis* L and *Ulva olivascens* Dangeard. Hidro-biologia 560:31–42

Walther GR, Post E, Convey P, Menzel A, Parmesan C, Beebee TJC et al (2002) Ecological responses to recent climate change. Nature 416:389–395

Wangberg I, Munthe J, Pirrone N et al (2001) Atmospheric mercury distribution in Northern Europe and in the Mediterranean region. Atmos Environ 35:3019–3025

Weatherhead EC, Andersen SB (2006) The search for signs of recovery of the ozone layer. Nature 441:39–45

Wiencke C, Gómez I, Pakker H, Flores-Moya A, Altamirano M, Hanelt D, Bischof K, Figueroa FL (2000) Impact of UV radiation on viability, photosynthetic characteristics and DNA on algal zoospores: implications for depth zonation. Mar Ecol Progr Ser 197:217–219

Williams SN, Schaefer SJ, Calvache ML, López D (1992) Global carbon dioxide emission to the atmosphere by volcanoes. Geochim Cosmochim Acta 56:1765–1770

Zepp RG, Erickson DJ, Paul ND, Sulzberger B (2007) Interactive effects of solar UV radiation and climate change on biogeochemical cycling. Photochem Photobiol Sci 6:286–300

Autochthonous Seagrasses

9

J. Lucas Pérez-Lloréns, Juan J. Vergara, Irene Olivé,
Jesús M. Mercado, Rafael Conde-Álvarez,
Ángel Pérez-Ruzafa, and Félix L. Figueroa

Abstract

Seagrasses are flowering plants that undergo their whole life cycle within shallow coastal habitats. All species share analogous architectural and growth patterns. They are modular plants composed of units arranged by a set of modules: a piece of rhizome, a bundle of leaves (shoot) attached to the rhizome and roots. Four species occur in the Mediterranean bioregion: the endemic *Posidonia oceanica*, the tropical *Cymodocea nodosa* and the temperates *Zostera marina* and *Z. noltii*. *Posidonia oceanica* is the largest one, with very slow growth rates and being considered the climax stage of Mediterranean subtidal bottoms. Meadows extend on 2.5–4.5 millions ha that is close to 25 % of the Mediterranean basin shallower than 50 m. *Cymodocea nodosa*, *Zostera marina* and, particularly, *Z. noltii* are smaller in size but fast growing as typical of the pioneering species. Seagrass meadows are among the most productive ecosystems on earth, providing important ecological services: nursery grounds, biofilters, water cleaners, coastline protectors and carbon sinks. However, despite its paramount importance there is widespread regression of such habitats. Although the four Mediterranean species have been assigned to the "Least Concern" category of the IUCN Red List, *P. oceanica* populations are experiencing the highest rate of decrease. Given the extremely slow growth rate of this species such losses are virtually irreversible. Direct and indirect (i.e., climate change) human activities affecting mostly to physical integrity of habitats, sediment and water quality, coastal sedimentary balance or species composition are argued to be the main drivers of seagrass decline in the Mediterranean Sea. European (and Mediterranean) countries, aware of the key important role that seagrasses play, have established management and conservation plans for these habitats. Thereafter, the aim of the present contribution is to present basic information about the biology and ecology of the Mediterranean seagrasses, the main threats facing these habitats, as well as to provide some information on the main conservation and management strategies.

J.L. Pérez-Lloréns (✉) • J.J. Vergara
Departamento de Biología, Facultad de Ciencias del Mar y Ambientales, Universidad de Cádiz, Campus de Excelencia Internacional del Mar (CEIMAR), 11510 Puerto Real, Cádiz, Spain
e-mail: joselucas.perez@uca.es; juanjose.vergara@uca.es

I. Olivé
CCMar – Center of Marine Sciences, Universidade do Algarve, Campus de Gambelas, 8005-139 Faro, Portugal
e-mail: iosamarra@ualg.pt

J.M. Mercado
Centro Oceanográfico de Málaga, Instituto Español de Oceanografía, Puerto Pesquero s/n, 29074 Fuengirola, Málaga, Spain
e-mail: jesus.mercado@ma.ieo.es

R. Conde-Álvarez
Estudios de Flora y Vegetación (EFYVE), 29007 Málaga, Spain
e-mail: rconde@efyve.com

Á. Pérez-Ruzafa
Departamento de Ecología e Hidrología, Facultad de Biología, Campus de Espinardo, Universidad de Murcia, 30100 Murcia, Spain
e-mail: angelpr@um.es

F.L. Figueroa
Departamento de Ecología, Universidad de Málaga, Campus Univ. de Teatinos s/n, 29071 Málaga, spain
e-mail: felix_lopez@uma.es

S. Goffredo and Z. Dubinsky (eds.), *The Mediterranean Sea: Its history and present challenges*,
DOI 10.1007/978-94-007-6704-1_9, © Springer Science+Business Media Dordrecht 2014

Keywords
Anthropogenic disturbances • Climate change • *Cymodocea nodosa* • Management • *Posidonia oceanica* • Seagrass • *Zostera marina* • *Zostera noltii*

Introduction

Seagrasses are flowering plants that endure their whole life cycle within shallow coastal habitats, extending from tropical to temperate latitudes. Even though they can thrive together with seaweeds, there are considerable differences between them from an evolutionary perspective. Therefore, since algae evolved in the sea around a billion of years ago, seagrasses evolved from terrestrial plants over the past 100 millions of years. Such "unequal" evolutionary trip resulted in a seagrass taxonomic diversity (60 species) much more lower than that of seaweeds (c.a., 9,000 species). For seagrasses, the life in water required an array of important adaptations such as (1) complex pollination mechanisms (involving spawning); (2) internal gas lacunae to transport oxygen to the roots and rhizomes and to maintain shoots in upright position when submersed; (3) chloroplasts located at the outermost layer of the leaves to improve gas exchanges and light capture; (4) strap-like, fibre-filled leaves with basal meristems, to withstand high hydrodynamics, and (5) extensive network of rhizomes and roots to assure an optimum anchorage in soft bottoms (Hemminga and Duarte 2000).

Besides to the low species richness, all seagrass species share analogous architectural patterns and growth dynamics. They are modular plants composed of units (ramets) arranged by a set of modules: a piece of rhizome, a bundle of leaves attached to it (shoot), and a root system. Additionally, ramets can bear flowers depending on the season. Due to their clonal nature, seagrass growth and spreading take place throughout the regular addition of the basic set of modules (Hemminga and Duarte 2000; Brun et al. 2006). The resulting meadow (either patchy or continuous) is usually composed by a reduced number of genetically diverse individuals, genets or clones. However, although clonal growth is the main way of meadow spreading and persistence, it does not create new, genetically distinct individuals, and thus, does not provide greater evolutionary potential or increased resilience to environmental change. In this sense, sexual reproduction is essential for initiating the formation of new clones and so it is also necessary for the development and maintenance of seagrass beds (Short et al. 2011).

Seagrass meadows occupy less than 0.15 % of the seabed (Charpy-Roubaud and Sournia 1990) but they are among the most productive ecosystems on earth (0.49 PgC year^{-1}, or c.a., 1.1 % of the total marine primary production) (Duarte and Chiscano 1999) providing important ecological functions or ecosystems services estimated at 25,400 € ha^{-1} year^{-1}, higher than, for example, coral reefs (8,200 € ha^{-1} year^{-1}), forests (1,300 € ha^{-1} year^{-1}) or croplands (108 € ha^{-1} year^{-1}) (Costanza et al. 1997, here recalculated to 2010 euros). Being aware of the caveats and limitations of such estimations, they highlight the importance of seagrasses. Among the principal ecosystems services are: (1) "promoters of biodiversity", since many organisms uses seagrasses (or its epiphytes) either as food, refuges or nursery ground (e.g., *Posidonia oceanica* meadows harbour 25 % of known Mediterranean species); (2) "biofilters" by stripping of nutrients and other contaminants from the water throughout foliar uptake; (3) "water clearers", since canopy enhances the settling probability of fine sediment particles increasing the water transparency; (4) "coastline protectors" since the belowground network of roots and rhizomes stabilizes sediment, besides the reduction of the erosive force of waves and tides by leaf baffling and leaf litter stacked on shore; (5) "carbon sinks", specially the long-lived-matt forming seagrasses such as the Mediterranean *P. oceanica*. In fact, seagrass meadows account up to 18 % (or 27.4–44 Tg C year^{-1}, with a maximum of 82) of all carbon storage ("blue carbon") in ocean sediments (Duarte and Cebrián 1996). Such carbon dioxide mitigation is especially valuable in the present (and future) situation of greenhouse gas increase causing climate change.

In addition to the goods and services provided by seagrasses, there is also ethnographic interest. Thus, since ancient times people from Mediterranean countries used these plants for different purposes. For example, dried leaves of *Posidonia oceanica* served as wrapping up material to carrying fragile pieces of glassware and ceramic (this is the origin of the name *Alga vitriariorum* as *P. oceanica* was known in ancient times); in the transportation of fresh fish from shoreline towns to inland cities; as building materials (roof insulation and adobes). Due to parasite-deterrent properties of seagrass leaves they were also used as material for stuffing pillows and mattresses (Pope Julius III popularized this use in the sixteenth century); mixed with lime and phosphates as a meal for feeding poultry; also in medicine *P. oceanica* was used to prevent respiratory diseases, alleviation of skin infections (e.g., acne) and leg pain by varicose veins; in some small villages of southern Spain the dried rhizomes of *P. oceanica* are still utilized as fuel to prepare traditional fish barbecues at the seashore (called *moragas*) (Pérez-Lloréns et al. 2012). Today, because of their sensitivity to water quality (transparency and nutrient concentrations), seagrasses are used as biological sentinels or "shore canaries" (Orth et al. 2006), for example as biological indicators in the

European Water Framework Directive (2000/60/CE) and Marine Strategy Framework Directive (2008/56/CE).

However, despite its paramount importance there is a worldwide regression of seagrass habitats (Orth et al. 2006; Short et al. 2011). Although there are somewhat contrasting estimates, a recent assessment indicates that about one-third of the global seagrass area has been already lost, and that losses are accelerating from less than 0.9 % $year^{-1}$ in the 1970s to more than 7 % $year^{-1}$ since 2000 (Waycott et al. 2009). In the Mediterranean, it is estimated that 46 % of the *Posidonia oceanica* meadows have undergone some reduction in range, density and/or extent, and 20 % have suffered acute regressions since 1970s (Díaz-Almela and Duarte 2008). Among the main causes, mostly anthropogenic (either direct or indirect), are: deterioration of light environment, physical disturbance, mechanical damage, eutrophication, introduction of invasive species and climate change (Duarte 2002). Many Mediterranean countries, aware of the key important role seagrasses plays, have established comprehensive and advanced monitoring programmes for seagrasses as well as management and conservations plans (Borum et al. 2004; Díaz-Almela and Duarte 2008).

Therefore, the aim of the present contribution is to provide basic information about the biology and ecology of the Mediterranean seagrasses, the main threats facing these habitats, and the principal conservation and management strategies.

Fig. 9.1 The seagrass *Posidonia oceanica* is a Mediterranean endemism. Meadows can growth as deep as 40–50 m in clear waters (Photograph: Juanma Ruiz)

Mediterranean Seagrass Species: Distribution, Biology and Ecology

Four seagrass species occur in the Mediterranean bioregion: the endemic *Posidonia oceanica*, the tropical *Cymodocea nodosa* and the temperates *Zostera marina* and *Z. noltii*.

Posidonia oceanica (Linnaeus) Delile

Posidonia oceanica (common name: Neptune grass) is a Mediterranean endemism and constitute one of the main mature (or climax) stages of Mediterranean subtidal bottoms. Available information for *P. oceanica* is higher than that for the other Mediterranean seagrass species. Meadows extend on 2.5–4.5 millions ha that is close to 25 % of the Mediterranean basin shallower than 50 m (Pasqualini et al. 1998). The western boundary is at the Alborán Sea where its growth is limited due to the influence of Atlantic water that is colder and less salty than the Mediterranean. The most westerly remnants are recorded 20 km east from the Gibraltar Strait on the Spanish coast (Estepona Bay), and 280 km towards Morocco coasts, near Sebkha-bou-Areg, and Chaffarine islands (Meinesz et al. 2009). Its presence in the north-western Mediterranean is widely recorded, whereas in the south-western and eastern Mediterranean is scarcer (Green and Short 2003). Nowadays, there is no *P. oceanica* in the eastern Levant Sea (Gobert et al. 2006) where the most north-eastern boundary would be close to the Kizilliman Marine Protected Area (Turkish coast, Celebi et al. 2007) and Alexandria (Egypt) on the south shore. Nevertheless, there are old records on the Syrian coast, but in 1976 the species was reported to be highly threatened (Mayhoub 1976).

Posidonia oceanica forms extensive underwater meadows on sandy (sometimes rocky) substrates, in clear open waters, from less than 1 m down to 25 m depth (Alborán Sea) or to 40 m (Ligurian or Aegean Seas) (Fig. 9.1). Light availability (i.e., water transparency) often limits the depth at which meadows occur (Duarte 1991a). Its mean compensation irradiance for photosynthesis (i.e., the light intensity at which oxygen production by photosynthesis compensates for its respiratory consumption) (c.a., 8 μmol photons m^{-2} s^{-1}) is lower than those reported for other Mediterranean species (Table 9.1). This characteristic and the large capacity to store and mobilize carbohydrates, allows *P. oceanica* to occur at deep locations. It is considered a stenohaline species

Table 9.1 Mean values of photosynthetic parameters, plant architectural traits, dynamic characteristics and meadow properties of the four Mediterranean seagrass species

	P. oceanica	*C. nodosa*	*Z. marina*	*Z. noltii*
Photosynthetic parameters				
Light-saturated photosynthetic rate (μmol O_2 g DW^{-1} h^{-1})	199.3	174.1	119.9	549.4
Compensation irradiance for photosynthesis (μmol m^{-2} s^{-1})	7.8	35.1	21.7	20.0
Saturation irradiance for photosynthesis (μmol m^{-2} s^{-1})	73.3	149.2	116.0	250.0
Architectural traits				
Leaf abundance (n° leaves $shoot^{-1}$)	6.1	3.3	4.2	3.2
Leaf length (cm)	75.2	34.6	39.7	10.8
Shoot weight (mg DW $shoot^{-1}$)	731.0	315.0	272.5	14.3
Horizontal rhizome internodal length (cm)	0.4	2.1	1.8	1.3
Vertical rhizome internodal length (mm)	1	1.4	–	–
Rhizome diameter (mm)	10	3.0	3.5	1.3
Fruit size (mm^3)	523.6	48.0	18.0	2.8
Dynamic features				
Shoot elongation rate (cm $shoot^{-1}$ day^{-1})	0.8	1.7	3.2	2.5
Horizontal rhizome elongation rate (cm $year^{-1}$)	4.2	40.2	40.9	54.8
Vertical rhizome elongation rate (cm $year^{-1}$)	1	1.4	–	–
Leaf plastochrone interval (day $leaf^{-1}$)	50.3	32.9	13.1	8.3
Leaf production rate (leaves $shoot^{-1}$ $year^{-1}$)	7.7	11.1	27.9	146.0
Horizontal rhizome production rate (rhizome internodes $year^{-1}$)	16.4	28.5	22.9	91.3
Leaf longevity (days)	302.8	45.0	51.4	25.0
Generation length (year)	35	3	1	1
Shoot recruitment rate (ln units $year^{-1}$)	0.2	1.4	1.5	1.4
Shoot mortality rate (ln units $year^{-1}$)	0.3	2.3	1.1	0.4
Patch/meadow properties				
Density (shoots m^{-2})	359	540	379	4945
Aboveground biomass (AB, g DW m^{-2})	675	485	272	135
Belowground biomass (BG, g DW m^{-2})	3785	500	147	88
AB/BG ratio	0.2	0.9	3.4	1.5
Leaf area index (LAI, m^2 m^{-2})	3.9	3.5	2.8	1.9
Aboveground production (gC m^{-2} $year^{-1}$)	392	446	568	875
Belowground production (gC m^{-2} $year^{-1}$)	43	78	199	643
Patch formation rate (n° patches ha^{-1} $year^{-1}$)	3	45	50	20

Data compiled from Bay (1984), Duarte (1991b), Pergent et al. (1994), Laugier et al. (1999), Hemminga and Duarte (2000), Brun et al. (2003b, 2006), Peralta et al. (2005), Lee et al. (2007), and Short et al. (2011)

supporting salinity values between 36.5 (e.g., Alborán Sea) and 39.7 (e.g., Cilician Sea), although it also occurs at lower salinities (e.g., from 21.5 to 28 in the Dardanelles Strait and in the Marmara Sea) or in hypersaline coastal lagoons (e.g., 39–44 in Farwà Lagoon on Libyan coast). However, low levels of carbon dioxide, rather than salinity, could be the limiting factor responsible for its absence in some coastal lagoons (Romero 2004). Its presence next to the mouth of large rivers (e.g., Rhône, Po or Nile) is limited by sediment and freshwater inputs. *Posidonia oceanica* withstand a rather wide range of temperatures, as inferred from its widespread latitudinal distribution (Green and Short 2003). The optimum temperatures range between 10 and 28 °C. Thus, its absence on a relatively narrow strip of the eastern Levant Sea (Lebanon and Israel coasts) has been associated to temperatures above the maximum range (Celebi et al. 2007). The nature of the substrate, as well as the hydrodynamics, can also limit the growth of *P. oceanica*. Meadows thrive on sandy or rocky bottoms, but are rare on muddy substrates because of the deterioration of light environment as well as the poor oxygenation of sediments that affects the plant physiology and the overall performance (e.g., presence of phytotoxins like sulfide). Hydrodynamics, either as accretion/erosion balance or as wave exposure, also influences the bed development. Thus, *P. oceanica* rarely occurs nearby river mouths or in confined waters. In sheltered bays, meadows can grow up to the water surface, developing fringing reefs (Fig. 9.2), but in more exposed waters canopy usually does not reach the water surface (Sánchez-Lizaso 2004).

Posidonia oceanica is a large, slow-growing rhizomatous plant possessing both horizontal (plagiotropic) and vertical (orthotropic) rhizomes. Rhizomes are arranged as a sequence

Fig. 9.2 A typical *Posidonia oceanica* fringing reef in shallow waters (Photograph: Juanma Ruiz)

Fig. 9.3 Detailed view of *Posidonia oceanica* shoots and orthotropic rhizomes fully covered by the fibrous rests of old leaf sheaths. The outermost leaves of the shoots (the oldest leaves) are covered by calcareous epiphytes (Photographs: Juanma Ruiz)

of segments (internodes) separated by nodes. Horizontal and vertical internodes are short (3.5–1 mm on average, respectively) and thick (up to 1 cm) revealing the slow growth of this species. Roots (up to 4 mm thick, and up to 40 cm long) are typically produced at the nodes of both kinds of rhizomes. Long (75 cm, on average) and wide (10 mm, on average) leaves (blade plus sheath) arranged in bundles on shoots (up to 8–10 leaves) arise from the nodes of the orthotropic rhizomes (Fig. 9.3). Leaf length undergoes seasonal variations being longer in summer (up to 1.4 m) and shorter in winter (20 cm, on average). Seasonal variations are also observed in biomass standing stocks and shoot density. On average, the aboveground biomass is 675 g DW m^{-2}, and the living belowground biomass is 3,785 g DW m^{-2} (Sánchez-Lizaso 2004). These values, specially the belowground ones, are much higher than those reported for the other Mediterranean seagrasses (Table 9.1). Besides seasonality there is also a bathymetric (and site-to-site) variation in biomass standing stock and shoot density, with higher values usually at shallow waters (more than 1,000 shoots m^{-2}) and lower ones at deeper waters (70 shoots m^{-2} or less). Meadows can be classified as a function of their densities into very sparse (50–150 shoots m^{-2}), dense (400–700 shoots m^{-2}) and very dense (700 shoots m^{-2}) (Giraud 1977).

Leaf longevity (303 days, on average), leaf plastochrone interval (i.e., elapsed time between the emergence of two successive leaves) (50 days $leaf^{-1}$) and generation time (35 years) are the highest among the Mediterranean seagrasses (Table 9.1). Senescent leaves are shed continuously throughout the year, but mainly in the autumn. Rhizomes are heavily lignified and easily recognizable by the compact fibrous rests of old, decaying leaf sheaths that remain attached to the rhizomes once the leaf blades have been lost (Fig. 9.3). The observed annual cyclic variation in the sheath thickness (and internodal length) has been used as a retrospective quantification method (known as "lepidocronology") to study the dynamics of the *P. oceanica* meadows (Duarte 2004).

Posidonia oceanica is a monoecius species, with male and female flowers in the same inflorescence. Flowering fluctuates highly depending on populations, years and sites, but use to be a rather infrequent event (less than 20 % of shoots) that happens in autumn. Massive flowering events have been recorded associated to climatic extremes such as extreme warm summers (Díaz-Almela et al. 2007, 2009). Inflorescences are big and produce large fruits (known as "sea olives") that shed the following spring (Figs. 9.4a, b). The buoyancy of seeds could contribute to the dispersal of the populations, but no direct observations are available to corroborate it. There is not a seed bank of *P. oceanica* since seeds germinate after 10 days without a dormancy period. Seedlings are rarely found reflecting that spreading of populations would occur mainly by vegetative growth of rhizomes. It would explain the low genetic variability of the *P. oceanica* populations both at local and at broader spatial scales (although new data based on microsatellite markers show genetic distinction between east–west population, at scale of Mediterranean Sea, as a whole (Procaccini et al. 2002)). Thus, the little investment and low success of sexual reproduction, together with the extremely slow clonal spread (see below) would explain the extraordinarily slow colonisation rate of *P. oceanica*. In fact, the time required for the patches to develop meadows has been estimated in several centuries (Sánchez-Lizaso 2004).

Clones colonize a 3-D space throughout rhizome spreading and branching. Growth of plagiotropic (4.2 cm $year^{-1}$, on average) and orthotropic (1 cm $year^{-1}$, on average) rhizomes, as well as branching rate (every 30 year, on average) is the lowest among the Mediterranean seagrasses (Duarte 1991b)

Fig. 9.4 (**a**) Detailed view of *Posidonia oceanica* inflorescence. (**b**) Fruit ("sea olive") (Photographs: Juanma Ruiz)

(Table 9.1). Besides the seasonal variation in growth rate (maximum in summer and minimum in winter, but smoother variations compared to the other Mediterranean seagrasses), the elongation rate of orthotropic rhizomes increases in response to burial resulting in a rise of the sea bottom. The network of interwoven fibrous live and dead rhizomes that decomposes slowly, mixed with sediment and organic debris can achieve several meters high and last for millennia. This bioconstruction is known as "matte" and the accumulation rate of such deposits has been estimated as 0.1–0.4 cm year^{-1} (Romero et al. 1994; Mateo et al. 1997). The study of the matte allows the reconstruction of the past environmental conditions (paleoreconstruction) of the Mediterranean Sea.

Posidonia oceanica-dominated ecosystems are very productive fixing c.a., 400 gC m^{-2} year^{-1}. Regardless of a large proportion (80 %) of the assimilated carbon is respired by the community itself, the net production of these ecosystems is about 72 gC m^{-2} year^{-1}, which is a net carbon fixation 60 times higher than that of the coastal bare sediments (Barrón et al. 2006). A significant proportion (42–62 %) of this fixed carbon is retained and buried in the matte for millennia (Larkum et al. 2006). Considering that *P. oceanica* meadows cover up to 4.5 millions ha in the Mediterranean (Pasqualini et al. 1998), such beds would sequester about 2 Tg C year^{-1}. Since there are no estimations on the amount of carbon withdrawn in other Mediterranean littoral and open-water habitats, it is not an easy task to attain accurate estimations on the importance of the seagrass beds on a whole Mediterranean basin. However, it must be considerable since nearly 50 % of the carbon sequestered in the whole ocean is buried in coastal vegetated habitats, and that all seagrass species account for 15 % of the total carbon buried in the ocean (Duarte et al. 2005). Then, *P. oceanica* beds fix and sequester a portion of atmospheric CO_2, providing an important ecosystem service, as is the regulation of the Earth´s climate.

Besides the carbon burial, the fixed carbon can be also exported to adjacent systems. After heavy autumn storms, coinciding with the main period of foliar renewal, a variable proportion of production (10–90 %, depending on local hydrodynamics) is exported as detached leaves and rhizome fragments and stacked on shore forming structures known as "banquettes" (and "*Posidonia* balls" or aegagropiles when constant rolling action of the sea shapes this material into balls) (Romero et al. 1994) (Figs. 9.5a, b). The stranded biomass supplies important amounts of sediments and nutrients to beaches especially in those locations where sediments are of biogenic origin. Additionally, this litter covers the sand of the beaches keeping it from erosion during heavy storms. As well, some of the detritus piled up on the shoreline remain in the water, increasing its viscosity, and thus, lessening the energy of the swell and the risk of coastline erosion (Marbà 2009).

Cymodocea nodosa (Ucria) Ascherson

Cymodocea nodosa (common name: Slender seagrass) is a warm water species widely distributed throughout the Mediterranean Sea, the eastern Atlantic, from south Portugal to Senegal and around the Canary Islands. It is a pioneering species that usually thrives in open sea (from shallow waters down to 30–40 m, depending on the water transparency), bays, small harbours and littoral lagoons (Fig. 9.6). This species prefers sandy or sand-muddy substrates tolerating a

Fig. 9.5 (**a**) Debris of *Posidonia oceanica* (detached shoots, rhizomes and roots) deposited on shore forming accumulations known as "banquettes". (**b**) Detail of the accumulations showing the distinctive "*Posidonia* balls", and fragments of rhizomes and shoots (Photographs: Juanma Ruiz)

Fig. 9.6 General view of a shallow *Cymodocea nodosa* meadow (Photograph: Ángel Pérez-Ruzafa)

wide range of nutrient concentrations, reducing conditions of the sediments and moderate levels of physical disturbance. Patches are also found in small sandy or gravel pockets that

Fig. 9.7 Mixed meadows of *Cymodocea nodosa* and the green seaweed *Caulerpa prolifera* are commonly found in shallow-sheltered bays (Photograph: Ángel Pérez-Ruzafa)

accumulate in crevices or small depressions on rocky flats. *Cymodocea nodosa* grows in warm environments with temperatures ranging from approximately 10 °C up to about 30 °C. Its sensitivity to temperature has been attributed to the tropical origin of this seagrass genus. It is also highly tolerant to salinity, withstanding fluctuations from 26 to 44. *Cymodocea nodosa* can occur either as monospecific stands (e.g., Urbinu lagoon, France) or as mixed meadows with *Zostera noltii* (e.g., Baia delle Saline, Italy), *Z. marina* (e.g., Venice Lagoon, Italy) or with the green alga *Caulerpa prolifera* (e.g., Mar Menor, Spain) (Fig. 9.7). In some locations it forms bands embracing the upper and lower limits of the *Posidonia oceanica* meadows (e.g., Natural Park of Cabo the Gata, Spain), colonizing also dead mattes of *P. oceanica* and enabling a further recolonization by this climax species (Marbà and Terrados 2004).

Cymodocea nodosa is a fast-growing rhizomatous plant that possesses vertical and horizontal rhizomes. Rhizomes are much more thinner (3.0 mm diameter, on average) than those of *Posidonia oceanica* (10 mm diameter, on average) and the orthotropic rhizomes lack the compact fibrous leaf sheath envelope (Fig. 9.8a). This species is easily distinguishable from *Zostera marina* by the long (2.1 cm on average) white to pink plagiotropic rhizome internodes and, mainly, by the presence of orthotropic rhizomes with slightly compressed and short internodes (1.4 mm, on average). Single, strongly branched roots (up to 3 mm thick and up to 40 cm long) appear at the nodes of both kinds of rhizomes. Leaves (34 cm long and 3 mm wide, on average) are arranged in bundles in shoots (3.3 leaves, on average) arising from nodes of horizontal and vertical rhizomes. The distinct lines identifying the nodes after leaf abscission are known as "leaf scars". Leaf length varies on season, depth and location. Leaf longevity and leaf plastochrone intervals depend on the time of leaf appearance but, on average, are 45–33 days leaf^{-1},

Fig. 9.8 (**a**) *Cymodocea nodosa* showing the characteristic horizontal (plagiotropic) pinkish and vertical (orthotropic) rhizomes and the single white roots arising from nodes. (**b**) Male plant showing flowers. (**c**) Seedling and female plant with a seed at the base of the shoot (Photographs: EDEA)

respectively (Table 9.1). Thus, leaf replacement dynamics (turn-over) is faster than in *P. oceanica* (Terrados and Marbà 2004).

Cymodocea nodosa is a dioecius species. Reproductive effort and success is subjected to temporal and spatial heterogeneity. Flowering (spring-summer) is quite rare occurring only in shoots older than 1 year, but is enhanced in response to burial. Besides, the spatial distribution and abundance of male and female clones could limit seed production, and in turn, the reproductive success in those sites where only clones of one sex occur. Female flowers produce two lenticular seeds that are considerably bigger (48 mm^3, on average) than those of *Zostera* spp. Seeds are formed at the base of the mother shoots and once detached they are rapidly buried into the sediment nearby the generative shoot because its negative-buoyancy building a rather ephemeral seed bank (7–9 months of dormancy period) (Figs. 9.8b, c). Although seed dispersal is rather limited, episodes of severe sediment dynamics can favour a long-distance dispersion. Seeds germinate from April to June of the next year (Terrados 1993). Patch formation rate in areas undergoing intense sexual reproduction has been estimated c.a., 90 clones ha^{-1} year^{-1}. However, clone mortality rate is about 50–70 % during the first year of life, hence, decreasing considerably the success of sexual reproduction (Duarte and Sand-Jensen 1990). Patches spread relatively fast since the growth rate of the horizontal rhizomes is among the highest of the Mediterranean seagrasses (40.2 cm year^{-1}, on average, but up to 200 cm year^{-1} has been recorded) (Duarte 1991b) (Table 9.1). The rapid space occupation resulting from fast clonal growth and the relatively high patch formation rate (45 patches ha^{-1} year^{-1}, on average) supports the colonizer role that *C. nodosa* plays during the succession process in the Mediterranean. The time span required for patches to develop meadows is estimated to be less than a decade. The vertical rhizomes of *C. nodosa* also extend, but at slower rates (1.4 cm year^{-1}, on average) than the horizontal ones. Thus, rhizome growth is plastic enough to allow this species to survive moderate environmental disturbances such as accretion/erosion episodes by increasing or decreasing the elongation rate of the vertical rhizomes, respectively (Duarte and Sand-Jensen 1990).

Growth, production, biomass standing stock and shoot density of *Cymodocea nodosa* beds undergo a marked seasonal pattern superposed to bathymetric and site-to-site variations. Growth rates vary highly throughout the seasonal cycle depending on temperature what it has been attributed to the tropical origin of this genus. Maximum values for growth (up to 6 cm shoot^{-1} day^{-1} for aboveground parts, and up to 0.6 cm day^{-1} for horizontal rhizomes), production (up to 844 gC m^{-2} year^{-1} for aboveground, and up to 144 gC m^{-2} year^{-1} for belowground), shoot standing stock biomass (up to 945 g DW m^{-2}) and shoot density (up to 2,000 shoots m^{-2}) are achieved during the spring-summer period. A considerable proportion of the net primary production (48–67 %) is exported to adjacent systems, a small amount (9 %) serves as food for herbivores and the remaining enters into the detritus food web.

Zostera marina Linnaeus

Zostera marina (common name: Eelgrass) presents a widespread and circumglobal distribution in northern latitudes: from artic waters to the Mediterranean Sea. It is abundant in the Baltic Sea, the North Sea and along the Atlantic coasts down to northern Spain. Further south it becomes less abundant and in the Mediterranean Sea it is considered a relict species. It forms small perennial isolated stands in shallow protected bays and coves of the northwestern Mediterranean (Spain, France, Italy) and Adriatic Sea extending from the intertidal to few meters depth (Fig. 9.9). However, denser

Fig. 9.9 Patch of *Zostera marina* at low tide. The light green shoots are flowering shoots (Photograph: EDEA)

Fig. 9.10 *Zostera noltii* meadow at low tide (Photograph: EDEA)

meadows occur regularly in coastal lagoons of the western Mediterranean such as the Thau lagoon (France) where it is often found with *Z. noltii* (Laugier et al. 1999) or in Venice Lagoon (Italy) where it forms mixed populations with *Cymodocea nodosa*. There are records of its presence in Tunisia and previously in the northern Aegean Sea (Boudouresque et al. 2009). Others old sightings as in Egypt or Syria (Mayhoub 1976) are doubtful due taxonomic confusion in the past. It has also been recorded in the Black Sea (Green and Short 2003). *Zostera marina* is mostly subtidal occurring down to 10–15 m depth (depending on water transparency) in sandy and muddy coastal bottoms of low to moderate wave exposure. It is a eurythermic (−1–25 °C) and euryhaline (5–35) species.

Zostera marina has only horizontal rhizomes (3.5 mm diameter, on average) with internodes (1.8 cm, long on average) that are white-green in recently produced ones to dark brown in the oldest. Two bundles of roots arise from the nodes. The roots are thin and long (up to 1 mm and 20 cm, respectively) covered by fine root hairs. Leaves (40 cm long and 6 mm wide, on average) are arranged in bundles in shoots (4.3 leaves, on average) arising from the nodes. Leaf length, biomass standing stock and shoot density are highly variable depending on season, depth and location. Thus, leaves up to 1.5 m long are observed in beds on soft sediments at intermediate depths. Leaf longevity and leaf plastochrone intervals are, on average 51.4–13.1 days leaf^{-1} respectively (Table 9.1) (Pérez-Lloréns 2004a).

Zostera marina is a monoecius species. Inflorescences are produced in shoots (known as flowering shoots) arising from long and thin stems (Fig. 9.9). Flowering (spring-early autumn) is frequent and produces thousands of seeds per square meter. Once the seeds (18 mm^3, on average) are mature the flowering shoots die and shed. Detached shoots float away from the bed contributing to a wider dispersal of seeds, although the majority of seeds drop within the mother meadow because of its negative buoyancy. The majority of patches (90–95 %) resulting from seed germination (50 patches ha^{-1} year^{-1}, on average) die within the first year and only few spread and persist, even several centuries or millennia (up to 3,000 years) in locations (Baltic Sea) where sexual reproduction is rare (Reusch et al. 1999). In general, the spread of patches in the Mediterranean is throughout clonal growth and the time required for the patches to develop meadows has been estimated to be less than a decade (Hemminga and Duarte 2000).

Zostera marina meadows follow a seasonal pattern in growth, production, biomass standing stock and shoot density that can be tuned by depth and location. Seasonality in growth is highly dependent on irradiance and less on temperature. Maximum values for growth (up to 5.7 cm shoot^{-1} day^{-1} for aboveground parts, with a mean leaf production rate of 28 leaves shoot^{-1} year^{-1}, and up to 0.25 cm day^{-1} for rhizomes), production (up to 1,985 gC m^{-2} year^{-1} for aboveground, and up to 657 gC m^{-2} year^{-1} for belowground), shoot standing stock biomass (up to 960 g DW m^{-2}) and shoot density (up to 2,100 shoots m^{-2}) are achieved during the spring-summer period (Pérez-Lloréns 2004a).

Zostera noltii Hornemann

Zostera noltii (common name: Dwarf eelgrass) is widespread in intertidal and shallow subtidal areas of the Atlantic coast of North America and Europe from the southern coast of Norway to the Mauritanian coast (Fig. 9.10). In the Mediterranean Sea it occurs in sheltered coasts such as bays, lagoons, small harbours or estuaries on sandy and muddy bottoms where can form monospecific or mixed meadows with *Cymodocea nodosa*, *Z. marina* or with the green seaweed *Caulerpa prolifera*. It is a eurythermic and euryhaline species with a wide range of salinity tolerance (25–51). Low

Fig. 9.11 (**a**) Plant of *Zostera noltii* showing the rhizome network, branching pattern and shoots. (**b**) Flowering shoot (Photographs: EDEA)

Fig. 9.12 *Zostera noltii* (*right*) is a pioneering species that use to colonize gaps after disturbance episodes in *Posidonia oceanica* meadows (*left*) (Photograph: José Antonio Rodríguez)

salinity improves germination and seedling development. Since it occurs at the intertidal or shallow subtidal (above 10 m depth) it is adapted to high irradiances (as the high saturating irradiance for photosynthesis denotes, Table 9.1), but can withstand relatively high turbidity episodes in estuarine water, as suggest its low minimum light requirement for growth (2 % of surface irradiance, Peralta et al. 2002). It is tolerant to the relatively anoxic conditions typical of the rich organic sediments where meadows thrive. The critical level of accretion or erosion tolerated by this species is extremely low due its small size and the presence of short vertical rhizomes (Pérez-Lloréns 2004b; Brun et al. 2005).

Architecture of *Zostera noltii* resembles that of *Z. marina* but reduced in the size of the modules. Rhizome internodes (3.5 mm diameter and 1.3 cm long, on average) are light green in the youngest parts and yellow or brown in the oldest ones. It has been reported recently the presence of short and thin vertical rhizomes in this species (Brun et al. 2005). Two bundles of thin roots (<1 mm) arise from rhizome nodes. Ribbon-shaped leaves (10.8 cm long and 1 mm wide, on average) are arranged in bundles in shoots (3.2 leaves, on average) arising from nodes. Each rhizome holds many shoots on short branches separated by rhizome internodes (Fig. 9.11a). As in the other seagrass species, leaf length, standing stock biomass and shoot density are highly variable depending on season, depth and location. Thus, leaves up to 65 cm long are observed in beds growing in deep areas. Leaf longevity and leaf plastochrone intervals vary depending on the recruitment season but on average, are 25.0–8.3 days leaf^{-1} respectively (Brun et al. 2003b; Peralta et al. 2005) (Table 9.1).

Zostera noltii is a monoecius species. Inflorescences are produced in flowering shoots arising from the rhizomes as the vegetative ones (Fig. 9.11b). Flowering represents less than 10 % of the shoots and can extend from March to November but usually varies among locations, since factors such as photoperiod, temperature, tidal amplitude and fluctuating salinity regimes control the flowering event. Seeds are very small (2.8 mm^3, on average) and are not likely to disperse far away because of its negative buoyancy. However, water currents may transport detached flowering shoots containing seeds over long distances. Seedling is also very infrequent (less than 5 % of patches are originated from seeds, on average, 20 patches ha^{-1} year^{-1}) (Table 9.1) suggesting that clonal growth instead sexual reproduction is the main way of meadows spreading (Brun et al. 2003a, b).

The high internode production (91.3 rhizome internodes year^{-1}, on average), elongation (54.8 cm year^{-1}, on average) and branching rates (at every rhizome node) lead to a rapid spread of patches. In fact, the time required for patches to develop meadows ranges from several months to a year. In addition, *Zostera noltii* also have the highest turnover rates among the Mediterranean seagrasses (e.g., leaf production rate is, on average, 146.0 leaves shoot^{-1} year^{-1}, and the leaf longevity 25.0 days), which is typical of colonising species allowing *Z. noltii* to cope with (or to recover after) considerable levels of disturbance. It is rather common that *Z. noltii* (or *Cymodocea nodosa*) colonizes gaps after disturbance episodes in *Posidonia oceanica* meadows (Fig. 9.12). As the other seagrass species, *Z. noltii* beds undergo clear seasonal cycles in growth, production, biomass standing stock and shoot density that can be altered by depth and location. Maximum values for growth (up to 3.5 cm shoot^{-1} day^{-1} for aboveground parts, and up to 0. 5 cm day^{-1} for rhizomes), production (up to 1,250 gC m^{-2} $year^{-1}$ for aboveground, and up to 1,280 gC m^{-2} $year^{-1}$for belowground), shoot biomass standing stock (up to 260 g DW m^{-2}) and shoot density (up to 10,000 shoots m^{-2}) are achieved during the spring-summer period. The annual foliar losses have been estimated in 420 gC m^{-2} $year^{-1}$ that accounts for near 75 % of the aboveground production (Marbà et al. 1996; Brun et al. 2003b; Peralta et al. 2005).

Disturbances in Seagrass Meadows

Despite the general agreement that seagrasses provide important ecosystem services and goods, being listed as priority habitats for conservation (e.g., Rio Biodiversity Convention, European Habitats Directive, European Water Framework Directive, National and Regional frameworks, etc.), there are growing evidences of a remarkable global regression (Orth et al. 2006; Short et al. 2011). Besides the natural disturbances affecting seagrass meadows, sometimes severely like the "wasting disease", a pathogen (slime mold) that caused the lost of many *Zostera marina* populations in the 1930s, direct or indirect human interventions are the principal drivers of the observed seagrass declines (Hemminga and Duarte 2000).

Anthropogenic Non-climatic Causes

Seagrass losses of 2,900 km^2 were reported from mid-1980s to mid-1990s as consequence of direct or indirect human pressures. An extrapolation of these numbers to unreported parts of the coasts would render c.a., 12,000 Km^2 of meadows probably lost (Short and Wyllie-Echeverria 1996). Recent estimations, based on a longer time observations (20 years), give values of about 33,000 km^2 (Spalding et al. 2003). Assuming a conservative value of global seagrass cover of 177,000 km^2 (Spalding et al. 2003), losses would range between 7 and 19 % of the documented seagrass areas (with values up to 65 % within inhabited areas, Lotze et al. 2006). However, reported losses probably represent a small fraction because global estimations of seagrass cover are very rough and many losses may remain unreported (Duarte et al. 2004).

Yearly, an estimated 100 million tourists visit the Mediterranean coasts, making it the European centre for international tourism, receiving around a third of all international tourists. About 60 % of the coastline is currently urbanized and up to more than 75 % in the regions with the most developed tourism industry, with harbours and ports occupying 1,250 km of the European Mediterranean coastline (French 1997). Such demographic pressure results in important disturbances of seagrass meadows. For example, it has been estimated that, on average, 46 % of the *Posidonia oceanica* meadows underwent some decline in range, density and/or coverage, and 20 % were severely deteriorated since the 1970s, and most dramatic large-scale losses happen in the northern Adriatic Sea where *P. oceanica* beds present at the beginning of the twentieth century have almost disappeared. However, it must be taken into account that some declines can be the consequence of long-term climate tendencies (e.g., the post-Last Glacial Maximum rise in sea-level, the Little Ice Age (LIA) cooling and the post-LIA warming) resulting in potential misinterpretation of the human influence in matte forming species as *P. oceanica* (Boudouresque et al. 2009).

Human activities affecting mostly to meadow physical integrity, sediment and water quality (including water transparency), coastal sedimentary balance or species composition are argued to be the main drivers of seagrass decline in the Mediterranean Sea (Boudouresque et al. 2009). Frequently, a single activity (e.g., coastal development) can involve both direct (e.g., mechanical damage) and indirect impacts (e.g., sediment unbalances, eutrophication, pollution, siltation, etc.) with synergistic effects on seagrass decline. In addition synergy can also emerge when several activities converge in a single location (Orth et al. 2006).

Mechanical Habitat Destruction

Direct habitat damage by land reclamation and port or infrastructure construction is a major cause of disturbance for seagrass meadows. For instance, the construction of Pointe-Rouge harbour (Marseille, southern France) directly destroyed 11 ha of a *Posidonia oceanica* meadow (Boudouresque et al. 2006).

Trawling is probably the most important and concerning activity causing large-scale destruction of *Posidonia oceanica* beds, especially that at deepest locations and with exposed rhizomes (González-Correa et al. 2005), since a standard trawler uproots 99,000–363,000 shoots h^{-1} (Martín et al. 1997). As an example, this fishing technique has been responsible for loss of the 12 % of the meadow extension in Corsica, almost 50 % in Alicante (Spain) or above 80 % in the Gulf of Gabe´s (Tunisia) (Boudouresque et al. 2006). Recovery of damaged beds might occur if trawling is forbidden, but the time span needed to recuperate the previous state may take almost 100 years given the very slow growth rate of *P. oceanica* (González-Correa et al. 2005).

Digging for clams can also harm seagrass meadows. It has been reported that fast-growing pioneering species like *Zostera noltii* are able to support certain threshold of disturbance. However, intensive clam harvesting practices (i.e., high frequency and intensity of disturbance, particularly during summer) may not allow the full recovery of *Z. noltii* meadows (Cabaço et al. 2005) (Fig. 9.13a).

Mooring can also affect meadows by uprooting shoots in those locations supporting high densities of small pleasure ships. It has been estimated that, on average, a single anchor

Fig. 9.13 Some direct and indirect human impacts on seagrass meadows. (**a**) Digging for clams. (**b**) Boat anchoring: the circular marks on intertidal seagrass meadows are produced by the drag of anchored boats due to their movement with flowing and ebbing tides. (**c**) Dense mat of the green algae *Chaetomorpha linum* growing on *Cymodocea nodosa-Zostera noltii* mixed meadow. The black colour of sediment denotes the anoxic conditions due to the loading of sediments with organic matter from algal die-off (Photographs: EDEA). (**d**) Habitat perturbations make *Posidonia oceanica* meadows more vulnerable to be invaded by alien species such as the green algae *Caulerpa racemosa* (Photograph: Juanma Ruiz)

uproots 34 shoots (Francour et al. 1999) or 68,000 shoots ha^{-1} (Corsica) (Boudouresque et al. 2006). If anchoring is in an intertidal domain, the movement of boats by ebbing and flowing tides and currents cause scars in intertidal populations of *Zostera noltii* or *Cymodocea nodosa* (Fig. 9.13b).

Dredging and sand withdrawal (e.g., for beach nourishment) are major causes of seagrass loss. Although the existence of a compact matte of *Posidonia oceanica* often keeps these areas from sand extraction it is not always the case, as for example, in Ischia (Italy) where 4 ha of meadow were directly destroyed by sediment pumping or, indirectly, by turbidity and burial by sand re-deposition (Flagella et al. 2006).

As just mentioned, meadow removal can led to a set of secondary effects like, increased turbidity (due to sediment resuspension), alteration of sediment dynamics (burial/erosion balance), and modification of physical processes (e.g., water currents) as a consequence of the sediment redistribution (Ruiz and Romero 2003). For example, nearby Toulon (south eastern France) the construction of artificial beaches buried 22 ha of *Posidonia oceanica* meadows, but sediments relocated during and after works buried 10 additional ha (Astier 1984). As consequence of sediment erosion/accretion episodes survival of shoots decreases. However, species with orthotropic rhizomes (*P. oceanica* and *Cymodocea nodosa*) can tolerate moderate burial since shoot recruitment is favoured because of the growth and branching of the orthotropic rhizomes are enhanced under such conditions. For example, *P. oceanica* beds can cope with burial rates of up to 4–5 cm $year^{-1}$. If sediment accretion exceeds such level bed decay accelerates, and accumulation of 14 cm result in 100 % shoot mortality (Manzanera et al. 1998). Analogously, orthotropic rhizome and leaf growth rates are kept at minimum when sediment is eroded. The seagrass response to sediment dynamics remains imprinted on rhizomes. It has been used for retrospective identification of burial/erosional events (Marbà and Duarte 1994). Fast-growing species lacking vertical rhizomes (*Zostera marina*) or with short and thin vertical rhizomes (*Z. noltii*) are most sensitive to burial. However, the high elongation rate and branching of the horizontal rhizomes allow a fast recovery after moderate burial events (Brun et al. 2007).

Eutrophication

Eutrophication (i.e., nutrient over-enrichment, especially nitrogen and phosphorus) observed in many coastal waters has been considered as a major cause of seagrass decline, especially in heavily developed parts of the world (Hemminga and Duarte 2000). Although currently eutrophication and pollution are not a generalized problem in the Mediterranean Sea (Jackson et al. 2006), some seagrass meadows are heavily affected because of localized human activities (e.g., domestic and industrial sewage outlets, agricultural runoff, aquaculture activities, etc.).

Eutrophication can harm seagrasses in two ways: directly because of the toxicity of some nutrients (e.g., ammonium) (Brun et al. 2002) or indirectly, by stimulating the growth of phytoplankton, epiphytes (micro- and/or macro-) or pleustophytic opportunistic macroalgae (Fig. 9.13c). Although the type of blooming algae will depend mostly on the water turnover rate (i.e., sheltered *vs* open locations), the overall consequence will be shading seagrasses beneath. Due to the high proportion of non- photosynthetic tissues (mostly roots and rhizomes), seagrasses require for survival and growth underwater irradiance values generally above 11 % of that incident on the water surface (Duarte 1991a) making them very vulnerable to deterioration of the light environment. However, despite the respiratory burden that belowground biomass represents, rhizomes are the main reservoir of soluble carbohydrates that are mobilized to meet whole-plant carbon demands for growth and survival under low light conditions (Hemminga and Duarte 2000). Accordingly, tolerance to prolonged shading is expected to be higher in *Posidonia oceanica* than in the other small-size Mediterranean species because of the largest capacity of *P. oceanica* to store carbohydrates (i.e., low AG/BG ratios, Table 9.1).

Besides the negative effects of light reduction on seagrass meadows, unfavourable biogeochemical alterations of the habitat arise: the loading of sediments with organic matter from algal die-off enhances the respiration of the benthic microbial community leading to anoxia. Such conditions are harmful to seagrasses not only because anoxia restricts starch (sucrose) translocation from rhizomes to shoots to meet the carbon demands under low light periods (Brun et al. 2003a) or impedes root respiration and nutrient uptake, but also because sulfide compounds in the surroundings of the rhizosphere may be toxic. Such reduced sediment conditions are long lasting even after the organic inputs stop, outspreading meadow deterioration (Hemminga and Duarte 2000).

Fish farms and other aquaculture developments are good examples of activities that are becoming increasingly in the shallow, sheltered coastal Mediterranean waters where seagrasses thrive, and are causing localized but significant impacts on meadows. Shading and high inputs of organic matter from fish cages lead to seagrass decline below and around fish cages, through processes comparable to those of the eutrophication outlined above (Ruiz et al. 2001). A comparative study addressing the effects of fish farming on *Posidonia oceanica* meadows across the Mediterranean Sea reported a negative impact on beds extending up to 200 m away from fish farms (Holmer et al. 2003), and, in Sicily, a negative impact on the orthotropic rhizome growth rate was detected as far as 1 km from a large fish farm installation. Since the resilience of this species to fish farm impact is overall very low (Marbà et al. 2006), the decline meadows due to the growing fish farming activities is currently rising (Jackson et al. 2006).

Introduced Species

The term 'introduced species' refers as any species introduced, intentionally or not, beyond its native range through human activities. In the last decades about 100 exotic macrophyte species have been introduced in the Mediterranean Sea mostly through shipping and aquaculture activities. At least five of them have an invasive behaviour to affect Mediterranean seagrass habitats: two Australian chlorophytes (*Caulerpa taxifolia* and *C. racemosa* var. *cylindracea*) and three Indo-Pacific rhodophytes (*Lophocladia lallemandii*, *Acrothamnion preissii* and *Womersleyella setacea*; Boudouresque et al. 2009).

The most well-known and studied case was the effect of *Caulerpa taxifolia* and *C. racemosa* var. *cylindracea* on *Posidonia oceanica* (e.g., Villèle and Verlaque 1995; Jaubert et al. 1999) and, to lesser extent, on *Zostera noltii* and *Cymodocea nodosa* meadows. Beyond the contrasting observations that literature brings on their influence on *P. oceanica* stands decline, it seems that habitat perturbations (e.g., trawling, eutrophication) makes seagrass beds more vulnerable to be invaded (Williams 2007) (Fig. 9.13d). It would explain that invasions have been mostly recorded in previously impacted and/or sparse meadows and not in dense and vigorous ones (Klein and Verlaque 2008).

The invasive rhodophyte *Lophocladia lallemandii* has been detected growing as dense mats of filaments on rhizomes and old leaves of *Posidonia oceanica* specimens occurring in small patches or at the periphery of denser meadows. Such mats can be so thick that produce mortality of the shoots (Ballesteros et al. 2007). *Acrothamnion preissii*, a new exotic rhodophyte that invades *P. oceanica* rhizomes has no evident effects on plants itself but it shifts most of the autochthonous rhizome epiphytes reducing the meadow's taxonomic diversity and habitat complexity. The rhodophyte *Womersleyella setacea* forms thick and long lasting turfs in several Mediterranean habitats, including *P. oceanica* beds. Mixed turfs of *A. preissii* and *W. setacea* may promote drastic declines in species richness of the understory assemblage in the invaded meadows (Piazzi et al. 2001).

Besides the cited observations, more surveys and experimental studies in invaded seagrass meadows would be necessary to quantify damages due to exotic species (Williams 2007). As abovementioned, dense *Posidonia oceanica* meadows seemed to cope with invasive species well, and no widespread losses have been reported. Nonetheless, competition can be a long-term ecological process and that short-term observations may be unsuccessful in detecting slow tendencies (Boudouresque et al. 2009).

Climate Change Potential Causes

Since the first Intergovernmental Panel on Climate Change report (IPCC 1990), numerous reports and scientific articles have addressed both the observed and the potential impacts of climatic change on species and their habitats. Although these studies documented changes that are clearly related to climate trends, they also revealed many challenges in predicting the outcome on species and ecosystems (Burkett et al. 2005). For example, climate change is thought to influence seagrass meadows by the rise in sea level (and associated coastal erosion and turbidity), the increased partial pressure of CO_2 and acidification of seawater, the elevated seawater temperature and the increasing frequency and strength of climatic extremes. However, it is important to remind that the ability to forecast the effects of the climate change depends largely on the uncertainties about (1) future emissions and concentration of greenhouse gases, (2) constructing scenarios of climate change, (3) sensitiveness of the climate change to perturbations and (4) influence of non-climatic drivers. In addition, there is an established, but incomplete evidence of critical ecosystem thresholds and the increasing likelihood of triggering non-linear responses (i.e., the output is not directly proportional to the input) once thresholds have been exceeded, resulting in abrupt changes and novel states (e.g., dystrophic crisis). Moreover, projecting the impacts of climate change on seagrass ecosystems is further complicated by a patchy knowledge of the interlinked temporal and spatial scales of ecosystem responses and the multiple (climatic *versus* non-climatic) drivers involved. Accordingly, predictions about the extent and direction of climate change on seagrass habitats will be associated with varying degrees of confidence. Thus, there is an acceptable level of confidence in predictions of global warming effects on dissolved inorganic carbon concentrations, sea level and on seagrass physiology; even in forecasting the effects of sea level rise on shoreline erosion and enhanced storm surges. However, there is a greater uncertainty to predict the effects of climate change on biotic interactions, wind patterns, and the frequency of extreme climatic events (IPCC 2007). It makes that forecasting the outcome of climate change on seagrass habitats becomes rather speculative.

CO_2 Increase and Acidification

It has been demonstrated that the anthropogenic CO_2 concentrations in the Mediterranean Sea are much higher than those of the Atlantic Ocean (Schneider et al. 2007). Accordingly, the expected pH drop in the Mediterranean would be relatively larger than that predicted globally (i.e., 0.5 pH units for the year 2100), resulting in CO_2 and HCO_3^- levels 50–6 % above those forecasted for the global ocean (IPCC 2007).

Ribulose-biphosphate carboxylase oxygenase (Rubisco) is the key enzyme involved in carbon fixation, using only CO_2 as substrate for carboxylation. In angiosperms, semi-saturation constants for CO_2 (30–250 μM) are higher than the CO_2 concentration in air-equilibrated seawater (10–15 μM, at pH 8.1). Furthermore, the CO_2 supply is often severely limited in dense seagrass canopies experiencing high biological activity (i.e., high C demand) under calm conditions (i.e., low hydrodynamics) (Enríquez and Rodríguez-Román 2006). To cope with the physical limitation of C mass transport, different carbon concentration mechanisms (CCMs) have evolved such as direct or indirect use of HCO_3^- (James and Larkum 1996) and/or the catalytic dehydration of HCO_3^- to CO_2 by the enzyme carbonic anhydrase (CA) (Invers et al. 1999). Despite such CCMs, and in contrast to many seaweed species, seagrasses are not fully saturated (i.e., 100 %) at the present CO_2 concentration in seawater (at pH 8.1), ranging from c.a., 91 % in *Cymodocea nodosa* to c.a., 60 % in *Zostera marina* with intermediate values for *Z. noltii* (c.a., 86 %) and *Posidonia oceanica* (c.a., 82 %) (Invers et al. 1999; Hellblom et al. 2001; Mercado et al. 2003). Thus, increased CO_2 and acidification would favour the competitive advantage of dense Mediterranean seagrass stands over seaweeds by rising photosynthesis and productivity and, probably, the depth limit of seagrass occurrence (Hemminga and Duarte 2000; Mercado 2011). However, it would be probably counterbalanced by the forecasted sea level rise and increased water turbidity (see below). The likely responses of seagrasses to climate change are forecasted mostly from physiological responses obtained in laboratory experiments (e.g., Zimmerman et al. 1997; Invers et al. 1999). However, present evidences that such physiological responses have led to observable changes in seagrass ecosystems are scarce, and other variables such as for example, the expected increase in water turbidity, temperature or nutrients (eutrophication) could hamper the effects of increased CO_2 and acidification on seagrass meadows dynamics.

Temperature

According to IPCC (2007) the global atmospheric temperature will raise by 1.1–6.4 °C by 2100, with a parallel increase of 0.053 °C $year^{-1}$ observed in the Mediterranean since 1970s, but not homogenous throughout the annual cycle (Brunet et al. 2007). Many biological processes are tightly controlled by temperature. Therefore, a change in the thermal regime (e.g., extreme temperatures, their extent, and seasonal rates of change) can directly affect several processes involved in seagrass growth and reproduction (e.g., respiration, photosynthesis, nutrient uptake, flowering and seed germination) as well as shifts in species composition, depending on the species-specific thermal tolerance ranges. At larger temporal and spatial scales changes in the biogeographic ranges are expected, with major effects suffered by those native populations of seagrasses existing nearby their boundaries of distribution.

It is well known that respiration increases more than photosynthesis as temperature rises. Thus, it is likely that global warming will result in higher respiration demands, that combined with reduced photosynthesis due to the forecasted light reduction (a combination of sea level rise and coastal erosion), will alter carbon balances compromising the maintenance of a positive carbon budget and, in turn, seagrass survival (Zimmerman and Alberte 1996). Mediterranean warming would benefit tropical species, such as *Cymodocea nodosa* or the newly established seagrass *Halophila stipulacea*, with higher photosynthetic and growth optimum temperatures than temperate species (*Zostera noltii, Z. marina*, *Posidonia oceanica*), which are likely to suffer a progressive regression. However, projecting the effects of increased temperature (and light availability) to the photosynthetic physiology of Mediterranean seagrasses, as well as to forecast the likely outcome driven by the climate change is highly speculative since effects appear to be rather species-specific (Enríquez et al. 2004) and because multiple (antagonistic, additive or synergistic) interactions with other human-related perturbations are expected.

A study carried out in several *Posidonia oceanica* meadows from the western Mediterranean revealed that elevated summer temperatures and persistent heat waves were associated with reductions in shoot growth and increases in shoot mortality and flowering intensity (Díaz-Almela et al. 2007, 2009). However, despite the observed massive flowering, the little success in seedling resulted in a sexual recruitment much more lower that the vegetative recruitment, and the total recruitment (sexual + vegetative) was, on average 4.5 times lower than mortality causing a progressive decline of those meadows. This study suggested that general regression of meadows could be enhanced by the Mediterranean warming trend, being such decline accelerated when seawater temperature rises above 28 °C. Moreover, enhanced temperature promotes microbial growth of sulphate-reducing bacteria and, consequently sulfide concentration in the sediment and anoxia. Under light-limited conditions with sulfide level rising toxicity in the sediment, seagrass carbon balance, growth and survival may be seriously harmed, as previously stated (Vaquer Sunyer and Duarte 2010).

Sea Level Rise, Coastal Erosion and Climatic Extremes

It is estimated that sea level in the Mediterranean will raise about 50–60 cm during the twenty-first century (IPCC 2007; Marcos et al. 2009). Moreover, precipitation trends in the Mediterranean basin during the last 50 years show an increase in both frequency and intensity of severe drought events, although the spatial distribution is not homogenous (Vicente-Serrano 2007). In the western Mediterranean basin the increase of drought events alternates with episodes of heavy rainfall and flooding (Millan et al. 2005). Therefore, sea level rise combined with climatic extremes will result in enhanced erosion of shores, increased water turbidity, modified currents and tidal ranges, altered salinity, as well as changes in sediment inputs and nutrient loadings to the coasts (Scavia et al. 2002) each of which could have major impacts on seagrass habitats. For example, rising sea levels will create new shallow areas for expansion allowing intertidal or shallow-subtidal populations of fast-growing pioneering species like *Zostera noltii* or *Cymodocea nodosa* to expand into these new inundated areas landward. However, the colonization of newly inundated areas upslope is only possible in zones free of shoreline constructions, which is not a common feature along the majority of Mediterranean coasts (Duarte 2002). On the contrary, meadows that currently occur at profundities close to its compensation depths (i.e., the depth at which photosynthesis equals respiration) will be probably hampered. Those species with orthotropic rhizomes (*Posidonia oceanica* and *C. nodosa*) would be able to maintain stands at similar depths, provided they can also trap particles to raise the seabed surface. However, increased background turbidity, due to coastal erosion, sediment resuspension or siltation, would clearly reduce the colonization depth. In addition, the UV-B/PAR ratio would decrease with increasing water turbidity, counteracting the reported beneficial role of UV-B radiations against stress conditions, since UV-B seems to be involved in the impairment and recovery of photosynthesis in *P. oceanica* (Figueroa et al. 2002). Also massive sediment movements associated to heavy storm events will affect meadows by burial or shoot uprooting (see "Anthropogenic non-climatic causes" section). Overall, the net effect of increasing frequency and strength of storm episodes on seagrasses is not clear and, as for the other climatic change drivers, rather speculative (Hemminga and Duarte 2000).

Interaction Between Non-climatic Anthropogenic and Climatic Drivers

As it has been previously stated, seagrass meadows decline can be the consequence of a combination of climatic and non-climatic causes. A good example was the regression and

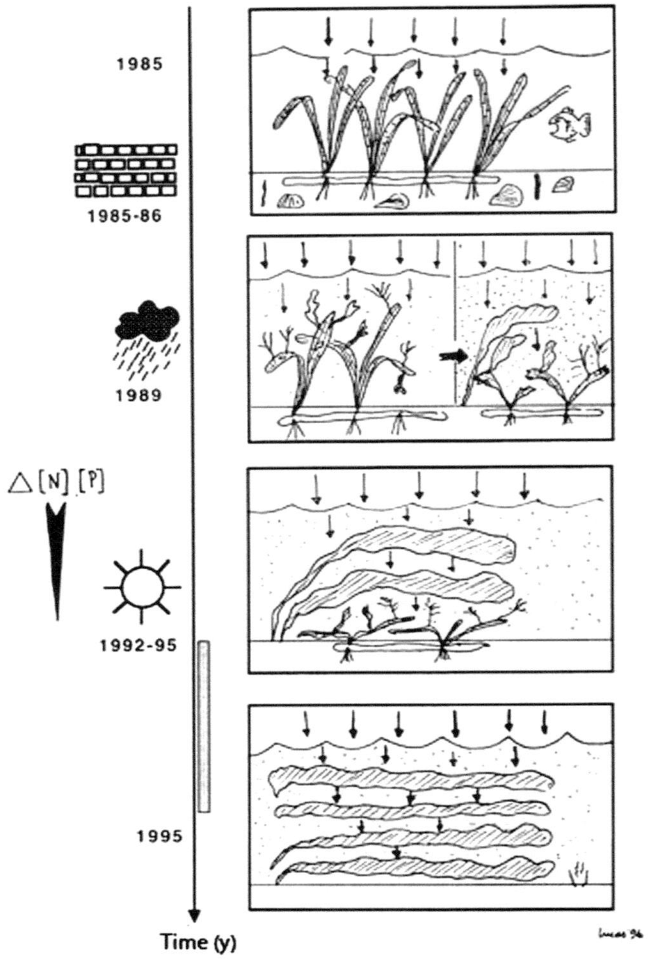

Fig. 9.14 Schematic sequence of events occurred at Palmones river estuary (southern Spain) due to human disturbances (dam construction, eutrophication) and extreme climatic episodes (flooding, drought) that resulted in a shift of species (Drawing: José Lucas Pérez-Lloréns)

the final loss of the *Zostera noltii* beds in Palmones river estuary (Algeciras Bay, southern Spain) close to Gibraltar Strait (Pérez-Lloréns and Niell 1993; Carreira et al. 1995) (Fig. 9.14). Briefly: in early 1980s intertidal mudflats of the estuary were fully covered by *Zostera noltii* stands and water quality and biodiversity were high. The construction and operation of a dam (mid 1980s) in the headwaters of the river reduced both the water flow (the area covered by water decreased by four in a decade) and the current speed, affecting nutrient exchange processes and sediment input to the estuary. In autumn 1989, recurrent and heavy rainfalls with flooding affected severely this location, specially the estuary and the seagrass meadows. Plants were uprooted by the opening of the dam or buried by large sediment loads from upstream. The intertidal mudflats were fully inundated by turbid water for weeks. The few surviving seagrass patches started to be epiphyted mostly by microalgae. A long lasting and extreme drought period (from 1992 to 1995) reduced drastically the freshwater inputs to the estuary. This reduction together with increasing nutrients loads (N, P) from agricultural and domestic runoff resulted in a widespread eutrophication that triggered massive proliferation of epiphytes and other opportunistic macroalgae (e.g., *Ulva*) that outcompeted seagrasses by shading and suffocation (Hernández et al. 1997). The system shifted from being seagrass-dominated to algal-dominated and underwent a dystrophic crisis, since dissolved oxygen dropped drastically due to the microbial remineralisation of the massive algal biomass die off, affecting not only to its structure and functioning, but also to the incomes of some local fishermen.

Management and Conservation of Mediterranean Seagrasses

According to Short et al. (2011), the four seagrass species occurring in Mediterranean waters can be assigned to the "Least Concern" category of the IUCN Red List. However, there are some differences in population trends. Thus, whereas *Cymodocea nodosa* populations seem to remain stable (with an annual distribution change estimated in 0.6 % year^{-1}), the populations of the other three species are decreasing, mostly due to human perturbations, at estimated rates of −1.4 % year^{-1} for *Zostera marina* and −5 % year^{-1} (and −10 % over the last 100 years) for *Posidonia oceanica* (Marbà et al. 2005). No estimations are available for *Zostera noltii*. Although for *P. oceanica* this estimation is more than twice the value of global rate of decline in seagrass ecosystems (2 % year^{-1}) (Orth et al. 2006), it does not meet the threshold for the "Threatened" category of the IUCN Red List (Short et al. 2011). However, as it has been aforementioned, given the extremely slow growth rate of *P. oceanica*, such losses are virtually irreversible. Declines of *P. oceanica* meadows have been reported in many parts of the Mediterranean, including pristine areas where no apparent human impacts exist. Such widespread regression could be due to the existence of a background decline, possibly related to general changes in the climate of the Mediterranean Sea (Duarte et al. 1999), or to the cumulative effects of natural and anthropogenic local processes (González-Correa et al. 2007). Taking into account all the Mediterranean species the decrease in extension possibly ranges between 0 and 10 % (considering as a baseline the beginning of the twentieth century) with a manifestly growing trend associated to human impacts (see Boudouresque et al. 2009, for further details). In this context there is an urgent need to develop and implement different measures of protection, management and/or restoration to counterbalance the deterioration of Mediterranean seagrass habitats.

Table 9.2 Metrics used to determine the ecological status of seagrass meadows and of water bodies on the basis of seagrass characteristics

Species	Characteristics	Reference
Posidonia oceanica (Spain)	Physiological: phosphorus, nitrogen and sucrose content, and ^{15}N and ^{34}S isotopic ratios in rhizomes. Copper, lead and zinc concentration in rhizomes	Romero et al. (2007)
	Individual morphological status: % leaves with necrosis and shoot leaf surface	
	Population status: meadow cover, shoot density and % plagiotropic rhizomes	
	Community: nitrogen content in epiphytes	
	Shoot density, shoot foliar surface, dead-matte cover, meadow cover, herbivore pressure, rhizome baring/burial, foliar necrosis, percentage of plagiotropic rhizomes, and leaf-epiphyte biomass	Fernández-Torquemada et al. (2008)
	Lower limit depth, lower limit type	López y Royo et al. (2009)
	Shoot density, shoot length	
Cymodocea nodosa (Spain)	Physiological: phosphorus, nitrogen, ^{15}N and ^{34}S isotopic ratios in rhizomes. Cadmium, copper and zinc concentration in rhizomes	Oliva et al. (2011)
	Individual morphological status: shoot size	
	Population status: root weight ratio	
	Community: epiphytes load	
Multispecies (UK)	Change in taxonomic composition from reference conditions	Foden and Brazier (2007)
Zostera marina	Shoot density	
Z. angustifolia	Spatial extent	
Z. noltii		
Multispecies (NL)	Taxonomic composition	de Jonge in Foden and Brazier (2007)
Z. angustifolia	Area (bed extent)	
Z. noltii	Coverage (bed density)	

Protection and Management Measures

Because of the growing concern about the progressive regression of Mediterranean seagrass habitats they have special protection in most of the laws and treaties like OSPAR or EU's Habitats Directive (Dir 92/43/CEE). *Posidonia oceanica* is explicitly considered by the EU Habitats Directive; however, in many cases (e.g., *Cymodocea nodosa* and *Zostera* spp.) such protection is still ambiguous and seagrass species are not protected by themselves, but either through indirect references as vegetated shallow stands, or by protecting their habitat as the case of "coastal lagoons" or "sandbanks which are slightly covered by seawater all the time". On the contrary, the European Water Framework Directive (WFD) considers seagrasses itself as an indicator of water quality (see below).

Setting management measures for conservation and recovering of seagrass meadows is a challenge and can be a daunting task. In Europe, the ecology, sensitivity, conservation and management requirements of *Zostera* spp. have been documented as part of the UK Marine SAC project and of the action plan process at a UK level (Davison and Hughes 1998). For the Mediterranean, there are also some international initiatives as the Action Plan for the Conservation of Marine Vegetation in the Mediterranean Sea (Pergent-Martini and Le Ravallec 2007). The first step for managing seagrass habitats is to develop adequate monitoring programs to control the regression or, optimistically, the recovery of these communities following implementation of management plans. In Europe, the WFD has meant a boost in the development of indicators to establish the ecological status of water bodies and the reference conditions in transitional and coastal waters. Different indices and methodologies based on seagrasses as biological indicators (since they are considered as "shore canaries", Orth et al. 2006) have been proposed (e.g., Orfanidis et al. 2003; Krause-Jensen et al. 2005; Foden and de Jong 2007; Foden and Brazier 2007; López y Royo et al. 2009). Most of the conventional indices only consider the presence, coverage or density of seagrasses as an indicator of good ecological status; however, new indices have been recently proposed based on physiological, morphological, and structural characteristics of individuals and meadows, such as the POMI index for *Posidonia oceanica* (Romero et al. 2007; Montefalcone 2009), the CYMOX index for *Cymodocea nodosa* (Oliva et al. 2011) or the ZoNI index for *Zostera noltii* (García-Marín et al. 2013) (Table 9.2).

Mapping seagrass extent and structure (patchiness, corridors, etc.) is one of the actions recorded in all management plans. Bionomic cartographies were common in the 1980s and early 1990s in the Mediterranean. However, such monitoring programs based on evaluation of cover and shoot density are not a reliable assessment of decline, as they have a high associated error and only detect changes when substantial damages on seagrass meadows occur (Duarte 2002). Remote sensing techniques, using aerial photography (Pasqualini et al. 1998), satellite images (Barillé et al. 2010) or acoustic methods (Freitas et al. 2008), can help to monitor long or medium term changes in extension, but do not provide information on shoot density or ecological status; it is also quite difficult to differentiate seagrasses from benthic macroalgae.

The most difficult, and costly issue for implementing management plans is to reduce nutrient and sediment discharges from both diffuse and point sources in surrounding watersheds (Orth et al. 2006). To keep nutrient and organic matter inputs in a minimum, urban and industrial sewages must be diverted to treatment plants to reduce nutrient loads. In addition, uncultivated soils along rivers and wetlands (that intercept agricultural nutrient runoffs and reduce siltation) must be also protected (Díaz-Almela and Duarte 2008). It is also imperative to prevent loading from fish farms, which must not be allowed in bays, and at a reasonable (at least 800 m) distance in open waters (Marbà et al. 2006). The undesirable effects of outlets from desalination (i.e., brine discharges) and power plants, which may increase salinity and seawater temperature in the vicinity of the meadows, must be also avoided (Sánchez-Lizaso et al. 2008). Dredging activities must be also prevented in the meadow or its vicinity.

Efficient management strategies may also include the protection of seagrass meadows with artificial reefs against trawling, the use of seagrass-friendly moorings, the wise management of seagrass litter stranded in the beach, and the control of invasive species such as *Caulerpa* spp. (Díaz-Almela and Duarte 2008). The establishment of "marine protected areas" is one of the main management measures. Most of these initiatives have been adopted as a fisheries management tool, but they are also very effective protecting biodiversity of fishes and invertebrates including genetic structure of the populations (Pérez-Ruzafa et al. 2006). However, little research has been done on the effects of these areas on seagrass populations. Marbà et al. (2002) reported that patch formation and patch growth rates of *Posidonia oceanica* in active colonizing areas increased after implementation of mooring regulations in the Cabrera National Park (Spain). Similarly, the leaf production tended to increase, and vertical rhizome growth to decrease, following the onset of regulation measures.

Preservation of genetic diversity can be a more complex task. The knowledge of the spatial patterns and scales of the genetic structure of most of the seagrass species is very scarce (Hemminga and Duarte 2000). The available information suggests that spatial heterogeneity can be very high, mainly in species inhabiting coastal lagoons and semi enclosed seas.

Restoration Measures

Another important management measure is to restore damaged seagrass meadows. Habitat restoration can be used to compensate for losses, not only for seagrasses itself, but also for fish, invertebrates, and other aquatic fauna inhabiting meadows. Habitat restoration modelling approaches, based on food-chain transfers, showed that restoration of seagrass beds would be more productive even than that of the salt marshes (McCay and Rowe 2003). Numerous restoration projects have been attempted, mostly at small scales (<1 ha) with a variety of techniques using either adult plants or seeds (Orth et al. 2006). Worldwide, the success of seagrass transplantation and restoration is around 30 % (Fonseca et al. 1998). In the Mediterranean, the first experiences with *Posidonia oceanica* date back to the 1990s (Molenaar et al. 1993; Piazzi et al. 1998) with positive results at small scale (70–100 % survival after 1–3 years). However, transplantation of meadow blocks (1 m^2) form donor sites showed little success (Sánchez-Lizaso et al. 2009). Genetic polymorphism favours the transplantation success in this species (Procaccini and Piazzi 2001), but the low growth rate of *P. oceanica* preclude that transplantation techniques can be implemented to recover large areas at short or medium term. In addition, seagrass transplantation is too costly to be implemented at large scale, which is specially a major drawback in developing countries (Duarte 2002).

In conclusion, seagrass ecosystems are among the most productive of the Biosphere, providing important ecosystems goods and services. Despite this paramount importance, they (including the seagrass meadows of the Mediterranean Sea) are in a vulnerable state showing worrying trends of decline. The main causes in the Mediterranean are the anthropogenic non-climatic drivers. It is expected that climate change will also affect these habitats. However, it is rather speculative to forecast the outcome of such effect because to the likely multiple interactions among different non-climatic and climatic drivers. A challenging issue to cope with the ongoing deterioration must be an effective management, including programs to promote public awareness, to increase the resilience, especially against the non-climatic causes, of these important Mediterranean habitats.

Acknowledgements The authors would like to thank Juanma Ruiz (IEO, Murcia), José Antonio Rodríguez and EDEA Research group (RNM-214) for providing the photos. The financial support throughout the research projects CTM2008-0012/MAR, CTM2011-24482 and CGL08-05407-C03-01 by the Ministry of Science and Innovation of Spain is thankful. This is contribution 2 from the CEIMAR Books & Monographs.

References

Astier JM (1984) Impacts des amenagements littoraux de la Rade de Toulon, lies aux techniques d'endinage, sur les herbiers a Posidonia oceanica. In: Boudouresque CF, Jeudy de Grissac A, Olivier J (eds) International workshop on Posidonia oceanica beds 1. GIS Posidonie Publisher, France, pp 255–259

Ballesteros E, Cebrian J, Alcoverro T (2007) Mortality of shoots of *Posidonia oceanica* following meadow invasion by the red alga *Lophocladia lallemandii*. Bot Mar 50:8–13

Barillé L, Robin M, Harin N, Bargain A, Launeau P (2010) Increase in seagrass distribution at Bourgneuf Bay (France) detected by spatial remote sensing. Aquat Bot 92:185–194

Barrón C, Duarte CM, Frankignoulle M, Borges AV (2006) Organic carbon metabolism and carbonate dynamics in a Mediterranean seagrass (*Posidonia oceanica*) meadow. Estuaries Coasts 29:417–426

Bay D (1984) A field study of the growth dynamics and productivity of *Posidonia oceanica* (L.) delile in Calvi bay, Corsica. Aquat Bot 20:43–64

Borum J, Duarte CM, Krause-Jensen D, Greve TM (2004) European seagrasses: an introduction to monitoring and management. Monitoring and Managing of European Seagrasses (M&MS) project, Hillerød

Boudouresque CF, Bernard G, Bonhomme P, Charbonnel E, Diviacco G, Meinesz A, Pergent G, Pergent-martini C, Ruitton S, Tunesi L (2006) Préservation et conservation des herbiers à *Posidonia oceanica*. Ramoge, Monaco

Boudouresque CF, Bernard G, Pergent G, Shili A, Verlaque M (2009) Regression of Mediterranean seagrasses caused by natural processes and anthropogenic disturbances and stress: a critical review. Bot Mar 52:395–418. doi:10.1515/bot.2009.057

Brun FG, Hernández I, Vergara JJ, Peralta G, Pérez-Lloréns JL (2002) Assessing the toxicity of ammonium pulses to the survival and growth of *Zostera noltii*. Mar Ecol Prog Ser 225:177–187

Brun FG, Hernández I, Vergara JJ, Pérez-Lloréns JL (2003a) Growth, carbon allocation and proteolytic activity in the seagrass *Zostera noltii* shaded by Ulva canopies. Funct Plant Biol 30:551–560

Brun FG, Pérez-Lloréns JL, Hernández I, Vergara JJ (2003b) Patch distribution and within-patch dynamics of the seagrass *Zostera noltii* hornem. In Los Toruños salt-sarsh, Cádiz Bay, Natural Park, Spain. Bot Mar 46:513–524

Brun FG, Hernández I, Vergara JJ, Pérez-Lloréns JL (2005) Evidence for vertical growth in the seagrass *Zostera noltii*. Bot Mar 48:446–450

Brun FG, Vergara JJ, Peralta G, García-Sánchez MP, Hernández I, Pérez-Lloréns JL (2006) Clonal building, simple growth rules and phylloclimate as key steps to develop functional-structural seagrass models. Mar Ecol Prog Ser 323:133–148

Brun FG, Cummaudo F, Olivé I, Vergara JJ, Pérez-Lloréns JL (2007) Clonal extent, apical dominance and networking features in the phalanx angiosperm *Zostera noltii* Hornem. Mar Biol 151:1917–1927

Brunet M, Jones PD, Sigró J, Saladié P, Aguilar E, Moberg A, Della-Marta PM, Liste RD, Walther A, López D (2007) Temporal and spatial temperature variability and change over Spain during 1850–2005. J Geophys Res 112, D12117. doi:10.1029/2006JD008249

Burkett VR, Wilcox DA, Stottlemyer R, Barrow W, Fagre D, Baron J, Price J, Nielsen JL, Allen CD, Peterson DL, Ruggerone G, Doyle T (2005) Nonlinear dynamics in ecosystem response to climatic, change: case studies and policy implications. Ecol Complex 2:357–394

Cabaço S, Alexandre A, Santos R (2005) Population-level effects of clam harvesting on the seagrass *Zostera noltii*. Mar Ecol Prog Ser 298:123–129

Carreira J, Clavero V, Gálvez JA, Guevara J, Hernández I, Lucena J, Niell FX, Peralta G, Pérez-Lloréns JL, Rodríguez R, Vergara JJ (1995) Eutrophication progresses in the estuaries sited at the end of small catchment areas. In: Rijstenbil J, Kamermans P, Nienhuis PH (eds) EUMAC synthesis report. NIOO, Yerseke

Celebi B, Gucu AC, Ok M, Serdar S, Akoglu E (2007) Survival of the *Posidonia oceanica* cuttings transplanted into the northeastern Levant Sea. Rapp Comm Int Mer Medit 38:446

Charpy-Roubaud C, Sournia A (1990) The comparative estimation of phytoplanktonic and microphytobenthic production in the oceans. Mar Microb Food Webs 4:31–57

Costanza R, d'Arge R, de Groot R, Farber S, Grasso M, Hannon B, Limburg K, Naeem S, O'Neill RV, Paruelo J, Raskin RG, Sutton P, van der Belt M (1997) The value of the world's ecosystem services and natural capital. Nature 387:253–260

Davison DM, Hughes DJ (1998) Zostera Biotopes (volume I). An overview of dynamics and sensitivity characteristics for conservation management of marine SACs, Scottish Association for Marine Science (UK Marine SACs Project). http://www.ukmarinesac.org.uk/publications.htm. Accessed 10 Dec 2011

Díaz-Almela E, Duarte CM (2008) Management of Natura 2000 habitats. 1120 *Posidonia beds (Posidonion oceanicae) European Commission. http://ec.europa.eu/environment/nature/natura2000/management/habitats/models_en.htm. Accessed 23 Nov 2011

Díaz-Almela E, Marbà N, Duarte CM (2007) Fingerprints of Mediterranean Sea warming in seagrass (*Posidonia oceanica*) flowering records. Glob Change Biol 13:224–235

Díaz-Almela E, Marbà N, Martínez R, Santiago R, Duarte CM (2009) Seasonal dynamics of *Posidonia oceanica* in Magalluf Bay (Mallorca, Spain): temperature effects on seagrass mortality. Limnol Oceanogr 54:2170–2182

Duarte CM (1991a) Seagrass depth limits. Aquat Bot 40:363–377

Duarte CM (1991b) Allometric scaling of seagrass form and productivity. Mar Ecol Prog Ser 77:289–300

Duarte CM (2002) The future of seagrass meadows. Environ Conserv 29:192–206

Duarte CM (2004) Las praderas de fanerógamas marinas. Las praderas de *Posidonia oceanica*. Algunos métodos y modelos en el estudio y la evaluación del estado de las praderas. Lepidocronología, crecimiento y edad de las praderas. In: Luque AA, Templado J (eds) Praderas y bosques marinos de Andalucía. Consejería de Medio Ambiente, Junta de Andalucía, Sevilla

Duarte CM, Cebrián J (1996) The fate of autotrophic production in the sea. Limnol Oceanogr 41:1758–1766

Duarte CM, Chiscano CL (1999) Seagrass biomass and production: a reassessment. Aquat Bot 65:159–174

Duarte CM, Sand-Jensen K (1990) Seagrass colonization – biomass development and shoot demography in *Cymodocea nodosa* patches. Mar Ecol Prog Ser 67:97–103

Duarte CM, Agustí S, Kennedy H, Vaqué D (1999) The Mediterranean climate as a template for Mediterranean marine ecosystems: the example of the northeast Spanish littoral. Prog Oceanogr 44: 245–270

Duarte CM, Marbà N, Santos R (2004) What may cause loss of seagrass? In: Borum J, Duarte CM, Krause-Jensen D, Greve TM (eds) European seagrasses: an introduction to monitoring and management. Monitoring and Managing of European Seagrasses (M&MS) Project, Hillerød

Duarte CM, Middelburg JJ, Caraco N (2005) Major role of marine vegetation on the oceanic carbon cycle. Biogeosciences 2:1–8

Enríquez S, Rodríguez-Román A (2006) Effect of water flow on the photosynthesis of three marine macrophytes from a fringing-reef lagoon. Mar Ecol Prog Ser 323:119–132

Enríquez S, Marbà N, Cebrián J, Duarte CM (2004) Annual variation in leaf photosynthesis and leaf nutrient content of four Mediterranean seagrasses. Bot Mar 47:295–306

Fernández-Torquemada Y, Díaz-Valdés M, Colilla F, Luna B, Sánchez-Lizaso JL, Ramos-Esplá AA (2008) Descriptors from Posidonia oceanica (L.) Delile meadows in coastal waters of Valencia, Spain, in the context of the EU Water Framework Directive. ICES J Mar Sci 65:1492–1497

Figueroa FL, Jiménez C, Viñegla B, Pérez-Rodríguez E, Aguilera J, Flores-Moya A, Atamirano A, Lebert M, Häder DP (2002) Effects of solar UV radiation on photosynthesis of the marine angiosperm *Posidonia oceanica* from southern Spain. Mar Ecol Prog Ser 230:59–70

Flagella S, Borriello I, Gambi MC, Buia MC (2006) Responses of *Posidonia oceanica* to environmental disturbance. Biol Mar Mediterr 13:215–219

Foden J, Brazier DP (2007) Angiosperms (seagrass) within the EU water framework directive: a UK perspective. Mar Pollut Bull 55:181–195

Foden J, de Jong DJ (2007) Assessment metrics for littoral seagrass under the European Water Framework Directive; outcomes of UK intercalibration with the Netherlands. Hydrobiologia 579: 187–197

Fonseca MS, Kenworthy WJ, Thayer GW (1998) Guidelines for the conservation and restoration of seagrasses in the United States and adjacent waters. NOAA Coastal Ocean Office, Silver Spring

Francour P, Ganteaume A, Poulain M (1999) Effects of boat anchoring in *Posidonia oceanica* seagrass beds in the Port-cros national Park (North-western Mediterranean Sea). Aquat Conserv 9: 391–400

Freitas R, Rodrigues AM, Morris EP, Pérez-Lloréns JL, Quintino V (2008) Single-beam acoustic ground discrimination of shallow water habitats: 50 kHz or 200 kHz frequency survey? Estuar Coast Shelf S 78:613–622

French PW (1997) Coastal and estuarine management. Routledge, London

García-Marín P, Cabaço S, Hernández I, Vergara JJ, Silva J, Santos R (2013) Multi-metric index based on the seagrass Zostera noltii (ZoNI) for ecological quality assessment of coastal and estuarine systems in SW Iberian Peninsula. Mar Pollut Bull 68:46–54

Giraud G (1977) Recensement des floraisons de *Posidonia oceanica* (L.) Delile en Méditerranée. Rapp Comm Int Mer Médit 24:127–130

Gobert S, Cambridg M, Velimirov G, Lepoint G, Bouquegnau JM, Dauvy P, Pergent-Martini C, Walker DA (2006) Biology of *Posidonia*. In: Larkum AWD, Orth RJ, Duarte CM (eds) Seagrasses: biology, ecology, and conservation. Springer, Dordrecht

González-Correa JM, Bayle-Sempere JT, Sánchez-Lizaso JL, Valle C, Sánchez-Jerez P, Ruíz JM (2005) Recovery of deep *Posidonia oceanica* meadows degraded by trawling. J Exp Mar Biol Ecol 320:65–76

González-Correa JM, Bayle-Sempere JT, Sánchez-Jerez P, Valle C (2007) *Posidonia oceanica* meadows are not declining globally. Analysis of population dynamics in marine protected areas of the Mediterranean Sea. Mar Ecol Prog Ser 336:111–119

Green EP, Short FT (2003) World atlas of seagrasses. University of California Press, London

Hellblom F, Beer S, Björk M, Axelsson L (2001) A buffer sensitive inorganic carbon utilisation system in *Zostera marina*. Aquat Bot 69:55–62

Hemminga MA, Duarte CM (2000) Seagrass ecology. Cambridge University Press, Cambridge

Hernández I, Peralta G, Pérez-Lloréns JL, Vergara JJ, Niell FX (1997) Biomass and dynamics of growth of *Ulva* species in Palmones river estuary. J Phycol 33:764–772

Holmer M, Pérez M, Duarte CM (2003) Benthic primary producers – a neglected environmental problem in Mediterranean maricultures. Mar Pollut Bull 46:1372–1376

Invers O, Perez M, Romero J (1999) Bicarbonate utilization in seagrass photosynthesis: role of carbonic anhydrase in *Posidonia oceanica* (L.) Delile and *Cymodocea nodosa* (Ucria) Ascherson. J Exp Mar Biol Ecol 235:125–133

IPCC (1990) Climate change. The IPCC impacts assessment. Australian Government Publishing Service, Canberra

IPCC (2007) Climate change 2007: synthesis report. Contribution of working groups I, II and III to the fourth assessment report of the intergovernmental panel on climate change. IPCC, Geneva

Jackson EL, Glanfield SC, Delacruz J, Langmead O, Arvanitidis C, Attrill MK, Tyler-Walters H, Mee L (2006) An objective approach to defining loss of seagrass in the Mediterranean: causal chain analysis and scenario predictions. Biol Mar Mediterr 13:240–244

James PL, Larkum AWD (1996) Photosynthetic inorganic carbon acquisition of *Posidonia australis*. Aquat Bot 55:149–157

Jaubert JM, Chisholm JRM, Ducrot D, Ripley HT, Roy L, Passeron-Seitre G (1999) No deleterious alterations in *Posidonia* beds in the bay of Menton (France) eight years after *Caulerpa taxifolia* colonization. J Phycol 35:1113–1119

Klein J, Verlaque M (2008) The *Caulerpa racemosa* invasion: a critical review. Mar Pollut Bull 56:205–225

Krause-Jensen D, Greve TM, Nielsen K (2005) Eelgrass as bioindicator under the European water framework directive. Water Resour Manag 19:63–75

Larkum AWD, Orth RJ, Duarte CM (2006) Seagrasses: biology, ecology and conservation. Springer, Dordrecht

Laugier T, Rigollet V, de Casabianca ML (1999) Seasonal dynamics in mixed eelgrass beds, *Zostera marina* L. and *Z. noltii* Hornem., in a Mediterranean coastal lagoon (Thau lagoon, France). Aquat Bot 63:51–69

Lee K-S, Park SR, Kim YK (2007) Effects of irradiance, temperature, and nutrients on growth dynamics of seagrasses: a review. J Exp Mar Biol Ecol 350:144–175

López y Royo C, Silvestri C, Salivas-Decaux M, Pergent G, Casazza G (2009) Application of an angiosperm-based classification system (BiPo) to Mediterranean coastal waters: using spatial analysis and data on metal contamination of plants in identifying sources of pressure. Hydrobiologia 633:169–179

Lotze HK, Lenihan HS, Bourque BJ, Bradbury RH, Cooke RG, Kay MC, Kidwell SM, Kirby MX, Peterson CH, Jackson JBC (2006) Depletion, degradation, and recovery potential of estuaries and coastal seas. Science 312:1806–1809

Manzanera M, Alcoverro T, Romero J (1998) The role of the remnant leaf sheaths (scales) attached to the *Posidonia oceanica* (L.) Delile rhizomes. J Exp Mar Biol Ecol 223:257–270

Marbà N (2009) Loss of seagrass meadows from the Spanish coasts: results of the praderas project. In: Duarte CM (ed) Global loss of coastal habitats: rates, causes and consequences. Fundación BBVA, Bilbao

Marbà N, Duarte CM (1994) Growth response of the seagrass *Cymodocea nodosa* to experimental burial and erosion. Mar Ecol Prog Ser 107:307–311

Marbà N, Terrados J (2004) Las praderas de fanerógamas marinas. Las praderas de *Cymodocea nodosa* Distribución y requerimientos ecológicos. In: Luque AA, Templado J (eds) Praderas y bosques marinos de Andalucía. Consejería de Medio Ambiente, Junta de Andalucía, Sevilla

Marbà N, Cebrián J, Enríquez S, Duarte CM (1996) Growth patterns of Western Mediterranean seagrasses: species-specific responses to seasonal forcing. Mar Ecol Prog Ser 133:203–215

Marbà N, Duarte CM, Holmer M, Martinez R, Basterretxea G, Orfila A, Jordi A, Tintoré J (2002) Effectiveness of protection of seagrass (*Posidonia oceanica*) populations in Cabrera National Park (Spain). Environ Conserv 29:509–518

Marbà N, Duarte CM, Díaz-Almela E, Terrados J, Álvarez E, Martínez R, Santiago R (2005) Direct evidence of imbalanced seagrass (*Posidonia oceanica*) shoot population dynamics along the Spanish Mediterranean. Estuaries 28:51–60

Marbà N, Santiago R, Díaz-Almela E, Álvarez E, Duarte CM (2006) Seagrass (*Posidonia oceanica*) vertical growth as an early indicator of fish farm-derived stress. Estuar Coast Shelf S 67:475–483

Marcos M, Tsimplis MN, Shaw AGP (2009) Sea level extremes in southern Europe. J Geophys Res 114:1–16

Martín MA, Sánchez-Lizaso JL, Ramos Esplá AA (1997) Cuantificación del impacto de las artes de arrastre sobre la pradera de *Posidonia oceanica* (L.) Delile, 1813. Publ Espec Inst Esp Oceanogr 23: 243–253

Mateo MA, Romero J, Pérez M (1997) Dynamics of millenary organic deposits resulting from the growth of the Mediterranean seagrass *Posidonia oceanica*. Estuar Coast Shelf S 44:103–110

Mayhoub H (1976) Recherches sur la végétation marine de la côte Syrienne. Etude expérimentale sur la morphogénèse et le développement de quelques espèces peu connues. PhD Thesis, University of Caen

McCay DPF, Rowe JJ (2003) Habitat restoration as mitigation for lost production at multiple trophic levels. Mar Ecol Prog Ser 264: 233–247

Meinesz A, Cirik S, Akcali B, Javel F, Migliaccio M, Thibaut T, Yuksek A, Procaccini G (2009) Posidonia oceanica in the Marmara Sea. Aquat Bot 90:18–22

Mercado JM (2011) Physiological basis for the use of seaweeds as indicators of anthropogenic pressures: the case of green tides. In: Kim S-K (ed) Handbook of marine macroalgae. Wiley-Blackwell, Hoboken

Mercado JM, Niell FX, Silva J, Santos R (2003) Use of light and inorganic carbon acquisition by two morphotypes of *Zostera noltii* Hornem. J Exp Mar Biol Ecol 297:71–84

Millán MM, Estrela MJ, Sanz MJ, Mantilla E, Martín M, Pastor F, Salvador R, Vallejo R, Alonso L, Gangoiti G, Ilardia JL, Navazo M, Albizuri A, Artíñano B, Ciccioli P, Kallos G, Carvalho RA, Andrés D, Hoff A, Werhahn J, Seufert G, Versino B (2005) Climatic feedbacks and desertification: the Mediterranean model. J Climate 18:684–701. doi:10.1175/jcli-3283.1

Molenaar H, Meinesz A, Caye G (1993) Vegetative reproduction in *Posidonia oceanica* – survival and development in different morphological types of transplanted cuttings. Bot Mar 36:481–488

Montefalcone M (2009) Ecosystem health assessment using the Mediterranean seagrass *Posidonia oceanica*: a review. Ecol Indic 9:595–604

Oliva S, Mascaró O, Llagostera I, Pérez M, Romero J (2011) Selection of metrics based on the seagrass *Cymodocea nodosa* and development of a biotic index (CYMOX) for assessing ecological status of coastal and transitional waters. Estuar Coast Shelf S 114:7–17. doi:10.1016/j.ecss.2011.08.022

Orfanidis S, Panayotidis P, Stamatis N (2003) An insight to the ecological evaluation index (EEI). Ecol Indic 3:27–33

Orth RJ, Carruthers TJB, Dennison WC, Duarte CM, Fourqurean JW, Heck KL, Hughes JAR, Kendrick GA, Kenworthy WJ, Olyarnik S, Short FT, Waycott M, Williams SL (2006) A global crisis for seagrass ecosystems. Bioscience 56:987–996

Pasqualini V, Pergent-Martini C, Clabaut P, Pergent G (1998) Mapping of *Posidonia oceanica* using aerial photographs and side scan sonar: application off the Island of Corsica (France). Estuar Coast Shelf S 47:359–367

Peralta G, Pérez-LLoréns JL, Hernández I, Vergara JJ (2002) Effects of light availability on growth, architecture and nutrient content of the seagrass *Zostera noltii* Hornem. J Exp Mar Biol Ecol 269: 9–26

Peralta G, Brun FG, Hernandez I, Vergara JJ, Pérez-Lloréns JL (2005) Morphometric variations as acclimation mechanisms in *Zostera noltii* beds. Estuar Coast Shelf S 64:347–356

Pérez-Lloréns JL (2004a) Las praderas de fanerógamas marinas. Las praderas de *Zostera marina*. In: Luque AA, Templado J (eds) Praderas y bosques marinos de Andalucía. Consejería de Medio Ambiente, Junta de Andalucía, Sevilla

Pérez-Lloréns JL (2004b) Las praderas de fanerógamas marinas. Las praderas de *Zostera noltii*. In: Luque AA, Templado J (eds) Praderas y bosques marinos de Andalucía. Consejería de Medio Ambiente, Junta de Andalucía, Sevilla

Pérez-Lloréns JL, Niell FX (1993) Seasonal dynamics of biomass and nutrient content in the intertidal seagrass *Zostera noltii* Hornem. From Palmones River estuary, Spain. Aquat Bot 46:49–66

Pérez-Lloréns JL, Hernández I, Bermejo R, Peralta G, Brun F, Vergara JJ (2012) Flora marina del litoral gaditano. Biología, ecología, usos y guía de identificación. Universidad de Cádiz, Cadiz

Pérez-Ruzafa A, González-Wangüemert M, Lenfant P, Marcos C, García-Charton JA (2006) Effects of fishing protection on the genetic structure of fish populations. Biol Conserv 129:244–255

Pergent G, Romero J, Pergent-Martini C, Mateo MA, Boudouresque CF (1994) Primary production, stocks and fluxes in the Mediterranean seagrass *Posidonia oceanica*. Mar Ecol Prog Ser 106:139–146

Pergent-Martini C, Le Ravallec C (2007) Guidelines for impact assessment on seagrass meadows. United Nations Environment Programme, Tunis

Piazzi L, Balestri E, Magri M, Cinelli F (1998) Experimental transplanting of *Posidonia oceanica* (L.) Delile into a disturbed habitat in the Mediterranean sea. Bot Mar 41:593–601

Piazzi L, Balata D, Cinelli F (2001) Incidence des Rhodophyceae exotiques *Acrothamnion preissii* et *Womersleyella setacea* sur le peuplement algal des rhizomes de *Posidonia oceanica*. In: Gravez V, Ruitton S, Boudouresque CF et al (eds) Fourth International workshop on *Caulerpa taxifolia*. GIS Posidonie, Marseille

Procaccini G, Piazzi L (2001) Genetic polymorphism and transplantation success in the mediterranean seagrass *Posidonia oceanica*. Restor Ecol 9:332–338

Procaccini G, Ruggiero MV, Orsini L (2002) Genetic structure and distribution of microsatellite population genetic diversity in *Posidonia oceanica* in the Mediterranean Basin. Bull Mar Sci 71:1291–1297

Reusch TBH, Borström C, Stam WT, Olsen JL (1999) An ancient eelgrass clone in the Baltic. Mar Ecol Prog Ser 183:301–304

Romero J (2004) Las praderas de fanerógamas marinas. La producción primaria y su destino. Características de los restos de la planta. In: Luque AA, Templado J (eds) Praderas y bosques marinos de Andalucía. Consejería de Medio Ambiente, Junta de Andalucía, Sevilla

Romero J, Pérez M, Mateo MA, Sala E (1994) The belowground organs of the Mediterranean seagrass *Posidonia oceanica* as a biogeochemical sink. Aquat Bot 47:13–19

Romero J, Martinez-Crego B, Alcoverro T, Pérez M (2007) A multivariate index based on the seagrass *Posidonia oceanica* (POMI) to assess ecological status of coastal waters under the water framework directive (WFD). Mar Pollut Bull 55:196–204

Ruíz JM, Romero J (2003) Effects of disturbances caused by coastal constructions on spatial structure, growth dynamics and photosynthesis of the seagrass *Posidonia oceanica*. Mar Pollut Bull 46: 1523–1533

Ruíz JM, Pérez M, Romero J (2001) Effects of fish farm loadings on seagrass (*Posidonia oceanica*) distribution, growth and photosynthesis. Mar Pollut Bull 42:749–760

Sánchez-Lizaso JL (2004) Las praderas de fanerógamas marinas. Las praderas de *Posidonia oceanica*. In: Luque AA, Templado J (eds) Praderas y bosques marinos de Andalucía. Consejería de Medio Ambiente, Junta de Andalucía, Sevilla

Sánchez-Lizaso JL, Romero J, Ruiz J, Gacia E, Buceta JL, Invers O, Fernández-Torquemada Y, Mas J, Ruiz-Mateo A, Manzanera M (2008) Salinity tolerance of the Mediterranean seagrass *Posidonia oceanica*: recommendations to minimize the impact of brine discharges from desalination plants. Desalination 221:602–607

Sánchez-Lizaso JL, Fernández-Torquemada Y, González-Correa JM (2009) Evaluation of the viability of *Posidonia oceanica* transplants associated with a marina expansion. Bot Mar 52:471–476. doi:10.1515/bot.2009.052

Scavia D, Field JC, Boesch DF, Buddemeier R, Cayan DR, Burkett V, Fogarty M, Harwell M, Howarth R, Mason C, Reed DJ, Royer TC,

Sallenger AH, Titus JG (2002) Climate change impacts on U.S. coastal and marine ecosystems. Estuaries 25:149–164

Schneider A, Wallace DWR, Kortzinger A (2007) Alkalinity of the Mediterranean Sea. Geophys Res Lett 34, L15608. doi:10.1029/2006gl028842

Short FT, Wyllie-Echeverria S (1996) Natural and human induced disturbance of seagrasses. Environ Conserv 23:17–27

Short FT, Polidoro B, Livingstone SR, Carpenter KE, Bandeira S, Bujang JS, Calupong HP, Carruthers TJB, Coles RG, Dennison WC, Erftemeijer PLA, Fortes MD, Freeman AS, Jagtap TG, Kamal AHM, Kendrick GA, Kenworthy WJ, La Nafie YA, Nasution IM, Orth RJ, Prathep A, Sanciangco JC, van Tussenbroek B, Vergara SG, Waycott M, Zieman JC (2011) Extinction risk assessment of the world's seagrass species. Biol Conserv 144: 1961–1971

Spalding M, Taylor M, Ravilious C, Short F, Green E (2003) Global overview: the distribution and status of seagrasses. In: Green E, Short F (eds) World atlas of seagrasses: present status and future conservation. University of California Press, Berkeley

Terrados J (1993) Sexual reproduction and sex banks of *Cymodocea nodosa* (Ucria) Ascherson meadows on the southeast Mediterranean coast of Spain. Aquat Bot 46:293–299

Terrados J, Marbà N (2004) Las praderas de fanerógamas marinas. Las praderas de *Cymodocea nodosa*. Características morfológicas. In: Luque AA, Templado J (eds) Praderas y bosques marinos de Andalucía. Consejería de Medio Ambiente, Junta de Andalucía, Sevilla

Vaquer-Sunyer R, Duarte CM (2010) Sulfide exposure accelerates hypoxia-driven mortality. Limnol Oceanogr 55:1075–1082. doi:10.4319/lo.2010.55.3.1075

Vicente-Serrano SM (2007) Evaluating the impact of drought using remote sensing in a Mediterranean, semi-arid region. Nat Hazards 40:173–208

Villèle X, Verlaque M (1995) Changes and degradation in a *Posidonia oceanica* bed invaded by the introduced tropical alga *Caulerpa taxifolia* in the North Western Mediterranean. Bot Mar 38:79–87

Waycott M, Duarte CM, Carruthers TJB, Orth RJ, Dennison WC, Olyarnik S, Calladine A, Fourqurean JW, Heck KL, Hughes AR, Kendrick GA, Kenworthy WJ, Short FT, Williams SL (2009) Accelerating loss of seagrasses across the globe threatens coastal ecosystems. Proc Natl Acad Sci 106:12377–12381. doi:10.1073/pnas.0905620106

Williams SL (2007) Introduced species in seagrass ecosystems: status and concerns. J Exp Mar Biol Ecol 350:89–110

Zimmerman RC, Alberte RS (1996) Effect of light/dark transition on carbon translocation in eelgrass *Zostera marina* seedlings. Mar Ecol Prog Ser 136:305–309

Zimmerman RC, Kohrs DG, Steller DL, Alberte RS (1997) Impacts of CO_2 enrichment on productivity and light requirements of eelgrass. Plant Physiol 115:599–607

Alien Benthic Algae and Seagrasses in the Mediterranean Sea and Their Connection to Global Warming

10

Razy Hoffman

Abstract

Nearly 140 alien benthic-algae taxa and one seagrass invader species were reported found in the Mediterranean Sea since the beginning of the twentieth century. Some caused major changes in the community structure and function of marine flora and fauna. Most non-indigenous species reached the Mediterranean due to anthropogenic activities. The opening of the Suez Canal, along with the increase of marine transportation and mariculture, enabled alien invasion from the Atlantic, Indian, and Pacific Oceans. Alien algae from the Atlantic Ocean probably invade the Mediterranean via the Straits of Gibraltar through cargo ships (through ballast water or attached to the hulls). Shellfish transfer is the most important vector for introducing alien seaweed from the Far East into the western Mediterranean. Global warming raised the surface water temperature, especially in the eastern Mediterranean Basin; therefore, it plays an important role in facilitating settlement establishment and the spread of alien algae and seagrasses from the tropical Atlantic, Indian, and Pacific Oceans in the Mediterranean. The control and reduction of marine invasions are also discussed.

Keywords

Mediterranean Sea • Alien species • Invasive species • Seaweed • Seagrass • Global warming • Introduction

Alien Invasion in the Mediterranean

More than 50 years ago, the English zoologist and ecologist Charles Sutherland Elton claimed that the spread of invasive species should be treated as one of the great historical convulsions in the world's fauna and flora (Elton 1958; Perrings 2010). Since then, invasion phenomena increased rapidly and invasive species became a major threat to fauna, flora, and even many ecosystems all over the world (Lowe et al. 2000). As studies indicated that in many cases invasion was not reversible, it is sometimes regarded as an irreversible type of pollution by certain scientists who study this phenomenon (Boudouresque and Verlaque 2005; Perrings 2010).

R. Hoffman (✉)
The George S. Wise Faculty of Life Sciences,
Tel-Aviv University, Tel-Aviv 69978, Israel
e-mail: razyho@hotmail.com

Biological invasion is the result of species introduction. An Introduced established species must meet four criteria: (1) it colonizes a new area where it did not previously occur; (2) the extension of its range is directly or indirectly linked to human activity; (3) there is no geographical continuity between its native area and the new area; and (4) new generations of the nonnative species are born *in situ* without human assistance, thus constituting self-sustaining populations (Boudouresque et al. 2005).

The significance of invasive species (also known as alien, exotic, nonindigenous, introduced, or nonnative species) in marine ecosystems worldwide has been highlighted and discussed intensively in recent years from ecological, environmental, and economic points of view (Wilcove et al. 1998; Dukes and Mooney 1999; Pimental et al. 2000; Davis et al. 2001; Streftaris et al. 2005). These invasive species often rapidly adapt to their new environments (Huey et al. 2000; Mooney and Cleland 2001), increase in abundance due to

S. Goffredo and Z. Dubinsky (eds.), *The Mediterranean Sea: Its history and present challenges*,
DOI 10.1007/978-94-007-6704-1_10, © Springer Science+Business Media Dordrecht 2014

high proliferation (Richardson et al. 2000; Kolar and Lodge 2001; Colautti and MacIsaac 2004), play conspicuous roles in the recipient ecosystems, and sometimes even displace keystone species (Boudouresque and Verlaque 2002) and, therefore, threaten native biological diversity (IUCN 2002). Some of these species can become economically harmful and sometimes even threaten human health (EPA 2001; Hoffman and Dubinsky 2010). The Success of invaders is, however, far from predictable (Richardson et al. 2000). Some studies indicated that only a small fraction of aliens become established, and generally nearly 1 % of these become pests (Williamson 1996).

About 6 % of the total species reported so far from the Mediterranean Sea are aliens and the overall inconceivable number of these non-indigenous species has almost reached 1,000, with 718 alien species reported from the eastern Mediterranean Sea, 328 from the western Mediterranean, 267 from the central Mediterranean, and 171 from the Adriatic Sea (Zenetos et al. 2010). The origins of these alien are mainly from the Indian, Pacific or Atlantic Oceans (Zenetos et al. 2005, 2008; Hoffman and Dubinsky 2010).

The Mediterranean Sea has experienced a massive invasion of hundreds of marine alien species [of Indo-Pacific origin], including Protozoa (Forminifera), algae, Polychaeta, Crustacea, Mollusca, Miscellaneous Invertebrata (Arthropoda Pycnogonida, Bryozoa, Chaetognatha, Chordata Ascidiacea, Cnidaria, Ctenophora, Echinodermata, Porifera, Sipuncula, Nematoda, and Platyhelminthes), and fish species, over a period of only 142 years since the opening of the Suez Canal in Egypt in November 1869 (Goren and Aronov 2002; Galil 2006a; Hoffman and Dubinsky 2010; Zenetos et al. 2010). This invasion is called Lessepsian migration and takes place mostly from the Red Sea to the Mediterranean. However, several species migrated in the opposite direction (anti-Lessepsian migration) (Golani and Ben-Tuvia 1995; Golani et al. 2002). The factors that facilitated the mainly northward transport to the Mediterranean were well described by Por (1971, 1978), who claimed that high salinity and high temperature (probably also as a result of global warming) in the Levant Basin of the Mediterranean increase the chances of success of Red Sea invaders. The fact that the Red Sea contains far more species when compared with the temperate fauna in the Mediterranean also contributes to the predominance of Red Sea organisms among Lessepsian migrants. The fact that Indo-Pacific organisms show more versatile adaptation to various ecological niches helps them compete successfully with Mediterranean species and occupy vacant ecological niches as well (Por 1971). It is now estimated that nearly 60 % of the alien species in the Mediterranean Sea passed through the Suez Canal (Zenetos et al. 2010).

Other reasons for the northern transport that support species migration are the winds and the northward water current leading from the southern end of the canal to its northern end (Hoffman and Dubinsky 2010). International maritime traffic also plays a very important role (as a transportation vector), conveying marine species all around the world through ships' ballast water (Nehring 2002, 2005; Streftaris et al. 2005) or attached to ships' hull fouling (Boudouresque and Verlaque 2010; Hoffman and Dubinsky 2010; Zenetos et al. 2010). Cargo ships pass through the Suez Canal and introduce aliens into the eastern Mediterranean.

The invasion of alien species to the western and central Mediterranean is a bit different. Although some alien species spread from the eastern to the central and western Mediterranean or arrived by ship directly from the Indo-Pacific, other alien species entered the western Mediterranean from the Atlantic Ocean through the Straits of Gibraltar. Moreover, many alien species in the central and especially the west Mediterranean arrived through shellfish transfer and mariculture (Boudouresque et al. 2011).

Alien Algae and Seagrasses in the Mediterranean Sea

During the twentieth century, the number of exotic alien marine macroalgal species found in the Mediterranean Sea nearly doubled every 20 years (Richardson et al. 2000; Boudouresque and Verlaque 2002). One hundred and thirty nine alien benthic alga and seagrass taxa from the Mediterranean Sea have been identified and reported so far (Table 10.1). Of these, 6 belong to the Cyanobacteria, 23 to the Chlorophyta, 81 to the Rhodophyta, 28 to Ochrophyta, and 1 is a seagrass species (*Halophila stipulacea*), belonging to the Tracheophyta phylum (Table 10.1). The highest numbers of these alien were reported from the western and eastern Mediterranean, with 92 and 78 taxa, respectively, while lower numbers were found in the central Mediterranean and Adriatic Sea, with 57 and 49 taxa, respectively (Zenetos et al. 2010; Table 10.1).

According to Streftaris and Zenetos (2006), 19 of these alien species have caused obvious biodiversity damage and are included in the list of the 100 worst invasive species in the Mediterranean (European Environmental Agency 2007), which means that 20 % of the worst alien species in the Mediterranean are marine macrophytes! However, the list of invasive seaweed species has grown and today (in 2012) we are talking about 21–29 species of marine macrophytes (seaweeds and seagrass) that should be regarded as invasive (see Table 10.1 and Zenetos et al. 2010). These invaders show negative effects on local Mediterranean marine flora and, therefore, threaten its diversity.

Rhodophyta phylum contributes the highest percentage (~60 %) of the reported alien algal taxa entering the Mediterranean (Fig. 10.1), however, distribution data for the indigenous Mediterranean seaweeds from the Chlorophyta,

Table 10.1 List of alien benthic algae and seagrasses reported to be found in the Mediterranean Sea

Alien taxon	Vector	Origin	Status
Cyanobacteria			
Ocillatoriaceae			
Lyngbya anomala (C.B.Rao) Umezaki & Watanabe[1]	?	IP[2]	?
Oscillatoria bonnemaisonii (P.L.Crouan & H.M.Crouan) P.L.Crouan & H.M.Crouan ex Gomont[3]	?	IP&At[2]	?
Oscillatoria miniata (Zanardini) Hauck ex Gomont[4]	?	IP[2]	?
Phormidiaceae			
Coleofasciculus chthonoplastes (Gomont) M.Siegesmund, J.R.Johansen & T.Friedl[3]	?	At&IP[2]	?
Microcoleus chthonoplastes (Mertens) Zanardini[3, 5]	?	At&IP[2]	?
Trichodesmium erythraeum Ehrenberg ex Gomont[4]	?	IP[2]	?
Chlorophyta			
Boodleaceae			
Boodlea composita (Harvey) F. Brand[6]	?	IP[2]	CA[6]
Caulerpaceae			
Caulerpa distichophylla Sonder[7]	SH?	IP[7]	IN[7]
C. mexicana Sonder ex Kützing[6, 7, 8]	SC[8]	IP[7]	ES[6, 7]
C. racemosa **var.** *cylindracea* (Sonder) Verlaque, Huisman & Boudouresque[7, 8]	AQ,SH[8]	IP[7]	IN[9]
C. racemosa **var.** lamourouxii f. requienii (Montagne) Weber-van Bosse[6, 7, 8]	SC[8]	IP[7]	ES[7, 8]
C. racemosa **var.** *turbinata* (J. Agardh) Eubank[7, 8]	SC[8]	IP[7]	QU[7]
C. scalpelliformis (Brown ex Turner) C. Agardh[6, 7, 8]	SC[8]	IP[7]	ES[7]
C. taxifolia (M.Vahl) C. Agardh[7, 8]	AQ[8]	IP[7]	IN[9]
Cladophoraceae			
Cladophora herpestica (Montagne) Kützing[7, 8]	SC[8]	IP[7]	ES[7]
C. hutchinsioides Hoek & Womersley[7]	?	IP[7]	ES[7]
C. patentiramea (Montagne) Kützing[7, 8]	SC,SH[8]	IP[7]	IN[7]
Codiaceae			
Codium arabicum Kützing[6, 10]	SH[10]	IP[10]	ES[10]
C. fragile **subsp.** *fragile* (Suringar) Hariot[7, 8]	SH,OY[8]	IP[7]	IN[9]
C. parvulum (Bory ex Audouin) P.C.Silva[6, 7, 10, 11]	SH[10]	IP[10]	IN[10, 11]
C. taylorii P.C.Silva[6, 7, 8]	SC	IP[7]	ES[6, 7, 8]
Dasycladaceae			
Batophora sp.[7]	?	At[7]	QU[7]
Neomeris annulata Dickie[6, 7, 8]	SC[8]	IP[7]	ES[7]
Derbesiaceae			
Derbesia boergesenii (Iyengar & Ramanathan) Mayhoub[7, 8]	SC[8]	IP[7]	CA[7]
D. rhizophora Yamada[7, 8]	OY[8]	IP[7]	ES[7]
Udoteaceae			
Boodleopsis sp.[6]	SC	IP[2]	CA[7]
Ulvaceae			
Ulva fasciata Delile[7, 8]	OY[8]?	IP[7]	ES[7]
U. pertusa Kjellman[7, 8]	OY[8]	IP[7]	IN[7]
Ulvaria obscura (Kützing) P.Gayral ex C, Bliding[7, 8]	OY[8]	IP[7]	ES[7]
Rhodophyta			
Acrochaetiaceae			
Acrochaetium codicola Børgesen[7, 8]	SH,OY[8]	IP&At[7]	ES[7]
A. robustum Børgesen[7, 8]	SC[8]	IP[7]	CA[7]
A. spathoglossi Børgesen[7, 8]	SC[8]	IP[7]	CA[7]
A. subseriatum Børgesen[7, 8]	SC[8]	IP[7]	CA[7]

(continued)

Table 10.1 (continued)

Alien taxon	Vector	Origin	Status
Areschougiaceae			
Sarconema filiforme (Sonder) Kylin[6, 7, 8, 12]	SC[8]	IP[7]	IN[6, 12]
S. scinaioides Børgesen[7, 8]	SC[8]	IP[7]	ES[7]
Bangiaceae			
Porphyra yezoensis Ueda[7, 8]	OY[8]	IP[7]	ES[7]
Bonnemaisoniaceae			
Asparagopsis armata Harvey[7, 8]	SH[8]	IP[7]	IN[9]
A. taxiformis (Delile) Trevisan de Saint-Léon[6, 7, 8, 12]	SC,SH[8]	IP[7]	IN[9]
Bonnemaisonia hamifera Hariot[7, 8]	SH[8]	IP[7]	ES[7]
Callithamniaceae			
Aglaothamnion feldmanniae Halos[7, 8]	SH[8]	At[7]	ES[7]
Caulacanthaceae			
Caulacanthus okamurae Yamada[7]	SH[8]	IP[7]	ES[7]
Ceramiaceae			
Acrothamnion preissii (Sonder) E.M.Wollaston[7, 8]	SH[8]	IP[7]	IN[9]
Antithamnion amphigeneum A.J.K.Millar[7, 8]	SH[8]	IP[7]	ES[7]
A. hubbsii E.Y.Dawson[7]	SH?	IP[7]	ES[7]
A. nipponicum Yamada & Inagak[7, 8]	OY[7]	IP[8]	ES[7]
Antithamnionella boergesenii (Cormaci & G.Furnari) Athanasiadis[7, 8]	SH?[8]	IP[7]	CA[7]
A. elegans (Berthold) J.H.Price & D.M.John[7, 8]	SH[8]	IP[8]	CA[7, 8]
A. spirographidis (Schiffner) E.M.Wollaston[7, 8]	SH[8]	IP[7]	ES[7]
A. sublittoralis (Setchell & Gardner) Athanasiadis[7, 8]	SH[8]	IP[7]	ES[7]
A. ternifolia (J.D.Hooker & Harvey) Lyle[7, 8]	SH[8]	IP[8]	ES[7]
Ceramium bisporum D.L.Ballantine[7, 8]	SH[8]	At[7]	CA[8]
C. strobiliforme G.W.Lawson & D.M.John[7, 8]	SH?[8]	At[7]	CA[8]
Corallinaceae			
Lithophyllum yessoense Foslie[7, 8]	OY[8]	IP[7]	IN[7]
Cystocloniaceae			
Hypnea anastomosans Papenfuss, Lipkin & P.C.Silva[7, 13]	SC[13]	IP[7]	ES[7]
H. cornuta (Kützing) J.Agardh[6, 7, 8, 12]	SC[8]	IP[8]	ES[7]
H. flagelliformis Greville ex J.Agardh[7, 8]	SC[8]	IP[7]	?
H. flexicaulis Yamagishi & Masuda[7]	SH?	IP[7]	ES[7]
H. spicifera (Suhr) Harvey[6, 14]	SC[14]	IP[2]	ES[6]
H. spinella (C. Agardh) Kützing[6, 7, 8]	SH?[8]	IP&At[7]	ES[7]
H. valentiae (Turner) Montagne[7, 8]	SC,SH[8]	IP[7]	ES[7]
Dasyaceae			
Dasya sessilis Yamada[7, 8]	OY[8]	IP[7]	ES[7]
Heterosiphonia japonica Yendo[7]	OY[8]	IP[7]	IN[7]
Delesseriaceae			
Apoglossum gregarium (E.Y.Dawson) M.J.Wynne[7, 8]	SH[8]	IP[7]	ES[7]
Nitophyllum stellato-corticatum Okamura[7, 8]	OY[8]	IP[7]	ES-IN[7, 8]
Galaxauraceae			
Galaxaura rugosa (J.Ellis & Solander) J.V.Lamouroux[6, 7, 8, 14, 15]	SH[9, 12]	IP&At[12]	IN[10]
Gigartinaceae			
Chondrus giganteus **f.** *flabellatus* Mikami[7, 8]	OY[8]	IP[7]	ES-IN[7, 8]
Goniotrichaceae			
Goniotrichiopsis sublittoralis G.M.Smith[7, 8]	SH[8]	IP[7]	ES[7]

(continued)

Table 10.1 (continued)

Alien taxon	Vector	Origin	Status
Gracilariaceae			
Gracilaria arcuata Zanardini[7,8]	SC[8]	IP[7]	CA[7]
G. vermiculophylla (Ohmi) Papenfuss[7]	OY[16]	IP[7]	IN[7]
Halymeniaceae			
Grateloupia asiatica S.Kawaguchi & H.W.Wang[7,8]	OY[8]	IP[7]	ES[7]
G. lanceolata (Okamura) Kawaguchi[7,8]	OY[8]	IP[7]	IN[17]
G. minima P.L.Crouan & H.M.Crouan[7]	OY?	IP[7]	ES[7]
Grateloupia patens (Okamura) S.Kawaguchi & H.W.Wang[7,8]	OY[8]	IP[7]	CA[7,8]
G. subpectinata Holmes[7,8]	OY[8]	IP[7]	ES[7]
G. turuturu Yamada[7,8]	OY[8]	IP[7]	IN[9]
Liagoraceae			
Ganonema farinosum (J.V.Lamouroux) K.C.Fan & Yung C.Wang[7,8,12]	SC[12]	IP[7]	ES[18]
Nemalion vermiculare Suringar[7,8]	OY[8]	IP[7]	ES[7]
Lomentariaceae			
Lomentaria flaccida Tanaka[7]	OY?	IP[7]	CA[7]
L. hakodatensis Yendo[7,8]	OY[8]	IP[7]	ES[7]
Phyllophoraceae			
Ahnfeltiopsis flabelliformis (Harvey) Masuda[7,8]	OY[8]	IP[7]	IN[17]
Plocamiaceae			
Plocamium secundatum (Kützing) Kützing[7,8]	?	IP&At[7]	ES[7]
Rhodomelaceae			
Acanthophora najadiformis (Delile) Papenfuss[6,7,8,12]	SC[8]	IP[7]	IN[6,12]
Chondria coerulescens (J. Agardh) Falkenberg[7,8]	OY[8]	At[7]	IN[17]
C. curvilineata F.S. Collins & Hervey[7,8]	SH[8]	At[7]	ES[7]
C. pygmaea Garbary & Vandermeulen[7,8]	SC[8]	IP[7]	ES[7]
Herposiphonia parca Setchell[7,8]	OY[8]	IP[7]	ES[7]
Laurencia caduciramulosa Masuda & Kawaguchi[7,8]	SH[8]	IP[7]	ES[7]
L. okamurae Yamada[7,8]	OY[8]	IP[7]	ES[7]
Lophocladia lallemandii (Montagne) F.Schmitz[7,8]	SC,SH[8]	IP[7]	IN[9]
Neosiphonia harveyi (J.W.Bailey) M.S.Kim, H.G.Choi, Guiry & G.W.Saunders[7,19]	SH[8]	IP[7]	IN[17]
Osmundea oederi (Gunnerus) G.Furnari[20,7]	OY[20],SH	At[7]	CA[7]
Polysiphonia atlantica Kapraun & J.N. Norris[7,8]	OY,SH[8]	At[7]	CA[7]
P. fucoides (Hudson) Greville[7,8]	FB[8]	At[7]	ES[7]
P. morrowii Harvey[7,8]	OY[8]	IP[7]	ES[7]
P. paniculata Montagne[7,8]	?	IP[7]	ES[7]
P. stricta (Dillwyn) Greville[7]	?	At[7]	ES[7]
Pterosiphonia tanakae S.Uwai & M.Masuda[7,8]	OY[8]	IP[7]	IN[17]
Symphyocladia marchantioides (Harvey) Falkenberg[7,8]	SH[8]	IP[7]	ES[7]
Womersleyella setacea (Hollenberg) R.E.Norris[7,8]	SH[8]	IP[7]	IN[9]
Rhodophysemataceae			
Rhodophysema georgei Batters[7,8]	OY[8]	IP&At[7]	CA[7]
Rhodymeniaceae			
Botryocladia madagascariensis G.Feldmann[7,8]	SH[8]	IP[7]	ES[7]
Chrysymenia wrightii (Harvey) Yamada[7,8]	OY[8]	IP[7]	ES[7]
Rhodymenia erythraea Zanardini[7,8]	SC,SH[8]	IP[7]	CA[7]

(continued)

Table 10.1 (continued)

Alien taxon	Vector	Origin	Status
Solieriaceae			
Agardhiella subulata (C.Agardh) Kraft & M.J.Wynne[7, 8]	OY[8]	IP&At[7]	ES[7]
Solieria dura (Zanardini) F.Schmitz[7, 8]	SC[8]	IP[7]	CA[7]
S. filiformis (Kützing) P.W.Gabrielson[7, 8]	?	At[7]	ES[7]
Stylonemataceae			
Goniotrichiopsis sublittoralis G.M.Smith[7, 8]	SH[8]	IP[7]	ES[7]
Wrangeliaceae			
Anotrichium okamurae Baldock[7, 8]	SH[8]	IP[7]	CA[7]
Griffithsia corallinoides (Linnaeus) Trevisan[7, 8]	OY[8]	IP&At[7]	ES[7]
Pleonosporium caribaeum (Børgesen) R.E.Norris[8, 21]	SH,OY[8]	At[8]	CA[8]
Ochrophyta			
Alariaceae			
Undaria pinnatifida (Harvey) Suringar[7, 8]	OY[8]	IP[7]	IN[9]
Chordaceae			
Chorda filum (Linnaeus) Stackhouse[7, 8]	OY[8]	IP&At[7]	ES[7]
Chordariaceae			
Acrothrix gracilis Kylin[7, 8]	OY[8]	IP&At[7]	ES[7]
Botrytella parva (Takamatsu) Kim[7, 8]	?	IP[7]	ES[7]
Cladosiphon zosterae (J. Agardh) Kylin[7, 22]	?	At[7]	ES[7]
Corynophlaea crispa (Harvey) Kuckuck[23]	?	At[2]	?
Halothrix lumbricalis (Kützing) Reinke[7, 8]	OY[8]	IP&At[7]	ES[7]
Leathesia marina (Lyngbye) Decaisne[7, 8]	OY[8]	IP&At[7]	ES[7]
Punctaria tenuissima (C. Agardh) Greville[7, 8]	OY[8]	At[7]	ES[7]
Microcoryne ocellata Strömfelt[24]	?	At[2]	?
Microspongium globosum Reinke[25]	?	At[2]	?
M. tenuissimum (Hauck) A.F. Peters[7]	?	At[7]	ES[7]
Sphaerotrichia firma (Gepp) A.D.Zinova[7, 8]	OY[8]	IP[7]	ES[7]
Desmarestiaceae			
Desmarestia viridis (O.F.Müller) J.V. Lamouroux[22]	?	IP&At[7]	IN[17]
Dictyotaceae			
Padina antillarum (Kützing) Piccone[7]	?	IP[7]	CA[7]
P. boergesenii Allender & Kraft[6, 7, 8]	SC[8]	IP[7]	IN[6]
P. boryana Thivy[6, 7, 8]	SC[8]	IP[7]	CA-ES[6,7]
Rugulopterix okamurae (E.Y.Dawson) Hwang, Lee & Kim[7, 8]	OY[8]	IP[7]	CA[7]
Spatoglossum variabile Figari & De Notaris[7, 8]	SC[8]	IP[7]	CA[7]
Stypopodium schimperi (Buchinger ex Kützing) Verlaque & Boudoresque[6, 7, 8, 10]	SC[8]	IP[7]	IN[10,9]
Ectocarpaceae			
Ectocarpus siliculosus **var.** *hiemalis* (P.L.Crouan & H.M.Crouan) Kuckuck[7, 8, 26]	?	At[7]	ES[7]
Fucaceae			
Fucus spiralis Linnaeus[7, 8]	FB[8]	At[7]	CA[7]
Laminariaceae			
Saccharina japonica (Areschoug) C.E.Lane, C.Mayes, Druehl & G.W.Saunders[7, 8]	OY[8]	IP[7]	CA[7]
Pylaiellaceae			
Pylaiella littoralis (Linnaeus) Kjellman[7, 22]	?	IP&At[7]	ES[7,17]
Sargassaceae			
Sargassum muticum (Yendo) Fensholt Setchell[7, 8]	OY[8]	IP[7]	IN[9]
Scytosiphonaceae			
Colpomenia peregrina Sauvageau[7, 8]	SH[8]	IP[7]	ES[7]
Compsoma saxicola (Kuckuck) Kuckuck[27]	?	At[2]	?
Scytosiphon dotyi M.J.Wynne[7, 8]	OY[8]	IP[7]	ES[7]

(continued)

Table 10.1 (continued)

Alien taxon	Vector	Origin	Status
Magnoliophyta			
Hydrocharitaceae			
Halophila stipulacea (Forsskål) Ascherson[7,8]	SC[22],SH?	IP[7]	IN[9]

[?] Means that the vector of introduction or the status of the alien species are unknown or questionable
[1]Reference: Kurt et al. (2001)
[2]Reference for distribution: Guiry and Guiry (2011)
[3]Reference: Taskin et al. (2001)
[4]Reference: Taskin et al. (2008)
[5]According to Guiry and Guiry (2011) this species is now regarded as *Coleofasciculus chthonoplastes* (Gomont) M.Siegesmund, J.R.Johansen & T.Friedl
[6]Hoffman et al. (submitted)
[7]Reference: Zenetos et al. (2010)
[8]Reference: Boudouresque and Verlaque (2010)
[9]Reference: Streftaris and Zenetos (2006)
[10]Reference: Hoffman et al. (2011)
[11]Reference: Israel et al. (2010)
[12]Reference: Hoffman and Dubinsky (2010)
[13]Lipkin (1972a) identified this species as *H. esperi* Bory de Saint-Vincent
[14]Reference: Aleem (1993)
[15]Reference: Hoffman et al. (2008a)
[16]Reference: Sfriso et al. (2010)
[17]Reference: Boudouresque et al. (2011)
[18]Based on Israeli national herbariums catalogs and 2011 unpublished surveys
[19]Boudouresque and Verlaque (2010) probably identified this species as *Polysiphonia harveyi* J.W.Bailey
[20]Reference: Serio et al. (2008)
[21]According to Wynne (2005) and Guiry and Guiry (2011) this species is now regarded as *Spongoclonium caribaeum* (Børgesen) M.J.Wynne
[22]Reference: Verlaque et al. (2011)
[23]Reference: Taskin (2006)
[24]Reference: Taskin et al. (2010)
[25]Reference: Ozturk et al. (2006)
[26]According to Guiry and Guiry (2011) this taxon is now regarded as *Ectocarpus siliculosus* (Dillwyn) Lyngbye
[27]Reference: Taskin et al. (2007)

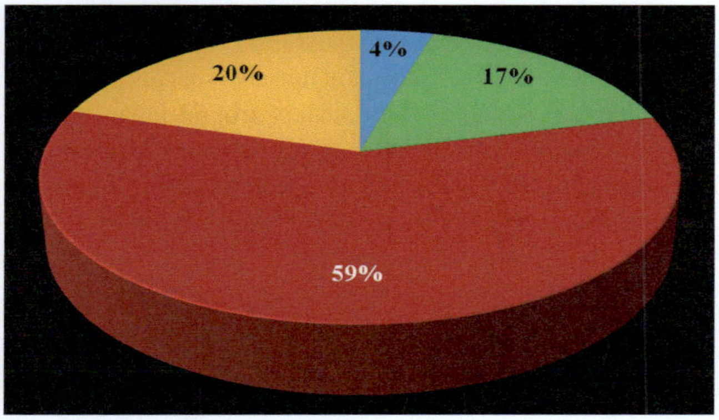

Fig. 10.1 Distribution of alien algae by phyla: *blue sector* represents Cyanobacteria, *green* – Chlorophyta, *red* – Rhodophyta, and *yellow* – Ochrophyta

Rhodophyta, and Ochrophyta phyla (Fig. 10.2a) are very close to the percentages established for the alien seaweed taxa found in the Mediterranean (Fig. 10.2b). Consequently, Rhodophyta does not appear to be more efficient in introduction events than the other two groups (Hoffman and Dubinsky 2010).

The vast majority of alien algae and seagrasses found in the Mediterranean so far are of natural Indo-Pacific origin and are naturally dispersed from there. The Atlantic Ocean contributed much less alien algae to the Mediterranean (Fig. 10.3). This picture is also clear when we compare the origins of alien seaweeds in the eastern (Fig. 10.4a), western (Fig. 10.4b), and central Mediterranean (Fig. 10.4c) and even the Adriatic Sea (Fig. 10.4d). Some alien algal species found in the Mediterranean are of Indo-Pacific and also Atlantic origin, as seen in Figs. 10.3 and 10.4. It is difficult to determine whether these species arrived from the Indo-Pacific or the Atlantic Oceans.

The main vectors that transport alien algae and seagrasses to the Mediterranean (Fig. 10.5) are shellfish transfer, shipping, and the Suez Canal (natural Lessepsian migration). However, a few alien seaweed species were released to the Mediterranean from aquariums or fishing baits. For 17 % of the alien algae, the vector of introduction is unknown. For 7 % of the alien algae, the vector of introduction is controversial (Zenetos et al. 2010 and Table 10.1).

The most famous invasive seaweeds in the Mediterranean Sea are probably *Caulerpa taxifolia* (killer algae), *Caulerpa*

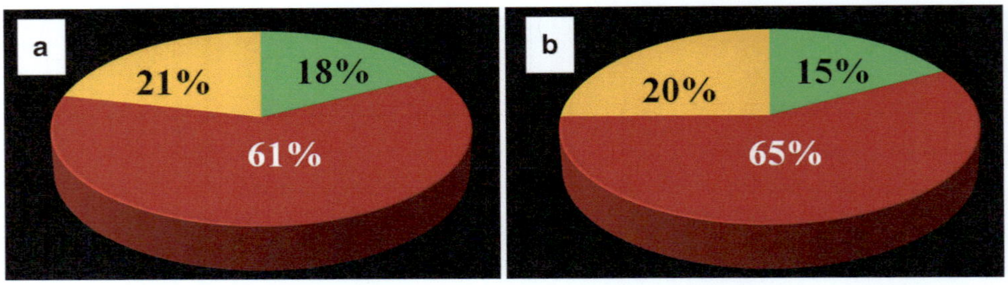

Fig. 10.2 Distribution of seaweeds by phyla: (**b**) native species in the Mediterranean (Boudouresque 2003; Zenetos et al. 2010) (**b**) alien taxa in the Mediterranean (Table 10.1). *Green* sector represents Chlorophyta, *red* – Rhodophyta, and *yellow* – Ochrophyta

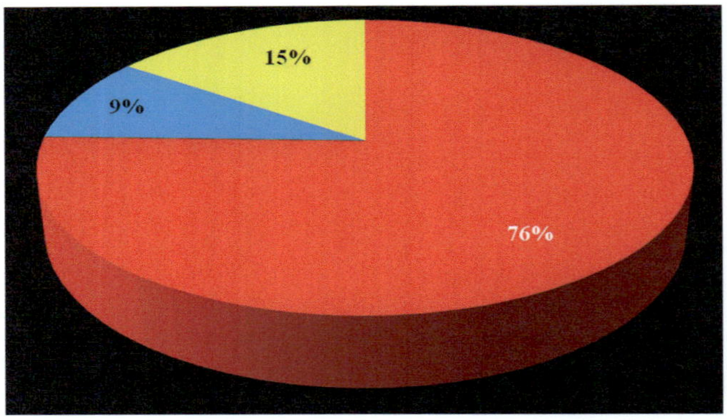

Fig. 10.3 General distribution of all alien algae reported to be found in the Mediterranean so far by their geographical origin. *Red sector* represents Indo-Pacific origin, *yellow* – Atlantic origin, and *blue* – Indo-Pacific and Atlantic origin (Modified from Zenetos et al. 2010 and Table 10.1)

racemosa var. *cylindracea* (Grape Caulerpa), *Codium fragile* subsp. *fragile* (Dead man's finger), the invasive strain of *Asparagopsis taxiformis* (Limu kohu), *Colpomenia peregrina*, *Sargassum muticum* (Jap weed), *Undaria pinnatifida* (Wakeme, Asian kelp), and *Womersleyella setacea*. These are regarded as the worst invasive (Streftaris and Zenetos 2006).

C. taxifolia (Chlorophyta), which was accidentally released to the Monaco Mediterranean shore from the Oceanographic Museum of Monaco public aquaria in 1984 (Meinesz 1999; Meinesz et al. 2001), spread rapidly to France, Italy, Spain, Croatia, Albania, Tunisia (Streftaris and Zenetos 2006; Zenetos et al. 2010), and probably to other Mediterranean shores. It is often called the "killer algae" since it is predominant in many underwater habitats, oppresses the local marine flora, and causes very serious ecological problems by dramatically decreasing biodiversity in native marine ecosystems (Gravez et al. 2001).

C. racemosa is a very old invader in the Mediterranean; however, this species has several subspecies and variants. *C. racemosa* var. *cylindracea* was first observed in the early 1990s in Libya. This taxon spread rapidly throughout most of the Mediterranean Sea, occupied a wide range of habitats, and developed extensive populations. Its rapid invasion of the Mediterranean deeply changed phytobenthic community structure (Piazzi et al. 2001) and decreased seaweed diversity (Gravez et al. 2001) at many locations. Studies of the effects of *C. racemosa* on local flora showed that the total percentage of cover and diversity of the indigenous flora and fauna was lower in invaded areas when compared with the control areas, where *C. racemosa* was absent (Piazzi and Cinelli 2003) and, therefore, it should be regarded as a serious threat to local marine flora, fauna, and ecosystems in the Mediterranean (Verlaque et al. 2004; Streftaris and Zenetos 2006).

S. muticum, *U. pinnatifida*, and *C. fragile* subsp. *fragile* were introduced into the Mediterranean through shellfish transfer (Boudouresque and Verlaque 2010). These three species (especially *S. muticum* and *C. fragile* subsp. *fragile*) also expand dispersal across the globe and became global threats to marine ecosystems all over the world.

S. muticum is a very successful invasive species (Critchley et al. 1990). It is native to the waters around Japan and the northwest Pacific, where it is an ecologically insignificant component of the marine flora (Norton 1977; Critchley 1983). During the last 50 years, this species invaded the east Pacific coast of North America and Mexico, the Mediterranean Sea (since 1980), the north Atlantic, and the North Sea as a consequence of the importation of cultivated Pacific oysters from Japan (Scagel 1956; Harries et al. 2007; Boudouresque and Verlaque 2010).

C. fragile subsp. *fragile* was first observed in the Mediterranean Sea in 1946 as *C. fragile* subsp. *tomentosoides* (van Goor) P.C. Silva (Boudouresque and Verlaque 2010; Zenetos et al. 2010). This species (and *S. muticum* too) also expands the dispersal of from Japan to the east Pacific, the north Atlantic and the North Sea, however; it has also been reported as being found in the south Pacific and South African shores (Chapman 1999). In some north Atlantic shores of America, this invader was found to grow at very high densities and damage the natural kelp forests and its ecosystem (Provan et al. 2005).

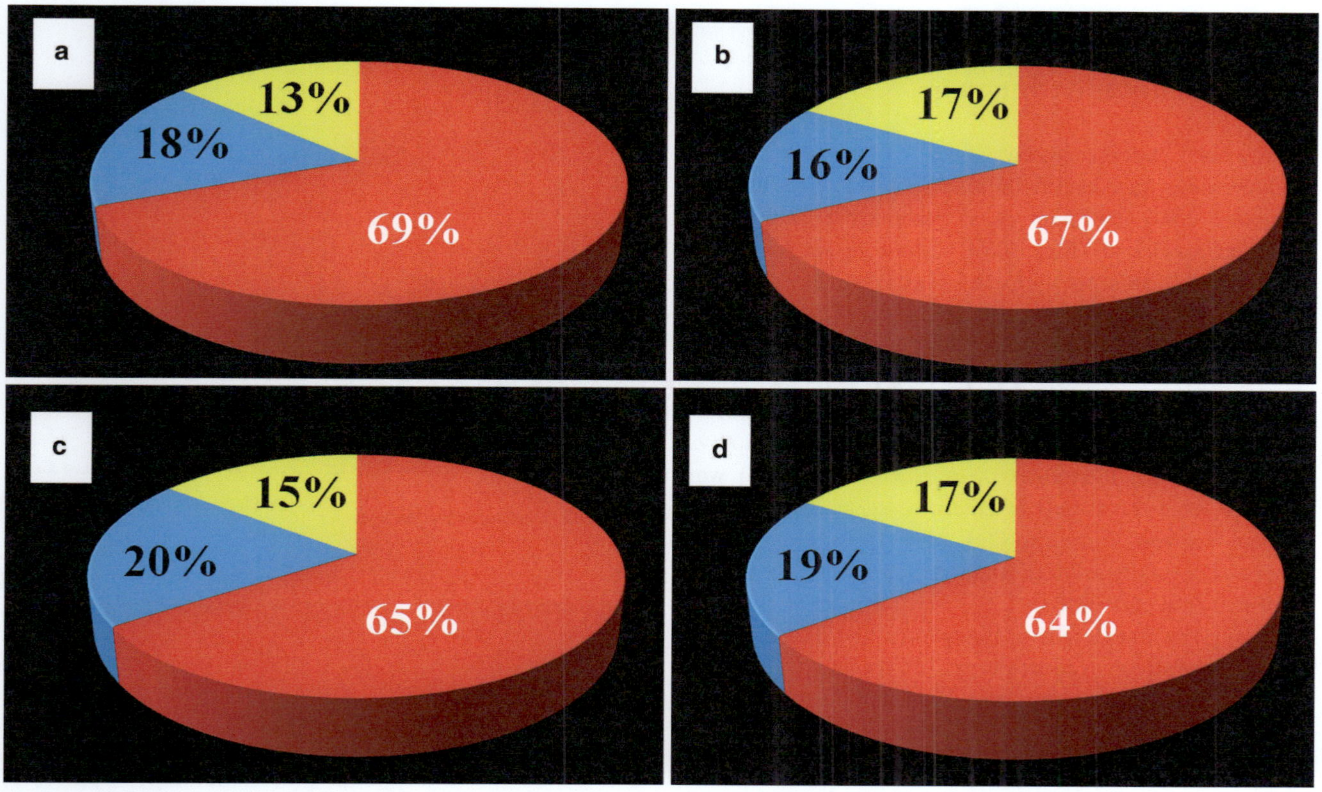

Fig. 10.4 (**a**, **b** and **c**) represent distribution in the eastern, western, and central Mediterranean, and (**d**) represents distribution in the Adriatic Sea (Modified from Zenetos et al. 2010 and Table 10.1)

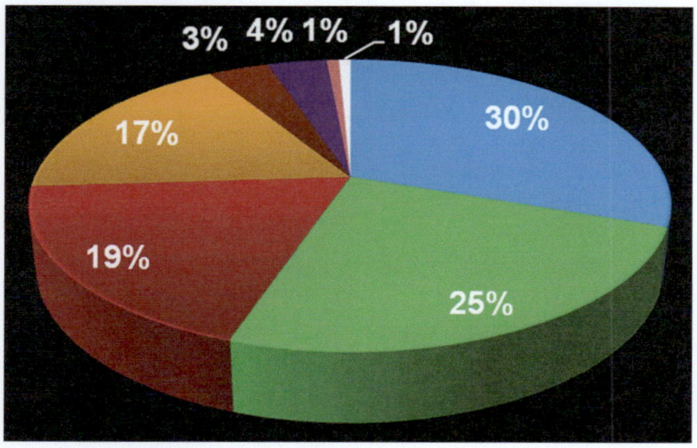

Fig. 10.5 Distribution of alien algae by vectors of introduction to the Mediterranean. *Blue* represents shellfish transfer, *green* – shipping, *red* – Suez Canal, *yellow* – unknown vector, *brown* – Suez Canal and/or shipping, *purple* – shellfish transfer and/or shipping, *pink* – marine aquarium, and *white* – fishing bait

Asian kelp (*U. pinnatifida*) was first reported seen in the Mediterranean Sea in 1971 at the Thau lagoon (France), which is identified as a hotspot to invaders introduced through shellfish transfer (Uwai et al. 2006; Boudouresque and Verlaque 2010). *U. pinnatifida* also invaded the south and east Pacific as well as the north Atlantic coasts (Uwai et al. 2006).

A. taxiformis was first reported seen in the Mediterranean at the end of the nineteenth century; however, in 1996, a new strain of this species was observed for the first time. This Indo-Pacific variant acted extremely invasive and spread rapidly to all parts of the Mediterranean Sea (Boudouresque and Verlaque 2010; Zenetos et al. 2010).

C. peregrine is an old-time invader since 1918 (Boudouresque and Verlaque 2010). This brown puffy (filled with air) seaweed, which probably arrived from the Pacific Ocean, is often called the oyster thief (also the nickname of *C. fragile* subsp. *fragile*) since it tends to grow on oyster shells, lifting the oysters floating away from the seabed and causing problems to oyster farmers, who lose some of their harvests.

W. setacea (such as *A. taxiformis*, *C. racemosa* var. *cylindracea* and *C. fragile* subsp. *fragile*) is found in all four basins of the Mediterranean (western, eastern, central, and also the Adriatic Sea) (Zenetos et al. 2010). This brown algae was first discovered in 1986 (Boudouresque and Verlaque 2010).

Alien Seaweeds and Their Impact in the Israeli Mediterranean

Twenty-one alien-seaweed taxa from the Israeli Mediterranean have been identified so far (6) (Hoffman and Dubinsky 2010, Hoffman et al. submitted; Table 10.1). *Caulerpa mexicana* (Fig. 10.6), *Caulerpa racemosa* var. *lamourouxii*, *Caulerpa scalpelliformis* (Fig. 10.7), *Codium taylorii* (Fig. 10.8), *Boodleopsis* sp., *Ulva fasciata*, *Sarconema filiforme* (Figs. 10.9 and 10.10), *Asparagopsis taxiformis*, *Hypnea cornuta* (Fig. 10.11), *Hypnea spinella* (Fig. 10.12), *Acanthophora najadiformis* (Fig. 10.13), *Padina boergesenii* (Fig. 10.14), *Stypopodium schimperi* (Fig. 10.15), and *Ganonema farinosum* (Fig. 10.16) were collected from the Israeli Mediterranean and stored at the national seaweed herbaria (Tel Aviv University and the Hebrew University of Jerusalem) in the twentieth century (Hoffman et al. submitted).

During the first decade of the twenty-first century, seven new alien seaweeds (constituting 33 % of the total aliens found) were found along the Mediterranean shores of Israel. Seaweed surveys (Hoffman et al. submitted) took place during the new millennium and checklists of the found taxa were prepared and compared to the catalogs of the Israeli national seaweed herbaria of the Mediterranean. Literature reports of macroalgal surveys from the Israeli Mediterranean (Edelstein 1960, 1962, 1964; Lipkin and Safriel 1971; Lundberg 1981, 1986, 1991, 1995, 1996) were also checked.

The new aliens found were: *Boodlea composita* (two observations at the same location), *Codium arabicum* (Fig. 10.17), *Codium parvulum* (Fig. 10.18), *Neomeris annulata*, *Hypnea spicifera*, *Galaxaura rugosa* (Fig. 10.19), and *Padina boryana* (Hoffman et al. submitted). *C. arabicum*, *C. parvulum*, and *G. rugosa* were recorded at the northern

Fig. 10.6 *Caulerpa Mexicana*

Fig. 10.7 *Caulerpa scalpelliformis*

Fig. 10.8 *Codium taylorii*

Fig. 10.9 *Sarconema filiforme* growing at the intertidal zone

Fig. 10.10 *Sarconema filiforme* often used as fishing bait at the north Israeli shore

Fig. 10.11 *Hypnea cornuta* growing in an intertidal pool

Israeli Mediterranean shore, particularly near the marine port at Haifa Bay (Fig. 10.20). The reports of *C. arabicum* and *C. parvulum* from in the bay are the first records of these alien species in the Mediterranean (Hoffman et al. 2011).

There are some *S. schimperi* populations along the Israeli coast; however, a large population is concentrated in Haifa Bay and the surrounding area. This species, which is well known as being invasive in the eastern Mediterranean basin, especially along the Levantine shores (Verlaque and Boudouresque 1991) in Syria (Mayhoub and Billard 1991)

Fig. 10.12 *Hypnea spinella* exposed during low tide

Fig. 10.13 *Acanthophora najadiformis*

Fig. 10.14 *Padina boergesenii*

Fig. 10.15 *Stypopodium schimperi*

Fig. 10.16 *Ganonema farinosum*

and Lebanon (Bitar et al. 2000, 2003), also spread to Egypt, Libya, and Turkey (Bitar et al. 2000), and was even reported to be found in the Aegean Sea (Cocito et al. 2000).

The appearance and concentration of *S. schimperi*, *G. rugosa*, *C. parvulum*, and *C. arabicum* in the Haifa Bay area indicated that the marine vegetation has changed since Edelstein described this site (Edelstein 1960, 1962, 1964) more than 50 years ago.

The conspicuous phenomena that clearly indicate proliferation of these alien seaweeds are enormous winter and spring drift events that occurred each year at the bay. During the years 2004–2008, several major algal blooms cast ashore thousands of tons of algal biomass (Israel et al. 2010; Hoffman et al. 2011). These drifts, which mainly contained fragmented *C. parvulum*, *G. rugosa*, and *S. schimperi* (Fig. 10.21), filled the seawater and rotted along the bay's beaches.

Observation and later calculation of alien and local seaweed abundance in the bay drift indicated that winter drifts

Fig. 10.17 *Codium arabicum* specimen drifted to shore at Haifa Bay

Fig. 10.18 *Codium parvulum*

Fig. 10.19 *Galaxaura rugosa*

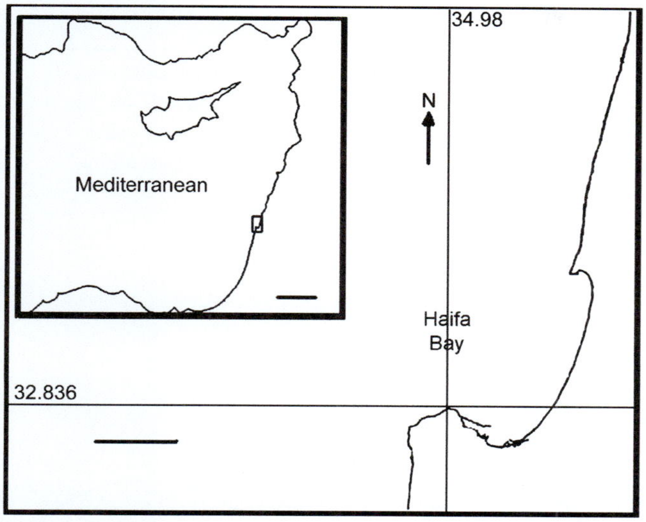

Fig. 10.20 Haifa Bay area (Insert map scale bars 100 km, main map scale bars 5 km)

Fig. 10.21 The three invaders seaweeds found in the Haifa Bay winter and spring drifts. From *right* to *left*, *Stypopodium schimpery*, *Galaxaura rugosa*, and *Codium parvulum*

contained ~80 % alien seaweeds (Fig. 10.22a), while in the spring, their share decreased to ~50 % (Fig. 10.22b) of the total (Hoffman et al. 2011). It seems that most alien seaweed biomass that drifted to the Haifa Bay shores during winter and spring came from the infralittoral and sublittoral zones. However, *C. parvulum*, *G. rugosa*, *S. schimperi*, and *C. arabicum* (the latter was the main alien in the summer drifts) were also found growing in small quantities in potholes and pools located in the bay intertidal zone (Hoffman et al. 2011).

The invasion and establishment of opportunistic alien seaweeds in Haifa Bay might be extremely harmful to the natural marine ecosystem for two reasons: (1) the replacement of local flora might lead to the extinction of some native seaweed species, e.g., the unexplained disappearance of *Halimeda tuna* (J. Ellis *et* Solander) J.V. Lamouroux (Hoffman et al. 2008b); this calcified green alga was very abundant in the Haifa Bay area over 50 years ago (Edelstein 1960, 1962; Edelstein and Komarovsky 1961); (2) alien Indo-Pacific seaweed proliferation might provide suitable settlement and establishment conditions for alien grazer species, which may displace native ones. In general, the invasion of alien species might cause the disruption of the local ecosystem structure and its energy and nutrient flows, resulting in its subsequent collapse (Hoffman et al. 2011).

Since becoming invasive is the climax of the process that begins with establishment and local spread, followed by increase in abundance (Richardson et al. 2000; Kolar and Lodge 2001), the aliens, *S. schimperi*, *G. rugosa*, *C. parvulum*, and *C. arabicum*, should be regarded and treated as invasive species in the Mediterranean, since they threaten the local marine flora, can reduce biodiversity, and have the potential for disrupting ecosystem function. Of these four species, *G. rugosa* and *C. arabicum* have the widest world dispersal among Indo-Pacific invaders, which may be indicative of their opportunistic capabilities and ecological flexibility. Therefore, it is quite possible that their dispersal might easily expand to other Mediterranean coastlines.

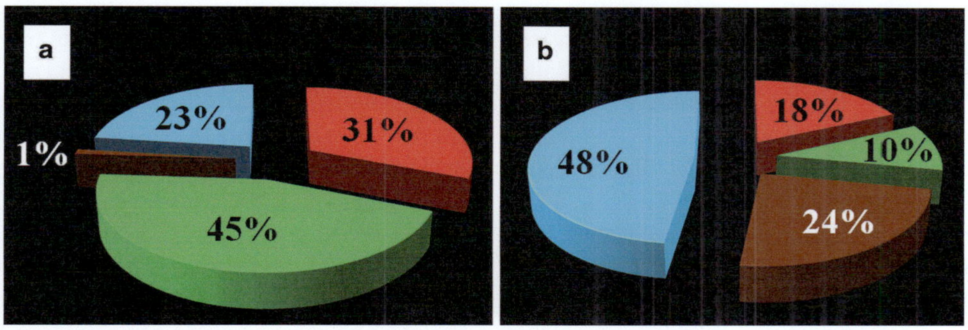

Fig. 10.22 Distribution pool (%) of seaweeds in the Haifa Bay drifts during winter (**a**) and spring (**b**) of 2005–2009. *Red* sector represents *G. rugosa*, *brown* – *S. schimperi*, *green* – *C. parvulum*, and *blue* – native algal species

Fig. 10.23 Alien Indo-Pacific seaweed invasion model for Haifa Bay. *Squares* represent the conditions that cause invasions; *circles* represent the factors that facilitate settlement and establishment

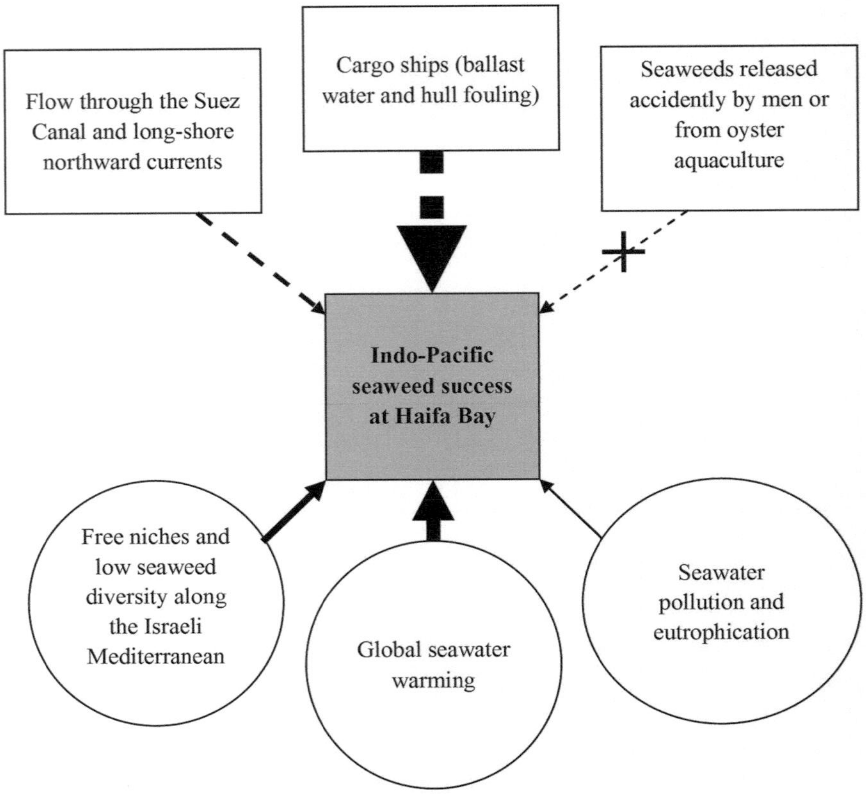

On the basis of the observation that showed that these new four invader seaweed species were mainly found at Haifa Bay (Hoffman and Dubinsky 2010; Hoffman et al. 2011, submitted), where many cargo ships anchor, and taking into consideration the weather conditions in the eastern Mediterranean Sea, a new model (Fig. 10.23) that shows all of the factors likely to cause and facilitate Indo-Pacific seaweed invasions in the region, was established. As industrialization proceeds, the concomitant growth of marine trade increases the invasion and dispersal of alien organisms worldwide, and particularly in the eastern Mediterranean Basin (Rilov and Crooks 2009). The tendency of most of the four newly reported Indo-Pacific invader seaweeds to concentrate in the Haifa Bay area strongly points to cargo ships as the main transport vector. These ships carry seaweed attached to their hulls or as fragments in their ballast water, and spill them after loading containers at port. Alien seaweeds can survive microscopic stages in the ballast water tanks of cargo ships that arrive at Mediterranean harbors (Flagella et al. 2007).

Seaweeds depend upon water-circulation patterns and waves to facilitate dispersal. As *C. arabicum*, *C. parvulum*, and *G. rugosa* were not found south of the Haifa Bay area, the flow through the Suez Canal probably had no effect on these invasions. However, their occurrence north of Haifa Bay indicates that the long-shore north-flowing currents may have had an effect on the northward spread following the first invasion (Hoffman et al. 2011).

Former Indo-Pacific invaders previously reported from the Israeli Mediterranean, and which became very common, such as *Acanthophora nayadiformis*, *Hypnea cornuta*, *Sarconema filiforme*, and *Codium taylori*, were also commonly found on the southern shores of Israel. Hence, there is a possibility that some of these species might have arrived at the Israeli Mediterranean in a northward flow through the Suez Canal and through long-shore north-flowing currents (Hoffman et al. 2011).

So far, there is no evidence that alien-seaweed species were accidentally released (as was the case of *C. taxifolia* in Monaco) or introduced to the Israeli Mediterranean coast through shellfish transfer since there are no oyster farms along these shores (Hoffman et al. 2011).

Global warming has serious effects on marine ecosystems (Belkin 2009), and might play a large role in the invasion process (Mooney and Hobbs 2000). Mediterranean sea-surface temperatures have increased by 2–3 °C in the southern and eastern Mediterranean since the 1980s (EEA 2007). Comparison of annual mean seawater-surface temperatures between the Israeli Mediterranean and the tropical Israeli Red Sea shows that these temperatures are very similar (24 and 23.5 °C, respectively). However, the mean sea-surface temperature amplitude in the Israeli Mediterranean is greater than that of the Red Sea (Hoffman et al. 2011). Therefore, the main factors facilitating the invasion of alien seaweed species of tropical origin into the Israeli Mediterranean shores might be the naturally high temperatures in the Levant Basin as well as the global-warming process. As sea-surface temperatures during winter rose, the conditions became ideal for the newly arriving tropical Indo-Pacific seaweed species to settle, establish communities, and extend their dispersal. The mean sea-surface temperature in Haifa Bay during February is nearly 20 °C, which is similar to tropical Red Sea surface temperatures (~21 °C); therefore, tropical Indo-Pacific seaweed invaders that arrive at the eastern Mediterranean Sea (such as those arriving from the Red Sea) find favorable temperatures in the recipient Israeli-Mediterranean marine ecosystem. These high temperatures may initially facilitate survival and settlement and later facilitate reproduction and dispersal (Hoffman et al. 2011).

Another important factor that makes the Haifa Bay shores a hotspot for alien seaweeds and facilitates their invasion is the low local macroalgal diversity. Although the Mediterranean Sea is considered rich in species considering its size (Rilov and Galil 2009), Israeli Levant macroalgal diversity is poor compared to that of other Mediterranean shores (on the basis of local published seaweed surveys and national herbaria). In addition, the Haifa Bay area is characterized by rich, diverse, and three-dimensional habitats present on the wide and long underwater reef.

In general, seawater near marine ports is chemically polluted (Ng and Song 2010), which sometimes makes environmental conditions less tolerable for some native species (Rilov and Crooks 2009). Therefore, seaweeds that grow near ports might be more tolerant than others to such pollution. Since Haifa Bay became highly polluted at the end of the twentieth century and at the beginning of the new millennium, some tolerant and opportunistic newcomer seaweeds might find this pollution to be a positive factor that gives them advantages over the less-tolerant local species, perhaps facilitating the success of invaders in the bay (Hoffman et al. 2011).

Halophila stipulacea – An Old-Timer Invasive Alien Seagrass in the Mediterranean

Nearly 60 seagrass species were reported found in marine ecosystems around the world. These are distributed in four families: Cymodoceaceae, Hydrocharitaceae, Posidoniaceae, and Zosteraceae [if we don't include Potagmogetonacea, which provides eight more species not regarded as seagrasses because they do not occur in oceanic water with consistently high salinity (Den Hartog 1970)], and eleven genera; however, only four of these (*Cymodocea nodosa*, *Posidonia oceanica* (endemic), *Zostera noltii*, and *Zostera marina*) are regarded as native Mediterranean species (Phillips and Meñez 1988; Lipkin et al. 2003; Procaccini et al. 2003; Borum and Greve 2004).

Fig. 10.24 *Halophila stipulacea* growing in aquarium in Israel

Halophila stipulacea (Forsskål) Ascherson (Hydrocharitaceae) (Fig. 10.24) is, so far, the only alien-seagrass species reported in the Mediterranean. This species, which mainly grows on sandy and muddy substrates along the eastern coast of Africa, originally has Indian Ocean and Red Sea distribution. The first record of this invasive species from the Mediterranean was in June 1894 (nearly two and a half decades after the opening of the Suez Canal), from the island of Rhodes (Fritsch 1895; Por 1971; Lipkin 1975a). Since then, this species spread to many coasts along the eastern basin of the Mediterranean, where it was also reported to be found in the Egyptian Mediterranean (Täckholm et al. 1941; Aleem 1962), Cyprus (Lipkin 1975b), Lebanon (Lipkin 1975b) Turkey (Alpinar 1987; Taskin et al. 2008), Greece and its surrounding islands (Politis 1930; Tsirika and Haritonidis 2005; Guiry and Guiry 2011), Albania (Tan and Mullaj 2000; Kashta et al. 2011), Malta (Verlaque 1994), Libya, and some shores in eastern Tunisia (Sghaier et al. 2011).

Although *H. stipulacea* was one of the first alien marine macrophytes reported in the Mediterranean since the opening of the Suez Canal, it did not spread fast and about 100 years after its first record from the eastern basin of the Mediterranean it was reported for the first time in the western Mediterranean (Procaccini et al. 2003), in the island of Sicily (Verlaque 1994; Procaccini et al. 1999; Di Martino et al. 2006). Later, it was also observed at the Tyrrhenian Sea in southwest Italy (Gambi et al. 2009).

Lately, *H. stipulacea* has also been reported for the first time in the Caribbean Sea in the west tropical Atlantic shores (Ruiz and Ballantine 2004) where it was found growing off the shores of Granada, Dominica, and St. Lucia (all parts of the Lesser Antilles) (Willette and Ambrose 2009). On the island of Dominica, it was found growing at seven locations and it was observed that this invasive alien species became the third most common seagrass in the area (an overall of six indigenous species were reported in this island previously) (Steiner et al. 2010).

H. stipulacea probably reached the Mediterranean through the Suez Canal; however, the fact that this species was mainly found in or near bays with ports, harbors, and marinas, such as the Port Said and Marsa Matrukh harbors in Egypt, Port of Sidon in Lebanon, Rhodes harbor, Cyprus harbors, Tobrouk Bay in Libya, the marina at Cap Monastir in Tunisia, Palinuro harbor in Italy, and several Caribbean bays, supports the assumption that the main transportation vector of this species is boats (Täckholm et al. 1941; Aleem 1962; Lipkin 1975a, b; Ruiz and Ballantine 2004; Gambi et al. 2009; Sghaier et al. 2011). According to Lipkin (personal communication), the main transportation vectors that spread *H. stipulacea* in the Mediterranean Sea are dredgers. These boats drag their nets in order to collect mollusks from the seabed. *H. stipulacea* shoots and other organisms are also caught in the nets and later released in the ports where the dredgers anchor. The fact that this seagrass was found all over the eastern basin of the Mediterranean except Israel supports this assumption. Owing to the political situation in the Middle East, dredgers from the Arabs countries that surround Israel are not allowed to approach the territorial Mediterranean seawater of Israel and, therefore, *H. stipulacea* did not reach the Israeli coast.

Today, *H. stipulacea* is considered one of the 100 worst invasive species in the Mediterranean. Certain studies that compared the general features of Mediterranean seagrass-species meadows in Italy (Table 10.2) showed that *H. stipulacea* has extremely high shoot density (especially when its beds are located in shallow waters) compared to the other four species. The leaf area and biomass of this invasive species were also found to be high (Buia et al. 2000; Procaccini et al. 2003).

Seagrass decline is a worldwide phenomenon and the global loss of seagrass from direct and indirect human impacts was estimated to be 33,000 km^2, or 18 % of the documented seagrass area during 20 years of population monitoring (Duarte et al. 2004). The main reason for seagrass-bed decline is seawater pollution, which increases eutrophication that leads to reduction in water clarity. As seagrasses have relatively high light requirements, the decrease in light penetration of coastal waters reduces their photosynthesis efficiency and makes them vulnerable (Duarte et al. 2004; Bernard et al. 2007).

Other reasons that contribute to seagrass decline include fishing, aquaculture, boating, anchoring, and coastal alteration.

Introduced alien species also play an important role in seagrass decline, especially along the Mediterranean coasts. Two alien exotic and toxic green seaweed species, *C. taxifolia* and *C. racemosa*, were reported to be a major threat to seagrass beds of the Mediterranean since they grow rapidly and compete with seagrasses for the soft-bottom substratum, which is favored by seagrasses (de Ville`le and Verlaque 1995; Ceccherelli and Cinelli 1997; Ceccherelli et al. 2000; Duarte et al. 2004; Pergent et al. 2008).

C. taxifolia grows rapidly, and largely appears to colonize areas devoid of seagrasses, but has also been reported to compete with *P. oceanica* off the shore of Monaco for space and resources, being able to damage its meadows (Duarte et al. 2004; Pergent et al. 2008). Moreover, this species drastically reduced biodiversity in seagrass beds (Streftaris and Zenetos 2006). The grape alga, *C. racemosa* var. *cylindracea*, easily invaded *C. nodosa* and *Z. noltii* beds and affected both soot and flower density of both seagrasses in their meadows (Ceccherelli and Campo 2002; Piazzi et al. 2005), but when it comes to *P. oceanica*, the Grape Caulerpa prefers dominating on its dead mattes, and no colonization was observed in dense meadows (Streftaris and Zenetos 2006).

No direct damage to the local seagrass meadows in the Mediterranean by *H. stipulacea* has been reported yet (Duarte et al. 2004); however, it seems that this invasive species prefers settling on dead mattes of endemic Mediterranean seagrass *P. oceanica* (Gambi et al. 2009), and therefore might prevent the future recovery of its mattes.

H. stipulacea also plays an important role in the process and the success of other alien invasions, as was the case with the invasion of the seaweed *C. pygmaea* (Zenetos et al. 2011) or the anaspidean mollusk *Syphonota geographica*, which is dietary depends on this seagrass (Mollo et al. 2008). *S. geographica* example indicates that the initial invasion of an exotic pest such as *H. stipulacea* might have paved the way for the subsequent invasion of trophic specialists that take advantage of niche opportunities (Mollo et al. 2008) and some of these invaders might also have negative effects on local Mediterranean flora and fauna.

Global climate changes might have a substantial long-term impact on seagrass ecosystems. The expected increase of surface-seawater temperatures in the Mediterranean may be a major threat to local seagrass populations, especially those living close to their low latitude border of distribution, such as *Z. marina*, *C. nodosa*, and in particular, the Mediterranean endemic seagrass *P. oceanica* (Duarte et al. 2004). *Z. marina* and *P. oceanica* have succeeded in colonizing cold temperate water; however, the rise of temperatures might push them to their distributional limits (Duarte et al. 2004), giving *C. taxifolia*, *C. racemosa*, and *H. stipulacea* advantages over the two indigenous seagrasses. Moreover, high water temperatures might increase the possibility of the invasion of other seagrasses from the Red Sea to the Mediterranean. Lipkin (1975c, 1977, 1979) reported seven species of seagrasses growing on the sandy or muddy bottom in the Gulf of Suez and Sinai Red Sea. Besides *H. stipulacea*, he also found *Halodule uninervis* (Forsskål) Ascherson, *Cymodocea rotundata* Ehrenberg and Hemprich ex

Table 10.2 General features of Mediterranean seagrass meadows (Buia et al. 2000; Procaccini et al. 2003)

	Posidonia oceanica shallow	*Posidonia oceanica* deep	*Cymodocea nodosa*	*Halophila stipulacea* shallow	*Halophila stipulacea* deep	*Zostera noltii*	*Zostera marina*
Density (shoots/m²)	700	160	925–1,925	19,728	13,000	269–1,246	216–1,093
Leaf area (m²/m²)	6.16–29	1.1–2.6	0.2–3.5	5	5.9	0.2–0.4	1.7–6.7
Leaf biomass gdw/m²	175–670	52–94	17–159	157.8		13–79	45–755

Ascherson, *Syringodium isoetifolium* (Ascherson) Dandy, *Thalassodendron ciliatum* (Forsskål) den Hartog, *Thalassia hemprichii* (Ehrenberg) Ascherson, and *Halophila ovalis* (R. Brown) J.D. Hooker. These tropical species are possible invaders to the Mediterranean through the Suez Canal due to global warming and the rise of seawater temperatures. About 40 years ago, Lipkin (1972a, b, 1977) already found *H. uninervis* growing associated with *H. stipulacea* on sandy and muddy bottoms (2–3 m deep) in the Great Bitter Lake, which is at the middle of the Suez Canal and about 60 km from its estuary to the Mediterranean Sea at Port Said and therefore it is quite possible that this species already reached the Mediterranean Sea, waiting to be discovered somewhere.

Shellfish Transfer and Alien Seaweed Invasion: A Study Case from the Gulf of Lion

Several studies conducted in the middle of the twentieth century caused Elton to make the surprising statement: "The greatest agency of all that spreads marine organisms to new quarters of the world must be the business of oyster cultures" (Elton 1958). This statement was strongly supported by recent studies concerning alien-seaweed species and the vectors that transport them into the Mediterranean Sea. Today, shellfish transfer is regarded as the most important vector for seaweed introduction into that sea (Maggs and Stegenga 1999; Verlaque 2001; Boudouresque and Verlaque 2010).

Shellfish transfer introduced 46 % of alien macrophytes, while shipping and the Suez Canal introduced 29 and 27 %, respectively. These results (the sum, including other vectors, is more than 100 %) also indicate that some alien macrophytes have been introduced several times by different vectors (Zenetos et al. 2010).

The Gulf of Lion is an excellent example of the contribution of shellfish aquaculture (oysters, clams, and mussels) to seaweed invasion. The Thau Lagoon (located at the Sète area on the France Mediterranean coast in the middle of the Gulf of Lion), which is the leading site of shellfish aquaculture in the Mediterranean Sea, is now also regarded as the main hotspot for alien seaweeds in the Mediterranean, with 58 alien seaweed species reported so far. These alien seaweeds, which mostly originated from the North Pacific Ocean, contributed 64 % west-Mediterranean alien seaweeds, and nearly 50 % of the total alien macrophytes of the Mediterranean (Verlaque et al. 2007; Zenetos et al. 2010; Boudouresque et al. 2011). According to Boudouresque et al. (2011), introduced seaweeds represent 32 % of the species diversity and 48–99 % of the seaweed biomass estimated on hard substrates at the Thau Lagoon.

The Thau Lagoon is not suitable for breeding the Japanese oyster *Crassostrea gigas* (Thunberg) (also called the Portuguese oyster), which is the main crop of the shellfish industry in the area and, therefore, the local *C. gigas* farms are completely dependent on importing new oysters and their spats from other locations. These farms have official permission to import spats and small *C. gigas* individuals produced in the Atlantic alone since 1977; however, oyster farmers imported billions of small *C. gigas* oysters since the 1970s at least through the 1990s from British Columbia, Japan and Korea located in the North Pacific Ocean (Boudouresque et al. 2011).

These evidences strongly point at *C. gigas* transfer as a primer vector of introduction, carrying alien seaweeds from the North Pacific Ocean as mature thalli or propagules attached to its shells or packing materials and also as a secondary dispersal of alien seaweeds in the temperate Gulf of Lion and other Mediterranean locations, such as Mar Piccolo and the Venice Lagoon in Italy.

Global Warming and Alien Seaweed Introduction in the Mediterranean

There is no doubt that the surface-water temperatures have increased rapidly in the Mediterranean Sea during recent decades, probably due to the global-warming process; however, its effects on invasion is still doubtable.

Generally, studies from around the world have indicated that there is a big connection between global warming and the spread of alien exotic species from tropical origins. Some studies proved that invasive species from warm origins have a superior capacity to tolerate increased temperatures compared to native species (Stachowicz et al. 2002; Ward and Masters 2007; Sorte et al. 2010).

Most of the alien species in the Mediterranean (especially in the eastern basin) are thermophiles. These species originate in warm water (Zenetos et al. 2010). However, several algologists and seaweed ecologists claim that it is too early to detect any quantitative or qualitative impacts of global warming on seaweed introduction; moreover, they showed that some alien seaweeds reported that were regarded as tropical actually originated in temperate seas (Boudouresque and Verlaque 2010).

Although global warming itself does not increase the number of species introduced into the Mediterranean Sea, it facilitates the settlement and establishment of alien species from tropical origins to warmer recipient marine ecosystems and, later, also facilitates their dispersal to nearby warm locations.

High average sea-surface water temperatures in the eastern basin of the Mediterranean, especially along the shores of Egypt, the Levant, Cyprus, and Turkey, welcome the tropical seaweeds and help them spread in the region. On the other hand, the western Mediterranean, which is still more temperate (colder than the eastern Mediterranean basin) is more welcoming to seaweeds coming from the colder seas of Japan, China, Korea, and even from the temperate shores of Australia (Boudouresque and Verlaque 2010).

How Should We Deal with the Alien Invasion in the Mediterranean Sea?

The best way to compete with the invasive-species phenomenon is through prevention, early detection, and rapid response (Williams and Grosholz 2008).

The natural migrations of alien species through the Suez Canal and the Straits of Gibraltar are fait accompli. Closing these sea crossings, stopping marine transportation, or even eliminating shellfish transfer to the Mediterranean in order to stop the invasion and introduction of alien marine creatures, are definitely not realistic and are out of the question. However, certain international actions should be taken in order to reduce or slow down the rate of introductions and invasions.

The first action that should be taken has educational value and is very important: to increase public awareness of the negative impacts of alien invasions (including invasion in the Mediterranean Sea) (ICES 2005). The public has the right to know the effects of aliens on the environment and the direct and indirect consequences and risks of massive invasions on us. These consequences and the possible ways to control them should be seriously discussed at schools and universities all around the Mediterranean, and the effect of marine and terrestrial invasions should be a main subject and an integral part of biology curricula.

The second action is to accept the recommendations proposed by marine scientists and act accordingly in order to reduce the invasion problem. It is shameful and very serious that Mediterranean countries that ratified international scientific conventions dealing with the solutions to this problem have not yet drafted any text of law to legally implement these conventions (Boudouresque and Verlaque 2005, 2010). Governments must adopt and implement conventions such as those published by the Food and Agriculture Organization of the United Nations (FAO 1996), the International Union for Conservation of Nature (IUCN 2002) the International Council for the Exploration of the Sea (ICES 2005), and the International Maritime Organization (IMO) (Tamelander et al. 2010), which recommended policies and actions that should take place in order to reduce marine-species introduction and their negative impacts. An excellent example of such a recommendation that should be implemented is the ones concerned with the reduction of alien invasion (especially alien seaweeds) due to mariculture and shellfish transfer. Verlaque et al. (2007) note a plan with seven comprehensive guidelines modified from the IUCN (Shine et al. 2000) and ICES (2005) guidelines in order to prevent future invasion by shellfish transfer. These guidelines are: (1) to increase the awareness of farmers concerning the risks associated with uncontrolled importation; (2) aquaculture should be based on native or local stock whenever possible. Import and transfer of livestock should be kept to a minimum, and the animals should be thoroughly inspected and quarantined for an appropriate observation period; (3) special attention should be paid during aquaculture trials with new exotic species (even with livestock from hatcheries); (4) non-native livestock to be introduced should be produced in hatcheries; (5) live products destined for consumption, processing, aquaria, or display should not be placed into the natural environment; (6) in the case of livestock transfers (including interregional ones), decontamination processes and/or quarantine, as proposed by the ICES, should be followed; and (7) efficient treatments (e.g., hot seawater for sterilization of oysters from epiphytes growing on its shell) to avoid introduction or secondary dispersal of exotic or native species should be carried out prior to each transfer, i.e., after the period of re-immersion preceding the transfer, and should be repeated upon arrival. These seven guidelines should be adopted by all Mediterranean countries that have aquaculture based on imported marine organisms (especially shellfish).

Shipping has universally been regarded as the single largest vector for the movement of aquatic alien species (Galil 2006b). A new convention formulated by the IMO focuses on ballast-water management strategies and recommends some new standards and procedures in order to prevent alien invasion via shipping, namely cargo ships. This convention had published recommendations designed to prevent the spread of alien species from ballast water origins (Tamelander et al. 2010).

IMO Regulation D-2, which deals with ballast-water performance standards, recommends that "Ships conducting ballast water management shall discharge less than 10 viable organisms per cubic meter greater than or equal to 50 μm in minimum dimension and less than 10 viable organisms per milliliter less than 50 μm in minimum dimension and greater than or equal to 10 μm in minimum dimension; and discharge of the indicator microbes shall not exceed the specified concentrations" (Tamelander et al. 2010).

In order to provide a new product to implement Regulation D-2, Siemens Water Technologies has developed and launched a new SiCURE™ Ballast Water Management System that provides a reliable, environmentally sound solution designed to protect against the proliferation of aquatic invasive species. This filter system is based on three pillars: (1) filters to remove or break larger organisms using a 40 μm weave wire screen; (2) hypochlorination (with NaOCl) using a concentric-tube electrode in order to oxidize and eliminate marine organisms that passed through the filters; and (3) proprietary control logic of the ballast water management system monitors the chlorine dose level necessary to provide the required efficacy (SiCURE™ 2012). Such a solution should be fixed by law all across the Mediterranean and every cargo ship that carries ballast water should be required to use an appropriate filtration system to minimize the spread of marine alien species in the Mediterranean Sea and the oceans.

In several regions, hull fouling is the dominant species-introduction vector and it is estimated that biofouling is responsible for the introduction of over two thirds of all non-native algal species globally. In 2001, the IMO set up the International Convention on the Control of Harmful Anti-fouling Systems on Ships (Anti-Fouling Convention) (Tamelander et al. 2010). This convention and the ones by ICES that followed during 2006, recommended emerging hull-fouling regulations and treatment options (ICES 2006), including seven guidelines with a general objective to minimize the number of unintentional species introductions associated with hull fouling: (1) to encourage the necessary research, development, and sharing of an adequate knowledge base to address the problems of hull-fouling mediated introductions of alien species in the Mediterranean; (2) to increase awareness of hull fouling as a major introduction vector; (3) to technically assist and advise the Mediterranean coastal states, if requested, to ratify the IMO AFS Convention; (4) to encourage the development and implementation of control efforts, such as hull cleaning measures; (5) to encourage the development of a framework for national legislation and regional cooperation to regulate the introduction of hull-fouling mediated species introductions, their eradication, and control; (6) to design a lead agency that would have a central responsibility within the government for coordinating the national response to the above issues; (7) to form a national task force to develop and implement the proposed guidelines. This national task force can be cross-sectoral and multidisciplinary (ICES 2006).

Other strategies on a national level published by the ICES were: (1) routine vessel monitoring to document the risk of species invasions in hull fouling; (2) identification of vessels that likely carry high-risk species in their hull fouling (risk assessment); (3) identification of ports that receive a large number of 'critical' vessels; (4) evaluation of hull treatment methods for 'critical' vessels; (5) make all dockyard and scrapyard operators aware that organisms removed from ship hulls should be collected and discharged safely on land; (6) strongly encourage marina operators to apply the proposed guidelines (ICES 2006).

Meetings such as the ICES Working Group on Introduction and Transfers of Marine Organisms (WGITMO), taking place in March 2012 in Lisbon, accompanied by International Oceanographic Commission (IOC) and IMO, and dealing with the progress achieved using new technologies and treatments against the alien invasion from ballast water and other ship-vector origins, are very important and encouraging, and it is expected that new regulations on these subjects will be published soon (ICES 2012).

The third action is to start a new global Mediterranean monitoring program focussing on the Suez Canal and the Strait of Gibraltar, in order to detect the new aliens that arrive naturally, and also focusing on some hotspot areas such as some of the North African shores, the Levantine-Basin coasts, harbors, and estuarine areas, which were not well studied so far. Monitoring the Suez Canal and its estuary near Port Said in order to detect and even prevent alien invasion to the Mediterranean is a very difficult and expensive mission that requires huge unending international efforts. A set of filters should be established through the canal in order to detect the new species that pass through it and reduce the infiltration of these alien species into the Mediterranean Sea.

Acknowledgments Razy Hoffman is a VATAT-supported post-doctoral fellow at the Steinhardt National Collections of Natural History, Tel Aviv University. I would like to thank Sharon Victor from the Mina and Everard Goodman Faculty of Life Sciences, Bar-Ilan University, for editing this chapter.

References

Aleem AA (1962) The occurrence of the seagrass *Halophila stipulacea* (Forsk.) Asch. On the west coast of Egypt. Bull Fac Sci 4:79–84, Alexandria

Aleem AA (1993) Marine algae of Alexandria, Egypt. University of Alexandria Publication, Alexandria

Alpinar K (1987) A new record for Turkish flora *Halophila stipulacea*. Istanbul Universitesi Eczacilik Fakultesi Mecmuasi 23:83–84

Belkin IM (2009) Rapid warming of large marine ecosystems. Prog Oceanogr 81:207–213

Bernard G, Boudouresque CF, Picon P (2007) Long term changes in *Zostera* meadows in the Berre lagoon (Provence, Mediterranean Sea). Estuar Coast Shelf S 73:617–629

Bitar G, Harmelin JG, Verlaque M, Zibrowius H (2000) Sur la flore marine benthique suppose lessepsienne de la côte Libanaise. Cas particulier de *Stypopodium schimperi*. In: Proceedings of the first Mediterranean symposium on marine vegetation, Ajaccio, France. Regional Activity Centre for Specially Protected Areas (RAC/SPA), Tunis

Bitar G, Dupuy de la Grandrive R, Foulquié M (2003) Second mission relating to the development of marine protected areas in Syrian coasts. Available on-line at:http://medmpa.rac-spa.org/pdf/Rapports/Syrie/Mission%20Report_October%202003.pdf. Viewed 1/15/12, 1–18 Aug 2003

Borum J, Greve TM (2004) The four European seagrass species. In: Borum J, Duarte CM, Krause-Jensen D, Greve TM (eds) European seagrasses: an introduction to monitoring and management. The M & MS project, Hillerod

Boudouresque CF (2003) The erosion of Mediterranean biodiversity. In: Rodríguez-Prieto C, Pardini G (eds) The Mediterranean Sea: an overview of its present state and plans for future protection. Servei de Publicacions de la Universitat de Girona, Girona

Boudouresque CF, Verlaque M (2002) Biological pollution in the Mediterranean Sea: invasive versus introduced macrophytes. Mar Pollut Bull 44:32–38

Boudouresque CF, Verlaque M (2005) Nature conservation, marine protected areas, sustainable development and the flow of invasive species to the Mediterranean Sea. Sci Rep Port-Cros Nation Park 21:29–54

Boudouresque CF, Verlaque M (2010) Is global warming involved in the success of seaweed introductions in the Mediterranean Sea? In: Israel A, Einav R, Seckbach J (eds) Seaweeds and their role in globally changing environments. Cellular origin, life in extreme habitats and astrobiology, vol 15. Springer Publishers, Dordrecht

Boudouresque CF, Ruitton S, Verlaque M (2005) Large-scale disturbances, regime shift and recovery in littoral systems subject to biological invasions. In: Velikova V, Chipev N (eds) Large-scale disturbances (regime shifts) and recovery in aquatic ecosystems: challenges for management towards sustainability. UNESCO-ROSTE/BAS Workshop on Regime Shifts. UNESCO Publication, Varna

Boudouresque CF, Klein J, Ruitton S, Verlaque M (2011) Biological invasions: the Thau lagoon, a Japanese biological island in the Mediterranean Sea. In: Ceccaldi HJ, Dekeyser I, Girault M, Stora G (eds) Global change: mankind-marine environment interactions. Proceedings of the 13th French-Japanese Oceanography Symposium. Springer, Dordrecht

Buia MC, Gambi MC, Zupo V (2000) Structure and functioning of Mediterranean seagrass ecosystems: an overview. Biol Mar Medit 7:167–190

Ceccherelli G, Campo D (2002) Different effects of *Caulerpa racemosa* on two co-accurring seagrasses in the Mediterranean. Bot Mar 45:71–76

Ceccherelli G, Cinelli F (1997) Short-term effects of nutrient enrichment of the sediment and interactions between the seagrass *Cymodocea nodosa* and the introduced green alga *Caulerpa taxifolia* in a Mediterranean bay. J Exp Mar Biol Ecol 217:165–177

Ceccherelli G, Piazzi L, Cinelli F (2000) Response of the non-indigenous *Caulerpa racemosa* (Forsskal) J. Agardh to the native seagrass *Posidonia oceanica* (L.) Delile: effect of density of shoots and orientation of edges of meadows. J Exp Mar Biol Ecol 243:227–240

Chapman AS (1999) From introduced species to invader: what determines variation in the success of *Codium fragile* ssp. *tomentosoides* (Chlorophyta) in the North Atlantic Ocean. Helgoländer Meeresun 52:277–289

Cocito S, Bianchi CN, Morri C, Peirano A (2000) First survey of sessile communities on subtidal rocks in an area with hydrothermal vents: milos Island, Aegean Sea. Hydrobiologia 426:113–121

Colautti RI, MacIsaac HJ (2004) A neutral terminology to define "invasive" species. Divers Distrib 10:135–141

Critchley AT (1983) *Sargassum muticum*: a taxonomic history including world wide and western pacific distributions. J Mar Biol Assoc UK 63:617–625

Critchley AT, Farnham WF, Yoshida T, Norton TA (1990) A bibliography of the invasive alga *Sargassum muticum* (Yendo) Fensholt (Fucales: Sargassaceae). Bot Mar 33:551–562

Davis MA, Thompson K, Grime JP, Charles S (2001) Elton and the dissociation of invasion ecology from the rest of ecology. Divers Distrib 7:97–102

de Ville`le X, Verlaque M (1995) Changes and degradation in a *Posidonia oceanica* bed invaded by the introduced tropical alga *Caulerpa taxifolia* in the north western Mediterranean. Bot Mar 38:79–87

Den Hartog C (1970) The seagrasses of the world. North Holland Publishing Company, Amsterdam

Di Martino V, Blundo MC, Tita G (2006) The Mediterranean introduced seagrass *Halophila stipulacea* in eastern Sicily (Italy): temporal variations of the associated algal assemblage. Vie Milieu 56:223–230

Duarte CM, Marba N, Santos R (2004) What may cause loss of seagrasses? In: Borum J, Duarte CM, Krause-Jensen D, Greve TM (eds) European seagrasses: an introduction to monitoring and management. The M & MS Project, Hillerod

Dukes JS, Mooney HA (1999) Does global change increase the success of biological invaders? Trends Ecol Evol 14:135–139

Edelstein T (1960) The biology and ecology of deep sea algae of the Haifa Bay area. PhD Dissertation, Hebrew University of Jerusalem (in Hebrew)

Edelstein T (1962) On the algal association and the ecology of the benthonic flora of the Haifa Bay. Rec Trav Stn Mar Endoume Bull 27:209–219

Edelstein T (1964) On the sublittoral algae of the Haifa Bay area. Vie Milieu 15:177–212

Edelstein T, Komarovsky B (1961) Epiphytic algae on *Halimeda tuna* f. *platydisca* (Dec.) Bart. In Haifa Bay. Bull Res Council Israel 10d:54–58

Elton CS (1958) The ecology of invasions by animals and plants. Methuen, London

EPA – United States Environmental Protection Agency (2001) http://www.epa.gov/gmpo/nonindig.html. Aquatic nuisance species. Annual Report

European Environmental Agency (2007) Halting the loss of biodiversity by 2010: proposal for a first set of indicators to monitor progress in Europe, Published by EEA, Technical Report N° 11/2007, Copenhagen

FAO – Food and Agriculture Organization of the United Nation (1996) Precautionary approach to capture fisheries and species introductions. Elaborated by the Technical Consultation on the Precautionary Approach to Capture Fisheries (Including Species Introductions). Lysekil, Sweden. FAO Technical Guidelines for Responsible Fisheries. No. 2. FAO, Rome. p 54. http://www.fao.org/DOCREP/003/W3592E/w3592e0a.htm#bm10, Viewed 5/12/2011, 6–13 June 1995

Flagella MM, Verlaque M, Soria A, Buia MC (2007) Macroalgal survival in ballast water tanks. Mar Pollut Bull 54:1395–1401

Fritsch C (1895) Ueber die Auffindung einer marinen Hydrocharidee im Mittelmeer. Verhandlungen der kk Zoologisch-Botanischen Gesellschaft Wien 45:104–106

Galil BS (2006a) The marine caravan-the Suez Canal and the Erithrean invasion. In: Gollasch S, Galil BS, Cohen AN (eds) Bridging divides, maritime canals as invasion corridors. Springer Publishers, Dordrecht

Galil BS (2006b) Shipwrecked – shipping impacts on the biota of the Mediterranean Sea. In: Davenport J, Davenport JL (eds) The ecology of transportation: managing mobility for the environment. Springer Verlag, Dordrecht

Gambi MC, Barbieri F, Bianchi CN (2009) New record of the alien seagrass *Halophila stipulacea* (Hydrocharitaceae) in the western Mediterranean: a further clue to changing Mediterranean Sea biogeography. Mar Biodivers Rec 2(1–7):e84. doi:10.1017/S175526720900058X

Golani D, Ben-Tuvia A (1995) Lessepsian migration and the Mediterranean fisheries of Israel. In: Armantrout NB (ed) Conditions of the world's aquatic habits. Proceedings of the world fish congress, theme 1. Oxford and IBH Publishing Company, New Delhi

Golani D, Orsi-Relini L, Massuti E, Quingnard JP (2002) 3: fishes. In: Briand F (ed) CIESM atlas of exotic species in the Mediterranean. CIESM Publishers, Monaco

Goren M, Aronov A (2002) First record of the Indo-Pacific parrot fish *Scarus ghobban* in the Eastern Mediterranean. Cybium 26:239–240

Gravez V, Ruitton S, Boudouresque CF, Meinesz A, Scabbia G, Verlaque M (2001) Fourth international workshop on *Caulerpa taxifolia*. GIS Posidonia Publ, Marseille

Guiry MD, Guiry GM (2011) AlgaeBase. World-wide electronic publication, National University of Ireland, Galway. http://www.algaebase.org; searched on 8 Oct 2011

Harries DB, Cook E, Donnan DW, Mair JM, Harrow S, Wilson JR (2007) The establishment of the invasive alga *Sargassum muticum* on the west coast of Scotland: rapid northwards spread and identification of potential new areas for colonisation. Aquat Invasions 2:367–377

Hoffman R, Dubinsky Z (2010) Invasive and Alien Rhodophyta in the Mediterranean and along the Israeli shores. In: Seckbach J, Chapman DJ (eds) Red algae in genome age. Cellular origin, life in extreme habitats and astrobiology, 13th edn. Springer Publishers, Dordrecht

Hoffman R, Iluz D, Kamenir Y, Dubinsky Z, Israel A, Pollution effects on benthic intertidal algal diversity, along the Israeli Mediterranean (submitted to Mar Pollut Bull)

Hoffman R, Israel A, Lipkin Y, Dubinsky Z, Iluz D (2008a) First record of two seaweeds from the Israeli Mediterranean *Galaxaura rugosa* (J Ellis & Solander) J.V. Lamouroux (Rhodophyta) and *Codium adhaerens* C. Agardh (Chlorophyta). Israel J Plant Sci 56:123–126

Hoffman R, Dubinsky Z, Israel A, Iluz D (2008b) The mysterious disappearance of *Halimeda tuna* from the intertidal zone along the Israeli Mediterranean. Israel J Ecol Evol 54:267–268

Hoffman R, Shemesh E, Ramot M, Dubinsky Z, Pinchasov-Grinblat Y, Iluz D (2011) First record of *Codium arabicum* Küt (Bryopsidales, Chlorophyta) in the Mediterranean Sea. Bot Mar 54:487–495

Huey RB, Gilchrist GW, Carlson ML, Berrigan D, Serra L (2000) Rapid evolution of a geographic cline in size in an introduced fly. Science 287:308–309

ICES – International Council for the Exploration of the Sea (2005) ICES code of practice on the introductions and transfers of marine organisms. International Council for the Exploration of the Seas, Copenhagen

ICES – International Council for the Exploration of the Sea (2006) Report of the ICES/IOC/IMO Working Group on Ballast and Other Ship Vectors (WGBOSV), 13–15 Mar 2006, Oostende. ACME:06

ICES – International Council for the Exploration of the Sea (2012) ICES Working Group on Introduction and Transfers of Marine Organisms (WGITMO) http://ices-usa.noaa.gov/working_groups/ACOM/WGITMO. Html Viewed 28 Feb 2012

Israel A, Einav R, Silva PC, Paz G, Chacana ME, Douek J (2010) First report of the seaweed *Codium parvulum* (Chlorophyta) in Mediterranean waters: recent blooms on the northern shores of Israel. Phycologia 49:107–112

IUCN – The World Conservation Union (2002) Policy recommendations. Sixth Meeting of the Conference of the Parties to the Convention on Biological Diversity (COP6), Hague. http://www.iucn.org/themes/pbia/wl/docs/biodiversity/cop6/invasives.doc 7–19 Apr 2002

Kashta L, Beqiraj S, Tilot V, Zuna V, Dodbiba E (2011) The first MPA in Albania, Sazani island-Karaburuni Peninsula, as a regional priority conservation area for marine biodiversity. Varstvo Narve 1:139–158

Kolar CS, Lodge DM (2001) Progress in invasion biology: predicting invaders. Trends Ecol Evol 16:199–204

Kurt O, Ozturk M, Taskın E, Aysel V (2001) Three new records for the marine algal flora of the Mediterranean Sea. In: Abstracts of 7th International Phycological Congress, Phycologia 40, supplement, ipc 7 Thessaloniki, 18–25 Aug 2001, p 68

Lipkin Y (1972a) Marine algal and sea-grass flora of the Suez Canal (the significance of this flora to the understanding of the recent migration through the Canal). Isr J Zool 21:405–446

Lipkin Y (1972b) Vegetation of the Bitter lakes in the Suez Canal water system. Isr J Zool 21:447–457

Lipkin Y (1975a) Halophila stipulacea, a review of a successful immigration. Aquat Bot 1:203–215

Lipkin Y (1975b) *Halophila stipulacea* in Cyprus and Rhodes, 1967–1970. Aquat Bot 1:309–320

Lipkin Y (1975c) A history, catalogue and bibliography of Red Sea seagrasses. Israel J Bot 24:89–105

Lipkin Y (1977) Seagrasses of Sinai and Israel. In: McRoy CP, Helfferich C (eds) Seagrass ecosystems: a scientific perspective. Marcel Dekker, Inc., New York

Lipkin Y (1979) Quantitative aspects of seagrass communities, particularly of those dominated by *Halophila stipulacea*, in Sinai (Northern Red Sea). Aquat Bot 7:119–128

Lipkin Y, Safriel U (1971) Intertidal zonation on rocky shores at Mikhmoret (Mediterranean, Israel). J Ecol 59:1–30

Lipkin Y, Beer S, Zakai D (2003) The seagrasses of the eastern Mediterranean and the Red Sea. In: Green EP, Short FT (eds) World atlas of seagrasses. University of California Press, Berkeley

Lowe S, Browne M, Boudjelas S, de Poorter M (2000) 100 of the world's worst invasive alien species. A selection from the Global Invasive Species Database. Published by The Invasive Species Specialist Group (ISSG), a specialist group of the Species Survival Commission (SSC) of the World Conservation Union (IUCN), Auckland

Lundberg B (1981) The algal vegetation at a platform shore at Mikhmoret. Proc Int Seaweed Symp 10:315–320

Lundberg B (1986) Variations in algal vegetation along the Mediterranean shore line of Israel as possible basis for planning of marine nature reserves. In: Dubinsky Z, Steinberger Y (eds) Environmental quality and ecosystem stability, vol III A/B. Bar-Ilan University Press, Ramat-Gan

Lundberg B (1991) Algal vegetation on the vermetid platforms, Habonim, Israel. Oebalia 17:493–507

Lundberg B (1995) Intertidal seaweeds in Israel, a field guide for the Mediterranean coasts. Supplement No. 3, Nature Conservation in Israel, Research and Surveys. Nature Reserves Authority, Jerusalem

Lundberg B (1996) Composition of the seaweed vegetation along the Mediterranean coast of Israel (report including data from 19 stations at 72 visits during the years 1973–1995). Nature Conservation in Israel, Research and Surveys, Supplement 3, Nature Reserves Authority, Jerusalem

Maggs CA, Stegenga H (1999) Red algal exotics on North Sea coasts. Helgoländer Meeresun 52:243–258

Mayhoub H, Billard C (1991) Contribution à la connaissance d'un *Stypopodium* (Dictyotales, Phaeophyceae) installé récemment sur les côtes syriennes. Cryptg Algol 12:125–136

Meinesz A (1999) Killer algae: the true tale of a biological invasion. University of Chicago Press, Chicago

Meinesz A, Belsher T, Thibaut T, Antolic B, Ben Mustapha K, Boudouresque C-F, Chiaverini D, Cinelli F, Cottalorda J-M, Djellouli A, El Abed A, Orestano C, Grau AM, Ivesa L, Jaklin A, Langar H, Massuti-Pascual E, Peirano A, Tunesi L, de Vaugelas J, Zavodnik N, Zuljevic A (2001) The introduced green alga *Caulerpa taxifolia* continues to spread in the Mediterranean. Biol Invasions 3:201–210

Mollo E, Gavagnin M, Carbone M, Castelluccio F, Pozone F, Roussis V, Templado J, Ghiselin MT, Cimino G (2008) Factors promoting marine invasions: a chemoecological approach. Proc Natl Acad Sci 105:4582–4586

Mooney HA, Cleland EE (2001) The evolutionary impact of invasive species. Proc Natl Acad Sci USA 98:5446–5451

Mooney HA, Hobbs RJ (2000) Invasive species in a changing world. Island Press, Washington, DC

Nehring S (2002) Biological invasions into German waters: an evaluation of the importance of different human-mediated vectors for nonindigenous macrozoobenthic species. In: Leppäkoski E, Gollasch S, Olenin S (eds) Invasive aquatic species of Europe: distribution, impacts and management. Kluwer, Dordrecht

Nehring S (2005) International shipping – a risk for aquatic biodiversity in Germany. In: Nentwig W, Bacher S, Cock MJW, Dietz H, Gigon A, Wittenberg R (eds) Biological invasions – from ecology to control, Neobiota 6. Pensoft Publishers, Sofia, pp 125–143

Ng AKY, Song S (2010) The environmental impacts of pollutants generated by routine shipping operations on ports. Ocean Coast Manage 53:301–311

Norton TA (1977) The growth and development of *Sargassum muticum* (Yendo) Fensholt. J Exp Mar Biol Ecol 26:41–53

Ozturk M, Wynne MJ, Taskin E (2006) First report of the *Microspongium globosum* Reinke (Phaeophyceae, Myrionemataceae) in the Mediterranean Sea. Nova Hedwigia 82:135–142

Pergent G, Boudouresque CF, Dumay O, Pergent-Martini C, Wyllie-Echeverria S (2008) Competition between the invasive macrophyte *Caulerpa taxifolia* and the seagrass *Posidonia oceanica*: contrasting strategies. BMC Ecol 8:20

Perrings C (2010) Elton and the economics of biological invasions. In: Richardson DM (ed) Fifty years of invasion ecology: the legacy of Charles Elton. Wiley-Blackwell, Oxford

Phillips RC, Meñez EG (1988) Seagrasses. Smithsonian Institution Press, Washington, DC

Piazzi L, Cinelli F (2003) Evaluation of benthic macroalgal invasion in the harbour area of the western Mediterranean Sea. Eur J Phycol 38:223–231

Piazzi L, Ceccherelli G, Cinelli F (2001) Threat to macroalgal diversity: effects of the introduced green alga *Caulerpa racemosa* in the Mediterranean. Mar Ecol Prog Ser 210:149–159

Piazzi L, Meinesz A, Verlaque M, Açali B, Antolic B, Argyrou M, Baltana D, Ballesteros E, Calvo S, Cinelli F, Cirik S, Cossu A, d'Archino R, Djellouli AS, Javel F, Lanfranco E, Mifsud C, Pala D, Panayotidis P, Peirano A, Pergent G, Petrocelli A, Ruitton S, Zuljevic A, Ceccherelli G (2005) Invasion of *Caulerpa racemosa* var. *cylindracea* (Caulerpales, Chlorophyta) in the Mediterranean Sea: an assessment of the spread. Cryptg Algol 26:189–202

Pimental D, Lach L, Zuniga R, Morrison D (2000) Environmental and economic costs of nonindigenous species in the United States. Bioscience 50:53–65

Politis J (1930) Plantes marines de la Grèce. Rapp PV Rèun Comm Int Explor Scient Met Mèditerranèe N S 5:195–205

Por FD (1971) 100 years of Suez Canal-century of Lessepsian migration-Retrospect and viewpoints. Syst Zool 20:138–159

Por FD (1978) Lessepsian migration: the influx of Red Sea biota into the Mediterranean by way of the Suez Canal. Springer, Berlin/Heidelberg/New York

Procaccini G, Acunto S, Fama P, Maltagliati F (1999) Structural, morphological and genetic variability in *Halophila stipulacea* (Hydrocharitaceae) populations in the western Mediterranean. Mar Biol 135:181–189

Procaccini G, Buia MC, Gambi MC, Perez M, Pergent G, Pergent-Martini C, Romero J (2003) The seagrasses of the western Mediterranean. In: Green EP, Short FT (eds) World atlas of seagrasses. Prepared by the UNEP World Conservation Monitoring Center. University of California Press, Berkeley

Provan J, Murphy S, Maggs CA (2005) Tracking the invasive history of the green alga *Codium fragile* ssp. *tomentosoides*. Mol Ecol 14:189–194

Richardson DM, Pysek P, Rejmanek M, Barbour MG, Panetta FD, West CJ (2000) Naturalization and invasion of alien plants: concepts and definitions. Divers Distrib 6:93–107

Rilov G, Crooks JA (2009) Marine bioinvasions: conservation hazards and vehicles for ecological understanding. In: Rilov G, Crooks JA (eds) Biological invasions in marine ecosystems. Ecological studies 204, vol 1. Springer, Berlin/Heidelberg

Rilov G, Galil B (2009) Marine bioinvasions in the Mediterranean sea-history, distribution and ecology. In: Rilov G, Crooks JA (eds) Biological invasions in marine ecosystems. Ecological studies 204, vol 31. Springer, Berlin/Heidelberg

Ruiz H, Ballantine DL (2004) Occurrence of the seagrass *Halophila stipulacea* in the tropical west Atlantic. Bull Mar Sci 75:131–135

Scagel RF (1956) Introduction of a Japanese alga, *Sargassum muticum* into the north-east Pacific. Fish Res Pap St Wash 1:49–58

Serio D, Petrocelli A, Cormaci M, Cecere E, Furnari G (2008) First record of *Osmundea oederi* (Gunnerus) G. gurnari comb. nov. (Rhodomelaceae, Rhodophyta) from the Mediterranean Sea. Cryptg Algol 29:119–127

Sfriso A, Maistro S, Andreoli C, Moro I (2010) First record of *Gracilaria vermiculophylla* (Gracilariales, Rhodophyta) in the Po Delta lagoons, Mediterranean Sea (Italy) (Note). J Phycol 46:1024–1027

Sghaier YR, Zakhama-Sraieb R, Benamer I, Charfi-Cheikhrouha F (2011) Occurrence of the seagrass *Halophila stypulacea* (Hydrocharitaceae) in the southern Mediterranean Sea. Bot Mar 54:575–582

Shine C, Williams N, Gündling L (2000) A guide to designing legal and institutional frameworks on alien invasive species. IUCN Gland, Cambridge/Bonn

SiCURE™ - http://www.water.siemens.com/en/products/chemical_feed_disinfection/hypochlorite_generation/Pages/sicure_ballast_water_management_system.aspx?stc=wwiis250128viewed27/2/2012

Sorte CJB, Williams SL, Zerebecki RA (2010) Ocean warming increases threat of invasive species in a marine fouling community. Ecology 91:2198–2204

Stachowicz JJ, Terwin JR, Whitlatch RB, Osman RW (2002) Linking climate change and biological invasions: ocean warming facilitates nonindigenous species invasions. Proc Natl Acad Sci USA 99:15497–15500

Steiner SCC, Macfarlane KJL, Price M, Willette DA (2010) The distribution of seagrasses in Dominica, Lesser Antilles. Rev Biol Trop 58:89–98

Streftaris N, Zenetos A (2006) Alien marine species in the Mediterranean – the 100 'worst invasives' and their impact. Mediterr Mar Sci 7:87–118

Streftaris N, Zenetos A, Papathanassiou E (2005) Globalisation in marine ecosystems: the story of non-indigenous marine species across European seas. Oceanogr Mar Biol Annu Rev 43:419–453

Täckholm V, Täckholm G, Drar M (1941) Flora of Egypt, vol I. Bull Fac Sci Egypt Univ 17:1–574

Tamelander J, Riddering L, Haag F, Matheickal J (2010) Guidelines for Development of National Ballast Water Management Strategies. GEF-UNDP-IMO GloBallast, London and IUCN, Gland. GloBallast Monographs No. 18

Tan K, Mullaj A (2000) *Parapholis filiformis* (Roth) C.E. Hubb., *Halophila stipulacea* (Forssk.) Asch., *Juncus subnodulosus* Schrank. Willdenowia 30:242

Taskin E (2006) First report of *Corynophlaea crispa* (Harvey) Kuckuck (Phaeophyceae, Corynophlaeaceae) in the Mediterranean Sea. Nova Hedwigia 82:217–225

Taskin E, Öztürk M, Kurt O, Aysel V (2001) Three new records for the marine algal flora of Turkey. Turk J Bot 25:245–248

Taskin E, Öztürk M, Kurt O, Öztürk M (2008) The check-list of the marine algae of Turkey. Ecem Kirtasiye, Manisa, Turkey

Taskin E, Kurt O, Cormaci M, Furnari G, Öztürk M (2010) Two brown algae from the eastern Mediterranean Sea: *Microcoryne ocellata* Strömfelt and *Corynophlaea flaccida* (C. agardh) Kützing. Fresen Environ Bull 19:892–896

Taşkın E, Öztürk M, Pedersen PM (2007) A rare brown alga in the Mediterranean Sea: *compsonema saxicola* (Kuckuck) Kuckuck (Phaeophyceae, Scytosiphonaceae). Rapp Comm Int Mer Médit 38:692

Tsirika A, Haritonidis S (2005) A survey of the benthic flora in the National Marine Park of Zakynthos (Greece). Bot Mar 48:38–45

Uwai S, Nelson W, Neill K, Wang WD, Aguilar-Rosas LE, Boo SM, Kitayama T, Kawai H (2006) Genetic diversity in *Undaria pinnatifida* (Laminariales, Phaeophyceae) deduced from mitochondria genes – origins and succession of introduced populations. Phycologia 45:687–695

Verlaque M (1994) Inventaire des plantes introduites en Méditerranée: origines et répercussions sur l'environnement et les activités humaines. Oceanol Acta 17:1–23

Verlaque M (2001) Checklist of the macroalgae of Tau Lagoon (Hérault, France): a hot spot of marine species introduction in Europe. Oceanol Acta 24:29–49

Verlaque M, Boudouresque CF (1991) *Stypopodium schimperii* (Dictyotales, Fucophyceae), algue de mer Rouge récemment apparue en Méditerranée. Cryptg Algol 12:195–211

Verlaque M, Afonso-Carrillo J, Gil-Rodriguez MC, Durand C, Boudouresque CF, Le Parco Y (2004) Blitzkrieg in a marine invasion: *Caulerpa racemosa* var. *Cylindracea* (Bryopsidales, Chlorophyta) reaches the Canary Islands (NE Atlantic). Biol Invasions 6:269–281

Verlaque M, Boudouresque CF, Mineur F (2007) Oyster transfers as a vector for marine species introductions: a realistic approach based on the macrophytes. CIESM Workshop Monographs, Monaco, 32:39–48

Verlaque M, Ruitton S, Boudouresque CF, Mineur F (2011) CIESM site. CIESM atlas of exotic macrophytes in the Mediterranean Sea. http://www.ciesm.org/atlas/appendix4.html. Searched on Nov 2011

Ward NL, Masters GL (2007) Linking climate change and species invasion: an illustration using insect herbivores. Glob Change Biol 13:1605–1615

Wilcove DS, Rothstein D, Dubow J, Phillips A, Losos E (1998) Quantifying threats to imperiled species in the United States. Bioscience 48:607–615

Willette DA, Ambrose R (2009) The distribution and expansion of the invasive seagrass *Halophila stipulacea* in Dominica, West Indies, with a preliminary report from St. Lucia. Aquat Bot 91:137–142

Williams SL, Grosholz ED (2008) The invasive species challenge in estuarine and coastal environments: marrying management and science. Estuar Coasts 31:3–20

Williamson M (1996) Biological invasions. Chapman and Hall, London

Wynne MJ (2005) A check-list of benthic marine algae of the tropical and subtropical western Atlantic: second revision. Beihefte zur Nova Hedwigia 129:1–152

Zenetos A, Çinar ME, Pancucci-Papadopoulou MA, Harmelin JG, Furnari G, Andaloro F, Bellou N, Streftaris N, Zibrowius H (2005) Annotated list of marine alien species in the Mediterranean with records of the worst invasive species. Mediterr Mar Sci 6:63–118

Zenetos A, Meriç E, Verlaque M, Galli P, Boudouresque CF, Giangrande A, Çınar ME, Bilecenoglu M (2008) Additions to the annotated list of marine alien biota in the Mediterranean with special emphasis on Foraminifera and parasites. Mediterr Mar Sci 9:119–165

Zenetos A, Gofas S, Verlaque M, Cinar M, Garcia Raso J, Bianchi C, Morri C, Azzurro E, Bilecenoglu M, Froglia C, Siokou I, Violanti D, Sfriso A, San Martin G, Giangrande A, Katagan T, Ballesteros E, Ramos-Esplà A, Mastrototaro F, Ocana O, Zingone A, Gambi M, Streftaris N (2010) Alien species in the Mediterranean Sea. A contribution to the application of European Union's Marine Strategy Framework Directive (MSFD). Part I. Spatial distribution. Mediterr Mar Sci 11:381–493

Zenetos A, Katsanevakis S, Poursanidis D, Crocetta F, Damalas D, Apostolopoulos G, Gravili C, Vardala-Theodorou E, Malaquias M (2011) Marine alien species in Greek seas: additions and amendments by 2010. Mediterr Mar Sci 12:95–120

The Zooplankton

11

Enric Saiz, Ana Sabatés, and Josep-Maria Gili

Abstract

The Mediterranean Sea bears one of the longest traditions in the study of zooplankton. Pioneer work focused more on taxonomy to suit the high diversity of zooplankton that inhabits the Mediterranean, and indeed very few ecosystems have such a deep and broad knowledge on zooplankton like the one gathered in the Mediterranean. Further research into ecological aspects encompassed more descriptive studies on distribution and seasonality patterns, to lead in current times to more functional and modelling approaches. In this chapter we present an overview of the seasonality and distribution patterns of zooplankton in the Mediterranean, with special emphasis on the driving mechanisms behind. We will focus essentially on the western basin, and provide an overview on the outcome of the studies conducted by the different schools of marine biologists and oceanographers established in the geographical areas surrounding the Catalan/Balearic Sea, the Ligurian Sea and the Thyrrenian Sea. The latitudinal gradient, the marked seasonality, and the particular productivity patterns in the Mediterranean provide a frame for a zooplankton community in which the succession of species and assemblages occurs with not drastic changes in the whole standing stocks. As in other seas, examples of diel, ontogenetic and seasonal vertical migration can be found in the Mediterranean; however, the presence of a deep chlorophyll maximum during the stratified period drives a distinct vertical distribution of epipelagic zooplankton. Mesoscale singularities such as density fronts and eddies, the presence of submarine canyons which modify the general circulation, and other local phenomena like riverine runoff are responsible for the enhancement of production and favours the aggregation of zooplankton, either by local increase phenomena or by passive accumulation. Finally, current challenges in zooplankton research in the Mediterranean, such as the long-term changes in relation to large-scale atmospheric forcing, the occurrence of gelatinous zooplankton proliferations and the changes in species distribution and presence of non-indigenous ones will be also discussed.

Keywords

Mediterranean Sea • Zooplankton • Diversity of zooplankton • Distribution and seasonality patterns • Western basin • Catalan/Balearic Sea • Ligurian Sea • Thyrrenian Sea • Zooplankton community • Diel • Ontogenetic and seasonal vertical migration • Deep chlorophyll maximum • Epipelagic zooplankton • Long-term changes • Large-scale atmospheric forcing • Gelatinous zooplankton • Non-indigenous species

E. Saiz • A. Sabatés • J.-M. Gili (✉)
Institut de Ciències del Mar, Consejo Superior de Investigaciones Científicas, Barcelona, Spain
e-mail: gili@icm.csic.es

S. Goffredo and Z. Dubinsky (eds.), *The Mediterranean Sea: Its history and present challenges*,
DOI 10.1007/978-94-007-6704-1_11,

Introduction

The general composition and diversity of the Mediterranean zooplankton is the consequence of a set of recent geological events occurred since the Messinian Salinity Crisis 5 M years ago, which resulted in extreme climatic changes caused by the combination of glaciations and interglacial periods. For that reason, it is difficult to know the original composition of the planktonic fauna in the Mediterranean. The present zooplankton in the Mediterranean is a mixture of faunas with diverse origin, which have experienced severe selective environmental pressures and followed different evolution processes (Pérès 1985; Fredj et al. 1992). Overall, the Mediterranean zooplankton can be considered as a true Atlantic fauna, composed mainly of subtropical species together with a component of clearly tropical origin and another one of northern species (combination of a minority of boreal origin and a mayor set of species from temperate and cold areas); in addition, an influence from oriental origin, especially in the Eastern Mediterranean, cannot be dismissed (Furnestin 1968, 1979; Bianchi and Morri 2000; Gómez 2006). Moreover, the recent discovery in deep habitats, such as the submarine canyons, of endemic species considered as Thetian relicts (Gili et al. 2000), reinforces the hypothesis that the deep Mediterranean basins are semi isolated habitats that contribute to the speciation processes (Fredj et al. 1992).

The study of zooplankton in the Mediterranean has a long tradition of more than two centuries. Pioneering work was carried out at the marine stations in Villefranche-sur-Mer (France) and Naples (Italy), which rendered excellent publications still in use, such as the *Manuel de Planctologie Méditerranéene* by Trégouboff and Rose (1957) and the collection of the *Fauna und Flora des Golfes von Neapel*, and from the Spanish side it deserves special consideration the *Introducción al estudio del plancton marino* by Massutí and Margalef (1950). The information gathered from the numerous studies carried out during the last 100 years originates mainly from the coastal and neritic environments, sampled along the different seasons of the year, but oceanic research on vessels has also been carried out with obvious temporal and geographical constraints; in addition, several long-term series of plankton collection are also available in the western Mediterranean, the Adriatic Sea and the Aegean Sea. Even if some specific taxonomic groups have received particular attention (e.g. copepods), the Mediterranean is perhaps one of the marine regions where a broad knowledge of practically all groups is available. Through time, different schools have developed and gathered the major part of specialists in the surrounding land (the south of France: Banyuls-sur-Mer, Marseille, and Villefranche-sur-Mer; Italy: Genova, Naples, and Trieste; Spain: Barcelona, Malaga and Mallorca), which have contributed to improve our understanding of the systematics, and the population and community ecology of the zooplankton in the western Mediterranean (e.g. synthesis works of Vives 1966; Furnestin 1968; Estrada et al. 1985; Gaudy 1985; Champalbert 1996). Recent monographic studies on the fauna and taxonomy of the Mediterranean zooplankton (Bouillon et al. 2004; Vives and Shmeleva 2006, 2010) and other more ecological reviews (Siokou-Frangou et al. 2010; Durrieu de Madron et al. 2011) are good examples of the contemporary dynamism of zooplanktonic and oceanographic research in the Mediterranean.

This chapter does not intend to be a complete analysis of the zooplankton ecology in the Mediterranean Sea. Focused mainly on the western Mediterranean, we will make an effort to summarize, in a descriptive and comprehensive manner, the major patterns of zooplankton distribution and seasonality in the area in relation to the environmental mechanisms behind, and will also appraise the variations that zooplankton might be experiencing under the current global threat of climate change.

The Physical Framework for Zooplankton

The Mediterranean Sea is located in the temperate zone of the Northern Hemisphere in a relatively narrow band of latitudes (30° to 45° N) (Fig. 11.1). This implies a certain latitudinal gradient but, essentially, a marked seasonal cycle. In the western basin, the surface temperature ranges from 12 to 13 °C in winter to 26 °C in summer, while in the eastern basin the surface temperature is approximately 16–17 °C in winter and 27 °C in summer. The deep Mediterranean waters (from 250 m to maximum depth) are very homogeneous and well oxygenated, with a temperature of about 12.8 °C and a salinity of about 38.2 (Durrieu de Madron et al. 2011).

The Mediterranean is a deep basin with typically narrow continental shelves, which account for less than 20 % of the total sea surface. Within the NW Mediterranean, however, some relatively wide continental shelves are found near the main river mouths (Gulf of Lions and Ebro shelf). Over most of the Mediterranean Sea annual evaporation exceeds rainfall and river runoff, so on average it behaves as a concentration basin (Lacombe et al. 1981; Hopkins 1985). As a result of the net nutrient export, the Mediterranean is globally considered as oligotrophic. As a whole, the biological production decreases from north to south and west to east and is inversely related to the increase in temperature and salinity (Béthoux et al. 1992; Siokou-Frangou et al. 2010). However, the complex coastline and bathymetry of the Mediterranean basin, added to a strong seasonality, circulation patterns and river runoff lead to a high environmental

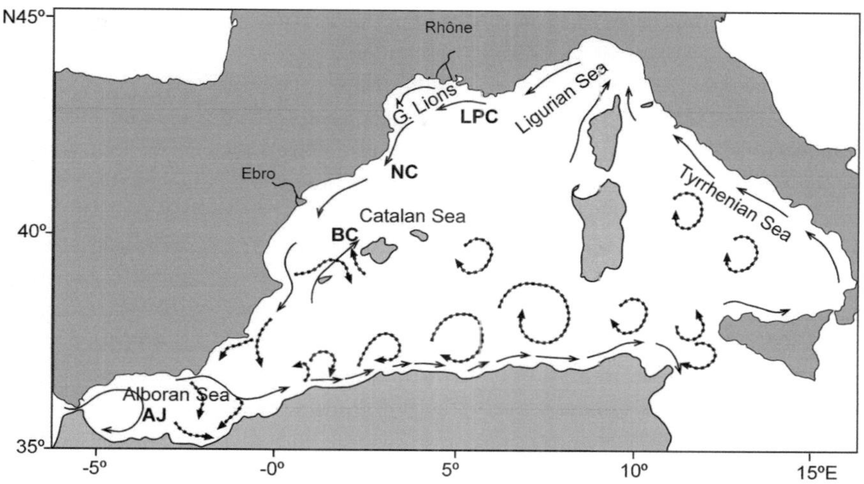

Fig. 11.1 Surface circulation of the western Mediterranean Sea (Redrawn from Millot and Taupier-Letage 2005). Acronyms: *LPC* Liguro-Provençal Current, *NC* Northern Current, *BC* Balearic Current, *AJ* Atlantic Jet

diversity at both regional and local scales, which may enhance production. The existence of these mechanisms would explain the moderate levels of primary production recorded, especially in the western basin, which sustain large fisheries and marine mammals communities (Estrada 1996; Coll et al. 2010).

The variety of habitats in the Mediterranean, together with a succession of colonization events, has shaped a remarkable diversity of species (Bianchi and Morri 2000). The latitudinal and longitudinal temperature gradients determine the distribution of the species. Thus, the species of subtropical origin are mainly found in the eastern basin and southern Mediterranean, where water temperature is higher than average. Cold-temperate species inhabit the northern areas (Gulf of Lions, Ligurian Sea, northern Adriatic) where water temperature is colder. Although the Mediterranean only represents 0.82 % in surface and 0.32 % in volume of the world's oceans, the Mediterranean Sea holds 4–18 % of all known marine species and has a high level of endemism (Bianchi and Morri 2000; Coll et al. 2010).

The general circulation of the Mediterranean is thermohaline and current patterns are cyclonic, adapted to the shape of the continental slopes (Millot 1999, 2005) (Fig. 11.1). The input flow of Atlantic Waters from the ocean follows the northern African coast. This current, and its geostrophic adjustment, are responsible for the existence of a strong thermohaline front (Tintoré et al. 1988; Prieur and Sournia 1994). However, part of this becomes unstable and forms a series of anticyclonic eddies (Millot 1985, 2005), which can be either trapped by the topography of the Alboran Sea or freely move around the Algerian basin and reach the Balearic Islands. The Balearic Islands can be considered the transitional region between the two main western Mediterranean subbasins: the Liguro-Provençal and the Algerian. Part of the Atlantic water flows across the Balearic channels forming the Balearic Current that follows the northern side of the Balearic Islands to the west coast of Corsica (Hopkins 1985; García et al. 1994). A surface front, which is not deeper than 200 m, associated with the Balearic Current, separates recent Atlantic water brought by the current from the resident waters of the centre of the northern basin (Salat 1995). The Balearic Current displays high variability, mainly driven by the fluctuations on the input from the Algerian basin through the Balearic channels (e.g. López-Jurado et al. 1995; Pinot et al. 2002; Millot 2005).

On the eastern side of the Ligurian basin the Atlantic water flow from the Balearic basin joins that from the Tyrrhenian Sea (Astraldi and Gasparini 1992) forming the Northern Current, which flows from the Ligurian to the Catalan Sea along the continental slope, adapted to the bathymetry and contouring the NW basin cyclonically (e.g. Allain 1960; Béthoux et al. 1988). Associated to the Northern Current there is a marked shelf-slope front, detectable down to a depth of around 400 m, again separating the relatively low salinity Atlantic water from the more saline open sea waters (Font et al. 1988). This front is strengthened by a surplus of buoyancy due to the local riverine runoff on the coastal side and by the strong evaporation due to the northern winds on the opposite side (Salat and Font 1987). The Northern Current has significant mesoscale activity, which leads to intense meanders, filaments and eddies that can develop and propagate along the Northern Current pathway (Millot 1990; Sammari et al. 1995; Flexas et al. 2002).

Driving Mechanisms for Plankton Production

The plankton productivity in the NW Mediterranean is influenced by a variety of physical factors, acting at both local and regional scales, whose effect in most cases depends on the seasonal cycle. The Mediterranean climate is characterized by a clear seasonal forcing in irradiance which, although subject to considerable interannual variability, because of variable cloud cover, leads to highly reproducible seasonal changes in surface temperature (Duarte et al. 1999). During the year, the upper layer of the water column, containing the photic zone, shifts between an interval of well-mixed water and a strongly stratified one. Winter cooling produces a breakdown of the thermocline and the vertical mixing bring nutrients from deep waters to the photic zone. The subsequent thermocline development in spring, due to increasing solar heating, prevents vertical motion, and then nutrient supply to the surface is interrupted. This cycle is common everywhere in mid-latitudes, but in the NW Mediterranean the stratification period is rather long, and the winter convection may reach the deepest layers (Lacombe et al. 1981). The alternation of stratified and mixing periods confers strong seasonality to primary production (Estrada et al. 1985).

The northern continental shelves (Gulf of Lions in the NW Mediterranean, but also the Northern Adriatic Sea and the Northern Aegean in the central and eastern sectors) are areas of intense vertical mixing and dense water formation in winter due to the strong and persistent continental winds (MEDOC Group 1970; Roether et al. 1996; Schott et al. 1996). The geographical area centered around 42°N 5°E receives the maximum wind stress (Reiter 1975; Jansá 1987) and is involved in deep convection during winter (Prieur et al. 1983). Fluctuations in the intensity of this phenomenon have been related to interannual variations in primary production, partly because stronger mixing would incorporate deeper and richer waters and partly because deep water spreading would occur over a larger area (Estrada 1996). These northwesterly winds are also responsible of a marked surface thermal front perpendicular to the coast, between 41° and 42° N, clearly identifiable in satellite images (López García et al. 1994).

Nevertheless, the stratified period is far from being as oligotrophic as often accepted. During that period, riverine runoff waters may spread over wide areas, due to the stratification, and can supply nutrients to the surface (Blanc et al. 1969; Ribera d'Alcalà et al. 2004). Although the contribution of this source to the total primary productivity is moderate (ca. 10–20 %, Salat et al. 2002) these freshwater inputs are very relevant because they help to locally maintain surface planktonic production in summer, and in addition they play a major role in balancing the water budget with the Atlantic. Continental runoff waters, mostly from the northern rivers, the Rhône and the Ebro, account for around 70 % of the total fresh water input into the western basin (Béthoux 1979).

There are other relevant mechanisms that enhance the productivity in the NW Mediterranean region that are not related to the water column structure. They are associated with the circulation and its mesoscale activity. Sills, changes in the slope orientation, canyons and some relatively shallow seamounts interact with the currents creating upward vertical components at locations that are typically upstream from these topographic features (Alvarez et al. 1996; Palanques et al. 2005). Current-frontal systems, like those of the Atlantic inflow in the Alboran Sea, or the Northern Current in the NW Mediterranean, show high mesoscale activity playing a decisive role in exchange processes between shelf and oceanic waters associated with instabilities or oscillations of the current (Wang et al. 1988; Millot and Taupier-Letage 2005). The generation of mesoscale eddies by frontal instabilities (Flexas et al. 2002; Taupier-Letage et al. 2003; Rubio et al. 2005) promotes vertical motions which support phytoplankton growth (Boucher et al. 1987; Prieur and Sournia 1994; Fielding et al. 2001) and also contribute to enhancing the productivity at the deep chlorophyll maximum level (Estrada 1985; Sabatés et al. 2004). Furthermore, these convergent fronts have been shown to be zones of accumulation of marine snow particles, possibly due to physical accumulation and coagulation of marine production (Gorsky et al. 2002; Stemmann et al. 2008).

Seasonal Patterns

The distinct seasonality of the climate and hydrographic features in the Mediterranean is a major factor determining the structure and functioning of the whole pelagic ecosystem (Duarte et al. 1999), and obviously influence the composition and succession patterns of the zooplankton assemblages. The long tradition on zooplankton research in the Mediterranean has provided an excellent collection of seasonal studies available in the literature, most of them mainly concerning the coastal or neritic areas; in occasions those studies besides the concomitant characterization of the physico-chemical features of the water column, also benefit from simultaneous information on phytoplankton and ciliate abundance and composition (e.g. Ribera d'Alcalà et al. 2004). These lower trophic levels are presumably relevant drivers of seasonality patterns due to their condition of major food items for zooplankton (Saiz and Calbet 2011) and fish larvae (Rossi et al. 2006).

Although located in temperate latitudes, the seasonality in the western Mediterranean differs from the typical cycles illustrated in books for the North Atlantic, in which a spring and autumn bloom drive zooplankton production (Kiørboe 1993). This dissimilarity transmits also into the upper trophic levels, as evidenced by the fact that in the Mediterranean different fish species spawn in succession over the entire year, so the larval peak production is not coupled to the seasonal late winter-early spring production

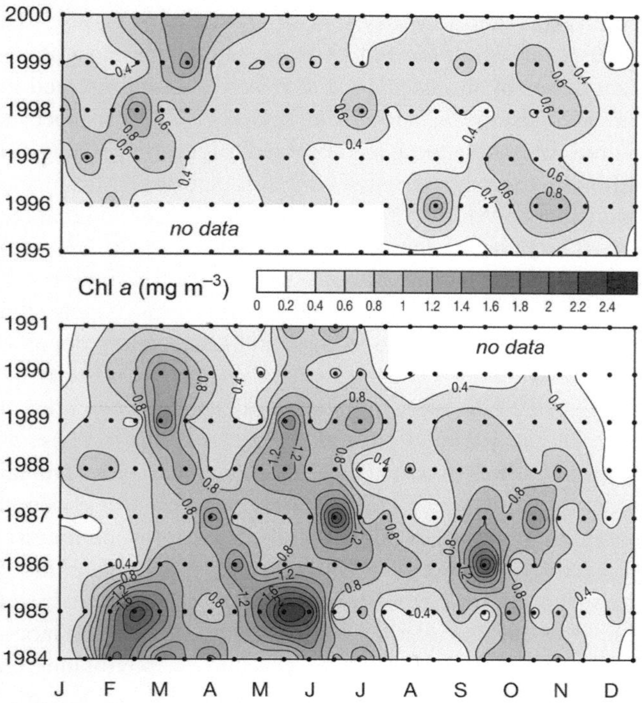

Fig. 11.2 Seasonal and interannual variability of chlorophyll *a* concentration (mg m^{-3}), averaged over the 0–60 m layer, at the station Marechiara off Naples for the period 1984–1991 and 1995–2000 (From Ribera d'Alcalà et al. (2004). With permission of the publisher)

bloom, as the match-mismatch hypothesis postulates (Cushing 1990).

In the Mediterranean water column seasonal stratification typically starts around April, and it is not completely destroyed until late in the year (November–December; Salat 1996; Ribera d'Alcalà et al. 2004). As mentioned before, during the stratified period the events of riverine runoff result in the formation of shallow haloclines spreading over wide areas, with the associated nutrient input, and can boost surface planktonic production over the wide areas they extend (Blanc et al. 1969; Ribera d'Alcalà et al. 2004).

The blooming of phytoplankton in late winter, prior to the thermal stratification of the water column, is a widespread event in the Mediterranean (see Duarte et al. 1999; Ribera d'Alcalà et al. 2004; D'Ortenzio and Ribera D'Alcalà 2009, and references therein) (Fig. 11.2). This winter chlorophyll peak occurs around February–March, and seems to be related to the seasonal presence in January–February of a high-pressure system in the Azores, providing stable water column and clear skies that permit the growth of phytoplankton. This late winter bloom is followed, at least in the northwestern sector and not so frequently in other areas, by a spring-summer bloom (May), confined to the surface layers due to the strong stratification. A third peak in autumn (October–November) is also frequent, that extends, like the winter one, over a deeper mixed layer. High interannual variability can be found in the timing and extent of the seasonal peaks, and in some years one of the peaks might be absent (Ribera d'Alcalà et al. 2004). Regarding ciliates, seasonal patterns are not always so clear and may differ among sites, for instance higher standing stocks are reported in winter in Blanes Bay (Spain) (Vaqué et al. 1997), and Villefranche-sur-Mer (France) (Bernard and Rassoulzadegan 1994), whereas at Naples (Italy) the minimal biomass is found in winter (Ribera d'Alcalà et al. 2004).

To truthfully depict the patterns of variation of zooplankton abundance, biomass and composition through the seasonal cycle, multiannual time-series have been also an essential instrument because allow to study recurrent patterns in zooplankton structure and succession (e.g. Mazzocchi and Ribera d'Alcalà 1995). Some significant differences can be found in the timing and peak abundances of zooplankton in the different areas of the Mediterranean, especially at the species level (e.g. Ribera d'Alcalà et al. 2004; Mazzocchi et al. 2007). Here we will illustrate the most common features of the seasonality scenario for the zooplankton in the coastal/neritic area of the western Mediterranean (Fig. 11.3), and will refer the reader to the original works for more detailed information. The knowledge on the seasonality of the zooplankton of open, deep waters in the Mediterranean basin is still scarce and major efforts are needed.

Many studies in the western Mediterranean describe typically 2–3 seasonal peaks of zooplankton (late winter or spring and late summer) over the year (Champalbert 1996; Calbet et al. 2001). As in the case of phytoplankton, the occurrence, amplitude and timing of peaks are quite variable among years (García-Comas et al. 2011). For instance Fernández de Puelles et al. (2007) reported that the standing stocks of zooplankton were highest in late winter-spring at the Mallorca channel, whereas in other studies, however, the winter standing stocks show the minimum values (e.g. Ribera d'Alcalà et al. 2004; Skovgaard and Saiz 2006). The general pattern followed by the mesozooplankton standing stocks seems to be essentially driven by the dynamics of the copepod community, which on an annual basis dominates the mesozooplankton (ca. 55–65 % of the abundance) and that during winter and spring achieve much higher contributions. At the species level, there is quite a lot of regularity in the seasonal patterns, and a seasonal succession of species assemblages, that appear recurrent and persistent every year, have been described (Mazzocchi and Ribera d'Alcalà 1995; Mazzocchi et al. 2011). Four key species (*Acartia clausi*, *Centropages typicus*, *Paracalanus parvus*, and *Temora stylifera*) seem to be the most abundant ones in the coastal/neritic Mediterranean, accounting for a large share of total copepod abundance from spring to autumn. They display a common succession pattern throughout the Mediterranean, although variations in the timing can be found among sites. First, *Acartia clausi* and *Centropages typicus* peak in

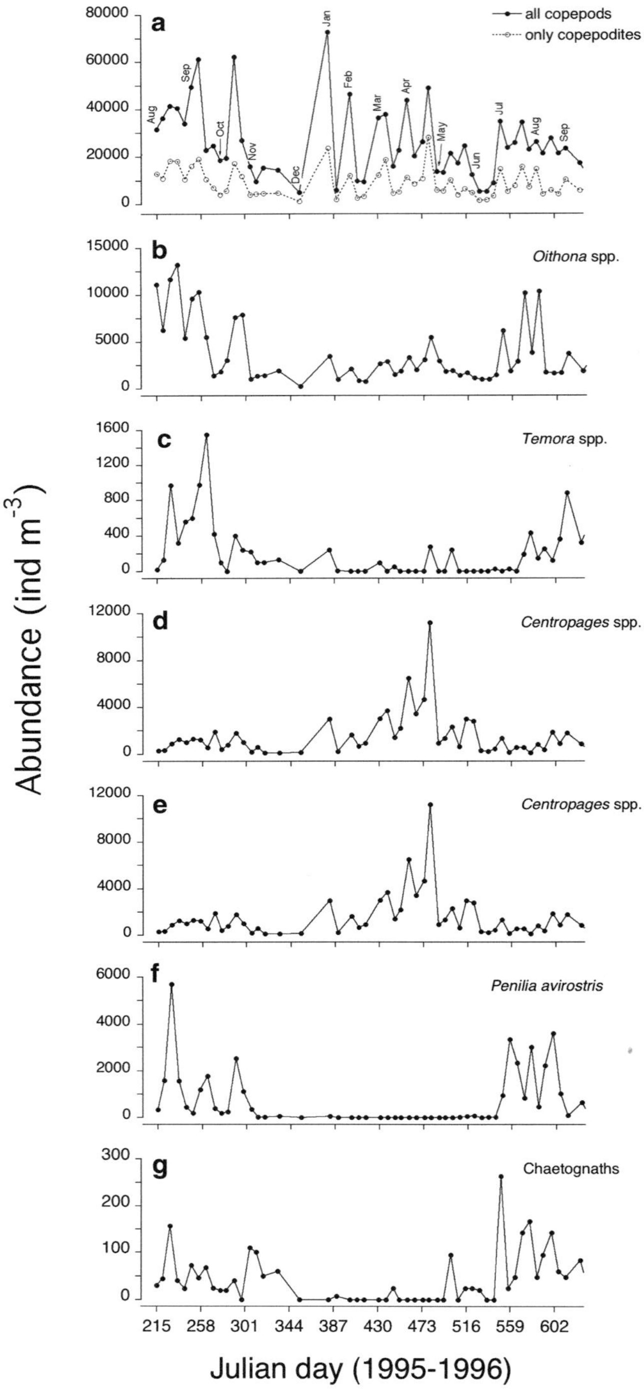

Fig. 11.3 Seasonal variations of several zooplankton groups off Blanes (Catalunya, Spain) (Redrawn and modified from Calbet et al. 2001)

spring-early summer; then *Paracalanus parvus* is important in summer and early autumn, and finally *Temora stylifera* is relevant in later summer-autumn. These populations are multivoltine, and according to Razouls (1974) *C. typicus* and *T. stylifera* have, respectively 7 and 5–6 generations per year in the neritic ecosystem of the Gulf of Lions. The genus *Clausocalanus* and *Oithona*, two of the most abundant genus in all oceans, are also very well represented in the Mediterranean, and a temporal succession of species within the genera has been observed, suggesting ecological differentiation among the congener species (Mazzocchi and Ribera d'Alcalà 1995). It is also worth noticing that the use of large mesh sizes in plankton nets (e.g. 200, 330 μm) has omitted from most studies a very relevant component of the zooplankton community, the smaller forms of metazoan zooplankton, namely copepod nauplii and young copepodites, and adults of small species as well (Calbet et al. 2001). This bias has an obvious influence on our present view of the seasonal patterns of the zooplankton assemblages. In terms of abundance and biovolume, on an annual basis the <200 μm fraction (mainly composed of copepod nauplii) comprised ca. 80 and 21 % of the total community, respectively (Calbet et al. 2001). From a functional point of view, the importance of this size fraction is very relevant due to the inverse relationship between body size and metabolic rates, and their grazing pressure on phytoplankton seems to be similar to the one exerted by the larger (>200 μm) zooplankton (Almeda et al. 2011). When such size fraction is taken into account, the small-sized copepod genus are much more prevalent, in particular *Oithona*, *Oncaea* and *Microsetella* (Calbet et al. 2001; Zervoudaki et al. 2006).

Cladocerans are also a very characteristic and relevant component of neritic zooplankton in the Mediterranean (Della Croce 1964; Vives 1966; Alcaraz 1970; Fernández de Puelles and Molinero 2008; Mazzocchi et al. 2011). They account for ca. 10–20 % of the mesozooplankton standing stock on an annual basis, and are mainly present in surface waters in late spring and, especially, in summer, when can be the major group in the mesozooplankton (Vives 1966; Alcaraz 1970; Fernández de Puelles and Molinero 2008; Mazzocchi et al. 2011). Due to their parthogenetic reproduction and the production of resting eggs, they are characterized by high population bursts (Egloff et al. 1997). The most abundant species is *Penilia avirostris*, which blooms in very stratified waters and can achieve very high numbers (Casanova 1965; Vives 1966; Alcaraz 1970; Thiriot 1972–1973; Calbet et al. 2002a; Atienza et al. 2006, 2008); this cosmopolitan filter feeder can graze efficiently on small flagellates (Atienza et al. 2006) and therefore is well adapted to oligotrophic conditions. As with copepods, seasonal peaks of abundance of *Penilia avirostris* also show some variation in timing throughout the Mediterranean: e.g. June–July in waters off Castellón, Spain (Vives 1966), July–August in Barcelona, Spain (Atienza et al. 2008), and September–October in Banyuls-sur-Mer, France (Thiriot 1972–1973). The other cladoceran species typically appear a bit earlier (Della Croce 1964;

Vives 1966; Alcaraz 1970; Calbet et al. 2001; Fernández de Puelles and Molinero 2008; Mazzocchi et al. 2011). With similar contribution to that of the cladocerans, appendicularians (ca. 10–17 %) are also an important component of neritic zooplankton (Calbet et al. 2001; Ribera d'Alcalà et al. 2004; Fernández de Puelles and Molinero 2008); although present throughout the year, Oikopleuridae seem to be more abundant in summer, whereas the Fritillaria are comparatively more important in autumn (Vives 1966; Fenaux 1968).

Other groups are less well represented in coastal and neritic waters, but also present their characteristic seasonal patterns (Champalbert 1996; Fernández de Puelles et al. 2007). Meroplanktonic larvae can occasionally be very important in coastal and confined areas, and peaks of occurrence are linked to the reproductive cycle of their benthic stages at different times of the year (Vives 1966; Calbet et al. 2001; Almeda et al. 2011). Doliolids are more abundant during the second half of the year (Vives 1966; Braconnot 1971), frequently in coincidence with periods when copepods are scarce and cladocerans (*Penilia avirostris*) are abundant (Ménard et al. 1997). The establishment of shallow haloclines in early summer and late autumn, as a consequence of storms and seasonal river swelling, and the occurrence of periods of calm weather (low winds) seem to favor the seasonal blooming of doliolids (Ménard et al. 1997); likely, these processes might be also relevant for other groups blooming in those periods. Chaetognaths are typically very scarce in winter and the maximum densities are found in August and September, in special for *Sagitta inflata*, the most abundant species in the northwestern Mediterranean (Vives 1966; Andréu 1990). Thaliacea are more common in late winter/early spring associated to the first primary production peak in the Mediterranean (Furnestin 1963; Vives 1966; Ménard et al. 1994). With regard to the Cnidaria, a seasonal maximum peak has been described for medusa and siphonophores in spring and early summer (Goy and Thiriot 1976; Ianora and Scotto di Carlo 1981; Gili et al. 1987; Gamulin and Kršinic 1993; Batistić et al. 2007). During this period of maximum abundance, the coastal communities are dominated by two species, the siphonophore *Muggiaea atlantica* and the medusa *Aglaura hemistoma,* whereas in the open sea, but with much lower abundance, dominate *Chelophyes appendiculata* and *Persa incolorata* (Berhaut 1969; Goy 1987); less abundant species that occur regularly throughout the year may exhibit minor peaks in autumn (Albertini-Berhaut 1971; Gili et al. 1987; Dallot et al. 1988; Andersen et al. 2001b). Not only local productivity peaks (Gili et al. 1988) but also physical factors favoring aggregation can be involved in such seasonal patterns (Graham et al. 2001).

Fish larvae also exhibit a pronounced seasonal variability in both abundance and number of species (Sabatés et al. 2007a), in obvious relation to the spawning cycles of the adults. Most neritic fish species (e.g. Sparidae, Labridae, Blennidae, Mullidae, Serranidae), as well as large migratory tuna, spawn during spring and summer; in winter, when species of relatively cold waters reproduce, the specific composition varies substantially and is less diverse, and larvae of Gadidae and Pleuronectidae are commonly found (Sabatés et al. 2007a). The most abundant commercial fish species, sardine and anchovy, have non-overlapping spawning periods, autumn-winter and spring-summer respectively, and are very well adapted to the productivity mechanisms characteristic of their respective spawning seasons; that is, vertical mixing on the shelf in winter and spreading of continental runoff at surface in spring-summer (Palomera et al. 2007). Their eggs and larvae dominate the ichthyoplanktonic fraction in neritic areas of the Mediterranean during these periods.

As a final remark, it is important to notice that although in the Mediterranean the adverse period is not too extreme compared to higher latitudes, still a certain number of species of zooplankton are not found in the water column year around, or only in very low numbers. The existence of resting stages that allow such species to spend the adverse period in the benthos is probably more common than usually acknowledged (Boero et al. 1996; Marcus and Boero 1998). Such life strategy is very apparent for some groups or species, like for instance the Cnidaria in which the polyps may become encysted as resting hydrorhizae and will recruit at the next favorable season (Boero and Bouillon 1993), or the cladocerans (Egloff et al. 1997). For copepods, the existence of resting and diapause eggs that may reside in the sediment of coastal and neritic environments is well known (Marcus 1996), but the number of studies on this subject in the Mediterranean is very limited (e.g. Ianora and Santella 1991; Marcus and Boero 1998; Belmonte and Pati 2007) and it is difficult to provide an assessment on their role in the population dynamics.

Vertical Structure

Different spatial and temporal scales frame the patterns of vertical distribution of zooplankton, as the result of the interaction of the behavior of the organisms (i.e. diel rhythms of vertical ascend at night and descend during the day, ontogenetic and seasonal migrations) with the physical forcing and physico-chemical-biological vertical structuring of the water column. These phenomena and patterns are common to all oceans and seas, but the marked seasonality in the Mediterranean, with alternating periods of mixing and stratification, strongly affects the vertical distribution of zooplanktonic organisms throughout the year cycle and shows some peculiarities worth mentioning here.

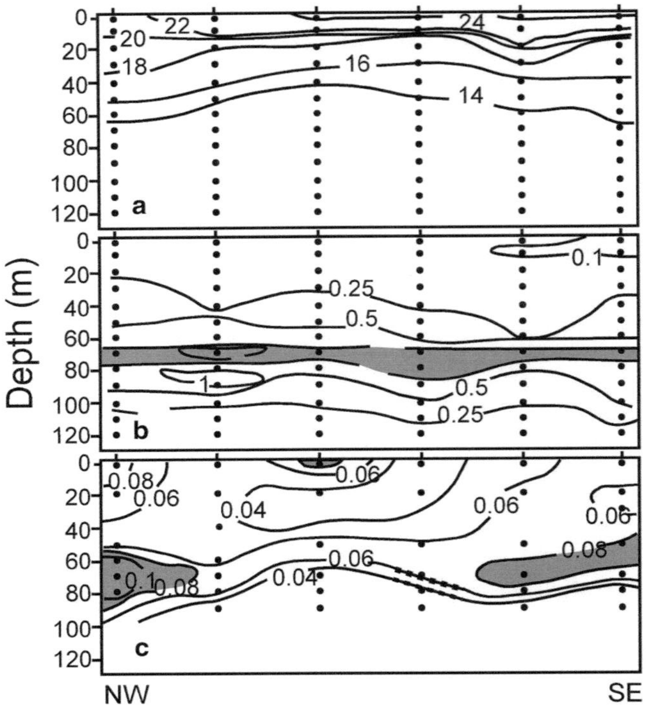

Fig. 11.4 Vertical profiles of (**a**) temperature (°C), (**b**) chlorophyll *a* concentration (µg chl *a* L^{-1}) and (**c**) mesozooplankton biomass (µg-at Nzoo L^{-1}) along a transect from Barcelona through the channel between Mallorca and Menorca, in the Catalan Sea, in summer. Concentrations >1 µg chl *a* L^{-1} and >0.08 µg-at Nzoo L^{-1}, respectively, are highlighted (Redrawn and modified from Alcaraz 1988)

Regarding the epipelagic realm, one of the major features of the vertical structure of the water column in the western Mediterranean is the stratification of the upper water column over a large portion of the year, resulting in the formation of a deep chlorophyll maximum which extends over large areas (Estrada et al. 1993) (Fig. 11.4). When the heat balance overrides vertical mixing, a thermocline starts to develop and deepens during summer down to deeper waters. During that period, nutrients are very scarce in the surface layers, and a deep chlorophyll maximum (DCM) develops associated with the thermocline or just below it, at depths where light is still sufficient for photosynthesis and nutrient inputs from the nutricline can warrant certain phytoplankton growth. It is important to consider that in the western Mediterranean the DCM is not only the result of a higher chlorophyll cell content at low light intensities, but can be the result of a cell maximum as well. The DCM can also correspond to a peak of primary production, although the major center of phytoplankton growth is in the upper layers (Estrada et al. 1985). In such an oligotrophic environment like the Mediterranean, especially during the stratification season, the presence of the DCM could be considered as an "oasis" to sustain secondary production. For this reason, the presence of the DCM during the long stratification period has relevant implications for the functioning and structuring of the pelagic food web in the Mediterranean. In addition the DCM seems to be also rich on ciliates, mainly the heterotrophic ones (Dolan and Marrasé 1995; Calbet et al. 2002a); ciliates are known as a preferred prey for the dominant zooplankter in marine systems, copepods (Saiz and Calbet 2011).

The vertical distribution of epipelagic mesozooplankton in the Mediterranean during summer is characterized by the presence of a zooplankton maximum at approximately the same depths that the DCM during the daytime (70–90 m), whereas at night, following the general pattern of vertical migration observed all over the world, zooplankton ascends to upper layers (25–50 m; Alcaraz 1985) (Fig. 11.4). As pointed out, the fact that epipelagic zooplankton remains during the day at the depth of the DCM and ascend at dusk to upper waters have important ecological implications, since zooplankton may feed not only at night but also during the day while at the DCM (Saiz and Alcaraz 1990). For this reason, the ascent of copepods at night during summer results in an upward flux of recycled nutrients (via zooplankton excretion) to the nutrient limited surface waters. Such recycling would, therefore counteract the loss of nutrients due to the downward vertical flux of particulated organic matter, and would help to sustain primary production in the surface waters. Estimates suggest that the contribution of regenerated nutrients to the phytoplankton requirements can increase 1.5 times due to the night ascend of zooplankton to the surface waters (Alcaraz 1988; Alcaraz et al. 2007). High densities of nauplii and copepodite stages of copepods have been reported associated to the DCM levels (Sabatés et al. 2007b, 2008; Olivar et al. 2010). In turn, larvae of some fish species, such as anchovy, have been found at the DCM during the daylight hours (the feeding period) benefiting from food aggregations at this level, ascending to the surface at night (Olivar et al. 2001; Sabatés et al. 2008) (Fig. 11.5).

This overlapping of the vertical distribution of the bulk of the epipelagic zooplankton biomass and the DCM seems to be a common feature that extends all over the Mediterranean during the stratification period (e.g. see Fig. 4 in Nowaczyk et al. 2011), although frequent exceptions, for instance of bimodal or multimodal distribution can also be found (Andersen et al. 2001b). At the species level such exceptions are not rare and should be interpreted in tight relation to the micro and fine scale structure of the water column and particular behavioral components (e.g. affinities to determined temperature and/or salinity, presence of potential prey or predators), as examples of niche segregation (e.g. vertical segregation of congeneric species of the genus *Clausocalanus*, Peralba and Mazzocchi 2004; see also a similar case for the eastern Mediterranean in Fragopoulu et al. 2001). Although waters above the DCM are nutrient poor, the maximum of phytoplankton productivity during the stratification period is typically well above the DCM, and other potential prey for

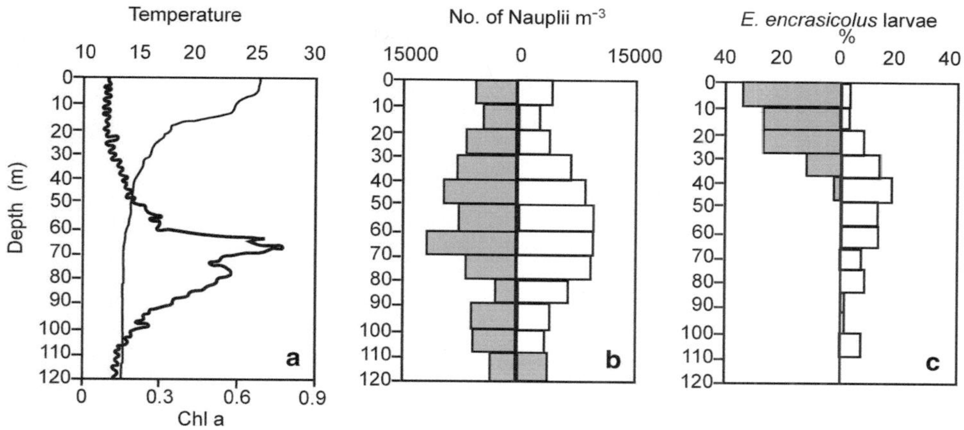

Fig. 11.5 Vertical distribution of (**a**) temperature (°C) and chlorophyll *a* (μg chl a L^{-1}), (**b**) nauplii stages of copepods, and (**c**) *Engraulis encrasicolus* larvae (> 8 mm SL) in the Catalan Sea (July 2003). *White* and *grey* bars are respectively day and night samples (Redrawn from Sabatés et al. 2008)

zooplankton (i.e. mixotrophic ciliates, Dolan and Marrasé 1995) could be more abundant above the DCM, therefore potentially providing a food supply for the zooplankters inhabiting those waters. Unfortunately, there is a lack of fine scale studies on the vertical distribution of zooplankton in the Mediterranean coupled to the distribution of potential prey/predators and hydrographic singularities that could provide further insights into the factors driving the vertical distribution of different zooplankton. At the smallest scales, there are evidences that zooplankton are capable to find and make profit from such patchiness in food distribution (Tiselius 1992; Saiz et al. 1993).

During the winter mixing period, in absence of strong vertical gradients and high algal biomass and primary productivity (Estrada et al. 1999; Morán and Estrada 2005), the vertical distribution of zooplankton abundance and biomass is more homogeneous along the water column (Vives 1966; Alcaraz et al. 2007), and the depth-dependent differentiation of zooplankton assemblages due vertical habitat partitioning observed in summer disappears (Ramfos et al. 2006). In this situation it has been reported that different species of the genus *Clausocalanus* did not show preferential depths distribution and were less segregated, presenting a more homogenous distribution (Peralba and Mazzocchi 2004). Nauplii and copepodite stages of copepods show a wide distribution along the first 100 m of the water column (Sabatés et al. 2007a; Olivar et al. 2010). Vertical distribution of sardine larvae, *Sardina pilchardus*, one the most abundant species during winter, also shows a wide distribution, in coincidence with their main food (Olivar et al. 2001; Sabatés 2004; Morote et al. 2010). Additionally, there seems to be seasonal variations in the intensity and extent of the diel vertical migration of zooplankton, which appears weaker from late summer to winter (Hure 1961).

Nowadays the study of the mesopelagic fauna has renewed a growing interest due to the acknowledge of the relevant role the mesopelagic zone has as a "filter" of the particulate organic matter between the epipelagial and the deep-sea and seafloor, and their significant contribution to export fluxes (Steinberg et al. 2000; Robinson et al. 2010). The western Mediterranean has a long tradition on studies of the deep zooplankton, including macroplankton and micronekton, although the degree of knowledge varies with the geographical area (i.e. Hure and Scotto di Carlo 1974; Vives 1978; Scotto di Carlo et al. 1984; Andersen and Sardou 1992; Andersen et al. 1992, 1998), and references therein). More detailed information has been gathered since the wide availability of multisampling nets since the early 1990s. There appear to be typically three zooplankton strata between 250 and 1,000 m depth during daytime, with ranges between 250–350, 400–550 and 700–800 m, inhabited with characteristic species assemblages with particular trophic relationships and life strategies (Gasser et al. 1998; Yoon et al. 2007). Different behavioral patterns can typically be found among the zooplankton inhabiting those deep waters (composed mainly of large copepods, but also accompanied by macroplankton and microneckton, e.g. euphausiids, pteropods, siphonophores and amphipods), ranging from strong diel migrators to weak or non-migrant species (Andersen and Sardou 1992; Andersen et al. 1998). The fish larvae of Myctophidae are another very relevant component of this community (e.g. Sabatés and Masó 1990; García Lafuente et al. 1998). This ecologically highly successful group of mesopelagic vertical migrators (Olivar et al. 2012) very likely plays a very significant role in the vertical carbon fluxes in the world oceans (Pakhomov et al. 1996), and exhibits high morphological diversity as an a strategy to optimize the utilization of trophic resources in the open-ocean habitat (Sabatés and Saiz 2000; Sabatés et al. 2003). It is also

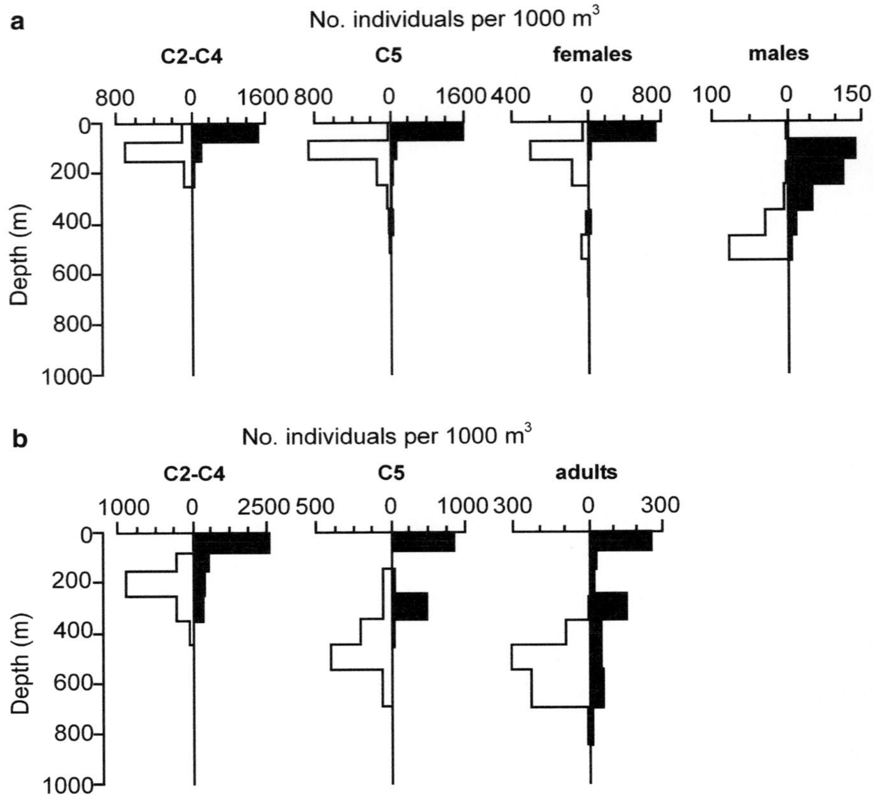

Fig. 11.6 Day (*white bars*) and night (*black bars*) vertical distribution of two mesopelagic copepod species ((**a**) *Neocalanus gracilis*, (**b**) *Euchaeta acuta*) in the northwestern Mediterranean, showing ontogenetic segregation (Redrawn and modified from Andersen et al. 2001a)

worth noticing that most studies dealing with mesopelagic zooplankton in the Mediterranean have employed very coarse mesh nets (i.e. 500 µm), but the very few studies with fine nets (50 µm) carried out in the eastern Mediterranean suggest that small copepods, especially from the poecilostomatoid genus *Oncaea*, can be also a significant and neglected component of those communities (Böttger-Schnack 1994).

Common patterns of ontogenetic vertical segregation of zooplankton (in which the younger stages inhabit less deep waters; i.e. Williams and Conway 1984; Uye et al. 1990), and of seasonal migration (in which the late developmental stages of zooplankton go though the unfavorable season in a resting or diapause state; overwintering: Conover and Huntley 1991; Hagen 1999; oversummering: Wang et al. 2003) found in other seas can also be found in the Mediterranean, although the number of studies dealing with this issue is also scarce (Fig. 11.6). (Andersen et al. 2001a) provides fine examples of the ontogenetic segregation of the day time vertical distribution of the copepods *Neocalanus gracilis*, *Pleuromamma abdominalis* and *Euchaeta acuta* in the Ligurian Sea at the DYFAMED station; at nighttime there seems to be overlapping, although vey likely a finer vertical resolution in the upper 100 m would show also some degree of segregation among stages. Regarding examples of seasonal migration and overwintering/oversummering strategies in the western Mediterranean, there are several reports of downward seasonal migration of large-sized copepods like *Eucalanus monachus*, *Eucalanus hyalinus* and *Calanus helgolandicus* (Hure and Scotto di Carlo 1974; Vives 1978; Gasser et al. 1998; Andersen et al. 2004). Typically these organisms descend to deeper waters (400–900 m) in summer and autumn, and rise to shallower waters in winter spring (Andersen et al. 2004). Such oversummering life strategy is likely related to the avoidance of the less productive season and the warmer surface waters during summer and autumn, and to exploit the peak of productivity in late winter-spring. In comparison with the Atlantic, the triggering of the seasonal descend seems to occur earlier in the year in the Mediterranean (Williams and Conway 1984; Andersen et al. 2001a). Gasser et al. (1998) reported high abundances of *Calanus helgolandicus* CV, with a large oil sac filling out most of their body in the deep waters (700–800 m) of the Ligurian Sea in June, as a signal of the oversummering stage.

Near-Bottom Zooplanktonic Communities

Although the near-bottom zooplankton dynamics is, in comparison with the pelagic one, much less known, their study has acquired an increasing interest in the last years due to its importance as a hot spot of diversity and to its key role in the trophic web (e.g. Cartes et al. 2010). The initial studies using sledges and dredges have been recently complemented with plankton nets attached to bottom trawls and have evidenced the existence of large swarms of euphausiids and mysids close to the bottom (e.g. Cartes et al. 1994; Cartes 1998). The supra-benthic fauna in those habitats, mainly composed of mysids, euphausiids, amphipods and isopods (Bellan-Santini 1990; Cartes et al. 1994; Macquart-Moulin 1998), seems to be well adapted to the variability generated by re-suspension processes caused by the currents near the bottom. Whereas certain groups like the mysids and the amphipods constitute a permanent component of this bottom-living fauna, others (e.g. euphausiids, isopods) perform vertical migrations and their dynamics is more linked to processes occurring in the water column, therefore playing an important role in the benthic-pelagic coupling, Macquart-Moulin 1998; Cartes et al. 2010). Under stratification conditions, in summer and autumn, swarms of the adult stages of the dominant macroplankton species can be found close to the bottom off the shelfbreak slope, likely in relation to the seasonally high abundance of their prey (e.g. the oversummering copepod *Calanus helgolandicus*) in those habitats (Cartes 1998; Cartes et al. 2010).

The study of planktonic organisms collected by sediment traps located near the bottom in the submarine canyons has permitted the discovery of a new deep zooplankton community (Gili et al. 2000). Gelatinous zooplankton, mainly jellies, Copepoda, Holothuria and Polycheta are the dominant groups in the traps and their abundance depend on both re-suspension processes near the bottom and the continuous flux of organic matter from the continental shelf, particularly during spring (Gili et al. 1998, 1999; Pagès et al. 2007; Sardá et al. 2009). The specific composition of the deep-sea canyon fauna highlights their relevance, in particular the jelly species, as an endemic component of the Mediterranean plankton (Casanova 1990) representative of relict species from the Thetis Ocean (Gili et al. 2000).

Horizontal Distribution Patterns at the Mesoscale

The spatial patterns of plankton biomass, productivity and species composition exhibit high variability as a result of the interaction, at a hierarchical spectrum of space and time scales, of the biological components with the physical oceanographic processes and environmental singularities. The existence, as a consequence of the environmental forcing, of spatial gradients in the local rate of population increase or decrease results in the development and persistence of biological spatial patterns (patchiness) in the ocean (Mackas et al. 1985). Although the Mediterranean Sea is considered globally an oligotrophic sea, a substantial amount of hydrographic features and biological heterogeneities can be found which, especially in the western basin, facilitate a certain increase in its potential fertility (Estrada 1996). As a result, large differences in the zooplankton communities between the eastern and western basins are evident, and a decrease in zooplankton abundance has been reported in relation to the strong eastward gradient in productivity and primary production (Mazzocchi et al. 1997; Dolan et al. 2002; Siokou-Frangou 2004; Minutoli and Guglielmo 2009). However, besides this basin scale trends, any attempt at ascertaining the major mesoscale spatial pattern in the Mediterranean is not an easy task, because most of the studies are local and synoptic data are scarce, and in addition the comparisons among areas and studies are flawed by the lack of standardized sampling methods and data treatment (Champalbert 1996). Zooplankton abundance largely depend on the size range of the organisms considered (Calbet et al. 2001; Nowaczyk et al. 2011), thus the use of different sampling mesh-size is a difficulty when comparing different zooplankton datasets. Sampling with finer mesh nets than the standard 200 µm, or with large bottles has revealed that biomass and abundance can increase by 2–20 fold (Andersen et al. 2001b; Calbet et al. 2001; Youssara and Gaudy 2001; Alcaraz et al. 2007).

Like in all oceans, a gradient of zooplankton abundance from coastal waters to the open sea is typically observed in the Mediterranean Sea (e.g. Vives et al. 1975; Champalbert 1996; Gaudy and Champalbert 1998), as a consequence of local eutrophication and river runoff. Physical and biological conditions are extremely variable over the shelf and coastal waters, where episodic pulses of production may occur (Denman and Powell 1984); in comparison, the oceanic region is characterized by a more steady production. As a result, mesozooplankton abundance in offshore waters oscillates within a narrow range and reveals lower seasonal variability than in coastal waters (Scotto di Carlo et al. 1984; Fernández de Puelles et al. 2003b); in addition, the changes in the composition of zooplankton assemblages along such environmental gradients must reflect the peculiarities in eco-physiological response and life history traits suitable to cope with the environmental variability (Runge 1988). Changes in copepod productivity have been observed in relation to the associated changes in the biomass and size spectrum of phytoplankton along this gradient (Saiz et al. 1999; Calbet et al. 2002b). In addition, in the northwestern Mediterranean continental runoff waters from the main rivers of the region, the Rhône and the Ebro, play an important role enhancing primary production of the shelf waters (Cruzado and Veláquez 1990; Salat et al. 2002; Diaz et al. 2008). Associated to these low salinity waters, high copepod densities have been recorded in the Gulf of Lions and near the Ebro delta (e.g. Razouls and Kouwenberg 1993; Sabatés et al. 2008) along with an increase in zooplankton biomass and feeding activity in the outflow plume of the Rhône River (Pagano et al. 1993). These rich zooplankton production areas have been reported to be important spawning grounds of anchovy in the northwestern Mediterranean, favoring larval survival as well as the diet and condition of the adults (Palomera et al. 2007; Banaru and Harmelin-Vivien 2009). Furthermore, over the shelf, topographic irregularities can greatly modify circulation producing complex plankton distributions (Alvarez et al. 1996). Hence, submarine canyons at the continental margin of the NW Mediterranean can interact with the Northern Current and modify the general circulation, generating topographically controlled up- and downwellings and affecting the shelf-slope water exchange (Alvarez et al. 1996; Durrieu de Madron et al. 1999; Palanques et al. 2005). All these processes favour high concentrations of zooplankton and fish larvae near the coast (Sabatés et al. 1989; Sabatés and Olivar 1996).

In the western basin, zooplankton abundance as well as the total biomass display a north–south decreasing gradient in relation to the decrease in primary production and increasing temperatures (D'Ortenzio and Ribera D'Alcalà 2009; Nowaczyk et al. 2011). Changes in the composition, phenology and distribution of zooplankton as a consequence of the latitudinal gradient in temperature between the northern and southern Mediterranean are expected, but comparative studies are scarce (Mazzocchi et al. 2007). In the Catalan coast, the northern sector, which is more directly influenced by strong northerly winds, is generally colder than central and southern parts. Thus, a surface thermal front roughly coincides with the limit of frequent northerly winds (López García et al. 1994). This thermal front has been reported to have a significant effect on zooplankton and fish larval distribution (Atienza 2006; Sabatés et al. 2009).

Numerous studies conducted in the western Mediterranean have highlighted the impact of mesoscale structures such as density fronts on the distribution and diversity of zooplankton, enhancing plankton patchiness and activity, and stimulating trophic transfers towards large predators (Boucher et al. 1987; Ibanez and Boucher 1987; Sabatés et al. 1989, 2004; Gorsky et al. 1991; Saiz et al. 1992; Alcaraz et al. 1994, 2007). The Northern Current (Ligurian Sea and Catalan Sea), the Balearic Current (Balearic Sea) and the Almeria–Oran front (Alboran Sea) constitute transitional zones with overall biological production enhanced by high phytoplankton biomass and primary production (Estrada 1985; Ibanez and Boucher 1987; Estrada and Margalef 1988; Prieur and Sournia 1994; Youssara and Gaudy 2001). Secondary producers have been also shown to be influenced by frontal structures in terms of biomass, productivity or specific diversity (Boucher et al. 1987; Gili et al. 1988; Sabatés et al. 1989; Alcaraz et al. 1994; Prieur and Sournia 1994; Seguin et al. 1994; Gaudy and Champalbert 1998; Youssara and Gaudy 2001; Fernández de Puelles et al. 2004). The higher biomass of zooplankton recorded at oceanic fronts is often explained by a physiological response of the organisms to the frontal environment. However, retention or accumulation of organisms at fronts may also depend on the interaction of the cross-frontal circulations with the differential motion of the organisms (sinking, floating and swimming; Owen 1981; Franks 1992).

In the Ligurian Sea several studies have emphasized the high variability of zooplankton populations linked to geostrophic frontal system (e.g. Boucher et al. 1987; Sournia et al. 1990; Pinca and Dallot 1995; Licandro and Ibanez 2000). The current organizes the hydrological structure of the region into three main distinct zones: a coastal-peripheral zone, a frontal zone, and an offshore central zone with a divergence near the offshore limits of the current jet and convergence inside the frontal zone (Boucher et al. 1987). Nevertheless, this basic hydrological organization may be complicated by other mesoscale structures, such as frontal meanders (Sournia et al. 1990; Prieur et al. 1993) and mesoscale eddies (Taupier-Letage and Millot 1986; Millot 1987), that play an important role in the distribution and concentration of zooplankton (Pinca and Dallot 1995). The main identified zones determine the structure and composition of zooplanktonic communities. High zooplankton densities, particularly copepods, have been reported in relation to the enhanced primary productivity associated with the frontal upward advection and to the secondary circulation (Boucher 1984; Boucher et al. 1987; Ibanez and Boucher 1987; Zakardjian and Prieur 1998) (Fig. 11.7). In addition, the frontal zone may act as a retention area where the enhanced feeding conditions for copepods, probably lead a strong cohort production (Molinero et al. 2008b), although some available information does not support higher copepod productivity (Saiz et al. 2007). Nevertheless, abundance variability of each species may show intense small-scale patchiness, even changing among their successive developmental stages. This arises from complex behavioral, feeding and reproductive strategies in response to environmental organization. This topic was particularly well illustrated by Boucher (1988) for the copepod *Calanus helgolandicus*.

In the Catalan Sea, the role of the Northern Current and its associated front on primary production (Estrada and Margalef 1988; Estrada et al. 1999) and zooplankton distribution and metabolism has also been extensively studied (e.g. Sabatés et al. 1989; Saiz et al. 1992; Alcaraz et al. 1994; Sabatés and Olivar 1996; Masó et al. 1998; Duró and Saiz 2000; Calbet et al. 2002a; Sabatés et al. 2004). In general, the spatial pattern of zooplankton biomass distribution is comparable to that observed by Boucher (1984) and Boucher et al. (1987) across the Ligurian Sea Front. High zooplanktonic biomass and fish larvae concentrations have regularly observed along the shelf-break in relation to the frontal convergence (Sabatés et al. 1989). The front may act as a barrier preventing dispersal of fish larvae towards the open sea (Sabatés and Olivar 1996). However, the patterns observed are subject to

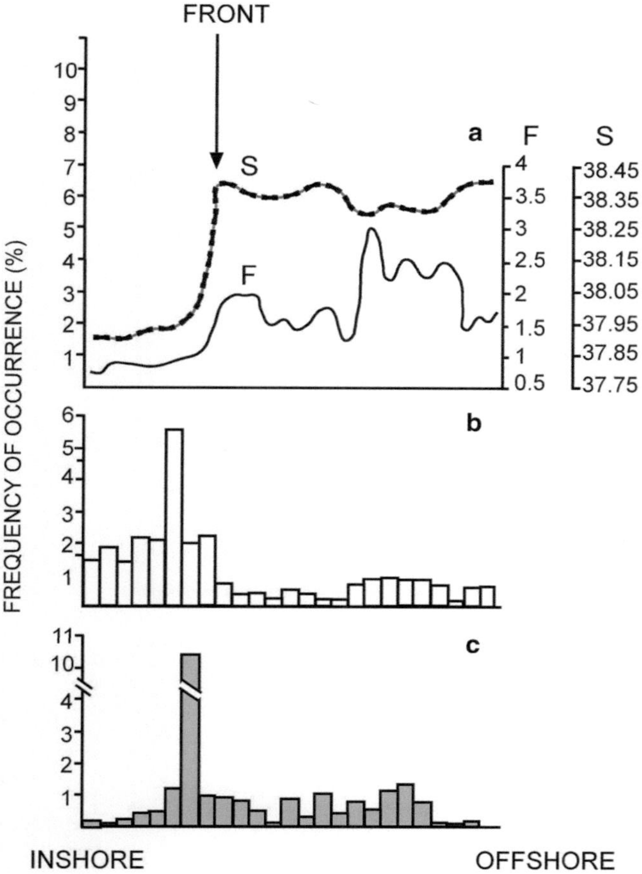

Fig. 11.7 Transect across the density front in the Ligurian Sea. (**a**) Salinity and fluorescence; (**b** and **c**) Relative abundance of, respectively, the calanoid copepods *Centropages typicus* and *Calanus helgolandicus* (Redrawn and modified from Boucher 1984)

considerable spatiotemporal variability, due to frontal mesoscale activity, which can show variations in intensity according to the seasonal variations in the location, strength and narrowness of the front (Sabatés and Olivar 1996; Sabatés et al. 2004). Such variability may reflect differently on zooplankton when considering state variables (i.e. biomass) or functional properties (metabolic activity) as a result of their different time scales (Calbet et al. 1996). Zooplanktonic biomass minima have been reported at the front with high zooplankton densities, particularly copepods, on its oceanic side (Alcaraz et al. 1994; Saiz et al. 1999; Calbet et al. 2002b; Sabatés et al. 2004), similarly to what was observed by Gaudy and Champalbert (1998) at the Ligurian Sea front (Fig. 11.8). These high zooplankton concentrations in the oceanic side of the front are related with peaks in phytoplanktonic biomass in that region (e.g. Estrada and Margalef 1988; Estrada et al. 1993) due to the existence of small convection cells along the front (Boucher et al. 1987) or episodes of cross-frontal ageostrophic circulation (Pinot et al. 1995); such coupling reflects on higher copepod grazing activity and predation pressure on primary producers (Saiz et al. 1992). Furthermore, biomass (organic carbon) and individual size (biomass) are more variable at the front than in the surrounding waters (Alcaraz et al. 2004, 2007). The highly dynamic frontal zone and the high instability induced there by intermittent, non-periodic, mechanical energy inputs would explain the observed variability (Alcaraz et al. 2007). The comparison of the repeated transects in a relatively short period of time, revealed that the front presented a meandering structure with changes in the position, direction and intensity of the current (Sabatés et al. 2004). Mechanisms of convergence/accumulation and divergence/dispersion of zooplankton and fish larvae were reported associated with the current's meandering flow pattern. Thus, the temporal scale at which the frontal system oscillates plays a key role in supplying nutrients to the photic zone on an intermittent basis but also exerts a decisive influence on processes affecting concentration/dispersal of zooplankton (Sabatés et al. 2004). Furthermore, is important to highlight the role of the Northern Current as a transport mechanism of planktonic organisms, in particular anchovy larvae from the spawning grounds of Gulf of Lions, towards the Catalan coast (Sabatés et al. 2007b). In addition to the direct transport by the current, anticyclonic eddies generated by instabilities of the current may entrap and aggregate the anchovy larvae, preventing their dispersion (Sabatés et al. 2007b) (Fig. 11.9).

The Alboran Sea, in the southern sector of the western Mediterranean Sea, is a highly energetic area. Inflow of Atlantic Waters and its geostrophic adjustment, are responsible for the existence of a strong thermohaline front (Tintoré et al. 1988; Prieur and Sournia 1994). The frontal system generates an area of cold, nutrient-rich waters where primary production is enhanced (Prieur and Sournia 1994; Videau et al. 1994). Higher zooplankton biomass and diversity are usually found in the Atlantic waters, as a consequence of the secondary production processes associated to the front, compared with surrounding waters (Dallot et al. 1988; Madin 1991; Seguin et al. 1994; Mills et al. 1996; Youssara and Gaudy 2001). In addition, maximum copepod egg production may occur near the front boundary, due to the existence of favorable feeding conditions (Youssara and Gaudy 2001). Nevertheless, the Atlantic Jet-frontal system has a strong variability at different time and spatial scales (Cheney and Doblar 1982; Parrilla 1984; Font et al. 2002). This time variability induces important changes in the distribution of phyto- and zooplankton. Rodríguez et al. (2001) showed that vertical velocities associated to mesoscale structures determined the phytoplankton size spectrum. Instabilities associated to the front are responsible for the subduction of phyto- and zooplankton to deep layers (Allen et al. 2001; Fielding et al. 2001; Youssara and Gaudy 2001; Andersen et al. 2004). In addition, changes in the Atlantic

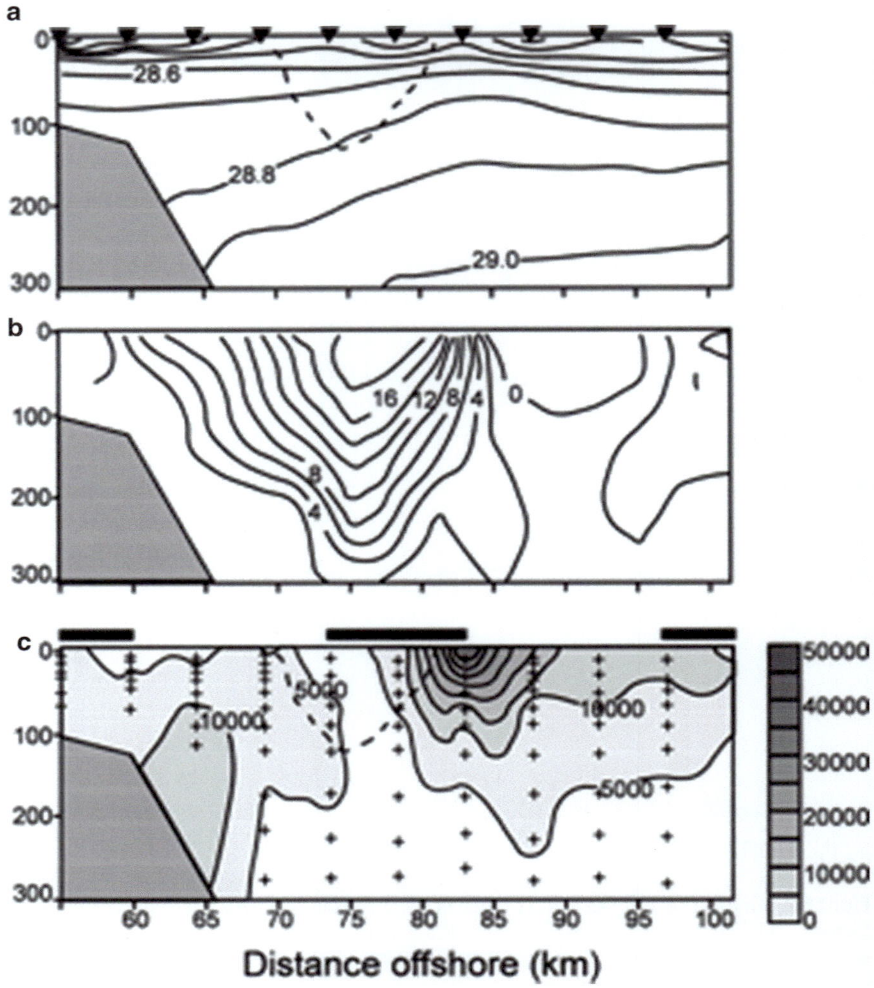

Fig. 11.8 Coastal-open sea transect across the Northern Current and its associated front, in the Catalan Sea (February, 1997). (**a**) Vertical distribution of density; (**b**) Geostrophic velocity (north–south) at 2 cm/s intervals; (**c**) Vertical distribution of copepods (No./100 m3). *Dotted line* in (a) and (c) encloses current velocities higher than 16 cm/s. *Dark segments* on *top* of (c) indicate night period (Redrawn and modified from Sabatés et al. 2004)

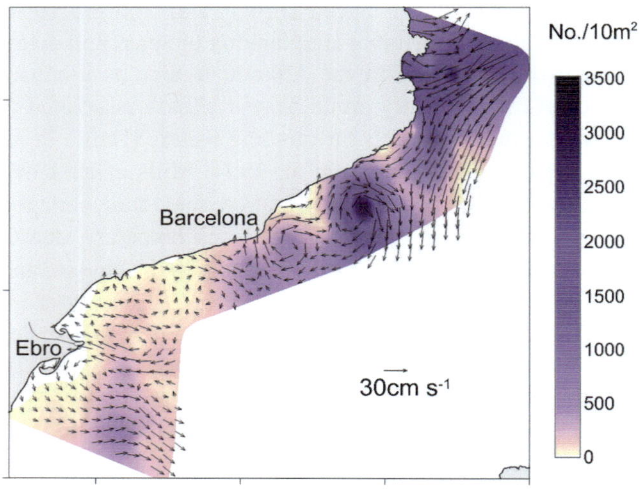

Fig. 11.9 Distribution of *Engraulis encrasicolus* larvae along the Catalan coast (July 2003) overlaid on the geostrophic current fields at 10 m, relative to 600 m. High concentrations of larvae are found in the northern area, associated to the intrusion of the Northern Current, and also inside the anticyclonic eddies originated by instabilities of the Northern Current (Redrawn and modified from Sabatés et al 2013)

Jet– frontal system involve changes in the vertical and horizontal distribution of fish larvae (Rubín 1997; García Lafuente et al. 1998; Vargas-Yáñez and Sabatés 2007). It has been suggested that the northwestern sector of the Alboran Sea is a favorable reproductive habitat for some commercial fish species such as the sardine, *Sardina pilchardus* (García et al. 1988; García Lafuente et al. 1998). Eggs and larvae of sardine are retained off the Málaga coast (i.e., in the geostrophic front area; Rubín et al. 1999) where the upwelling processes (Sarhan et al. 2000) favors high phytoplankton and zooplankton concentrations (Rodríguez et al. 1982; Minas et al. 1991).

Long-Term Changes

The distinctive physical, climatic and ecological features of the Mediterranean make it a very sensitive area to climate change (Lejeusne et al. 2010; Calvo et al. 2011). It is widely accepted that the Mediterranean is already one of the most impacted seas in the world, since climate change interacts synergistically with many other anthropogenic stressors such as fishing, eutrophication, pollution, and introduced exotic species (Turley 1999; Lejeusne et al. 2010). Long-term changes in the hydrographic and meteorological conditions, with warming water, less rainfall, lower wind-speed and progressive acidification have been reported in the Mediterranean (e.g. Béthoux et al. 1999; Rixen et al. 2005; Vargas-Yáñez et al. 2010). These changes appear to be related to atmospheric patterns, such as the NAO, and other mid latitude teleconnection patterns (Vignudelli et al. 1999; Tsimplis and Josey 2001; Rixen et al. 2005). Climate change, through long-term temperature increase, has been demonstrated to affect the boundaries of biogeographic regions, with some warm-water species extending their ranges and colonizing new areas where they were previously absent (Bianchi and Morri 2000; Sabatés et al. 2009; Philippart et al. 2011).

In the Mediterranean, climate change is undoubtedly affecting also the basic biology and ecology of planktonic organisms and the ecosystem functioning (e.g. Molinero et al. 2008a; Conversi et al. 2010; Lejeusne et al. 2010; Martín et al. 2012). However, the magnitude and extension of these changes, and whether or not a common trend can be found is still not clear. This concern about the effects of climate change in the pelagic ecosystem has provided a renovated interest on the use of time-series monitoring studies in order to describe the long-term, natural patterns of zooplankton abundance and composition, and assess their relation to large-scale atmospheric processes and global change. Hence, a recent burst of studies has appeared in an effort to provide new insights on the inter-annual variations in zooplankton standing stocks, assemblage composition and phenological traits in the Mediterranean (e.g. Molinero et al. 2008c; Fernández de Puelles et al. 2009; García-Comas et al. 2011; Mazzocchi et al. 2012), and relate them to basin-scale, regional and local climate forcing related to global change. Due to the key position of zooplankton in pelagic trophic webs, changes in zooplankton communities at decadal and longer-time scales can effect the functioning of the pelagic food web, and may result in severe repercussion for fisheries and other human-related activities (i.e., outburst of gelatinous zooplankton and the human use of beach and swimming activities at sea).

In the western Mediterranean three zooplankton time-series can be considered the most relevant ones, due to their temporal coverage and to the quantity and relevance of the research derived from them: namely the Point B station of Villefranche-sur-Mer in the Ligurian Sea (started in 1966), the Marechiara Station off Naples in the Thyrrenian Sea (started in 1995), and the Baleares Station off Mallorca Island, in the Catalan/Balearic Sea (started in 1994). Here we will summarize the more recent findings derived from their analyses, and refer the reader to the original works and some very recent reviews (Siokou-Frangou et al. 2010; Berline et al. 2012) for more comprehensive information. When doing such exercise, it is important to take into consideration the diversity of environments, sampling strategies (frequency, nets employed,…), taxonomic resolution, etc., encompassed in those time-series, which often can make difficult a straightforward comparison. These issues will be taken into consideration afterward.

The Time Series at Station Point B off Villefranche-sur-Mer

The station Point B is located at the entrance of the Bay of Villefranche-sur-Mer at ca. 80 m depth, on a narrow coastal shelf of the Ligurian Sea influenced by the Northern Current and by a submarine canyon nearby. This is the longest zooplankton time-series in the Mediterranean, sampled weekly (five times a week) since 1966 (Fig. 11.10). Plankton was collected with vertical hauls performed with a Juday-Bogorov net (330 μm mesh size), coupled with additional samples for macroplankton taken with a 690-μm Régent net. The 330 μm-mesh series was discontinued in 2003, but a new collection series started in 1995 using a 200-μm WP-2, which better covered the size range of copepods. In their thorough analyses of zooplankton data during the 1966–1993 period at station Point B, targeted into a few relevant copepod species and several jellyfish and chaetognaths, Molinero et al. (2005a, 2008a, 2008c) found that the effects of local climate variability, especially above certain threshold, translated into changes in the winter water temperature and the degree of water column stratification, which eventually resulted in an increase on jellyfish populations during the mid-late 1980s. According to these authors, the higher frequency of jellyfish outbreaks in the northwestern Mediterranean may significantly alter the functioning of the pelagic ecosystem, evidenced by significant decreases in both copepod (by predation) and chaetognaths (by competition for prey) standing stocks in the Ligurian Sea since the late 1980s (Molinero et al. 2005a, 2008a, c). Such changes may cause a disruption of the functioning of the planktonic ecosystem, affecting both directly and indirectly fish recruitment and dynamics. The climatological and hydrological forcing at regional and local scales responsible for those effects on the zooplankton in the Ligurian Sea appeared to be linked, in last instance, to the long-term changes in the North Atlantic climate. A further analysis of the same Point B series of samples by García-Comas et al. (2011), including ten more years of data (up to 2003) and encompassing not only some target species, but the whole standing stocks of different broad zooplankton groups (e.g. copepods, chaetognaths, appendicularians, etc.) (Fig. 11.10), showed that the previously observed patterns did not hold completely when the study period was prolonged. Even when comparing the same series of years, the increasing trend of jellyfish abundance from the late 1980s, reported by Molinero, differed

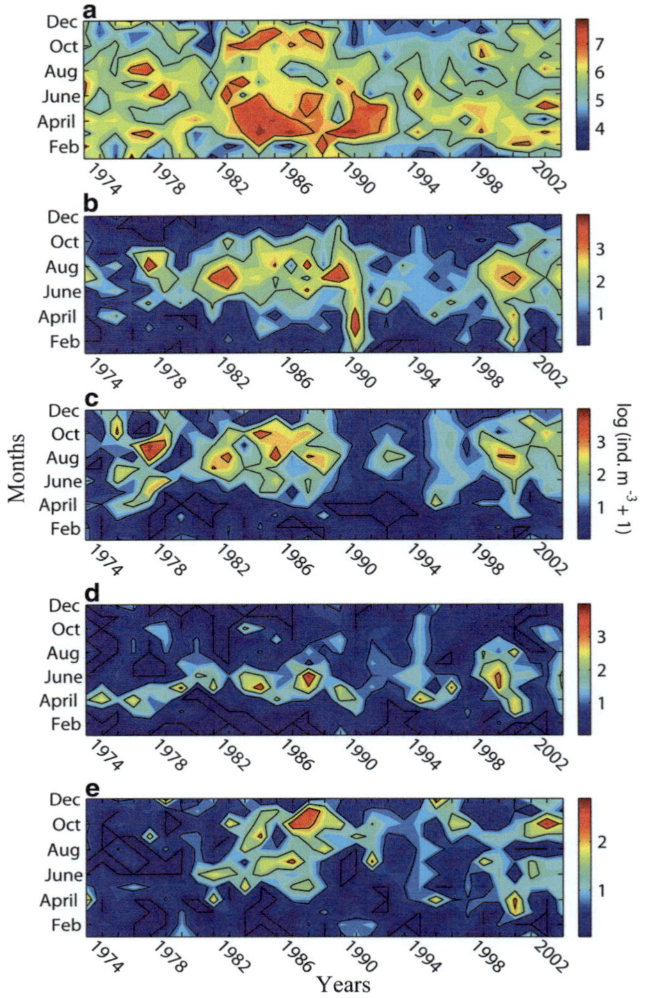

Fig. 11.10 Seasonal and interannual variability of several zooplankton groups at the station Point B off Villefranche-sur-Mer from 1974 to 2003. (**a**) Copepods; (**b**) Decapod larvae; (**c**) Chaetognaths; (**d**) Siphonophores; (**e**) Jellyfish (From García-Comas et al. (2011). With permission of the publisher)

from the observations of García-Comas et al. (2011) with a decrease in total medusae stocks from 1989. In this last study, all the considered zooplankton groups showed similar long-term patterns over the 30-year period of study, and evidenced a certain synchrony with that of environmental factors, following almost-decadal periods (García-Comas et al. 2011). It appears, then, that the trends observed by Molinero for some target species and the proposed regime shift in 1987 for the Mediterranean in synchrony with changes in the Atlantic ocean and the Baltic and Black Seas (Molinero et al. 2008c; Conversi et al. 2010) are challenged when more years are included in the analysis and the bulk of zooplankton standing stocks are considered (García-Comas et al. 2011). Furthermore, the influence of the North Atlantic climate, as represented by the NAO index, on the local hydrological and climatological factors was only evident until the early 1990s. García-Comas et al. (2011) suggested that the link between climate forcing and zooplankton, would be mediated not by a raise of water temperature and thermal stratification, as proposed by Molinero, but by episodes of strong winter mixing, associated to cold and low precipitation (higher salinity) periods, which would augment the nutrient input in the photic zone enhancing phytoplankton productivity, therefore through bottom-up effects increase zooplankton stocks. Contrarily, the recent analysis of the parallel time series taken with a 200-μm WP-2 at Point B during the period 1995–2005 by Vandromme et al. (2011), which also include phytoplankton (chlorophyll) biomass data, seems to support a top-down control of zooplankton on phytoplankton.

The Time Series at Station Marechiara Off Naples

The station Marechiara (off Naples, Italy) is located at a depth of ca. 75 m, in a coastal site hydrodynamically very active, influenced by the land runoff from the nearby populated area, and by intrusions of oligotrophic offshore waters, similar to the open Tyrrhenian Sea (Ribera d'Alcalà et al. 2004; Mazzocchi et al. 2011). Sampling started on 1984 (biweekly sampled), was discontinued temporally between 1991 and 1995, and then continued since 1995 (weekly). Zooplankton is collected by vertical tows with a Nansen net provided with 200-μm mesh size. Mazzocchi et al. (2011, 2012) present the more recent update and analysis of long-term changes in mesozooplankton standing stocks and composition at the site (samples analyzed until 2006). Mesozooplankton standing stocks show relatively low variability over the years (average annual mean: 1,700 ind m^{-3}), with a period of negative anomalies between 1995 and 2002. At the species level, the most abundant zooplankton species displayed high interannual variability (Fig. 11.11), but no remarkable long-term trend could be evidenced, with a few exceptions (e.g. decreased abundance of the copepods *Acartia clausi* and *Centropages typicus* in summer). Despite the extensive variability in hydrographic features and autotrophic communities evidenced at the study site over the years, robust and persisting (over time) zooplankton species assemblages can be distinguished in association with the seasonal cycles of hydrographic and biological features. Although in some cases interannual variations of those assemblages could be associated to environmental long-term trends (e.g. a significant warming trend in summer temperature, a decreasing trend in the autotrophic community), no clear, common patterns were found and the copepod and zooplankton assemblages were notably persistent. It appears that the variability associated

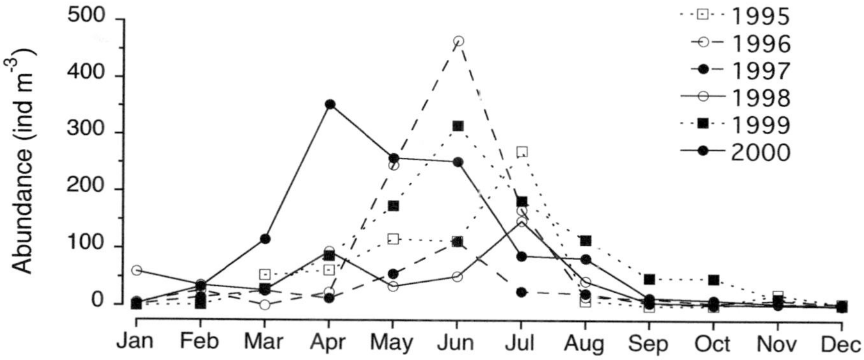

Fig. 11.11 Phenological variation in the occurrence of the calanoid copepod *Centropages typicus* at the station Marechiara off Naples (Redrawn and modified from Mazzocchi et al. 2007)

to changes during the seasonal cycle is much larger than the interannual variations so far observed in the monitoring series.

The Time Series at the Baleares Station off Mallorca

The zooplankton time series at the Baleares station, located at the Mallorca channel in the Catalan/Balearic Sea consists of samples taken every 10 days since 1994 with a Bongo-20 fitted with 250 μm mesh size (Fernández de Puelles et al. 2007); an additional 5-year period (1994–1999) included two other stations sampled monthly across the shelf and additional finer mesh size (Fernández de Puelles et al. 2003a, b). The Mallorca channel is a very relevant area for the meridional water exchange in the western Mediterranean, and is subject to intense mesoscale variability and water masses mixing as the result of being exposed to intrusions of water masses from either the Northern Current (saltier and colder), or from recent, fresher and warmer, Atlantic waters (Pinot et al. 2002). This particular location acts as an excellent observatory to assess how large-scale climate patterns affecting the northern hemisphere and in particular the North Atlantic (i.e. NAO) can act as drivers of regional meteorological variability and hydrographic patterns in the Catalan/Balearic Sea, and therefore influence the north western Mediterranean zooplankton communities (Fernández de Puelles et al. 2007; Fernández de Puelles and Molinero 2008). At seasonal scale, during winter the surface waters of the Mallorca channel mainly show the characteristics of the Northern Current waters, whereas in summer and autumn the warmer and less saline recent Atlantic waters exert a stronger influence (Pinot et al. 2002). Such changes reflect in the standing stocks of zooplankton in the area, in particular copepods, that happen to be higher during the first part of the year, when the Northern Current waters, more productive, have stronger influence (Fernández de Puelles et al. 2003b). The high interannual variability in zooplankton abundance in the Mallorca channel appears to be driven by the large-scale atmospheric forcing in the North Atlantic Ocean, which may reflect in the alternation on the intensity of the influence of Northern Current and Atlantic waters in the area, but also on differential local cooling processes during mild versus cold winters (Fernández de Puelles et al. 2003b). These effects are evidenced by a strong negative relationship between water temperature and zooplankton standing stocks (also conspicuous when dealing with particular taxonomic groups), in special for those periods with strong anomalies (>1 SD) in the North Atlantic climatological indices considered (Fernández de Puelles et al. 2007; Fernández de Puelles and Molinero 2008).

General Remarks

A comprehensive picture of the long-term changes in the zooplankton communities in the western Mediterranean as a whole, and of the influence that large-scale atmospheric processes and global change might have on them, as can be deducted after reading the descriptions above, is not a simple issue. Several considerations have to be taken into account when drawing general conclusions. For instance, their extension in time and the presence of gaps in the sampling are important factors to ponder, because often the number of years (samples) processed is still too short for proper and robust assessment of interannual variability in relation with large-scale atmospheric forcing. Furthermore, in a recent and thorough effort by Berline et al. (2012), which includes also the zooplankton time-series in the Adriatic and Eastern Mediterranean, the authors failed to find significant correlations between climate indices and local temperature or zooplankton abundance, and concluded that probably the monitoring sites are too strongly influenced by local forcing (either climatic and/or anthropogenic) to evidence larger-scale forcing, although synchronous behaviors can be observed among monitoring sites. The peculiarities of the site (i.e. the Baleares station located in neritic waters in the Mallorca channel, main area for meridional water transport in the western Mediterranean; the Point B station near Villefranche-sur-Mer, close to a submarine canyon; the Marechiara station, in a neritic area close to a populated city like Naples) certainly must be considered in order to disentangle the hidden patterns in the typically variable biological data and to extrapolate the observations to other areas. As Mazzocchi et al. (2012) noticed, one should be cautious when coastal sites are used to assess the impact of climate change because these highly variable environments are inhabited by zooplankton species well adapted to cope with variable environmental conditions, and therefore less prone to show dramatic changes. Unfortunately, long-term monitoring sites in open deep waters, free of coastal influences and less sensitive to changes in circulation patterns, are expensive in terms of cost and human-power and are lacking in the Mediterranean.

Another source of uncertainty in the analysis of the time-series data is the taxonomic resolution employed. In this regard, Berline et al. (2012) pointed out that the use of broad taxonomic groups in their study (e.g. copepods, cladocerans, chaetognaths) instead of the species level could have easily masked changes in species composition, phenology, etc. This is consistent with the fact that when standing stocks of broad taxonomic groups are considered, the ranges of variability, either interannual (e.g. Baleares station: in the period 1994–2003, the annual mean abundance of copepods ranged from 454 to 655 ind m^{-3}, with a coefficient of variation of 12 %, Fernández de Puelles et al. 2007), or seasonal (e.g. Baleares station: in the period 1994–2003, the monthly mean abundance of copepods ranged from 347 to 723 ind m^{-3}, with a coefficient of variation of 23 %, Fernández de Puelles et al. 2007) appear rather moderate. At the single species level, Molinero et al. (2005b) reported phenological and abundance changes during the period 1966–1993 at the Point B station in Villefranche-sur-Mer for two significant components of the copepod assemblages in the western Mediterranean, the species *Centropages typicus* and *Temora stylifera*, which could be related to large-scale climate forcing (NAO index). However, in a more ambitious study Mazzocchi et al. (2007) focused on the same species *Centropages typicus* but encompassing several observational sites in both the western and the eastern Mediterranean, only at two locations (Marechiara off Naples and Point B off Villefranche-sur-Mer) the long-term phenological changes in *Centropages typicus* populations could be related to the NAO index, and the response differed between sites. The large size of the Mediterranean basin, encompassing several degrees of longitude and latitude, and its location under the influence of mid-latitude and tropical regimes, is probably responsible for the diversity in patterns observed and the degree of influence of long-term atmospheric forcing related, which would be linked not only to the North Atlantic climate but to other climate regimes as well (Lionello et al. 2006; Martín et al. 2012; Mazzocchi et al. 2012). The range of variation in environmental factors than can be coped with is much narrower when considered at the species level that when the whole community is considered, and hence changes due to climate variations should be first detected at the species level. Mazzocchi et al. (2011) also cautions that although dominant species might not show conspicuous signals of long-term changes, the presence or absence of rare species might be a better proxy to first evidence subtle changes in zooplankton communities. This is the case for the Gulf of Naples, where the disappearance of rare copepod species such as *Acartia margalefi* and *Paracartia latisetosa* (which have disappeared in other areas of the Mediterranean as well), and the more frequent intrusion of offshore copepods at the monitoring site have been reported (Mazzocchi et al. 2012). In this regard, the recent implementation of automatic and semiautomatic zooplankton counting, sizing and identifying methods and technologies (Alcaraz et al. 2003; Grosjean et al. 2004; Gorsky et al. 2010) has helped to speed up the processing of large number of samples and standardize data (e.g. García-Comas et al. 2011; Vandromme et al. 2011), although at the cost of a much lower taxonomic resolution that would be required to detect such changes in rare species. Nevertheless, in order to assess the magnitude of the alterations in the whole planktonic food-web structure and dynamics due to climate change and the implications for the export of carbon to upper trophic levels, the whole assemblage or community might be more relevant (García-Comas et al. 2011).

Jellyfish Blooms and Changes in Fauna

Increases in jellyfish populations have been reported for many pelagic marine ecosystems worldwide (Mills 2001; Purcell 2005) and the Mediterranean is not an exception (Benovic and Lucic 2001; Boero et al. 2008a). Although proliferations of jellyfish, and other gelatinous zooplankton, are a natural phenomenon in coastal and oceanic Mediterranean waters (CIESM 2008) they seem to have become more frequent over the past few years. Many hypotheses have been proposed to explain these blooms throughout the world, which are also applicable to the Mediterranean: climate change, eutrophication, overexploitation, aquaculture and species invasions (Purcell et al. 1999, 2007; Mills 2001). Proliferations of gelatinous zooplankton may therefore be indicative of considerable changes in the functioning of marine ecosystems and have a wide range of ecological implications, including alteration of plankton assemblages through both, top-down and bottom-up effects (Mills 2001; Purcell 2005).

Qualitative scientific records of jellyfish populations and proliferations date from more than 200 years ago in the Mediterranean (Goy et al. 1989; Kogovsek et al. 2010), and despite their heterogeneity and scarcity these data suggest a

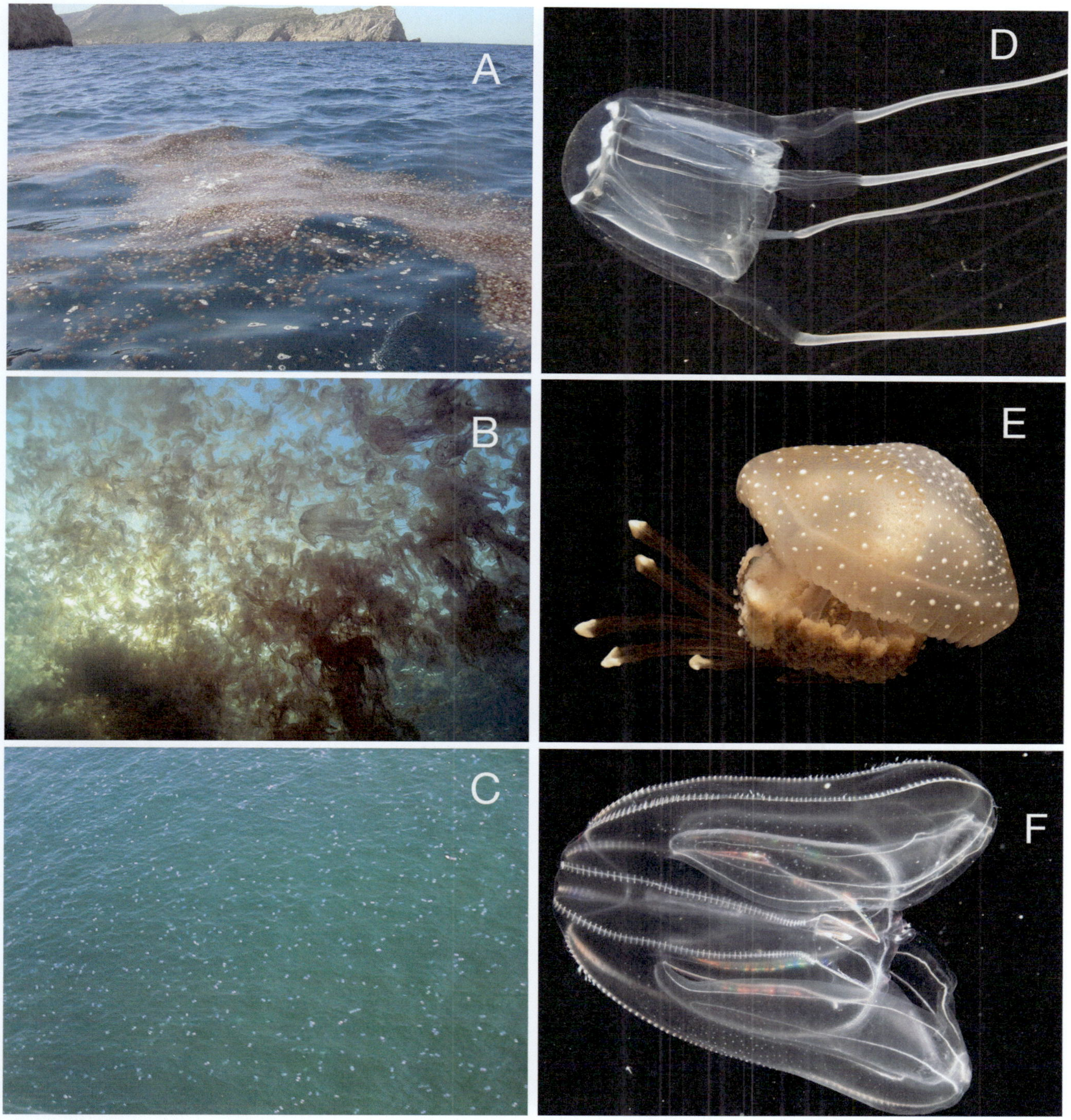

Fig. 11.12 *Left panels*: Three examples of jellyfish proliferations. Swarms of *Pelagia noctiluca* in surface waters of the Balearic Islands in (**a**) June and (**b**) July and (**c**) swarm of *Rhizostoma pulmo* in the Catalan Coast in July. *Right panels*: Three examples of recent invasive species in the Mediterranean (**d**) the jelly *Carybdea marsupialis*, (**e**) the jelly *Phyllorhiza punctata* and (**f**) the ctenophore *Mnemiopsis leidyi* (Photographs of David Diaz (a), Giampiero Mancini (b), Agencia Catalana de l'Aigua (c), Eduardo Obis (d–f))

periodicity in the blooms. The most spectacular events are those of *Pelagia noctiluca* (Fig. 11.12). High densities of *P. noctiluca* have been observed regularly over the past 12 years in the Adriatic and in the whole Mediterranean (Malej and Malej 2004). Goy et al. (1989) reported a periodic occurrence of *P. noctiluca* blooms related to climatic factors, such as low rainfall, high temperatures and high atmospheric pressure. The analysis of data of recent decades, however, suggests that the intensity, timing, and distribution of *P. noctiluca* population outbreaks seem to be driven by recent climatic and hydrological conditions in the Mediterranean (Daly Yahia et al. 2010; Licandro et al. 2010). The common jellyfish *Aurelia aurita* and *Rhizostoma pulmo* are regularly present in the Adriatic Sea, but recently their proliferation has become frequent, especially in the northern coasts (Malej and Malej 1992; Ramšak and Stopar 2007). A similar phenomenon occurs in the Catalan coast where large jellyfish swarms at the beaches caused a great number of swimmers stung (Gili et al. 2010). The presence and arrivals of jellies along the Catalan beaches have been explained in base to local climatic conditions linked to changes in wind and rain regimes (Rubio and Muñoz 1997; Gili and Pagès 2005).

Recent works pointed out that long-term variability of gelatinous zooplankton is strongly affected by climate fluctuations. From the study of the few existing plankton data series and semi quantitative multiyear records in the northwestern Mediterranean, Molinero et al. (2008a) showed that increases of gelatinous plankton were associated with positive phases of the NAO after 1980, with warm and stable water column providing a suitable environmental conditions for their development (see section "Long-term changes" for further discussion). The recent study by Licandro et al. (2012), from the same area, showed that the standing stock of calycophoran siphonophores did not show any significant change during nearly three decades, with the annual maximum in spring. Nevertheless, major changes in the community composition were reported since the middle 1980s, with the decrease of the dominant species *Muggiaea kochi* allowing the increase of the congeneric *Muggiaea atlantica* and *Chelophyes appendiculata*. These community shifts were related to hydroclimatic changes, being salinity, temperature and water stratification the main factors associated with these changes. These results are in line with previous observations of Riera et al. (1986) and Buecher (1999).

Besides the direct effects of climate change, increases in gelatinous zooplankton have also been correlated with overfishing (Lynam et al. 2004; Purcell 2012). Jellyfish and fish interact both as predators and competitors of each other (Brodeur et al. 1999; Pagès 2001), as jellyfish and other gelatinous plankton are consumers of fish eggs and larvae (Lynam et al. 2005; Sabatés et al. 2010). Most of the energy that once circulated through fishes is now processed by other predators, such as pelagic gelatinous carnivores, that fill a gap within the trophic system by the decrease in fish stocks (Mills 2001; Purcell 2012). Other important evidences of the human impact on jelly outbreaks have been demonstrated in coastal lagoons in the Mediterranean (Benovic et al. 2000; Pagès 2001) and other coastal regions related (Purcell et al. 1999). In these areas, the intensive use of fertilizers due to the increment of agricultural activities have resulted in eutrophication, which has favored some species with endosymbiotic algae, such as *Cotylorhiza tuberculata*. Moreover, the increase of available hard substrates (e.g. damns, artificial reefs, shells from bivalve aquaculture) have enhanced the chance for suitable planula settlement and subsequent ephyra production (Pagès 2001; Holst and Jarms 2007). In general, the proliferation of artificial structures associated with the exponential increase of shipping, aquaculture and other coastal industries and defense structures, could, thus, be a major driver of the global rise of jellyfish blooms in Mediterranean coastal areas (Duarte et al. 2013). Thus, a progressive increase in the presence of jellyfish swarms during the next years in the Catalan coast has been proposed as consequence of the interaction between environmental changes and anthropogenic pressures (Calvo et al. 2011).

The long-term water temperature increase has also affected the boundaries of biogeographic regions, with some thermophilic species extending their ranges and colonizing new areas where they were previously absent (Laubier 2003). The northward migration of species with an affinity for warm waters has been reported in different Mediterranean regions for fish and benthic organisms (Francour et al. 1994; Astraldi et al. 1995; Bianchi and Morri 2000; Sabatés et al. 2006). The increase of thermophilic biota in the Mediterranean Sea would involve changes in both indigenous (meridionalization) and non-indigenous (tropicalization) species (Boero et al. 2008b).

Regarding plankton, strong biogeographical shifts in copepod assemblages in relation to the water temperature increase have been observed in the North Atlantic, involving a northward extension of warm water species and a decrease in the number of cold water species (Beaugrand et al. 2002). However, in the Mediterranean such clear examples for plankton are still scarce. Very likely the changes will be first detected in the northern coldest regions and common species will be less sensitive due to their wide distribution in the whole basin. For instance, in the Northern Adriatic, changes in the specific composition of copepods have been reported in relation to the temperature increase. The relative abundance of southern species, i.e. *Diaixis pygmoea, Oithona similis* and *Oithona nana,* has increased in the last years while, on the contrary, relic cold-water species, such as *Pseudocalanus elongatus,* have decreased (Fonda Umani and Conversi 2008). Moreover, Mazzocchi et al. (2012) reported that rare copepod species, for instance *Acartia margalefi* and *Paracartia latisetosa*, have changed drastically their presence in the Thyrrenian Sea and other parts of the Mediterranean, but the mechanism behind these changes are not clear. More clear examples, in parallel with the northwards expansion of pelagic fish species (i.e. *Sardinella aurita* and *Pomatomus saltatrix*) due to sea warming, can be found in the ichthyoplankton (Sabatés et al. 2009, 2012) (Fig. 11.13).

Other changes in zooplankton biota in the Mediterranean are related to the increase of non-indigenous invasive species, phenomenon known as tropicalization. The availability

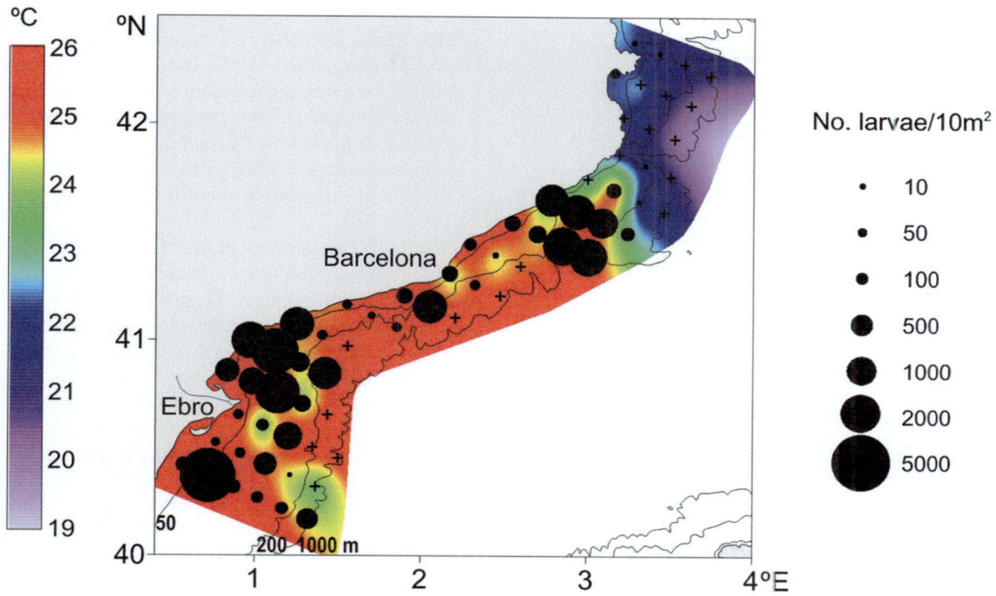

Fig. 11.13 Distribution of *Sardinella aurita* larvae along the Catalan coast (July 2003) overlaid on sea surface temperature. The northern limit of the larval distribution is bounded by the thermal front associated to the intrusion of the Northern Current (Redrawn and modified from Sabatés et al. 2009)

of proper conditions for both tropical and temperate species makes the Mediterranean a perfect sea for biological invasions (Boero et al. 2009). In addition, the seasonality in the Mediterranean is a key factor that facilitates the coexistence of many species (Coma et al. 2000) including those species that resist or remain as resting stages or in dormancy (Boero et al. 1996). Today, more than 500 non-indigenous species are listed from the Mediterranean Sea (Galil 2007), mostly benthic and of tropical and subtropical origin (Zenetos et al. 2010). The best known invasive species are the Lessepsian ones, entering from the Red Sea to the Mediterranean by the Suez Canal (e.g. Kovalev 2006, and references therein). In some cases, these Lessepsian species have become problematic, for example the Sciphomeduse *Rhopilema nomadic,* forming stable populations in the eastern basin (Spanier and Galil 1991). In addition to the entrance of new species from both Suez and Gibraltar (the Atlantic flow never stopped), the Mediterranean is also experiencing the transport of exotic species by ships, both as fouling or in the ballast waters (Lotan et al. 1992). The appearance of the calanoid copepod *Acartia tonsa* (Gaudy and Viñas 1985) and the cyclopoid copepod *Oithona davisae* (Saiz et al. 2003) in the Mediterranean are likely examples of such transport. Another recent example of the introduction of foreign species is the presence and increase in the western Mediterranean of the invasive ctenophore, *Mnemiopsis leidyi* (Fuentes et al. 2009) (Fig. 11.12). The success of the introduced *M. leidyi* has been mainly attributed to the reduced competition due to overfishing of zooplanktivorous fish populations and the lack of predators (Siapatis et al. 2008). Another important invasive species is the Scyphomedusae *Phyllorhiza punctata* with a relative well-documented history of invading tropical and subtropical environments over the last years (Galil et al. 1990; Boero et al. 2009). The proliferation of the cubomedusa *Carybdea marsupialis* in the western Mediterranean Sea have been related to the increase in water temperature and to the enhanced chances for polyp settlement in artificial substrata (Bordehore et al. 2011).

In summary, the historical process of species distribution changes, migrations, colonization, and invasions has given rise to a puzzle of species that conform the actual Mediterranean biodiversity. It is evident that the Mediterranean has always been changing (both physically and biologically). Nevertheless, nowadays, like most marine ecosystems worldwide, the Mediterranean is highly impacted by several human-mediated threats. These major disturbances, in addition to species introductions and climate change, may severely impact the natural balance of planktonic communities and, therefore, have repercussions to the whole ecosystem.

Acknowledgments The authors are indebted to the Editors for their invitation to participate in this book. We acknowledge the support of grants CTM2009-08783, CTM2010-18874 and CTM2011-23480. We specially thank Drs. Miquel Alcaraz and Stamatina Isari for their helpful comments and suggestions on the manuscript, and Drs. García-Comas, Mazzocchi and Ribera d'Alcalà for kindly providing access to original figures and data.

References

Albertini-Berhaut J (1971) Distribution des trachyméduses *Persa incolorata* McGrady 1857, *Aglaura hemistoma* Péron et Lesueur 1809 dans le golfe de Marseille. Bull Soc Zool Fr 95:677–687

Alcaraz M (1970) Ciclo anual de los cladóceros en aguas de Castellón (Mediterráneo occidental). Invest Pesq 34:281–290

Alcaraz M (1985) Vertical distribution of zooplankton biomass during summer stratification in the western Mediterranean. In: Gibbs PE (ed) Proceedings of the 19th European marine biology symposium, Plymouth, Devon, 16–21 Sept 1984. Cambridge University Press, Cambridge

Alcaraz M (1988) Summer zooplankton metabolism and its relation to primary production in the western Mediterranean. Oceanol Acta 9(Sp Issue):185–191

Alcaraz M, Saiz E, Estrada M (1994) Excretion of ammonia by zooplankton and its potential contribution to nitrogen requirements for primary production in the Catalan Sea (Nw Mediterranean). Mar Biol 119:69–76

Alcaraz M, Saiz E, Calbet A, Trepat I, Broglio E (2003) Estimating zooplankton biomass through image analysis. Mar Biol 143: 307–315

Alcaraz M, Calbet A, Irigoien X, Trepat I, Saiz E (2004) Short-term variability of zooplankton biomass and size structure in the northwestern Mediterranean. Rapp Comm Int Expl Sci Mer Mediterr 37:306

Alcaraz M, Calbet A, Estrada M, Marrase C, Saiz E, Trepat I (2007) Physical control of zooplankton communities in the Catalan Sea. Prog Oceanogr 74:294–312

Allain C (1960) Topographie dynamique et courants généraux dans le bassin occidental de la méditerranée (Golfe du Lion, Mer catalane, Mer d'Alboran et ses abords, secteur à l'est de la Corse). Rev Trav Inst Peches Marit 24:121–145

Allen SE, Vindeirinho C, Thomson RE, Foreman MGG, Mackas DL (2001) Physical and biological processes over a submarine canyon during an upwelling event. Can J Fish Aquat Sci 58:671–684

Almeda R, Calbet A, Alcaraz M, Saiz E, Trepat I, Arin L, Movilla J, Saló V (2011) Trophic role and carbon budget of metazoan microplankton in northwest Mediterranean coastal waters. Limnol Oceanogr 56:415–430

Alvarez A, Tintoré J, Sabatés A (1996) Flow modification and shelf-slope exchange induced by a submarine canyon off the northeast Spanish coast. J Geophys Res 101:12043–12055

Andersen V, Sardou J (1992) The diel migrations and vertical distributions of zooplankton and micronekton in the northwestern Mediterranean Sea. 1. Euphausiids, mysids, decapods and fishes. J Plankton Res 14:1129–1154

Andersen V, Sardou J, Nival P (1992) The diel migrations and vertical distributions of zooplankton and micronekton in the northwestern Mediterranean Sea. 2. Siphonophores, hydromedusae and pyrosomids. J Plankton Res 14:1155–1169

Andersen V, Francois F, Sardou J, Picheral M, Scotto M, Nival P (1998) Vertical distributions of macroplankton and micronekton in the Ligurian and Tyrrhenian Seas (northwestern Mediterranean). Oceanol Acta 21:655–676

Andersen V, Gubanova A, Nival P, Ruellet T (2001a) Zooplankton community during the transition from spring bloom to oligotrophy in the open NW Mediterranean and effects of wind events. 2. Vertical distributions and migrations. J Plankton Res 23:243–261

Andersen V, Nival P, Caparroy P, Gubanova A (2001b) Zooplankton community during the transition from spring bloom to oligotrophy in the open NW Mediterranean and effects of wind events. 1. Abundance and specific composition. J Plankton Res 23:227–242

Andersen V, Devey C, Gubanova A, Picheral M, Melnikov V, Tsarin S, Prieur L (2004) Vertical distributions of zooplankton across the Almeria-Oran frontal zone (Mediterranean Sea). J Plankton Res 26:275–293

Andréu P (1990) Los Quetognatos de las costas de la Península Ibérica. PhD thesis, University of Barcelona

Astraldi M, Gasparini G (1992) The seasonal characteristics of the circulation in the North Mediterranean Basin and their relationship with the atmospheric-climatic conditions. J Geophys Res 97:9531–9540

Astraldi M, Bianchi CN, Gasparini GP, Morri C (1995) Climatic fluctuations, current variability and marine species distribution: a case study in the Ligurian Sea (north-west Mediterranean). Oceanol Acta 18:139–149

Atienza D (2006) Ecology of the marine cladoceran *Penilia avirostris*. PhD thesis, Universitat Politècnica de Catalunya

Atienza D, Calbet A, Saiz E, Alcaraz M, Trepat I (2006) Trophic impact, metabolism, and biogeochemical role of the marine cladoceran *Penilia avirostris* and the co-dominant copepod *Oithona nana* in NW Mediterranean coastal waters. Mar Biol 150:221–235

Atienza D, Saiz E, Skovgaard A, Trepat I, Calbet A (2008) Life history and population dynamics of the marine cladoceran *Penilia avirostris* (Branchiopoda: Cladocera) in the Catalan Sea (NW Mediterranean). J Plankton Res 30:345–357

Banaru D, Harmelin-Vivien M (2009) Trophic links and riverine effects on food webs of pelagic fish of the north-western Black Sea. Mar Freshw Res 60:529–540

Batistić M, Jasprica N, Carić M, Lučić D (2007) Annual cycle of the gelatinous invertebrate zooplankton of the eastern south Adriatic coast (NE Mediterranean). J Plankton Res 29:671–686

Beaugrand G, Reid PC, Ibañez F, Lindley JA, Edwards M (2002) Reorganization of north Atlantic marine copepod biodiversity and climate. Science 296:1692–1694

Bellan-Santini D (1990) Mediterranean deep-sea amphipods: composition, structure and affinities of the fauna. Prog Oceanogr 24: 275–287

Belmonte G, Pati AC (2007) Hatching rate and diapause duration in eggs of *Paracartia latisetosa* (Copepoda: Calanoida). J Plankton Res 29(Suppl 1):i39–i47

Benovic A, Lucic D (2001) Jellyfish outbreaks: natural cycle or stress response effect? In: Briand F (ed) Gelatinous zooplankton outbreaks: theory and practice, vol 14. CIESM Workshop Monographs, Monaco

Benovic A, Lucic D, Onofri V (2000) Does change in an Adriatic hydromedusan fauna indicate an early phase of marine ecosystem destruction? Mar Ecol 21:221–231

Berhaut J (1969) Variations mensuelles d'abondance de la trachyméduse *Liriope tetraphylla* (Chamisso et Eysenhard, 1821) dans le golfe de Marseille. Bull Mus Hist nat Paris 40:1222–1231

Berline L, Siokou-Frangou I, Marasović I, Vidjak O, Fernández de Puelles ML, Mazzocchi MG, Assimakopoulou G, Zervoudaki S, Fonda-Umani S, Conversi A, Garcia-Comas C, Ibanez F, Gasparini S, Stemmann L, Gorsky G (2012) Intercomparison of six Mediterranean zooplankton time series. Prog Oceanogr 97–100:76–91

Bernard C, Rassoulzadegan F (1994) Seasonal variations of mixotrophic ciliates in the northwest Mediterranean Sea. Mar Ecol Prog Ser 108:295–301

Béthoux J (1979) Budgets of the Mediterranean Sea – their dependance on the local climate and on the characteristics of the Atlantic waters. Oceanol Acta 2:157–163

Béthoux J, Prieur L, Bong J (1988) Le courant Ligure au large de Nice. Oceanol Acta 9(Sp Issue):59–67

Béthoux J, Morin P, Madec C, Gentili B (1992) Phosphorus and nitrogen behaviour in the Mediterranean Sea. Deep-Sea Res Part A 39:1641–1654

Béthoux J, Gentili B, Morin P, Nicolas E, Pierre C, Ruiz-Pino D (1999) The Mediterranean Sea: a miniature ocean for climatic and environmental studies and a key for the climatic functioning of the north Atlantic. Prog Oceanogr 44:131–146

Bianchi C, Morri C (2000) Marine biodiversity of the Mediterranean Sea: situation, problems and prospects for future research. Mar Pollut Bull 40:367–376

Blanc F, Leveau M, Szekielda K (1969) Effects eutrophiques au débouché d'un grand fleuve (Rhône). Mar Biol 3:233–242
Boero F, Bouillon J (1993) Zoogeography and life cycle patterns of Mediterranean hydromedusae (Cnidaria). Biol J Linn Soc 48: 239–266
Boero F, Belmonte G, Fanelli G, Piraino S, Rubino F (1996) The continuity of living matter and the discontinuities of its constituents: do plankton and benthos really exist? Trends Ecol Evol 11:177–180
Boero F, Bouillon J, Gravili C, Miglietta MP, Parsons T, Piraino S (2008a) Gelatinous plankton: irregularities rule the world (sometimes). Mar Ecol Prog Ser 356:299–310
Boero N, Féral JP, Azzurro E, Cardin V, Riedel B, Despalatovic M, Munda I, Moschella P, Zaouali J, Fonda Umani S, Theocharis A, Wiltshire K, Briand F (2008b). In: Briand F (ed) Climate warming and related changes in Mediterranean marine biota, CIESM workshop monographs 35. CIESM, Monaco
Boero F, Putti M, Trainito E, Prontera E, Piraino S, Shiganova T (2009) Recent changes in Western Mediterranean Sea biodiversity: the establishment of *Mnemiopsis leidyi* (Ctenophora) and the arrival of *Phyllorhiza punctata* (Cnidaria). Aquat Invasions 4:675–680
Bordehore C, Fuentes V, Atienza D, Barberá C, Fernandez-Jover D, Roig M, Acevedo-Dudley MJ, Canepa AJ, Gili JM (2011) Detection of an unusual presence of the cubozoan *Carybdea marsupialis* in shallow beaches at the locality of Denia (SW Mediterranean). Mar Biodivers Rec 4:1–6
Böttger-Schnack R (1994) The microcopepod fauna in the Eastern Mediterranean and Arabian Seas: a comparison with the Red Sea fauna. Hydrobiologia 292:271–282
Boucher J (1984) Localization of zooplankton populations in the Ligurian marine front: role of ontogenic migration. Deep-Sea Res Part A 31:469–484
Boucher J (1988) Space-time aspects in the dynamics of plankton stages. In: Rothschild B (ed) Toward a theory on biological-physical interactions in the world oceans. Kluwer, Amsterdam
Boucher J, Ibanez FI, Prieur L (1987) Daily and seasonal variations in the spatial distribution of zooplankton populations in relation to the physical structure in the Ligurian Sea Front. J Mar Res 45:133–173
Bouillon J, Medel MD, Pagès F, Gili JM, Boero F, Gravili C (2004) Fauna of the Mediterranean Hydrozoa. Sci Mar 68(Suppl 2):5–438
Braconnot JC (1971) Contribution à l'étude biologique et écologique des tuniciers pélagiques salpides et doliolides. I. Hydrologie et écologie des salpides. Vie Milieu 22:257–286
Brodeur RD, Mills CE, Overland JE, Walters GE, Schumacher JD (1999) Evidence for a substantial increase in gelatinous zooplankton in the Bering Sea, with possible links to climate change. Fish Oceanogr 8:296–306
Buecher E (1999) Appearance of *Chelophyes appendiculata* and *Abylopsis tetragona* (Cnidaria, Siphonophora) in the Bay of Villefranche, northwestern Mediterranean. J Sea Res 41:295–307
Calbet A, Alcaraz M, Saiz E, Estrada M, Trepat I (1996) Planktonic herbivorous food webs in the Catalan Sea (NW Mediterranean): temporal variability and comparison of indices of phyto-zooplankton coupling based on state variables and rate processes. J Plankton Res 18:2329–2347
Calbet A, Garrido S, Saiz E, Alcaraz M, Duarte CM (2001) Annual zooplankton succession in coastal NW Mediterranean waters: the importance of the smaller size fractions. J Plankton Res 23:319–331
Calbet A, Broglio E, Saiz E, Alcaraz M (2002a) Low grazing impact of mesozooplankton on the microbial communities of the Alboran Sea: a possible case of inhibitory effects by the toxic dinoflagellate *Gymnodinium catenatum*. Aquat Microb Ecol 26:235–246
Calbet A, Saiz E, Alcaraz M (2002b) Copepod egg production in the NW Mediterranean: effects of winter environmental conditions. Mar Ecol Prog Ser 237:173–184
Calvo E, Simó R, Coma R, Ribes M, Pascual J, Sabatés A, Gili JM, Pelejero C (2011) Effects of climate change on Mediterranean marine ecosystems: the case of the Catalan Sea. Clim Res 50:1–29
Cartes JE (1998) Dynamics of the bathyal benthic boundary layer in the northwestern Mediterranean: depth and temporal variations in macrofaunal-megafaunal communities and their possible connections within deep-sea trophic webs. Prog Oceanogr 41:111–139
Cartes J, Sorbe J, Sarda F (1994) Spatial distribution of deep-sea decapods and euphausiids near the bottom in the northwestern Mediterranean. J Exp Mar Biol Ecol 179:131–144
Cartes JE, Fanelli E, Papiol V, Zucca L (2010) Distribution and diversity of open-ocean, near-bottom macroplankton in the western Mediterranean: analysis at different spatio-temporal scales. Deep-Sea Res Part I 57:1485–1498
Casanova JP (1965) *Penilia avirostris* Dana, indicateur d'eaux diluées. Rev Trav Inst Peches Marit 29:197–204
Casanova J (1990) Les chaetognathes bentho-planctoniques en Méditerranée. Rapp Comm Int Explor Mer Mediterr 32:208
Champalbert G (1996) Characteristics of zooplankton standing stock and communities in the Western Mediterranean Sea: relations to hydrology. Sci Mar 60:97–113
Cheney RE, Doblar RA (1982) Structure and variability of the Alborán Sea frontal system. J Geophys Res 87:585–594
CIESM (2008) Climate warming and related changes in Mediterranean marine biota, vol 35. CIESM workshop monographs, Monaco
Coll M, Piroddi C, Steenbeek J, Kaschner K, Lasram FBR, Aguzzi J, Ballesteros E, Bianchi CN, Corbera J, Dailianis T (2010) The biodiversity of the Mediterranean Sea: estimates, patterns, and threats. PLoS One 5:e11842
Coma R, Ribes M, Gili JM, Zabala M (2000) Seasonality in coastal benthic ecosystems. Trends Ecol Evol 15:448–453
Conover RJ, Huntley M (1991) Copepods in ice-covered seas – distribution, adaptations to seasonally limited food, metabolism, growth patterns and life cycle strategies in polar seas. J Mar Syst 2:1–41
Conversi A, Umani SF, Peluso T, Molinero JC, Santojanni A, Edwards M (2010) The Mediterranean Sea regime shift at the end of the 1980s, and intriguing parallelisms with other European basins. PLoS One 5:e10633
Cruzado A, Veláquez ZR (1990) Nutrients and phytoplankton in the Gulf of Lions, northwestern Mediterranean. Cont Shelf Res 10:931–942
Cushing DH (1990) Plankton production and year-class strength in fish populations: an update of the match/mismatch hypothesis. Adv Mar Biol 26:249–293
D'Ortenzio F, Ribera D'Alcalà M (2009) On the trophic regimes of the Mediterranean Sea: a satellite analysis. Biogeosciences 6:139–148
Dallot S, Goy J, Carré C (1988) Peuplements de carnivores planctoniques gélatineux et structures productives en Méditerranée occidentale. Oceanol Acta 9(Sp Issue):193–209
Daly Yahia MN, Batistic M, Lucic D, Fernández de Puelles ML, Licandro P, Malej A, Molinero JC, Siokou-Frangou I, Zervoudaki S, Prieto L, Goy J, Daly Yahia-Kéfi O (2010) Are the outbreaks of *Pelagia noctiluca* (Forsskål,1775) more frequent in the Mediterranean basin? In: Gislason A, Gorsky G (eds) Proceedings of the joint ICES/CIESM workshop to compare zooplankton ecology and methodologies between the Mediterranean and the north Atlantic (WKZEM), ICES cooperative research report 300, ICES, Copenhagen
Della Croce N (1964) Distribuzione e biologia del cladocero marino *Penilia avirostris* Dana. Bull Inst Océanogr Monaco 62:1–16
Denman KL, Powell TM (1984) Effects of physical processes on planktonic ecosystems in the coastal ocean. Oceanogr Mar Biol Ann Rev 22:125–168
Diaz F, Naudin J-J, Courtier C, Rimmelin P, Oriol L (2008) Biogeochemical and ecological functioning of the low-salinity water lenses in the region of the Rhone River freshwater influence, NW Mediterranean Sea. Cont Shelf Res 28:1511–1526
Dolan JR, Marrasé C (1995) Planktonic ciliate distribution relative to a deep chlorophyll maximum: Catalan Sea, NW Mediterranean, June 1993. Deep-Sea Res Part I 42:1965–1987

Dolan JR, Claustre H, Carlotti F, Plouvenez S, Moutin T (2002) Microzooplankton diversity: relationships of tintinnid ciliates with resources, competitors and predators from the Atlantic Coast of Morocco to the Eastern Mediterranean. Deep-Sea Res Part I 49:1217–1232

Duarte CM, Agustí S, Kennedy H, Vaqué D (1999) The Mediterranean climate as a template for Mediterranean marine ecosystems: the example of the northeast Spanish littoral. Prog Oceanogr 44:245–270

Duarte CM, Pitt K, Lucas C, Purcell J, Uye S, Robinson K, Brotz L, Decker MB, Sutherland KL, Malej A, Madin L, Mianzan H, Gili JM, Fuentes V, Atienza D, Pagès F, William W, Graham M, Condon R (2013) Global ocean sprawl as a trojan horse for jellyfish blooms. Front Ecol Environ 11:91–97

Duró A, Saiz E (2000) Distribution and trophic ecology of chaetognaths in the western Mediterranean in relation to an inshore-offshore gradient. J Plankton Res 22:339–361

Durrieu de Madron X, Radakovitch O, Heussner S, Loÿe-Pilot MD, Monaco A (1999) Role of the climatological and current variability on shelf-slope exchanges of particulate matter. Evidence from the Rhône continental margin (NW Mediterranean). Deep-Sea Res Part I 46:1513–1538

Durrieu de Madron X, Guieu C, Conan P, Cossa D, D'Ortenzio F, Estournel C, Gazeau F (2011) Marine ecosystems' responses to climatic and anthropogenic forcings in the Mediterranean by 'The MERMEX Group'. Prog Oceanogr 91:97–166

Egloff DA, Fofonoff PW, Onbé T (1997) Reproductive biology of marine Cladocerans. Adv Mar Biol 31:79–167

Estrada M (1985) Deep phytoplankton and chlorophyll maxima in the western Mediterranean. In: MoraitouApostolopoulou M, Kiortsis V (eds) Mediterranean marine ecosystems. Plenum Press, New York

Estrada M (1996) Primary production in the northwestern Mediterranean. Sci Mar 60(Suppl 2):55–64

Estrada M, Margalef R (1988) Supply of nutrients to the Mediterranean photic zone along a persistent front. Oceanol Acta 9(Sp Issue): 133–142

Estrada M, Vives F, Alcaraz M (1985) Life and the productivity of the open sea. In: Margalef R (ed) Key environment: Western Mediterranean. Pergamon Press, Oxford

Estrada M, Marrasé C, Latasa M, Berdalet E, Delgado M, Riera T (1993) Variability of deep chlorophyll maximum characteristics in the Northwestern Mediterranean. Mar Ecol Prog Ser 92:289–300

Estrada M, Varela RA, Salat J, Cruzado A, Arias E (1999) Spatio-temporal variability of the winter phytoplankton distribution across the Catalan and North Balearic fronts(NW Mediterranean). J Plankton Res 21:1–20

Fenaux R (1968) Quelques aspects de la distribution verticale chez les appendiculaires en Méditerranée. Cah Biol Mar 9:23–29

Fernández de Puelles ML, Molinero JC (2008) Decadal changes in hydrographic and ecological time-series in the Balearic Sea (western Mediterranean), identifying links between climate and zooplankton. ICES J Mar Sci 65:311–317

Fernández de Puelles ML, Grás D, Hernández-León S (2003a) Annual cycle of zooplankton biomass, abundance and species composition in the neritic area of the Balearic Sea, western Mediterranean. Mar Ecol 24:123–139

Fernández de Puelles ML, Pinot JM, Valencia J (2003b) Seasonal and interannual variability of zooplankton community in waters off Mallorca island (Balearic Sea, western Mediterranean): 1994–1999. Oceanol Acta 26:673–686

Fernández de Puelles ML, Valencia J, Jansá J, Morillas A (2004) Hydrographical characteristics and zooplankton distribution in the Mallorca channel (western Mediterranean): spring 2001. ICES J Mar Sci 61:654–666

Fernández de Puelles ML, Alemany F, Jansá J (2007) Zooplankton time-series in the Balearic Sea (western Mediterranean): variability during the decade 1994–2003. Prog Oceanogr 74:329–354

Fernández de Puelles ML, Lopez-Urrutia Å, Morillas A, Molinero JC (2009) Seasonal variability of copepod abundance in the Balearic region (western Mediterranean) as an indicator of basin scale hydrological changes. Hydrobiologia 617:3–16

Fielding S, Crisp N, Allen J, Hartman M, Rabe B, Roe H (2001) Mesoscale subduction at the Almeria-Oran front: part 2. Biophysical interactions. J Mar Syst 30:287–304

Flexas M, Durrieu de Madron X, Garcia M, Canals M, Arnau P (2002) Flow variability in the Gulf of Lions during the MATER HFF experiment (March-May 1997). J Mar Syst 33:197–214

Fonda Umani S, Conversi A (2008) Possible climate impacts on the northern Adriatic pelagic ecosystem. In: Briand F (ed) Climate warming and related changes in Mediterranean marine biota, CIESM workshop monographs 35, Monaco

Font J, Salat J, Tintoré J (1988) Permanent features of the circulation in the Catalan Sea. Oceanol Acta 9(Sp Issue):51–57

Font J, Rousseau S, Shirasago B, García-Gorriz E, Haney R (2002) Mesoscale variability in the Alboran Sea: synthetic aperture radar imaging of frontal eddies. J Geophys Res 107:3059

Fragopoulu N, Siokou-Frangou I, Christou E, Mazzocchi M (2001) Patterns of vertical distribution of Pseudocalanidae and Paracalanidae (Copepoda) in pelagic waters (0 to 300 m) of the eastern Mediterranean Sea. Crustaceana 74:49–68

Francour P, Boudouresque C, Harmelin J, Harmelin-Vivien M, Quignard J (1994) Are the Mediterranean waters becoming warmer? Information from biological indicators. Mar Pollut Bull 28:523–526

Franks PJS (1992) Phytoplankton blooms at fronts: patterns, scales, and physical forcing mechanisms. Rev Aquat Sci 6:121–137

Fredj G, Bellan-Santini D, Meinardi M (1992) Etat des connaissances sur la faune marine méditerranéenne. Bull Inst Océanogr Monaco 9(Special Issue):133–145

Fuentes VL, Atienza D, Gili JM, Purcell JE (2009) First records of *Mnemiopsis leidyi* A. Agassiz 1865 off the NW Mediterranean coast of Spain. Aquat Invasions 4:671–674

Furnestin M (1963) Les Chaetognathes du grupe Serratodentata en Méditerranée. Rapp P-v Reun Comm Int Explor Sci Mer Mediterr 17:631–634

Furnestin M-L (1968) Le zooplancton de la Mediterranée (Bassin occidental). Essai de synthèse. J Cons perm int Explor Mer 32:25–69

Furnestin M (1979) Aspects of the zoogeography of the Mediterranean plankton. In: van der Spoel S, Pierrot-Bults A (eds) Zoogeography and diversity of plankton. Bunge, Utrecht

Galil B (2007) Loss or gain? Invasive aliens and biodiversity in the Mediterranean Sea. Mar Pollut Bull 55:314–322

Galil B, Spanier E, Ferguson W (1990) The Scyphomedusae of the Mediterranean coast of Israel, including two Lessepsian migrants new to the Mediterranean. Zool Meded 64:95–105

Gamulin T, Kršinic F (1993) On the occurrence of Calycophorae (Siphonophora) in the southern Adriatic and Tyrrhenian Sea: a comparison of the annual cycles off Dubrovnik and Naples. J Plankton Res 15:855

García A, Rubín JP, Rodríguez JM (1988) La distribución de las áreas de puesta y cría de la sardina (*S. pilchardus*) en el sector noroccidental costero del mar de Alboran, en Marzo de 1982. Complementado con información adicional de parámetros biológicos y físico-químicos. Inf Téc Inst Esp Oceanogr 59:3–24

García E, Tintoré J, Pinot JM, Font J, Manriquez M (1994) Surface circulation and dynamics of the Balearic Sea. Coast Estuar Stud 46:73–91

García-Comas C, Stemmann L, Ibanez F, Berline L, Mazzocchi MG, Gasparini S, Picheral M, Gorsky G (2011) Zooplankton long-term changes in the NW Mediterranean Sea: decadal periodicity forced by winter hydrographic conditions related to large-scale atmospheric changes? J Mar Syst 87:216–226

García Lafuente J, Cano N, Vargas M, Rubín JP, Hernández-Guerra A (1998) Evolution of the Alboran Sea hydrographic structures during July 1993. Deep-Sea Res Part I 45:39–65

Gasser B, Payet G, Sardou J, Nival P (1998) Community structure of mesopelagic copepods (>500 mu m) in the Ligurian sea (western Mediterranean). J Mar Syst 15:511–522

Gaudy R (1985) Features and peculiarities of zooplankton communities from the western Mediterranean. In: Moraitou-Apostopoulos M, Kiortsis V (eds) Mediterranean marine ecosystems. Plenum, New York

Gaudy R, Champalbert G (1998) Space and time variations in zooplankton distribution south of Marseilles. Oceanol Acta 21: 793–802

Gaudy R, Viñas MD (1985) Premiére signalisation en Méditerranée du copépode pélagique *Acartia tonsa*. Rapp Comm Int Explor Sci Mer Mediterr 29:227–229

Gili J, Pagès F (2005) Les proliferacions de meduses. Jellyfish blooms. Butll Soc Hist Nat Balear 48:9–22

Gili JM, Pagès F, Vives F (1987) Distribution and ecology of a population of planktonic cnidarians in the western Mediterranean. In: Bouillon J, Boero F, Cicogna F, Cornelius PFS (eds) Modern trends in the systematics, ecology, and evolution of hydroids and hydromedusae. Clarendon, Oxford

Gili J, Pagès F, Sabatés A, Ros J (1988) Small-scale distribution of a cnidarian population in the western Mediterranean. J Plankton Res 10:385–401

Gili JM, Bouillon J, Pagès F, Palanques A, Puig P, Heussner S (1998) Origin and biogeography of the deep-water Mediterranean Hydromedusae including the description of two new species collected in submarine canyons of northwestern Mediterranean. Sci Mar 62:113–134

Gili JM, Bouillon J, Pagès F, Palanques A, Puig P (1999) Submarine canyons as habitats of prolific plankton populations: three new deep-sea Hydroidomedusae in the western Mediterranean. Zool J Linn Soc 125:313–329

Gili JM, Pages F, Bouillon J, Palanques A, Puig P, Heussner S, Calafat A, Canals M, Monaco A (2000) A multidisciplinary approach to the understanding of hydromedusan populations inhabiting Mediterranean submarine canyons. Deep-Sea Res Part I 47:1513–1533

Gili JM, Fuentes V, Atienza D, Lewinsky I (2010) Report of the Medusa project. Technical report, vol 8. Agència Catalana de l'Aigua, Generalitat de Catalunya, Barcelona

Gómez F (2006) Endemic and Indo-Pacific plankton in the Mediterranean Sea: a study based on dinoflagellate records. J Biogeogr 33:261–270

Gorsky G, da Silva L, Dallot S, Laval P, Braconnot J, Prieur L (1991) Midwater tunicates: are they related to the permanent front of the Ligurian Sea(NW Mediterranean)? Mar Ecol Prog Ser 74:195–204

Gorsky G, Prieur L, Taupier-Letage I, Stemmann L, Picheral M (2002) Large particulate matter in the western Mediterranean I. LPM distribution related to mesoscale hydrodynamics. J Mar Syst 33:289–311

Gorsky G, Ohman MD, Picheral M, Gasparini S, Stemmann L, Romagnan J-B, Cawood A, Pesant S, Garcia-Comas C, Prejger F (2010) Digital zooplankton image analysis using the ZooScan integrated system. J Plankton Res 32:285–303

Goy J (1987) Summer submergence of Persa incolorata Melrady, 1857 (Cnidaria, Hydromedusa) in the Mediterranean. Ann Inst Oceanogr 63:47–55

Goy J, Thiriot A (1976) Conditions estivales dans la divergence de Méditerrannée nord-occidentale-II-macroplancton et micronecton-Étude qualitative et estimation quantitative des cnidaires et des euphausiacés. Ann Inst Oceanogr 52:33–44

Goy J, Morand P, Etienne M (1989) Long-term fluctuations of *Pelagia noctiluca* (Cnidaria, Scyphomedusa) in the western Mediterranean Sea. Prediction by climatic variables. Deep-Sea Res Part A 36:269–279

Graham WM, Pages F, Hamner WM (2001) A physical context for gelatinous zooplankton aggregations: a review. Hydrobiologia 451:199–212

Grosjean P, Picheral M, Warembourg C, Gorsky G (2004) Enumeration, measurement, and identification of net zooplankton samples using the ZOOSCAN digital imaging system. ICES J Mar Sci 61:518–525

Hagen W (1999) Reproductive strategies and energetic adaptations of polar zooplankton. Invertebr Reprod Dev 36:25–34

Holst S, Jarms G (2007) Substrate choice and settlement preferences of planula larvae of five Scyphozoa (Cnidaria) from German Bight, North Sea. Mar Biol 151:863–871

Hopkins T (1985) Physics of the sea. In: Margalef R (ed) Key environment: Western Mediterranean. Pergamon Press, Oxford

Hure J (1961) Migration journalière et distribution saisonnière verticale du zooplancton dans la région profonde de l'Adriatique. Acta Adriat 9:1–59

Hure J, Scotto di Carlo B (1974) New patterns of diurnal vertical migration of some deep-water copepods in the Tyrrhenian and Adriatis Seas. Mar Biol 28:179–184

Ianora A, Santella L (1991) Diapause embryos in the neustonic copepodAnomalocera patersoni. Mar Biol 108:387–394

Ianora A, Scotto di Carlo B (1981) The distribution and annual cycles of Siphonophora Calycophora in the Gulf of Naples and adjacent waters. Arch Oceanogr Limnol 20:51–65

Ibanez F, Boucher J (1987) Anisotropie des populations zooplanctoniques dans la zone frontale de Mer Ligure. Oceanol Acta 10:205–216

Jansá A (1987) Distribution of the mistral: a satellite observation. Meteorol Atmos Phys 36:201–214

Kiørboe T (1993) Turbulence, phytoplankton cell size, and the structure of pelagic food webs. Adv Mar Biol 29:1–72

Kogovsek T, Bogunovic B, Malej A (2010) Recurrence of bloom-forming scyphomedusae: wavelet analysis of a 200-year time series. Hydrobiologia 645:81–96

Kovalev A (2006) On the problem of Lessepsian migrations of zooplanktonic organisms. Mediterr Mar Sci 7:67–71

Lacombe H, Gascard J, Gonella J, Bethoux J (1981) Response of the Mediterranean to the water and energy fluxes across its surface, on seasonal and interannual scales. Oceanol Acta 4:247–255

Laubier L (2003) Changement et vulnérabilité des peuplements marins côtiers. C R Geosci 335:561–568

Lejeusne C, Chevaldonné P, Pergent-Martini C, Boudouresque CF, Pérez T (2010) Climate change effects on a miniature ocean: the highly diverse, highly impacted Mediterranean Sea. Trends Ecol Evol 25:250–260

Licandro P, Ibanez F (2000) Changes of zooplankton communities in the Gulf of Tigullio (Ligurian Sea, western Mediterranean) from 1985 to 1995. Influence of hydroclimatic factors. J Plankton Res 22:2225–2253

Licandro P, Conway D, Daly Yahia MN, Fernández de Puelles ML, Gasparini S, Hecq J, Tranter P, Kirby R (2010) A blooming jellyfish in the northeast Atlantic and Mediterranean. Biol Lett 6:688–691

Licandro P, Souissi S, Ibanez F, Carré C (2012) Long-term variability and environmental preferences of calycophoran siphonophores in the Bay of Villefranche (north-western Mediterranean). Prog Oceanogr 97–100:152–163

Lionello P, Malanotte-Rizzoli P, Boscolo R, Alpert P, Artale V, Li L, Luterbacher J, May W, Trigo R, Tsimplis M (2006) The Mediterranean climate: an overview of the main characteristics and issues. Dev Earth Environ Sci 4:1–26

López García MJ, Millot C, Font J, García-Ladona E (1994) Surface circulation variability in the Balearic Basin. J Geophys Res 99:3285–3296

López-Jurado J, Garcia Lafuente J, Cano Lucaya N (1995) Hydrographic conditions of the Ibiza channel during November 1990, March 1991 and July 1992. Oceanol Acta 18:235–243

Lotan A, Ben-Hillel R, Loya Y (1992) Life cycle of *Rhopilema nomadica*: a new immigrant scyphomedusan in the Mediterranean. Mar Biol 112:237–242

Lynam CP, Hay SJ, Brierley AS (2004) Interannual variability in abundance of north Sea jellyfish and links to the North Atlantic Oscillation. Limnol Oceanogr 49:637–643

Lynam CP, Heath MR, Hay SJ, Brierley AS (2005) Evidence for impacts by jellyfish on north Sea herring recruitment. Mar Ecol Prog Ser 298:157–167

Mackas DL, Denman KL, Abbott MR (1985) Plankton patchiness: biology in the physical vernacular. Bull Mar Sci 37:652–674

Macquart-Moulin C (1998) Gut repletion during diel vertical migration in the benthopelagic crustacean *Eurydice truncata* Norman, 1868 (Isopoda, Cirolanidae). J Plankton Res 20:817–829

Madin LP (1991) Distribution and taxonomy of zooplankton in the Alboran Sea and adjacent western Mediterranean, a literature survey and field guide. Woods Hole Oceanographic Institution, Technical report WHOI-91-26

Malej A, Malej M (1992) Population dynamics of the jellyfish *Pelagia noctiluca* (Forsskal, 1775). In: Colombo G, Ferrara I (eds) Marine eutrophication and population dynamics. Olsen & Olsen, Fredensborg

Malej A, Malej M (2004) Invasion of the jellyfish *Pelagia noctiluca* in the northern Adriatic: a non-success story. In: Dumont H, Shiganova TA, Niermann U (eds) Aquatic invasions in the Black, Caspian, and Mediterranean Seas. Kluwer Academic Publishers, New York

Marcus NH (1996) Ecological and evolutionary significance of resting eggs in marine copepods: past, present, and future studies. Hydrobiologia 320:141–152

Marcus NH, Boero F (1998) Minireview: the importance of benthic-pelagic coupling and the forgotten role of life cycles in coastal aquatic systems. Limnol Oceanogr 43:763–768

Martín P, Sabatés A, Lloret J, Martin-Vide J (2012) Climate modulation of fish populations: the role of the Western Mediterranean Oscillation (WeMO) in sardine (*Sardina pilchardus*) and anchovy (*Engraulis encrasicolus*) production in the north-western Mediterranean. Clim Change 110:925–939

Masó M, Sabatés A, Olivar MP (1998) Short-term physical and biological variability in the shelf-slope region of the NW Mediterranean during the spring transition period. Cont Shelf Res 18:661–675

Massutí M, Margalef R (1950) Introducción al estudio del plancton marino. Patronato Juan de la Cierva de Investigación Técnica (CSIC), Sección de Biología Marina, Barcelona

Mazzocchi MG, Ribera d'Alcalà M (1995) Recurrent patterns in zooplankton structure and succession in a variable coastal environment. ICES J Mar Sci 52:679–691

Mazzocchi M, Christou E, Fragopoulu N, Siokou-Frangou I (1997) Mesozooplankton distribution from Sicily to Cyprus (eastern Mediterranean): I. General aspects. Oceanol Acta 20:521–535

Mazzocchi MG, Christou ED, Capua ID, Fernández dP ML, Fonda-Umani S, Molinero JC, Nival P, Siokou-Frangou I (2007) Temporal variability of *Centropages typicus* in the Mediterranean Sea over seasonal-to-decadal scales. Prog Oceanogr 72:214–232

Mazzocchi MG, Licandro P, Dubroca L, Di Capua I, Saggiomo V (2011) Zooplankton associations in a Mediterranean long-term time-series. J Plankton Res 33:1163

Mazzocchi MG, Dubroca L, García-Comas C, Di Capua I, Ribera d'Alcalà M (2012) Stability and resilience in coastal copepod assemblages: the case of the Mediterranean long-term ecological research at Station MC (LTER-MC). Prog Oceanogr 97–100:135–151

MEDOC Group (1970) Observation of formation of deep water in the Mediterranean Sea, 1969. Nature 227:1037–1040

Ménard F, Dallot S, Thomas G, Braconnot J (1994) Temporal fluctuations of two Mediterranean salp populations from 1967 to 1990. Analysis of the influence of environmental variables using a Markov chain model. Mar Ecol Prog Ser 104:139–152

Ménard F, Fromentin JM, Goy J, Dallot S (1997) Temporal fluctuations of doliolid abundance in the bay of Villefranche-sur-Mer (northwestern Mediterranean Sea) from 1967 to 1990. Oceanol Acta 20:733–742

Millot C (1985) Some features of the Algerian current. J Geophys Res 90:7169–7176

Millot C (1987) The structure of mesoscale phenomena in the Ligurian Sea inferred from the DYOME experiment. Ann Geophys 5B:21–30

Millot C (1990) The Gulf of Lions' hydrodynamics. Cont Shelf Res 10:885–894

Millot C (1999) Circulation in the western Mediterranean Sea. J Mar Syst 20:423–442

Millot C (2005) Circulation in the Mediterranean Sea: evidences, debates and unanswered questions. Sci Mar 69(Suppl 1):5–21

Millot C, Taupier-Letage I (2005) Additional evidence of LIW entrainment across the Algerian subbasin by mesoscale eddies and not by a permanent westward flow. Prog Oceanogr 66:231–250

Mills CE (2001) Jellyfish blooms: are populations increasing globally in response to changing ocean conditions? Hydrobiologia 451:55–68

Mills C, Pugh P, Harbison G, Haddock S (1996) Medusae, siphonophores and ctenophores of the Alboron Sea, southwestern Mediterranean. Sci Mar 60:145–163

Minas HJ, Coste B, LeCorre P, Minas M, Raimbault P (1991) Biological and geochemical signatures associated with the water circulation through the Strait of Gibraltar and in western Alboran Sea. J Geophys Res 96:8755–8771

Minutoli R, Guglielmo L (2009) Zooplankton respiratory Electron Transport System (ETS) activity in the Mediterranean Sea: spatial and diel variability. Mar Ecol Prog Ser 381:199–211

Molinero JC, Ibanez F, Nival P, Buecher E, Souissi S (2005a) North Atlantic climate and northwestern Mediterranean plankton variability. Limnol Oceanogr 50:1213–1220

Molinero JC, Ibañez F, Souissi S, Chifflet M, Nival P (2005b) Phenological changes in the northwestern Mediterranean copepods *Centropages typicus* and *Temora stylifera* linked to climate forcing. Oecologia 145:640–649

Molinero JC, Casini M, Buecher E (2008a) The influence of the Atlantic and regional climate variability on the long-term changes in gelatinous carnivore populations in the northwestern Mediterranean. Limnol Oceanogr 53:1456–1467

Molinero JC, Ibanez F, Souissi S, Bosc E, Nival P (2008b) Surface patterns of zooplankton spatial variability detected by high frequency sampling in the NW Mediterranean. Role of density fronts. J Mar Syst 69:271–282

Molinero JC, Ibanez F, Souissi S, Buecher E, Dallot S, Nival P (2008c) Climate control on the long-term anomalous changes of zooplankton communities in the northwestern Mediterranean. Glob Change Biol 14:11–26

Morán XAG, Estrada M (2005) Winter pelagic photosynthesis in the NW Mediterranean. Deep-Sea Res Part I 52:1806–1822

Morote E, Olivar MP, Villate F, Uriarte I (2010) A comparison of anchovy (*Engraulis encrasicolus*) and sardine (*Sardina pilchardus*) larvae feeding in the northwest Mediterranean: influence of prey availability and ontogeny. ICES J Mar Sci 67:897–908

Nowaczyk A, Carlotti F, Thibault-Botha D, Pagano M (2011) Distribution of epipelagic metazooplankton across the Mediterranean Sea during the summer BOUM cruise. Biogeosciences 8: 2159–2177

Olivar MP, Salat J, Palomera I (2001) Comparative study of spatial distribution patterns of the early stages of anchovy and pilchard in the NW Mediterranean Sea. Mar Ecol Prog Ser 217:111–120

Olivar MP, Emelianov M, Villate F, Uriarte I, Maynou F, Álvarez I, Morote E (2010) The role of oceanographic conditions and plankton availability in larval fish assemblages off the Catalan coast (NW Mediterranean). Fish Oceanogr 19:209–229

Olivar M, Bernal A, Molí B, Peñaa M, Balbín R, Castellón A, Miquel J, Massutí E (2012) Vertical distribution, diversity and assemblages of mesopelagic fishes in the western Mediterranean. Deep-Sea Res Part I 62:53–69

Owen RW (1981) Patterning of flow and organisms in the larval anchovy environment. In: Sharp GD (ed) Workshop on the effects of environmental variation on the survival of larval pelagic fishes, Intergovernmental Oceanographic Commission, Workshop report 28. IOC/UNESCO, Paris

Pagano M, Gaudy R, Thibault D, Lochet F (1993) Vertical migrations and feeding rhythms of mesozooplanktonic organisms in the Rhone River plume area (north-west Mediterranean Sea). Estuar Coast Shelf Sci 37:251–269

Pagès F (2001) Past and present anthropogenic factors promoting the invasion, colonization and dominance by jellyfish of a Spanish coastal lagoon. In: Briand F (ed) Gelatinous zooplankton outbreaks: theory and practice, CIESM workshop series 14. CIESM, Monaco

Pagès F, Martín J, Palanques A, Puig P, Gili JM (2007) High occurrence of the elasipodid holothurian *Penilpidia ludwigi* (von Marenzeller, 1893) in bathyal sediment traps moored in a western Mediterranean submarine canyon. Deep-Sea Res Part I 54:2170–2180

Pakhomov E, Perissinotto R, McQuaid C (1996) Prey composition and daily rations of myctophid fishes in the Southern Ocean. Mar Ecol Prog Ser 134:1–14

Palanques A, García-Ladona E, Gomis D, Martín J, Marcos M, Pascual A, Puig P, Gili JM, Emelianov M, Montserrat S, Guillén J, Tintoré J, Segura M, Jordi A, Ruiz S, Basterretxea G, Font J, Blasco D, Pagès F (2005) General patterns of circulation, sediment fluxes and ecology of the Palamós (La Fonera) submarine canyon, northwestern Mediterranean. Prog Oceanogr 66:89–119

Palomera I, Olivar MP, Salat J, Sabatés A, Coll M, García A, Morales-Nin B (2007) Small pelagic fish in the NW Mediterranean Sea: an ecological review. Prog Oceanogr 74:377–396

Parrilla G (1984) Mar de Alboran. Situación del giro anticiclónico en Abril de 1980. Bol Inst Esp Oceanogr 1:106–113

Peralba Ä, Mazzocchi MG (2004) Vertical and seasonal distribution of eight *Clausocalanus* species (Copepoda: Calanoida) in oligotrophic waters. ICES J Mar Sci 61:645–653

Pérès J (1985) History of the Mediterranean biota and the colonization of the depths. In: Margalef R (ed) Key environments: Western Mediterranean. Pergamon Press, Oxford

Philippart C, Anadón R, Danovaro R, Dippner J, Drinkwater K, Hawkins S, Oguz T, O'Sullivan G, Reid P (2011) Impacts of climate change on European marine ecosystems: observations, expectations and indicators. J Exp Mar Biol Ecol 400:52–69

Pinca S, Dallot S (1995) Meso- and macrozooplankton composition patterns related to hydrodynamic structures in the Ligurian Sea (Trophos-2 experiment, April-June 1986). Mar Ecol Prog Ser 126:49–65

Pinot J-M, Tintoré J, Lopez-Jurado JL, Fernandez de Puelles ML, Jansa J (1995) Three-dimensional circulation of a mesoscale eddy/front system and its biological implications. Oceanol Acta 18:389–400

Pinot JM, López-Jurado J, Riera M (2002) The CANALES experiment (1996–1998). Interannual, seasonal, and mesoscale variability of the circulation in the Balearic Channels. Prog Oceanogr 55:335–370

Prieur L, Sournia A (1994) " Almofront-1"(April-May 1991): an interdisciplinary study of the Almeria-Oran geostrophic front, SW Mediterranean Sea. J Mar Syst 5:187–204

Prieur L, Bethoux J, Bong J, Tailliez D (1983) Particularités hydrologiques et formation d'eau profonde dans le bassin Liguro-Provencal en 1981–1982. Rapp Comm Int Mer Mediterr 28:51–53

Prieur L, Copin-Montegut C, Claustre H (1993) Biophysical aspects of 'Almofront-l', an intensive study of a geostrophic frontal jet. Ann Inst Oceanogr 69:71–86

Purcell JE (2005) Climate effects on formation of jellyfish and ctenophore blooms: a review. J Mar Biol Assoc U K 85:461–476

Purcell JE (2012) Jellyfish and ctenophore blooms coincide with human proliferations and environmental perturbations. Ann Rev Mar Sci 4:1.1–1.27

Purcell JE, Malej A, Benovic A (1999) Potential links of jellyfish to eutrophication and fisheries. In: Malone TC, Malej A, Harding LW, Smodlaka N, Turner RE (eds) Ecosystems at the land-sea margin: drainage basin to coastal sea, coastal and estuarine studies,55. American Geophysical Union, Washington, DC

Purcell JE, Uye SI, Lo WT (2007) Anthropogenic causes of jellyfish blooms and their direct consequences for humans: a review. Mar Ecol Prog Ser 350:153–174

Ramfos A, Isari S, Somarakis S, Georgopoulos D, Koutsikopoulos C, Fragopoulu N (2006) Mesozooplankton community structure in offshore and coastal waters of the Ionian Sea (eastern Mediterranean) during mixed and stratified conditions. Mar Biol 150:29–44

Ramšak A, Stopar K (2007) Dispersal ecology and phylogeography of Scyphomedusae in the Mediterranean Sea. MarBEF Newsl Autumn 2007:20–21

Razouls C (1974) Variations annuelles quantitatives de deux espèces de copépodes planctoniques *Centropages typicus* et *Temora stylifera* de la région de Banyuls: cycles biologiques et estimation de la production. III. Dynamique des populations et calcul de la production. Cah Biol Mar 15:51–88

Razouls C, Kouwenberg J (1993) Spatial distribution and seasonal variation of mesozooplankton biomass in the Gulf of Lions (northwestern Mediterranean). Oceanol Acta 16:393–401

Reiter ER (1975) Handbook for forecasters in the Mediterranean; weather phenomena of the Mediterranean basin; part 1. General description of the meteorological processes. DTIC document

Ribera d'Alcalà M, Conversano F, Corato F, Licandro P, Mangoni O, Marino D, Mazzocchi MG, Modigh M, Montresor M, Nardella M, Saggiomo V, Sarno D, Zingone A (2004) Seasonal patterns in plankton communities in a pluriannual time series at a coastal Mediterranean site (Gulf of Naples): an attempt to discern recurrences and trends. Sci Mar 68(suppl1):1–266

Riera T, Gili JM, Pagès F (1986) Estudio cuantitativo y estacional de dos poblaciones de cnidarios planctónicos frente a las costas de Barcelona (Mediterráneo occidental): Ciclos entre 1966–67 y 1982–83. Misc Zool 10:23–32

Rixen M, Beckers J, Levitus S, Antonov J, Boyer T, Maillard C, Fichaut M, Balopoulos E, Iona S, Dooley H (2005) The western Mediterranean deep water: a proxy for climate change-art. no. L12608. Geophys Res Lett 32:NIL_47–NIL_50

Robinson C, Steinberg DK, Anderson TR, Arístegui J, Carlson CA, Frost JR, Ghiglione JF, Hernández-León S, Jackson GA, Koppelmann R (2010) Mesopelagic zone ecology and biogeochemistry - a synthesis. Deep-Sea Res Part II 57:1504–1518

Rodríguez J, García A, Rodríguez V (1982) Zooplanktonic communities of the divergence zone in the northwestern Alboran Sea. Mar Ecol 3:133–142

Rodríguez J, Tintoré J, Allen JT, Blanco JM, Gomis D, Reul A, Ruiz J, Rodríguez V, Echevarria F, Jiménez-Gómez F (2001) Mesoscale vertical motion and the size structure of phytoplankton in the ocean. Nature 410:360–363

Roether W, Manca BB, Klein B, Bregant D, Georgopoulos D, Beitzel V, Kovaƒçevifá V, Luchetta A (1996) Recent changes in eastern Mediterranean deep waters. Science 271:333–335

Rossi S, Sabatés A, Latasa M, Reyes E (2006) Lipid biomarkers and trophic linkages between phytoplankton, zooplankton and anchovy (*Engraulis encrasicolus*) larvae in the NW Mediterranean. J Plankton Res 28:551–562

Rubín JP (1997) El ictioplancton estival en el mar de Alborán y estrecho de Gibraltar: composición cualitativa–cuantitativa y distribución espacial (julio 1993). Publ Espec Inst Esp Oceanogr 24:27–41

Rubín JP, Cano N, Prieto L, García CM, Ruíz J, Echevarría F, Corzo A, Gálvez JA, Lozano F, Alonso-Santos JC, Escánez J, Juárez A, Zabala L, Hernández F, García Lafuente J, Vargas M (1999) La estructura del ecosistema pelágico en relación con las condiciones oceanográficas y topográficas en el golfo de Cádiz, estrecho de Gibraltar y mar de Alborán (sector noroeste), en julio de 1995. Inf Téc Inst Esp Oceanogr 175:1–73

Rubio P, Muñoz JM (1997) Predicción estival del riesgo de blooms de *Pelagia noctiluca* (litoral central catalán). In: Creus Novau J (ed) Situaciones de riesgo climático en España. Instituto Pirenaico de Ecología, Jaca

Rubio A, Arnau PA, Espino M, Flexas MM, Jord*f*Å G, Salat J, Puigdef*f*Åbregas J, S-Arcilla A (2005) A field study of the behaviour of an anticyclonic eddy on the Catalan continental shelf (NW Mediterranean). Prog Oceanogr 66:142–156

Runge JA (1988) Should we expect a relationship between primary production and fisheries? The role of copepod dynamics as a filter of trophic variability. Hydrobiologia 167/168:61–71

Sabatés A (2004) Diel vertical distribution of fish larvae during the winter-mixing period in the northwestern Mediterranean. ICES J Mar Sci 61:1243–1252

Sabatés A, Masó M (1990) Effect of a shelf-slope front on the spatial distribution of mesopelagic fish larvae in the western Mediterranean. Deep-Sea Res Part A 37:1085–1098

Sabatés A, Olivar P (1996) Variation of larval fish distributions associated with variability in the location of a shelf-slope front. Mar Ecol Prog Ser 135:11–20

Sabatés A, Saiz E (2000) Intra-and interspecific variability in prey size and niche breadth of myctophiform fish larvae. Mar Ecol Prog Ser 201:261–271

Sabatés A, Gili J, Pagès F (1989) Relationship between zooplankton distribution, geographic characteristics and hydrographic patterns off the Catalan coast (western Mediterranean). Mar Biol 103:153–159

Sabatés A, Bozzano A, Vallvey I (2003) Feeding pattern and the visual light environment in myctophid fish larvae. J Fish Biol 63:1476–1490

Sabatés A, Salat J, Masó M (2004) Spatial heterogeneity of fish larvae across a meandering current in the northwestern Mediterranean. Deep-Sea Res Part I 51:545–557

Sabatés A, Martín P, Lloret J, Raya V (2006) Sea warming and fish distribution: the case of the small pelagic fish, *Sardinella aurita*, in the western Mediterranean. Glob Change Biol 12:2209–2219

Sabatés A, Olivar MP, Salat J, Palomera I, Alemany F (2007a) Physical and biological processes controlling the distribution of fish larvae in the NW Mediterranean. Prog Oceanogr 74:355–376

Sabatés A, Salat J, Palomera I, Emelianov M, Fernández de Puelles M, Olivar MP (2007b) Advection of anchovy (*Engraulis encrasicolus*) larvae along the Catalan continental slope (NW Mediterranean). Fish Oceanogr 16:130–141

Sabatés A, Zaragoza N, Grau C, Salat J (2008) Vertical distribution of early developmental stages in two coexisting clupeoid species, *Sardinella aurita* and *Engraulis encrasicolus*. Mar Ecol Prog Ser 364:169–180

Sabatés A, Salat J, Raya V, Emelianov M, Segura-Noguera M (2009) Spawning environmental conditions of *Sardinella aurita* at the northern limit of its distribution range, the western Mediterranean. Mar Ecol Prog Ser 385:227–236

Sabatés A, Pagès F, Atienza D, Fuentes V, Purcell JE, Gili JM (2010) Planktonic cnidarian distribution and feeding of *Pelagia noctiluca* in the NW Mediterranean Sea. Hydrobiologia 645:153–165

Sabatés A, Martín P, Raya V (2012) Changes in life history traits in relation to climate change: bluefish (*Pomatomus saltatrix*) in the northwestern Mediterranean. ICES J Mar Sci 69:1000–1009

Sabatés A, Salat J, Raya V, Emelianov M (2013) Role of mesoscale eddies in shaping the spatial distribution of the coexisting Engraulis encrasicolus and Sardinella aurita larvae in the northwestern Mediterranean. J Mar Syst 111–112:108–119

Saiz E, Alcaraz M (1990) Pigment gut contents of copepods and deep phytoplankton maximum in the western Mediterranean. J Plankton Res 12:665–672

Saiz E, Calbet A (2011) Copepod feeding in the ocean: scaling patterns, composition of their diet and the bias of estimates due to microzooplankton grazing during incubations. Hydrobiologia 666:181–196

Saiz E, Rodriguez V, Alcaraz M (1992) Spatial distribution and feeding rates of *Centropages typicus* in relation to frontal hydrographic structures in the Catalan Sea (western Mediterranean). Mar Biol 112:49–56

Saiz E, Tiselius P, Jonsson PR, Verity P, Paffenhöfer GA (1993) Experimental records of the effects of food patchiness and predation on egg production of *Acartia tonsa*. Limnol Oceanogr 38:280–289

Saiz E, Calbet A, Irigoien X, Alcaraz M (1999) Copepod egg production in the western Mediterranean: response to food availability in oligotrophic environments. Mar Ecol Prog Ser 187:179–189

Saiz E, Calbet A, Broglio E (2003) Effects of small-scale turbulence on copepods: the case of *Oithona davisae*. Limnol Oceanogr 48:1304–1311

Saiz E, Calbet A, Atienza D, Alcaraz M (2007) Feeding and production of zooplankton in the Catalan Sea (NW Mediterranean). Prog Oceanogr 74:313–328

Salat J (1995) The interaction between the Catalan and Balearic currents in the southern Catalan Sea. Oceanol Acta 18:227–234

Salat J (1996) Review of hydrographic environmental factors that may influence anchovy habitats in northwestern Mediterranean. Sci Mar 60(Suppl 2):21–32

Salat J, Font J (1987) Water mass structure near and offshore the Catalan coast during the winters of 1982 and 1983. Ann Geophys (B Terr Planet Phys) 5:49–54

Salat J, Garcia MA, Cruzado A, Palanques A, Arín L, Gomis D, Guillén J, de León A, Puigdefàbregas J, Sospedra J, Velásquez ZR (2002) Seasonal changes of water mass structure and shelf slope exchanges at the Ebro Shelf (NW Mediterranean). Cont Shelf Res 22:327–348

Sammari C, Millot C, Prieur L (1995) Aspects of the seasonal and mesoscale variabilities of the northern current in the western Mediterranean Sea inferred from the PROLIG-2 and PROS-6 experiments. Deep-Sea Res Part I 42:893–917

Sardá R, Gil J, Taboada S, Gili J (2009) Polychaete species captured in sediment traps moored in northwestern Mediterranean submarine canyons. Zool J Linn Soc 155:1–21

Sarhan T, García Lafuente J, Vargas M, Vargas JM, Plaza F (2000) Upwelling mechanisms in the northwestern Alboran Sea. J Mar Syst 23:317–331

Schott F, Visbeck M, Send U, Fischer J, Tramma L, Desaubies Y (1996) Observations of deep convection of the Gulf of Lions, northern Mediterranean, during the winter of 1991/92. J Phys Oceanogr 26:505–524

Scotto di Carlo B, Ianora A, Fresi E, Hure J (1984) Vertical zonation patterns for Mediterranean copepods from the surface to 3000 m at a fixed station in the Tyrrhenian Sea. J Plankton Res 6:1031–1056

Seguin G, Errhif A, Dallot S (1994) Diversity and structure of pelagic copepod populations in the frontal zone of the eastern Alboran sea. Hydrobiologia 292:369–377

Siapatis A, Giannoulaki M, Valavanis VD, Palialexis A, Schismenou E, Machias A, Somarakis S (2008) Modelling potential habitat of the invasive ctenophore *Mnemiopsis leidyi* in Aegean Sea. Hydrobiologia 612:281–295

Siokou-Frangou I (2004) Epipelagic mesozooplankton and copepod grazing along an east–west transect in the Mediterranean Sea. Rapp Comm Int Explor Sci Mer Mediterr 37:439

Siokou-Frangou I, Christaki U, Mazzocchi M, Montresor M, Ribera DA (2010) Plankton in the open Mediterranean Sea: a review. Biogeosciences 7:1543–1586

Skovgaard A, Saiz E (2006) Seasonal occurrence and role of protistan parasites in coastal marine zooplankton. Mar Ecol Prog Ser 327:37–49

Sournia A, Brylinski J-M, Dallot S, Le Corre P, Leveau M, Prieur L, Froget C (1990) Fronts hydrologiques au large des côtes françaises: Les sites-ateliers du programme Frontal. Oceanol Acta 13:413–438

Spanier E, Galil BS (1991) Lessepsian migration: a continuous biogeographical process. Endeavour 15:102–106

Steinberg DK, Carlson CA, Bates NR, Goldthwait SA, Madin LP, Michaels AF (2000) Zooplankton vertical migration and the active transport of dissolved organic and inorganic carbon in the Sargasso Sea. Deep-Sea Res Part I 47:137–158

Stemmann L, Prieur L, Legendre L, Taupier-Letage I, Picheral M, Guidi L, Gorsky G (2008) Effects of frontal processes on marine aggregate dynamics and fluxes: an interannual study in a permanent geostrophic front (NW Mediterranean). J Mar Syst 70:1–20

Taupier-Letage I, Millot C (1986) General hydrodynamical features in the Ligurian Sea inferred from the DYOME experiment. Oceanol Acta 9:119–131

Taupier-Letage I, Puillat I, Millot C, Raimbault P (2003) Biological response to mesoscale eddies in the Algerian basin. J Geophys Res 108:10.1029

Thiriot A (1972–1973) Les cladocères de Méditerranée occidentale III.- Cycle et répartition a Banyuls-sur-Mer (Golfe du Lion). synthèse des années 1965–1969. Vie Milieu 23:243–295

Tintoré J, Violette PEL, Blade I, Cruzado A (1988) A study of an intense density front in the eastern Alboran Sea: the Almeria Oran front. J Phys Oceanogr 18:1384–1397

Tiselius P (1992) Behavior of *Acartia tonsa* in patchy food environments. Limnol Oceanogr 37:1640–1651

Trégouboff G, Rose M (1957) Manuel de Planctologie Méditerranéene. CNRS, Paris

Tsimplis MN, Josey SA (2001) Forcing of the Mediterranean Sea by atmospheric oscillations over the north Atlantic. Geophys Res Lett 28:803–806

Turley CM (1999) The changing Mediterranean Sea - a sensitive ecosystem? Prog Oceanogr 44:387–400

Uye S, Huang C, Onbe T (1990) Ontogenetic diel vertical migration of the planktonic copepod *Calanus sinicus* in the Inland Sea of Japan. Mar Biol 104:389–396

Vandromme P, Stemmann L, Berline L, Gasparini S, Mousseau L, Prejger F, Passafiume O, Guarini J, Gorsky G (2011) Inter-annual fluctuations of zooplankton communities in the Bay of Villefranche-sur-Mer from 1995 to 2005 (northern Ligurian Sea, France). Biogeosciences 8:3143–3158

Vaqué D, Blough HA, Duarte CM (1997) Dynamics of ciliate abundance, biomass and community composition in an oligotrophic coastal environment (NW Mediterranean). Aquat Microb Ecol 12:71–83

Vargas-Yáñez M, Sabatés A (2007) Mesoscale high-frequency variability in the Alboran Sea and its influence on fish larvae distributions. J Mar Syst 68:421–438

Vargas-Yáñez M, Moya F, García-Martínez MC, Tel E, Zunino P, Plaza F, Salat J, Pascual J, López-Jurado JL, Serra M (2010) Climate change in the western Mediterranean Sea 1900–2008. J Mar Syst 82:171–176

Videau C, Sournia A, Prieur L, Fiala M (1994) Phytoplankton and primary production characteristics at selected sites in the geostrophic Almeria–Oran front system SW Mediterranean Sea. J Mar Syst 5:235–250

Vignudelli S, Gasparini GP, Astraldi M, Schiano ME (1999) A possible influence of the north Atlantic oscillation on the circulation of the western Mediterranean Sea. Geophys Res Lett 26:623–626

Vives F (1966) Zooplancton nerítico de las aguas de Castellón (Mediterráneo occidental). Invest Pesq 30:49–166

Vives F (1978) Sobre la distribución de *Calanus helgolandicus*, Claus (Copepoda, Calanoida) en el Mediterráneo occidental. Res Exp Cient B/O Cornide 7:83–92

Vives F, Shmeleva A (2006) Crustacea, Copépodos Marinos I. Calanoida. Fauna Ibérica, vol 29. Museo Nacional de Ciencias Naturales, CSIC, Madrid

Vives F, Shmeleva A (2010) Crustacea, Copépodos marinos II. Non Calanoida. Fauna Ibérica, vol 33. Museo Nacional de Ciencias Naturales, CSIC, Madrid

Vives F, Santamaría G, Trepat I (1975) El zooplancton de los alrededores del estrecho de Gibraltar en Junio-Julio de 1972. Res Exp Cient B/O Cornide 4:7–100

Wang DP, Vieira MEC, Salat J, Tintore J, La Violette PE (1988) A shelf/slope frontal filament off the northeast Spanish coast. J Mar Res 46:321–332

Wang R, Zuo T, Wang K (2003) The Yellow Sea Cold Bottom Water - an oversummering site for *Calanus sinicus* (Copepoda, Crustacea). J Plankton Res 25:169–183

Williams R, Conway D (1984) Vertical distribution, and seasonal and diurnal migration of *Calanus helgolandicus* in the Celtic Sea. Mar Biol 79:63–73

Yoon W, Nival P, Choe S, Picheral M, Gorsky G (2007) Vertical distribution and nutritional behaviour of *Cyclothone braueri*, *Nematoscelis megalops*, *Meganyctiphanes norvegica* and *Salpa fusiformis* in the NW Mediterranean mesopelagic zone ICES CM 2007/F:03. Paper presented at the ICES Annual Science Conference 2007, 17–21 Sept 2007, Helsinki

Youssara F, Gaudy R (2001) Variations of zooplankton in the frontal area of the Alboran Sea (Mediterranean Sea) in winter 1997. Oceanol Acta 24:361–376

Zakardjian B, Prieur L (1998) Biological and chemical signs of upward motions in permanent geostrophic fronts of the western Mediterranean. J Geophys Res 103:27849–27866

Zenetos A, Gofas S, Velarque M, Çinar ME, García Raso JE, Bianchi CN, Morri C, Azzurro E, Bilecenoglu M, Froglia C, Siokou I, Violanti D, Sfriso A, San Martín G, Giangrande A, Katagan T, Ballesteros E, Ramos-Esplá A, Mastrototaro F, Ocaña O, Zingone A, Gambi MC, Streftarus N (2010) Alien species in the Mediterranean Sea by 2010. A contribution to the application of European Union's Marine Strategy Framework Directive (MSFD). Part I. Spatial distribution. Mediterr Mar Sci 11:381–493

Zervoudaki S, Nielsen T, Christou E, Siokou-Frangou I (2006) Zooplankton distribution and diversity in a frontal area of the Aegean Sea. Mar Biol Res 2:149–168

Zoobenthos

12

Josep-Maria Gili, Rafael Sardá, Teresa Madurell, and Sergi Rossi

Abstract

Among Mediterranean marine life, benthic communities are possibly the most peculiar in terms of richness and endemic species. The distribution and structure of benthic fauna (sessile and vagile) is driven by environmental gradients that change with season, depth, the type of substrate and the interaction between organisms. The combined action of these multiple factors results in a high variety of assemblages and communities. This chapter focuses on the benthic fauna, with the aim to provide a broad description of the hard and soft bottom communities and the general trends of their characteristics and variability. Special emphasis is given to the ecological strategies of the fauna that inhabit in complex benthic ecosystems. The pressures and impacts on these benthic fauna and ecosystems, from alien species invasions, warming effects, ocean acidification and other direct and indirect human perturbations are also illustrated.

Keywords

Alien species invasions • Benthic communities • Benthic fauna • Complex benthic ecosystems • Ecological strategies • Endemic species • Environmental gradients • Hard and soft bottom communities • Human perturbations • Mediterranean marine life • Ocean warming and acidification • Sessile fauna • Vagile fauna

Introduction

The Mediterranean marine environment is characterized by a set of environmental factors that define its biota, especially in shallow communities. Extreme reduction of tides, transparent and blue waters, relatively high salinity and mean water temperature together with an evident seasonality with four recognizable seasons are the identity signals of this temperate sea. These particular features renders the Mediterranean a very singular environment, and given the small extension of the Mediterranean (0.82 % of world oceans), with a very high marine biodiversity (between the 4 and 18 % of all marine species known) (Bianchi and Morri 2000). Among Mediterranean marine life, benthic communities are possibly the most peculiar in terms of richness and originality.

Environmental Gradients

In marine Mediterranean benthic environments, the distribution of fauna is ruled by environmental gradients defined as: humidity (increasing from the atmosphere to the water); action-wave related factors (maximal at level zero and decreasing in both shoreward and seawater direction); water temperature and light intensity (decreasing with depth); etc.

J.-M. Gili (✉) • T. Madurell
Institut de Ciències del Mar, Consejo Superior de Investigaciones Científicas, Barcelona, Spain
e-mail: gili@icm.csic.es

R. Sardá
Institut d'Estudis Avançats de Blanes, Consejo Superior de Investigaciones Científicas, Blanes, Spain

S. Rossi
Institut de Ciència i Tecnologia Ambientals, Barcelona, Spain

S. Goffredo and Z. Dubinsky (eds.), *The Mediterranean Sea: Its history and present challenges*,
DOI 10.1007/978-94-007-6704-1_12, © Springer Science+Business Media Dordrecht 2014

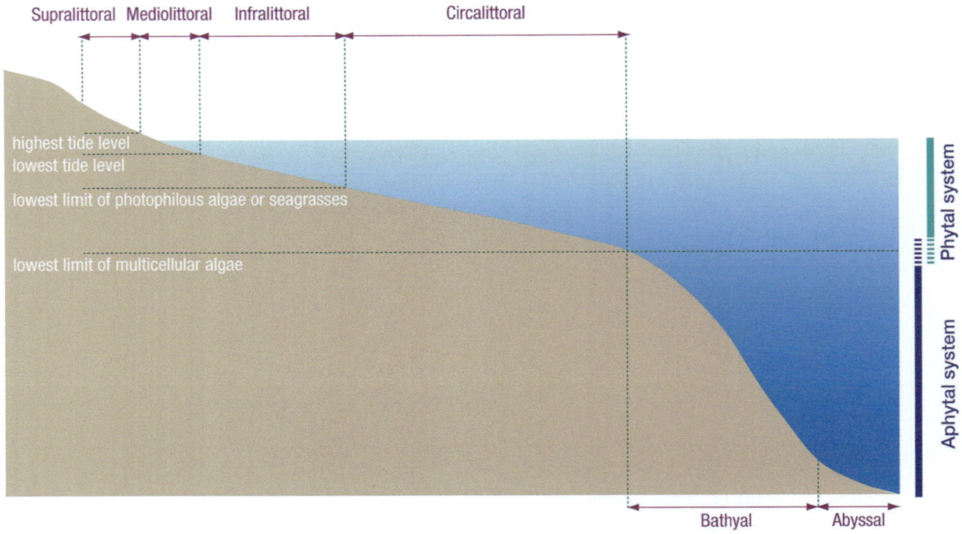

Fig. 12.1 Main benthic zonation of the Mediterranean Sea (Modified from Pérès 1982)

(Ros et al. 1985) (Fig. 12.1). Environmental changes are generally sharper along the vertical axis than along the horizontal plane especially in shallow areas. Other factors that change with depth such as the light spectrum, the water movement, the grain size of soft sediment (decreasing with depth), and the amount of hard substrates that are higher in shallow and shelf zones (Riedl 1971), play also an important role influencing the biota distribution.

The gradient on the vertical axis is often combined with a patchy heterogeneity in the horizontal plane caused mainly by the inclination, exposure to light, texture, type of substrate, orientation to predominant currents, etc. Concurrently, along these vertical gradients in shallow environments the plant-animal relationship changes with decreasing light. The macro-algae negatively impact on most of the sessile invertebrates by overgrowing in the shallower areas, or generating a canopy effect. These algae create a habitat for small and mobile fauna that finds shelter and food. All these factors generate diversity, which is enhanced by the increasing structural role of the organisms themselves, as plants and algae in the photic zone, the sessile animals in the aphotic zone. The variety of assemblages and communities results from the combined action of these multiple factors.

Mediterranean hard and soft-bottom communities were described from a qualitative and quantitative point of view on the studies of French pioneers in this field (Laborel 1961; Pérès and Picard 1964; Picard 1965; Laubier 1966; Bellan-Santini 1969; Guille 1971; Pérès 1982) and more recent reviewers (Ros et al. 1984, 1985; Fredj and Laubier 1985; Pérès 1985; Bellan-Santini et al. 1994; Bianchi et al. 2004; Ballesteros 2006). These classical papers identify the main coastal and shelf communities found in the western Mediterranean. Different assemblages are described by the presence of key species following two main environmental parameters, bathymetry and substrate or sediment granulometry. In shallower areas the overlap of both parameters often occurs due to the presence of mixed sediments. In deeper habitats, bathymetry and granulometry are correlated due to a general decrease of hydrodynamism with depth, and communities generally extend over wider areas. No similar detailed community descriptions exist for the eastern Mediterranean and most papers (Karakassis and Eleftheriou 1997; Simboura et al. 2000, 2005; Simboura and Zenetos 2002; Chintiroglou et al. 2005) follow the previous mentioned classical works. A detailed and actual lists of the Mediterranean benthic communities for the establishment of national inventories of natural sites of conservation interest can be found in the synthesis of RAC/SPA (2006), a revised list is given by Fraschetti et al. (2008).

Spatial Trends and Variability

Zoobenthic Assemblages and Communities on Hard Substrata

Supralittoral

The supralittoral zone is never, or only very rarely, immersed. This habitat requires a relatively high degree of humidity that is supplied by the wave spray. Immersions occur mainly during heavy storms, though under certain conditions may be irregular because of the small tidal range. The upper part of this zone is only wetted very few times a year, whereas the lower part is almost continually splashed. The living community of this zone is homogeneous on a wide scale and has a very simple composition. The severe environmental conditions have selected a few successful animal strategies. In the upper part of vertical rocky zones are common the herbivore littorinid gastropods (*Melaraphe neritoides*), cirripeds (*Chtamalus depressus*) and some detritivore mobile isopods (*Ligia italica*) (Pérès and Picard 1964; Ros et al. 1984).

Mediolittoral

The mediolittoral assemblages require or tolerate immersion and are adapted to the periodic ebb and flow of the intertidal zone. In the upper part of this zone humectation is due to the wave action and immersion is rare. In the lower part submersions are more frequent. The upper part is colonized by species of the supralittoral zone such as the very common molluscs strongly adhered to the rocks *Patella rustica* and decapoda (*Pachygrapsus marmoratus*) in crevices and sheltered areas. In the lower part animals can dominate the space (e.g. mussel belts, *Mytillus edulis*). On the rocky coast, cirripeds occupy the upper zone, often closely packed. On the lower region of rocky coasts live less tolerant animals, mussels and gastropods that cling to the wave-battered algal belts or hide into the crevices of encrusting algae belts. In areas with rich and dense communities of encrusting seaweeds many sessile species resist the wave action such as foraminifera (*Miniacina miniacea*), cnidarians (*Actinia equina*), sponges (*Hallichondria* spp.), hydrozoans (*Paracoryne huvei*) and many briozoans, polychaetes, crustaceans (amphipods, isopods and some decapoda such as *Eriphia spinifrons*) and molluscs (Pérès 1967; Bouillon 1975; Ros et al. 1985; Morri et al. 1990).

Infralittoral

The infralittoral zone is characterized by an upper limit where species cannot endure emergence and by a lower limit marked by the disappearance of photophilic algae and marine phanerogams. The extension of this zone is highly variable, depending on substrate exposure, light penetration and turbidity. On rocky substrata the dominant species of fauna are associated to photophilic algal talus and phanerogams plants. Between the bases of the algal talus there is an encrusting animal community composed of skeletal material of polychaete worms, bryozoans and gastropods. Other encrusting species, sponges (*Myxilla rosacea*, *Hymedesmia* spp, *Phorbas topsenti*) and colonial ascidians, compete with the algae for the substrata. Sedentary species strongly attached to the rocky substrate such as molluscs (*Acanthochiton fascicularis*) are common, although crustaceans and members of others groups also occur (Bellan-Santini 1969; Ros et al. 1984).

A wide variety of species spend part of their life inside the rock crevices or in the interstices left or formed by others species. Examples of these include many polychaetes (*Nereis* spp, *Lepidonotus* spp, *Spirobranchus* sp, etc.), crustaceans (amphipods and isopods), molluscs (*Jujubinus gravinae*, etc.), ophiurids (*Ophiotrix fragilis* and *Amphipholis squamata*) and also species that cling to the under face of boulders (*Coscinasterias tenuispina* or *Asterina gibbosa*). The complexity of this animal community increases with the endobiont species on sponges and ascidians.

There are other groups of species that move from one algal turf to another or adhering to the fronds or to sessile animals. The most common are amphipods (*Dexamine* spp, *Hyale* spp, etc.), but also isopods (*Cymodoce truncta*) together with picnogonids, molluscs, prosobranchs and opistobranchs gastropods and free-living polychaetes (e.g. *Platinereis dumerilii).* The algal fronds have also an important epiphytic community of sessile organisms such as hydrozoans with caprellids amphipods associated, bryozoans and encrusting tunicates.

On the algal talus there are other animals that compete for the space and develop large sizes such as cnidarians (*Aiptasia diaphana*, *Anemonia sulcata*, *Balanophyllia regia*) many hydrozoans and bryozoans (*Pentapora ottomulleriana*, *Turbicellepora magnicostata*). These species are chiefly substrate-occupiers, which compete strongly for the space. A second group of species, sessile molluscs (*Spondylus gaederopus*, *Ostrea edulis*) and tunicates (*Microcosmus sabatieri*, *Pyura dura*, etc.), are mainly substrate-suppliers that contribute to the spatial structuring of infralittoral communities. A final important group is the vagile fauna with species of different sizes that are highly mobile and linked in varying ways to the photophilic algae talus. Most of them are predators and include decapods (e.g. *Pilummus hirtellus*, *Alpheus dentipes*), molluscs (*Octopus vulgaris* and many gastropods), echinoderms (e.g. *Paracentrotus lividus*, *Echinaster sepositus*) and many fishes of the families Gobiidae and Bleniidae (Ros et al. 1985).

Circalittoral

The circalittoral zone extends from the lower level of photophilic algae and sea-grasses to the end of the continental shelf. Fixed and colonial animals (sponges, cnidarians, bryozoans, tunicates) and red calcareous algae predominate on hard bottoms. The algae growing on coarse gravelly substrates coalesce the calcareous sediments to form a continuous, organogenous substrate on which the coralligenous community develops. This assemblage is the most structured and species-rich in the Mediterranean benthos. In general, the currents in the circalittoral zone are steady, though occasionally strong, and water is constantly cool. The plants are no longer the main contributors to the biomass, although its role is important as builders of the coralligenous (Ros et al. 1984; Uriz et al. 1993; Ballesteros 2006). The main variables delimitating these communities are changes in the relief, slope and substrata features in rocky habitats, grain size of the sediments, inter-specific competition and the alternation between hard and soft bottoms.

Three structural levels can be found in the circalittoral communities. The basal habitat with many species intermixes with algae in a reciprocal encrusting process. These species contribute to the community in two ways, as builders and bioeroders. One example (as a bioeroder) are the sponge species of the genus *Cliona*, which penetrate carbonate based organisms (Rosell et al. 1999; Cerrano et al. 2001; Rosell and Uriz 2002). The lower level is dominated by a high diversity of massive, erect and reptant species, such as sponges, cnidarians, bryozoans and tunicates, while the higher level contain the most visible and representative species. Their abundance in each habitat characterizes the so-called *facies*, being the contributors to the three-dimensional structure of these bottoms. Gorgonians (e.g. *Paramuricea clavata, Corallium rubrum* or *Eunicella singularis*), antipatharians (*Antipathella subpinnata*) (Bo et al. 2008, 2009), stylasterids (*Errina aspera*) (Salvati et al. 2010) and sponges (e.g. *Axinella polypoides*) are examples of this higher structural level (Fig. 12.2a, b, c). Other important components of the circalittoral communities are those species that move around the lower and higher levels (crustaceans, echinoderms or polychaetes). There is also a fauna specialized in occupying the higher levels as sessile epibionts or temporary residents. The coralligenous community is the more emblematic and developed of the circalittoral habitats, being a hot spot diversity habitat in the Mediterranean Sea (Laubier 1966; Fredj and Laubier 1985; Gili and Ros 1985a; Ballesteros et al. 1993; Ballesteros 2006).

Another very peculiar community in the Mediterranean circalittoral bottoms are the submarines caves. The fauna inhabiting in caves have a clear connection with the coralligenous but also with communities from deep waters (Riedl 1966; Uriz et al. 1992, 1993; Bianchi and Morri 1994; Bianchi et al. 1996). In this dark and cool habitats there is a continuously impoverishment of the particle concentration and the benthic cover from the open section to the darker parts of the caves (Gili et al. 1986; Martí et al. 2004) (Fig. 12.3). The walls are dominated by suspension feeders (sponges, cnidarians and bryozoans) and vagile fauna characterized by very specialized species of crustaceans and fishes (Bibiloni et al. 1989; Gili and Ballesteros 1991) (Fig. 12.2d).

Coralligenous assemblages are also found on soft isolated rocks or other rocky substrata (Fig. 12.2e). These latter assemblages can reach deeper areas, predominantly between 100 and 200 m depth. This particular community represents one of the most diverse assemblages in the Mediterranean shelf and slope even though these habitats have been destroyed by bottom trawling during decades. These rocky substrates are often covered with a silty sediment layer, and the sessile species have to penetrate through this sediment layer to attach to the hard substrate. The predominant species are suspension feeders such as gorgonians, alcyonarians, black corals, sponges and molluscs (True 1970; Gili and Ros 1985b; Uriz et al. 1992, 1993; Ballesteros 2006; Bo et al. 2011a, b).

Bathyal

Hard bottom communities in the bathyal domain are represented mainly by the cold-water coral community (Pérès 1985; Zibrowius and Taviani 2005; Bo et al. 2012). The deep-sea or cold-water coral community is found on hard substrates of the bathyal zone, between 200 and 1,000 m depth (Fig. 12.2f). These cold-water corals are only found when the slope is steep enough to expose hard substrates as vertical cliffs. The dominant species are the reef forming corals *Lophelia pertusa* and *Madrepora oculata*, which appear as scattered clumps along the slope. There is a rich associated fauna, the permanent species that live around or above the big boulders and rocks, and the temporal inhabitants such fishes and crustaceans that use this habitat as food source and nursery (Freiwald and Roberts 2005; Costello et al. 2005; Orejas et al. 2009; D'Onghia et al. 2010).

Zoobenthic Assemblages and Communities on Soft-Bottoms

Infralittoral

Infralittoral exposed soft-bottoms habitats are mostly sandy, with average grain sands depending on whether they are located in moderate or highly exposed areas. The predominant community, *the Beach sandy community*, usually covers bottoms from 5 to 7 m depth being colonized by several species of bivalves of the genus *Donax*, *Tellina* and *Lentidium.* Following the *Beach sandy community*, the so-called *Fine well-sorted sand community* (Pérès and Picard 1964) is the predominant one. This community has been recently named *Littoral Fine sands community* (Colombini et al. 2003; Labrune et al. 2007). Where high currents occur, fine sands are replaced by coarse sands and fine gravels, and the community changes to the *Littoral Coarse sands community.* This community contains a large diversity of bivalve species being *Spisula subtruncata* the most abundant one followed by polychaetes (*Ditrupa arietina*, *Owenia fusiformis*, and small spionids and paraonids). Under certain circumstances, these communities are densely colonized by seagrasses (mainly *Cymodocea nodosa* and *Posidonia oceanica*) increasing spatial heterogeneity and therefore microhabitat availability. This process leads to an increase in diversity and abundance that causes a clear differentiation in the community structure. Finally, the *Littoral sandy mud community* is found in the deeper part of the infralittoral. In this zone a change in species composition is observed due to low current waters, which relates with more fine granulometry in the composition of sediment. Bivalves such as *Loripes lacteus* and *Tapes* spp. and crustaceans like *Pestarella tyrrhena* are among the characteristic species.

Infralittoral non-exposed soft-bottom habitats, on the other hand, are mostly dominated by silty sands and clays. This type of environments are generally found near river

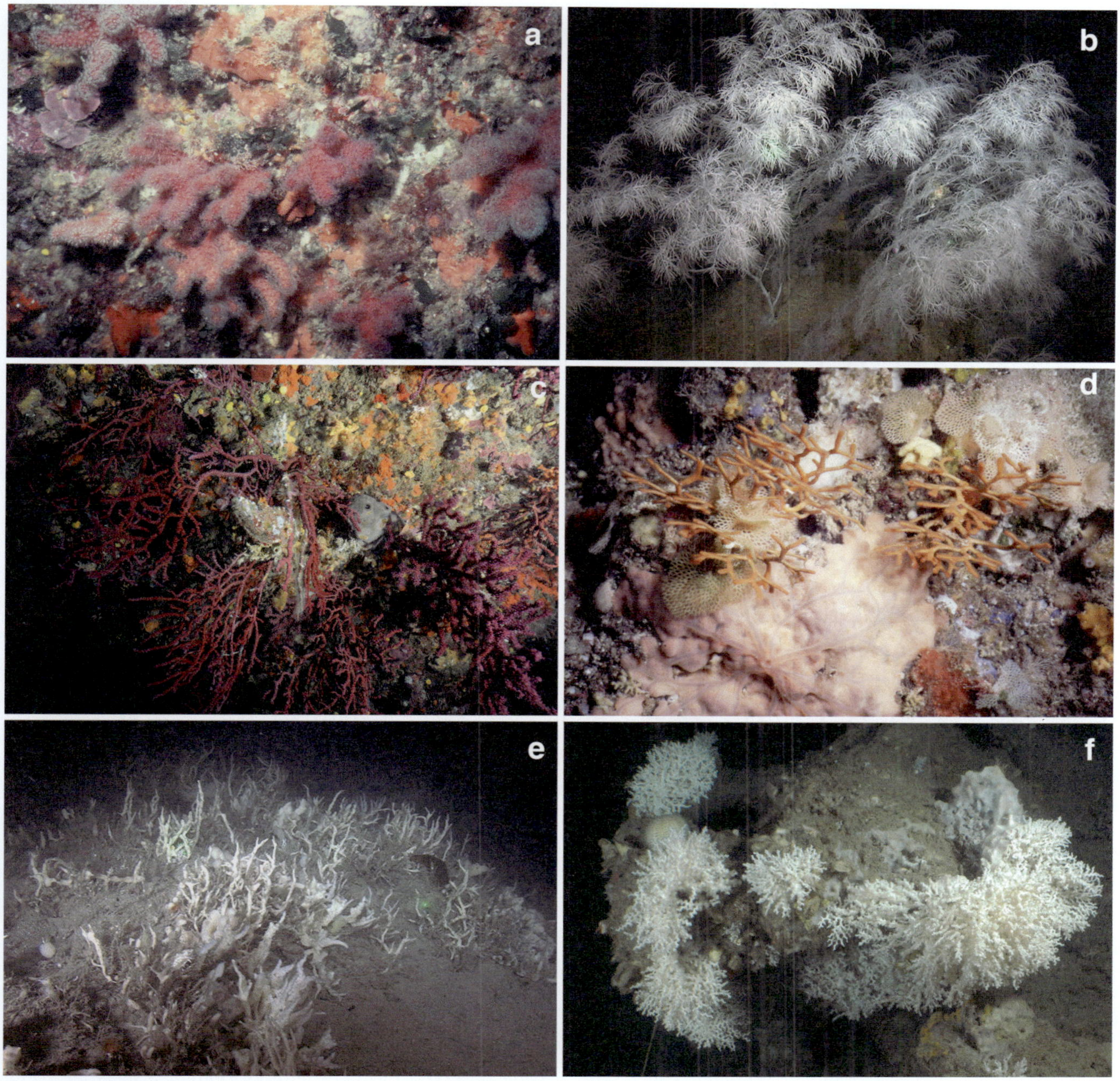

Fig. 12.2 Several examples of rocky bottom communities dominated by sessile organisms in the western Mediterranean: (**a**) circalittoral community dominated by the alcyonarian *Alcyonium acaule* at 20 m depth in the Medes Islands (NW Mediterranean) (foto by J.M. Gili); (**b**) circalittoral community dominated by the black coral *Antipathella subpinnata* at 120 m depth in Son Bou Canyon in Menorca Channel (NW Mediterranean); (**c**) coralligenous community dominated by the gorgonian *Paramuricea clavata* at 45 m depth in the Cap de Creus (NW Mediterranean) (Photo by Aldo Ferrucci); (**d**) circalittoral community located in the entrance of submarine caves in the NE Mallorca Island (NW Mediterranean) at 15 m depth dominated by the bryozoans *Smittina cervicorn*is and *Retepora* sp. And the sponge *Spirastrella cuntatrix* (Photo by J.M. Gili); (**e**) community of deep rocky habitats at the end of the continental shelf of the Cap de Creus (NW Mediterranean) at 110 m depth, dominated by the sponge *Desmacidon fruticosum*; (**f**) community of depth cold-water corals dominated by *Madrepora oculata* at 200 m depth located at the Cap de Creus submarine canyon (NW Mediterranean) (Photographs **b**, **e** and **f** by JAGO-ICM-GEOMAR)

mouths and/or in deltas and coastal lagoon environments. The distribution of the species, assemblages and communities reflects the environmental gradients (i.e. granulometry, salinity and anoxic conditions) that can be found inside the lagoon, and are mostly related with their communication with the open sea. Bivalve species such as *Cerastoderma edule* and *Scrobiculari plana*, polychates such as *Streblospio shrubsolii* and *Hediste diversicolor* and crustaceans such as

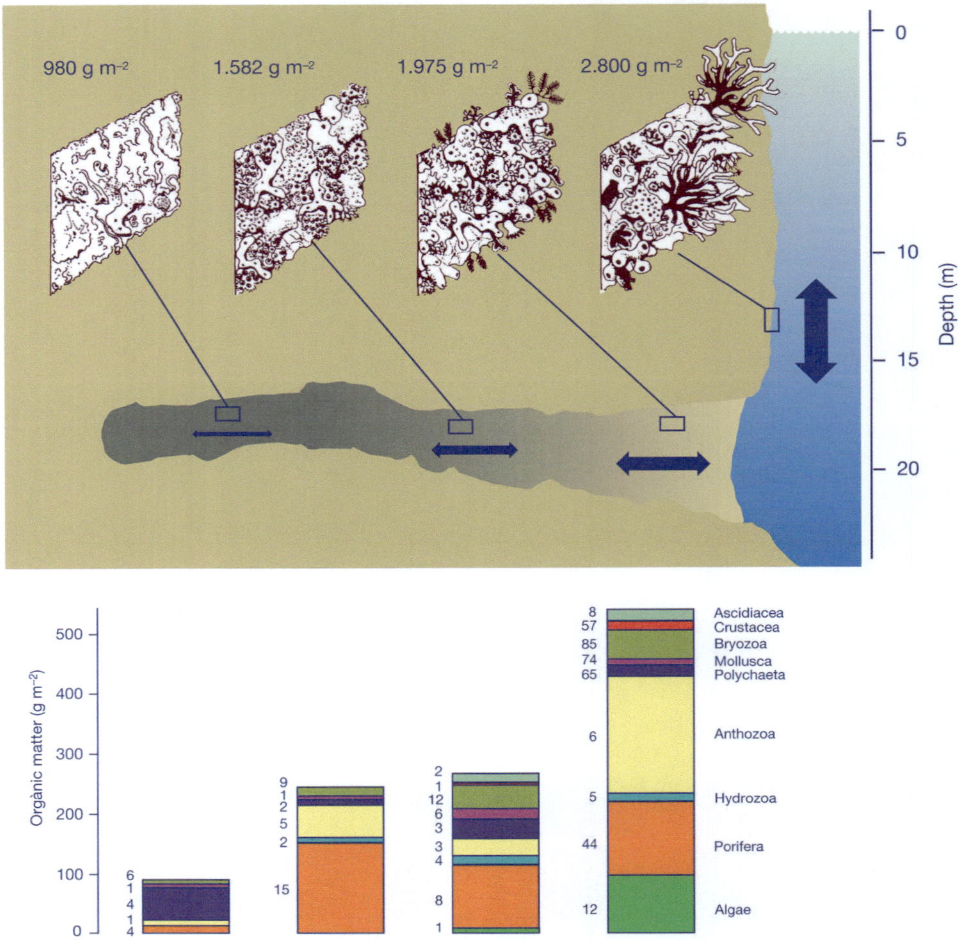

Fig. 12.3 Schematic representation of four Mediterranean circalittoral communities distributed from a vertical wall outside a cave to a horizontal one inside the cave. The number at the *top* of each representation is the community biomass in dry weight. The width and direction of *arrows* roughly indicate the direction and intensity of water flow. The numbers on the side of each bar indicate the number of species within each group (Figure modified from Gili and Coma 1998)

Cyathura carinata and *Microdeutopus gryllotalpa* are some of the most important species.

An extreme environment in shallow waters is the hydrothermal vents. In the Mediterranean they occur in shallow depths of up to 115 m. The dominant sediment, called ooze, is mud with a high percentage of organic remains. These areas rely on primary production and benthic fauna is not distinct from the surrounding areas (De Biasi et al. 2004) except for the abundant gastropod *Cyclope neritea* in the bacterial mat areas (Dando et al. 1995).

Circalittoral

Circalittoral soft-bottom environments present a more complex situation (Somaschini et al. 1998). Pérès and Picard (1964) and Picard (1965) reported the presence of different circalittoral assemblages. On the other hand, Pérès (1967) and Salen Picard (1981, 1982) described many degradation detritic facies. And finally Guille (1971) observed many transition zones among these assemblages, identifying only one circalittoral soft-bottom community with many sub-communities or facies. Other transitional zones, mostly between sandy and muddy bottoms, have been observed by other authors (Fèbvre-Chevalier 1969; Gambi and Fresi 1981; Falciai et al. 1983; Fresi et al. 1983; Zavodnik et al. 1985; Gravina 1986). All these aforementioned classifications were obtained basically with the help of dredges and box corers devices. Recent observations made with remote operated vehicles (ROV's) or small submarines have allowed to make direct observations on these soft-bottoms environments and discern also the fragile epifaunal species, which are usually observed in patches. Thus, by visual inspection of these habitats we can now classify all the communities found in circalittoral shelf environments, into four main large communities with many different assemblages and/or facies inhabiting them (Sardá et al. 2012).

Fig. 12.4 Characteristic assemblages and zoobenthic communities on soft-bottoms: continental shelf detritic communities in the Cap the Creus (NW Mediterranean) at 120 m depth (**a**) facies with *Leptometra phalangium* at; (**b**) facies dominated by gorgonians *Eunicella* sp. and sponges (photographs a and b by NEMO-ICM-Gavin Newman); (**c**) Maërl bed community at 52 m depth from Cabrera Island (NW Mediterranean) (photo by David Diaz); (**d**) communities of bathyal muds, facies with *Funiculina quadrangularis* in the Menorca channel (NW Mediterranean) at 140 m depth (Photo by JAGO-ICM-GEOMAR)

In the upper part of the circalittoral domain, the *Littoral sandy mud community* can also be found, although with a higher number of deeper species in the faunistic composition. *Continental shelf detritic communities* are the second large group of assemblages that can be found. The relative fraction of muds in their sediment composition determine different assemblages, from sandy detritic environments to more muddy. The presence of epifaunal species such as gorgonians, pennatulaceans (*Pennatula rubra*, *Pteroides spinosum),* alcionaria (*Alcyonium palmatum),* ceriantharia *(Cerianthus membranaceus),* crinoidea *(Leptometra phalangium*), ofiuroidea (*Ophiothrix quinquemaculata),* and even terebelid polychaetes, determines the different facies of the same community (Fig. 12.4a, b). On continental shelf bottoms that meet the continental slope or submarine canyons, the proportion of mud decreases, dominating the gravel and rock substrata. In this zone, the present epifauna is more diverse with the presence of sponges and echinoderms. Finally, the *Platform mud community* is the largest community that can be found occupying most part of the circalittoral environment. The composition is mainly silty sand sediments and clay, which vary between the terrigenous coastal mud in the upper part to the deep muds in the deeper part. The fauna is characterized by the endofaunal components with a large fraction being small polychaete species of genus such as *Tharyx*, *Monticellina*, *Prionospio* and *Lumbrineris* (Guille 1965; Desbruyères et al. 1972; Gili et al. 1987).

A very specific community that normally colonizes detritic environments in the upper part of the circalittoral is the well-known *Maërl bed community* (Fig. 12.4c). Maërl beds are mainly composed of free-living non-geniculate coralline algae (Corallinacea: Rhodophyta) that have an algal core and sandy granules or shells with a more or less regular and concentric growth around them. These structures develop and accumulate on soft-bottoms, influenced by laminar currents and gives a rich and diverse benthic community (Ballesteros 1994; Peña and Bárbara 2008).

Deep-Sea

Bathyal and Abyssal

Our knowledge on bathyal and abyssal soft-bottom communities is still scarce and few papers dare to work at a community level. The prevalent division presented by Pérès (1985) divides the slope communities into three horizons, the upper slope horizon extending to 400–500 m, the middle slope horizon that can extend until 1,200–1,400 m, characterized by compact muds and a high diversity, and the lower slope horizon that goes into the abyssal plains. Despite the reviews by Pérès 1985; Fredj and Laubier (1985) and Laubier and Emig (1993) quantitative data on faunal composition and ecological knowledge are generally scarce, although new studies on deep-sea habitats have lead to valuable new knowledge on deep Mediterranean benthos (Boury-Esnault et al. 1994; Uriz and Rosell 1990; Tselepides et al. 2000; Kröncke et al. 2003; Cartes et al. 2004; Ramirez-Llodra et al. 2008). Although it is accepted that Mediterranean deep-sea biodiversity should be high as shown by comprehensive biodiversity studies (Bianchi and Morri 2000; Coll et al. 2010; Danovaro et al. 2010), there is still few data to describe the different deep-sea communities. To present, bathyal facies are defined according to the compaction of the sediment. Sandy muds are quite frequent in the upper horizon whereas bathyal and abyssal muds constitute the largest bottom environments in the Mediterranean. Some of the characteristic species giving name to those facies include the cnidarians *Isidella elongata* and *Funicula quandrangularis* (Fig. 12.4d), the echinoderm *Bryopsis lyrifera*, the gasterop *Apporhais seressianus* and the sponges *Thenea muricata and Pheronema grayi*. Fishes and crustaceans (mainly decapods and peracarids) are particularly abundant (Cartes et al. 2004; Sardà et al. 2004; Galil 2004). The communities inhabiting these bottoms can be different to those found in world oceans due to a shallower average depth, high deep-water turnover rates, higher temperature and extreme oligotrophy with low nutrient concentrations, especially in the eastern part. All these key issues together are forecasting also a unique large percentage of endemic species.

From the continental shelf to the abyssal plain, soft bottoms environments are located on the continental slopes and/or submarine canyons. Organic matter originating in the water column and on the continental shelf flows along the canyon axis and sinks together with inorganic particulate matter. As a consequence, organic and inorganic debris, and large accumulations of sediments can be observed on the floor of submarine canyons (Canals et al. 2006). These deposits are responsible for distinct species assemblages and higher faunal densities and biomass in comparison to nearby non-canyon. In addition, canyons can hold littoral species carried down by the adventive inputs, be responsible for local upwelling and occasionally the presence of benthic species in the water column, acting as a source and reservoir of endemic species (Gili et al. 1998, 2000; Sardá et al. 2010). The environmental conditions in submarine canyons may vary greatly from one location to another. Physical conditions such as strong bottom currents are responsible of resuspension, and seasonal changes in the particle rain may result in a habitat heterogeneity that could be comparable to that of tropical forests or rocky coasts (Grassle 1989).

Cold Seeps

Cold seep biological communities relying on chemosynthesis and associated to mud volcanoes and faults have been relatively recently discovered in the southeastern Mediterranean Sea, south of Crete and Turkey (in the Mediterranean ridge and Anaximander mountains) and north of Egypt at the Nile Seep-Sea Fan at depths ranging from 500 to 2,000 m. Cold seep habitats are extremely heterogeneous in structure and composition, linked to the different reduced microhabitats and substrate types in carbonate crusts and sediments. Faunal assemblages associated with Mediterranean cold seeps are still relatively unknown. Bivalves and tubeworms associated with endosymbiotic and chemoautotrophic bacteria characterize the structure of the faunal assemblages of cold seeps, as well as numerous gastropods. Described symbiont-bearing species include the polychaetes *Lamellibrachia anaximandri, Siboglinidae monilifera*, the bivalves *Myrtea amorpha, Lucinoma kazani, Idas modiolaeformis, Thyasira striata* and *Isorropodon perplexum* (Salas and Woodside 2002; Olu-Le Roy et al. 2004; Werne et al. 2004; Duperron et al. 2008; Ritt et al. 2010). Other megafaunal species of exceptional size compared with their background counterparts include the sponges (*Rhizaxinella pyrifera*) and crabs (*Chaceon mediterraneus*) (Olu-Le Roy et al. 2004).

The benthic cartography of the Mediterranean benthic habitats according to EUNIS (European Nature Information System) habitat classification can be found at http://jncc.defra.gov.uk/page-5020 (Fig. 12.5).

Ecological Strategies

Most benthic Mediterranean bottoms are considered complex habitats and thus with many ecological niches. Due to substrata features, environmental factors and the role of the organisms themselves, the structural and dynamic complexity of sea bottoms enables the presence of many different habitats and organisms with diverse ecological strategies. Besides its precise role in the community, species fall into a bionomic or ecological continuum among the so-called *r* and *K* strategies (Pianka 1970; Margalef 1974). According to Hiscock and Mitchell (1980) any species may be summarized into the following four strategies: long-lived species

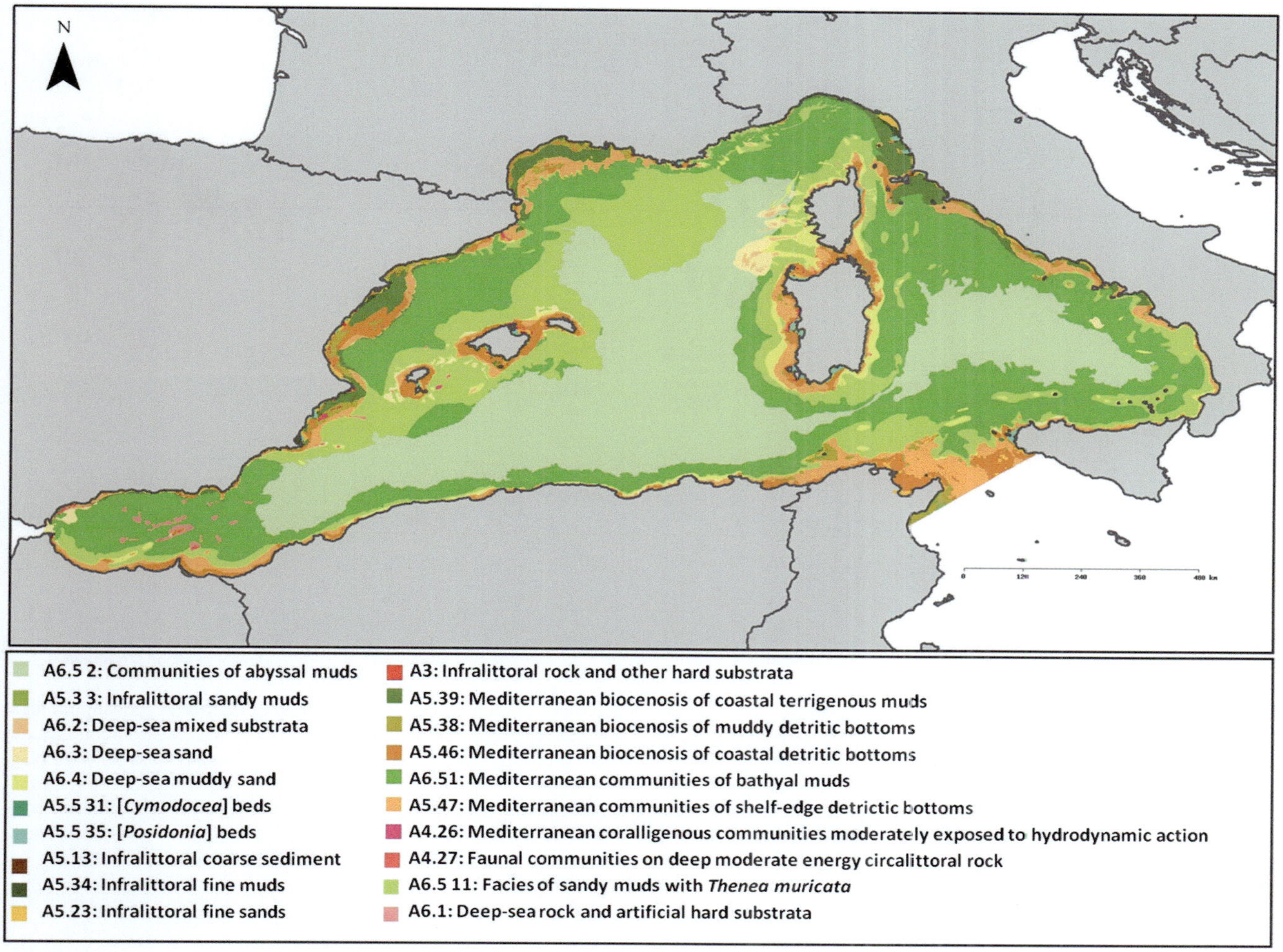

Fig. 12.5 Benthic cartography (EUSeaMap) from the Northwestern Mediterranean following EUNIS classification (Adapted from http://jncc.defra.gov.uk/page-5020)

(moderate to big size, which stabilize and characterize the community; e.g. the bio-constructors); seasonal species (appear only at certain times of the year but have an important impact or role in the communities); opportunistic species (high reproductive potential, they can temporally monopolize the space); and species that respond to small alterations in the environmental conditions (usually with a known and variable temporally structural significance).

In general, both ecological strategies *r* and *k*, follow clear trends and are very representative of Mediterranean zoobenthos. *R*-strategist represents the opportunistic, fugitive, pioneer and generalist species. They exploit the lack of organization and environmental stability, with a high offspring production; the vast majority inevitably disappear. They are highly adaptable and frequent in unstable and stressed environments. On the other hand, the *k*-strategist exemplifies the specialists or strategists. They can stabilize their populations in a given environment, and produce few very protected offspring that are well adapted to ecological stable environments. In sessile modular species they also can reproduce asexually.

In the Mediterranean zoobenthos the ecological strategies can be summarized in the following trends:

It the upper littoral levels, the benthic strategies favoured are those that tend to counteract the mechanical stress associated with these shallow habitats. The balance between high illumination rates results in productive and dense algae communities that support a diverse and rich fauna. Because of the seasonal variations in algal cover, and the high environmental stress, most species have high reproduction rates, production and fast turnover. There are also diverse animal strategies colonizing the hard substrates, the algae or plant talus, especially by the vagile species.

At intermediate levels such as the circalittoral, the current stress is not mechanical but refers to light availability. Light decline reduces plant capacity to occupy the space, and therefore the competition with animals, especially sessile fauna, is maximal. The supply of food is adequate because of the continuous hydrodynamic regime and the oversupply of plankton and benthic production from the upper layer communities. The degree of organization is high. Many species are builders,

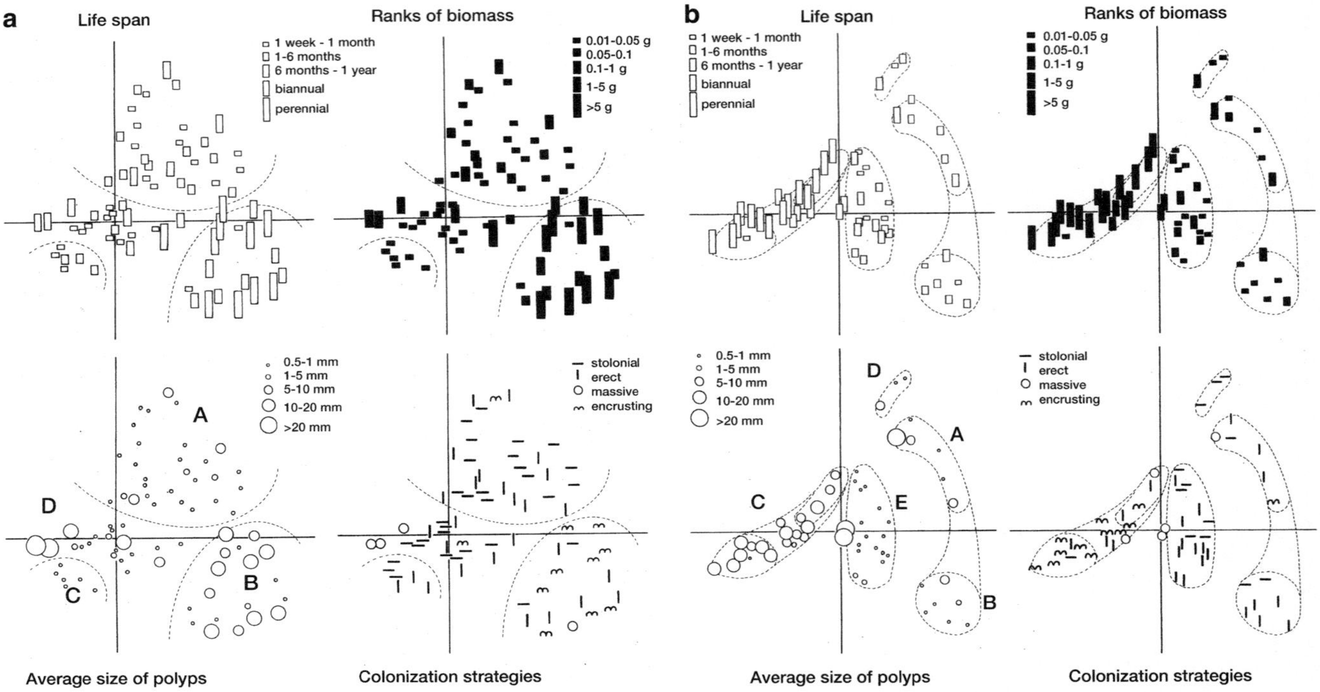

Fig. 12.6 Two examples of ecological strategies present in the different communities from an extensive survey in the Medes Islands (**a**) and along the Catalan coast (**b**) (NW Mediterranean). The 104 benthic cnidarian species considered have been ordered by means of a Principal Component Analysis (Figure modified from Gili et al. 1989, see this paper for more information and species list). (**a**) *A* mediolittoral community and assemblages of photophilic algae; *B* circalittoral community mainly coralligenous assemblages; *C* assemblages of species growing on algae and phanerogams communities; *D* species of wide distribution in rocky shallow communities. (**b**) *A* mediolittoral community; *B* same assemblage but species forming dense populations; *C* circalittoral communities, coralligenous and submarine caves assemblages; *D* mediolittoral community mainly phanerogams medows; *E* supralittoral and mediolittoral shallow communities dominated by species with photophilic algae

having high production rates and investing in three-dimensional structures. They have moderate to slow growth rates and a slow turnover. At deeper levels (e.g. the deep circalittoral, dark caves or shelf-break communities) the stress is greater due to high sedimentation and the scarce food supply. In these environments only animal assemblages can survive, mainly *k*-strategists, and their adaptation is often determined by the acquisition of far-reaching specializations (Fig. 12.6).

In shallow and deep benthic communities there is a convergent trend among the different groups towards a few adaptive options, which are themselves diverse according each particular community (Zabala and Ballesteros 1989). The most suitable strategy for spatial competition is the most common trophic guild, the filter feeding. This strategy may be found through solitary or colonial sessile organization, with or without a skeleton, and ultimate growth to a maximal size, being most of them *k*-strategists. They have long life cycles, efficient capabilities for filtering and defence, and a large capacity for covering the ground. At this ecological model the advantages of colonial vs. solitary strategies must be also considered, as an alternative and complementary way for successional colonization and population stability (Jackson 1977). At the other extreme are the individual organisms, vagile and of limited growth, with short life cycles, high reproductive capacities, with planktonic and long-lived larvae, clearly *r*-strategists. They are minor competitors and prefer unstable surface levels. They opportunistically cover any area that is left open or, when occupying more stable substrates, show strong spatial and temporal fluctuations (Ros et al. 1985; Gili et al. 1989).

Temporal Variability

Seasonal variations are common phenomena in all marine ecosystems. In the Mediterranean, these changes are controlled by climate driven processes and anthropogenic forcing (Duarte et al. 1999). Life cycles of marine organisms show marked seasonal patterns in growth, reproduction and abundance. In the water column, these seasonal cycles are directed by factors, such as photoperiod, light intensity, food availability, oxygen, salinity and temperature. All these factors influence the energy expenditure in marine organisms both planktonic and benthonic. It is therefore a challenge to integrate the effects of multiple factors, as well as their interactive effects, to determine seasonality in the life cycles of marine organisms.

Plankton and Benthic-Pelagic Coupling

The composition and dynamics of plankton communities display strong seasonal changes (Longhurst 1998). In the Mediterranean, periods of water stratification and lower levels of water motion that reduce the biological activity follow high productive seasons. In temperate seas, spring season is considered a key period because it concentrates the majority of the total annual primary and secondary production (Margalef 1985). In autumn or early winter a second short productive period also occurs, although the primary productivity drops (Ribera d'Alcalá et al. 2004) and the seston becomes highly refractory (Grémare et al. 1997; Rossi et al. 2003), having a very low availability for a wide range of organisms. From the end of winter and beginning of spring the primary production is high (Estrada 1996) and many zooplanktonic groups display their maximal peaks of abundance (Sabatés et al. 1989; Calbet et al. 2001). This general tendency throughout the water column has also been observed in close bottom planktonic communities (Ribes et al. 1999; Rossi and Gili 2005). Planktonic communities receive a major input of detrital and dissolved organic matter from river run-off and from benthic debris coming from macroalgae and phanerogams, which influence the biological production (Thomsen and van Weering 1998). Besides the organic input, the near-bottom water layer is also subject to highly variable turbulence that influences the plankton and seston dynamics (Wainright 1990; Rossi and Gili 2009). Recent data on suspension feeders from benthic communities corroborate this seasonal patterns for planktonic communities (Coma et al. 2000).

Boero et al. (1996) and Marcus and Boero (1998) transferred to ecology a host of biological research on the life cycles of many planktonic species with benthic resting stages. Marine systems function based on pulses of production during which some species become very abundant for short periods. Phytoplankton blooms, followed by zooplankton blooms are the engine of all marine systems. Especially in coastal waters, many planktonic species spend the adverse season in the benthos, as resting stages. The pulses, thus, are based on the hatching of the resting stages and on its match with favorable conditions (Boero 1994). The resting stage banks of the benthos are the real potential for the expression of biodiversity in the plankton. Pati et al. (1999) suggested that the predation of the meiofauna on the benthic resting stages of plankton might play a keystone role in regulating the diversity of plankton pulses. The role of resting stages in benthic-pelagic coupling might well be enhanced by the down- and upwelling currents generated by marine canyons (Della Tommasa et al. 2000). Many gelatinous plankters do have benthic stages and the dynamics of their populations rely on benthic processes (Boero et al. 2008). These crucial aspects of benthic pelagic coupling are still vastly unexplored and represent a new frontier in marine ecology.

In shallow waters, it is proved that the seasonal environmental variability both in the water column and near-bottom are closely linked and determine biological processes. Sedimentation, re-suspension, vertical mixing and nutrient fluxes conduct the coupling between plankton and benthic systems. At the same time, benthic organisms have a great influence on the dynamics of water column providing nutrients coming from regeneration processes in shallow and estuarine areas (Valiela 1995). However, in deeper communities the benthic-pelagic coupling is seasonally asymmetrical with surface events, although ecological processes are still linked with the biological production in surface layers (Gori et al. 2012).

Seasonality in the deep-sea may be less pronounced, but can be easily evidenced by the increased seasonal influx of detritus into these areas. With the exception of some extreme environments such as the cold seeps, found in the eastern Mediterranean region, most deep ecosystems in the Mediterranean depend completely on alloctonous organic and inorganic inputs (Cartes et al. 2004). Danovaro et al. (1999) reported mass fluxes at equal depths being up to two orders of magnitude higher in the Western Mediterranean (Gulf of Lions) than in the Eastern Mediterranean (Cretan sea). The 10 % of the carbon in surface waters is exported to 1,000 m depth in the Western Mediterranean, but only 2–3 % in the Eastern Mediterranean, and the bacterial densities are four times higher in the former than in the latter. The same authors also reported different efficiencies in the transfer of organic matter to the deep-sea between the west and the east, 10 % and 1 % respectively. This has deep implications in terms of benthic-pelagic coupling. The general conclusion from the research carried out in these environments, mainly in the NW Mediterranean deep waters (Cartes et al. 2001, 2008; Fanelli and Cartes 2004;), indicates that seasonal peaks in primary production or particulate organic matter translate into seasonal peaks of secondary production in the deep-sea.

Algae Dominated Communities and Sea-Grass Communities

Most habitats in temperate regions exhibit seasonal peaks in the standing stock of algae in late spring throughout summer and with minima in winter (Murray and Littler 1984). In most communities, these seasonal patterns involve changes in community structure (Mann and Lazier 2006), and strongly influence functional processes (Boudouresque 1971). There is also a shift in the time of production and biomass peaks, which take place in spring in shallow sublittoral algal communities, and in summer in the deeper assemblages. However this scenario changes when the communities suffer the effect of fish and urchin predation, which modify benthic communities and affect the composition, abundance and dynamics of shallow sublittoral algal communities. At high densities and during the production phase of the algae, sea urchins display an overgrazing activity that reduce the algal biomass because of the patchy distribution and denudation by the urchin, especially on erect talus. In extreme cases of predation, grazers reduce totally the erect stratum of fleshy algae, resulting in a monotonous assemblage of encrusting algae and in a reduction of algal communities seasonality (Velarque 1987; Sala et al. 1998a, b). In the Mediterranean fishes are not the dominant predators and have less effect on algal abundance and seasonal variability than sea urchins (Velarque 1990). However, overgrazing the populations of fleshy erect macroalgae by fishes facilitates the increase in biomass of chemically defended or less competitive species assemblages, which disguise the natural community patterns (Sala and Boudouresque 1997). In seagrasses, the seasonal changes in meadow structure, shoot density or leaf surface and biomass, have an influence on the species composition and abundances of associated vagile fauna like polychaetes and fish, which are in turn influences by the ecology and life cycle of single species like recruitment and migrations (Francour 1997; Gambi et al. 1998).

Vagile Fauna

The sessile components of benthic communities are associated with the morphological and physical characteristics of the seabed. These features determine seasonal periods of growth alternated with periods of inactivity and dormancy (Coma et al. 2000). In contrast, much of the fauna that lives close to the bottom exhibits varying degrees of mobility, from crawling gastropod molluscs, starfishes and sea urchins to highly mobile mysidacea and fish. These organisms perform seasonal migrations through a vertical gradient, avoiding high temperatures in summer and looking for food during winter in shallower habitats (San Vicente and Sorbe 2003; Bellan-Santini et al. 1994). Another group that moves seasonally are the bottom dwelling fishes and their migrations are linked to certain habitat types, and even to different regional areas (García-Rubies and Macpherson 1995). Such variation in habitat use can be attributed to different life history stages. Juveniles burrow into the substratum to evade predation, but the adults use a wider range of habitats such as big boulders or caves (Planes et al. 2000; Macpherson and Raventos 2005).

On the other hand, larval stages of many invertebrate and fish species use transitory habitats as an area for shelter and feeding. Many of these habitats occur in hard bottom communities associated to the sessile species that cover temporally the substrata. The most common types of habitats are macro-algal and sea-grass beds where species find also refuge (Bruno and Bertness 2001). Occasionally, communities dominated by animals as bryozoans, gorgonians, corals and polychaetes play a similar role and have a great strategic importance for species that inhabit them (Bellan-Santini et al. 1994; Pederson and Peterson 2002). During periods of high growth of the foliage and thalli, meadows of macro-algae provide shelter and protection to different stages of many marine species that are most vulnerable to predation. Such habitats are essential in explaining the life cycles of species and also the seasonal and inter-annual variations in populations that synchronize their larval stages with the factors affecting the opportunities to colonize ephemeral habitats.

In deeper soft-bottoms environments of the continental shelves, there is a clear lack of knowledge in the temporal dynamics One of the few studies that have been carried out in the Mediterranean on this topic (de Juan and Cartes 2011) showed a marked seasonal variability linked to the high phytoplankton production in spring. The springtime production supplies organic matter to the benthos, inducing a higher infaunal abundance. There was also a reduction in the infaunal crustacean abundance and species richness in late summer. These changes were attributed to two factors, the establishment of the thermocline (which could limit the availability of organic matter to the seabed), and the high water temperatures (which caused the production of phytoplankton exudates and accumulation of mucilaginous aggregates over the seabed).

The low food input into the deep-sea results in scarce food resources, high food partitioning, highly diversified diets, and very complex trophic webs. It can also be observed a trend towards a decrease of feeding intensity with depth for decapods and fish, which suggests a reduction on the metabolic activity with increasing depth (Carrasón and Cartes 2002; Fanelli and Cartes 2004).

Animal-Dominated Communities

There are great differences in the temporal changes in the structure and dynamics between algal-dominated and animal-dominated communities. Shallow algal communities show the largest temporal changes, though it is difficult to establish a clear seasonal pattern. When the cover area is partitioned between algae and animals, they both find favourable conditions for growing in the spring to summer period for algae, and autumn to winter for animals (Garrabou et al. 2002). In deeper communities, dominated by animals, few structural seasonal changes have been observed and these might be detected only in the physiological parameters of animal populations such as reproduction periods or slowly

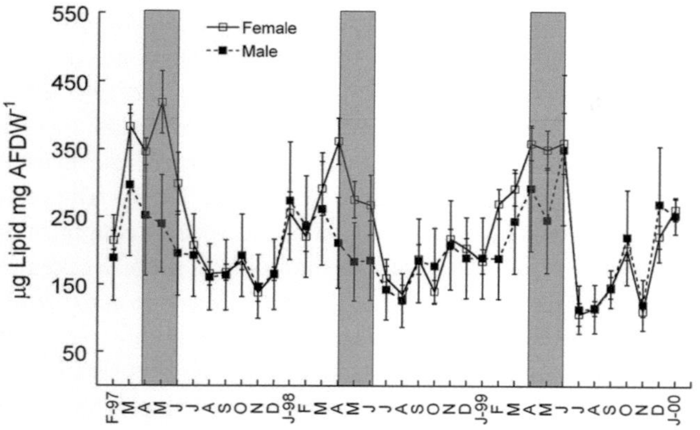

Fig. 12.7 Three years cycle of the lipid contents in the gorgonian *Paramuricea clavata*. Shadow areas indicate reproduction period (Figure modified from Rossi et al. 2006a). The marked seasonality of food quality and availability is reflected in the energy storage of mediterranean benthic suspension feeders. Food constraints and reproductive traits are reflected in summer (reproduction and aestivation process), and in autumn (low quality seston). Reproduction success is tightly related with the energy storage capability

growth patterns (Turon and Becerro 1992; Garrabou et al. 2002; Blanquer and Agell 2008; De Caralt et al. 2008). In these animal-dominated communities the abundance of ephemeral organisms and the trophic ecology of perennial ones are closely related, in response to summer and late autumn food shortages of the water column (Rossi et al. 2006a). Ingesta and assimilation may vary seasonally with seston concentration and quality, depending on the feeding strategy (Coma et al. 2000; Jordana et al. 2001; Rossi and Gili 2005).

In the Mediterranean, the seasonal variation of environmental factors such as temperature, food availability and photoperiod can cause shifts in the resource allocation of benthic organisms (Caswell 1989). From an energetic point of view, organisms change the seasonal pattern of energy invested in reproduction, growth and maintenance influencing the food storage (Rossi et al. 2006a) (Fig. 12.7). In general, the main energy output coincides with the reproductive period at the end of the most successful feeding period, especially spring and early autumn (Coma et al. 1998). The seasonal variation in the biochemical composition of the tissue of an organism may be interpreted as a record of the water column productivity fluctuations in the benthic community. It has been shown in gorgonians that the effect of seasonal fluctuations in seston quantity and quality were apparent in the protein, carbohydrate, and lipid tissue concentration values (Rossi et al. 2006a; Gori et al. 2007; Rossi and Tsounis 2007). Other approaches include the shift of the resource allocation into defensive mechanisms by species that grow in high competitive communities. In sponges, higher competitive pressure in the shaded environment results in increased investment in defensive/supportive (mineral and organic) structures and a lower investment in somatic growth and reproductive output of larvae (Uriz et al. 1995).

Another seasonal strategy related to the energetic constraints is a predictable pattern of temporal dormancy. In the Mediterranean, summer dormancy predominates, in contrast to cold temperate seas, which are characterized by winter dormancy (Sardá et al. 1999; Garrabou 1999; Coma et al. 2000; Betti et al. 2012; Di Camillo et al. 2012). The physiological changes related with this resting state help the organisms to survive in adverse weather conditions and food scarcity. The plasticity of many sessile invertebrates allows them to withstand net energy deficits by digesting themselves until additional food becomes available (Gili and Coma 1998). Biochemical composition in Mediterranean gorgonians (Rossi et al. 2006a; Rossi and Tsounis 2007) shows periodic summer minimum energy storage of lipids and proteins, as well as a drop in late autumn-early winter. These observations agree with the expected physiological changes associated to summer and late autumn energy shortage (Grémare et al. 1997; Rossi et al. 2003), and with an annual pattern of investment in growth and reproduction of gorgonian species. Seasonal dormancy phenomena have been related not only with energetic constraints but also with physiological adaptations to exceptional changes in water temperature ranges (Sulak et al. 2007). The presence of resting stages is a common mechanism in marine organisms, which can remain inactive for long periods, acting as biodiversity reservoirs. The importance of life cycle dynamics in structuring marine communities contributes to explain seasonal fluctuations not only in benthic but also in phyto- and zooplanktonic communities (Boero et al. 1996).

Tendencies

The Mediterranean Sea is one of the most important biodiversity hotspots of the world with many different species and natural resources (Bianchi and Morri 2000; Coll et al. 2010). However, it is increasingly compromised by human activities, something that is accentuated by its semi-enclosed configuration, which restricts its communication with the open ocean. Human pressures on this environment are increasing habitat and species destruction to become a priority issue for policy-makers. In fact, Mediterranean Sea has been identified as a hotspot of sea warming effects, ocean acidification, alien species invasion and other direct and indirect human perturbations, being considerably more impacted than other places of the world (Durrieu de Madron et al. 2011). Increasing demographic and economic developments make the general panorama of the Mediterranean benthic communities complex, since several factors are synergically acting at population and community levels.

Directly Man-Induced Pressures

Fishing Activity

The fishing activity in the Mediterranean is characterized by the multi-specificity of catches and the absence of large single stocks, especially in the demersal regime (Farrugio et al. 1993). Although artisanal fleets and gears constitute the most important fisheries in the Mediterranean, nowadays there is a tendency in the development of semi-industrial fleets, mainly of trawl fishing. Five thousand two hundred and seventy four trawlers operate in Mediterranean waters, fully exploiting the continental shelves and upper slopes of the basin (FAO 2011). Bottom trawling shows the least degree of selectivity with the widest-range impact on different organisms of the ecosystem due to habitat destruction and modification (Demestre et al. 2000), and displays the largest impacts on some demersal target species and non-target groups (i.e. small organisms discarded and juveniles of vulnerable species such as rays and demersal sharks). At present, trawl fisheries constitute probably the most acute direct problem for soft-bottom and detritic habitats, but was an evident source of destruction in other communities like *Posidonia oceanica* or *Cymodocea nodosa*. Together with species like hake (*Merluccius merluccius*) or the anglerfish (*Lophius piscatorius)* and other demersal fishes, the most common catches of bottom trawlers are crustaceans. Except for the Norwegian lobster (*Nephrops norvegicus*), the scientific assessments carried out on the red shrimp (*Aristeus antennatus*), the giant red shrimp (*Aristeomorpha foliacea*) and the pink shrimp (*Parapenaeus longirostris*) are giving alarms for overexploitation and the highly damaging effect on the accompanying fauna and surrounding environments (General Fisheries Commission of the Mediterranean 2012).

Common illegal practices in Mediterranean trawling include using nets with a smaller mesh opening than the permitted, fishing in prohibited zones and seabeds and using engines with a higher power than officially declared. Their catches also include abundant by-catch and discards. Though theoretically it would be simple to analyze the intensity of trawling in the Mediterranean using the track of the fleet with positioning systems, the reality is that data does not exist or it is not available. Studies in the North Sea showed that the area trawled by these vessels each year is equivalent to the size of the entire sea (Leth and Kuijpers 1996), and this figure could be similar for the continental shelves in the Mediterranean. The impact of trawling on the ecosystem can be categorized into two basic factors: (a) the selectivity of fishing techniques, with regard to the target species and to catching young fish and other sea animals that do not correspond to the target species, and (b) the habitat destruction, physical and biological consequences of the fishing arts on the bottom (Fig. 12.8a, b).

The selectivity depends on the mesh size used and its shape. The larger the size of the mesh opening of the net, the greater the selectivity, whereas diamond shapes generally tend to result in poorer selectivity than square shaped ones. However, other reports have demonstrated that the selectivity of fishing techniques depends more on the target species (behavior, life cycle, etc.) than on the size of mesh used (e.g. young mullet, Tokaç and Tosunoglu 1996). Although selectivity depends on the shape of the species caught, it is also highly affected by the large amounts of rubbish taken in the catch. A survey conducted on a Norway lobster trawler fishery in the Mediterranean showed that one piece of rubbish was brought up per six Norway lobsters (Ragonese et al. 1994). Rubbish is becoming an alarming problem in fisheries.

The physical impact of bottom trawling on ecosystems is also very high, reducing the complexity of benthic communities (Auster 1998). All the components involved in trawling have the capability to affect the seabed, the net, the chains, the weights and, especially, the doors. These can go various centimeters into the seabed (up to 30 cm depth) depending on the sediment and the technique used, damaging benthic ecosystems (Collie et al. 2000; de Juan et al. 2007, 2011). All trawl components show a huge impact on soft-bottom environments yielding less species selectivity than other fishing techniques, damaging the substrate and its sessile species, and increasing its slow recuperation. The reduction of invertebrates (echinoderms, polychaetes and molluscs) in trawled zones reach up to 65 % when compared with areas where this kind of fishing has not taken place (Bergman and Hup 1992).

Trawling can create also similar effects to eutrophication, increasing the presence of opportunist species in benthic environments. The reduction of biomass in these zones, when compared with an area where trawling has not taken place, can be ten times greater, especially with regard to fish species. Additionally, trawling discards can give rise to episodes of anoxia on the seabed, increasing the mortality of target species (Browder 1981) and causing changes in the structure and composition of these species (Chan and Liew 1986). The effect of discards can be noted on other taxonomic groups of fauna as marine birds. Discards may support a demographic explosion of the most opportunistic species to the detriment of other, more vulnerable ones (Garthe and Hüppop 1998), producing also changes in habitats, behavior and reproduction. Trawling also affects the water column and the ecosystem in general, by changes in biogeochemistry and the nutrient effluents caused by the resuspension of sediments, nutrients and pollutants, as well as damage to the invertebrates that are responsible for irrigating, compacting and oxygenating the seabed (Schwinghamer et al. 1996). The increase in turbidity can also influence the photosynthetic capacity of plants, with the resultant decrease in their distribution and productivity in the photic zone (Caddy 2000).

In general, trawling on soft-bottom habitats affects the most to long-living epibenthic species and less the infaunal

Fig. 12.8 Examples of perturbations: (**a**) trawl marks and (**b**) gosh net entangling adult lobsters *Palinurus elephas* (Photographs **a** and **b** by NEMO-ICM-Gavin Newman) as examples of direct human perturbations by fishing in the Cap de Creus continental shelf (NW Mediterranean) at 110 m depth; mass mortalities events affecting *Paramuricea clavata* at 20 m depth in Corsica (Central Mediterranean) (**c**) healthy colony and (**d**) a colony showing almost complete loss of tissue, the denuded axis have been colonized by different epibiontic species (Photographs **c** and **d** by MedRecover Marine Conservation Research Group); (**e**) mucilage aggregations on gorgonian *Paramuricea clavata* at 20 m depth, Montgrí Coast (NW Mediterranean); (**f**) invasive species *Oculina patagonica,* colony found at 15 m depth in Medes Islands (NW Mediterranean) (Photographs **e** and **f** by Núria Teixidó)

ones. Consequently the oldest and longest living species are the scarcest in trawled areas, while younger, more short-lived species are the most abundant (Tuck et al. 1998). The effects of trawling in the long-term could also favor small passive filter-feeders over larger active filter-feeder species. Effects of trawling are as important as those produced by strong natural perturbations (Pusceddu et al. 2005). Trawling may have a very high impact on important and unique habitats of long-term formation, seagrass beds, maërl habitats, and deep coral reefs. In deep-sea, the impact of trawling on species assemblages could be even greater. Because these areas are not regularly subjected to natural changes, as frequently as in shallower waters, are more vulnerable to artificially induced changes (Langton and Auster 1999).

Nutrient Increase

Disease and mucilage formation are phenomena that are directly attributed to nutrient increase in the Mediterranean Sea. The mucilage phenomenon has been affecting different areas of the North-Western Mediterranean during the last decades (Mistri and Ceccherelli 1996a; Giuliani et al. 2005) (Fig. 12.8e). The first record was in 1991 in the Tyrrhenian Sea (Innamorati et al. 1993), but it has been spreading in other areas during the last 20 years (Pusceddu et al. 2009). This phenomenon, in which different algae form a compact bloom that covers benthic communities (Giuliani et al. 2005), has a direct effect on suspension feeders, covering during a prolonged time their feeding structures. Entire populations can be affected, being the extent of the injuries negatively correlated with the size of the colonies (Mistri and Ceccherelli 1996a). However, as in the previously described hot waves mass mortalities, the recovery of the colonies were also observed after 2 years of the event (Mistri and Ceccherelli 1996b). An indirect effect of hypoxia (and even anoxia) is also observed because of the oxygen depletion by bacterial respiration (Schiaparelli et al. 2003). Is not clear which is the origin of the mucilage, neither if is related with local contamination (high nutrient concentration), but it is certain that has effects on the pelagic and benthic functioning (Pusceddu et al. 2009).

Another phenomenon related with the microbiological or viral attack in hard bottom suspension feeders has been described in sponges (Gaino et al. 1992). The disease apparently produced by bacteria, made profound alterations on the skeleton with bacterial damage fibres (Cebrian et al. 2011). Commercial sponge beds were heavily damaged, especially in shallower areas, and because of its intensive harvesting, were locally depleted (Pronzato and Manconi 2008).

Harvesting

Commercial sponges and red coral are the most affected animal species by direct impact on hard bottom substrates (Pronzato and Manconi 2008; Tsounis et al. 2010). *Spongia*, *Hyppospongia*, and *Corallium rubrum*, have been exploited for thousands of years. However, the most intense harvesting period is centred in the last two centuries. Sponges also suffered a devastating disease that dropped dramatically the landings and the commercialization of these species (Pronzato and Manconi 2008). Other sponge species have been recently identified as animals of pharmacological interest (like *Dysidea avara*), threatening the resource to extract the target metabolites (Pronzato et al. 2000). Red coral has also been harvested in a completely unsustainable manner (Tsounis et al. 2007), because no biological parameters are considered in the actual fishery models. The intense harvesting and mass mortalities may collapse this resource at a local level in the first 50 m depth, due to a synergic effect (Garrabou et al. 2001; Santangelo et al. 2007). Even if there is a real effort to recover both harvested suspension feeders organisms (Pronzato et al. 2000; Bramanti et al. 2007; Linares et al. 2012; Benedetti et al. 2011), there is still a long way to manage properly these overharvested species. Other important harvesting activities that may cause severe direct or indirect impact on rocky benthic invertebrate assemblages and their habitat as in the case of the edible sea urchin *Paracentrotus lividus* (Sala et al. 1998a, b; Pais et al. 2007) and the rock-boring date mussel *Lithophaga lithophaga* (Fanelli et al. 1994; Guidetti et al. 2003).

Local Impacts

Local impacts on suspension feeders due to the direct action of humans have also important consequences. Sala et al. (1996) found that the bryozoan *Pentapora fascialis* was more exposed to SCUBA diver's action in frequented than in non-frequented areas. The same perturbation (dead or partially injured colonies) or the lack of populations recovery were found in a long term monitoring of the gorgonian *Paramuricea clavata* and red coral *Corallium rubrum* (Coma et al. 2004; Linares et al. 2012). SCUBA diving may be re-directed to a more respectful and appropriate diving behaviour (Bramanti et al. 2011). However, simple actions like bubbling near the colonies during the spawning period may cause the loss of sexual products before the fecundation process occurs, and therefore damaging the entire population (Tsounis et al. 2012).

Other not so evident damages in suspension feeders assemblages are caused by line fishing, which may affect more than 30 % of the gorgonians in highly frequented areas by amateurs or professionals (Bavestrello et al. 1997; Bramanti et al. 2011). Lines denude the axis of the animals, which favours the development of epibiont aggregates. Mechanical stress has also been recorded, damaging colonies that can be partially or totally broken. These kind of injuries have an effect on the gonadal output and in the energy storage capability of gorgonians that may produce less potential recruits (Tsounis et al. 2012). Other negative local impacts are those performed on sea-grasses by the anchoring of boats from recreational boating that causes mechanical damage and habitat loss (Francour et al. 1999; Milazzo et al. 2004; Lloret et al. 2008).

Natural and/or Indirectly Man-Induced Pressures

Increasing Temperature

The temperature increase observed in recent decades in the ocean has also been observed in the Mediterranean Sea. Two significant mass mortalities due to a heat wave (1999 and

2003) have been detected and their impact in hard bottom communities quantified (Cerrano et al. 2000; Perez et al. 2000; Garrabou et al. 2009). In these mass mortality events, the so-called animal forest were affected in the first 10–50 m depth in a wide range (more than 1,000 km of coast in the second heat wave), partially or totally killing gorgonians, corals, sponges or bryozoans. Affected colonies also recorded extensive attacks of microorganisms (protozoan and fungi) (Cerrano et al. 2000) (Fig. 12.8c,d). The anomalous high temperatures, the energetic constrains associated with water column stability conditions and pathogens developments, were the most probable causes of mortality (Bally and Garrabou 2007; Coma et al. 2009; Vezzulli et al. 2010). Other mass mortality events have been reported in the Mediterranean (Bavestrello et al. 1994; Harmelin and Marinopoulos 1994; Maldonado et al. 2010; Cebrian et al. 2011), although in previous outbreaks the number of affected species and/or spatial scale concerned was lower than in the above-mentioned 1999 and 2003 mass mortalities.

Several time series have examined the effects of the heat waves on different organisms especially in the coralligenous communities. Linares et al. (2008) showed the warming effect on the reproductive output of the gorgonian *Paramuricea clavata*, being the most affected colonies those with less gonadal production. Bramanti et al. (2005) observed a drop in the recruitment of shallow red coral patches, probably a consequence of the partial or total mortality of the colonies (Garrabou et al. 2001). Alternatively, not all the areas recovered in the same way after the heat wave. Unlike in the Port Cros National Park (Linares et al. 2005), *Paramuricea clavata* had high recruitment rates and tissue recover in the Eastern Ligurian Sea (Cupido et al. 2008, 2009), showing that population dynamics of gorgonians may differ with local environmental factors (Bramanti et al. 2009). Also the autoecology of the species is a key factor to understand the capacity to recover, being *Eunicella singularis* the species that showed the best recovery in the above-mentioned area (Fava et al. 2010). Recently, genetic studies demonstrated significant genetic structure between populations at small spatial scales (tens and hundreds of meters), suggesting that gene flow is highly restricted (Ledoux et al. 2010; Mokhtar-Jamaï et al. 2011). These results indicate that recovery of populations from large perturbations is very likely only based in local processes.

Ocean Acidification

Another less understood but probable damage suffered by the benthic communities due to the climate change in the Mediterranean sea will be the ocean acidification effects. Ocean acidification is predicted to impact all areas of the oceans and affect a variety of marine organisms (Durrieu de Madron et al. 2011). The diversity of responses among species prevents clear predictions about the impact of acidification at the ecosystem level (Hall-Spencer et al. 2008; Kroeker et al. 2011).

Organisms like polychaetes or amphipods seem to be non-affected by the future acidification panorama, but decapods or gastropods may suffer severe impacts on its calcareous structures (Kroeker et al. 2011). Other calcareous organisms may be extremely affected by future ocean acidification, as shown by Lombardi et al. (2011) in the bryozoan *Schizoporella errata*. In this suspension feeder, the putative defensive polymorphs (avicularia) were significantly fewer, and retarded growth of zooidal basal and lateral walls was evident at low pH. Other studies show the effects of the chemical balance alteration in micronutrients essential for the physiological needs of bryozoans. In *Myriapora truncata*, prolonged low pH exposition increases the enrichment of Mg in outer layers, enhancing the skeleton vulnerability to dissolution. Alternatively, the symbiotic anthozoan *Cladocora caespitosa* seems to be non-affected by pCO_2 increase. So, the common belief that calcification rates will be affected by ocean acidification in all calcifier organisms may not be the common rule (Rodolfo-Metalpa et al. 2010). In fact, much work is needed to better understand the potential impact of pCO_2 increase, not only in the physiology of organisms but also at the community and between organism interaction levels.

Easterly Wind-Storms

Easterly wind-storms occur frequently and have high intensity in the Western Mediterranean Sea (Mendoza et al. 2011). The immediate effect of these storms is a direct mechanic perturbation on the benthic fauna and flora. Pluri-annual algal canopies like *Cystoseira zosteroides* may be heavily affected by autumn storms, but their recovery depends upon local population size structure, mortality and recruitment (Navarro et al. 2011). The changes produced in the algal and sessile animal composition, as well as in the abundance of vagile organism, may be crucial for the survivorship of higher trophic levels.

But there is also another less evident effect of easterly wind-storms that influences benthic communities. Storms affect resuspension modifying the composition of the organic fraction of settling particles, decreasing the labile fraction of particulate organic matter (Grémare et al. 1997; Rossi et al. 2003). The storm may result in a transitory increase of the abundance of fine particles at the water-sediment interface, coating these particles with highly refractory matter (Grémare et al. 2003). Even if in the deep coastal shelf and in submarine canyons such storms may mobilize particles suitable for benthic organisms (Puig et al. 2001; Sardà et al. 2009), in near coastal shallow areas can give rise to a shortage in food availability. The decrease in food availability because of high resuspension and heavy river runoff particle transport rises the food stress level in passive suspension feeders (Rossi et al. 2006b).

Biological Invasions

Biological invasions in marine habitats represent one of the main factors of human-induced global changes (Occhipinti-Ambrogi and Savini 2003). The Mediterranean Sea has been subjected to introductions of non-indigenous species by ship traffic and aquaculture several centuries ago, but it has been during the last hundred years when these impacts have been accelerating (Galil 2000). Besides the Suez channel opening in Egypt. Also, the proliferation of artificial hard structures has been a key factor for the spreading of many species (Bulleri and Airoldi 2005), as well as the transformation of benthic areas by different direct or indirect human intrusions (Coma et al. 2011). An updated checklist of marine alien species within each subregion of the Mediterranean Sea, along with their acclimatization status and origin has been recently provided by Zenetos et al. (2010). A total of 955 alien species are known in the Mediterranean, the vast majority of them have been introduced in the Eastern Mediterranean (718), less in the Western Mediterranean (328) and Central Mediterranean (267), and least in the Adriatic (171), being this numbers underestimated. Excluding microalgae, for which knowledge is still insufficient, aliens have increased notably the total species richness of the Mediterranean Sea by 5.9 %. A total of 134 alien species are classified as invasive or potentially invasive (108 are present in the Eastern Mediterranean, 75 in the Central Mediterranean, 53 in the Adriatic and 64 in the Western Mediterranean). Invasive species presence indicates that they have spread, are spreading or have demonstrated their potential to spread elsewhere, and have an adverse effect on biological diversity, ecosystem functioning, socio-economic values and/or human health in invaded regions. A large portion of these invasive species are benthic species.

One of the most studied invasions of hard bottom suspension feeders is the *Oculina patagonica* case (Fig. 12.8f). This species is a new immigrant from the Southwest Atlantic to the Mediterranean Sea, which has now a widespread distribution even in the eastern Mediterranean (Fine et al. 2001). Coma et al. (2011) observed a longterm series of expansion in this symbiotic anthozoan, as well as the abundance of *Paracentrotus lividus*, the main grazer of macroalgae in hard bottom substrates. Their results show that part of the increasing presence of *Oculina patagonica* may be explained by the increase of sea urchin abundance, which creates a barren rock suitable for the settling of the alien species (Coma et al. 2011). Sponges have also been reported to invade the Mediterranean. The calcareous sponge *Paraleucilla magna* is proliferating in the western Mediterranean since 10 years ago (Guardiola et al. 2012) and its origin remains unknown. In general, the first stages of colonization and the capacity to survive during the settling process have been observed as key factors to better understand alien species success on hard bottom substrates (Rius et al. 2009a, b).

In summary, Mediterranean benthic communities have been perturbed and changed by human activities since the beginning of the civilization. With the progressive increase in human population in the coastal zone, the impact has increased especially on the coast but also along the continental shelf. Today, signs of disturbances caused by man activities can be seen in almost all regions of the Mediterranean Sea. Both trawling and pollution drastically affect the loss of diversity and benthic habitat reduction. The anthropogenic effects must be added to those produced by global warming that affect especially the shallow populations and may favor the proliferation of invasive species. The conservation of the Mediterranean requires urgent measures such as marine protected areas and a proper change in its management, based on rigorous scientific knowledge (Rield 1980; Sala 2004). Benthic communities, especially those dominated by animals, have been unfairly ignored in management plans and conservation all around the world (Arntz et al. 1999).

Acknowledgments The authors are indebted to the Editors for their invitation to participate in this book. We acknowledge the support of projects LIFE+INDEMARES "Inventory and designation of marine Natura 2000 areas in the Spanish seas". S. Rossi was supported by the research program "Ramon y Cajal" from the Spanish Ministry of Science and Education. We are also grateful to Jordi Corbera for his help with the illustrations. We specially thank Dr. Giorgio Bavestrello, Dr. Ferdinando Boero, Dr. Lorenzo Bramanti, Dr. Joaquim Garrabou, Dr. Giorgos Tsounis and Dr. Iousune Uriz, for reviewing the manuscript and for their valuable comments and suggestions.

References

Arntz WE, Gili JM, Reise K (1999) Unjustifiably ignored: reflections on the role of benthos in marine ecosystems. In: Gray JS, Ambrose W, Szaniawska A (eds) Biogeochemical cycling and sediment ecology. Kluwer Academic Publishers, Dordrecht

Auster PJ (1998) A conceptual model of the impacts of fishing gear on the integrity of fish habitats. Conserv Biol 12:1198–1203

Ballesteros E (1994) The deep-water *Peyssonnelia* beds from the Balearic Islands (western Mediterranean). Mar Ecol 15:233–253

Ballesteros E (2006) Mediterranean coralligenous assemblages: a synthesis of present knowledge. Oceanogr Mar Biol Annu Rev 44:123–195

Ballesteros E, Zabala M, Uriz MJ, Garcia-Rubiés A, Turon X (1993) El Benthos: les communities. In: Alcover JA, Ballesteros E, Fornós JJ (eds) Història Natural de l'Arxipèlag de Cabrera. CSIC-Edit Moll, Mallorca

Bally M, Garrabou J (2007) Thermodependent bacterial pathogens and mass mortalities in temperate benthic communities: a new case of emerging disease linked to climate change. Glob Change Biol 13:2078–2088

Bavestrello G, Bertone S, Cattaneo-Vietti R, Cerrano C, Gaino E, Zanzi D (1994) Mass mortality of *Paramuricea clavata* (Anthozoa, Cnidaria) on Portofino promontory cliffs, Ligurian Sea, Mediterranean Sea. Mar Life 4:15–19

Bavestrello G, Cerrano C, Zanzi D, Cattaneo-Vietti R (1997) Damage by fishing activities to the gorgonian coral *Paramuricea clavata* in the Ligurian Sea. Aquat Conserv Mar Freshw Ecosyst 7:253–262

Bellan-Santini D (1969) Contribution à l'étude des peuplements infralittoraux sur substrat rocheux. Rec Trav Sta Mar Endoume 63:1–293

Bellan-Santini D, Lacaze JC, Poizat C, Pérès JM (1994) Les biocénoses marines et littorales de Méditerranée, synthèse, menaces et perspective. Mus Nat Hist Nat Paris Coll Patrim Nat 19:1–246

Benedetti A, Bramanti L, Tsounis G, Faimali M, Pavanello G, Rossi S, Gili JM, Santangelo G (2011) Applying cathodically polarised substrata to the restoration of a high value coral. Biofouling 27:799–809

Bergman MJN, Hup M (1992) Direct effects of beamtrawling on macrofauna in sandy sediment in the southern North Sea. ICES J Mar Sci 49:5–11

Betti F, Bo M, Di Camillo CG, Bavestrello G (2012) Life history of *Cornularia cornucopiae* (Anthozoa: Octocorallia) on the Conero Promontory (North Adriatic Sea). Mar Ecol 33:49–55

Bianchi CN, Morri C (1994) Studio bionomico comparativo di alcune grotte marine sommerse: definizione di una scala di confinamento. Inst Ital Speleol Mem 6 Ser II:107–123

Bianchi CN, Morri C (2000) Marine biodiversity of the Mediterranean Sea: situation, problems and prospects for future research. Mar Pollut Bull 40:367–376

Bianchi CN, Cattaneo-Vietti R, Cinelli F, Morrri C, Pansini M (1996) The biological study of the submarine caves of the Mediterranean: present knowledge and perspectives. Boll Mus Inst Biol Univ Genova 60–61:41–69

Bianchi CN, Pronzato R, Cattaneo-Vietti R et al (2004) Hard bottoms. Biol Mar Mediterr 11:185–215

Bibiloni MA, Uriz MJ, Gili JM (1989) Sponge communities in three submarine caves of the Balearic Islands (western Mediterranean: adaptations and faunistic composition). Mar Ecol 10:317–334

Blanquer A, Agell G (2008) Hidden diversity in sympatric species. Adjusting life-history dynamics to share substrate. Mar Ecol Prog Ser 371:109–115

Bo M, Tazioli S, Bavestrello G (2008) *Antipathella subpinnata* (Antipatharia, Myriopathidae) in Italian seas. Ital J Zool 75:185–195

Bo M, Bavestrello G, Canese S, Giusti M, Salvati E, Angiolillo M, Greco S (2009) Characteristics of a black coral meadow in the twilight zone of the central Mediterranean Sea. Mar Ecol Prog Ser 397:53–61

Bo M, Bavestrello G, Canese S, Giusti M, Angiolillo M, Cerrano C, Salvati E, Greco S (2011a) Coral assemblage off the Calabrian Coast (South Italy) with new observations on living colonies of *Antipathes dichotoma*. Ital J Zool 78:231–242

Bo M, Bertolino M, Borghini M, Castellano M, Covazzi Harriague A, Di Camillo CG, Gasparini GP, Misic C, Povero P, Pusceddu A, Schroeder K, Bavestrello G (2011b) Characteristics of the mesophotic megabenthic assemblages of the vercelli seamount (north Tyrrhenian Sea). PLoS One 6:e16357

Bo M, Bertolino M, Bavestrello G, Canese S, Giusti M, Angilolillo M, Pansini M, Taviani M (2012) Role of deep sponge grounds in the Mediterranean Sea: a case study in southern Italy. Hydrobiologia 687:163–177

Boero F (1994) Fluctuations and variations in coastal marine environments. Mar Ecol 15:3–25

Boero F, Belmonte G, Fanelli G, Piraino S, Rubino F (1996) The continuity of living matter and the discontinuities of its constituents: do plankton and benthos really exist? Trends Ecol Evol 11:177–180

Boero F, Bouillon J, Gravili C, Miglietta MP, Parsons T, Piraino S (2008) Gelatinous plankton: irregularities rule the world (sometimes). Mar Ecol Prog Ser 356:299–310

Boudouresque CF (1971) Contribution à l'étude phytosociologique des peuplements algaux des côtes varoises. Végétatio 22:83–184

Bouillon J (1975) Sur la reproduction et l'écologie de *Paracoryne huvei* Picard (Tubularoidea-Athecata-Hydrozoa-Cnidaria). Arch Biol 86:45–96

Boury-Esnault N, Pansini M, Uriz MJ (1994) Spongiares bathyaux de la mer d'Alboran et du Golge Ibero-marocain. Mem Mus Nat Hist Nat Paris 160:1–174

Bramanti L, Magagnini G, De Maio L, Santangelo G (2005) Recruitment, early survival and growth of the Mediterranean red coral *Corallium rubrum* (L 1758), a 4-year study. J Exp Mar Biol Ecol 314:69–78

Bramanti L, Rossi S, Tsounis G, Gili JM, Santangelo G (2007) Recruitment and early survival of red coral on settlement plates: some clues for demography and restoration. Hydrobiologia 580:219–224

Bramanti L, Iannelli L, Santangelo G (2009) Mathematical modelling for conservation and management of gorgonian corals: young and olds, could they coexist? Ecol Model 220:2851–2856

Bramanti L, Vielmini I, Rossi S, Stolfa S, Santangelo G (2011) Approaching recreational scuba divers to emblematic species conservation: the case of red coral (*Corallium rubrum*). J Nat Conserv. 19:312–318

Browder JA (1981) Use of energy flow model to evaluate alternative harvesting strategies in a multispecies fishery. In: Proceedings of international symposium on energy ecology modeling, Louisville

Bruno JF, Bertness MD (2001) Habitat modification and facilitation in benthic marine communities. In: Bertness MD, Gaines SD, Hay ME (eds) Marine community ecology. Sinauer Associates, Sunderland

Bulleri F, Airoldi L (2005) Artificial marine structures facilitate the spread of a nonindigenous green alga, *Codium fragile* ssp. *tomentosoides*, in the north Adriatic Sea. J Appl Ecol 42:1063–1072

Caddy JF (2000) Marine catchment basin effects versus impacts of fisheries on semi-enclosed seas. ICES J Mar Sci 57:628–640

Calbet A, Garrido S, Saiz E, Alcaraz M, Duarte CM (2001) Annual zooplankton succession in coastal NW Mediterranean waters: the importance of the smaller size fractions. J Plankton Res 23:319–331

Canals M, Puig P, Durrieu de Madron X, Heussner S, Palanques A, Fabrés J (2006) Flushing submarine canyons. Nature 444:354–357

Carrasón M, Cartes JE (2002) Trophic relationship in a Mediterranean deep-sea fish community: partition of food resources, dietary overlap and connections within the benthic boundary layer. Mar Ecol Prog Ser 241:41–55

Cartes JE, Maynou F, Morales-Nin B, Massutí E, Moranta J (2001) Trophic structure of a bathyal benthopelagic boundary layer community south of the Balearic Islands (southwestern Mediterranean). Mar Ecol Prog Ser 215:23–35

Cartes JE, Maynou F, Sardà F, Company JB, Lloris D, Tudela S (2004) The Mediterranean deep-sea ecosystems: an overview of their diversity, structure, functioning and anthropogenic impacts. In: The Mediterranean deep-sea ecosystems: an overview of their diversity, structure, functioning and anthropogenic impacts, with a proposal for conservation. IUCN/Málaga WWF, Rome

Cartes JE, Madurell T, Fanelli E, López-Jurado JL (2008) Dynamics of suprabenthos and zooplankton around Mallorca (Balearic Islands, NW Mediterranean): influence of environmental variables and effect on higher trophic levels. J Mar Syst 71:316–335

Caswell H (1989) Life-history strategies. In: Cherrett JM (ed) Ecological concepts. Blackwell Scientific Publications, Oxford

Cebrian E, Uriz MJ, Garrabou J, Ballesteros E (2011) Sponges mass mortalities in a warming Mediterranean Sea: are cyanobacteria-harboring species worst off? PLoS One 6:e20211. doi:10.1371/journal.pone.0020211

Cerrano C, Bavestrello G, Bianchi CN, Cattaneo-Vietti R, Bava S, Morganti C, Morri C, Picco P, Sara G, Sciaparelli S, Siccardi A, Spogna F (2000) A catastrophic mass-mortality episode of gorgonians and other organisms in the Ligurian Sea (northwestern Mediterranean), summer 1999. Ecol Lett 3:284–293

Cerrano C, Bavestrello G, Bianchi CN, Calcinai B, Cattaneo-Vietti R, Morri C, Sarà M (2001) The role of sponge bioerosion in the Mediterranean coralligenous accretion. In: Faranda FM, Guglielmo G, Spezie G (eds) Mediterranean ecosystems: structures and processes. Springer, Milano

Chan EH, Liew HC (1986) Characteristics of an exploited tropical shallow-water demersal fish community in Malaysia. In: Maclean JL, Dizon LB, Hosillos LV (eds) First Asian fish forum. Asian Fish Society, Manila

Chintiroglou C, Antoniadou C, Vafidis D, Kotsoubas D (2005) A review on the biodiversity of hard substrate invertebrate communities in the Aegean Sea. Mediterr Mar Sci 6:51–62

Coll M, Piroddi C, Steenbeek J et al (2010) The biodiversity of the Mediterranean Sea: estimates, patterns, and threats. PLoS One 5:e11842. doi:10.1371/journal.pone.0011842

Collie JS, Hall SJ, Kaiser MJ, Poiner IR (2000) A quantitative analysis of fishing impacts on shelf-sea benthos. J Anim Ecol 69:785–798

Colombini I, Fallaci M, Milanesi F, Scapini F, Chelazzi L (2003) Comparative diversity analysis in sandy littoral ecosystems of the western Mediterranean. Estuar Coast Shelf Sci 58:93–104

Coma R, Ribes M, Gili JM, Zabala M (1998) An energetic approach to the study of life-history traits of two modular colonial benthic invertebrates. Mar Ecol Prog Ser 162:89–103

Coma R, Ribes M, Gili JM, Zabala M (2000) Seasonality in coastal benthic ecosystems. Trends Ecol Evol 15:448–453

Coma R, Pola E, Ribes M, Zabala M (2004) Long-term assessment of temperate octocoral mortality patterns, protected vs. unprotected areas. Ecol Appl 14:1466–1478

Coma R, Ribes M, Serrano E, Jiménez E, Salat S, Pascual J (2009) Global warming-enhanced stratification and mass mortality events in the Mediterranean. Proc Natl Acad Sci U S A 106:6176–6181

Coma R, Serrano E, Linares C et al (2011) Sea urchins predation facilitates coral invasion in a marine reserve. PLoS One 6:e22017. doi:10.1371/journal.pone.0022017

Costello MJ, McCrea M, Freiwald A, Lundälv T, Jonsson L, Bett BJ, van Weering TCE, de Hass H, Roberts JM, Allen D (2005) Role of cold-water *Lophelia pertusa* coral reefs as fish habitat in the NE Atlantic. In: Freiwald A, Roberts JM (eds) Cold-water corals and ecosystems. Springer, Berlin/Heidelberg/New York

Cupido R, Cocito S, Sgorbini S, Bordone A, Santangelo G (2008) Response of a gorgonian (*Paramuricea clavata*) population to mortality events: recovery or loss? Aquat Conserv 18:984–992

Cupido R, Cocito S, Barsanti M, Sgorbini S, Periano A, Santangelo G (2009) Unexpected long-term population dynamics in a canopy-forming gorgonian coral following mass mortality. Mar Ecol Prog Ser 394:195–200

D'Onghia G, Maiorano P, Sion L, Giove A, Capezzuto F, Carlucci R, Tursi A (2010) Effects of deep-water coral banks on the abundance and size structure of the megafauna in the Mediterranean Sea. Deep-Sea Res II 57:397–411

Dando PR, Hughes JA, Thiermann F (1995) Preliminary observations on biological communities at shallow vents in the Aegean Sea. In: Parson LM, Walker CL, Dixon DR (eds) Hydrothermal vents and processes, vol 87, Geological society special publication. Geological Society, London, pp 303–317

Danovaro R, Dinet A, Duineveld G, Tselepides A (1999) Benthic response to particulate fluxes in different trophic environments: a comparison between the Gulf of Lions-Catalan Sea (western Mediterranean) and the Cretan Sea (eastern Mediterranean). Prog Oceanogr 44:287–312

Danovaro R, Company JB, Corinaldes C et al (2010) Deep-sea biodiversity in the Mediterranean: the known, the unknown and the unknowable. PLoS One 5:e11832. doi:10.1371/journal.pone.0011832

De Biasi AM, Bianchi CN, Aliani S, Cocito S, Peirano A, Dando PR, Morri C (2004) Epibenthic communities in a marine shallow area with hydrothermal vents (Milos Island, Aegean Sea). Chem Ecol 20:89–105

De Caralt S, Uriz MJ, Wijffels R (2008) Grazing, differential size-class dynamics and survival of the Mediterranean sponge *Corticium candelabrum*. Mar Ecol Prog Ser 360:97–106

de Juan S, Cartes J (2011) Influence of environmental factors on the dynamics of macrobenthic crustaceans on soft-bottoms of the Ebro Delta continental shelf (northwestern Mediterranean). Sci Mar 75:691–700

de Juan S, Thrush SF, Demestre M (2007) Functional changes as indicators of trawling disturbance on a benthic community located in a fishing ground (NW Mediterranean Sea). Mar Ecol Prog Ser 334:117–129

de Juan S, Demestre M, Sánchez P (2011) Exploring the degree of trawling disturbance by the analysis of benthic communities ranging from a heavily exploited fishing ground to an undisturbed area in the NW Mediterranean. Sci Mar 75:507–516

Della Tommasa L, Belmonte G, Palanques A, Puig P, Boero F (2000) Resting stages in a submarine canyon: a component of shallow-deep-sea coupling? Hydrobiologia 440:249–260

Demestre M, Sanchez P, Kaiser MJ (2000) The behavioural response of benthic scavengers to otter trawling disturbance in the Mediterranean. In: Kaiser MJ, de Groot SJ (eds) The effects of fishing on non-target species and habitats. Blackwell Scientific, Oxford

Desbruyères D, Guille A, Ramos J (1972) Bionomie benthique du plateau continental de la cote catalane espagnole. Vie Milieu 23:335–363

Di Camillo CG, Coppari M, Bartolucci I, Bo M, Betti F, Bertolino M, Calcinai B, Cerrano C, De Grandis G, Bavestrello G (2012) Temporal variations in growth and reproduction of *Tedania anhelans* and *Chondrosia reniformis* in the north Adriatic Sea. Hydrobiologia 687:299–313

Duarte CM, Agustí S, Kennedy H, Vaqué D (1999) The Mediterranean climate as a template for Mediterranean marine ecosystems: the example of the northeast Spanish littoral. Prog Oceanogr 44:245–270

Duperron S, Halary S, Lorion J, Sibuet M, Gaill F (2008) Unexpected co-occurrence of six bacterial symbionts in the gills of the cold seep mussel *Idas* sp. (Bivalvia: Mytilidae). Environ Microbiol 10:433–445

Durrieu de Madron X, Guieu C, Sempéré R et al (2011) Marine ecosystems' responses to climatic and anthropogenic forcings in the Mediterranean by the MerMex group. Prog Oceanogr 91:97–166

Estrada M (1996) Primary production in the northwestern Mediterranean. Sci Mar 60:55–64

Falciai L, Castagnolo L, Focardi S, Gambi MC, Spadini V, Zampi M, Della Seta G, Minervini R (1983) Organismi bentonici di fondi mobili. In: IRSA, L'esperimento Tevere. Influenza di un fiume sull'ecosistema marino prospiciente la sua foce. Quad Ist Ric Acque 66:217–230

Fanelli E, Cartes JE (2004) Feeding habits of Pandalid shrimps in the Alboran Sea (SW Mediterranean): influence of biological and environmental variables. Mar Ecol Prog Ser 280:227–238

Fanelli G, Riraino S, Belmonte G, Geraci S, Boero F (1994) Human predation along Apulian rocky coasts (SE Italy): desertification caused by *Lithophaga lithophaga* (Mollusca) fisheries. Mar Ecol Prog Ser 110:1–8

FAO (2011) General Fisheries Comission for the Mediterranean task 1 statistical bulletin. A synopsis of information on fleet segments and operational units by GFCM geographical sun-areas. Available online at: http://www.gfcm.org/gfcm/topic/17106/en. Accessed 2 Aug 2012

Farrugio H, Oliver P, Biagi F (1993) An overview of he history, knowledge, recent and future research trends in Mediterranean fisheries. In: Lleonart J (ed) Northwestern Mediterranean fisheries, vol 57, Scientia marina. Instituto de Ciencias del Mar, C.S.I.C, Barcelona, pp 105–119

Fava F, Bavestrello G, Valisano L, Cerrano C (2010) Survival, growth and regeneration in explants of four temperate gorgonian species in the Mediterranean Sea. Ital J Zool 77:44–52

Fèbvre-Chevalier C (1969) Étude bionomique des substrats meubles dragables du golfe de Fos. Tethys 1:421–476

Fine M, Zibrowius H, Loya Y (2001) *Oculina patagonica*: a non-lessepsian scleractinian coral invading the Mediterranean Sea. Mar Biol 138:1195–1203

Francour P (1997) Fish assemblages of *Posidonia oceanica* beds at Port-Cros (France, NW Mediterranean): assessment of composition and long-term fluctuations by visual census. Mar Ecol 18:157–173

Francour P, Ganteaume A, Poulain M (1999) Effects of boat anchoring in *Posidonia oceanica* seagrass beds in the Port-Cros National Park (north-western Mediterranean Sea). Aquat Conserv Mar Freshw Ecosyst 9:391–400

Fraschetti S, Terlizzi A, Boero F (2008) How many habitats are there in the sea (and where)? J Exp Mar Biol Ecol 366:109–115

Fredj G, Laubier L (1985) The deep Mediterranean benthos. In: Moraitou-Apostopoulou M, Kiortsis V (eds) Mediterranean marine ecosystems. Plenum Press, New York

Freiwald A, Roberts JM (2005) Cold-water corals and ecosystems. Springer, Berlin/Heidelberg/New York

Fresi E, Gambi MC, Focardi S, Bargagli R, Baldi F, Falciai L (1983) Benthic communities and sediment types: a structural analysis. Mar Ecol 4:101–121

Gaino E, Pronzato R, Corriero G, Paola Buffa P (1992) Mortality of commercial sponges: incidence in two Mediterranean áreas. Ital J Zool 59:79–85

Galil BS (2000) A sea under siege-alien species in the Mediterranean. Biol Invasions 2:177–186

Galil BS (2004) The limit of the sea: the bathyal fauna of the Levantine sea. Sci Mar 68:63–72

Gambi MC, Fresi E (1981) Ecology of soft-bottom macrbenthos along the coast of southern Tuscany (Parco Naturale della Maremma). Rapp Comm Intern Mer Mediterr 27:123–125

Gambi MC, Conti G, Bremec CS (1998) Polychaete distribution, diversity and seasonality related to seagrass cover in shallow soft bottoms of the Tyrrhenian Sea (Italy). Sci Mar 62:1–7

García-Rubies A, Macpherson E (1995) Substrate use and temporal pattern of recruitment in juvenile fishes of the Mediterranean littoral. Mar Biol 124:35–42

Garrabou J (1999) Life history traits of two Mediterranean coelenterates: *Alcyonium acaule* and *Parazoanthus axinellae*, with emphasis on growth. Mar Ecol Prog Ser 178:193–204

Garrabou J, Perez T, Sartoretto S, Harmelin JG (2001) Mass mortality event in red coral *Corallium rubrum* populations in the Provence region (France, NW Mediterranean). Mar Ecol Prog Ser 217:263–272

Garrabou J, Ballesteros E, Zabala M (2002) Structure and dynamics of North-western Mediterranean rocky benthic communities along a depth gradient. Estuar Coast Shelf Sci 55:493–508

Garrabou J, Coma R, Bensoussan N et al (2009) Mass mortality in north western Mediterranean rocky benthic communities: effects of the 2003 heat wave. Glob Change Biol 15:1090–1103

Garthe S, Hüppop O (1998) Foraging success, klepto-parasitism and feeding techniques in scavenging seabirds; does crime pay? Helgoländer Meeresun 52:187–196

General Fisheries Commission of the Mediterranean-GFCM (2001) 26th session of the Scientific Advisory Committee (SAC) of the General Fishery Commission for the Mediterranean (GFCM), Chania

General Fisheries Commission of the Mediterranean-GFCM (2012) 14th session of the Scientific Advisory Committee (SAC) Report of the working group on stock assessment of demersal species, Chania

Gili JM, Ballesteros E (1991) Structure of cnidarian populations in Mediterranean sublittoral benthic communities as a result of adaptation to different environmental conditions. Oecol Aquat 10:243–254

Gili JM, Coma R (1998) Benthic suspension feeders: their paramount role in littoral marine food webs. Trends Ecol Evol 13:316–321

Gili JM, Ros JD (1985a) Estudio cuantitativo de tres poblaciones circalitorales de cnidarios bentónicos. Investig Pesq 49:323–352

Gili JM, Ros JD (1985b) Study and cartography of the benthic communities of Medes Islands (NE Sapin). Mar Ecol 6:219–238

Gili JM, Riera T, Zabala M (1986) Physical and biological gradients in a submarine cave on the western Mediterranean coast (north-east Spain). Mar Biol 90:291–297

Gili JM, Ros JD, Pagès F (1987) Types of bottoms and benthic Cnidaria from the trawling grounds (littoral and bathyal) off Catalonia (NE Spain). Vie Milieu 37:85–98

Gili JM, Murillo J, Ros JD (1989) The distribution pattern of benthic Cnidarians in the western Mediterranean. Sci Mar 53:19–35

Gili JM, Bouillon J, Pagès F, Palanques A, Puig P, Heussner S (1998) Origin and biogeography of the deep-water Mediterranean Hydromedusae including the description of two new species collected in submarine canyons of northwestern Mediterranean. Sci Mar 62:113–134

Gili JM, Pagès F, Bouillon J, Palanques A, Puig P, Heussner S, Calafat A, Canals M, Monaco A (2000) A multidisciplinary approach to the understanding of hydromedusan populations inhabiting Mediterranean submarine canyons. Deep-Sea Res I 47:1513–1533

Giuliani S, Virno Lamberti C, Sonni C, Pellegrini D (2005) Mucilage impact on gorgonians in the Tyrrhenian Sea. Sci Total Environ 353:340–349

Gori A, Linares C, Rossi S, Coma R, Gili JM (2007) Spatial variability in reproductive cycle of the gorgonians *Paramuricea clavata* and *Eunicella singularis* (Anthozoa, Octocorallia) in the western Mediterranean Sea. Mar Biol 151:1571–1584

Gori A, Viladrich N, Gili JM, Kotta M, Cucio C, Magni L, Rossi S (2012) Reproductive cycle and trophic ecology in deep versus shallow populations of the Mediterranean gorgonian *Eunicella singularis*. Coral Reefs 31:823–837

Grassle JF (1989) Species diversity in deep-sea communities. Trends Ecol Evol 4:12–15

Gravina MF (1986) Analisi della distribuzione dei Policheti nei fondi mobili costieri delle Cinque Terre (Liguria). Boll Mus Inst Biol Univ Genova 52:197–212

Grémare A, Amouroux JM, Charles F, Dinet A, Riaux-Gobin C, Baudart J, Medernach L, Bodiou JY, Vétion G, Colomines JC, Albert P (1997) Temporal changes in the biochemical composition and nutritional value of the particulate organic matter available to surface deposit-feeders: a two year study. Mar Ecol Prog Ser 150:195–206

Grémare A, Amouroux JM, Cauwet G et al (2003) The effects of a strong winter storm on physical and biological variables at a shelf site in the Mediterranean. Oceanol Acta 26:407–419

Guardiola M, Frotscher J, Uriz MJ (2012) Genetic structure and differentiation at a short-time scale of the introduced calcareous sponge *Paraleucilla magna* to the western Mediterranean. Hydrobiologia. doi:10.1007/s10750-011-0948-1

Guidetti P, Frachetti S, Terlizzi A, Boero F (2003) Distribution patterns of sea urchins and barrens in shallow Mediterranean rocky reefs impacted by the illegal fishery of the rock-boring mollusc *Lithophaga lithophaga*. Mar Biol 143:1135–1142

Guille A (1965) Exploration en soucoupe plongeante Cousteau de l'entrée nord-est de la baie de Rosas (Espagne). Bull Inst Oceanogr 65:1–12

Guille A (1971) Bionomie benthique du plateau continental de la côte catalane francaise II. Les communautés de la macrofaune. Vie Milieu 21:149–280

Hall-Spencer JM, Rodolfo-Metalpa R, Martin S et al (2008) Volcanic carbon dioxide vents show ecosystem effects of ocean acidification. Nature 454:96–99

Harmelin JG, Marinopoulos J (1994) Population structure and partial mortality of the gorgonian *Paramuricea clavata* (Risso) in the North-western Mediterranean (France, Port Cros Island). Mar Life 4:5–13

Hiscock K, Mitchell R (1980) The description and classification of sublittoral epibenthic ecosystems. In: Price JH, Irvine DEG, Farnham WF (eds) The shore environment, 2 ecosystems. Academic, London

Innamorati M, Raddi E, Buzzichelli S, Melley A, Balzi M (1993) Le mucillaggini nel Mar Tirreno. Biol Mar 1:23–26

Jackson JBC (1977) Competition on marine hard substrata: the adaptative significance of solitary and colonial strategies. Am Nat 3:743–767

Jordana E, Grèmare A, Lantoine F, Courties C, Charles F, Amoroux JM, Vètion G (2001) Seasonal changes in the grazing of coastal picoplankton by the suspension-feeding polychaete *Ditrupa arietina* (O.F. Muüller). J Sea Res 46:245–259

Karakassis J, Eleftheriou A (1997) The continental shelf of Crete: structure of macrobenthic communities. Mar Ecol Prog Ser 160:185–196

Kroeker KJ, Micheli F, Gambi MC, Martz TR (2011) Divergent ecosystem responses within a benthic marine community to ocean acidification. Proc Nat Acad Sci 108:14515–14520

Kröncke I, Turkay M, Fiege D (2003) Macrofauna communities in the eastern Mediterranean Sea. Mar Ecol 24:193–216

Laborel J (1961) Contribution à l'étude des peuplements benthiques sciaphiles sur substrat rocheux en Méditerranée. Rec Trav Sta Mar Endoume 33:117–173

Labrune C, Grémare A, Amouroux JM, Sardá R, Gil J, Taboada S (2007) Assessment of soft-bottom polychaete assemblages in the Gulf of Lions (NW Mediterranean) based on a mesoscale survey. Estuar Coast Shelf Sci 71:133–147

Langton RW, Auster PJ (1999) Marine fishery and habitat interactions: to what extent are fisheries and habitat interdependent? Bull Am Fish Soc 24:14–21

Laubier L (1966) Le coralligène des Albères. Monographie bióceno-tique. Ann Inst Oceanogr Paris 43:139–316

Laubier L, Emig CC (1993) La faune benthique profonde de Méditerrannée. In: Della Croce FR (ed) Symposium Mediterranean Seas 2000. Istituto de Scienze ambientale marina, S. Margherita Ligure, pp 397–428

Ledoux JB, Mokhtar-Jamai K, Roby C, Feral JP, Garrabou J, Aurelle D (2010) Genetic survey of shallow populations of the Mediterranean red coral (*Corallium rubrum* (Linnaeus, 1758)): new insights into evolutionary processes shaping current nuclear diversity and implications for conservation. Mol Ecol 19:675–690

Leth JO Kuijpers A (1996) Effects on the seabed sediment from beam trawling in the North Sea. In: ICES annual science conference, Reykjavík

Linares C, Coma R, Diaz D, Zabala M, Hereu B, Dantart L (2005) Immediate and delayed effects of a mass mortality event on gorgonian population dynamics and benthic community structure in the NW Mediterranean Sea. Mar Ecol Prog Ser 305:127–137

Linares C, Coma R, Zabala M (2008) Effects of a mass mortality event on gorgonian reproduction. Coral Reefs 27:27–34

Linares C, Garrabou J, Hereu B, Diaz D, Marschal C, Sala E, Zabala M (2012) Assessing the effectiveness of marine reserves on unsustainably harvested long-lived sessile invertebrates. Conserv Biol 26:88–96

Lloret J, Zaragoza N, Caballero D, Riera V (2008) Impacts of recreational boating on the marine environment of Cap de Creus (Mediterranean Sea). Ocean Coast Manag 51:749–754

Lombardi C, Gambi MC, Vasapollo C, Taylor P, Cocito S (2011) Skeletal alterations and polymorphism in a Mediterranean bryozoan at natural CO2 vents. Zoomorphology 130:135–145

Longhurst A (1998) Ecological geography of the sea. Academic, San Diego

Macpherson E, Raventos N (2005) Settlement patterns and psotsettlement survival in two Mediterranean littoral fishes: influences of early-life traits and environmental variables. Mar Biol 148:167–177

Maldonado M, Sánchez-Tocino L, Navarro C (2010) Recurrent disease outbreaks in corneous demosponges of the genus *Ircinia*: epidemic incidence and defense mechanisms. Mar Biol 157:1577–1590

Mann KH, Lazier JRN (2006) Dynamics of marine ecosystems. Biological-physical interactions in the oceans, 3rd edn. Blackwell Publishing, Malden

Marcus N, Boero F (1998) Production and plankton community dynamics in coastal aquatic systems: the importance of benthic-pelagic coupling and the forgotten role of life cycles. Limnol Oceanogr 43:763–768

Margalef R (1974) Ecologia. Omega, Barcelona

Margalef R (1985) Introduction to the Mediterranean. In: Margalef R (ed) Key environments. Western Mediterranean. Pergamon Press, Oxford

Martí R, Uriz MJ, Ballesteros E, Turon X (2004) Benthic assemblages along two Mediterranean caves: species diversity and coverage as a function of abiotic parameters and geographic distance. J Mar Biol Assoc UK 84:557–572

Mendoza ET, Jimenez JA, Mateo J (2011) A coastal storms intensity scale for the Catalan Sea (NW Mediterranean). Nat Hazards Earth Syst Sci 11:2453–2462

Milazzo M, Badalamenti F, Ceccherelli G, Chemello R (2004) Boat anchoring on *Posidonia oceanica* beds in a marine protected area (Italy, western Mediterranean): effect of anchor types in different anchoring stages. J Exp Mar Biol Ecol 299:51–62

Mistri M, Ceccherelli VU (1996a) Damage and partial mortality in the gorgonian *Paramuricea clavata* in the strait of Messina (Tyrrhenian Sea). Mar Life 5:43–49

Mistri M, Ceccherelli VU (1996b) Effects of a mucilage event on the Mediterranean gorgonian *Paramuricea clavata*. I-Short term impacts at the population and colony levels. Ital J Zool 63:221–230

Mokhtar-Jamaï K, Pascual M, Ledoux JB, Coma R, Feral JP, Garrabou J, Aurelle D (2011) From global to local genetic structuring in the red gorgonian *Paramuricea clavata*: the interplay between oceanographic conditions and limited larval dispersal. Mol Ecol 20:3291–3305

Morri C, Castelli A, Diviacco G, Mori M, Bianchi CN (1990) Zonazione di comunità bentiche lungo l'estuario della magra (mar ligure orientale). Atti Soc Tosc Sci Nat Mem Ser B 97:311–327

Murray SN, Littler MM (1984) Analysis of seaweed communities in a disturbed rocky intertidal environment near Whites Point, Los Angeles, California, USA. Hydrobiologia 116/117:374–382

Navarro L, Ballesteros E, Linares C, Hereu B (2011) Spatial and temporal variability of deep-water algal assemblages in the northwestern Mediterranean: the effects of an exceptional storm. Estuar Coast Shelf Sci 95:52–58

Occhipinti-Ambrogi A, Savini D (2003) Biological invasions as a component of global change in stressed marine ecosystems. Mar Pollut Bull 46:542–551

Olu-Le Roy K, Sibuet M, Fiala-Medioni A, Gofas S, Salas C, Mariotti A, Foucher J-P, Woodside J (2004) Cold seep communities in the deep eastern Mediterranean Sea: composition, symbiosis and spatial distribution on mud volcanoes. Deep Sea Res I 51:1915–1936

Orejas C, Gori A, Lo Iacomo C, Puig P, Gili JM, Dale MRT (2009) Cold-water corals in the Cap de Creus canyon (north-western Mediterranean): spatial distribution, density and anthropogenic impact. Mar Ecol Prog Ser 397:37–51

Pais A, Chessa LA, Serra S, Ruiu A, Meloni G, Donno Y (2007) The impact of commercial and recreational harvesting for Paracentrotus lividus on shallow rocky reef sea urchin communities in Northwestern Sardinia, Italy. Estuar Coast Shelf Sci 73:589–597

Pati AC, Belmonte G, Ceccherelli UV, Boero F (1999) The inactive temporary component: an unexplored fraction of meiobenthos. Mar Biol 134:419–427

Pederson EJ, Peterson MS (2002) Bryozoans as ephemeral estuarine habitat and a larval transport mechanism for mobile benthos and young fishes in the north-central Gulf of Mexico. Mar Biol 140:935–947

Peña V, Bárbara I (2008) Maërl community in the north-western Iberian Peninsula: a review of floristic studies and long-term changes. Aquat Conserv Mar Freshw Ecosyst 18:339–366

Pérès JM (1967) The Mediterranean benthos. Oceanogr Mar Biol Annu Rev 5:449–533

Pérès JM (1982) Zonation. In: Kinne O (ed) Marine ecology. Wiley, Chichester

Pérès JM (1985) History of the Mediterranean biota and the colonization of the depths. In: Margalef R (ed) Key environments. Western Mediterranean. Pergamon Press, Oxford

Pérès JM, Picard J (1964) Nouveau Manuel de Bionomie Benthique de la Mer Méditerranée. Rec Trav St mar Endoume 31:5–137

Perez T, Garrabou J, Sartoretto S, Harmelin JG, Francour P, Vacelet J (2000) Mortalité massive d'invertébrés marins: un événement sans précédent en Méditerranée nord-occidentale. C R Acad Sci III 323:853–865

Pianka ER (1970) On *r*- and *K*- selection. Am Nat 104:592–597

Picard J (1965) Recherches qualitatives sur les biocenoses marines des substrats dragables de la région marseilaise. Rec Trav St Mar Endoume 35:3–160

Planes S, Galzin R, Garcia-Runies A, Goni R, Harmelin JG, Le Diréach L, Lenfant P, Quetglas A (2000) Effects of marine protected areas on recruitment processes with special reference to Mediterranean littoral ecosystems. Environ Conserv 27:126–143

Pronzato R, Manconi R (2008) Mediterranean commercial sponges: over 5000 years of natural history and cultural heritage. Mar Ecol 29:146–166

Pronzato R, Cerrano C, Cubeddu T, Magnino G, Manconi R, Pastore S, Sarà A, Sidri M (2000) The millennial history of commercial sponges: from harvesting to farming and integrated aquaculture. Biol Mar Mediterr 7:132–143

Puig P, Palanques A, Guillén J (2001) Near-bottom suspended sediment variability caused by storms and near-inertial internal waves on the Ebro mid continental shelf (NW Mediterranean). Mar Geol 178:81–93

Pusceddu A, Grémare A, Escoubeyrou K, Amouroux JM, Fiordelmondo C, Danovaro R (2005) Impact of natural (storm) and anthropogenic (trawling) sediment resuspension on particulate organic matter in coastal environments. Cont Shelf Res 25:2506–2520

Pusceddu A, Fonda Umani S, Danovaro R (2009) Marine mucilage: a spreading oceans' influenza? In: Proceedings of the 2009 ASLO meeting, Nice

RAC/SPA (2006). Classification of benthic marine habitat types for the Mediterranean region. Available online at http://www.rac-spa.org/sites/default/files/doc_fsd/lchm_en.pdf. Accessed 2 Aug 2012

Ragonese S, Rizzo P, Giusto GB (1994) Domestic wastes and the *Nephrops norvegicus* fishery in the Strait of Sicily. Biol Mar Mediterr 1:309–310

Ramirez-Llodra E, Ballesteros M, Company JB, Dantart L, Sardà S (2008) Spatio-temporal variations of biomass and abundance in bathyal non-crustacean megafauna in the Catalan Sea (northwestern Mediterranean). Mar Biol 153:297–309

Ribera d'Alcalá M, Conversano F, Corato F, Licandro P, Mangoni O, Marino D, Mazzocchi MG, Modigh M, Montresor M, Nardella M, Saggiorno V, Sarno D, Zingone A (2004) Seasonal patterns in plankton communities in a pluriannual time series at a coastal Mediterranean site (Gulf of Naples): an attempt to discern recurrences and trends. Sci Mar 68:65–83

Ribes M, Coma R, Gili JM (1999) Seasonal variations of particulate organic carbon, dissolved organic carbon and the contribution of the microbial communities to the live particulate organic carbon in a shallow near-bottom ecosystem at the Northwestern Mediterranean Sea. J Plankton Res 21:1077–1100

Riedl R (1966) Biologie der Meereshöhlen. Paul Parey, Hamburg

Riedl R (1971) Water movement. Animals. In: Kinne O (ed) Marine ecology. Wiley, Chichester

Rield R (1980) Marine ecology – a century of changes. Mar Ecol 1:3–46

Ritt B, Sarrazin J, Caprais JC, Noël P, Gauthier O, Pierre C, Henry P, Desbruyères D (2010) First insights into the structure and environmental setting of cold-seep communities in the Marmara Sea. Deep Sea Res I 57:1120–1136

Rius M, Turon X, Marshall DJ (2009a) Non-lethal effects of an invasive species in the marine environment: the importance of early life-history stages. Oecologia 159:873–882

Rius M, Pineda MC, Turon X (2009b) Population dynamics and life cycle of the introduced ascidian *Microcosmus squamiger* in the Mediterranean Sea. Biol Invasions 11:2181–2194

Rodolfo-Metalpa R, Martin S, Ferrier-Pagès C, Gattuso JP (2010) Response of the temperate coral *Cladocora caespitosa* to mid- and long-term exposure to pCO2 and temperature levels projected for the year 2100 AD. Biogeosciences 7:289–300

Ros JD, Olivella I, Gili JM (1984) Els Sistemes Naturals de les illes Medes. Arxius de la Secció de Ciències, 73 Institut d'Estudis Catalans, Barcelona

Ros JD, Romero J, Ballesteros E, Gili JM (1985) Diving in blue water. The benthos. In: Margalef R (ed) Key environments. Western Mediterranean. Pergamon Press, Oxford

Rosell D, Uriz MJ (2002) Excavating and endolithic sponge species (Porifera) from the Mediterranean: species descriptions and identification key. Org Divers Evol 1:1–31

Rosell D, Uriz MJ, Martin D (1999) Infestation by excavating sponges on the oyster (*Ostrea edulis*) populations of the Blanes littoral zone (north western Mediterranean). J Mar Biol Assoc UK 79:409–413

Rossi S, Gili JM (2005) Temporal variation and composition of near-bottom seston features in a Mediterranean coastal area. Estuar Coast Shelf Sci 65:385–395

Rossi S, Gili JM (2009) Near bottom phytoplankton and seston: importance in the pelagic-benthic coupling processes. In: Kersey WT, Munger SP (eds) Marine phytoplankton. Nova Science Publishers Inc, New York

Rossi S, Tsounis G (2007) Temporal and spatial variation in protein, carbohydrate, and lipid levels in *Corallium rubrum* (anthozoa, octocorallia). Mar Biol 152:429–439

Rossi S, Grémare A, Gili JM, Amouroux JM, Jordana E, Vétion G (2003) Biochemical characteristics of settling particulate organic matter at two north-western Mediterranean sites: a seasonal comparison. Estuar Coast Shelf Sci 58:423–434

Rossi S, Gili JM, Coma R, Linares C, Gori A, Vert N (2006a) Seasonal cycles of protein, carbohydrate and lipid concentrations in *Paramuricea clavata*: (Anthozoa, Octocorallia): evidences for summer-autumn feeding constraints. Mar Biol 149:643–651

Rossi S, Snyder MJ, Gili JM (2006b) Protein-carbohydrate-lipid levels and HSP70-HSP90 (stress protein) expression over an annual cycle of a benthic suspension feeder: useful tools to detect feeding constraints in a benthic suspension feeder. Helgol Mar Res 60:7–17

Sabatés A, Gili JM, Pagès F (1989) Relationship between zooplankton distribution, geographic characteristics and hydrographic patterns off the Catalan coast (western Mediterranean). Mar Biol 103:153–159

Sala E (2004) The past and present topology and structure of Mediterranean subtidal rocky-shore food webs. Ecosystems 7:333–340

Sala E, Boudouresque CF (1997) The role of fishes in the organization of a Mediterranean sublittoral community. 1. Algal communities. J Exp Mar Biol Ecol 212:25–44

Sala E, Garrabou J, Zabala M (1996) Effects of diver frequentation on Mediterranean sublittoral populations of the bryozoan *Pentapora fascialis*. Mar Biol 126:451–459

Sala E, Boudouresque CF, Harmelin-Vivien M (1998a) Fishing, trophic cascades, and the structure of algal assemblages: evaluation of an old but untested paradigm. Oikos 82:425–439

Sala E, Ribes M, Hereu B, Zabala M, Alvà V, Coma R, Garrabou J (1998b) Temporal variability in abundance of the sea urchins *Paracentrotus lividus* and *Arbacia lixula* in the northwestern Mediterranean: comparison between a marine reserve and an unprotected area. Mar Ecol Prog Ser 168:135–145

Salas C, Woodside J (2002) *Lucinoma kazani* n. sp. (Mollusca: Bivalvia): evidence of a living benthic community associated with a cold seep in the eastern Mediterranean Sea. Deep Sea Res I 49:991–1005
Salen-Picard C (1981) Evolution d'un peuplement de vase terrigène côtière soumis à des rejets de dragages, dans le golfe de Fos. Téthys 10:83–88
Salen-Picard C (1982) Nouvelles données sur la biocenose des fonds detritiques envases. In: 28 Congres Assemblée Pleniere CIESM, Cannes
Salvati E, Angiolillo M, Bo M, Bavestrello G, Giusti M, Cardinali A, Puce S, Spaggiari C, Greco S, Canese S (2010) The population of *Errina aspera* (Hydrozoa: Stylasteridae) of the Messina Strait (Mediterranean Sea). J Mar Biol Assoc UK 90:1331–1336
San Vicente C, Sorbe JC (2003) Biology of the suprabenthic mysid *Schistomysis assimilis* (Sars, 1877) on Creixell beach, Tarragona (nortwestern Mediterranean). Bol Inst Esp Oceanogr 19:391–406
Santangelo G, Bramanti L, Iannelli M (2007) Population dynamics and conservation biology of the over-exploited Mediterranean red coral. J Theor Biol 244:416–423
Sardá R, Pinedo S, Martin D (1999) Seasonal dynamics of macroinfaunal key species inhabiting shallow soft-bottom in the bay of Blanes (NW Mediterranean). Acta Oecolog 20:315–326
Sardà F, D'Onghia G, Politou C-Y, Tselepides A (2004) Mediterranean deep-sea biology. Monogr Sci Mar 68(Suppl 3):1–204
Sardá R, Diedrich J, Tintoré J, Cormier R (2010) The ecosystem-based management system: a formal tool for the management of the marine environement. Littoral 14001. doi:10.1051/litt/201114001
Sardá R, Gil J, Taboada S, Gili JM (2009) Polychaete species captured in sediment traps moored in northwestern Mediterranean submarine canyons. Zool J Linn Soc 155:1–21
Sardá R, Rossi S, Martí X, Gili JM (2012) Marine benthic cartography of the Cap de Creus (NE Catalan Coast, Mediterranean Sea). Sci Mar 76:159–171
Schiaparelli S, Castellano M, Povero P, Sartoni G, Cattaneo-Vietti R (2003) A benthic mucilage event in North-western Mediterranean Sea and its possible relationships with the summer 2003 European heatwave: short term effects on littoral rocky assemblages. Mar Ecol 28:341–353
Schwinghamer P, Guigne JY, Siu WC (1996) Quantifying the impact of trawling on benthic habitat structure using high resolution acoustics and chaos theory. Can J Fish Aquat Sci 53:288–296
Simboura N, Zenetos A (2002) Benthic indicators to use in ecological quality classification of Mediterranean soft bottom marine ecosystems, including a new biotic index. Mediterr Mar Sci 3–2:77–111
Simboura N, Nicolaidou A, Thessalou-Legaki M (2000) Polychaete communities of Greece: an ecological overview. Mar Ecol 21:129–144
Simboura N, Panayotidis P, Papathanassiou E (2005) A synthesis of the biological quality elements for the implementation of the European Water Framework Directive in the Mediterranean ecoregion: the case of Saronikos Gulf. Ecol Indic 5:253–266
Somaschini A, Martini N, Gravina MF, Belluscio A, Corsi F, Ardizzioni GD (1998) Characterization and cartography of some Mediterranean soft-bottom benthic communities (Ligurian Sea, Italy). Sci Mar 62:27–36
Sulak KJ, Brooks RA, Randall MT (2007) Seasonal refugia and trophic dormancy in Gulf sturgeon: test and refutation of the thermal barrier hypothesis. Am Fish Soc Symp 56:19–49
Thomsen L, van Weering T (1998) Spatial and temporal variability of particulate matter in the benthic boundary layer at the NW European Continental Margin (Goban Spur). Prog Oceanogr 42:61–76
Tokaç A, Tosunoglu Z (1996) Selection parameters for red mullet (*Mullus barbatus* L., 1758.). Istanb J Aquat Prod Spec Issue 1999: 517–534
True MA (1970) Étude quantitative de quatre peuplements sciaphiles sur substrat rocheux dans la région marseillaise. Bull Inst Oceanogr Monaco 69:1–48
Tselepides A, Papadopoulou N, Podaras D, Plaiti W, Koutsoubas D (2000) Macrobenthic community structure over the continental margin of Crete (South Aegean Sea, NE Mediterranean). Prog Oceanogr 46:401–428
Tsounis G, Rossi S, Gili JM, Arntz WE (2007) Red coral fishery in the Costa Brava (NW Mediterranean): case study of the overharvested precious coral. Ecosystems 10:975–986
Tsounis G, Rossi S, Grigg R, Santangelo G, Bramanti L, Gili JM (2010) The exploitation and conservation of precious corals. Oceanogr Mar Biol Annu Rev 48:161–212
Tsounis G, Martínez L, Bramanti L, Viladrich N, Martínez A, Gili JM, Rossi S (2012) Effects of human impact on the reproductive effort and allocation of energy reserves in the Mediterranean octocoral *Paramuricea clavata*. Mar Ecol Prog Ser 449:161–172
Tuck ID, Hall SJ, Robertson MR, Armstrong E, Basford DJ (1998) Effects of physical trawling disturbance in a previously unfished sheltered Scottish sea loch. Mar Ecol Prog Ser 162:227–242
Turon X, Becerro MA (1992) Growth and survival of several ascidian species from Northwestern Mediterranean. Mar Ecol Prog Ser 82:235–247
Uriz MJ, Rosell D (1990) Sponges from bathyal depths (1000–1700 m) in the western Mediterranean Sea. J Nat Hist 24:374–391
Uriz MJ, Rosell D, Martin D (1992) Sponge population of the Cabrera archipelago (Balearic Islands). Characteristics, distribution, and abundance of the most representative species. Mar Ecol 13: 101–117
Uriz MJ, Zabala M, Ballesteros E, Garcia-Rubiés A, Turon X (1993) El Benthos: les coves. In: Alcover JA, Ballesteros E, Fornós JJ (eds) Història Natural de l'Arxipèlag de Cabrera. CSIC-Edit Moll, Mallorca
Uriz MJ, Turon X, Becerro MA, Galera J, Lozano J (1995) Patterns of resource allocation to somatic, defensive, and reproductive functions in the Mediterranean encrusting sponge *Crambe crambe* (Demospongiae, Pecilosclerida). Mar Ecol Prog Ser 124:159–170
Valiela I (1995) Marine ecological processes. Springer, New York
Velarque M (1987) Contribution a l'etude du phytobenthos d'un ecosystem photophile thermophile en Mediterranee occidentale. These, Univiversitée d'Aix-Marseille II
Velarque M (1990) Relations entre *Sarpa salpa* (Linnaeus, 1758) (Téléodtéen, Sparidae), les autres poisons brouters et le phytobenthos algal méditerranée. Oceanol Acta 13:373–388
Vezzulli L, Previati M, Pruzzo C, Marchese A, Bourne DG, Cerrano C, the VivrioSEa Consortium (2010) Vibrio infections triggering mass mortality events in a warming Mediterranean Sea. Environ Microbiol 12:2007–2019
Wainright SC (1990) Sediment-to-water fluxes of particulate material and microbes by resuspension and their contribution to the planktonic food web. Mar Ecol Prog Ser 62:271–281
Werne JP, Haese RR, Zitter T et al (2004) Life at cold seeps: a synthesis of biogeochemical and ecological data from Kazan mud volcano, eastern Mediterranean Sea. Chem Geol 205(1093):367–390
Zabala M, Ballesteros E (1989) Estrategias dependientes de la superficie y flujo de energia en comunidades bentónicas marinas o, porque no hay corales en el Mediterráneo. Sci Mar 53:3–17
Zavodnik D, Vidakovic J, Amoureu L (1985) Contribution to the sediment macrofauna in the area of Rovinj (north Adriatic Sea). Cah Biol Mar 26:431–444
Zenetos A, Gofas S, Verlaque M et al (2010) Alien species in the Mediterranean Sea by 2010. A contribution to the application of European Union's Marine Strategy Framework Directive (MSFD). Part I. Spatial distribution. Mediterr Mar Sci 11:381–493
Zibrowius H, Taviani M (2005) Remarkable sessile fauna associated with deep coral and other calcareous substrates in the Strait of Sicily, Mediterranean Sea. In: Freiwald A, Roberts JM (eds) Cold-water corals and ecosystems. Springer, Berlin/Heidelberg/New York

Foraminifera

13

Anna Sabbatini, Caterina Morigi, Maria Pia Nardelli, and Alessandra Negri

Abstract

Foraminifera are eukaryotic unicellular microorganisms inhabiting all marine environments.

The study of these protists has huge potential implications and benefits. They are good indicators of global change and are also promising indicators of the environmental health of marine ecosystems. Nevertheless, much remains to be learnt about foraminiferal ecology.

In this chapter we intend to introduce the main issues in the study of foraminifera in the Mediterranean Sea and the state-of-the-art developments in the study of these organisms. The aims of this chapter are: (**1**) to provide a brief history of the study of foraminifera and (**2**) to review recent developments in the study of modern foraminifera, particularly as they apply to Mediterranean faunas. Our intention is to describe the development of the use of foraminiferal assemblages in Mediterranean applied ecological studies up to their possible use as bio-indicator for the monitoring of marine ecosystems.

Keywords

Benthic foraminifera • Hard and soft-shelled taxa • Methodology • Ecology • Bioindicator • Mediterranean Sea

Introduction

Foraminifera were first described and illustrated in the eighteenth and nineteenth centuries (von Linnaeus 1758; Fichtel and Moll 1798; De Montfort 1808), but it was the French naturalist Alcide d'Orbigny who established a firm foundation for their study. In his 1826 work "Tableau Méthodique de la Classe des Céphalopodes", d'Orbigny made the same mistake as earlier authors in describing these microscopic shells as those of minute cephalopods. He believed that the granuloreticulopods of living specimens were tiny tentacles, and also, like von Linnaeus (1758), recognized that many of the planispiral shells resembled *Nautilus* (Lipps et al. 2011). However, d'Orbigny's major contribution was to establish these organisms as a distinct order ("Foraminifères") because the chambers had apertures (*foramina*) that ensured the communication between different parts of the test instead of the siphons found in typical

A. Sabbatini (✉) • A. Negri
Dipartimento di Scienze della Vita e dell'Ambiente, Università Politecnica delle Marche, Via Brecce Bianche, 60131 Ancona, Italy
e-mail: a.sabbatini@univpm.it; a.negri@univpm.it

C. Morigi
Dipartimento di Scienze della Terra, Università di Pisa, Via S. Maria 53, 56126 Pisa, Italy
e-mail: morigi@dst.unipi.it

M.P. Nardelli
Dipartimento di Scienze della Vita e dell'Ambiente, Università Politecnica delle Marche, Via Brecce Bianche, 60131 Ancona, Italy

Laboratoire des Bio-Indicateurs Actuels et Fossiles (BIAF), CNRS, UMR6112, LPGN, LUNAM Université, Université d'Angers, 2 Boulevard Lavoisier, 49045 Angers CEDEX 01, France
e-mail: pia.nardelli@gmail.com; mariapia.nardelli@univ-angers.fr

S. Goffredo and Z. Dubinsky (eds.), *The Mediterranean Sea: Its history and present challenges*, DOI 10.1007/978-94-007-6704-1_13,

cephalopods. D'Orbigny's (1826) ascribed 544 species to his Order Foraminifères; 335 were new species names that were published without descriptions. Since then, these minute organisms have become interesting objects of study by both geologists and biologists (Cifelli 1990).

For many years, fossil foraminiferal faunas have been used to date sediments and to reconstruct paleo-environments (Haq and Boersma 1998). The resulting knowledge of climatic and oceanographic changes in the geological past has been used to predict and model future environmental change (Sen Gupta 2002). However, in recent years there has been renewed interest among the scientific community in foraminifera as indicators of modern global change, i.e. global warming, rising sea levels and loss of biodiversity (Hillaire-Marcel and de Vernal 2007; Ingels et al. 2012) and as bioindicators of environmental health (Hallock et al. 2003; Gooday et al. 2009; Frontalini and Coccioni 2011). This last applied use of foraminifera involves the study of the existing faunas and their actual ecology. The application of biochemical and molecular techniques, normally applied to other organisms, are making scientists aware that foraminifera can be used in ecological studies and/or protocols for biomonitoring programs, in the same way as the larger metazoan meiofaunal and macrofaunal assemblages, and with similar or even better results.

The aims of this chapter are: (1) to provide a brief history of the study of foraminifera and (2) to review recent developments in the study of modern foraminifera, particularly as they apply to Mediterranean faunas. Our intention is to describe the development of the use of foraminiferal assemblages in Mediterranean applied ecological studies up to their possible use as bio-indicator for the monitoring of marine ecosystems. We highlight (1) the heterogeneous nature of the existing body of data that cannot be easily integrated to produce an overall synthesis of foraminiferal community parameters and (2) the consequent need for a standardised methodology. The latter will lead to an improved understanding of the role of foraminifera in the functioning of Mediterranean Sea ecosystems and aspects of their biology that are beyond the scope of environmental studies.

What Are Foraminifera?

Foraminifera are single-celled eukaryotic organisms (protists) with tests (shells) that are present as fossils in the sediments of the last 545 million years, as well as in modern oceans. Tests may be made of organic material, sand grains or other particles cemented together ('agglutinated'), or crystalline $CaCO_3$ (calcite or aragonite). The hard tests are commonly divided into chambers that are added during growth, although the simplest forms are open tubes or hollow spheres.

Fully grown individuals range in length size from about 100 µm or less to almost 20 cm. Some have a symbiotic relationship with algae. They consume food ranging from dissolved organic molecules, bacteria, diatoms and other single-celled algae, to small animals such as copepods. They catch their food with a highly mobile network of thin pseudopodia (called reticulopodia) that extend from one or more apertures in the shell. Foraminifera also use their pseudopodia for multiple other functions including locomotion, respiration and test building.

Fig. 13.1 Benthic and planktonic foraminiferal assemblages of deep sediment from tropical Atlantic Ocean (From Morigi 1999) (*Scale bar*=700 µm)

There are an estimated 4,000 benthic species of foraminifera living in the world's oceans today (Murray 2007) on and in the sediment, on rocks and on macroalgae at the sea bottom, while only 40 species are planktonic (Hemleben et al. 1989). Foraminifera are found in all marine environments, from the intertidal to the deepest ocean trenches, and from the tropics to the poles, from brackish to hyper-saline waters. Recent studies suggest that they are present in freshwater and even in terrestrial habitats (Meisterfeld et al. 2001; Holzmann and Pawlowski 2002; Holzmann et al. 2003). Foraminifera are among the most abundant shelled organisms in many marine environments (Hayward et al. 2011). A cubic centimetre of sediment may yield hundreds of living individuals, and many more dead shells. In some habitats their shells are an important component of the sediment. For example, the pink sands of some beaches get much of their colour from the pink to red-colour shells of a particular species of foraminifer. In regions of the deep ocean far from land, the bottom sediment is often made up almost entirely of the shells of foraminifera (Fig. 13.1).

How Foraminifera Can Be Used in Applied Science

The study of fossil foraminifera has many practical applications beyond expanding our knowledge of the diversity of life. In particular they are useful in the field of stratigraphy, paleoenvironmental reconstruction, and oil exploration. The earliest foraminifera occurred in the Precambrian-Cambrian boundary, 545 million years ago (McIlroy et al. 1994) and they show a continuous evolutionary development up to present time, so that different species are found at different times. This, together with the fact that they are abundant in all marine environments, and easy to collect, even from deep oil wells, explains why they have been extensively used for dating marine sedimentary rocks. Foraminifera are sensitive to environmental conditions and often associated with particular environmental settings (Haq and Boersma 1998; Sabbatini et al. 2002; Panieri 2005, 2006a, b; Gooday et al. 2010). This allows paleontologists to use foraminiferal fossils to reconstruct environments in the geological past. In this way, foraminifera have been used to map the former planetary distributions of the tropics, locate ancient shorelines, and track global ocean temperature changes during the ice ages (Hillaire-Marcel and de Vernal 2007). To reconstruct ancient environments, paleontologists employ metrics that are similar to those used to characterize modern assemblages. These include species diversity, the relative numbers of planktonic and benthic species, the ratios of different shell types, and shell chemistry (Murray 1991; Hillaire-Marcel and de Vernal 2007). One widely used set of proxies for environmental conditions in ancient oceans is provided by the ratios of stable isotopes present in the shell carbonate. Although modified by physiological ("vital") effects, these ratios reflect the chemistry of the water in which foraminifera grew. For example, in 1955, Cesare Emiliani discovered that the ratio of stable oxygen isotopes depends on the water temperature, because warmer water tends to evaporate off more of the lighter isotopes. Since then, measurements of stable oxygen isotopes in planktonic and benthic foraminiferal shells from hundreds of deep-sea cores worldwide have been used to map past surface and bottom-water temperatures (Rohling and Cooke 2002). These data helps us understand how climate and ocean currents have changed in the past and may change in the future.

Many geologists work as biostratigraphers and use foraminifera extracted from drill cuttings recovered from oil wells to date sediments and reconstruct past environments. The stratigraphic analyses made using foraminifera as descriptors is so precise that these fossils are even used to direct sideways drilling within an oil-bearing horizon in order to increase well productivity. Since the 1920's the oil industry has been an important employer of paleontologists specialised in the study of these microscopic fossils. As a result of their potential "economical significance", foraminifera are better known for their spectacular fossil record than for their variety and abundance in modern marine environments. But, at the same time, curiosity-driven research, and the need to understand the present in order to interpret the past, has recently propelled paleontologists to learn more about the ecology of these protists.

Ecology of Benthic Foraminifera

Ecological studies of modern foraminifera (especially benthic taxa) started in the 1950s (Phleger 1960) and have increased over the past 60 years. Nevertheless, the main interest of the scientific community in this group remains focused on the use of their excellent fossil record to understand marine environmental changes in the geological (and historical) past. This leads to the paradox that the paleoecology of fossil foraminifera (based on some pioneer studies performed by geologist in the second half of the nineteenth century), is often better known than the ecology of modern species. Little is known about life cycles and lifestyles of most species of foraminifera. Reasonably complete life cycles are known for fewer than 30 of the 4,000 extant species. The few species that have been studied show a rich diversity of foraminiferal life cycles (i.e. involving alternating generations, apogamic, binary fission, different type of gametes and mode of fertilization) and a wide range of behaviours and diets. The classical life-cycle (i.e. gametogamy) in *Elphidium crispum* has been shown to be environmentally sensitive (Myers 1943). The whole life-cycle (both sexual and asexual phases) is completed in 1 year in temperate regions such as the Mediterranean basin.

These relatively large, shell-bearing protists typically constitute half or more of the deep-sea meiobenthos and are often an important constituent of the larger (>300 μm) macrofaunal size fraction as well. Together with bacteria, they are key players in the functioning of deep-sea benthic ecosystems. Some benthic species burrow actively through sediment at speeds up to 1 cm per hour, while others attach themselves to the surface of rocks or marine macroalgae. Many species feed at a low trophic level and play a crucial role in the long-term processing of fresh, photosynthetically-produced organic material that is transported to the ocean-floor as rapidly-sinking aggregates (Gooday 1993; Gooday et al. 2008). At least in some environmental setting, they collectively, process the same amount of labile organic matter as bacteria, although their biomass is a tiny fraction of that of bacteria (Moodley et al. 2002). Foraminifera are abundant enough to be an important part of the marine food chain, and their predators include scaphopods, isopods, marine snails, sand dollars and small fishes (Lipps 1983).

Studies conducted during the last decades have led to a better understanding of the biology of modern foraminifera (Le Cadre and Debenay 2006; Bentov et al. 2009; de Nooijer et al. 2009). However this group is scarcely used in monitoring studies because a standardization of protocols has not been achieved to date. Recently Schönfeld (2012) tackled this issue and reviewed the development of field and laboratory methods, their constraints and consequences for faunal and data analyses evidencing that much work remain to do.

The most important requirement is to discriminate between living and dead assemblages (Murray 2000). Over the last 20 years, new methods have been developed for this purpose, each one having a different degree of accuracy and based on a different rationale. Rose Bengal (RB), a non-vital stain that binds proteins and other macromolecules, is still the most widely used in ecological studies to recognize presumably dead (unstained) foraminifera from their living (stained) counterparts (Walton 1953; Murray and Bowser 2000). However, it does not discriminate between viable and recently dead organisms. Therefore, it becomes very important to effectively recognize the cell metabolism. In this context, the *Fluorescent In Situ Hybridization technique* (FISH), complementary to the CellTracker Green method (Bernhard et al. 2006), represents a new and useful approach to identify living cells possessing an active metabolism and also able to discriminate their grade of activity (Borrelli et al. 2011).

The Early Works on Mediterranean Foraminifera

The earliest work on Mediterranean foraminifera, dating from the late eighteenth and early nineteenth centuries, was descriptive and focused on taxonomic inventories (Soldani 1789, 1795; d'Orbigny 1826). During the last century, local faunal assemblages or selected species from the Western and Eastern Mediterranean were described; these included the works of Buchner (1940) and Hofker (1960) in the Gulf of Naples, Colom (1974) in the Balearic Sea, Le Calvez and Le Calvez (1958) in the Gulf of Lyon; Cherif (1970) in the Aegean Sea (Miliolacea), Fornasini (1902, 1904, 1905, 1906a, b) and Wiesner (1923) in the Adriatic Sea. Studies of benthic foraminiferal distributions in the Mediterranean started in the late 1860s when Jones and Parker (1860) proposed a synoptical table of the fossil and Recent species and varieties of benthic foraminifera (littoral to intertidal) from the Tyrrhenian, Adriatic and Levantine basins. More recent investigations, starting in the 1950s, have extended from shallow water down to abyssal (4,500 m) depths.

Parker (1958), Todd (1958), Blanc-Vernet (1969) and Colom (1974) were the first to conduct qualitative studies of modern bathyal benthic foraminiferal fauna in the Mediterranean Sea. In particular, Parker (1958) studied the distribution of 110 benthic and 18 planktonic species in the Eastern Mediterranean and Aegean Sea and described four bathymetric faunal boundaries for the benthic species. Blanc-Vernet (1969) investigated living benthic foraminifera from the Aegean Sea, off Rhodes, Crete and Peloponnesus, along the coast of Marseille and Corsica and described their biogeographic, seasonal and habitat-specific distribution. Parisi (1981) worked on samples from bathyal depths (1,003–3,593 m) in the Tyrrhenian Sea and Straits of Sicily. Bizon and Bizon (1983) reported on the geographic and bathymetric distribution of species down to 2,000 m off Marseille, Corsica, and in the Ligurian Sea. Two studies have analysed samples from both the Eastern and Western Mediterranean. Cita and Zocchi (1978) worked in the Alboran, Balearic, Tyrrhenian, Ionian, and Levantine basins (166–4,625 m), while Cimerman and Langer (1991) provided a comprehensive review of the distribution and morphology of benthic foraminifera from numerous localities in the Adriatic Sea and from various sample stations in the Tyrrhenian Sea.

In general, earlier studies have focused on restricted areas. For instance, Albani and Serandrei Barbero (1982, 1990), Serandrei Barbero et al. (1989) and Albani et al. (1991) worked on recent benthic foraminifera in the Venice Lagoon (Northern Adriatic Sea) and recognized areas characterized by similar hydrographic conditions basing on these faunas. Sgarrella et al. (1983) studied modern benthic foraminifera from the Gulf of Policastro in the southern Tyrrhenian Sea in order to determine the influence of fresh-water discharge on the assemblages. In an important study covering a much wider geographical area, Jorissen (1987, 1988) analysed the distributions of benthic foraminiferal taphocoenoses found in 285 grab samples and piston-core tops from the Adriatic Sea. For the first time, he correlated the distribution and the morphology of these organisms to environmental parameters, such as the input of nutrients and suspended load from Italian rivers (mainly from the Po outflow) and the surface circulation responsible for the transport and distribution of these products to the bottom.

The relatively few studies of modern planktonic foraminifera in the Mediterranean include those of Blanc-Vernet (1969), Cifelli (1974), Thunell (1978), Blanc-Vernet et al. (1979), Vénec-Peyré (1990). Of particular note is the later work of Pujol and Vergnaud-Grazzini (1995) which is the most accurate study so far of the distribution of living planktonic foraminifera along a NW-SE transect across the Mediterranean Sea. Their observations indicated that geographical distributions and living depths are related to regional hydrography and productivity of the Mediterranean basins.

The Last 30 Years of Efforts in the Study of Benthic Foraminifera

In the past 30 years, research in this field has increased greatly, prompted by the need to understand modern foraminiferal distributions in order to interpret marine environmental changes in the historical past. This led to an increased

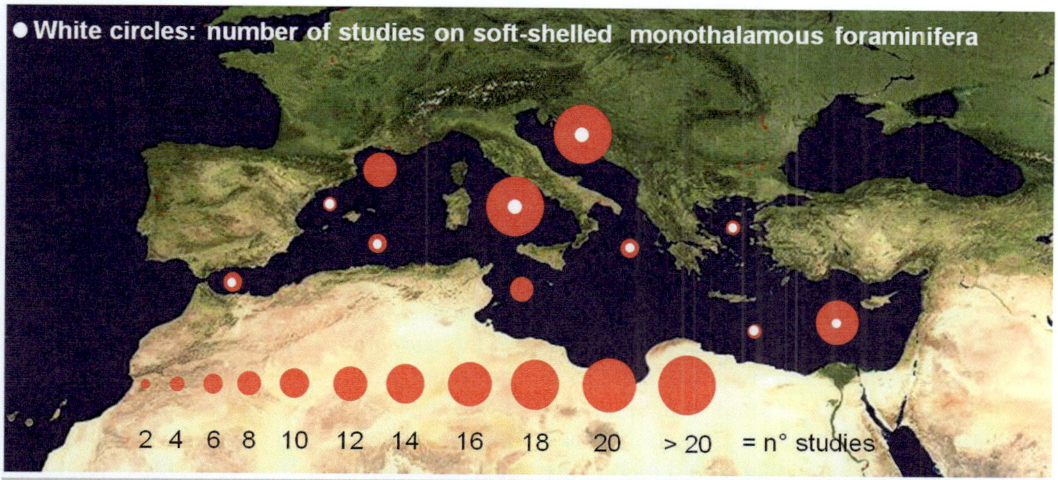

Mediterranean regions (n° studies)	Unstained	Total assemblage RB stained + Dead	Living assemblage RB stained	Total number	Soft-shelled (monothalamous taxa)
		RB = Rose Bengal			
Alboran Sea		3	2	5	1
Algerian sub-basin		4	2	6	1
Balearic Sea		2	(1)	3	1
Ligurian-Provencal sub-basin	2	1	8 (1)	12	
Tyrrhenian Sea	16	11	4 (2)	33	3
Sicily Channel	2	4	1	7	
Ionian Sea	1	2	3	6	1
Adriatic Sea	13	2	16 (3)	34	4
Aegean Sea			3	3	2
Cretan Passage		2	2	4	2
Levantine sub-basin	2	3	11	16	2
western Mediterranean	1			1	
eastern Mediterranean	3			3	1
	40	34	59	133	18

Fig. 13.2 Geographical distribution of total foraminiferal studies in the Mediterranean Sea. The *red circle* diameter is proportional to the number of total foraminiferal studies and the *white circle* indicate the number of studies on soft-shelled monothalamous taxa. The number between parenthesis in the column of the living assemblage indicates experimental studies where foraminifera are observed alive. *RB* Rose Bengal

emphasis on trying to understand the ecological requirements of modern foraminifera. In addition, benthic foraminifera have emerged as reliable indicators of the state of marine environments, in particular in shallow-water settings (Gooday et al. 2009; Balsamo et al. 2010, 2012; Frontalini and Coccioni 2011). In order to address these aims, different approaches were used, including the study of both unstained assemblages (i.e. the total assemblage comprising live and dead individuals without differentiating them) and living (Rose Bengal stained = RB stained) assemblages (Fig. 13.2).

The question of whether total or living assemblages best reflect the average environmental conditions is extensively debated by researchers (Murray 1982; Bergamin et al. 2003). For instance, Scott and Medioli (1980) assessed the validity of using the total (RB stained and dead) assemblage in ecological studies. They found that the high seasonal variability of the living (RB stained) assemblage may be attributed to seasonal climatic changes rather than changes in the prevailing marine environment. Murray (1982, 2000), however, argued that ecological studies must be based on the living assemblage, analysed over a period of time, in order to determine the relationships between living and dead assemblages. Alve and Murray (1994) found that, due to post-mortem processes influencing the dead (unstained) assemblage, such as dissolution of calcareous tests or transport, only results based on the living assemblage are reliable. Murray and Bowser (2000) emphasized that the main problem with total assemblages is that data on living assemblages (biocoenoses, not influenced by taphonomic changes) are combined with those on dead assemblages (tanathocoenoses or even taphocoenoses modified by taphonomic processes). In addition, the proportion of live and dead tests is influenced by several factors such as the thickness of sampled sediment layer, temporal variations of standing crop and the sedimentation rate. From this discussion it is clear that living assemblages, although certainly autochthonous, are affected by substantial temporal changes due to the high irregular foraminiferal life cycles and patchily distributed populations. Consequently, only samples collected during different seasons of the year can be considered to reflect the overall environment. On the other hand, total assemblages are affected by post-mortem processes, but they have the advantage that they represent the average environmental conditions during the time span corresponding to the deposition of the sediment sample. This approach is simpler, more practical and less costly and therefore may be preferred in environments where taphonomic processes are limited and autochthonous/allochthonous specimens can be recognized.

Various authors have used these different approaches to investigate modern Mediterranean benthic foraminiferal faunas during the last 30 years. During the 1980s and 1990s, a number of researchers described the relationships between the distribution of the unstained (live and dead individuals without differentiating them) benthic foraminiferal assemblages and the main environmental variables, i.e. oxygen, temperature, salinity, organic matter and grain size (Table 13.1, Fig. 13.3a). In the Adriatic Sea, Albani and Serandrei Barbero (1982, 1990), Albani et al. (1984, 1991, 1998, 2007, 2010) and Serandrei Barbero et al. (1989, 1999) thoroughly described modern benthic foraminifera on the continental shelf of the northern basin and the lagoon of Venice, interpreting them as indicators of different environmental settings, from intertidal to shallow water. Other studies based on unstained samples were conducted in the Tyrrhenian Sea and the Strait of Sicily. In these areas, where the human impact due to the presence of major ports (Naples and Augusta harbours) and industries (Bagnoli Bay) is high, work on benthic foraminifera has focused principally on possible links between pollution and assemblage characteristics, including changes in density and biodiversity, sensitive species, deformation of the shell (Bergamin et al. 2005; Ferraro and Lirer 2006; Ferraro et al. 2006, 2009; Di Leonardo et al. 2007; Valenti et al. 2008; Romano et al. 2008, 2009a; Carboni et al. 2009).

Other recent studies that use benthic foraminifera for environmental characterization have analysed the total fauna (Rose Bengal stained + dead) (Table 13.1, Fig. 13.3b). The majority of sampling sites are located in the Tyrrhenian Sea where authors anticipated that the presence of heavy metals would drive changes in benthic foraminiferal assemblages and cause test deformations (Bergamin et al. 2009; Cherchi et al. 2009; Romano et al. 2009b; Aloulou et al. 2012; Caruso et al. 2011). This approach was used by Coccioni (2000) in the Adriatic Sea and Samir and El-Din (2001) in the Levantine basin (Fig. 13.3b). Studies based on live plus dead assemblages have also addressed foraminiferal distributions (Donnici and Serandrei Barbero 2002; Buosi et al. 2012). De Rijk et al. (1999, 2000) analysed the distribution of Recent benthic foraminifera along a west–east bathyal and abyssal transect in the Mediterranean and their relation to the organic matter flux to the seafloor. Other papers document the impact of different environmental parameters (physical or chemical) on foraminiferal assemblages. For example, Milker et al. (2009) examined the influence of temperature on the distribution of modern shallow-water faunas, whereas Carboni et al. (2004) and Frezza and Carboni (2009) describe assemblages in the Tyrrhenian Sea influenced by the outflow of the river Ombrone, and Panieri (2005, 2006b) described the adaptation of benthic foraminifera to extreme environments (i.e. hydrothermal vent).

Finally, ecological studies of live (RB stained) assemblages have focused on their distribution and diversity, as well as their utility in biomonitoring (Table 13.1, Fig. 13.3c). Studies performed either in shallow areas or at deep sites (Fig. 13.4) have ranged from the description of foraminiferal microhabitats within the first 7 cm of sediment at a single shallow site close to the Po outlets (Barmawidjaja et al. 1992) to the spatial micro-distributions in the shallow subtidal zone in the northernmost Adriatic Sea (Hohenegger et al. 1993). Fontanier et al. (2008) compared samples from the Gulf of Lion slope (343–1,987 m) and one site located at 920 m in the Lacaze-Duthier Canyon, while Mojtahid et al. (2009, 2010) and Goineau et al. (2011, 2012) explored environmental control on live benthic foraminifera in a river-dominated shelf setting in their study of the Rhône prodelta (15–100 m). Pancotti (2011) conducted the only existing study of live assemblages in samples from the Eastern to Western Mediterranean. Her data provided new insights into foraminiferal diversity in the Mediterranean deep-sea, in particular, an apparent east-to-west increase in species richness corresponding to the productivity gradient, as well as indicating future research directions regarding factors controlling and threatening deep-sea biodiversity (Danovaro et al. 2010).

A number of authors have addressed the temporal variation (seasonal and/or inter-annual) of foraminiferal faunas, in terms of density and biodiversity, in relation to changes over time in key environmental parameters (i.e. oxygen, grain size, organic matter) (Soetaert et al. 1991; Jorissen et al. 1992; Pranovi and Serandrei Barbero 1994; Donnici et al. 1997; Schmiedl et al. 2000; Jannink 2001; Serandrei Barbero et al. 2003; Duijnstee et al. 2004; Panieri 2006b; Lampadariou et al. 2009; Sabbatini et al. 2010, 2012; Frontalini et al. 2011). A few papers consider the use of live (RB stained) foraminifera as environmental pollution indicators. Among these, Bergamin et al. (2003), Frontalini and Coccioni (2008), Coccioni et al. (2009), Frontalini et al. (2009, 2010), Buosi et al. (2010) used the FAI index (Foraminiferal Abnormality Index) to detect, on the basis of foraminiferal test morphology, the degree of ecosystem contamination in the central Adriatic and along the Italian coast of the Tyrrhenian Sea. Also of note are the studies of Yanko et al. (1999) describing the response of benthic foraminifera to heavy metal pollution along Mediterranean coast of Israel.

Others authors (e.g. Arieli et al. 2011) evaluated the potential long-term effect of rising sea-surface temperature caused by a thermal pollution from a power station on living benthic foraminifera, while Hyams-Kaphzan et al. (2009) and Elshanawany et al. (2011) explored the effects of anthropogenic eutrophication in the Eastern Mediterranean Sea.

Unfortunately, the numerous foraminiferal studies conducted in the Mediterranean have utilised different

Table 13.1 Methodological features of foraminiferal studies in the Mediterranean Sea listed for Unstained, Total (Rose Bengal stained + dead) and living (Rose Bengal stained) assemblages

Area	Sampling tecnique	Replicates	Thickness (cm)	Standardization	Staining	Wet vs Dry	Sieve-size fraction (μm)	Soft vs Hard	Author(s)
Adriatic Sea	n.a.	No	Surficial	No	Unstained	Dry	>125	Hard	Albani and Serandrei Barbero (1982)
Adriatic Sea	Grab sample	No	0–3; 0–4	[a]	Unstained	[a]	n.a.	[a]	Albani and Serandrei Barbero (1990)
Adriatic Sea	Hand-scraping	No	0–1; 1–2	No	Unstained	Dry	n.a.	Hard	Albani et al. (1984)
Adriatic Sea	Grab sample	No	0–3; 0–5	No	Unstained	Dry	n.a.	Hard	Albani et al. (1991)
Adriatic Sea	Grab sample	No	Surficial	No	Unstained	Dry/floatations in CCl_4	>63	Hard	Albani et al. (1998)
Adriatic Sea	Grab sample	No	0–5; 0–7	No	Unstained	Dry	>125	Hard	Albani et al. (2007)
Adriatic Sea	Grab sample	No	0–5; 0–7	No	Unstained	Dry/floatations in CCl_4	>125	Hard	Albani et al. (2010)
Tyrrhenian Sea	Core	No	0–3; 0–5	g^{-1} Dry sediment	Unstained	Dry	>125	Hard	Bergamin et al. (2005)
Tyrrhenian Sea	Grab sample	No	0–2	No	Unstained	Dry	>63	Hard	Carboni et al. (2009)
Adriatic Sea	Core	No	0–40		Unstained	[a]	>63	Hard	Chierici et al. (1962)
Tyrrhenian Sea	Box-core	Yes	n.a.	g^{-1} Dry sediment	Unstained	Dry	>90	Hard	Di Leonardo et al. (2007)
Tyrrhenian Sea	Box-core	No	0–1; 0–3	No	Unstained	Dry	>90	Hard	Ferraro and Lirer (2006)
Tyrrhenian Sea	Vibro-core	Yes	0–20	No	Unstained	Dry	>125	Hard	Ferraro et al. (2006)
Tyrrhenian Sea	Vibro-core	No	0–20	g^{-1} Dry sediment	Unstained	Dry	>90	Hard	Ferraro et al. (2009)
Levantine basin	Box-core	[a]	[a]	[a]	Unstained	[a]	[a]	[a]	Hyams-Kaphzan et al. (2008)
Adriatic Sea	Grab sample/ piston core	No	n.a.	n.a	Unstained	Dry	150–595	Hard	Jorissen (1987, 1988)
Tyrrhenian Sea, Sicily Channel	Gravity core	No	0–1	no	Unstained	Dry	>63	Hard	Parisi (1981)
Eastern Mediterranean Sea	Piston/gravity core	No	[a]	[a]	Unstained	Dry	>150	Hard	Parker (1958)
Tyrrhenian Sea	Grab sample	No	Surficial	g^{-1} Dry sediment	Unstained	Dry	>125	Hard	Romano et al. (2008)
Tyrrhenian Sea	Grab sample	No	0–2	g^{-1} Dry sediment	Unstained	Dry	>63	Hard	Romano et al. (2009b)
Adriatic Sea	Grab sample	No	0–3; 0–4	[a]	Unstained	[a]	n.a.	[a]	Serandrei Barbero et al. (1989)
Adriatic Sea	Grab sample	No	Surficial	No	Unstained	Dry	>63	Hard	Serandrei Barbero et al. (1999)
Tyrrhenian Sea	Grab sample	No	0–3; 0–4	No	Unstained	Dry	>63	Hard	Valenti et al. (2008)
Strait of Sicily	Grab sample	No	0–1; 0–2	50 cm^3	RB + Dead	Dry	>63	Hard	Aloulou et al. (2012)
Tyrrhenian Sea	Grab sample	No	0–2	g^{-1} Dry sediment	RB + Dead	Dry	>63	Hard	Bergamin et al. (2009)

(continued)

Table 13.1 (continued)

Area	Sampling tecnique	Replicates	Thickness (cm)	Standardization	Staining	Wet vs Dry	Sieve-size fraction (μm)	Soft vs Hard	Author(s)
Tyrrhenian Sea	Grab sample	No	0–1	g^{-1} Dry sediment	RB + Dead	Dry	>50	Hard	Buosi et al. (2012)
Tyrrhenian Sea	Grab sample	No	n.a.	g^{-1} Dry sediment	RB + Dead	Dry	>125	Hard	Carboni et al. (2004)
Tyrrhenian Sea	Grab sample	No	0–4	g^{-1} Dry sediment	RB + Dead	Dry	>63	Hard	Caruso et al. (2011)
Tyrrhenian Sea	Hand coring	Yes	0–2	10 cm^3	RB + Dead	Dry/floatation $ZnBr^2$	63–125; >125	Hard	Cherchi et al. (2009)
Adriatic Sea	Grab sample	No	n.a.	[a]	RB + Dead	[a]	>125	[a]	Coccioni (2000)
Mediterranean Sea	Box-core	No	0–2	No	RB + Dead	Dry	>125	Hard	de Rijk et al. (1999)
Mediterranean Sea	Box-core	No	0–2	No	RB + Dead	Dry	>125	Hard	de Rijk et al. (2000)
Adriatic Sea	Grab sample	No	0–7	No	RB + Dead	Dry	>125	Hard	Donnici and Serandrei Barbero (2002)
Tyrrhenian Sea	Grab sample	No	0–5; 0–7	No	RB + Dead	Dry	>125	Hard	Frezza and Carboni (2009)
Western Mediterranean Sea	Grab sample/box core	No	0–1; 0–2	10 cm^3	RB + Dead	Dry	>125	Hard	Milker et al. (2009)
Tyrrhenian Sea	Box-core/gravity core	No	0–1	10 cm^3	RB + Dead	Dry/floatation sodium polytungstate	> 63	Hard	Panieri (2005)
Tyrrhenian Sea	Scuba diving	Yes	0–2	10 cm^3	RB + Dead	Dry/floatation sodium polytungstate	>63	Hard; Soft only qualitative	Panieri (2006b)
Tyrrhenian Sea	Grab sample	No	0–2	g^{-1} Dry sediment	RB + Dead	Dry	>63	Hard	Romano et al. (2009a)
Eastern Mediterranean Sea	Grab sample	No	0–1; 0–2	g^{-1} Dry sediment	RB + Dead	Dry	>63; >125 geochemistry	Hard	Samir and El-Din (2001)
Tyrrhenian Sea	Grab sample	No	n.a.	No	RB + Dead	Dry	>106	Hard	Sgarrella et al. (1983)
Levantine basin	Hand-scraping	Yes	surficial	g^{-1} Dry sediment	RB	Dry	63–1000	Hard	Arieli et al. (2011)
Adriatic Sea	Box-core	No	0–2	g^{-1} Dry sediment	RB	Dry	>63	Hard	Barmawidjaja et al. (1992)
Tyrrhenian Sea	Grab sample	No	n.a.	g^{-1} Dry sediment	RB	Dry	>125	Hard	Bergamin et al. (2003)
Aegean Sea	Grab sample	No	0–1	g^{-1} Dry sediment	RB	Dry	>63	Hard	Bergin et al. (2006)
Tyrrhenian Sea	n.a.	No	surficial	g^{-1} Dry sediment	RB	Dry	>63	Hard	Buosi et al. (2010)
Adriatic Sea	Box-core	No	0–10	cm^2	RB	Dry	63–150; >150	Hard	De Stigter (1996)
Adriatic Sea	Grab sample	No	Surficial	200 cm^3	RB	Dry/floatations in CCl_4	>63	Hard	Donnici et al. (1997)
Adriatic Sea	Sediment core	Yes	0–2	no	RB	Wet	>63	Hard	Duijnstee et al. (2004)

Eastern Mediterranean Sea	Grab sample	No	0–3	50 cm^3	RB	Dry	>63	Hard	Elshanawany et al. (2011)
Western Mediterranean Sea	Multi-core	Yes	0–10	50 cm^3	RB	Wet	>150	Hard	Fontanier et al. (2008)
Adriatic Sea	Grab sample	No	0–2	50 cm^3	RB	Dry	>63	Hard	Frontalini and Coccioni (2008)
Tyrrhenian Sea	Grab sample	No	n.a.	50 cm^3	RB	Dry	>63	Hard	Frontalini et al. (2009)
Tyrrhenian Sea – Adriatic Sea	Grab sample	No	n.a.	50 cm^3	RB	Dry	>63	Hard	Frontalini et al. (2010)
Adriatic Sea	Grab sample	No	0–5	50 cm^3	RB	Dry	>63	Hard	Frontalini et al. (2011)
Western Mediterranean Sea	Multi-core	No	0–5	100 cm^2	RB	Wet	63–150; >150	Hard	Goineau et al. (2011)
Western Mediterranean Sea	Multi-core	Yes	0–5	100 cm^2	RB	Dry	63–150; >150	Hard	Goineau et al. (2012)
Adriatic Sea	Core	Yes	0–5	90 cm^3	RB	Dry	>63	Hard	Hohenegger et al. (1993)
Levantine basin	Box-core	No	0–10	10 cm^3	RB	Wet	63–2000	Hard	Hyams-Kaphzan et al. (2009)
Levantine basin	Box-core	No	0–10	50 cm^3	RB	Wet	63–150; 150–595	Hard; Soft only qualitative	Jannink (2001)
Adriatic Sea	Box-core	No	n.a.	g^{-1} dry sediment	RB	Dry	>63	Hard	Jorissen et al. (1992)
Aegean and Levantine Sea	Multi-core	Yes	0–10	10 cm^2	RB	Wet	>32	Soft+Hard	Lampadariou et al. (2009)
Western Mediterranean Sea	Multi-core	No	0–5	100 cm^2	RB	Wet and dry	63–150; >150	Hard	Mojtahid et al. (2009)
Western Mediterranean Sea	Multi-core	No	0–5	100 cm^2	RB	Wet and dry	>150	Hard	Mojtahid et al. (2010)
Adriatic Sea	Grab sample	Yes	0–2	10 cm^2	RB	Wet	>63	Soft+Hard	Nardelli (2012)
Mediterranean Sea	Box-core	Yes	0–5	10 cm^2	RB	Wet	>63	Soft+Hard	Pancotti (2011)
Adriatic Sea	Scuba diving	No	0–1	no	RB	Wet	>63	Hard	Panieri (2006a)
Adriatic Sea	Grab sample	No	0–3; 0–4	[a]	RB	[a]	>63	[a]	Pranovi and Serandrei Barbero (1994)
Adriatic Sea	Box-core	No	0–2	20 ml	RB	Wet	>63	Soft+Hard	Pucci (2006)
Adriatic Sea	Box-core	No	0–2	10 cm^2	RB	Wet	>63	Soft+Hard	Sabbatini et al. (2010)
Adriatic Sea	Grab sample	Yes	0–1	10 cm^2	RB	Wet	>63	Soft+Hard	Sabbatini et al. (2012)
Western Mediterranean Sea	Multi-core	No	0–10	g^{-1} dry sediment	RB	Dry	>125	Hard	Schmiedl et al. (2000)
Adriatic Sea	Grab sample/ box-core	No	0–10	[a]	RB	[a]	>125	[a]	Serandrei Barbero et al. (2003)
Western Mediterranean Sea	Box-core	No	0–20	10 cm^2	RB	Wet/ludox	500; 250; 38	Xenophyophores	Soetaert et al. (1991)

n.a. not available

[a]Indicate data not available; informations are referred to Frontalini and Coccioni (2011), Appendix A. Supplementary data. Frontalini and Coccioni (2011)

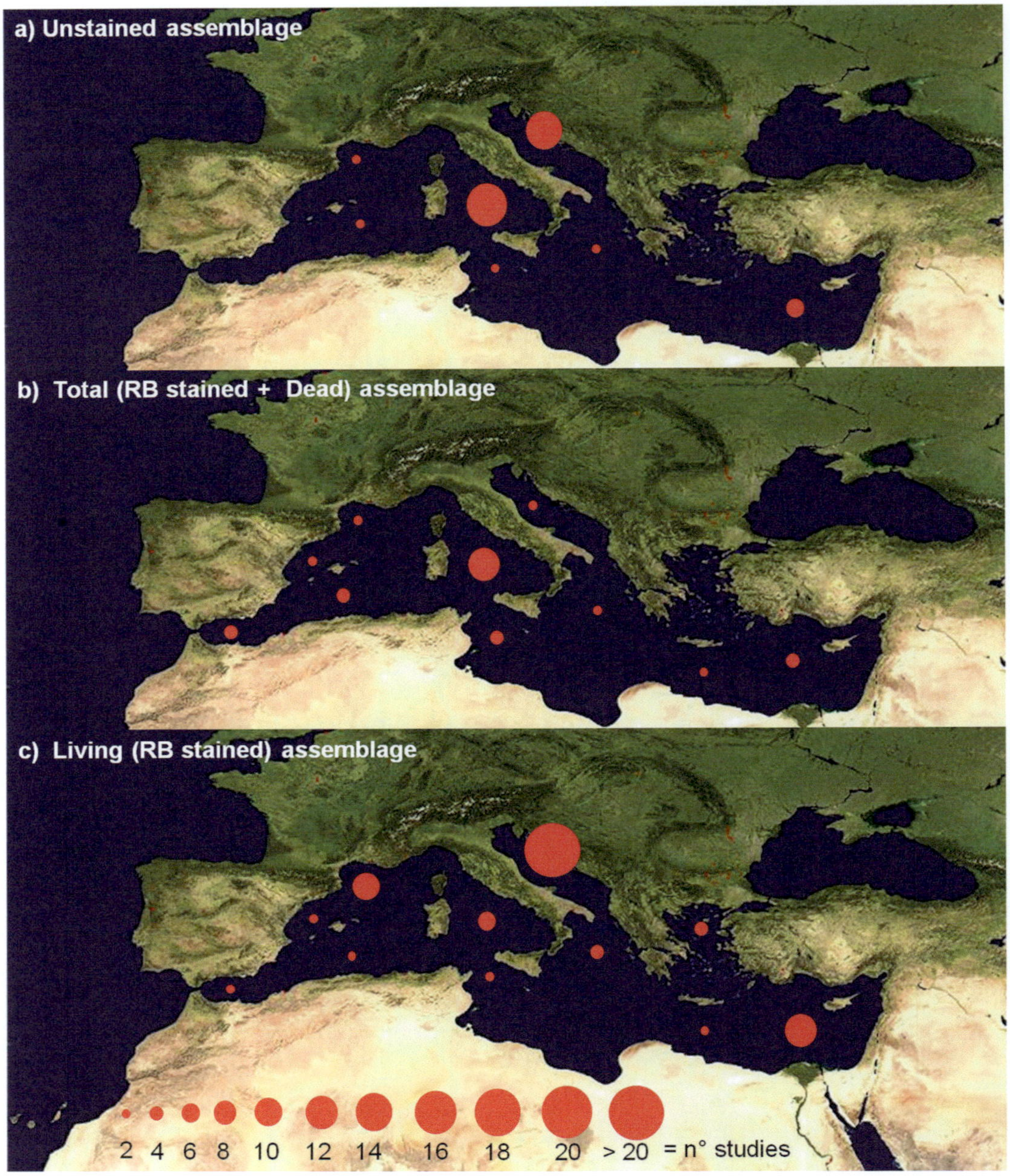

Fig. 13.3 Geographical distribution of total foraminiferal studies in the Mediterranean Sea divided for Unstained (**a**), Total (Rose Bengal stained + dead) (**b**) and living (Rose Bengal stained) assemblages (**c**). The *red circle* diameter is proportional to the number of studies on foraminifera. In the living assemblages, experimental studies where foraminifera are observed alive are included

methodologies (Balsamo et al. 2010, 2012; Frontalini and Coccioni 2011). This hampering the comparison of different studies and therefore the possibility to gather consistent data on biodiversity and abundance trends, or on the impact of a particular pollutant, or pollution in general, on the foraminifera. The problems are compounded by differences in staining and sampling methodologies, and the fact that an important part of the foraminiferal fauna is often neglected. In the following paragraph we will examine these two important issues (Table 13.1).

Problems in the Sampling Methodology

A variety of sampling gears has been used to collect material for the study of foraminifera (Murray 1991; Schönfeld 2012; Schönfeld et al. 2012). Earlier studies were based on samples obtained using grabs, gravity cores, or piston cores, which do not retain the surface sediment where living foraminifera are concentrated (Massiotta et al. 1976; Jorissen 1987; Parisi 1981) (Table 13.1). Even some recent investigations have

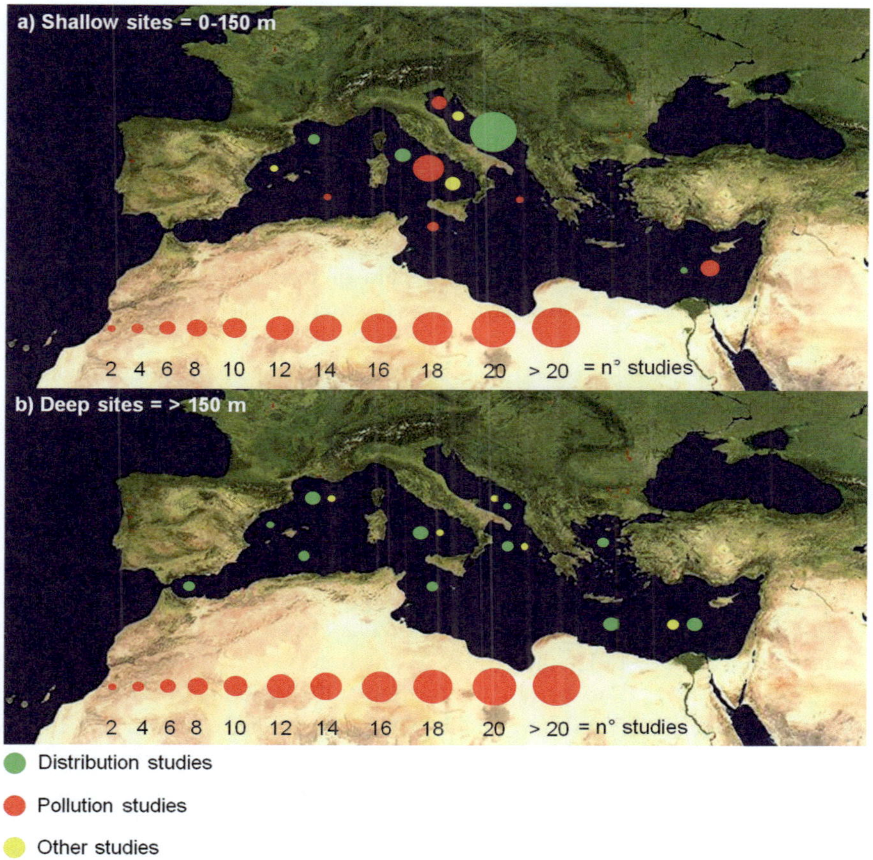

Fig. 13.4 Bathymetrical distribution of foraminiferal studies in (**a**) shallow sites (0–150 m) and (**b**) deep sites (>150 m). Studies typology (*coloured circles*) are shown: foraminiferal fauna distribution and ecology (*green circles*), pollution and biomonitoring (*red circles*) and other studies (*yellow circles*). The circle diameter is proportional to the number of studies on foraminifera. In the last category, we included papers on taxonomy, time series analyses, geochemistry and experimental laboratory

been based on samples taken using grabs, due to problems in sampling in harbour areas and the unavailability of a box corer or multi-corer. In some cases the first few centimetres of sediment are removed and in others, subsamples are taken using plexiglas tubes (Donnici et al. 1997; Bergamin et al. 2003, 2009; Cherchi et al. 2009; Coccioni et al. 2009; Frontalini et al. 2010; Aloulou et al. 2012; Caruso et al. 2011; Elshanawany et al. 2011) (Table 13.1). However, most modern studies have employed box cores (de Stigter 1996; Soetaert et al. 1991; Barmawidjaja et al. 1992; Jorissen et al. 1992; de Rijk et al. 1999, 2000; Jannink 2001; Serandrei Barbero et al. 2003; Ferraro et al. 2006; Di Leonardo et al. 2007; Hyams-Kaphzan et al. 2009) (Table 13.1) or hydraulically-damped multiple corers (Schmiedl et al. 2000; Fontanier et al. 2008; Mojtahid et al. 2009, 2010; Goineau et al. 2011, 2012).

There is also a considerable variety in the subsampling procedure. Sample from grabs or box corers are often limited to the first few centimetres of sediment (0–1, 0–2 cm up to 0–5 and 0–7 cm) in both distributional and pollution studies. Several authors studied only the top 2 cm of sediment (Albani et al. 1984; Barmawidjaja et al. 1992; Panieri 2005, 2006a, b; Bergin et al. 2006; Romano et al. 2008, 2009a, b; Bergamin et al. 2009; Carboni et al. 2009; Sabbatini et al. 2010; Aloulou et al. 2012; Arieli et al. 2011) (Table 13.1). Others consider the first 3, 4, 5 or 7 cm of sediment (Frontalini et al. 2011) or even 20 cm (Ferraro et al. 2006, 2009) as one unit, thereby mixing the different levels without considering the living depth of individual species. Other authors, however, have addressed the important issue of the vertical distribution of foraminiferal species in the sediment. Generally, the first 2 cm are sub-sampled every 0.5 cm and levels below 2 cm are sub-sampled every cm. Only a few authors have examined sediment layers down to 10 cm depth (de Stigter 1996; Schmiedl et al. 2000; Jannink 2001; Fontanier et al. 2008; Hyams-Kaphzan et al. 2009). The studies of Hohenegger et al. (1993), Pancotti (2011), Pucci et al. (2009), Mojtahid et al. (2009, 2010) and Goineau et al. (2011, 2012) were limited to the first 5 cm.

Another important problem concerns sample replication, which provides statistically useful information on the small-scale density and biodiversity variability of faunal assemblages in terms of density and diversity. Although this is standard practice in metazoan meiofaunal and macrofaunal research, the use of replicated samples is still fairly rare in studies of foraminiferal distributions (Hohenegger et al. 1993; Duijnstee et al. 2004; Fontanier et al. 2008; Lampadariou et al. 2009; Pancotti 2011;

Goineau et al. 2012), as well as in biomonitoring studies (Ferraro et al. 2006; Panieri 2006b; di Leonardo et al. 2007; Cherchi et al. 2009; Arieli et al. 2011). Sieve mesh size (Table 13.1) is another crucial variable that strongly influences assemblage composition. In the Mediterranean the following meshes have been used: 32, 38, 63, 90, 125, 150, 595 and 1,000 μm. A final point to consider is that in many geological-oriented studies, specimens are not identified to species level but grouped together as genera or morphogroup, making impossible to analyse the full extent of the assemblage diversity.

The Hard vs. Soft Shelled Foraminifera Issue

Few authors have included soft-shelled monothalamous species in their study of Mediterranean foraminifera: Soetaert et al. (1991) in the Gulf of Lions; Moodley et al. (1997), Pucci (2006), Pancotti (2011), Nardelli (2012) and Sabbatini et al. (2010, 2012) in the Adriatic Sea; Hatziyanni (1999), Lampadariou et al. (2009) in the Eastern Mediterranean Sea. Only Pucci (2006), Pancotti (2011), Nardelli (2012) and Sabbatini et al. (2010, 2012) have studied this rarely-studied component in terms of abundance and species diversity and in relation to the environmental setting (Fig. 13.5). Instead, some studies (Bizon and Bizon 1983; de Rijk et al. 2000; Fontanier et al. 2008) only report counts for selected species of soft-shelled monothalamous foraminifera. All other authors have confined their investigations to hard-shelled species and therefore have not encompassed the full range of foraminiferal biodiversity in the Mediterranean (Fig. 13.6). As reported in the previous paragraphs, early studies did not consider treatment with Rose Bengal and therefore yielded total assemblages, that is, a mixture of live and dead tests. Other studies instead considered foraminiferal specimens stained with Rose Bengal to distinguish between alive and dead organisms at the time of collection. In addition, most analyses are based on the dry picking of individuals, but in some cases, (Table 13.1; e.g., Jannink 2001; Duijnstee et al. 2004; Panieri 2006a; Hyams-Kaphzan et al. 2009) the foraminifera were picked out from sample residues in water. This technique instead allows the evaluation of all the foraminifera, including the soft-shelled monothalamous forms with delicate organic or agglutinated walls that shrink and disappear when dried.

Soft-shelled monothalamous foraminifera are often an important component of benthic fauna in both shallow and deep-water settings (Gooday 2002) and ignoring them would lead to underestimating the real variability of foraminiferal abundance and diversity. In the deep sediments of the Mediterranean Sea the soft-shelled monothalamous foraminifera account for up to almost 30 % of the entire assemblage both in the western and eastern basin (Pancotti 2011). In the shallow northern Adriatic Sea, this component ranges from 20 to 60 % of the living (RB stained) assemblage (Sabbatini et al. 2010); it can reach even 80 % of relative abundance in shallow waters of the central Adriatic (Nardelli 2012).

Unfortunately, soft-shelled monothalamous foraminifera are time consuming to extract, and largely undescribed. Moreover, they have little fossilization potential and therefore they are often ignored because they are not useful in paleoecological and geological studies. Also in comparison to the many workers on foraminifera in general, there are few specialists on soft-shelled monothalamous foraminifera.

Nevertheless, there are some scattered, early records of soft-walled allogromiids from the Mediterranean Sea. Notable among these is the paper by Grüber (1884), who described several species from coastal waters of the Bay of Naples. These included *Craterina mollis*, later established as the type species of the genus *Allogromia* by Rhumbler (1904). Other examples are from Huxley (1910) who reports *Shepheardella taeniformis* from the Bay of Naples, and earlier Dujardin (1835) who described the gromiid *Gromia oviformis* (a close relative of the foraminifera) based on material from the NW Mediterranean coast and elsewhere. In more recent years, there have been few species-level studies of soft-shelled, monothalamous foraminifera from the Mediterranean. They include Nyholm's (1951) description of an allogromiid-like protist, *Marenda nematodes*, from the Catalan coast. He distinguished the new species from free-living nematodes, which it closely resembles, by its slow movements when irritated by the light of the microscope.

Comprehensive studies of "entire" foraminiferal assemblages (i.e. including both soft and hard-shelled forms) are a recent development (Pucci 2006; Sabbatini et al. 2010, 2012; Pancotti 2011; Nardelli 2012). Pucci (2006) studied the biodiversity of benthic foraminifera along a shallow transect from the Po outflow to the central Adriatic Sea. Based on the results obtained in the period between May and June 2004, the coast between Goro (near the mouth of the river Po) and Cattolica (Central Adriatic coast) was divided into three areas with different foraminiferal assemblages linked to physical-chemical parameters (chlorophyll, oxygen, temperature and turbidity) and specific grain sizes. Pucci (2006) also reported qualitative data on soft-shelled monothalamous taxa, indicating that they were rather uncommon (6 % of all the stained foraminifera in the samples). However, they were distributed across all 14 transects along the Adriatic area from Cattolica to Goro, and were most abundant in the northern transects in front of the Comacchio region. The relative abundance of soft-shelled monothalamous foraminifera reached 65 % at one station located near the coast at 5 m water depth off Comacchio. Most of the soft-shelled monothalamous species were undescribed and there was a relatively high abundance of small, thin wall specimens.

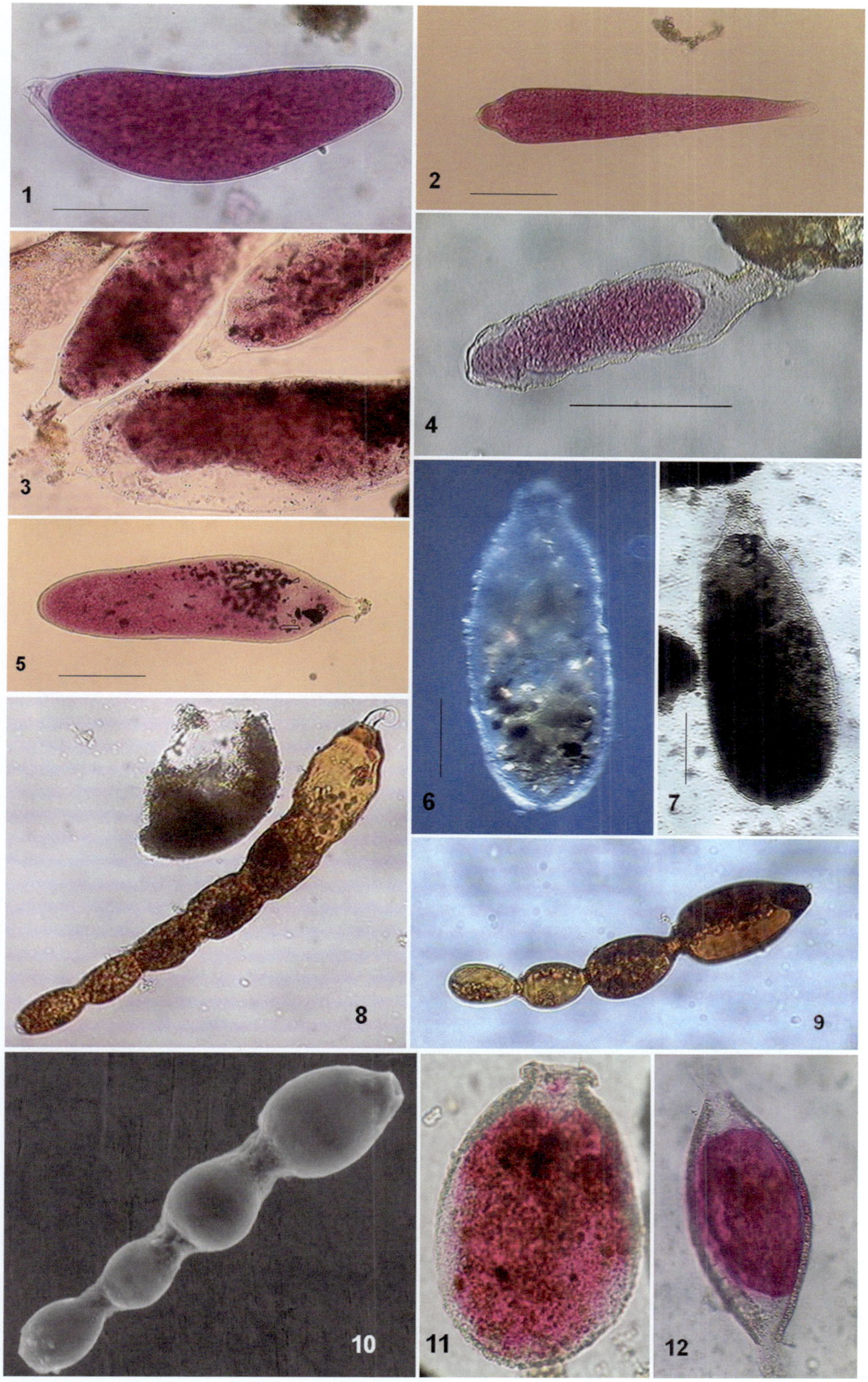

Fig. 13.5 Foraminifera from the Mediterranean Sea: monothalamous taxa with an organic test; unless otherwise stated, all are Rose Bengal stained specimen viewed under the optical microscope: **1.** *Goodayia* sp., magnification 400X. **2.** *Micrometula* sp. magnification 200X. **3.** *Gloiogullmia* sp., magnification 200X. **4.** *Vellaria* sp., magnification 400X. **5.** *Psammophaga*-like type 2, magnification 200X. **6.** *Psammophaga-like* type 1, live specimen, magnification 200X. **7.** *Psammophaga*-like type 1 (different specimen from 6), phase contrast image of a live specimen, magnification 200X. Specimens 1, 2, 3 and 5 are collected from the northern Adriatic Sea, water depth between 8 and 20 m, while specimens 4, 6, 7 are from the central Adriatic Sea, water depth between 10 and 15 m. **8.** *Nodellum membranaceum*, magnification 200X. **9.** *Resigella* sp. magnification 400X. **10.** *Resigella sp.* (same specimen as in 9), Scanning Electron Micrograph, magnification 600X. **11.** *Saccamminid* sp.8, magnification 400X. **12.** *Vanhoeffenella gaussi*, magnification 400X. Specimens 8 and 9 were collected from the western Mediterranean Sea, water depth between ~2,500–2,600 m, specimen 10 is from the eastern Mediterranean Sea, water depth ~4,300 m and specimen 11 is from the central Mediterranean Sea, water depth ~3,900 m. Scale bar = 100 μm; if different, the scale is indicated in the figure

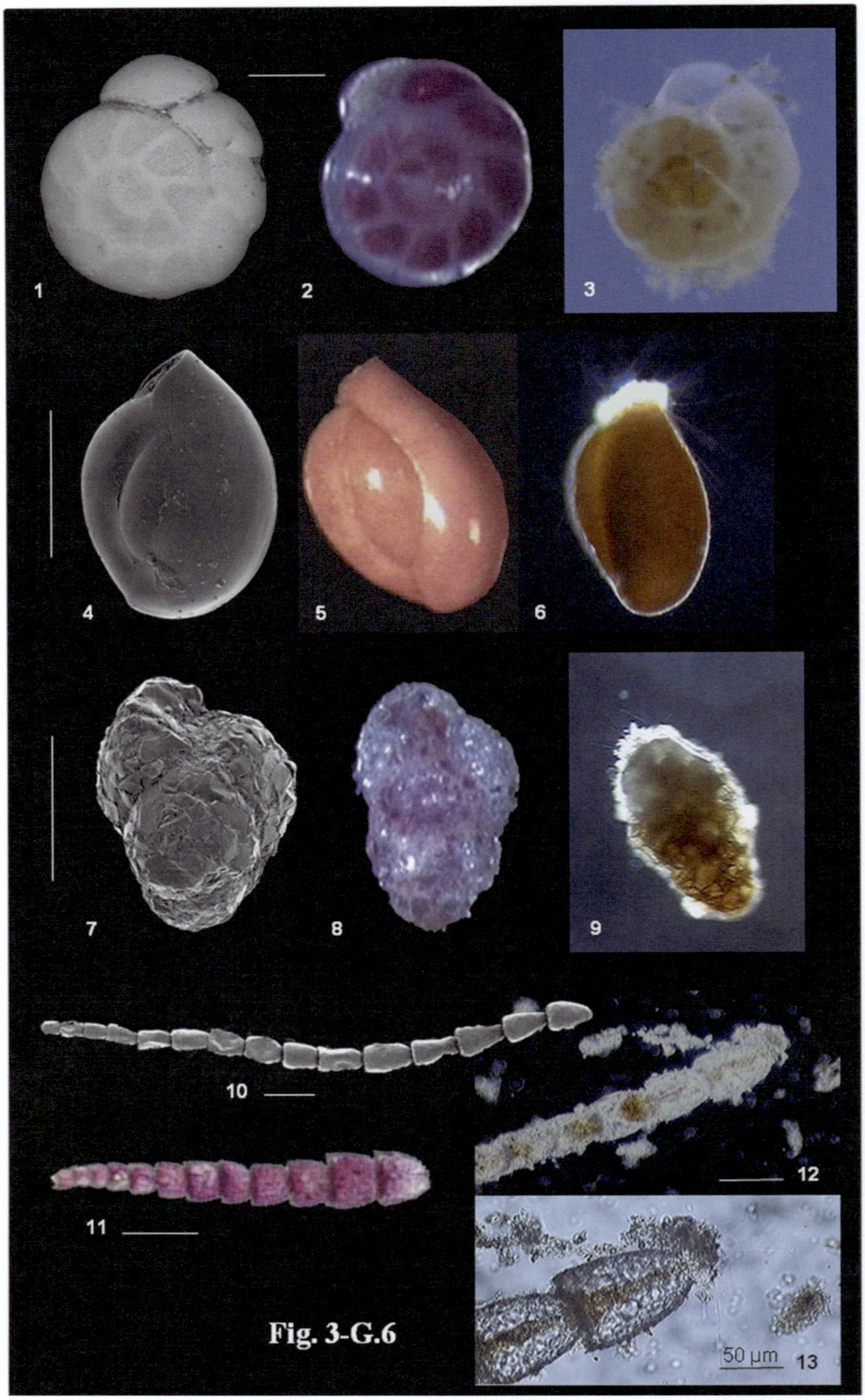

Fig. 13.6 Foraminifera from the Mediterranean Sea: hard-shelled polythalamous taxa with calcareous (perforated or imperforated) and agglutinated tests. Images are of dead (unstained) specimens photographed by Scanning Electron Microscopy, 'living' specimens stained with Rose Bengal and live specimens with pseudopods photographed under a light microscope with either reflected or transmitted light. All specimens are from the central Adriatic Sea, water depth 10–15 m. **1–3.** *Ammonia parkinsoniana* (calcareous species with perforated test), three individuals of this species are illustrated; **1.** Scanning Electron Micrograph. **2.** Rose Bengal stained specimen, reflected light.

Sabbatini et al. (2010) investigated the foraminiferal faunas, including the soft-shelled monothalamous component, along a shallow bathymetric transect in the Gulf of Trieste. The distribution of foraminiferal species was a function of differences in water depth, granulometry and distance from fresh water sources and other chemical and physical parameters (temperature, salinity and dissolved oxygen). The absolute and relative abundance of soft-shelled monothalamous foraminifera decreased with distance from the coast (and the nutrient source, the Isonzo River). All the soft-shelled monothalamous taxa found were new for the North Adriatic waters, undescribed at the species level and, in most cases, even at the genus level. Similar taxa are also abundant in deep waters. A regional-scale study extending from the western to the eastern part of the deep Mediterranean basin (Pancotti 2011) revealed that soft-shelled monothalamous foraminifera, the vast majority of them undescribed, represent at least 50 % of the assemblage at depths >1,500 m.

Benthic foraminifera make an important contribution to meiofaunal biomass. In some areas (the Algerian-Provençal and the Levantine basins), their biomass is comparable to that of the metazoan meiofauna. Sabbatini et al. (2012) analysed relationships between foraminiferal communities and trophic status in coastal sediments, revealing that temporal (seasonal) variability in the quantity and composition of the food sources is responsible of the variability of foraminiferal assemblages. These authors also suggested that soft-shelled monothalamous foraminifera (allogromiids *sensu lato*) respond to the nutritional quality of sedimentary organic matter rather than to its quantity. Nardelli (2012) described the occurrence of soft-shelled monothalamous foraminifera in a shallow water hydrocarbon seepage from the central Adriatic Sea; the soft-shelled component is particularly dominant (80 % of the entire foraminiferal assemblages) at the proximal station of the hydrocarbon seep influenced by the presence of high concentration of volatile aliphatic compounds.

The studies reviewed above emphasize the importance of soft-shelled monothalamous foraminifera and their potential in biomonitoring studies even if they cannot provide information on past ecosystems because they do not fossilize. The soft-shelled taxa must be taken into account in order to achieve a comprehensive taxonomic and ecological overview of foraminiferal assemblages. Their study can add information of importance in biomonitoring studies, particularly in shallow-water ecosystems where they can account for >50 % of living foraminifera. Moreover, ecological studies of some key "allogromiid" species in coastal environments could lead to the recognition of sensitive or tolerant species with the potential to be used as bioindicators in the same way as hard-shelled foraminifera.

The Last Frontier in the Study of Foraminifera

The last frontier in the study of benthic foraminifera is the experimental approach. Laboratory experiments make it possible to evaluate foraminiferal responses to changes in one or more chemical-physical parameters under controlled conditions, either at the level of the whole fauna (in micro- or mesocosms) or of one or a few selected species (in culture). The results obtained in the laboratory could represent a model, albeit simplified, of ecosystem functioning, and can be tested *in situ*. Duijnstee (2001) conducted laboratory experiments to explore how marine snow events (causing anoxia) influenced foraminiferal growth, reproduction and survival. Comparison of community structure in stressed situations and less stressed conditions can provide information on how the different species will respond to oxygen stress. This is very important, because oxygen availability is often considered to be the most important variable determining the structure of benthic communities in environments with high nutrient loads, as in the Adriatic Sea. Ernst (2002) examined this issue further in microcosm experiments aimed at assessing the separate effect of the oxygen concentration and organic flux on benthic foraminiferal assemblages.

In the Adriatic Sea, Pucci et al. (2009) conducted mesocosm experiment to evaluate the survival of benthic foraminifera under hypoxic conditions, a potential source of stress, especially in eutrophic and shallow environments subjected to pollution from industrial activity. In anoxic sediments, the upward migration of foraminiferal species could be caused by decreasing oxygen concentrations in deeper sediment layers but also by changes in the distribution and availability of trophic resources at different sediment levels. In this context, Heinz et al. (2001) described the response of benthic foraminifera from the Gulf of Taranto (Ionian Sea) and Gulf of Lions (Ligurian-Provençal sub-basin) to simulate phytoplankton pulses under laboratory conditions.

Fig. 13.6 (continued) **3.** Live specimen, transmitted light. **4–6.** *Triloculina rotunda* (calcareous species with imperforated test), three individuals are illustrated; **4.** Scanning Electron Micrograph. **5.** Rose Bengal stained specimen, reflected light. **6.** Live specimen with pseudopods, phase contrast, length 470 μm, magnification 100X. **7–9.** *Eggerella scabra* (species with agglutinated test), three individuals are illustrated; **7.** Scanning Electron Micrograph. **8.** Rose Bengal stained specimen, reflected light. **9.** Live specimen with pseudopods, phase contrast, magnification 100X. **10–13.** *Leptohalysis scottii* (species with agglutinated test); **10.** Scanning Electron Micrograph. **11.** Rose Bengal stained specimen, transmitted light. **12.** Phase contrast image of live specimen. **13.** Phase contrast image of live specimen; detail of the aperture and pseudopods (same specimen described in 12). Scale bar = 100 μm; if different, the scale is indicated in the figure

The use of benthic foraminiferal assemblages for the assessment of the quality of marine ecosystems has grown recently because of the high potential and of these organisms as monitoring tools (Schönfeld 2012; Schönfeld et al. 2012). Foraminifera respond rapidly to environmental changes, are relatively easy and cheap to sample and have an excellent fossil record, which can provide some information about the pre-impact conditions of the environment under scrutiny. However, many aspects of their biology remain far from clear, yet. Unlike other taxa already used for biomonitoring and ecotoxicological purposes, little is known in detail about how different kinds of impact affect their ecology and biology, including growth, death and reproduction rates, mechanisms of defence, intra- and inter-specific relationships. Moreover, the effects of chemical pollution on the biomineralization processes of calcareous species are poorly understood. To explore some of these issues through experimental studies, Nardelli (2012) performed a series of experiments on a miliolid species (*Triloculina rotunda*) aiming to investigate the effects of zinc contamination on its growth, survival, reproduction and cellular ultrastructure. The experimental species proved to be highly tolerant to zinc, in terms of survival (<50 % of deaths until 10 mg/l of zinc). This is probably due to its ability of this species to bioaccumulate the metal, as evidenced by ultrastructural observations at transmission microscopy (TEM). On the other hand, effects on growth rates (stop or delay of growth) were already observed at a zinc concentration of 0.1 mg/l and they seem to influence metal incorporations rates into the shell. In fact at the zinc concentration of 0.1 mg/l corresponded to a decrease of zinc incorporation rates into calcite, possibly as a consequence of cellular disease. The study also demonstrated that zinc, by itself, is not able to cause test deformations, as previously hypothesized by several authors (e.g. Sharifi et al. 1991; Samir and El-Din 2001; Romano et al. 2008; Madkour and Ali 2009). The work of Nardelli (2012), in which the cytology, biogeochemistry, and ecology of foraminiferal species were examined under controlled conditions, offers a promising approach to improving our knowledge of aspects of foraminiferal biology that are beyond the scope of environmental studies. Unfortunately, a considerable research effort is still required to further develop the culturing protocols necessary to improve this kind of experiments.

Concluding Remarks

The aim of this chapter has been to introduce the main issues in the study of foraminifera in the Mediterranean Sea and the state-of-the-art developments in the study of these organisms. The study of these protists has huge potential implications and benefits. They are good indicators of global change and are also promising indicators of the environmental health of marine ecosystems. Nevertheless, much remains to be learnt about foraminiferal ecology. We stress here that the study of the ecology of foraminifera has been often hampered to date by inconsistent methodologies which have yielded an equally inconsistent body of data that cannot be easily integrated to produce an overall synthesis of community parameters. In the future, researchers will need to focus on specific topics and apply similar methodologies to improve our understanding of the role of foraminifera in the functioning of both present and past Mediterranean Sea ecosystems.

Acknowledgements The authors gratefully acknowledge Andrew J Gooday (National Oceanography Centre) and Antonio Pusceddu (DiSVA, Polytechnic University of Marche) for constructive discussions and for critical assessment of the manuscript.

References

Albani AD, Serandrei Barbero R (1982) A foraminiferal fauna from the lagoon of Venice. Mar Micropaleontol 25:187–217

Albani AD, Serandrei Barbero B (1990) I foraminiferi della Laguna e del Golfo di Venezia. Mem Sci Geol 42:271–341

Albani AD, Favero I, Serandrei Barbero R (1984) Benthonic foraminifera as indicators of intertidal environments. Geo-Mar Lett 4:43–47

Albani AD, Favero I, Serandrei Barbero R (1991) The distribution and ecological significance of recent foraminifera in the lagoon south of Venice. Rev Española Micropaleontol 23:129–143

Albani AD, Favero I, Serandrei Barbero R (1998) Distribution of sediment and benthic foraminifera in the Gulf of Venice, Italy. Estuar Coast Shelf Sci 46:251–265

Albani AD, Favero I, Serandrei Barbero R (2007) Foraminifera as ecological indicators in the Lagoon of Venice, Italy. Ecol Indic 7:239–253

Albani AD, Donnici S, Serandrei Barbero R, Rickwood PC (2010) Seabed sediments and foraminifera over the Lido Inlet: comparison between 1983 and 2006 distribution patterns. Cont Shelf Res 30:847–858

Aloulou F, EllEuch B, Kallel M (2012) Benthic foraminifera assemblages as pollution proxies in the Northern Coast of Gabes Gulf, Tunisia. Environ Monit Assess 184:777–795

Alve E, Murray J (1994) Ecology and taphonomy of benthic foraminifera in a temperate mesotidal inlet. J Foraminifer Res 24:18–27

Arieli RN, Almogi-Labin A, Abramovich S, Herut B (2011) The effect of the thermal pollution on benthic foraminiferal assemblages in the Mediterranean shoreface adjacent to Hadera power plant (Israel). Mar Pollut Bull 62:1002–1012

Balsamo M, Albertelli G, Ceccherelli VU, Coccioni R, Colangelo MA, Curini-Galletti M, Danovaro R, D'Addabbo R, De Leonardis C, Fabiano M, Frontalini F, Gallo M, Gambi C, Guidi L, Moreno M, Pusceddu A, Sandulli R, Semprucci F, Todaro MA, Tongiorgi P (2010) Meiofauna of the Adriatic Sea: present knowledge and future perspectives. Chem Ecol 26:45–63

Balsamo M, Semprucci F, Frontalini F, Coccioni R (2012) Meiofauna as a tool for marine ecosystem biomonitoring. In: Cruzado A (ed) Marine ecosystem. InTech, Rijeka

Barmawidjaja DM, Jorissen F, Puskaric S, Van der Zwann GJ (1992) Microhabitat selection by benthic foraminifera in the northern Adriatic Sea. J Foraminifer Res 22:297–317

Bentov S, Brownlee C, Erez J (2009) The role of seawater endocytosis in the biomineralization process in calcareous foraminifera. Proc Natl Acad Sci U S A 106:21500–21504

Bergamin L, Romano E, Gabellini M, Ausili A, Carboni MG (2003) Chemical-physical and ecological characterisation in the environmental project of a polluted coastal area: the Bagnoli case study. Mediterr Mar Sci 4:5–20

Bergamin L, Romano E, Celia M (2005) Pollution monitoring of Bagnoli Bay (Tyrrhenian Sea, Naples Italy): a chemical-physical and ecological approach. Aquat Ecosyst Health 8:293–302

Bergamin L, Romano E, Finoia MG, Venti F, Bianchi J, Colasanti A, Ausili A (2009) Benthic foraminifera from the coastal zone of Baia (Naples, Italy): assemblage distribution and modification as tools for environmental characterization. Mar Pollut Bull 59:234–244

Bergin F, Kucuksezgin F, Uluturhan E, Barut IF, Meric E, Avsar N, Nazik A (2006) The response of benthic foraminifera and ostracoda to heavy metal pollution in Gulf of Izmir (Eastern Aegean Sea). Estuar Coast Shelf Sci 66:368–386

Bernhard JM, Ostermann DR, Williams DS, Blanks JK (2006) Comparison of two methods to identify live benthic foraminifera: a test between Rose Bengal and Cell Tracker Green with implications for stable isotope paleoreconstructions. Paleoceanography 21:PA4210. http://onlinelibrary.wiley.com/doi/10.1029/2006PA001290/abstract

Bizon G, Bizon JJ (1983) Les foraminifères des sediments profonds. Pêtrole et techniques 302:59–94

Blanc-Vernet L (1969) Contribution à l'Etude des Foraminifères de Méditerranée. Relation entre la microfaune et le sédiment. Biocénoses actuelles, thanatocénoses pliocènes et quaternaires. Récueil des Travaux de la Station Marine d'Endoume 64:1–251

Blanc-Vernet L, Clairefond P, Orsolini P (1979) La Mer Pelagienne. Les foraminifères Annales de l'Université de Provence 6:171–209

Borrelli C, Sabbatini A, Luna GM, Nardelli MP, Sbaffi T, Morigi C, Danovaro R, Negri A (2011) Technical note: determination of the metabolically active fraction of benthic foraminifera by means of *Fluorescent In Situ Hybridization* (FISH). Biogeosciences 8:2075–2088

Buchner P (1940) Die Lagenen des Golfes von Neapel und der marinen Ablagerungen auf Ischia (Beiträge zur Naturgeschichte der Insel Ischia 1). Nova Acta Leopoldina, Neue Folge 9:363–560

Buosi C, Frontalini F, Da Pelo S, Cherchi A, Coccioni R, Bucci C (2010) Foraminiferal proxies for environmental monitoring in the polluted lagoon of Santa Gilla (Cagliari, Italy). Present Environ Sustain Dev 4:91–104

Buosi C, du Châtelet A, Cherchi A (2012) Benthic foraminiferal assemblages in the current-dominated Strait of Bonifacio (Mediterranean Sea). J Foraminifer Res 42:39–55

Carboni MG, Frezza V, Bergamin L (2004) Distribution of recent benthic foraminifers in the Ombrone River Basin (Tuscany continental shelf, Tyrrhenian Sea Italy): relations with fluvial run off. In: Coccioni R, Galeotti S, Lirer F (eds) Proceedings in the 1st Italian meeting on environmental micropaleontology. Grzybozsky Foundation Special Publication, Urbino

Carboni MG, Succi MC, Bergamin L, Di Bella L, Frezza V, Landini B (2009) Benthic foraminifera from two coastal lakes of southern Latium (Italy). Preliminary evaluation of environmental quality. Mar Pollut Bull 59:268–280

Caruso A, Cosentino C, Tranchina L, Brai M (2011) Response of benthic foraminifera to heavy metal contamination in marine sediments (Sicilian coasts, Mediterranean Sea). Chem Ecol 27:9–30

Cherchi A, Da Pelo S, Ibba A, Mana D, Buosi C, Floris N (2009) Benthic foraminifera response and geochemical characterization of the coastal environment surrounding the polluted industrial area of Portovesme (South-Western Sardinia, Italy). Mar Pollut Bull 59:281–296

Cherif OH (1970) Die Miliolacea der West-Küste von Naxos (Griechenland) und ihre Lebensbereiche. Dissertation, University of Clausthal

Chierici MA, Busi MT, Cita MB (1962) Contribution à une étude écologique des foraminifères dans la Mer Adriatique. Rev Micropaleontol 2:123–142

Cifelli R (1974) Planktonic foraminifera from the Mediterranean and adjacent Atlantic waters (Cruise 49 of the Atlantis II, 1969). J Foraminifer Res 4:171–183

Cifelli R (1990) Foraminiferal classification from d'Orbigny to Galloway. Cushman Found Foramineral Res Spec Publ 27:1–88

Cimerman F, Langer M (1991) Mediterranean foraminifera. Slovenska Akademija Znanosti Umetnosti, Ljubljana

Cita MB, Zocchi M (1978) Distribution patterns of benthic foraminifera on the floor of the Mediterranean Sea. Oceanol Acta 1:445–462

Coccioni R (2000) Benthic foraminifera as bioindicators of heavy metal pollution – a case study from the Goro Lagoon (Italy). In: Martin RE (ed) Environmental micropaleontology: the application of microfossils to environmental geology. Kluwer Academic/Plenum Publishers, New York

Coccioni R, Frontalini F, Marsili A, Mana D (2009) Benthic foraminifera and trace element distribution: a case-study from the heavily polluted lagoon of Venice (Italy). Mar Pollut Bull 59:257–267

Colom G (1974) Foraminiferos ibéricos. Introducción al estudio de las especies bentónicas recientes. Investig Pesq 38:1–245

d'Orbigny A (1826) Tableau Métodique de la Classe des Céphalopodes. Ann Sci Nat 7:245–314

Danovaro R, Company JB, Corinaldesi C, D'Onghia G, Galil B, Gambi C, Gooday AJ, Lampadariou N, Luna GM, Morigi C, Karine O, Polymenakou P, Ramirez-Llodra E, Sabbatini A, Sardà F, Sibuet M, Tselepides A (2010) Deep-Sea biodiversity in the Mediterranean Sea: the known, the unknown and the unknowable. PLoS One 5:e11832. http://www.plosone.org/article/info%3Adoi%2F10.1371%2Fjournal.pone.0011832

De Montfort PD (1808) Conchyliologie systématique et classification méthodique des Coquilles. Schoell, Paris

de Nooijer LJ, Toyofuku T, Kitazato H (2009) Foraminifera promote calcification by elevating their intracellular pH. Proc Natl Acad Sci U S A 106:15374–15378

de Rijk S, Troelstra SR, Rohling EJ (1999) Benthic foraminiferal distribution in the Mediterranean Sea. J Foraminifer Res 29:93–103

de Rijk S, Jorissen F, Rohling EJ, Troelstra SR (2000) Organic flux on bathymetric zonation of Mediterranean benthic foraminifera. Mar Micropaleontol 40:151–166

de Stigter HC (1996) Recent and fossil benthic foraminifera in the Adriatic Sea: distribution patterns in relation to organic carbon flux and oxygen concentration at the seabed. Geologica Ultraiectina 144:1–254

Di Leonardo R, Bellanca A, Capotondi L, Cundy A, Neri R (2007) Possible impacts of Hg and PAH contamination on benthic foraminiferal assemblages: an example from the Sicilian coast, central Mediterranean. Sci Total Environ 388:168–183

Donnici S, Serandrei Barbero R (2002) The benthic foraminiferal communities of the northern Adriatic continental shelf. Mar Micropaleontol 44:93–123

Donnici S, Serandrei Barbero R, Taroni G (1997) Living benthic foraminifera in the Lagoon of Venice (Italy). Populations dynamics and its significance. Micropaleontology 43:440–454

Duijnstee I (2001) Experimental ecology of foraminifera. Dissertation, University of Utrecht

Duijnstee I, de Lugt I, Vonk Noordegraaf H, van der Zwaan B (2004) Temporal variability of foraminiferal densities in the northern Adriatic Sea. Mar Micropaleontol 50:125–148

Dujardin F (1835) Recherche sur les organismes inférieurs. Ann Sci Nat Paris Zool 2:343–377

Elshanawany R, Ibrahim MI, Milker Y, Schmiedl G, Badr N, Kholeif SEA, Zonneveld KAF (2011) Anthropogenic impact on benthic foraminifera, Abu-Qir Bay, Alexandria, Egypt. J Foraminifer Res 41:326–348

Emiliani C (1955) Pleistocene temperatures. J Geol 63:538–578

Ernst SR (2002) An experimental study on the proxy value of benthic foraminifera. Geologica Ultraiectina 220:1–157

Ferraro L, Lirer F (2006) Morphological variations of benthonic foraminiferal tests from Holocene sediments of the Punta Campanella shelf (south Tyrrhenian Sea). In: Coccioni R, Marsili A (eds) Proceedings of the second and third Italian meetings on environmental micropaleontology. Grzybozsky Foundation Special Publication, Urbino

Ferraro L, Sprovieri M, Alberico I, Lirer F, Prevedello L, Marsella E (2006) Benthic foraminifera and heavy metals distribution: a case study from the Naples Harbour (Tyrrhenian Sea, southern Italy). Environ Pollut 142:274–287

Ferraro L, Sammartino S, Feo ML, Rumolo P, Salvagio Manta D, Ennio Marsella E, Sprovieri M (2009) Utility of benthic foraminifera for biomonitoring of contamination in marine sediments: a case study from the Naples Harbour (Southern Italy). J Environ Monit 11:1226–1235

Fichtel L, Moll JPC (1798) Testacea Microscopica Aliaque Minuta ex Generibus Argonauta et Nautilus ad Naturam Delineata et Descripta. Pichler, Wien

Fontanier C, Jorissen FJ, Lansard B, Mouret A, Buscail R, Schmidt S, Kerherve´ P, Buron F, Zaragosi S, Hunault G, Ernoult E, Artero C, Anschutz A, Rabouille C (2008) Live foraminifera from the open slope between Grand Rhone and Petit Rhone Canyons (Gulf of Lions, NW Mediterranean). Deep-Sea Res I 55:1532–1553

Fornasini C (1902) Sinossi metodica dei foraminiferi sin qui rinvenuti nella sabbia de Lido di Rimini. Memorie della Regia Accademia de le Scienze de l'Istituto di Bologna 5:205–212

Fornasini C (1904) Illustrazione di specie orbignyane di Miliolidi istituite nel 1826. Memorie della Regia Accademia de le Scienze de l'Istituto di Bologna 6:1–17

Fornasini C (1905) Illustrazione di specie orbignyane di Miliolidi istituite nel 1826. Memorie della Regia Accademia de le Scienze de l'Istituto di Bologna 6:1–14

Fornasini C (1906a) Illustrazione di specie orbignyane di "Rotalidi" istituite nel 1826. Memorie della Regia Accademia de le Scienze de l'Istituto di Bologna 6:61–70

Fornasini (1906b) Illustrazione di specie orbignyane di Nodosaridi, di Rotalidi et d'altri foraminiferi. Memorie della Regia Accademia de le Scienze de l'Istituto di Bologna 6:41–54

Frezza V, Carboni MG (2009) Distribution of recent foraminiferal assemblages near the Ombrone river mouth (Northern Tyrrhenian Sea, Italy). Rev Micropaleontol 52:43–66

Frontalini F, Coccioni R (2008) Benthic foraminifera for heavy metal pollution monitoring: a case study from the central Adriatic Sea coast of Italy. Estuar Coast Shelf Sci 76:404–417

Frontalini F, Coccioni R (2011) Benthic foraminifera as bioindicators of pollution: a review of Italian research over the last three decades. Rev Micropaleontol 54:115–127

Frontalini F, Buosi C, Da Pelo S, Coccioni R, Cherchi A, Bucci C (2009) Benthic foraminifera as bio-indicators of trace element pollution in the heavily contaminated Santa Gilla lagoon (Cagliari, Italy). Mar Pollut Bull 58:858–877

Frontalini F, Coccioni R, Bucci C (2010) Benthic foraminiferal assemblages and trace element contents from the lagoons of Orbetello and Lesina. Environ Monit Assess 170:245–260

Frontalini F, Semprucci F, Coccioni R, Balsamo M, Bittoni P, Covazzi-Harriague A (2011) On the quantitative distribution and community structure of the meio and macrofaunal communities in the coastal area of the Central Adriatic Sea (Italy). Environ Monit Assess 180:325–344

Goineau A, Fontanier C, Jorissen FJ, Lansard B, Buscail R, Mouret A, Kerhervé P, Zaragosi S, Ernoult E, Artéro C, Anschutz P, Metzger E, Rabouill C (2011) Live (stained) benthic foraminifera from the Rhône prodelta (Gulf of Lion, NW Mediterranean): environmental controls on a river-dominated shelf. J Sea Res 65:58–75

Goineau A, Fontanier C, Jorissen FJ, Buscail R, Kerhervé P, Cathalot C, Pruski AM, Lantoine F, Bourgeois S, Metzger E, Legrand E, Rabouille C (2012) Temporal variability of live (stained) benthic foraminiferal faunas in a river-dominated shelf – faunal response to rapid changes of the river influence (Rhone prodelta, NW Mediterranean). Biogeosciences 9:1367–1388

Gooday AJ (1993) Deep-sea benthic foraminiferal species which exploit phytodetritus: characteristic features and controls on distribution. Mar Micropaleontol 22:187–205

Gooday AJ (2002) Organic-walled allogromiids: aspects of their occurrence, diversity and ecology in marine habitats. J Foraminifer Res 32:384–399

Gooday AJ, Nomaki H, Kitazato H (2008) Modern deep-sea benthic foraminifera: a brief review of their biodiversity and trophic diversity. In: Austin WEN, James RH (eds) Biogeochemical controls on palaeoceanographic environmental proxies. Geological Society, London

Gooday AJ, Jorissen F, Levin LA, Middelburg JJ, Naqvi SWA, Rabalais NN, Scranton M, Zhang J (2009) Historical records of coastal eutrophication-induced hypoxia. Biogeosciences 6:1707–1745

Gooday AJ, Bett BJ, Escobar E, Ingole B, Levin LA, Neira C, Raman AV, Sellanes J (2010) Habitat heterogeneity and its influence on benthic biodiversity in oxygen minimum zones. Mar Ecol 31:125–147

Grüber A (1884) Die Protozoen des Hafens von Genua. Nova Acta der kaiserlich-Leopoldinisch-Carolinische. Deutschen Akademie der Naturforscher 46:475–539

Hallock P, Lidz BH, Cockey-Burkhard EM, Donnelly KB (2003) Foraminifera as bioindicators in coral reef assessment and monitoring: the foram index. Environ Monit Assess 81:221–238

Haq BU, Boersma A (1998) Introduction to marine micropaleontology. Elsevier, Amsterdam

Hatziyanni E (1999) Ecology of benthic meiofauna and foraminifera. Dissertation, University of Crete

Hayward BW, Cedhagen T, Kaminski M, Gross O (2011) World Modern Foraminifera database. http://www.marinespecies.org/foraminifera. Accessed 22 Aug 2012

Heinz P, Schmiedl G, Kitazato H, Hemleben C (2001) Response of deep-sea benthic foraminifera from the Mediterranean sea to simulated phytoplankton pulses under laboratory conditions. J Foraminifer Res 31:210–227

Hemleben C, Spindler M, Anderson OR (1989) Modern planktonic foraminifera. Springer, New York

Hillaire-Marcel C, de Vernal A (2007) Proxies in Late Cenozoic Paleoceanography. Elsevier Science, Amsterdam

Hofker J (1960) Foraminiferen aus dem Golf von Neapel. Paläontol Z 34(3/4):233–262

Hohenegger J, Piller WE, Baal C (1993) Horizontal and vertical spatial microdistribution of foraminifers in the shallow subtidal Gulf of Trieste, Northern Adriatic Sea. J Foraminifer Res 23:79–101

Holzmann M, Pawlowski J (2002) Freshwater foraminifera from lake Geneva: past and present. J Foraminifer Res 32:344–350

Holzmann M, Habura A, Giles H, Bowser S, Pawlowski J (2003) Freshwater foraminiferans revealed by analysis of environmental DNA samples. J Eukaryot Microbiol 50:135–139

Huxley T (1910) Note on *Shepheardella taeniformis* Siddall. Zool Anz 36:124–125

Hyams-Kaphzan O, Almogi-Labin A, Sivan D, Benjamini C (2008) Benthic foraminifera assemblage change along the southeastern Mediterranean inner shelf due to fall-off of Nile-derived siliciclastics. N Jahrb Geol Palaeontol 248:315–344

Hyams-Kaphzan O, Almogi-Labin A, Benjamini C, Herut B (2009) Natural olygotrophy vs. pollution-induced eutrophy on the SE Mediterranean shallow shelf (Israel): environmental parameters and benthic foraminifera. Mar Pollut Bull 58:1888–1902

Ingels J, Vanreusel A, Brandt A, Catarino AI, David B, De Ridder C, Dubois P, Gooday AJ, Martin P, Pasotti F, Robert H (2012) Possible effects of global environmental changes on Antarctic benthos: a synthesis across five major taxa. Ecol Evol 2:453–485

Jannink NT (2001) Seasonality, biodiversity and microhabitats in benthic foraminiferal communities. Geologica Ultraiectina 203:1–192

Jones TR, Parker WK (1860) On the rhizopodal fauna of the Mediterranean, compared with that of the Italian and some other Tertiary deposits. Q J Geol Soc 16:292–307
Jorissen FJ (1987) The distribution of benthic foraminifera in the Adriatic Sea. Mar Micropaleontol 12:21–48
Jorissen FJ (1988) Benthic foraminifera from the Adriatic Sea; principles of phenotypic variations. Utrecht Micropaleontol Bull 37:1–174
Jorissen FJ, Barmawidjaj DM, Puskaric S, van der Zwaan GJ (1992) Vertical distribution of benthic foraminifera in the northern Adriatic Sea: the relation with the organic flux. Mar Micropaleontol 19:131–146
Lampadariou N, Tselepides A, Hatziyanni E (2009) Deep-sea meiofaunal and foraminiferal communities along a gradient of primary productivity in the eastern Mediterranean Sea. Sci Mar 73:337–345
Le Cadre V, Debenay JP (2006) Morphological and cytological responses of *Ammonia* (Foraminifera) to copper contamination: implication for the use of foraminifera as bioindicators of pollution. Environ Pollut 143:304–317
Le Calvez J, Le Calvez Y (1958) Répartition des Foraminifères dans la Baie de Villefranche. I Miliolidae. Ann Inst Oceanogr 35(3):159–234
Lipps JH (1983) Biotic interactions in benthic foraminifera. In: Trevesz MJS, McCall PL (eds) Biotic interactions in recent and fossil benthic communities. Plenum Press, New York
Lipps JH, Finger KL, Walker SE (2011) What should we call the foraminifera? J Foraminifer Res 41:309–313
Madkour HA, Ali MY (2009) Heavy metals in the benthic foraminifera from the coastal lagoons, Red Sea, Egypt: indicators of anthropogenic impact on environment (case study). Environ Geol 58:543–553
Massiotta R, Cita MB, Mancuso M (1976) Benthonic foraminifers from bathyal depths in the Eastern Mediterranean. Marit Sediments Spec Publ 1:251–262
McIlroy D, Green OR, Brasier MD (1994) The world's oldest foraminiferans. Microsc Anal 147:13–15
Meisterfeld R, Holzmann M, Pawlowski J (2001) Morphological and molecular characterization of a new terrestrial allogromiid species: *Edaphoallogromia australica* gen. et spec. nov. (Foraminifera) from Northern Queensland (Australia). Protist 152:185–192
Milker Y, Schmiedl G, Betzler C, Römer M, Jaramillo-Vogel D, Siccha M (2009) Distribution of recent benthic foraminifera in shelf carbonate environments of the Western Mediterranean Sea. Mar Micropaleontol 73:207–229
Mojtahid M, Jorissen FJ, Lansard B, Fontanier C, Bombled B, Rabouille C (2009) Spatial distribution of live benthic foraminifera in the Rhône prodelta: faunal response to a continental–marine organic matter gradient. Mar Micropaleontol 70:177–200
Mojtahid M, Jorissen FJ, Lansard B, Fontanier C (2010) Microhabitat selection of benthic foraminifera in sediments off the Rhone river mouth (NW Mediterranean Sea). J Foraminifer Res 40:231–246
Moodley L, van der Zwaann GJ, Herman PMJ, Kempers L, van Breugel P (1997) Differential response of benthic meiofauna to anoxia with special reference to foraminifera (Protista: Sarcodina). Mar Ecol Prog Ser 158:151–163
Moodley L, Middelburg JJ, Boschker HTS, Duineveld GCA, Pel R, Herman PMJ, Heip CHR (2002) Bacteria and foraminifera: key players in a short-term deep-sea benthic response to phytodetritus. Mar Ecol Prog Ser 236:23–29
Morigi C (1999) Clima e produttività oceanica nel tardo Quaternario: analisi statistica delle associazioni a foraminiferi nell'Atlantico tropicale. Dissertation, University of Bologna
Murray JW (1982) Benthic foraminifera: the validity of living, dead or total assemblages for the interpretation of palaeoecology. J Micropalaeontol 1:137–140
Murray JW (1991) Ecology and palaeoecology of benthic foraminifera. Wiley/Longman Scientific and Technical, New York/Harlow
Murray JW (2000) The enigma of the continued use of total assemblages in ecological studies of benthic foraminifera. J Foraminifer Res 30:244–245
Murray JW (2007) Biodiversity of living benthic foraminifera: how many species are there? Mar Micropaleontol 64:163–176
Murray JW, Bowser SS (2000) Mortality, protoplasm decay rate, and reliability of staining techniques to recognize 'living' foraminifera: a review. J Foraminifer Res 30:66–70
Myers EH (1943) Life activities of foraminifera in relation to marine ecology. Proc Am Philos Soc 86:439–459
Nardelli MP (2012) Response of benthic foraminifera to pollution through experimental and *in situ* studies: advances in biological aspects and tools for future application in biomonitoring. Dissertation, Polytechnic University of Marche
Nyholm K-G (1951) A new monothalamous foraminifer. *Marenda nematoides* n. gen. N. sp. Contrib Cushman Found Foraminifer Res 2:95
Pancotti I (2011) Variazioni longitudinali, batimetriche e biogeografiche di abbondanza, biomassa e diversità della meiofauna a foraminiferi bentonici nel Mediterraneo profondo e Atlantico. Dissertation, Polytechnic University of Marche
Panieri G (2005) Benthic foraminifera from a recent, shallow-water hydrothermal environment in the Aeolian Arc (Tyrrhenian Sea). Mar Geol 218:207–229
Panieri G (2006a) Foraminiferal response to an active methane seep environment: a case study from the Adriatic Sea. Mar Micropaleontol 61:116–130
Panieri G (2006b) The effect of shallow marine hydrothermal vent activity on benthic foraminifera (Aeolian Arc, Tyrrhenian Sea). J Foraminifer Res 36:3–14
Parisi E (1981) Distribuzione dei foraminiferi bentonici nelle zone batiali del Tirreno e del Canale di Sicilia. Riv Ital Paleontol 87:293–328
Parker WK (1958) Eastern Mediterranean foraminifera. Rep Swed Deep-Sea Exped 8:217–283
Phleger FB (1960) Ecology and distribution of recent foraminifera. The Johns Hopkins Press, Baltimore
Pranovi F, Serandrei Barbero R (1994) Benthic communities of northern Adriatic areas subject to anoxic conditions. Memorie di Scienze Geologiche 46:79–92
Pucci F (2006) Ecologia dei foraminiferi bentonici: risposta sperimentale alle condizione di anossia e applicazione alla ricostruzione paleo ambientale. Dissertation, Polytechnic University of Marche
Pucci F, Geslin E, Barras C, Morigi C, Sabbatini A, Negri A, Jorissen FJ (2009) Survival of benthic foraminifera under hypoxic conditions: results of an experimental study using the Cell Tracker Green method. Mar Pollut Bull 59:336–351
Pujol C, Vergnaud-Grazzini C (1995) Distribution patterns of live planktic foraminifera as related to regional hydrography and productive systems of the Mediterranean Sea. Mar Micropaleontol 25:187–217
Rhumbler R (1904) Systematische Zusammenstallung der recenten Reticulosa (Nuda + Foraminifera) I Teil. Archiv für Protistenkunde 3:181–294
Rohling EJ, Cooke S (2002) Stable oxygen and carbon isotopes in foraminiferal carbonate shelles. In: Sen Gupta BK (ed) Modern foraminifera. Kluwer Academic Publishers, London
Romano E, Bergamin L, Finoia MG, Carboni MG, Ausili A, Gabellini M (2008) Industrial pollution at Bagnoli (Naples, Italy): benthic foraminifera as a tool in integrated programs of environmental characterisation. Mar Pollut Bull 56:439–457
Romano E, Bergamin L, Finoia MG, Magno MC, Mercatali I, Ausili A, Gabellini M (2009a) The effects of human impact on benthic foraminifera in the Augusta harbour (Sicily, Italy). In: Moksness E, Dahl E, Støttrup JG (eds) Integrated coastal zone management. Wiley-Blackwell Publishing, London

Romano E, Bergamin L, Ausili A, Pierfranceschi G, Maggi C, Sesta G, Gabellini M (2009b) The impact of the Bagnoli industrial site (Naples, Italy) on sea-bottom environment. Chemical and textural features of sediments and the related response of benthic foraminifera. Mar Pollut Bull 59:245–256

Sabbatini A, Morigi C, Negri A, Gooday AJ (2002) Soft-shelled benthic foraminifera from a hadal site (7,800 m water depth) in the Atacama Trench (SE Pacific): preliminary observations. J Micropalaeontol 21:131–135

Sabbatini A, Bonatto S, Gooday AJ, Morigi C, Pancotti I, Pucci F, Negri A (2010) Modern benthic foraminifes at Northern shallow sites of Adriatic Sea and soft-walled, monothalamous taxa: a brief overview. Micropaleontology 59:359–376

Sabbatini A, Bonatto S, Bianchelli S, Pusceddu A, Danovaro R, Negri A (2012) Foraminiferal assemblages and trophic state in coastal sediments of the Adriatic Sea. J Mar Syst 105:163–174. doi:10.1016/j.jmarsys.2012.07.009

Samir AM, El-Din AB (2001) Benthic foraminiferal assemblages and morphological abnormalities as pollution proxies in two Egyptian bays. Mar Micropaleontol 41:193–227

Schmiedl G, de Bovée F, Buscail R, Charrière B, Hemleben C, Medernach L, Picon P (2000) Trophic control of benthic foraminiferal abundance and microhabitat in the bathyal Gulf of Lions, western Mediterranean Sea. Mar Micropaleontol 40:167–188

Schönfeld J (2012) History and development of methods in recent benthic foraminiferal studies. J Micropalaeontol 31:53–72

Schönfeld J, Alve E, Geslin E, Jorissen F, Korsun S, Spezzaferri S, Members of the FOBIMO group (2012) The FOBIMO (FOraminiferal BIo-MOnitoring) initiative – towards a standardised protocol for soft-bottom benthic foraminiferal monitoring studies. Mar Micropalaeontol 94–95:1–13

Scott DB, Medioli FS (1980) Living vs total foraminiferal populations: their relative usefulness in paleoecology. J Paleontol 54:814–831

Sen Gupta BK (2002) Modern foraminifera. Kluwer Academic Publishers, London

Serandrei Barbero R, Albani A, Favero V (1989) Distribuzione dei foraminiferi recenti nella laguna a Nord di Venezia. Boll Soc Geol Ital 108:279–288

Serandrei Barbero R, Carbognin L, Taroni G, Cova E (1999) Distribution of recent benthic foraminifera in the southern basin of the Venice lagoon (Italy): statistical evaluation of taxa significance. Micropaleontology 45:99–111

Serandrei Barbero R, Morisieri M, Carbognin L, Donnici S (2003) An inner shelf foraminiferal fauna and its response to environmental processes (Adriatic Sea, Italy). Revista Española de Micropaleontología 35:241–264

Sgarrella F, Barra B, Improta A (1983) The benthic foraminifers of the Gulf of Policastro (Southern Tyrrhenian Sea, Italy). Bollettino della Società dei Naturalisti in Napoli 92:67–114

Sharifi AR, Croudace TW, Austin RL (1991) Benthonic foraminiferids as pollution indicators in Southampton water, Southern England, UK. J Micropalaeontol 10:109–113

Soetaert K, Heip C, Vincx M (1991) The meiobenthos along a Mediterranean deep-sea transect off Calvi (Corsica) and in an adjacent canyon. Mar Ecol 12:227–242

Soldani A (1789) Testaceographiae ac Zoophytographiae parvae et microscopicae. Tomus Primus. Senis, Firenze

Soldani A (1795) Testaceographiae ac Zoophytographiae parvae et microscopicae. Tomus Primus pars tertia. In: Molini G, Rossi FS (eds) Testaceographiae ac Zoophytographiae parvae et microscopicae. Senis, Firenze

Thunell RC (1978) Distribution of recent planktonic foraminifera in surface sediments of the Mediterranean Sea. Mar Micropaleontol 3:147–173

Todd R (1958) Foraminifera from western Mediterranean deep-sea cores. Rep Swed Deep-Sea Exped 8:167–215

Valenti D, Tranchina L, Brai M, Caruso A, Cosentino C, Spagnolo B (2008) Environmental metal pollution considered as noise: effects on the spatial distribution of benthic foraminifera in two coastal marine areas of Sicily (southern Italy). Ecol Model 213:449–462

Vénec-Peyré MT (1990) Contribution of foraminifera to the study of recent sedimentation in the gulf of Lions (western Mediterranean Sea). Cont Shelf Res 10:869–883

von Linnaeus C (1758) Systema naturae, sive regna tria natura systematice proposita per classes, ordines, genera et species. Holmiae, Laurentii Salvii, Stockholm

Walton WR (1953) Techniques for recognition of living foraminifera. Contrib Cushman Found Foraminifer Res 3:56–60

Wiesner H (1923) Die miliolideen der östlichen Adria: Privately printed, Prag-Bubenec, 113 p

Yanko V, Arnold AJ, Parker WC (1999) Effects of marine pollution on benthic foraminifera. In: Sen Gupta BK (ed) Modern foraminifera. Kluwer Academic Publishers, London

Mediterranean Corals Through Time: From Miocene to Present

14

Agostina Vertino, Jarosław Stolarski,
Francesca R. Bosellini, and Marco Taviani

Abstract

Stony corals, especially scleractinians, are a recurrent component of the benthic fauna of the Mediterranean basin and its Mesozoic-to-Cenozoic precursors. Both morphological and geochemical features of coral skeletons place these organisms among the most important natural paleoarchives of the Mediterranean geological history. The present day low diversity of the Mediterranean scleractinian fauna (25 genera and only 33 species) strikingly contrasts with its high diversity in the Early-Middle Miocene (over 80 genera and hundreds of species). The decline in coral richness has occurred since the late Middle Miocene onwards. This impoverishment trend was not linear, but abrupt in shallow-water environments during and immediately after the Late Miocene and more gradual since the Pliocene onwards. At the end of the Miocene, the Mediterranean coral fauna underwent a drastic modification that led to the disappearance of almost all zooxanthellate corals and the well-established shallow-water coral-reef province. However, the generic diversity of azooxanthellate and deep-water corals did not undergo significant modifications, that were instead much stronger at the end of the Pliocene and of the Pleistocene. Indeed, before the Calabrian stage, all remnant Indo-Pacific-like azooxanthellate genera disappeared and a clear NE Atlantic affinity was established, whereas at the Pleistocene – Holocene boundary, there was a clear reduction in psychrospheric deep-water taxa. The causes that led to the impoverishment of the Mediterranean coral fauna diversity are complex and not all fully understood. However, there is a clear link between the coral diversity decrease, the gradual northward shift outside the tropical belt of the Mediterranean region, and the major climate modifications on a global scale during the last 20 million years.

Keywords

Corals • Scleractinia • Mediterranean • Neogene • Pleistocene • Recent • Climate

A. Vertino (✉)
Department of Earth and Environmental Sciences,
University of Milano-Bicocca, Piazza della Scienza 4,
20126, Milan, Italy
e-mail: agostina.vertino@unimib.it

J. Stolarski
Institute of Paleobiology, Polish Academy of Sciences,
Twarda 51/55, 00-818 Warsaw, Poland
e-mail: stolacy@twarda.pan.pl

F.R. Bosellini
Department of Chemical and Geological Sciences,
University of Modena and Reggio Emilia,
L.go S.Eufemia 19, 41100 Modena, Italy
e-mail: francesca.bosellini@unimore.it

M. Taviani
Institute of Marine Sciences, National Research Council
(ISMAR-CNR), Bologna, via Gobetti 101, 40129 Bologna, Italy

Biology Department, Woods Hole Oceanographic Institution,
266 Woods Hole Road, Woods Hole, MA 02543, USA
e-mail: marco.taviani@bo.ismar.cnr.it

S. Goffredo and Z. Dubinsky (eds.), *The Mediterranean Sea: Its history and present challenges*,
DOI 10.1007/978-94-007-6704-1_14, © Springer Science+Business Media Dordrecht 2014

Introduction

"Coral" is a popular term used to refer to marine animals in the class Anthozoa (phylum Cnidaria) that typically form colourful and fascinating colonies. More properly, the locution "stony corals" (Cairns et al. 1999) is instead used to indicate those cnidarians that form calcified skeletons and that belong to both the class Anthozoa (order Scleractinia and subclass Octocorallia *partim*) and Hydrozoa (family Stylasteridae).

This paper is mostly focused on the stony coral representatives of the order Scleractinia, whose evolutionary history dates back to the Paleozoic (Stolarski et al. 2011), though it is mostly known from its sudden Mesozoic appearance about 240 Ma (Roniewicz and Morycowa 1993). Scleractinians produce an aragonite skeleton, with some possible exceptions in the Cretaceous (Stolarski et al. 2007), and comprise the main reef-building organisms of both shallow and deep modern seas. They have long been used for reconstructing the history of oceans and seas, mainly because they can build large carbonate structures ("reefs") which are well preserved in the geological record, and because most extant reef-building scleractinian species are restricted to tropical shallow marine waters (Veron 1995, 2000) thus having a distinct climatic significance (Rosen 1999). Indeed reef-building zooxanthellate scleractinians, i.e., living in symbiosis with the unicellular algae zooxanthellae (e.g., Stanley and van de Schootbrugge 2009), thrive in the modern euphotic zone of the circum-global latitude belt comprised between 35–40°N and about 35°S where the lowest mean annual sea-surface temperature is around 16 °C (Rosen 1999). Azooxanthellate scleractinians, deprived of photosymbionts, are almost as rich in species as the zooxanthellate ones (711 *vs* 777 species, Roberts et al. 2009) and also include frame-building taxa which function as primary engineers of extensive carbonate bioconstructions. However they are much less known than the tropical cousins because they thrive in dark, cold (mostly 4–8 °C) and generally deep (up to > 1,000 m) oceanic waters (e.g. Freiwald et al. 2004).

Zooxanthellate (z-) and azooxanthellate (az-) corals have been a recurrent component of the benthic fauna of the Mediterranean basin and its Mesozoic-to-Cenozoic precursors. Being particularly sensitive to environmental modifications, they have undergone several changes in both diversity, reef-building capacity and geographic distribution through time. However, despite the spectacular record in Cenozoic deposits of peri-Mediterranean regions, the history of Mediterranean corals (in particular azooxanthellate ones) is still poorly understood.

This contribution provides a critical overview of Miocene to Recent Mediterranean corals, with special emphasis on the az-scleractinian genera distribution through space and time (Figs. 14.1–14.3). Our inferences are based on literature review and analysis of fossil and modern coral specimens held in the collections of several institutions (Museum of Paleontology and Botanical Garden, University of Modena and Reggio Emilia, IPUM; "G. Gemmellaro" Museum of Palaeontology and Geology, University of Palermo, MPG-P; Museum of Paleontology, University of Catania, MP-C; Museum of Natural History, Turin University, MN-T; Museum of Paleontology, Sapienza University of Rome, MP-R; Department of Earth Sciences, University Milano-Bicocca, DES-M; Institute of Marine Sciences – CNR, Bologna; Muséum National d'Histoire Naturelle, Paris; Smithsonian National Museum of Natural History, Washington D.C., NMNH). We present and briefly discuss the most relevant aspects related to the presence in the Mediterranean basin of both zooxanthellate and azooxanthellate coral taxa during about the last 20 million years. The distinction between z- and az-coral genera presented in this paper follows the criteria proposed by Wilson and Rosen (1998) and Perrin and Bosellini

Fig. 14.1 Map showing the Mediterranean Sea and the location of the Miocene to Pleistocene outcrops containing the fossil azooxanthellate scleractinian corals (az-corals) mentioned in the text. Colouring of stratigraphic units (*colours of stars*) follows the Commission for the Geological Map of the World (http://www.ccgm.org; see also http://www.stratigraphy.org)

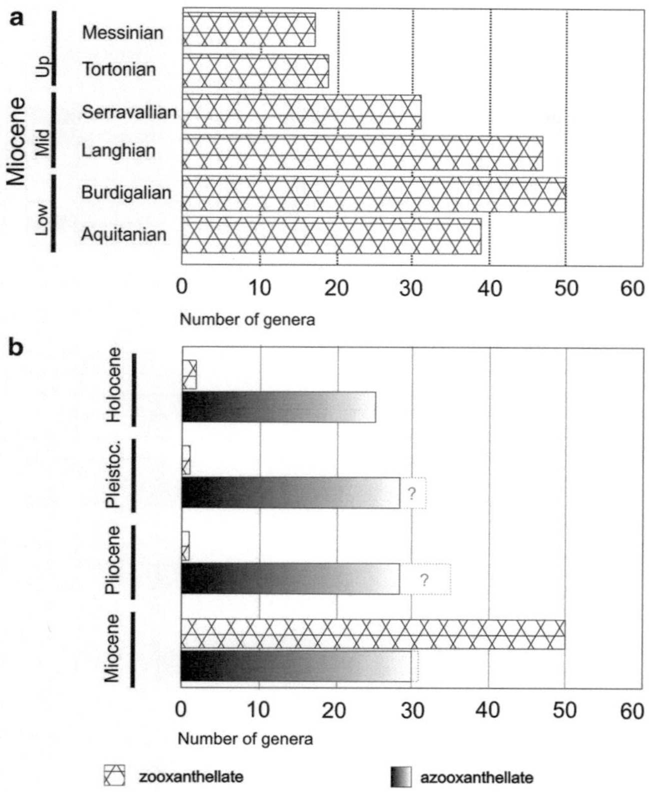

Fig. 14.2 Abundance of genera of zooxanthellate scleractinian corals during the Miocene (**a**) and of both zooxanthellate and azooxanthellate scleractinian corals from Miocene to Holocene (**b**). *White* histograms with *grey dotted lines* and *question marks* indicate taxa mentioned in previous works and whose occurrence needs to be verified (Fig. 14.2a modified after Bosellini and Perrin 2008)

(2012) for fossil corals. Although zooxanthellae are not preserved in the fossil record as body fossils (Stanley and Helmle 2010), we infer with some confidence their former presence in the coral tissues based upon actualistic assumption made at the generic systematic level.

Our work aims to highlight the importance of fossil corals as a natural archive of information on the history of the Mediterranean basin, and to foster the need for more studies related to both taxonomic and stratigraphic aspects of Neogene-Pleistocene coral deposits from peri-Mediterranean regions. In fact, most information on Neogene corals, and in particular azooxanthellate ones, dates back to the first half of the twentieth century, therefore both the stratigraphy of coral-bearing deposits and the taxonomy at the species level is in need of a comprehensive review. Some recent overviews of Oligo-Miocene Mediterranean z-corals (Bosellini and Perrin 2008; Perrin and Bosellini 2012), of Pleistocene az-scleractinians (Vertino 2003), and this study represent a first step towards critical revision of fossil Mediterranean corals, but there is still a long road ahead.

Mediterranean Coral Fauna in the Miocene

Early-Middle Miocene

Aquitanian z-corals are documented from a number of localities (e.g., Chevalier 1961; Perrin and Bosellini 2012), but their diversity seems to be modest. Az-corals from the same stage are instead virtually unknown except for the occasional records of *Desmophyllum*, *Phyllangia* and some rhizangiid corals (Chevalier 1961; Oosterban 1988; Cahuzac and Chaix 1993, 1996).

The Burdigalian-Langhian seems to have been a particularly favorable time for corals. They have been reported from several localities (review in Chevalier 1961; Bosellini and Perrin 2008; Perrin and Bosellini 2012) and in northern Italy alone (mostly in the Turin Hills deposits) over 70 scleractinian genera, including z- and az-corals, have been recorded. The Burdigalian shallow-water bioconstructions were of moderate lateral extension, with Porites and Tarbellastraea as the most common genera associated with *Favites*, *Montastraea*, *Favia*, *Solenastrea*, *Siderastrea*, *Thegioastrea*, *Caulastraea*, *Goniopora* and *Stylophora* (Perrin and Bosellini 2012). These corals were found in shallow-water mixed carbonate-siliciclastic sediments as well as in mass-gravity dominated deposits (e.g., Turin Hills localities), mixed with deeper-water az-corals. Langhian z-coral deposits document a Middle-Miocene latitudinal expansion of true coral reefs and reef complexes, followed by a contraction during the Serravallian, when z-corals apparently disappear from distinct areas such as the Aquitaine Basin, Catalonia, Corsica, Sardinia, Gulf of Suez and Central Paratethys (Perrin et al. 1998; Perrin and Bosellini 2012). These local environmental deteriorations are reflected in a strong decrease of z-coral genera between Langhian and Serravallian (Bosellini and Perrin 2008; Fig. 14.2a).

The Early-Middle Miocene also represents a flourishing time interval for Mediterranean az-corals. So far 30 genera and 7 subgenera can be certainly ascribed to this time period (Figs. 14.2–14.3), and the occurrence of further taxa (mentioned in nineteenth and early twentieth century literature) needs to be verified. Early-Middle Miocene az-corals have been recorded from several peri-Mediterranean regions (Fig. 14.1). However, the most diverse fauna examined so far pertains to Burdigalian-Langhian deposits from northern Italy (Turin Hills), where exceptionally large morphotypes, such as *Balanophyllia* specimens over 10 cm in length (Zuffardi-Comerci 1932; Chevalier 1961; A. Vertino pers. obs.), are rather common. More than half of the Early-Middle Miocene az-scleractinian genera are colonial and morphologically indistinguishable from taxa living today in (i) infralittoral (e.g., *Astrangia*, *Cladangia*, *Cladocora*, Fig. 14.4c, d, *Culicia*, *Madracis* "*pharensis*-like"), (ii)

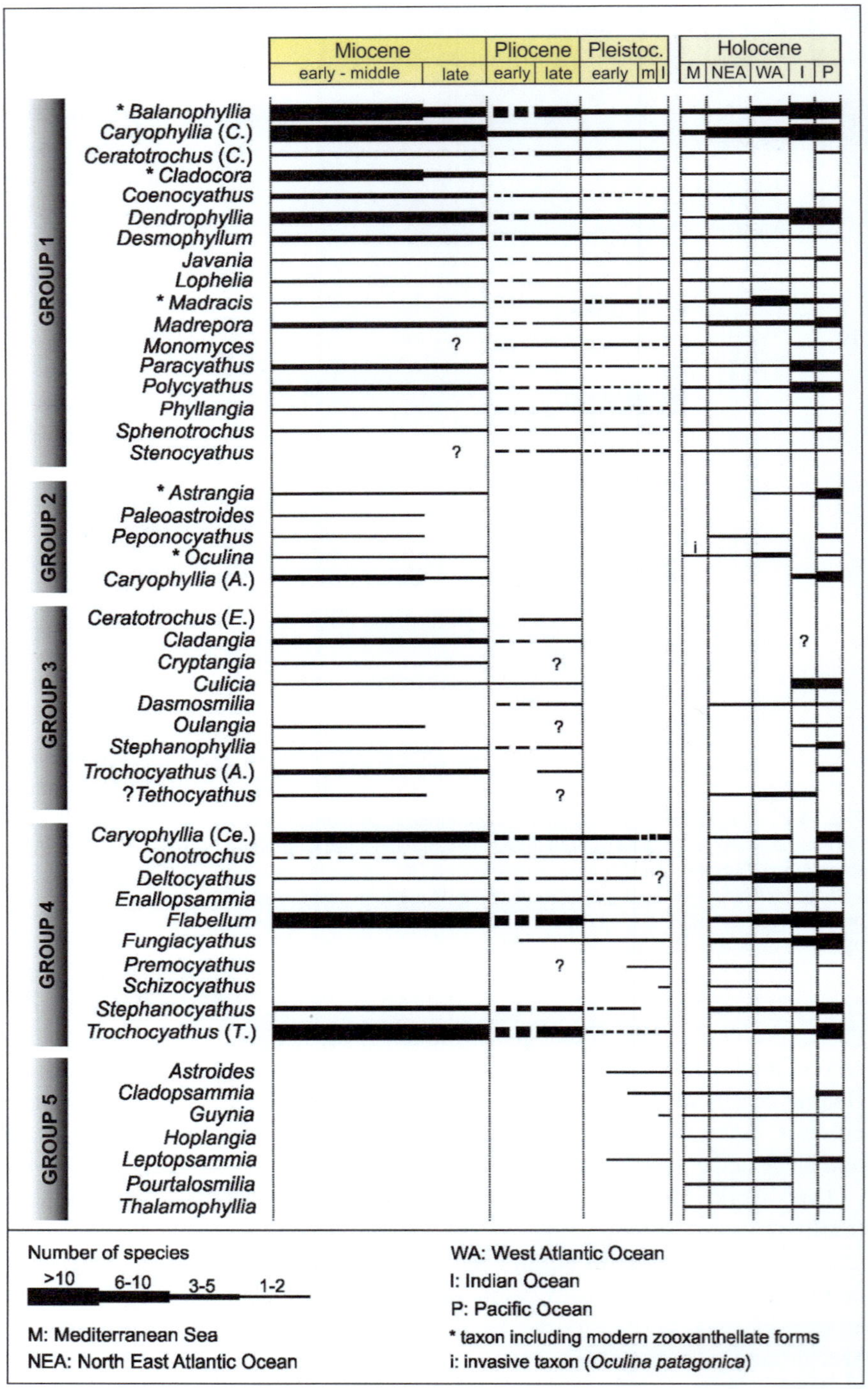

Fig. 14.3 List of Miocene to Recent Mediterranean azooxanthellate scleractinian genera and subgenera recognized so far. Taxa assigned to five main groups in line with their presence in the Mediterranean through time. *Group 1*: present since the Miocene or at least the Early Pliocene; *group 2*: disappeared after the Miocene; *group 3*: disappeared after the Pliocene; *group 4*: disappeared after the Pleistocene; *group 5*: currently living in the Mediterranean and not known before the Pleistocene or Holocene. Present distribution (and symbiotic status) of corals in the Mediterranean Sea, East–west Atlantic, Indian and Pacific oceans as stated in Zibrowius 1980; Vertino 2003; Roberts et al. 2009. Extant genera and subgenera, with exception of *Caryophyllia* (*Ceratocyathus*), according to Cairns and Kitahara 2012. Stratigraphy after the International Chronostratigraphic Chart 2012 (http://www.stratigraphy.org). *Dotted lines* and *question marks* refer to fossil occurrences documented only in ancient literature and in need of taxonomic confirmation.

circalittoral to epibathyal (e.g., *Coenocyathus*, *Dendrophyllia*, Fig. 14.4e, f, *Polycyathus*) and (iii) preferentially or exclusively bathyal environments (e.g., *Lophelia*, *Madrepora*, Fig. 14.4h, *Enallopsammia*). Most solitary corals are instead typical of lower circalittoral to bathyal environments (such as *Desmophyllum*, Fig. 14.4a, b) with exceptions of some species belonging to the genera *Balanophyllia* and *Caryophyllia* that are morphologically close to shallow-water species. Specimens resembling the typical azooxanthellate frame-building species *Lophelia pertusa* and *Madrepora oculata* (Fig. 14.4h) are relatively common but present only as isolated fragments. Among the genera and subgenera known so far, the only extinct ones are *Paleoastroides*, *Cryptangia* and *Ceratotrochus* (*Edwardsotrochus*), although the last genus has been recorded in the Mediterranean throughout the Pliocene

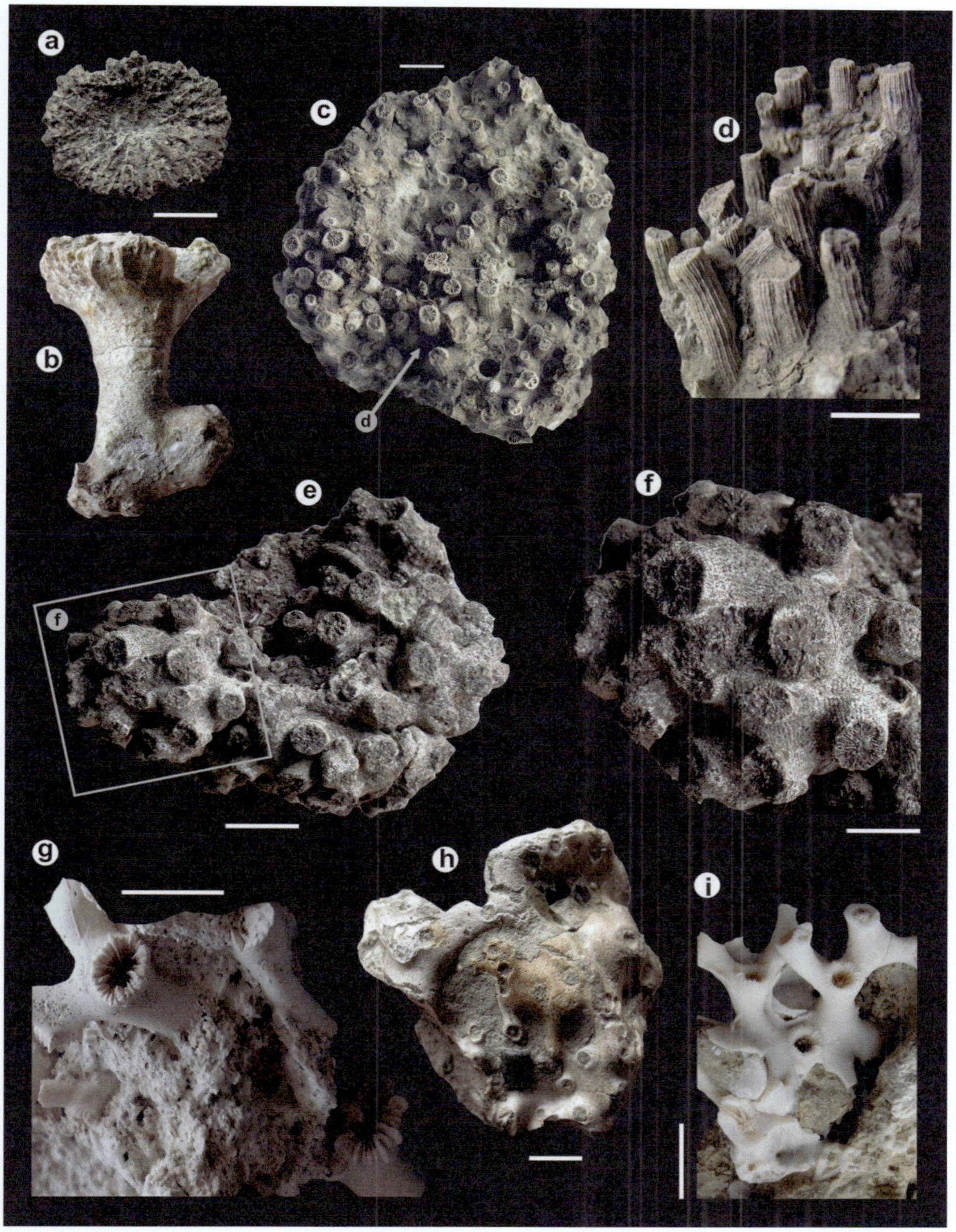

Fig. 14.4 Fossil examples of scleractinian corals persistent in the Mediterranean from Miocene to Recent (group 1, see Fig. 14.3): *Desmophyllum* cf. *dianthus* (**a**, **b**), Baldissero, Turin Hills, Miocene, MNH-T/10369; *Cladocora* cf. *caespitosa* (**c**, **d**), "Argille Azzurre Fm", Castell'Arquato, Pliocene, IPUM/16925 (specimen figured in Montanaro 1931, Tav IV, Fig. 14.6); *Dendrophyllia* sp. (**e**, **f**), Rio Batteria, Turin Hills, Miocene, MNH-T/10360; *Lophelia pertusa* (**g**), Gravitelli, Messina, Pleistocene, MP-C/VIPlC3; *Madrepora oculata* (**h**), Sciolze, Turin Hills, Miocene, MNH-T/10072, (**i**), La Montagna, Messina, Pleistocene, MP-C/VIPlC4. Scale bars equal 10 mm, only in (**e**) equals 20 mm

(Figs. 14.3; 14.5c, f). Almost all Early-Middle Miocene az-scleractinian genera still exist in modern Indian and/or Pacific waters, around 70 % in the NE Atlantic and only 50 % in the modern Mediterranean (Fig. 14.3). It is noteworthy that the following six taxa *Caryophyllia* (*Acanthocyathus*), *Conotrochus*, *Culicia*, *Oulangia*, *Trochocyathus* (*Aplocyathus*) (14.5a, b) *Stephanophyllia*, (Fig. 14.5e) are today exclusive to Indo-Pacific regions and that most of the very common and diverse genera of Early-Middle Miocene deposits, such as *Balanophyllia*, *Dendrophyllia*, *Flabellum* (14.5h–j), *Trochocyathus* (*Trochocyathus*) are among the most diverse genera in

Fig. 14.5 Fossil examples of Miocene to Pleistocene scleractinian corals extinct in the Mediterranean. *Trochocyathus* (*Aplocyathus*) *armatus*, (**a**, **b**), Grangia, Turin Hills, Miocene, MNH-T/10689; *Ceratotrochus* (*Edwardsotrochus*) *duodecimcostatus*, (**c**, **f**), Civiciano, Siena basin, Pliocene, DES-M/VIPlC1; *Trochocyathus multispinosus* (**d**), Tortona, Miocene, IPUM/32352; *Stephanophyllia elegans* (**e**), Tortona, Miocene, IPUM/16851; *Deltocyathus italicus*, Tortona, Miocene, IPUM/32390; *Flabellum avicula* (**h**, **i**), Chero, Piacenza, Pliocene, IPUM/16937; *Flabellum vaticani* (**j**), Cava Lanciani, Vaticano, Rome, Pliocene, MP-R/3638 (very similar to modern *F. knoxi* specimens, held in the Smithsonian NMNH and collected in the south Pacific at c. 420 m depth); *Stephanocyathus elegans* (**k**), Messina, Pleistocene, MPG-P/547. Scale bars equal 10 mm in (**a**–**g**, **k**) and 20 mm in (**h**–**j**)

modern Indo-Pacific waters (Fig. 14.3). Among the extant genera, only *Peponocyathus* seems to disappear from the proto-Mediterranean before the Tortonian stage.

Late Miocene

Z-coral reefs were abundant and well developed during the late Tortonian and Messinian in several peri-Mediterranean regions (southeastern Spain, the Balearic Islands, Italy, Sicily, Algeria, and Morocco), in particular the western Mediterranean (Martín and Braga 1994; Brachert et al. 1996; Braga et al. 1996; Martín et al. 2010). However, with respect to the Serravallian, the northern limit of Tortonian z-coral assemblages is shifted to the south (Perrin and Bosellini 2012) and the reef-building scleractinians are much less diverse than in the past reaching typically two to four genera (mostly *Tarbellastrea*, *Porites*, *Siderastrea* and *Solenastrea*) per locality. The reef belt shrinks further during the Messinian when very low-diversity reefs (typically dominated by the genera *Porites* and *Siderastrea*) are especially frequent in the western part of the Mediterranean basin (review in Bosellini and Perrin 2008 and Perrin and Bosellini 2012). From west to east they occurred in: Morocco (Ben-Moussa et al. 1989; Saint Martin and André 1992; Saint Martin and Cornée 1996; Rouchy et al. 2003), Spain (Esteban and Giner 1980; Rheinhold 1995; Esteban et al. 1978; Braga 1996; Sánchez-Almazo et al. 2001; Vennin et al. 2004; Martín et al. 2010), Baleares (Pomar 1991), Algeria (Saint Martin 1996), peninsular Italy (Bossio et al. 1996; Bosellini et al. 2001, 2002; Bosellini 2006; Romano et al. 2007), Sicily and Pelagian islands (Catalano 1979; Pedley and Grasso 1994; Pedley 1996; Buccheri et al. 1999; Moissette et al. 2002), Malta (Saint Martin and André 1992), Crete (Brachert et al. 2006), Cyprus (Follows 1992), and Turkey (Karabiyikoğlu et al. 2005). In general, the subtropical Messinian reef corals formed under mesotrophic to oligotrophic conditions (Bosellini et al. 2002; Romano et al. 2007) at an estimated temperature of around 16–17 °C (Bosellini and Perrin 2008). Despite their low diversity, the framebuilding capacity of z-corals in this period was particularly high, with the formation of true and large reef structures (including quite many fringing reef complexes) within both pure carbonate and mixed carbonate-siliciclastic settings (Pomar and Hallock 2007; Perrin and Bosellini 2012). Corals display a large suite of growth habits, up to aberrant forms. Reefs differ greatly even in the same geographic location both horizontally and vertically showing distinct ecological succession along a paleobathymetric gradient, suggested to be partly controlled by variations in sea-level and related changes in hydrodynamic energy and light conditions (e.g., Pomar 1991; Bosellini et al. 2001, 2002; Vennin et al. 2004) or by sediment supply (Rheinhold 1995). For any event, the further deterioration of Messinian seawater caused a stop to such coral growth in the Mediterranean with last 'reefs' being represented by microbialites (Esteban and Giner 1980; Riding et al. 1991a, b; Moissette et al. 2002; Roveri et al. 2009) and *Halimeda* and vermetid constructions (e.g., Braga et al. 1996; Bosellini et al. 2002). Interestingly small bioconstructions, up to 3 m high and up to 10 m across, commonly associated with oolitic bars and large microbial carbonate mounds were still present in the Messinian post-evaporitic Terminal Complex (Sorbas Member; Western Mediterranean) and characterized by only *Porites* (Esteban and Giner 1977; Riding et al. 1991a, b; Calvet et al. 1996; Braga and Martin 2000; Roveri et al. 2009; Martín et al. 2010). The final act of the Messinian Salinity Crisis (MSC) seems to have swept away most if not all stenoecious marine biota from the Mediterranean (Taviani 2002, 2004, and references therein). However, the main cause of the disappearance of the Mediterranean z-coral reef province at the end of the Miocene appears to be only secondarily linked with the MSC. Indeed, the cause of the diversity decrease of Mediterranean z-coral reefs, very evident after the Serravallian (Bosellini and Perrin 2008) and throughout the Late Miocene (Martín et al. 2010), has been mostly related to (i) the closure of the open marine seaway through Middle East (Bosellini and Perrin 2008) and (ii) the decreasing seawater surface temperature linked with the gradual northward shift of the Mediterranean region outside the tropical belt (e.g., Rosen 1999; Perrin 2002; Bosellini and Perrin 2008; Martín et al. 2010).

Concerning Late Miocene a–z corals, most occurrences are found in northern Apennine localities, such as Tortona, Stazzano, S. Agata, Monte Baranzone, Montegibbio (De Angelis 1894; Simonelli 1896; Montanaro 1929; Zuffardi-Comerci 1937; Chevalier 1961; Corselli 2001), and in Sicily (Cafici 1883; Alemagna 1936) and pertain to the Tortonian stage. They are particularly rich in solitary forms such as *Balanophyllia*, *Ceratotrochus multispinosus* (Fig. 14.5d), *Ceratotrochus* (*Edwardsotrochus*) *duodecimcostatus*, *Deltocyathus italicus* (Fig. 14.5g), *Flabellum* and *Stephanophyllia* (Fig. 14.5e). Some Middle Miocene genera have not been recorded so far in Late Miocene deposits (Fig. 14.3) but this is possibly an artefact resulting from the scantiness of suitable outcrops of Tortonian to Messinian age. An exceptional case is represented by the lower Messinian taphocoenosis at Carboneras in the Spanish Betic Cordillera (Barrier et al. 1991). The benthic assemblage here is quite diverse (calcified sponges, cnidarians, molluscs and stalked crinoids) and contains numerous scleractinians (14 species at least, belonging to 13 genera) including Mediterranean extant genera such *Madrepora*, *Dendrophyllia*, *Desmophyllum*, *Caryophyllia*, *Javania*, *Balanophyllia* and *Flabellum*.

Truly cold-water coral bioconstructions (*Lophelia*-*Madrepora* dominated), comparable to the modern ones, have not been recorded so far in Late Miocene peri-Mediterranean outcrops. However, framestones and/or rudstones dominated by azooxanthellate colonial corals (*Dendrophyllia* and/or *Oculina*) have been identified in southern Italy (Mastandrea et al. 2002).

Coral Responses to a Cooling Mediterranean: From the Pliocene to the Glacial Ages

The end of the Miocene corresponds to the disappearance of the tropical/subtropical coral reef province and almost all zooxanthellate coral genera from the Mediterranean (Chevalier 1961; Rosen 1999; Bosellini and Perrin 2008; Martín et al. 2010; Perrin and Kiessling 2010; Perrin and Bosellini 2012 and reference therein). On a global scale, the beginning of the Pliocene marks a strong decrease in coral diversity and, for tropical shallow-water coral reefs, the establishment of a latitudinal belt comparable to that in the Holocene (Perrin and Kiessling 2010).

Pliocene

The Pliocene shallow-water coral fauna is less diverse with respect to the Miocene, lacking almost all zooxanthellate colonial corals with the only certain exception of the species *Cladocora caespitosa* (e.g., Aguirre and Jiménez 1998; Dornbos and Wilson 1999). Montanaro (1931) mentions *Siderastraea* and *Plesiastraea* from the Pliocene of northern Italy ("Astiano di Riorzo", Emilia Romagna region, Italy) but this occurrence needs to be verified. Typical shallow-water (< 50 m water depth) Pliocene taxa are colonial *Cladangia*, *Cladopsammia*, *Culicia*, *Madracis* and *Policyathus* and some species of solitary *Balanophyllia* and *Caryophyllia*. However the majority of the scleractinian genera known so far from Pliocene deposits are solitary forms preferentially or exclusively living in modern circalittoral to bathyal environments (De Angelis 1893, 1894, 1895; Osasco 1895; Simonelli 1895, 1896; Zuffardi-Comerci 1927, 1937; Montanaro 1931; Dieni and Omenetto 1960; Chevalier 1961; Russo 1980; Corselli 1997). In particular, the marine Piacenzian clayey deposits from northern Italy (Argille Azzurre Fm: Montanaro 1931; Zuffardi-Comerci 1937; Spadini 2012 pers. comm.) host abundant specimens of the genus *Flabellum*, the subgenus *Edwardsotrochus* (Fig. 14.5c, f) and dendrophylliids that may reach very large size as their Miocene counterparts. Very well preserved specimens of *Flabellum* (*F. siciliense*) and *Stephanophyllia* occur also in coarser Pliocene sediments, such as the bioclastic sands and calcarenites of Altavilla Milicia in Sicily (MPl 5 in age according to Ruggieri and D'Arpa 2004). The presence of deep-water frame-building genera, such as the colonial *Lophelia* and *Madrepora* (known in the nineteenth and early twentieth century literature as *Lophohelia* and *Amphihelia*, *Diplohelia*, respectively), the solitary gregarious *Desmophyllum* and some dendrophylliids (*Enallopsammia*, *Dendrophyllia*), is ascertained in the middle Pliocene from northern Italy (unpublished data), but their occurrence in several other Italian outcrops (e.g. Seguenza 1864, 1880; De Angelis 1894; Simonelli 1895, 1896; Montanaro 1931; Zuffardi-Comerci 1932; Chevalier 1961) needs to be corroborated with careful stratigraphic and taxonomic studies. Moreover deep-sea coral-bearing deposits dominated by *Madrepora*, *Lophelia*, *Desmophyllum* are reported as "Upper Pliocene" from several outcrops in southern Italy (e.g. Barrier 1984, 1987; Roux et al. 1988; Barrier et al. 1989) and Rhodes (Hanken et al. 1996). However, some of the Italian deposits have been recently ascribed to the Calabrian stage (Vertino 2003). Notably, the colonial az-scleractinian *Coenocyathus* is recorded from a Pliocene deep-water chemoherm in the Stirone river, northern Apennines (Monegatti et al. 2001).

The Indo-Pacific affinity recognized for the Miocene coral fauna is still discernible in Mediterranean Pliocene assemblages, though the number of species of several genera appears to be reduced (Fig. 14.3). Indeed, taxa living today exclusively in Indo-Pacific waters, such as *Culicia*, *Stephanophyllia*, and less commonly *Caryophllia* (*Acantocyathus*) and *Trochocyathus* (*Aplocyathus*), are known from several Pliocene Italian deposits and also from southern Spain and France, northern Africa, Greece (e.g., Chevalier 1961; Chaix and Cahuzac 2005; Vertino 2003; Martinell and Domènech 2009). Moreover the genera *Balanophyllia*, *Dendrophyllia*, *Trochocyathus* and *Flabellum* are still diverse in Pliocene outcrops and include species very similar to those in the modern Indo-Pacific (e.g., Fig. 14.5j).

Pleistocene

Similarly to what has been observed for shallow-water molluscs by Monegatti and Raffi (2010), climate variations during and at the end of the Pliocene seem to have had a strong impact on the diversity of both shallow- and deep-water az-corals. The data collected so far implies that the Pleistocene coral fauna is much more similar to the modern NE Atlantic than to the Pliocene one. Before the beginning of the Pleistocene (i) the typical "Indo-Pacific" genera and subgenera (*Culicia*, *Stephanophyllia*, *Acantocyathus* and *Aplocyathus*) seem to disappear in the Mediterranean and (ii) the diversity of dendrophylliids and of the very common Mio-Pliocene genera *Flabellum* and *Trochocyathus* remarkably decreased (Fig. 14.3). All genera and almost all species known from Pleistocene deposits of peri-Mediterranean regions live (or are very similar to species living) today in the NE Atlantic Ocean. Actually the genera *Astroides*, *Cladopsammia*, *Guynia*, *Leptopsammia* and *Schizocyathus* have not been recorded so far in Mediterranean deposits older than Pleistocene. However, they could have been misidentified and, regarding small-sized species such as *Guynia* and *Schizocyathus*, overlooked in Mio-Pliocene deposits.

The Pleistocene scleractinians known so far from the Mediterranean can be placed into four groups according to their environmental distribution: (1) Infralittoral, photophilous, e.g., *Cladocora caespitosa* (Bernasconi et al. 1997; Titschack et al. 2008; Peirano et al. 2004, 2009; Antonioli et al. 2009; Drinia et al. 2010; references therein), *Balanophyllia europaea* (e.g. Barrier et al. 1990) and sciaphilous, e.g., *Astroides calycularis* (Zibrowius 1995), *Cladopsammia rolandi*; (2) Infralittoral to upper circalittoral; preferentially on overhangs and caves above 50 m water depth, e.g., *Caryophyllia inornata*, *Hoplangia durotrix*, *Leptopsammia pruvoti*; (3) Lower circalittoral to bathyal, e.g., *Caryophyllia clavata*, *C. smithii*, *Ceratotrochus magnaghi*, *Paracyathus pulchellus* (Vertino 2003 and reference therein); (4) Preferentially or exclusively bathyal, e.g., *Caryophyllia coronata*, *Caryophyllia* (*Ceratocyathus*) *elegans*, *Desmophyllum dianthus*, *Flabellum messanense*, *Fungiacyathus* aff. *fragilis*, *Lophelia pertusa* (Fig. 14.4g), *Madrepora oculata* (Fig. 14.4i), *Stephanocyathus elegans* (Fig. 14.4k) (Di Geronimo 1979; Vertino 2003; Di Geronimo et al. 2005; Titschack and Freiwald 2005; Taviani et al. 2011a, b). In the Calabrian stage (*G. cariacoensis* Zone, Vertino 2003), as well as in the Late Pleistocene (Remia and Taviani 2005; McCulloch et al. 2010; Taviani et al. 2011a, b), bathyal corals displaying a high frame-building potential (above all *L. pertusa*, *M. oculata*, *D. dianthus*) created extensive carbonate bioconstructions aligned with extant "cold-water coral reefs" from the modern ocean (e.g. Freiwald et al. 2004; Roberts et al. 2009). The most striking example of cold-water coral communities in the Early Pleistocene Mediterranean is represented by the spectacular coral framestones and rudstones cropping out along the margins of the Messina Strait (southern Italy, Seguenza 1864, 1873–1877, 1880; Di Geronimo 1987; Barrier et al. 1989, 1996; Zibrowius 1987, 1991; Roux et al. 1988; Di Geronimo et al. 2005; Vertino 2003). The abundance of still-submerged fossil cold-water corals (CWC), mostly ascribed to the Late Pleistocene (either predating or postdating the last glacial maximum: Taviani and Colantoni 1984; Taviani et al. 2005b; McCulloch et al. 2010; Angeletti and Taviani 2011; Taviani et al. 2011a, b) implies that better conditions than those today did exist in the very recent past of the Mediterranean for the settlement and maintenance of Atlantic-type frame-building CWC.

Present Coral Fauna: NE Atlantic Relict

The last remarkable modification that led to the modern composition of the Mediterranean coral fauna occurred at the end of the Pleistocene. The investigations carried out so far on Holocene benthic communities show that nine scleractinian genera (*Conotrochus*, *Deltocyathus*, *Enallopsammia*, *Flabellum*, *Fungiacyathus*, *Pemocyathus*, *Schizocyathus*, *Stephanocyathus*, *Trochocyathus*), the subgenus *Caryophyllia* (*Ceratocyathus*) and over 15 species (Vertino 2003) disappeared at the end of the Pleistocene (Fig. 14.3). Interestingly, most of these Pleistocene taxa are morphologically indistinguishable from species that live today in very deep and cold waters of the NE Atlantic (e.g. *Fungiacyathus fragilis*, *Caryophyllia sarsiae*, *C. ambrosia*, *Premocyathus polymorphus*, *Aulocyathus atlanticus*, *Stephanocyathus elegans*, *Flabellum alabastrum*, *Schizocyathus fissilis*, *Balanophyllia caryophylloides*, *Enallopsammia rostrata*).

In total 25 genera and 33 species live today in the Mediterranean Sea, most of them on hard substrates of the continental shelf (Zibrowius 1980). Only *Sphenotrochus andrewianus*, typical species of the "biocoenosis of coarse-grained sands and fine gravels under bottom currents (SGCF)", and the morphotypes "clavus" and "vermiformis" of the two circalittoral to epibathyal species *Caryophyllia smithii* and *Stenocyathus vermiformis*, respectively, are restricted to mobile substrates. The modern shallow-water coral fauna is very similar to the Pleistocene one; in fact, all infralittoral Pleistocene photophilous and sciafilous species are still present in modern Mediterranean waters, although the geographic distribution and/or development of some colonial species appears to be highly reduced. For instance, *Astroides calycularis*, at present known only in rocky coastal areas of the southwestern Mediterranean, had a wider geographic distribution during Pleistocene interglacial periods (when the superficial water temperature was probably higher) being found as fossil in several localities of the northern Mediterranean (Zibrowius 1995). Moreover *C. caespitosa* (Fig. 14.6b), the only zooxanthellate bioconstructor coral of the modern Mediterranean, forms still extensive banks near the Tunisian coast, in the Aegean (Laborel 1961; Zibrowius 1980; Peirano et al. 1998) and Adriatic Sea (Kružić et al. 2012), but seemed to be much more flourishing in warmer periods of the Pliocene and Pleistocene. *Oculina patagonica* represents another remarkable difference between the modern infralittoral Mediterranean coral fauna and Pleistocene one. This is indeed a

Fig. 14.6 Examples of modern (living) Mediterranean scleractinian corals. Solitary, gregarious, azooxanthellate *Desmophyllum dianthus* (**a**); colonial zooxanthellate *Cladocora caespitosa* (**b**); colonial azooxanthellate *Lophelia pertusa* (**c**), *Madrepora oculata* (**d**, *white*; corals with *red* polyps represent *Desmophyllum*) and *Dendrophyllia cornigera* (**e**, **f**). With exception of *Cladocora*: (Porto Cesareo, 11.5 m depth) all photographed specimens were collected in the bathyal environment (at ca. 600 m water depth) of the Mediterranean Sea during the M70-1 cruise, 2006, from the SML province (**a**, **c**) and the Bari Canyon (**d–f**). Scale bars equal 10 mm in (**a–c**), 20 mm in (**d**)

non-native invasive zooxanthellate species of recent introduction in the Mediterranean (60–70 years?) which today is widely known throughout the basin: Italy (Liguria), France (Marseille), Spain (Algeciras to Catalonia, Alboran, Cabrera and Columbretes), Algeria and Tunisia, Egypt (Alexandria area), Israel, Lebanon, Turkey, Greece (Zibrowius 1992; Sartoretto et al. 2008).

Apart from the exceptions mentioned above, the diversity of infralittoral to circalittoral and circalittoral to epibathyal corals (ecological groups "2" and "3" described in the previous paragraph) seems to be unchanged from the Pleistocene onwards. Instead, as mentioned above, the modern bathyal coral community is notably different than the Pleistocene one, lacking all cold stenothermic species which live today in the neighbouring NE Atlantic bathyal bottoms at temperature lower than 10 °C (Vertino 2003 and references therein). The extensive *Lophelia*-dominated bioconstructions, which characterized the Pleistocene Mediterranean and are the main components of the still growing NE Atlantic banks and giant mounds (comprehensive review in Freiwald et al. 2004; Roberts et al. 2009), underwent a strong regression at the end of the Pleistocene. Nevertheless, well-developed frame-building CWC communities are still present on Atlantic-like mounds, in terms of extension and biodiversity, of some locations of the western Mediterranean (Alboran Sea) (Lo iacono et al. in press). Moreover, bathyal *Madrepora*-dominated communities still thrive in the Ionian Sea, on the top and upper flanks of the small Santa Maria di Leuca (SML) mounds (e.g., Taviani et al. 2005a; Savini and Corselli 2010; Vertino et al. 2010a; Fig. 14.6a, c), as well as along flanks of canyons, fault escarpments and overhangs in several other localities of the Mediterranean Sea (Freiwald et al. 2009; Orejas et al. 2009; Gori et al. 2013; Sanfilippo et al. 2012; Fig. 14.6d–f).

Finally, it is worth mentioning the role as "secondary building" that the genus *Dendrophyllia* has been playing since at least the Early Miocene in the (proto-) Mediterranean. This genus is present in the modern basin with two species, *D. ramea* and *D. cornigera*, which differ in geographical and depth distribution (Zibrowius 1980, 2003; Salomidi et al. 2010). In the Mediterranean *D. ramea* is commonly observed around 30–40 m water depth whereas *D. cornigera* has a much wider bathymetric and geographic distribution. Indeed, the latter species has been recorded alive from around 80 up to 730 m water depth and it is rather common in bathyal communities (Zibrowius 1980; Freiwald et al. 2009; Mastrototaro et al. 2010; Vertino et al. 2010a; Salomidi et al. 2010), locally associated to the frame-building corals *M. oculata* and *L. pertusa*. *D. cornigera* can be considered as a community-structuring species because it can build very large colonies and, at certain places, its broken branches can create peculiar rubble facies which host diverse epi- and endofauna communities (Rosso et al. 2010; Vertino et al. 2010a; Taviani et al. 2011a, b). Likely, most of the *D. cornigera* rubble facies which characterize today the bathyal seafloor of several Mediterranean regions dates back to the Late Pleistocene implying, as already pointed out by Blanc et al. (1959) and Zibrowius (1980), a greater development of this species in the recent geological past of the basin. Remarkably, comparable dendrophylliid-dominated facies have been recorded so far both in Miocene (Mastandrea et al. 2002; Bosellini et al. 1999) and Early Pleistocene outcrops from southern Italy (Vertino 2003).

Corals Other than Scleractinians

Calcified corals also include anthozoans in the subclass Octocorallia and hydrozoans of the family Stylasteridae that secrete a calcitic skeleton. Their fossil history is comparatively poor with respect to scleractinians. Nevertheless, they are consistently cited in the paleontological literature. Bathyal representatives of the octocoral family Isididae (mostly the genus *Keratoisis*) have been recorded from the Miocene (e.g. De Angelis and Neviani 1897; Barrier et al. 1991; Zuffardi-Comerci 1932), Pliocene (Simonelli 1895) and Plio-Pleistocene (e.g. Seguenza 1864; Di Geronimo 1979, 1987; Fois 1990; Barrier et al. 1996; Vertino 2003; Di Geronimo et al. 2005) of the Mediterranean basin. Moreover, octocoral holdfasts referred to as Primnoidae have been recognized in the Miocene coral collection from the Turin Hills (Museum of Natural History, Turin University), in Plio-Pleistocene outcrops from southern Italy (Di Geronimo 1979; Grasshoff and Zibrowius 1983; unpublished data) and submerged deposits from the eastern Mediterranean of presumable Pleistocene age. Among octocorals, the genus *Corallium* has also been recorded several times in Miocene to Pleistocene outcrops of peri-Mediterranean regions (e.g., Zuffardi-Comerci 1927, 1932; Chevalier 1961; Barrier et al. 1991; Vertino et al. 2010b). Finally, a rich stylasterid fauna (14 species in at least 8 genera: *Calyptopora, Conopora, Crypthelia, Distichopora, Lepidopora, Pliobothrus, Stenohelia, Stylaster*) characterizes the lower Messinian taphocoenosis of Carboneras (Barrier et al. 1991), and *Errina* and *Stenohelia* have been collected from the Pleistocene of southern Italy (Di Geronimo et al. 2005). It is worth remarking that, with exception of the sytlasterid *Errina*, all coral genera mentioned in this paragraph are absent in the modern Mediterranean. Similarly to what has happened to scleractinians, Mediterranean octocorals and stylasterids underwent an abrupt diversity decrease in shallow waters at the end of the Miocene and in deep-waters at the end of the Pleistocene.

Corals as Palaeoceanographic Archives

In recent years a growing attention has been granted to certain corals as potential palaeoceanographic archives because of their encoding of seawater signals (temperature, nutrients, pH) within their carbonate skeleton. One of the reasons for the current success in oceanographic reconstructions relies upon the original aragonite composition of the exoskeleton in scleractinians that permits precise age assessment using U-series dating. Corals can be used to decipher oceanographic signals at various time scales from decadal to millennia (e.g., Schrag and Linsley 2002; Stirling and Andersen 2009; Thompson et al. 2011). Regarding specifically the Mediterranean basin, this technique has been applied thus far to both shallow- and deep-water Pleistocene to Holocene scleractinians using stable (oxygen, carbon, boron) and radiogenic (strontium and neodymium) isotopes and/or trace elements (Montagna et al. 2005, 2011a, b; Silenzi et al. 2005; López Correa et al. 2010; Trotter et al. 2011). In the shallow water environment, the prime candidate has been the zooxanthellate colonial coral *Cladocora caespitosa* that appears to be a reliable archive of seawater surface temperature (Peirano et al. 2004; Silenzi et al. 2005; Montagna et al. 2007, 2008a, b, 2011a), pH (Trotter et al. 2011) and pre- and post-bomb surface radiocarbon. Simultaneously, growing attention has been devoted to the geochemical screening of bathyal taxa to determine their reliability as oceanographic recorders. The solitary coral *Desmophyllum dianthus* has been shown to store decipherable information in its aragonitic exoskeleton (Montagna et al. 2005). In particular, it has been demonstrated that the P/Ca ratio of the skeletal aragonite of this quasi-cosmopolitan species is linearly correlated to the ambient seawater phosphorus concentration (Montagna et al. 2006; Anagnostou et al. 2011). Nonetheless, most of the geochemical signals are modified by the coral physiology during the biomineralization processes, complicating the paleoclimate reconstruction (Meibom et al. 2008; Brahmi et al. 2012). Culturing experiments and geochemical studies at fine scale resolution of Mediterranean corals are required to provide reliable evidence for these interpretations.

Regarding intermediate bathymetric situations in the water column (basically lower circalittoral to epibathyal benthic habitats), not much attention has been paid in evaluating such corals in the Mediterranean (i.e. *Dendrophyllia* spp, and *Caryophyllia* spp.). Geochemical applications to calcitic corals are still rather rare, although some attention has been given to fossil Isididae and both recent and fossil *Corallium* (Vielzeuf et al. 2008; Montagna et al. 2011b).

Current geochemical research on Mediterranean corals targets the Li/Mg ratio (Montagna et al. 2009) and neodymium isotopes (^{143}Nd/^{144}Nd: Montagna et al. 2011b) as reliable proxies for seawater temperature independent from 'vital effects' and water mass circulation, respectively.

Biodiversity of Tomorrow

What factors may shape the future Mediterranean coral diversity? Ocean acidification poses a direct threat to marine calcifiers, above all aragonitic corals, including those in the Mediterranean (Montagna et al. 2008a). On the other hand, recent studies on both shallow and deep water corals seem to document a certain resilience of scleractinians (at least azooxanthellate ones) to increasing temperature (Caroselli et al. 2012) and decreasing seawater pH, indicating that no extinction is foreseen in the near future (Rodolfo-Metalpa et al. 2011; Maier et al. 2012; McCulloch et al. 2012a, b).

Non-native invasion by coral species (e.g., *Oculina patagonica*) has already affected the aboriginal Mediterranean fauna. The warming trend of Mediterranean seawater might pave the road to the coming and successful establishment of true tropical corals, likely from the Red Sea as Lessepsian migrants. This issue has been considered by Por (2009) who makes the point that the Levantine Basin is already presenting a thermal status conducive to the maintenance of scleractinians such *Stylophora pistillata* and *Siderastrea savignyiana*, an argument conforming to the concept of the Godot basin introduced by Taviani 2002; Oliverio and Taviani 2003. However, the hydrological conditions prevailing in the Suez maritime connection between the two seas, and the relatively short larval life of corals seems to limit this migration (Por 2009). Although in the near future some subtropical-tropical corals may eventually settle in the Mediterranean, it is arguable that the expected cooling imported by the next ice age would be so detrimental as to cause their extinction (Chap. 16 by Sabelli and Taviani, this volume).

Discussion and Final Remarks

Data collected so far have led us to categorize the known Miocene to Recent az-scleractinian genera and subgenera in five main groups (Fig. 14.3). The first one includes the so-called "persistent" taxa, those that have lived in the Mediterranean since the Miocene or, at least, the Early Pliocene (some examples shown in Fig. 14.4). Almost all members of this group today show a very wide geographic distribution, and many of them have undergone a clear diversity decline at species level since the late Miocene. Interestingly, the most diverse Miocene genera (dendrophylliids and *Caryophyllia*) correspond to the most diverse ones of the Indo-Pacific Oceans. Groups 2, 3, 4 comprise the "extinct" taxa, which disappeared from the Mediterranean at the end of the Miocene, Pliocene or Pleistocene (some examples are shown in Fig. 14.5). Group 2 includes genera and subgenera that seem to disappear at the end of the Miocene, with exception of the recent alien species *Oculina patagonica*. Group 3 comprises taxa that were extinct in Mediterranean waters at the end of the Pliocene whereas to group 4 belong the taxa (mostly bathyal) which disappeared at the end of the Pleistocene. Finally, in group 5 we have gathered those genera that have been almost unknown in the fossil record so far or recorded only not earlier than the Pleistocene Calabrian stage.

On the whole, the Mediterranean scleractinian diversity has clearly diminished since the Late Miocene. This trend was not linear but abrupt in shallow-water environments during and immediately after the Late Miocene and more gradual from the Pliocene onwards. Indeed, at the end of the Miocene, the Mediterranean coral fauna underwent a drastic modification that led to the disappearance of the well established shallow-water coral-reef province. The generic diversity of azooxanthellate and deep-water corals did not undergo evident modifications, that were instead much stronger at the end of the Pliocene and of the Pleistocene (Fig. 14.3).

Summarizing, we can assert with confidence that the Mediterranean coral fauna composition has undergone three major modifications since the late Miocene:

- Miocene – Pliocene boundary: (a) disappearance of the "Mediterranean z-coral reef province" and of almost all z-genera; (b) species diversity decrease of most az-coral genera.
- Pliocene – Pleistocene boundary (or at least before the Calabrian stage) (a) diversity decrease at the genus level, (b) disappearance of Indo-Pacific-like taxa; (c) clear outline of Atlantic affinity.
- Pleistocene – Holocene boundary: strong diversity decrease at the genus level with disappearance of deep-

water taxa living today at very low temperature (< 10 °C) in the neighbouring NE Atlantic.

The causes of these three major changes and of the general coral diversity decrease since the Serravallian are not fully understood. However, there seems to be a clear link between the Mediterranean coral diversity demise and the major climate modifications on a global scale during the last 20 million years. In particular, the decrease of z-coral genera since the late Middle Miocene (Bosellini and Perrin 2008) seems to correspond with the gradual cooling and re-establishment of a major ice-sheet on Antarctica after the late Middle Miocene climatic optimum (Zachos et al. 2001). Moreover, the differences between the Piacenzian (Late Pliocene) and Calabrian (Early Pleistocene) az-coral diversity (Fig. 14.3) could be related to the main oceanographic modifications that were driven by the major expansion of the Northern Hemisphere ice sheets at ca. 2.7 Ma (Shackleton et al. 1984; Dwyer et al. 1995). It is worth remarking that the decrease in Mediterranean coral diversity from the late Miocene on has been also ascribed to oceanographic modifications triggered by the gradual northward shift outside the tropical belt of the Mediterranean region (e.g. Rosen 1999; Perrin 2002; Bosellini and Perrin 2008; Martín et al. 2010).

The disappearance of the stenothermic bathyal corals at the end of the Pleistocene is still under debate; however, it has been generally ascribed to the shift from psychrosphere to thermosphere conditions of the Mediterranean Sea as a consequence of the transition from glacial to interglacial age. Indeed, the extensive development of Mediterranean frame-building CWC communities as well as the presence of psychrospheric bathyal species (scleractinian and other groups of organisms) has been often related to glacial periods (Corselli 2001; Taviani et al. 2005b). Nevertheless, the absence of abundant cold-water corals in the Mediterranean during the Last Glacial Maximum (McCulloch et al. 2010) implies that, during glacial periods, bathyal corals did not thrive in this basin if other important environmental conditions, such as nutrition, oxygenation (Fink et al. 2012), low sedimentation rate, were not optimal.

As mentioned above this contribution represents a first step towards a better understanding of the evolutionary history of Mediterranean corals. In order to make a proper use of the extraordinary Cenozoic record of the peri-Mediterranean regions and of the corals as paleoenvironmental indicators, paleontological data needs to be combined with updated regional geological and stratigraphical information. Furthermore, taxonomic review at the species level is essential to improve our knowledge of the Mediterranean coral fauna and, as a consequence, the geological history of this basin.

Acknowledgements This article is a contribution to the EU CoralFISH (contract n° 213144), Hermione (contract n° 226354), and Coconet (contract n° 287844) programmes, to the BIOMAP project of Regione Puglia, to the RITMARE project and to "Dote ricercatori" (FSE, Regione Lombardia). The contribution of AV was facilitated by the ESF COCARDE-ERN network. ISMAR-CNR Bologna scientific contribution n. 1780. We warmly thank Enzo Burgio, Stephen Cairns, Italo Di Geronimo, Daniele Ormezzano, Antonietta Rosso, Paolo Serventi and Helmut Zibrowius for their kind support with the fossil and Recent Scleractinian collections mentioned in the text and for constructive discussions. Thanks are due to Valerio Spadini for providing information on corals from the Pliocene Basin of Siena and to Luca Pedriali for supplying the *Ceratotrochus* (*E.*) *duodecimcostatus* specimen presented in Fig. 14.5. Moreover, we sincerely thank Captains, Officers, Crew and Colleagues aboard R/V *Urania* (during the CNR cruises CORTI, GECO, CORSARO, MARCOS, DECORS; 2003–2011), R/V Universitatis (during the APLABES, 2003–2005, and CoralFISH-MAGIC, 2010, cruises) and R/V *Meteor* (during the M70-1 cruise, 2006) for their collaboration in the study of modern and fossil (submerged) Mediterranean deep-sea corals. Constructive criticism on the manuscript by Nancy Budd, Paolo Montagna and Mark Wilson is gratefully acknowledged.

References

Aguirre J, Jiménez AP (1998) Fossil analogues of present-day *Cladocora caespitosa* coral banks: sedimentary setting, dwelling community, and taphonomy (Late Pliocene, W Mediterranean). Coral Reefs 17:203–213

Alemagna C (1936) Nuove ricerche sul sistema miocenico della Sicilia sud-orientale. Zuccarello e Izzi, Catania

Anagnostou E, Sherrell RM, Gagnon A, LaVigne M, Field MP, McDonough WF (2011) Seawater nutrient and carbonate ion concentrations recorded as P/Ca, Ba/Ca, and U/Ca in the deep-sea coral *Desmophyllum dianthus*. Geochim Cosmochim Acta 75:2529–2543

Angeletti L, Taviani M (2011) Entrapment, preservation and incipient fossilization of benthic predatory molluscs within deep-water coral frames in the Mediterranean Sea. Geobios 44:543–548

Antonioli F, Ferranti L, Fontana A, Amorosi A, Bondesan A, Braitenberg C, Dutton A, Fontolan G, Furlani C, Lambeck K, Mastronuzzi G, Monaco C (2009) Holocene relative sea-level changes and vertical movements along the Italian and Istrian coastlines. Quat Int 206:102–133

Barrier P (1984) Évolution tectono-sédimentaire pliocène et pléistocène du détroit de Messine. (Italie). Dissertation, University of Marseille-Luminy

Barrier P (1987) Stratigraphie des dépôts pliocènes et quaternaires du Détroit de Messine. Doc et Trav IGAL 11:59–81

Barrier P, Di Geronimo I, Montenat C, Roux M, Zibrowius H (1989) Présence de faunes bathyales atlantiques dans le Pliocène et le Pléistocène de Méditerranée (détroit de Messine, Italie). Boll Soc Géol France 5:787–796

Barrier P, Di Geronimo I, Zibrowius H, Raisson F (1990) Faune Sénégalienne du paléoescarpement du Capo Vaticano (Calabre Meridionale). Implications néotectoniques. In: Robba E (ed) Proceedings of 4th Symposium Ecol Paleoecol Benthic Communities. Museo Regionale Sci Nat, Torino

Barrier P, Zibrowius H, Lozouet P, Montenat C, Ott D'Estevou P, Serrano F, Soudet HJ (1991) Une faune de fond dur du bathyal supérieur dans le Miocène terminal des Cordillères Bétiques (Carboneras, SE, Espagne). Mésogée 51:3–13

Barrier P, Di Geronimo I, La Perna R, Rosso A, Sanfilippo R, Zibrowius H (1996) Taphonomy of deep-sea hard and soft bottom communities: the Pleistocene of Lazzàro (southern Italy). In: Meléndez Hevia G, Blasco Sancho MF, Pérez Urresti I (eds) Communication

Reunion de Tafonomia y Fosilización. Institucion Fernando El Catolico, Zaragoza

Ben-Moussa A, El Hajjaji K, Pouyet S, Demarq G (1989) Les mega-faunes marines du Messinien de Melilla (nord-est Maroc). Palaeogeogr Palaeoclimatol 73:51–60

Bernasconi MP, Corselli C, Carobene L (1997) A bank of scleractinian coral *Cladocora caespitosa* in the Pleistocene of the Crati valley (Calabria, southern Italy): growth versus environmental conditions. Bol Soc Paleontol Ital 36:53–61

Blanc JJ, Pérès JM, Picard J (1959) Coraux profonds thanathocoenoses quaternaires en Méditerranée. Coll Internat CNRS 83:185–192

Bosellini F (2006) Biotic changes and their control on Oligocene-Miocene reefs: a case study from the Apulia Platform margin (southern Italy). Palaeogeogr Palaeoclimatol 241:393–409

Bosellini F, Perrin C (2008) Estimating Mediterranean Oligo-Miocene sea-surface temperatures: an approach based on coral taxonomic richness. Palaeogeogr Palaeoclimatol 258:71–88

Bosellini A, Bosellini F, Colalongo ML, Parente M, Russo A, Vescogni A (1999) Stratigraphic architecture of the Salento Coast from Capo d'Otranto to S. Maria di Leuca (Apulia, Southern Italy). Riv Ital Paleontol 105:397–416

Bosellini F, Russo A, Vescogni A (2001) Messinian reef-building assemblages of the Salento Peninsula (southern Italy): palaeobathy-metric and palaeoclimatic significance. Palaeogeogr Palaeoclimatol 175:7–26

Bosellini F, Russo A, Vescogni A (2002) The Messinian reef complex of the Salento Peninsula (southern Italy): stratigraphy, facies and paleoenvironmental interpretation. Facies 47:91–112

Bossio A, Esteban M, Mazzanti R, Mazzei R, Salvatorini G (1996) Rosignano reef complex (Messinian), Livornesi Mountains, Tuscany, central Italy. In: Franseen EK, Esteban M, Ward WC, Rouchy J-M (eds) Models fror carbonate stratigraphy from Miocene reef complexes of the Mediterranean region. Society for Sedimentary Geology, Tulsa

Brachert TC, Betzler C, Braga JC, Martin JM (1996) Record of climatic change in neritic carbonates: turnovers in biogenic associations and depositional modes (Upper Miocene, southern Spain). Geol Rundsch 85:327–337

Brachert TC, Reuter M, Kroeger K, Felis T, Lohmann G, Micheels A, Fassoulas C (2006) Porites corals from Crete (Greece) open a window into Late Miocene (10 Ma) seasonal and interannual climate variability. Earth Planet Sci Lett 245:81–94

Braga JC (1996) Geometries of reef advance in response to relative sea-level changes in a Messinian (uppermost Miocene) fringing reef (Cariatiz reef, Sorbas Basin, SE Spain). Sediment Geol 107:61–81

Braga JC, Martin JM (2000) Subaqueous siliciclastic stromatolites: a case history from Late Miocene beach deposits in the Sorbas Basin of SE Spain. In: Riding RE, Awamil SM (eds) Microbial sediments. Springer, Berlin

Braga JC, Martin JM, Riding R (1996) Internal structure of segment reefs: *Halimeda* algal mounds in the Mediterranean Miocene. Geology 24:35–38

Brahmi C, Kopp C, Domart-Coulon I, Stolarski J, Meibom A (2012) Skeletal growth dynamics linked to trace-element composition in the scleractinian coral. Geochim Cosmochim Acta 99:146–158

Buccheri G, Renda P, Morreale C, Sorrentino G (1999) Il Tirreniano dell'isola di Lampedusa (Arcipelago Pelagiano, Agrigento, Italia). Le successioni di cala Maluk e Cala Uccello. Boll Soc Geol Ital 118:361–373

Cafici I (1883) La formzioen miocenica nel territorio di Licodia-Eubea (Provincia di Catania). Atti R Accad Lincei Sci Fis Nat 14:1–39

Cahuzac B, Chaix C (1993) Les faunes de coraux (Anthozoaires Scléractiniaires) de la façade atlantique française au Chattien et au Miocène. Ciênc Terra 12:57–69

Cahuzac B, Chaix C (1996) Structural and faunal evolution of Chattian–Miocene reefs and corals in western France and the northeastern Atlantic Ocean. In: Franseen EK, Esteban M, Ward WC, Rouchy JM (eds) Models for carbonate stratigraphy: from Miocene reef complexes of Mediterranean regions. Society for Sedimentary Geology, Tulsa

Cairns SD (1999) Species richness of recent Scleractinia (Appendix: list of extant stony corals). Atoll Res Bull 459:1–12

Cairns SD, Kitahara MV (2012) An illustrated key to the genera and subgenera of the Recent azooxanthellate Scleractinia (Cnidaria, Anthozoa), with an attached glossary. Zookeys 227:1–47

Calvet F, Zamarreno I, Valles D (1996) Late Miocene reefs of the Alicante-Elche Basin, southeast Spain. In: Franseen EK, Esteban M, Ward WC, Rouchy JM (eds) Models for carbonate stratigraphy: from Miocene reef complexes of Mediterranean regions. Society for Sedimentary Geology, Tulsa

Caroselli E, Mattioli G, Levy O, Falini G, Dubinsky Z, Goffredo S (2012) Inferred calcification rate of a Mediterranean azooxanthel-late coral is uncoupled with sea surface temperature along an 8° latitudinal gradient. Front Zool 9:32

Catalano R (1979) Scogliere ed evaporiti messiniane in Sicilia. Modelli genetici ed implicazioni strutturali. Lav Ist Geol Univ Palermo 18:1–21

Chaix C, Cahuzac B (2005) Le genre *Culicia* (Scléractiniaire): systé-matique, écologie et biogéographie au Cénozoïque. Eclogae Geol Helv 98:169–187

Chevalier JP (1961) Récherches sur les Madréporaires et les formations récifales Miocene de la Méditerranée occidentale. Mém Soc Géol Fr 40:1–562

Corselli C (1997) Il mare in fondo al pozzo: i fossili pliocenici del Pozzo Bagoderi di Malnate. Quaderni del sistema museale Alta Valle Olona, Varese

Corselli C (2001) Change and diversity: the Mediterranean deep corals from the Miocene to the present. In: Faranda FM, Guglielmo L, Spezie G (eds) Mediterranean ecosystems: structures and processes. Springer Italia, Genova

De Angelis G (1893) I zoantari fossili dei dintorni di Roma. Boll Soc Geol Ital 12:1–25

De Angelis G (1894) I Corallari dei terreni terziari dell'Italia settentrio-nale, Collezione Michelotti, Museo Geologico della Regia Università di Roma. Atti R Accad Lincei Sci Fis Nat 1:164–280

De Angelis G (1895) Descripción de los Antozoos fósiles pliocénicos de Cataluña. Mem R Accad Cienc Artes Barcelona 3:1–25

De Angelis G, Neviani A (1897) Corallari e briozoi neogenici. Boll Soc Geol Ital 15:571–594

Di Geronimo I (1979) Il Pleistocene in facies batiale di Valle Palione (Grammichele, Catania). Boll Malacol 15:85–156

Di Geronimo I (1987) Bionomie des peuplements benthiques des sub-strats meubles et rocheux plio-quaternaires du Détroit de Messine. Doc Trav IGAL 11:153–169

Di Geronimo I, Messina C, Rosso A, Sanfilippo R, Sciuto F, Vertino A (2005) Enhanced biodiversity in the deep: early Pleistocene coral communities from southern Italy. In: Freiwald A, Roberts JM (eds) Cold-water corals and ecosystems. Springer, Berlin/Heidelberg

Dieni J, Omenetto P (1960) Studio di una macrofauna del Pliocene infe-riore di Orosei. Riv Ital Paleontol 66:605–618

Dornbos SQ, Wilson MA (1999) Paleoecology of a Pliocene coral reef in Cyprus: recovery of a marine community from the Messinian Salinity Crisis. Neues Jahrb Geol P-A 213:103–118

Drinia H, Koskeridou E, Antonarakou A, Tzortzaki E (2010) Benthic foraminifera associated with the zooxanthellate coral *Cladocora caespitosa* in the Pleistocene of the Kos Island (Aegean Sea, Greece): sea level changes and palaeoenvironmental conditions. Bull Geol Soc Gr 43:613–619

Dwyer GS, Cronin TM, Baker PA, Raymo ME, Buzas JS, Correge T (1995) North Atlantic deepwater temperature change during late Pliocene and late Quaternary climatic cycles. Science 270: 1347–1351

Esteban M, Giner J (1977) El arrecife de Santa Pola. In: Salas R (ed) Primer Seminario Práctico de Asociaciones Arrecifales y Evaporíticas. Universitat de Barcelona, Barcelona

Esteban M, Giner J (1980) Messinian coral reefs and erosion surfaces in Cabo de Gata (Almeria, SE Spain). Acta Geol Hisp 15:97–104

Esteban M, Calvet F, Dabrio C, Baron A, Giner J, Pomar L, Salas R, Permanyer A (1978) Aberrant features of the Messinian coral reefs, Spain. Acta Geol Hisp 13:20–22

Fink H, Wienberg C, Hebbeln D, McGregor H, Taviani M, Freiwald A (2012) Oxygen control on Holocene cold-water coral development in the Mediterranean Sea. Deep-Sea Res I 62:89–96

Fois E (1990) Stratigraphy and palaeogeography of the Capo Milazzo area (NE Sicily, Italy): clues to the evolution of the southern margin of the Tyrrhenian Basin during the Neogene. Palaeogeogr Palaeoclimatol 78:87–108

Follows EJ (1992) Patterns of reef sedimentation and diagenesis in the Miocene of Cyprus. Sediment Geol 79:225–253

Freiwald A, Fosså JH, Grehan AJ, Koslow JA, Roberts JM (2004) Cold-water coral reefs. UNEP-WCMC Biodivers Ser 22:1–84

Freiwald A, Beuck L, Rüggeberg A, Taviani M, R/V Meteor M70-1 participants (2009) The white coral community in the central Mediterranean revealed by ROV surveys. Oceanography 22:58–74

Gori A, Orejas C, Madurell T et al (2013) Bathymetrical distribution and size structure of cold-water coral populations in the Cap de Creus and Lacaze-Duthiers canyons (northwestern Mediterranean). Biogeosciences 10:2049–2060

Grasshoff M, Zibrowius H (1983) Kalkkrusten auf Achsen von Hornkorallen, rezent und fossil. Senckenb Marit 15:111–145

Hanken NM, Bromley RG, Miller J (1996) Plio-Pleistocene sedimentation in oscillating grabens north-east Rhodes, Greece. Geol J 31:271–296

Karabiyikoğlu M, Tuzcu S, Çiner A, Deynoux M, Örçen S, Hakyemez A (2005) Facies and environmental setting og the Miocene coral reefs in the late-orogenic fill of the Antaya Basin, western Taurides, Turkey: implications for tectonic control and sea-level changes. Sediment Geol 173:345–371

Kružić P, Sršen P, Benković L (2012) The impact of seawater temperature on coral growth parameters of the colonial coral *Cladocora caespitosa* (Anthozoa, Scleractinia) in the eastern Adriatic Sea. Facies 58:477–491

Laborel J (1961) Le concretionnement algal "coralligène" et son importance géomorphologique en Méditerranée. Recueil des Travaux de la Station Marine d'Endoume 23:37–60

López Correa M, Montagna P, Vendrell B, McCulloch M, Taviani M (2010) Stable isotopes ($\delta^{18}O$ & $\delta^{13}C$), trace and minor element compositions of Last Glacial to Recent scleractinians and bivalves at Santa Maria di Leuca deep-water coral province, Ionian Sea (2010). Deep-Sea Res II 57:471–486

Lo Iacono C, Gràcia E, Ranero CR, Emelianov M, Huvenne V, Bartolomé R, Booth-Rea G, Prades J, The MELCOR Cruise Party (in press) The West Melilla Mounds, Eastern Alboran Sea: morphological characterization and paleo-environmental context. Deep-Sea Res II

Maier C, Watremez P, Taviani M, Weinbauer MG, Gattuso JP (2012) On board experiments to determine calcification rates and the effect of ocean acidification on Mediterranean cold-water corals. Proc R Soc B-Biol Sci 279:1716–1723

Martín JM, Braga JC (1994) Messinian events in the Sorbas Basin in southeastern Spain and their implications in the recent history of the Mediterranean. Sediment Geol 90:257–268

Martín JM, Braga JC, Sánchez-Almazo IM (2010) Temperate and tropical carbonate-sedimentation episodes in the Neogene Betic basins (southern Spain) linked to climatic oscillations and changes in Atlantic-Mediterranean connections: constraints from isotopic data. In: Mutti M, Piller WE, Betzler C (eds) Carbonate systems during the Oligocene-Miocene climatic transition. Wiley, Oxford

Martinell J, Domènech R (2009) Commensalism in the fossil record: *Eunicid polychaete* bioerosion on Pliocene solitary corals. Acta Palaeontol Pol 54:143–154

Mastandrea A, Muto F, Neri C, Papazzoni CA, Perri E, Russo F (2002) Deep-water coral banks: an example from the "Calcare di Mendicino" (Upper Miocene, northern Calabria, Italy). Facies 47:27–42

Mastrototaro F, D'Onghia G, Corriero G et al (2010) Biodiversity of the white coral ecosystem off Cape Santa Maria di Leuca (Mediterranean Sea): an update. Deep-Sea Res II 57:412–430

McCulloch M, Taviani M, Montagna P, López Correa M, Remia A, Mortimer G (2010) Proliferation and demise of deep-sea corals in the Mediterranean during the Younger Dryas. Earth Planet Sci Lett 298:143–152

McCulloch M, Falter J, Trotter J, Montagna P (2012a) Coral resilience to ocean acidification and global warming through pH up-regulation. Nat Clim Change 2:623–627

McCulloch M, Trotter J, Montagna P, Falter J, Dunbar R, Freiwald A, Forsterra G, López Correa M, Maier C, Rüggeberg A, Taviani M (2012b) Resilience of cold-water scleractinian corals to ocean acidification: Boron isotopic systematics of pH and saturation state up-regulation. Geochim Cosmochim Acta 87:21–34

Meibom A, Cuif J-P, Mostefaoui S, Dauphin Y, Houlbreque F, Meibom K, Dunbar R (2008) Compositional variations at ultra-structure length-scales in coral skeleton. Geochim Cosmochim Acta 72:1555–1569

Moissette P, Saint Martin J-P, André J-P, Pestrea S (2002) L'association microbialite-bryozoaires dans le Messinien de Sicile et de Sardaigne. Geodiversitas 24:611–623

Monegatti P, Raffi S (2010) The Messinian marine molluscs record and the dawn of the eastern Atlantic biogeography. Palaeogeogr Palaeocl 297:1–11

Monegatti P, Raffi S, Roveri M, Taviani M (2001) One day trip in the outcrops of Castell'Arquato Plio-Pleistocene Basin: from the Badland of Monte Giogo to the Stirone River. University of Parma, Parma

Montagna P, López Correa M, Rüggeberg A, McCulloch M, Rodolfo-Metalpa R, Ferrier-Pagès C, Freiwald A, Goldstein S, Henderson G, Mazzoli C, Russo S, Silenzi S, Taviani M, Trotter J (2009) Li/Mg ratios in shallow and deep-sea coral exoskeleton as a new temperature proxy. AGU Fall meeting, San Francisco, 14–18 December 2009

Montagna P, McCulloch M, Taviani M, Remia A, Rouse G (2005) High resolution trace and minor element compositions in deep-water solitary scleractinian corals (*Desmophyllum dianthus*) from the Mediterranean Sea and the Great Australian Bight. In: Freiwald A, Roberts JM (eds) Cold-water corals and ecosystems. Springer, Berlin/Heidelberg

Montagna P, McCulloch M, Taviani M, Mazzoli C, Vendrell B (2006) Phosphorus in cold-water corals as a proxy for seawater nutrient chemistry. Science 312:1788–1791

Montagna P, McCulloch M, Mazzoli C, Silenzi S, Odorico R (2007) The non-tropical coral *Cladocora caespitosa* as the new climate archive for the Mediterranean Sea: high-resolution (weekly) trace element systematics. Quat Sci Rev 26:441–462

Montagna P, McCulloch M, Mazzoli C, Taviani M, Silenzi S (2008a) High-resolution geochemical records from cold-water corals: proxies for paleoclimate and paleoenvironmental reconstructions and the role of coral physiology. CIESM workshop monographs 36, pp 55–60

Montagna P, Silenzi S, Devoti S, Mazzoli C, McCulloch M, Scicchitano G, Taviani M (2008b) High-resolution natural archives provide new tools for climate reconstruction and monitoring in the Mediterranean Sea. Rend Lincei-Sci Fis Nat 19:121–140

Montagna P, Taviani M, Silenzi S, McCulloch M, Mazzoli C, Goldstein S (2011a) Marine climate archives and geochemical proxies: a review and future investigations on the Mediterranean Sea. In: Brugnoli E, Cavarretta G, Mazzola S et al (eds) Marine research at CNR. Dipartimento Terra e Ambiente – Consiglio Nazionale delle Ricerche, Roma

Montagna P, López Correa M, Goldstein S, McCulloch MT, Freiwald A, Taviani M, Trotter JA, Raddatz J (2011b) Neodymium isotopic composition of gorgonian corals as reliable tool to reconstruct water mass circulation. Mineral Mag Goldschmidt Conf Abstracts: 1491

Montanaro E (1929) Coralli tortoniani di Montegibbio (Modena). Boll Soc Geol Ital 48:107–127

Montanaro E (1931) Coralli pliocenici dell'Emilia. Palaeontogr Ital 31:63–91

Oliverio M, Taviani M (2003) The eastern Mediterranean Sea: tropical invasions and niche opportunities in a "Godot Basin". Biogeographia 24:313–327

Oosterban AFF (1988) Early Miocene corals from the Aquitaine basin (SW France). Meded Werkgr Tert Kwart Geol 25:247–284

Orejas C, Gori A, Lo Iacono C, Puig P, Gili JM, Dale MRT (2009) Cold-water corals in the Cap de Creus canyon, northwestern Mediterranean: spatial distribution, density and anthropogenic impact. Mar Ecol Prog Ser 397:37–51

Osasco E (1895) Di alcuni corallari pliocenici del Piemonte e della Liguria. Atti R Accad Sci Torino 31:225–238

Pedley M (1996) Miocene reefs distribution and their association in the central Mediterranean region: an overview. In: Franseen EK, Esteban M, Ward WC, Rouchy J-M (eds) Models for carbonate stratigraphy from Miocene reef complexes of the Mediterranean region. Society for Sedimentary Geology, Tulsa

Pedley M, Grasso M (1994) A model for the Late Miocene reef tripolaceous associations of Sicily and its relevance to aberrant coral growth-forms and reduced biological diversity within the Paleomediterranean. Geol Mediterr 21:109–121

Peirano A, Morri C, Mastronuzzi G, Bianchi CN (1998) The coral *Cladocora caespitosa* (Anthozoa, Scleractinia) as a bioherm builder in the Mediterranean Sea. Mem Descr Carta Geol Ital 52:59–74

Peirano A, Morri C, Bianchi CN et al (2004) The Mediterranean coral *Cladocora caespitosa*: a proxy for past climate fluctuations. Glob Planet Change 40:195–200

Peirano A, Kružić P, Mastronuzzi G (2009) Growth of Mediterranean reef of *Cladocora caespitosa* (L.) in the Late Quaternary and climate inferences. Facies 55:325–333

Perrin C (2002) Tertiary: the emergence of modern reef ecosystems. In: Kiessling W, Flügel E (eds) Phanerozoic reef patterns. Society for Sedimentary Geology, Tulsa

Perrin C, Bosellini F (2012) Palaeobiogeography of scleractinian corals: changing patterns during the Oligocene-Miocene climatic transition in the Mediterranean. Earth Sci Rev 111:1–24

Perrin C, Kiessling W (2010) Latitudinal trends in Cenozoic reef patterns and their relationship to climate. In: Mutti M, Piller WE, Betzler C (eds) Carbonate systems during the Oligocene-Miocene climatic transition. Wiley, Oxford

Perrin C, Plaziat JC, Rosen BR (1998) The Miocene coral reefs and reef corals of the SW Gulf of Suez and NW Red Sea: distribution, diversity and regional environmental controls. In: Purser BH, Bosence DWJ (eds) Sedimentation and tectonics in rift basins, Red Sea - Gulf of Aden. Chapman and Hall, London

Pomar L (1991) Reef geometries, erosion surfaces and high-frequency sea-level changes, upper Miocene reef complex, Mallorca, Spain. Sedimentology 38:243–269

Pomar L, Hallock P (2007) Changes in coral-reef structure through the Miocene in the Mediterranean province: adaptive vs. environmental influence. Geology 35:899–902

Por F (2009) Tethys returns to the Mediterranean: success and limits of tropical re-colonization. BioRisk 3:5–19

Remia A, Taviani M (2005) Shallow-buried Pleistocene *Madrepora-coral* mounds on a muddy continental slope, Tuscan Archipelago, NE Tyrrhenian Sea. Facies 50:419–425

Rheinhold C (1995) Guild structure and aggradation pattern of Messinian *Porites* patch reefs: ecological succession and external environmental control (San Miguel de Salinas Basin, SE Spain). Sediment Geol 97:157–175

Riding R, Braga JC, Martin JM (1991a) Oolite stromatolites and thrombolites, Miocene, Spain: analogues of Recent giant Bahamian examples. Sediment Geol 71:121–127

Riding R, Martin JM, Braga JC (1991b) Coral-stromatolite reef framework, Upper Miocene, Almeria, Spain. Sedimentology 38: 799–818

Roberts JM, Wheeler A, Freiwald A, Cairns S (2009) Cold-water corals: the biology and geology of deep-sea coral habitats. Cambridge University Press, Cambridge

Rodolfo-Metalpa R, Houlbrèque F, Tambutté E, Boisson F, Baggini C, Patti FP, Jeffree R, Fine M, Foggo A, Gattuso J-P, Hall-Spencer JM (2011) Coral and mollusc resistance to ocean acidification adversely affected by warming. Nat Clim Change 1:308–312

Romano C, Neri C, Russo A, Russo F, Stolarski J (2007) Le biofacies e la storia diagenetica delle biocostruzioni del Miocene superiore, affioranti lungo le coste tirreniche dell'Italia meridionale. Geol Rom 40:77–96

Roniewicz E, Morycowa E (1993) Evolution of the Scleractinia in the light of microstructural data. Cour Forsch-Inst Senckenberg 164:233–240

Rosen BR (1999) Paleoclimatic implications of the energy hypothesis from Neogene corals of the Mediterranean region. In: Agusti J, Rook L, Andrews P (eds) The evolution of Neogene terrestrial ecosystems in Europe. Cambridge University Press, Cambridge

Rosso A, Vertino A, Di Geronimo I, Sanfilippo R, Sciuto F, Di Geronimo I, Violanti D, Corselli C, Taviani M, Mastrototaro F, Tursi A (2010) Hard- versus soft-bottom thanatofacies from the Santa Maria di Leuca deep-water coral province, Recent Mediterranean. Deep-Sea Res II 57:360–379

Rouchy JM, Pierre C, Et-Touhami M, Kerzazi K, Caruso A, Blanc-Valleron MM (2003) Late Messinian to early Pliocene paleoenvironmentaal changes in the Melilla basin (NE Morocco) and their relation to Mediterranean evolution. Sediment Geol 163:1–27

Roux M, Barrier P, Di Geronimo I, Montenat C (1988) Découverte de Crinoïdes pédonculés bathyaux d'origine atlantique dans le Pliocène supérieur et le Pleistocène moyen méditerranéen: conséquences biogéographiques. C R Acad Sci Paris 307:259–364

Roveri M, Gennari R, Lugli S, Manzi V (2009) The Terminal Carbonate Complex: the record of sea-level changes during the Messinian salinity crisis. GeoActa 8:63–77

Ruggieri G, D'Arpa C (2004) Pliocene ostracofauna of the Altavilla section (Altavilla Milicia, Sicily). Systematic remarks. Boll Soc Paleontol Ital 43:159–180

Russo A (1980) The psychrospheric coral fauna from the lower Pliocene of northern Italy. Acta Palaeont Pol 25:614–617

Saint Martin JP (1996) Messinian coral reefs of western Orania, Algeria. In: Franseen EK, Esteban M, Ward WC, Rouchy J-M (eds) Models for carbonate stratigraphy from Miocene reef complexes of the Mediterranean region. Society for Sedimentary Geology, Tulsa

Saint Martin JP, André J (1992) Les construction coralliennes de la plateforme carbonate messinienne de Malte (Méditerranée centrale). Geol Mediterr 19:145–163

Saint Martin JP, Cornée JJ (1996) The Messinian reef complex of Melilla, northeastern Rif, Morocco. In: Franseen EK, Esteban M, Ward WC, Rouchy J-M (eds) Models for carbonate stratigraphy from Miocene reef complexes of the Mediterranean region. Society for Sedimentary Geology, Tulsa

Salomidi M, Zibrowius H, Issaris Y, Milionis K (2010) *Dendrophyllia* in Greek waters, Mediterranean Sea, with the first record of *D. ramea* (Cnidaria, Scleractinia) from the area. Mediterr Mar Sci 11:189–194

Sánchez-Almazo IM, Spiro B, Braga JC, Martín JM (2001) Constraints of stable isotope signatures on the depositional palaeoenvironments of Upper Miocene reef and temperate carbonates in the Sorbas basin, SE Spain. Palaeogeogr Palaeoclimatol 175:153–172

Sanfilippo R, Vertino A, Rosso A, Beuck L, Freiwald A, Taviani M (2012) Serpula aggregates and their role in deep-sea coral communities in the southern Adriatic. Facies doi:10.1007/s10347-012-0356-7

Sartoretto S, Harmelin J-G, Bachet F, Bejaoui N, Zibrovius H (2008) The alien coral *Oculina patagonica* De Angelis 1908 (Cnidaria; Scleractinia) in Algeria and Tunisia. Aquat Invasions 3:173–180

Savini A, Corselli C (2010) High-resolution bathymetry and acoustic geophysical data from Santa Maria di Leuca Cold Water Coral province (northern Ionian Sea-Apulian continental slope). Deep-Sea Res II 57:326–344

Schrag DP, Linsley BK (2002) Corals, chemistry, and climate. Science 296:277–278

Seguenza G (1864) Disquisizioni paleontologiche intorno ai Corallarii fossili delle rocce terziarie del distretto di Messina. Mem R Acc Sci Torino 21:399–560

Seguenza G (1873, 1877) Studi stratigrafici sulla formazione pliocenica dell'Italia meridionale. Boll Reg Comit Geol Ital 4:29–45, 84–103, 131–153, 213–230, 280–301, 345–357 (1873); 8:7–27, 91–99, 359–367 (1877)

Seguenza G (1880) La formazione terziaria nella provincia di Reggio (Calabria). Atti R Accad Lincei Sci Fis Nat 6:1–446

Shackleton NJ, Backman J, Zimmerman H et al (1984) Oxygen isotope calibration of the onset of ice-rafting and history of glaciation in the north Atlantic region. Nature 307:620–623

Silenzi S, Bard E, Montagna P, Antonioli F (2005) Isotopic records in a non-tropical coral (*Cladocora caespitosa*) from the Mediterranean Sea: evidence of a new high-resolution climate archive. Glob Planet Change 49:94–120

Simonelli V (1895) Gli Antozoi pliocenici del Ponticello di Savena presso Bologna. Paleont Ital 1:149–168

Simonelli V (1896) Antozoi neogenici del Museo Parmense. Paleont Ital 2:185–201

Stanley GD, Helmle KP (2010) Middle Triassic coral growth bands and their implication for photosymbiosis. Palaios 25:754–763

Stanley GD, van de Schootbrugge B (2009) The evolution of the coral-algal symbiosis. In: Van Oppen JH, Lough JM (eds) Coral bleaching: patterns processes, causes and consequences. Springer, Berlin

Stirling C, Andersen M (2009) Uranium-series dating of fossil coral reefs: extending the sea-level record beyond the last glacial cycle. Earth Planet Sci Lett 284:269–283

Stolarski J, Meibom A, Przeniosło R, Mazur M (2007) A Cretaceous scleractinian coral with a calcitic skeleton. Science 318:92–94

Stolarski J, Kitahara VM, Miller D, Cairns SD, Mazur M, Meibom A (2011) The ancient evolutionary origins of Scleractinia revealed by azooxanthellate corals. BMC Evol Biol 11:316. doi:10.1186/1471-2148-11-316

Taviani M (2002) The Mediterranean benthos from late Miocene up to present: ten million years of dramatic climatic and geologic vicissitudes. Biol Mar Mediterr 9:445–463

Taviani M (2004) Shaping the biogeography of the Mediterranean basin: one geologist's perspective. Biogeogeographia 24:15–22

Taviani M, Colantoni P (1984) Paléobiocenoses profondes a scléractiniaires sur l'escarpement de Malte-Syracuse (Mer Méditerranée): leur structure, leur âge et leur signification. Rev Inst Fr Pétr 39:547–552

Taviani M, Corselli C, Freiwald A, Malinverno E, Mastrototaro F, Remia A, Savini A, Tursi A (2005a) First geo-marine survey of living cold-water *Lophelia* reefs in the Ionian Sea (Mediterranean basin). Facies 50:409–417

Taviani M, Freiwald A, Zibrowius H (2005b) Deep-coral growth in the Mediterranean Sea: an overview. In: Freiwald A, Roberts M (eds) Deep-water corals & ecosystems. Springer, Berlin

Taviani M, Angeletti L, Antolini B, Ceregato A, Froglia C, López Correa M, Montagna P, Remia A, Trincardi F, Vertino A (2011a) Geo-biology of Mediterranean deep-water coral ecosystems. In: Brugnoli E, Cavarretta G, Mazzola S et al (eds) Marine research at CNR. Dipartimento Terra e Ambiente – Consiglio Nazionale delle Ricerche, Roma

Taviani M, Vertino A, López Correa M, Savini De Mol B, Remia A, Montagna P, Angeletti L, Zibrowius H, Alves T, Salomidi M, Ritt B, Henry P (2011b) Pleistocene to Recent scleractinian deep-water corals and coral facies in the eastern Mediterranean. Facies 57:579–603

Thompson WG, Curran HA, Wilson MA, White B (2011) Sea-level oscillations during the last interglacial highstand recorded by Bahamas corals. Nat Geosci 4:684–687

Titschack J, Freiwald A (2005) Growth, deposition, and facies of Pleistocene bathyal coral communities from Rhodes, Greece. In: Freiwald A, Roberts M (eds) Deep-water corals & ecosystems. Springer, Berlin

Titschack J, Nelson CS, Beck T, Freiwald A, Radtke U (2008) Sedimentary evolution of a Late Pleistocene temperate red algal reef (Coralligène) on Rhodes, Greece: correlation with global sea-level fluctuations. Sedimentology 55:1747–1776

Trotter J, Montagna P, McCulloch M, Silenzi S, Reynaud S, Mortimer G, Martin S, Ferrier-Pagès C, Gattuso J-P, Rodolfo-Metalpa R (2011) Quantifying the pH "vital effect" in the temperate zooxanthellate coral *Cladocora caespitosa*: validation of the boron seawater pH proxy. Earth Planet Sci Lett 303:163–173

Vennin E, Rouchy J-M, Chaix C, Blanc-Valleron M-M, Caruso A, Rommevau V (2004) Paleoecological constraints on reef-coral morphologies in the Tortonian-early Messinian of the Lorca Basin, SE Spain. Palaeogeogr Palaeoclimatol 213:163–185

Veron JEN (1995) Corals in space and time: the biogeography and evolution of the Scleractinia. University of New South Wales Press, Sydney

Veron JEN (2000) Corals of the world. Australian Institute of Marine Science, Townsville

Vertino A (2003) Sclerattiniari plio-pleistocenici ed attuali del Mediterraneo. Dissertation, University of Messina

Vertino A, Savini A, Rosso A, Di Geronimo I, Mastrototaro F, Sanfilippo R, Gay G, Etiope G (2010a) Benthic habitat characterization and distribution from two representative sites of the deep-water SML Coral Province (Mediterranean). Deep-Sea Res II 57:380–396

Vertino A, Zibrowius H, Taviani M, Rocca M (2010b) Fossil coralliidae in the Mediterranean basin. In: Bussoletti E, Cottingham D, Bruckner A et al (eds) Red coral, science management and trade: lessons from the Mediterranean. NOAA, Silver Spring

Vielzeuf D, Garrabou J, Baronnet A, Grauby O, Marschal C (2008) Nano and macroscale biomineral architecture of red coral (*Corallium rubrum*). Am Mineral 93:1799–1815

Wilson MEJ, Rosen BR (1998) Implications of paucity of corals in the Paleogene of SE Asia: plate tectonics or centres of origin. In: Hall R, Holloway JD (eds) Biogeography and geological evolution of SE Asia. Backhuys Publishers, Leiden

Zachos J, Pagani M, Sloan L, Thomas E, Billups K (2001) Trends, rhythms, and aberrations in global climate 65 Ma to present. Science 292:686–693

Zibrowius H (1980) Les Scléractiniaires de la Méditerranée et de l'Atlantique nord-oriental. Mém Inst Océanogr Monaco 11:284

Zibrowius H (1987) Scléractiniaires et Polychètes Serpulidae des faunes bathyales actuelle et plio-pléistocène de Méditerranée. Doc et Trav IGAL 11:255–257

Zibrowius H (1991) Les Scléractiniaires du Miocène au Pléistocène de Sicilie et de Calabre de Giuseppe Seguenza (1864–1880) (Cnidaria: Anthozoa) Atti Acad Peloritana Pericolanti. Cl Sci Mat Fis Nat 67:159–179

Zibrowius H (1992) Ongoing modification of the Mediterranean marine fauna and flora by the establishment of exotic species. Mésogée 51:83–107

Zibrowius H (1995) The "southern" *Astroides calycularis* in the Pleistocene of the northern Mediterranean – an indicator of climatic changes (Cnidaria, Scleractinia). Geobios 28:9–16

Zibrowius H (2003) The "white coral community", canyon and seamount faunas of the deep Mediterranean Sea. Project report for the preparation of a Strategic Action Plan for the Conservation of Biological Diversity in the Mediterranean region (SAP BIO). http://sapbio.rac-spa.org/aeng.pdf. Accessed 21 Sept 2012

Zuffardi-Comerci R (1927) Faunetta di corallari pliocenici dell'isola di Rodi. Atti R Acad Sci Torino 63:231–237

Zuffardi-Comerci R (1932) Corallari-Zoantari fossili del Miocene della "Collina di Torino". Paleont Ital 23:86–132

Zuffardi-Comerci R (1937) Contributo dato dai corallari – durante i periodi geologici – alla formazione dei terreni calcarei d'Italia. Riv Ital Paleontol Stratigr 43:9–35

Mediterranean Coral Population Dynamics: A Tale of 20 Years of Field Studies

15

Erik Caroselli and Stefano Goffredo

Abstract

In contrast with the number of studies on tropical species, analyses of the variation of growth parameters with environmental variables in temperate areas are very scarce. The growth of only four species of scleractinians has been analyzed in natural Mediterranean populations: *Cladocora caespitosa* (Family: Faviidae), *Balanophyllia europaea* and *Leptopsammia pruvoti* (Family: Dendrophylliidae), and *Caryophyllia inornata* (Family: Caryophylliidae). Notwithstanding the importance of obtaining information on coral population dynamics, the first study on a Mediterranean scleractinian dates back less than 10 years. To date, field analyses of the relationships between environmental parameters and growth and population dynamics of Mediterranean scleractinians are available only for two solitary dendrophyliid species: *B. europaea* and *L. pruvoti*. Available literature on growth and population dynamics of natural scleractinian populations of the Mediterranean Sea is reviewed in the present work. As general trends, it seems that: (1) solitary species have a definite growth pattern, in contrast with colonial species; and (2) symbiotic species are more sensitive to increasing temperatures and more vulnerable to global warming. Knowledge on the growth and population dynamics of Mediterranean species has significantly increased in the last 20 years, and more effort is still required to gather the necessary information for protecting a hotspot of biodiversity like the Mediterranean Sea from the threats of global environmental change.

Keywords

Temperate areas • Growth parameters • Environmental variables • Mediterranean scleractinians • *Cladocora caespitosa* • *Balanophyllia europaea* • *Leptopsammia pruvoti* • *Caryophyllia inornata* • Coral population dynamics • Dendrophyliid species • Natural scleractinian populations • Mediterranean Sea • Definite growth pattern • Symbiotic species • Global warming • Zooxanthellate corals • Azooxanthellate corals

E. Caroselli
Marine Science Group, Department of Biological, Geological and Environmental Sciences, Alma Mater Studiorum – University of Bologna, Via Selmi 3, 40126 Bologna, Italy

S. Goffredo (✉)
Department of Biological, Geological and Environmental Sciences, Alma Mater Studiorum – University of Bologna, Via Selmi 3, 40126 Bologna, Italy
e-mail: s.goffredo@unibo.it

Introduction

Temperature and irradiance variations associated with latitude have major consequences on global coral distribution patterns (Kleypas et al. 1999). Latitude is the main factor influencing the variation of light and temperature (Kain 1989), which are strongly linked to coral growth, physiology, and demography (Kleypas et al. 1999; Lough and Barnes 2000). In general, coral growth decreases with increasing latitude, to a limit

S. Goffredo and Z. Dubinsky (eds.), *The Mediterranean Sea: Its history and present challenges*,
DOI 10.1007/978-94-007-6704-1_15, © Springer Science+Business Media Dordrecht 2014

beyond 30°N and 30°S, where coral reef development no longer occurs (Kinsey and Davies 1979). Coral "growth" is a composite of the three related characters of calcification rate, skeletal density, and linear extension rate (calcification = skeletal density x linear extension; Lough and Barnes 2000; Carricart-Ganivet 2004). The measurement of all the three growth components is fundamental when assessing the effect of the environment on coral growth, since none of the three is a perfect predictor of the other two (Dodge and Brass 1984). The three variables have been analyzed in the tropical genera *Porites* (Lough and Barnes 2000; Cooper et al. 2008; Tanzil et al. 2009), *Montastraea* (Carricart-Ganivet 2004) and *Diploastrea* (Cantin et al. 2010), where their variation has been linked to changes in temperature and irradiance associated with latitude. In contrast with the number of studies on tropical species, analyses of the variation of growth parameters with environmental variables in temperate areas are very scarce (e.g. Howe and Marshall 2002). The three growth parameters of only one scleractinian species have been studied in the Mediterranean Sea along a natural temperature gradient (Goffredo et al. 2009).

Since the first investigations, the characteristics of scleractinian population structure and dynamics have been related to environmental conditions and to the symbiosis with unicellular algae (Grigg 1975; Loya 1976; Buddemeier and Kinzie III 1976; Bablet 1985; Hughes and Jackson 1985). It is now commonly accepted that the demographic traits of coral populations may reveal relationships between the organisms and their environment, and can be used to assess habitat stability and suitability (Grigg 1975; Bak and Meesters 1998; Meesters et al. 2001; Goffredo et al. 2004, 2008). Moreover, information such as population turnover are useful when planning strategies for reef restoration on degraded or damaged coastal areas (Connell 1973; Rinkevich 1995; Chadwick-Furman et al. 2000; Goffredo and Chadwick-Furman 2003; Knittweis et al. 2009). Notwithstanding their importance, few studies have quantified population dynamics parameters of scleractinian corals, partly because of the processes of fragmentation, fusion and partial colony mortality, which cause corals of similar size to be of widely different ages, thus distorting the age-size relationships (Hughes and Jackson 1985; Babcock 1991). The scarce studies on population dynamics of scleractinian corals were reviewed in 1973, describing their growth and survivorship (Connell 1973). Since then, demography has been studied for some species in the Red Sea (Loya 1976; Chadwick-Furman et al. 2000; Goffredo and Chadwick-Furman 2003; Glassom and Chadwick 2006; Guzner et al. 2007), Pacific (Fadlallah 1983; Nozawa et al. 2008; Roth et al. 2010; Knittweis et al. 2009; Thresher et al. 2011), Caribbean (Hughes and Jackson 1985; Johnson 1992; Meesters et al. 2001; Vermeij 2006; Goffredo and Lasker 2006, 2008; Green et al. 2008; Vermeij and Sandin 2009; Edmunds 2010), Great Barrier Reef (Babcock 1991), Southwestern Atlantic (Lins de Barros and Pires 2006), and the Mediterranean (Goffredo et al. 2004, 2008, 2010; Shenkar et al. 2005; Teixido et al. 2011; Caroselli et al. 2012). Replication, growth and death of the modules can be used to model the growth of modular individuals (Harper 1977), and studies of modular growth have often focused on plasticity of form and the complexity of both individual colony growth and population dynamics of these organisms (Hughes and Jackson 1985; Babcock 1991; Hughes 1989).

In certain species whose individuals rarely fragment or fuse, and where partial colony mortality can be recognized by growth anomalies, it is possible to reliably determine coral age (Babcock 1991; Chadwick-Furman et al. 2000). The growth and dynamics of modular organisms that match these prerequisites can be analyzed using age-based models applied to colony morphology (Grigg 1975; Hughes 1989; Goffredo and Lasker 2006). In some solitary corals, age estimates can easily be obtained by counting the externally visible growth bands (Chadwick-Furman et al. 2000; Goffredo and Chadwick-Furman 2003). Growth band analysis has been used to determine the age of colonial scleractinian and gorgonian corals (Knuston et al. 1972; Logan and Anderson 1991; Goffredo and Lasker 2006), and in solitary forms (Chadwick-Furman et al. 2000; Goffredo and Chadwick-Furman 2003; Goffredo et al. 2004, 2008, 2010; Caroselli et al. 2012). Hence, for some species growth and population dynamic models based on age can be applied to describe demographic characteristics (Grigg 1984; Ross 1984; Chadwick-Furman et al. 2000; Goffredo and Chadwick-Furman 2003; Goffredo et al. 2004, 2008, 2010; Caroselli et al. 2012).

Growth

Only four species of scleractinians have been analyzed in natural Mediterranean populations: *Cladocora caespitosa* (Family: Faviidae), *Balanophyllia europaea* and *Leptopsammia pruvoti* (Family: Dendrophylliidae), and *Caryophyllia inornata* (Family: Caryophylliidae). Available literature is shown in Table 15.1.

The first scleractinian whose growth has been measured in a natural Mediterranean population is *C. caespitosa*. The growth of single polyps of this endemic, common, colonial and zooxanthellate coral, which is able to form wide bioherms (Zibrowius 1982; Shuhmacher and Zibrowius 1985; Schiller 1993b; Kružić and Požar-Domac 2003), has been measured in the Adriatic Sea (Table 15.1; Fig. 15.1), where the colonies have exhibited linear growth rates of 2.8–6.2 mm year^{-1} (Schiller 1993a; Kružić and Požar-Domac 2002). Even if with a high variability, similar values (1.3–6.1 mm year^{-1}) have been obtained by analyzing the annual growth bands on X-ray prints of polyps from the Ligurian and

Table 15.1 Available literature on the growth of natural populations of Mediterranean scleractinians

Reference	Species	Study site
Schiller (1993a)	*Cladocora caespitosa*	1
Peirano et al. (1999)	*C. caespitosa*	2
Peirano et al. (2001)	*C. caespitosa*	2
Kružić and Požar-Domac (2002)	*C. caespitosa*	3
Goffredo et al. (2004)	*Balanophyllia europaea*	4
Peirano et al. (2004)	*C. caespitosa*	2
Goffredo et al. (2007)	*B. europaea*, *Leptopsammia pruvoti*	4, 5, 6, 7, 8, 9
Goffredo et al. (2008)	*B. europaea*	4, 5, 6, 7, 8, 9
Goffredo et al. (2009)	*B. europaea*	4, 5, 6, 7, 8, 9
Goffredo et al. (2010)	*L. pruvoti*	4
Caroselli et al. (2011)	*B. europaea*, *L. pruvoti*	4, 5, 6, 7, 8, 9
Teixido et al. (2011)	*L. pruvoti*, *Caryophyllia inornata*	10, 11
Kersting and Linares (2012)	*C. caespitosa*	12
Caroselli et al. (2012)	*L. pruvoti*	4, 5, 6, 7, 8, 9

For each reference, the studied species and the codes of the study sites used in Fig. 15.1 are indicated

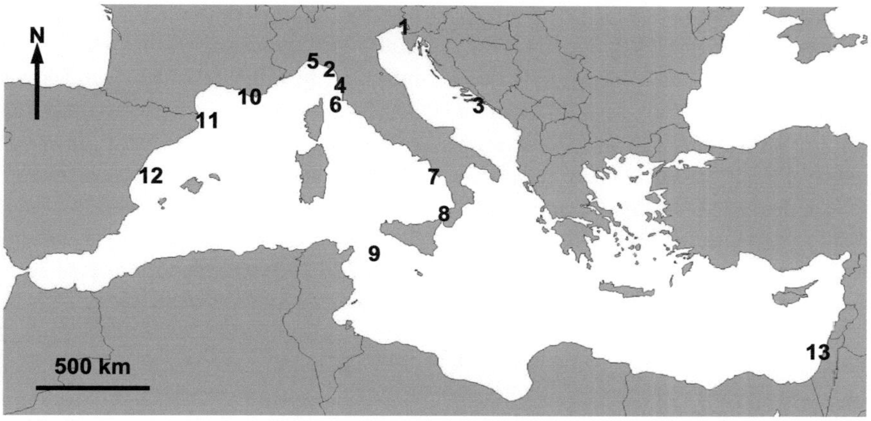

Fig. 15.1 Map of the Mediterranean Sea. The sites where field studies on growth and population dynamics of Mediterranean scleractinians available in the literature have been performed, are indicated. *1*: Bay of Piran, Northern Adriatic Sea, Slovenia (45° 31′ N, 13° 33′ E); *2*: La Spezia region, Ligurian Sea, Italy (44° 05′ N, 9° 45′ E); *3*: Veliko Jezero, Southern Adriatic Sea, Croatia (42° 46′ N, 17° 21° E); *4*: Calafuria, Ligurian Sea, Italy (43° 27′ N, 10° 21′ E); *5*: Genova, Ligurian Sea, Italy (44° 20′ N, 9° 8′ E); *6*: Elba Island, Northern Tyrrhenian Sea, Italy (42° 45′ N, 10° 24′ E); *7*: Palinuro, Central Tyrrhenian Sea, Italy (40° 2′ N, 15° 16′ E); *8*: Scilla, Southern Tyrrhenian Sea, Italy (38° 1′ N, 15° 38′ E); *9*: Pantelleria Island, Strait of Sicily, Italy (36° 45′ N, 11° 57′ E); *10*: Riou Archipelago, Balearic Sea, France (43° 10′ N, 5° 23′ E); *11*: Medes Islands, Balearic Sea, Spain (42° 3′ N, 3° 13′ E); *12*: Columbretes Islands, Balearic Sea, Spain (39° 53′ N, 0° 41′ E); *13*: Sdot Yam, Levantine Sea, Israel (39° 29′ N, 34° 53 E)

Adriatic Seas, and from the Columbretes Islands in Spain (Table 15.1; Fig. 15.1) (Peirano et al. 1999; Kružić and Požar-Domac 2002; Kersting and Linares 2012). The species has been found to deposit one high density band in winter, and one low density band in summer (Peirano et al. 1999), confirming the trend previously suggested for temperate corals (Highsmith 1979). The summer low density band in *C. caespitosa* is produced when the density of the deposited skeleton and the linear extension, and consequently also the calcification rate (Lough and Barnes 2000; Carricart-Ganivet 2004), are lower than in winter months, while the opposite occurs for the high density band (Peirano et al. 1999, 2005). The same pattern may be true for all temperate corals, but there are several clues that the banding pattern may be related not only to temperature, but also to light and/or other factors such as the heterotrophy/autotrophy budget during the year (Miller 1995; Peirano et al. 1999, 2005). It seems that *C. caespitosa* is well adapted to winter conditions, characterized by low solar irradiance and high energy supply, while the summer appears as a period of stress and starvation due to the lower nutrient amounts and lower photosynthetic efficiency of the zooxanthellae caused by the excessive irradiance and temperature (Peirano et al. 2005). The calcification capability of this species (0.004–12.8 kg $CaCO_3$ m^{-2} $year^{-1}$), even if very variable among sites, depths and colonies, comprises values comparable with the calcification potential of tropical corals, making this species a major carbonate producer among Mediterranean organisms (Peirano et al. 2001). Moreover, fossil specimens have proved to record in their skeleton information on the environmental conditions experienced in the past, making this species a high-resolution proxy for analyzing the changes in Mediterranean climate from short (annual) to long (millennial) timescales (Peirano et al. 2004, 2009; Silenzi et al. 2005). Today, *C. caespitosa* is sensitive to thermal and irradiance stress. Mass mortality events of the species, and of other Mediterranean benthic organisms, have been observed in NW Mediterranean during summer periods characterized by prolonged temperature values above the normal ones (Rodolfo-Metalpa et al. 2000; Cerrano et al. 2000; Garrabou et al. 2001).

The first study on a Mediterranean Dendrophylliid has been performed on the endemic, zooxanthellate, solitary coral *B. europaea* in the Calafuria population on the Italian coasts (Table 15.1; Fig. 15.1; Goffredo et al. 2004). Growth analysis has been performed both by measuring actual growth increments through time in the field, and by dating corals by counting the annual growth bands on computerized tomography (CT) scans (Kenter 1989; Logan and Anderson 1991; Bosscher 1993; Goffredo et al. 2004). The normal X-ray technique commonly used for tropical massive species (Barnes and Lough 1989) proved to be uneffective in the small solitary species usually found in Mediterranean rocky shores, as well as in some tropical corals with complex growth forms (Cantin et al. 2010). For *B. europaea* growth data obtained by the field measurements and by CT scans have been fitted to the von Bertalanffy age-length growth function, which is characterized by a decreasing linear growth rate as coral age increases (von Bertalanffy 1938; see Pauly 1984 and Sparre et al. 1989 for derivation of the von Bertalanffy curve parameters). Differently from the common trend of colonial corals, whose growth is usually assumed as constant throughout the entire lifespan, studied species of Mediterranean solitary corals have a decreasing growth rate as they grow old (Goffredo et al. 2004, 2008, 2010; Caroselli et al. 2012). While growth assessment based on field measurements is rather accurate, it considers growth increments on a relatively small timescale (months). Instead, growth band count accuracy is limited by the resolution of the scan, but it considers the growth rate during the entire lifespan of the animal. The growth curves resulting from field measurements and from CT scan analysis were very similar, thus indicating that the two growth assessment methods were comparable (Goffredo et al. 2004). This preliminary analysis has shown to be fundamental, since further analyses of growth in different populations (Goffredo et al. 2008) may be performed by using only the CT scan technique, thus saving funds and time that would be required for in situ measurements in distant locations. *B. europaea* was the first temperate scleractinian species whose all growth components (skeletal density, linear extension rate and calcification rate) have been measured (Goffredo et al. 2009). When analyzed along a latitudinal gradient of temperature spanning 7 ° of latitude, populations of this species exhibited a decrease of skeletal density (from 0.0020 to 0.0009 g mm^{-3};−53.3 %), and calcification rate (from 0.0029 to 0.0010 g mm^{-2} $year^{-1}$), while the average linear extension rate remained constant (1.16 mm $year^{-1}$), with increasing temperature (Table 15.1; Fig. 15.1; Goffredo et al. 2007, 2008, 2009). A fine-scale analysis of the components of skeletal density (bulk density, micro-density and porosity, *sensu* Bucher et al. 1998) indicated that the observed decrease of skeletal bulk density with temperature (Goffredo et al. 2007) is due to an increase in the porosity of the skeleton, while the density of the calcium carbonate composing it remains constant (Caroselli et al. 2011). Even if this species is zooxanthellate, temperature has always shown a higher effect on the variation of biological parameters than irradiance did (Goffredo et al. 2007). The main hypothesis to explain these negative effects of temperature on growth of *B. europaea* states that the photosynthesis of the zooxanthellae may be inhibited by high temperatures (Howe and Marshall 2002; Al-Horani 2005) causing a cascading inhibition effect on the calcification process (Gattuso et al. 1999; Al-Horani et al. 2005), which would result in a lower calcification rate and more porous skeletons (Goffredo et al. 2007, 2008, 2009; Caroselli et al. 2011). However, this hypothesis is yet to be tested, and other explanations related to differences in nutrients availability between study sites may contribute to the observed trends (Goffredo et al. 2007, 2008, 2009; Caroselli et al. 2011). Today, there is concern for the future of this endemic species at the light of the expected increase in global seawater temperatures (Goffredo et al. 2008, 2009), and, as well as for *C. caespitosa*, mass mortality events of the species have been observed in NW Mediterranean during summer periods characterized by prolonged temperatures values above the normal ones (Rodolfo-Metalpa et al. 2000; Cerrano et al. 2000).

Analyses on the closely related solitary dendrophylliid *L. pruvoti* in the same populations have yielded very different results. The main ecological difference between the two species is the lack of a symbiotic relationship with zooxanthellae in *L. pruvoti*, which results in a distinct habitat separation with respect to *B. europaea* (Zibrowius 1980). In fact, even if they share part of their bathimetric distribution (0–40 m for *B. europaea*, 0–70 m for *L. pruvoti*), *L. pruvoti* lives in the shadows on the vaults of caves and crevices, while *B. europaea* colonizes open and illuminated habitats, with a peak of population density around 6 m depth (Zibrowius 1980; Goffredo et al. 2004). Field and growth bands measurements have shown that *L. pruvoti* has a definite growth pattern that can be modeled on the von Bertalanffy growth function, with a decreasing growth rate with coral age, rising the possibility that this is a general characteristic of solitary coral forms (Chadwick-Furman et al. 2000; Goffredo and

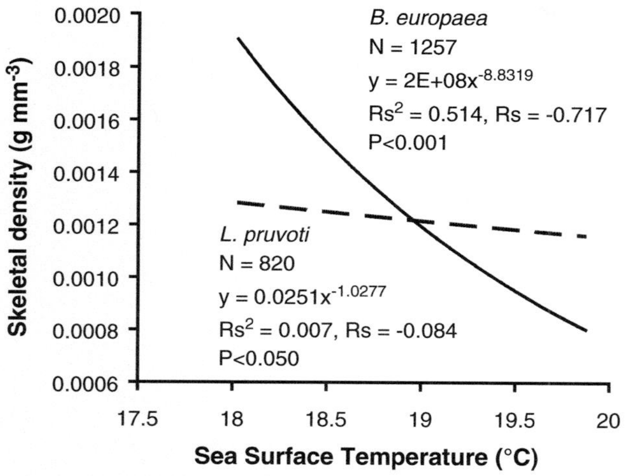

Fig. 15.2 *Balanophyllia europaea* and *Leptopsammia pruvoti.* Variation in skeletal density with sea surface temperature on a latitudinal gradient spanning 7 ° of latitude along the western Italian coasts. *Rs^2* Spearman's determination coefficient, *Rs* Spearman's correlation coefficient, *N* number of individuals (Derived by Goffredo et al. 2007)

Chadwick-Furman 2003; Goffredo et al. 2004, 2008, 2010; Caroselli et al. 2012), even if also some gorgonians have the same growth pattern (Grigg 1974; Goffredo and Lasker 2006). Even if, from a statistical point of view, the skeletal density is significantly correlated with temperature in populations arranged along a gradient of environmental conditions spanning 7° of latitude, from a biological point of view the difference between the extremes is very low (from 0.0013 to 0.0012 g mm^{-3};−7.7 %), especially if compared with the sharp decline in skeletal density with temperature of *B. europaea* (from 0.0019 to 0.0008 g mm^{-3};−57.9 % Fig. 15.2; Goffredo et al. 2007). The growth rates of populations are homogeneous along the latitudinal gradient, with an average value of about 0.9 mm $year^{-1}$ (derived from Caroselli et al. 2012) which is very similar to the estimated growth rate measured during a 5 years monitoring of polyps in French and Spanish waters (0.7 mm $year^{-1}$; Teixido et al. 2011). The homogeneous growth rates suggest a higher tolerance of this species to increasing temperature, with respect to the zooxanthellate *B. europaea* (Goffredo et al. 2010; Caroselli et al. 2012). This is confirmed also by the analyses of bulk density, micro-density, which is the density of the calcium carbonate crystals composing the skeleton (Bucher et al. 1998), and porosity of the skeletons in the populations along the latitudinal gradient. Bulk density and porosity of *L. pruvoti* are homogeneous, while micro-density increases significantly with temperature indicating that this species may even benefit from increasing temperature (Caroselli et al. 2011).

Finally, linear extension rate has been measured in the solitary azooxanthellate Caryophyllid *Caryophyllia inornata*, monitored for 25 years in two sites in Spanish and French waters (Teixido et al. 2011). The recorded values (0.68 mm $year^{-1}$ on average) are similar to the ones of the solitary azooxanthellate *L. pruvoti*, but lower than the zooxanthellate solitary *B. europaea* and extremely lower than the colonial *C. caespitosa*, confirming the hypothesis that zooxanthellae photosynthesis plays a major role in enhancing the calcification capabilities of their hosts (Gattuso et al. 1999; Al-Horani et al. 2005).

Population Dynamics

Notwithstanding the importance of obtaining information on coral population dynamics, the first study on a Mediterranean scleractinian dates back less than 10 years, and has been performed on the solitary endemic species *B. europaea* (Goffredo et al. 2004). The clear dependency between size and age in this species allowed to determine the age distributions in populations located along a natural gradient of temperature spanning 7 ° of latitude (Goffredo et al. 2004, 2008). By applying population dynamics models commonly used in coral and fish management (Pauly 1984; Sparre et al. 1989; Chadwick-Furman et al. 2000; Goffredo and Chadwick-Furman 2003; Goffredo and Lasker 2008; Knittweis et al. 2009), it has been possible to assess the effect of environmental parameters variation on some important demographic parameters such as population abundance and structure stability, instantaneous rate of mortality, average age, percentage of immature individuals, age at maximum biomass, and average age of biomass (Goffredo et al. 2004, 2007, 2008). With increasing seawater temperature, populations of this species are characterized by a decrease of population abundance (from 50 to 4 individuals per square meter;−91.7 %), structure stability (from 0.935 to 0.423;−54.8 %) and percentage of immature individuals (from 56.9 % to 31.4 %;−44.0 %) (Goffredo et al. 2007, 2008). All these parameters are related to the reproductive biology of the species, then it has been hypothesised that the same inhibition of zooxanthellae photosynthesis that would affect skeletal growth at higher seawater temperatures may also cause a decrease of energy availability for reproduction, and for the zooxanthellate larvae of *B. europaea* (Goffredo and Zaccanti 2004; Goffredo et al. 2008). In fact, in zooxanthellate corals, most of the energy required for gametogenesis comes from photosynthesis (Rinkevich 1989; Carlon 2002; Baird and Marshall 2002), thus a reduction of photosynthetic efficiency with increasing temperature may explain the observed negative effects on population abundance, stability and presence of young individuals (Goffredo et al. 2008).

As support to the above hypothesis, the same analyses have been performed for populations of the azooxanthellate *L. pruvoti* in the same sites (Goffredo et al. 2010; Caroselli et al. 2012). As expected, this species has an homogeneous population abundance across the 7 ° of latitude studied (10,155 individuals per square meter, SD=5,741), while differences in demographic parameters seem related to local factors and are not dependent on seawater temperature variations (Goffredo et al. 2010; Caroselli et al. 2012). The fact

Table 15.2 Available literature on population dynamics of natural populations of Mediterranean scleractinians

Reference	Species	Study site
Goffredo et al. (2004)	*Balanophyllia europaea*	4
Shenkar et al. (2005)	*Oculina patagonica*	13
Goffredo et al. (2007)	*B. europaea*, *Leptopsammia pruvoti*	4, 5, 6, 7, 8, 9
Goffredo et al. (2008)	*B. europaea*	4, 5, 6, 7, 8, 9
Goffredo et al. (2010)	*L. pruvoti*	4
Teixido et al. (2011)	*L. pruvoti* and *Caryophyllia inornata*	10, 11
Caroselli et al. (2012)	*L. pruvoti*	4, 5, 6, 7, 8, 9

For each reference, the studied species and the codes of the study sites used in Fig. 15.1 are indicated

that the azooxanthellate *L. pruvoti* does not present negative effects on growth and demographic characteristics related to high temperatures, while the zooxanthellate *B. europaea* is strongly and negatively affected, suggests that the symbiotic condition may indeed play a major role in determining this different response. In one population of *L. pruvoti* (Calafuria; Table 15.2; Fig. 15.1), by combining the information from growth and population dynamics (Goffredo et al. 2010) with the ones on reproductive biology (Goffredo et al. 2005, 2006), it has been possible to estimate the larval mortality (98 %) of this species at Calafuria (Goffredo et al. 2010). Performing the same analysis on all the populations of *B. europaea* and *L. pruvoti* along the latitudinal gradient will allow to further clarify the reproductive processes most affected by seawater temperature and likely to cause the observed trends in abundance and stability of the populations. This is currently being performed by the EU 7th Framework Program project "Corals and Global Warming: The Mediterranean versus the Red Sea" (CoralWarm; www.coralwarm.eu).

An analysis of size frequencies of the colonial zooxanthellate *Oculina patagonica* in one population on the Israeli coast (Table 15.2; Fig. 15.1) has discovered that in this species the mortality related to bleaching events is higher in large colonies than in smaller ones (Shenkar et al. 2005). This result has been considered quite surprising, since mortality is usually inversely related to colony size (Hughes and Jackson 1980; Johnson et al. 1995). It has been hypothesised that large colonies may have a higher probability to be infected by the causative agent of bleaching in this species (the bacterium *Vibrio shiloi*; Kushmaro et al. 1996), confirming the usefulness of population structure analysis to assess the relationships between species and their environment (Bak and Meesters 1998).

A recent long term population dynamics analysis on several benthic Mediterranean species in French and Spanish waters includes *L. pruvoti* (monitored for 5 years) and the Caryophylliid solitary azooxanthellate *Caryophyllia inornata* (monitored for 25 years; Table 15.2; Fig. 15.1; Teixido et al. 2011). *C. inornata* has showed a higher mortality rate (9.2 % year^{-1}) than *L. pruvoti* (less than 3.4 % year^{-1}), with peaks up to 40 % year^{-1} during the 25 years monitored (Teixido et al. 2011). From a population perspective, the higher mortality of *C. inornata* has been balanced by a higher recruitment rate (32 recruits year^{-1} against less than 3 recruits year^{-1} in *L. pruvoti*), but the survival among recruits is only 50 % (Teixido et al. 2011). For both species, the mean lifespan has been estimated as 25–30 years, a higher value than previously reported for *L. pruvoti* at Calafuria (13 years; Goffredo et al. 2011) but similar to Palinuro and Pantelleria (26 years) and lower than in Elba, Scilla and Genova (44, 47 and 73 years, respectively; derived by Caroselli et al. 2012). This indicates that even if the growth of *L. pruvoti* is quite stable in the Mediterranean locations where it has been studied, its population dynamics traits may be extremely variable, probably due to local environmental differences (Teixido et al. 2011; Caroselli et al. 2012).

Available literature on population dynamics of Mediterranean scleractinians studied in the field is reported in Table 15.2.

Environmental Controls

Climate change is the defining environmental, economic, and social issue of our time, and it is now certain that the rapid increase in CO_2 concentration in the atmosphere since the nineteenth century industrial revolution is significantly changing the physical and chemical environment of the planet (Hoegh-Guldberg 2011). Global climatic changes are accelerating, and the average surface temperature of the Earth is likely to increase by 1.1–6.4 °C by the end of the twenty-first century (Solomon et al. 2007). Climate change is having higher and faster effects on marine communities than on terrestrial ones (Richardson and Poloczanska 2008). The magnitude of temperature warming is expected to be higher in temperate areas than in tropical ones (Solomon et al. 2007), and the Mediterranean Sea is likely to be one of the regions mostly affected by the ongoing warming trend and by an increase in extreme events (Lejeusne et al. 2010), thus representing a natural focus of interest for research. The Mediterranean is already one of the most impacted

Table 15.3 Available literature on the effects of the environment on growth and/or population dynamics of natural populations of Mediterranean scleractinians

Reference	Species	Study site
Goffredo et al. (2007)	*Balanophyllia europaea* and *Leptopsammia pruvoti*	4, 5, 6, 7, 8, 9
Goffredo et al. (2008)	*B. europaea*	4, 5, 6, 7, 8, 9
Goffredo et al. (2009)	*B. europaea*	4, 5, 6, 7, 8, 9
Caroselli et al. (2011)	*B. europaea* and *L. pruvoti*	4, 5, 6, 7, 8, 9
Caroselli et al. (2012)	*L. pruvoti*	4, 5, 6, 7, 8, 9

For each reference, the studied species and the codes of the study sites used in Fig. 15.1 are indicated

seas, due to its central position as the cradle of civilization in antiquity, and as a contemporary hub of oil and commodities shipping (Queguiner 1978). Global temperature increase is one of the greatest threats for coral and coral reefs survival (Hughes et al. 2003). Recently, the coralligenous community, one of the most diverse in the Mediterranean (~1,666 species; Ballesteros 2006), where suspension feeders are dominant, has been strongly affected by several mass mortality events related to high temperatures (Cerrano et al. 2000; Perez et al. 2000; Rodolfo-Metalpa et al. 2000; Coma et al. 2009; Garrabou et al. 2009). However, to date, field analyses of the relationships between environmental parameters and growth and population dynamics of Mediterranean scleractinians are available only for two solitary dendrophyliid species: *B. europaea* and *L. pruvoti* (Table 15.3; Fig. 15.1; Goffredo et al. 2007, 2008, 2009; Caroselli et al. 2011, 2012).

The zooxanthellate *B. europaea* is a Mediterranean endemic species which is likely to be negatively affected by seawater warming, since increasing temperature lowers its population abundance, its skeletal density (Goffredo et al. 2007), by increasing its skeletal porosity (Caroselli et al. 2011), and lowers its calcification rate (Goffredo et al. 2009). Moreover, in warmer waters populations are less stable and show a progressive deficiency of young individuals, so that there is concern for the future of this endemic species (Goffredo et al. 2008). This negative effect of temperature increase seems to be related to the symbiosis with zooxanthelle, whose photosynthesis is likely to be depressed at high temperatures, causing negative effects on the growth and reproductive traits of this species (Goffredo et al. 2007, 2008, 2009; Caroselli et al. 2011). However, this hypothesis still has to be tested. *L. pruvoti*, instead, seems to be quite tolerant to the same temperature range experienced by *B. europaea*, since none of its biological traits studied until now in the same sites, time interval and using the same methods, results negatively affected by temperature increase (Goffredo et al. 2007; Caroselli et al. 2011, 2012). It is even possible that *L. pruvoti* may benefit from increasing temperature, since it increases the density of the calcium carbonate crystals (micro-density; Bucher et al. 1998) composing its skeleton (Caroselli et al. 2011). The higher tolerance of *L. pruvoti*, relative to *B. europaea*, may indeed rely on the absence of symbionts, and thus the lack of an inhibition of host physiological processes by the heat-stressed zooxanthellae. This may be a general difference between symbiotic and asymbiotic corals. In fact, also the zooxanthellate *C. caespitosa* is sensitive to high temperatures and exhibited mass mortality events during anomalous summer temperature increases (Peirano et al. 2005; Rodolfo-Metalpa et al. 2000; Cerrano et al. 2000; Garrabou et al. 2001), and its photosynthesis responds negatively to the long term exposure to temperatures recorded during recent warm summers (Rodolfo-Metalpa et al. 2006). In contrast, growth of the azooxanthellate *L. pruvoti* is quite stable in the Mediterranean, and its population dynamics, even if highly variable, is not related to temperature (Teixido et al. 2011; Caroselli et al. 2012). To verify if zooxanthellate corals are more tolerant to the increasing temperature than azooxanthellate ones, further studies on other species are needed, as well as studies on the photosynthetic efficiency of the symbionts at different temperatures. This may help predicting the possible species shifts that are likely to occur as a consequence of global climatic change. Knowledge on the growth and population dynamics of Mediterranean species has significantly increased in the last 20 years, and more effort is still required to gather the necessary information for protecting a hotspot of biodiversity like the Mediterranean Sea from the threats of global environmental change.

Conclusion

The present work reviewed the available literature on growth and population dynamics of natural scleractinian populations of the Mediterranean Sea, which has considerably increased during the last 20 years. As general trends, it seems that: (1) solitary species have a definite growth pattern, in contrast with colonial species; and (2) symbiotic species are more sensitive to increasing temperatures, and more vulnerable to global warming. Future work on different species/locations will be able to verify if these general trends are confirmed in other scleractinian species both in the Mediterranean Sea and in other basins.

References

Al-Horani FA (2005) Effects of changing seawater temperature on photosynthesis and calcification in the scleractinian coral *Galaxea fascicularis*, measured with O_2, Ca^{2+} and pH microsensors. Sci Mar 69:347–354

Al-Horani FA, Ferdelman T, Al-Moghrabi SM, de Beer D (2005) Spatial distribution of calcification and photosynthesis in the scleractinian coral *Galaxea fascicularis*. Coral Reefs 24:173–180

Babcock RC (1991) Comparative demography of three species of scleractinian corals using age- and size-dependent classifications. Ecol Monogr 6:225–244

Bablet JP (1985) Report on the growth of a scleractinian. In: Proceedings of 5th international coral reef symposium 4, Tahiti, pp 361–365

Baird AH, Marshall PA (2002) Mortality, growth and reproduction in scleractinian corals following bleaching on the Great Barrier Reef. Mar Ecol Prog Ser 237:133–141

Bak RPM, Meesters EH (1998) Coral population structure: the hidden information of colony size-frequency distributions. Mar Ecol Prog Ser 162:301–306

Ballesteros E (2006) Mediterranean coralligenous assemblages: a synthesis of the present knowledge. Oceanogr Mar Biol Annu Rev 44:123–195

Barnes DJ, Lough JM (1989) The nature of skeletal density banding in scleractinian corals: fine banding and seasonal patterns. J Exp Mar Biol Ecol 2:119–134

Bosscher H (1993) Computerized tomography and skeletal density of coral skeletons. Coral Reefs 12:97–103

Bucher DJ, Harriott VJ, Roberts LG (1998) Skeletal micro-density, porosity and bulk density of acroporid corals. J Exp Mar Biol Ecol 228:117–136

Buddemeier RW, Kinzie RA III (1976) Coral growth. Oceanogr Mar Biol Annu Rev 14:183–225

Cantin NE, Cohen AL, Karnauskas KB, Tarrant AM, McCorkle DC (2010) Ocean warming slows coral growth in the central Red Sea. Science 329:322–325

Carlon DB (2002) Production and supply of larvae as determinants of zonation in a brooding tropical coral. J Exp Mar Biol Ecol 268:33–46

Caroselli E, Prada F, Pasquini L, Nonnis Marzano F, Zaccanti F, Falini G, Levy O, Dubinsky Z, Goffredo S (2011) Environmental implications of skeletal micro-density and porosity variation in two scleractinian corals. Zoology 114:255–264

Caroselli E, Zaccanti F, Mattioli G, Falini G, Levy O, Dubinsky Z, Goffredo S (2012) Growth and demography of the solitary scleractinian coral *Leptopsammia* pruvoti along a sea surface temperature gradient in the Mediterranean Sea. PLoS One 7:e37848

Carricart-Ganivet JP (2004) Sea surface temperature and the growth of the West Atlantic reef-building coral *Montastraea annularis*. J Exp Mar Biol Ecol 302:249–260

Cerrano C, Bavestrello G, Bianchi CN, Cattaneo-Vietti R, Bava S, Morganti C, Morri C, Picco P, Sara G, Schiaparelli S, Siccardi A, Sponga F (2000) A catastrophic mass-mortality episode of gorgonians and other organisms in the Ligurian Sea (north-western Mediterranean), summer 1999. Ecol Lett 3:284–293

Chadwick-Furman NE, Goffredo S, Loya Y (2000) Growth and population dynamic model of the reef coral *Fungia granulosa* Kluzinger, 1879 at Eilat, northern Red Sea. J Exp Mar Biol Ecol 249:199–218

Coma R, Ribes M, Serrano E, Jimenez E, Salat J, Pasqual J (2009) Global warming-enhanced stratification and mass mortality events in the Mediterranean. Proc Natl Acad Sci U S A 106:6176–6181

Connell JH (1973) Population ecology of reef building corals. In: Jones OA, Endean R (eds) Biology and geology of coral reefs, vol. II: biology 1. Academic, New York

Cooper TF, De'ath G, Fabricius KE, Lough JM (2008) Declining coral calcification in massive *Porites* in two nearshore regions of the northern Great Barrier Reef. Glob Change Biol 14:529–538

Dodge RE, Brass GW (1984) Skeletal extension, density and calcification of the reef coral, *Montastraea annularis*: St. Croix, U.S. Virgin Islands. Bull Mar Sci 34:288–307

Edmunds PJ (2010) Population biology of *Porites astreoides* and *Diploria strigosa* on a shallow Caribbean reef. Mar Ecol Prog Ser 418:87–104

Fadlallah YH (1983) Population dynamics and life history of a solitary coral, *Balanophyllia elegans*, from central California. Oecologia 58:200–207

Garrabou J, Perez T, Sartoretto S, Harmelin JG (2001) Mass mortality in red coral *Corallium rubrum* populations in the Provence region (France, NW Mediterranean). Mar Ecol Prog Ser 217:263–272

Garrabou J, Coma R, Bensoussan N, Bally M, Chevaldonne P, Cigliano M, Diaz D, Harmelin JG, Gambi MC, Kersting DK, Ledoux JB, Lejeusne C, Linares C, Marschal C, Perez T, Ribes M, Romano JC, Serrano E, Teixido N, Torrents O, Zabala M, Zuberer F, Cerrano C (2009) Mass mortality in the NW Mediterranean rocky benthic communities: effects of the 2003 heat wave. Glob Change Biol 15:1090–1103

Gattuso JP, Allemand D, Frankignoulle M (1999) Photosynthesis and calcification at cellular, organismal and community levels in coral reefs: a review on interactions and control by carbonate chemistry. Am Zool 39:160–183

Glassom D, Chadwick NE (2006) Recruitment, growth and mortality of juvenile corals at Eilat, northern Red Sea. Mar Biol 142: 411–418

Goffredo S, Chadwick-Furman NE (2003) Comparative demography of mushroom corals (Scleractinia, Fungiidae) at Eilat, northern Red Sea. Mar Biol 142:411–418

Goffredo S, Lasker HR (2006) Modular growth of a gorgonian coral generates predictable patterns of colony and population growth. J Exp Mar Biol Ecol 336:221–229

Goffredo S, Lasker HR (2008) An adaptive management approach to an octocoral fishery based on the Beverton-Holt model. Coral Reefs 27:751–761

Goffredo S, Zaccanti F (2004) Laboratory observations of larval behavior and metamorphosis in the Mediterranean solitary coral *Balanophyllia europaea* (Scleractinia, Dendrophylliidae). Bull Mar Sci 74:449–458

Goffredo S, Mattioli G, Zaccanti F (2004) Growth and population dynamics model of the Mediterranean solitary coral *Balanophyllia europaea* (Scleractinia, Dendrophylliidae). Coral Reefs 23:433–443

Goffredo S, Radetić J, Airi V, Zaccanti F (2005) Sexual reproduction of the solitary sunset cup coral *Leptopsammia pruvoti* (Scleractinia, Dendrophylliidae) in the Mediterranean. 1. Morphological aspects of gametogenesis and ontogenesis. Mar Biol 147:485–495

Goffredo S, Airi V, Radetić J, Zaccanti F (2006) Sexual reproduction of the solitary sunset cup coral *Leptopsammia pruvoti* (Scleractinia, Dendrophylliidae) in the Mediterranean. 2. Quantitative aspects of the annual reproductive cycle. Mar Biol 148:923–932

Goffredo S, Caroselli E, Pignotti E, Mattioli G, Zaccanti F (2007) Variation in biometry and population density of solitary corals with environmental factors in the Mediterranean Sea. Mar Biol 152:351–361

Goffredo S, Caroselli E, Mattioli G, Pignotti E, Zaccanti F (2008) Relationships between growth, population structure and sea surface temperature in the temperate solitary coral *Balanophyllia europaea* (Scleractinia, Dendrophylliidae). Coral Reefs 27:623–632

Goffredo S, Caroselli E, Mattioli G, Pignotti E, Dubinsky Z, Zaccanti F (2009) Inferred level of calcification decreases along an increasing temperature gradient in a Mediterranean endemic coral. Limnol Oceanogr 54:930–937

Goffredo S, Caroselli E, Mattioli G, Zaccanti F (2010) Growth and population dynamic model for the non-zooxanthellate temperate solitary coral *Leptopsammia pruvoti* (Scleractinia, Dendrophylliidae). Mar Biol 157:2603–2612

Goffredo S, Vergni P, Reggi M, Caroselli E, Sparla F, Levy O, Dubinsky Z, Falini G (2011) The skeletal organic matrix from Mediterranean coral Balanophyllia europaea influences calcium carbonate precipitation. PLoS One 6:e22338

Green DH, Edmunds PJ, Carpenter RC (2008) Increasing relative abundance of *Porites astreoides* on Caribbean reefs mediated by an overall decline in coral cover. Mar Ecol Prog Ser 359:1–10

Grigg RW (1974) Growth rings: annual periodicity in two gorgonian corals. Ecology 55:876–881

Grigg RW (1975) Age structure of a longevous coral: a relative index of habitat suitability and stability. Am Nat 109:647–657

Grigg RW (1984) Resource management of precious corals: a review and application to shallow water reef building corals. P S Z N I Mar Ecol 5:57–74

Guzner B, Novoplansky A, Chadwick NE (2007) Population dynamics of the reef-building coral *Acropora hemprichii* as an indicator of reef condition. Mar Ecol Prog Ser 333:143–150

Harper JL (1977) Population biology of plants. Academic, London

Highsmith RC (1979) Coral growth rates and environmental control of density banding. J Exp Mar Biol Ecol 37:105–125

Hoegh-Guldberg O (2011) The impact of climate change on coral reef ecosystems. In: Dubinsky Z, Stambler N (eds) Coral reefs: an ecosystem in transition. Springer Science+Business Media BV, Dordrecht

Howe SA, Marshall AT (2002) Temperature effects on calcification rate and skeletal deposition in the temperate coral, *Plesiastrea versipora* (Lamarck). J Exp Mar Biol Ecol 275:63–81

Hughes RN (1989) A functional biology of clonal animals. Chapman and Hall, New York

Hughes TP, Jackson JBC (1980) Do corals lie about their age-some demographic consequences of partial mortality, fission and fusion. Science 209:713–715

Hughes TP, Jackson JBC (1985) Population dynamics and life histories of foliaceous corals. Ecol Monogr 55:141–166

Hughes TP, Baird AH, Bellwood DR, Card M, Connolly SR, Folke C, Grosberg R, Hoegh-Guldberg O, Jackson JBC, Kleypas J, Lough JM, Marshall P, Nyström M, Palumbi SR, Pandolfi JM, Rosen B, Roughgarden J (2003) Climate change, human impacts, and the resilience of coral reefs. Nature 301:929–933

Johnson KG (1992) Population dynamics of a free-living coral: recruitment, growth and survivorship of *Manicina areolata* (Linnaeus) on the Caribbean coast of Panama. J Exp Mar Biol Ecol 164:171–191

Johnson KG, Budd AF, Stemann TA (1995) Extinction selectivity of neogene Caribbean reef corals. Paleobiology 21:52–73

Kain JM (1989) The seasons in the subtidal. Br Phycol J 24:203–215

Kenter JAM (1989) Applications of computerized tomography in sedimentology. Mar Geotechnol 8:201–211

Kersting D-K, Linares C (2012) *Cladocora caespitosa* bioconstructions in the Columbretes Islands Marine Reserve (Spain, NW Mediterranean): distribution, size structure and growth. Mar Ecol 33:427–436. doi:10.1111/j.1439-0485.2011.00508.x

Kinsey DW, Davies PJ (1979) Carbon turnover, calcification and growth in coral reefs. In: Trudinger PA, Swaine DJ (eds) Biogeochemical cycling of mineral forming elements. Elsevier, Amsterdam

Kleypas JA, McManus JW, Menez AB (1999) Environmental limits to coral reef development: where do we draw the line? Am Zool 39:146–159

Knittweis L, Jompa J, Richter C, Wolff M (2009) Population dynamics of the mushroom coral *Heliofungia actiniformis* in the Spermonde Archipelago, South Sulawesi, Indonesia. Coral Reefs 28:793–804

Knuston DW, Buddemeier RW, Smith SV (1972) Coral chronometers: seasonal growth bands in reef corals. Science 177:270–272

Kružić P, Požar-Domac A (2002) Skeleton growth rates of coral bank of *Cladocora caespitosa* (Anthozoa, Scleractinia) in Veliko jezero (Mljet National Park). Period Biol 104:123–129

Kružić P, Požar-Domac A (2003) Banks of the coral *Cladocora caespitosa* (Anthozoa, Scleractinia) in the Adriatic Sea. Coral Reefs 22:536

Kushmaro A, Loya Y, Fine M, Rosenberg E (1996) Bacterial infection and coral bleaching. Nature 380:396

Lejeusne C, Chevaldonné P, Pergent-Martini C, Boudouresque CF, Pérez T (2010) Climate change effects on a miniature ocean: the highly diverse, highly impacted Mediterranean Sea. Trends Ecol Evol 25:250–260

Lins de Barros MM, Pires DO (2006) Colony size-frequency distributions among different populations of the scleractinian coral *Siderastrea stellata* in southwestern Atlantic: implications for life history patterns. Braz J Oceanogr 54:213–223

Logan A, Anderson IH (1991) Skeletal extension growth rate assessment in corals, using CT scan imagery. Bull Mar Sci 49:847–850

Lough JM, Barnes DJ (2000) Environmental controls on growth of the massive coral *Porites*. J Exp Mar Biol Ecol 245:225–243

Loya Y (1976) The red sea coral *Stylophora pistillata* is an r strategist. Nature 259:478–480

Meesters WH, Hilterman M, Kardinaal E, Keetman M, de Vries M, Bak RPM (2001) Colony size–frequency distributions of scleractinian coral populations: spatial and interspecific variation. Mar Ecol Prog Ser 209:43–54

Miller MW (1995) Growth of a temperate coral: effects of temperature, light, depth, and heterotrophy. Mar Ecol Prog Ser 122:217–225

Nozawa Y, Tokeshi M, Nojima S (2008) Structure and dynamics of a high-latitude scleractinian coral community in Amakusa, southwestern Japan. Mar Ecol Prog Ser 358:151–160

Pauly D (1984) Fish population dynamics in tropical waters: a manual for use with programmable calculators. International Center for Living Aquatic Resources Management, Manila

Peirano A, Morri C, Bianchi CN (1999) Skeleton growth and density pattern of the temperate zooxanthellate scleractinian *Cladocora caespitosa* from the Ligurian Sea (NW Mediterranean). Mar Ecol Prog Ser 185:195–201

Peirano A, Morri C, Bianchi CN, Rodolfo-Metalpa R (2001) Biomass, carbonate standing stock and production of the Mediterranean coral *Cladocora caespitosa* (L.). Facies 44:75–80

Peirano A, Morri C, Bianchi CN, Aguirre J, Antonioli F, Calzetta G, Carobene L, Mastronuzzi G, Orrù P (2004) The Mediterranean coral *Cladocora caespitosa*: a proxy for past climate fluctuations? Glob Planet Change 40:195–200

Peirano A, Abbate M, Cerrati G, Difesca V, Peroni C, Rodolfo-Metalpa R (2005) Monthly variations in calix growth, polyp tissue, and density banding of the Mediterranean scleractinian *Cladocora caespitosa* (L.). Coral Reefs 24:404–409

Peirano A, Kružić P, Mastronuzzi G (2009) Growth of Mediterranean reef of *Cladocora caespitosa* (L.) in the Late Quaternary and climate inferences. Facies 55:325–333

Perez T, Garrabou J, Sartoretto S, Harmelin JG, Francour P et al (2000) Mass mortality of marine invertebrates: an unprecedented event in the northwestern Mediterranean. C R Acad Sci Paris, Ser III 323:853–865

Queguiner J (1978) The Mediterranean as a maritime trade route. Ocean Manag 3:179–189

Richardson AJ, Poloczanska ES (2008) Under-resourced, under threat. Science 320:1294–1295

Rinkevich B (1989) The contribution of photosynthetic products to coral reproduction. Mar Biol 101:259–263

Rinkevich B (1995) Restoration strategies for coral reefs damaged by recreational activities—the use of sexual and asexual recruits. Restor Ecol 3:241–251

Rodolfo-Metalpa R, Bianchi CN, Peirano A, Morri C (2000) Coral mortality in NW Mediterranean. Coral Reefs 19:24

Rodolfo-Metalpa R, Richard C, Allemand D, Ferrier-Pages C (2006) Growth and photosynthesis of two Mediterranean corals, *Cladocora caespitosa* and *Oculina patagonica*, under normal and elevated temperatures. J Exp Biol 209:4546–4556

Ross MA (1984) A quantitative study of the stony coral fishery in Cebu, Philippines. P S Z N I Mar Ecol 5:75–91

Roth L, Koksal S, van Woesik R (2010) Effects of thermal stress on key processes driving coral-population dynamics. Mar Ecol Prog Ser 411:73–87

Schiller C (1993a) Ecology of the symbiotic coral *Cladocora caespitosa* (L.) (*Faviidae*, *Scleractinia*) in the bay of Piran (Adriatic Sea): I. Distribution and biometry. P S Z N I Mar Ecol 14:205–219

Schiller C (1993b) Ecology of the symbiotic coral *Cladocora caespitosa* (*Faviidae*, *Montastreinae*) in the bay of Piran: II. Energy budget. P S Z N I Mar Ecol 14:221–238

Shenkar N, Fine M, Loya Y (2005) Size matters: bleaching dynamics of the coral *Oculina patagonica*. Mar Ecol Prog Ser 294:181–188

Shuhmacher H, Zibrowius H (1985) What is hermatypic? A redefinition of ecological groups in corals and other organisms. Coral Reefs 4:1–9

Silenzi S, Bard E, Montagna P, Antonioli F (2005) Isotopic and elemental records in a non-tropical coral (*Cladocora caespitosa*): discovery of a new high-resolution climate archive for the Mediterranean Sea. Glob Planet Change 49:94–120

Solomon S, Qin D, Manning M, Marquis M, Averyt K, Tignor MMB, Le Roy Miller Jr H, Chen Z (2007) Climate Change 2007: the physical science basis. Contribution of Working Group I to the fourth assessment report of the Inter-governmental Panel on Climate Change. Cambridge University Press, Cambridge

Sparre P, Ursin E, Venema SC (1989) Introduction to tropical fish stock assessment. FAO fisheries technical paper, Rome

Tanzil JTI, Brown BE, Tudhope AW, Dunne RP (2009) Decline in skeletal growth of the coral *Porites lutea* from the Andaman Sea, South Thailand between 1984 and 2005. Coral Reefs 28:519–528

Teixido N, Garrabou J, Harmelin JG (2011) Low dynamics, high longevity and persistence of sessile structural species dwelling on Mediterranean coralligenous outcrops. PLoS One 6:e23744

Thresher RE, Adkins J, Thiagarajan N (2011) Modal analysis of the deep-water solitary scleractinian, *Desmophyllum dianthus*, on SW Pacific seamounts: inferred recruitment periodicity, growth and mortality rates. Coral Reefs 30:1063–1070

Vermeij MJA (2006) Early life-history dynamics of Caribbean coral species on artificial substratum: the importance of competition, growth and variation in life-history strategy. Coral Reefs 25:59–71

Vermeij MJA, Sandin SA (2009) Density-dependent settlement and mortality structure the earliest life phases of a coral population. Ecology 89:1994–2004

Von Bertalanffy L (1938) A quantitative theory of organic growth (inquiries on growth laws II). Hum Biol 10:181–213

Zibrowius H (1980) Les scléractiniaires de la Méditeranée et de l'Atlantique nord-oriental. Mem Inst Oceanogr (Monaco) 11:1–284

Zibrowius H (1982) Taxonomy in ahermatypic scleractinian corals. Paleontogr Am 54:80–85

The Making of the Mediterranean Molluscan Biodiversity

16

Bruno Sabelli and Marco Taviani

Abstract

The present Mediterranean molluscan biodiversity, comprising more than 2,000 species, was shaped by the multiple events connected with the geodynamic and climatic evolution of this basin in the last million years coupled with the process of biological evolutionary processes on the global scale. Climate, hydrology, basin physiography, and connections via seaways were and are the main forcing factors tuning the type of molluscs making the Mediterranean diversity at any geological instant, resulting in (often cyclical) taxonomic additions and subtractions. However, this basic motif operating in the Mediterranean since its formation as an independent basin over millions of years, has been seriously perturbed by the deliberate or accidental introduction of non-indigenous (alien or non native) species in the last decades. The taxonomic addition of mostly warm-water species is taking place at an unprecedented pace with respect to even the dramatic changes that have marked glacial-interglacial transitions, thus making the present day Mediterranean Sea a planetary biogeographic *unicum*.

Keywords

Mediterranean • Mollusca • Benthos • Diversity • Evolution • Biogeography • Threats

Introduction

The Mediterranean Sea is considered as a biodiversity hotspot of exceptional planetary value (Coll et al. 2010). With a present estimate of ca. 17,000 marine species it accounts for as much as 7.5 % of total marine diversity (Coll et al. 2010; Danovaro et al. 2010). Such comparably high richness when contrasted with the relatively small surface area of the basin (ca. 2.5 million km^2, 0,7 % of the global ocean surface) is the net result of the basin's complex geological history, mid-latitude geographic position, hydrological regimes, varied coastal (rocky shores, sandy beaches, lagoons, islands etc.) and submarine topography (shelves, bathyal plains, canyons, seamounts, active submarine volcanoes, anoxic bottoms, mud volcanoes etc.), wide bathymetric range (the mean depth is about 1,525 m, and the deepest point is the Matapan Deep System or Matapan–Vavilov Deep in the Hellenic Trench at ca. 5,121 m), and the past and present regional variability of climate within the basin (Taviani 2002; Sardà et al. 2004; Danovaro et al. 2010) (Fig. 16.1).

The Mollusca is one of the most species-rich phyla inhabiting the Mediterranean waters from brackish lagoons to bathyal depths. For instance, among macro-invertebrates, molluscs outnumber all other taxonomic groups (Relini 2008, 2010). The Mediterranean molluscan fauna is considered as

B. Sabelli (✉)
Department of Biological, Geological and Environmental Sciences, University of Bologna, via Selmi 3, 40126 Bologna, Italy
e-mail: bruno.sabelli@unibo.it

M. Taviani
ISMAR-CNR, Bologna, via Gobetti 101, 40129 Bologna, Italy

Biology Department, Woods Hole Oceanographic Institution, 266 Woods Hole Road, Woods Hole, MA 02543, USA
e-mail: marco.taviani@bo.ismar.cnr.it

S. Goffredo and Z. Dubinsky (eds.), *The Mediterranean Sea: Its history and present challenges*,
DOI 10.1007/978-94-007-6704-1_16, © Springer Science+Business Media Dordrecht 2014

Fig. 16.1 Snapshots of Mediterranean mollusc biodiversity. The established present situation (**a**) the quintessential Mediterranean endemic bivalve *Pinna nobilis* (Linnaeus 1758) embedded in coarse sandy sediment at 10 m within a seagrass meadow, Montecristo island, Tyrrhenian Sea (© E. Lipparini 2009). The recent past (**b**) a shell assemblage dominated by the pectinid *Pseudamussium peslutrae*, one of the most common bivalves in the Quaternary Mediterranean basin, CNR Cruise SETE of R/V *Urania*, southern Adriatic Sea, ca.–350 m. The on-going change (**c**) screen from an industrial plant in Augusta, Sicily heavily biofouled by the non-indigenous invasive mussel *Brachidontes pharaonis* (© M. Faimali 2008)

the best known in the world (Oliverio 2003), numbering more than 2,000 species (Coll et al. 2010), although this figure is being steadily increased by the addition of new taxa, including introduced species. Interestingly but not surprisingly, even among the non-indigenous species, Mollusca play a leading role accounting for almost one fifth of exotic taxa presently known from the Mediterranean (Zenetos et al. 2010; CIESM 2012).

In this chapter we discuss the present day Mediterranean molluscan biodiversity, tracing its history and evaluating its potential developments in the mid term. One major advantage of using molluscs for such an analysis is the phylum's high biodiversity and ecological adaptability promoting colonization of virtually all marine environments. Another is the occurrence in most member species of a durable mineral skeleton, the shell, which notoriously permits an easier evaluation of taxonomy (therefore diversity and biogeography) through time at various temporal scales. By large, the focus of our discussion is centred upon shell-bearing benthic molluscs that also permits a comparison with the fossil record.

From Past to Present

The Mediterranean marine biodiversity as presently known is the result of many interacting factors of both regional and global scale. The present situation is but the very last of the many "diversity snapshots" configuring at any temporal instant the biological load of this basin. Today's biodiversity is intimately linked but not equal to past and future changes. Various studies have tried to describe and explain the history of the marine biota of the Mediterranean in light of its very complex and dynamic geologic and climatic evolution and the reader should refer to these papers for details (Pérès 1989; Por and Dimentman 1989; Bianchi and Morri 2000; Corselli 2001; Taviani 2002, with references).

Most Authors agree upon the pivotal event represented by the late Miocene Messinian Salinity Crisis (CIESM 2008; Roveri et al. 2008). In short, a widely shared view is that the marine stenoecious Mediterranean marine fauna was annihilated within a hydrologicaly-compromised basin, and then completely renewed by the following Pliocene inundation from the Atlantic Ocean at 5.33 million years (Hsü et al. 1977; Taviani 2004; Monegatti and Raffi 2010, with references therein). Some Authors contend, however, that this Mediterranean reflooding might have taken place already in the terminal Miocene during the so-called Lago-mare event (Carnevale et al. 2008; Ligios et al. 2012). At any event, the post-Messinian crisis Mediterranean marine molluscan biodiversity basically evolved from this wave of invaders. On the other hand, this view is not unanimously accepted and is contrasted on some aspects as it does not explain all observations. For instance, Por (2009), on the basis of records of fossil fish and of certain benthic species (e.g., echinoids: Néraudeau et al. 2001) in the westernmost part of the basin, admits the permanence of trans-Messinian marine stocks in coastal areas, beside the presence of supposedly Tethyan relics such as the present endemic seagrass *Posidonia oceanica* (L) Delile 1813, to the Mediterranean. Tethyan relics, basically marine organisms echoing the times when the proto-Mediterranean was still in direct connection with the Indo-Pacific region, may include molluscs both in the Plio-Pleistocene and recent Mediterranean fauna (e.g., Ponder 1989; La Perna 2005, 2006; La Perna and D'Abramo 2011). However, the significance of Tethyan relics has been discussed by Taviani (2002) that opposes the view that these faunal elements did survive permanently inside the Mediterranean.

We can probably accept that at least a substantial part of the post-crisis Mediterranean benthic fauna is of Atlantic origin and that the core of the Mediterranean mollusc diversity in the Pliocene is represented by Miocene species (Raffi 1993), most of which hypothetically to have necessarily survived the crisis outside the basin in the Atlantic (Sabelli and Taviani 1984; Jones 1984; Taviani 2004). Monegatti and Raffi (2010) argue that most of the paleoendemic molluscan fauna inhabiting the pre-Messinian proto-Mediterranean basin vanished before the salinity crisis and that the molluscan fauna replenishing the Mediterranean in the early Pliocene was shaped in the Miocene of the adjacent Atlantic Ocean. This Miocene core was later modified by the addition of neoendemics (putatively) evolved within the basin (Oliverio 1995b; Le Renard et al. 1996; Pedriali and Robba 2001; Forli et al. 2002, 2004; Landau et al. 2004, and many others).

In any case, within 3 million years this Pliocene fauna was exposed to the dramatic oceanographic and climatic modifications associated with the change from the tropical-subtropical situation of the Pliocene (Zanclean-Piacenzian, from ca. 5.33 to 2.59 Ma) to the cooler times of the Pleistocene starting with the Gelasian (2.59–1.80 Ma), which had important repercussions on the marine biota (Taviani 2002; Monegatti and Raffi 2010). These events have been documented to have strongly impacted the molluscan diversity of the Mediterranean causing for instance the selective extinction within this basin of a number of taxa of warm-water affinity inhabiting the continental margin (Le Renard et al. 1996; Monegatti and Raffi 2001, 2010; La Perna et al. 2002, 2004; Ceregato et al. 2007). The deep water mollusc fauna does not necessarily conform completely to the trend observed in shallower water environments although, at least some bivalves, such as *Acharax doderleini* Mayer 1861, seems to be disappeared in the Piacenzian (Taviani et al. 2011a). As suggested by diversity in family Nuculidae, the Plio-Pleistocene Mediterranean mollusc fauna contained putative endemic shallow water taxa up to the Early Pleistocene, some of which did not survive further into the basin (La Perna 2007)

It has been repeatedly observed that many Neogene Mediterranean gastropods only differ from their Pleistocene to Recent counterparts in the typology of their embryonic shell (Solsona and Martinell 1999). This is the case for instance of *Neverita olla* (de Serres, 1829) vs *N. josephinia* Risso 1826 (Pedriali and Robba 2009), *Pagodula vaginata* (De Cristofori and Jan 1832) vs *P. echinata* (Kiener 1842) (La Perna 1996), *Cyclope migliorinii* (Bevilacqua 1928) vs *C. neritea* Linnaeus 1758 (Gili and Martinell 2000), some Terebridae (Bouchet 1981) and many others. This rather mysterious trend has not yet been satisfactory explained apart from a generic attribution to climatic driven causes. The real taxonomic significance of such chronological pairs still needs to be properly addressed and fully understood.

Ultimately, as the Pleistocene progressed, the Mediterranean achieved a great part of its present status of a temperate marine basin at least as regards seawater temperature. The onset of cyclical glacial ages on Earth caused the periodic entry inside the basin of benthic elements, mostly molluscs, from the adjacent Eastern Atlantic Ocean (e.g., Gignoux 1913; Pelosio and Raffi 1973; Raffi 1986; Malatesta and Zarlenga 1988; Taviani et al. 1991; Bellomo 1993; Marabini et al. 1995). With respect to the Mediterranean mollusc fauna, these periodical cold times of intense colonization of its coastal habitats by such Atlantic Boreo-Celtic species (referred to in the literature as "cold" or "northern guests"), i.a. *Arctica islandica* (Linnaeus 1767), *Modiolus modiolus* Linnaeus 1758, *Pseudamussium peslutrae* (Linnaeus 1771), *Mya truncata* Linnaeus 1758, *Panomya norvegica* (Spengler 1793), *Buccinum undatum* Linnaeus 1758, *Buccinum humphreysianum* Bennet 1824, *Neptunea contraria* (Linnaeus 1771), profoundly impacted the structure of the basin's biota for the longest duration of the post-Gelasian Quaternary. The last expansion of these north-eastern Atlantic molluscs took place during the last glacial epoch (Mars 1958; Colantoni 1973; Colantoni et al. 1975; Taviani 1978; Domènech and Martinell 1980, 1982; Curzi et al. 1984; Taviani et al. 1991), after which they (temporarily) withdrew from the Mediterranean except *B. humphreysianum* and *P. peslutrae* (see below). Shorter-lasting temperature reversals (interglacials) allowed the colonisation of the Mediterranean by subtropical elements the best known examples of which is *Strombus* and the associate 'Senegalese' fauna of the last interglacial (also known as "warm guests" in the literature) comprising a cohort of subtropical species (Gignoux 1913; Issel 1914; Ruggieri and Buccheri 1968; Bellomo 1996; Taviani 2002 with references). More in detail, the *Strombus* fauna includes a number of extant West African taxa, such as *Persististrombus latus* (Gmelin 1791) (better known in the literature as *Strombus bubonius* Lamarck 1822), *Polinices lacteus* (Guilding 1834), *Monoplex trigonus* (Gmelin 1791), *Gemophos viverratus* (Kiener, 1834), *Conus ermineus* Born 1778, *Brachidontes puniceus* (Gmelin 1791), and many others that apparently only once reached and established themselves successfully in the Mediterranean, during the short time of the penultimate interglacial (ca. 125 ka BP: Marine Isotope Stage 5e, commonly recorded in the Mediterranean stratigraphy as the Tyrrhenian Stage: Antonioli et al. 2008). This could have also been the time when *Patella ferruginea* Gmelin 1791, spread in the Mediterranean where it apparently acclimatized successfully up to present, although only in the western basin.

As for the deep sea, it is noteworthy that during the Pliocene (at least from the Piacenzian) and part of the Pleistocene, bottom water temperature kept the Mediterranean basin within psycrospheric conditions (Di Geronimo and La Perna 1996, 1997; Rosso and Di Geronimo 1998; Corselli 2001; La Perna 2003, 2004, 2007; La Perna et al. 2004). Such cooler than present temperature favoured the establishment of a diverse molluscan fauna in the Pliocene and up to the Sicilian stage in the Pleistocene decreasing in diversity from then onwards with the establishment of warm homothermal conditions (Raffi and Taviani 1984; Di Geronimo and La Perna 1997; La Perna 2004; Tabanelli 2008). Glacial times seem to be more propitious in the last 700 ka for temporal increases in the deep sea benthic diversity as documented by Late Glacial submerged subfossil taphocoenoses containing various Atlantic species no longer living at present in the Mediterranean (Taviani 1974; Raffi and Taviani 1984; Taviani et al. 2011b).

In summary, what we see today in the Mediterranean Sea is the (temporary) result of various filters. Above all, the geodynamic evolution of the basin controls through time the opening and closure of seaways at various times and the establishment of a variety of physiographic conditions, the global climate with its regional responses influenced land and sea (Taviani 2002), and both factors acted together in driving the oceanographic properties of Mediterranean water masses that sustain its benthic fauna.

Table 16.1 Comparison between shells-only vs living mollusc records from different Mediterranean localities

Locality	Dead	Living	Ratio
Secche Tor Paterno (Latium)[a]	223	134	0.60
Gulf of Trieste[b]	481	319	0.66
El Maresma (Barcelona)[c]	332	213	0.64
Punta Pioppeto (Procida Is., Naples)[d]	405	152	0.38
Gulf of Gabes (Tunisia)[e]	400	145	0.36
Gulf of Valencia (only gastropods)[f]	359	127	0.35
Tremiti Is. (Apulia)[g]	388	96	0.25
Gulf of Manfredonia (Apulia)[h]	84	8	0.1

Data from:
[a]PG Albano (personal communication)
[b]Vio and De Min (1996)
[c]Peñas et al. (2009)
[d]Soppelsa et al. (2007)
[e]Cecalupo et al. (2008)
[f]Oliver Baldovì (2007)
[g]Mazziotti et al. (2008)
[h]Panetta et al. (2007)

The Marine Molluscs of the Mediterranean: How Many Species at Present?

The term biodiversity usually refers to the main three levels of genes, species and ecosystems (Wilson 1988), and is here limited to the species level (species diversity or species richness). Molluscs represent the 18 % of the entire marine fauna of the Mediterranean sea, they are only surpassed by crustaceans (19 %) but exceed the annelids (10 %) representing the third most diverse marine invertebrate taxon (Coll et al. 2010). The most recent catalogues of the marine Mollusca species inhabiting the Mediterranean Sea list more than 2,100 species (see Coll et al. 2010). Mediterranean molluscs include members of all known existing classes, i.e. Caudofoveata 9 species (0.4 %), Solenogastres 29 species (1.4 %), Monoplacophora 1 species (0.04 %: no living individuals collected so far), Polyplacophora 30 species (1.4 %), Gastropoda 1,558 species (73 %), Bivalvia 430 species (20 %), Scaphopoda 14 species (0.7 %) and Cephalopoda 65 species (3 %). Molluscs are mainly benthic, but some gastropods are holoplanktic: essentially Heteropoda (14 species) and Thecosomata (49 species), while cephalopods are mostly nektonic (45 species). Holoplanktic molluscs are not treated here and the reader may refer to the monograph on the eastern Mediterranean by Janssen (2012) and to Janssen and Peijnenburg (Chap. 20, this volume) for the recentmost update.

Besides depending upon the authorial concept of species, any number given is far from being definitive and the tally is in need of constant updating mostly because of the steady addition of novel taxa through deliberate or unintentional human action. Furthermore, some species are only known from their mineralized exoskeleton (the shell), rather than from living individuals (Table 16.1). This fact introduces a potential bias (that is, an over-estimation) in assessing the present biodiversity especially as regards deep water taxa and holoplanktonic species which could well represent, in some cases, pre-modern remains. This holds particularly true for deep sea molluscs whose empty shells may at time even pertain to former late Pleistocene communities (e.g., Taviani 1974, 1978; Taviani and Colantoni 1979; Di Geronimo and

Bellagamba 1985; Taviani and Taviani 1986; Bouchet and Taviani 1989; Bonfitto et al. 1994; Pons-Moya and Pons 2000; Panetta et al. 2003).

With respect to the shelled molluscs, three contributions deserve special attention. Monterosato (1878) published the first complete compendium by listing 989 species. About a century later Piani (1980) published a list of 1,502 species, a figure which has now risen to 1667 due to the work of Coll et al. (2010) representing an average increase of about five species/year.

The discrepancies apparent from a simple comparison of recently produced lists underlies the current uncertainness in defining species in Mollusca. For example, the most recent compilations by Coll et al. (2010) and Cossignani and Ardovini (2011) report 1,221 vs. 1,341 shelled gastropods and 430 vs. 441 bivalves, respectively. Such substantial differences derive from the various criteria accepted by different authors as to as to which entities are valid species. Alpha taxonomy of shell bearing molluscs was, and still is, primarily based on shell morphology and colour pattern, and occasionally on morphometric analysis (Carvajal-Rodriguez et al. 2005, 2006; Criscione et al. 2009; Criscione and Patti 2010). It is always important to keep in mind that shelled molluscs have for millennia captured human attention for their aesthetic value (Taviani and Fiocchi 1997). This attitude, similarly for Lepidoptera (butterflies) or Coleoptera (beetles), has direct implications for the taxonomy of the Phylum that may be both positive and negative. The positive aspect is that fascination with this group has resulted in a vast number of workers, a large proportion of whom are amateurs, actively searching for shells, thus making enormous quantities of data and material available for study. The other side of the coin might be a sort of distressed superficiality or incompleteness in defining species based upon hasty subjective evaluation of shell features only (the frenzy for "discovering" new species as evoked by Ghisotti 1974a). In the last decades taxonomy benefitted by the routinely application of advanced technologies such as scanning electron microscopy that permit study of morphology at a great level of detail representing a substantial step forwards in assessing species, especially in morphologically complex groups. An excellent case of this is the family Triphoriidae where SEM images of larval stages (Thiriot-Quivrieux and Rodriguez-Babio 1975; Bouchet 1984), coupled with the radular morphology and of the colour patterns of the living mollusc, resulted in the identification of a number of Atlanto-Mediterranean species and genera out of what was previously considered a single very variable species *Triphora perversa* with its variants (Bouchet and Guillemot 1978; Bouchet 1984, 1995). In fact, besides the shell, the radula is often used to delimit taxa at species level in Gastropoda but its value is somewhat controversial primarily because of its variability in relation to ontogeny (Warén 1990), sex (Fretter 1984; Mutlu 2004), and phenotypic plasticity (Padilla 1998). The integration of shell and soft part morphologic features has been used for unravelling the systematics of complex groups such as for example Marginellidae and Cerithiidae (Gofas 1992; Garilli and Galletti 2006).

A further issue with strong repercussions on gastropod taxonomy is the occurrence of species pairs characterized by possessing an identical teleconch but distinct protoconchs (size, sculpture and number of whorls). The protoconch is assumed to reflect contrasting developmental pathways, i.e. direct and/or lecithotrophic vs. planktotrophic and this bears on taxonomic evaluation (see Oliverio 1996 and bibliography therein). The current view is that such mollusc pairs that only differ in shell protoconch type could be considered as distinct specific entities (Bouchet 1989; Solsona and Martinell 1999; Mariottini et al. 2009). The alternative hypothesis postulated by some authors that the same mollusc species could present two different kind of development (poecilogony) is still to be demonstrated (Hoagland and Robertson 1988; Bouchet 1989). However, some authors admit to this possibility attesting the uncertainness still reigning on this crucial issue (Cumming 1993; Gibson and Chia 1995; Chester 1996; Krug 2007). The adoption of the protoconch morphology (= developmental strategy) as diagnostic in separating species has in the last 30 years resulted in establishing many new taxa of Mediterranean gastropods solely or almost entirely based upon this criterion. The most emblematic case in point is offered by the species-rich family Rissoidae. Before the 1960s, the family included 98 nominal species, but this figure has now reached 154; many of these additions are based upon species pairs as described before. Similar considerations apply to other families as well (i.e. Nassariidae, Conidae, Pyramidellidae). Some species have also been split primarily by their very different egg capsules, such as the gastropods *Nassarius cuvierii* (Payraudeau, 1826) and *N. unifasciatus* (Kiener, 1834) (Moreno and Templado 1994).

The isozyme electrophoresis approach, although scantily adopted by European malacologists, has contributed to clarifying the status of some Mediterranean taxa, especially when integrated with the more traditional descriptors just outlined. For examples, it has been possible to resolve such controversial taxa as the bivalves *Chamelea gallina* (Linnaeus 1758) and *C. striatula* (Da Costa 1778) (Backeljau et al. 1994) and, among the gastropods, *Nassarius reticulatus* (Linnaeus 1758) and *N. nitidus* (Jeffreys, 1867) (Sanjuan et al. 1997), species within the Rissoidae (Oliverio 1984; Colognola et al. 1986; Munksgaard 1990) and Cerithiidae (Boisselier-Dubayle and Gofas 1999).

To date, the growing application of molecular techniques is bringing fresh breath to the alpha taxonomy of Mediterranean molluscs. Sequencing of DNA or RNA represents the most recent approach widely and successfully used for supraspecific and phylogenetic recontructions (e.g., Schiaparelli et al. 2005; Plazzi et al. 2011; Donald et al. 2012; Sharma et al. 2012). The application of bio-molecular methods for species-level definition is gaining momentum but still presents its own problems; nevertheless, such methods have revealed sibling species in bivalves (the *Mactra corallina* group: Guarniero et al. (2010), and resolved debated morphospecies in Polyplacophora (*Acanthochitona crinita* (Pennant 1777)-*A. oblonga* (Leloup 1981)): Bonfitto et al. 2011) and in the gastropod genera *Ocinebrina* (Crocetta et al. 2012) and *Cerithiopsis* (Modica et al. 2013). The intertidal habitat-engineering vermetid gastropod *Dendropoma petraeum* (Monterosato 1884) has been shown to likely contain four criptic species at least (Calvo et al. 2009), and the existence of one possible cryptic species in the eastern Mediterranean has also been proposed for the limpet *Patella rustica* Linnaeus 1758 (Sá-Pinto et al. 2010). Interestingly two main lineages within the common brackish water bivalve *Cerastoderma glaucum* (Poiret 1789), one Aegean-Ponto-Caspian, the other central-western Mediterranean-Boreo Atlantic, diverged considerably, possibly suggesting an incipient speciation (Nikula and Väinölä 2003). Finally, the genetic screening of a non-native bivalve of unknown original provenance known for long as *Anadara demiri* Piani 1981 has finally identified this widespread invasive taxon as being the common east American arcid *Anadara transversa* Say, 1822 (Albano et al. 2009).

The combination of multiple descriptors (genetics, shell morphology, anatomy, egg masses and ecological function) is probably the best approach thus far to try categorizing a species. As regards Mediterranean molluscan diversity, some points strictly linked to the foregoing discussion need to be explored further: the role and meaning of endemic, deep-sea molluscs and non-indigenous species.

Endemics

The Mediterranean marine fauna is considered to host a large number of endemic taxa which also include molluscs (Bianchi and Morri 2000; Coll et al. 2010, with references therein). To the best of our knowledge, no synoptic review dealing specifically with the Mediterranean endemic malacofauna as a whole has been published to date. This topic has been partly discussed by Gofas (1998) while treating the malacofauna around the Gibraltar Strait. Most of such endemics are distributed in shallow waters, while only a few have been recorded from bathyal depths (see below). Endemic molluscs total several tens of putative species, and belong largely to the Gastropoda (iconography in Cossignani and Ardovini 2011). A full discussion of the taxonomic status of all species to date only reported from the Mediterranean basin is beyond the scope of this paper. Here we prefer to limit ourselves to some emblematic cases. Perhaps the best known is *Pinna nobilis* Linnaeus 1758, the largest Mediterranean bivalve, considered an endangered species under strict protection (Council Directive 92/43/EEC 1992) since its populations appear in decline (Vicente 1990).

Up to the present, many ‘endemic’ species have been recognized as such based solely upon distinct morphological features of their shells. This is, for example, the case of the ‘historic’ endemic trochids *Gibbula spratti* (Forbes 1844) from the Aegean Sea and *G. albida* (Gmelin 1791) from the Adriatic (and in the Black Sea where it was originally described as *G. euxinica* Andrjewski 1937). The three ascertained extant species within *Clathromangelia* are non planktotrophic endemics of which *C. granum* is distributed in the central and eastern Mediterranean, while the other two (*C. strigillata* Pallary 1904 and *C. loiselieri* Oberling 1970) are localized in the Aegean-Levant and the Gabés basin (Oliverio 1995b). It equally applies to the many Rissoidae (especially taxa in the genus *Alvania*); Muricidae (e.g., *Muricopsis cevikeri* Houart 2000), Fasciolariidae (*Fusinus* spp. and others, most of which have been described in the last decades (i.e., Houart 2000; Buzzurro and Russo 2007)). Supplementary developmental and ecological information additional to shell morphology is provided only in some cases (e.g., *Rissoa auriscalpium*-*R. italiensis*: Russo and Patti 2005). The application of electrophoresis lends support to specific status of the bivalve *Chamelea gallina* (Linnaeus 1758) relative to its prevalently Atlantic couterpart *C. striatula* (da Costa 1778) making this abundant species Mediterranean endemics (Backeljau et al. 1994).

An interesting condition showing the operational relevance in systematics of integrating more descriptors is offered by one of the most classic of Mediterranean endemics, the sublittoral trochid *Gibbula nivosa* (A. Adams, 1851) only known from Malta (a rare case of punctiform geographic range: Evans et al. 2011), and considered as probably extinct (Schembri et al. 2007) until the recent finding of live populations (Evans et al. 2010). Although established on its peculiar shell morphology and coloration on-going molecular research provides full support for its specific status (Barco et al. 2013). Such an integrated approach would surely solve the many taxonomic problems of the presumed endemic fauna of semi-confined marine coastal settings in the Mediterranean Sea, like the Venice lagoon (home to for example the gastropods *Calliostoma virescens* Coen 1833 and *Ocinebrina ingloria* (Crosse 1865)), the Ganzirri and Pantano del Faro marine lakes near Messina (home to the gastropod *Nassarius tinei* (Maravigna 1840)), and above all the Gulf of Gabés, on the north African shore of the Mediterranean. The latter situation has been described as a potential species factory (Sabelli and Taviani 1980) because of the high proportion of apparently exclusive species inhabiting this Tunisian site (Ghisotti 1972a, b, c). The recent work by Cecalupo et al. (2008) overall lists 584 species, of which 33 are considered as endemics, that is to say about the 7 % a value similar to the 6 % suggested by Sabelli and Taviani (1980). These putative neoendemics all have a sibling species in the Mediterranean from which they likely derived.

The Mediterranean deep sea also includes some ascertained mollusc endemic molluscs, especially those associated with hydrocarbon/sulphide seepage sites (treated below).

An intriguing case of a very localized and putative endemic-rich site that merits to be studied by integrating available taxonomic techniques, are the volcanic "caves" along the Sicilian coast (Palazzi and Villari 2001) said to host 5 exclusive species, some not formally described (*Puncturella picciridda, Monophorus* sp., *Muricopsis* sp., *Neolepton discriminatum, Lucinoma spelaeum*). Another endemic gastropod (*Skeneoides digeronimoi* La Perna 1998) is reported from a volcanic 'cave' at Ustica island in the Tyrrhenian Sea (La Perna 1998b), and an arcid bivalve (*Asperarca magdalenae*) has been described from the Gymnasium calcareous cave off southeast Sicily and as a Pleistocene fossil also from Sicily (La Perna 1998a).

Most such endemic taxa established solely on their shell morphology are currently accepted as valid (WoRMS 2012). Clearly, the final acceptance of the specific status of morphologically different entities regarded as endemics to the Mediterranean, urgently requires a comparative and in-depth screening of their genetics.

The Deep Sea

In general, the deep-sea benthic fauna of the Mediterranean Sea (200–4,000 m: Danovaro et al. 2010) is considered to be an impoverished annex of the deep Atlantic Ocean, further characterized by a longitudinal gradient of decreasing biodiversity from the Strait of Gibraltar eastwards (Fredj and Laubier 1985; Bouchet and Taviani 1992a; Sardà et al. 2004; Tecchio et al. 2011). At present, the supposedly marked taxonomic depauperation of the Eastern Mediterranean vs the Western Mediterranean deep sea fauna has been re-dimensioned when considering prokaryota and eukarya all together (Danovaro et al. 2010) but still seems to hold true for megabenthos (Tecchio et al. 2011), and it also applies to molluscs which show a comparably higher diversity in the west, especially in the Alboran and Gibraltar areas (Bouchet and Taviani 1992a,b; Salas 1996), a trait shared with other taxonomic groups as well (Abelló et al. 2002; Arvanitidis et al. 2002; Danovaro et al. 2007).

Based upon deep-sea gastropod distribution Bouchet and Taviani (1992a) have proposed that at least part of the Mediterranean deep sea benthos derives from planktotrophic larval influxes from the Atlantic Ocean entering the Mediterranean *via* the surface Atlantic water (MAW = Modified Atlantic Water, Font et al. 1998), thus breaching the otherwise insurmountable Gibraltar sill. Relying upon empirical data showing a prominence of planktotrophic gastropods and concomitant eastward decrease in diversity, these authors hypothesize therefore the existence of pseudopopulations (i.e. not self-maintaining in the basin) in the Mediterranean deep-sea benthos. While this is a possible dispersal and settling mechanism for some species, as admitted by Bouchet and Taviani (1992a) this concept does not apply to the deep sea benthos as a whole. In fact, Sardà et al. (2004, with references therein) make the point that various benthic organisms in the deep Levantine basin consist of self-maintaining viable autochthonous populations (see also Koutsoubas et al. 2000 regarding more specifically molluscs). La Perna (2004) observes that the Mediterranean deep water fauna of protobranchiate bivalve is partly inherited from the past and therefore is a permanent and viable resident component, without excluding the potential presence of neo-endemics as a result on intrabasinal evolution. In this respect, only a few species of deep sea molluscs are at present considered endemic to the Mediterranean (i.e. *Yoldiella ovulum, Yoldiella wareni, Ledella marisnostri*: La Perna 2004). In fact, the most consistent number of endemics at bathyal depths pertain to specialized molluscs inhabiting chemosynthetic cold seep habitats. Many data on deep sea molluscs traditionally derive from sampling of soft-sediment habitats and accordingly mostly reflect life in the relatively monotonous muddy bathyal bottoms. In the recent years however, substantial improvements in submarine technology had led to the exploration of many habitats previously either poorly known or totally unsampled. A case in point are the deep water coral banks, canyon heads, seamounts and steep cliffs (e.g., Zibrowius and Taviani 2005; Freiwald et al. 2009; Mastrototaro et al. 2010). Research on such hard bottoms is steadily providing important information on remarkable and elusive molluscan life as well, including giant oysters (Gofas et al. 2007), large sessile bivalves (López-Correa et al. 2005) and rare gastropods (Taviani et al. 2009). Even more remarkable and productive in terms of upgrading our knowledge on deep-sea biodiversity is the on-going research on cold seep chemosynthetic habitats, such as mud volcanoes and pockmarks releasing hydrogen sulphide and hydrocarbons. These habitats are widespread in the Mediterranean deep sea and host an exclusive mollusc fauna largely made up of endemic species (Taviani 2004; Taviani et al. 2011a). These taxa include mainly bivalves, most of which associated with bacterial symbionts, e.g., "*Solemya*" sp., *Idas modiolaeformis* (Sturany 1896), *Lucinoma kazani* Salas and Woodside 2002, *Myrtea amorpha* (Sturany 1896), *Thyasira striata*

(Sturany 1896), and subordinately gastropods, e,g. *Lurifax vitreus* Warén and Bouchet 2001, *Clelandella myriamae* Gofas 2004 (Corselli and Basso 1996; Salas and Woodside 2002; Gofas 2004; Olu-le Roy et al. 2004; Rodrigues et al. 2011; Taviani et al. 2011a; Chap. 5 by Taviani, this volume). It should be noticed that many of these Mediterranean endemics were described long ago (Sturany 1896) and have thus been included in the classic Mediterranean catalogues of molluscs. However, most of these species have yet to be studied genetically to ascertain their species status as distinct species compared to closely related Atlantic taxa.

Non-indigenous (Alien or Non-native) Species

The recent conquest of the Mediterranean Sea by an increasing *armada* of non-native species is an unprecedented event in the biogeography of the Mediterranean, because of the number of species involved (Zibrowius 1991; CIESM 2002; Leppäkoski et al. 2002; Zenetos et al. 2003; Dumont et al. 2004; Galil 2008; Zenetos et al. 2010; Occhipinti-Ambrogi et al. 2011; Chap. 22 by Goren, this volume; Chap. 27 by Galil, this volume). Probably no other large marine basin on Earth could claim to house in excess of 209 species of exotic origin, of which 109 established, as does the Mediterranean Sea (Zenetos et al. 2010). Some of these species are ecologically (and some economically: Garaventa et al. 2012) impacting the Mediterranean, locally assuming a key role in the marine costal ecosystems (Streftaris and Zenetos 2006; Galil 2007). The highest number of ascertained alien molluscs in the Mediterranean have come from the Red Sea *via* the Suez Canal and are so-called Lessepsian migrants (Gofas and Zenetos 2003). The latest estimate records 117 such species (Zenetos et al. 2010), all of them exclusively occurring in shallow water habitats (Chemello and Oliverio 1996). In addition to Lessepsians, other Indo-West Pacific immigrants have reached the Mediterranean possibly through ship-mediated transport as it has been postulated for the gastropod *Cantharus tranquebaricus* (Gmelin 1791) (Mienis 2003), and the strombid *Conomurex persicus* (Swainson 1821) (see Zibrowius 2002) sometimes uncorrectly cited as *C. decorus* (Röding 1798) (Moolenbeck and Dekker 1993). As for a few other tropical elements, *C. persicus* is spreading quite successfully in the eastern Mediterranean (Oliverio 1995a). Remarkably, it represents a renewed situation with *Strombus* s.l. forming part of the Mediterranean marine fauna as seen up to the Pliocene and once again in the last interglacial, providing an argument in favour of the concept of Godot basins (Taviani 2002; Oliverio and Taviani 2004). The arrival of Atlantic tropical aliens is on the contrary a rare phenomenon. An example is the gastropod *Mitrella psilla* (Duclos 1846) found in the Bay of Tunis and thought to have been accidentally introduced by shipping (Antit et al. 2010). Besides such tropical elements, many other molluscan taxa now form part of the Mediterranean fauna, often reaching the status of stable reproducing populations in the basin (Gofas and Zenetos 2003; Occhipinti-Ambrogi et al. 2011). Major dispersal pathways for such molluscs are deliberate, such as aquaculture (UNEP/MAP-RAC/SPA 2008), or unintentional anthropic introduction (CIESM 2002). This phenomenon is far from being concluded as new taxa are continuously introduced (i.a. Kabasakal et al. 2005; Louizidou et al. 2007; Ovalis and Zenetos 2007; Tzomos et al. 2011; Öztürk 2012), and some acclimatize themselves in the Mediterranean (Fabbri and Landi 1999; Keppel et al. 2012), while already-established non-native species steadily expand their range within the basin (e.g., Russo 1999; Russo and Mel 2002; Öztürk and Poutiers 2005; Zenetos et al. 2005; Çeviker and Albayrak 2006; Öztürk et al. 2007; Sarà et al. 2008; Tzomos et al. 2011; Crocetta 2011; Tsiakkiros and Zenetos 2011; Turk and Furlan 2011). Beside the case of *Strombus* discussed before, some non-tropical aliens appear to be particularly successful even while confronting the long-established autochthonous mollusc fauna. An emblematic example is presented by the arcid bivalve *Anadara kagoshimensis* (Tokunaga 1906), an invasive species mainly recorded in the Mediterranean literature as *Scapharca cornea* (Reeve 1844) or *Anadara inaequivalvis* (Bruguière 1789). First spotted in clean and well-sorted sands of fully marine coastal waters in the Adriatic Sea (e.g., Ghisotti 1973; Ghisotti and Rinaldi 1976) and in the Black Sea (Gomoiu 1981) surprisingly it soon successfully invaded very confined brackish and muddy waters in northern Adriatic lagoons (Poluzzi et al. 1981; Taviani et al. 1985; Poluzzi and Taviani 1986). Nowadays, it is a common and well-established faunal element of many coastal and lagoonal systems in the Mediterranean (e.g., Occhipinti-Ambrogi et al. 2011). A rather similar case is that of the Japanese clam *Ruditapes philippinarum*, originally introduced for aquaculture purposes in the 1980s (Bodoy et al. 1981; Lazzari and Rinaldi 1994; Breber 1985, 2002; Cesari and Pellizzato 1985), and expanding quite successfully in the Mediterranean from coastal marine settings to confined lagoons (Zamouri-Langar et al. 2006). Similar considerations could be applied to other non-tropical species (some of which are very invasive) such as the gastropods *Crepidula fornicata* (Linnaeus 1758) and *Rapana venosa* (Valenciennes 1846), and the bivalves *Arcuatula senhousia* (Benson in Cantor 1842), *Limnoperna securis* (Lamarck 1819), *Crassostrea gigas* (Thunberg 1793) and *Mya arenaria* (Linnaeus 1758) (Ghisotti 1974b; Stora 1976; Clanzig 1989; Cachia 1981; Madhioub and Zaouali 1988; Hoenselaar and Hoenselaar 1989; Zibrowius 1991; Blanchard 1997; Porcheddu et al. 1999; Zenetos et al. 2005; Crocetta 2011).

Biogeographic Considerations

Multiple factors influence the biogeography of marine benthos, including historical legacy, physical barriers and seaways, oceanographic vectors (currents), and dispersal mechanisms. The Mediterranean Sea is a distinct province included in the Temperate Northern Atlantic realm (Spalding et al. 2007). The overall biogeographic complexity of this region is evident when considering that the Mediterranean province itself is subdivided into as many as seven ecoregions that are west to east: Alboran Sea, Western Mediterranean, Ionian Sea, Tunisian Plateau/Gulf of Sidra, Adriatic Sea, Levantine Sea, Aegean Sea (Spalding et al. 2007).

As a consequence, the total benthic molluscan stock inhabiting the Mediterranean is unevenly distributed in the basin (see Table 16.2). It should be said that the geographic range of individual molluscan taxa only seldom, if ever, strictly coincides with the boundary of any ecoregion. In fact, there is a substantial number of species whose range is almost basinwide, although each region seems characterized by its own peculiarities.

To provide examples, the Alboran Sea, a region much influenced by the Atlantic fauna, contains many species that are localized or endemic in that general area, some expanding on both sides (Atlantic and Mediterranean) of the Gibraltar Strait (Gofas 1998; Gofas et al. 2011a, b). This area is to date one of the best known in the entire Mediterranean Sea. It has been receiving attention from malacologists beginning from the pioneering works of Pallary (1902, 1920). Van Aartsen et al. (1984) list around 380 species, an estimate substantially upgraded by Peñas et al. (2006) who listed 603 species found around the island of Alboran. Finally, the recent monographs by Gofas et al. (2011a, b) provide a synoptic and critical overview of all molluscs known from Andalusia which is largely located in the Alboran Sea ecoregion. Some 30 molluscs inhabit a very restricted geographic range (1 to 2° in latitude) centred on the Strait of Gibraltar and could be considered endemics to this area (Gofas 1998; Peñas et al. 2006). Besides, about 40 tropical Atlantic species have spread into the Alboran Sea but decrease in number along a west to east gradient with about 25 reaching up to Algeria and five into the westernmost part of the Mediterranean (Gofas and Zenetos 2003). As much as 11 % of the Alboran mollusc fauna appears to be exclusive of this Mediterranean ecozone and 4 % are endemic.

The far distant northern Adriatic hosts some remarkable endemics as well, as described above (e.g. *Calliostoma virescens* Coen 1933), or species that are preferentially distributed in that sector of the basin influenced by cooler waters and reduced salinity, like *Gibbula albida* (Gmelin 1791), *Gibbula adriatica* (Philippi 1844), and *Littorina saxatilis* (Olivi 1792). Furthermore, *Alvania schwartziana* Brusina 1866, *Trivia multilirata* (Sowerby G. B. II 1870) and *Ocinebrina ingloria* (Crosse 1865) also appear to be limited to this sub-basin. It is noteworthy that all these species, apart from *T. multilirata*, are non planktotrophic or are direct developers like *L. saxatilis*. This latter presents a puzzling disjunct distribution (Atlantic Ocean and Venice lagoon: Cesari 1994). However, biomolecular evidence seems to support a recent, possibly anthropic northern Adriatic introduction in historical times (Panova et al. 2011).

Table 16.2 Comparison among check lists of shelled gastropods and bivalves, plotting Mediterranean vs selected localities

	Gastropoda	Bivalvia
Mediterranean Sea[a]	1,221	430
Andalucia[b*]	668	275
Italy[c]	835	315
Greece[d]	539	230
Turkey[e]	380	206
Israel[f]	408	193

Data from:
[a]Coll et al. (2010)
[b]Gofas et al. (2011a, b)
[c]Cattaneo Vietti and Giovine (2008), Oliverio (2008), and Schiaparelli (2008)
[d]Tenekidis (1989)
[e]Demir (2003)
[f]Barash and Danin (1992)
*The Andalucia checklist has been selected in the absence of recent compilations on the Spanish Mediterranean malacofauna as a whole, and also because of its substantial peculiarity with respect to the northern Iberian molluscan fauna

Noticeably, some warm temperate species are modally distributed along the south-western Mediterranean coasts, although reaching up to eastern Sicilian waters (e.g. *Bivetiella cancellata* (Linnaeus 1767)) or Sardinia and the Tuscan Archipelago (e.g. *Patella ferruginea*). On the other hand, cold temperate species, possibly leftovers from the last glacial times, display a localized Mediterranean distribution, such as, *Pseudamussium peslutrae* Linnaeus 1771 (=*P. septemradiatum* (Müller 1776)), in the deep Alboran Sea (Salas 1996), and *Buccinum humphreysianum* Bennet 1824, extending as north as the Gulf of Lion (Casamor and Ghisotti 1968).

One of the leading physical factors governing the distribution of molluscs in the Mediterranean, and therefore their biogeography, is obviously hydrography. Firstly, on the basinal scale the current regime (summarized in Font et al. 1998) is offering the avenues for dispersal. Secondly, seawater attributes such as salinity and temperature set additional constraints on the successful establishment of any given taxon. Salinity for instance selects the marine benthic molluscs capable of living within euryaline seawater bodies such as lagoons. Brackish lagoons do not seem to host endemic species but molluscan assemblages nevertheless differ biogeographically inside the Mediterranean basin (Bedulli and Sabelli 1990). Temperature may act to impede or favour the dispersal of taxa within the basin even in the face of propitious current directions, as seen in the sub-tropical non-native species of the Levantine basin.

Biologically, the type of development and the length of larval life are of paramount importance in determining the biodiversity achievable within the Mediterranean Sea. A few papers address specifically the issue of larval life length related to the distribution of molluscs (Bouchet 1981, 1989; Oliverio 1996). The latter compares the larval strategies of gastropods found in *Posidonia* beds along a west–east transect from Spain to Levantine Turkey observing an increase in the ratio of non planktotrophic to planktotrophic species. This trend is interpreted as being directly dependent on the evolution of different larval developmental pathways.

The developmental strategy in Mollusca is so relevant to the biogeography and biodiversity of the Mediterranean as to deserve to be treated in some detail.

Developmental Strategies and Their Biogeographic Consequences

All marine molluscs develop inside an egg envelope in a first larval form (known as pericalymma in protobranch bivalves), and in a trochophore-like larva in gastropods and non-protobranch bivalves (Page 2009). The following larval stage, the veliger, is in most cases a swimming one, a strategy enabling in principle a widespread dispersion in benthic species of reduced vagility. Veligers in various species have different lifespans ranging from a few hours to a week in lecithotrophic larvae, and from one to several weeks or even years in planktotrophic larvae (Strathmann and Strathmann 2007). The shell keeps memory of the type of development. Thus, the occurrence of only one or two kinds of protoconch (prodissoconch in bivalves) in shells is generally informative about direct/lecithotrophic, or planktotrophic development. The planktotrophic larval stage is securing a dispersal over long distances as well documented for instance by amphiatlantic species among deep water gastropods (e.g. Bouchet and Warén 1980, 1985; Taviani and Taviani 1986).

Obviously, great caution should be applied when trying inferring the actual dispersal capability of any given species, based solely upon indirect information as larval shell morphology. In fact, the effectiveness of dispersal is constrained primarily by the requirements of the metamorphosing larva. It seems that the more a veliger swims away from the parental habitat, the more difficult is for them finding a suitable place for metamorphosis. Remarkably, a wide geographic area characterized by little if any habitat discontinuity in its extension has been documented to be settled successfully by "direct" developers or short swimming larvae in just a few generations. Examples of such are provided by Foighil (1989) for the bivalve genus *Lasea*, and by Scheltema and Williams (2009) for the family Nuculidae. These observations support the "paradox of Rockall", an island in the northeastern Atlantic whose littoral zone is uniquely inhabited by benthic invertebrates lacking a planktonic larva (Johanesson 1988). In this island, only the brooding *Littorina saxatilis* (Olivi 1792) is present while the planktonic *L. littorea* Linnaeus 1758, is absent, in spite that they normally co-exist along both the western and eastern continental coasts.

Regarding constraints to dispersion, another case of some interest is presented by the gastropod *Cerithium scabridum* Philippi 1848, an alien species firstly reported from the Levantine basin about 140 years ago (Keller 1883). This locally invasive species possesses a planktotrophic larval phase lasting 45–60 days (Ayal and Safriel 1982) in principle providing a potential wide dispersal capability. Adults exhibit unspecialized feeding habits, occurring on all kind of substrata in intertidal or shallow water of fully marine and brackish habitats and displaying a high level of genetic variability (Lavie and Nevo 1986). All together such characteristics should enable an almost unlimited distribution within the Mediterranean but this does not seem to be the case. In fact, *Cerithium scabridum* exhibits a continuous distribution within the Levantine basin but a rather patchy distribution in the southern Mediterranean with local populations often rich in individuals. To account for this spotty distribution, Gofas and Zenetos (2003) hypothesized that this species was likely introduced by shipping, first in Sicily and later in Tunisia. Recent findings of dead shells within harbours at Vibo Valentia, Tyrrhenian Sea (Crocetta et al. 2009) and of a living population in Otranto, Apulia (Albano and Trono 2008) lend support to such a hypothesis.

The swimming or buoyancy capability of larvae and their resulting passive drift with currents is not the only mode of dispersal. Molluscs lacking a planktonic larval stage can be dispersed by mucous threads (Martel and Chia 1991), as propagules: young crawling stages or even sexually mature adults in the case of small sized species (e.g. Sullivan 1948; Bradley and Cooke 1959; Sellmer 1967 for bivalves and Newell 1964; Anderson 1971; Little and Nix 1976; Levinton 1979 for gastropods). A peculiar dispersal mechanism has been described in *Mytilus edulis* Linnaeus 1758, whose post-larvae are able to remain pelagic by secreting a gas bubble into the mantle cavity (Nelson 1925). Tidal currents spread young specimens sometimes over tens of kilometers (Beukema and Vlas 1989). Noticeably, a basic circadian periodicity in migratory rhythms of drifting juvenile molluscs has been observed (Armonies 1992). Frequent drifting excursions enhance rafting opportunities thus favouring long distance dispersal (Martel and Chia 1991). Floating material (plastic debris, algae, logs and other) is usually concentrated by wind and currents into drift lines that can then travel hundred kilometers (Birkeland 1971), becoming available to floating animals within a few hours or days (Highsmith 1985). This mechanism can be more effcient than normally envisaged. A high proportion of widely distributed Antarctic benthic invertebrate taxa brood or produce egg capsules from which benthic juveniles hatch, preventing in principle a large diffusion: On the contrary, many such taxa are widely distributed so that it has been envisaged that algal rafting likely plays a role as dispersal mechanism (Dell 1972; Simpson 1977; Pearse 1979; Picken 1980). Furthermore, Dell (1972) posits that a brooding female is a better colonist than a larva.

The classic passive transport by wading birds (Rees 1965; Green and Figuerola 2005) or by other migrant marine animals like turtles (Oliverio et al. 1992), is by now evolved to include humans as a significant dispersal mechanism of overwhelming importance.

About 15 % of the exotic molluscs currently found in the Mediterranean basin have been introduced by shipping/ballast waters or aquaculture, and for three species, i.e. *Gibbula albida*, *Siphonaria pectinata* (Linnaeus 1758) and *Perna perna* (Linnaeus 1758), an intra-Mediterranean transfer has been reported (Gofas and Zenetos 2003). Moreover, this kind of transfer is also hypothesized for other alien species: *Cerithium scabridum*, *Rapana venosa* (Valenciennes 1846) and *Fulvia fragilis* (Forskål in Niehbur 1775) (Crocetta et al. 2008).

If dispersal mechanism represents the main source of construction of the past and present Mediterranean biodiversity, ecology constitutes a constraint to the diversity and distribution of molluscs mostly affecting larval settlement and metamorphosis or juvenile life. As shown by Zenetos (1997) there is a substantial uniformity in the diversity of marine Bivalvia in a geographic area as large as Greece, but ecological patterns, such as depth and degree of confinement, nevertheless play here a major role in the distribution of individual species. Under this respect, Table 16.3 presents a comparison among shelled gastropods, bivalves and their ratio in several Mediterranean localities with different attributes. The role of ecological factors in shaping distribution of marine organisms is well exemplified by the noteworthy transformation of bottom communities in the Adriatic Sea. In about 50 years the variation in sedimentation rate, organic load and flocculation resulted in the substitution of the *Turritella* community described by Vatova (1949) with a *Corbula-Amphiura filiformis* community (Scaccini 1967; Schiaparelli et al. 2000; Chiantore et al. 2001), with rarefaction and possible local extinction of some of the species associated with the former. Another striking case is oyster reefs (*Ostrea edulis* Linnaeus 1758) much more abundant in the Mediterranean coastal areas in the recent past and at present in strong recession with few exceptions (Airoldi and Beck 2007).

Table 16.3 Comparison among shelled gastropods, bivalves and their ratio in several Mediterranean localities with different attributes

Locality	Gastropoda	Bivalvia	Gastr/biv
Mediterranean[a]	1,221	430	2.8
Barcelona[b]	213	105	2
Valencia[c]	359	184	2
Andalucia[d]	668	275	2.4
Algeciras[e]	266	110	2.4
Alboran[f]	454	136	3.3
Tuscany[g]	426	214	2
Procida (Naples)[h]	288	105	2.7
Tor Paterno (Rome)[i]	124	29	4.3
Scilla (Calabria)[j]	369	87	4.2
Adriatic[k]	511	254	2
Gulf of Trieste[l]	294	172	1.7
N Adriatic (tegnue)[m]	124	68	1.8
off Venice lagoon[n]	121	82	1.5
Venice lagoon[o]	78	86	0.9
Emilia Romagna[p]	68	115	0.6
Rovigno[q]	199	142	1.4
Cattaro[r]	164	146	1.1
Tremiti Islands[s]	275	113	2.4
Gulf of Manfredonia[t]	30	51	0.6
Greece[u]	539	230	2.3
Turkey[v]	380	206	1.8
Ciprus[w]	442	172	2.6
Israel[x]	408	193	2.1
Malta[y]	609	242	2.5
Gulf of Gabes[z]	390	157	2.5

Data from:
[a]Coll et al. (2010)
[b]Peñas et al. (2009)
[c]Oliver Baldovì (2007) and Tamayo Goya (2008)
[d]Gofas et al. (2011a, b)
[e]van Aartsen et al. (1984)
[f]Peñas et al. (2006)
[g]Terreni (1981)
[h]Soppelsa et al. (2007)
[i]PG Albano unpublished
[j]Vazzana (2010)
[k]Coen (1937)
[l]Vio and De Min (1996)
[m]Casellato and Stefanon (2008)
[n]Cesari and Mizzan (1994)
[o]Cesari (1994)
[p]Rinaldi (1991)
[q]Coen and Vatova (1932)
[r]Stjepeević and Parenzan (1980)
[s]Mazziotti et al. (2008)
[t]Panetta et al. (2007)
[u]Tenekidis (1989)
[v]Demir (2003)
[w]Öztürk et al. (2004)
[x]Barash and Danin (1992)
[y]Cachia et al. (2004)
[z]Cecalupo et al. (2008)

Threats to the Mediterranean Mollusc Diversity

There is a widespread perception that in the current Anthropocene (*sensu* Crutzen and Stoermer 2000) the marine biodiversity is under attack on a global scale often because of direct or indirect human actions (Gray 1997; Hoegh-Guldberg and Bruno 2010). Among the main causes often considered by scientists as serious threats to marine biodiversity at all levels from species to ecosystems we can list: (1) global warming, (2) ocean acidification, (3) habitat loss, (4) pollution, and (5) overfishing (Gray 1997; Sala and Knowlton 2006). In this scenario, the Mediterranean is not exception (Airoldi and Beck 2007; Claudet and Fraschetti 2010; UNEP 2010; Coll et al. 2011).

Global Warming

The present trend of rising temperature as a result of increasing levels of greenhouse gases in the atmosphere is generating a growing concern about its pernicious impact on marine ecosystems. According to projections, global warming may potentially depressing habitat complexity, rearranging trophic webs, fostering new ecological assemblages, favouring exotic newcomers, and introducing diseases (Hoegh-Guldberg and Bruno 2010). In the Mediterranean, an increase of seawater temperature is already causing visible alterations in the marine biota (Danovaro et al. 2004; Peirano et al. 2005; Bianchi 2007; CIESM 2008; Parravicini et al. 2009; Por 2009; Coll et al. 2010; Faimali and Taviani 2011; Chap. 17 by Albano, this volume). The macroscopic evidence regarding molluscs is surely the amazing increase in the number of exotic species, both Lessepsians or accidentally-introduced, steadily added to the autochtonous molluscs fauna of the basin and by large represented by warm-affinity organisms (Zibrowius 1991; Galil 2008, 2009; Occhipinti-Ambrogi et al. 2011). Although such species-increment appears as the most evident outcome of tropicalization, it is the influence of rising seawater temperature on Mediterranean habitats as a whole that deserves attention. In fact, many important habitats, as for example seagrass beds house various exclusive or preferentially associating autochtonous molluscs (e.g., Pérès and Picard 1964; Terlizzi and Russo 1995; Beqiraj et al. 2008; Gofas et al. 2011a, b; Albano and Sabelli 2012) that may respond to thermal variations by reduction, displacement or eradication.

Ocean Acidification

Ocean acidification (OA), induced by uptake by seawater of anthropogenic CO_2, might adversely impact the calcification potential of carbonate shelled organisms (Orr et al. 2005), and have important repercussions on other physiological functions and ultimately on the trophic web (Fabry et al. 2008; Rossoll et al. 2012; Seibel et al. 2012). Therefore, OA is clearly a potential factor controlling the Mediterranean mollusc fauna. Aragonitic holoplanktonic species for instance, such as thecosomatous pteropods, are supposedly at risk from increasingly acidified water masses (Comeau et al. 2009, 2012). Tank experiments on larvae of the Mediterranean *Cavolinia inflexa* (Leseur 1813) exposed to decreasing seawater pH documented a strong reduction in shell secretion (Comeau et al. 2010). Moreover, ocean acidification has been shown to alter other metabolic processes beside calcification. However, in face of such negative scenarios, a recent metabolic study on Pacific pteropods which include also taxa represented in the present-day Mediterranean fauna (*Hyalocylis striata* (Rang 1828), *Clio pyramidata* Linnaeus 1767, *Diacavolinia longirostris* (Blainville 1821), and *Creseis virgula* Rang 1828) seem to demonstrate an individual resilience to augmented acidification (Maas et al. 2012), as also other calcifiers and non-calcifying organisms (Nielsen et al. 2010; McCulloch et al. 2012a, b). Regarding benthic molluscs, data from naturally acidified waters do not straightforwardly lend support that they are detrimental to the calcification capacity of Mediterranean shelled molluscs (Rodolfo-Metalpa et al. 2011). Much still needs to be learned about the mid and long term effects of OA on organisms with carbonate skeletons, but it is unlikely that it could be considered, on the short term at least, as a serious threat to Mediterranean molluscs

Habitat Loss

Habitat degradation, and in some cases complete loss, induced by human activity is for sure one of the most serious, if not the most important, menaces to marine biodiversity (Airoldi and Beck 2007; Coll et al. 2010). Human-induced habitat fragmentation is documented for the important endemic habitat-forming *Posidonia oceanica* that already suffers from competition by invasive species (Peirano et al. 2005; Airoldi and Beck 2007; Montefalcone et al. 2010a, b). It is likely that the decline and the possible loss of local stands of this seagrass might result in the rarefaction or even extinction of those molluscan taxa associated with it, which are often restricted to specific geographic sectors of the Mediterranean Sea. Fishing malpractices (including overfishing) can, in some cases, be considered as causing habitat loss through inducing sensitive changes both directly in the ecological structure of biological communities (species removal, trophic web disruption, etc.) and indirectly in the physical environment (Coll et al. 2010). For instance, the over-trawling of sectors of the Adriatic continental shelf is stressing the exploitable (as well as the co-occurring non-commercial) living stocks (e.g., *Chamelea gallina*) with local decline in populations inhabiting soft sediments, as well as impacting the recruitment through unnatural stirring and ploughing of the sediment (Romanelli et al. 2009). Another case in point centred on molluscs is the now-illegal collection through rock-hammering or rock blasting of edible boring molluscs (such as the date-mussel *Lithophaga lithophaga* Linnaeus 1758) that has caused considerable and long-lasting ecological damage on limestone coasts (Russo and Cicogna 1991; Fanelli et al. 1994; Bianchi and Morri 2000; Fraschetti et al. 2001; Guidetti and Boero 2004; Devescovi et al. 2005; Rovere et al. 2009).

Pollution

Habitat degradation due to pollution is another factor adversely impacting marine ecosystems from coastal lagoons down to bathyal depths (WWF/IUCN 2004; Coll et al. 2010). It encompasses a variety of phenomena, from the release at sea of dangerous and toxic substances to oil spills and solid waste dumping (Valavanidis and Vlachogianni 2011 with references therein). At the species level, chemicals may interfere with body functions generally in a negative way (including lethality) and cause a decrease in the viability of organisms due to histo-pathological alterations in their organs, chromosomal aberrations and breaks, DNA damage (Bresler et al. 1999, 2003), or the biochemistry of the gonads during the spawning phase (Hamdani and Soltani-Mazouni 2011). Biotoxins and heavy metals may cause physiological stress in bivalves, as seen, for example, in *Mytilus galloprovincialis* Lamarck 1819, from the Albanian coast (Sadikaj et al. 2009). At least one organotin pollutant, tributyltin (TBT), has been documented to act directly on molluscs eventually driving a decline in populations (e.g. the muricid *Stramonita haemastoma* Linnaeus 1767, Terlizzi et al. 2001). TBT is a xenobiotic used (but now banned in many countries) as a biocide in marine antifouling paints and wood preservatives; it is recognized as inducing an alteration in the reproductive system of prosobranch gastropods known as imposex (Braber 1970). Imposex has been demonstrated in several Atlanto-Mediterranean species including several muricids and nassariids (Gibbs et al. 1990; Oehlmann et al. 1996, 1998; Garaventa et al. 2007; Lemghich and Benajiba 2007; Wirzinger et al. 2007; Rodríguez et al. 2009). As clearly demonstrated by Oehlmann et al. (1996) in *Ocinebrina aciculata* (Lamarck 1822), TBT not only induces the development of an additional male sexual organs (mainly the penis and/or vas deferens) in females, sterilization and sex change, but also damages of capsule gland (inhibition of breeding) and the occlusion of vulva by vas deferens tissue (a barrier to copulation) (Bryan et al. 1986) even if recent findings about TBT effect, pointed out that, at least in *Hexaplex trunculus* Linnaeus 1758, it could be not the only imposex cue (Garaventa et al. 2006). Sewage related pollution has been documented to negatively affect rocky shore assemblages (Terlizzi et al. 2002), including the structuring of molluscan assemblages (Terlizzi et al. 2005). Changes in soft bottom molluscan assemblages has been also quantified as a results of offshore gas platforms (Terlizzi et al. 2008). There is however no strong evidence that pollution is really threatening extinction of any Mediterranean mollusc, although some concern exists for the endangered *Patella ferruginea* as, for example, in the populations in the Ceuta harbour (Espinosa et al. 2007).

Final Remarks

The present Mediterranean diversity is transient. Biodiversity *per se* is in a sense ephemeral, and the durability and resilience of its components are accordingly transitory. As the natural historical evolution of our planet teaches, every diversity of every single place at any particular time is permanently liable to be partly or totally jeopardized by a variety of factors thus vanishing, evolving or adapting. The present (mollusc) marine diversity is thus only the result of a collection of evolving diversities without solution of continuity accompanying the highly complex history of this basin.

The Mediterranean is by now in an ultra-dynamic and accelerated situation of aggrading diversity especially in response to an alien expansion that has no equal in its geological past. But is this present trend going to set the pace for this basin on a longer time scale? The answer is probably negative as soon as we consider the global paleoclimatic behaviour as understood to date. In fact, according to the theory that cyclity of Quaternary ice ages is astronomically tuned by precessional forcing at ca. 11 kyrs periodicity, the current interglacial (Holocene or Marine Isotope Stage 1) should by now be approaching its end (Kukla et al. 1972). Because of the uncertainties linked to increasing atmospheric carbon dioxide concentrations, however, this termination could be deferred by many centuries (Tzedakis et al. 2012). At some time in the future, however, a new ice age will impose a drastic climatic deterioration over the Earth which in turn will in all likeliness dictate once again the typology of marine ecosystems of the Mediterranean basin. The expected depression of the temperature will seal the fate of the hundreds of Lessepsian and other ‘warm water’ alien species presently inhabiting the Mediterranean and causing so much concern at present (Oliverio and Taviani 2004). Thus, the evocative prophecy by Por (2009) of a Mediterranean Sea returning to the warm days of Tethys times might in fact collide with a major change in climatic direction governed by astronomical causes. In such a scenario temperate invasive species could in principle be spared by climatic deteriorations as those routinely experienced by the Mediterranean in the Pleistocene.

Acknowledgments This paper benefitted over the years of enlightening discussion with a number of professional and amatorial malacologists. Alessandro Ceregato (ISMAR-CNR Bologna), Francesca Garaventa (ISMAR-CNR Genova), Serge Gofas (University of Malaga), Marco Faimali (ISMAR-CNR Genova), Rafael La Perna (University of Bari), Marco Oliverio (University La Sapienza, Rome), Patrick Schembri (University of Malta), José 'Pepe' Templado (Museo Nacional de Ciencias Naturales-CSIC, Madrid), Antonio Terlizzi (University of Lecce) kindly revised the ms providing many constructive suggestions. This article contributes to the EU Hermione (contract n° 226354) and Coconet (contract n° 287844) programmes. ISMAR-CNR Bologna scientific contribution n. 1769.

Bibliography

Abelló P, Carbonell A, Torres P (2002) Biogeography of epibenthic crustaceans on the shelf and upper slope of the Iberian Peninsula Mediterranean coasts: implications for the establishment of natural management areas. Sci Mar 66(Suppl 2):183–198

Airoldi L, Beck MW (2007) Loss, status and trends for coastal marine habitats of Europe. Oceanogr Mar Biol Annu Rev 45:345–405

Albano PG, Sabelli B (2012) The molluscan assemblages inhabiting the leaves and rhizomes of a deep water *Posidonia oceanica* settlement in the central Tyrrhenian Sea. Sci Mar 76(4):721–732

Albano PG, Trono D (2008) Record of the alien species *Cerithium scabridum* Philippi, 1848 (Gastropoda: Cerithiidae) from Otranto, southern Adriatic Sea. Boll Malcol 44:1–4

Albano PG, Rinaldi E, Evangelisti F, Kuan M, Sabelli B (2009) On the identity and origin of *Anadara demiri* (Bivalvia: Arcidae). J Mar Biol Assoc UK 89:1289–1298

Anderson A (1971) Intertidal activity, breeding and the floating habit of *Hydrobia ulvae* in the Ythan estuary. J Mar Biol Assoc UK 51: 423–437

Antit M, Gofas S, Azzouna A (2010) A gastropod from the tropical Atlantic becomes an established alien in the Mediterranean. Biol Invasions 12:991–994

Antonioli F, Deino A, Ferranti L, Keller J, Marabini S, Mastronuzzi G, Negri A, Piva A, Vai GB, Vigliotti L (2008) Lo studio della sezione "Il Fronte" per la definizione del Piano Tarantiano (Puglia, Italy). Il Quat 20:31–34

Armonies W (1992) Migratory rhythms of drifting juvenile molluscs in tidal waters of the Wadden Sea. Mar Ecol Prog Ser 83:197–206

Arvanitidis C, Bellan G, Drakopoulos P, Valavanis V, Dounas C, Koukouras A, Eleftheriou A (2002) Seascape biodiversity patterns along the Mediterranean and the Black sea: lessons from the biogeography of benthic polychaetes. Mar Ecol Prog Ser 244:139–152

Ayal Y, Safriel UN (1982) r-curves and the cost of the planktonic stage. Am Nat 119:391–401

Backeljau T, Bouchet P, Gofas S, de Bruyn L (1994) Genetic variation, systematics and distribution of the venerid clam *Chamelea gallina*. J Mar Biol Assoc UK 74:211–223

Barco A, Evans J, Schembri PJ, Taviani M, Oliverio M (2013) Testing the applicability of DNA barcoding for mediterranean species of top-shells (Gastropoda, Trochidae, Gibbula s.l.). Mar Biol Res 9:785–793

Barash A, Danin Z (1992) Fauna Palestina. Mollusca I. annotated list of Mediterranean molluscs of Israel and Sinai. The Israel Academy of Sciences and Humanities, Jerusalem

Bedulli D, Sabelli B (1990) E' possibile una zoogeografia delle lagune mediterranee attraverso la distribuzione dei Molluschi? Oebalia 16(Suppl):133–141

Bellomo E (1993) Presenza di "Ospiti boreali" (Mollusca, Gastropoda) nel Pleistocene di Contrada Case Alte (Villa S. Giovanni). Boll Soc Paleontol Ital 32:393–399

Bellomo E (1996) Predation traces on specimens of *Polinices lacteus* (Guilding, 1831), (Gastropoda, Prosobranchia, Naticidae) from last interglacial assemblage of Bovetto (Reggio Calabria, Southern Italy). Boll Soc Paleontol Ital 3(Spec):9–17

Beqiraj S, Kashta L, Kuci M, Rasemi D, Mato X, Gace A (2008) Benthic macrofauna of *Posidonia oceanica* meadows in the Albanian coast. Nat Montenegr 7:55–69

Beukema JI, de Vlas J (1989) Tidal-current transport of thread-drifting postlarval juveniles of the bivalve *Macoma balthica* from the Wadden Sea to the North Sea. Mar Ecol Prog Ser 52:193–200

Bianchi CN (2007) Biodiversity issues for the forthcoming tropical Mediterranean Sea. Hydrobiologia 580:7–12

Bianchi CN, Morri C (2000) Marine biodiversity of the Mediterranean Sea: situation, problems and prospects for future research. Mar Pollut Bull 40:367–376

Birkeland C (1971) Biological observations on Cobb Seamount. Northwest Sci 45:193–199

Blanchard M (1997) Spread of the slipper limpet *Crepidula fornicata* (L. 1758) in Europe. Current state and consequences. Sci Mar 61(Suppl 2):109–118

Bodoy A, Maître-Allain T, Riva A (1981) Croissance comparée de la palourde européenne *Ruditapes decussatus* et de la palourde japonaise *Ruditapes philippinarum* dans un écosystème artificiel méditerranéen. Vie Mar 2:39–51

Boisselier-Dubayle MC, Gofas S (1999) Genetic relationships between marine and marginal marine populations of *Cerithium* species from the Mediterranean Sea. Mar Biol 135:671–682

Bonfitto A, Oliverio M, Sabelli B, Taviani M (1994) A quaternary deep-sea marine molluscan assemblage from East Sardinia (Western Tyrrhenian Sea). Boll Malacol 11:305–309

Bonfitto A, Dell'Angelo B, Evangelisti F, Sabelli B (2011) The genus *Acanthochitona* (Mollusca: Polyplacophora) in the Mediterranean Sea: morphological and molecular data. Sci Mar 75:171–180

Bouchet P (1981) Evolution of larval development in Eastern Atlantic Terebridae (Gastropoda), Neogene to Recent. Malacologia 21:363–369

Bouchet P (1984) Les Triphoridae de Mediterranée et du proche Atlantique (Mollusca, Gastropoda). Lav Soc Ital Malacol 21: 5–58

Bouchet P (1989) A review of poecilogony in Gastropods. J Molluscan Stud 55:67–78

Bouchet P (1995) Nouvelles observations sur la systematique des Triphoridae de Mediterranée et du proche Atlantique. Boll Malacol 31:205–220

Bouchet P, Guillemot H (1978) The *Triphora perversa*-complex in Western Europe. J Molluscan Stud 44:344–356

Bouchet P, Taviani M (1989) Atlantic deep-sea gastropods in the Mediterranean: new findings. Boll Malacol 2:137–148

Bouchet P, Taviani M (1992a) The Mediterranean deep-sea fauna: pseudopopulations of Atlantic species? Deep Sea Res Part I Oceanogr Res Pap 39:169–184

Bouchet P, Taviani M (1992b) La colonizzazione dei bacini a soglia. Il caso del Mar Mediterraneo. Lav Soc Malacol Ital 24:120–130

Bouchet P, Warén A (1980) Revision of the North-East Atlantic bathyal and abyssal Turridae (Mollusca: Gastropoda). J Molluscan Stud 8(Suppl):1–119

Bouchet P, Warén A (1985) Revision of the northeast Atlantic bathyal and abyssal neogastropoda (excluding Turridae). Boll Malacol 2(Suppl):123–296

Braber SJM (1970) The occurrence of a penis-like outgrowth behind the right tentacle in spent females of *Nucella lapillus* (L). Proc Malacol Soc Lond 39:231–233

Bradley WH, Cooke P (1959) Living and ancient populations of the clam *Gemma gemma* in a Maine coast tidal flat. Fish Bull 137:305–334

Breber P (1985) L'introduzione e l'allevamento in Italia dell'arsella del Pacifico *Tapes decussatus*, Reeve (Bivalvia, Veneridae). Oebalia 9:675–680

Breber P (2002) Introduction and acclimatisation of the Pacific carpet clam, *Tapes philippinarum*, to Italian waters. In: Leppäkoski E, Gollasch S, Olenin S (eds) Invasive aquatic species of Europe – distribution, impact and management. Kluwer, Dordrecht

Bresler V, Bissinger V, Abelson A, Dizer H, Sturm A, Kratke R, Fishelson L, Hansen P-D (1999) Marine molluscs and fish as biomarkers of pollution stress in littoral regions of the Red Sea, Mediterranean Sea and North Sea. Helgoland Mar Res 53:218–243

Bresler V, Abelson A, Fishelson L, Feldstein T, Rosenfeld M, Mokady O (2003) Marine molluscs in environmental monitoring. I. Cellular and molecular responses. Helgoland Mar Res 57:157–165

Bryan GW, Gibbs PE, Hummerstone LG, Burt GR (1986) The decline of the gastropod *Nucella lapillus* around South-West England: evidence for the effect of trybutiltin from antifouling paints. J Mar Biol Assoc UK 66:611–640

Buzzurro G, Russo P (2007) *Fusinus* del mediterraneo – Mediterranean *Fusinus*. Grafiche ATA, Milano

Cachia C (1981) Notes on some uncommon species of molluscs from the Maltese islands. Boll Malacol 17:291–294

Cachia C, Mifsud C, Sammut PM (2004) The marine mollusca of the Maltese Islands. Part four. The classes: Caudofoveata, Solenogastres, Bivalvia, Scaphopoda and Cephalopoda. Backhuys, Leiden

Calvo M, Templado J, Oliverio M, Machordom A (2009) Hidden Mediterranean biodiversity: molecular evidence for a cryptic species complex within the reef building vermetid gastropod *Dendropoma petraeum* (Mollusca: Caenogastropoda). Biol J Linn Soc 96:898–912

Carnevale G, Longinelli A, Caputo D, Barbieri M, Landini W (2008) Did the Mediterranean marine reflooding preceded the Mio-Pliocene boundary? Paleontological and geochemical evidence from upper Messinian sequences of Tuscany, Italy. Paleogeogr Paleoclimatol Paleoecol 257:81–105

Carvajal-Rodriguez A, Conde-Padin P, Rolan-Alvarez E (2005) Decomposing shell form into size and shape by geometric morphometric methods in two sympatric ecotypes of *Littorina saxatilis*. J Molluscan Stud 71:313–318

Carvajal-Rodriguez A, Guerra-Varela J, Fernandez B, Rolan E, Rolan-Alvarez E (2006) An example of application of geometric morphometric tools to the morphological diagnosis of two sibling species in *Nassarius* (Mollusca, Prosobranchia). Iberus 24:81–88

Casamor L, Ghisotti F (1968) *Buccinum humphreysianum* Bennet, 1825. Schede malacologiche del Mediterraneo. Toffoletto, Milano

Casellato S, Stefanon A (2008) Coralligenous habitat in the northern Adriatic Sea: an overview. Mar Ecol 29:321–341

Cattaneo Vietti R, Giovine F (2008) Opisthobranchia. Biol Mar Med 15(Suppl):279–295

Cecalupo A, Buzzurro G, Mariani M (2008) Contributo alla conoscenza della malacofauna del Golfo di Gabès (Tunisia). Quad Civ Staz Idrobiol Milano 31:1–175

Ceregato A, Raffi S, Scarponi D (2007) The circalittoral/bathyal paleocommunities in the Middle Pliocene of Northern Italy: the case of the *Korobkovia oblonga–Jupiteria concava* paleocommunity type. Geobios 40:555–572

Cesari P (1994) I Molluschi della Laguna di Venezia. Arsenale, Venezia

Cesari P, Mizzan L (1994) Dati sulla malacofauna marina costiera del Veneziano. Boll Mus Civ St Nat Venezia 43:179–190

Cesari P, Pellizzato M (1985) Molluschi pervenuti in laguna di Venezia per apporti volontari o casuali. Acclimazione di *Saccostrea commercialis* (Iredale & Roughely, 1933) e di *Tapes philippinarum* (Adams & Revve, 1850). Boll Malacol 21:237–274

Çeviker D, Albayrak S (2006) Three alien molluscs from Iskenderun Bay (SE Turkey). Aquat Invasions 1:76–79

Chemello R, Oliverio M (1996) Lessepsian migrations: a theoretical "island-jumping" model. Biol Mar Med 3:444–446

Chester CM (1996) The effect of adult nutrition on the reproduction and development of the estuarine nudibranch, *Tenellia adspersa* (Nordmann, 1845). J Exp Mar Biol Ecol 198:113–130

Chiantore M, Bedulli D, Cattaneo-Vietti R, Schiaparelli S, Albertelli G (2001) Long-term changes in the Mollusc-Echinoderm assemblages in the north and coastal middle Adriatic Sea. Atti Ass Ital Oceanogr Limn 14:63–75

CIESM (2002) Alien marine organisms introduced by ships in the Mediterranean and Black seas. CIESM, Monaco

CIESM (2008) The Messinian salinity crisis from mega-deposits to microbiology. In: Briand F (ed) A consensus report. CIESM, Monaco

CIESM (2012) Atlas of exotic species in the Mediterranean Sea. http://www.ciesm.org/online/atlas/index.htm. Accessed 27 Sept 2012

Clanzig S (1989) Introduction récente d'invertébré dans les lagunes Mediterranéennes du Languedoc-Roussilon (France). Bull Soc Zool Fr 114:151–152

Claudet J, Fraschetti S (2010) Human-driven impacts on marine habitats: a regional m eta-analysis in the Mediterranean Sea. Biol Conserv 143:2195–2206

Coen G (1937) Nuovo saggio di una Sylloge Molluscorum Adriaticorum. Mem Regio Com Talass Ital 240:1–173

Coen G, Vatova A (1932) Malacofauna Arupinensis. Thalassia 1:1–53

Colantoni P (1973) A glacial mollusc fauna from Baronie seamount (off Eastern Sardinia). Rapp Comm Int Mer Médit 21:897–900

Colantoni P, Noto P, Taviani M (1975) Prime datazioni assolute di una fauna fossile a *Pseudamussium septemradiatum* dragata nel basso Adriatico. G Geol 40(Supplt 3):133–140

Coll M, Piroddi C, Steenbeek J, Kashner K, Lasram FBR, Aguzzi J, Ballesteros E, Bianchi CN, Corbera J, Dailinis T, Danovaro R, Estrada M, Froglia C, Galil BS, Gasol JM, Gerwagen R, Gil J, Guilhaumon F, Kesner-Rayes K, Kitsos MS, Koukouras A, Lampradariou N, Laxamana E, Lopez Fe de la Cuadra CM, Lotze HK, Martin D, Mouillot D, Oro D, Raichevich S, Rius-Barile J, Saiz-Salinas JI, San Vincente C, Somot S, Templado J, Turon X, Vafidis D, Villanueva R, Voultsiadou E (2010) The biodiversity of the Mediterranean Sea: estimates, patterns, and threats. PLoS One 5:1–36

Coll M, Piroddi C, Albouy C, Lasram FBR, Cheung WWL, Christensen V, Karpouzi VS, Guilhaumon F, Mouillot D, Paleczny M, Palomares ML, Steenbeek J, Trujillo P, Watson R, Pauly D (2011) The Mediterranean Sea under siege: spatial overlap between marine biodiversity, cumulative threats and marine reserves. Glob Ecol Biogeogr 21:465–480

Colognola R, Masturzo P, Russo G, Scardi M, Vinci D, Fresi E (1986) Biometric and genetic analysis of the marine rissoid *Rissoa auriscalpium* and its ecological implications. Mar Ecol 7:265–285

Comeau S, Gorsky G, Jeffree R, Teyssié JL, Gattuso JP (2009) Impact of ocean acidification on a key Arctic pelagic mollusc (*Limacina helicina*). Biogeosciences 6:1877–1882

Comeau S, Gorsky G, Alliouane S, Gattuso JP (2010) Larvae of the pteropod *Cavolinia inflexa* exposed to aragonite undersaturation are viable but shell-less. Mar Biol 157:2341–2345

Comeau S, Gattuso JP, Nisumaa AM, Orr J (2012) Impact of aragonite saturation state changes on migratory pteropods. Proc R Soc B Biol Sci 279:732–738

Corselli C (2001) Change and diversity: the Mediterranean deep corals from the Miocene to the Present. In: Faranda FM, Guglielmo L, Spezie G (eds) Mediterranean ecosystems: structures and processes. Springer Italia, Genova

Corselli C, Basso D (1996) First evidence of benthic communities based on chemosynthesis in the Napoli mud volcano (Eastern Mediterranean). Mar Geol 132:227–239

Cossignani T, Ardovini R (2011) Malacologia Mediterranea. L'Informatore Piceno, Ancona

Council Directive 92/43/EEC (1992) http://eur-lex.europa.eu/LexUriServ/LexUriServ.do?uri=CELEX:31992L0043:EN:html. Accessed 10 Oct 2012

Criscione F, Patti FP (2010) Similar shells are not necessarily a reliable guide to phylogeny: *Rissoa guerinii* Récluz, 1843, and *Rissoa lia* (Monterosato, 1884) (Caenogastropoda: Rissoidae): a case study. Nautilus 124:117–128

Criscione F, Scuderi D, Patti FP (2009) Revising alpha-taxonomy in shelled gastropods: the case of *Rissoa panhormensis* Verduin, 1985 (Caenogastropoda: Rissoidae). Nautilus 123:303–312

Crocetta F (2011) Marine alien Mollusca in the Gulf of Trieste and neighbouring areas: a critical review and state of knowledge (updated in 2011). Acta Adriat 52:247–260

Crocetta F, Bonomolo G, Albano PG, Barco A, Houart R, Oliverio M (2012) The status of the northeastern Atlantic and Mediterranean small mussel drills of the *Ocinebrina aciculata* complex (Mollusca: Gastropoda: Muricidae), with the description of a new species. Sci Mar 76:177–189

Crutzen PJ, Stoermer EF (2000) The "Anthropocene". IGBP Newsl 41:17–18

Cumming RL (1993) Reproduction and variable larval development of an ectoparasitic snail, *Turbonilla* sp. (Pyramidellidae, Opisthobranchia), on cultured giant clams. Bull Mar Sci 52:760–771

Curzi P, D'Onofrio S, Roveri M, Taviani M (1984) Core ADS 74–24: a window on the latest Quaternary history of the Adriatic Sea. G Geol 46(Suppl 3):61–77

Danovaro R, Dell'Anno A, Pusceddu A (2004) Biodiversity response to climate change in a warm deep sea. Ecol Lett 7:821–828

Danovaro R, Gambi C, Lampadariou N, Tselepides A (2007) Deep-sea nematode biodiversity in the Mediterranean basin: testing for longitudinal, bathymetric and energetic gradients. Ecography 31:231–244

Danovaro R, Company JB, Corinaldesi C, D'Onghia G, Galil B, Gambi C, Gooday AJ, Lampadariou N, Luna GM, Morigi C, Olu K, Polymenakou P, Ramirez-Llodra E, Sabbatini A, Sardà F, Sibuet M, Tselepides A (2010) Deep-sea biodiversity in the Mediterranean Sea: the known, the unknown, and the unknowable. PLoS One 5:e11832

Dell RK (1972) Antarctic benthos. In: Russell FS, Yonge M (eds) Advances in marine biology, vol 10. Academic, London

Demir M (2003) Shells of Mollusca collected from the seas of Turkey. Turk J Zool 27:101–140

Devescovi M, Ozretic B, Iveša L (2005) Impact of date mussel harvesting on the rocky bottom structural complexity along the Istrian coast (Northern Adriatic, Croatia). J Exp Mar Biol Ecol 325:134–145

Di Geronimo I, Bellagamba M (1985) Malacofaune dei dragaggi BS 77–1 e BS 77–2 (Sardegna nord orientale). Boll Soc Paleontol Ital 24:111–129

Di Geronimo I, La Perna R (1996) *Bathyspinula excisa* (Philippi 1844) (Bivalvia, Protobranchia): a witness of the Plio-Quaternary history of the deep Mediterranean benthos. Riv Ital Paleontol Stratigr 102:105–118

Di Geronimo I, La Perna R (1997) Pleistocene bathyal molluscan assemblages from Southern Italy. Riv Ital Paleontol Stratigr 103:389–426

di Maria Monterosato TA (1878) Enumerazione e sinonimia delle conchiglie mediterranee. Giorn Sci Nat Econ Palermo 13:61–115

Domènech R, Martinell J (1980) Contribución al conocimiento de la fauna malacológica sumergida del Würm del litoral gerundense. Com Prim Congr Nac Malac, Madrid

Domènech R, Martinell J (1982) Fauna malacòlogica submergida del Würm del litoral Gironí. Descriptiva i sistematica. Bull Inst Cat Hist Nat 48:31–60

Donald KM, Preston J, Williams ST, Reid DG, Winter D, Alvarez R, Buge B, Hawkins SJ, Templado J, Spencer HG (2012) Phylogenetic relationships elucidate colonization patterns in the intertidal grazers *Osilinus* Philippi, 1847 and *Phorcus* Risso, 1826 (Gastropoda: Trochidae) in the northeastern Atlantic Ocean and Mediterranean Sea. Mol Phylogenet Evol 62:35–45

Dumont HJ, Shiganova TA, Niermann U (eds) (2004) Aquatic invasions in the Black, Caspian and Mediterranean Seas. Kluwer, Dordrecht

Espinosa F, Guerra-García JM, García-Gómez JC (2007) Sewage pollution and extinction risk: an endangered limpet as a bioindicator? Biodivers Conserv 16:377–397

Evans J, Borg JA, Schembri PJ (2010) Rediscovery of live *Gibbula nivosa* (Gastropoda: Trochidae). Rapp Proc verb Réun CIESM 39:507

Evans J, Borg JA, Schembri PJ (2011) Distribution, habitat preferences and behaviour of the critically endangered Maltese top-shell *Gibbula nivosa* (Gastropoda: Trochidae). Mar Biol 158:603–611

Fabbri R, Landi L (1999) Nuove segnalazioni di molluschi, crostacei e pesci esotici in Emilia-Romagna e prima segnalazione di *Corbicula fluminea* (O.F. Müller, 1774) in Italia. Quad Studi Nat Romagna 12:9–20

Fabry VJ, Seibel BA, Feely RA, Orr JC (2008) Impacts of ocean acidification on marine fauna and ecosystem processes. ICES J Mar Sci 65:414–432

Faimali M, Taviani M (2011) Una rivoluzione su scala planetaria. Darwin 45:84–89

Fanelli G, Piraino S, Belmonte G, Geraci S, Boero F (1994) Human predation along Apulian rocky coasts (SE Italy): desertification caused by *Lithophaga lithophaga* (Mollusca) fisheries. Mar Ecol Prog Ser 110:1–8

Foighil DO (1989) Planktotrophic larval development is associated with a restricted geographic range in *Lasaea,* a genus of brooding, hermaphroditic bivalves. Mar Biol 103:349–358

Font J, Millot C, Salas J, Julià A, Chic O (1998) The drift of modified Atlantic water from the Alboran Sea to the eastern Mediterranean. Sci Mar 62:211–216

Forli M, Dell'Angelo B, Ciappelli F, Taviani M (2002) A new species of Haliotidae (Mollusca, Vetigastropoda) in the Italian Pliocene. Boll Malacol 38:149–154

Forli M, Dell'Angelo B, Montagna P, Taviani M (2004) A new large *Patella* (Mollusca: Archaeogastropoda) in the Pliocene of the Mediterranean Basin. Boll Malacol 40:1–9

Fraschetti S, Bianchi CN, Terlizzi A, Fanelli G, Morri C, Boero F (2001) Spatial variability and human disturbance in shallow subtidal hard substrate assemblages: a regional approach. Mar Ecol Prog Ser 212:1–12

Fredj G, Laubier L (1985) The deep Mediterranean benthos. In: Moraitou-Apostopulou M, Kiortsis V (eds) Mediterranean marine ecosystems. NATO Conf Ser Plenum Press, New York

Freiwald A, Beuck L, Rüggeberg A, Taviani M, R/V Meteor M70-1 Participants (2009) The white coral community in the central Mediterranean revealed by ROV surveys. Oceanography 22:58–74

Fretter V (1984) Prosobranchs. In: Wilbur KM (ed) The Mollusca, vol 1, Reproduction. Academic, New York

Galil B (2007) Seeing red: alien species along the Mediterranean coast of Israel. Acquat Invasions 2:281–312

Galil BS (2008) Alien species in the Mediterranean Sea – which, when, where, why? Hydrobiologia 597:105–116

Galil BS (2009) Taking stock: inventory of alien species in the Mediterranean sea. Biol Invasions 11:359–372

Garaventa F, Faimali M, Terlizzi A (2006) Imposex in pre-pollution times. Is TBT to blame? Mar Pollut Bull 52:696–718

Garaventa F, Centanni E, Pellizzato F, Faimali M, Terlizzi A, Pavoni A (2007) Imposex and accumulation of organotin compounds in populations of *Hexaplex trunculus* (Gastropoda, Muricidae) from the Lagoon of Venice (Italy) and Istrian Coast (Croatia). Mar Pollut Bull 54:602–625

Garaventa F, Corrà C, Piazza V, Giacco E, Greco G, Pane L, Faimali M (2012) Settlement of the alien mollusc *Brachidontes pharaonis* in a Mediterranean industrial plant: bioassays for antifouling treatment optimization and management. Mar Environ Res 76:90–96

Garilli V, Galletti L (2006) Taxonomical characters for distinguishing *Cerithium lividulum* Risso, 1826, and *C. renovatum* Monterosato, 1884 (Gastropoda, Caenogastropoda, Cerithiidae). Basteria 70:109–122

Ghisotti F (1972a) Le conchiglie del golfo di Gabès (prima parte). Conchiglie 8:63–89

Ghisotti F (1972b) Le conchiglie del golfo di Gabès (seconda parte). Conchiglie 8:101–113

Ghisotti F (1972c) Le conchiglie del golfo di Gabès (terza parte). Conchiglie 8:133–144

Ghisotti F (1973) *Scapharca* cfr. *cornea* (Reeve), ospite nuova del Mediterraneo. Conchiglie 9:68

Ghisotti F (1974a) La frenesia della nuova specie. Conchiglie 10:65–70

Ghisotti F (1974b) *Rapana venosa* (Valenciennes) nuova ospite adriatica? Conchiglie 10:125–126

Ghisotti F, Rinaldi E (1976) Osservazioni sulla popolazione di *Scapharca*, insediatasi in questi ultimi anni su un tratto del litorale romagnolo. Conchiglie 12:183–195

Gibbs PE, Bryan GW, Pascoe PL, Burt GR (1990) Reproductive abnormalities in female *Ocenebra erinacea* (Gastropoda) resulting from tributyltin-induced imposex. J Mar Biol Assoc UK 70:639–656

Gibson GD, Chia F-S (1995) Developmental variability in the poecilogonous opisthobranch *Haminaea callidegenita*: life-history traits and effects of environmental parameters. Mar Ecol Prog Ser 121:139–155

Gignoux M (1913) Les formations marines pliocénes et quaternaires de l'Italie du Sud e de la Sicile. Ann Univ Lyon 36(ns 1):1–693

Gili C, Martinell J (2000) Phylogeny, speciation and turnover. The case of Mediterranean gastropods of the genus *Cyclope* Risso, 1826. Lethaia 33:236–250

Gofas S (1992) Le genre *Granulina* (Marginellidae) en Méditerranée et dans l'Atlantique Oriental. Boll Malacol 28:1–26

Gofas S (1998) Marine molluscs with a very restricted range in the Strait of Gibraltar. Divers Distrib 4:255–266

Gofas S (2004) Geographical differentiation in *Clelandella* (Gastropoda: Trochidae) in the northeastern Atlantic. J Molluscan Stud 71:133–144

Gofas S, Zenetos A (2003) Exotic molluscs in the Mediterranean Basin: current status and perspectives. Oceanogr Mar Biol Annu Rev 41:237–277

Gofas S, Freiwald A, López Correa M, Remia A, Salas C, Taviani M, Wisshak M, Zibrowius H (2007) Oyster beds in the deep sea. World Congress of Malacology, Antwerp

Gofas S, Moreno D, Salas C (2011a) Moluscos marinos de Andalucia, vol 1. Servicio de Publicaciones/Interscambio Cientifico Universidad de, Malaga

Gofas S, Moreno D, Salas C (eds) (2011b) Moluscos marinos de Andalucia, vol 2. Servicio de Publicaciones e Interscambio Cientifico/Universidad de Malaga, Malaga

Gomoiu M-T (1981) Some problems concerning actual ecological changes in the Black sea. Cercet Mar IRCM 14:109–127

Gray JS (1997) Marine biodiversity: patterns, threats and conservation needs. Biodivers Conserv 6:153–175

Green AJ, Figuerola J (2005) Recent advances in the study of long-distance dispersal of aquatic invertebrates via birds. Divers Distrib 11:149–156

Guarniero I, Plazzi F, Bonfitto A, Rinaldi A, Trentini M, Passamonti M (2010) The bivalve mollusc *Mactra corallina*: genetic evidence of existing sibling species. J Mar Biol Assoc UK 90:633–644

Guidetti P, Boero F (2004) Desertification of Mediterranean rocky reefs caused by date-mussel, *Lithophaga lithophaga* (Mollusca: Bivalvia), fishery: effects on adult and juvenile abundance of a temperate fish. Mar Pollut Bull 48:978–982

Hamdani A, Soltani-Mazouni N (2011) Changes in biochemical composition of the gonads of *Donax trunculus* L. (Mollusca, Bivalvia) from the Gulf of Annaba (Algeria) in relation to reproductive events and pollutiom. Jordan J Biol Sci 4:149–156

Highsmith RF (1985) Floating and algal rafting as potential dispersal mechanisms in brooding invertebrates. Mar Ecol Prog Ser 25:169–179

Hoagland KE, Robertson R (1988) An assessment of poecilogony in marine invertebrates: phenomenon or fantasy? Biol Bull 174:109–125

Hoegh-Guldberg O, Bruno JF (2010) The impact of climate change on the world's marine ecosystems. Science 328:1523–1528

Hoenselaar HJ, Hoenselaar J (1989) *Musculista senhousia* (Benson in Cantor, 1842) in the western Mediterranean (Bivalvia, Mytilidae). Basteria 53:73–76

Houart R (2000) New species of Muricidae (Gastropoda) from the northeastern Atlantic and the Mediterranean Sea. Zoosystema 22:459–469

Hsü KJ, Montadert L, Bernoulli D, Cita MB, Erickson A, Garrison RE, Kidd RB, Mèlierés F, Müller C, Wright R (1977) History of the Messinian salinity crisis. Nature 267:399–403

Issel A (1914) Lembi fossiliferi quaternari e recenti nella Sardegna meridionale. Acc Naz Lincei 23(Suppl 5):759–770

Janssen AW (2012) La Quaternary to Recent holoplanktonic Mollusca (Gastropoda) from bottom samples of the eastern Mediterranean Sea: systematics, morphology. Boll Malacol 48(Suppl 9):1–105

Johanesson K (1988) The paradox of Rockall: why is a brooding gastropod (*Littorina saxatilis*) more widespread than one having a planktonic larval dispersal stage (*L. littorea*)? Mar Biol 99: 507–513

Jones CC (1984) Messinian refugia: evidence of some Venerinae bivalves. Ann Geol Pays Hell 22:69–77

Kabasakal H, Karhan SU, Kabasakal E, Kalakan E, Buzzurro G (2005) First record of *Strombus (Doxander) vittatus vittatus* Linnaeus, 1758 from the Mediterranean Sea, with a brief review of strombids in the Levant Basin. Nach Ersten Vorarlb Malakol Gesellsch 13:63–66

Keller C (1883) Die Fauna im Suez-Kanal und die Diffusion der mediterranen und erythräischen Thierwelt. Neue Denkschr algem schweiz Ges gesammt Naturwiss 29:1–39

Keppel E, Sigovini M, Tagliapietra D (2012) A new geographical record of *Polycera hedgpethi* Er. Marcus, 1964 (Nudibranchia: Polyceridae) and evidence of its established presence in the Mediterranean Sea, with a review of its geographical distribution. Mar Biol Res 8(10):969–981

Koutsoubas D, Tselepides A, Eleftheriou A (2000) Deep sea molluscan fauna of the Cretan Sea (Eastern Mediterranean): faunal, ecological and zoogeographical remarks. Senkenb Mar 30:85–98

Krug PJ (2007) Poecilogony and larval ecology in the gastropod genus *Alderia*. Am Malacol Bull 23:99–111

Kukla G, Mattheus RK, Mitchell JM (1972) The end of the present interglacial. Quat Res 2:261–269

La Perna R (1996) Phyletic relationships and ecological implications between *Pagodula vaginata* (De Cristofori & Jan, 1832) and *Pagodula echinata* (Kiener, 1840) (Gastropoda, Muricidae). Boll Soc Paleontol Ital 35:81–92

La Perna R (1998a) On *Asperarca* Sacco, 1898 (Bivalvia, Arcidae) and two new Mediterranean species. Boll Malacol 33:11–18

La Perna R (1998b) A new Mediterranean *Skeneoides* (Gastropoda: Skeneidae) from a shallow-water cave. J Conchol 36:21–27

La Perna R (2003) The Quaternary deep-sea protobranch fauna from the Mediterranean: composition, depth-related distribution and changes. Boll Malacol 39:17–34

La Perna R (2004) The identity of *Yoldia micrometrica* Sequenza, 1877 and three new deep-sea protobranchs from the Mediterranean (Bivalvia). J Nat Hist 38:1045–1057

La Perna R (2005) A gigantic deep-sea Nucinellidae from the tropical West Pacific (Bivalvia: Protobranchia). Zootaxa 881:1–10

La Perna R (2006) Taxonomy of the family Neilonellidae (Bivalvia, Protobranchia): Miocene and Plio-Pleistocene species of *Pseudoneilonella* Laghi, 1896 from Italy. Veliger 49:196–208

La Perna R (2007) Revision of the Nuculidae (Bivalvia: Protobranchia) from the Cerulli Irelli collection (Pleistocene, Mediterranean). Palaeontogr Ital 91:109–140

La Perna R, D'Abramo M (2011) The genus *Nemocardium* Meek 1876 in the Plio-Pleistocene of Italy (Bivalvia, Cardiidae). J Conchol 40:559–568

La Perna R, Landau B, Marquet R (2002) Species of *Granulina* (Gastropoda, Marginellidae) from the Pliocene of Málaga (southern Spain) with description of four new species. Cainozoic Res 1: 111–120

La Perna R, Ceregato A, Tabanelli C (2004) Mediterranen Pliocene protobranchs: the genera *Jupiteria* Bellardi, 1875, *Ledella* Verril & Bush, 1897 and *Zealeda* Marwick, 1924 (Mollusca, Bivalvia). Boll Malacol 40:25–36

Landau B, Beu A, Marquet R (2004) The early Pliocene Gastropoda (Mollusca) of Estepona, Southern Spain. Part 5: Tonnoidea, Ficoidea. Palaeontos 5:35–102

Lavie B, Nevo E (1986) Genetic diversity of marine gastropods: contrasting strategies in *Cerithium rupestre* and *Cerithium scabridum* in the Mediterranean Sea. Mar Ecol Prog Ser 28:99–103

Lazzari G, Rinaldi E (1994) Alcune considerazioni sulla presenza di specie extramediterranee nelle lagune salmastre di Ravenna. Boll Malacol 30:195–202

Le Renard J, Sabelli B, Taviani M (1996) On *Candinia* (Sacoglossa: Juliidae), a new fossil genus of bivalved gastropods. J Paleontol 70:230–235

Lemghich I, Benajiba MH (2007) Survey of imposex in prosobranch molluscs along the northern Mediterranean coast of Morocco. Ecol Indic 7:209–214

Leppäkoski E, Gollasch S, Olenin S (eds) (2002) Invasive aquatic species of Europe – distribution, impact and management. Kluwer, Dordrecht

Levinton JS (1979) The effect of density upon deposit-feeding populations: movement, feeding and floating of *Hydrobia ventrosa* Montagu (Gastropoda: Prosobranchia). Oecologia 43:27–39

Ligios S, Anadón P, Castorina F, D'Amico C, Esu D, Gliozzi E, Gramigna P, Mola M, Monegato G (2012) Ostracoda and Mollusca biodiversity and hydrochemical features in Lat Miocene brackish basins of Italy. Geobiology 45:351–367

Little C, Nix W (1976) The burrowing and floating behaviour of the gastropod *Hydrobia ulvae*. Estuar Coast Mar Sci 4:537–544

López Correa M, Freiwald A, Hall-Spencer J, Taviani M (2005) Distribution and habitats of Acesta excavata (Bivalvia: Limidae) with new data on its shell ultrastructure. In: Freiwald A, Roberts M (eds) Cold-water corals & ecosystems. Springer, Berlin

Louizidou P, Thessalou-Legaki M, Zenetos A (2007) First record of the Red Sea immigrant *Chrysallida fischeri* (Hornung and Memod, 1925) (Gastropoda: Pyramidellidae) from Greek waters. Aquat Invasions 2:95–98

Maas AE, Wishner KF, Seibel BA (2012) The metabolic response of pteropods to acidification reflects natural CO_2-exposure in oxygen minimun zone. Biogeosciences 9:747–757

Madhioub MN, Zaouali J (1988) Captage de l'huître creuse *Crassostrea gigas* dans le lac d'Ichkeul, Tunisie. Bull Inst Natl Sci Tech Oceanogr Pêche Salambó 15:47–60

Malatesta A, Zarlenga F (1988) Northern guests in the Pleistocene Mediterranean sea. Geol Rom (1986) 25:91–154

Marabini S, Taviani M, Vai GB, Vigliotti L (1995) Yellow sand facies with *Arctica islandica*: low stand signature in an early Pleistocene Front-Apennine Basin. G Geol 57(Suppl 3):259–275

Mariottini P, Smriglio C, Di Giulio A, Oliverio M (2009) A new fossil conoidean from the Pliocene of Italy, with comments on the *Bela menkhorsti* complex (Gastropoda: Conidae). J Conchol 40:5–14

Mars P (1958) Les faunes malacologiques quaternaires "froides" de Méditerranée. Le gisement du cap de Creus. Vie Milieu 9:67–89

Martel A, Chia F-S (1991) Drifting and dispersal of small bivalves and gastropods with direct development. J Exp Mar Biol Ecol 150:131–147

Mastrototaro F, D'Onghia G, Corriero G, Matarrese A, Maiorano P, Panetta P, Gherardi M, Longo C, Rosso A, Sciuto F, Sanfilippo R, Gravili C, Boero F, Taviani M, Tursi A (2010) Biodiversity of the white coral bank off Cape Santa Maria di Leuca (Mediterranean Sea): an update. Deep Sea Res Part II Trop Stud Oceanogr 57:412–430

Mazziotti C, Agamennone F, Tisselli M (2008) Checklist della malacofauna delle Isole Tremiti (Medio Adriatico). Boll Malacol 44(5–8):71–86

McCulloch M, Falter J, Trotter J, Montagna P (2012a) Coral resilience to ocean acidification and global warming through pH up-regulation. Nat Clim Change 2(8):623–627

McCulloch M, Trotter J, Montagna P, Falter J, Dunbar R, Freiwald A, Forsterra G, López Correa M, Maier C, Rüggeberg A, Taviani M (2012b) Resilience of cold-water scleractinian corals to ocean acidification: Boron isotopic systematics of pH and saturation state upregulation. Geochim Cosmochim Ac 87:21–34

Mienis HK (2003) *Cantharus tranquebaricus*: first records of another Indian ocean species in the eastern Mediterranean. Triton 8:11

Modica MV, Mariottini P, Prkíc J, Oliverio M (2013) DNA-barcoding of sympatric species of ectoparasitic gastropods of the genus *Cerithiopsis* (Mollusca: Gastropoda: Cerithiopsidae) from Croatia. J Mar Biol Assoc UK 93:1059–1065

Monegatti P, Raffi S (2001) Taxonomic diversity and stratigraphic distribution of mediterranean Pliocene bivalves. Paleogeogr Paleoclimatol Paleoecol 297:1–11

Monegatti P, Raffi S (2010) The Messinian marine molluscs record and the dawn of the eastern Atlantic biogeography. Paleogeogr Paleoclimatol Paleoecol 165:171–193

Montefalcone M, Albertelli G, Morri C, Bianchi CN (2010a) Patterns of wide-scale substitution within meadows of the seagrass *Posidonia oceanica* in NW Mediterranean Sea: invaders are stronger than natives. Aquat Conserv 20:507–515

Montefalcone M, Parravicini V, Vacchi M, Albertelli G, Ferrari M, Morri C, Bianchi CN (2010b) Human influence on seagrass habitat fragmentation in NW Mediterranean Sea. Estuar Coast Shelf Sci 86:292–298

Moolenbeck RG, Dekker H (1993) On the identity of *Strombus decorus* and *Strombus persicus*, with description of *Strombus decorus masirensis* n. ssp. and a note on *Strombus fasciatus*. Vita Marina 42:3–10

Moreno D, Templado J (1994) El completo de especies "*Nassarius cuvierii–N. unifasciatus*" (Gastropoda, Nassariidae) en el SE de España. Iberus 12:33–47

Munksgaard C (1990) Electrophoretic separation of morphologically similar species of the genus *Rissoa* (Gastropoda; Prosobranchia). Ophelia 31:97–104

Mutlu E (2004) Sexual dimorphisms in radula of *Conomurex persicus* (Gastropoda: Strombidae) in the Mediterranean Sea. Mar Biol 145:693–698

Nelson TC (1925) On the occurrence and food habits of ctenophores in New Jersey inland coastal waters. Biol Bull 48:92–111

Néraudeau D, Goubert E, Lacour D, Rouchy J-M (2001) Changing biodiversity of Mediterranean irregular echinoids from the Messinian to the Present-Day. Paleogeogr Paleoclimatol Paleoecol 175:43–60

Newell R (1964) Some factors controlling the upstream distribution of *Hydrobia ulvae* (Pennant). (Gastropoda. Prosobranchia). Proc Zool Soc Lond 142:85–106

Nielsen LT, Jakobsen HH, Hansen PJ (2010) High resilience of two coastal plankton communities to twenty-first century seawater acidification: Evidence from microcosm studies. Mar Biol Res 6:542–555

Nikula R, Väinölä R (2003) Phylogeography of *Cerastoderma glaucum* (Bivalvia: Cardiidae) across Europe: a major break in the Eastern Mediterranean. Mar Biol 143:339–350

Occhipinti-Ambrogi A, Marchini A, Cantone G, Castelli A, Chimenz C, Cormaci M, Froglia C, Furnari G, Gambi MC, Giaccone G, Giangrande A, Gravili C, Mastrototaro F, Mazziotti C, Orsi-Relini L, Piraino S (2011) Alien species along the italian coasts: an overview. Biol Invasions 13:215–237

Oehlmann J, Fioroni P, Stroben E, Markert B (1996) Tributyltin (TBT) effects on *Ocinebrina aciculata* (Gastropoda: Muricidae): imposex develpment, sterilization, sex change and population decline. Sci Total Environ 188:205–223

Oehlman J, Stroben E, Schulte-Oehlman U, Bauer B (1998) Imposex development in response to TBT pollution in Hinia incrassate (Strōm, 1768) (Prosobranchia, Stenoglossa). Aquat Toxicol 43:239–260

Oliver Baldovì JD (2007) Catàlogo de los Gasteròpodos testàceos marinos de la parte Sur del Golfo de Valencia (Espana). Iberus 25:29–61
Oliverio M (1984) Developmental vs. genetic variation in two Mediterranean rissoid complexes. J Molluscan Stud 60:461–465
Oliverio M (1995a) The status of the living mediterranean *Strombus*, or: what is a lessepsian migrant? Notiz CISMA 16:35–40
Oliverio M (1995b) The systematics of the radula-less gastropod *Clathromangelia* (Caenogastropoda, Conoidea). Zool Scr 24:193–201
Oliverio M (1996) Contrasting developmental strategies and speciation in N. E. Atlantic Prosobranchs: a preliminary analysis. In: Taylor J (ed) Origin and evolutionary radiation of the Mollusca. Oxford University Press, Oxford
Oliverio M (2003) The Mediterranean molluscs: the best known malacofauna of the world … so far. Biogeographia 24:195–208
Oliverio M (2008) Gastropoda prosobranchia. Biol Mar Med 15(Suppl):235–278
Oliverio M, Taviani M (2004) The Eastern Mediterranean Sea: tropical invasions and niche opportunities in a "Godot Basin". Biogeographia 24:313–318
Oliverio M, Gerosa G, Cocco M (1992) First record of *Pinctada radiata* (Bivalvia, Pteriidae) epibiont on the loggerhead sea turtle *Caretta caretta* (Chelonia, Cheloniidae). Boll Malacol 28:149–152
Olu-Le Roy K, Sibuet M, Fiala-Médioni A, Gofas S, Salas C, Mariotti A, Foucher JP, Woodside J (2004) Cold seep communities in the deep eastern Mediterranean Sea: composition, symbiosis and spatial distribution on mud volcanoes. Deep Sea Res Part I Oceanogr Res Pap 51:1915–1936
Orr JC, Fabry VJ, Aumont O, Bopp L, Doney SC et al (2005) Anthropogenic ocean acidification over the twenty-first century and its impact on calcifying organisms. Nature 437:681–686
Ovalis P, Zenetos A (2007) On the establishment of two more alien molluscs (*Chama aspera* Reeve, 1846 and *Chama asperella* Lamarck, 1819) in the eastern Mediterranean. Mediterr Mar Sci 8:97–100
Öztürk B (2012) A new alien gastropod *Pseudorhaphitoma iodolabiata* (Hornung & Mermod, 1928) (Mangeliidae, Mollusco) in the Mediterranean Sea. Mediterr Mar Sci 13:30–32
Öztürk B, Poutiers J-M (2005) Fulvia fragilis (Bivalvia: Cardiidae): a lessepsian mollusc species from Izmir Bay (Aegean Sea). J Mar Biol Assoc UK 85:351–356
Öztürk B, Buzzurro G, Benli A (2004) Marine molluscs from Cyprus: new data and checklist. Boll Malacol 39:49–78
Öztürk B, Önen M, Dogan A (2007) Addition to the knowledge of Turkish mollusc fauna. Rapp Comm Int Mer Médit CIESM 38:566
Padilla D (1998) Inducible phenotypic plasticity of the radula in Lacuna (Gastropoda: Littorinidae). Veliger 41:201–204
Page LR (2009) Molluscan larvae: pelagic juveniles or slowly metamorphosing larvae? Biol Bull 216:216–225
Palazzi S, Villari A (2001) Molluschi e brachiopodi delle grotte sottomarine del Taorminese. La Conchiglia 32(Suppl a 297):1–56
Pallary P (1902) Liste des mollusques testacés de la baie de Tanger. J Conchyl 50:1–39
Pallary P (1920) Exploration scientifique du Maroc organisée par la Societé de Géographie de Paris et continuée par la Scocieté des Sciences Naturelles du Maroc. Deuxiéme fascicule. Malacologie. Larose Paris and Institut Scientifique Chérifen Rabat
Panetta P, Mastrototaro F, Matarrese A, Tursi A (2003) Tanatocenosi a molluschi presenti nelel acque adriatiche albanesi. Biol Mar Med 10:597–601
Panetta P, Mastrototaro F, Matarrese A (2007) Malacofauna dei fondi incoerenti del Golfo di Manfredonia. Boll Malacol 43:57–61
Panova M, Blakeslee AMH, Miller AW, Mäkinen T, Ruiz GM, Johannesson K, André C (2011) Glacial history of the North Atlantic marine snail *Littorina saxatilis*, inferred from distribution of mitochondrial DNA lineages. PLoS One 6:e17511
Parravicini V, Guidetti P, Morri C, Montefalcone M, Donato M, Bianchi CN (2009) Consequnces of sea water temperature anomalies on a Mediterranean submarine cave ecosystem. Estuar Coast Shelf Sci 86:276–282
Pearse JS (1979) Polyplacophora. In: Giese AC, Pearse JS (eds) Reproduction of marine invertebrates, vol V. Academic, New York
Pedriali L, Robba E (2001) *Euspira magenesi*, a new species of the Naticidae (Gastropoda) from the Pliocene of Italy. Riv Ital Paleontol Stratigr 107:483–487
Pedriali L, Robba E (2009) A revision of the Pliocene naticids of northern and central Italy. III. The subfamilies Poliniceinae and Sininae. Riv Ital Paleontol Stratigr 15:371–429
Peirano A, Damasso V, Montefalcone M, Morri C, Bianchi CN (2005) Effects of climate, invasive species and antropogenic impacts on the growth of the seagrass *Posidonia oceanica* (L.) Delile in Liguria (NW Mediterranean Sea). Mar Pollut Bull 50:817–822
Pelosio G, Raffi S (1973) Considerazioni sul limite Plio-Pleistocene nella serie del T. Crostolo (Preappennino reggiano). Aten Parm Acta Nat 9:39–67
Peñas A, Rolan E, Luque AA, Templado J, Moreno D, Rubio F, Salas C, Sierra A, Gofas S (2006) Moluscos marinos de la isla de Alboran. Iberus 24:23–151
Peñas A, Rolan E, Almera J (2009) Fauna malacologica de un fondo detrìtico fangoso en El Maresma, Barcelona (nordeste de la Penìnsula Ibérica). Iberus 27:19–56
Pérès JM (1989) History of the Mediterranean biota and the colonization of the depths. In: Por FD, Dimentman D (eds) The legacy of Tethys. Kluwer, Dordrecht
Pérès JM, Picard J (1964) Nouveau manuel de bionomie benthique de la mer Méditerranée. Rec Trav Sta Mar Endoume 47:3–137
Piani P (1980) Catalogo dei molluschi conchiferi viventi nel Mediterraneo. Boll Malacol 16:113–224
Picken GB (1980) Reproductive adaptations of Antarctic benthic invertebrates. Biol J Linn Soc 14:67–75
Plazzi F, Ceregato A, Taviani M, Passamonti M (2011) A molecular phylogeny of bivalve mollusks: ancient radiations and divergences as revealed by mitochondrial genes. PLoS One 6:e27147
Poluzzi A, Taviani M (1986) Analisi quantitativa delle associazioni a molluschi di due sacche (Scardovari e Canarin) del delta meridionale del fiume Po. Mem Soc Geol It 27:403–410
Poluzzi A, Sabelli B, Taviani M (1981) Auto-sinecologia dei molluschi dei fondi mobili del delta settentrionale del Po (estate 1980). Boll Soc Paleont It 20:169–178
Ponder WF (1989) Mediterranean Cingulopsidae, a relict eastern Tethyan fauna (Gastropoda, Cingulopsidae). Boll Malacol 25:85–90
Pons-Moya J, Pons GX (2000) Molluscs de fons fangosos batials del Coll de Mallorca (SW de Mallorca, Illes Baleares, Mediterrània Occidental). Boll Soc Hist Nat Baleares 43:105–110
Por FD (2009) Tethys returns to the Mediterranean: success and limits of tropical re-colonization. BioRisk 3:5–19
Por FD, Dimentman C (1989) The legacy of Tethys. Kluwer, Dordrecht
Porcheddu AS, Francour P, Soltan D (1999) Considerazioni sul ritrovamento di una populazione di *Mya arenaria* L., 1758 negli stagni di Berri e di Vaïne (France meridionale). Boll Malacol 34: 167–171
Raffi S (1986) The significance of marine boreal molluscs in the early Pleistocene faunas of the Mediterranean area. Paleogeogr Paleoclimatol Paleoecol 52:267–289
Raffi S (1993) L'origine della fauna attuale del mar Mediterraneo: il caso dei Bivalvi. XIX Seminario sulla evoluzione biologica e i grandi problemi della biologia. Faune attuali e faune fossili. Accademia nazionale dei Lincei. Contributi del centro Linceo interdisciplinare "Beniamino Segre" Rome 86:189–207
Raffi S, Taviani M (1984) An outline of the late Tertiary to Recent history of the Mediterranean deep-sea mollusk faunas. Ann Géol Pays Hellén 32:257–261

Rees WJ (1965) The aerial dispersal of Mollusca. Proc Malacol Soc Lond 36:269–282
Relini G (ed) (2008) Checklist della flora e della fauna dei mari italiani. I Parte. Biol Mar Med 15(Suppl 1):1–385
Relini G (ed) (2010) Checklist della flora e della fauna dei mari italiani/checklist of the flora and fauna in italian seas. Parte II. Biol Mar Med 17(Suppl 1):387–828
Rinaldi E (1991) Le conchiglie della costa romagnola. Essegi, Ravenna
Rodolfo-Metalpa R, Houlbrèque F, Tambutté E, Boisson F, Baggini C, Patti FP, Jeffree R, Fine M, Foggo A, Gattuso J-P, Hall-Spencer JM (2011) Coral and mollusc resistance to ocean acidification adversely affected by warming. Nat Clim Change 1:308–312
Rodrigues CF, Duperron S, Gaudron SM (2011) First documented record of a living solemyid bivalve in a pockmark of the Nile Deep-sea Fan (eastern Mediterranean Sea). Mar Biodivers Rec. doi:10.1017/S175526721100008X
Rodríguez JG, Tueros I, Borja Á, Franco J, Alonso JIG, Garmendia JM, Muxica I, Sariego C, Valencia V (2009) Butyltin compounds, sterility and imposex assessment in *Nassarius reticulatus* (Linnaeus, 1758), prior to the 2008 European ban on TBT antifouling paints, within Basque ports and along coastal areas. Cont Shelf Res 29:1165–1173
Romanelli M, Cordisco CA, Giovanardi O (2009) The long-term decline of the *Chamelea gallina* L. (Bivalvia: Veneridae) clam fishery in the Adriatic Sea: is a synthesis possible? Acta Adriat 50:171–205
Rosso A, Di Geronimo I (1998) Deep-sea Pleistocene Bryozoa of southern Italy. Geobios 30:303–317
Rossoll D, Bermúdez R, Hauss H, Schulz KG, Riebesell U, Sommer U, Winder M (2012) Ocean acidification-induced food quality deterioration constrains trophic transfer. PLoS One 7:e34737
Rovere A, Bellati S, Parravicini V, Firpo M, Morri C, Bianchi CN (2009) Abiotic and biotic links work two ways: effects on the deposit at the cliff foot induced by mechanical action of date mussel harvesting (*Lithophaga lithophaga*). Estuaries Coasts 32:333–339
Roveri M, Manzi V, Gennari R, Iaccarino S, Lugli S (2008) Recent advancements in the Messinian stratigraphy of Italy and their Mediterranean-scale implications. Boll Soc Paleontol Ital 47:71–85
Ruggieri G, Buccheri G (1968) Una malacofauna tirreniana dell'isola di Ustica (Sicilia). Geol Rom 7:27–58
Russo P (1999) Nuovi dati sulla distribuzione geografica di *Strombus (Conomurex) persicus* Swainson, 1821. Boll Malacol 34:103–104
Russo GF, Cicogna F (1991) The date mussel (*Lithophaga lithophaga*) a "case" in the Gulf of Naples. In: Boudouresque CF, Avon M, Gravez V (eds) Les Espèces Marines à Protéger en Méditerranée. GIF Posidonie publ, Marseille
Russo P, Mel P (2002) Ritrovamento di *Musculista senhousia* (Benson in Cantor, 1842) nella laguna di Venezia. La Conchiglia 34:53–55
Russo GF, Patti FP (2005) Early life history of two closely related gastropods, *Rissoa auriscalpium* and *Rissoa italiensis* (Caenogastropoda: Rissoidae). Mar Biol 147:429–437
Sabelli B, Taviani M (1980) Reflexions sur les endemiques (mollusques marins) du Golfe de Gabes (Tunisie). CIESM, Cagliari
Sabelli B, Taviani M (1984) The paleobiogeographic distribution of the Mediterranean benthic molluscs and the Messinian salinity crisis or where did the molluscs go? Ann Géol Pays Hellén 32:263–269
Sadikaj R, Panariti E, Arapi D (2009) Biotoxin levels in bivalve molluscs in Albania and the related health impact. Maced J Anim Sci 1:205–211
Sala E, Knowlton N (2006) Global marine biodiversity trends. Ann Rev Environ Resour 31:93–122
Salas C (1996) The Bivalves from off the Southern Iberian Peninsula collected by the FAUNA I and BALGIM expeditions. Haliotis 25:33–100
Salas C, Woodside J (2002) *Lucinoma kazani* n. sp. (Mollusca: Bivalvia): evidence of a living benthic community associated with a cold seep in the eastern Mediterranean Sea. Deep Sea Res Part I Oceanogr Res Pap 49:991–1005
Sanjuan A, Pérez-Losada M, Rolàn E (1997) Allozyme evidence for cryptic speciation in sympatric populations of *Nassarius* spp. (Mollusca: Gastropoda). J Mar Biol Assoc UK 77:773–784
Sá-Pinto A, Baird SJE, Pinho C, Alexandrino P, Branco M (2010) A three-way contact zone between forms of *Patella rustica* (Mollusca: Patellidae) in the central Mediterranean Sea. Biol J Linn Soc 100:154–169
Sarà G, Romano C, Mazzola A (2008) A new Lessepsian species in the western Mediterranean (*Brachidontes pharaonis* Bivalvia: Mytilidae): density, resource allocation and biomass. Mar Biodivers Rec. doi:10.1017/S175526720600087X
Sardà F, Calafat A, Mar Flexas M, Tselepides A, Canals M, Espino M, Tursi A (2004) An introduction to Mediterranean deep-sea biology. Sci Mar 68(Suppl 1):7–38
Scaccini A (1967) Dati preliminari sulel zoocenosi bentoniche e sulla biomassa in una zona dell'alto e medio Adriatico. Note Lab Biol Mar Pesca Fano 2:25–56
Scheltema RS, Williams IP (2009) Reproduction among protobranch bivalves of the family Nuculidae from sublittoral, bathyal, and abyssal depths off the New England coast of North America. Deep Sea Res Part II Trop Stud Oceanogr 56:1835–1846
Schembri PJ, Borg JA, Deidun A, Knittweis L, Mellado Lopez T (2007) Is the endemic Maltese Topshell *Gibbula nivosa* extinct? Rapp Proc Verb Réun CIESM 38:592
Schiaparelli S (2008) Bivalvia. Biol Mar Med 15(Suppl):296–314
Schiaparelli S, Chiantore M, Cattaneo-Vietti R, Bedulli D, Albertelli G (2000) The autoecology of some mud-dwelling species helps understanding changes in benthic communities of the northern Adriatic Sea. Atti Ass Ital Oceanogr Limn 13:195–205
Schiaparelli S, Barucca M, Olmo E, Boyer M, Canapa A (2005) Phylogenetic relationships within Ovulidae (Gastropoda: Cypraeoidea) based on molecular data from the 16S rRNA gene. Mar Biol 147:411–420
Seibel BA, Maas AE, Dierssen HM (2012) Energetic plasticity underlies a variable response to ocean acidification in the pteropod, *Limacina helicina antarctica*. PLoS One 7:e30464
Sellmer GP (1967) Functional morphology and ecological life history of the gem clam, *Gemma gemma* (Eulamellibranchia: Veneridae). Malacologia 5:137–223
Sharma PP, González VL, Kawauchi GY, Andrade SC, Guzmán A, Collins TM, Glover EA, Harper EM, Healy JM, Mikkelsen PM, Taylor JD, Bieler R, Giribet G (2012) Phylogenetic analysis of four nuclear protein-encoding genes largely corroborates the traditional classification of Bivalvia (Mollusca). Mol Phylogenet Evol 65: 64–74
Simpson RD (1977) The reproduction of some littoral molluscs from Macquarie Island (sub-Antarctic). Mar Biol 44:125–142
Solsona M, Martinell J (1999) Protoconch as a taxonomic tool in Gastropoda systematics. Application in the Pliocene Mediterranean Naticidae. Geobios 32:409–419
Soppelsa O, Crocetta F, Fasulo G (2007) I molluschi marini di Punta Pioppeto (Isola di Procida – Campania). Boll Malacol 43:21–32
Spalding MD, Fox HE, Allen GR, Davidson N, Ferdaña AA, Finlayson M, Halpern BS, Jorge MA, Lombana A, Luorie SA, Martin KD, McManus E, Molnar J, Recchia CA, Robertson J (2007) Marine ecoregions of the world: a bioregionalization of coastal and shelf areas. Bioscience 57:573–583
Stjepeević J, Parenzan P (1980) Il golfo delle Bocche di Cattaro – condizioni generali e biocenosi bentoniche con carta ecologica delle sue due baie interne: di Kotor (Cattaro) e di Risan (Risano). Studia Marina 9–10:3–146
Stora G (1976) Evolution des peuplements benthiques d'un étang marin soumis à un effluent d'eaux douces. Bull Ecol 7:275–285
Strathmann MG, Strathmann RR (2007) An extraordinarily long larval duration of 4.5 years from hatching to metamorphosis for teleplanic veligers of *Fusitriton oregonensis*. Biol Bull 213:152–159

Streftaris N, Zenetos A (2006) Alien marine species in the Mediterranean – the 100 'worst invasives' and their impact. Med Mar Sci 7:87–118

Sturany R (1896) Zoologische Ergebnisse VII. Mollusken I (Prosobranchier und Opisthobranchier; Scaphopoden; Lamellibranchier) gesammelt von S.M. Schiff "Pola" 1890–1894. Denkschr Kaiserl Akad Wissensch Math-Naturwiss Cl 63:1–36

Sullivan CM (1948) Bivalve larvae of Malpeque Bay, P. E. I. Bull Fish Res Bd Can 77:1–58

Tabanelli C (2008) Associazioni di paleocomunità batiali a molluschi bentonici nel Pliocene della Romagna. Metodologie ed indicazioni per nuove tematiche. Quad Studi Nat Romagna 26:1–80

Tamayo Goya JC (2008) Catàlogo de los bivalvos marinos del sector central del Golfo de Valencia (España). Iberus 26:69–80

Taviani M (1974) Nota sul ritrovamento di cinque specie di Molluschi Gastropoda, Prosobranchia poco conosciuti o nuovi per le acque del Mediterraneo. Quad Civ Staz Idrobiol Milano 5:39–50

Taviani M (1978) Associazioni a Molluschi pleistoceniche-attuali dragate nell'Adriatico meridionale. Boll Zool 45:297–306

Taviani M (2002) The Mediterranean benthos from late Miocene up to present: ten million years of dramatic climatic and geologic vicissitudes. Biol Mar Med 9:445–463

Taviani M (2004) Shaping the biogeography of the Mediterranean basin: one geologist's perspective. In: Marine biogeography of the Mediterranean Sea: patterns and dynamics of biodiversity, vol 24, Biogeographia. Nuova Immagine Ed, Siena, pp 15–22

Taviani M, Colantoni P (1979) Thanatocoenoses würmiennes associées aux coraux blancs. Rapp Proc Verb Réun CIESM 25:141–142

Taviani M, Fiocchi C (1997) L'Uomo e la conchiglia. In: Endrizzi L, Marzatico F (eds) Ori delle Alpi. Quad Sez Arch Castello Buonconsiglio Monum Collez Provinc Trento 6:143–149

Taviani M, Taviani N (1986) Remarks on *Coralliophila richardi* (P. Fischer, 1882) (Gastropoda, Prosobranchia). Boll Malacol 22:65–72

Taviani M, Poluzzi A, Taviani N, Calamaras G (1985) *Scapharca inaequivalvis*; a recent invader of the southern lagoons in the Po river delta plain. Rapp Comm intern Mer Med CIESM 29:179–181

Taviani M, Bouchet P, Metivier B, Fontugne M, Delibrias G (1991) Intermediate steps of southwards faunal shifts testified by last glacial submerged thanatocoenoses in the Atlantic Ocean. Paleogeogr Paleoclimatol Paleoecol 86:331–338

Taviani M, Angeletti L, Freiwald A, Dimech M, Harasewych MG, Mifsud C, Oliverio M (2009) Coralliophilinae (Gastropoda: Muricidae) associated with deep-water coral banks in the Mediterranean. Nautilus 123:106–112

Taviani M, Angeletti L, Ceregato A (2011a) Chemosynthetic bivalves of the family Solemyidae (Bivalvia, Protobranchia) in the Neogene of the Mediterranean Basin. J Paleontol 85:1067–1076

Taviani M, Vertino A, López Correa M, Savini De Mol B, Remia A, Montagna P, Angeletti L, Zibrowius H, Alves T, Salomidi M, Ritt B, Henry P (2011b) Pleistocene to Recent scleractinian deep-water corals and coral facies in the Eastern Mediterranean. Facies 57:579–603

Tecchio S, Ramírez-Llodra E, Sardà F, Company JB, Palomera I, Mechó A, Pedrosa-Pàmies R, Sanchez-Vidal A (2011) Drivers of deep Mediterranean megabenthos communities along longitudinal and bathymetric gradients. Mar Ecol Prog Ser 439:181–192

Tenekidis NE (1989) On a collection of shells from the Greek seas. Protopata, Athens

Terlizzi A, Russo GF (1995) Variabilità strutturale del popolamento a molluschi di alcune praterie di *Posidonia oceanica* del Mediterraneo. Biol Mar Med 2:423–426

Terlizzi A, Fraschetti S, Gianguzza P, Faimali M, Boero F (2001) Environmental impact of antifouling technologies: state of the art and perspectives. Aquat Conserv Mar Freshw Ecosyst 11:311–317

Terlizzi A, Fraschetti S, Guidetti P, Boero F (2002) The effects of sewage discharge on shallow hard substrate sessile assemblages. Mar Pollut Bull 44:544–550

Terlizzi A, Scuderi D, Fraschetti S, Anderson MJ (2005) Quantifying effects of pollution on biodiversity: a case study of highly diverse molluscan assemblages in the Mediterranean. Mar Biol 148: 293–305

Terlizzi A, Bevilacqua S, Scuderi D, Fiorentino D, Guarnieri G, Giangrande A, Licciano M, Felline S, Fraschetti S (2008) Effects of offshore platforms on soft bottom macro-benthic assemblages: a case study in a Mediterranean gas field. Mar Pollut Bull 56:1303–1309

Terreni G (1981) Molluschi conchiferi del mare antistante la costa toscana (Gastropoda, Scaphopoda, Amphineura, Bivalvia, Cephalopoda). Benvenuti e Cavaciocchi, Livorno

Thiriot-Quievreux C, Rodriguez-Babio C (1975) Etude des protoconques de quelques prosobranches de la région de roscoff. Cah Biol Mar 16:135–148

Tsiakkiros L, Zenetos A (2011) Further additions to the alien mollusc fauna along the Cypriot coast: new opisthobranchia species. Acta Adriat 52:115–124

Turk T, Furlan B (2011) New records of Indo-Pacific and Atlantic mollusc species (Opisthobranchia) in the Eastern Mediterranean and Adriatic Sea. Ann Se Hist Nat 21:5–10

Tzedakis PC, Channell JET, Hodell DA, Kleiven HF, Skinner LC (2012) Determining the natural length of the current interglacial. Nat Geosci 5:138–141

Tzomos T, Chartosia N, Christodoulou M, Kitsos M-S (2011) New records and range expansion of lessepsian migrants in the Levantine and Aegean Seas. Mar Biodivers Rec 3:e10

UNEP (2010) The Mediterranean Sea biodiversity: state of the ecosystems, pressures, impacts and future priorities. UNEP-MAP RAC/SPA, Tunis

UNEP/MAP-RAC/SPA (2008) Guidelines for controlling the vectors of introduction into the Mediterranean of non-indigenous and invasive marine species. RAC/SPAC, Tunis

Valavanidis A, Vlachogianni T (2011) Marine litter: man-made solid waste pollution in the Mediterranean Sea and coastline. Abundance, composition and sources identification. http://chem-tox-ecotox.org/wp/?p=749. Accessed 27 Sept 2012

van Aartsen JJ, Menkhorst HPM, Gittenberger E (1984) The marine Mollusca of the Bay of Algeciras, Spain, with general notes on *Mitrella*, Marginellidae and Turridae. Basteria 1(Suppl 2):135

Vatova A (1949) La fauna bentonica dell'Alto e Medio Adriatico. Nova Thal 1:1–110

Vazzana A (2010) La malacofauna del circalitorale di Scilla (Stretto di Messina). Boll Malacol 46:65–74

Vicente N (1990) Estudio ecológico y protección del molusco lamelibranquio *Pinna nobilis* L. 1758 en la costa mediterránea. Iberus 9:269–279

Vio E, De Min R (1996) Contributo alla conoscenza dei Molluschi marini del golfo di Trieste. Atti Mus Civ St Nat Trieste 47:173–233

Warén A (1990) Ontogenetic changes in the trochoidean (Archaeogastropoda) radula, with some phylogenetic interpretations. Zool Scr 19:179–187

Wilson EO (ed) (1988) Biodiversity. National Academic Press, New York

Wirzinger G, Vogt C, Bachmann J, Hasenbank M, Liers C, Stark C, Ziebart S, Oehlmann J (2007) Imposex of the netted whelk *Nassarius reticulatus* (Prosobranchia) in Brittany along a transect from a point source. Cah Biol Mar 48:85–94

WoRMS (2012) World register of marine species. http://www.marinespecies.org/aphia.php?p=taxdetails&id=565481. Accessed 29 Sept 2012

WWF/IUCN (2004) The Mediterranean deep-sea ecosystems: an overview of their diversity, structure, functioning and anthropogenic impacts, with a proposal for conservation. IUCN, Malaga

Zamouri-Langar N, Chouba L, Jarboui O, Mirabet R (2006) Presence du bivalve Veneridae *Ruditapes philippinarum* (Adams & Reeve, 1850) sur le littoral sud de la Tunisie. Bull Inst Natn Scien Tech Mer Salambô 33:23–27

Zenetos A (1997) Diversity of marine bivalvia in Greek waters: effects of geography and environment. J Mar Biol Assoc UK 77:463–472
Zenetos A, Gofas S, Russo G, Templado J (2003) CIESM atlas of exotic species in the Mediterranean, vol 3, Mollusca. CIESM, Monaco
Zenetos A, Koutsoubas D, Vardala-Theodoru E (2005) Origin and vectors of introduction of exotic molluscs in greek waters. Belg J Zool 135:279–286
Zenetos A, Gofas S, Verlaque M, Cinar ME, Raso EG, Bianchi CN, Morri C, Azzurro E, Bilecenoglu M, Froglia C, Siokou I, Violanti D, Sfriso A, San Martin G, Giangrande A, Katagan T, Ballesteros E, Ramos-Esplá A, Mastrototaro F, Ocana O, Zingone A, Gambi MC, Streftaris N (2010) Alien species in the Mediterranean Sea by 2010 A contribution to the application of European Union's Marine Strategy Framework Directive (MSFD). Part I. Spatial distribution. Mediterr Mar Sci 11:381–493
Zibrowius H (1991) Ongoing Modification of the Mediterranean marine fauna and flora by the establishment of exotic species. Mésogée 51:83–107
Zibrowius H (2002) Assessing scale and impact of ship-transported alien fauna in the Mediterranean. CIESM Workshop Monogr 20:63–68
Zibrowius H, Taviani M (2005) Remarkable sessile fauna associated with deep coral and other calcareous substrates in the Strait of Sicily, Mediterranean Sea. In: Freiwald A, Roberts M (eds) Deep-water corals & ecosystems. Springer, Berlin

Recent Changes in the Distribution of Autochthonous Marine Molluscs in the Mediterranean Sea

17

Paolo G. Albano

Abstract

The changes in the distribution of two autochthonous molluscs in the Mediterranean Sea are described from the early nineteenth century to date. *Echinolittorina punctata* (Gastropoda: Littorinidae) had until the end of the twentieth century a distribution limited to the western Mediterranean (southern Spain and from Morocco to northern Tunisia, a few localities in southern Sicilia) and to the eastern Mediterranean (from Egypt to Syria). At the end of the twentieth century the range started to expand to all along the coast of Sicilia and northward along the Italian peninsula. Today, the species is found north of Roma. *Eastonia rugosa* (Bivalvia: Mactridae) used to have a distribution limited to the western Mediterranean, not exceeding the Siculo-Tunisian sill. During the end of the twentieth century and in the 2000s, the species was recorded from the Gulf of Gabès and from progressively northern localities in Italy. Today the species is found north of Livorno, in Toscana. The distributional changes of a few more Mediterranean molluscs is described in less detail, but wishes to highlight potential further descriptors of change of the biodiversity of the Mediterranean Sea.

Keywords

Autochthonous molluscs • Climate warming forcing • *Eastonia rugosa* • *Echinolittorina punctata* • Human introduction • Mediterranean molluscs • Mediterranean Sea • Molluscan fauna • Monitoring programmes

Abbreviations

ANSP	Academy of Natural Sciences of Philadelphia
LACM	Los Angeles County Museum
MCZR	Museo Civico di Zoologia, Roma
MZUB	Museo di Zoologia dell'Università di Bologna
NMR	Natuurhistorisch Museum Rotterdam
SBMNH	Santa Barbara Museum of Natural History

P.G. Albano (✉)
Dipartimento di Scienze Biologiche, Geologiche e Ambientali, University of Bologna, c/o Prof. B. Sabelli, via Selmi 3, 40126 Bologna, Italy
e-mail: pgalbano@gmail.com

Introduction

The changes in the distribution of species are becoming one of the most relevant issues in works on the biodiversity of the Mediterranean Sea (Bianchi and Morri 1993, 1994; Bianchi 2007; CIESM 2008; UNEP-MAP-RAC/SPA 2008; Lejeusne et al. 2009; Coll et al. 2010). Whether driven by morphological modifications of the Basin (e.g. the opening of the Suez Canal), exploitation of marine species, bioinvasions or climate change, basic knowledge on the past and present distribution of species is essential to understand the status of marine biota and cause-effect interactions.

The objective of this chapter is to describe the changes of the range of two species of molluscs: *Echinolittorina punctata* (Gmelin 1791) (Gastropoda: Littorinidae) and *Eastonia rugosa* (Helbling 1779) (Bivalvia: Mactridae). Both species are autochthonous in the Mediterranean Sea: they have a

S. Goffredo and Z. Dubinsky (eds.), *The Mediterranean Sea: Its history and present challenges*,
DOI 10.1007/978-94-007-6704-1_17, © Springer Science+Business Media Dordrecht 2014

wide distribution on the north-western African coasts of the Atlantic Ocean and used to have a predominantly southern distribution in the Mediterranean Sea.

Both species have undergone a wide extension of their ranges in the last decades. Lacking major monitoring programmes, this phenomenon is here documented on the basis of records from the literature, public and private collections. This is of course a work in continuous progress. As long as new data are collected the pattern is expected to become clearer and resolution higher. I have considered herein data collected until December 15th, 2011. However, the present results seem robust enough for describing the main directions of change of the species distribution. The description below has a wide view on the Mediterranean Sea and then focuses on what has happened on the Italian coastline in the last decades. It is along Italy that the most surprising and best documented range extensions happened recently.

Mapping the Distribution: Retrieving Data from the Literature and Collections

Data used to describe the temporal variation of the distribution was obtained from two different sources: the literature and collections, both private and public. Table 17.1 details the data set.

Bibliographic references have been divided into primary and secondary. Primary references are those that give original data on collecting events, previously unpublished. Primary references are the most valuable ones, giving first hand and usually more detailed data. Secondary data usually summarize the known information, retrieved from previously published primary references or other sources. The latter usually have a more general approach to the description of the distribution. However, their value is in the ability to give a glimpse of the distributions at the publication time. Especially in the early literature, it is often difficult to find all primary references and collection records. Secondary references are an adequate surrogate of primary sources until these become available. However, citations in the text will be preferably given for primary references.

Table 17.1 Data set used to describe the distribution of the species

	Echinolittorina punctata	*Eastonia rugosa*
Primary references	146 [32]	108 [37]
Secondary references	87	34
Collections	317	72
Total	550	231

Between square brackets there is the number of primary references considered for reporting the absence of the species

Public collections were inspected both directly and remotely. In the latter case, every record which was outside the expected range at the time of collecting was checked by direct contact with curators, often asking for photographs of the specimen to check the identification.

Data from older collections sometimes lack the collection date of the specimens. To place in time these cases, the date of death of the collector has been considered. If the collector clearly indicated that specimens came from other collectors (it is often the case in the Monterosato collection, for example) then the date of death of the source of specimens was considered.

Private collections were mainly investigated remotely. Several collectors sent their data and observations and, again, records outside the expected range were checked by photographs.

Absence records were also collected, because in the case of regional works the record of absences helped in describing the patterns of distribution.

Echinolittorina punctata and *Eastonia rugosa*: Two Stories of "Success"

Research on these two species was motivated by the sudden finding of populations in unexpected places. My first encounter with *Eastonia rugosa* dates back to December 2004 on the coasts to the north of Roma, on the Tyrrhenian Sea (Albano 2006). *Echinolittorina punctata* showed up while beachcombing the shores south of Gallipoli, in the south-eastern Italian region Puglia, on the Ionian Sea (Albano and Trono 2008). There were not consistent previous findings of the species in these areas and common knowledge in the malacological Italian community expected them much southerly. Since these two encounters, research started and it was evident that on one side those populations were stable because findings were confirmed later and population density had increased, and on the other side it was possible to delineate a progressive distribution extension which is described here under.

Despite that the core of observations was around the Italian coasts, the research of species records was carried out on the entire Mediterranean Sea and neighbouring Atlantic Ocean in order to place what is happening in Italy in a wider biogeographic context. The range of the distribution in the Mediterranean Sea is shown 50 years by 50 years, starting from 1850.

Range Extension of *Echinolittorina punctata*

E. punctata (Fig. 17.1a, b) is a littorinid gastropod that lives on hard substrates from the littoral fringe to the upper eulittoral. Its taxonomic position has been recently reviewed by

Fig. 17.1 (**a**, **b**) *Echinolittorina punctata* (Gmelin 1791), height 7.7 mm, Penisola di Magnisi, north of Siracusa, Sicilia (legit S. Palazzi, September 2008). (**c**, **d**) *Melarhaphe neritoides* (Linnaeus 1758) height 9.2 mm, Fiumicino (Roma) (legit F. De Santis, February 2009). (**e**, **f**) *Littorina saxatilis* (Olivi 1792) height 5.4 mm, Venezia (legit P.G. Albano, May 2011). (**g**) *Echinolittorina punctata* (Gmelin 1791) egg capsule, diameter 0.3 μm, Catania (Spawned in laboratory, July 2008)

Reid (2011). Among the most important synonyms in the context of this review, *Turbo siculus* Brugnone 1850 refers to populations found in south-eastern Sicilia, *Litorina syriaca* Philippi 1847 refers to populations found in Syria. *Litorina tigrina* d'Orbigny 1842 was used for some Spanish populations. However, since the early twentieth century the nomenclature has remained stable.

In the Mediterranean Sea there is little risk of misidentification with other species. The littoral rocks where the species is found host another littorinid, *Melarhaphe neritoides* (Linnaeus 1758) (Fig. 17.1c, d). This species is generally smaller and with a different colour pattern. However, encrusted and not fully mature specimens of *E. punctata* may be (and have been) misidentified with *M. neritoides*. As an example, Lanfranco (1973a) published the first record of the species from Malta, but he underlined that "specimens of this periwinkle have been in my collection for several years without my noticing". The third littorinid living in the Mediterranean Sea, *Littorina saxatilis* (Olivi 1792), shows some phenotypes with spotted shell (Fig. 17.1e, f), but it has a completely disjointed distribution, being restricted to the North Adriatic Sea and Gulf of Gabés (whose population was named *Littorina nervillei* Dautzenberg 1893).

The dispersal of the species is guaranteed by pelagic egg capsules (Fig. 17.1g). The development time is not known, but the congeneric tropical species *E. hawaiiensis* (Rosewater and Kadolsky 1981) hatches after 3 days and metamorphosis occurs after 21–25 days (Struhsaker and Costlow 1968); a similar duration of planktotrophic development can be predicted for all species from their uniform larval shells (Williams and Reid 2004).

In 1850, *Echinolittorina punctata* was recorded from Malaga (McAndrew 1851 as *Littorina tigrina* d'Orbigny 1842, a synonym of *Littoraria nebulosa* (Lamarck 1822)) and Syria (Philippi 1847, where the species was described as *Litorina syriaca*). The species was also recorded from Augusta, in south-eastern Sicilia (20 km north of Siracusa), by Aradas and Benoit (1870) and Monterosato (1872). This record is based on the material collected by Brugnone who introduced the new taxon *Turbo siculus* in 1850. The finding by Brugnone may be the very first in Sicilia because the species was not recorded by Costa (1829) and Philippi (1836, 1844). Specimens collected by Brugnone at Pozzallo (the south-easternmost tip of Sicilia) are now in the Monterosato collection at the Museo Civico di Zoologia in Roma (MCZR) (lots 21697 and 21718). Unfortunately, the date of collection is not present on the labels. Despite that it is not explicitly recorded, the species may have well been present at that time on most of the coasts of the southern Mediterranean Sea between Spain and Syria. Forbes (1844) did not cite the species from the Aegean Sea, however this area is characterized until recently by a lack of proper monitoring and information in relation to the extension of its coastline.

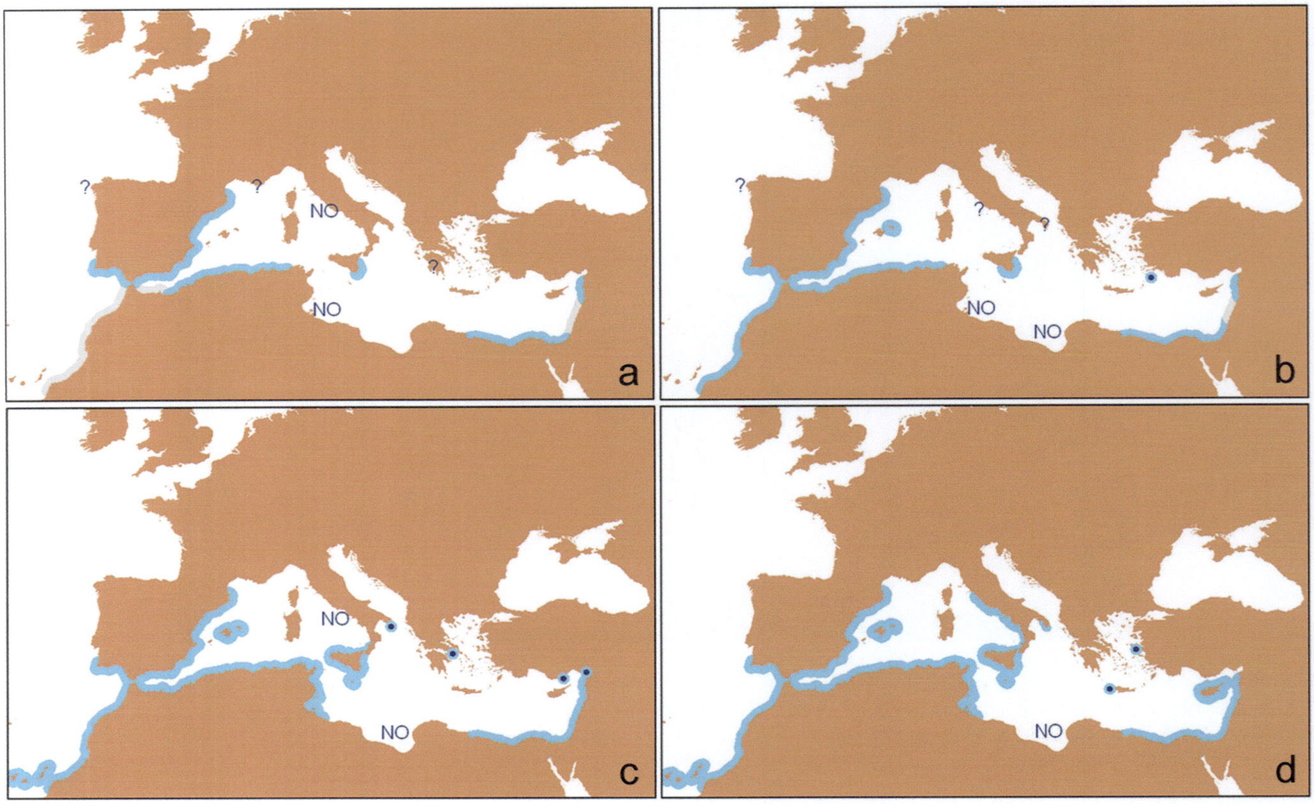

Fig. 17.2 Distribution of *Echinolittorina punctata* in 1900 (**a**), 1950 (**b**), 2000 (**c**) and today (**d**). *Sky blue* areas are the distribution reconstructed from bibliography and collections. *Grey* areas are expected distribution not confirmed by bibliography and collections probably due to lack of sampling in the area. *Blue* points are single records of interest because in poorly sampled areas or because the first records of successive range extensions. "*NO*" means the species was not recorded by regional works in the area at the time. Question marks, "*?*", indicate areas where doubtful records were reported

In 1900 (Fig. 17.2a), the species was recorded from southern Portugal by McAndrew (1854) and Weinkauff (1883) and the Mediterranean coasts of Spain (McAndrew 1857; Hidalgo 1867; Monterosato 1878; Weinkauff 1883). The species was cited for Algeria by Weinkauff (1868), Monterosato (1877a), Pallary (1900) and Egypt by Weinkauff (1868). Several records from Syria are mainly from secondary references (Weinkauff 1868, 1883; Monterosato 1875, 1878) however a lot without the collecting date is present in the Monterosato collection (lot 21696). The Monterosato collection also contains a lot from Algeria (lot 21718) coming from the Tiberi collection, again without the collecting date. The distribution in south-eastern Sicilia was reported by Monterosato (1872, 1875) probably on the basis of the lots cited above and still preserved in his collection. Records from Asturias and Galicia (McAndrew 1854) are not considered reliable by Reid (2011). Locard (1886) reported the species from southern France, on the basis of Michaud's material. This record was not confirmed by Dautzenberg (1881) for Cannes nor by Bucquoy et al. (1882–1898) for the Roussillon. A lot from Greece (no details about the locality) is present in the ANSP collection (lot 18577) collected by M. Conemenos; little is known about his biography but it is likely that this collector was active mainly during the second half of the nineteenth century. It is important to evidence that at this time there are no reliable records from peninsular Italy and the Gulf of Gabés.

In 1950 (Fig. 17.2b) most of the range is confirmed. Pallary (1920) recorded it from several localities in Morocco and indeed several specimens are present in the Monterosato collection with Pallary's label (MCZR, lot 21699). Praus Franceschini (1906) recorded the species from Algeria and several lots are present in the Monterosato collection in MCZR (specimens from Alger collected by Joly in lot 21781, a hundred specimens from Alger in lot 21718, further specimens from Oran in lots 21706 and 2178) and in the Los Angeles County Museum (LACM) (lot 60078, collected by R. Tremper in Algiers). Remarkably, the species was still not recorded from the Gulf of Gabés (Pallary 1904, 1906) and Libya (Monterosato 1917a, b). Pallary did not record the species from Tunis either (1914). However, the species was present in Egypt (Pallary 1912a; Tomlin 1927; Moazzo 1939) and specimens are present in the Monterosato collection from Alexandria (MCZR, lot 21697 and 21706). In the

Middle East the species was recorded from Syria (Pallary 1912b, 1919; Monterosato collection at MCZR lot 21706) and Israel (Santa Barbara Museum of Natural History SBMNH, lot 367006). The Spanish distribution was confirmed (Hidalgo 1910, 1917; Monterosato collection lot 21697, specimens from Malaga and Villanueva y Geltrù) and extended to Menorca (Hidalgo 1917). In Italy, the species was recorded from Catania (Sicilia) by Patané (1946) and is present in the Monterosato collection from other localities in Sicilia: Ragusa (lot 21715 and 21718), Pozzallo and Scoglitti (unnumbered lots). At this time the species was not as yet recorded from Malta. However, Lanfranco (1973a) discovered the species among his lots of *Melarhaphe neritoides* and therefore misidentification may have been the cause of the absence of previous records. The species is however rare in Malta even today (C. Mifsud, personal communication, November 2011). Two doubtful records are from Italy. The first is by Praus Franceschini (1906) who recorded specimens from Tricase, on the Adriatic shores of Puglia in south-eastern Italy. This record could not be checked since most of the Praus Franceschini collection in Napoli was destroyed during World War II and these lots are missing (N. Maio, personal communication, October 2011). This record was never confirmed later. The second is by Brunelli (1928) who cited the species from Santa Marinella (Roma) the present day northernmost locality in the Tyrrhenian Sea. Brunelli was not a malacologist (he cites the species while describing the bioerosion of littoral rocks) and it was not possible to check the identification. However, the species was not recorded any more from these Italian regions until the 2000s (Albano 2010) and it seems unlikely that the species was not intercepted by the many collectors who surveyed the coasts so close to Roma, where a lively community of collectors reside. Another interesting record is from Rhodes (Academy of Natural Sciences of Philadelphia, ANSP, 18579) because so few are the records even today from the eastern Mediterranean Sea. This record is supposed to be from the early twentieth century, probably before World War I (P. Callomon, personal communication, November 2011).

In the next decades (Fig. 17.2c), most of the distribution was confirmed. New records were from Mallorca (Bosch and Moreno 1983), several localities in Israel (Lipkin and Safriel 1971; Palant and Fishelson 1968; Nordsieck 1972; Noy et al. 1987; Nevo and Lavie 1989), Lebanon (Nuwayhid et al. 1985), Turkey (from Tasuçu and Iskenderun, A. Lugli, private collection; Yumurtalik, M. Tisselli, private collection; Antalya, Natuurhistorisch Museum Rotterdam, NMR993000033293), Greece (Keratià, Nicolay and Angioy 1988) and Cyprus (Ozturk et al. 2004). However, the most remarkable records are from Tunisia, Malta and Italy. In Tunisia, the species is recorded from Kelibia in 1973 (B. Sabelli collection at the Museo di Zoologia dell'Università di Bologna, MZUB), Cap Serrat in 1987 (A. Lugli collection), Tabarka in 1991 (Williams and Reid 2004) and from several localities in the Gulf of Gabés: Jerba in 1994 (D. Viola collection), 1995 (NMR993000033315), several other localities in 1998 (Enzenross and Enzenross 2001; NMR SNSD4280) and in 2000 (I. Nofroni collection, he observed abundant populations). In this context, the populations recorded by Antit et al. (2008) along the Tunisian coasts fit in the expected distribution of the species at that time. In Malta there was the first formal record of the species in 1973 (Cachia 1973; Lanfranco 1973b). In Italy, the distribution previously restricted to the south-eastern tip of Sicilia extended all over the island and beyond. The species was recorded from Porto Palo in 1975 and Marinella di Selinunte in 1978 (south-western Sicilia) (ENEA database 1998), Messina (north-eastern Sicilia) in 1980 (NMR993000039672) and 1994 (NMR993000025145). A particular event in the history of this species concerns its distribution on the north-western shores of Sicilia: D'Anna (1986) reported that a Sicilian collector, Vittorio Emanuele Orlando, had introduced specimens of *E. punctata* collected on the southern shores of the island to localities in the Gulf of Castellammare, west of Palermo (Cala Rossa near Terrasini, Trappeto and Balestrate) in 1972–1973. D'Anna then went back to the same localities at the beginning of the 1980s and found conspicuous populations of *E. punctata* in the area, claiming they were the descendants of those introduced specimens. However, the species was not found by S. Palazzi (personal communication, April 2011), a skilful malacologist who knew well the area, during his surveys at the beginning of the 1980s nor do we have any record from local collectors in the 1970s. Whether this introduction was actually successful and played a role in the subsequent range extension in the Tyrrhenian Sea is still to be ascertained. However, the species was then further found at Carini (Palermo) in 1983, Torre Pozzillo (Palermo) in 1994 (NMR993000025147) and is still present today from Trappeto to Carini (pers. obs.). In 1997, the first reliable record for peninsular Italy was given for Zambrone, Vibo Valentia (Calabria) (D'Anna 1997), and in 1999 there was the first record for the Ionian coasts of Puglia, in south-eastern Italy: a dead shell from Santa Caterina (Lecce).

These last records from unusual Italian localities were just the beginning of a prodigious extension to other areas of peninsular Italy and of the enlargement of populations (Fig. 17.2d). The species was recorded from the coasts of Salerno in 2001 (D'Anna 2001), 2002 (A. Margelli, personal communication, March 2008) and 2004 (Soppelsa et al. 2004). Several records were then published for Lazio where the species was recorded in the southern sectors in 2004 (Soppelsa et al. 2004) and then from 2007 at Fiumicino and northward to Santa Marinella a few kilometers south of Civitavecchia (Albano 2010). In 2007 a rich population was then found on the Ionian coasts of Puglia, in the area of Gallipoli (Albano and Trono 2008). Since then, the species

Fig. 17.3 *Eastonia rugosa* (Helbling 1779). (**a–d**) width 68.5 mm, Siracusa, Sicilia (legit W. De Mattia, August 1998). (**b–e**) Width 61.5 mm, Foglino (Roma) (Meli collection, MCZR). (**c–f**) Width 56.8 mm, Anzio (Roma) (Meli collection, MCZR)

spread along the coast and is now found in several localities with stable populations.

In the second half of the twentieth century the species colonized the entire island of Sicilia and in the last 15 years, the species spread northward along peninsular Italy for 600 km.

Range Extension of *Eastonia rugosa*

E. rugosa (Fig. 17.3a–d) is a mactrid bivalve that inhabits sublittoral soft substrates. No recent works are available about its taxonomy and nothing is known about its reproduction and development. The species has been cited in the older literature as *Eastonia locardi* d'Oliveira 1896 or *Spisula lamarckii* Gray 1837, but its nomenclature is stable since the beginning of the twentieth century. In the past, the species had sometimes been placed in the genus *Lutraria*.

There are also little chances of misidentification in the Mediterranean Sea. However, specimens of *Petricola lithophaga* (Retzius 1788) (Bivalvia: Veneridae; family placement according to WoRMS database (accessed on December 27th, 2011)) have been misidentified with young *Eastonia rugosa*, probably due to the superficial similarity of the white colour and valve sculpture of strong radial ribs. However, the external ligament of *P. lithophaga* readily distinguishes it from *Eastonia rugosa*, which has an internal ligament. Moreover, *P. lithophaga* is an endolithic species. This misidentification has happened for a specimen from Corse kept at the NMR (993000021751). Moreover, also records in the literature from unusual habitats are doubtful, such as the record of a "juvenile" specimen from a precoralligenous substrate at 18 m depth in Bagheria (Palermo) in 1973 (Orlando 1979).

One of the difficulties when dealing with this species is that it is easy to collect only when beached and therefore usually dead and without soft parts. However, valves stranded on a beach do not necessarily imply a living population, but may come from sub-fossil and fossil deposits off the coast. This problem has to be taken into consideration even in the case of sublittoral samplings when dead specimens are found. Unfortunately, it is rare that authors specify the state of collected specimens.

The known distribution of the species in 1850 is limited to the Iberian Peninsula (Vigo, Faro, Gibraltar, Malaga) and La Goulette in north Tunisia (McAndrew 1851). However, the species was probably present all along the north-eastern African coasts because in 1900 (Fig. 17.4a) the species was recorded from the Canary Islands (McAndrew 1854) eastward to Morocco (McAndrew 1857), Algeria (Weinkauff 1862; Monterosato 1877a). Further records were again from the Iberian Peninsula: Asturias and Galicia, Algarve (McAndrew 1854, 1857), Cadiz (MCZR 188012, 195319) and Andalucia (McAndrew 1854, 1857; Hidalgo 1870). The species was not recorded from southern France (Bucquoy et al. 1882–1898; Locard 1886, 1891), Corse (Payraudeau 1826), Sicilia (Costa 1829; Philippi 1836, 1844), Malta (Caruana 1867), Gulf of Gabès (Dautzenberg 1883), Greece (Forbes 1844).

An unusual record in this context was from the Tyrrhenian coasts of Italy, in Lazio. Meli (1896) reported specimens collected at the Anzio harbor in 1881 and Foglino near Nettuno in 1893. These specimens are still preserved in the MCZR (Fig. 17.3e, c–f, b–e respectively). Both specimens consist of loose valves, without any trace of ligament and periostracum, and would fit the hypothesis of fossil material washed ashore (Fig. 17.3b–d). Indeed, the same author reports that *Eastonia rugosa* is found in

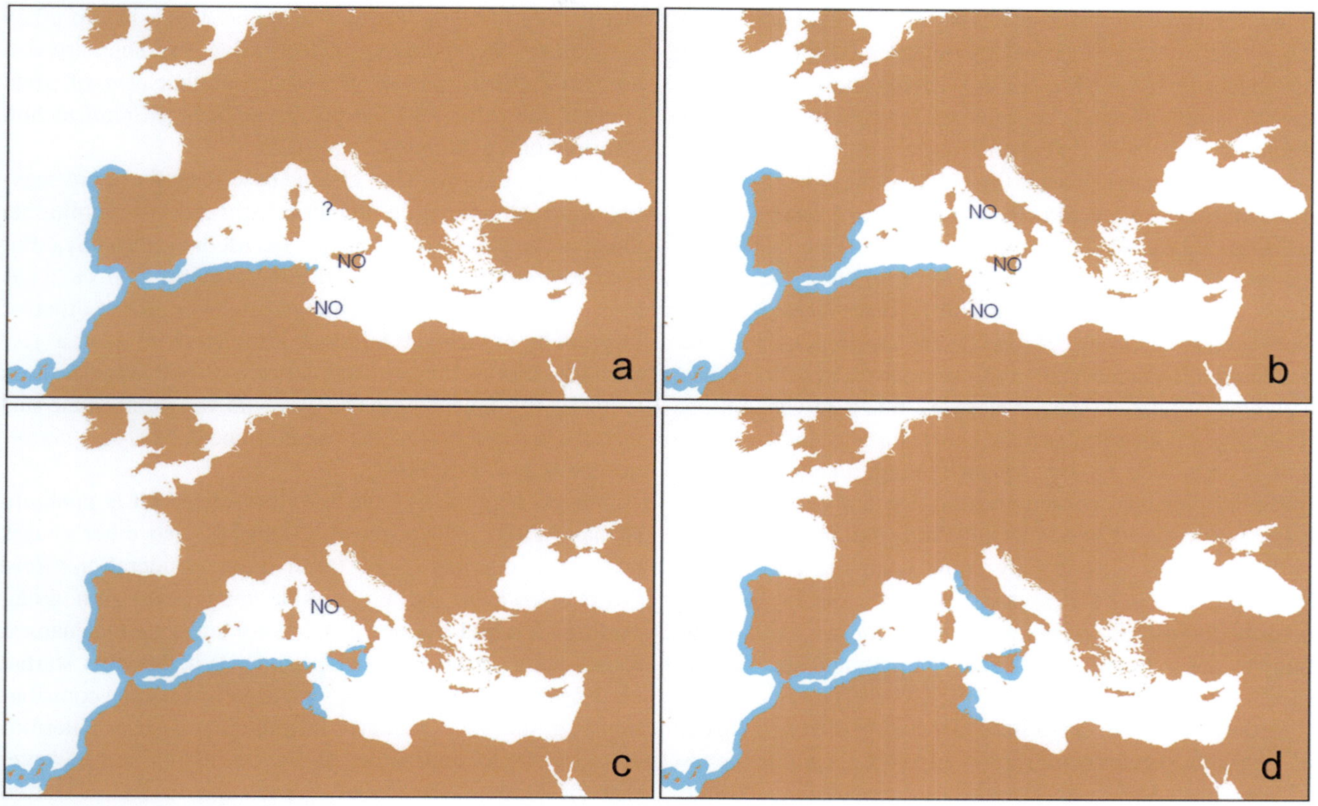

Fig. 17.4 Distribution of *Eastonia rugosa* in 1900 (**a**), 1950 (**b**), 2000 (**c**) and today (**d**). *Sky blue* areas are the distribution reconstructed from bibliography and collections. "*NO*" means the species was not recorded by regional works in the area at the time. Question marks, "*?*", indicate areas where doubtful records were reported

quaternary sands in Foglino and in the sands below the train station of Anzio (just 400 m from the coastline). In the same work, Meli reported the finding by the collector Biagio Donati of several specimens, still retaining the ligament, in the harbour muds of Civitavecchia. He then reported that Donati had thought that these specimens were brought in by fishermen and that he did not ever see living individuals. A few years before, Monterosato (1877b) studied the shells of the Civitavecchia harbor in the Donati collection but he did not cite this species which should have certainly caught his attention.

At the beginning of the twentieth century (Fig. 17.4b), the species was again reported from many Atlantic and western Mediterranean localities: several places in Morocco (Pallary 1920), Algeria (Pallary 1900, Monterosato collection lot 19763), Portugal, Gibraltar and Spain (Hidalgo 1917) while it was still not recorded from the Gulf of Gabès (Pallary 1904, 1906). It was absent from the eastern Mediterranean Sea: Lybia (Monterosato 1917a, b), Egypt (Pallary 1912a), Syria (Pallary 1912b). No further specimens were recorded from Italy.

During the second half of the twentieth century, the species began the colonization of Italy (Fig. 17.4c). In 1972 a single valve was found at Marina di Ragusa and in 1977 the species was abundant on the same beach, complete and living specimens were collected (Orlando 1979). At the beginning of the 1970s the species was found also in Siracusa (B. Sabelli collection at MZUB). Then several other findings were reported: in 1980 at Scoglitti (Ragusa) and Marzamemi (Siracusa) (Grasso 1981), in 1982 at Donnalucata (Ragusa), at the Anapo estuary (Siracusa) and Acitrezza (Catania) (ENEA database 1998), in 1983 at Messina (Micali and Giovine 1983), in 1984 at Licata (Agrigento) (ENEA database 1998). Subsequent records confirmed the distribution around Sicilia: Punta Braccetto (Ragusa) in 1994 (NMR 993000049271), Siracusa in 1995 (D. Viola collection) and 1998 (P.G. Albano collection). At this time the species was not recorded from Lazio (Anonymous 1986) and Toscana (Terreni 1981, 1983). Abundant populations were recorded in northern Tunisia in the Gulf of Tunis (Zaouali 1971, B. Sabelli collection at MZUB) but in the same period the species was not found in the Gulf of Gabès (B. Sabelli, P. Piani, personal communication). A first record from Djerba was in 1995 (NMR993000036123) then in 1998 the species was reported from several localities in the Gulf of Gabès (Enzenross and Enzenross 2001).

As for *Echinolittorina punctata*, at the beginning of the twenty-first century the species range extended fast along the Italian coastline (Fig. 17.4d). Living individuals of the species were informally reported on internet forums as collected in 2003 from Ladispoli (Roma) (by I. Nofroni) and Marina di Cerveteri (Roma) (by A. Antonini), in 2004 again from Marina di Cerveteri (Roma) and Palo Laziale (Roma) (Albano 2006). In 2005 the species was recorded from Marina di Cecina (Livorno) (by D. Ferri) and again from Lazio in Anzio (La Valle et al. 2007). It was later recorded from several localities in Lazio (F. De Santis, personal communication, December 2011) to nowadays. In 2007, a new area was colonized: specimens were found in Calambrone (Pisa) (by S. Galatolo) and the existence of a stable population is confirmed by subsequent findings to nowadays (L. Innocenti, personal communication, December 2011). This is the northernmost Italian locality known to the present day. Further records continuously come from the cited Italian localities, meaning that populations are stable in the range described. Still, the species is not recorded from Malta (maybe due to lack of suitable habitat) and the Eastern Mediterranean Sea.

From the first record in Italy in the early 1970s to the end of 1990s the species colonized most of Sicilia and in the last 15 years, the species has extended its distribution to a further 800 km northward.

Descriptors of Change

Which Are the Drivers of Change?

The ranges described above should be read in two different ways depending on the temporal range considered. During the nineteenth century and the first half of the twentieth century researchers started to describe the Mediterranean fauna and the distributional gaps should be seen as a lack of sampling and information rather than true distributional gaps. However, approximately since 1950 distributional records started to be more frequent and spread over the Mediterranean basin due to an increased workforce both among professional researchers and amateur collectors.

Of course, this is not uniform all along the Mediterranean coastline. The western Mediterranean European coasts enjoyed and still enjoy a more intense research effort than the north African coasts (with the sole exception of the Gulf of Gabès that, due to its many endemics, attracted and still attract attention). The extreme eastern Mediterranean Sea started to be carefully studied during the second half of the twentieth century due to the many lessepsian migrants which enriched its fauna. However, only Israel may have a good knowledge of its marine fauna, while many other places, like the long Turkish coastline which lies in a key position for the spreading of lessepsian migrants, are still not satisfactorily known. Greece is another area of great interest due to its position but it has been difficult to find information on its fauna.

Anyway, the pattern of change described above are striking, especially along the Italian coastline, whose molluscan fauna is one of the best known in the Mediterranean Sea. For both *Echinolittorina punctata* and *Eastonia rugosa* it has been possible to describe a range extension in the order of magnitude of several hundred kilometers in just a few decades. Due to the southern affinity of both species, it is tempting to connect these changes with variations in the surface sea temperature of the Mediterranean Sea, as a consequence of global warming.

Indeed, hypotheses have been formulated for *E. punctata* (Albano 2010; Albano et al. 2010) but also other causes have been claimed to be taken into consideration (Reid 2011). However, the progressive range extension along north Sicilia and the peninsular Italy suggests that natural forces have driven the phenomenon. Whether this started entirely for natural causes or with an active human contribution (e.g. the V.E. Orlando's introduction west of Palermo) is still to be ascertained. It has to be highlighted however that the range extension of *E. punctata* in Sicilia started along the southern and eastern coasts well before this event and that the newly discovered population in Puglia is certainly not connected with it.

The present-day patchy distribution of *Eastonia rugosa* is more intriguing. Despite that it is distributed along most of Sicilia, the species is not recorded yet for Calabria and Campania (a coastline almost 400 km long). It is however rather continuously distributed in Lazio and Toscana. Such distributional gaps may raise the doubt that the range extension could be driven by human factors like maritime traffic or export of fish products as suggested by Zaouali (2008). The distribution of *E. rugosa* may also be driven by environmental conditions like sediment granulometry and influx of fresh water as observed by other authors (e.g. Orlando 1979).

Of course all the hypotheses on the active human introduction of these species would change the status from autochthonous to alien in the newly colonized sectors (sensu Pyšek et al. 2009). In any case, until the active role of humans is demonstrated, the condition of these species is better described as autochthonous species extending their range.

What Are Other Molluscs Doing?

Other species of molluscs have been proposed as descriptors of change of the biodiversity of the Mediterranean Sea under climate warming forcing. Bianchi and Morri (1993) recorded as warm-water species reported from the Ligurian Sea *Patella ferruginea* Gmelin 1791 (Gastropoda: Patellidae) and *Charonia lampas* (Linnaeus 1758) (Gastropoda: Ranellidae). The CIESM project "Tropical signals" enlisted among its macrodescriptors of warm water affinity: *Dendropoma* spp. (Gastropoda: Vermetidae), *Erosaria spurca* (Linnaeus 1758) and *Luria lurida* (Linnaeus 1758) (Gastropoda: Cypraeidae), *Stramonita haemastoma* (Linnaeus 1767) (Gastropoda: Muricidae), *Perna perna* (Linnaeus 1758) (Bivalvia: Mytilidae) and *Pinna rudis* Linnaeus 1758 (Bivalvia: Pinnidae).

Some of these species, however, had a distribution in the colder areas of the Mediterranean Sea already in the past. *Charonia lampas* was cited by Bucquoy et al. (1882–1898) for the Roussillon and for the Adriatic Sea by Brusina (1866) and Coen (1937). *Luria lurida* was recorded at the end of the nineteenth century north to Livorno and in the Adriatic Sea (Monterosato 1897). *Stramonita haemastoma* is recorded from southern France by Bucquoy et al. (1882–1898).

Some other species could be more difficult to monitor because of their cryptic habitat, like *Pinna rudis*. Despite that this species was observed at unusually high latitudes in the Tyrrhenian Sea in recent times (Oliverio 1997) and may therefore give a signal of change, its habit to settle in crevices or at the bottom of walls on hard substrates and its usually low density makes it a difficult target for monitoring. Moreover, the recent records cited above in the Tyrrhenian basin reported the species for shallower waters than usual, a trend still to be interpreted in terms of thermophily. Indeed, in the northern parts of its range it also inhabits sublittoral bottoms below 50 m, rendering it an even more difficult target for sampling.

Patella ferruginea is an endangered species with a former wide distribution in the western Mediterranean Sea. It is considered a species in regression due to over-collecting and therefore of conservation concern (Scotti and Chemello 2000). The possibility to disentangle the effects of human predation from climatological drivers coupled with its rarity discourage its use as descriptor of change.

Among the species which may be useful in tracking changes there probably is *Erosaria spurca*. It is usually a shallow subtidal species, recorded from Sicilia, Spain and several northern African localities (Monterosato 1897). Some other records are scattered in the twentieth century literature in various locations in the Tyrrhenian Sea and it has been recently found by divers in Liguria (B. Sabelli, personal communication, December 2011). Revising its distribution records may help to better define whether this species had a range extension or not in the Mediterranean Sea in the last decades.

Dendropoma spp. represent a species complex comprising at least four cryptic species with allopatric distribution (Calvo et al. 2009). These gregarious species form dense aggregations cemented by crustose coralline algae (*Neogoniolithon brassicaflorida*) along the lower midlittoral fringe (from 0 m down to 6 m) on the rocky shores of the warmest areas of the Mediterranean Sea (Laborel 1987). They are considered indicators of sea-level and sea-surface temperature changes on a secular scale, recording changes in SST in terms of relative $\delta^{18}O$ changes of carbonate (Antonioli et al. 1999; Silenzi et al. 2004). This species complex has a mainly southern distribution in the Mediterranean Sea, but there is lacking information on whether it is extending its range in the last decades. This complex lacks a pelagic larval stage (Calvo et al. 2009), therefore any distributional response to warming may be slower than other species with pelagic larval stage. Indeed, the value of these species as SST descriptors hypothesized by Antonioli et al. (1999) is for phenomena in the order of magnitude of a thousand years: the oldest platforms built by *Dendropoma* are reported from the south-eastern Mediterranean Sea (6,000 years BP) while the platforms of southern Italy are younger (2,350 years BP) and this is interpreted as a northward migration related to the long term effect of the Holocene sea surface temperature warming.

Perna perna is a mussel with a wide geographic distribution outside the Mediterranean Sea ranging from the Atlantic coasts of South America to several places around Africa both on the Atlantic and Indian oceans. On the basis of a molecular phylogeny of the genus *Perna*, Wood et al. (2007) did not regard the species *P. picta* (Linnaeus 1758) from North Africa as a separate species but as a synonym of *P. perna*. In the Mediterranean Sea it is historically reported from southern Spain, Algeria, Malta (Monterosato 1878). Its first record for Sicilia is by Buccheri and Palisano (1976) on the basis of populations found a few kilometers east of Palermo. The first record for the eastern Mediterranean Sea is by Barash and Danin (1992) for Israel. Giannuzzi-Savelli et al. (2001) showed specimens from Calabria but did not specify whether they were from its Tyrrhenian or Ionian coasts. *Perna* species are broadcast spawners and have a pelagic larval stage of 2–4 weeks (Wood et al. 2007) and have therefore a great dispersal ability. In South Africa, competition with the locally alien *Mytilus galloprovincialis* Lamarck 1819 has been noted (Hanekom 2008) resulting in vertical habitat partitioning (Rius and McQuaid 2006; Zardi et al. 2006; Nicastro et al. 2010a, b). However, in the Mediterranean Sea the two species have been observed coexisting, equally sharing the available space (Zaouali 1973; Buccheri and Palisano 1976). The distribution of the species in the basin and its trend may deserve further research.

Back to *Echinolittorina punctata* and *Eastonia rugosa*

At present *Echinolittorina punctata* and *Eastonia rugosa* represent the two best documented cases of range extensions of autochthonous molluscs in the Mediterranean Sea. Despite that further species may be added soon if research continues on this topic, it is worthwhile to highlight the interesting features of these two species in particular.

Echinolittorina punctata is extremely easy to observe in its habitat, allowing non destructive sampling techniques and the easy evaluation of presence, population density, population dynamics etc. It is a species which is poorly present in the fossil record due to its hard substrate habitat. On the contrary, *Eastonia rugosa* is an infaunal bivalve and therefore is easily observable only when washed up on the beach. It may be sampled directly due to its shallow water habitat, but certainly with higher temporal and financial costs than *E. punctata*. However, its habitat favours fossilization and indeed there already are several records in palaeontological literature. Both species are macroscopic and can be easily identified even in the field. Moreover, their taxonomy is stabilized.

Some aspects of their distribution history still deserve understanding, but these two species and hopefully a few more in the near future may become a pool of species able to describe the changes happening in the biodiversity of the Mediterranean Sea.

Acknowledgements Massimo Appolloni, Museo Civico di Zoologia in Roma allowed access to the collections under his responsibility (Meli, Monterosato, Rigacci, Settepassi). The following curators helped in sending information on lots in the collections under their responsibility: Paul Callomon, Academy of Natural Sciences, Philadelphia, USA; Lindsey Groves, Natural History Museum of Los Angeles, USA; Frans Slieker, Natural History Museum Rotterdam, The Netherlands.

Mirco Travaglini and Stefania Brunetti, Library of the Department of Experimental Evolutionary Biology, University of Bologna kindly helped tracing hard to find literature.

A great number of people sent data about their findings to me and I wish to thank the whole shell collectors community for their support.

Marco Oliverio, Dipartimento di Biologia e Biotecnologie "Charles Darwin", La Sapienza University, Roma, Bruno Sabelli, Dipartimento di Biologia Evoluzionistica Sperimentale of the University of Bologna, and Constantine Mifsud, Malta, made valuable comments on a first draft of the manuscript.

Maps were drawn using the coastline data downloadable from the OpenStreetMap website (http://planet.openstreetmap.org/).

References

Albano PG (2006) Ritrovamenti presso la spiaggia di Palo Laziale (Roma). Not Soc Ital Malacol 24:21–22

Albano PG (2010) Further northward extension of the distribution of *Echinolittorina punctata* (Mollusca: Gastropoda) along the Italian Tyrrhenian coastline: new stations in Lazio. Cah Biol Mar 51:201–204

Albano PG, Trono D (2008) On the occurrence of *Echinolittorina punctata* (Gmelin, 1791) (Gastropoda: Littorinidae) in Puglia, South-Eastern Italy. Boll Malacol 44:123–126

Albano PG, Sabelli B, Adani M, Pinardi N (2010) The termophilous species *Echinolittorina punctata* as a new descriptor of tropicalization in the Mediterranean Sea – first data. Biol Mar Mediterr 17:90–93

Anonymous (1986) Elenco dei molluschi conchiferi del Lazio. La Conchiglia, 18 (Suppl): 1–16

Antit M, Gofas S, Azzouna A (2008) New records of upper shore Mollusca for the Tunisian coast: newcomers or overlooked? Mar Biodivers Rec 1:e99

Antonioli F, Chemello R, Improta S, Riggio S (1999) *Dendropoma* intertidal reef formations and their paleoclimatological significance, NW Sicily. Mar Geol 161:155–170

Aradas A, Benoit L (1870) Conchigliologia vivente marina della Sicilia e delle isole che la circondano. Stabilimento Tipografico C. Galatola, Catania

Barash A, Danin Z (1992) Annotated list of Mediterranean molluscs of Israel and Sinai. Fauna palaestina, Mollusca I. The Israel Academy of Sciences and Humanities, Jerusalem

Bianchi CN (2007) Biodiversity issues for the forthcoming tropical Mediterranean Sea. Hydrobiologia 580:7–21

Bianchi CN, Morri C (1993) Range extensions of warm-water species in the northern Mediterranean: evidence for climatic fluctuations? Porcup Newsl 5:156–159

Bianchi CN, Morri C (1994) Southern species in the Ligurian Sea (northern Mediterranean): new records and a review. Boll Mus Ist Biol Univ Genova 58–59:181–197

Bosch M, Moreno I (1983) First record of *Littorina punctata* (Gmelin, 1791) (Gastropoda: Prosobranchia) on the island of Mallorca (Spain). Veliger 25:401–403

Brugnone GA (1850) Un nuovo Turbo siciliano. L'Armonia, Giornale politico, scientifico e letterario per la Sicilia 1(14):4

Brunelli G (1928) Sulla natura biofisica della erosione foveolare delle arenarie nella costa Tirrena. Atti R Accad Lincei; Mem cl Sci Fis Mat Nat Ser VI 8:423–424

Brusina S (1866) Contribuzione pella fauna dei molluschi dalmati. Verh k-k Zool-bot Ges Wien 16:1–134

Buccheri G, Palisano G (1976) Nuovi dati sulla distribuzione geografica di *Perna (Perna) picta* (Born, 1780) e considerazioni sistematiche sulla specie. Conchiglie 12:143–156

Bucquoy E, Dollfus G-F, Dautzenberg P (1882–1898) Les mollusques marins du Rousillon. J.-B. Bailliere and Fils, Paris

Cachia C (1973) Further notes on marine shelled Mollusca. Malt Nat 1:30–32

Calvo M, Templado J, Oliverio M, Machordom A (2009) Hidden Mediterranean biodiversity: molecular evidence for a cryptic species complex within the reef building vermetid gastropod *Dendropoma petraeum* (Mollusca: Caenogastropoda). Biol J Linn Soc 96:898–912

Caruana AA (1867) Enumeratio ordinata molluscorum gaulo-melitensium. British Press, Malta

CIESM (2008) Climate warming and related changes in Mediterranean marine biota. CIESM workshop monographs 35. Monaco

Coen G (1937) Nuovo saggio di una Sylloge Molluscorum Adriaticorum. R Com Talassogr Ital, Mem CCXL:VI-173

Coll M, Piroddi C, Steenbeek J, Kaschner K, Ben Rais Lasram F et al (2010) The biodiversity of the Mediterranean Sea: estimates, patterns, and threats. PLoS One 5(8):e11842. doi:10.1371/journal.pone.0011842

Costa OG (1829) Osservazioni zoologiche intorno ai testacei dell'Isola di Pantelleria. Tip. della Minerva, Napoli

D'Anna G (1986) Studio sulla distribuzione di *Littorina punctata* (Gmelin) e *Littorina neritoides* (L.) nel Golfo di Castellammare (Sicilia Nord-Occidentale). Lav Soc Ital Malacol 22:39–50

D'Anna G (1997) Segnalazione di *Littorina punctata* (Gmelin, 1791) per Zambrone (Vibo Valentia). Not Soc Ital Malacol 15:20

D'Anna G (2001) Presenza di *Nodilittorina punctata* (Gmelin, 1791) sulle coste del basso Salentino. La Conchiglia 33(300):53–54

d'Oliveira MP (1896) Eastonia locardi, n. sp. Revista de Sciencias Naturaes e Sociaes 4:32–33

d'Orbigny AD (1842) Mollusques, Echinodermes, Foraminifères et Polypiers recueillis aux Iles Canaries par MM. Webb et Berthelot et décrits par Alcide d'Orbigny. Mollusques. Béthune, Paris

Dautzenberg P (1881) Liste de coquilles recueillies à Cannes par MM. E. and Ad. Dollfus. Feuille Jeunes Nat 11(129):117–121
Dautzenberg P (1883) Liste des coquilles du Golfe de Gabès. J Conchyliol 31:289–330
Dautzenberg P (1893) Description d'une nouvelle espèce du genre Littorina provenant des côtes de la Tunisie. J Conchyliol 41:35–36
ENEA Università di Parma, Società Italiana di Malacologia (1998) Censimento della malacofauna marina delle coste italiane. http://www.santateresa.enea.it/wwwste/malaco/home.htm. Accessed 30 Nov 2011
Enzenross L, Enzenross R (2001) Untersuchungen über das Vorkommen mariner Mollusken in tunesischen Gewässern. Schr Malakozool 17:45–62
Forbes E (1844) Report on the Mollusca and Radiata of the Aegean Sea, and on their distribution, considered as bearing on geology. Report of British Association for the Advancement of Science (1843), pp 130–193
Giannuzzi-Savelli R, Pusateri F, Palmeri A, Ebreo C, Coppini M, Margelli A, Bogi C (2001) Atlante delle conchiglie marine del Mediterraneo, vol VII. Edizioni Evolver, Roma
Gmelin JF (1791) Caroli a Linnaei Systema Naturae per Regna Tria Naturae, Editio Decima Tertia, Aucta Reformata. Tome 1, Pars 6 (Vermes). G.E. Deer, Leipzig
Grasso S (1981) Ritrovamento di molluschi rari o poco noti lungo i litorali della Sicilia meridionale. Boll Malacol 17:49–50
Gray JE (1837) A synoptical catalogue of the species of certain tribes or genera of shells contained in the collection of the British Museum and the author's cabinet. Magazine of Natural History NS 1:370–376
Hanekom N (2008) Invasion of an indigenous *Perna perna* mussel bed on the south coast of South Africa by an alien mussel *Mytilus galloprovincialis* and its effect on the associated fauna. Biol Invasions 10:233–244
Helbling GS (1779) Beyträge zur Kenntniß neuer und seltener Konchylien, aus einigen Wienerischen Sammlungen. Abhandlungen einer Privatgesellschaft in Böhmen, zur Aufnahme der Mathematik, der vaterländischen Geschichte, und der Naturgeschichte 4:102–131
Hidalgo JG (1867) Catalogue des mollusques testacés marins des cotes de l'Espagne et des iles Baléares – suite. J Conchyliol 15:357–426
Hidalgo JG (1870) Moluscos marinos de Espana, Portugal y las Baleares. Imprenta de Miguel Ginesta, Madrid
Hidalgo JG (1910) Catalogo de los moluscos testaceos marinos de la costa y bahia de Cadiz. Rev R Acad Cienc exactas, fis nat 9:776–803, 865–886, 946–978
Hidalgo JG (1917) Fauna Malacologica de España, Portugal y las Baleares. Trab Mus Nac Cienc Nat, Zool 30:1–752
La Valle P, Vani D, Liboni A, Smriglio C (2007) *Eastonia rugosa* (Helbling 1799) (Bivalvia, Mactridae) occurring along the Latium coasts. Basteria 71:93–95
Laborel J (1987) Marine biogenic constructions in the Mediterranean. Sci Rep Port-Cross Natl Park 13:97–126
Lamarck J-B (1819) Histoire naturelle des animaux sans vertèbres, vol 6(1). Paris.
Lamarck J-B (1822) Histoire naturelle des animaux sans vertèbres, vol 6(2), vol 7. Paris.
Lanfranco G (1973a) The periwinkles of Malta. Br Shell Collector's Club Newsl 8:2
Lanfranco G (1973b) Notes on some sedentary invertebrates in Maltese waters. Malt Nat 4:26–29
Lejeusne C, Chevaldonné P, Pergent-Martini C, Boudouresque CF, Pérez T (2009) Climate change effects on a miniature ocean: the highly diverse, highly impacted Mediterranean Sea. Trends Ecol Evol 25:250–260. doi:10.1016/j.tree.2009.10.009
Linnaeus C (1758) Systema Naturae per regna tria naturae, secundum classes, ordines, genera, species, cum characteribus, differentiis, synonymis, locis. Editio decima, reformata. Laurentius Salvius, Holmiae
Linnaeus C (1767) Systema naturae sive regna tria naturae, secundum classes, ordines, genera, species, cum characteribus, differentiis, synonymis, locis. Editio duodecima. Laurentii Salvii, Holmiae
Lipkin Y, Safriel U (1971) Intertidal zonation on rocky shores at Mikhmoret (Mediterranean, Israel). J Ecol 59(1):1–30
Locard A (1886) Prodrome de Malacologie Française. Catalogue Général des Mollusques vivants de France. H. Georg, Lyon/J.-B. baillière et Fils, Paris
Locard A (1891) Les coquilles marines des cotes de France. Description des Familles, genres et espèces. J.-B. Bailliere and Fils, Paris
McAndrew R (1851) Notes on the distributions and range of depth of Mollusca and other marine animals observed on the coasts of Spain, Portugal, Barbary, Malta and southern Italy in 1849. Report of British Association for the Advancement of Science 20, pp 264–304
McAndrew R (1854) On the geographical distribution of testaceous Mollusca in the north Atlantic and neighbouring seas. Proc Lit Philos Soc Liverp 8:8–56
McAndrew R (1857) Report on the Marine testaceous Mollusca of the North-east Atlantic and neighbouring seas, and the physical conditions affecting their development. Report of the British Association for the Advancement of Science 26, pp 101–158
Meli R (1896) Sulla *Eastonia rugosa* Chemn. (*Mactra*) ritrovata vivente e fossile nel littorale di Anzio e Nettuno. Bull Soc Malacol Ital 20:45–65
Micali P, Giovine F (1983) Elenco dei molluschi rinvenuti nello Stretto di Messina. Comune di Messina, Messina
Moazzo PG (1939) Mollusques testacés marins du canal de Suez. Mém Inst Egypte 38:1–283
Monterosato T (1872) Notizie intorno alle conchiglie mediterranee. Ufficio Tipografico di Michele Amenta, Palermo
Monterosato T (1875) Nuova rivista delle conchiglie mediterranee. Atti Acc Palermitana Sci Lett Arti, Sez II 5:1–50
Monterosato T (1877a) Note sur quelques coquilles provenant des cotes d'Algerie. J Conchyliol 25:24–29
Monterosato T (1877b) Notizie sulle conchiglie della rada di Civitavecchia. Ann Mus Civ Stor Nat Genova 9:407–428
Monterosato T (1878) Enumerazione e sinonimia delle conchiglie mediterranee. Giornale di Scienze Naturali ed Economiche 13:61–115
Monterosato T (1897) Sur les *Cypraea* de la Méditerranée. J Conchyliol 45(3):153–165
Monterosato T (1917a) Molluschi delle coste Cirenaiche raccolti dall'Ing. Camillo Crema. R Com Talassogr Ital 106:1–14
Monterosato T (1917b) Molluschi viventi e quaternari raccolti lungo le coste della Tripolitania dall'Ing. Camillo Crema. Boll Soc Zool Ital S III 4:1–28
Nevo E, Lavie B (1989) Selection of allozyme genotypes of two species of marine gastropods (genus *Littorina*) in experiments of environmental stress by nonionic detergent and crude oil-surfactant mixtures. Genet Sel Evol 21:295–302
Nicastro KR, Zardi GI, McQuaid CD (2010a) Differential reproductive investment, attachment strength and mortality of invasive and indigenous mussels across heterogeneous environments. Biol Invasions 12:2165–2177
Nicastro KR, Zardi GI, McQuaid CD, Stephens L, Radloff S, Blatch GL (2010b) The role of gaping behaviour in habitat partitioning between coexisting intertidal mussels. BMC Ecol 10:17
Nicolay K, Angioy M (1988) Tavole sinottiche di conchiologia mediterranea ed europea. Tav. XXXV: Fam. Littorinidae Gray, 1840. Parte I. La Conchiglia 20(226–227):28–29
Nordsieck F (1972) Marine Gastropoden aus der Shiqmona-Bucht in Israel. Arch Molluskenkd 102(4–6):227–245

Noy R, Lavie B, Nevo E (1987) The niche-width variation hypothesis revisited: genetic diversity in the marine gastropods *Littorina punctata* (Gmelin) and *L. neritoides* (L.). J Exp Mar Biol Ecol 109:109–116
Nuwayhid MA, Evans DL, Young LGL (1985) Defensive behaviour in littoral snails from Lebanon. Malacol Rev 18:9–14
Oliverio M (1997) Records of *Pinna rudis* from the Tuscan Archipelago. Boll Malacol 32:151–152
Olivi G (1792) Zoologia Adriatica ossia Catalogo ragionata degli Animali del Golfo e delle Laguna di Venezia; preceduto da una Dissertazione sulla Storia fiscia e naturale del Golfo; e accompagnato da Memorie, ed Osservazioni di Fiscia Storia naturale ed Economia. Bassano
Orlando VE (1979) Presenza nella Sicilia sud-orientale di *Eastonia rugosa* (Helbling, 1779) (Mollusca Bivalvia). Nat Sicil, S IV 3:39–44
Ozturk B, Buzzurro G, Avni Benli H (2004) Marine molluscs from Cyprus: new data and checklist. Boll Malacol 39:49–78
Palant B, Fishelson L (1968) *Littorina punctata* (Gmelin) and *Littorina neritoides* (L.), (Mollusca, Gastropoda) from Israel: ecology and annual cycle of genital system. Isr J Zool 17:145–160
Pallary P (1900) Coquilles marines du littoral du Départment d'Oran. J Conchyliol 48:211–422
Pallary P (1904) Addition à la faune malacologique du Golfe de Gabès. J Conchyliol 52:212–248
Pallary P (1906) Addition à la faune malacologique du Golfe de Gabès. 3e article. J Conchyliol 54:77–124
Pallary P (1912a) Catalogue des mollusques du littoral méditerranéen de l'Egypte. Mém Inst Egyptien 7:69–200
Pallary P (1912b) Liste des mollusques marins des cotes de la Syrie. Feuille des Jeunes Naturalistes Ser 5 42:171–174
Pallary P (1914) Liste des mollusques du Golfe de Tunis. Bull Soc Hist Nat Afr Nord 6:12–27
Pallary P (1919) Enumeration des mollusques marins des cotes de la Syrie. Bull Soc Hist Nat Afr Nord 10:166–176
Pallary P (1920) Exploration scientifique du Maroc organisée par la Société de Géographie de Paris et continuée par la Société des Sciences Naturelles du Maroc, Deuxième fascicule. Malacologie. Larose, Rabat/Paris
Patané L (1946) Sulla biologia di *Littorina punctata* (Gm.). Boll Soc Ital Biol Sper 22:928–929
Payraudeau BC (1826) Catalogue descriptif et méthodique des annelides et des mollusques de l'Ile de Corse. Imprimerie J. Tatsu, Paris
Philippi RA (1836) Enumeratio molluscorum Siciliae cum viventium tum in tellure tertiaria fossilium quae in itinere suo observavit. Berolini
Philippi RA (1844) Enumeratio molluscorum Siciliae cum viventium tum in tellure tertiaria fossilium quae in itinere suo observavit. Volumen secundum. Halis Saxonum
Philippi RA (1847) Abbildungen und beschreibungen neuer oder wenig gekannter conchylien, unter mithülfe mehrer deutscher conchyliologen. Vol 2. T. Fischer Cassel
Praus Franceschini C (1906) Elenco delle conchiglie del Golfo di Napoli e del Mediterraneo esistenti nel Museo Zoologico di Napoli. Annuario Mus Zool R Univ Napoli NS 2:1–68
Pyšek P, Hulme PE, Nentwig W (2009) Glossary of the main technical terms used in the handbook. In: DAISIE handbook of alien species in Europe. Springer, Berlin/Heidelberg
Reid DG (2011) The genus *Echinolittorina* Habe, 1956 (Gastropoda: Littorinidae) in the eastern Atlantic Ocean and Mediterranean Sea. Zootaxa 2974:1–65
Retzius AJ (1788) Venus lithophaga descripta. Memoires de l'Academie Royale des Sciences (Turin), 11–14
Rius M, McQuaid CD (2006) Wave action and competitive interaction between the invasive mussel *Mytilus galloprovincialis* and the indigenous *Perna perna* in South Africa. Mar Biol 150:69–78
Rosewater J, Kadolsky D (1981) Rectifications in the nomenclature of some Indo-Pacific Littorinidae II. Proc Biol Soc Wash 94:1233–1236
Scotti G, Chemello R (2000) I molluschi marini mediterranei degni di protezione: stato delle conoscenze e forme di tutela. Boll Malacol 36:61–70
Silenzi S, Antonioli F, Chemello R (2004) A new marker for sea surface temperature trend during the last centuries in the temperate areas: vermetid reef. Glob Planet Change 40:105–114
Soppelsa O, Crocetta F, Fasulo G (2004) Prime segnalazioni di *Nodilittorina punctata* (Gmelin, 1791) per il golfo di Salerno, il golfo di Napoli ed i litorali laziali (Tirreno centro-meridionale). Boll Soc Nat Napoli NS 2:41–44
Struhsaker JW, Costlow JD (1968) Larval development of *Littorina picta* (Prosobranchia, Mesogastropoda), reared in the laboratory. Proc Malacol Soc Lond 38:153–160
Terreni G (1981) Molluschi conchiferi del mare antistante la costa toscana (Gastropoda, Scaphopoda, Amphineura, Bivalvia, Cephalopoda). Tip. Benvenuti and Cavaciocchi, Livorno
Terreni G (1983) Molluschi conchiferi del mare antistante la costa toscana (Gastropoda, Scaphopoda, Amphineura, Bivalvia, Cephalopoda). Emendatio ed Addenda. Quad Mus St Nat Livorno 4:107–125
Tomlin JRB (1927) Cambridge Expedition to the Suez Canal, 1924. Report on the Mollusca (Amphineura, Gastropoda, Scaphopoda, Pelecypoda). Transactions of the Zoological Society of London 22, pp 291–320
UNEP-MAP-RAC/SPA (2008) Impact of climate change on biodiversity in the Mediterranean Sea. RAC/SPA Edit, Tunis
Weinkauff HC (1862) Catalogue des coquilles marines recueillies sur les cotes de l'Algerie. J Conchyliol 10:301–371
Weinkauff HC (1868) Die Conchylien des Mittelmeeres, ihre geographische und geologische Verbreitung. Vol 2: Mollusca Cephala. Verlag von Theodor Fischer Cassel
Weinkauff HC (1883) Catalog der gattung Litorina Férussac. Jahrb Dtsch Malakozool Ges nebst nachrichtsblatt 10:213–227
Williams ST, Reid DG (2004) Speciation and diversity on tropical rocky shores: a global phylogeny of snails of the genus *Echinolittorina*. Evolution 58:2227–2251
Wood AR, Apte S, MacAvoy ES, Gardner JPA (2007) A molecular phylogeny of the marine mussel genus *Perna* (Bivalvia: Mytilidae) based on nuclear (ITS1and2) and mitochondrial (COI) DNA sequences. Mol Phylogenet Evol 44:685–698
Zaouali J (1971) Note sur la presence de *Standella* (*Eastonia*) *rugosa* Gmelin (Eulamellibranchia, Mactridae) dans le Golfe de Tunis. Bull Inst Oceanogr Peche Salammbô 2:95–97
Zaouali J (1973) Note sur la presence de *Perna perna* L. (=*Mytilus africanus* Chemnitz) dans la région de Bizerte (Tunisie). Bull Inst Océanogr Péche Salammbo 2:637–642
Zaouali J (2008) Méditerranée: mixité ethnique et coexistence pacifique? Cas du golfe de Tunis. In: Briand F. (ed) Climate warming and related changes in Mediterranean marine biota. CIESM workshop monographs 35, pp 89–98
Zardi GI, Nicastro KR, McQuaid CD, Rius M, Porri F (2006) Hydrodynamic stress and habitat partitioning between indigenous (*Perna perna*) and invasive (*Mytilus galloprovincialis*) mussels: constraints of an evolutionary strategy. Mar Biol 150: 79–88

18 Advances in Predicting the Impacts of Global Warming on the Mussels *Mytilus galloprovincialis* in the Mediterranean Sea

Basile Michaelidis, Hans-O. Pörtner, Inna Sokolova, and Lars Tomanek

Abstract

Mussels *Mytilus galloprovincialis* have been used as model bivalves to study the impacts of global warming on their physiological performance in Thermaikos Gulf, North Greece. The studies have been conducted under laboratory and field conditions for more than 6 years and focused on the biochemical, metabolic, physiological and energetic responses of *M. galloprovincialis* to increases in the ambient temperature. Here we summarize the findings concerning the responses of *mussels* to environmental temperature, present an integrated model of their physiological performance during thermal stress and discuss these findings in the light of the predicted temperature changes in the Thermaikos Gulf from the regional climate trends and the mean global temperature projections for the period 1990–2100 based on IS92 emission scenarios of the Intergovernmental Panel for Climate Change (IPCC). Our findings indicate that mussels in Thermaikos Gulf currently face the temperatures close to their upper thermal limits, especially during the summer, and thus are likely vulnerably to any further increase in the temperature such as expected during the global warming.

Keywords

Acclimation at different temperatures • Acute heat and chronic cold stress • Antioxidant response • Biochemical mechanisms • Climate changes • Critical temperatures • Depletion of energy reserves • Energetics of *Mytilus galloprovincialis* • Energy allocation to reproductive organs • Feeding ingestion and absorption rates • Food availability, Global warming • Heat stress in intertidal organisms • Hypoxia • Intergovernmental Panel for Climate Change • Lethal temperature • Mediterranean ecosystems • Membrane damage • Multiple stressors • Mussel farming • Mussel populations • Mussels' body temperature • *Mytilus galloprovincialis* • Ocean acidification • Ocean circulation patterns • Pathogens and harmful algal blooms • Physiology • Pollution • Protein synthesis for stress defense • Proteomic • Sea level rise • Survival of mussels • Temperature extremes • Temperature-dependent growth • Thermal biology • Thermal intolerance • Thermal limits • Thermal stress • Threshold temperatures • Up-regulation of oxidative stress proteins

B. Michaelidis (✉)
Laboratory of Animal Physiology, Department of Zoology, School of Biology, Faculty of Science, Aristotle University of Thessaloniki, Thessaloniki 54124, Greece
e-mail: michaeli@bio.auth.gr

H.-O. Pörtner
Integrative Ökophysiologie, Alfred-Wegener-Institut für Polar- und Meeresforschung, Postfach 120161, D-27515 Bremerhaven, Germany

I. Sokolova
Department of Biology, University of North Carolina at Charlotte, Charlotte, NC 28223-0001, USA

L. Tomanek
Department of Biological Sciences, Center for Coastal Marine Sciences and Environmental Proteomics Laboratory, California Polytechnic State University, San Luis Obispo, CA 93407-0401, USA

S. Goffredo and Z. Dubinsky (eds.), *The Mediterranean Sea: Its history and present challenges*, DOI 10.1007/978-94-007-6704-1_18,

Introduction

Increasing greenhouse gas concentrations are expected to lead to substantial atmospheric warming during this century. Mean global temperature projections for the period 1990–2100 based on IS92 emission scenarios of the Intergovernmental Panel for Climate Change (IPCC) range from an increase by 1.5 °C above pre-industrial, via a most likely change by 2.5 °C to a high degree of warming by 4.5 °C (IPCC 2007). As a consequence of the atmospheric warming, the oceans are expected to warm which may have serious impacts on marine ecosystems. Coastal marine systems are among the most ecologically and socio-economically vital on the planet and there is a strong scientific consensus that these marine ecosystems, along with the resources and services they provide, are threatened by anthropogenic climate change (Harley et al. 2006). Based on the IPCC assessment (2007), the following abiotic and biotic elements of climate change are likely to impact marine organisms:

1. Global warming: There is agreement that our planet (including the oceans) will be heated by at least 1.1 °C during this century and that the increase can reach up to 3° C. In the intertidal and coastal zones, the degree of local warming and the amplitude of thermal fluctuations may even exceed this global predicted increase (Helmuth et al. 2002, 2006).
2. Ocean acidification occurs due to the uptake of anthropogenic CO_2 into the oceans and is considered one of the strongest potential threats to marine organisms within the next decades (Gattuso and Hansson 2011).
3. Sea level rise will profoundly influence river deltas, increase saline water intrusion and bring about major biotic changes. The IPCC (2007) has estimated that oceans will rise by 0.1–1 m over this century, while more recent estimates project an increase of 1.5–2 m (IPCC 2007).
4. Changes in ocean circulation patterns driven by climate change will affect ocean productivity, impact phyto- and zooplankton biomasses and alter food webs (Richardson 2008; Gambaiani et al. 2009).
5. Outbreaks of infectious diseases and parasitism are expected to become more frequent (Harley et al. 2006).

Climate change science has developed rapidly over the last few decades as an effort by the scientific community to understand not only how natural and managed marine ecosystems have responded historically to changes in climate, but also to develop methods to measure and predict ongoing and future impacts of global climate change (Parmesan 2006; Helmuth et al. 2006). However, the responses of marine ecosystems to the combined effects of climate change and anthropogenic activities remain poorly understood. Moreover, one important sector of the coastal economy and ecology – aquaculture – has received little attention in the context of climate change research.

Changes in environmental variables have both direct and indirect impacts on marine organisms including those which are exploited commercially. Direct effects of environmental variables including climate change involve changes in behaviour and physiology, including growth, reproductive capacity and mortality. Indirect effects alter the productivity, structure and composition of the marine system on which these organisms depend for food and shelter. These changes can create physiological (growth, development, reproduction, disease) and ecological (organic and inorganic nutrient cycles, predation and competition by invasive species) effects on cultured organisms and require operational adjustments in aquaculture (such as species selection, site selection and grow-out technologies).

Understanding and predicting the potential effects of global warming on marine organisms are complicated by two major issues: (1) the interactive effects of multiple stressors that simultaneously affect marine organisms in their habitats; (2) local spatio-temporal variability in the environment superimposed on the changes on larger global and regional scales. These aspects must be taken into account when developing explanatory and/or predictive models for the future climate-driven changes in marine ecosystems. The effects of multiple stressors may arise in one of two principle ways: (1) the impact of one factor is either strengthened or weakened by variation in another factor and (2) the combined influence of two stressors pushes an individual or population beyond a critical threshold that would not be reached via variation in a single stressor. The situation may be further complicated in the field because changes in species physiological performance and fitness occur not only as the result of "acute" extremes of a stressor, which kill organisms over periods of days and weeks, but also as the result of slower, more chronic, cumulative stressors (of longer duration, but typically smaller magnitude) that reduce growth, health, and reproduction. The detection of changes in natural ecosystems may thus require long-term monitoring of population sizes, reproduction and growth. Furthermore, identifying and predicting marine ecosystem response to climate change relies on an understanding of the physiological mechanisms that drive these changes (Helmuth et al. 2006). Therefore, both long term sublethal ("chronic") and short duration ("acute") stressors contribute to ecological responses to climate change.

Climate changes are likely to be spatially and temporally heterogeneous, as will be the subsequent impact on organisms and ecosystems (Helmuth and Hofmann 2001). Therefore, microhabitats are important in governing physiological responses and performance, and are also influenced by many larger scale factors, including the hydrological status of the marine systems and the influence of pollutants, winds, currents (Helmuth et al. 2006). Assessing and forecasting the direct and indirect effects of climate change (i.e. ecological forecasting) mandates that we understand the roles of

(a) habitat heterogeneity in driving local micro-climate, (b) the effect of the organism's characteristics (morphology, colour, etc.) on modifying how micro-climatic conditions affect their physiology; (c) the impacts of those signals on physiological responses (and ultimately fitness); (d) the capacity for organisms to acclimate to those signals at varying temporal scales; (e) the indirect effects of physiological responses on interspecific interactions and (f) the role of dispersal in maintaining connectivity between organisms living at nested spatial scales (Helmuth et al. 2010). Studies of physiological responses (from molecular to whole organism levels) in response to the changes in the physical factors under the laboratory and field conditions are critical for ecological forecasting because they provide us with the knowledge of how environmental signals are translated into physiological performance and fitness of marine organisms (Helmuth 2009; Helmuth et al. 2010).

Physiological traits and underlying biochemical mechanisms are important in setting stress tolerance limits of a species and can be directly linked to the phenomena observed in the field such as temperature-induced changes in species abundance (Pörtner 2002a, b; Pörtner and Knust 2007; Somero 2010, 2012). Moreover, physiological approaches combined with the new "-omics" technologies (genomics, transcriptomics, and proteomics) may enable diagnoses of the "state of health" of natural populations exposed to changes in abiotic conditions and yield insights into the nature of sublethal stress at the cellular level (Tomanek 2011; Somero 2012). Consequently, physiology is central to understanding past and current responses and to predicting future responses of organisms to global climate change (Helmuth et al. 2010; Pörtner 2010; Somero 2010). Within the biological sciences, communication among physiologists, geneticists, population biologists and community ecologists will help provide a more comprehensive picture of biological change. Climatologists and oceanographers can refine our understanding of where and how climate change will impact coastal systems and help integrate information on physiological function with knowledge of temporal and spatial patterns in the physical environment (Pörtner 2002a; Somero 2005).

Oxygen- and Capacity-Limited Thermal Tolerance (OCLTT): An Integrative Model to Explain Reduction of Physiological Performance During Thermal Stress

Physiological ecologists use performance curves (Pörtner 2002a, b, 2010) to define the complex relationships between an organism's responses (such as growth, reproduction, and survival) and environmental factors (e.g. temperature, CO_2, hypoxia, currents, wind). Performance curves describe both the conditions under which an organism can function and its physiological limits. Importantly, performance curves are species-specific and can change over the course of an organism's lifetime as a function of developmental stage, size, age, exposure to stress, and field acclimatization. As applied to thermal tolerance, performance curves assist with integration and analysis of physiological responses to environmental temperature and of underlying molecular and biochemical mechanisms that are important in setting thermal limits and species boundaries. Recent studies in marine fish and invertebrates have shown that the limits of the thermal tolerance window are characterized by the onset of internal (systemic) hypoxia (hypoxemia) despite fully oxygenated waters, leading to anaerobic metabolism at both cold and warm temperature extremes. These observations contributed to the development of the concept of oxygen- and capacity-limited thermal tolerance (OCLTT) in animals (Pörtner 2001, 2002a, b; Pörtner et al. 2004, 2005a, b). This concept suggests that first level of thermal intolerance at low and high temperature extremes in animals is a reduction in whole organism aerobic scope at low and high borders of the thermal envelope, so-called pejus thresholds (*Tp*). This reduction is caused by a limited capacity of the oxygen supply mechanisms (ventilation, circulation) to cover an animal's temperature-dependent oxygen demand, resulting in hypoxemia which develops beyond the pejus temperature, *Tp*. With continued cooling or warming, aerobic scope finally disappears at low or high critical threshold temperatures (*Tc*), where transition to anaerobic mitochondrial metabolism and progressive insufficiency of cellular ATP provision occurs. Beyond *Tc*, onset of anaerobic metabolism also co-occurs with the cellular heat shock response (HSR) (Pörtner 2010). Recent studies reported a close relationship between the onset of anaerobiosis and the HSR in bivalves (Gracey et al. 2008; Anestis et al. 2010a, b). In *Modiolus barbatus* a shift to anaerobic metabolism and the HSR is observed after acclimation to temperatures beyond 26° C (Anestis et al. 2008). A possible link between anaerobiosis and the onset of the HSR is provided through the production of reactive oxygen species (ROS) during thermal stress (Heise et al. 2003, 2006) that can denature proteins, especially the cytoskeleton, and trigger the production of heat shock proteins (Tomanek and Zuzow 2010).

The OCLTT concept, as a general principle, explains how aerobic scope is limited by insufficient oxygen supply at both sides of the thermal window (in temperate and tropical ectotherms) and sets the performance window in animals, with an optimum close to the upper pejus temperature (e.g. Pörtner and Knust 2007; Pörtner and Farrell 2008). The resulting temperature-dependent performance curve in animals provides a mechanistic explanation for the thermal performance curve used traditionally in evolutionary analyses of thermal biology (review in: Angiletta 2009). Moreover, the OCLTT concept allows understanding the physiological mechanisms that shape and limit performance. It successfully integrates mechanisms at whole-organism and tissue levels with the underlying biochemical and genomic mechanisms. In addition, it provides a matrix for the integration of the impacts of various stressors (including food limitation, ocean acidification, pollution and others) on temperature dependent performance (Pörtner 2010). Although the conceptual model of OCLTT indicates that thermal limitation sets in at high organisational levels (i.e. the level of oxygen supply), biochemical stress mechanisms mediate the consequences of oxygen limitation at the cellular level (Pörtner 2002a, b; Somero 2012). Importantly, once sufficient oxygen delivery resumes after return to non-stressful conditions, the costs of cellular damage repair and establishment of cellular homeostasis determine the physiological costs that diminish organism's performance.

An important implication of the OCLTT model is the prediction that multiple environmental stressors will result in reduced aerobic scope and thereby narrow the window of thermal tolerance of marine organisms, depending on the species' overall tolerance range. Among "natural" stressors expected during the climate change, elevated CO_2 concentrations (ocean acidification) and hypoxia cause a narrowing of thermal windows as seen in crustaceans (Melzner et al. 2007; Pörtner and Farrell 2008; Walther et al. 2009). Elevated CO_2 levels also affected the metabolic response to temperature increase in a bivalve (eastern oysters *Crassostrea virginica*) leading to elevated basal energy demand (Beniash et al. 2010; Lannig et al. 2010), and thereby, increasing sensitivity to temperature stress. Synergistic effects of metal pollution and elevated temperature also lead to reduced aerobic scope. Increased sensitivity to temperature stress in organisms exposed to metals (and *vice versa)* has been demonstrated in marine invertebrates including bivalves (Cherkasov et al. 2006; Lannig et al. 2006; Sokolova and Lannig 2008; Ivanina et al. 2009). It is likely that other environmental stressors (e.g. toxins, UV radiation or disease) that increase metabolic costs for cellular homeostasis, detoxification and damage repair will have similar effects sensitizing marine organisms, including mussels, to thermal stress. Further investigations are needed to determine the synergistic effects of multiple stressors on aerobic scope and the thermal tolerance window of *M. galloprovincialis*, especially in the basins with high anthropogenic pressure and variable environmental conditions such as the Thermaikos Gulf and the Mediterranean Sea in general.

Subtidal and Intertidal Mussels as Models for Studying the Effects of Global Warming

During the last decades the effects of climate change have been intensively investigated in intertidal ecosystems (Somero 2002; Harley et al. 2006; Helmuth et al. 2006, 2010; Helmuth 2009). Intertidal organisms including mussels inhabit an interface between aquatic and terrestrial habitats where they are exposed to extreme physical conditions during low tides (Newell 1979). They experience body temperatures that exceed the temperature of the surrounding air and regularly approach sublethal thermal limits (Helmuth 1999; Helmuth and Hofmann 2001; Tomanek and Sanford 2003). Slow or immobile organisms such as mussels that reside higher in the intertidal are more likely to experience prolonged thermal and desiccation stresses and greater temperature extremes than are organisms lower in the intertidal, and are unable to migrate to avoid these stresses (Hofmann and Somero 1995; Roberts et al. 1997; Halpin et al. 2002).

The extreme variability and vertical gradient of environmental parameters including temperature in the intertidal zone provides an opportunity to test relevant hypotheses by comparing intertidal and subtidal (sub)populations of the same species. In subtidal mussels, the climate signal is mediated through temperature and physicochemistry of the water. This contrasts the situation in intertidal mussels subjected to air exposure, oxygen deficiency, dehydration and more extreme temperatures. Therefore, the climate signal, its relevant components and its effects should be clearer and the signal-to-noise ratio higher in the subtidal animals than in the intertidal ones. The mussel, *Mytilus galloprovincialis* is an ideal model species for such comparisons (Anestis et al. 2007; Ioannou et al. 2009).

There are several important parameters commonly used to estimate the impacts of thermal stress on the mussels' physiological performance. At the whole-organism level, body temperature is important for understanding and interpreting the physiological responses of mussels in the field because it can strongly vary among individuals in different microhabitats and significantly deviate from air or water temperature. Body temperature can be measured by temperature probes inserted between the valves of live mussels or using biomimetic sensors ('*robomussels*'). "Robomussels" are temperature data loggers that mimic the thermal characteristics of an individual mussel and can record temperature data at 10-min intervals for up to 7 months (Helmuth et al. 2010). Integrated physiological indicators such as scope for growth (SFG) and condition index (CI) can be used to determine how thermal stress affects the ability of mussels to use the available energy from the ingested food for somatic or gonadal growth and basal maintenance (including health) and assess the temperature-induced energy stress and its consequences for the organism's fitness (Anestis et al. 2010b). Physiology-based ecological models such as the dynamic energy budget (DEB) model can then be used to describe and predict population growth depending on the rates at which the organism assimilates and utilises energy for maintenance, growth and reproduction depending on the state of the organism and of its environment (Ren and Ross 2005; Van der Veer et al. 2009).

At the cellular level, a common metric used to assess the response to thermal stress in mussels is the expression of heat shock proteins (e.g., Hsp70) (Hofmann 2005; Tomanek 2008, 2010). The differential expression of inducible Hsp70 proteins is positively correlated with seasonal temperature changes and with the vertical position in the intertidal and subtidal zones (Tomanek and Somero 1999; Somero 2002; Anestis et al. 2007, 2008; Ioannou et al. 2009; Tomanek 2010; Somero 2012). Biochemical markers, such as RNA/DNA ratios, metabolic markers including enzymatic activities and metabolites can also be used to assess protein synthesis and metabolic profiles in response to thermal stress (Dahlhoff 2004). Oxidative stress is a common hallmark of temperature-induced physiological stress (Abele et al. 2002; Heise et al. 2003, 2006) and can be assessed using the activities of key antioxidant enzymes, non-enzymatic antioxidants, and biomarkers of oxidative damage to DNA, proteins and lipids (Lesser and Kruse 2004; Tomanek and Zuzow 2010; Tomanek 2011; Fields et al. 2012; Somero 2012). Recently, new techniques in genomics, transcriptomics and metabolomics (Gracey and Cossins 2003; Hofmann and Place 2007; Hofmann and Gaines 2008; Tomanek and Zuzow 2010; Somero 2012) that allow an integral view of the respective cellular and physiological phenotypes have opened new doors for measuring the responses of mussels to their physical environment in both the laboratory and field.

Climate Change and Ecological Pressures in the Thermaikos Gulf-Mediterranean Sea

The Mediterranean Sea has been identified as a "*hotspot*" for climate change highly vulnerable to its effects (Jeftic et al. 1992; Nicholls and Hoozemans 1996; Gambaiani et al. 2009; Calvo et al. 2011; Durrieu de Madron et al. 2011). In the Mediterranean Sea, the global climate change is expected to alter marine biodiversity and productivity, trigger changes in trophic interactions, cause toxic algal blooms and propagation of thermophilic species (Gambaiani et al. 2009). Regionally, the Thermaikos Gulf in South Greece is considered one of the most vulnerable systems with the greatest sensitivity and the lowest adaptability to climate change effects. The Thermaikos Gulf is a part of a protected estuarine area (under the Ramsar convention) but it supports active environmentally-friendly mussel farming compatible with the protected status of the area. It is the largest source of farmed mussels (*M. galloprovincialis*) in Greece, representing about the 88 % of the country's cultivated sea area and 80–85 % of the total national production. Mussel farming has been in this area for more than 25 years providing employment to a significant proportion of the local population and yielding the mussel production in excess of the 12,000 t/year. Mussel populations in the Thermaikos Gulf are subtidal and remain submersed throughout the year.

Analysis of meteorological data since 1950 indicates a tendency of air temperatures to increase in the Thermaikos Gulf and the Mediterranean Sea (Vargas-Yánẽz et al. 2008, 2010). In an effort to predict how climate change might influence the mussels' population in Thermaikos Gulf, our research during the last 6 years has focused on the physiological performance and molecular, biochemical and metabolic stress responses of mussels *M. galloprovincialis* to increasing ambient temperature. Here we present and discuss our findings and provide an overview of the model of physiological performance of *M. galloprovincialis* during thermal stress that emerged from the recent laboratory and field studies.

Temperature: A Key Factor Shaping Performance of *M. galloprovincialis*

Linking Whole Organism to Cellular Thermal Limits Via Heat Shock, Metabolic and Behavioural Responses

Heat Shock Response (HSR) of *M. galloprovincialis* to Thermal Stress

Heat shock proteins (Hsps) are highly evolutionarily conserved molecular chaperones that refold proteins denatured from a variety of insults including thermal stress (Feder and Hofmann 1999). Hsps prevent the aggregation of heat-damaged proteins and facilitate their renaturation following a heat shock, thus playing an important role in thermotolerance (Parsell et al. 1993; Parsell and Lindquist 1993). From an ecological perspective the threshold temperatures inducing expression of Hsps may determine extreme thermal limits of an organism. It has been proposed that this threshold is related to the species distribution boundaries in an ecosystem (Somero 2002; Hofmann 2005; Tomanek 2008, 2010).

The minimum heat-shock induction temperature (T_{on}), the temperature of the maximal response (T_{max}), and the shut-off temperature (T_{off}) appear to be species specific, and reflect the evolutionary history of a species (Dietz and Somero 1992; Tomanek and Somero 1999). The temperatures inducing HSR also exhibit considerable plasticity as a function of developmental stage, acclimation temperature, and season (Dietz and Somero 1992; Hofmann and Somero 1995; Chapple et al. 1998; Roberts et al. 1997; Tomanek and Somero 1999; Buckley and Hofmann 2001). Moreover, microclimate and physical conditions in habitats are of great importance in influencing the HSR and thermotolerance of marine organisms (Somero 2002. Helmuth and Hofmann 2001; Tomanek 2008, 2010). The mechanisms setting the threshold temperatures remain to be explored.

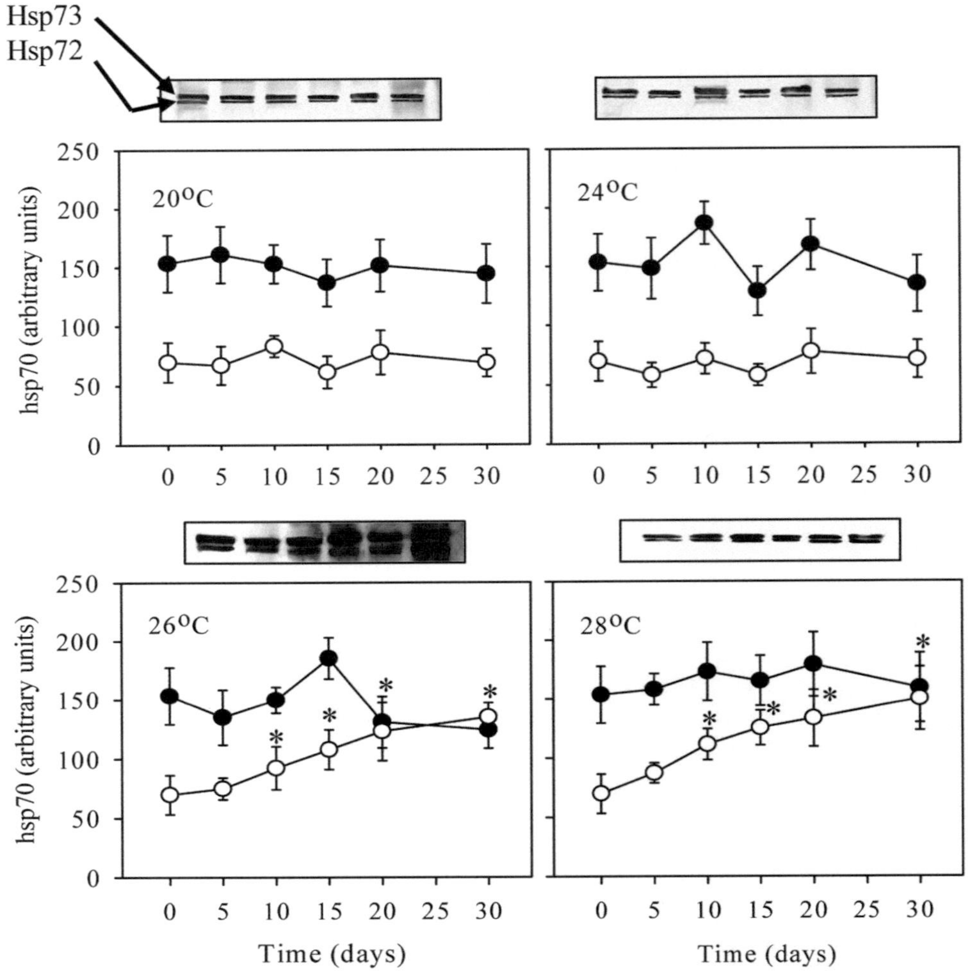

Fig. 18.1 Levels of Hsp70 in the mantle of submersed *Mytilus galloprovincialis* during acclimation to different water temperatures. Tissue extracts were subjected to SDS-PAGE and immunoblotted for Hsp70. Representative immunoblots are shown for each acclimation temperature. Blots were quantified by laser-scanning densitometry. Values are means ± SE, n = 5 preparations from different animals. Inducible isoform Hsp72, *open circles*; constitutive isoform Hsp73, *closed circles*. *P < 0.05 compared with the control (0 days) (Anestis et al. 2007). The results show a gradual increase in the levels of inducible form Hsp72 especially after acclimation of *Mytilus galloprovincialis* at 26° C

Our recent studies focused on the threshold temperatures of HSR induction as a marker of the upper thermal tolerance limit in *M. galloprovincialis*. Threshold temperatures (T_{on}) inducing expression of Hsp70 and Hsp90 in the mantle and posterior adductor muscle (PAM) were determined during acclimation of the mussels to 20, 24, 26, or 28 °C for 30 days. Two Hsp70 isoforms were found in the mussel tissues at all temperatures: an inducible Hsp72 that showed a gradual increase during 30 days of acclimation at temperatures beyond 26 °C (Fig. 18.1), and a constitutive Hsp73 that did not vary in response to the temperature. Hsp90 expression showed a similar thermal response profile to that of Hsp72. Analysis of the proteomic changes in gill tissue during temperature acclimation in *M. galloprovincialis* and one of its more heat-sensitive congeners, *M. trossulus*, showed that both species adjusts several metabolic enzymes (such as ubiquinol cytochrome c reductase), and cytoskeletal proteins (including tubulin) during laboratory acclimation to 19° C. However, *M. galloprovincialis* shows up-regulation of oxidative stress proteins and molecular chaperones during acclimation to 13 and 7 °C, possibly due to adjustments in membrane composition during cold exposure (Fields et al. 2012). Proteomic changes to acute heat stress show similar changes in oxidative stress proteins and molecular chaperones in *M. galloprovincialis* (Tomanek and Zuzow 2010).

A T_{on} of 26 °C for HSR in *M. galloprovincialis* determined in laboratory studies (Anestis et al. 2007) is also confirmed by the field surveys of Hsp70 and Hsp90 in the mantle and posterior adductor muscle (PAM) tissues of mussels between February and August 2008. In a subtidal population of mussels, HSR occurs in two peaks, the first from February to mid April and the second between mid July and August, in parallel to seasonal temperature extremes (Fig. 18.2). Specifically, the expression of the inducible isoform Hsp72 and Hsp90 showed a minor peak in April and a dramatic increase in summer months (June–August) (Fig. 18.2). The summer peak of Hsp72 and Hsp90 expression coincided with the increase in the water temperature (with rapid warming starting in June and reaching 28–29 °C between mid July and August) suggesting that ambient temperature is the major abiotic factor triggering the expression of Hsps during summer time. Similarly, cold extreme temperatures were interpreted to cause HSP expression in February to April, which is rather correlated with the reproductive activity of mussels (Ioannou et al. 2009).

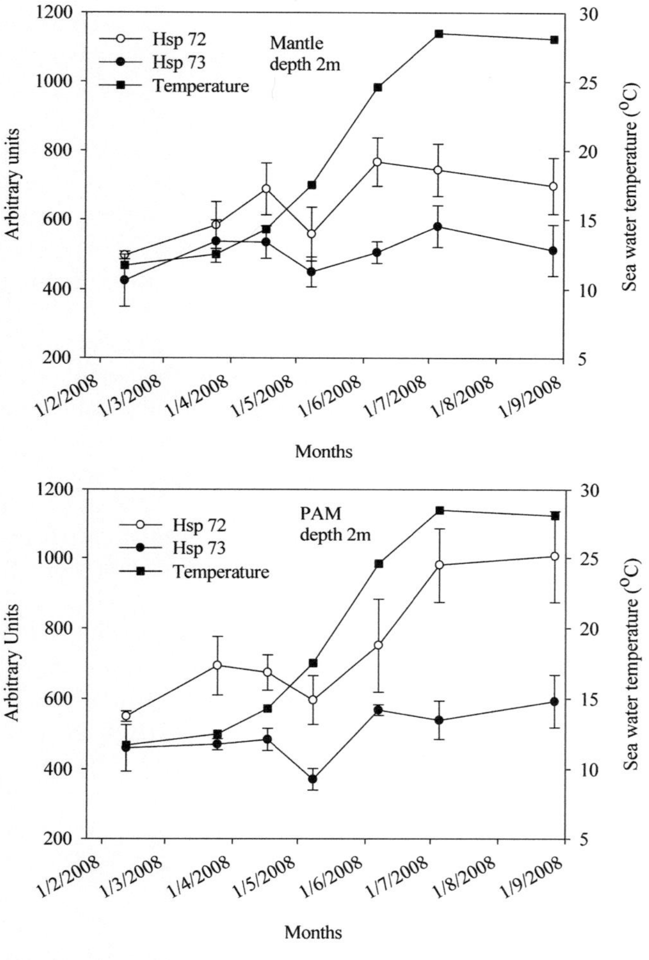

Fig. 18.2 Seasonal changes in the levels of Hsp70 (inducible isoform Hsp72, *open circles*; constitutive isoform Hsp73, *closed circles*) in the mantle and posterior adductor muscle (PAM) of submersed *Mytilus galloprovincialis* during acclimatization at Thermaikos Gulf. Tissue extracts were subjected to SDS-PAGE and immunoblotted for Hsp70. *$P<0.05$ compared with the control (0 days). The results indicated two peak of expression one early spring and the second mid summer. The first is probably correlated with the reproductive activity of mussels, while the second is closely correlated with the elevation of seawater temperature (for further discussion see Ioannou et al. 2009)

The seasonal pattern of HSR in submersed *M. galloprovincialis* is in line with that reported by Tomanek (2008) and Barua and Heckathorn (2004) for rocky intertidal mussels. These authors reported that a relationship between the induction temperature for the HSR and temperature-dependent growth is a step function, such that around the optimal growth temperature an acclimation of HSR induction temperatures would be unnecessary. Beyond this range a modification of morphology, physiology, or biochemistry (e.g. changes in cell membrane composition, etc.) occurs during thermal acclimation. Such changes would alter the sensitivity to high temperatures and result in a change of the HSR induction temperature (Tomanek and Somero 2002). Shifting HSR induction temperatures during early warming might explain the small HSR. The shift of the induction temperature is probably limited, and beyond some upper limits there is no further change in the HSR induction temperatures. Tomanek (2008) reported that rocky intertidal animals that experience highly variable thermal conditions start synthesizing Hsps below the highest temperatures that they experience. Also, Tomanek (2008) pointed out that although eurythermal species such as mussels are poised to cope with thermal extremes under the present-day climate conditions, their physiological limits are already close to the highest body temperatures they currently experience and they have a limited capacity to modify these limits through acclimation, thus making them particularly vulnerable to any further increase in temperature.

Our findings indicate that thermoprotection of protein integrity by Hsps is diminished in *M. galloprovincialis* in Thermaikos Gulf when mussels are exposed to temperatures beyond of 28–29 °C (Anestis et al. 2007; Ioannou et al. 2009). Similar to our results, Hofmann and Somero (1995) showed that the threshold induction temperature for Hsp70 in the gills of *M. galloprovincialis* is about 26 °C. Buckley et al. (2001) also reported that the threshold HSR induction temperature was about 26 °C in the mussels *M. trossulus* acclimated at 20–23 °C. Recent proteomic analysis showed an HSR induction threshold to be 28 °C for Hsp70 but may be as low 24 °C for small Hsps in *M. trossulus* (Tomanek and Zuzow 2010). Notably, the levels of a constitutive Hsp70 isoform did not increase in warm field acclimatized mussels (Buckley et al. 2001). The differential expression of inducible and constitutive isoforms of Hsp70 has been discussed by Buckley et al. (2001) and characterized at the proteomic level (Tomanek and Zuzow 2010). It has been suggested that accumulation of the inducible Hsp70 might act as a buffer against subsequent heat stress, conferring increased thermotolerance in gradually warming environments. Our results are in consistent with the above assumption showing a progressive increase in the inducible isoform of Hsp70 in the PAM and mantle of *M. galloprovincialis* in summer (Anestis et al. 2007; Ioannou et al. 2009) and indicating a negative feedback loop with higher Hsp levels inhibiting the onset of the HSR.

Notably, the temperature extremes in the intertidal zone often reach beyond the HSP induction temperature of *M. galloprovincialis* suggesting that the survival of mussels under these conditions must be dependent on the relatively short-term exposure and well known passive tolerance mechanisms exploited during air exposure. However, HSP induction temperatures are also surpassed in submersed mussels of Thermaikos Gulf, indicating a high degree of thermal specialization to just within the seasonal temperature range. The physiological reasons of thermal specialization have been discussed as related to energy economy (Pörtner 2006).

Correlation Between Intermediary Metabolism and HSR in Thermally Stressed *M. galloprovincialis*

The HSR in response to increased warming demands additional energy expenditure to cover the cost of protein synthesis for stress defence. Energy is required for the activation of transcription of heat-shock genes, the synthesis of Hsps, and the ATP-dependent chaperoning function by Hsps. Furthermore, proteomic changes during heat stress in metabolic enzymes suggest a trade-off between the production of ROS and energy metabolism (Tomanek and Zuzow 2010). The induction and functions of the Hsp70 family are dependent on the availability of ATP (Mallouk et al. 1999). Models for the inducible Hsp72 expression propose that under normal cellular ATP concentrations, Hsps are not effective as chaperones and that decreased contents of ATP, ATP/ADP ratio and energy charge may be conducive to Hsp72 induction (Palleros et al. 1991; Kabakov and Gabai 1994). Moreover, proteolytic degradation and replacement of damaged proteins is an ATP-demanding process which is compounded by the energy expended to synthesize and maintain pools of Hsps. Many denatured proteins undergo a process known as ubiquitination in which a small protein, ubiquitin, is bound covalently to the denatured target protein. Ubiquitination requires ATP, as does the subsequent proteolysis of the ubiquitinated protein through non-lysosomal proteolytic pathways. Hofmann and Somero (1995) showed that the levels of ubiquitinated proteins are higher in heat-stressed mussels, but subunit isoforms of the proteasome, the ubiquitin-dependent protease, are down-regulated with heat stress in *M. galloprovincialis*, suggesting that the actual degradation occurs during recovery (Tomanek and Zuzow 2010). One reason for this may be that ATP synthesis is impaired during the heat stress in intertidal organisms due to anaerobiosis and metabolic depression, and thus forces all energetically costly repair processes, including protein homeostasis, to be delayed until return to non-stressful conditions (Hofmann and Somero 1996a, b).

A key aspect of the cellular stress response is the modulation of major pathways of energy metabolism, and the induction of key enzymes of energy metabolism, which may provide the reducing and energy equivalents needed for stress-related cell functions (Kültz 2005). The duration of thermal stress will determine the extent of heat damage and consequently the extent of energy expenditure and the rate of ATP production for energy-demanding processes. Our studies on the changes in activity of a key metabolic enzyme, pyruvate kinase (PK) in *M. galloprovincialis* during acclimation/acclimatization to increasing temperatures are consistent with this hypothesis (Fig. 18.3). PK controls the flux of phospho*enol*pyruvate (PEP) to succinate during anaerobiosis and its modification to a less active form contributes to metabolic depression (Storey and Storey 1990; Brooks and Storey 1997). Consequently, it is used as an indicator of shift from aerobic to anaerobic metabolism. No change in PK activity occurred at temperatures lower than 24 °C. However, warming beyond 24 °C and up to 26 °C reduced PK activity in the mantle and PAM of mussels. The relative activity of the enzyme (V_o/V_{max}) decreased from 0.95 to 0.57 within 15 days of acclimation, indicating a shift of PK toward a less active form and induction of metabolic depression (Anestis et al. 2007). In contrast, further warming beyond 26 °C caused a reactivation of PK activity from the mantle and PAM. Within the first 10 days of exposure to 26 or 28 °C PK values returned to levels similar to those observed in the mussels acclimated at 24 °C. Thereafter, the activity and V_o/V_{max} of PK from mantle reached levels above control values, indicating activation of the enzyme and reactivation of glycolysis (Fig. 18.3) (Anestis et al. 2007; Ioannou et al. 2009). Overall, the time course of glycolytic reactivation is closely correlated with that of HSR induction suggesting that the energy-demanding process of Hsp synthesis is partly met by activation of glycolytic ATP production in the tissues of *M. galloprovincials* at temperatures beyond 26 °C. These observations also match the induction of Hsps in field mussels exposed to temperatures above 26 °C (Ioannou et al. 2009). A schematic presentation of the profile of glycolytic flow in the tissues of *M. galloprovincialis* during acclimation at different temperatures is given in the Fig. 18.4.

Warming and Oxidative Stress in *M. galloprovincialis*

Oxidative stress (a misbalance between cellular pro-oxidants and antioxidants) is a common hallmark of environmental (including thermal) stress response in marine organisms. Reactive oxygen and nitrogen species (ROS and RNS, respectively) are generated as by-products of cellular metabolism. ROS are the major cellular pro-oxidants (predominantly generated in mitochondria) that can react with and damage macromolecules including proteins, lipids and nucleic acids (Halliwell and Gutteridge 2007). Enzymatic

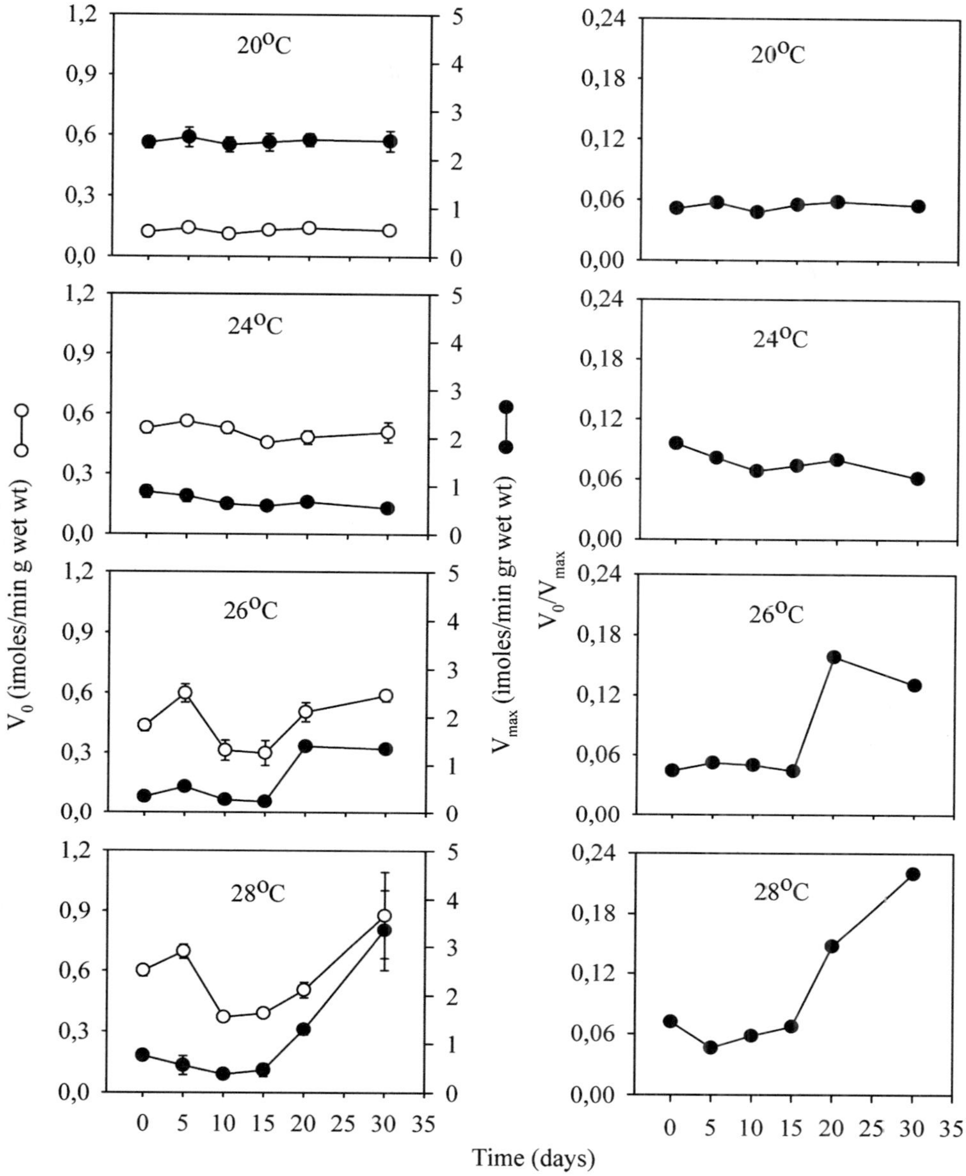

Fig. 18.3 Activity of pyruvate kinase (PK) from the mantle of *Mytilus galloprovincialis* during acclimation to different water temperatures. Activity (μmol/min per g per wt) was determined at 2 mM (V_{max}) and 0.05 mM (V_o). Values are means ± SE, n = 5 preparations from different animals. V_{max} (*closed circles*) and V_o (*open circles*) are given at *left*, and the ratio V_o/V_{max} is given in the *right* (Anestis et al. 2007)

(such as superoxide dismutase, catalase and glutathione peroxidase) and non-enzymatic (including glutathione, vitamins E and C and metallothioneins) antioxidants detoxify and scavenge ROS and RNS preventing them from reacting with important cellular molecules (Halliwell and Gutteridge 2007). Environmental stress can shift the dynamic equilibrium between the production and removal of ROS and RNS by either increasing the rate of their production, inhibition and/or depletion of antioxidants, or combination of the two mechanisms (Lushchak 2010).

Environmental temperature modulates ROS production as well as expression and activity of antioxidant defence mechanisms in marine bivalves, including mussels (review in: Abele and Puntarulo 2004). The antioxidative defence system also acts as a defence system supporting passive tolerance against temperature extremes, together with the HSR and anaerobic metabolism (Pörtner 2002a, b, 2010). The antioxidant response of *M. galloprovincialis* to temperature acclimatization is complex and varies with the year, location, environmental variability in pollution or oxygen levels as well as among different antioxidants exploited (Petrović et al. 2004; Santovito et al. 2005; Box et al. 2007; Pisanelli et al. 2009; Vidal-Liñán et al. 2010). In *M. galloprovincialis* from the Adriatic Sea, activities of two key

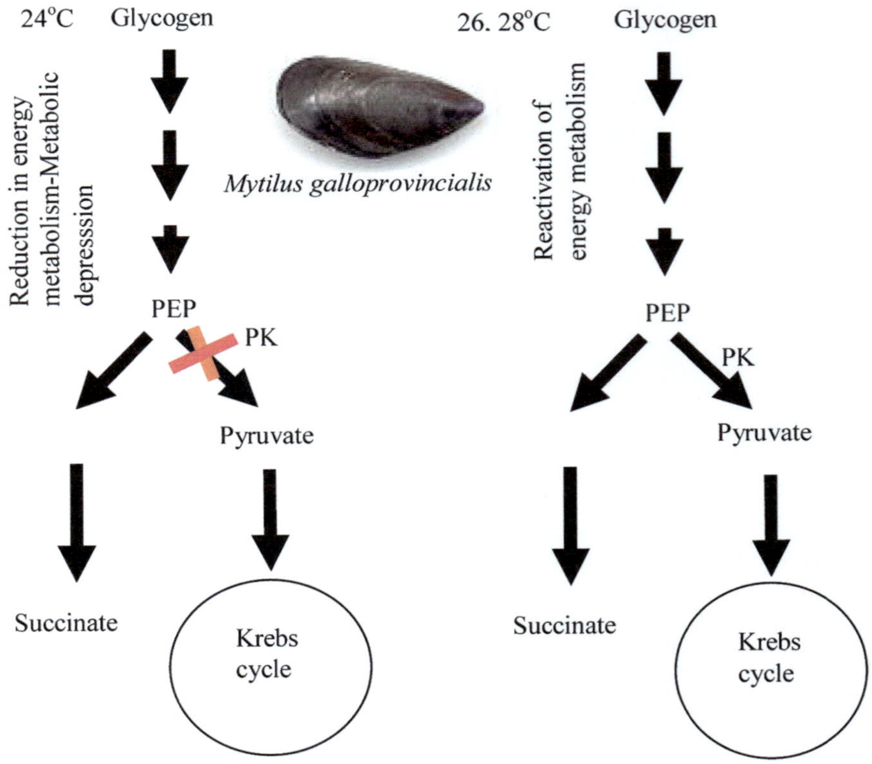

Fig. 18.4 Schematic presentation of the profile of glycolytic flow in the tissues of *Mytilus galloprovincialis* during acclimation at different temperatures. The model is based on the data presented in the Fig. 18.3 and adapted from Anestis et al. (2007). It is shown that the glycolytic flow closely correlates with the acclimation temperature. When *Mytilus galloprovincialis* is acclimated at 24° C, PK activity is reduced indicating a metabolic depression and the glycolytic flow ends in an anaerobic product succinate (Storey and Storey 1990; Brooks and Storey 1997). At acclimation temperatures higher than 26° C, PK activity is reactivated, resulting in increase of glycolytic rate and ATP turnover. These changes in the glycolytic rates coincide with the initiation of Hsps expression suggesting that the energy demand for Hsps synthesis during thermal stress is met by activation of intermediary metabolism. PEP, phospho*enol*pyruvate; PK, pyruvate kinase

antioxidant enzymes (catalase and glutathione reductase) and total antioxidant capacity (TAOC) were elevated in spring as a possible anticipatory response to summer stress (Bocchetti and Regoli 2006). In summer, the activities of these enzymes were suppressed hand-in-hand with a partial depletion of cellular glutathione, higher activity of glutathione peroxidase and decreased TAOC indicating an oxidative challenge and a high burden on the glutathione-dependent antioxidant system (Bocchetti and Regoli 2006; Bocchetti et al. 2008). High levels of oxidative stress in *M. galloprovincialis* in summer were supported by a marked decrease in lysosomal stability indicative of membrane damage (Bocchetti and Regoli 2006). Proteomic analyses of gill tissue of *M. galloprovincialis* exposed to acute heat and chronic cold stress showed increased abundance of a number of oxidative stress proteins exposed to acute heat and chronic cold stress (Tomanek and Zuzow 2010). These studies suggest a trade-off between levels of oxidative stress and energy metabolism. Notably, additional stressors (such as pollution or ocean acidification) may synergistically enhance the negative effects of thermal stress on the cellular redox balance leading to increased oxidative stress and cellular damage in bivalves (including mussels) from anthropogenically impacted areas such as the Mediterranean Sea (Lannig et al. 2006; Verlecar et al. 2007; Sokolova and Lannig 2008; Tomanek 2011).

Correlating Molecular and Metabolic Responses of *M. galloprovincialis* with Behavioural Patterns Under Progressive Thermal Stress

The concomitant changes in the HSR and metabolic patterns in response to a temperature change in mussels match with the behaviour of mussels, specifically the movement of valves that regulates the rate of water filtration and food intake. The duration of valve opening and the mean fraction of time during which valves remained open or closed were estimated from the recorded traces of shell movements in mussels acclimated for 10 days at different temperatures (7, 14, 18, 20, 24, 26, 28 and 30 °C). The behavioural response of mussels to increasing acclimation temperatures was biphasic. In the lower temperature range (7–16 °C) the period during which mussels were kept the valves open increased with increasing acclimation temperature. Thereafter, however, the mean period of valve opening declined until

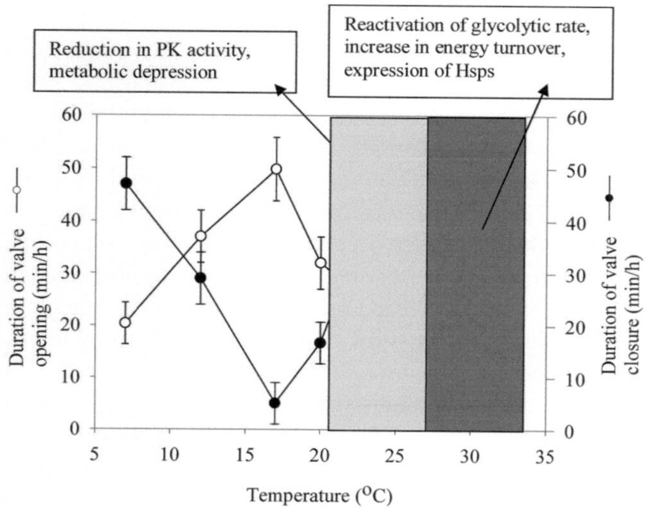

Fig. 18.5 Mean periods of valve closure or opening during acclimation to different water temperatures. Values are means ± SE, n = 5 (Adapted from Anestis et al. 2007). Acclimation at 20–24° C leads to more prolonged valve closure correlated with the reduction in PK activity and depression of metabolic rate (cf Fig. 18.4). In contrast, at higher temperatures valves remain open longer with a concomitant activation of PK activity and increase in glycolytic rate. This behaviour may help mussels to increase the oxygen consumption and food intake rates to meet the energy demand for the synthesis of stress proteins and to defend against the thermal stress

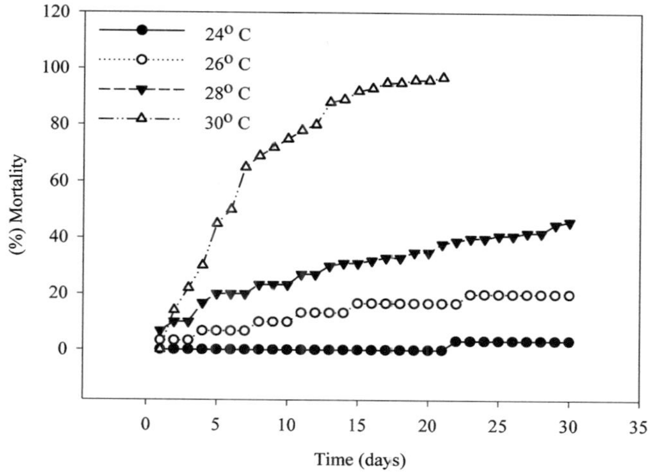

Fig. 18.6 Effect of water temperature on the mortality of *Mytilus galloprovincialis* during 30 days of acclimation to different temperatures (Anestis et al. 2007). Mussels' mortality increased markedly during acclimation of mussels at the temperatures above 26° C. These data correspond to the maximum thermal tolerance limits of *Mytilus galloprovincialis* in Thermaikos Gulf

beyond 26 °C there again was a trend to increase the period of valve opening with increasing temperatures (Fig. 18.5).

This behavioural pattern of opening and closure of valves can help to explain the observed changes in HSR and metabolic response of mussels during acclimation to different temperatures. Thus, *M. gallprovincialis* acclimated to 24 °C displayed extended periods of valve closure compared to controls (Fig. 18.5) and these behavioural changes are paralleled with the observed reduction in the PK activity status (Fig. 18.3), indicating low glycolytic rate and reduced rates of energy turnover during moderate warming. This agrees with previous reports indicating that metabolic depression accompanies prolonged valve closure in bivalves (Anestis et al. 2007). In this context, temperature-induced anaerobiosis has been repeatedly reported for marine invertebrates including mollusks (Sommer et al. 1997). Reactivation of metabolism beyond 26 °C is consistent with extended gaping observed at high temperatures, which might reflect enhanced oxygen demand at extreme temperatures. In fact, an increase in the rate of oxygen consumption has been reported for *M. edulis* exposed to temperatures higher than 24 °C (Anestis et al. 2007 and references therein). The relative roles of the transient depression of metabolic rate and the reactivation of metabolic rate upon warming remain to be further explored. Enhanced mortality beyond 26 °C would suggest that the reactivation of metabolism is an emergency response not effective in the long-term protection of the mussels.

Integrating the Responses of *M. galloprovincialis* (from Molecules to Behaviour, Energetics to Mortality) to Increased Ambient Temperature

The temperature tolerance range of an organism can be defined as the range of temperatures where body growth is positive. The optimum temperature is thought to be a reflection of balance between the positive (rate-enhancing) and negative (e.g. temperature-induced disturbance of macromolecular function and/or ingestion rate) effects of temperature on metabolism (Van der Veer et al. 2009). Although the energy demand of the HSR is accompanied by reactivation of energy metabolism when *M. galloprovincialis* is acclimated to temperatures beyond 26° C, mortality increased gradually with the rise of acclimation temperature (Fig. 18.6). Thus, mortality was 5 % within 5 days and 20 % after 30 days of exposure to 26 °C. Mussel mortality increased further upon exposure to 28 and 30 °C, when 80 % of the animals died within 15 days of acclimation (Anestis et al. 2007). The data demonstrate that *M. galloprovincialis* cannot survive seawater temperatures at and beyond 26 °C for extended periods of time (Fig. 18.6). According to unpublished observations, 30–55 % of the mussels die in Thermaikos Gulf, when daytime temperature of surface seawater rises up to 28 °C, emphasizing that mussels live at the upper limit of their acclimation capacity as outlined above. Similarly, field studies on farmed *M. galloprovincialis* in Alfacs Bay, Spain showed that during summer and when seawater temperature rises beyond of 25 °C in summer, mussels decrease their feeding ingestion and absorption rates and efficiencies (Galimany et al. 2011). Similar to *M. galloprovincialis*, *M. edulis* displays an upper lethal temperature between 26 and

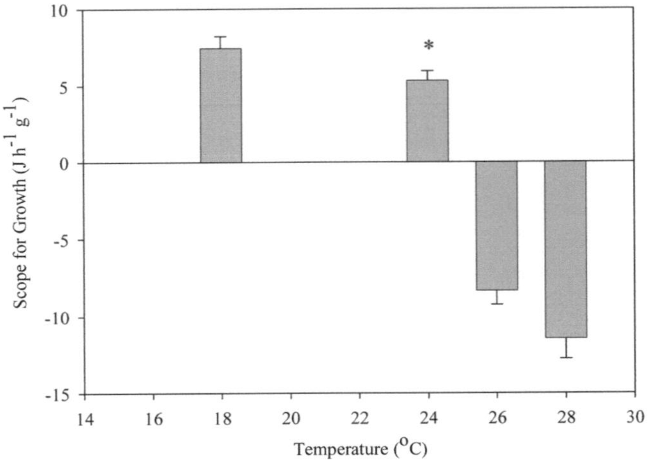

Fig. 18.7 Changes in scope for growth (SFG) by *Mytilus galloprovincialis* calculated after 30 days of acclimation at different temperatures. Values are means ± SEM, n = 15. The data indicate a marked reduction of ability of mussels to gain energy from the ingested food when acclimated at temperatures higher than 26° C (Anestis et al. 2010b)

28 °C. Mussels exposed to such high water temperatures are not able acquire sufficient food to meet energy demand which would explain the high summer mortality rates (Ramón et al. 2007).

The OCLTT concept (Pörtner 2010) draws a link between cellular (HSR, anaerobic metabolism) and whole organism (overall metabolism, behavioural) responses to temperature, with limits at a whole organism level set narrower than at the molecular level (Pörtner 2002a). Thus, elevated mortality of heat-exposed mussels can result from a mismatch between functional capacity and oxygen supply, which is mirrored in a mismatch between energy demand and energy availability. During exposure to thermal stress, the elevated energy demand for stress defence may contribute to the reduction of oxygen supply to tissues. Also, energy demand cannot be met by energy gain through the intake and catabolism of food any more. Our studies showed that during acclimation to temperatures beyond 26 °C, *M. galloprovincialis* gradually losses its ability to assimilate the consumed food and to maintain a positive energy balance. Scope for growth (SFG) becomes negative in mussels acclimated to 28 °C (Fig. 18.7), mainly due to the temperatures-induced inhibition of the clearance rate (Fig. 18.8) (Anestis et al. 2010b). As demonstrated in other studies on mussels (i.e. Sobral and Widdows 1997; Smaal and Widdows 1994) clearance rate (i.e. food acquisition) is the most sensitive parameter affecting the SFG during exposure of mussels to high ambient temperatures. For example, filtration rate increases in *M. edulis* when

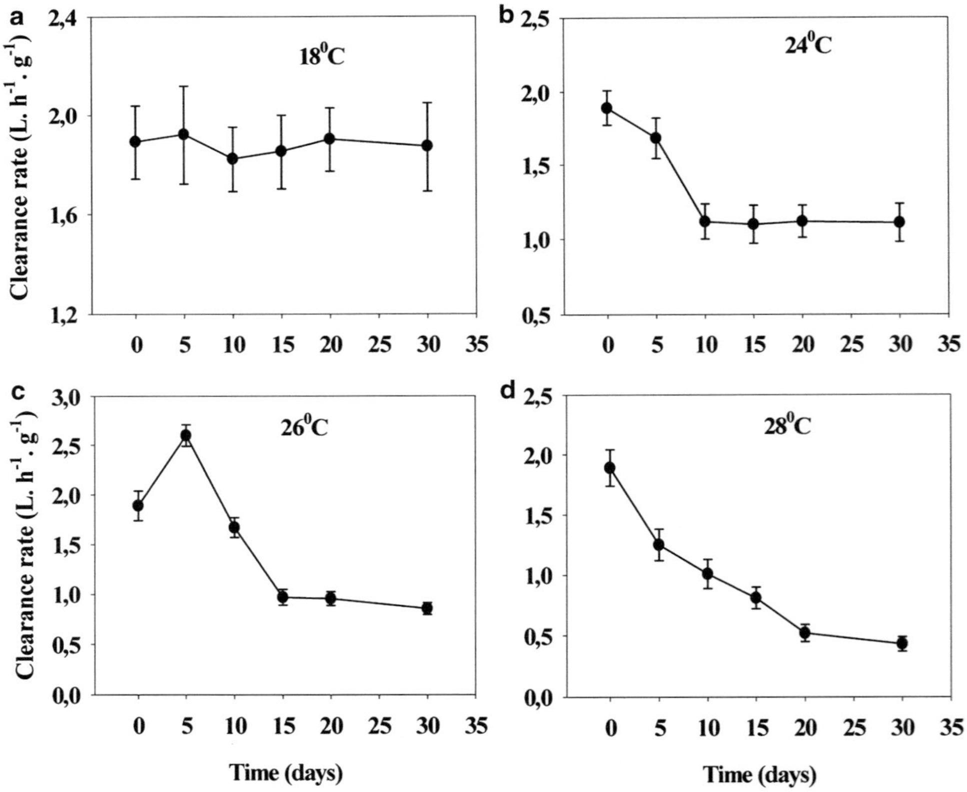

Fig. 18.8 Effects of water temperature on the clearance rate of *Mytilus galloprovincialis* during 30 days of acclimation. Values are means ± SEM, N = 15. Animals acclimated at 18° C were used as controls, and the values obtained at other temperatures were compared to those controls (zero day at each panel, * P<0.05)

acclimationtemperatures rise up to 15–17 °C, while at temperatures beyond 25 °C, filtration rate is significantly reduced (Gonzalez and Yevich 1976; Bayne et al. 1976; Widdows 1973; Schulte 1975). Moreover, Bayne et al. (1976) showed an intermediate dependence of filtration rate on temperature up to 22 °C for *M. californianus*, which declined at 26 °C. In *M. californianus* the SFG was highest at 17–22 °C and declined at 26 °C. This is consistent with the fact that temperate *Mytilus* spp. are well known for their ability to tolerate low temperatures whereas very high temperatures (generally > 22 °C) can have serious detrimental effects on clearance rate and energy gain from ingested food (Seed and Suchanek 1992). Similar to *Mytilus spp.*, other mollusc species show similar responses to temperature with respect to clearance rates. For example Resgalla et al. (2007) observed that under chronic conditions, *Perna perna* is capable of compensating the clearance rate between 15 and 30 °C, but with a tendency for this rate to increase up to 25 °C and be inhibited at 25 °C. Sobral and Widdows (1997) also reported that by increasing temperature beyond 20 °C resulted in lower clearance rates, leading to a marked reduction of SFG in an infaunal bivalve *Ruditapes decussatus*. They also reported that high temperatures (above 27 °C) are stressful to the clams, as shown by the low and negative values of SFG.

Similarly, Tremblay et al. (1998a) reported decreased food availability and very low measures of absorption efficiency resulting in negative values of SFG in mussels in August. The long-term depletion of energy reserves at high temperatures may contribute to increased mussel mortality. For mussels (*M. edulis*) from the British Isles, thermal stress, particularly exposure to temperatures over 20 °C, decreases energetic reserves and induces cellular autophagy (Lowe et al. 1982). Tremblay et al. (1998a, b) demonstrated that in mid-August mussels from both resistant and susceptible stocks at the Magdalen Islands had negative values of SFG and very low values of O:N ratio, indicating protein catabolism during thermal stress. The period during which mass mortality of mussels normally occurred (August), was associated with elevated activities of lysosomes, indicating high autophagy in tissues as part of the protein degrading activity.

Multiple Stressors Affecting Performance of *M. galloprovincialis*

We have focused our analysis on temperature tolerance but are well aware that climate change comes with several factors changing in the marine environment. As discussed in the respective sections several of these factors interfere with thermal tolerance, by enhancing thermal sensitivity. One of these is the factor CO_2 causing ocean acidification, a process presently exacerbating in the oceans (Caldeira and Wickett 2005). These factors acting synergistically with temperature extremes have been proposed be understood most comprehensively by analysing them on a matrix of temperature effects. In that way, by narrowing the thermal window of a species, these factors modulate the principle effect of temperature on large scale biogeography of marine species and are integrated in the present picture (Pörtner 2010), considering the projection that ocean acidification will become effective over the decades to come. Two special examples are discussed here, the role of parasitism and of food availability in shaping stress resistance, with the general insight that parasitism and food limitation act in similar ways, by reducing resistance to other stressors including temperature extremes.

Parasitism and Scope for Growth (SFG): Effects of Marteiliosis

Global warming may accelerate the growth and distribution of pathogens and harmful algal blooms in marine environments, affecting health of aquatic organisms as well as humans (Harvell et al. 1999; IPCC 2007). In bivalves including mussels, several intracellular protozoan parasites can cause diseases resulting in increased morbidity and mortality of wild and cultured populations. In mussels, protozoan infections (including marteiliosis caused by *Marteilia* sp.) can lead to glycogen depletion (indicated by emaciation and discoloration of the digestive gland) which in turn leads to the cessation of growth, a poor condition index and disruption of the gametogenic cycle indicative of energy deficiency (Figueras and Montes 1988; Figueras et al. 1991; Robert et al. 1991; Villalba et al. 1993a). At the tissue level, marteiliosis leads to formation of granulocytomas, tissue necrosis in the digestive gland (Villalba et al. 1993b) and increases in the haemocytes in the haemolymph (Carballal et al. 1998). While sublethal effects of marteliosis are predominantly linked with activation of immune response, cellular damage and energy deficit, bivalve mortalities appear to be associated with disruption of the digestive tubule epithelia due to sporulation of the parasite (Villalba et al. 1993b; Karagiannis and Angelidis 2007).

Marteilia sp. is prevalent in the Thermaikos Gulf and is likely able to complete its life cycle within this area (Angelidis et al. 2001; Virvilis et al. 2003; Rayyan et al. 2006; Virvilis and Angelidis 2006). The histopathological examination of the infected mussels showed t plasmodia of *Marteilia* sp. in the epithelial cells of the stomach. Sporangia of *Marteillia* sp. were found in the epithelial cells of the digestive tubules, where sporulation occurred. In wild and cultured mussel populations the prevalence of the parasite varies depending on the culture technique and is higher in the mussels cultured “on table” compared to those grown on long lines (Karagiannis and Angelidis 2007).

Seasonal data suggests that elevated seawater temperature favours growth and proliferation of *Marteilia* sp. in the Thermaikos Gulf with significantly higher parasite prevalence detected in July compared to October (Karagiannis and Angelidis 2007).

In a recent study (Anestis et al. 2010b), we showed that the presence of *Mertelia* spp. impacts the energetics of *M. galloprovincialis*. Specifically, the food clearance rate of the infected mussels was significantly higher than that of the non-infected mussels by about 16 %. However, the infected mussels exhibited a marked decrease in the ability to absorb energy from the ingested food as shown by the lower food assimilation efficiency (Ae) in infected mussels. Infection did not affect respiration and excretion rates which remained at the same levels in the healthy and infected mussels. It is likely that elevated clearance rate and food intake is a compensatory mechanism to offset the low food absorption efficiency in the infected mussels. However, this compensation is incomplete as indicated by the lower SFG values in infected mussels.

Food Availability and Stress Defence

Food availability and quality are major factors affecting growth rates and survival of mussels. For example, mussels' growth rates are often positively correlated with concentrations of chlorophyll a during spring and summer months on the west coast of North America (Menge et al. 1997, 2002). Gardner (2000) also showed that low seston quality may limit mussels' distribution by reducing growth rate and SFG. Local variations in productivity are known to drive community structure (Menge et al. 1997, 2008) and affect species richness (MacLeod et al. 2004). The importance of bottom-up effects (such as food limitation or low levels of local productivity) in regulating community structure in freshwater, terrestrial and pelagic ocean systems has long been recognized (Dahlhoff and Menge 1996; Menge et al. 1999, 2002; Menge 2000; McQuaid and Lindsay 2000; Helson et al. 2007). Smith et al. (2006) provided evidence that productivity losses associated with regional climate change may be linked to declines in biodiversity in intertidal mussel communities, when abundances of most species fell and rare species disappeared.

Quantification of the role of energetic costs in limiting the maximal height in the intertidal zone at which a species can occur is of great importance and represents an important challenge for future investigations (Somero 2002). Recent studies suggest that the abundance of mussels on New Zealand rocky shores may decrease in the face of increasing aerial temperatures predicted under global climate change scenarios, drastically altering intertidal community structure (Petes et al. 2007). Although the impact of climate change can simply occur through shifts in the temperature regime

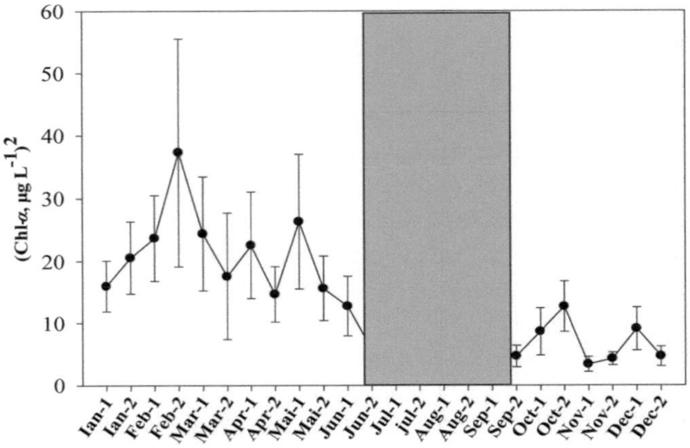

Fig. 18.9 Annual variation of chlorophyll-a (*Chl-α*) concentrations (μg·L^{-1}); mean square (x^2) *Chl-α* values and standard deviation in one sampling station at the inner part of Thermaikos Gulf at fortnightly intervals from 10 years data (2000–2009). Similar to the rest of the Mediterranean Sea (Coma et al. 2009), Chl-*α* levels are lowest in the Thermaikos Gulf during summer time

(McCarty 2001; Parmesan 2006), evidence suggests that interactions between temperature and other factors including energy shortage are likely to have a widespread influence on mussels' community dynamics (Harley et al. 2006; Parmesan 2006). OCLTT together with the energy shortage phenomenon would support an integrative view to understand how anomalous climatic conditions may induce mass mortality of some benthic species (Coma and Ribes 2003; Coma et al. 2009).

Availability of adequate particulate food is a consistent growth stimulus for bivalves under current farming conditions (Navarro et al. 1991; Saxby 2002). In the Adriatic Sea the period of fastest growth of *M. galloprovincialis* coincides with the period of highest primary productivity (Marasovič et al. 2005), suggesting that the higher growth rate may be a result of increased food availability. An important factor influenced by food availability and quality is the condition index (CI). The bivalve CI is widely employed in environmental monitoring programmes as it integrates physiological responses to stress with changes in somatic growth. This factor seems to be influenced by the reproductive state of mussels and by the temperature of seawater (Sasikumar and Krishnakumar 2011). Condition index showed spatial and temporal variations with higher values during fall and winter, and lower values during spring and summer (Peharda et al. 2007). The growth of mussels in the Thermaikos Gulf appears rather lower compared to other areas of the Mediterranean Sea due to the low quantity of available food in the mussel cultivations. Moreover, the lowest CI values are found in the Thermaikos Gulf mussels during mid summer (Galinou-Mitsoudi S, unpublished data). The pattern of changes in the content of chlorophyll-a (*Chl-a*) in the Thermaikos Gulf for the period 2000–2009 is shown in Fig. 18.9. Each point in the Fig. 18.9 expresses the mean value of *Chl-a* for each month during the period 2000–2009. *Chl-a* content was highest between February and April, and lowest during the warm summer months (Katsikatsou et al. 2012). Food availability (*Chl-a*) in the Ther-maikos Gulf is similar to that in the rest of the Mediterranean Sea (Coma et al. 2009).

Integration of the availability of food and stress responses of *M. galloprovincialis* indicate that periods of low availability of food are associated with a loss of aerobic scope. *M. galloprovincialis* has to face the concominant reduction in food and the gradual elevation of ambient temperature from spring to summer. Both factors may lead to constraints in energy allocation to reproductive organs and further exacerbation of heat stress. Energy shortage can involve oxidative stress as indicated by field studies (Regoli et al. 2004; Lesser et al. 2010), and transcriptomic and proteomic analyses (Tomanek and Zuzow 2010; Lockwood et al. 2010; Tomanek 2011). This cellular response can impair the ability of stressed organisms to grow and reproduce (Petes et al. 2007, 2008; Fearman and Moltschaniwskyja 2010), resulting in changes of community structure and richness (Menge et al. 1997, 2008; MacLeod et al. 2004; Smith et al. 2006).

Correlating Physiological Traits and Responses with Regional Oceanographic and Climatological Patterns

Even small changes in climatic variables often produce significant changes in the abundance of marine species because of direct and indirect effects of the climate. Direct effects are linked to changes in water temperature and salinity which can affect the organism's survival, reproduction and biotic interactions. Indirect effects are linked to hydrodynamic variations affecting ocean circulation, stratification of water, mixing patterns and consequently nutrient availability for marine organisms (Kawasaki et al. 1991). As the world warms, climate will change in the Mediterranean region, making the Mediterranean basin a vulnerable region (Nicholls and Hoozemans 1996; Calvo et al. 2011). Recent evidence points to a large-scale warming of the Mediterranean basin (Béthoux et al. 1990; Astraldi et al. 1995) and the changing biodiversity in response (Francour et al. 1994). Although the relationship between weather and oceanographic conditions is not straightforward, the long-term studies on seawater temperature conditions show

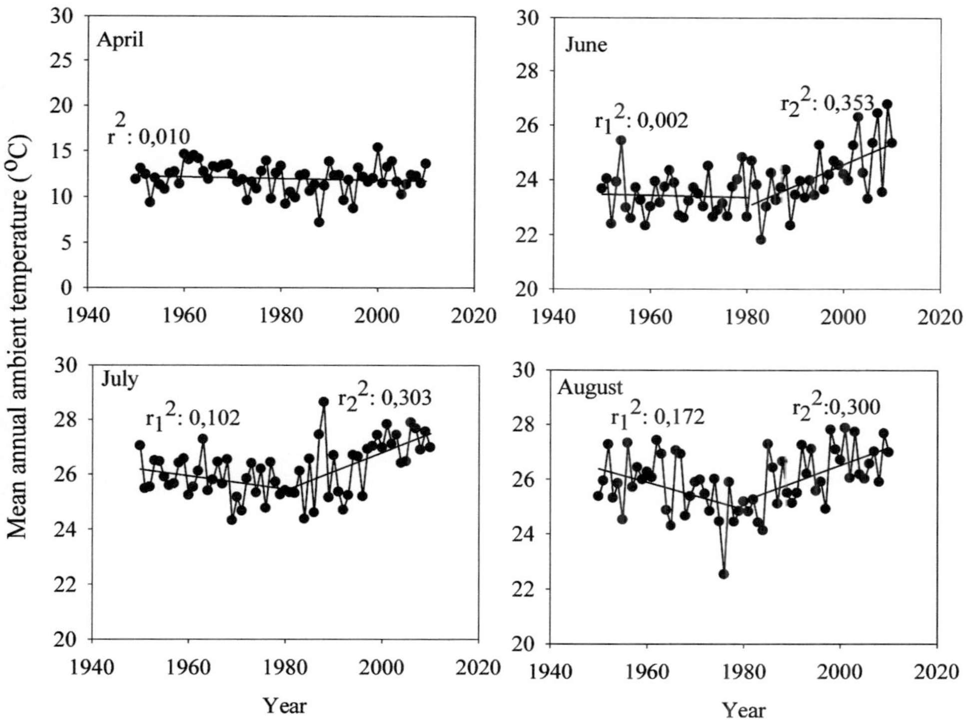

Fig. 18.10 Annual variation in ambient air temperature in the area of Thermaikos Gulf. The analysis of meteorological data since 1950 indicates increasing air temperatures in the Themaikos Gulf especially after 1980. The climate trends are in line with the mean global temperature projections for the period 1990–2100 based on IS92 emission scenarios of the Intergovernmental Panel for Climate Change (IPCC)

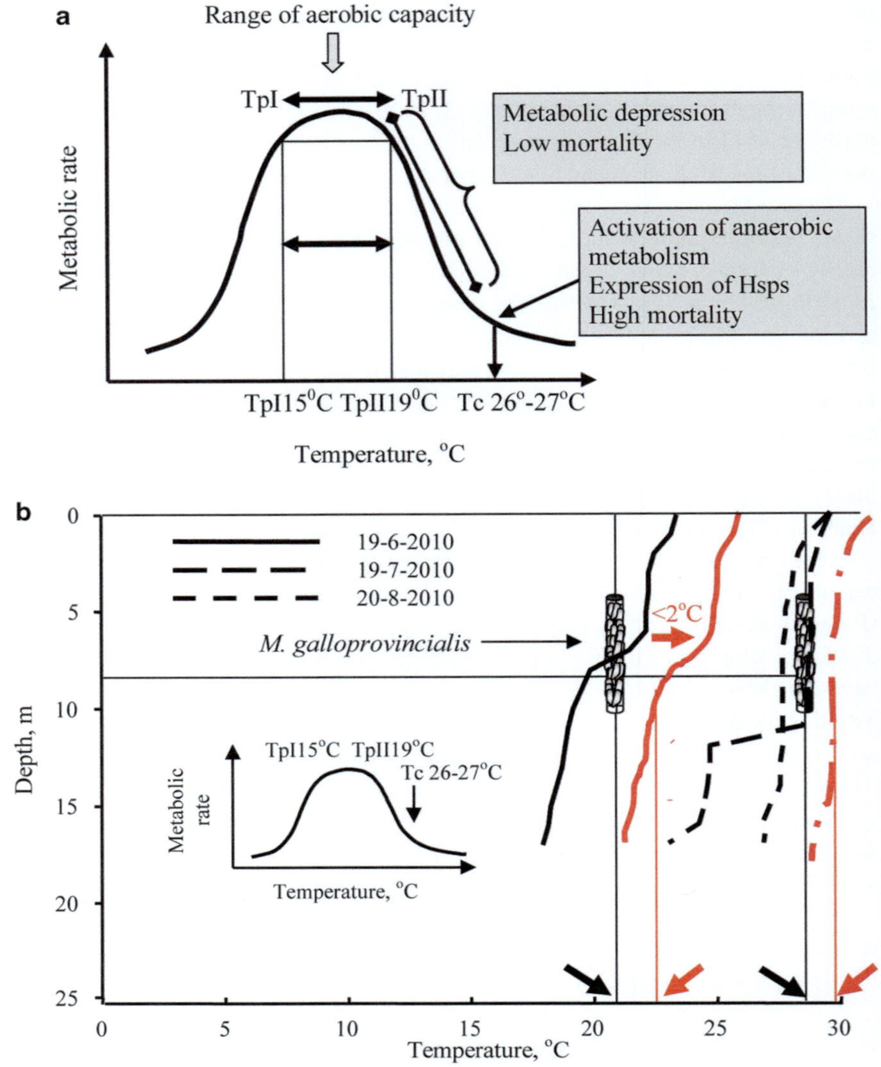

Fig. 18.11 (**a**) Thermotolerance model for *Mytilus galloprovincialis* based on Pörtner (2001) and our studies on metabolic (PK), biochemical (Hsps) and energetic indicators and temperature-dependent mortality of *Mytilus galloprovincialis*. The mussels exhibit their maximum aerobic capacity between 14 and 19° C, and experience thermal stress at temperatures at and beyond 26° C, with critical temperature (*Tc*) around 27–28° C. TpI and TpII, pejus temperatures (Pörtner 2001). (**b**) Changes in seawater temperature in Thermaikos Gulf based on regional climate trends (see Fig. 18.10) and the mean global temperature projections for the period 1990–2100 based on IS92 emission scenarios of the Intergovernmental Panel for Climate Change (IPCC) (IPCC 2007). These data suggest that prolonged exposure of *Mytilus galloprovincialis* to sublethal temperatures is not excluded in the future indicating impacts on the mussels' populations

a distinct warming trend of coastal waters worldwide including the Mediterranean Sea (Vargas-Yáñez et al. 2008, 2010; Calvo et al. 2011).

The richness of microclimates in the Mediterranean (ranging from climate conditions similar to those of the Northern Sea in the Adriatic to almost tropical conditions in the Eastern Mediterranean) makes any prediction of large spatial scales difficult. Indeed, most effects of climate change on marine biodiversity have so far been identified only on regional scales. In contrast, several recent models have also demonstrated the potential of the global or regional climate change to cause reductions in biodiversity and local extinctions (Roberts et al. 2002; Thomas et al. 2004). The regional climate prediction is not an insoluble problem but it is characterized by uncertainty due to the inherent unpredictabilities of external forcing – both anthropogenic and natural – and of the climate system itself (Mitchell and Hulme 1999), which must be accepted an incorporated into the global climate change analyses and projections. The reliability of climate change projections generally decreases from the global to the regional and further to the local scale. The problem for the Mediterranean lies in the fact that marked regional differences in the rate of change are expected and that a wide disagreement exists between the patterns of change projected by various models.

In the intertidal zone, environmental stress is likely driven by multiple factors, including temperature, desiccation, and reduced feeding time. In the field, the mussels' body temperature is determined by a complex series of interactions between the organism and its environment, leading to higher body temperatures due to solar radiation (Helmuth and Hofmann 2001; Helmuth et al. 2006). The absolute tidal height of an organism on the shore, the amount of wave splash that it receives, and the local tidal cycle all interact to determine the timing and duration of exposure to terrestrial conditions at low tide (Helmuth et al. 2006). Consequently the mechanisms linking climate change and intertidal marine ecosystems are complex and incompletely understood. However, the data from our studies in conjunction with the climate trends in the region of Thermaikos Gulf permit us to predict certain trends for the subtidal mussel populations. Analysis of meteorological data since 1950 indicates a tendency of air temperatures to increase in the Thermaikos Gulf (Fig. 18.10). Moreover, records of sea temperatures for 10 years show that the permanently submersed mussels face increasingly high temperatures during summer (Fig. 18.11b). Seawater temperatures in the Thermaikos Gulf fluctuate between 26 and 28° C during June and August (Anestis et al. 2008). Thus, taking into account the seasonal changes in temperature of seawater, it seems that *M. galloprovincialis* in this region can temporarily experience and tolerate temperatures near or beyond its upper critical temperatures (*Tc*) (Fig. 18.11a). However, when mussels encounter water temperatures higher than 25 °C for extended periods, they experience tissue damage and undergo heat-induced mass mortality when day temperatures of surface seawater rise up to 28 °C. Moreover, the detection frequency of the pathogenic protozoan was significantly higher at the July sampling compared to the October sampling indicating that warming promotes growth of *Marteilia* sp. in the Thermaikos Gulf (Karagiannis and Angelidis 2007). This suggests a possible outbreak of marteiliosis within the next decades of warming with potentially a devastating impact on the populations of *M. galloprovincialis*. In a broader ecological and ecotoxicological context, parasitism by *Marteilia* sp. can also impair the mussels' resistance to other types of stressors such as pollution, hypoxia and ocean acidification.

Taking into consideration the thermal limits of *M. galloprovincialis,* we conclude that under the present day conditions, *M. galloprovincialis* lives close to its incipient lethal temperatures during summer. Accordingly, an increase of air temperature by 2–4 °C will threaten the populations of *M. galloprovincialis* in the region of Thermaikos Gulf pushing them beyond their upper thermal limits (Fig. 18.11b). Moreover, energetic constraints due to high metabolic activity and prolonged low levels of food availability should be explored as potential co-determinant of mussels' mortality. For comparison, recent investigations in a clam showed that warming in late summer results in a significant decrease in growth rate (body mass and shell length increase) and an increase in mortality rate. The addition of planktonic food dampens the negative effect of warming on the growth rates. This suggests that one reason for the negative growth effect of temperature increase in late summer might lie in a negative energetic balance caused by an enhanced metabolic rate at limited food levels (Weitere et al. 2009) and the associated reduction of thermal tolerance. Similar relationships may hold true for the mussel *M. galloprovincialis*. Given the important ecological role of this species in Mediterranean ecosystems, declines in mussel populations will have serious implications for biodiversity and ecosystem structure, as well as for the economy of the region.

References

Abele D, Puntarulo S (2004) Formation of reactive species and induction of antioxidant defence systems in polar and temperate marine invertebrates and fish. Comp Biochem Physiol – Part A 138:405–415

Abele D, Heise K, Pörtner HO, Puntarulo S (2002) Temperature dependence of mitochondrial function and production of reactive oxygen species in the intertidal mud clam *Mya arenaria*. J Exp Biol 205:1831–1841

Anestis A, Lazou A, Pörtner HO, Michaelidis B (2007) Behavioral, metabolic, and molecular stress responses of marine bivalve *Mytilus galloprovincialis* during long-term acclimation at increasing ambient temperature. Am J Physiol Regul Integr Comp Physiol 293:R911–R921

Anestis A, Pörtner HO, Lazou A, Michaelidis B (2008) Metabolic and molecular stress responses of sublittoral bearded horse mussel *Modiolus barbatus* to warming sea water: implications for vertical zonation. J Exp Biol 211:2889–2898

Anestis A, Pörtner HO, Michaelidis B (2010a) Anaerobic metabolic patterns related to stress responses in hypoxia exposed mussels *Mytilus galloprovincialis*. J Exp Mar Biol Ecol 394:123–133

Anestis A, Karayiannis D, Angelidis P, Staikou A, Michaelidis B (2010b) Response of *Mytilus galloprovincialis* (L.) to increasing seawater temperature and to marteiliosis: metabolic and physiological parameters. Comp Biochem Physiol A 156:57–66

Angelidis P, Virvilis C, Photis G, Chollet B, Berthe F (2001) First report of Marteilia disease of the flat oyster Ostrea edulis, in the gulf of Thessaloniki, Greece. In: 10th international conference on 'Diseases of Fish and Shellfish' EAFP, 10–14 September, Dublin

Angiletta MJ (2009) Thermal adaptation: a theoretical and empirical synthesis. Oxford University Press, Oxford

Astraldi M, Bianchi CN, Gasparini GP, Morri C (1995) Climatic fluctuations, current variability and marine species distribution: a case study in the Ligurian Sea (north-west Mediterranean). Oceanol Acta 18:139–149

Barua D, Heckathorn AS (2004) Acclimation of the temperature set-points of the heat-shock response. J Therm Biol 29:185–193

Bayne LB, Bayne JC, Carefoot CT, Thompson JR (1976) The physiological ecology of *Mytilus californianus* Conrad 1. Metabolism and energy balance. Oecologia (Berl) 22:211–228

Beniash E, Ivanina A, Lieb NS, Kurochkin I, Sokolova IM (2010) Elevated levels of carbon dioxide affect metabolism and shell formation in oysters *Crassostrea virginica* (Gmelin). Mar Ecol Prog Ser 419:95–108

Béthoux JP, Gentili B, Raune J, Tailliez D (1990) Warming trend in the western Mediterranean deep water. Nature 347:660–662

Bocchetti R, Regoli F (2006) Seasonal variability of oxidative biomarkers, lysosomal parameters, metallothioneins and peroxisomal enzymes in the Mediterranean mussel *Mytilus galloprovincialis* from Adriatic Sea. Chemosphere 65:913–921

Bocchetti R, Lamberti CV, Pisanelli B, Razzetti EM, Maggi C, Catalano B, Sesta G, Martuccio G, Gabellini M, Regoli F (2008) Seasonal variations of exposure biomarkers, oxidative stress responses and cell damage in the clams, *Tapes philippinarum*, and mussels, *Mytilus galloprovincialis*, from Adriatic Sea. Mar Environ Res 66:24–26

Box A, Sureda A, Galgani F, Pons A, Deudero S (2007) Assessment of environmental pollution at Balearic Islands applying oxidative stress biomarkers in the mussel *Mytilus galloprovincialis*. Comp Biochem Physiol C 146:531–539

Brooks SPJ, Storey KB (1997) Glycolytic controls in estivation and anoxia: a comparison of metabolic arrest in land and marine molluscs. Comp Biochem Physiol A 118:1103–1114

Buckley AB, Owen ME, Hofmann GE (2001) Adjusting the thermostat: the threshold induction temperature for the heatshock response in intertidal mussels (genus *Mytilus*) changes as a function of thermal history. J Exp Biol 204:3571–3579

Caldeira K, Wickett ME (2005) Ocean model predictions of chemistry changes from carbon dioxide emissions to the atmosphere and ocean. J Geophys Res 110(C9):C09S04

Calvo E, Simó R, Coma R et al (2011) Effects of climate change on Mediterranean marine ecosystems: the case of the Catalan Sea. Clim Res 50:1–29

Carballal MJ, Villalba A, Lopez C (1998) Seasonal variation and effects of age, food availability, size, gonadal development, and parasitism on the hemogram of *Mytilus galloprovincialis*. J Invertebr Path 72:304–312

Chapple JP, Smerdon GR, Berry RJ, Hawkins AJS (1998) Seasonal changes in stress-70 protein levels reflect thermal tolerance in the marine bivalve *Mytilus edulis* L. J Exp Mar Biol Ecol 229:53–68

Cherkasov AS, Biswas PK, Ridings DM, Ringwood AH, Sokolova IM (2006) Effects of acclimation temperature and cadmium exposure on cellular energy budgets in a marine mollusk *Crassostrea virginica*: linking cellular and mitochondrial responses. J Exp Biol 209:1274–1284

Coma R, Ribes M (2003) Seasonal energetic constraints in Mediterranean benthic suspension feeders: effects at different levels of ecological organization. Oikos 101:205–215

Coma R, Ribes M, Serrano E, Jiménez E, Salat J, Pascualc J (2009) Global warming-enhanced stratification and mass mortality events in the Mediterranean. PNAS 106:6176–6181

Dahlhoff EP (2004) Biochemical indicators of stress and metabolism: applications for marine ecological studies. Annu Rev Physiol 66:183–207

Dahlhoff EP, Menge BA (1996) Influence of phytoplankton concentration and wave exposure on the ecophysiology of *Mytilus californianus*. Mar Ecol Prog Ser 144:97–107

Dietz TJ, Somero GN (1992) Species- and tissue-specific synthesis patterns for heat-shock proteins HSP70 and HSP90 in several marine teleost fishes. Physiol Zool 66:863–880

Durrieu de Madron X, Guieu C, Sempéré R, Conan P, Cossa D, D'Ortenzio F et al (2011) Marine ecosystems' responses to climatic and anthropogenic forcings in the Mediterranean. Prog Oceanogr 91:97–166

Fearman J, Moltschaniwskyja NA (2010) Warmer temperatures reduce rates of gametogenesis in temperate mussels, *Mytilus galloprovincialis*. Aquaculture 305:120–125

Feder ME, Hofmann GE (1999) Heat-shock proteins, molecular chaperones, and the stress response: evolutionary and ecological physiology. Annu Rev Physiol 61:243–282

Figueras AJ, Montes J (1988) Aber disease of edible oysters caused by *Marteilia refringens*. In: Fisher WS (ed) Disease processes in marine bivalve molluscs. American Fisheries Society, Special Publication, Bethesda, pp 38–46

Figueras AJ, Jardon CF, Caldas JR (1991) Diseases and parasites of rafted mussels (*Mytilus galloprovincialis* Lmk): preliminary results. Aquaculture 99:17–33

Francour P, Boudouresque CF, Harmelin JG, Harmelin-Vivien ML, Quignard JP (1994) Are the Mediterranean waters becoming warmer?Information from biological indicators. Mar Pollut Bull 28(9):523–526

Galimany E, Ramón M, Ibarrola I (2011) Feeding behavior of the mussel *Mytilus galloprovincialis* (L.) in a Mediterranean estuary: a field study. Aquaculture 314:236–243

Gambaiani DD, Mayol P, Isaac SJ, Simmonds MP (2009) Potential impacts of climate change and greenhouse gas emissions on Mediterranean marine ecosystems and cetaceans. J Mar Biol Assoc UK 89(1):179–201

Gardner JPA (2000) Where are the mussels on Cook Strait (New Zealand) shores? Low seston quality as a possible factor limiting multi-species distributions. Mar Ecol Prog Ser 194:123–132

Gattuso JP, Hansson L (2011) Ocean acidification: background and history. In: Gattuso JP, Hansson L (eds) Ocean acidification. Oxford University Press, Oxford, pp 1–20

Gonzalez JG, Yevich P (1976) Responses of an estuarine population of the blue mussel *Mytilus edulis* to heated water from a steam generating plant. Mar Biol 34:177–189

Gracey AY, Cossins AR (2003) Application of microarray technology in environmental and comparative physiology. Annu Rev Physiol 65:231–259

Gracey AY, Chaney ML, Boomhower PJ, Tyburczy RW, Connor K, Somero GN (2008) Rhythms of gene expression in a fluctuating intertidal environment. Curr Biol 18:1501–1507

Halliwell B, Gutteridge JMC (2007) Free radicals in biology and medicine. Oxford University Press, Oxford/New York

Halpin PM, Sorte CJ, Hofmann GE, Menge BA (2002) Patterns of variation in levels of Hsp70 in natural rocky shore populations from microscales to mesoscales. Integr Comp Biol 42:815–824

Harley CDG, Hughes AR, Hultgren KM et al (2006) The impacts of climate change in coastal marine systems. Ecol Lett 9:228–241

Harvell CD, Kim K, Burkholder JM et al (1999) Emerging marine diseases – climate links and anthropogenic factors. Science 285:1505–1510

Heise K, Puntarulo S, Pörtner HO, Abele D (2003) Production of reactive oxygen species by isolated mitochondria of the Antarctic bivalve *Laternula elliptica* (King and Broderip) under heat stress. Comp Biochem Physiol C 134:79–90

Heise K, Puntarulo S, Nikinmaa M, Abele D, Pörtner HO (2006) Oxidative stress during stressful heat exposure and recovery in the North Sea eelpout (*Zoarces viviparus*). J Exp Biol 209:353–363

Helmuth B (1999) Thermal biology of rocky intertidal mussels: quantifying body temperatures using climatological data. Ecology 80:15–34

Helmuth B (2009) From cells to coastlines: how can we use physiology to forecast the impacts of climate change? J Exp Biol 212:753–760

Helmuth B, Hofmann GE (2001) Microhabitats, thermal heterogeneity, and patterns of physiological stress in the rocky intertidal zone. Biol Bull 201:374–384

Helmuth B, Harley CDG, Halpin PM, O'Donnell M, Hofmann GE, Blanchette CA (2002) Climate change and latitudinal patterns of intertidal thermal stress. Science 298:1015–1017

Helmuth B, Mieszkowska N, Moore P, Hawkins SJ (2006) Living on the edge of two changing worlds: forecasting the responses of rocky intertidal ecosystems to climate change. Ann Rev Ecol Evol Syst 37:373–404

Helmuth B, Broitman RB, Yamane L, Gilman ES, Mach K, Mislan KAS, Denny WM (2010) Organismal climatology: analyzing environmental variability at scales relevant to physiological stress. J Exp Biol 213:995–1003

Helson JG, Pledger S, Gardner APJ (2007) Does differential particulate food supply explain the presence of mussels in Wellington Harbour (New Zealand) and their absence on neighbouring Cook Strait shores? Estuarine Coastal Shelf Sci 72:223–234

Hofmann GE (2005) Patterns of Hsp gene expression in ectothermic marine organisms on small to large biogeographic scales. Integr Comp Biol 45:247–255

Hofmann GE, Gaines SD (2008) New tools to meet new challenges: emerging technologies for managing marine ecosystems for resilience. Bioscience 58(1):43–52

Hofmann GE, Place SP (2007) Genomics-enabled research in marine ecology: challenges, risks and pay-offs. Mar Ecol Prog Ser 332:249–255

Hofmann GE, Somero GN (1995) Evidence for protein damage at environmental temperatures: seasonal changes in levels of ubiquitin conjugates and hsp70 in the intertidal mussel *Mytilus trossulus*. J Exp Biol 198(1509):1518

Hofmann GE, Somero GN (1996a) Interspecific variation in thermal denaturation of proteins in the congeneric mussels *Mytilus trossulus* and *M. galloprovincialis*: evidence from the heat-shock response and protein ubiquitination. Mar Biol 126:65–75

Hofmann GE, Somero GN (1996b) Protein ubiquitination and stress protein synthesis in *Mytilus trossulus* occurs during recovery from tidal emersion. Mol Mar Biol Biotechnol 5:175–184

Ioannou S, Anestis A, Pörtner HO, Michaelidis B (2009) Seasonal patterns of metabolism and the heat shock response (HSR) in farmed mussels *Mytilus galloprovincialis*. J Exp Mar Biol Ecol 381:136–144

IPCC (2007) Climate change 2007: the physical science basis. Contribution of Working Group I to the fourth assessment report of the Intergovernmental Panel on Climate change [Solomon S, Qin D, Manning D, Chen Z, Marquis M, Averyt KB, Tignor M, Miller HL (eds)]. Cambridge University Press, Cambridge/New York, pp 996

Ivanina AI, Taylor C, Sokolova IM (2009) Effects of elevated temperature and cadmium exposure on stress protein response in eastern oysters *Crassostrea virginica* (Gmelin). Aquat Toxicol 91:245–254

Jeftic L, Milliman JD, Sestini G (1992) Climate change and the Mediterranean. Edward Arnold, London

Kabakov EA, Gabai V (1994) Heat-shock proteins maintain the viability of ATP-deprived cells: what is the mechanism? Trends Cell Biol 4:193–195

Karagiannis D, Angelidis P (2007) Infection of cultured mussels *Mytilus galloprovincialis* by the protozoan *Marteilia sp.* in the Thermaikos Gulf (N Greece). Bull Eur Assoc Fish Pathol 27(4):131–141

Katsikatsou M, Anestis A, Pörtner HO, Vratsistas A, Aligizaki K, Michaelidis B (2012) Field studies and projections of climate change effects on the bearded horse mussel *Modiolus barbatus* in the Gulf of Thermaikos, Greece. Mar Ecol Prog Ser 449:183–196

Kawasaki T, Tanaka S, Toba Y, Taniguchi A (1991) Long–term variability of pelagic fish populations and their environment. Pergamon Press, Oxford/New York/Beijing/Frankfort/Seoul/Sydney/Tokyo, p 402

Kültz D (2005) Molecular and evolutionary basis of the cellular stress response source. Annu Rev Physiol 67:225–257

Lannig G, Flores JF, Sokolova IM (2006) Temperature dependent stress response in oysters, *Crassostrea virginica*: pollution reduces temperature tolerance in oysters. Aquat Toxicol 79:278–287

Lannig G, Eilers S, Pörtner HO, Sokolova IM, Bock C (2010) Impact of ocean acidification on energy metabolism of oyster, *Crassostrea gigas* – changes in metabolic pathways and thermal response. Mar Drugs 8(8):2318–2339

Lesser PM, Kruse AV (2004) Seasonal temperature compensation in the horse mussel, *Modiolus modiolus*: metabolic enzymes, oxidative stress and heat shock proteins. Comp Biochem Physiol A 137:495–504

Lesser PM, Bailey AM, Merselis GD, Morrison RJ (2010) Physiological response of the blue mussel *Mytilus edulis* to differences in food and temperature in the Gulf of Maine. Comp Biochem Physiol A 156:541–551

Lockwood LB, Sanders GJ, Somero GN (2010) Transcriptomic responses to heat stress in invasive and native blue mussels (genus *Mytilus*): molecular correlates of invasive success. J Exp Biol 213:3548–3558

Lowe DM, Moore MN, Bayne BL (1982) Aspects of gametogenesis in the marine mussel *Mytilus edulis* L. J Mar Biol Assoc UK 62:133–145

Lushchak VI (2010) Environmentally induced oxidative stress in aquatic animals. Aquat Toxicol 101:13–30

MacLeod CD, Hauser N, Peckham H (2004) Diversity, relative density and structure of the cetacean community in summer months east of Great Abaco, Bahamas. J Mar Biol Assoc UK 84:469–474

Mallouk Y, Vayssier-Taussat M, Bonventre JV, Polla BS (1999) Heat shock protein 70 and ATP as partners in cell homeostasis. Int J Mol Med 4:463–474

Marasovič I, Nincevic L, Kusflpilic G, Marinovic S, Marinov S (2005) Long-term changes of basic biological and chemical parameters at two stations in the middle Adriatic. J Sea Res 54:3–14

McCarty JP (2001) Ecological consequences of recent climate change. Conserv Biol 15:320–331

McQuaid CD, Lindsay TL (2000) Effect of wave exposure on growth and mortality rates of the mussel *Perna perna*: bottom-up regulation of intertidal populations. Mar Ecol Prog Ser 206:147–154

Melzner F, Mark FC, Pörtner HO (2007) Role of blood-oxygen transport in thermal tolerance of the cuttlefish, *Sepia officinalis*. Integr Comp Biol 47:645–655

Menge BA (2000) Top-down and bottom-up community regulation in marine rocky intertidal habitats. J Exp Mar Biol Ecol 250:257–289

Menge BA, Daley BA, Wheeler PA, Strub PT (1997) Rocky intertidal oceanography: an association between community structure and near-shore phytoplankton concentration. Limnol Oceanogr 42:57–66

Menge BA, Daley BA, Lubchenco J, Sanford E, Dahlhoff E, Halpin PM, Hudson G, Burnaford JL (1999) Top-down and bottom-up regulation of New Zealand rocky intertidal communities. Ecol Monogr 69(3):297–330

Menge BA, Olson AM, Dahlhoff EP (2002) Environmental stress, bottom-up effects, and community dynamics: integrating molecular-physiological and ecological approaches. Integr Comp Biol 42:892–908

Menge BA, Chan F, Lubchenco J (2008) Response of a rocky intertidal ecosystem engineer and community dominant to climate change. Ecol Lett 11:151–162

Mitchell DT, Hulme M (1999) Predicting regional climate change: living with uncertainty. Prog Phys Geogr 23:57–78

Navarro E, Iglesias JIP, Perez Camacho A, Labarta U, Beiras R (1991) The physiological energetics of mussels (*Mytilus galloprovincialis* Lmk) from different cultivation rafts in the Ria de Arosa (Galicia, N.W. Spain). Aquaculture 94:197–212

Newell RC (1979) Biology of intertidal animals, 3rd edn. Marine Ecological Surveys, Faversham

Nicholls RJ, Hoozemans FMJ (1996) The Mediterranean: vulnerability to coastal implication of climate change. Ocean Coast Manag 31:105–132

Palleros DR, Welch WJ, Fink AL (1991) Interaction of hsp70 with newly synthesized proteins: effects of temperature and nucleotides on the kinetics of binding. Proc Natl Acad Sci USA 88:5719–5723

Parmesan C (2006) Ecological and evolutionary responses to recent climate change. Annu Rev Ecol Evol Syst 37:637–669

Parsell DA, Lindquist S (1993) The function of heat-shock proteins in stress tolerance: degradation and reactivation of damaged proteins. Annu Rev Genet 27:437–496

Peharda M, Župan I, Bavčević L, Frankić A, Tin Klanjšček T (2007) Growth and condition index of mussel *Mytilus galloprovincialis* in experimental integrated aquaculture. Aquac Res 38:1714–1720

Petes EL, Menge AB, Murphy DG (2007) Environmental stress decreases survival, growth, and reproduction in New Zealand mussels. J Exp Mar Biol Ecol 351:83–91

Petes EL, Menge AB, Harris IA (2008) Intertidal mussels exhibit energetic trade-offs between reproduction and stress resistance. Ecol Monogr 78:387–402

Petrović S, Semenčić L, Ozretić B, Ozretić M (2004) Seasonal variations of physiological and cellular biomarkers and their use in the biomonitoring of north adriatic coastal waters (Croatia). Mar Pollut Bull 49:713–720

Pisanelli B, Benedetti M, Fattorini D, Regoli F (2009) Seasonal and inter-annual variability of DNA integrity in mussels *Mytilus galloprovincialis*: a possible role for natural fluctuations of trace metal concentrations and oxidative biomarkers. Chemosphere 77:1551–1557

Pörtner HO (2001) Climate change and temperature dependent biogeography: oxygen limitation of thermal tolerance in animals. Naturwissenschaften 88:137–146

Pörtner HO (2002a) Climate change and temperature dependent biogeography: systemic to molecular hierarchies of thermal tolerance in animals. Comp Biochem Physiol 132A:739–761

Pörtner HO (2002b) Physiological basis of temperature dependent biogeography: tradeoffs in muscle design and performance in polar ectotherms. J Exp Biol 205:2217–2230

Pörtner HO (2006) Climate dependent evolution of Antarctic ectotherms: an integrative analysis (EASIZ, SCAR). Deep Sea Res II 53:1071–1104

Pörtner HO (2010) Oxygen- and capacity-limitation of thermal tolerance: a matrix for integrating climate-related stressor effects in marine ecosystems. J Exp Biol 213:881–893

Pörtner HO, Farrell AP (2008) Physiology and climate change. Science 322:690–692

Pörtner HO, Knust R (2007) Climate change affects marine fishes through the oxygen limitation of thermal tolerance. Science 315:95–97

Pörtner HO, Langenbuch M, Reipschläger A (2004) Biological impact of elevated ocean CO_2 concentrations: lessons from animal physiology and earth history. J Oceanogr 60:705–718

Pörtner HO, Lucassen M, Storch D (2005a) Metabolic biochemistry: its role in thermal tolerance and in the capacities and in the capacities of physiological and ecological function. Fish Physiol 22:79–118

Pörtner HO, Langenbuch M, Michaelidis B (2005b) Synergetic effects of increased CO_2, temperature and hypoxia on marine animals. From earth history to global change. J Geophys Res-Oceans 110(C9): art. no. C09S10 SEP 23

Ramón M, Fernández M, Galimany E (2007) Development of mussel (*Mytilus galloprovincialis*) seed from two different origins in a semi-enclosed Mediterranean Bay (N.E. Spain). Aquaculture 264:148–159

Rayyan A, Damianidis P, Antoniadou C, Chintiroglou CC (2006) Protozoan parasites in cultured mussels *Mytilus galloprovincialis* in the Thermaikos gulf (north Aegean Sea, Greece). Dis Aquat Organ 70:251–254

Regoli F, Cerrano C, Chierici E, Chiantore MC, Bavestrello G (2004) Time-course variations of oxyradical metabolism, DNA integrity and lysosomal stability in mussels, *Mytilus galloprovincialis*, during a field translocation experiment. Aquat Toxicol 68:167–178

Ren JS, Ross AH (2005) Environmental influence on mussel growth: a dynamic energy budget model and its application to the greenshell mussel *Perna canaliculus*. Ecol Model 189:347–362

Resgalla C Jr, Brasil ES, Laitano KS, Reis Filho RW (2007) Physioecology of the mussel *Perna perna* (Mytilidae) in southern Brazil. Aquaculture 270:464–474

Richardson AJ (2008) In hot water: zooplankton and climate change. ICES J Mar Sci 65:279–295

Robert R, Borel M, Pichot Y, Trut G (1991) Growth and mortality of the European oyster *Ostrea edulis* in the Bay of Arcachon (France). Aquat Living Resour 4:265–274

Roberts DA, Hofmann GE, Somero GN (1997) Heat shock protein expression in *Mytilus californianus*: acclimatization (seasonal and tidal-height comparisons) and acclimation effects. Biol Bull 192:309–320

Roberts CM, McClean CJ, Veron JEN et al (2002) Marine biodiversity hotspots and conservation priorities for tropical reefs. Science 295:1280–1284

Santovito G, Piccinni E, Cassini A, Irato P, Albergoni V (2005) Antioxidant responses of the Mediterranean mussel, *Mytilus galloprovincialis*, to environmental variability of dissolved oxygen. Comp Biochem Physiol C 140:321–329

Sasikumar G, Krishnakumar KP (2011) Aquaculture planning for suspended bivalve farming systems: the integration of physiological response of green mussel with environmental variability in site selection. Ecol Indic 11:734–740

Saxby SA (2002) A review of food availability, sea water characteristics and bivalve growth performance at coastal culture sites in temperate and warm temperate regions of the world. Fish Res Rep No 132, pp 42

Schulte HE (1975) Influence of algal concentration and temperature on the filtration rate of *Mytilus edulis*. Mar Biol 30:331–341

Seed R, Suchanek TH (1992) Population and community ecology of Mytilus. In: Gosling E (ed) The mussel *Mytilus*: ecology, physiology, genetics and culture. Elsevier, Amsterdam, pp 87–157

Smaal AS, Widdows J (1994) The scope for growth of bivalves as an integrated response parameter in biological monitoring. In: Kramer KJM (ed) Biomonitoring of coastal waters and estuaries. CRC Press, Boca Raton, pp 247–262

Smith RJ, Fong P, Ambrose FR (2006) Dramatic declines in mussels bed community: response to climate change? Ecol Lett 87:1153–1161

Sobral P, Widdows J (1997) Effects of elevated temperatures on the scope for growth and resistance to air exposure of the clam *Ruditapes decussatus* (L.), from southern Portugal. Sci Mar 61:163–171

Sokolova IM, Lannig G (2008) Interactive effects of metal pollution and temperature on metabolism in aquatic ectotherms: implications of global climate change. Clim Res 37:181–201

Somero GN (2002) Thermal physiology and vertical zonation of intertidal animals: optima, limits, and costs of living. Integr Comp Biol 42:780–789

Somero GN (2005) Linking biogeography to physiology: evolutionary and acclimatory adjustments of thermal limits. Front Zool 2. http://www.frontiersinzoology.com/content/2-1/1

Somero NG (2010) The physiology of climate change: how potentials for acclimatization and genetic adaptation will determine 'winners' and 'losers'. J Exp Biol 213:912–920

Somero GN (2012) The physiology of global change: linking patterns to mechanisms. Annu Rev Mar Sci 4:2.1–2.23

Sommer A, Klein B, Pörtner HO (1997) Temperature induced anaerobiosis in two populations of the polychaete worm *Arenicola marina* (L.). J Comp Physiol 167B:25–35

Storey KB, Storey JM (1990) Metabolic rate depression and biochemical adaptation in anaerobiosis, hibernation and aestivation. Q Rev Biol 65:145–174

Thomas CD, Cameron A, Green RE et al (2004) Extinction risk from climate change. Nature 427:145–148

Tomanek L (2008) The importance of physiological limits in determining biogeographical range shifts due to global change: the heat-shock response. Physiol Biochem Zool 81:709–717

Tomanek L (2010) Variation in the heat shock response and its implication for predicting the effect of global climate change on species' biogeographical distribution ranges and metabolic costs. J Exp Biol 213:971–979

Tomanek L (2011) Environmental proteomics: changes in the proteome of marine organisms in response to environmental stress, pollutants, infection, symbiosis, and development. Annu Rev Mar Sci 3:14.1–14.27

Tomanek L, Sanford E (2003) Heat-shock protein 70 (Hsp70) as a biochemical stress indicator: an experimental field test in two congeneric intertidal gastropods (Genus: *Tegula*). Biol Bull 205:276–284

Tomanek L, Somero GN (1999) Evolutionary and acclimation-induced variation in the heat-shock responses of congeneric marine snails (genus tegula) from different thermal habitats: implications for limits of thermotolerance and biogeography. J Exp Biol 202:2925–2936

Tomanek L, Somero GN (2002) Interspecific- and acclimationinduced variation in levels of heat-shock proteins 70 (hsp70) and 90 (hsp90) and heat-shock transcription factor-1 (HSF1) in congeneric marine snails (genus *Tegula*): implications for regulation of *hsp* gene expression. J Exp Biol 205:677–685

Tomanek L, Zuzow JM (2010) The proteomic response of the mussel congeners *Mytilus galloprovincialis* and *M. trossulus* to acute heat stress: implications for thermal tolerance limits and metabolic costs of thermal stress. J Exp Biol 213:3559–3574

Tremblay R, Myrand B, Sevigny JM, Guderley H (1998a) Bioenergetic and genetic parameters in relation to susceptibility of blue mussels, *Mytilus edulis* (L) to summer mortality. J Exp Mar Biol Ecol 221:27–58

Tremblay R, Myrand B, Guderley H (1998b) Temporal variation of lysosomal capacities in relation to susceptibility of mussels, *Mytilus edulis*, to summer mortality. Mar Biol 132:641–649

Van der Veer WH, Cardoso JFMF, Peck AM, Kooijman LMS (2009) Physiological performance of plaice *Pleuronectes platessa* (L.): a comparison of static and dynamic energy budgets. J Sea Res 62:83–92

Vargas-Yánẽz M, Garcia MJ, Salat J, Garcia-Martinez MC, Pascual J, Moya F (2008) Warming trends and decadal variability in the western Mediterranean shelf. Glob Planet Change 63:177–184

Vargas-Yánẽz M, Zunino P, Benali A, Delpy M, Pastre F, Moya F, Carmen García-Martínez M-C, Tel E (2010) How much is the western Mediterranean really warming and salting?. J Geophys Res 115:C04001, 12 PP

Verlecar XN, Jena KB, Chainy GBN (2007) Biochemical markers of oxidative stress in *Perna viridis* exposed to mercury and temperature. Chem Biol Interact 167:219–226

Vidal-Liñán L, Bellas J, Campillo JA, Beiras R (2010) Integrated use of antioxidant enzymes in mussels, *Mytilus galloprovincialis*, for monitoring pollution in highly productive coastal areas of Galicia (NW Spain). Chemosphere 78:265–272

Villalba A, Mourelle SG, Carballal MJ, Lopez MC (1993a) Effects of infection by the protistan parasite *Marteilia refringens* on the reproduction of cultured mussels *Mytilus galloprovincialis* in Galicia (NW Spain). Dis Aquat Organ 17:205–213

Villalba A, Mourelle SG, Carballal MJ, Lopez MC, Azevedo C (1993b) Marteiliasis affecting cultured mussels *Mytilus galloprovincialis* of Galicia NW Spain: I. Etiology, phases of the infection, and temporal and spatial variability in prevalence. Dis Aquat Organ 16:61–72

Virvilis C, Angelidis P (2006) Presence of the parasite *Marteilia* sp. in the flat oyster (*Ostrea edulis* L) in Greece. Aquaculture 25:1–5

Virvilis C, Angelidis P, Photis G (2003) Presence of the parasite *Marteilia* sp. in the shellfish of the Thermaikos Gulf in northern Greece. Bull Eur Assoc Fish Pathol 23:157–161

Weitere M, Vohmann A, Schulz N et al (2009) Linking environmental warming to the fitness of the invasive clam *Corbicula fluminea*. Glob Chang Biol 15:2838–2851

Widdows J (1973) The effects of temperature on the metabolism and activity of *Mytilus edulis* L. Neth J Sea Res 7:387–398

Holoplanktonic Mollusca: Development in the Mediterranean Basin During the Last 30 Million Years and Their Future

19

Arie W. Janssen and Katja T.C.A. Peijnenburg

Abstract

A short introduction is given to the systematics and morphology of pelagic Gastropoda (Mollusca: Heteropoda and Pteropoda), also explaining some details of the animals' particular way of life. The fossil record of these invertebrates is discussed briefly, followed by an overview of existing literature, focussing on the Mediterranean Basin. Developments during the Mediterranean Cenozoic are explained stage by stage and the numerical occurrences of species within the basin are correlated with palaeo-temperatures, generally with very good results. In the same context illustrations are added of index species or otherwise interesting taxa. The very intensively studied Quaternary occurrences are referred to briefly, citing other papers recently published. A preliminary Mediterranean biozonation scheme, recently developed and based on fossil Pteropoda, is referred to and schematically included in one of the illustrations. In the last part, we discuss possible future developments of holoplanktonic molluscs, resulting on the one hand from increased knowledge of the systematics, phylogeny and biogeography due to recent interest in this group and the application of modern powerful morphometric and molecular techniques. On the other hand, there exist major concerns about the future of this group resulting from the (combined) effects of global warming and ocean acidification.

Keywords:

Systematics • Morphology • Pelagic Gastropoda • Mollusca • Heteropoda • Pteropoda • Fossil records • Mediterranean basin • Mediterranean Cenozoic • Palaeo-temperatures • Mediterranean biozonation scheme • Fossil Pteropoda • Holoplanktonic molluscs • Phylogeny • Biogeography • Global warming • Ocean acidification

A.W. Janssen (✉)
Naturalis Biodiversity Center (NCB), P.O. Box 9517,
2300 RA Leiden, The Netherlands

Zeeweg 161, 2224 CE Katwijk, The Netherlands
e-mail: ariewjanssen@gmail.com

K.T.C.A. Peijnenburg
Naturalis Biodiversity Center (NCB), P.O. Box 9517,
2300 RA Leiden, The Netherlands

Institute for Biodiversity and Ecosystem Dynamics,
University of Amsterdam, P.O. Box 94248,
1090 GE Amsterdam, The Netherlands
e-mail: K.T.C.A.Peijnenburg@uva.nl

S. Goffredo and Z. Dubinsky (eds.), *The Mediterranean Sea: Its history and present challenges*,
DOI 10.1007/978-94-007-6704-1_19, © Springer Science+Business Media Dordrecht 2014

Introduction

Within the immense group of invertebrate animals united in the phylum Mollusca, the group of the Gastropoda, – snails – is the most well-known and by far the most rich in species. Gastropods find habitation in a wide range of environments, including land, fresh- and seawater, occurring from high elevations in mountains to abyssal depths in the ocean. Virtually all of the many thousands of known species live on solid surface, on land as well as in aquatic settings. In most cases they stay in contact with their 'terra-firma' by means of a special organ, called 'the foot', a very muscular structure suitable for locomotion, usually facilitated by mucus secretion as a lubrication agent. As gastropods seemingly creep on their stomach, this means of propulsion is the origin of the latinized word (of greek origin) 'gastropoda': from *γαστήρ* (*gastér*, stem: *gastr-*=stomach) and *πούς* (*poús*, stem: *pod-*=foot).

Gastropods are best known by their shells, which with their attractive shapes and colours are popular collector's items. This is especially true for tropical marine genera, including Cypraeidae, Conidae, Muricidae, Strombidae, Mitridae, Olividae etc., to such an extent that nowadays many countries maintain laws prohibiting collecting or export of shells, even when found dead on beaches. Traditionally gastropod systematics have not only been based on the anatomy of the soft parts, but also on characteristics of the shell morphology.

Among marine gastropods, however, some relatively small groups have developed a completely different, so-called pelagic or planktonic way of life, and live their lives free of contact with the sea bottom. A similar way of life is adopted by many other vertebrate and invertebrate animal groups, for many of them only during parts of their life cycle, e.g. during their larval stages. Molluscs living their entire life in the water column, from egg to maturity and death, are called 'holoplanktonic'. Their exclusively aragonitic shells – if present – may reach the sea bottom only after death or even be dissolved before reaching the bottom if this is situated below the aragonite compensation depth (Berger 1978).

Among the gastropods several systematically remote groups have independently developed a holoplanktonic way of life. The most familiar are those known as 'pteropods' and 'heteropods', nowadays obsolete but still commonly used names for what are formally called 'Thecosomata' and 'Gymnosomata' (for pteropods) and 'Littorinimorpha, Pterotracheoidea' (for heteropods) (Bouchet and Rocroi 2005). Also a few species of the Nudibranchia have a similar way of life. The gastropod family Janthinidae, the quite spectacular species of which live at or close to the surface of ocean waters, never near the bottom, belong to the pleuston, a special group of holoplanktonic organisms dependent on the water/air-interface.

A strong interest in the anatomy and systematics of pelagic gastropods (among others) developed from early in the nineteenth century onwards, partly because much material became available, collected during the many famous ocean expeditions executed in those days, resulting in abundant papers by authors such as Abildgård, de Blainville, delle Chiaje, Cuvier, de Férussac, Gegenbaur, Krohn, Lesueur, d'Orbigny, Péron, Poli, Rang, Souleyet, etc. An extensive summary of early research is given in Bronn and Keferstein (1862–1866). Numerous data from later research to modern times, as well as illustrations of the pelagic gastropod groups, are summarized from earlier literature in the beautiful book by Lalli and Gilmer (1989).

Systematics and General Morphology

Heteropoda

The superfamily Pterotracheoidea, nowadays included in the 'clade' Littorinimorpha, is subdivided into three families: Atlantidae, Carinariidae and Pterotracheidae. Detailed information on these heteropods can be found in Roger R. Seapy's (2011) pages on the Tree of Life website. Modifications to pelagic life in these families are a transformation of the original gastropod foot into a relatively large, vertical swimming fin, a reduction of the shell-wall thickness and the presence of surprisingly well-developed eyes. Soft parts are largely transparent and usually show internal organs without anatomical dissection. The animals are carnivorous and grab their prey by means of a long proboscis in which the radula is situated.

The family Atlantidae in the modern concept includes about 22 living species, the largest of which has a shell diameter of 1 cm only. They live predominantly in the epipelagic zone showing a day/night-rhythm: near the surface at darkness, at greater depth in daytime (negative fototaxis).

In the genus *Atlanta*, with about 19 or 20 Recent species, the entirely aragonitic, glassy transparent shells are coiled dextrally in a flat spiral, completely visible from the apical side of the shell, of which the last whorl is extended by a flattened, extremely thin but double-walled calcareous keel, considerably enlarging the shell's surface. The animal can withdraw completely into its shell, and seal the aperture with a cartilaginous plate, called the operculum. Identification of most species is rather difficult and in many cases is hardly possible using the shell alone; anatomical characteristics like eye-type, radula or operculum are frequently necessary for a reliable recognition of the species. In the genera *Protatlanta* and *Oxygyrus*, each with a single Recent species, anatomical features resemble the genus *Atlanta* closely, but show a

Fig. 19.1 *Carinaria lamarckii* de Blainville, 1817, living specimen photographed by scuba diving in the Florida Current (Courtesy of Ronald W. Gilmer)

partial reduction of shell calcification; in the former the keel surrounding the body whorl is entirely cartilaginous, in the latter the last half of the bodywhorl and the keel are cartilaginous. The *Oxygyrus* shell has a nautiloid shape, with its initial whorls covered by subsequent volutions.

Carinariidae (Fig. 19.1) are a group in which three Recent genera are distinguished: *Carinaria*, *Cardiapoda* and *Pterosoma*. The animals can be quite large (*Carinaria cristata* Linné, 1767 reaches a body size of 68 cm; Seapy 2011) and have reduced shells. In *Carinaria*, with five Recent species, and *Pterosoma* (one species) the shell is cap-shaped, consisting of a very small, dextrally coiled embryonal shell (protoconch), followed by a strongly widening, very fragile adult shell part (teleoconch) that covers the visceral organs. The animal cannot withdraw into the shell. In *Cardiapoda* the shell is strongly reduced, hardly more than a microscopic coiled protoconch, embedded within the soft parts.

In the Pterotracheidae two genera, monotypic *Firoloida*, and *Pterotrachea* with three species, are distinguished. The body length of the largest species reaches more than 30 cm (Seapy 2011), but the adults lack a shell, which is only present during the larval stage and is shed during metamorphosis. Such microscopic (less than 1 mm), dextrally coiled pterotracheid larval shells collected from bottom sediments cannot yet reliably be related to the known adult forms (Janssen 2012a). Loss of the shell in the adult stage and a streamlined body enabling rapid swimming movements make these heteropods among the molluscs best adapted to pelagic life (Seapy 2011).

Heteropoda prefer relatively high water temperatures and are therefore mainly distributed in tropical and subtropical water masses. They are predominantly known from plankton catches and bottom samples, and are hardly ever found washed ashore on beaches.

Pteropoda-Thecosomata (Fig. 19.2)

Pteropods have also developed special organs as adaptations to pelagic life. The original foot is transformed into two 'wings' or parapodia, which has given them the popular name of 'papillon de mer' or 'sea butterfly'. Systematically, the clade Thecosomata is subdivided into Euthecosomata and Pseudothecosomata. In Euthecosomata two superfamilies: Limacinoidea and Cavolinioidea are distinguished, all species of which have a shell. The animals feed in a very special, rather spectacular way by producing a mucus web (Gilmer 1972, 1974; Gilmer and Harbison 1986) many times larger than their own body size, in which microplankton is caught (phytoplankton, protozoans etc.) and subsequently ingested together with the web. Species in these groups are protandric hermaphrodites (Lalli and Wells 1978), changing their gender from male to female in the course of life. Apart from 'normal' sexual reproduction, a very curious asexual reproduction by strobilation or schizogamy has been described for the species *Clio pyramidata* Linné, 1767 (van der Spoel 1973, 1979; Pafort-van Iersel and van der Spoel 1986), that, however, by some authors was explained as a preservation artifact.

In the Euthecosomata some 60–65 Recent species are now distinguished (quite a few of them still insufficiently characterized), subdivided into two superfamilies: Limacinoidea, with sinistrally coiled shells, and Cavolinioidea, with uncoiled, conical or bilaterally symmetrical shells. Most Euthecosomata are either epi- (0 – c. 200 m water depth) or mesopelagic (200–1,000 m), but some species are bathypelagic and live in water depths of 1,000 m or more. A negative fototaxis, as in heteropods, is known for a large number of species. Most species show a preference for warmer environments and are distributed in tropical and subtropical realms. In colder water the species diversity is reduced, but individuals frequently occur in dense populations and form together with 'krill' (euphausid crustaceans) an important component of the food source of fishes and balaenid whales.

In Recent Limacinoidea only one family, Limacinidae, is recognised, comprising three genera (*Heliconoides*, *Limacina* and *Thielea*) and seven Recent species. In one species, *Heliconoides inflata* (d'Orbigny, 1834), brood protection occurs (Lalli and Wells 1973).

Recent Cavolinioidea are subdivided into four families, viz. Creseidae (6 species), Cuvierinidae (5 or 6 species), Cliidae (10 or 11 species; Fig. 19.3) with respectively conical, bottle-shaped or pyramidal shells, and Cavoliniidae (35–40 species?), where the shell is more or less bulbous in shape with partly separated dorsal and ventral parts. Cavoliniidae nowadays include three Recent genera, *Cavolinia*, *Diacavolinia* and *Diacria*. Many new taxa have been introduced during the last decades, especially in these last two genera partly based on variations in colour pattern of the shell (e.g. Bontes and van der Spoel 1998) or sometimes

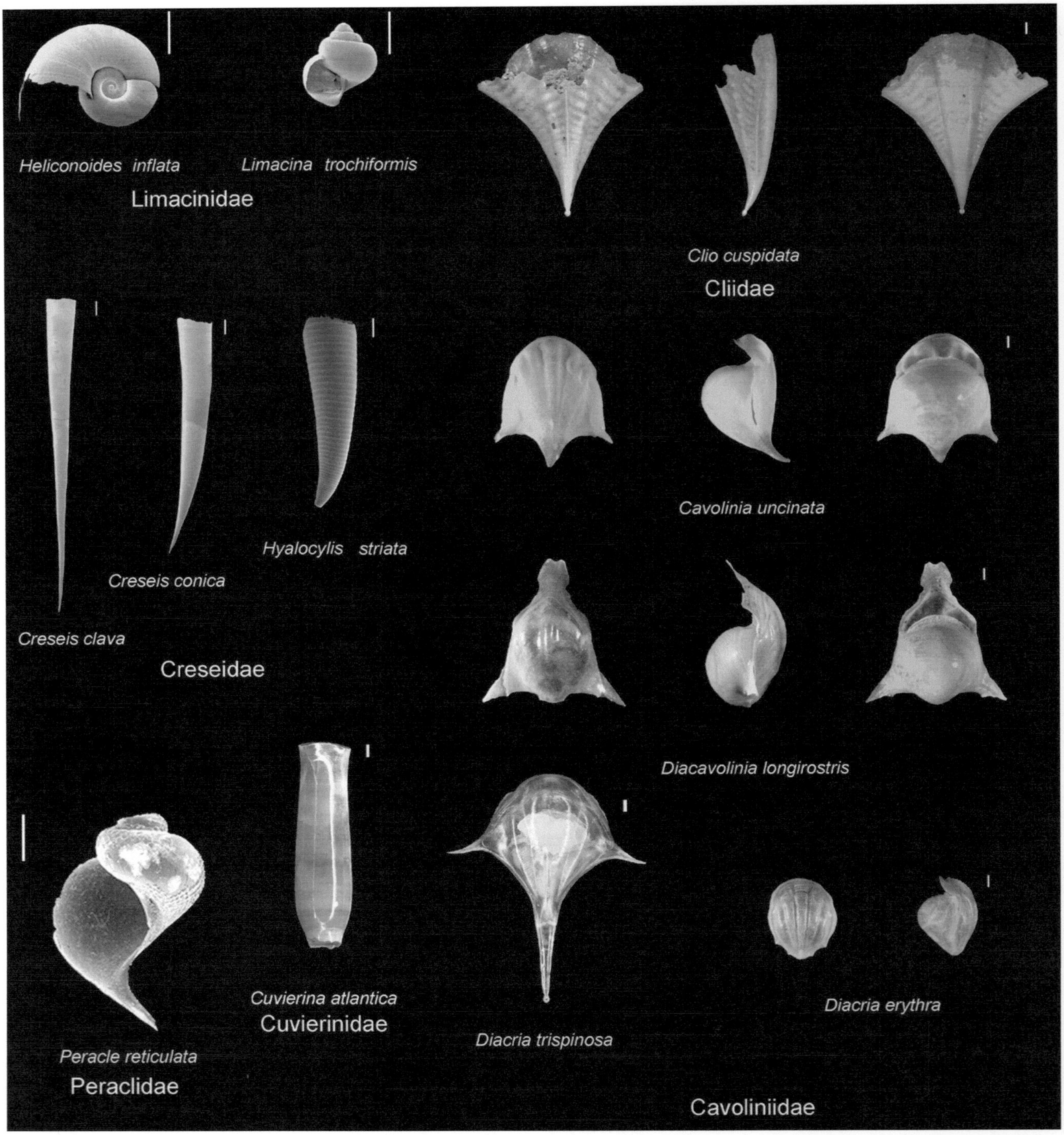

Fig. 19.2 Examples of Recent Thecosomata shells (*Cuvierina atlantica* and *Diacria trispinosa*, courtesy Alice Burridge and Christine Hörnlein; other figures after Janssen 2007, 2012a). Bar lengths represent 0.5 mm

very small differences in shell morphology (e.g. van der Spoel et al. 1993), the taxonomic value of which still has to be ascertained.

In the Pseudothecosomata, differing in some anatomical respects from Euthecosomata, some species are shelled throughout life and others only have a shell during the larval stage, but in the adult stage are provided with an internal cartilaginous-gelatinous 'pseudoconch'. About 20 species are distinguished in the superfamily Cymbulioidea, belonging to three families. Only in the Peraclidae, including a single

Fig. 19.3 Living specimen of *Clio pyramidata* Linné, 1767 f. *lanceolata* (Lesueur 1813); living specimen from the northern Atlantic Ocean. Specimen size c. 15 mm (Courtesy Russell R. Hopcroft)

Fig. 19.4 Living specimen of the gymnosome pteropod *Clione limacina* (Phipps 1774) from Canada Basin (Arctic Ocean), imaged in the lab. Specimen size approximately 8 mm (Courtesy Russel R. Hopcroft, University of Alaska)

genus *Peracle* with seven Recent species, is a shell present in the adult stage. Cymbuliidae (10 Recent species, several of them with as yet unacknowledged validity) have a minuscule shell during the premetamorphosis stage. Such a larval shell has not yet been recorded from the Desmopteridae, of which two or three species are known.

Pteropoda-Gymnosomata (Fig. 19.4)

Although the Gymnosomata ('sea angels'), a group of (in the adult state) shell-less pteropods probably includes some 50 living species they remain relatively little known. As they demonstrate strongly different anatomical features from those of thecosomes, for a very long time they have been considered as systematically only distantly related. Recently, however, molecular and renewed anatomical research has demonstrated that they are probably sister groups, and it was proposed to unite them (again!) with the formal name of Pteropoda (Klussmann-Kolb and Dinapoli 2006). Gymnosomes live in open ocean settings and can only be obtained by careful plankton trawling. They are hardly ever found in shallow water and virtually never wash ashore on beaches. Preservation is difficult, as specimens caught alive tend to deform strongly and collapse obscuring their characters when preserved in alcohol or formaldehyde.

Krohn (1860) had already described the larvae and their shells of several gymnosomatous molluscs, but was not able to match them to any of the adult species. Lalli and Conover (1973, 1976) were the first to describe the minute larval shells (protoconchs) of some identified species of gymnosomatous pteropods by successfully reproducing specimens under laboratory conditions from eggs to larvae and keeping them alive beyond metamorphosis. Despite the fact that such protoconchs had been collected earlier from bottom samples (Fig. 19.5) they were not recognised as gymnosome shells but erroneously taken for protoconchs of *Clio* species. After publication of Lalli and Gilmer's papers this was rectified for Mediterranean and Red Sea observations by Grecchi (1987). Several more types of gymnosomatous larval shells have been found (van der Spoel and Diester-Haass 1976; Kunz 1996; Janssen 2007, 2012a, b), but none of them can yet be reliably related to one of the known species.

One of the more abundant gymnosomes, the colder water species *Clione limacina* (Phipps, 1774) (Fig. 19.4), apparently easily kept under laboratory conditions, has since the 1980s been the subject of numerous papers on neurophysiology, appearing in periodicals like Experimental Brain Research, Neurophysiology, Journal of Comparative Physiology, Journal of Experimental Physiology, etc., as well as in various Russian journals (for an introduction see the website of Bullock 2011).

Fig. 19.5 Examples of unidentified Recent Gymnosomata larval shells (After Kunz 1996 in Janssen 2007). Bar length represents 0.1 mm

The Fossil Record

General

Obviously, only shelled species can be expected to occur as fossils. Holoplanktonic molluscs usually have an extensive horizontal distribution along climate zones, influenced by ocean currents and many groups show a rapid development. Such features make them virtually perfect index fossils for biostratigraphical purposes. But, because of their usually reduced shell-wall thickness, pteropods and heteropods are extremely fragile which severely reduces their fossilization potential. Their large size compared to other pelagic fossils used in biostratigraphy (planktonic foraminifera, calcareous nannoplankton, dinoflagellates) as well as their aragonitic nature, however, make them less practical instruments in biostratigraphy, as large samples are necessary to obtain a fair impression of an assemblage. Especially in boreholes, from which only restricted quantities of sediment are usually available, that feature hampers a regular usage of these fossils. Aragonite is a rather easily dissolved modification of calcite ($CaCO_3$) and such shells tend to disappear soon from sediments during diagenesis. However, one major advantage of using pteropods for biostratigraphy over the other holoplanktonic organisms frequently used for this purpose is that their fragile shells do not survive significant transportation and are rarely reworked from one sedimentary unit into another. Pteropods occur only as reworked elements when they are preserved as internal moulds of calcite, pyrite or phosphorite (Curry 1965; Janssen and Little 2010), which indeed is also frequently the case in the Mediterranean area (Janssen 2012b).

Heteropoda

The oldest known heteropods date back to the Jurassic. Bandel and Hemleben (1987) recognised three small (<5 mm) species, two of which belong to the carinariid genus *Coelodiscus* and one to *Pterotrachea*, a genus at that time only known by Recent representatives. Another, as yet unnamed *Pterotrachea* species was recently described from Eocene (Ypresian-Lutetian) deposits in Uzbekistan (Janssen et al. 2011).

The family Bellerophinidae, nowadays (Bouchet and Rocroi 2005) included in the Pterotracheoidea, is exclusively known from Mesozoic occurrences. Because of a nautiloid shape the genus *Bellerophina* (Fig. 19.6) has been considered to be a cephalopod, although the shell more closely resembles the Recent atlantid *Oxygyrus* in shape.

The genus *Brunonia*, the systematic position of which has been enigmatic for a long time, was considered to represent a new subfamily Brunoninae in the Carinariidae by Dieni (1990). *Brunonia* was first described by Yokoyama (1890) from Cretaceous rocks in Japan, but has a wide distribution along the former Tethys area, inclusive of the Mediterranean (Sardinia). Its place in the Carinariidae was accepted by Bouchet and Rocroi (2005).

Three heteropods have been described from the Eocene of France. *Eoatlanta spiruloides* (Lamarck, 1804) of the Paris Basin Lutetian age, is a planispiral species of which the uncarinated bodywhorl is detached from the foregoing whorl. That species, however, seems to be unrelated to the Heteropoda and most probably represents a benthic gastropod. Erroneously identified with the same name, another species was recorded from the Danian of Faxe (Denmark) by Ravn (1933) and Rosenkrantz (1960), but that occurrence

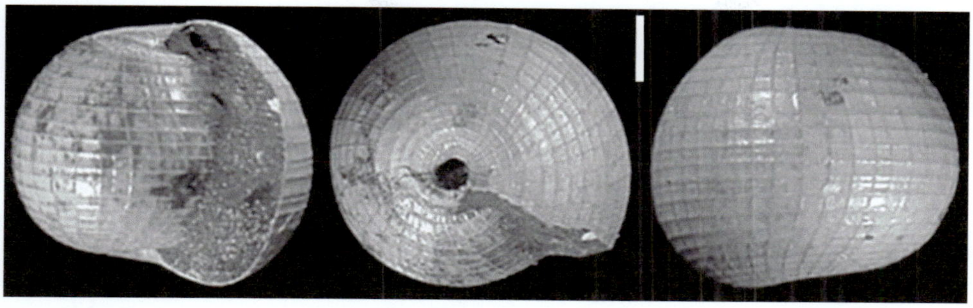

Fig. 19.6 *Bellerophina minuta* (J. Sowerby 1814) from the Lower Gault of Folkestone, UK; apertural (*left*), apical (*middle*) and dorsal (*right*) views, showing almost planispiral coiling (Photograph Jim Craig†. Courtesy of Steven Tracey). Bar represents 1 mm

equally seems to represent a benthic species and is not a heteropod.

Cossmann (1902) introduced *Carinaria mirabilis* from the Lutetian of the Loire Basin and a very similar species, *Atlanta arenularia* Gougerot and Braillon, 1965, was described from the Paris Basin Bartonian. These two latter species differ in various respects from *Carinaria* as well as from *Atlanta* and rather seem to represent a intermediary form between these two genera.

More typical species of the Atlantidae and Carinariidae are known from Oligocene and younger rocks. Also several species of *Protatlanta* are described, however these differ somewhat from the Recent type species. A new genus might eventually be necessary for these taxa.

Fig. 19.7 Mass occurrence of the oldest pteropod known: *Heliconoides mercinensis* (Watelet and Lefèvre 1885), in calcitic internal mould preservation; late Paleocene (Selandian) Fur Formation, Cementstone 'C-Blok', from Mors (Jylland, Denmark) (NBC collection, Leiden, The Netherlands). Bar length 5 mm

Pteropoda

Paleozoic genera like *Tentaculites*, *Conularia*, *Hyolithes* etc. were formerly included in the Pteropoda (e.g. Barrande 1867), although soon Pelseneer (1888) had expressed doubts. Several of these indeed resemble the Recent representatives closely, even to such an extent that they were sometimes included in genera based on Recent type species (e.g. *Styliola*; Barrande 1867). Nowadays, however, such fossils are no longer considered to represent pteropods or even molluscs, but rather to be worms, brachiopods, or cnidarians (e.g. McKinney et al. 1995; Sendino et al. 2011).

Pteropods have also been recorded from Mesozoic rocks (Troelsen 1938; Carozzi 1954), but these too are now considered to belong to other fossil groups. Blanckenhorn (1889, 1934) described pteropods from the Cretaceous of Syria and Turkey but these formations were later (Avnimelech 1945) recognised as being of Miocene age.

The oldest real pteropod currently known is a limacinid, *Heliconoides mercinensis* (Watelet and Lefèvre, 1885), originally described from the early Eocene (Ypresian) of the Paris Basin, but known to be present already in latest Paleocene rocks of the North Sea Basin (Fig. 19.7) and Alabama USA (Janssen and King 1988; Janssen 2010).

If indeed the species *Heliconoides mercinensis* is 'the first' pteropod and the ancestor of all later pteropods, development of this animal group must have been almost explosive: by the early Eocene (Ypresian) many different species, belonging to two families, Limacinidae and Creseidae, were already present, recorded from north and central America (Collins 1934; Hodgkinson et al. 1992), the North Sea (Curry 1965; Janssen and King 1988), Paris (Dollfus and Ramond 1885) and Aquitaine (Curry 1982; Cahuzac and Janssen 2010) basins, from Uzbekistan (Janssen et al. 2011) and from Iran (publication in prep.).

An evolution from sinistrally coiled to straight-shelled pteropods was implicitly suggested by Curry (1965) and Curry and Rampal (1979). This phenomenon seems to be confirmed by several pteropod species present during the Ypresian and Lutetian, demonstrating successive stages of despiralisation in the limacinid genus *Altaspiratella*

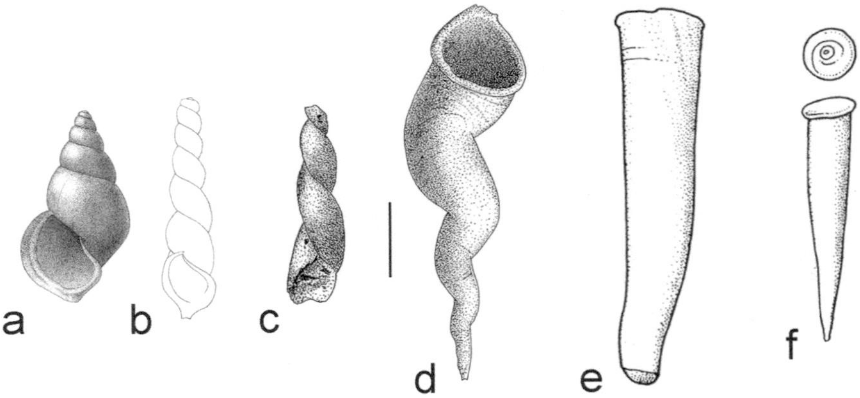

Fig. 19.8 Possible development of Limacinidae (genus *Altaspiratella*) into Creseidae (genera *Camptoceratops* and *Euchilotheca*). (**a**): *Altaspiratella bearnensis* (Curry 1982); (**b**): *A. multispira* (Curry 1982); (**c**): *A.* sp. nov.; (**d**): *Camptoceratops priscus* (Godwin-Austen 1882); (**e**): *Euchilotheca elegans* Harris, 1894; (**f**): *E. succincta* (Defrance 1828) (**a**, **b**, **d**) from the Ypresian of Gan, SW France; (**c**) from the latest Ypresian/earliest Lutetian of Iran; (**e**) from the Ypresian of Bracklesham and (**f**) from the Lutetian of Parnes, France (Both after Curry 1965. Courtesy of Malacological Society of London). Bar length represents 1 mm

(Fig. 19.8a–c), subsequently leading to creseid genera like *Camptoceratops* (Fig. 19.8d) and *Euchilotheca* (Fig. 19.8e, f). There are, however, some discrepancies: the genus *Camptoceratops* is apparently already present before the most elongate limacinid. This might be a consequence of as yet incompletely known vertical ranges.

From the Oligocene period relatively few pteropod species are known. Several of the Eocene occurrences become extinct (with *Heliconoides*, *Limacina*, *Creseis* and *Praehyalocylis* as survivors). During the Rupelian the first certain species of the genera *Clio* and *Vaginella* appeared and starting with the late Oligocene (Chattian) a renewed rapid development took place, culminating during the middle Miocene. Several new genera came into existence, but disappeared already before the Mio-Pliocene transition. Two species starting in the late Oligocene, *Heliconoides inflata* (d'Orbigny, 1834) and *Styliola subula* (Quoy and Gaimard, 1827), are very long-ranging and still survive today. Also several species from the Miocene period survived into the Pliocene and are present in the currently living fauna. A strong reduction in numbers of species is seen during the various stages of the Pleistocene, especially during the colder intervals, with a noticeable revival again during the Holocene.

Fig. 19.9 Luigi Bellardi (1818–1889) (Courtesy Biblioteca Accademia delle Scienze di Torino)

Fossil Holoplanktonic Mollusca in the Mediterranean Basin

Early Research

By far the greater part of older literature is based on material from Italy. The first fossil pteropod ever described from the Mediterranean is *Cuvierina astesana* (Rang, 1829), from the Pliocene near Asti (Piemonte, northern Italy). Bronn (1831) listed four pteropod species (one of which belongs to the Scaphopoda, however) from Italy. Several further species were named by Bonelli in a catalogue of the molluscs in the Torino museum, probably written up long before 1840 but his manuscript was never published. Still, the nomenclaturally invalid names given by Bonelli remained in use for a long time in Italian literature.

Further holoplanktonic molluscs from northern Italy were described by authors like Sismonda (1842, 1847), Michelotti (1847), Pictet (1855), Mayer (1868) and others, but it was Luigi Bellardi (1873) (Fig. 19.9), in the first volume of the famous monographic series 'I molluschi dei terreni terziari del Piemonte e della Liguria' who concentrated data from earlier literature and adequately introduced a number of new species. Of the 23 species recorded by Bellardi 21 are pteropods and two are heteropods (Carinariidae). Most of these species originate from Miocene rocks of the 'Colli Torinesi', a few are from other localities in northern Italy. Most of the samples studied by Bellardi are still available in the Turin 'Dipartimento di Scienze della Terra' (Janssen 1995). Bellardi introduced nine species as new, but nowadays he also is the author of several other species, as he validated a number of taxa unofficially named by Bonelli.

Giuseppe Seguenza (1867 and several later papers) recorded pteropods and heteropods from the Plio-Pleistocene of southern Italy and Sicily. In 1878 Nicola Tiberi published an overview in French, predominantly based on earlier literature, of all cephalopods, pteropods and heteropods from Italy known as fossils as well as from the living fauna. Two years later an Italian version of that paper (Tiberi 1880) was published. This author did not appreciate Bellardi's 1873 paper as much as we do nowadays. Concerning two pteropod species introduced by Bonelli/Bellardi, '*Balantium sulcosum*' and '*B. calix*' (both of which we still consider to be valid species in the genus *Clio*), Tiberi remarks (1878: 73) 'Il est évident que si cette espèce, ainsi que la précédente, ne sont que des anomalies de la *Cleodora pyramidata*, elles ne sont dues qu'à des fantaisies d'auteurs qui se plaisent à tout propos à multiplier les espèces sans aucun nécessité'. This remark demonstrates a lack of knowledge rather than a sound criticism.

In 1896 Simonelli introduced two new pteropod species from Pliocene clays at Sivizzano near Parma. Another paper on Miocene pteropods is by Audenino (1899), who described pteropods, among which were several new taxa, from Monte dei Cappuccini, in the Turin Hills. Most of Audenino's material is still available in Torino. One species in his paper, introduced as '*Limacina Formae*', is nowadays included in the genus *Sphaerocina* Jung, 1971, of which it is the type species, and separated from all other pteropods in the family Sphaerocinidae (Janssen and Maxwell *in* Janssen 1995).

Only a few papers appeared in the early years outside northern Italy. Bourguignat (1868) introduced three pteropod species, unfortunately still enigmatic, from the Pliocene of Algeria. Simonelli (1895) introduced a remarkable pteropod species, *Cavolinia cookei*, from Malta. Other pteropods from Malta were recorded, usually identified to genus level only (*Hyalea*, *Vaginella*), by Cooke (1893, 1896). Blanckenhorn's papers on pteropods from Asia minor are already mentioned above. Verri and de Angelis d'Ossat published three papers (1899, 1900, 1901) on Miocene deposits in Umbria, recording also the pteropods from these rocks, but they unfortunately gave no descriptions or illustrations.

Twentieth Century Research

In 1901 Blanckenhorn described a new pteropod from the Pliocene of Egypt under the name of *Hyalaea angusticostata* Blanckenhorn, 1901, a taxon nowadays synonymised with the Recent species *Cavolinia tridentata* (Forskål 1775). Again most later papers on fossil holoplanktonic molluscs appeared on the basis of Italian material. Rafaello Bellini (1905) summarized the existing knowledge, inclusive of the Recent pteropods, uncritically listing no less than 50 species. The illustrations in his paper are poor, frequently even incorrectly drawn copies from earlier papers and in many cases more confusing than clarifying the identification of species.

A pteropod assemblage consisting of 15 species from the Miocene of the Gargano peninsula was properly described by Giuseppe Checchia-Rispoli in 1921, introducing eight taxa as new, and estimating the age as Langhian. The same fauna was discussed by Giuseppe Sirna (1968) who described another two new species, but differed in opinion concerning the age of the Gargano deposit. Later the same Gargano assemblage and its age were discussed by other authors (see below). A single Miocene pteropod, *Diacria sangiorgii*, was extensively described by Franco Scarsella (1934) from the Tortonian of Fornaci d'Appennino. From Miocene rocks at Montegibbio (near Modena) Dieci (1962) recorded six pteropod species, indicating a Langhian age. In the same paper the author gave an overview of known Italian pteropod species of Miocene and Pliocene age, listing no less than 52 species, some of which are now considered synonyms.

In the 1960s Siebrecht van der Spoel of the University of Amsterdam (Netherlands) started a long series of papers on Recent holoplanktonic molluscs, culminating in two important monographs (1967, 1976) and many other papers, partly in collaboration with his students and other co-workers. As is clear from the title of van der Spoel's (1967) PhD thesis ('a group with remarkable developmental stages') a rather revolutionary life cycle for pteropods was advocated, leading to renewed discussions on pteropod systematics and biology. During the same time interval Ronald Gilmer (1972, 1974) made his spectacular scuba diving observations on pteropod feeding behavior.

Possibly initiated or, at least, stimulated by the interest in the living pelagic mollusc assemblages and the many papers published on that subject, a renewed awareness of fossil pteropod biostratigraphy materialized in the 1970s at the Milano University's geology department and resulted in a series of papers by Elio Robba, the first one appearing in 1971, followed by a number of other papers, partly together with other authors, discussing pteropod assemblages, mainly from northern Italy. In these papers biostratigraphical interpretations are obtained by correlation with other microfossil disciplines, such as planktonic foraminifers and/or calcareous nannoplankton. In this way Robba (1971) described 'associazioni a pteropodi' of Langhian age from the Cessole Formation, then in 1972 early Miocene assemblages (for a part even late Oligocene in age, see below) of the Langhe area and later in 1977 Serravallian assemblages from the same part of northern Italy. In that period also DiGeronimo (1974) published a paper on pteropods and heteropods from the Miocene of the Vetto d'Enza area in Reggio Emilia.

Robba and Spano (1978) described pteropods from the middle Miocene of southern Sardinia, followed in 1983 by Carlo Spano's paper on pteropods from the 'Miocene inferiore' of northern Sardinia. In 1979 Robba, together with Giulio Pavia of the Torino university, published on pteropods from the 'Messinian' locality of Tetti Borelli in the Turin hills, the material of which was revised by Janssen (1995, 1999b) who corrected the age to Tortonian. D'Alessandro et al. (1979) published a voluminous paper on the Neogene deposits of Gargano, paying much attention to the holoplanktonic molluscs, thereby revising the earlier papers of Checchia-Rispoli and Sirna. Soon afterwards contacts between Assunta D'Alessandro (University of Bari) and Elio Robba led to a new paper (1981) in which not only the Gargano pteropods were drastically revised, but which also included descriptions of an important occurrence of holoplanktonic molluscs from the Lecce area (localities Melpignano and Cursi, Salento, Puglia), leading to the conclusion that both the Gargano and Lecce assemblages were reworked from originally Langhian deposits into rocks of Serravallian/Tortonian age (with which conclusion, as far as Gargano is concerned, the first author of this paper does not agree). Robba's activities with fossil pelagic molluscs ended with a paper by Bernasconi and Robba (1982) discussing paleobiogeography by attempting to unravel the dispersion routes of pteropods through time. Nowadays, 30 years later, with so much more data available on systematics, distribution and occurrence, many of their conclusions are in fact overruled or in need of reassessment.

DiGeronimo et al. (1981) described the fauna from clays of the Tellaro Formation outcropping near Vittoria in Sicily and compared it with that of the Blue Clay Formation in Malta, inferring a similar age for both. With the kind help of Italo DiGeronimo the first author was able to collect several times at one of the Sicilian localities (Poggio Musenna), which resulted in a paper (Janssen 1999b) denying the correlation on the basis of pteropod distributions and suggesting a Tortonian age for the Sicilian and a Serravallian for the Maltese clays.

Numerous papers were published, by a variety of authors, on the pelagic molluscs recorded from Pleistocene to Holocene Mediterranean rocks, largely obtained from bottom samples and cores from offshore localities. For details see Janssen (2012a).

Cenozoic Pelagic Mollusc Assemblages of the Mediterranean Correlated with Palaeo-Temperatures

From the literature, e.g. distribution maps in van der Spoel (1967, 1976) and Bé and Gilmer (1977), it is clear that at present the distribution of holoplanktonic molluscs largely agrees with climatic zones, influenced by ocean currents. Tropical, bi-subtropical and bipolar etc. distribution patterns are distinguished. It may be supposed that the same has been true for the past. A sufficiently detailed Cenozoic palaeotemperature curve was recently published by Antón Uriarte in his Earths Climate History website who allowed us to use this curve here. In Fig. 19.10 the stages since the late Oligocene are given with their absolute age. The curve represents δ ^{18}O values reflecting global palaeotemperatures. Next to the curve estimated numbers of existing Mediterranean species are given per stage (in green). Numbers per stage, however, do not mean that those species have been present during the complete interval. Some may have disappeared or else arrived, in the course of the stage. Numbers in most cases are estimates as there are numerous uncertainties, either about the correct identification of species, or about the precise age of the rocks from which the material was extracted. Also it should be realized that occurrence or absence of species may be influenced by other factors than temperature, such as ocean currents, sea level changes, salinity, absence or presence of connections with the Atlantic and/or Pacific, and evolutionary development of taxa.

Eocene (55.8–33.9 Ma) (Ages Referred to are According to Gradstein et al. 2004)

Although pteropods from the Eocene are well-known from several parts of the world (North Sea Basin, Aquitaine Basin, USA, Canada, central Asia, Japan, Australia) hardly any Mediterranean occurrence is known, in spite of the fact that deposits of that age are widely represented (e.g. Mancin et al. 2003). Sacco (1899: 366) refers to the genera *Balantium* and *Vaginella* from Eocene deposits of the 'Appennino romagnolo', without any further details. Both genera, however, are unknown from Eocene rocks. A single unpublished occurrence of Eocene (Bartonian) pteropods is known from Egypt.

Oligocene (33.9–23.03 Ma)

Rupelian

The only known occurrence of supposed Rupelian pteropods in the Mediterranean is at Arcugnano (Vicenza, Italy) from where a single example of an unidentified limacinid is present

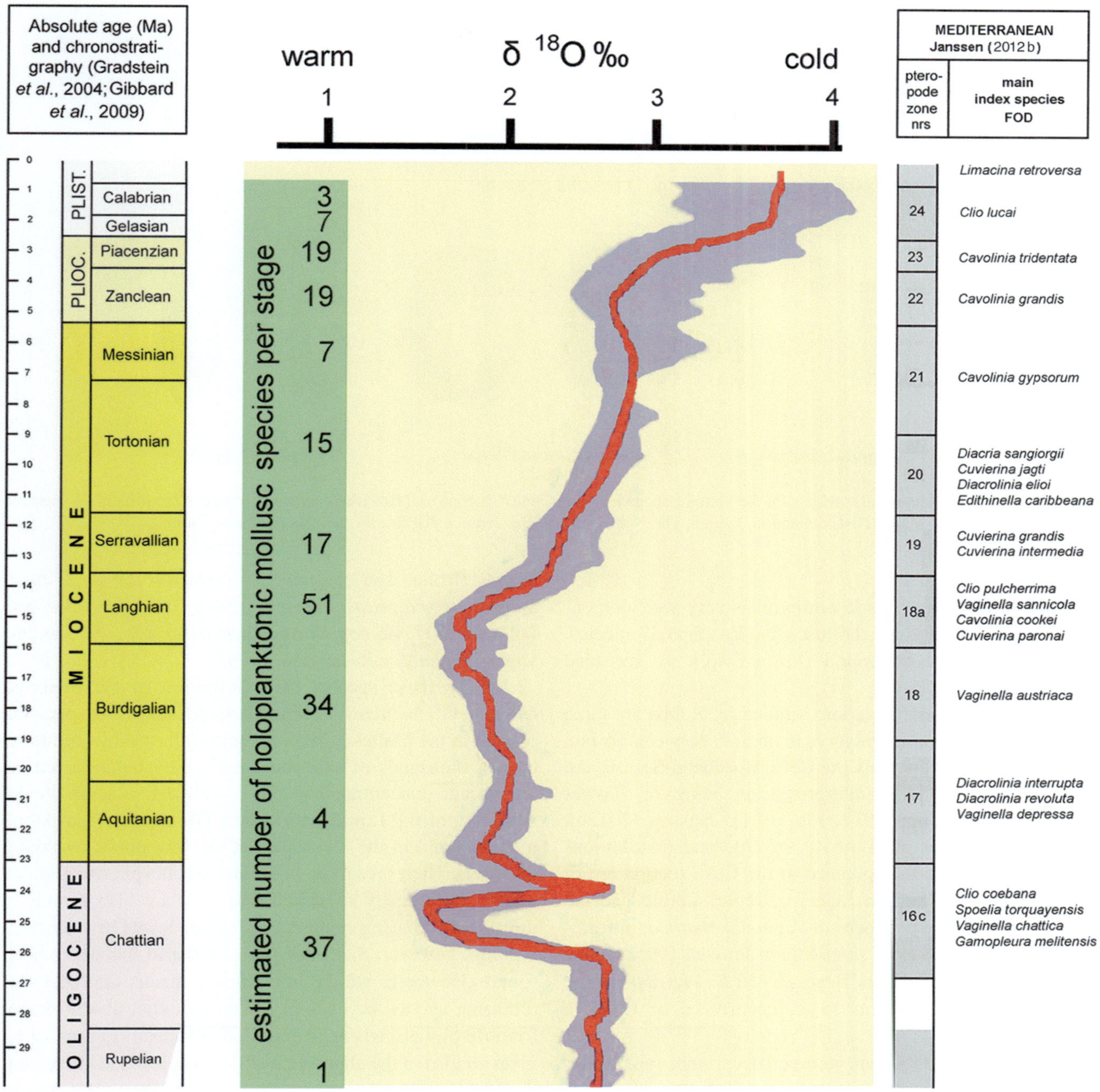

Fig. 19.10 Palaeotemperature curve for the interval Chattian-Pleistocene (After Uriarte 2011), with absolute age and stages (*left-hand columns*), estimated number of holoplanktonic mollusc species per stage (in *green*) and Pteropod Zones for the Mediterranean with index species (*right-hand columns*) (After Janssen 2012b)

in the NBC collection, without any additional data. We have personally sampled the complete section at Massignano (near Ancona, Italy) in which the GSSP for the Rupelian was designated (Premoli Silva and Graham Jenkins 1993), but not one mollusc was found in the samples. The single record for the Rupelian almost certainly underestimates the number of species actually living in the palaeo-Mediterranean at that time. From other regions, North Sea Basin, Aquitaine Basin, several more species are known, and since the basin was at that time in open connection with both the Atlantic and Indian oceans it may be supposed that more species were actually present. The low number recorded here probably reflects the very restricted availability of pteropod bearing deposits of Rupelian age from deeper waters in the area.

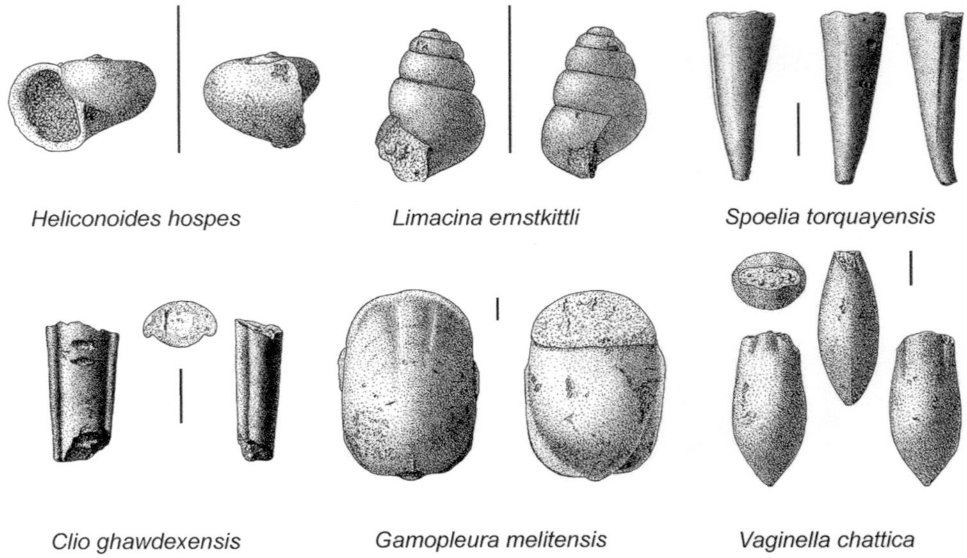

Fig. 19.11 Thecosomatous index fossils for the Mediterranean late Oligocene (Chattian), all from the Lower or the base of the Middle Globigerina Limestone Member of Malta (NBC collection, Leiden, The Netherlands; After Janssen 2012b). Bar lengths represent 1 mm

Chattian

The late Oligocene climatic optimum is very well demonstrated by holoplanktonic molluscs. No less than 37 species, including just one heteropod (*Atlanta* sp.) are recorded (Fig. 19.10).

The oldest Italian formations studied by Robba are three sections in the Langhe province, in an area between Torino, Cuneo and Genova, namely the Ceva, Rocchetta-Seriole and Valle Uzzone sections, interpreted as being of 'Lower Miocene' age (Robba 1977, Fig. 19.1). Janssen (2012b), however, concludes on the basis of pteropod distributions that zones 1 and 2 distinguished in the Ceva section are of late Oligocene (Chattian) age. A single sample (Torino collection) with 22 specimens of *Vaginella chattica* (unfortunately misidentified as *V. sannicola* in Janssen 1995) from a locality indicated as 'Villa Besozzi, Val S. Martino' in the Colli Torinesi, also points to an occurrence of Chattian sediments.

No further late Oligocene pteropod occurrences are known from Italy but many are known in the Maltese archipelago (Fig. 19.11). For a long time the Lower Globigerina Limestone Member in these islands was considered to be of early Miocene (Aquitanian) age (Felix 1973 and references therein, Pedley et al. 1977; Janssen 2004), but is nowadays considered, on the basis of planktonic foraminifera and calcareous nannoplankton, to be of Chattian age (Foresi et al. 2008). This point of view is now fully confirmed by pteropods as well. The pteropods even indicate clearly (Janssen 2012b) that the basal part of the Middle Globigerina Limestone Member of Malta also is of Chattian age, as demonstrated by the absence of index fossils for the Miocene and the continued occurrence of species (Fig. 19.11) like *Heliconoides hospes* (Rolle, 1861), *Limacina ernstkittli* Janssen, 2012, *Spoelia torquayensis* Janssen, 1990, *Clio ghawdexensis* Janssen, 2004, *Gamopleura melitensis* Janssen, 1995 and/or *Vaginella chattica* R. Janssen, 1979.

Among these species the distribution of the pteropod *Gamopleura melitensis* is particularly puzzling. The species is present in the Maltese Chattian rocks in incredible numbers; tens of thousands of specimens can easily be collected, as they occur concentrated in the so-called C 1 level on the boundary of the Lower and Middle Globigerina Limestone members and in the lower part of the limestones overlying that level. Their preservation as internal phosphoritic moulds makes them easily reworked and they are also found, in upwards decreasing numbers, in higher levels as well. It is striking, however, that this species hitherto has never been found elsewhere, which undoubtedly means that rocks of the same age as the ones in Malta are either absent or not fossiliferous elsewhere in the Mediterranean. Figure 19.12 gives an idea of the abundance of this species. It is interesting that another Chattian species, *Gamopleura maxwellii*, closely resembling *G. melitensis*, was recently described from New Zealand by Grebneff et al. (2011).

Miocene (23.03–5.33 Ma)

Aquitanian

The dramatic drop in temperature near the end of the Chattian may be one of the reasons for the very restricted number of species (four only) for the Aquitanian. Pteropods from the earliest Miocene rocks are only known from Italy and have not been demonstrated with certainty yet in Malta or elsewhere.

Fig. 19.12 *Gamopleura melitensis* Janssen, *in* Rehfeld and Janssen 1995, paratype sample (in phosphoritic internal mould preservation) from the type locality Baħrija 1 (Malta), base of Middle Globigerina Limestone Member, phosphorite level C 1 (Chattian) (NBC collection, Leiden, The Netherlands; After Janssen 2012b)

Unfortunately most available Italian material supposed to be of Aquitanian age is from older collections with insufficient or unreliable data, which reduces an enumeration of Aquitanian pteropods to a short list. The species *Heliconoides inflata* (d'Orbigny, 1834), *Limacina valvatina* (Reuss 1867), *Gamopleura taurinensis* (Michelotti, 1847) and *Vaginella depressa* Daudin, 1800 were definitely present during that time. The real number of species was probably higher.

Burdigalian

The gradually increasing temperature during the Burdigalian coincides with a higher number of species, of which 34 are recorded. During this stage several new genera and species occur. Of the genus *Diacrolinia*, already present during the Late Aquitanian elsewhere (SW France), two species are found in the Mediterranean: *D. interrupta* (Bellardi, 1873) and *D. revoluta* (Bellardi, 1873). Several other species already present in the Aquitanian rocks continued during the Burdigalian, among which were *Gamopleura taurinensis* and *Vaginella depressa*.

It is interesting to note that the development from *Vaginella depressa* to its successor species *V. austriaca* (at approximately 19 Ma), easily demonstrated in the Mediterranean and agreeing with observations in the Central Paratethys, and the Aquitaine and North Sea basins, did not take place in the Pacific realm (Japan, Australia, New Zealand). In the Pacific *V. depressa* continued until the Langhian/Serravallian (Janssen 2006; Cahuzac and Janssen 2010) most probably because an exchange of pteropod populations between the Mediterranean and the Indo-Pacific was interrupted by the middle Miocene closure of the Suez isthmus as a result of large scale tectonic activity (Harzhauser et al. 2007).

Langhian

Near the end of the Burdigalian and during the Langhian another climatic optimum is clear from the curve in Fig. 19.10. The number of holoplanktonic gastropods increases to the unrivalled number of 51, which is still an underestimate as a number of taxa only known by their larval shells were not included in the count, as they may or may not have belonged to some of the identified species. The high number of taxa may partly also be explained by a rapid evolution within the basin, but also by increasing immigration from the Atlantic: several of the species are also known to occur in the Caribbean Basin.

The pteropod assemblage changes considerably during the Langhian stage by the sometimes remarkably abundant appearance of new species: *Cuvierina paronai* Checchia-Rispoli, 1921, *Clio pulcherrima* (Mayer, 1868), *Cavolinia cookei* Simonelli, 1895, *C. pycna* Jung, 1971, *Diacrolinia aurita* (Bellardi, 1873), *Edithinella varanica* Sirna, 1968, *Vaginella gibbosa* Audenino, 1899, *V. lapugyensis* Kittl, 1886, *V. sannicola* Janssen, 1990, *Sphaerocina formai* (Audenino, 1899) and several other taxa. Assemblages of this age are known from northern Italy, Gargano, Salento (but there reworked in younger deposits), Malta (Fig. 19.13) and Cyprus. The rapid development of the genus *Vaginella* is especially striking. In some of the younger Langhian rocks (top levels of the Upper Globigerina Limestone Member) of the Maltese archipelago, specimens of that genus, mainly *V. austriaca* and its forma *acutissima* Audenino, 1899 are represented by millions of specimens. *Sphaerocina formai* does not survive the Langhian/Serravallian boundary, but other Sphaerocinidae species are known from younger deposits elsewhere.

Serravallian

A gradual lowering of temperature through the Serravallian, Tortonian and Messinian is reflected in the number of species recognised from this interval. Compared to the Langhian there is a drastic reduction in species, decreasing from 17 during Serravallian times to only 7 in the Messinian. Serravallian rocks are known from northern Italy (Robba 1977), from Turkey (Erünal-Erentöz 1958; Janssen 1999a) and from Malta. Several species continue from the Langhian (*Cavolinia paronai*, *Edithinella varanica*, *Vaginella austriaca*, *V. lapugyensis*) and some new species appear: *Cuvierina curryi* Janssen, 2005, *C. grandis* D'Alessandro and Robba, 1981, *C. intermedia* (Bellardi, 1873) and *Diacrolinia larandaensis* Janssen 1999a.

Tortonian

Pteropod assemblages of Tortonian age are known from Sicily (Tellaro Formation), northern Italy (Tetti Borelli, S. Agata Fossili Formation, Montaldo Member), Cyprus (Pakhna Formation, Maronia Marlstone) and possibly Malta

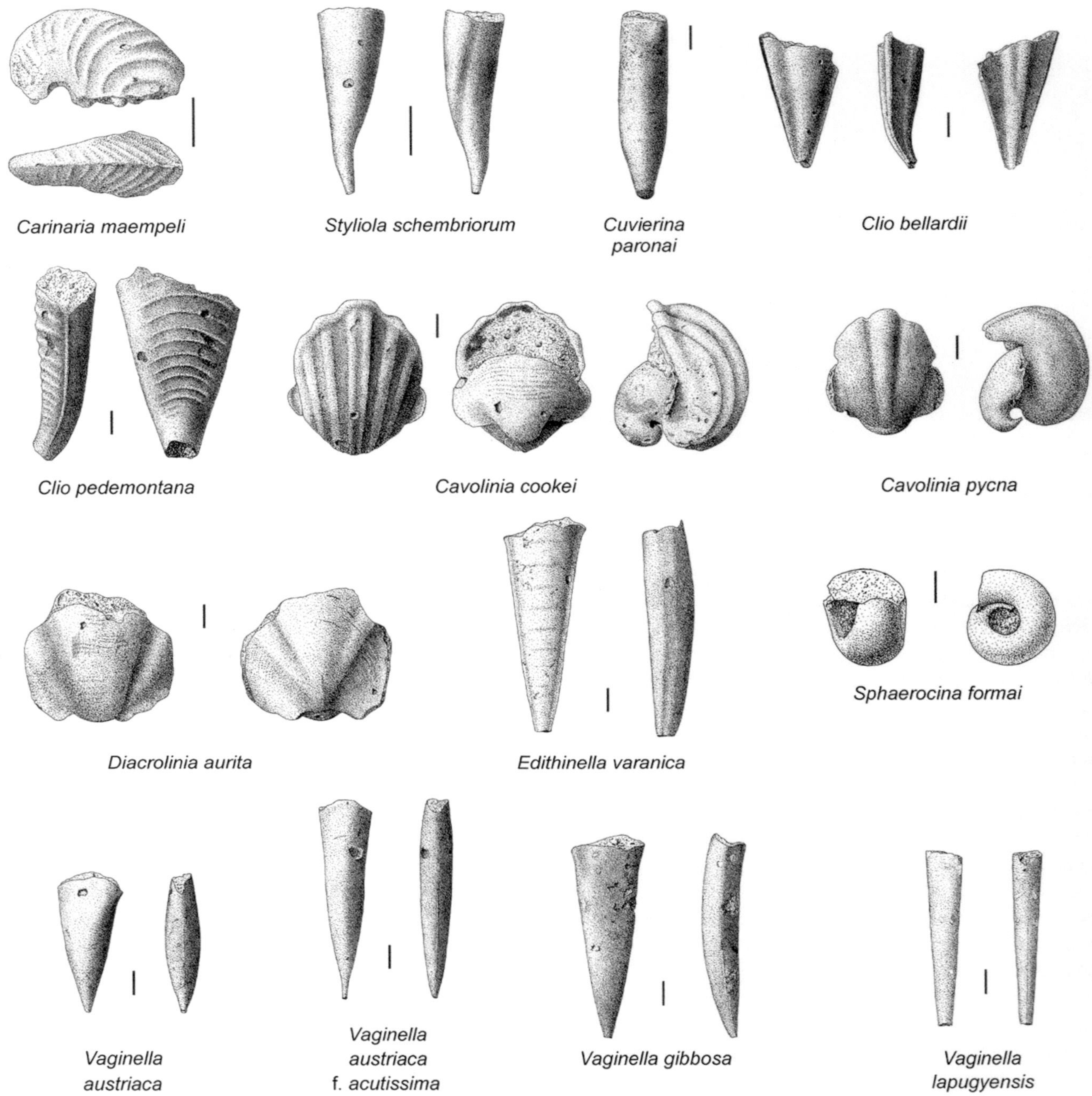

Fig. 19.13 Important fossil species for the Mediterranean Langhian. All illustrated specimens are from the Upper Globigerina Limestone Member of Malta (NBC collection, Leiden, The Netherlands; After Janssen 2012b). Bar lengths represent 1 mm

(Greensand Formation). Approximately 15 species are known, among which *Cuvierina inflata* (Bellardi, 1873), *C. jagti* Janssen, 1995, *Clio giulioi* Janssen, 1995, *Diacria sangiorgii* Scarsella, 1934, *Diacrolinia elioi* Janssen, 1995 and *Edithinella caribbeana* (Collins, 1934) may be considered index species for a Tortonian age. The occurrence of some of these species in internal mould preservation in the Maltese Greensand Formation must be reworked if those rocks indeed are of Messinian age, as supposed by Mazzei (1985) and Jacobs et al. (1996).

Messinian

From the pre-evaporitic Messinian period seven species of holoplanktonic molluscs are known from northern Italy (e.g. Biglini, Moncucco Torinese, Sant'Agata Fossili; Sturani 1973, 1978; Sturani and Sampò 1973; Janssen 1995). An unpublished sample from the Monte del Casino section at Borgo-Tossignano (Emilia-Romagno) was collected December 1989 by Marco Taviani and is kept in the Bologna museum. Most important pteropod species for this interval are *Bowdenatheca jamaicensis* Collins, 1934, *Cavolinia gypsorum* (Bellardi, 1873) and *Diacria trispinosa* (de Blainville, 1821). Irene Zorn (1997) recorded these pteropod species and the heteropod *Carinaria lamarckii* de Blainville, 1817 from the Heraklion Basin of Crete.

During the later Messinian the 'salinity crisis' (5.75–5.32 Ma) obliterated almost all marine life in the Mediterranean (Clauzon et al. 1996; de Lange and Krijgsman 2010) resulting from the Gibraltar isthmus which largely or completely disconnected the Mediterranean from the Atlantic. Probably the entire holoplanktonic mollusc population in the Mediterranean became extinct at this time.

Fig. 19.14 Topotypical specimen of *Clio guidottii* Simonelli 1896 from Zanclean clays at Sivizzano near Traversetolo (Italy, Parma province) (NBC collection). Bar length represents 1 cm

Pliocene (5.33–2.59 Ma)

Zanclean

Supposedly the complete Pliocene heteropod and pteropod assemblage immigrated into the Mediterranean after the reopening of the Strait of Gibraltar. Also favored by improving climatic conditions 19 holoplanktonic mollusc species are recorded from this interval. Apart from several species continuing after the Miocene a number of new species appear, among which are *Striolimacina imitans* (Gabb, 1873), *Creseis clava* (Rang, 1828), *Cuvierina astesana* (Rang, 1829), *Clio braidensis* (Bellardi, 1873), *C. guidottii* Simonelli, 1896 (Fig. 19.14), *Cavolinia grandis* (Bellardi, 1873) and *Diacria italica* Grecchi, 1982. All these species are known from various northern Italian deposits, one of the most rewarding localities being Quattro Castella (Monticelli) in Reggio Emilia, but that fauna (NBC collections) has not yet been published. Sediments exposed in a former claypit at Puget-sur-Argens (SE France, Var) dated as Zanclean on the basis of planktonic foraminifera yielded material rich in holoplanktonic gastropods, in which, however, the most important index species seem to be absent (Janssen, 2012c).

Piacenzian

Pteropod assemblages yielding the same number of species as in the Zanclean are also known from several localities in northern Italy. The main characteristic is the replacement of the pteropod *Cavolinia grandis* by *C. tridentata*, one of the largest *Cavolinia* species still living in the Mediterranean. A rich variety of species was encountered in rocks of, according to the pteropods, Piacenzian age, in SE Spain, locality Estepona. Janssen (2004) described 18 species, three of which are heteropods, and 15 of which are pteropods, among which the species *Cavolinia tridentata* is convincingly present. Two other cavoliniids are worth mentioning: *Cavolinia floridana* (Collins, 1934), at the time only known by its holotype described from the Choctawatchee Formation in Florida USA. The species appeared to be quite common at Estepona. A previously undescribed cavoliniid species, introduced as *C. landaui* Janssen, 2004, was also frequent in Estepona. Neither species is, however, yet known from elsewhere in the Mediterranean.

Pleistocene and Holocene (2.59 Ma – Recent)

During the Gelasian a reduction to seven species may be correlated with a drop in temperature that continues in the later Pleistocene, but some 'warmer water' species are still present: *Heliconoides inflata*, *Styliola subula*, *Clio pyramidata* f. *lanceolata*, *Diacria trispinosa* and an unidentified *Cavolinia* sp. are common occurrences, as e.g. in the Torrente Stirone outcrop near San Nicomede (Parma).

In clays of Calabrian age, outcropping in the left bank of the Enza river, at San Polo d'Enza (Reggio Emilia), the number of species further reduced to three only, one of which was a commonly occurring large *Clio* species that appeared to be new and was described as *C. lucai* Janssen, 2000

Fig. 19.15 Holotype of *Clio lucai* Janssen, 2000, from Calabrian clays exposed in the left bank of Enza River at San Polo d'Enza (Parma, Italy) (NBC collections, Leiden). Bar length is 1 cm

(Fig. 19.15). A single specimen is also known from the Gelasian of the Torrente Stirone.

An overwhelming number of papers has been published on holoplanktonic molluscs from younger pleistocene Mediterranean rocks. Most of these are, in part at least, dedicated to the striking occurrence of a real cold water pteropod species, *Limacina retroversa* (Fleming, 1823), a species nowadays confined to the northern Atlantic Ocean. This species is frequently found to be present in large numbers in many Mediterranean bottom samples and cores and is considered to be characteristic for colder intervals. By comparing its numerical abundance to supposed 'warm water' species like *Heliconoides inflata*, *Limacina trochiformis* (d'Orbigny, 1834), *Cavolinia inflexa* (Lesueur, 1813), and/or *Diacria trispinosa* several authors have tried to construct palaeotemperature curves for the Pleistocene. For a listing of such papers see Janssen (2012a).

During the later Pleistocene cyclical periods with alternating cold (glacial) and warm (interglacial) periods influenced the holoplanktonic mollusc assemblages strongly. As detailed data for the Mediterranean Pleistocene are difficult to extract and to correlate from the abundant literature we only refer here, as examples, to some figures given in Janssen (2012a) for a gravity core in the eastern Mediterranean (taken W of Gávdhos, at coordinates 34°58'4"N 23°44'8"E), ranging from the latest Saalian onwards. The core was sampled centimeter-wise and datings were done on planktonic foraminifera. From the single available sample of Saalian age 19 species were recorded. In samples of the overlying Eemian interval initially the number decreases, but rises in the middle part to 27 (in one of the 1 cm samples more than 2,000 specimens are present), to fall again to only 4 species per sample in the highest Eemian samples. During the cold Weichselian the number of species drops strongly, usually with only a few (4–6) species present per sample (and specimen numbers of rarely more than 30 per sample), among which unidentified *Atlanta* and the pteropods *Limacina retro-ver-sa* (Fleming, 1823) and *Clio pyramidata* L., 1767 are frequently present, especially so in the higher parts of this interval. In the last, Holocene interval (with a sediment thickness of only 23 cm) the number of species varies from 10 to 22 in the uppermost sample, whereas the number of specimens, up to 800 per cm interval, may be considered quite high.

As a connection to the eastern seas through the Suez isthmus was already closed during the Miocene, migration of pteropods and heteropods, following climatic trends, took place through the Strait of Gibraltar. Immigration of new species occurs also at the present time. As evaporation of Mediterranean waters exceeds the influx of water (e.g. from rivers), a permanent supply takes place by inward currents through the Strait of Gibraltar and several Atlantic species are found occasionally in the western part of the Mediterranean, but do not find suitable environmental circumstances to reproduce and do not belong to the permanent fauna. This fact makes it difficult to estimate the number of species permanently living at present in the Mediterranean Basin. A fair approach for Thecosomata was published by Corselli and Grecchi (1990), who accepted only 13 species as continously living in the basin. The number of holoplanktonic mollusc species in the Mediterranean decreases from west to east, but some species are strikingly more common in the Levantine Basin than in the Western Basin, e.g. *Hyalocylis striata* (Rang 1829) and *Cavolinia gibbosa* (d'Orbigny, 1834), the latter even considered to represent a new, locally adapted species, *C. gibboides*, by Rampal (2002).

Pteropod Biozonation

The first biostratigraphical zonation based on Pteropoda was developed for the Cenozoic of the North Sea Basin by Janssen and King (1988) and, revised by several later authors, was used as a basis for a similar zonation for the Aquitaine Basin (Cahuzac and Janssen 2010). A preliminary biozonation, for the Mediterranean basin, largely relying on the earlier two systems, is published in Janssen (2012b) and a summary is added herein on the right hand side of Fig. 19.10, together with an enumeration of the most important index fossils. The imperfection of this zonation is evident from the fact that the zone boundaries almost all coincide with stage boundaries, which is unlikely to be correct. Further detailed research will be necessary to ultimately clarify and refine a more realistic subdivision.

The Future

Several aspects may influence how holoplanktonic mollusc populations in the world's oceans, and particularly the number of recognized species in a certain area, will change in the future. On the one hand, there is the increase in knowledge of the systematics and biogeography of holoplanktonic molluscs due to recent interest in this group and the application of modern powerful morphometric and molecular techniques. On the other hand, there exist major concerns about the future of thecosome pteropods because of their vulnerability to rising CO_2 and associated ocean acidification. We will discuss both aspects below, with particular reference to the Mediterranean Sea.

Marine zooplankton have long been viewed as slowly evolving taxa with widespread, panmictic populations (e.g. Angel 1993; Norris 2000). However, recent molecular studies of marine zooplankton have challenged this view by revealing the presence of numerous morphologically cryptic species (e.g. Knowlton 2000; Goetze 2010) and by showing that ecological and environmental barriers can play an important role in separating populations (e.g. Goetze 2003; Peijnenburg et al. 2004; Peijnenburg & Goetze (in press)). Even within the Mediterranean Sea, significant population genetic structuring has been reported, for instance, separating eastern and western basin populations of planktonic copepods (Papadopoulos et al. 2005; Yebra et al. 2011) and chaetognaths (Peijnenburg et al. 2006).

Thus far holoplanktonic gastropods have received little attention from molecular biologists even though the group has considerable potential for integrated molecular and palaeontological study of evolution. A recent molecular barcoding study by Jennings et al. (2010), part of the Census of Marine Life (http://www.coml.org/), was the first to publish a large dataset for pteropods and heteropods sampled from six ocean regions across the globe (although the Mediterranean was not sampled). These authors sequenced a portion of the cytochrome c oxidase I subunit gene (COI, or standard mitochondrial barcoding gene) for 115 individuals of 41 species. Jennings et al. (2010) found high phylogenetic support for a monophyletic Pteropoda (similar to Klussmann-Kolb and Dinapoli 2006) and a monophyletic Heteropoda suggesting that both groups represent single radiations into the pelagic realm. The authors further suggest that a taxonomic revision of holoplanktonic gastropods will be required, however, the lack of resolution of their molecular marker (COI) at intermediate nodes precluded conclusions about the evolutionary history within either group.

Molecular barcoding studies can be very useful in facilitating identification of described species, reveal potential cryptic variation within them, and may allow detection of new or undescribed species, especially in groups with incomplete or unclear taxonomies (e.g. Bucklin et al. 2010; Goetze 2010). However, it is crucial that species identifications are doublechecked by taxonomic experts, morphological data are collected and accessible, and large numbers of individuals per species are screened (typically >50 per species according to Goetze 2010). Jennings et al. (2010) report that for most species for which multiple subspecies or 'formae' were collected, the variants were genetically so distinct that they could be regarded as different species. Examples include *Limacina helicina helicina* (Phipps 1774) and *L. helicina antarctica* Woodward, 1854 (see also Hunt et al. 2010) and *Creseis virgula* (Rang 1828) and *C. conica* Eschscholtz, 1829. Although sample sizes were very small, Jennings et al. (2010) suggest that for some species no significant population structure was found between ocean basins, e.g. *Clione limacina*, *Clio recurva* (Children, 1823), *Diacria trispinosa*, whereas for other species significant geographic structure was detected, e.g. *Cavolinia uncinata* (d'Orbigny, 1834), *Clio cuspidata* (Bosc, 1802), *Cuvierina columnella* (Rang 1827) and *Diacria major* (Boas 1886). However, a major flaw of this study is that the authors seemed to have failed to consult the most recent taxonomic literature. For example, Janssen (2005) described how only two Recent *Cuvierina* species are found in the Atlantic, viz. *C. atlantica* Bé, MacClintock & Currie, 1972 and *C. cancapae* Janssen, 2005, and two different species in the Indian ocean, viz. *C. columnella* and *C. urceolaris* (Mörch, 1850). Considerable confusion can arise if genetic barcodes cannot be connected to correct taxonomic names or when taxonomic identifications are doubtful. This should be avoided by consulting taxonomic experts from the beginning of a barcoding project. Nevertheless, we can expect from these initial results that the number of recognized species within holoplanktonic molluscs will probably increase in the near future as more detailed morphological and molecular data becomes available.

The other major development is the question how planktonic molluscs will respond to rising atmospheric carbon dioxide (CO_2) and associated global warming and ocean acidification in the future. Ocean acidification is the process resulting from the uptake of atmospheric CO_2 by the oceans, causing a lowering of pH and calcium carbonate ($CaCO_3$) saturation state in surface waters and has been nicknamed the 'evil twin' of global warming (e.g. Doney et al. 2009). This process is of particular interest because it impacts shell-forming marine organisms, and pteropods have been identified as extremely vulnerable due to their thin shells of aragonite, a highly soluble form of calcium carbonate (Feely et al. 2004; Orr et al. 2005; Fabry et al. 2008; Comeau et al. 2010a). From pre-industrial levels, contemporary surface ocean pH has dropped on average by about 0.1 pH units and additional declines of 0.2 and 0.3 pH units will occur over the twenty-first century unless human CO_2 emissions are drastically reduced (Orr et al. 2005). Surface ocean $CaCO_3$ saturation states are declining everywhere, but the greatest and quickest changes are experienced at the poles. Of particular concern is the fact that rates of change in global ocean pH are unprecedented, a factor 30–100 times faster than temporal changes in the recent geological past (Pelejero et al. 2010; Doney 2010). Moreover, the most recent past ocean acidification event occurred during the Paleocene-Eocene Thermal Maximum (55.8 Ma) which is at about the same time of the oldest pteropod fossils known thus far.

How pteropods will respond to future ocean acidification is still largely unknown. However, laboratory experiments have shown a decrease in $CaCO_3$ precipitation rates at lower pH (Comeau et al. 2010a, b). Additionally, a recent study by Wall-Palmer et al. (2011) examining the distribution and preservation of pteropods over the last 250,000 years in marine sediment cores from the Caribbean Sea indicated that pteropod calcification is closely linked to global changes in pCO_2 and pH similar to findings for foraminifers (e.g. Barker and Elderfield 2002). Hence these studies show that euthecosome pteropods can be affected by conditions predicted for the twenty-first century. However, field evidence of the impact of ocean acidification does not exist (yet) and intraspecific variability in pteropod responses to ocean acidification may exist, as has been reported for coccolithophores (Langer et al. 2009). Therefore, to end with a more optimistic note, there may be enough variability (evolutionary potential) present in current pteropod populations to allow adaptation to future ocean conditions. In fact, in a recent review on zooplankton evolution, Peijnenburg and Goetze (in press) argue that zooplankton may be well poised for evolutionary responses to global change.

How the Mediterranean holoplanktonic mollusc fauna will be affected by climate change is unknown, but changes in zooplankton species assemblages have already been reported (Conversi et al. 2009, 2010; Molinero et al. 2008; de Madron et al. 2011). According to Lejeusne et al. (2010) the Mediterranean is a sea under siege. Sea temperatures are steadily increasing, extreme climatic events and related disease outbreaks are becoming more frequent, faunas are shifting and invasive species are spreading. Furthermore, Touratier and Goyet (2011) estimated that waters in the Mediterranean are already acidified (−0.14 to −0.05) with waters from the eastern basin less impacted than waters from the western basin. Compared to the acidification of the world ocean's surface waters, the Mediterranean Sea appears to be one of the regions that are most impacted by acidification (de Madron et al. 2011).

A direct consequence of warming is an increase in the abundance of thermotolerant species and a decrease in 'cold' stenothermal species. This process is sometimes referred to as 'tropicalisation' or 'meridionalisation' of the Mediterranean (Bianchi 2007, Lejeusne et al. 2010). In terms of the number of pteropod species living in the Mediterranean, an increase in temperature could be regarded as favourable as most currently living species prefer warm waters. We could expect that increasing temperatures may lead to the permanent settlement of new species entering the Mediterranean through the Gibraltar Strait. Species such as *Cuvierina atlantica*, *Diacria danae* van Leyen and van der Spoel, 1982, *D. trispinosa*, and *Cavolinia uncinata*, which are abundant on the Atlantic side of Gibraltar and already encountered in the Mediterranean as wandering specimens, may be expected to establish permanent populations in the Mediterranean in the near future. In fact, this process is already ongoing as demonstrated with several examples from other animal groups by Por (2009).

Another source of warm water species entering the Mediterranean are the so-called 'Lessepsian' migrants, coming from the Red Sea through the Suez Canal (e.g. Godeaux and Toulemont 1990; Por 2009). This source was already demonstrated for numerous organisms, benthic molluscs included. For the holoplanktonic molluscs, this route is probably less favourable because of the shallow Suez Canal (only 24 m deep). However, the presence of apparently Recent, or at least very well preserved specimens of *Cavolinia uncinata*, hardly ever recorded from the eastern Mediterranean, is surprising. They were found in just one sample, Me25-25 Kg1 taken off the Nile delta, 32°00.57'N 31°53.24'E (Senckenberg Museum collection, Frankfurt am Main, Germany; Janssen 2012a) which is close to the Suez Canal. *Cavolinia uncinata* is a common species in the northern part of the Red Sea (Singer 1994; Janssen 2007) and therefore we think this may be the first Lessepsian pteropod species ever reported.

Little doubt is left that climate change is affecting marine biodiversity worldwide, however how it will affect marine ecosystems is unclear. The most difficult task will be to understand how numerous biotic and abiotic factors interact. The Mediterranean Sea is one of the most impacted seas in the world, and it is known that climate change tends to interact synergistically with other disturbances (e.g. Lejeusne et al. 2010). Holoplanktonic molluscs, and particularly euthecosome pteropods, could serve as important bio-indicators of changes in ocean chemistry. However, for this group to serve as indicator species, we first need to know more about their biodiversity, phylogenetic relationships and biogeography before they become critically impacted.

References

Angel MV (1993) Biodiversity of the pelagic ocean. Conserv Biol 7:760–772

Audenino L (1899) Pteropodi miocenici del Monte dei Cappucini in Torino. Bull Soc Malac Ital 20:97–114

Avnimelech M (1945) Revision of fossil Pteropoda from southern Anatolia, Syria and Palestine. J Paleontol 19:637–647

Bandel K, Hemleben C (1987) Jurassic heteropods and their modern counterparts (Planktonic Gastropoda, Mollusca). N Jb Geol Paläont Abh 174:1–22

Barker S, Elderfield H (2002) Foraminiferal calcification response to glacial-interglacial changes in atmospheric CO_2. Science 297:833–836

Barrande J (1867) Système Silurien du centre de la Bohême 1. Recherches paléontologiques 3. Classe des mollusques. Ordre des Ptéropodes. Barrande, Prague-Paris

Bé AWH, Gilmer RW (1977) A zoogeographic and taxonomic review of euthecosomatous Pteropoda. In: Ramsay ATS (ed) Oceanic micropaleontology 1. Academic, London

Bellardi L (1873) I molluschi dei terreni terziari del Piemonte e della Liguria, 1. Cephalopoda, Pteropoda, Heteropoda, Gastropoda (Muricidae et Tritonidae). Mem r Accad Sci Torino 27:1–264

Bellini R (1905) Les ptéropodes des terrains tertiaires en quaternaires d'Italie. Ann Soc R Zool Malacol Belg 40:23–45

Berger WH (1978) Deep sea carbonate: pteropod distribution and the aragonite compensation depth. Deep Sea Res 25:447–452

Bernasconi MP, Robba E (1982) The thecosomatous pteropods: a contribution towards the Cenozoic Tethyan paleobiogeography. Boll Soc Paleont Ital 21:211–222

Bianchi CN (2007) Biodiversity issues for the forthcoming tropical Mediterranean Sea. Hydrobiologia 580:7–21

Blainville [HMD] (1821) Hyale, *Hyaloea* (Malacoz.). Dict Sci Nat 22:65–83

Blanckenhorn M (1889) Pteropodenreste aus der oberen Kreide nord-syriens und aus dem hessischen Oligocän. Z d geol Gesellsch 61:593–602

Blanckenhorn M (1934) Die Bivalven der Kreideformation von Syrien-Palaestina, nebst einem ergänzenden Anhang über Brachiopoden, Gastropoden und Pteropoden und einem Überblick über die gesammte Molluskenfauna. Palaeontographica A81:161–296

Boas JEV (1886) Spolia atlantica. Bidrag til pteropodernes. Morfologi og Systematik samt til Kundskaben om deres geografiske Udbredelse. Vidensk Selsk Skr 6 (naturvidensk math Afd 4):1–231

Bontes B, van der Spoel S (1998) Variation in the *Diacria trispinosa* group, new interpretation of colour patterns and description of *D. rubecula* n. sp. (Pteropoda). Bull Zool Mus Univ Amsterdam 16:77–84

Bosc LAG (1802) Histoire naturelle des coquilles, contenant leur description, les mœurs des animaux qui les habitent et leurs usages 2. Deterville, Paris

Bouchet P, Rocroi J-P (2005) Classification and nomenclator of gastropod families. Malacologia 47:1–397

Bourguignat JR (1868) Études géologique et paléontologique des hauts plateaux de l'Atlas entre Boghar et Tiharet. In: Souvenirs d'une exploration scientifique dans le Nord de l'Afrique. Challamel ainé, Paris, pp 1–35

Bronn HG (1831) Italiens Tertiär-Gebilde und deren organische Einschlüsse, 3. Uebersicht der fossilen Ueberreste in den tertiären subapenninischen Gebirgen. Groos, Heidelberg

Bronn HG, Keferstein W (1862–1866) Die Klassen und Ordnungen des Thier-Reichs, wissenschaftlich dargestellt in Wort und Bild, 3. Malacozoa, 2. Kopftragende Weichthiere (Malacozoa cephalophora). C.F. Winter, Leipzig/Heidelberg

Bucklin A, Steinke D, Blanco-Bercial L (2010) DNA barcoding of marine metazoa. Annu Rev Mar Sci 3:471–508

Bullock TH (2011) Neuroethology of zooplankton. http://cogprints.org/127/1/Plankton.htm. Accessed Dec 2011

Cahuzac B, Janssen AW (2010) Eocene to Miocene pteropods (Gastropoda, Euthecosomata) from the Aquitaine Basin, SW France. Scripta Geol 141:1–193

Carozzi A (1954) L'organisme 'C' J. Favre (1927) est une *Vaginella* portlandienne. Arch Sci Soc Phys Hist Nat Genève 7:107–111

Checchia-Rispoli G (1921) I pteropodi del Miocene garganico. Boll R Com Geol Ital 48:1–28

Children JG (1823) Lamarck's genera of shells. Q J Sci Lit Arts 15:216–258

Clauzon G, Suc J-P, Gautier F, Berger A, Loutre M-F (1996) Alternate interpretation of the Messinian salinity crisis: controversy resolved? Geology 24:363–366

Collins RL (1934) A monograph of the American tertiary pteropod mollusks. Johns Hopkins Univ Stud Geol 11:137–234

Comeau S, Gorsky G, Alliouane S, Gattuso JP (2010a) Larvae of the pteropod *Cavolinia inflexa* exposed to aragonite undersaturation are viable but shell-less. Mar Biol 157:2341–2345

Comeau S, Jeffree R, Teyssie JL, Gattuso JP (2010b) Response of the arctic pteropod *Limacina helicina* to projected future environmental conditions. PLoS One 5(6):e11362. doi:10.1371/journal.pone.0011362

Conversi A, Peluso T, Fonda-Umani S (2009) Gulf of Trieste: a changing ecosystem. J Geophys Res Oceans 114:C03S90. doi:10.1029/2008JC004763

Conversi A, Umani SF, Peluso T et al (2010) The Mediterranean sea regime shift at the end of the 1980s, and intriguing parallelisms with other European basins. PLoS One 5(5):e10633. doi:10.1371/journal.pone.0010633

Cooke JH (1893) The marls and clays of the Maltese islands. Q J Geol Soc 49:117–128

Cooke JH (1896) Notes on the Globigerina limestone of the Maltese islands. Geol Mag NS4:502–511

Corselli C, Grecchi G (1990) Considerazioni sui Thecosomata attuali del Bacino Mediterraneo. Lav Soc Ital Malacol 24:120–130

Cossmann M (1902) Mollusques eocéniques de la Loire Inférieur 2(2). Bull Soc Sci nat Ouest France 1:55–209

Curry D (1965) The English Palaeogene pteropods. Proc Malacol Soc London 36:357–371

Curry D (1982) Ptéropodes eocènes de la tuilerie de Gan (Pyrénées-Atlantiques) et de quelques autres localités du SW de la France. Cah Micropal 4:35–44

Curry D, Rampal J (1979) Shell microstructure in fossil thecosome pteropods. Malacologia 18:23–25

D'Alessandro A, Robba E (1981) Pteropodi neogenici della Puglia (Italia meridionale). Riv Ital Paleont Strat 86:605–698

D'Alessandro A, Laviano A, Ricchetti G, Sardella A (1979) Il Neogene del Monte Gargano. Boll Soc Paleont Ital 18:9–116

Dieci G (1962) Pteropodi langhiani di Rio delle Bàgole (Montegibbio, Appennino settentrionale modenese). Boll Soc Paleontol Ital 1:31–42

Defrance [JLM] (1828) *Vaginella*. (Foss.). Dict Sci Nat 56:427

de Lange GJ, Krijgsman W (2010) Messinian salinity crisis: a novel unifying shallow gypsum/deep dolomite formation mechanism. Mar Geol 275:273–277

de Madron XD, Guieu C, Sempere R et al (2011) Marine ecosystems' responses to climatic and anthropogenic forcings in the Mediterranean. Progr Oceanogr 91:97–166

Dieni I (1990) *Brunonia annulata* (Yokoyama, 1890) (Carinariidae, Mesogastropoda) nel Cretaceo inferiore della Sardegna. Boll Soc Paleont Ital 29:43–51

DiGeronimo I (1974) Molluschi pelagici in livelli di marne mioceniche presso Vetto (R. Emilia). Conchiglie 10:181–192

DiGeronimo I, Grasso M, Pedley HM (1981) Palaeoenvironment and palaeogeography of Miocene marls from southeast Sicily and the Maltese islands. Palaeogeogr Palaeoclimatol Palaeoecol 34:173–189

Dollfus V, Ramond G (1885) Liste des ptéropodes du terrain tertiaire parisien. Ann Soc R Malacol Belg 20:38–44

Doney SC (2010) The growing human footprint on coastal and open-ocean biogeochemistry. Science 328:1512–1516

Doney SC, Fabry VJ, Feely RA, Kleypas JA (2009) Ocean acidification: the other CO2 problem. Annu Rev Mar Sci 1:169–192

d'Orbigny A (1834) Voyage dans l'Amérique méridionale (le Brésil, la république orientale de l'Uruguay, la République Argentine, la Patagonie, la République du Chili, la République de Bolivia, la République du Pérou), exécuté pendant les années 1826, 1827, 1828, 1829, 1830, 1831, 1832 et 1833. Bertrand & Strasbourg, Levrault, Paris

Erünal-Erentöz L (1958) Mollusques du Néogène des bassins de Karaman, Adana et Hatay (Turquie). Publ Inst Ét Rech Min Turq 4:1–232

Fabry VJ, Seibel BA, Feely RA, Orr JC (2008) Impacts of ocean acidification on marine fauna and ecosystem processes. Ices J Mar Sci 65:414–432

Feely RA, Sabine CL, Lee K et al (2004) Impact of anthropogenic CO2 on the CaCO3 system in the oceans. Science 305:362–366

Fleming J (1823) On a reversed species of *Fusus*, (*Fusus retroversus*). Mem Wernerian Nat Hist Soc 4:498–500

Felix R (1973) Oligo-Miocene stratigraphy of Malta and Gozo (PhD-thesis Rijksuniv Utrecht). Meded Landbouwhogesch Wageningen 73–20:1–104

Foresi LM, Mazzei R, Salvatorini G, Donia F (2008) Biostratigraphy and chronostratigraphy of the Maltese Lower Globigerina Limestone Member (Globigerina Limestone Formation): new preliminary data based on calcareous plankton. Boll Soc Paleont Ital 46:175–181

Forskål P *in* Niebuhr C (1775) Descriptiones animalium avium, amphibiorum, piscium, insectorum, vermium; quae in itinere Orientali observavit Petrus Forskål, prof. Haun. Post mortem autoris edidit Carsten Niebuhr. Adjuncta est materia medica Kahirina atque tabula Maris Rubri geographica. Möller, Hauniae

Gabb WM (1873) Description of some new genera of Mollusca. Proc Acad Nat Sci Phila 1872:270–274

Gibbard PL, Head MJ, Walker MJC, Subcommission on Quaternary Stratigraphy (2009) Formal ratification of the Quaternary System/Period and the Pleistocene Series/Epoch with a base at 2.58 Ma. J Quaternary Sci 25:96–102

Gilmer RW (1972) Free floating mucus webs: a novel feeding adaptation for the open ocean. Science 176:1239–1240

Gilmer RW (1974) Some aspects of feeding in thecosomatous pteropod mollusks. J Exp Mar Biol Ecol 15:127–144

Gilmer RW, Harbison GR (1986) Morphology and field behavior of pteropod molluscs: feeding methods in the families Cavoliniidae, Limacinidae and Peraclididae (Gastropoda: Thecosomata). Mar Biol 91:47–57

Godeaux J, Toulemont A (1990) A propos des migrations lessepsiennes. Bull Inst Océanogr Monaco numéro spécial 7:103–107

Godwin-Austen HH (1882) On a fossil species of *Camptoceras*, a freshwater mollusk from the Eocene of Sheerness-on-Sea. Q J Geol Soc Lond 38:218–221

Goetze E (2003) Cryptic speciation on the high seas; global phylogenetics of the copepod family Eucalanidae. Proc R Soc Lond B Biol Sci 270:2321–2331

Goetze E (2010) Species discovery in marine planktonic invertebrates through global molecular screening. Mol Ecol 19:952–967

Gougerot L, Braillon J (1965) Découverte d'une nouvelle espèce d'*Atlanta* (Hétéropode) dans le Bartonien de Barisseuse (Oise). Bull Soc Géol France 7:302–303

Gradstein FM et al (2004) A geologic time scale 2004. Cambridge University Press, Cambridge (39 authors)

Grebneff A, Janssen AW, Lee DE (2011) A new euthecosomatous gastropod *Gamopleura maxwelli* sp. n. (Gastropoda, Cavolinioidea, Cavoliniidae) from the Late Oligocene of southern New Zealand. N.Z. J Geol Geophys 54:69–74

Grecchi G (1987) Revisione tassonomica degli esemplari fossili e subfossili di *Clio polita* rinvenuti in Mediterraneo e in Mar Rosso. Boll Malacol 23:297–303

Harzhauser M, Kroh A, Mandi O, Piller WE, Goehlich U, Reuter M, Berning B (2007) Biogeographic responses to geodynamics: a key study all around the Oligo-Miocene Tethyan seaway. Zool Anz 246:241–256

Hodgkinson KA, Garvie CL, Bé AWH (1992) Eocene euthecosomatous Pteropoda (Gastropoda) of the Gulf and eastern coasts of North America. Bull Am Paleontol 103:5–62

Hunt B, Strugnell J, Bednarsek N et al (2010) Poles apart: the "bipolar" pteropod species *Limacina helicina* is genetically distinct between the Arctic and Antarctic oceans. PLoS One 5(e9835):9831–9834

Jacobs E, Weissert H, Shields G (1996) The Monterey event in the Mediterranean: a record from shelf sediments of Malta. Paleoceanography 11:717–728

Janssen AW (1995) Systematic revision of holoplanktonic Mollusca in the collections of the 'Dipartimento di Scienze della Terra' at Torino, Italy. Mus reg Sci nat Torino Monogr 17:1–233

Janssen AW (1999a) Notes on the systematics, morphology and biostratigraphy of fossil holoplanktonic Mollusca, 4. A collection of euthecosomatous pteropods from the Miocene of the Karaman Basin, Turkey. Basteria 63:11–15

Janssen AW (1999b) Notes on the systematics, morphology and biostratigraphy of fossil holoplanktonic Mollusca, 6. Biostratigraphical interpretation of an assemblage from Poggio Musenna (Sicily, Italy) in comparison to northern Italian and Maltese localities. Basteria 63:111–120

Janssen AW (2004) Fossils from the lower Globigerina Limestone Formation at Wardija, Gozo (Miocene, Aquitanian), with a description of some new pteropod species (Mollusca, Gastropoda). Cent Mediterr Nat 4:1–33

Janssen AW (2005) Development of Cuvierinidae (Mollusca, Euthecosomata, Cavolinioidea). During the Cainozoic: a non-cladistic approach with a re-interpretation of recent taxa. Basteria 69:25–72

Janssen AW (2006) Notes on the systematics, morphology and biostratigraphy of fossil holoplanktonic Mollusca, 17. On the status of some pteropods (Gastropoda, Euthecosomata) from the Miocene of New Zealand, referred to as species of *Vaginella*. Basteria 70:71–83

Janssen AW (2007) Holoplanktonic Mollusca (Gastropoda) from the Gulf of Aqaba, Red Sea and Gulf of Aden (Late Holocene-Recent). Veliger 49:140–195

Janssen AW (2010) Pteropods (Mollusca, Euthecosomata) from the Early Eocene of Rotterdam (The Netherlands). Scripta Geol Sp Issue 7:161–175

Janssen AW (2012a) Late quaternary to recent holoplanktonic Mollusca (Gastropoda) from bottom samples of the eastern Mediterranean Sea: systematics, morphology. Boll Malacol 49(suppl 9):1–105

Janssen AW (2012b) Systematics and biostratigraphy of holoplanktonic mollusca from the Oligo-Miocene of the Maltese Archipelago. Mus Reg Sci Nat Torino Bollettino 28:197–601

Janssen AW (2012c) Early Pliocene heteropods and pteropods (Mollusca, Gastropoda) from Le Puget-sur-Argens (Var), France. Cainoz Res 9:145–187

Janssen AW, King C (1988) Planktonic molluscs (Pteropods). In: Vinken R et al (eds) The northwest European Tertiary Basin. Results of the International Geological Correlation Programme Project no. 124. Geol Jb. A100:356-368

Janssen AW, Little CTS (2010) Holoplanktonic Mollusca from the Miocene of Cyprus. Palaeont 53:1111–1145
Janssen AW, King C, Steurbaut E (2011) Notes on the systematics, morphology and biostratigraphy of fossil holoplanktonic Mollusca, 21. Early and middle Eocene (Ypresian-Lutetian) holoplanktonic Mollusca (Gastropoda) from Uzbekistan. Basteria 75:71–93
Jennings RM, Bucklin A, Ossenbrügger H, Hopcroft RR (2010) Species diversity of planktonic gastropods (Pteropoda and Heteropoda) from six ocean regions based on DNA barcode analysis. Deep Sea Res II 57:2199–2210
Klussmann-Kolb A, Dinapoli A (2006) Systematic position of the pelagic Thecosomata and Gymnosomata within Opisthobranchia (Mollusca, Gastropoda) – revival of the Pteropoda. J Zool Syst Evol Res 44:118–129
Knowlton N (2000) Molecular genetic analyses of species boundaries in the sea. Hydrobiologia 420:73–90
Krohn A (1860) Beiträge zur Entwicklungsgeschichte der Pteropoden und Heteropoden. Engelmann, Leipzig
Kunz A (1996) Schalenmorphologische Merkmale der Ontogenesestadien ausgewählter Pteropoden (Gastropoda: Heterostropha). BSc thesis Univ Hamburg (Geol-Paläont) (unpublished)
Lalli CM, Conover RJ (1973) Reproduction and development of *Paedoclione doliiformis*, and a comparison with *Clione limacina* (Opisthobranchia, Gymnosomata). Mar Biol 19:13–22
Lalli CM, Conover RJ (1976) Microstructure of the veliger shells of gymnosomatous pteropods (Gastropoda: Opisthobranchia). Veliger 18:237–240
Lalli CM, Gilmer RW (1989) Pelagic snails. The biology of holoplanktonic gastropod mollusks. Stanford University Press, Stanford
Lalli CM, Wells FE (1973) Brood protection in an epipelagic the cosomatous pteropod, *Spiratella* (‘*Limacina*’) *inflata* (D’Orbigny). Bull Mar Sci 23:933–941
Lalli CM, Wells FE (1978) Reproduction in the genus *Limacina* (Opisthobranchia: Thecosomata). J Zool London 186:95–108
Langer G, Nehrke G, Probert I, Ly J, Ziveri P (2009) Strain-specific responses of *Emiliania huxleyi* to changing seawater carbonate chemistry. Biogeosciences 6:2637–2646
Lejeusne C, Chevaldonne P, Pergent-Martini C, Boudouresque CF, Perez T (2010) Climate change effects on a miniature ocean: the highly diverse, highly impacted Mediterranean Sea. Trends Ecol Evol 25:250–260
Lesueur [JA] (1813) Mémoire sur quelques espèces d’animaux mollusques et radiaires recueillis dans la Méditerranée, près de Nice. Nouv Bull Sci Soc Philomath 3:281–285
Mancin N, Pirini C, Bicchi E, Ferrero E, Valleri G (2003) Middle Eocene to middle Miocene planktonic foraminiferal biostratigraphy for internal basins (monferrato and northern Apennines, Italy). Micropaleontology 49:341–358
Mayer C (1868) Description de coquilles fossiles des terrains tertiaires supérieurs (suite). J Conchyl 8:102–112
Mazzei R (1985) The Miocene sequence of the Maltese islands: biostratigraphic and chronostratigraphic references based on nannofossils. Atti Soc Tosc Sci Nat A92:165–197
McKinney FK, Delvolve J-J, Sobiera J (1995) *Conularia* sp. from the Pyrénées: further support for scyphozoan affinities of the Conularida. Lethaia 28:229–236
Michelotti G (1847) Description des fossiles des terrains miocènes de l’Italie septentrionale. Natuurk Verhand Holl Maatsch Wetensch Haarlem 3:1–408
Molinero JC, Ibanez F, Souissi S et al (2008) Climate control on the long-term anomalous changes of zooplankton communities in the Northwestern Mediterranean. Glob Chang Biol 14:11–26
Mörch OAL (1850) Catalogus conchyliorum quae reliquit C.P. Kierulf, Md. Dr., nunc publica auctione X Decembris MDCCCL Hafniae Dividenda. Typis Trieri, Hafniae
Norris RD (2000) Pelagic species diversity, biogeography, and evolution. Paleobiology 26:236–258
Orr JC, Fabry VJ, Aumont O et al (2005) Anthropogenic ocean acidification over the twenty-first century and its impact on calcifying organisms. Nature 437:681–686
Pafort-van Iersel T, van der Spoel S (1986) Schizogamy in the planktonic opisthobranch *Clio* - a previously undescribed mode of reproduction in the Mollusca. Int J Invert Reprod Dev 10:43–50
Papadopoulos LN, Peijnenburg KTCA, Luttikhuizen PC (2005) Phylogeography of the calanoid copepods *Calanus helgolandicus* and *C. euxinus* suggests Pleistocene divergences between Atlantic, Mediterranean, and black Sea populations. Mar Biol 147:1353–1365
Pedley HM, House MR, Waugh H (1977) The geology of Malta and Gozo. P Geologists’ Assoc 87:325–341
Peijnenburg KTCA, Breeuwer JAJ, Pierrot-Bults AC, Menken SBJ (2004) Phylogeography of the planktonic chaetognath *Sagitta setosa* reveals isolation in European seas. Evol 58:1472–1487
Peijnenburg KTCA, Fauvelot C, Breeuwer AJ, Menken SBJ (2006) Spatial and temporal genetic structure of the planktonic *Sagitta setosa* (Chaetognatha) in European seas as revealed by mitochondrial and nuclear DNA markers. Mol Ecol 15:3319–3338
Peijnenburg KTCA, Goetze E (in press) High evolutionary potential of marine zooplankton. Ecology and Evolution
Pelejero C, Calvo E, Hoegh-Guldberg O (2010) Paleo-perspectives on ocean acidification. Trends Ecol Evol 25:332–344
Pelseneer P (1888) Report on the Pteropoda collected by H.M.S. Challenger during the years 1873-1876, 3. Anatomy. Rep Sci Res Voy HMS Challenger 23:1–97
Pictet FJ (1855) Traité de paléontologie ou histoire naturelle des animaux fossiles considérés dans leurs rapports zoologiques et géologiques 3, 2nd edn. Baillière, Paris
Phipps CJ (1774) A voyage towards the North Pole undertaken by his Majesty’s command 1773. W. Bowyer, J. Nichols, London
Por FD (2009). Tethys returns to the Mediterranean: success and limits of tropical re-colonization. In: Krupp F, Musselman JL, Kotb MMA, Weidig I (eds) Environment, biodiversity and conservation in the Middle East. Proceedings of the First Middle Eastern Biodiversity Congress, Aqaba, Jordan, 20–23 Oct 2008. BioRisk 3:5–19
Premoli Silva I, Graham Jenkins D (1993) Decision on the Eocene-Oligocene boundary stratotype. Episodes 16:379–382
Quoy [JRC], Gaimard [JP] (1827) Observations zoologiques faites à bord de l’Astrolabe, en mai 1826, dans le détroit de Gibraltar (suite et fin). Description des genres biphore, carinaire, hyale, flèche, cléodore, anatife et briarée. Ann Sci Nat 10:225–239 (atlas)
Rang [PCAL] (1827) Description de deux genres nouveaux (*Cuvieria* et *Euribia*) appartenant à la classe des ptéropodes. Ann Sci Nat 12:320–329
Rang [PCAL] (1828) Notice sur quelques mollusques nouveaux appartenant au genre cléodore, et établissement et monographie du sous-genre créseis. Ann Sci Nat 13:302–319
Rang [PCAL] (1829) Description de cinq espèces de coquilles fossiles appartenant à la classe des ptéropodes. Ann Sci Nat 16:492–499
Rampal J (2002) Biodiversité et biogéographie chez les Cavoliniidae (Mollusca, Gastropoda, Opisthobranchia, Euthecosomata). Régions faunistiques marines Zoosystema 24:209–258
Ravn JPJ (1933) Étude sur les pélécypodes et gastropodes daniens du calcaire de Faxe. Kong Dansk Vidensk Selsk Skr, Naturvidensk Mathem Afd 9:1–71
Reuss AE (1867) Die fossile Fauna der Steinsalzablagerung von Wieliczka in Galizien. Sitz-Ber math-naturw Cl k Akad Wissensch Wien 55:17–182
Robba E (1971) Associazioni a pteropodi della Formazione di Cessole (Langhiano). Riv ital Paleontol S 77:19–126
Robba E (1972) Associazioni a pteropodi nel Miocene inferiore delle Langhe (Piemonte). Riv ital Paleontol S 78:467–524

Robba E (1977) Pteropodi serravalliani della Langhe (Piemonte). Riv ital Paleontol S 83:575–640

Robba E, Spano C (1978) Gasteropodi pelagici nel Miocene medio del Campidano meridionale (Sardegna). Riv ital Paleontol S 84:751–796

Rolle F (1861) Über einige neue oder wenig gekannte Mollusken-Arten aus Tertiär-Ablagerungen. Sitz-Ber math-naturw Cl k Akad Wissensch Wien 44:205–224

Rosenkrantz A (1960) Danian Mollusca from Denmark. International Geological Congress, Report on the twenty-first session, Norden, pp. 193–198.

Scarsella F (1934) Di una nuova specie di pteropodo del Miocene appenninico. Boll Soc Geol Ital 53:177–181

Seapy .R (2011). Pterotracheoidea. http://tolweb.org/Pterotracheoidea/27801. Accessed Nov 2011

Sendino C, Zágoršek K, Vyhlasová Z (2011) The aperture and its closure in an Ordovician conulariid. Acta Palaeont Pol 56:659–663

Simonelli V (1895) Sopra un nuovo pteropode del Miocene di Malta. Boll Soc Geol Ital 14:19–24

Simonelli V (1896) Sopra due nuovi pteropodi delle Argille di Sivizzano nel Parmense. Boll Soc Geol Ital 15:182–191

Singer BS (1994) An explosion of the pelagic population in the Gulf of Aqaba, January '94. Levantina 80:8–10

Sirna G (1968) Pteropodi del Miocene di S. Nicola Varano (Gargano) con descrizione di due nuove specie. Geol Romana 7:417–430

Sismonda A (1842). Synopsis methodica animalium invertebratorum Pedemontii fosilum. Torino

Sismonda A (1847) Synopsis methodica animalium invertebratorum Pedemontii fosilum, Torino (2nd edn)

Sowerby J (1814) The mineral conchology of Great Britain; or coloured figures and descriptions of those remains of testaceous animals or shells, which have been preserved at various times and depths in the earth, 1:97–178

Spoel S van der (1967) Euthecosomata, a group with remarkable developmental stages (Gastropoda, Pteropoda). PhD thesis, University Amsterdam. J. Noorduijn, Gorinchem

Sturani C (1973) A fossil eel (*Anguilla* sp.) from the Messinian of Alba (Tertiary Piedmontese Basin). Palaeoenvironmenta; and palaeogeographic implications. In: Drooger C (ed) Messinian events in the Mediterranean. North Holland Publ. Co., Amsterdam/London

Sturani C (1978) Messinian facies in the Piedmont Basin. In: Catalano R, Ruggieri G, Sprovieri R (eds). Messinian evaporites in the Mediterranean. Mem Soc Geol Ital16:11–25

Sturani C, Sampò M (1973) Il messiniano inferiore in facies diatomitica nel bacino terziario Piemontese. Mem Soc Geol Ital 12:335–358

Tiberi N (1878) Mollusques marins d'Italie. Céphalopodes, ptéropodes, hétéropodes vivant de la Méditerranée et fossiles du terrain tertiaire d'Italie. Mém Soc Roy Malacol Belgique 13:52–84

Tiberi N (1880) Cefalopodi, pteropodi, eteropodi viventi nel Mediterraneo e fossili nel terreno terziario italiano con aggiunte e correzioni. Bull Soc Malacol Ital 6:5–49

Touratier F, Goyet C (2011) Impact of the eastern Mediterranean transient on the distribution of anthropogenic CO2 and first estimate of acidification for the Mediterranean Sea. Deep-Sea Res I 58:1–15

Troelsen J (1938) Pteropoden-Reste aus dem oberen Senon dänemarks. Medd Dansk Geol Foren 9:183–185

Uriarte A (2011) Earth's climate history. http://web.me.com/uriarte/Earths_Climate/Earths_Climate_History.html. Accessed Nov 2011

van der Spoel S (1973) Strobilation in a mollusc; the development of aberrant stages in *Clio pyramidata* Linnaeus, 1767 (Gastropoda, Pteropoda). Bijdr Dierk 43:202–215

van der Spoel S (1976) Pseudothecosomata, Gymnosomata and Heteropoda (Gastropoda). Bohn, Scheltema and Holkema, Utrecht

van der Spoel S (1979) Strobilation in a pteropod (Mollusca, Opisthobranchia). Malacologia 18:27–30

van der Spoel S, Diester-Haass L (1976) First records of fossil gymnosomatous protoconchae (Pteropoda, Gastropoda). Bull Zoöl Mus Univ Amsterdam 5:85–88

van der Spoel S, Bleeker J, Kobayasi H (1993) From *Cavolinia longirostris* to twenty-four *Diacavolinia* taxa, with a phylogenetic discussion (Mollusca, Gastropoda). Bijdr Dierk 62: 127–166

Verri A, de Angelis d'Ossat G (1900) Secondo contributo allo studio del Miocene nell'Umbria. Boll Soc Geol Ital 19:240–279

Verri A, De Angelis d'Ossat G (1901) Terzo contributo allo studio del Miocene nell'Umbria. Boll Soc Geol Ital 20:1–23

Verri A, de de Angelis d'Ossat G (1899) Contributo allo studio del Miocene nell'Umbria. Rend R Accad Lincei 8:543–550

Wall-Palmer D, Hart MB, Smart CW, Sparks RSJ, Le Friant A, Boudon G, Deplus C, Komorowski JC (2011) Pteropods from the Caribbean Sea: variations in calcification as an indicator of past ocean carbonate saturation. Biogeosciences 9:309–315

Watelet A, Lefèvre T (1885) Note sur des ptéropodes du genre *Spirialis* découverts dans le Bassin de Paris (1880). Ann Soc Malacol Belg 15:100–103

Yebra L, Bonnet D, Harris RP, Lindeque PK, Peijnenburg KTCA (2011) Barriers in the pelagic: population structuring of *Calanus helgolandicus* and *C. euxinus* in European waters. Mar Ecol Progr Ser 428:135–149

Yokoyama M (1890) Versteinerungen aus der japanischen Kreide. Palaeontographica 36:159–202

Zorn I (1997) Hololanktonic gastropods from the early messinian of the Heraklion Basin (Crete, Greece). Contrib Tert Quat Geol 34:31–45

Ocean Acidification and Warming Effects on Crustacea: Possible Future Scenarios

20

Awantha Dissanayake

Abstract

Continuously-increasing concentrations of atmospheric carbon dioxide CO_2, primarily through the burning of fossil fuels, are rapidly increasing the oceanic concentrations of CO_2 and leading to the phenomenon of ocean acidification. Evidence to date on the effects of altered seawater chemistry on the biota is growing, yet is in its infancy. Evidence of effects is limited mostly to fish, molluscs and echinoderms, yet there is a growing body of evidence of effects of ocean acidification on the Crustacea. Our predictive ability on physiological effects and the potential ecosystem level effects is currently limited. By posing fundamental questions, the answers may lie in implementing mechanistic-level studies in order to elucidate organism physiological limits and species' potential to adapt to future oceanic conditions.

Keywords

Ocean acidification • Climate change • Physiology • Temperature • pH • pCO_2 • Organism performance

Introduction

Climatic change is driven by the anthropogenic-forced physical drivers; elevated temperature and carbon dioxide (CO_2), which are causing increasing warming of the earth's oceans, elevating oceanic CO_2 concentrations (hypercapnia), and thus, acidification of the oceans. Since the beginning of the industrial revolution (*ca.* 1,750), atmospheric levels of carbon dioxide (CO_2) have been steadily increasing. Present day levels of atmospheric CO_2 (380 ppm) are greater than pre-industrial levels (280 ppm) (Feely et al. 2004). The increase in anthropogenic CO_2, attributed to the burning of fossil fuels (IPCC 2001), has been rising at a far greater rate than previously recorded in the Earth's history (Pearson and Palmer 2000). Under current models, continuously increasing production of CO_2 emissions (under the 'business as-usual scenario') are predicted to raise the current atmospheric concentrations to 540–980 ppm by the turn of the century, and *ca.* 2,000 ppm by the year 2300 (Caldeira and Wickett 2003). Oceans are regarded as natural 'sinks' for carbon and account for 30–50 % of the atmospheric carbon that has been emitted over the last 250 year (Feely et al. 2004; Siegenthaler and Sarmiento 1993).

The modern age of anthropogenic global climate change is thus, known as the 'Anthropocene' (Crutzen and Stoermer 2000), where ocean acidification is known as the other CO_2 problem (Sabine et al. 2004). Global climate change also includes ocean warming, and the synergistic effect of both factors (elevated CO_2 and temperature) may have significant repercussions to animal physiology (Pörtner and Farrell 2008). Climate change biology, and ocean acidification research, in particular, is a rapidly developing scientific discipline, but is still in its infancy. There is an emerging awareness that mechanistic understanding is one of the research priorities for elucidating physiological effects and thus, projecting potential ecological and ecosystem

A. Dissanayake (✉)
Institute for East China Sea Research, Nagasaki University, Taira-machi 1551-7, Nagasaki 851-2213, Japan
e-mail: awanthadissanayake@gmail.com

S. Goffredo and Z. Dubinsky (eds.), *The Mediterranean Sea: Its history and present challenges*, DOI 10.1007/978-94-007-6704-1_20,

restructuring and its effect(s) on biological productivity (Peck et al. 2009). Elucidating possible climate change impacts on marine biota is of paramount importance, because generally, invertebrates are more sensitive to hypercapnia than fish (Ishimatsu and Dissanayake 2010). However, crustaceans are an important invertebrate group, especially, in economic terms, for example fisheries and aquaculture. Still, our knowledge on impacts of ocean climate change on this group is particularly poor (Ishimatsu and Dissanayake 2010; Whiteley 2011). The purpose of this review is to outline the effects of ocean acidification and ocean warming on the crustacean phylum, due to the changing climate. The present discussion will not delve into specific effects on each of the groups or species in turn, rather highlighting the recent evidence, major developments and future possibilities.

As ocean acidification is purported to affect calcifying organisms, specific groups have preferentially been used as physiological models to elucidate physiological effects of altered pCO_2 seawater chemistry at various stages of the life-cycles, such as bivalve molluscs (Gazeau et al. 2007; Range et al. 2011), corals (Fine and Tchernov 2007; Gattuso 1998) and echinoderms (reviewed in Ishimatsu and Dissanayake 2010; Kroeker et al. 2011). The reason the aforementioned groups have been preferentially chosen over the Crustacea maybe due to two important physiological reasons; firstly, the altered seawater chemistry due to pCO_2, pH, and carbonate chemistry directly and indirectly affects the external shells/skeletons via dissolution and altered calcification rates. Secondly, the vulnerability of the aforementioned groups may also be linked to lack of ability to buffer internal changes associated with elevated CO_2 concentrations in external aquatic environment. Crustacean models in particular, have been relatively under represented. Crustaceans, however, constitute an important link in coastal and open-water ecosystems, as prey items for higher trophic levels e.g. fish, as well as humans. Specifically, for the Mediterranean, the Crustacea form a large significant portion of the marine capture fisheries (~ 85–90 % of the 1.8 million tonnes total in 2006; FAO 2008). Crustacean species are ubiquitously found in the marine environment, distributed latitudinally from the poles to temperate and tropical latitudes and occupy a range of habitats, from the coastal shallow-water intertidal and benthos, to coastal open-waters and especially to deep-water benthic environments, where the conditions are characterised as relatively stable with respect to physical variables (Kennett and Ingram 1995; Pörtner et al. 2004). With respect to ocean acidification evidence, much of the focus has been directed towards the brachyuran decapods (Spicer et al. 2007; Walther et al. 2009). However, ocean acidification has also been evaluated in species representative of various trophic levels, such as the secondary producers, e.g. copepods, sessile species, such as barnacles, as well as the amphipods.

Climate change factors, such as ocean warming and acidification, both resultant from increasing anthropogenically-emitted CO_2, are proposed to have a strong impact on aquatic organisms. The physiological evidence regarding ocean acidification and Crustacea is wide-ranging in terms of study duration, from the short-term (hours to days) and mid-term (days to months); the ecological long-term consequences are as yet unknown. In order to elucidate physiological and ecological, (thus ecophysiological) impacts in organisms, the mechanisms that give rise to changes in physiological function (i.e. diagnostic evidence) will aid our understanding of the potential ecological consequences (i.e. prognosis). The current review will attempt to summarise evidence to date and surmise the possible ecological long-term consequences and thus, such body of preliminary evidence serves to outline the scientific direction in which to direct our understanding.

Summary of Physiological Impacts

Although, there is a relative lack of knowledge and ocean acidification is an emergent topic in climate change biology, the body of evidence of physiological effects in Crustacea from the last 8 years, emanate largely from laboratory studies, conducted over short- and mid-term time scales. However, mechanistic studies conducted in previous decades served to elucidate the physiological mechanisms for alteration of the *internal milieu* due to hypercapnia (elevated pCO_2) in crustaceans (Cameron 1978; Cameron and Iwama 1987), but also in fish (Iwama and Heisler 1991; Toews et al. 1983). It is likely that changes in seawater chemistry will affect the internal physiological functioning of marine organisms, such as the acid–base balance (Raven et al. 2005). Thus, organisms that live in aquatic environments face the problem of maintaining a constant internal environment (i.e. acid–base balance in extracellular fluid) that enables cells to function efficiently and is independent of the external environment (Rankin and Davenport 1981; Seibel and Walsh 2003). Previous mechanistic studies evaluating short-term exposure durations, often tested acid–base imbalances at higher pCO_2 concentrations (>10,000 µatm CO_2 or ~ 1 kPa) compared to predicted future ocean acidification scenarios (*ca.* 1,000 µatm for end of the century or 1,900 µatm for year 2400) (Caldeira and Wickett 2003); and have demonstrated that acid–base balance is closely associated with ionic regulation as both internal processes are linked (Batterton and Cameron 1978; Cameron 1985, 1978; Cameron and Iwama 1987).

Mechanisms of acid–base regulation require electroneutral ion exchange with the external environment, occurring across gill epithelia where H^+ and HCO_3^- ions are used as counter-ions for sodium (Na^+) and chloride (Cl^-) (HCO_3^-/Cl^- and H^+/Na^+) (Cameron et al. 1983; Mantel and Farmer 1983; Wheatly and R. P. Henry 1992). Ionic and osmotic regulation is defined as 'maintenance of ion concentrations (ionic) and total particle concentration (osmotic) in a body fluid differing to that of the external medium (Robertson 1949). It is generally regarded that efficient ionic/osmotic regulators are efficient acid–base regulators, and any acid–base alterations are more likely to be regulated, and thus, compensated in strong ionic regulators compared to weak ionic regulators (Whiteley 2011). A limited number of studies have evaluated acid–base balance effects of environmentally-relevant ocean acidification exposure scenarios, with the general conclusion that acid–base alterations are compensated for by hypercapnic buffering of bicarbonate (HCO_3^-). Dissanayake et al. (2010) demonstrated that OA exposure (0.3 kPa) in two Palaemonid species (*Palaemon elegans* and *P. serratus*) resulted in initial acidosis in haemolymph (after 5–14 days), however, complete and 'over' compensation was resultant after prolonged exposure after 30 days. Palaemonids have previously been shown to be efficient hyper- and hypo-ionic/osmoregulators, especially in both low and high saline waters, respectively (Freire et al. 2003; Panikkar 1941). Two species of brachyuran decapods *Cancer magister and Necora puber* considered to be relatively-weak ionic regulators were also able to compensate haemolymph acid–base alterations, and due to concomitant bicarbonate buffering; exposed to 1 kPa CO_2 within 24 h for *C. magister* (Pane and Barry 2007) and 0.2 kPa CO_2 for *N. puber* (Spicer et al. 2007). Continued exposure in *N. puber* demonstrated that limits existed in the bicarbonate buffering mechanism; after 16 days where concentrations reached 27 mmol l^{-1}. Increased hypercapnia exposure (2 kPa CO_2 which equates to pH 6.05) resulted in haemolymph acidosis and subsequent mortalities after 4–5 days, despite a high bicarbonate pool (*ca.* 55 mmol l^{-1}). Although, a medium-term exposure (30 days) demonstrated contradictory results, whereby, crabs were able to compensate for previously-observed acid–base alterations with a lower bicarbonate pool (Small et al. 2010). Such differences demonstrate that regulatory processes may have energetic repercussions, if regulation is achieved. Ion regulatory changes may result from an up-regulation of Na^+/K^+-ATPase, the enzyme responsible for ion exchange, as shown previously for palaemonids and other decapods (Mendonça et al. 2007; Freire et al. 2008; Ituarte et al. 2008; Masui et al. 2009). Carbonic anhydrase also plays a role in supplying counter-ions (H^+ and HCO_3^-) for ionic exchange through the hydration of respired CO_2 and plays an important role in CO_2 excretion and acid–base balance (Henry and Cameron 1983; McMahon et al. 1984). It could be hypothesised that under hypercapnia exposure any upregulation in carbonic anhydrase (and also Na^+/K^+-ATPase, as mentioned previously) could serve to compensate for hypercapnic changes in acid–base balance as a result of active ion transport (Henry and Wheatly 1992; Seibel and Walsh 2003). Bicarbonate buffering therefore mainly occurs, as previously mentioned, via ion exchange with the external medium, with around 90 % 'sourced' from seawater and the remainder internally, either enzyme up-regulation or via exoskeletal/carapace dissolution (Cameron 1985).

There is, however, evidence where bicarbonate buffering does not occur. The burrowing shrimp *Upogebia deltaura*, an important ecosystem engineer in benthic systems, when exposed to hypercapnia scenarios representative of coastal ocean acidification (35 d to ca. 1,400 μatm; pH 7.64 and ca. 2,700 μatm; pH 7.35) and those associated with geological CO_2 sequestration leaks (>14,000 μatm, pH 6.71); results showed that *U. deltaura* appears to be tolerant to levels predicted for the 2100 ocean acidification scenario. At higher levels, however, shrimps experienced extracellular acidosis but no difference in haemolymph bicarbonate [HCO_3^-], suggesting little or no buffering capacity, and there was no observable evidence of other physiological costs with regards to metabolism, osmotic regulation, calcification, growth and activity. At the highest scenario, significant differences in activity were observed prior to 100 % mortality. Thus, suggesting that in event of a CO_2 sequestration leakage event, the activity, and therefore, potentially, the ability of these sediment bio-turbators may be comprised, resultant in loss of this important ecosystem engineer in benthic environments (Donohue et al. 2012).

It is generally held that organism lifestyle and traits may have repercussions on the degree of ocean acidification effects. Slow moving species may be more susceptible compared to more motile species; generally, slow-moving species have higher magnesium [Mg^{2+}] concentrations within body fluids compared to faster-moving species, such as pelagic crustaceans (Robertson 1949, 1953). Internal magnesium concentrations are, thus, widely regarded as an indicator of activity, as magnesium acts on the neuro-muscular junctions and hence why increased Mg, (usually administered as MgCl or $MgSO_4$) has been widely used as an anaesthetic (see crustacean examples of internal [Mg2+] in Burton, 1995; and reviewed in Morritt and Spicer 1993). Magnesium is also a co-factor in enzyme systems, such as the transfer of phosphate groups e.g. ATP (Morritt and Spicer 1993). If alteration in magnesium regulation occurs, resulting in elevated internal [Mg] could lead to a decrease in activity, and repercussion for ecologically-relevant behaviours and thus, species fitness. Penaeid prawns exposed to an ocean acidification scenario (0.2 kPa CO_2; pH 7.48 for 10 days) displayed elevated [Mg^{2+}] within the haemolypmph; penaeids are active swimmers and thus, maintain low magnesium concentrations (Wittmann et al.

2010). Any potential increase in [Mg^{2+}], thus, breakdown or alteration of iono-regulation, could cause a narcotising effect on the individual and potentially affect overall behavioural activity, such as reduced swimming performance, as previously observed in this species with hypercapnia exposure (Dissanayake and Ishimatsu 2011).

Further evidence of disruption of iono-regulation is observed in hermit crabs exposed to reduced seawater pH (pH 6.80; ~12,000 μatm CO_2) where elevated chloride concentrations [Cl^-] were found in haemolymph (de la Haye et al. 2012), corroborating observations previously reported for Palaemonids (Dissanayake et al. 2010). A reduced response to prey odour, such as antennule flicking rate, described as the 'sniffing response' in crustaceans, and less successful in locating the odour source, as well as an overall decline in locomotory activity compared to those in normocapnic seawater (de la Haye et al. 2011, 2012). High CO_2 exposure caused hermit crabs to be less active, owing to possible metabolic depression and a trade-off in maintainance functions (acid–base balance) versus behaviour, although, a statistically-significant correlation was not observed. Although, great strides have been taken to elucidate ocean acidification effects upon Crustacea, it is apparent further studies are needed to further evaluate the connection between energy chanelling for up-regulation of maintainance functions and the need to engage in ecologically-relevant behaviours, such as foraging in the face of altered seawater conditions; the consequences of which could be an overall reduction in organisms' fitness.

Growth, Reproduction and Developmental Effects

In terms of ocean acidification effects on growth and development, most studies have been evaluated on early life-stages and not yet linked developmental effects to the relevant stages of maturity, such as reproductive maturity. Marine calenoid copepods have been model species in which to study effects of elevated CO_2 on growth, reproduction and development, due to relatively short generation times, fast development of discrete larval stages, and ecological-importance as secondary producers in marine food webs. Contrasting evidence has showed, however, that a disparity exists between species, thus, highlighting differences between species tolerance and sensitivity, which makes predictive modelling of ecosystems effects far from straightforward (reviewed in Whiteley 2011). Kurihara et al. (2004) demonstrated in *Acartia steurei* and *A. erythraea*, that under elevated CO_2 conditions (at levels far greater than forecasted under future OA scenarios; > 2,000 μatm CO_2), the production of eggs, egg hatching rate and nauplii mortality all decreased with increasing CO_2. In *A. tsuensis,* however, no significant effects were observed for any developmental parameters, including survival, body size, developmental time of initial copepod generation. The development from eggs to mature adults in subsequent two generations of females acclimated under these high CO_2 conditions (2,380 μatm CO_2), also did not exhibit any negative effects, thus, highlighting a discrepancy even within a genus of Calenoid copepods. Mayor et al. (2007) demonstrated in the ecologically-important copepod *Calanus finmarchicus* that under very high CO_2 levels (8,000 μatm CO_2) a reduction in viability of eggs produced, however, did not affect the energetic content of the adult females or the daily reproduction rate. It is currently known that energetic consequences for the viability of eggs and energy content of female *C. finmarchicus* is dependant on the availability of food during the spring bloom. Under limited food conditions, a 'sacrificial reproductive strategy' is postulated, whereby female copepods divert energy resources by protein catabolism to increase egg production (Mayor et al. 2009). Under end of the century ocean warming and acidification scenarios (i.e. 2 or 4°°C and 1,000 μatm CO_2; pH 7.73), effects on *C. helgolandicus* were observed, whereby the acclimation temperature and the batch of eggs used may have been the driving factors for apoptosis or programmed cell death observed in eggs and nauplii (Mayor et al. 2012). The viability of eggs and larval development has also bee studied in antarctic krill, *Euphausia superba*, with no effects observed at 1,000 μatm CO_2, however, at 2,000 μatm CO_2, significant developmental effects were observed whereby development was halted at the gastrula stage (2 days) in 90 % of the embryos, resulting in 100 % unsuccessful hatching rates of the embryos (Kawaguchi et al. 2010).

Developmental effects in early-life stages have also been evaluated in the European lobster *Homarus gammarus* (Arnold et al. 2009), the barnacles, *Amphibalanus amphitrite* (McDonald et al. 2009) and *Semibalanus balaniodes* (Findlay et al. 2009). No developmental effects were observed in the early stages of development in the European lobster, from Zoea I to IV stages after 28 days of exposure to 1,200 μatm CO_2. However, the carapace dry mass and both calcium and magnesium content was significantly depressed at stage IV. Developmental effects and net decline in calcification in important post-larval stages, along with a reduced shell mass of developing larval lobsters may affect recruitment success in this economically-important species (Arnold et al. 2009). No differences were reported for cyprid size, nauplii survival, growth from juvenile to adult or egg production in *Amphibalanus amphitrite* exposed to pH 7.4 (McDonald et al. 2009). However, Findlay et al. (2009) demonstrated that hatching rates were delayed in *Semibalanus balaniodes* by 19 days under high CO_2 conditions (pH 7.7; 900 μatm CO_2). Growth effects in adults has only been established in the Palaemonid prawn *Palaemon pacificus*, whereby average growth was observed to be suppressed in prawns exposed to high CO_2 levels; 8 % less compared to adults after 6 weeks exposure in adult females exposed to pH 7.64; 1,900 µatm CO_2 (Kurihara et al. 2008).

From the current dearth of knowledge on effects on growth, reproduction and development, it is apparent much progress needs to be made to identify species' tolerances and 'tipping' points in sensitivity. However, factors that influence viability and thus, progeny survival and success of future generations, may actually be context-dependent, as well as species-dependant (Mayor et al. 2012). These authors advocate that the variability in developmental responses is linked to the acclimation capacity of the parental generation. Thus, highlighting, the importance of acclimation and the 'knock-on' effects on the darwinian fitness of the species will have huge ramifications, if we are to increase our predicative abilities on the consequences of future ocean acidification and warming scenarios.

Synergistic Effects of Ocean Acidification and Other Environmental Factors

There are a growing number of ocean acidification studies to date, not just focussing on crustaceans. There is also a lack of evidence of synergistic effects of OA and other environmental variables; however, there are now emergent studies reporting on the synergy of OA and temperature. Temperature is the most obvious abiotic factor to study climate change effects on marine organisms, as forecasted climate scenarios depend on both climate change drivers; ocean warming, due to temperature and ocean acidification, due to increased carbon dioxide (IPCC 2007).

Temperature

Current ecosystem changes are generally seen to be driven by temperature effects on organism functioning (Pörtner and Farrell 2008), rather than the impact of increased CO_2 diffusion and subsequent altered seawater chemistry of the oceans, as evidenced at a major international conference (PICES- *Climate Change Effects on Fish and Fisheries,* Sendai, Japan 2010), where the vast majority of studies presented were regarding temperature effects on organisms. In order to accurately predict the effects of climate change on marine ecosystems and the organisms that inhabit it, providing the ecosystem goods and services that we require, more studies focussing on the synergistic effects of abiotic factors (and combinations there of) are needed. The paradigm proposed by Pörtner and Farrell (2008) represents the thermal window of aerobic performance of an organism, where optimal limits of organism performance occurs (Pörtner and Farrell 2008). Any deviations from optimum, either towards colder temperatures or warmer temperatures, result in 'pejus' (turning worse) or critical limits. Aerobic thermal windows vary over the life-cycle of an organism and also between species from differing thermal niches, e.g. polar, temperate or tropical species (Pörtner and Farrell 2008). An example of decreased organism aerobic performance was primarily demonstrated in fish, the eelpout *Zoarces viviparus*, as a mismatch between the demand for oxygen and the capacity of oxygen supply to tissues. Crustacean examples include decapods exposed to hypercapnia (1,000 µatm CO_2 for *Cancer pagurus*; 710–3,000 µatm for *Hyas araneus*), a reduction in both blood oxygen partial pressure and cardiac activity was recorded with increasing temperature in both species. An example of temperature manipulation alone in the spider crab *Maja squinado*, demonstrated the same effects, i.e. limited capacity for ventilation and circulation at extreme temperatures causes insufficient oxygen supply. These examples seem to suggest a shift in the oxygen carrying capacity of blood and concomitant decreased thermal tolerance in these benthic decapod crustaceans. The repercussions of which suggest a narrowing of the 'thermal window' i.e. thermal tolerance to elevated temperature (Frederich and Pörtner 2000; Metzger et al. 2007; Walther et al. 2009). Although, it is argued that such a paradigm of decreased organism performance maybe thought of limiting in terms of failing to consider thermally-induced phenotypic plasticity or thermal acclimatisation (Franklin and Seebacher 2009; see examples within). It is proposed that organisms are not specialised to their environment and may display phenotypic plasticity in response to altered conditions. The capacity for displaying such plasticity in phenotypic responses may be the key for species survival, however, they may be limits to an organism's capability to display phenotypic plasticity. It is also counter-argued that while some species display conserved limited thermal acclimation capabilities beyond present habitat temperatures, some on the otherhand do not. However, variable specialisation according to thermal acclimation requires further testing using important darwinian fitness parameters, such as growth, reproduction, and development (see Pörtner et al. 2009, and examples therein).

Climatic change and variability may exert effects upon biota in both singly and/or in combination with other physical and chemical factors. Factors include environmental factors such as temperature, aeolian-driven vertical mixing, changes in salinity, oxygen and pH. Effects upon organisms are classed as direct effects and may be exerted on physiology, development, reproduction, behaviour and ultimately, survival. Indirect effects are exerted via ecosystem processes and trophic level effects. Indirect effects, may ultimately, alter the trophic structure of ecosystems by altering abundance of predators, prey and competitor species (see Brander 2010 for examples). Current investigations into biological effects of climate variability have yielded regime shifts in fish species. In the North Pacific, for example, analysis of long-term data sets have revealed regime shifts in species abundance to oscilate between Japanese anchovy (*Engraulis japonicus*) and Japanese sardine (*Sardinops melanostictus*) (Takasuka et al. 2007, 2008). Differences in spawning thermal optima between the species could explain the cause of the recorded regime shifts, with spawning temperature patterns displaying warm/eurythermal Japanese anchovy (mean 22 °C) and cool/stenothermal Japanese sardine (16–17 °C) in the western North Pacific. Increases in sea surface temperatures, in general, are purported to be higher in Northern marine systems than southern systems, and especially, enclosed systems such as the Mediterranean (Philippart et al. 2011). If climate change affects regional and global biological production, e.g. fisheries, trophic structure and species composition may be altered. Altered species compositions of open systems have been well-documented with regard to fish species, especially for the North Sea (Perry et al. 2005). Open systems, such as the North Sea may experience a shift from Northern to warmer Southern species. Relatively-closed systems, such as the Mediterranean are predicted to suffer from loss of endemic species and experience invasions of non-indigenous species (Occhipinti-ambrogi 2007; Philippart et al. 2011), where the 'tropicalisation' of species is purported to occur in the future (Lejeusne et al. 2010). Globally, it is known the ocean warming will have a strong impact on global patterns of biodiversity, however, the effects of ocean acidification are less clear and currently subject to speculation (Tittensor et al. 2010)

Hypoxia – 'Dead'Zones

Dead zones or oxygen minimum zones are areas of low oxygen or hypoxia and defined as dissolved oxygen (DO) levels of below 2 mg/L O_2 (Diaz and Rosenberg 2008). However, the current perceived threshold DO levels may be underestimating the effects of hypoxia, as levels greater than 2 mg/L of O_2/L are responsible for mass mortality events observed in crustacean and fish species (Vaquer-Sunyer and Duarte 2008). The formation of dead zones is thought to have exponentially increased since the 1960s, from *ca.* 20 to currently in excess of 400, and may have serious ecosystem consequences such as decreased biological productivity (Zhang et al. 2010). The formations of dead zones are driven by increases in primary production and coastal eutrophication arising from riverine run-off (Diaz and Rosenberg 2008) and are likely to increase as CO_2 concentrations increase (Brewer and Peltzer 2009). The accumulation of particulate organic matter (POC) direct from riverine inputs may exacerbate microbial activity and enhanced consumption of dissolved oxygen in the benthos; such a process may explain the formation of hypoxic waters (Diaz and Rosenberg 2008; Rabouille et al. 2008; Wang 2009), with coastal hypoxia due to anthropogenic factors being the prime example in the Adriatic, specifically regarding the Mediterranean (reviewed in Zhang et al. 2010).

Currently, there are coastal locations where low oxygen levels occur naturally, due to localised patterns of circulation processes and natural respiration. The oxygen levels found here are often in combination with low pH conditions and under-saturated with regard to calcite or aragonite, producing corrosive, acidic waters, posing a threat to the local biota (Feely et al. 2010). Such physiologically-stressful seawater conditions, which potentially may be exacerbated by anthropogenic loading, including ocean acidification, regional coastal land-use and associated nutrient enrichment will pose serious threats to organisms that may be living close to their physiological limits; the extent of which are currently unknown (Pörtner and Farrell 2008).

Well-established synergistic physiological effects of hypoxia and elevated CO_2 are the reduced immune functions of decapod crustaceans. Several studies have shown that the combination of both hypoxia and hypercapnia results in the reduced innate ability of decapods to neutralise the threat of invading pathogens, e.g. bacteria compared to normoxic conditions, therefore, altering the normal role of circulating haemocytes in the removal of an invading pathogen. Holman et al. (2004) observed a significant increase in the number of *Vibrio campbellii* that remained in the hemolymph of *Callinectes sapidus,* the blue crab, following hypercapnic hypoxia exposure (pCO_2 = 1.8 kPa; pO2 = 4 kPa). Burgents et al. (2005) also observed a reduction in the immune function of *Litopenaeus vannamei.* Recent evidence has also demonstrated immune suppression in a economically-important European species the Norway lobster, *Nephrops norvegicus,* whereby under ocean acidification, a reduction in 50 % carrying-capacity of the haemocytes in the circulatory system was observed. Subsequently, the phagocytic-capability of these haemocytes was compromised in this species (Hernroth et al. 2012). The resultant effect, as demonstrated in three economically-important fishery species, is the decreased resistance to invading pathogens.

Current levels of such hypoxic, low pH water due to accumulation of metabolic CO_2 are found as low as pH 7.6 in the bottom water of Lower St. Lawrence Estuary (Mucci et al. 2011) and pH 7.4–7.5 e.g. in Puget Sound (USA) (Feely et al. 2010). The best estimates of the 'ventilation age' i.e. time taken for mixing of these waters with the surface mixed layers is an average of 16 years, indicating the poor mixing and high accumulation capacity for hypoxia in these deep waters (Mucci et al. 2011). Evidence of coastal hypoxic zones are not only restricted to the 'press exposures' in deep waters, but are further evidenced in shallow-water episodic events. Recent studies record 'pulse events' of intrusions of low pH, hypoxic waters in Monterey Bay (USA), sourced from deep, off-shore waters toward near-shore environments, as low as 17 m (Booth et al. 2012). Regarding the Mediterranean, the carbon chemistry in this area is poorly understood, however, it has been recently shown that acidification has occurred in open shallow waters and deep waters, from the Balearic Sea to the Levantine Basin, in the range of 0.05–0.14 pH units, since the industrial era (Touratier and Goyet 2011).

Conclusion and Future Priorities

Despite the evidence to date on the potential impacts of ocean acidification on marine Crustacea, as well as other organisms, our predictive ability on the long-term consequences remains poorly understood. Our predictive ability on effects on marine organisms are based on modelling approaches. Current models e.g. predicting effects on changes in species ranges are either correlative or mechanistic. Mechanistic models, which determine future responses, incorporate mechanistics effects, ecophysiological adaptations, and evolutionary processes, and actually go beyond simple correlative approaches (Peck et al. 2009). Two studies conducted on model crustaceans have incorporated physiological effects from laboratory exposures, as well incorporating modeling approaches to predict the long-term consequences to the species fitness. Findlay et al. (2010), as previously mentioned, demonstrated that hatching rates were delayed in *Semibalanus balaniodes* by 19 days under high CO_2 conditions. However, further evaluation by these authors demonstrated that under future ocean warming and ocean acidification scenarios, each factor will be a significant driving force for the survival of future progenies of this barnacle species. A critical temperature of 13 °C has been identified, whereby below this temperature, ocean acidification (pH 7.8) has a significant effect on survival of larvae, than that of elevated temperatures alone. Model predictions show that the combined effect of ocean warming and acidification, would result in extinction in populations at its the southern range limit, compared to ocean warming scenario alone, predicted to be years 2077–2098, respectively (Findlay et al. 2010). Comparatively, a multi-generational model based on a two generation laboratory study of the effects of ocean acidification (pH 7.67) on the reproductive output of female copepods, *Tisbe battaglai* demonstrated a gradual decline in the production of nauplii forecasted over the next 100 years (equivalent to 2,430 generations). Further effects included changes in growth, cuticle composition and led the authors to conclude that resource allocation occurs preferentially towards reproductive output, at the expense of maternal somatic growth (Fitzer et al. 2012). These mechanistic models depend on the incorporation of the data on species capacities to resist and adapt to change, and thus, improve our capacity to predict effects of future oceanic conditions. Along with such examples of modeling approaches, a better understanding of the interaction of phenotypic plasticity, genetic evolution and population growth can only be reached by confronting studies of natural populations to experimental evolution in the laboratory (Chevin et al. 2012).

The potential for physiological effects of global climate change and ocean acidification on Crustacea will be no more apparent than in the Mediterranean. The Mediterranean Sea, whilst described as a 'miniature ocean' and a biodiversity 'hotspot' (Coll et al. 2010) and has also been proposed as a giant mesocosm of the world's oceans, where multiple anthropogenic impacts have been observed (Lejeusne et al. 2010). A predicted scenario in the Mediterranean Sea is considered to be decreased rainfall and wind, warmer surface waters and a prolonged stratification period of surface waters (Calvo et al. 2011). The cumulative effects, as discussed within this review, will more likely continue due to anthropogenic impacts from a growing human population, rising from current levels of seven billion to predicted levels of nine billion by 2050 (Smith 2011), and there is an ever-increasing concern of attempting to feed the global population by sourcing protein from the marine environment, either from marine capture fisheries or aquaculture. Current predictions state that to meet the world's food demands, the current capacity of fish production needs to be increased by 50 % (Rice and Garcia 2011). Thus, the dependence on protein for human consumption from the sea will increase placing further importance on the preservation of sustainable fisheries and aquaculture practices.

Ocean acidification as a global phenomenon is widely accepted and the modern rate of pH change/acidification compared to previous events in the earth's history is cause for concern (Pearson and Palmer 2000). Recently, research efforts have gathered momentum to elucidate the potential effects on marine organisms, from the organism level to that of ecosystems. In summary, physiological studies can, therefore, aid our predictions of identifying species currently living at their physiological limits and identifying those species which are best suited to adapt (Somero 2010). A recommended starting point is identifying the thermal tolerances of organisms, which requires a multidisciplinary and integrative approach; employing molecular biology, physiology, ecology to evolutionary approaches (Madeira et al. 2012). The goal is to ultimately, predict effects at a biological, social and economical level. Several research priorities have been outlined with regards to ocean acidification to further increase the current paucity in knowledge; firstly, how will acidification interact with other stressors, secondly, outlining species tolerances and ability to adapt to future oceanic conditions, at region scales and global scales, what the ramifications are to species interactions and ecosystem function and finally, improving our monitoring and predictive abilities (Garrard et al. 2012).

References

Arnold KE, Findlay H, Spicer JI, Daniels CL, Boothroyd D (2009) Effect of CO_2-related acidification on aspects of the larval development of the European lobster, *Homarus gammarus* (L.). Biogeoscience 6:1747–1754. doi:10.5194/bgd-6-3087-2009

Batterton CV, Cameron JN (1978) Characteristics of resting ventilation and response to hypoxia, hypercapnia and emersion in the blue crab *Callinectes sapidus*. J Exp Zool 203:403–418. doi:10.1002/jez.1402030308

Booth JAT, McPhee-Shaw EE, Chua P, Kingsley E, Denny MP, Roger B, Steven J, Zeidberg LD, Gilly WF (2012) Natural intrusions of hypoxic, low pH water into nearshore marine environments on the California coast. Cont Shelf Res 45:108–115. doi:10.1016/j.csr.2012.06.009

Brander K (2010) Impacts of climate change on fisheries. J Marine Syst 79:389–402. doi:10.1016/j.jmarsys.2008.12.015

Brewer PG, Peltzer E (2009) Limits to marine life. Science 324:347–348. doi:10.1126/science.1170756

Burgents JE, Burnett KG, Burnett LE (2005) Effects of hypoxia and hypercapnic hypoxia on the localization and the elimination of *Vibrio campbellii* in *Litopenaeus vannamei*, the Pacific white shrimp. Biol Bull 208:159–168

Caldeira K, Wickett ME (2003) Anthropogenic carbon and ocean pH. Nature 425:365. doi:10.1038/425365a

Calvo E, Simó R, Coma R et al (2011) Effects of climate change on Mediterranean marine ecosystems: the case of the Catalan Sea. Climate Res 50:1–29. doi:10.3354/cr01040

Cameron JN (1978) Effects of hypercapnia on blood acid–base status, NaCl fluxes, and trans-gill potential in freshwater blue crabs, *Callinectes sapidus*. J Comp Physiol B 123:137–141

Cameron JN (1985) Compensation of hypercapnic acidosis in the aquatic blue crab, *Callinectes sapidus*: the predominance of external seawater over carapace carbonate as the proton sink. J Exp Biol 114:197–206

Cameron JN, Iwama GK (1987) Compensation of progressive hypercapnia in channel catfish and blue crabs. J Exp Biol 133:183–197

Cameron JN, Mangum CP, Bliss DE (1983) Environmental adaptations of the respiratory system: ventilation, circulation and oxygen transport. In: Vernberg J, Vernberg WB (eds) The biology of crustacea: environmental adaptations. Academic, New York

Chevin L-M, Collins S, Lefèvre F (2012) Phenotypic plasticity and evolutionary demographic responses to climate change: taking theory out to the field. Funct Ecol. doi:10.1111/j.1365-2435.2012.02043.x

Coll M, Piroddi C, Steenbeek J et al (2010) The biodiversity of the Mediterranean Sea: estimates, patterns, and threats. PLoS One 5:e11842. doi:10.1371/journal.pone.0011842

Crutzen PJ, Stoermer EF (2000) The "anthropocene". Global Change Newslett 41:17–18

de la Haye KL, Spicer JI, Widdicombe S, Briffa M (2011) Reduced sea water pH disrupts resource assessment and decision making in the hermit crab *Pagurus bernhardus*. Anim Behav 82:495–501. doi:10.1016/j.anbehav.2011.05.030

de la Haye KL, Spicer JI, Widdicombe S, Briffa M (2012) Reduced pH sea water disrupts chemo-responsive behaviour in an intertidal crustacean. J Exp Mar Biol Ecol 412:134–140. doi:10.1016/j.jembe.2011.11.013

Diaz RJ, Rosenberg R (2008) Spreading dead zones and consequences for marine ecosystems. Science 321:926–929. doi:10.1126/science.1156401

Dissanayake A, Ishimatsu A (2011) Synergistic effects of elevated CO2 and temperature on the metabolic scope and activity in a shallow-water coastal decapod (*Metapenaeus joyneri*) (Crustacea: Penaeidae). ICES J Mar Sci 68:1147–1154. doi:10.1093/icesjms/fsq188

Dissanayake A, Clough R, Spicer JI, Jones MB (2010) Effects of hypercapnia on acid–base balance and osmo-/iono-regulation in prawns (Decapoda: Palaemonidae). Aquat Biol 11:27–36. doi:10.3354/ab00285

Donohue P, Calosi P, Bates A et al (2012) Impact of exposure to elevated pCO 2 on the physiology and behaviour of an important ecosystem engineer, the burrowing shrimp *Upogebia deltaura*. Aquat Biol 15:73–86. doi:10.3354/ab00408

FAO (2008) The state of world fisheries and aquaculture. FAO, Rome

Feely RA, Sabine CL, Lee K et al (2004) Impact of anthropogenic CO2 on the CaCO3 system in the oceans. Science 305:362–366. doi:10.1126/science.1097329

Feely RA, Alin SR, Newton J et al (2010) The combined effects of ocean acidification, mixing, and respiration on pH and carbonate saturation in an urbanized estuary. Estuar Coast Shelf S 88:442–449. doi:10.1016/j.ecss.2010.05.004

Findlay HS, Kendall MA, Spicer JI, Widdicombe S (2009) Future high CO2 in the intertidal may compromise adult barnacle *Semibalanus balanoides* survival and embryonic development rate. Mar Ecol Prog Ser 389:193–202. doi:10.3354/meps08141

Findlay HS, Burrows MT, Kendall MA et al (2010) Can ocean acidification affect population dynamics of the barnacle Semibalanus balanoides at its southern range edge? Ecology 91:2931–2940. doi:dx.doi.org/10.1890/09-1987.1

Fine M, Tchernov D (2007) Scleractinian coral species survive and recover from decalcification. Science 315:1811. doi:10.1126/science.1137094

Fitzer SC, Caldwell GS, Close AJ et al (2012) Ocean acidification induces multi-generational decline in copepod naupliar production with possible conflict for reproductive resource allocation. J Exp Mar Biol Ecol 418–419:30–36. doi:10.1016/j.jembe.2012.03.009

Franklin CE, Seebacher F (2009) Adapting to climate change. Science 323:876

Frederich M, Pörtner HO (2000) Oxygen limitation of thermal tolerance defined by cardiac and ventilatory performance in spider crab, *Maja squinado*. Am J Physiol- Reg Int Comp Physiol 279:R1531–R1538

Freire CA, Cavassin F, Rodrigues EN et al (2003) Adaptive patterns of osmotic and ionic regulation, and the invasion of fresh water by the palaemonid shrimps. Comp Biochem Phys A 136:771–778. doi:dx.doi.org/10.1016/j.cbpb.2003.08.007

Freire CA, Onken H, Mcnamara JC (2008) A structure–function analysis of ion transport in crustacean gills and excretory organs. Biochemistry 151:272–304

Garrard SL, Hunter RC, Frommel AY et al (2012) Biological impacts of ocean acidification: a postgraduate perspective on research priorities. Mar Biol. doi:10.1007/s00227-012-2033-3

Gattuso J (1998) Effect of calcium carbonate saturation of seawater on coral calcification. Global Planet Change 18:37–46. doi:10.1016/S0921-8181(98)00035-6

Gazeau F, Quiblier C, Jansen JM et al (2007) Impact of elevated CO_2 on shellfish calcification. Geophys Res Lett 34:1–5. doi:10.1029/2006GL028554

Henry BYRP, Cameron JN (1983) The role of carbonic anhydrase in respiration, ion regulation and acid–base balance in the aquatic crab *Callinectes sapidus* and the terrestrial crab *Gecarcinus lateraus*. J Exp Biol 223:205–223

Henry RP, Wheatly MG (1992) Interaction of respiration, ion regulation, and acid–base balance in the everyday life of aquatic crustaceans. Am Zool 32:407–416. doi:10.1093/icb/32.3.407

Hernroth B, Sköld HN, Wiklander K (2012) Simulated climate change causes immune suppression and protein damage in the crustacean *Nephrops norvegicus*. Fish Shellfish Immunol 33(5):1095–1101. doi:10.1016/j.fsi.2012.08.011

Holman JD, Burnett KG, Burnett LE (2004) Effects of hypercapnic hypoxia on the clearance of *Vibrio campbellii* in the Atlantic blue crab, *Callinectes sapidus* Rathbun. Biol Bull 206:188–196

IPCC (2001) Climate Change 2001: the scientific basis. Contribution of Working Group I to the third assessment report of the Intergovernmental Panel on Climate Change

IPCC (2007) Climate change 2007: the physical science basis. Contribution of working group I to the fourth assessment report of the intergovernmental panel on climate change. Cambridge University Press, Cambridge

Ishimatsu A, Dissanayake A (2010) Life threatened in acidic coastal waters. In: Ishimatsu A, Lie HJ (eds) Coastal environmental and ecosystem issues of the east china Sea. Nagasaki University, Tokyo

Ituarte BR, López Mañanes A, Spivak E, Anger K (2008) Activity of Na^+K^+-ATPase in a freshwater shrimp' *Palaemonetes argentinus* (Caridea, Palaemonidae): ontogenetic and salinity-induced changes. Aquat Biol 3:283–290. doi:10.3354/ab00089

Iwama BYGK, Heisler N (1991) Acid–base regulation during environmental hypercapnia in the rainbow trout (*Oncorhynchus mykiss*). J Exp Biol 18:1–18

Kawaguchi S, Kurihara H, King R et al (2010) Will krill fare well under southern ocean acidification? Biol Lett 7:288–291. doi:10.1098/rsbl.2010.0777

Kennett JP, Ingram BL (1995) A 20, 000-year record of ocean circulation and climate change from the Santa Barbara basin. Nature 377:510–514. doi:10.1038/377510a0

Kroeker KJ, Micheli F, Gambi MC, Martz TR (2011) Divergent ecosystem responses within a benthic marine community to ocean acidification. Proc Natl Acad Sci 108:14515–14520. doi:10.1073/pnas.1107789108

Kurihara H, Shimode S, Shirayama Y (2004) Effects of raised CO2 concentration on the egg production rate and early development of two marine copepods (*Acartia steueri* and *Acartia erythraea*). Mar Pollut Bull 49:721–727. doi:dx.doi.org/10.1016/j.marpolbul.2004.05.005

Kurihara H, Matsui M, Furukawa H et al (2008) Long-term effects of predicted future seawater CO_2 conditions on the survival and growth of the marine shrimp *Palaemon pacificus*. J Exp Mar Biol Ecol 367:41–46. doi:10.1016/j.jembe.2008.08.016

Lejeusne C, Chevaldonné P, Pergent-Martini C et al (2010) Climate change effects on a miniature ocean: the highly diverse, highly impacted Mediterranean Sea. Trends Ecol Evol 25:250–260. doi:10.1016/j.tree.2009.10.009

Madeira D, Narciso L, Cabral HN, Vinagre C (2012) Thermal tolerance and potential impacts of climate change on coastal and estuarine organisms. J Sea Res 70:32–41. doi:10.1016/j.seares.2012.03.002

Mantel LH, Farmer LL (1983) Osmotic and ionic regulation. In: Mantel LH (ed) The biology of crustacea. Internal anatomy and physiological regulation. Academic, New York

Masui DC, Mantelatto FLM, Mcnamara JC et al (2009) Na +, K+-ATPase activity in gill microsomes from the blue crab, *Callinectes danae*, acclimated to low salinity: novel perspectives on ammonia excretion. Comp Biochem Phys A 153:141–148. doi:10.1016/j.cbpa.2009.01.020

Mayor D, Matthews C, Cook K et al (2007) CO_2-Induced acidification affects hatching success in *Calanus finmarchicus*. Mar Ecol Prog Ser 350:91–97. doi:10.3354/meps07142

Mayor DJ, Anderson TR, Pond DW, Irigoien X (2009) Egg production and associated losses of carbon, nitrogen and fatty acids from maternal biomass in *Calanus finmarchicus* before the spring bloom. J Marine Syst 78:505–510. doi:10.1016/j.jmarsys.2008.12.019

Mayor DJ, Everett NR, Cook KB (2012) End of century ocean warming and acidification effects on reproductive success in a temperate marine copepod. J Plankton Res 34:258–262. doi:10.1093/plankt/fbr107

McDonald MR, McClintock JB, Amsler CD et al (2009) Effects of ocean acidification over the life history of the barnacle *Amphibalanus amphitrite*. Mar Ecol Prog Ser 385:179–187. doi:10.3354/meps08099

McMahon BR, Burnett LE, Fur PL (1984) Carbon dioxide excretion and carbonic anhydrase function in the Red Rock Crab *Cancer productus*. J Comp Physiol B 154:371–383. doi:10.1007/BF00684444

Mendonça NN, Masui DC, McNamara JC et al (2007) Long-term exposure of the freshwater shrimp macrobrachium olfersii to elevated salinity: effects on gill (Na+, K+)-ATPase [alpha]-subunit expression and K+-phosphatase activity. Comp Biochem Phys A 146:534–543. doi:10.1016/j.cbpa.2006.01.019

Metzger R, Sartoris FJ, Langenbuch M, Pörtner HO (2007) Influence of elevated CO2 concentrations on thermal tolerance of the edible crab *Cancer pagurus*. J Therm Biol 32:144–151. doi:dx.doi.org/10.1016/j.jtherbio.2007.01.010

Morritt D, Spicer JI (1993) A brief re-examination of the function and regulation of extracellular magnesium and its relationship to activity in crustacean arthropods. Comp Biochem Physiol A Physiol 106:19–23

Mucci A, Starr M, Gilbert D, Sundby B (2011) Acidification of lower St. Lawrence Estuary bottom waters. Atmos Ocean 49:206–218. doi:10.1080/07055900.2011.599265

Occhipinti-ambrogi A (2007) Global change and marine communities: alien species and climate change. Mar Pollut Bull 55:342–352. doi:10.1016/j.marpolbul.2006.11.014

Pane EF, Barry JP (2007) Extracellular acid–base regulation during short-term hypercapnia is effective in a shallow-water crab, but ineffective in a deep-sea crab. Mar Ecol Prog Ser 334:1–9. doi:10.3354/meps334001

Panikkar NK (1941) Osmoregulation in some palaemonid prawns. J Mar Biol Assoc UK 25:317–359

Pearson PN, Palmer MR (2000) Atmospheric carbon dioxide concentrations over the past 60 million years. Nature 406:695–699. doi:10.1038/35021000

Peck LS, Clark MS, Morley SA et al (2009) Animal temperature limits and ecological relevance: effects of size, activity and rates of change. Funct Ecol 23:248–256. doi:10.1111/j.1365-2435.2008.01537.x

Perry AL, Low PJ, Ellis JR, Reynolds JD (2005) Climate change and distribution shifts in marine fishes. Science 308:1912–1915. doi:10.1126/science.1111322

Philippart CJM, Anadón R, Danovaro R et al (2011) Impacts of climate change on European marine ecosystems: observations, expectations and indicators. J Exp Mar Biol Ecol 400:52–69. doi:10.1016/j.jembe.2011.02.023

Pörtner H-O, Farrell AP (2008) Physiology and climate change. Science 322:690–692. doi:10.1126/science.1163156

Pörtner HO, Langenbuch M, Reipschläger A (2004) Biological impact of elevated ocean CO2 concentrations: lessons from animal physiology and earth history. J Oceanogr 60:705–718

Pörtner H-O, Farrell AP, Knust R et al (2009) Adaptating to climate change – response. Science 323:876–867

Rabouille C, Conley DJ, Dai MH et al (2008) Comparison of hypoxia among four river-dominated ocean margins: the Changjiang (Yangtze),

Mississippi, Pearl, and Rhone rivers. Cont Shelf Res 28:1527–1537. doi:dx.doi.org/10.1016/j.csr.2008.01.020

Range P, Chícharo MA, Ben-hamadou R et al (2011) Calcification, growth and mortality of juvenile clams *Ruditapes decussatus* under increased pCO2 and reduced pH: variable responses to ocean acidification at local scales ? J Exp Mar Biol Ecol 396:177–184. doi:10.1016/j.jembe.2010.10.020

Rankin CJ, Davenport J (1981) Animal osmoregulation. Blackie, Glasgow

Raven J, Caldeira K, Elderfield H et al (2005) Ocean acidification due to increasing atmospheric carbon dioxide. The Royal Society, London

Rice JC, Garcia SM (2011) Fisheries, food security, climate change, and biodiversity: characteristics of the sector and perspectives on emerging issues. ICES J Mar Sci 68:1343–1353. doi:10.1093/icesjms/fsr041

Robertson JD (1949) Ionic regulation in some marine invertebrates. J Exp Biol 26:182–200

Robertson JD (1953) Further studies on ionic regulation in marine invertebrates. J Exp Biol 30:277–296

Sabine CL, Feely RA, Gruber N et al (2004) The oceanic sink for anthropogenic CO_2. Science 305:367–371. doi:10.1126/science.1097403

Seibel BA, Walsh PJ (2003) Biological impacts of deep-sea carbon dioxide injection inferred from indices of physiological performance. J Exp Biol 206:641–650. doi:10.1242/jeb.00141

Siegenthaler U, Sarmiento JL (1993) Atmospheric carbon dioxide and the ocean. Nature 365:119–125. doi:10.1038/365119a0

Small D, Calosi P, White D et al (2010) Impact of medium-term exposure to CO_2- enriched seawater on the physiological functions of the velvet swimming crab *Necora puber*. Aquat Biol 10:11–21. doi:10.3354/ab00266

Smith K (2011) We are seven billion. Nat Climat Change 1:331–336. doi:10.1038/nclimate1235

Somero GN (2010) The physiology of climate change: how potentials for acclimatization and genetic adaptation will determine “winners” and “losers”. J Exp Biol 213:912–920. doi:10.1242/jeb.037473

Spicer J, Raffo A, Widdicombe S (2007) Influence of CO_2-related seawater acidification on extracellular acid–base balance in the velvet swimming crab *Necora puber*. Mar Biol 151:1117–1125. doi:10.1007/s00227-006-0551-6

Takasuka A, Oozeki Y, Aoki I (2007) Optimal growth temperature hypothesis: Why do anchovy flourish and sardine collapse or vice versa under the same ocean regime? Can J Fish Aquat Sci 64:768–776. doi:10.1139/f07-052

Takasuka A, Oozeki Y, Kubota H, Lluch-Cota SE (2008) Contrasting spawning temperature optima: Why are anchovy and sardine regime shifts synchronous across the North Pacific? Prog Oceanogr 77:225–232. doi:10.1139/F09-051

Tittensor DP, Mora C, Jetz W et al (2010) Global patterns and predictors of marine biodiversity across taxa. Nature 466:1098–1101. doi:10.1038/nature09329

Toews DP, Holeton GF, Heisler N (1983) Regulation of the acid–base status during environmental hypercapnia in the marine teleost fish *Conger conger*. J Exp Biol 107:9–20

Touratier F, Goyet C (2011) Impact of the Eastern Mediterranean Transient on the distribution of anthropogenic CO_2 and first estimate of acidification for the Mediterranean Sea. Deep-Sea Res I 58:1–15. doi:10.1016/j.dsr.2010.10.002

Vaquer-Sunyer R, Duarte CM (2008) Thresholds of hypoxia for marine biodiversity. Proc Natl Acad Sci 105:15452–15457. doi:10.1073/pnas.0803833105

Walther K, Sartorius FJ, Bock C, Pörtner HO (2009) Impact of anthropogenic ocean acidification on thermal tolerance of the spider crab *Hyas araneus*. Biogeosci Disc 6:2837–2861. doi:10.5194/bg-6-2207-2009

Wang B (2009) Hydromorphological mechanisms leading to hypoxia off the Changjiang estuary. Mar Environ Res 67:53–58, dx.doi.org/10.1016/j.marenvres.2008.11.001

Wheatly MG, Henry RP (1992) Extracellular and intracellular acid–base regulation in crustaceans. J Exp Zool 263:127–142. doi:10.1002/jez.1402630204

Whiteley N (2011) Physiological and ecological responses of crustaceans to ocean acidification. Mar Ecol Prog Ser 430:257–271. doi:10.3354/meps09185

Wittmann A, Held C, Pörtner H, Sartoris F (2010) Ion regulatory capacity and the biogeography of Crustacea at high southern latitudes. Polar Biol 33:919–928. doi:10.1007/s00300-010-0768-1

Zhang J, Gilbert D, Gooday AJ et al (2010) Natural and human-induced hypoxia and consequences for coastal areas: synthesis and future development. Biogeosci 7:1443–1467. doi:10.5194/bg-7-1443-2010

Bryozoan Constructions in a Changing Mediterranean Sea

21

Chiara Lombardi, Paul D. Taylor, and Silvia Cocito

Abstract

Bryozoans in the Mediterranean Sea are recognized as bioconstructional framework builders, both primary builders that construct frameworks alone or in combination with other organisms, or secondary builders that play various functional roles. Ten bryozoan species or complexes of species in the Mediterranean are responsible for providing habitats for diverse species and assemblages, thus playing important roles in promoting biodiversity and habitat heterogeneity. Four habitat-forming bryozoans respond to climatic changes (global warming and ocean acidification) by altering their colony growth, zooidal morphology and development, skeletal mineralogy and geochemistry. Under conditions of reduced pH, these species reallocate energy resources within the colony by regulating zooid size, proportion of polymorphs, number of zooidal generations, colony growth rate, and investment in the organic components involved in biomineralizational processes. Mediterranean 'bryo-constructions' are suitable 'ecosystem models' to be monitored in the context of climate change, especially ocean acidification and warming.

Keywords

Bioconstructional framework builders • Biodiversity • Biomineralizational processes • Bryo-constructions • Bryozoan species • Bryozoans • Climatic changes • Colony growth • Geochemistry • Global warming • Habitat heterogeneity • Habitat-forming bryozoans • Mediterranean Sea • Ocean acidification • Primary builders • Secondary builders • Skeletal mineralogy • Zooid size • Zooidal morphology

Introduction

Bioconstructors are marine organisms responsible for building complex biogenic structures through the aggregation of their calcareous skeletons (Laborel 1987). Based on two life strategies, gregariousness, where conspecific individuals settle and grow close to each other, and coloniality, in which asexual reproduction produces aggregates of conjoined clonal zooids, framework builders may be highly dynamic through their own skeletal growth processes as well as the forces of physical and biological destruction (Laborel 1987). Biogenic calcareous structures vary greatly in size, ranging from a few cms to many kilometers, and in complexity, forming composite structures that can be distributed globally (Cocito 2004; Wood et al. 2012).

The biomineralized skeletons of bioconstructing organisms may be highly resistant through geological time, resulting in relatively complete fossil records that are amenable to studies of long-term evolutionary changes, including patterns of species loss and appearance, such as the introduction of 'alien' species. Complex and well-developed calcareous skeletons of biogenic structures play important roles in the carbon cycle,

C. Lombardi (✉) • S. Cocito
Marine Environment and Sustainable Development Unit, ENEA, Via Forte Santa Teresa, 19032 Pozzuolo di Lerici, La Spezia, Italy
e-mail: chiara.lombardi@enea.it

P.D. Taylor
Department of Earth Sciences, Natural History Museum, Cromwell Road, London SW7 5BD, UK

S. Goffredo and Z. Dubinsky (eds.), *The Mediterranean Sea: Its history and present challenges*,
DOI 10.1007/978-94-007-6704-1_21, © Springer Science+Business Media Dordrecht 2014

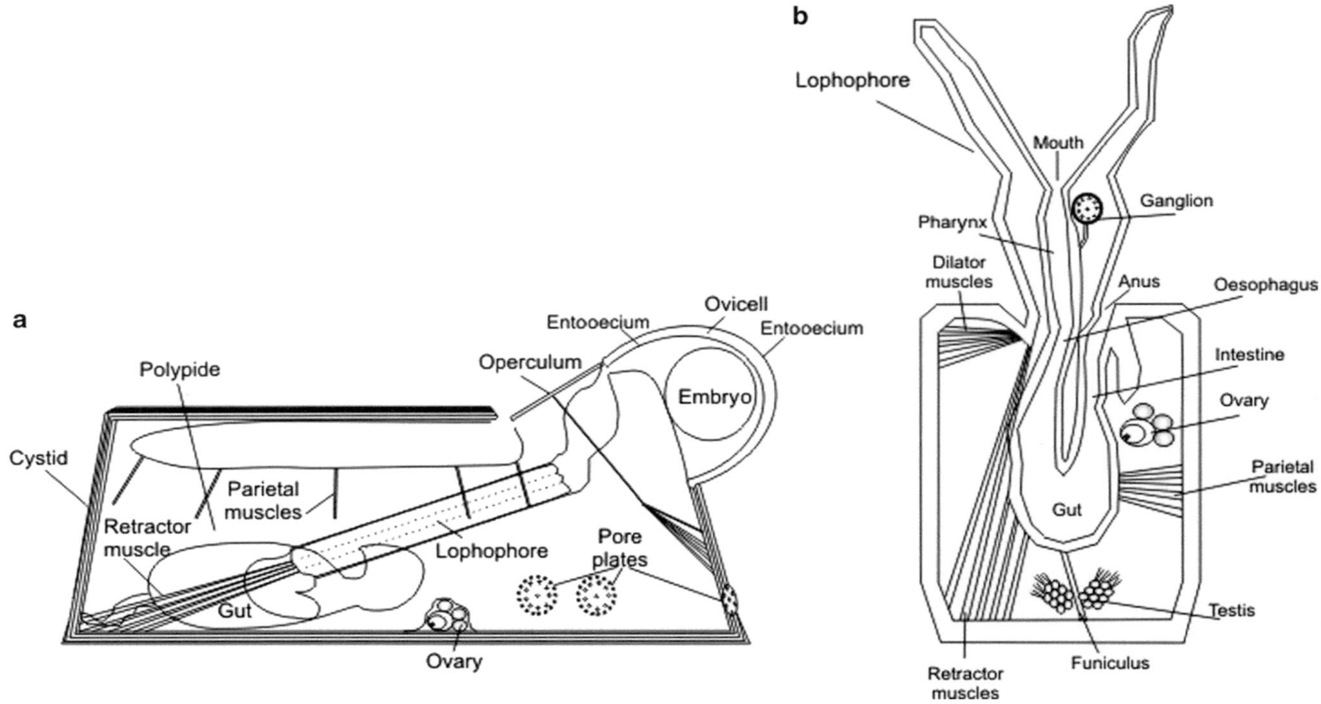

Fig. 21.1 Diagrammatic representations of bryozoan zooid structures. (**a**) Ascophoran cheilostome with lophophore retracted (Adapted from Hayward and Ryland 1999) (**b**) Generalized bryozoan zooid with lophophore extended

both as carbon sources and sinks. Furthermore, they constitute excellent 'target species' and 'target assemblages' for studying the effects of global warming and ocean acidification.

Constructional Bryozoa in the Mediterranean Sea

Among organisms making bioconstructions, bryozoans have been widely recognized as important habitat-forming organisms (Cocito 2004; Taylor and Wilson 2003; Wood et al. 2012). After their first appearance in the Early Ordovician, this phylum expanded across freshwater, brackish and marine ecosystems.

The most phylogenetically basal class of bryozoans is Phylactolaemata, which includes entirely freshwater animals without biomineralized skeletons (Waeschenbach et al. 2012). Two classes - Stenolaemata and Gymnolaemata – form a sister-grouping of predominantly marine species usually having calcareous skeletons. With a rich fossil record extending back to the Early Ordovician, Stenolaemata comprises four extinct (Trepostomata, Cystoporata, Cryptostomata, Fenestrata) and one extant order (Cyclostomata), all well calcified. Gymnolaemata consist of the paraphyletic order Ctenostomata, which is entirely soft-bodied but recorded as borings back to the Early Ordovician (Taylor and Ernst 2004), and Cheilostomata, today the dominant group of bryozoans, an order of species with calcareous skeletons which first appeared in the Late Jurassic (Taylor 1994).

Bryozoan colonies comprise between a few to many thousands of conjoined, clonal modules (zooids), each ranging from 0.1 to a little over 1 mm in length. Although genetically identical, the zooids within a colony differ morphologically according to their developmental stage (ontogeny), the developmental stage of the colony when they were budded (astogeny), and as a result of functional differentiation (polymorphism). Each zooid is composed of a polypide and a cystid. The polypide comprises a U-shaped gut and associated muscles, and an inverted cone- or bell-shaped ring of tentacles, the lophophore, which protrudes through an external opening (primary orifice) in the skeleton for feeding. During feeding, the cilia on the tentacles beat to create a water current that drives phytoplankton towards the mouth (Fig. 21.1a) located at the confluence of the tentacles. A U-gut shaped forms a loop suspended in the fluid-filled body cavity (coelom or pseudocoelom) from the mouth to the anus which opens outside the ring of tentacles. The cystid comprises essentially the outer body walls of the zooid, including cuticle, mineralized skeleton and peritoneum (Hayward and Ryland 1999; McKinney and Jackson 1989; Taylor 2005) (Fig.21.1b).

Formation of the biomineralized bryozoan skeleton occurs between an outer cuticle and secretory epithelial cells (Tavener-Smith and Williams 1972; Lutaud 1987). Even though biomineralization processes are still poorly known, the mineralogy of both recent and fossil bryozoans has been quite extensively documented (see Poluzzi and Sartori 1974; Smith et al. 2006; Taylor et al. 2009). Most species have entirely calcitic skeletons, but some are aragonitic or bimineralic (i.e. mixtures of calcite and aragonite) (Smith et al. 2006; Taylor et al. 2008). In calcitic and bimineralic species, the calcite phase can have a variable content of $MgCO_3$. Around 30 % of bryozoan species have low-Mg calcite skeletons (0–4 mol% $MgCO_3$), the majority have intermediate-Mg calcite skeletons (4–8 wt% $MgCO_3$), whereas high-Mg calcite skeletons (>8 wt% $MgCO_3$) have been found in a few species (Smith et al. 2006). This mineralogical variety, coupled with varied zooidal and colonial growth strategies as well as the presence of diverse ultrastructural fabrics, contributes to the great complexity of bryozoan skeletons.

Bryozoan colonies can range from simple linear chains of zooids, to highly integrated colonies with determinant growth and coordinated behaviour (Hageman et al. 2003; Jackson 1985). They reproduce both sexually and asexually. Sexual reproduction results in the formation of new colonies by larvae which swim away from the parent colony and settle on a hard substrate elsewhere. Asexual reproduction allows colony growth by asexual budding of new individuals at the peripheral growing edge and in some instances plays a role in the clonal propagation of colonies through fragmentation and regrowth.

Growth strategies are extremely complex in bryozoan colonies (Hageman et al. 1998, 2003) and the theoretical range of possible architectures adopted by colonies is huge. Colony shapes present among bryozoans include lobes, sheets, discs, cylinders (solid or hollow), nodules, spheroids or ellipsoids, cones and domes (solid or hollow). Differently shaped colonies favour the formation of specific bioconstructions. Furthermore, growth orientation relative to the substrate, the arrangement of zooidal series and of zooidal frontal surfaces, dimensions and bifurcations of structural units, and secondary calcification processes also contribute to the form of bioconstructions. Calcified encrusting colonies can be composed of unilaminar or multilaminar sheets of zooids, while erect colonies show varying frequencies of bifurcations. Different types of connections between structural units in the colony (e.g. fused structural units forming 'net-shaped' colonies, extrazooidal connections, cuticular tubes, calcified connections) are responsible for determining the shapes and sizes of colonies (Hageman et al. 1998, 2003).

The ability of bryozoans to form long-lasting carbonate structures makes them important carbonate-producers (McKinney and Jackson 1989; Taylor and Allison 1998), with a significant role in the carbon cycle. Bryozoan skeletons reflect a combination of organic and environmentally regulated processes (Lombardi et al. 2008b; Lowenstam and Weiner 1989; Smith and Key 2004; Smith et al. 2006; Taylor et al. 2008). Biomineral precipitation depends to varying degrees on environmental factors. For example, aragonite and magnesium calcite deposition are related to seawater temperature and chemistry (Morse et al. 1997, 2007; Stanley and Hardie 1998; Montañez 2002; Skinner and Jahren 2003; Feeley et al. 2004; Smith and Key 2004; Taylor et al. 2009; Kuklinski and Taylor 2009). Nevertheless, taxon-specific and ontogenetic factors also contribute to mineralogical variations seen within colonies, within species, among colony-forms and within families (Poluzzi and Sartori 1974; Smith et al. 1998; Smith and Key 2004). Isotopically, bryozoan skeletons are generally in equilibrium with seawater (Lowenstam and Weiner 1989; Bone and James 1993; Crowley and Taylor 2000; Smith and Key 2004), although in a few instances disequilibrium may result from kinetic effects or possibly symbiosis (Crowley and Taylor 2000).

Biomineralization processes acting at genotypic, phenotypic and ecophenotypic levels allow some bryozoan species to adopt appropriate morphological responses to environmental changes, giving them utility as 'bioindicator' organisms (McKinney and Jackson 1989). For example, zooid size may correlate with food resources, growth rates (Okamura 1987; Hunter and Hughes 1994), salinity, oxygen concentration, and extreme hydrodynamic conditions (Okamura 1985; O'Dea and Okamura 1999). Importantly, culture experiments and field observations have revealed a consistent negative correlation between zooid size and ambient temperature (Menon 1972; Hunter and Hughes 1994; Okamura et al. 2011).

The Mediterranean benthic fauna contains approximately 480 species of bryozoans (Rosso 2003). Most of these species have calcareous skeletons and some contribute to bioconstructions which vary in their spatial extents, structures and taxonomic compositions, being generally monospecific in high-energy, shallow-water environments but paucispecific in open, deep-water habitats. Cryptic microhabitats often shelter bryozoan-dominated assemblages containing numerous species (Harmelin 1985). Mediterranean bryozoan bioconstructions are usually characterized by encrusting multilaminar sheets which are able to form dome-like colonies, wrapping other organisms or inert substrates (Cigliano et al. 2007; Cocito et al. 2012; Kocak et al. 2002; Poluzzi and Coppa 1991), and also by large erect arborescent colonies with multilaminar and/or robust branches (Cocito 2004; Cocito et al. 2004; Lombardi et al. 2008a; McKinney and Jackson 1989; Novosel 2005)

Strongly calcified marine bryozoans are important producers of modern temperate carbonate sediment, both in the Mediterranean (Fornos and Ahr 2006; Bayhan et al. 2001) and elsewhere (e.g. James and Bone 2010). Bryozoan-rich deposits show a marked change in latitudinal distribution through geological time: whereas post-Palaeozoic examples were almost exclusively formed outside the tropics, such deposits were common in the tropics during the Palaeozoic

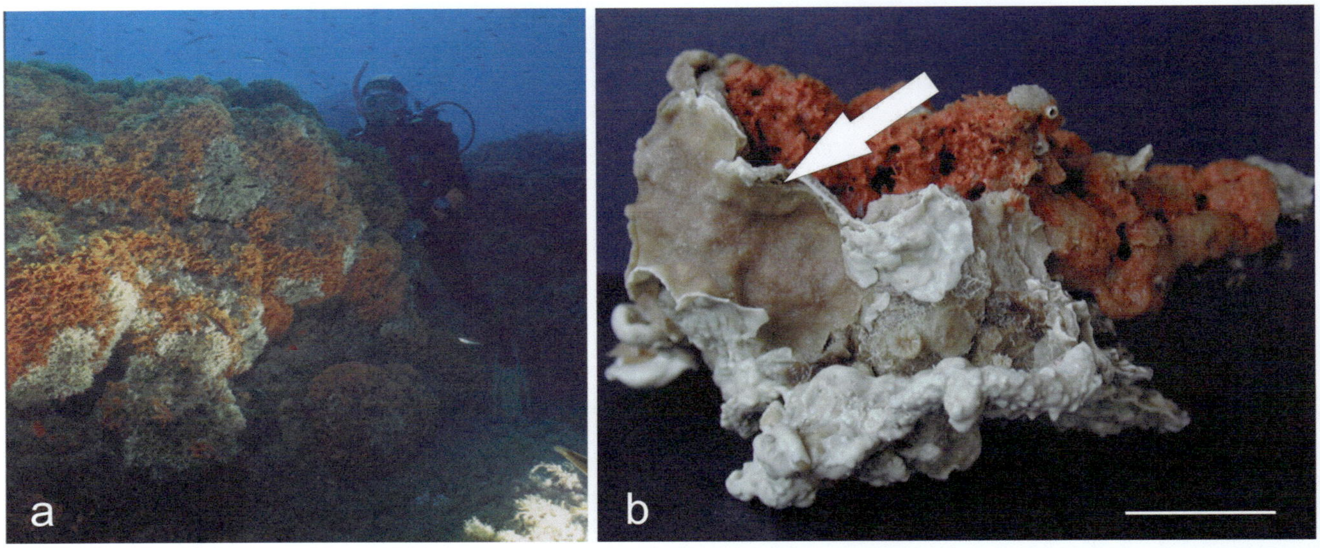

Fig. 21.2 (**a**) Primary framework constructed by erect, rigid colonies of the orange bryozoan *Pentapora fascialis* at Tinetto Island (Gulf of La Spezia, Ligurian Sea). (**b**) Encrusting bryozoans (arrowed) acting as binders within a secondary framework sampled at S.M. di Leuca (Apulian coast, Ionian Sea) (scale bar = 3 cm)

(Taylor and Allison 1998). A good example of bryozoan bioconstructions in the fossil record is provided by the abundant bryozoan nodules (bryoliths) that occur in the Early Pleistocene Cape Arkhangelos Calcarenite of Rhodes (Greece). These bryoliths, composed almost exclusively of multilaminar colonies of the scan cheilostome *Calpensia nobilis*, nucleated on pebbles or bivalve shells, and contain intergrown bivalves and serpulid worms, together with other rarer skeletal organisms (Moissette et al. 2010).

Constructional Roles

The manner in which bryozoans occur in benthic bioconstructions reflects their bioconstructional roles. Large bryozoans were designated as 'frame-builders' by Duncan (1957) but without any strict definition of the term. Wood et al. (2012) used the term 'frame-building' for species that regularly grow to ≥50 mm in three dimensions, a size considered necessary for the development of complex habitats for macro-invertebrates.

Calcareous bryozoans have the potential to form bioconstructions, both as primary builders, by constructing a framework either alone or with other organisms (mainly algae, serpulids, corals and other bryozoans), or by playing secondary roles according to different functional categories. Assignment of organisms to particular functional groups is mainly based on Fagerstrom's (1991) classification which took into consideration bryozoan growth-form and habit. *Primary frame-builders* are considered to be erect, large and well-skeletonized organisms that provide most of the volume and rigidity to the framework, as well as multilayered encrusting organisms able to form dense crusts (Fig. 21.2). These adapt to the roughness of the substrate by folding and producing overlapping layers, and sometimes by incorporating other organisms. If the overgrown organisms are soft-bodied they perish after overgrowth. Cocito et al. (2000) defined the process of 'ghost frame-building' as a peculiar form of bioimmuration, i.e. the organic overgrowth of sessile organisms by encrusting animals with mineralized skeleton (Taylor 1990). In contrast to bioimmuration which usually produces a precise mould, the exact shape of the soft-bodied organism is not preserved in the resulting structure.

Pentapora spp., *Reteporella* spp., *Smittina cervicornis* (Pallas), *Myriapora truncata* (Pallas), *Adeonella* spp. and *Schizoretepora serratimargo* (Hincks) are erect, large and well-skeletonized Mediterranean species that grow on shaded rocks and loose substrates where they may function as primary frame-builders. Examples of encrusting primary frame-builders are *Schizoporella* spp., *Schizomavella* spp., *Pentapora ottomüllleriana* (Moll), *Schizobrachiella sanguinea* (Norman), *Rhynchozoon* spp., *Calpensia nobilis* (Esper) and *Reptadeonella violacea* (Johnston). These are perennial species with fast and continuous growth on various types of substrates, such as living organisms, organogenic constructions and rocks. A prerequisite for the formation of well-structured bioconstructions, displayed by many species of bryozoans, is the persistence of dense assemblages, which in turn depends on continuous larval settlement, rapid clonal growth, often with fragmentation and partial mortality (Jackson 1977). Species characterized by extreme longevity and large size can create structures with permanent attachments to extensive substrates that are considerably elevated above the substrate and possess cavities and thus high spatial heterogeneity.

Secondary frame-builders include other encrusting bryozoans, which expand and unite the components of the framework and the settling sediment, thus belonging to the functional category of 'binders'. Other erect, non-skeletal to poorly-skeletonized organisms act as 'bafflers', reducing current velocity on the surface of the framework, enhancing sediment deposition and cavity filling ('dwellers') within the framework (Fig. 21.3). Calcareous encrustations of secondary frame-builders contribute to the reinforcement or strengthening of the overall framework.

Ecological Roles

By generating physical structures, thereby modifying the habitat and its structural complexity, bryozoan bioconstructions provide living space for other organisms over time scales of years to decades (Smith et al. 2001). Quantitative changes in the amount of available living space occur as the bioconstruction grows. Habitat structure has been considered an important determinant of the number, identity and abundance of species within communities through the principle of 'biological habitat provision' (Thompson et al. 1996). The surface of bioconstructions serves as a substrate for a variety of organisms, from bacteria to algae, hydrozoans, sponges, other bryozoans and invertebrates, and these substrata persist over ecological time, potentially allowing ecological successions to occur. Cavities inside the structure offer protection from predation, wave scour and high light irradiance, thus providing a better habitat for many organisms. Moreover, the physical structure effects local water flow, influencing settlement and recruitment, increasing the variety of suitable feeding sources and altering the balance of biotic interactions including predation and competition. Thus, the presence of habitat-forming bryozoans can facilitate colonization by more or different species. This associated fauna is an important characteristic of bryozoan habitats. It may encourage the bryozoans by providing a stable substratum for further growth or by 'welding' branches together, enhancing the integrity of the structure. The increase in size of bioconstructions induces modifications not only in the composition of biota associated but also in the shape and growth rate of these structures (Cocito 2004).

Although the term 'habitat-former' ('habitat-forming organism') has been widely used (Bruno et al. 2003), its application depends on scale. Processes generated by bryozoan habitat-formers can be apparent from centimetric to metric scales. Here we limit habitat-forming bryozoans to those dominating metres of seafloor or any type of biotic or artificial substrate and contributing significantly to habitat complexity.

In the Mediterranean about ten bryozoan species or combinations of multiple bryozoan species have been reported as habitat-forming (Fig. 21.3, Table 21.1). *Pentapora fascialis* (Pallas) has been reported from 15 Italian sites, from the western Ligurian Sea to the Sicily Channel, and is locally abundant at three sites: Alassio Bay (Ligurian Sea), Formiche Islands (Tyrrhenian Sea) and Scoglitti (Sicily Channel) (Lombardi et al. 2008a). It attains a cover of 78 % at Tinetto Island (Ligurian Sea) (Cocito and Ferdeghini 2001). The same species was reported from other localities in southeastern Spain (Medes Islands, Columbretes Islands) (Sala et al. 1996; Templado et al. 2002) and southern France (Marseille) (Harmelin and Capo 2001). Large colonies (mean diameter 65 cm, up to 1 m) of *P. fascialis* living around freshwater springs in the northeastern Adriatic were found to be associated with upwelling currents caused by karstic freshwater springs (Cocito et al. 2004; Novosel et al. 2004). The encrusting bryozoan *Schizoporella errata* (Waters) grows in harbours and on shallow rocky bottoms, forming sizeable buildups in Spain and Italy (Maluquer 1985; Cocito et al. 2000) where colonies up to 40 cm in width have been recorded. The articulated erect species *Cellaria salicornioides* (Lamouroux) forms 'meadows' off Croatia (McKinney and Jaklin 2000; Novosel 2005), described as one of the largest unbroken areas of bryozoans known, more than 100 × 100 m in extent and attaining 80–90 % bryozoan cover. Through successive phases of regeneration and overgrowth, multilayered encrusting colonies of *Calpensia nobilis* form thick cylinders on every kind of hard substrate, particularly on rhizomes of the seagrass *Posidonia oceanica* in southern Spain and in Ischia Strait, Naples (Romero Colmenero and Sanchez Lizaso 1999; Cigliano et al. 2007).

Other common examples of monopolization of rocks by bryozoans are provided by mixed assemblages of encrusting cheilostome bryozoans comprising *Schizomavella* spp., *Schizoporella* spp. and *Parasmittina* spp. The area covered by these assemblages as single patches can reach 0.5 m^2 on vertical or slightly overhanging walls, especially of boulders (Harmelin J-G pers. comm.). Artificial reefs deployed in 2007 in Marseille (France) for the restoration of an active artisanal fishery in a site where *Posidonia* beds were previously destroyed (Charbonnel et al. 2011), today host large, dense and abundant colonies of cheilostomes, mainly *P. fascialis* and *Turbicellepora avicularis* (Hincks), that participate greatly in the 3-D structure of the living cover of the substrate (Harmelin J-G 2012).

Habitat-forming bryozoans provide living space for diverse species and assemblages. Understanding the effect of habitat-forming bryozoans on diversity can allow a better appreciation of the fundamental ecological relationships between habitat complexity and diversity, and is also useful for management purposes. Single or multiple bryozoan species can contribute to habitat complexity at any one site, sometimes in association with other frame-building taxa

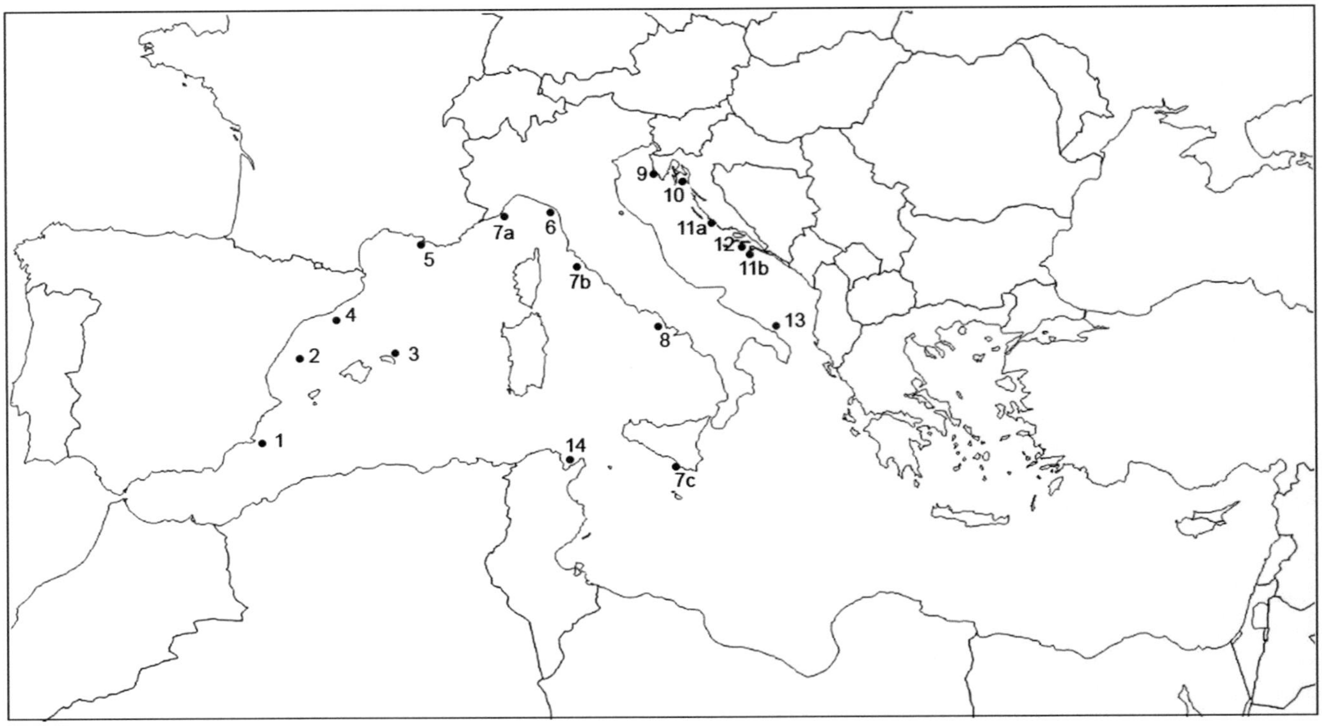

Fig. 21.3 Occurrences of habitat-forming bryozoans in the Mediterranean Sea. Site numbers correspond to those in Table 21.1

Table 21.1 Occurrences of habitat-forming bryozoans in the Mediterranean Sea

Site	Location	Species	References
1	Tabarca Island and Cabo de Palos, S Spain	*Calpensia nobilis*	Romero Colmenero and Sanchez Lizaso (1999)
2	Columbretes Islands, SE Spain	*Pentapora fascialis, Reteporella* sp.	Templado et al. (2002)
3	Mahón Harbour, Menorca, SE Spain	*Schizoporella errata*	Maluquer (1985)
4	Medes Islands, SE Spain	*Pentapora fascialis*	Sala et al. (1996)
5	Marseille, S France	Multiple	Harmelin and Capo (2001)
6	Tino Island, Ligurian Sea, NW Italy	*Schizoporella errata, Pentapora fascialis*	Cocito et al. (2000) and Cocito and Ferdeghini (2001)
7	(a) Alassio Bay (Ligurian Sea), (b) Formiche Islands (Tyrrhenian Sea), (c) Scoglitti (Sicily Channel)	*Pentapora fascialis*	Lombardi et al. (2008a)
8	Ischia Strait, Tyrrhenian Sea, Italy	*Calpensia nobilis*	Cigliano et al. (2007)
9	Istrian coast, Croatia, NE Adriatic	Multiple	Hayward and McKinney (2002), McKinney and Jaklin (2000) and Novosel 2005
10	Senj Archipelago, Croatia, NE Adriatic	*Pentapora fascialis*	Cocito et al. (2004) and Novosel et al. (2004)
11	(a) Jabuka Island and Bisevo Island, Central E Adriatic, (b) Lastovo Island, SE Adriatic	Multiple	Novosel (2005)
12	Korčula Island, Croatia, SE Adriatic,	*Pentapora fascialis, Schizobrachiella sanguinea*	Novosel (2005)
13	Apulian coasts, SW Adriatic	Multiple	Sarà (1969) and Ferdeghini et al. (2001)
14	Gulf of Tunis, Gulf of Gabes, Tunisia	Multiple	Mustapha et al. (2002)

Site numbers correspond to those of Fig. 21.3. Multiple species refers to >2 species

(molluscs, sponges, etc.) (Morgado and Tanaka 2001). Small, epibiotic bryozoans are among the most species-rich and frequently occurring animals living directly on habitat-forming bryozoan colonies. For example, colonies of *Pentapora fascialis* hosted 27 species of bryozoans (Ferdeghini and Cocito 1999), 30 species of bryozoans grew on *Cellaria salicornioides* (McKinney and Jaklin 2000), whereas the encrusting species *Schizoporella errata* supported five species of bryozoans (Maluquer 1985; Cocito et al. 2000). Some of the richest inter-phyletic associations include 19 species of molluscs found living on *P. fascialis* (Ferdeghini and Cocito 1999), 84 species associated with *P. fascialis* colonies in the Ligurian Sea (Ferdeghini and Cocito 1999), and 58 epibiont species on *C. salicornioides* branches (McKinney and Jaklin 2000).

Even though bryozoan-dominated habitat is widespread and occurs throughout the marine environment, little information is available on the ecology of habitat-forming bryozoans. Two conditions have been reported as essential to enable colonization by bryozoans: the existence of suitable hard (e.g. rocks, shells) or firm (e.g. marine plants) substrata, and an adequate supply of plankton as a food resource (Taylor 2000). Conversely, high rates of sedimentation and/or disturbance, and stagnant, low-oxygen conditions are unfavourable. Living on rocky outcrops, particularly on raised surfaces, can reduce the risk of sedimentation and also improve access to faster moving water outside the benthic boundary layer. Habitat-forming bryozoans often occur in channels or around headlands where water movement is fast. Dense meadows of *Cellaria fistulosa* (Linnaeus) and *C. salicornioides* were found to be located on a sediment–floored plain swept by strong bottom currents, whereas meadows of *Margaretta cereoides* densely covered rocky plateau with vigorous hydrodynamics (Novosel 2005). At two sites in the Mediterranean, *Schizoporella errata* occupied the majority of substrate provided by wharf piles where water current flow is permanent (Maluquer 1985).

Responses to Climate Change

Within the perspective of climate change – ocean acidification (OA) and global warming (GW) – occurring in the Mediterranean Sea, the ability of bryozoans to create permanent carbonate structures with varied biomineralogies makes this group a good candidate for exploring the effects of OA and GW. As a natural sink for CO_2, oceans absorb about a quarter of anthropogenic CO_2 emissions (Sabine et al. 2004), with consequences on the geochemical balance of the oceans. CO_2 reacts with water to form carbonic acid, which leads to an increase in hydrogen ions and a consequent decline in pH and also generates a decrease in the carbonate ion (CO_3^{2-}). As with all marine biomineralizers secreting calcareous skeletons, bryozoans metabolise carbonate ions in conjunction with organic material (proteins, carbohydrates, lipids) to form their skeletons. The concentration of carbonate ions in the ocean largely determines whether there is dissolution or precipitation of the two main biomineral polymorphs of $CaCO_3$, aragonite and calcite, including Mg-rich calcite (Smith 2009). OA does not occur in isolation, but rather in concert with other changes, such as GW, that may have synergistic, antagonistic, additive or neutral effects. Furthermore, not only biomineralization processes but also other important physiological processes, such as reproduction, growth and metabolic responses, can be affected by OA and GW.

An example of an ecophenotypically plastic Mediterranean cheilostome used for environmental inference is *Pentapora fascialis* in which the influence of temperature on colony growth, zooid size and skeletal carbonate mineralogy and geochemistry has been studied. Colonies have well-defined bands ('summer' and 'winter' growth bands; Patzold et al. 1987; Lombardi et al. 2006) interpreted as responses to seasonal changes that effect carbonate deposition (Fig. 21.4a). Although several factors have been hypothesized to cause growth band formation in bryozoans (Stebbing 1971; Winston 1983; Barnes 1995), seawater temperature may be the main cause in species of *Pentapora* (Lombardi et al. 2006, 2008b; Knowles et al. 2010). Smaller zooids are present in summer than winter growth bands, corresponding to the 'size-temperature' rule (Atkinson 1994). Observations on mineralogical differences between winter and summer bands not only corroborate the effect of seasonal changes of seawater temperature on bryozoan mineralogy (Lowenstam 1954; Crowley and Taylor 2000; Smith and Key 2004), but also show the rapidity with which ambient temperature impacts the biomineralization process. An increase in aragonite and in calcite with higher Mg content characterizes those parts of the skeleton formed in the summer, as previously observed in other bimineralic bryozoan species (Poluzzi and Sartori 1973; Rucker and Carver 1969). This underlines the potential value of *Pentapora* as an indicator of thermal conditions (Lombardi et al. 2006; O'Dea 2005; Knowles et al. 2010) (Table 21.2).

Other experimental studies have revealed the effect of GW and OA on three Mediterranean constructional cheilostomes (Rodolfo-Metalpa et al. 2010; Lombardi et al. 2011a, b, c). Transplant experiments were conducted in a natural CO_2 vent area at Ischia Island (Naples), located on the south side of Castello Aragonese (40° 43.84′ N, 13° 57.08′ E) between 0.5 and 3 m. The peculiar gas composition (90–95 % CO_2, 3–6 % N_2, 0.6–0.8 % O_2, and 0.2–0.8 % CH_4, emitted at a rate of about 1.4×10^6 l days^{-1} at room temperature and atmospheric pressure; Hall-Spencer et al. 2008) and pH gradient (extremely low: 6.6 ± 0.5; low pH: 7.7 ± 0.3; normal pH: 8.1 ± 0.1; Kroeker et al. 2012) make this area suitable as a natural laboratory for studying the impact of OA on Mediterranean marine organisms and communities. Three of the major Mediterranean habitat-forming bryozoans, *Myriapora truncata*, *Schizoporella errata* and *Calpensia nobilis*, were collected by scuba diving in rocky areas far from the volcanic vents (May – September 2009 and June – September 2010) and successively transplanted in cages at 3–4 m depth along a 200 m transect for different periods of time before recovery (see Rodolfo-Metalpa et al. 2010; Lombardi et al. 2011a, b, c).

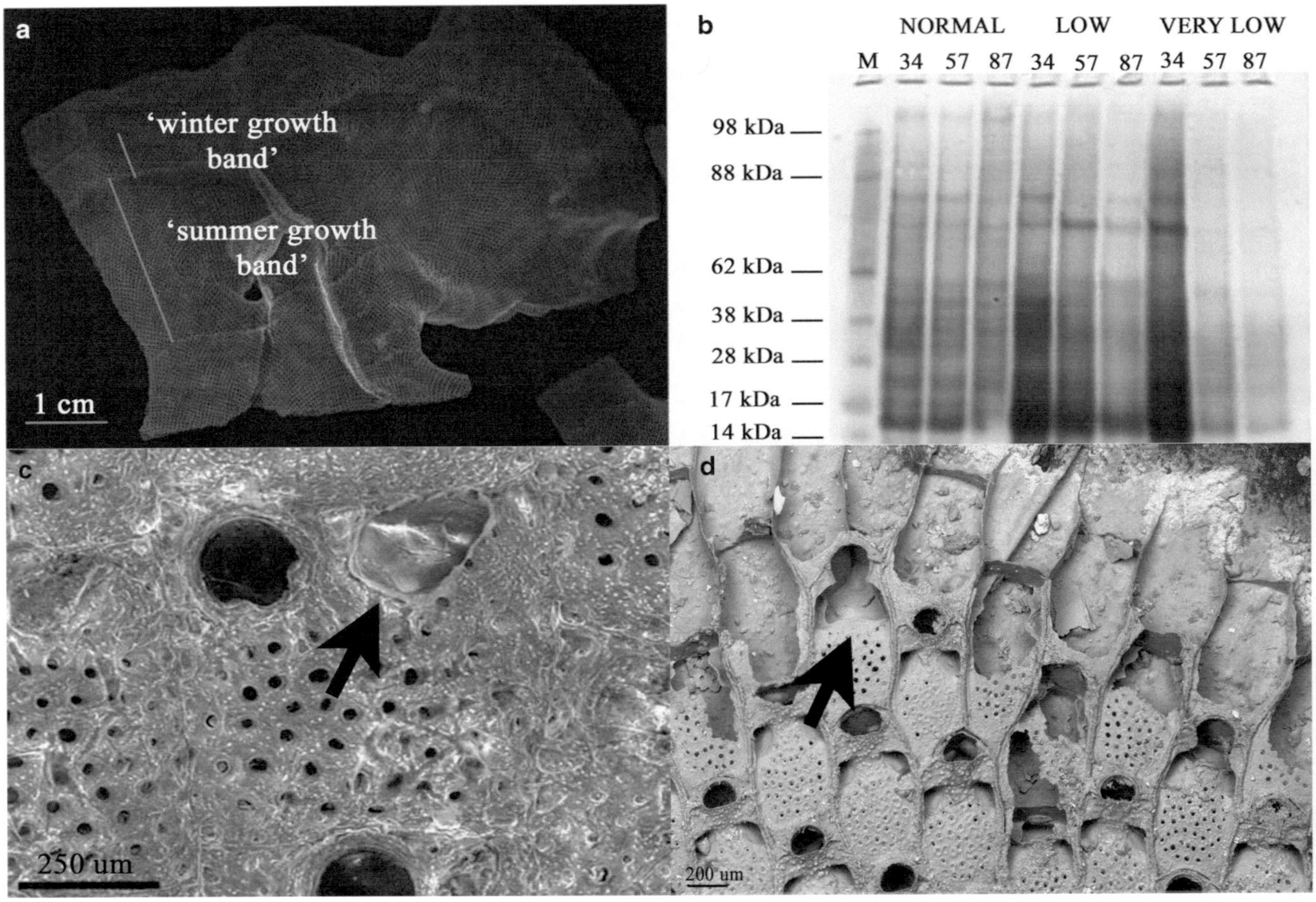

Fig. 21.4 (**a**) *Pentapora fascialis*. X-ray radiography of a lamina showing a "winter" and a "summer" growth band; (**b**) *Myriapora truncata*. Silver stained SDS PAGE gel for protein analysis. Molecular weight marker, samples exposed to normal, low and extreme low pH conditions for 34, 57 and 87 day periods; (**c**) *Schizoporella errata*. Avicularian 'bud' (*arrowed*) in colony grown in a low pH site; (**d**) *Calpensia nobilis*. Reduced number of zooid generation and key-hole shaped apertures (*arrowed*) at the growing edge from a colony grown at a low pH site

Table 21.2 Mineralogy (calcitic, aragonitic, *HMC*=High Mg Calcite), environmental variables, structure/compounds investigated and responses of some bryozoan species

Species	Mineralogy	Variable	Structure/compound investigated	Responses	References
Pentapora fascialis	Bimineralic mainly aragonitic	t°C	Colony growth Zooid size Mineralogy and geochemistry	Reduced growth in cold water Larger zooids in cold water Higher wt% aragonite in warm water	Lombardi et al. (2006, 2008b)
Myriapora truncata	Calcitic (HMC)	t°C pH	Net calcification Colony growth Zooidal growth Skeletal walls Mineralogy and geochemistry Cuticle and polypide Protein	Halted calcification Stop growth Decrease in Mg Investment in skeletal protection: increase of external cuticle thickness Change in protein concentrations	Rodolfo-Metalpa et al. (2010) and Lombardi et al. (2011a, b)
Schizoporella errata	Bimineralic mainly aragonitic	pH	Zooidal growth Skeletal walls and structures Polymorphs	Reduction of colony growth Reallocation of energy resources within the colony: reduction of defensive polymorphs	Lombardi et al. (2011c) and Lombardi (unpublished)
Calpensia nobilis	Bimineralic, mainly calcitic	pH	Colony growth Zooidal growth Skeletal wall and structure	Reduction of colony growth Reduction of number of generations of partly formed zooids at growing edge	Lombardi et al. (unpublished)

In *M. truncata*, net calcification, mineralogy and geochemistry, zooid formation, organic tissues and content of protein involved in calcification differed according to temperature and pH (Rodolfo-Metalpa et al. 2010; Lombardi et al. 2011a, b). After the 45-day experiment, net calcification rate for colonies exposed to warm temperature (25–28 °C) and low pH (mean pH 7.7) were the same as those growing in normal conditions. However, a more prolonged period of deployment (128 days) in high temperatures halted calcification at both normal and low pH sites, and all transplants died when high temperatures were combined with extremely low pH sites (Rodolfo-Metalpa et al. 2010). Colonies remained alive at the low and extremely low pH sites during the 45-day experiment but corrosion was very striking after 128 days, with colonies from the extremely low pH site (mean pH 6.6) showing significant loss of skeleton and thus greatly reduced calcification compared to those from normal pH sites. Compared to the control, these colonies also had lower levels of Mg (mean 8 wt% versus 9.5 wt% Mg) within their skeletons. As Mg content was higher in the outer layers of the skeletal walls in colonies from the normal pH site, corrosion of these layers may explain the lower Mg level found in colonies exposed to lower pH conditions (Lombardi et al. 2011a). The enrichment of Mg in outer layers of the skeleton may enhance the vulnerability of *M. truncata* to dissolution given that the solubility of calcite increases with Mg content. Nevertheless, established colonies of *M. truncata* may be resilient to the levels of ocean acidification predicted for the next 200 years. This could be the result of the presence of a thick organic cuticle which after a month in low pH increased in thickness compared to colonies from normal pH conditions, suggesting a protective role against dissolution of the high-Mg calcite skeleton (Lombardi et al. 2011b). At the low and extremely low pH sites, no zooids were budded, pointing to reallocation of energy resources from producing new zooids to protection. A significant change in the protein profile and expression displayed by samples from low and extremely low pH sites suggests that *M. truncata* makes an initial attempt to overcome the decrease in pH by up-regulating protein production but eventually exhausts the biochemical energy needed to maintain this rate of protein production, leading to eventual death (Lombardi et al. 2011b) (Fig. 21.4b, Table 21.2).

Another example of a Mediterranean bryozoan reallocating energy resources when growing in acidic conditions is the bimineralic species *Schizoporella errata*. Colonies showed fewer putative defensive polymorphs (avicularia) in low pH, the avicularia often appearing as incomplete 'buds' (Fig. 21.4b), as well as retarded growth of the basal and lateral walls of the autozooids at the extremely low pH site. The lower proportion of complete avicularia suggests a switch in resource allocation away from defense. In addition, corrosion of the skeleton was observed in both new and old zooids at the low pH site, and feeding zooids were slightly smaller but had larger orifices for the protrusion of feeding lophophores. These findings corroborate previous studies demonstrating the potential vulnerability of this species to OA, while also showing the possibility of the bryozoan to respond by adjusting resource allocation among zooids of different types (Lombardi et al. 2011c) (Table 21.2).

In a third bryozoan, the bimineralic *Calpensia nobilis*, transplanted colonies showed regular growth and zooid development when exposed to low pH conditions even for 6 weeks. However, periods of exposure exceeding 6 weeks induced modification in zooidal development within colonies, with apparent corrosion of the narrow gymnocyst forming the edge of the orifice. In addition, the extent of the zone of zooids with developing skeletons was reduced from the normal four generations (Lombardi, unpublished.) (Fig. 21.4c, Table 21.2).

Conclusions

Although the importance of bryozoans as frame-builders and habitat-formers is still poorly known in a global context, in places such as in the Mediterranean Sea there is a growing body of evidence showing the importance of these biomineralizing animals in forming biogenic constructions, enhancing marine biodiversity, providing and storing carbon, and responding to climatic changes (Cocito 2004; Novosel 2005; Wood et al. 2012).

The regional environment of the Mediterranean Sea is a useful model to understand the effects of climate change on the complexity of the physical and ecological interactions in small and complex basins, and thus on biodiversity (http://medsea-project.eu). In spite of numerous studies reporting detrimental effects of global changes on single species, both in the laboratory and in the field, very little is known about the effects at ecosystem and community levels (Kroeker et al. 2012). Because bryozoan carbonate bioconstructions provide habitats suitable for other organisms, the richness of these associated faunas, the interactions established between the host bryozoan and associated organisms, and the ability of bryozoans to respond to environmental changes, Mediterranean 'bryo-constructions' are suitable ecosystem models which can be monitored in the context of climate changes occurring in the Mediterranean Sea. The coastal bioconstructor species that were used in transplant experiments proved to be more vulnerable to the effects of ocean acidification when the water was warmest, indicating that ocean acidification will probably exacerbate the benthic mass mortality events that have been recorded with increasing frequency in the warming Mediterranean Sea. Differential responses of benthic bioconstructors to decreased pH and elevated seawater temperatures can lead to substantial changes in community structure, loss of biodiversity, trigger shifts in ecosystem function, impact marine food webs and, especially when combined with changes in salinity and nutrients, substantially alter the productivity of the sea.

It is not known whether there will be enough time for these benthic bioconstructors to adapt to survive the rapid rate changes predicted for the future. Our understanding of the processes that underlie observed effects on ecosystems and biogeochemistry is still rudimentary, as is our ability to forecast future impacts. There is an urgent need to develop tools to assess and quantify these impacts across the entire range of biological responses, from subcellular regulation to ecosystem reorganization, and from short-term physiological acclimation to evolutionary adaptation.

Acknowledgments We would like to thank J.-G. Harmelin (Marseille), M. Novosel (Zagreb) and L. Mizzan (Venice) for providing information and literature.

References

Atkinson D (1994) Temperature and organism size—a biological law for ectotherms? Adv Ecol Res 25:1–57

Barnes DKA (1995) Seasonal and annual growth in erect species of Antarctic bryozoans. J Exp Mar Biol Ecol 188:181–198

Bayhan E, Ergin M, Temel A, Keskin S (2001) Sedimentology and mineralogy of superficial bottom deposits from Aegean-Canakkale-Marmare transition (Eastern Mediterranean): effects of marine and terrestrial factors. Mar Geol 175:297–315

Bone Y, James NP (1993) Bryozoans as carbonate sediment producers on the cool-water Lancepede Shelf, southern Australia. Sediment Geol 86:247–271

Bruno JF, Stachowicz JJ, Bertness MD (2003) Inclusion of facilitation into ecological theory. Trends Ecol Evol 18:119–125

Charbonnel E, Carnus F, Ruitton S, Le Direac'h L, Harmelin J-G, Beurois J (2011) Artificial reefs in Marseille: from complex natural habitats to concepts of efficient artificial reef design. In: Ceccaldi HJ, Dekeyser I, Girault M, Stora G (eds) Global change: mankind-marine environment interactions. Springer, Dordrecht

Cigliano M, Cocito S, Gambi MC (2007) Epibiosis of *Calpensia nobilis* (Esper)(Bryozoa: Cheilostomida) on *Posidonia oceanica* (L.) Delile rhizomes: effects on borer colonization and morpho-chronological features of the plant. Aquat Bot 86:30–36

Cocito S (2004) Bioconstruction and biodiversity: their mutual influence. Sci Mar 68:137–144

Cocito S, Ferdeghini F (2001) Carbonate standing stock and carbonate production of the bryozoan *Pentapora fascialis* in the north-western Mediterranean. Facies 45:25–30

Cocito S, Ferdeghini F, Morri C, Bianchi CN (2000) Patterns of bioconstruction in the cheilostome bryozoan *Schizoporella errata*: the influence of hydrodynamics and associated biota. Mar Ecol Prog Ser 192:153–161

Cocito S, Novosel M, Novosel A (2004) Carbonate bioformations around underwater freshwater springs in the north-eastern Adriatic Sea. Facies 50:13–17

Cocito S, Lombardi C, Ciuffardi F, Gambi MC (2012) Colonization of Bryozoa on seagrass *Posidonia oceanica* 'mimics': biodiversity and recruitment pattern over time. Mar Biodiv 42:189–201

Crowley SF, Taylor PD (2000) Stable isotope composition of modern bryozoan skeletal carbonate formation from the Otago Shelf, New Zealand. NZ J Mar Freshwat Res 34:331–351

Duncan H (1957) Bryozoans. Geol Soc Am Mem 67:783–799

Fagerstrom JA (1991) Reef-building guilds and a checklist for determining guild. Coral Reefs 10:47–52

Feeley RA, Sabine CL, Lee K, Berelson W, Kleypas J, Fabry VJ, Millero FJ (2004) Impact of anthropogenic CO_2 on the $CaCO_3$ system in the oceans. Science 305:362–366

Ferdeghini F, Cocito S (1999) Biologically generated diversity in two bryozoan buildups. Biol Mar Mediterr 6:191–197

Ferdeghini F, Cocito S, Azzaro L, Sgorbini S, Cinelli F (2001) Bryozoan bioconstructions in the coralligenous formations of S.M. Leuca (Apulia, Italy). Biol Mar Mediterr 8:238–245

Fornos JJ, Ahr WM (2006) Present-day temperate carbonate sedimentation on the Balearic Platform, western Mediterranean: compositional and textural variation along a low-energy textural variation along a low energy isolated ramp. Geol Soc London 255:71–84

Hall-Spencer JM, Rodolfo-Metalpa R, Martin S, Fine M, Turner SM, Tedesco D, Buia M-C (2008) Volcanic carbon dioxide vents show ecosystem effects of ocean acidification. Nature 454:96–99

Hageman SJ, Bock PE, Bone Y, McGowran B (1998) Bryozoan growth habits: classification and analysis. J Paleo 72:418–436

Hageman SJ, Lukasik J, McGowran B, Bone Y (2003) Paleoenvironmental significance of *Celleporaria* (Bryozoa) from modern and tertiary cool-water carbonates of southern Australia. Palaios 18:510–527

Harmelin J-G (1985) Bryozoan dominated assemblages in Mediterranean cryptic environments. In: Nielsen C, Larwood GP (eds) Bryozoa: Ordovician to recent. Olsen and Olsen, Fredensborg

Harmelin JG, Capo S (2001) Effects of sewage on bryozoan diversity in Mediterranean rocky bottoms. In: Wyse Jackson PN, Buttler CJ, Spencer Jones ME (eds) Bryozoan studies. A.A. Balkema Publishers, Lisse

Hayward PJ, McKinney FK (2002) Northern Adriatic Bryozoa from the vicinity of Rovinj, Croatia. Bull Am Mus Nat Hist 270:1–139

Hayward PJ, Ryland JS (1999) Cheilostomatous Bryozoa, part 2: Hippothooidea–Celleporoidea. In: Barnes RSK, Crothers JH (eds) Synopses of the British Fauna (New Series). Field Studies Council, Shrewsbury

Hunter E, Hughes RN (1994) The influence of temperature, food ratio and genotype on zooid size in *Celleporella hyalina* (L.). In: Hayward PJ, Ryland JS, Taylor PD (eds) Biology and palaeobiology of bryozoans. Olsen and Olsen, Fredensborg

Jackson JBC (1977) Competition on marine hard substrata: the adaptive significance of solitary and colonial strategies. Am Nat 111:743–747

Jackson JBC (1985) Distribution and ecology of clonal and aclonal benthic invertebrates. In: Jackson JBC, Buss LW, Cook RE (eds) Population biology and evolution of clonal organisms. Yale University Press, New Haven

James NP, Bone Y (2010) Neritic carbonate sediments in a temperate realm. Springer, Dordrecht

Knowles T, Leng MJ, Williams M, Taylor PD, Okamura B (2010) Interpreting seawater temperature range using oxygen isotopes and zooid size variation in *Pentapora foliacea* (Bryozoa). Mar Biol 157:1171–1180

Kocak F, Balduzzi A, Benli HA (2002) Epiphytic bryozoan community of *Posidonia oceanica* (L.) Delile meadow in the northern Cyprus (Eastern Mediterranean). Indian J Mar Sci 31:235–238

Kroeker KJ, Micheli F, Gambi MC (2012) Ocean acidification causes ecosystem shifts via altered competitive interactions. Nat Clim Chan. doi:10.1038/nclimate1680

Kuklinski P, Taylor PD (2009) Mineralogy of Arctic bryozoan skeleton in a global context. Facies 55:333–342

Laborel J (1987) Marine biogenic constructions in the Mediterranean. A review. Sci Rep Port-Cros Natl Park 13:97–126

Lombardi C, Cocito S, Occhipinti-Ambrogi A, Hiscock K (2006) The influence of temperature on zooid size and growth rate in *Pentapora fascialis* (Bryozoa: Cheilostomata). Mar Biol 149:1103–1109

Lombardi C, Cocito S, Occhipinti-Ambrogi A, Porter JS (2008a) Distribution and morphological variation of colonies of the bryozoan *Pentapora fascialis* (Bryozoa: Cheilostomata) along the western coast of Italy. J Mar Biol Ass UK 88:711–717

Lombardi C, Cocito S, Hiscock K, Occhipinti-Ambrogi A, Setti M, Taylor PD (2008b) Influence of seawater temperature on growth bands, mineralogy and carbonate production in a bioconstructional bryozoan. Facies 54:333–342

Lombardi C, Rodolfo-Metalpa R, Cocito S, Gambi MC, Taylor PD (2011a) Structural and geochemical alterations in the Mg calcite bryozoan *Myriapora truncata* under elevated seawater pCO_2 simulating ocean acidification. Mar Ecol 32:211–222

Lombardi C, Rodolfo-Metalpa R, Silvia C, Gambi MC, Taylor PD (2011b) Effects of ocean acidification on growth, organic tissue and protein profile of the Mediterranean bryozoan *Myriapora truncata*. Aquat Biol 13:251–262

Lombardi C, Cocito S, Gambi MC, Cisterna B, Flach F, Taylor PD, Keltie K, Freer A, Cusack M (2011c) Effects of ocean acidification on growth, organic tissue and protein profile of the Mediterranean bryozoan *Myriapora truncata*. Aquat Biol 13:251–262

Lowenstam HA (1954) Environmental relations of modification compositions of certain carbonate secreting marine invertebrates. Proc Natl Acad Sci USA 40:39–48

Lowenstam HA, Weiner S (1989) On biomineralization. Oxford University Press, New York

Lutaud G (1987) The construction of cuticle and early stages of calcification in the zooidal bud of malacostegans. In: Ross JRP (ed) Bryozoa: present and past. Western Washington University, Bellingham

Maluquer P (1985) Algunas consideraciones sobre la fauna asociada a las colonias de *Schizoporella errata* (Waters, 1878) del puerto de Mahón (Menorca, Baleares). Publ Dept Zool Univ Barcelona 11:23–28

McKinney FK, Jackson JBC (1989) Bryozoan evolution. Chicago University Press, Chicago

McKinney FK, Jaklin A (2000) Spatial niche partitioning in the *Cellaria* meadow epibiont association, northern Adriatic Sea. Cah Biol Mar 41:1–17

Menon NR (1972) Heat tolerance, growth and regeneration in the three North Sea bryozoans exposed to different constant temperatures. Mar Biol 15:1–11

Moissette P, Cornée JJ, Koskeridou E (2010) Pleistocene rolling stones or large bryozoan nodules in a mixed silicilastic-carbonate environment (Rhodes, Greece). Palaios 25:24–39

Montañez IP (2002) Biological skeletal carbonate records changes in major ion chemistry of paleo-oceans. Proc Natl Acad Sci USA 99:15852–15854

Morgado EH, Tanaka MO (2001) The macrofauna associated with the bryozoan *Schizoporella errata* (Walters) in southeastern Brazil. Sci Mar 65:173–181

Morse JW, Wang Q, Tsio MY (1997) Influence of temperature and Mg: Ca ratio on $CaCO_3$ precipitates from seawater. Geology 25:85–87

Morse JW, Arvidson RS, Luttge A (2007) Calcium carbonate formation and dissolution. Chem Rev 107:342–381

Mustapha BK, Komatsu T, Hattour A, Sammari CH, Zarrouk S (2002) Bionomie des étages infra et circalittoral du golfe de Gabès. Bull Inst Natl Sci Tech Mer Salammbô 29:1–16

Novosel M (2005) Bryozoans of the Adriatic Sea. Denisia 16:231–246

Novosel M, Olujic G, Cocito S, Pozar-Domac A (2004) Submarine freshwater springs in the Adriatic Sea: a unique habitat for the bryozoan *Pentapora fascialis*. In: Moyano HI, Cancino JM, Wyse Jackson PN (eds) Bryozoan studies. A.A. Balkema Publishers, Lisse

O'Dea A (2005) Zooid size parallels contemporaneous oxygen isotopes in a large colony of *Pentapora foliacea* (Bryozoa). Mar Biol 146:1075–1081

O'Dea A, Okamura B (1999) Influence of seasonal variation in temperature, salinity and food availability on module size and colony growth of the estuarine bryozoan *Conopeum seurati*. Mar Biol 135:581–588

Okamura B (1985) The effects of ambient flow velocity, colony size, and upstream colonies on the feeding success of Bryozoa. II *Conopeum reticulum* (Linnaeus), an encrusting species. J Exp Mar Biol Ecol 89:69–80

Okamura B (1987) Seasonal change in zooids size and feeding activity in epifaunal colonies of *Electra pilosa*. In: Ross JP (ed) Bryozoa: present and past. Western Washington University, Bellingham

Okamura B, O'Dea A, Knowles T (2011) Bryozoan growth and environmental reconstruction by zooid size variation. Mar Ecol Prog Ser 430:133–146

Patzold J, Ristedt H, Wefer G (1987) Rate of growth and longevity of a large colony of *Pentapora foliacea* (Bryozoa) recorded in their oxygen isotopes profiles. Mar Biol 96:535–538

Poluzzi A, Coppa MG (1991) Zoarial strategies to win substratum space in *Calpensia nobilis* (Esper). In: Bigey FP (ed) Bryozoaires actuels et fossiles: Bryozoa Living and Fossil. Société des Sciences Naturelles de l'Ouest de la France, Nantes

Poluzzi A, Sartori R (1973) Carbonate mineralogy of some Bryozoa from Talbot shoal (Strait of Sicily, Mediterranean). Gior Geol 39:11–25

Poluzzi A, Sartori R (1974) Report on the carbonate mineralogy of bryozoans. In: Pouyet S (ed) Bryozoa. Documents des Laboratoires de Géologie, Lyon

Rodolfo-Metalpa R, Lombardi C, Cocito S, Hall-Spencer J, Gambi MC (2010) Effects of ocean acidification and high temperatures on the bryozoan *Myriapora truncata* at natural CO_2 vents. Mar Ecol 31:447–456

Romero Colmenero L, Sanchez Lizaso JL (1999) Effects of *Calpensia nobilis* (Esper 1796) (Bryozoa: Cheilostomida) on the seagrass *Posidonia oceanica* (L.). Delile. Aquat Bot 62:217–223

Rosso A (2003) Bryozoan diversity in the Mediterranean sea. Biogeographia 24:227–250

Rucker JB, Carver RE (1969) A survey of the carbonate mineralogy of cheilostome Bryozoa. J Paleontol 43:791–799

Sabine CL, Feeley RA, Gruber N, Key RM, Lee K, Bullistar JL, Wanninkhof R, Wong CS, Wallace DWR, Tilbrook B, Millero FJ, Peng T-H, Kozyr A, Ono T, Rios AF (2004) The oceanic sink for anthropogenic CO_2. Science 305:367–371

Sala E, Garrabou J, Zabala M (1996) Effects of diver frequentation on Mediterranean sublittoral populations of the bryozoan *Pentapora fascialis*. Mar Biol 126:451–459

Sarà M (1969) Research on coralligenous formations: problems and perspectives. Pubbl Staz Zool Napoli 37:124–134

Skinner HCW, Jahren AH (2003) Biomineralization. In: Schlesinger WH (ed) Treatise on geochemistry. Elsevier, Amsterdam

Smith AM (2009) Bryozoans as southern sentinels of ocean acidification: a major role for a minor phylum. Mar Freshw Res 60:475–482

Smith AM, Key MM Jr (2004) Controls, variation, and a record of climate change in detailed stable isotope record in a single bryozoan skeleton. Quat Res 61:123–133

Smith AM, Nelson CS, Spencer HG (1998) Skeletal carbonate mineralogy of New Zealand bryozoans. Mar Geol 151:27–46

Smith AM, Stewart B, Key MM Jr, Jamet CM (2001) Growth and carbonate production by *Adeonellopsis* (Bryozoa: Cheilostomata) in Doubtful Sound, New Zealand. Palaeogeo Palaeoclim Palaeoecol 175:201–210

Smith AM, Key MM Jr, Gordon DP (2006) Skeletal mineralogy of bryozoans: taxonomic and temporal patterns. Earth Sci Rev 78:287–306

Stanley SM, Hardie LA (1998) Secular oscillations in carbonate mineralogy of reef-building and sediment producing organisms driven by tectonically forced shifts in seawater chemistry. Palaeogeo Palaeoclim Palaeoecol 144:3–19

Stebbing ARD (1971) Growth of *Flustra foliacea* (Bryozoa). Mar Biol 9:267–273

Tavener-Smith R, Williams A (1972) The secretion and structure of the skeleton of living and fossil Bryozoa. Philos Trans R Soc Lond B-Biol Sci 264:97–160

Taylor PD (1990) Preservation of soft-bodied and other organisms by bioimmuration: a review. Paleontol J 33:1–17

Taylor PD (1994) An early cheilostome bryozoan from the Upper Jurassic of Yemen. N Jb Geol Paläont Abh 191:331–344

Taylor PD (2000) Origin of the modern bryozoan fauna. In: Culver SJ, Rawson PF (eds) Biotic Response to Global Change. The Last 145 Million Years. Cambridge University Press, Cambridge

Taylor PD (2005) Bryozoans. In: Selley RC, Cocks LRM, Plimer IR (eds) Encyclopaedia of geology. Elsevier, Amsterdam

Taylor PD, Allison PA (1998) Bryozoan carbonates in space and time. Geology 26:459–462

Taylor PD, Ernst A (2004) Bryozoans. In: Webby BD, Paris F, Droser ML, Percival IG (eds) The great Ordovician biodiversification event. Columbia University Press, New York

Taylor PD, Wilson MA (2003) Palaeoecology and evolution of marine hard substrate communities. Earth Sci Rev 62:1–103

Taylor PD, Kudryavtsev AB, Schopf JW (2008) Calcite and aragonite distributions in the skeletons of bimineralic bryozoans as revealed by Raman spectroscopy. Invertbr Biol 127:87–97

Taylor PD, James NP, Bone Y, Kuklinski P, Kyser TK (2009) Evolving mineralogy of cheilostome bryozoans. Palaios 24:440–452

Templado J, Calvo M, Garcia Carrascosa AM, Boisset F, Jimenez J (2002) Flora y fauna de la reserve marina de las Islas Columbretes. Museo Nacional Ciencias Naturales, Madrid

Thompson RC, Wilson BJ, Tobin ML, Hill AS, Hawkins AJ (1996) Biologically generated habitat provision and diversity of rocky shore organisms at a hierarchy of spatial scale. J Exp Mar Biol Ecol 202:73–84

Waeschenbach A, Taylor PD, Littlewood DTJ (2012) A molecular phylogeny of bryozoans. Molecular Phylogenetics Evoln 62:718–735

Winston JE (1983) Patterns of growth, reproduction and mortality in bryozoans from the Ross Sea, Antarctica. Bull Mar Sci 33:688–702

Wood ACL, Probert PK, Rowden AA, Smith AM (2012) Complex habitat generated by marine bryozoans: a review of its distribution, structure, diversity, threats and conservation. Aquat Conservat Mar Freshwat Ecosyst 22:547–563

The Fishes of the Mediterranean: A Biota Under Siege

22

Menachem Goren

Abstract

This chapter summarizes and presents an overall view of the various aspects of the native Mediterranean fish diversity. The Mediterranean Sea is a land-locked sea, comprising ca. 0.7 % of the hydrosphere area hosting ca. 5 % of the world's marine fish species and supplies ca. 1 % of the global marine catch. Eight percent of the fish species in the Mediterranean have been categorized as regionally threatened.

The total number of native marine fish species in the Mediterranean (May 2012) is 580 belonging to 151 families. The classes Myxini and Holocephali are represented in the Mediterranean by a single family and a single species each; The Cephalaspidomorphi by a single family and with two species; the Elasmobranchii (cartilaginous fish) by 24 families and 81 species; the Actinopterygii (bony fishes) is represented in the Mediterranean by 124 families and 495 species.

The total number of reported deep-sea fishes (below 1,000 m) is 60, belonging to 33 families.

The near-shore regions along the European coast, from Spain to Italy (including the Adriatic Sea), are the richest parts of the sea with 360–460 fish species. The near-shore poorest parts are most of Egyptian and eastern Libyan coasts with 200–270 species.

Keywords:

Mediterranean • Biodiversity • Marine fishes • Nanism • Deep sea fishes

Introduction

The Mediterranean Sea is a land-locked sea, comprising ca. 0.7 % of the hydrosphere area and hosting 7 % of the world marine species (Bianchi and Mori 2000; Coll et al. 2012). Its native fish comprise only ca. 5.1 % of the world's marine fish species (calculated from Bianchi and Mori 2000). The fish constitute a significant component of the ecological system of the Mediterranean and have huge economic value. Consequently, they have attracted the attention of both scientists and of the public since ancient times as evident from archeological findings in Greece, Italy and Egypt (Vargas et al. 2010; Giner 2010; Boylan 2012).

This chapter summarizes and presents an overall view of the various aspects of the native Mediterranean fish diversity. In order to accommodate the vast amount of information gathered during the preparation of the chapter to spatial limitations of this book, I have tried to minimize the number of tables and references, and direct the reader to recent published articles that cover the various issues.

When dealing with the biota of the Mediterranean, we need to be aware that it is continuously changing due to a combination of anthropogenic pressures such as overfishing, habitat degradation, water warming and the massive invasion of alien species, especially in the eastern part of the sea. These pressures have severely affected the fish diversity in the Mediterranean. As evident from Abdul Malak et al. (2011)

M. Goren (✉)
Department of Zoology, George S. Wise Faculty of Life Sciences, Tel Aviv University, Tel Aviv 69978, Israel
e-mail: gorenm@post.tau.ac.il

S. Goffredo and Z. Dubinsky (eds.), *The Mediterranean Sea: Its history and present challenges*, DOI 10.1007/978-94-007-6704-1_22,

an evaluation of the conservation status of the Mediterranean marine fishes reveals that 8 % of the evaluated species have been categorized as regionally threatened (critically endangered, endangered or vulnerable). These figures, although alarming, do not reflect the fundamental changes in the biodiversity of the Mediterranean and the role played by the alien species in effecting this change. Despite the tremendous impact of the alien species on Mediterranean biodiversity they were excluded from this chapter and will be dealt separately.

The Origin of the Mediterranean Fishes

The Mediterranean Sea has a long and an interesting history. It goes back to ca. 250 million years ago, when the first version of the Tethys Sea (the precursor of the Mediterranean) was formed by continental movement (Tang 2012; Por 1989). The desiccation of the Mediterranean during the Messinian Salinity Crisis, about 5.5–6 million years ago (Hsü et al. 1973, 1977; Krijgsman et al. 1999; Duggen et al. 2003; Garcia-Castellanos et al. 2009) probably led to the complete extinction of the ancient Tethyan biota (Bouchet and Taviani 1992).

The post-Messinian period saw the establishment of a connection with the Atlantic, which filled the Mediterranean basin with north-east Atlantic water and biota, enabling continuous exchange of biota between the two water bodies through the Straits of Gibraltar. This remained the sole connection between the Mediterranean biota and external marine biota until the opening of the Suez Canal in 1869. Thus, the native biota in the Mediterranean is of Atlantic origin. The partial isolation of the Mediterranean and the unique local conditions promoted the speciation of endemic species that today comprise a significant proportion of the Mediterranean biota. A higher resolution of the biogeographic affiliation of the biota can be found Bianchi and Mori (2000).

The exchange of biota between the Mediterranean and the north east Atlantic is an ongoing process, in which over the past 50 years, 34 fish species of Atlantic origin have been reported for the first time in the Mediterranean (Golani et al. 2002; Heemstra et al. 2011), and this probably reflects similar changes in other marine groups.

The possibility of the existence of relicts from the pre Messinian period has been suggested by several scientists (Benson 1976; Por and Dimentman 1985; Por 1989; Vacelet et al. 1989; Boury-Esnault et al. 1992 and others). Although there is "a strongly debated question is whether the Tethyan marine fauna disappeared completely during this period or survived in scattered areas with normal salinity" (Boury-Esnault et al. 1992), there are some indications that support this hypothesis (Por 1989; Vacelet et al. 1989; Boury-Esnault et al. 1992). However, there is no indication that any of the Mediterranean marine fish is one of these relicts.

How Many Native Fish Inhabit the Mediterranean?

The issue of how many native fish species inhabit in the Mediterranean has been dealt with by many authors, especially since the mid-1980s, when digital databases became available for scientists. Most of the studies arrived at similar estimates. Fredj and Maurin (1987) listed 638 species (including 43 of Indo-Pacific origin); Bianchi and Mori (2000) 613 species; Quignard and Tomasini (2000) 664; Coll et al. (2010, Table 22.1) 650 fish species, including 116 aliens; Coll et al. (2012) 625 species and Abdul Malak et al. (2011) listed 519 native species. Some of these articles provide a list of species, some just numbers.

In order to attend an up-to-date list of the native Mediterranean fish species, I critically evaluated and compiled relevant available data. The main sources were as follow: (a) fishbase - A Global Information System on Fishes (Froese and Pauly 2012a; hereafter Fishbase); (b) IUCN "overview on Mediterranean fishe" (Abdul Malak et al. 2011; hereafter IUCN); and (c) Quignard and Tomasini (2000). In addition, special attention given to the lists of Frej and Maurin (1987), Bilecenoglu et al. (2002), and Golani (2005). An extract from Fishbase (May, 2012) yields a list of 717 fish species in the Mediterranean, 580 of which are classified as "native" or "endemic"; IUCN lists 519 native species; Quignard and Tomasini counted 570 native species. In order to obtain the best possible estimate, and to update the list of native fish species, I have compared the lists of native species extracted from these sources. Each discrepancy between lists was investigated using additional literature sources. Finally, I checked the systematic status of each species in the Catalog of Fishes (CAS) – on-line version 15 March 2012 (Eschmeyer 2012). Although some species were classified in Fishbase as "questionable" regarding their validity, they are included in the statistics presented here since CAS considers them valid species. A different problem involved the definition of an alien vs. native species. In this chapter the criteria of the Mediterranean Science Commission (CIESM 2012) were employed. CIESM considers as alien those species of north-east Atlantic origin that were recorded in the Mediterranean after 1960.

The total number of native marine fish species in the Mediterranean, as of May 2012, is 580 belonging to 151 families (a full list of Mediterranean species and their systematic position is given in Table 22.1). The average number of species per family is 3.82. The family diversity of the

Table 22.1 List of Mediterranean fishes, 2012

Cephalaspidomorphi
Petromyzontidae
Lampetra fluviatilis (Linnaeus, 1758)
Petromyzon marinus (Linnaeus, 1758)
Holocephali
Chimaeridae
Chimaera monstrosa (Linnaeus, 1758)
Myxini
Myxinidae
Myxine glutinosa (Linnaeus, 1758)
Actinopterygii
Balistidae
Balistes capriscus (Gmelin, 1789)
Acanthuridae
Acanthurus monroviae (Steindachner, 1876)
Acipenseridae
Acipenser sturio (Linnaeus, 1758)
Huso huso (Linnaeus, 1758)
Alepisauridae
Alepisaurus ferox (Lowe, 1833)
Alepocephalidae
Alepocephalus rostratus (Risso, 1820)
Ammodytidae
Ammodytes tobianus (Linnaeus, 1758)
Gymnammodytes cicerelus (Rafinesque, 1810)
Anguillidae
Anguilla anguilla (Linnaeus, 1758)
Apogonidae
Apogon imberbis (Linnaeus, 1758)
Argentinidae
Argentina sphyraena (Linnaeus, 1758)
Glossanodon leioglossus (Valenciennes, 1848)
Atherinidae
Atherina boyeri (Risso, 1810)
Atherina hepsetus (Linnaeus, 1758)
Atherina presbyter (Cuvier, 1829)
Aulopidae
Aulopus filamentosus (Bloch, 1792)
Bathysauridae
Bathysaurus mollis (Günther, 1878)
Batrachoididae
Halobatrachus didactylus (Bloch and Schneider, 1801)
Belonidae
Belone belone (Linnaeus, 1761)
Belone svetovidovi (Collette and Parin, 1970)
Tylosurus acus (Lacepède, 1803)
Berycidae
Beryx decadactylus (Cuvier, 1829)
Blenniidae
Aidablennius sphynx (Valenciennes, 1836)
Blennius ocellaris (Linnaeus, 1758)
Coryphoblennius galerita (Linnaeus, 1758)
Hypleurochilus bananensis (Poll, 1959)
Lipophrys adriaticus (Steindachner and Kolombatovic, 1883)
Lipophrys canevae (Vinciguerra, 1880)
Lipophrys dalmatinus (Steindachner and Kolombatovic, 1883)
Lipophrys pholis (Linnaeus, 1758)
Lipophrys trigloides (Valenciennes, 1836)
Microlipophrys nigriceps (Vinciguerra, 1883)
Parablennius gattorugine (Brünnich, 1768)
Parablennius incognitus (Bath, 1968)
Parablennius pilicornis (Cuvier, 1829)
Parablennius rouxi (Cocco, 1833)
Parablennius sanguinolentus (Pallas, 1811)
Parablennius tentacularis (Brünnich, 1768)
Parablennius zvonimiri (Kolombatovic, 1892)
Salaria basilisca (Valenciennes, 1836)
Salaria pavo (Risso, 1810)
Scartella cristata (Linnaeus, 1758)
Bothidae
Arnoglossus grohmanni (Bonaparte, 1837)
Arnoglossus imperialis (Rafinesque, 1810)
Arnoglossus kessleri (Schmidt, 1915)
Arnoglossus laterna (Walbaum, 1792)
Arnoglossus rueppelii (Cocco, 1844)
Arnoglossus thori (Kyle, 1913)
Bothus podas (Linnaeus, 1758)
Bramidae
Brama brama (Bonnaterre, 1788)
Bythitidae
Bellottia apoda (Giglioli, 1883)
Cataetyx alleni (Byrne, 1906)
Cataetyx laticeps (Koefoed, 1927)
Grammonus ater (Risso, 1810)
Callanthiidae
Callanthias ruber (Rafinesque, 1810)
Callionymidae
Callionymus fasciatus (Valenciennes, 1837)
Callionymus lyra (Linnaeus, 1758)
Callionymus maculatus (Rafinesque, 1810)
Callionymus pusillus (Delaroche, 1809)
Callionymus reticulatus (Valenciennes, 1837)
Callionymus risso (LeSueur, 1814)
Synchiropus phaeton (Günther, 1861)
Caproidae
Capros aper (Linnaeus, 1758)
Carangidae
Alectis alexandrina (Geoffroy Saint Hilaire, 1817)
Campogramma glaycos (Lacepède, 1801)
Caranx crysos (Mitchill, 1815)
Caranx hippos (Linnaeus, 1766)
Caranx rhonchus (Geoffroy Saint Hilaire, 1817)
Decapterus macarellus (Cuvier, 1833)
Elagatis bipinnulata (Quoy and Gaimard, 1825)
Lichia amia (Linnaeus, 1758)
Naucrates ductor (Linnaeus, 1758)
Pseudocaranx dentex (Schneider, 1801)
Seriola dumerili (Risso, 1810)
Trachinotus ovatus (Linnaeus, 1758)
Trachurus mediterraneus (Steindachner, 1868)

(continued)

Table 22.1 (continued)

Trachurus picturatus (Bowdich, 1825)
Trachurus trachurus (Linnaeus, 1758)
Carapidae
Carapus acus (Brünnich, 1768)
Echiodon dentatus (Cuvier, 1829)
Echiodon drummondii (Thompson, 1837)
Centracanthidae
Centracanthus cirrus (Rafinesque, 1810)
Spicara maena (Linnaeus, 1758)
Spicara smaris (Linnaeus, 1758)
Centriscidae
Macroramphosus scolopax (Linnaeus, 1758)
Centrolophidae
Centrolophus niger (Gmelin, 1788)
Hyperoglyphe perciformis (Mitchill, 1818)
Schedophilus medusophagus (Cocco, 1839)
Schedophilus ovalis (Valenciennes, 1833)
Cepolidae
Cepola macrophthalma (Linnaeus, 1758)
Chaetodontidae
Chaetodon hoefleri (Steindachner, 1881)
Chlopsidae
Chlopsis bicolor (Rafinesque, 1810)
Chlorophthalmidae
Chlorophthalmus agassizi (Bonaparte, 1840)
Citharidae
Citharus linguatula (Linnaeus, 1758)
Clinidae
Clinitrachus argentatus (Risso, 1810)
Clupeidae
Alosa agone (Scopoli, 1786)
Alosa alosa (Linnaeus, 1758)
Alosa fallax (Lacepède, 1803)
Sardina pilchardus (Walbaum, 1792)
Sardinella aurita (Valenciennes, 1847)
Sardinella maderensis (Lowe, 1838)
Sprattus sprattus (Linnaeus, 1758)
Congridae
Ariosoma balearicum (Delaroche, 1809)
Conger conger (Linnaeus, 1758)
Gnathophis mystax (Delaroche, 1809)
Coryphaenidae
Coryphaena equiselis (Linnaeus, 1758)
Coryphaena hippurus (Linnaeus, 1758)
Cottidae
Taurulus bubalis (Euphrasen, 1786)
Cynoglossidae
Symphurus ligulatus (Cocco, 1844)
Symphurus nigrescens (Rafinesque, 1810)
Cyprinodontidae
Aphanius fasciatus (Nardo, 1827)
Aphanius iberus (Valenciennes, 1846)
Dactylopteridae
Dactylopterus volitans (Linnaeus, 1758)
Echeneidae
Echeneis naucrates (Linnaeus, 1758)
Remora australis (Bennett, 1840)
Remora brachyptera (Lowe, 1839)
Remora osteochir (Cuvier, 1829)
Remora remora (Bennett, 1840)
Engraulidae
Engraulis encrasicolus (Linnaeus, 1758)
Epigonidae
Epigonus constanciae (Giglioli, 1880)
Epigonus denticulatus (Dieuzeide, 1950)
Epigonus telescopus (Risso, 1810)
Microichthys coccoi (Ruppell)
Microichthys sanzoi (Sparta, 1950)
Evermannellidae
Evermannella balbo (Risso, 1820)
Exocoetidae
Cheilopogon exsiliens (Linnaeus, 1771)
Cheilopogon heterurus (Rafinesque, 1810)
Exocoetus obtusirostris (Günther, 1866)
Exocoetus volitans (Linnaeus, 1758)
Hirundichthys rondeletii (Valenciennes, 1846)
Hirundichthys speculiger (Valenciennes, 1847)
Gadidae
Gadiculus argenteus (Guichenot, 1850)
Merlangius merlangus (Linnaeus, 1758)
Micromesistius poutassou (Risso, 1826)
Trisopterus luscus (Linnaeus, 1758)
Trisopterus minutus (Lacepède, 1800)
Gasterosteidae
Gasterosteus aculeatus (Linnaeus, 1758)
Gempylidae
Ruvettus pretiosus (Cocco, 1829)
Gobiesocidae
Apletodon dentatus (Facciolà, 1887)
Apletodon incognitus (Hofrichter and Patzner, 1997)
Diplecogaster bimaculata (Bonnaterre, 1788)
Gouania willdenowi (Risso, 1810)
Lepadogaster candolii (Risso, 1810)
Lepadogaster lepadogaster (Risso, 1810)
Lepadogaster purpurea (Bonnaterre, 1788)
Opeatogenys gracilis (Canestrini, 1864)
Gobiidae
Aphia minuta (Risso, 1810)
Buenia affinis (Iljin, 1930)
Buenia jeffreysii (Günther, 1867)
Chromogobius quadrivittatus (Steindachner, 1863)
Chromogobius zebratus (Kolombatovic, 1891)
Corcyrogobius liechtensteini (Kolombatovic, 1891)
Crystallogobius linearis (Düben, 1845)
Deltentosteus collonianus (Risso, 1820)
Deltentosteus quadrimaculatus (Valenciennes, 1837)
Didogobius bentuvii (Miller, 1966)
Didogobius schlieweni (Miller, 1993)

(continued)

Table 22.1 (continued)

Didogobius splechtnai (Ahnelt and Patzner, 1995)
Favonigobius melanobranchus (Fowler, 1934)
Gammogobius steinitzi (Bath, 1971)
Gobius ater (Bellotti, 1888)
Gobius auratus (Risso, 1810)
Gobius bucchichi (Steindachner, 1870)
Gobius cobitis (Pallas, 1811)
Gobius couchi (Miller and El Tawil, 1974)
Gobius cruentatus (Gmelin, 1789)
Gobius fallax (Sarato, 1889)
Gobius gasteveni (Miller, 1974)
Gobius geniporus (Valenciennes, 1837)
Gobius kolombatovici (Kovacic and Miller, 2000)
Gobius niger (Linnaeus, 1758)
Gobius paganellus (Linnaeus, 1758)
Gobius roulei (de Buen, 1928)
Gobius strictus (Fage, 1907)
Gobius vittatus (Vinciguerra, 1883)
Gobius xanthocephalus (Heymer and Zander, 1992)
Gobiusculus flavescens (Fabricius, 1779)
Knipowitschia caucasica (Berg, 1916)
Knipowitschia panizzae (Verga, 1841)
Lebetus guilleti (Le Danois, 1913)
Lesueurigobius friesii (Malm, 1874)
Lesueurigobius sanzi (de Buen, 1918)
Lesueurigobius suerii (Risso, 1810)
Millerigobius macrocephalus (Kolombatovic, 1891)
Odondebuenia balearica (Pellegrin and Fage, 1907)
Pomatoschistus bathi (Miller, 1982)
Pomatoschistus canestrinii (Ninni, 1883)
Pomatoschistus knerii (Steindachner, 1861)
Pomatoschistus marmoratus (Risso, 1810)
Pomatoschistus microps (Nordmann, 1840)
Pomatoschistus minutus (Pallas, 1770)
Pomatoschistus norvegicus (Collett, 1902)
Pomatoschistus pictus (Malm, 1865)
Pomatoschistus quagga (Heckel, 1837)
Pomatoschistus tortonesei (Miller, 1969)
Pseudaphya ferreri (de Buen and Fage, 1908)
Speleogobius trigloides (Zander and Jelinek, 1976)
Thorogobius ephippiatus (Lowe, 1839)
Thorogobius macrolepis (Kolombatovic, 1891)
Vanneaugobius dollfusi (Brownell, 1978)
Vanneaugobius pruvoti (Fage, 1907)
Zebrus zebrus (Risso, 1826)
Zosterisessor ophiocephalus (Pallas, 1814)
Gonostomatidae
Cyclothone braueri (Jespersen and Tåning, 1926)
Cyclothone microdon (Günther, 1878)
Cyclothone pygmaea (Jespersen and Tåning, 1926)
Gonostoma denudatum (Rafinesque, 1810)
Haemulidae
Parapristipoma humile (Bowdich, 1825)
Parapristipoma octolineatum (Valenciennes, 1833)
Plectorhinchus mediterraneus (Guichenot, 1850)
Pomadasys incisus (Bowdich, 1825)
Hemiramphidae
Hyporhamphus picarti (Valenciennes, 1846)
Heterenchelyidae
Panturichthys fowleri (Ben Tuvia, 1953)
Ipnopidae
Bathypterois dubius (Vaillantv 1888)
Bathypterois grallator (Goode and Bean, 1886)
Istiophoridae
Istiophorus albicans (Latreille, 1804)
Makaira nigricans (Lacepède, 1802)
Tetrapturus albidus (Poey, 1860)
Tetrapturus belone (Rafinesque, 1810)
Tetrapturus georgii (Lowe, 1841)
Kyphosidae
Kyphosus saltatrix (Linnaeus, 1758)
Labridae
Acantholabrus palloni (Risso, 1810)
Centrolabrus trutta (Lowe, 1834)
Coris julis (Linnaeus, 1758)
Ctenolabrus rupestris (Linnaeus, 1758)
Labrus bergylta (Ascanius, 1767)
Labrus merula (Linnaeus, 1758)
Labrus mixtus (Linnaeus, 1758)
Labrus viridis (Linnaeus, 1758)
Lappanella fasciata (Cocco, 1833)
Symphodus bailloni (Valenciennes, 1839)
Symphodus cinereus (Bonnaterre, 1788)
Symphodus doderleini (Jordan, 1890)
Symphodus mediterraneus (Linnaeus, 1758)
Symphodus melanocercus (Risso, 1810)
Symphodus melops (Linnaeus, 1758)
Symphodus ocellatus (Linnaeus, 1758)
Symphodus roissali (Risso, 1810)
Symphodus rostratus (Bloch, 1797)
Symphodus tinca (Linnaeus, 1758)
Thalassoma pavo (Linnaeus, 1758)
Xyrichtys novacula (Linnaeus, 1758)
Lampridae
Lampris guttatus (Brünnich, 1788)
Liparidae
Eutelichthys leptochirus (Tortonese, 1959)
Paraliparis murieli (Matallanas, 1984)
Lobotidae
Lobotes surinamensis (Bloch, 1790)
Lophiidae
Lophius budegassa (Spinola, 1807)
Lophius piscatorius (Linnaeus, 1758)
Lophotidae
Lophotus lacepede (Giorna, 1809)
Lotidae
Gaidropsarus biscayensis (Collett, 1890)
Gaidropsarus granti (Regan, 1903)

(continued)

Table 22.1 (continued)

Gaidropsarus mediterraneus (Linnaeus, 1758)
Gaidropsarus vulgaris (Cloquet, 1824)
Molva dypterygia (Pennant, 1784)
Molva macrophthalma (Rafinesque, 1810)
Molva molva (Linnaeus, 1758)
Lutjanidae
Lutjanus jocu (Bloch and Schneider, 1801)
Luvaridae
Luvarus imperialis (Rafinesque, 1810)
Macrouridae
Coelorinchus caelorhincus (Risso, 1810)
Coelorinchus labiatus (Koehler, 1896)
Coryphaenoides guentheri (Vaillant, 1888)
Coryphaenoides mediterraneus (Giglioli, 1893)
Hymenocephalus italicus (Giglioli, 1884)
Nezumia aequalis (Günther, 1878)
Nezumia sclerorhynchus (Valenciennes, 1838)
Trachyrincus scabrus (Rafinesque, 1810)
Merlucciidae
Merluccius merluccius (Linnaeus, 1758)
Microstomatidae
Microstoma microstoma (Risso, 1810)
Nansenia iberica (Matallanas, 1985)
Nansenia oblita (Facciolà, 1887)
Molidae
Mola mola (Linnaeus, 1758)
Ranzania laevis (Pennant, 1776)
Moridae
Eretmophorus kleinenbergi (Giglioli, 1889)
Gadella maraldi (Risso, 1810)
Guttigadus latifrons (Holt and Byrne, 1908)
Lepidion guentheri (Giglioli, 1880)
Lepidion lepidion (Giglioli, 1880)
Mora moro (Risso, 1810)
Physiculus dalwigki (Kaup, 1858)
Rhynchogadus hepaticus (Facciolà, 1884)
Moronidae
Dicentrarchus labrax (Linnaeus, 1758)
Dicentrarchus punctatus (Bloch, 1792)
Mugilidae
Chelon labrosus (Risso, 1826)
Liza aurata (Risso, 1810)
Liza ramada (Risso, 1827)
Liza saliens (Risso, 1810)
Mugil cephalus (Linnaeus, 1758)
Oedalechilus labeo (Cuvier, 1829)
Mullidae
Mullus barbatus (Linnaeus, 1758)
Mullus surmuletus (Linnaeus, 1758)
Pseudupeneus prayensis (Cuvier, 1829)
Muraenesocidae
Cynoponticus ferox (Linnaeus, 1758)
Muraenidae
Anarchias euryurus (Lea, 1913)
Gymnothorax unicolor (Delaroche, 1809)
Muraena helena (Linnaeus, 1758)
Myctophidae
Benthosema glaciale (Reinhardt, 1837)
Ceratoscopelus maderensis (Lowe, 1839)
Diaphus holti (Tåning, 1918)
Diaphus metopoclampus (Cocco, 1829)
Diaphus rafinesquii (Cocco, 1838)
Diogenichthys atlanticus (Tåning, 1928)
Electrona risso (Cocco, 1829)
Gonichthys cocco (Cocco, 1829)
Hygophum benoiti (Cocco, 1838)
Hygophum hygomii (Lütken, 1892)
Lampanyctus crocodilus (Risso, 1810)
Lampanyctus pusillus (Risso, 1810)
Lobianchia dofleini (Zugmayer, 1911)
Lobianchia gemellarii (Cocco, 1838)
Myctophum punctatum (Rafinesque, 1810)
Notoscopelus bolini (Nafpaktitis, 1975)
Notoscopelus elongatus (Costa, 1844)
Notoscopelus kroyeri (Malm, 1861)
Symbolophorus veranyi (Moreau, 1888)
Nemichthyidae
Nemichthys scolopaceus (Richardson, 1848)
Nettastomatidae
Facciolella oxyrhyncha (Gilbert, 1890)
Nettastoma melanurum (Rafinesque, 1810)
Saurenchelys cancrivora (Peters, 1864)
Nomeidae
Cubiceps capensis (Smith, 1845)
Cubiceps gracilis (Lowe, 1843)
Notacanthidae
Notacanthus bonaparte (Risso, 1840)
Polyacanthonotus rissoanus (Filippi and Verany, 1859)
Ophichthidae
Apterichtus anguiformis (Peters, 1877)
Apterichtus caecus (Linnaeus, 1758)
Dalophis imberbis (Delaroche, 1809)
Echelus myrus (Linnaeus, 1758)
Mystriophis crosnieri (Blache, 1971)
Ophichthus ophis (Linnaeus, 1758)
Ophichthus rufus (Rafinesque, 1810)
Ophisurus serpens (Linnaeus, 1758)
Ophidiidae
Benthocometes robustus (Goode and Bean, 1886)
Ophidion barbatum (Linnaeus, 1758)
Ophidion rochei (Müller, 1845)
Parophidion vassali (Risso, 1810)
Ostraciidae
Acanthostracion notacanthus (Bleeker, 1863)
Acanthostracion quadricornis (Linnaeus, 1758)
Lactophrys trigonus (Linnaeus, 1758)
Paralepididae
Arctozenus risso (Bonaparte, 1840)
Lestidiops jayakari (Boulenger, 1889)
Lestidiops pseudosphyraenoides (Ege, 1918)

(continued)

Table 22.1 (continued)

Lestidiops sphyrenoides (Risso, 1820)
Lestidium atlanticum (Borodin, 1928)
Paralepis coregonoides (Risso, 1820)
Paralepis speciosa (Bellotti, 1878)
Sudis hyalina (Rafinesque, 1810)
Peristediidae
Peristedion cataphractum (Linnaeus, 1758)
Phosichthyidae
Ichthyococcus ovatus (Cocco, 1838)
Vinciguerria attenuata (Cocco, 1838)
Vinciguerria poweriae (Cocco, 1838)
Phycidae
Phycis blennoides (Brünnich, 1768)
Phycis phycis (Linnaeus, 1766)
Pleuronectidae
Platichthys flesus (Linnaeus, 1758)
Pleuronectes platessa (Linnaeus, 1758)
Polynemidae
Galeoides decadactylus (Bloch, 1795)
Polyprionidae
Polyprion americanus (Schneider, 1801)
Pomacentridae
Chromis chromis (Linnaeus, 1758)
Pomatomidae
Pomatomus saltatrix (Linnaeus, 1766)
Priacanthidae
Priacanthus arenatus (Cuvier, 1829)
Regalecidae
Regalecus glesne (Ascanius, 1772)
Scaridae
Sparisoma cretense (Linnaeus, 1758)
Sciaenidae
Argyrosomus regius (Asso, 1801)
Sciaena umbra (Linnaeus, 1758)
Umbrina canariensis (Valenciennes, 1843)
Umbrina cirrosa (Linnaeus, 1758)
Umbrina ronchus (Valenciennes, 1843)
Scomberesocidae
Scomberesox saurus (Walbaum, 1792)
Scombridae
Acanthocybium solandri (Cuvier, 1832)
Auxis rochei (Risso, 1810)
Auxis thazard (Lacepède, 1800)
Euthynnus alletteratus (Rafinesque, 1810)
Katsuwonus pelamis (Linnaeus, 1758): 297
Orcynopsis unicolor (Geoffroy Saint Hilaire, 1817)
Sarda sarda (Bloch, 1793): 44 48
Scomber colias (Gmelin, 1789)
Scomber scombrus (Linnaeus, 1758)
Scomberomorus tritor (Cuvier, 1832)
Thunnus alalunga (Bonnaterre, 1788)
Thunnus thynnus (Linnaeus, 1758)
Scophthalmidae
Lepidorhombus boscii (Risso, 1810)
Lepidorhombus whiffiagonis (Walbaum, 1792)
Scophthalmus maximus (Linnaeus, 1758)
Scophthalmus rhombus (Linnaeus, 1758)
Zeugopterus regius (Bonnaterre, 1788)
Scorpaenidae
Pontinus kuhlii (Bowdich, 1825)
Scorpaena azorica (Eschmeyer, 1969)
Scorpaena elongata (Cadenat, 1943)
Scorpaena loppei (Cadenat, 1943)
Scorpaena maderensis (Valenciennes, 1833)
Scorpaena notata (Rafinesque, 1810)
Scorpaena porcus (Linnaeus, 1758)
Scorpaena scrofa (Linnaeus, 1758)
Scorpaena stephanica (Cadenat, 1943)
Scorpaenodes arenai (Torchio, 1962)
Sebastidae
Helicolenus dactylopterus (Delaroche, 1809)
Serranidae
Anthias anthias (Linnaeus, 1758)
Epinephelus aeneus (Geoffroy Saint Hilaire, 1817)
Epinephelus caninus (Valenciennes, 1843)
Epinephelus costae (Steindachner, 1878)
Epinephelus marginatus (Lowe, 1834)
Hyporthodus haifensis (Ben Tuvia, 1953)
Mycteroperca rubra (Bloch, 1793)
Serranus atricauda (Günther, 1874)
Serranus cabrilla (Linnaeus, 1758)
Serranus hepatus (Linnaeus, 1758)
Serranus scriba (Linnaeus, 1758)
Serrivomeridae
Serrivomer brevidentatus (Roule and Bertin, 1929)
Soleidae
Bathysolea profundicola (Vaillant, 1888)
Buglossidium luteum (Risso, 1810)
Dicologlossa cuneata (Moreau, 1881)
Microchirus azevia (de Brito Capello, 1867)
Microchirus boscanion (Chabanaud, 1926)
Microchirus ocellatus (Linnaeus, 1758)
Microchirus variegatus (Donovan, 1808)
Monochirus hispidus (Rafinesque, 1814)
Pegusa impar (Bennett, 1831)
Pegusa lascaris (Risso, 1810)
Solea aegyptiaca (Chabanaud, 1927)
Solea solea (Linnaeus, 1758)
Synapturichthys kleinii (Risso, 1827)
Sparidae
Boops boops (Linnaeus, 1758)
Dentex dentex (Linnaeus, 1758)
Dentex gibbosus (Rafinesque, 1810)
Dentex macrophthalmus (Bloch, 1791)
Dentex maroccanus (Valenciennes, 1830)
Diplodus annularis (Linnaeus, 1758)
Diplodus bellottii (Steindachner, 1882)
Diplodus cervinus (Lowe, 1841)
Diplodus puntazzo (Cetti, 1777)
Diplodus sargus (Linnaeus, 1758)

(continued)

Table 22.1 (continued)

Diplodus vulgaris (Geoffroy Saint Hilaire, 1817)
Lithognathus mormyrus (Linnaeus, 1758)
Oblada melanura (Linnaeus, 1758)
Pagellus acarne (Risso, 1826)
Pagellus bellottii (Steindachner, 1882)
Pagellus bogaraveo (Brünnich, 1768)
Pagellus erythrinus (Linnaeus, 1758)
Pagrus auriga (Valenciennes, 1843)
Pagrus caeruleostictus (Valenciennes, 1830)
Pagrus pagrus (Linnaeus, 1758)
Sarpa salpa (Linnaeus, 1758)
Sparus aurata (Linnaeus, 1758)
Spondyliosoma cantharus (Linnaeus, 1758)
Sphyraenidae
Sphyraena sphyraena (Linnaeus, 1758)
Sphyraena viridensis (Cuvier, 1829)
Sternoptychidae
Argyropelecus hemigymnus (Cocco, 1829)
Argyropelecus olfersii (Cuvier, 1829)
Maurolicus muelleri (Gmelin, 1788)
Valenciennellus tripunctulatus (Esmark, 1871)
Stomiidae
Bathophilus nigerrimus (Giglioli, 1884)
Borostomias antarcticus (Lönnberg, 1905)
Chauliodus sloani (Schneider, 1801)
Stomias boa (Risso, 1810)
Stromateidae
Stromateus fiatola (Linnaeus, 1758)
Synaphobranchidae
Dysomma brevirostre (Facciola, 1887)
Syngnathidae
Hippocampus guttulatus (Cuvier, 1829)
Hippocampus hippocampus (Linnaeus, 1758)
Minyichthys sentus (Dawson, 1982)
Nerophis lumbriciformis (Jenyns, 1835)
Nerophis maculatus (Rafinesque, 1810)
Nerophis ophidion (Linnaeus, 1758)
Syngnathus abaster (Risso, 1826)
Syngnathus acus (Linnaeus, 1758)
Syngnathus phlegon (Risso, 1826)
Syngnathus taenionotus (Canestrini, 1871)
Syngnathus tenuirostris (Rathke, 1837)
Syngnathus typhle (Linnaeus, 1758)
Synodontidae
Synodus saurus (Linnaeus, 1758)
Tetragonuridae
Tetragonurus cuvieri (Risso, 1810)
Tetraodontidae
Ephippion guttifer (Bennett, 1831)
Lagocephalus lagocephalus (Linnaeus, 1758)
Trachichthyidae
Aulotrachichthys sajademalensis (Kotlyar, 1979)
Hoplostethus mediterraneus (Cuvier, 1829)
Trachinidae
Echiichthys vipera (Cuvier, 1829)
Trachinus araneus (Cuvier, 1829)
Trachinus draco (Linnaeus, 1758)
radiatus (Cuvier, 1829)
Trachipteridae
Trachipterus trachypterus (Gmelin, 1789)
Zu cristatus (Bonelli, 1820)
Trichiuridae
Lepidopus caudatus (Euphrasen, 1788)
Trichiurus lepturus (Linnaeus, 1758)
Triglidae
Chelidonichthys cuculus (Linnaeus, 1758)
Chelidonichthys lucerna (Linnaeus, 1758)
Chelidonichthys obscurus (Bloch and Schneider, 1801)
Eutrigla gurnardus (Linnaeus, 1758)
Lepidotrigla cavillone (Lacepède, 1801)
Lepidotrigla dieuzeidei (Audoin in Blanc and Hureau, 1973)
Trigla lyra (Linnaeus, 1758)
Trigloporus lastoviza (Brünnich, 1768)
Tripterygiidae
Tripterygion delaisi (Cadenat and Blache, 1970)
Tripterygion melanurum (Guichenot, 1850)
Tripterygion tripteronotum (Risso, 1810)
Uranoscopidae
Uranoscopus scaber (Linnaeus, 1758)
Xiphiidae
Xiphias gladius (Linnaeus, 1758)
Zeidae
Zeus faber (Linnaeus, 1758)
Zoarcidae
Melanostigma atlanticum (Koefoed, 1952)
Elasmobranchii
Alopiidae
Alopias superciliosus (Lowe, 1839)
Alopias vulpinus (Bonnaterre, 1788)
Carcharhinidae
Carcharhinus amboinensis (Müller and Henle, 1839)
Carcharhinus brachyurus (Günther, 1870)
Carcharhinus brevipinna (Müller and Henle, 1839)
Carcharhinus limbatus (Müller and Henle, 1839)
Carcharhinus longimanus (Poey, 1861)
Carcharhinus melanopterus (Quoy and Gaimard, 1824)
Carcharhinus obscurus (Lesueur, 1818)
Carcharhinus plumbeus (Nardo, 1827)
Prionace glauca (Linnaeus, 1758)
Centrophoridae
Centrophorus granulosus (Schneider, 1801)
Cetorhinidae
Cetorhinus maximus (Günnerus, 765)
Dalatiidae
Dalatias licha (Bonnaterre, 1788)
Dasyatidae
Dasyatis centroura (Mitchill, 1815)
Dasyatis marmorata (Steindachner, 1892)
Dasyatis pastinaca (Linnaeus, 1758)
Dasyatis tortonesei (Capapé, 1975)

(continued)

Table 22.1 (continued)

Pteroplatytrygon violacea (Bonaparte, 1832)
Taeniura grabata (Geoffroy Saint Hilaire, 1817)
Echinorhinidae
Echinorhinus brucus (Bonnaterre, 1788)
Etmopteridae
Etmopterus spinax (Linnaeus, 1758)
Gymnuridae
Gymnura altavela (Linnaeus, 1758)
Hexanchidae
Heptranchias perlo (Bonnaterre, 1788)
Hexanchus griseus (Bonnaterre, 1788)
Hexanchus nakamurai (Teng, 1962)
Lamnidae
Carcharodon carcharias (Linnaeus, 1758)
Isurus oxyrinchus (Rafinesque, 1810)
Lamna nasus (Bonnaterre, 1788)
Myliobatidae
Mobula mobular (Bonnaterre, 1788)
Myliobatis aquila (Linnaeus, 1758)
Pteromylaeus bovinus (Geoffroy Saint Hilaire, 1817)
Rhinoptera marginata (Geoffroy Saint Hilaire, 1817)
Odontaspididae
Carcharias taurus (Rafinesque, 1810)
Odontaspis ferox (Risso, 1810)
Oxynotidae
Oxynotus centrina (Linnaeus, 1758)
Pristidae
Pristis pectinata (Latham, 1794)
Pristis pristis (Linnaeus, 1758)
Rajidae
Dipturus batis (Linnaeus, 1758)
Dipturus oxyrinchus (Linnaeus, 1758)
Leucoraja circularis (Couch, 1838)
Leucoraja fullonica (Linnaeus, 1758)
Leucoraja melitensis (Clark, 1926)
Leucoraja naevus (Müller and Henle, 1841)
Raja africana (Bloch and Schneider, 1801)
Raja asterias (Delaroche, 1809)
Raja brachyura (Lafont, 1871)
Raja clavata (Linnaeus, 1758)
Raja miraletus (Linnaeus, 1758)
Raja montagui (Fowler, 1910)
Raja polystigma (Ogilby, 1910)
Raja radula (Delaroche, 1809)
Raja rondeleti (Bougis, 1959)
Raja undulata (Lacepède, 1802)
Rostroraja alba (Lacepède, 1803)
Rhinobatidae
Glaucostegus halavi (Forsskål, 1775)
Rhinobatos cemiculus (Geoffroy Saint Hilaire, 1817)
Rhinobatos rhinobatos (Linnaeus, 1758)
Scyliorhinidae
Galeus atlanticus (Vaillant, 1888)
Galeus melastomus (Rafinesque, 1810)
Scyliorhinus canicula (Linnaeus, 1758)
Scyliorhinus stellaris (Linnaeus, 1758)
Somniosidae
Centroscymnus coelolepis (Barbosa du Bocage and de Brito Capello, 1864)
Somniosus rostratus (Risso, 1826)
Sphyrnidae
Sphyrna lewini (Griffith and Smith, 1834)
Sphyrna mokarran (Rüppell, 1837)
Sphyrna zygaena (Linnaeus, 1758)
Squalidae
Squalus acanthias (Linnaeus, 1758)
Squalus blainville (Risso, 1826)
Squalus megalops (Macleay, 1881)
Squalus uyato (Rafinesque, 1810)
Squatinidae
Squatina aculeata (Cuvier, 1829)
Squatina oculata (Bonaparte, 1840)
Squatina squatina (Linnaeus, 1758)
Torpedinidae
Torpedo marmorata (Risso, 1810)
Torpedo nobiliana (Bonaparte, 1835)
Torpedo torpedo (Linnaeus, 1758)
Triakidae
Galeorhinus galeus (Linnaeus, 1758)
Mustelus asterias (Cloquet, 1821)
Mustelus mustelus (Linnaeus, 1758)
Mustelus punctulatus (Risso, 1827)

Date extracted from Froese and Pauly (2012a), Eschmeyer (2012), Abdul Malak et al. (2011), and slightly modified by the author of this chapter

ichthyofauna of the Mediterranean is 1.9175 (equation as in Goren 1993; Quignard and Tomasini 2000). These numbers are much lower than those of the Red Sea and other tropical seas (Goren 1993). The average number of species per family in the Red Sea is 6.93 (calculated from Golani and Bogorodsky 2010) and family diversity in the Red Sea is 4.156 (Goren 1993).

The richest family in the Mediterranean is the Gobiidae, with 57 species. The other rich families (represented in the Mediterranean by more than ten species) are listed in Table 22.2. The distribution pattern of the species among the families is shown in Table 22.3. As evident from this table, the great majority of families are represented in the Mediterranean by only one to three species.

Systematic Analysis

Following the classification of the CAS, there are five fish classes in the Mediterranean.

Table 22.2 List of the richest fish families in the Mediterranean

Class	Family	No. of species
Actinopterygii	Gobiidae	57
Actinopterygii	Sparidae	23
Actinopterygii	Labridae	21
Actinopterygii	Blenniidae	20
Actinopterygii	Myctophidae	19
Elasmobranchii	Rajidae	17
Actinopterygii	Carangidae	15
Actinopterygii	Soleidae	13
Actinopterygii	Scombridae	12
Actinopterygii	Syngnathidae	12
Actinopterygii	Serranidae	11

Table 22.3 The frequency distribution of the Mediterranean species among the families

No. of families with 11 or more species	11
No. of families with six to ten species	15
No. of families with five species	6
No. of families with four species	12
No. of families with three species	18
No. of families with two species	26
No. of families with one species	63
Total no. of families	151

Table 22.4 The Mediterranean elasmobranch families and their species richness

Family	Number of species
Rajidae	17
Carcharhinidae	9
Dasyatidae	6
Myliobatidae	4
Scyliorhinidae	4
Squalidae	4
Triakidae	4
Hexanchidae	3
Lamnidae	3
Rhinobatidae	3
Sphyrnidae	3
Squatinidae	3
Torpedinidae	3
Alopiidae	2
Odontaspididae	2
Pristidae	2
Somniosidae	2
Centrophoridae	1
Cetorhinidae	1
Dalatiidae	1
Echinorhinidae	1
Etmopteridae	1
Gymnuridae	1
Oxynotidae	1

Class **Myxini**, represented in the Mediterranean by a single family – Myxinidae, and a single species (*Myxine glutinosa*). The fish is found occasionally in the western Mediterranean but is rare in the central Mediterranean. No record of this species is known from the Levantine basin.

Class **Cephalaspidomorphi**, represented in the Mediterranean by a single family – Petromyzontidae, and with two species (*Lampetra fluviatilis* and *Petromyzon marinus*). Both species are eurihalins, spending most of their life in freshwater habitats and partly in marine habitats, close to estuaries (Fishbase).

Class **Holocephali**, represented in the Mediterranean by a single family – Chimaeridae, and a single species (*Chimaera monstrosa*). This species is found frequently in the western part of the Mediterranean at depths of 350–700 m, whereas in the eastern Mediterranean the fish is rare and found at depths of 1,000 m and below (Goren and Galil 1997).

Class **Elasmobranchii** (cartilaginous fish) with 10 orders in the Mediterranean, 24 families and 81 species. Most cartilaginous families are represented in the Mediterranean by a few species; however, three families (Rajidae, Carcharhinidae and Dasyatidae) are much richer in species (17, 8 and 6 species, respectively) (Table 22.4). The average number of species per family is 3.4. Family diversity for Mediterranean elasmobrachs is 1.2319.

Endemism

Five elasmobranch species (6 %) are endemic (species that are also found in the Black Sea are not considered as endemic to the Mediterranean): *Leucoraja melitensis*, *Raja polystigma*, *Raja radula*, *Raja rondeleti*, and *Dasyatis tortonesei*.

Habitat Analysis

Analysis of the distribution of the Mediterranean elasmobranch species among the marine habitats (following Fishbase classification) reveals that 42 % are demersal species, ca. 20 % are bathydemersal, with the benthopelagic and reef-associated species comprising ca. 11 % each, and the pelagic species comprising 16 % of the elasmobranches. This means that 86 % of species are dependent, to a certain degree, on the benthos for their food supply; whereas only 16 % depend solely on the pelagic organisms as a source of food.

Trophic Level

Analysis of the trophic level of the Mediterranean elasmobranches, based on data from Fishbase (for methods see Froese and Pauly 2012b) reveals that most fish (68 %)

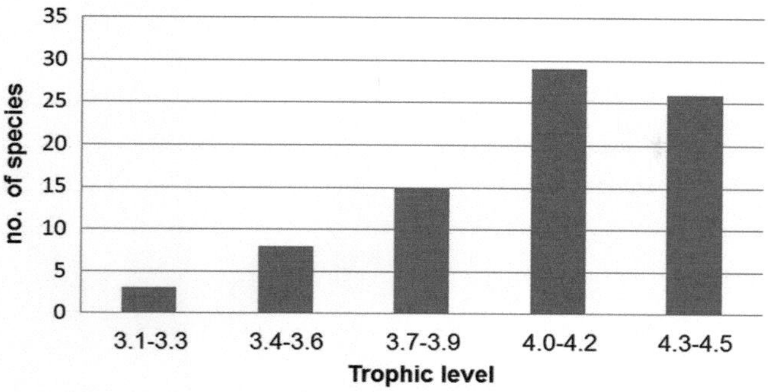

Fig. 22.1 The trophic level of the Mediterranean elasmobranches

are ranked as top predators, with a trophic level of 4 and above (Fig. 22.1).

Class **Actinopterygii** (bony fishes). This class is represented in the Mediterranean by 20 orders, 124 families and 495 species. As mentioned above, the richest family is the Gobiidae. The families with more than ten species are listed in Table 22.2. The average number of species per family is ca. 4.0. Family diversity of the bony fish is 1.8194.

Endemism

Forty-nine of the bony fishes (ca. 10 %) are endemic to the Mediterranean (species found in the Black Sea are not considered as endemic) (Table 22.5). This number is much lower than the 114 endemic bony fishes counted by Fredj and Maurin (1987) and than the 80 species noted by Costello et al. (2010). We can attribute the wide gap between the estimates mainly to differences in the definition of endemic species. Fredl and Mourin, and probably also Costelo et al. included in their calculations fishes from the Black Sea. An extract of endemic species from Fishbase revealed 43 species. One of these species, *Acipenser naccarii*, is found also in the Black Sea and thus is not included in the list of endemic species in Table 22.5. Others, *Caelorinchus mediterraneus*, *Chromogobius zebratus*, *Gobius kolombatovici*, *Kuhlia rubens*, *Rhynchogadus hepaticus* and *Syngnathus taenionotus*, are incorrectly categorized in Fishbase as "native" instead of "endemic", although their distribution is limited to the Mediterranean. The species included here as endemic. *Eretmophorus gouania* and *Willdenowi kleinenbergi* included here as endemic although in Fishbase these species are regarded as Mediterranean-Atlantic species. However, I could not find any reliable source to confirm their presence outside the Mediterranean.

Habitat Analysis

Analysis of the distribution of the bony fish species among the marine habitats (following Fishbase classification) reveals that 47.5 % species are demersal; ca. 12 % are benthopelagic; ca. 10 % are benthopelagic and reef-associated (each); ca 6 % are pelagic-neritic; and ca. 8 % are pelagic.

Table 22.5 The endemic bony fishes of the Mediterranean

Buenia affinis
Caelorinchus mediterraneus
Chromogobius zebratus
Corcyrogobius liechtensteini C
yclothone pygmaea D
idogobius bentuvii D
idogobius schlieweni D
idogobius splechtnai
Eretmophorus kleinenbergi
Eutelichthys leptochirus
Gammogobius steinitzi
Gobius geniporus
Gobius kolombatovici
Gobius strictus
Gobius vittatus
Gouania willdenowi
Grammonus ater
Kuhlia rubens
Lepidion lepidion
Lesueurigobius suerii
Microichthys coccoi
Microichthys sanzoi
Microlipophrys nigriceps M
illerigobius macrocephalus N
ansenia iberica N otoscopelus
elongatus Odondebuenia
balearica Opeatogenys
gracilis Ophichthus rufus
Panturichthys fowleri
Paralepis speciosa
Paraliparis murieli
Parophidion vassali
Pomatoschistus bathi
Pomatoschistus canestrinii
Pomatoschistus knerii
Pomatoschistus quagga
Pomatoschistus tortonesei
Pseudaphya ferreri
Rhynchogadus hepaticus S
alaria basilisca S peleogobius
trigloides Symphodus
doderleini Symphodus
melanocercus Syngnathus
taenionotus Syngnathus
taenionotus Tetrapturus
belone Thorogobius
macrolepis Tripterygion
melanurum

Trophic Analysis

Analysis of the trophic level of the Mediterranean bony fish, based on data from Fishbase, reveals that ca. 4 % feed mostly on plants, ca. 37 % are omnivores, 37 % are predators, and 22 % are top predators (Fig. 22.2).

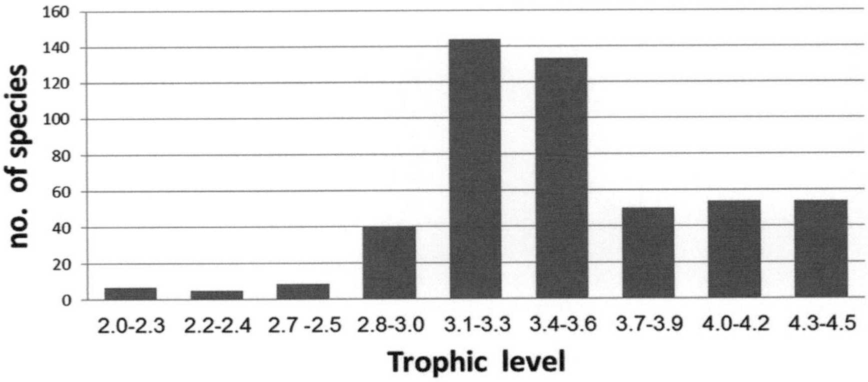

Fig. 22.2 The trophic level of the Mediterranean bony fishes

West vs. East: Spatial Distribution of Native Fish in the Mediterranean

The distribution of the species among the various parts of the Mediterranean is uneven. Ben-Tuvia (1983), Por (1989), Fredj and Laubier (1985), Fredj and Maurin (1987), Goren and Galil (1997), Quignard and Tomasini (2000) and others have described a decline in the number of native species, invertebrates and fish, from the north-west Mediterranean toward its south-east. Close examination of the data revealed that the parts of the sea with most species are the near-shore regions along the European coast from Spain to Italy (including the Adriatic Sea). The numbers of fish species in these regions vary from 359 to 460 (Quignard and Tomasini 2000; Coll et al. 2010). In general, the number of fish species in the western Mediterranean is over 500 while the number of species in the eastern part is much lower (Coll et al. 2010). Goren and Galil (1997) estimated the number of fish species in the Levantine basin as comprising 63 % of the Mediterranean fishes. The lower number of species in the letter region is also typical for various invertebrate groups (Coll et al. 2010).

An analysis of the native fish species in the Levantine basin (based on Bilecenoglu et al. 2002; Golani 2005; Golani et al. 2006a; and the fish collection of the Zoological Museum of Tel-Aviv University) revealed that the Levantine basin hosts 345 native fish species (ca. 60 % of the total Mediterranean native fishes) belonging to 116 families (ca. 74 % of families of native species).

Some of the richer families (e.g. as Sparidae, Blenniidae, Dasyatidae, Apogonidae, Serranidae) are nearly equally represented in both parts of the Mediterranean, whereas others (such as Gobiidae, Labridae, Rajidae) are poorly represented in the east.

The major relevant difference between the west and the east that can explain the lower number of species in the east is that of the higher water temperature, especially during the summer. Coll et al. (2010, Fig. 1b) show a clear temperature gradient from north-west to south-east. The temperatures in the south-east Mediterranean are much higher than those in the north-west. Considering the temperate Atlantic origin of the native Mediterranean fishes, this gradient can explain the poverty of species of the eastern Mediterranean (Por 1989). However, an additional reason might be the low habitat complexity in the south-east Mediterranean. The moderate soft-bottom slope of the continental shelf and scarcity of sea-weed habitats is reflected in a relatively low habitat complexity, which can impact species richness in the east. The impact of habitat complexity on species richness is discussed by Harmelin-Vivien et al. (1994) and Harmelin-Vivien et al. (2005). With the recent significant rise in water temperature increased significantly (EEA 2007) we can expect certain native species to limit their distribution to the cooler waters, a phenomenon that will probably be masked by the expansion of alien species.

Deep-Water Fishes

There are various definitions for the term "deep sea fishes". Danovaro et al. (2010) consider the deep-water biota as those that dwell in water deeper than 200 m. Golani et al. (2006b) consider the fishes collected at 250–300 m as deep sea fishes. Others consider the deep sea as the zone below 1,000 fathoms (ca. 1,800 m, Fishbase). In this chapter we follow Trujillo and Thurman (2011) and classify deep-sea fishes as fishes found below 1,000 m.

Although the Mediterranean Sea is the cradle of western civilization, and the biota of its shallow waters have been studied since the dawn of maritime civilization, the depth of the sea had remained almost unknown until the end of the nineteenth and the early twentieth century.

The pioneers in deep-sea sampling in the Mediterranean were Captain Magnaghi, hydrographer of the Italian navy and Professor Giglioli, who took samples from between the islands of Corsica and Sardinia from 1881 to 1883, on board the Washington (Stefanescu et al. 1992b) followed shortly after by the POLA expedition (1890–1893), the DANA expedition (1908–1910), and the THOR expedition (1921–1922), all of which studied the deep water biota of the Mediterranean (Galil and Goren 1994). Due to technical limitations at that time, little information regarding the deep-sea fishes was obtained by these expeditions. Since the 1950s, the deep-sea ichthyofauna of the western Mediterranean has been studied, as shown in Frej and Maurin (1987) and Stefanescu et al. (1992a). However, the ichthyofauna of the eastern Mediterranean remained almost unknown until the 1990s, when Galil and Goren studied this biota in the Levantine basin and increased the number of reported fish species from 7 to 39 (Galil and Goren 1994; Goren and Galil 1997, 2002; Galil 2004; Goren et al. 2006).

As of May 2012, the total number of reported deep-sea fishes in the Mediterranean is 60 (ca. 10 % of the native fishes), belonging to 33 families. No alien species were reported from this habitat. The average number of species per family is ca. 1.8. The family diversity at these depths is 1.371, relatively low compared to the entire ichthyofauna of the Mediterranean, and probably reflecting the lower complexity of the habitats at these depths. An extract of the deep water fishes from Fishbase revealed 70 deep-sea species, but since the majority of the fish are found also in the Atlantic, it is not clear whether the information regarding their depth range relates to the deep Mediterranean waters or to the Atlantic.

The richest family was found to be the Myctophidae with 11 species followed by the Macrouridae with seven species. The Bythitidae and Moridae are represented in the deep-sea with three species each. All the other 28 families are represented by a single or two species. Many of the Mediterranean deep-sea fish are also found in shallower water. Only two of the deep-sea species, *Lepidion lepidion* and *Cyclothone pygmaea*, are endemic to the Mediterranean. The low proportion of endemism (3.3 %) in the depths led Bouchet and Taviani (1992) to suggest that "much of this Mediterranean deep-sea fauna consists of reproductively sterile pseudopopulations that are constantly derived through larval inflow from Atlantic mother populations obtain on the fishes of deep sea". This hypothesis, however, has never been proven, although Roldán et al. (1998), who studied population genetic structure of the European hake, *Merluccius merluccius*, suggested a "... likelihood of some gene flow between Atlantic and Mediterranean populations...".

A comparison between the western and eastern deep-sea fish diversity reveals the number of deep-sea fish species in the west to be 42, belonging to 23 families (1.83 species per family, family diversity 1.229) whereas in the east there are 39 species belonging to 23 families (1.70 species per family, family diversity 1.199). The overlap in species composition of the two basins is nearly 50 %. The similarity on the families' composition as well as the family diversity is high.

The two most abundant species in the east are *Nezumia sclerorhynchus* and *Bathypterois dubius*, comprising 29.5 and 24.8 % of the total catch, respectively (Goren and Galil, unpublished data). In the west the most abundant species are *Bathypterois dubius* and *Lepidion lepidion*, comprising 39.0 and 23.4 % of the total catch, respectively (Stefanescu et al. 1992b). Regarding the biomass of fish, Stefanescu et al. (1992b: 197) noted that "the Mediterranean deep-sea ichthyofauma is very impoverished in comparison with the adjacent Atlantic Basin fauna". A comparison with the eastern Mediterranean revealed it to be much poorer than the west (Goren and Galil 2002) leading them to called the Levantine deep-sea habitat "a deep-sea desert". Ungaro et al. (2001) who studied the deep-water fishes of the Adriatic, also reported a very low biomass of fish. they found that at the depths below 1,000 m the most abundant fish were *Nezumia sclerorhynchus* and *Lepidion lepidion*.

Nanism

Many east-Mediterranean animals, fish and invertebrates, are smaller than their conspecifics in the western Mediterranean and in the Atlantic. This phenomenon was termed by Por (1989) "Levantine nanism". Sonin et al. (2007), who summarized published data as well as personal communications from marine experts, reported that Levantine nanism was observed in sponges, sipunculids crustaceans, coelenterates, polychaetes and fish. They compared body sizes between similar aged *Mullus barbatus* off the Israeli coast and the Sicilian coast, and found significant differences between the two populations. The adult fish (4–5 years old) in Sicily were 20–25 % longer than the Israeli fishes.

My own findings have also revealed that some fishes belonging to the families Spraidae, Mullidae and Lophidae are much smaller in the Levant than in the western Mediterranean and north-eastern Atlantic.

In a comparison carried out of populations of the grouper *Mycteroperca rubra* off the Tunisian and Israeli coasts no differences were found in size/age between the two sites (Aronov and Goren 2008). In a comparison between populations of the grouper *Epinaphalus marginatus* in Israel, Tunis, Egypt and France, fish up to the age of 3 years did not differ in size, however, fishes older than 4 years were slightly longer in the western populations than in the eastern ones (Aronov 2002).

Since the publication by Bergmann (1847) on the suggested relationship between temperature and body size of animals, many scientists have attempted to posit an hypothesis or a mathematical model that would explain the phenomenon (see Atkinson and Sibly 1997). The ongoing debate and the continuous study of the relationship between temperature and size is an indication that this complex issue still is far from being solved (e.g. Atkinson and Sibly 1997; Angilletta and Dunham 2003; Kingsolver and Huey 2008). The suggested hypothesis to explain the "Levantine nanism" has linked it to the limited food resources and/or the higher temperatures in the east, which trigger early sexual maturity that consequently halts the animal's growth (Wirssubski 1953; Por 1989). The relationship between body size, temperature and sea productivity, in marine biota, thus need to be examined on a large scale, based a comprehensive comparative study of fish communities in various geographic ranges, including the Red Sea and the Indian Ocean, and under different ecological conditions.

Fishery in the Mediterranean

Besides being an attraction for 200 million tourists annually (Coll et al. 2010), and a route for to 280 million tons of cargo (in 2006; Reynaud 2009), the Mediterranean, like the entire hydrosphere, is a source of food and income to the peoples living along its shores. An FAO (2012) publication, "An overview: major trends and issues", reports that the global annual landing from 2001 to 2009 remained around 90 million tons whereas the production of mariculture increased during this period from nearly 12 million tons in 2001 to 17.5 million tons in 2009. In 2007 the global per capita consumption of fish was estimated at 17.8 kg, with fish accounting for 16.6 % of the global population's intake of animal protein and 6.4 % of all protein consumed.

The marine fishery in the Mediterranean (including the Marmara Sea and excluding the Black Sea, data extracted in May 2012), during the period 2000–2008, reports an average annual landing of ca. 900,000 t of fish (ca. 1 % of the global marine catch). Analysis of the data shows that the fish group "herrings, sardines, anchovies" was the dominant group in the catch, with an average of nearly 390,000 t (ca. 44 % of the total). "Miscellaneous pelagic fishes" contributed nearly 116,750 t (13 % of the total) and the group "tunas, bonitos, billfishes" added 87,250 t (ca. 10 %). In total, the pelagic species that comprise ca. 8.0 % of bony species in the Mediterranean, comprise ca. 67 % of the total catch. The average annual catch of nonpelagic bony fish (categorized as "cods, hakes, haddocks", "flounders, halibuts, soles", "miscellaneous coastal fishes" and "miscellaneous demersal fishes") is ca. 25 % of the total catch, unidentified fish comprise ca 7 % of the total, and the group "sharks, rays, chimaeras" contributes ca. 1 % of the catch.

The richest part of the Mediterranean for fishery is the Balearic Sea with an annual average of ca. 248,000 t (28 % of the total Mediterranean catch), followed by the Ionian Sea, with an average of ca 156,000 t (ca.17 %) while the Adriatic and the Aegean contribute 10 and 11 %, respectively. The average Levantine catch is ca. 73,000 t (8 %). The Israeli average catch during the period 2000–2008 was 2,950 t (ca 4 % of the Levantine catch, calculated from the annual report of the Israeli Fishery Department (Shapiro 2007)).

From 1950 until the mid 1980s the total catch for the entire Mediterranean has increased gradually from 400,000 t to the present level (Aquarone et al. 2009). Analysis of FAO (2012) data reveal that that during the period 2000–2008 in most regions the annual catch remained within 10–15 % of the average and no-long term significant trend can be identified. The only region that showed a slight gradual increase in the catch was that of the Levant. This alleged stability in the catch should not lead to the false conclusion that fishery in the Mediterranean has reached its optimal exploitation (or sustainable level). The Mediterranean is heavily exploited and needs a comprehensive fishery management that will be based on a thorough understanding of the food web (see Coll and Libralato 2012), the biology of the various species, and the behavior of the Mediterranean as an ecological system.

Acknowledgements I am grateful to Naomi Paz for editing the manuscript.

References

Abdul Malak D, Livingstone SR, Pollard D, Polidoro BA, Cuttelod A, Bariche M, Bilecenoglu M, Carpenter KE, Collette BB, Francour P, Goren M, Kara MH, Massutí E, Papaconstantinou C, Tunesi L (2011) Overview of the conservation status of the marine fishes of the Mediterranean Sea. International Union for Conservation of Nature and Natural Resources, Malaga

Angilletta MJ, Dunham AE (2003) The temperature-size rule in ectotherms: simple evolutionary explanations may not be general. Am Nat 162:332–342

Aquarone MC, Adams S, Misud P (2009) IV-7 Mediterranean Sea: LNE #26. http://www.lme.noaa.gov/LMEWeb/LME_Report/lme_26.pdf . Accessed 17 Aug 2012

Aronov A (2002) Comparative study of the ecology of three groupers (Epinephelinae, Serranidae) at the shallow rocky habitats of the Israeli Mediterranean coast. Dissertation, Tel Aviv University

Aronov A, Goren M (2008) Ecology of the mottled grouper (*Mycteroperca Rubra*) in the eastern Mediterranean. Elect J Ichthyol 4:43–55

Atkinson D, Sibly M (1997) Why are organisms usually bigger in colder environments? Making sense of the life history puzzle. Trends Ecol Evol 12:235–239

Benson RH (1976) Changes in the ostracodes of the Mediterranean with the Messinian salinity crisis. Palaeogeo Palaeoclimat Palaeoecol 20:147–170

Ben-Tuvia A (1983) The Mediterranean Sea. B. Biological aspects. In: Ketchum BH (ed) Estuaries and enclosed seas, ecosystems of the world. Elsevier, Amsterdam

Bergmann C (1847) Ueber die Verhältnisse der Wärmeökonomie der Thiere zu ihrer Größe. Göttinger Studien 1:595–798

Bianchi CN, Mori C (2000) Marine biodiversity of the Mediterranean Sea: situation, problems and prospects for future research. Mar Pollut Bull 40:367–376

Bilecenoglu M, Taskavak E, Mater S, Kaya M (2002) Checklist of the marine fishes of Turkey. Zootaxa 113:1–194

Bouchet P, Taviani M (1992) The Mediterranean deep-sea fauna: pseudopopulations of Atlantic species? Deep-Sea Res 39:169–184

Boury-Esnault N, Pansini M, Uriz MJ (1992) A new *Discorhabdella* (Porifera, Demospongiae), a new Tethyan relict of pre-Messinian biota? J Nat Hist 26:1–7

Boylan M (2012) Aristotle: biology. Internet encyclopedia of philosophy. http://www.iep.utm.edu/aris-bio/. Accessed 17 Aug 2012

CIESM (2012) http://www.ciesm.org/online/atlas/index.htm. Accessed 23 Aug 2012

Coll M, Libralato S (2012) Contributions of food web modelling to the ecosystem approach to marine resource management in the Mediterranean Sea. Fish Fish 13:60–88

Coll M, Piroddi C, Steenbeek J, Kaschner K, Ben Rais Lasram F et al (2010) The biodiversity of the Mediterranean Sea: estimates, patterns, and threats. PLoS One 5:e11842

Coll M, Piroddi C, Albouy C, Ben Rais Lasram F, Cheung WWL, Christensen V, Karpouzi J, Guilhaumon V, Mouillot SF, Paleczny D, Palomares M, Steenbeek ML, Trujillo P, Watson R, Pauly D (2012) The Mediterranean Sea under siege: spatial overlap between marine biodiversity, cumulative threats and marine reserves. Global Ecol Biogeo 21:465–480

Costello MJ, Coll M, Danovaro R, Halpin P, Ojaveer H et al (2010) A census of marine biodiversity knowledge, resources, and future challenges. PLoS One 5:e12110

Danovaro R, Company JB, Corinaldesi C, D'Onghia G, Galil B et al (2010) Deep-sea biodiversity in the Mediterranean Sea: the known, the unknown, and the unknowable. PLoS One 5:e11832

Duggen S, Hoernle K, van den Bogaard P, Rüpke L, Morgan JP (2003) Deep roots of the Messinian salinity crisis. Nature 422:602–606

EEA (2007) Europe's environment, the fourth assessment. 5. Marine and coastal environment. http://www.eea.europa.eu/publications/state_of_environment_report_2007_1 Accessed 17 Aug 2012

Eschmeyer WN (2012) Catalog of fishes-electronic version. http://research.calacademy.org/research/ichthyology/catalog/fishcatmain.asp. Accessed 17 June 2012

FAO (2012) Fisheries and aquaculture information and statistics service 23/05/2012. http://www.fao.org/fishery/statistics/en. Accessed 17 Aug 2012

Fredj G, Laubier L (1985) The deep Mediterranean benthos. In: Moraito-Apostolopoulu M, Kiortsis V (eds) Mediterranean marine ecosystems. Plenum Press, New York

Fredj G, Maurin C (1987) Les poissons dans la banque de donnés MEDIFAUNA. Application à l'étude des caractaristiques de la faune ichtyologique Méditerranéenne. Cybium 11:220–333

Froese R, Pauly D (2012a) FishBase. World wide web electronic publication. http://www.fishbase.org (version 05/2012). Accessed 17 June 2012

Froese R, Pauly D (2012b) Food webs and trophic levels in: FishBase. World wide web electronic publication. http://www.fishbase.org/FishOnLine/English/FOL_FishAsPart.htm#Food_webs_and_trophic_levels. Accessed 17 Aug 2012

Galil BS (2004) The limit of the sea: the bathyal fauna of the Levantine Sea. Sci Mar 68(Suppl 3):63–72

Galil BS, Goren M (1994) The deep sea Levantine fauna – new records and rare occurrences. Senckenb Marit 25:41–52

Garcia-Castellanos D, Estrada F, Jiménez-Munt I, Gorini C, Fernàndez M, Vergés J, De Vicente R (2009) Catastrophic flood of the Mediterranean after the Messinian salinity crisis. Nature 462: 778–781

Giner CA (2010) Fishing nets in the ancient world: the historical and archaeological evidence. In: Bekker-Nielsen T, Bernal Casasola D (eds) Proceedings of the international workshop on "Ancient nets and fishing gear in classical antiquity. A first approach". Aarhus University Press, Aarhus

Golani D (2005) Checklist of the Mediterranean fishes of Israel. Zootaxa 947:1–90

Golani D, Bogorodsky SV (2010) The fishes of the Red Sea – reappraisal and updated checklist. Zootaxa 2463:1–135

Golani D, Orsi-Relini L, Massuti E, Quignard JP (2002) CIESM atlas of exotic species in the Mediterranean (updated by E Azzurro 2009) http://www.ciesm.org/atlas/appendix1.html. Accessed 23 Aug 2012

Golani D, Öztürk B, Başusta N (2006a) Fishes of the eastern Mediterranean. Turkish Marine Research Foundation, Istanbul

Golani D, Gökoglu M, Güven O (2006b) Two new records of deep-water fish species from the eastern Mediterranean. Mar Biodivers Rec 1:e46

Goren M (1993) Statistical aspects of the Red Sea ichthyofauna. Isr J Zool 39:293–298

Goren M, Galil BS (1997) New records of deep-sea fishes from the Levant Basin and a note on the deep-sea fishes of the Mediterranean. Isr J Zool 43:197–203

Goren M, Galil BS (2002) Records of *Cataetyx laticeps* and *Ophidion barbatum* in the eastern Mediterranean, with comments on the deep ichthyofauna. Cybium 26:150–152

Goren M, Mienis H, Galil BS (2006) Not so poor – new records for the deep sea fauna of the Levant Sea, eastern Mediterranean. Mar Biodivers Rec 1:2–5

Harmelin-Vivien ML, Harmelin JG, Francour P (1994) Influence de quelques facteurs du milieu sur le peuplement de poissons des prairies a`Caulerpa taxifolia a` Monaco. In: Boudouresque CF, Meinez A, Gravez V (eds) First international workshop on *Caulerpa taxifolia*. GIS Posidonie, Marseille

Harmelin-Vivien ML, Bitar G, Harmelin J-G, Monestiez P (2005) The littoral fish community of the Lebanese rocky coast (eastern Mediterranean Sea) with emphasis on Red Sea immigrants. Biol Invasions 7:625–637

Heemstra P, Aronov A, Goren M (2011) First record of the Atlantic island grouper *Mycteroperca fusca* (Lowe, 1838) in the Mediterranean Sea. Mar Biodivers Rec 3:e92

Hsü KJ, Ryan WBF, Cita MB (1973) Late Miocene desiccation of the Mediterranean. Nature 242:240–244

Hsü KJ, Montadert L, Bernoulli D, Cita MB, Garrison RE, Kidd RB, Mèlierés F, Müller C (1977) History of the Mediterranean salinity crisis. Nature 267:399–403

Kingsolver JG, Huey RB (2008) Size, temperature, and fitness: three rules. Evol Ecol Res 10:251–268

Krijgsman W, Hilgen FJ, Raf I, Sierro FJ, Wilsonk DS (1999) Chronology, causes and progression of the Messinian salinity crisis. Nature 400:652–654

Por FD (1989) The legacy of Tethys. An aquatic zoogeography of the Levant. Kluwer Academic Publishers, Dordrecht

Por FD, Dimentman C (1985) Continuity of Messinian biota in the Mediterranean basin. In: Stanley PJ, Wezel FC (eds) Geological evolution of the Mediterranean basin. Springer, Berlin

Quignard JP, Tomasini JA (2000) Mediterranean fish diversity. Biol Mar Medit 7:1–66

Reynaud C (2009) The components of maritime transport in the Mediterranean economy and territory. Territory and Transport Mediterranean: 255–260 http://www.iemed.org/anuari/2009/aarticles/a255.pdf. Accessed 17 Aug 2012

Roldán MI, García-Marín JL, Utter FM, Pla C (1998) Population genetic structure of European hake, *Merluccius merluccius*. Heredity 81:327–334

Shapiro J (ed) (2007) The fisheries and aquaculture of Israel 2007 (Hebrew and English). http //www.fishery.moag.gov.il/NR/rdonlyres/5EB9733E-D5BB-4F81-810D-C51805EA8142/0/Shnaton_2007.pdf. Accessed 23 Aug 2012

Sonin O, Spanier E, Levi D, Patti B, Rizzo P, Andreoli MG (2007) Nanism (dwarfism) in fish: a comparison between red mullet *Mullus barbatus* from the southeastern and the central Mediterranean. Mar Ecol Prog Ser 343:221–228

Stefanescu C, Lloris D, Rucabado J (1992a) Deep-living demersal fishing fishes in the Catalan Sea (western Mediterranean) below a depth of 1,000 m. J Nat Hist 26:197–213

Stefanescu C, Rucabado J, Lloris D (1992b) Depth-size trends in western Mediterranean demersal deep-sea fishes. Mar Ecol Prog Ser 81:205–213

Tang CM (2012) "Tethys Sea." Encyclopaedia Britannica. Encyclopaedia Britannica Online Academic Edition. Encyclopædia Britannica Inc, Web http://www.britannica.com/EBchecked/topic/588887/Tethys-Sea. Accessed 17 Aug 2012

Trujillo AP, Thurman HV (2011) Essentials of oceanography, 10th edn. Prentice Hall, Boston

Ungaro N, Marano G, Rivas G (2001) Notes on ichthyofauna of the deep basin of the Southern Adriatic Sea. Sarsia 86:153–156

Vacelet J, Boury-Esnault N, Zibrovius N (1989) Unexpected deep-water records of calcareous sponges (Calcarea). Deep-Sea NewsLett 15:24–25

Vargas G, Alberto E, Florido del Corral D (2010) The origin and development of tuna fishing nets (almadrabas). In: Bekker-Nielsen T, Bernal Casasola D (eds) Proceedings of the international workshop on "Ancient nets and fishing gear in classical antiquity. A first approach". Aarhus University Press, Aarhus

Wirssubski A (1953) On the biology and biotope of the red mullet. Bull Sea Fish Res St Israel 7:1–20

Physiological Responses of Marine Animals Towards Adaptation to Climate Changes

23

Elena Fabbri and Enrico Dinelli

Abstract

According to climatic models, the Mediterranean basin will be one of the regions most affected by the ongoing warming trend and increase in extreme events. The Mediterranean is already one of the most impacted seas in the world, where climate change interacts with many other stressors. Coastal lagoons, in particular, represent critical areas for their importance in terms of land use, economic relevance and anthropogenic pressure, and are the main objects of our analysis. A concept emerged in recent studies on climate changes, suggesting that the only environmental signals that matter to an organism are those that the organism experiences. Thus, animal responses may be very different from those expected at a large-scale, and the impacts of climate change can be different according to a number of local/organismal conditions. The present contribution is focused on the effects of climate change-driven factors on animal physiology, considering that physiology bridges the gap between mechanistic molecular understanding and the larger scale ecosystem responses. Indeed, adaptive responses to large-scale perturbations, such as climate change, affect all biological levels but they initially take place at the cellular and individual levels, and are then integrated and translated to upper levels of biological organization. The geochemical features that may influence animals responses are also addressed.

Keywords:

Adaptive responses • Animal physiology • Animal responses • Climate change • Coastal lagoons • Ecosystem responses • Geochemical responses large-scale perturbations • Mechanistic molecular responses • Mediterranean basin • Physiology

Introduction

The biosphere is comprised of a large array of organisms that vary in their abilities to resist or respond to environmental challenges. Climate change on its turn encompasses changes in numerous environmental factors including pH, water level, salinity and temperature (Helmuth 2009) and a number of related changes, e.g. oxygen and food availability, that ultimately modify organism performances and capability of adaptation (Monaco and Helmuth 2011). Several recent reviews are available in the literature addressing the issue at the molecular (e.g. Peck 2011; Somero 2010), ecological (Wernberg et al. 2011), evolutionary (Hoffmann and Sgrò 2011) and previsional (The MerMex Group 2011) levels, and further excellent contributions are provided in different chapters of this book. This review in particular aims at underlining some peculiar aspects of impacts on the physiology of aquatic fauna living in Mediterranean coastal lagoons. Our approach is far from being exhaustive, and for a wider perspective concerning the keystone role of physiology in

E. Fabbri (✉) • E. Dinelli
Department of Biological, Geological, and Environmental Sciences, University of Bologna, Via Selmi 3, 40100 Bologna, Italy
e-mail: elena.fabbri@unibo.it

S. Goffredo and Z. Dubinsky (eds.), *The Mediterranean Sea: Its history and present challenges*,
DOI 10.1007/978-94-007-6704-1_23,

marine climate change research we recommend the reading of recent reviews by Monaco and Helmuth (2011) and Doney et al. (2012).

According to climatic models, the Mediterranean basin will be one of the regions most affected by the ongoing warming trend and by an increase in extreme events. There are reasons to believe that the Mediterranean is already one of the most impacted seas in the world, since climate change interacts synergistically with many other impacts (Vargas-Yáñez et al. 2010). Indeed, the natural balance of Mediterranean coast ecosystems are severely impacted by major disturbances, which may cause the extensive loss of biodiversity. These include habitat destruction and alien species introductions, overfishing, pollution, phenomena tightly related with the increasing of anthropic pressure due to urbanization on the northwestern shores and outstanding growth population on the southern and eastern shores. The Mediterranean basin is therefore a focus of interest for research on climate change effects.

It appears that all parts of the Mediterranean will be equally affected by global climate changes; indeed, different climatic conditions are involved in the Mediterranean basin, so responses are expected to vary. A further concept emerged in recent studies on climate changes, suggesting that the only environmental signals that matter to an organism are those that the organism experiences (Kearney 2006). The physiological niche of an animal is driven by the interaction of the organism morphology, size and behavior with its local habitat, and responses may be very different from those expected at a large-scale (Helmuth et al. 2011). Extreme examples have been reported (Denny et al. 2011) showing that two species may have similar performances, although they experience very different levels of stress under identical field conditions because the predator (*Mytilus*) is able to maintain a lower body temperature with respect to the prey (*Pisaster*). Therefore, the impacts of climate change can be different according to a number of local/organismal conditions, and in this sense, coastal lagoons represent very peculiar habitats, often different from each other.

Here we will focus on the effects of climate change-driven factors on animal physiology, considering that physiology provides a key contribution bridging the gap between mechanistic molecular understanding and the larger scale ecosystem responses. Although scientific research on climate changes impact is facing an enormously wide task, we believe that much can be learned through integrated approaches dealing with the mechanisms underlying the biochemical/physiological consequences of climate changes at organism scales. This statement is based on the fact that organism responses to environmental changes represent a continuum array from the biochemical (i.e. enzyme activation/inhibition and gene expressions) and the physiological flexibility (i.e. protein expressions, metabolic, cardiovascular and respiratory adaptation) which take place within seconds, minutes or hours, to the gene frequency variations and the evolutionary changes based on new fixed genes or gene combination through selection which ultimately alter populations and communities in the range of years or centuries, if we exclude microbial systems (Peck 2011).

Acclimatization and genetic adaptation are predicted to be the most important processes allowing species survival to climate change, with species with very long generation time strongly depending on physiological flexibility through acclimatization processes to survive until genetic adaptation will become effective (Somero 2010). We must consider that several climate change-related phenomena may take place faster and be rather strong, though transient. This is true for rapid temperature shifts, decreasing salinities as a consequence of storms, flooding events and mobilization of pollutants. Biochemical/physiological responses become therefore significant for a major number of species, and indeed crucial to survive changes.

Coastal lagoon ecosystems were and are still undergoing heavy human impacts (Lotze et al. 2006). Large concentrations of contaminants were accumulated in sediments of coastal lagoons due to past industrial activity, thus one of the issues to be faced is the fate of these contaminants with changing environmental conditions, including those related to climate change (Noyes et al. 2009). Changes in temperature and salinity, for example, imply further changes e.g. in oxygen availability or food intake, and also modify contaminant bioavailability, metabolism and excretion. Climate change can affect in a direct and indirect way the bioavailability of chemical compounds with a possible scenario in which the levels of contaminants in the tissues of aquatic organisms increase. These effects are probably more relevant in fragile water bodies characterized by shallow waters and a low water-volume to surface-area ratio, as the coastal lagoons, where biogeochemical cycles and water balances are modified quickly by changes in temperature, precipitation patterns and rising sea level. Toxicokinetics may be altered as well, for example, different uptake or detoxification rates may be induced by different temperature, salinity, oxygen, etc., that in turn influence food levels. On the other way around, the physiological state of animals may be important. For instance, food scarcity modifies fat, protein, and carbohydrate content in an organism, which may influence the partition of chemicals over the different compartments. It is well known that organic chemicals accumulated in fat increase their body concentrations and cause severe toxicity after long starvation periods (Verreault et al. 2008).

Considering that the processes which govern the fate of contaminants in sediments are complex interactions with the biogeochemical cycles of major redox and biogenic elements (C, O, P, S, Si, Fe, etc.), the geochemical features of coastal lagoons that may influence biological responses are considered in this review.

Climate Change in Coastal Lagoon Environments

Coastal lagoons represent critical areas for their importance in terms of land use (i.e., protected areas, agriculture), economic importance (i.e. aquaculture, fishing, tourism, industrial activities) and anthropogenic pressure (i.e. industrial settlements, large cities, effluent discharge) that can lead to accumulation of contaminants, organic matter and nutrients from various sources (Newton et al. 2003; Viaroli et al. 2007; Brito et al. 2012a). They represent environments separated from the sea by some kind of barrier (more or less stable), generally with limited exchange with the marine environment. Lagoons are characterized by high amounts of organic matter (Chapman 2012) which can be affected by severe eutrophication problems (Lloret et al. 2008) and are considered to be one of the most fragile marine environment in relation to climate change and its effects on the marine environment (Fig. 23.1).

The increase in temperature, the thermal forcing of the system, has caught much attention and its occurrence on ocean temperature are reported in many areas (Philippart et al. 2011) and also in the Mediterranen Sea (The MerMex Group 2011) and can be reasonably expected to be more significant in restricted coastal areas with limited circulation, such as coastal lagoons. Sea level rise will be one of the first effect of climate change (IPCC 2007a, b; Gosling et al. 2011) and to the eustatic phenomenon additional local factors such as natural or anthropogenically induced subsidence could

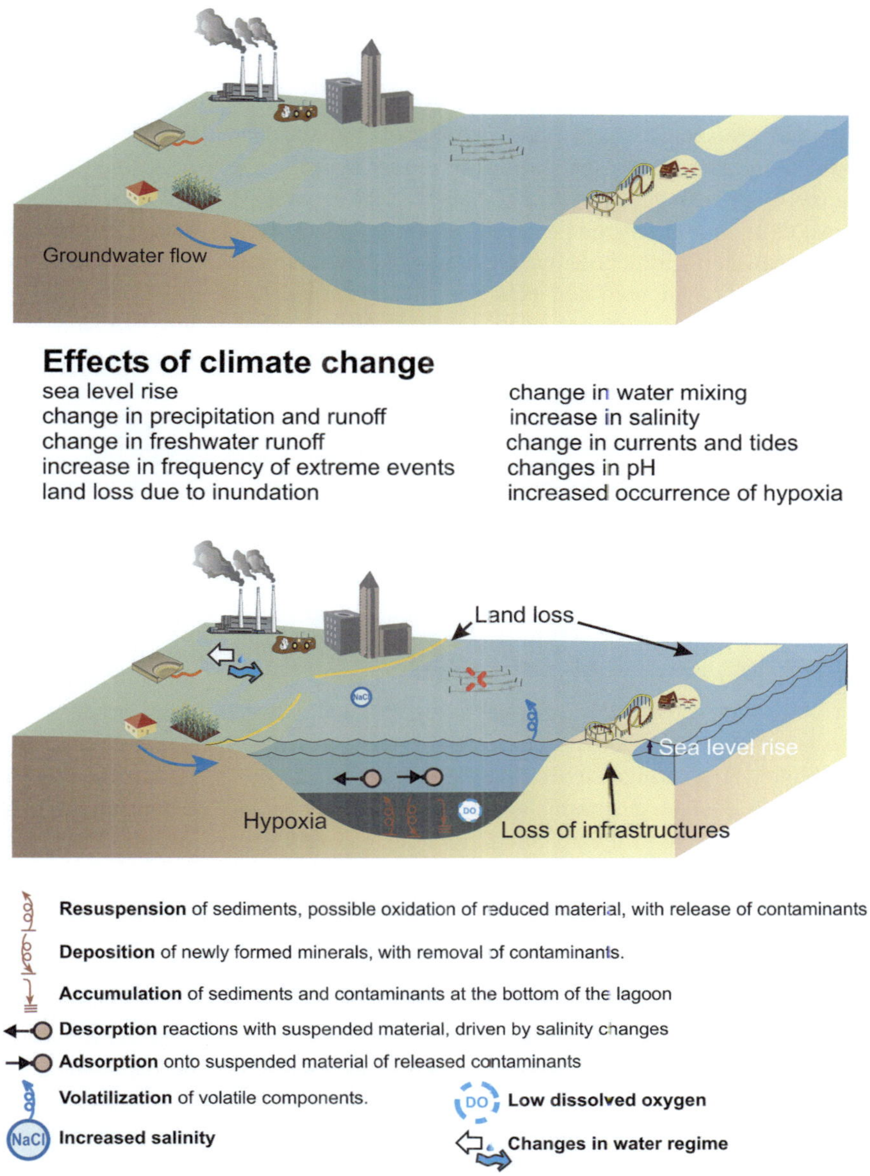

Fig. 23.1 Conceptual model of the effect that sea level rise induced to climate change could have on a coastal lagoon environment. Some of the effects could be present or not according to site specific features, whereas actions taken to counteract or limit the inundation are not considered

increase the relative variation in coastal areas (Nicholls 2002; FitzGerald et al. 2008) that only partially and locally could be matched by increase in sediment supply (Day et al. 2011). According to the estimated change in sea level considered in the adopted models, i.e. 0.18–0.38 m for an approximately 2 °C world, and 0.26–0.59 m for an approximately 4 °C world (IPCC 2007a, b) there could be a loss of coastal wetlands in the order of 5–20 % only due to eustatic processes (Nicholls 2004). In many natural conditions coastal lagoons can react to sea level rise by migrating landward, flooding dryland areas, but if such evolution is not possible for natural geomorphologic condition or for the presence of protection structures, the persistence of this environment is at risk.

Among the foreseen scenarios for middle latitude settings due to global warming there are changes in the hydrological cycle and a decrease in precipitation in southern Europe (IPCC 2007a, b). Extreme events, such as storms, are also predicted by IPCC (2007a, b) and they will increase the flooding risk (e.g., Nicholls 2002, 2007; Zanchettin et al. 2007) of the lagoon systems and their surroundings. Among side effects, sediment resuspension and flooding of contaminated land could be enhanced by these modifications in the hydrological regime, contributing to a rapid degradation of the ecosystem.

Another issue that is emphasized for the marine ecosystem is seawater acidification which is an effect of increased CO_2 dissolution in seawater, that has already produced a decrease in 0.1 pH units compared to pre-industrial values (Gattuso and Hansson 2011; Orr 2011). However in coastal zones the trends might be not so clear for the overlapping of other processes interfering with the carbonate equilibrium, basically related to biogeochemical cycles active in coastal areas. Nutrient fluxes are a very important controlling factor, since a decrease can modify the metabolic balance from net autotrophy to net heterotrophy, leading to a pH decrease (Gazeau et al. 2011) while an opposite trend results during eutrophication episodes. However microbial degradation occurring at the bottom and within the sediment can lead to acidification of bottom waters. In addition coastal waters may be more sensitive to riverine input, producing local effects different from global trends.

Salinity is a further environmental parameter affected by climate change in coastal lagoons, particularly in the light of the change in precipitation and temperature for the Mediterranean area (IPCC 2007a, b). Representing transitional environments, coastal lagoons are by definition zones with marked salinity gradients so that the possible scenarios could be of extreme concentration during dry periods, often leading to precipitation of evaporative salts.

If the above mentioned effects are a general forcing for the coastal ecosystem, we must consider that coastal lagoons represent a sink for many kind of contaminants transported from rivers, from atmospheric deposition or released by activities within the area. Either organic and inorganic contaminants accumulate in sediments, which are subject to recycling and can be sometimes concentrated also in organisms which could be of concern for humans feeding on these resources (Schiedek et al. 2007). Sediments are generally fine grained, except for zones where currents are more intense, enriched in organic matter, anoxic at shallow depth which means slow decomposition processes of organic material and, for inorganic contaminants, stability of forms associated to sulfides and organic matter. One problem that occurs even nowadays, is associated to sediment resuspension; this is induced by storm and wave action that can produce pulsating inputs of contaminated sediments in the water column leading to possible release of contaminants from the solid particles, mostly related to changes in the redox state. Contaminants can be transformed in new forms, sometimes more bioavailable, that extends the potential targets to a wide number of species (Roberts 2012, and references therein). Although not widely investigated, it is worth noting that the presence of further components, such as suspended particulates or colloids (mostly iron and manganese hydroxides) may limit the time that mobile forms spend in the environment, thus creating a pulsating flux of contaminants for short periods.

Flooding can inundate land nearby to coastal lagoons, which can be contaminated thus causing remobilization of contaminants, through resuspension, dissolution or exchange reactions. Thus, contaminants introduced into the marine environments are new or in greater quantities compared to previous situations. These will include for examples new types of pesticides, due to modifications in farming practices or to change in crop cultivation related to variation in climatic conditions. Moreover, changes in the precipitation amount and regime could modify the timing of contribution of contaminants to the coastal environment leading to different effects on biota (Schiedek et al. 2007). Critical for pesticides are for example the variations in temperature regime that influence their solubility in water, making them more volatile, increase their degradation through microbial activity in the sediments and increase their hydrolysis in water, making them either more or less bioreactive.

Changes in salinity may have various impacts: some studies have shown an increasing metal uptake by diverse aquatic organisms at reduced salinities (Hall and Anderson 1995; Wright 1995; Lee et al. 1998). Salinity controls also the solubility of many PAHs (Ramachandran et al. 2006). Thus, alteration of abiotic parameters will also regulate the extent of exposure of aquatic organisms to toxic substances. Hypoxia or anoxia episodes might represent environmental issues that could be a side effect of increased nutrient fluxes to the coastal environment for changes in precipitation regime increased development of water column stratification leading to eutrophication episodes. These are expected to affect the environmental quality of the ecosystem inducing changes in the benthic community composition (Schiedek et al. 2007).

Climate Change and Animal Physiology

Adaptive responses to large-scale perturbations, such as climate change, affect all biological levels. They initially take place at the cellular and individual levels, but are subsequently integrated and translated to upper levels. Most aquatic organisms are ectotherms, which makes temperature an important environmental factor controlling physiological processes. Warming is expected to differentially affect the warm and cold-temperate fish and shellfish that utilize a coastal area. Species with more southerly distributions and temperature tolerances will likely benefit, but investment in growth and reproduction may be reduced by some cold temperate species near the upper limits of thermal tolerances. Fish embryos are generally less able to compensate for changes in temperature and tend to have narrower temperature tolerances than other life stages; of course thermal tolerances of embryos and early larval stages are particularly important in determining shifts in species distributions (Rombough and Moroz 1997). Higher water temperatures during winter, in particular, may have positive effects on some species; on the opposite, higher temperatures may decrease the areal extent of the habitats for cold-temperate species both directly and in combination with low dissolved oxygen. Increasing temperatures may also facilitate the successful northward expansion of non-native species (Stachowicz et al. 2002) as well as of pathogens (Cook et al. 1998). Warmer winters appear to have already increased infections in oyster populations in Atlantic Coast estuaries (i.e. Cook et al. 1998). In addition, warming often enhances pollutant impacts. As temperature influences all kinds of physiological processes, an interaction between temperature and toxicants is easily predictable when they act on these processes. Among numerous examples, higher temperature-induced-mercury methylation (Booth and Zeller 2005) was suggested as a possible mechanism to increase mercury uptake in fish (Bambrick and Kjellström 2004).

The toxicity of common pollutants (i.e. heavy metals) to fish generally increases at higher temperatures (Carere et al. 2011), due to a greater production of bioactivated free radicals, that are more toxic than the parent compound, or to the increased uptake of the original toxicant. It was proposed that thermal stress acts on neural membranes, often accompanied by tissue hypoxia (Rosas and Ramirez 1993). Relevant studies on marine vertebrate and invertebrate species proposed the concept of oxygen and capacity-limited thermal tolerance (Pörtner and Knust 2007) to explain the mechanism that regulates both thermotolerance windows and thermal optimal values of animals. According to this view, detrimental temperatures bring about insufficient oxygen supply and transport to tissues, which coupled with high baseline oxygen demand at elevated temperatures likely reduce the organism performance (Angilletta 2009). As exposure to toxicants may increase the metabolic oxygen demand, higher temperatures will therefore potentiate the toxic effect.

Temperature may also influence toxicokinetics of compounds by affecting metabolic rate or feeding activity of aquatic vertebrates and invertebrates as well as toxin uptake, elimination, and detoxication rates (Donker et al. 1998; Smit and Van Gestel 1997). Other physiological features of animals are modified, for example, by the induction of proteins protective against heat or its consequences, such as oxidative stress. Indeed, animals will develop a stress response, and shift energy from growth and reproduction to stress protein synthesis (i.e. heat shock proteins; Piano et al. 2002; Fabbri et al. 2008) or enzymes activities (i.e. antioxidant defences). These and other proteins such as metallothioneins, which act as antioxidant scavengers and at the same time as important metal chelators influence animal sensitivity to toxicants (Piano et al. 2004). The heat shock protein response is among the most investigated physiological pathways in response to global changes (Somero 2012), as well discussed also in this book.

Temperature may alter physical factors such as partition coefficients or diffusion rates, in that way affecting toxicokinetics. Interestingly, modifications in the partitioning of toxicants over different body parts may result in their higher accumulation in the target tissue, leading to lower effect concentrations (Van Wezel and Jonker 1998). On the other hand, increased diffusion rates may favour higher toxicant uptake rates. Physiological processes altered by temperature may influence toxicokinetics as well. In ectotherms, an increase of ambient temperature leads to a higher metabolic rate. High food or nutrient levels may also raise animal metabolism, generally inducing higher toxicant uptake rates. Elimination and detoxication mechanisms may counteract the thermal effect on uptake rate, as the rates of these processes tend to increase as well. However, increases in toxicity when temperature raises are widely reported, while this is not true for elimination and detoxication rates.

On these basis, we expect increased toxicity at increasing temperatures when both stressors act on metabolic processes, such as energy metabolism or respiration. Further, we expect increased toxicant concentrations in tissues at high temperatures, due to elevated uptake rates. Rosas and Ramirez (1993) found a decreased thermal tolerance in prawn (*Macrobrachium rosenbergii*) exposed to cadmium and chromium. Jacobson et al. (1997) found increased copper sensitivity of the mussel *Actinonaias pectorosa* at elevated temperatures. Persoone et al. (1989) observed a similar response to increased temperature in *Daphnia magna* exposed to potassium dichromate. Some hypothesis were made by these Authors to justify the observed decreases in thermal tolerance in animals exposed to pollution. For example, neural membranes exposed to temperature driven changes and concomitant decrease in oxygen concentration would be more vulnerable to stress. It has also to be considered that the partitioning over storage lipid and water is modified by temperature, thus at lower temperatures more lipophilic contaminants have to be accumulated in the storage lipids before the critical concentration in the membrane lipids is reached, leading to cell toxicity. The solubility of compounds into the water is temperature dependent, therefore changes in water temperature influence the bioavailability of toxicants. It is also expected that rising temperatures increase the toxic effects of chemicals acting on metabolism, for example, on oxygen uptake and distribution. Higher toxicant uptake rates will follow, with an increased internal toxicant concentrations at elevated temperatures.

Salinity fluctuations are intrinsic features of estuaries and coastal lagoons, although possible also in freshwater habitats. Increasing frequency and magnitude of flood events at coastal sites due to climate change could lead for example to longer and more frequent periods of reduced salinity. It is well known that physiological processes at the basis of osmoregulation are different in the two environments. The internal medium of freshwater organisms (both invertebrates as well as vertebrates) is hypertonic relative to their environment. Thus, water tends to flow into the body due to osmotic pressure and animals passively loose ions. The regulation of the ionic balance mainly occurs by active uptake of Na^+ and Cl^- ions from the water, and much energy is addressed to this function. In contrast, the blood of marine vertebrates and some invertebrate classes is hypotonic with respect to the water. Therefore, the osmotic pressure causes passive inflow of ions and loss of water. To compensate for these processes, fish drink seawater and the excess of ions are excreted by the chloride cells through a complex array of mechanisms and energy costs. The ability to tolerate a range of salinities is a key adaptation for estuarine and brackish water fish. Nevertheless, adaptation is costly thus a certain amount of energy must be allocated in this direction, and there are considerable individual variation in the tolerance limits within many species, affected in part by size but also by the variability in a number of interacting factors, including oxygen uptake, acid-base balance, thermal tolerance, feeding, etc.

Recent studies on mussels (Hamer et al. 2008) showed that mussel acclimation to reduced salinities resulted in decline calculated on the basis of *M. galloprovincialis* tissue weight and shell sizes, although the condition index was similar. Oxygen consumption rate of mussels increased considerably to about 51 and 65 % in lower saline concentrations (28 and 18 psu) compared to control mussels (37 psu). DNA integrity was negatively impacted by reduced salinity acclimation, which also resulted in a significantly higher p38MAPK phosphorylation, indication of stress and potential alteration of homeostatic responses. A period of reduced salinity comparable to a moderate flood event negatively affected the immune function of *M. edulis* and caused significant changes in the metabolic fingerprint of mussel blood, i.e. the set of small-molecule metabolites such as metabolic intermediates, hormones and other signalling molecules, and secondary metabolites (Bussell et al. 2008). Circulating haemocyte concentration, percentage of eosinophilic haemocytes and level of phagocytosis were reduced in seawater when salinity was maintained for 2 days at the 50 % of that of seawater. Changes in immunological variables in response to reduced salinity could be a consequence of many processes including reduced movement of the haemocytes due to osmotic effects or enhanced infiltration of haemocytes into the tissues (Cajaraville et al. 1996). Alternatively, at reduced salinity, mussels close their valves. This behavior, coupled with the energetic cost of adaptation, may lead to an increase in autophagy (Sadok et al. 1997). Such evidence indicate that changes due to decreased salinity are expected to increase the susceptibility of mussels to other stresses.

The interaction between pollution and salinity is rather complex, because salinity may act both on physiological processes as well as on the chemical itself. As a major issue, salinity influences the chemical speciation of metals. A rise in salinity increases the degree of metal complexation, decreasing bioavailability and decreasing toxicity because the free ionic form of metals is more toxic (Bury et al. 1999; Karen et al. 1999). In addition, the increase in ionic strength at higher salt concentrations decreases the activity of toxic compounds. Metal toxicity was often found to increase at low salinities; increased bioavailability is correlated to such effect, as competition of the metals with calcium and magnesium at uptake sites decreases with lower salinity. It is also true that the increase in toxicity at lower salinity is not similar for all species, depending on biochemical interactions, different modes of toxic action or pathways of toxicant entry (Bury et al. 1999; Karen et al. 1999). For example, the above cited differences in osmoregulation in fresh and saltwater organisms obviously determine species-specific accumulation and excretion of toxicants. It is generally thought that increased physiological costs for osmoregulation, due to low salinity, decrease the overall fitness of organisms that become less capable to endure toxic stress. From this statement, it might be expected that the lowest toxicity is at the isosmotic point of organisms, where the exchange of water and ions with the ambient water is the least and energy can be eventually driven to maintain physiological homeostasis.

However, higher salinity water is not free of problems. In freshwater, metals mainly exist in the dissociated and most available form, while at higher salinity metals are bound to particles. The assumption of particulate matters, including algae, together with bound toxicants represent indeed a powerful vehicle of contaminant intake through the diet. The amount of compounds accumulated by living animals is also determined by the residence time of the chemical.

Salinity may influence this process, as for example the degradation half-life of an organophosphate insecticide is higher in saline- relative to fresh- waters. Interestingly, also toxicity of organophosphate insecticides increased at rising salinity, while no clear relationship was observed for other pesticides.

Variations in environmental factors may also change toxic effect concentrations by altering the sensitivity of target organisms, as reported for temperature already several years ago (Fisher and Wadleigh 1985). On the other hand, animals are able to regulate the internal toxicant concentrations, either by increased elimination, detoxication, or storage in specific body parts. Because these regulatory mechanisms cost energy, less energy becomes available for other processes since assimilated energy is first allocated to maintenance, and a second step addressed to growth or reproduction (Smit and Van Gestel 1997). In line with such an energy demand, Geyer et al. (1993) observed less negative effects in animals with high fat content relative to animals with low fat content. This seems to indicate that well-fed animals exposed to toxicants can pay the energy costs for coping with toxic effects more easily than those held at scarce food conditions. We may also consider that the interaction between environmental factors and toxicants is not constant in natural systems, as environmental conditions, toxicant concentrations and the physiological state and metabolic needs of organisms may change in time. Nevertheless, various data indicate that organisms living near their environmental tolerance limits are more susceptible to temperature increases (Sorte et al. 2011) and additional stressors including toxicants; at this regards, concentrations that are sublethal to organisms living at optimal environmental conditions may become lethal at suboptimal conditions.

Dissolved Oxygen

Low dissolved oxygen affects growth, mortality, distributions, and food web interactions of a wide range of organisms. Seasonal hypoxia leading to benthic animals mortality typically occurs in shallow waters with hypoxic or anoxic episodes in bottom water shoreward (Tyler and Targett 2007). In addition to directly increasing mortality, hypoxia may have strong effects on the ecosystem and alter behavioral and physiological responses of animals that in turn modify their trophic interactions (Breitburg et al. 2001). For example, increases in summer temperatures and increased anoxia or hypoxia may exclude fish species from benthic feeding grounds and bioenergetically favorable cool deep-water environments (Coutant 1985; Secor and Gunderson 1998). In a simulation on the combined effects of hypoxia and temperature, Niklitschek and Secor (2005) observed that even a small temperature increase (1 °C) during the summer could eliminate suitable habitats for juvenile sturgeons. Warming will increase the extent and severity of hypoxia effects on aquatic vertebrates by affecting dissolved O_2 concentrations, but also by increasing the organism oxygen requirements (e.g., Shimps et al. 2005).

It is a matter of fact that aquatic habitats affected by hypoxia or even anoxia have recently expanded. Climate change contributed the problem due to increased precipitation and temperature bringing nutrient-rich, fresh and relatively warm water to sensitive areas (Harley et al. 2006). Hypoxia is often caused by increased loads of organic matter and various organism groups respond differently to increased total organic carbons and sulfide. For benthic groups that are favoured by high TOC and less sensitive to hypoxia and sulfide (e.g., annelids and echinoderms), the effects of contaminants will be reduced while for arthropods, which are more sensitive to total organic carbons and hypoxia, the effects of contaminants will be increased (Lenihan et al. 2003). Furthermore, reduced feeding rates during hypoxic events are documented (Ripley and Foran 2007) and starved organisms are more sensitive to contaminants. Thus, changes in benthic community composition due to hypoxia and concomitant pollution are expected.

Changes in pH

Like the open ocean, coastal lagoons are subjected to anthropogenic acidification as a result of increasing CO_2 concentrations in the atmosphere (Blackford 2010). Coastal lagoons are also subjected to other human-induced perturbations, in particular eutrophication, as well as to strong biogeochemical and hydrological activities. Consequently, the few time-series data available reveal changes in the carbonate chemistry more complex than in the open ocean (Gazeau et al. 2011). The rate of net calcification of mussels (*Mytilus edulis*) and oysters (*Crassostrea gigas*) decreased linearly with increasing CO_2 in short-term manipulations (Gazeau et al. 2007). That calcifying species are more at risk from the effects of ocean acidification than noncalcifying ones is intuitively attractive, however it is not fully proved. For example, longer-term experiments on mussels and oysters showed no effect or even increased calcification rates (Ries et al. 2009) at slightly lower pH. Changes in calcification rates, either positive or negative, have been measured in sea urchins (Ries et al. 2009) and brittlestars (Wood et al. 2008).

Calcification is not the only process impacted by acidification in bivalves, which in fact increases metabolic rates and lowers immune responses (Andersson et al. 2009). Oysters growth declined linearly with decreasing pH levels, while no pH effects were observed for mussels. Further described potential alterations concern acid base balance (sea urchins: Miles et al. 2007; sea stars: Hernroth et al. 2011), reduction of immune response (Hernroth et al. 2011), and increase in respiration rate (Wood et al. 2008).

A down regulation of genes involved in calcification, cellular stress response, metabolism and apoptosis were reported in *Strongylocentrotus* larvae raised in low pH seawater (Todgham and Hofmann 2009; O'Donnell et al. 2010). Larvae from intertidal *P. lividus* seem to be more resistant to acidification than those of species collected from subtidal sites, suggesting that sea urchins living in the stressful intertidal zone may be adapted or acclimatized to pH stress. In their study Clark et al. (2009) observed that the larvae of the Antarctic sea urchin *Sterechinus neumayeri* were the least affected by low pH compared to tropical and temperate sea urchin species; *S. neumayeri* would be adapted to higher CO_2 conditions in the Polar waters and, therefore, may have greater capacity to acclimatize to lowered seawater pH.

Decreased swimming speed was observed in sperm of urchins (Havenhand et al. 2008) and reduced sperm flagellar motility in holothurians (Morita et al. 2010) exposed to acidification, indicating sensitivity of the gametes to elevated CO_2. A decrease in pH of seawater lowers the internal pH of sperm, leading to a decrease in sperm motility (Christen et al. 1986). The regulation of internal pH of sperm is controlled by a Na^+/K^+-ATPase pump and Na^+/H^+ antiporters located on the plasma membrane of spermatozoa (Gatti and Christen 1985). In spermatozoa already subjected to pH stress in the gonads, the transmembrane proteins may be more effective or expressed in a greater number, due to either acclimatization or natural selection. Since the fertilization rate of sea urchins is affected by sperm motility (Havenhand et al. 2008), the fertilization rate of acclimatized/selected sperm is expected to be higher. The mechanisms at the basis of this phenomenon are not fully understood, and the potential ability of some species to cope with seawater acidification deserves further investigation. On the contrary, Havenhand and Schlegel (2009) did not found pH effects on sperm motility in *Crassostrea gigas*. In other studies, clams were exposed to sediments collected in three locations in the Gulf of Cadiz and to contaminated particles derived from an accidental mining spill in Spain (Riba Lopez et al. 2010) The concentration of metals, with the exception of Cd, within clams significantly increased when sediment pH was lowered by one or two units. Indeed, acidification increases metal availability. As a matter of fact, the behavior of heavy metals is complex and is determined by several parameters (e.g., pH, concentration of organic matter, redox potentials).

Bivalves play an important role on the functioning of coastal marine ecosystems since they link primary productivity with upper trophic levels. In many coastal areas bivalves are also important economic resources. In recent studies, involving the direct manipulation of the concentration of CO_2 in seawater, significant effects of acidification were reported for different species and life-stages of bivalves: reduced calcification (Gazeau et al. 2007; Kurihara et al. 2009; Miller et al. 2009); reduced shell growth (Berge et al. 2006; Michaelidis et al. 2005); and increased mortality (Talmage and Gobler 2009). More recent studies however (Range et al. 2011) performed on clams reared in water from the Rio Formosa lagoon (Portugal) showed no effect of pH in the range from 8.2 to 7.5 on mortality, growth and calcification. In the light of these and further observations, it is becoming clear that also the biological responses to water acidification will be species-specific and therefore much more variable and complex than previously hypothesized (Miller et al. 2009). The responses may be variable at local scales, which emphasizes the danger of extrapolating results from a few model species or from one region to another. In particular, local response patterns will be more relevant in coastal lagoons and estuarine waters. Organisms inhabiting these areas are naturally exposed to greater variations in terms of seawater carbonate chemistry which is likely to affect their sensitivity to future acidification phenomena (Gazeau et al. 2010).

Laboratory investigations widely contributed to the understanding of animal responses to environmental changes. They also fed mathematical models to explore ecosystems under different climate change scenarios and provided information useful to plan subsequent field experiments. However they usually represent an incomplete approach to draw conclusions about the biological impacts of climate change since usually based on single environmental variables, while interactions between organisms and the surrounding changing environment is much more complex. Additive, antagonistic or synergistic effects of contaminant mixtures and physical changes on aquatic organisms are indeed expected. Meta-analysis from a relevant number of studies reviewed by Crain et al. (2008) suggested overall cumulative effects of multiple stressors, that are likely to be worse than expected based on the independent impacts of each stressor. However, in some cases, stressors either ameliorated one another, or one stressor had such a large effect the addition of a second stressor had no additional impact.

Besides the question of how animal species living in coastal lagoons will be affected by climate change, a further answer deserves responses, that is how the quality of these environments may be efficiently monitored in a climate change scenario. Indeed, a future challenge will be to anticipate the risks and develop ways to mitigate the combined effects of climate change and pollution.

Coastal lagoons are highly vulnerable to urban pressure, tourism, agriculture and coastal dynamics (Rodríguez-Rodríguez 2007; Durán et al. 2009). Coastal lagoons are also highly visible suppliers of major ecosystem services not only in their surroundings, but also at great distances (i.e., food provision). These areas are also used for extensive and intensive aquaculture activities, mainly shellfish farms, and are designated as sites of biodiversity relevance. Together with estuaries, coastal lagoons provide more ecosystem services and societal benefits than other aquatic systems. Since climate change impacts must be taken into account when implementing the Environmental Impact Assessment (EIA), Strategic Environmental Assessment (SEA) Directives, and spatial planning policies (Chapman 2012) to adequately protect these vulnerable environments, it is essential to monitor the changes produced both in the short- and long-term by climate change driven factors. Thus, information need to be highlighted also to conservation bodies and government agencies, and some updated tools for the assessment of organism health status under a climate change scenario will be suggested. Testable predictions carried out using mechanistic approaches (Helmuth et al. 2005) may provide useful information to decision makers, and potentially elucidate important principles by which climate changes affect organisms and ecosystems. They may also play an important role in communication with end-users (either inhabitants or stakeholders getting profits form coastal lagoons). To a large extent, chemico-physical data are collected and archived with little or no thought given to their potential biological applications (Helmuth et al. 2010). Similarly, many physiological studies are carried out with only a limited ecological context, and viceversa (Denny and Helmuth 2009). Examples of cross-disciplinary research when exploring the ecological impacts of climate change are advisable, but rare (Denny and Helmuth 2009; Hofmann et al. 2010).

A suite of indicators needed to address climate change issues and determine possible causes of ecosystem changes comprises biological indicators related to organisms (behavioural aspects, state of the body, etc.), populations (recruitment, mortality, numbers and geographical distribution) and community structure (biodiversity, structure and functioning) (Philippart et al. 2011). These mainly refer to open ocean, although they could be adapted to coastal environments; however, some more rapidly responsive indicators should be applied to follow in an acceptable rate the alterations induced by climate change in combination with additional chemical or physical stressors. The quality assessment and monitoring of coastal lagoons has always been difficult because of their intrinsic high variability in term of chemical and physical characteristics, which limits the powerness of the analytical chemistry and ecotoxicological approaches. Since chemicals significantly affect physiological processes in animals, alternative tools for biomonitoring were recently recommended, using sentinel organisms and biomarkers (Viarengo et al. 2007). These are in line with the importance of considering physiological mechanisms when forecasting the effects of global change on organisms and ecosystems, as recently emphasized (e.g. Pörtner and Farrell 2008). Biomarkers are biochemical, physiological or histological indices of either exposure to, or effect of, pollutants (Viarengo et al. 2007). They provide qualitative and semi-quantitative information of the nature of the insult, and gives the unique contribution to identify the deleterious effect of combination of stressors. In our opinion, such approach also represents an accurate accounting for cumulative impacts of environmental stress and their interactions at the biological level, useful to collect early information on alterations in progress. Fast evaluation can enable the activation of measures for pollutant reduction before the damage has reached the entire community. Many international programs were carried out in order to standardize the measurement of biomarkers and correlate molecular and cellular responses to those of individuals and populations (Viarengo et al. 2007). Expert systems able to integrate biomarker data are now available to translate biological data into realistic and objective evaluation of the changes in organism physiology induced by pollutants (Dagnino et al. 2007).

The stress response of aquatic species is being examined also at a finer scale, i.e. at the genomic or proteomic level (Somero 2010). Some caveats must be taken into accounts when the -omic methodologies are employed to this purpose, including the circadian gene expression patterns that reflect, for example, biosynthetic activities; in case of intertidal species, also gene and protein expression periodicities driven by tidal cycles must be considered. Nevertheless, -omic methodologies are providing extremely useful suggestions towards the development of new and more helpful biomarkers.

The Pialassa Baiona: An Interesting Case Study

We had the chance to perform a combined geochemical-biological investigations within the Pialassa Baiona, which proved to be an excellent study system for evaluating climate change effects combined with pollution. The Pialassa Baiona (Fig. 23.2) is a brackish lagoon consisting of small, shallow ponds and deeper artificial channels connected to the North Adriatic Sea through the harbour of Ravenna (Italy). The area also receives freshwater from inland through some drainage channels, which cross cultivated lands. Further freshwater inputs represented by wastewaters coming from the industrial area are located in the southernmost part of the lagoon. Continuous deposition and remobilisation of fine-grained sediments are promoted by a weak water dynamic. Fine-grained sediments represent a very important component to study because they represent the fraction preferentially associated with pollutants and the portion of sediment filtered by mussels (Donnini et al. 2007).

Until the '70s the lagoon was impacted by poorly treated industrial wastewaters, produced mainly by chemical plants (Fabbri et al. 2006). Several chemical and geochemical monitoring surveys indicated that the lagoon is affected by a

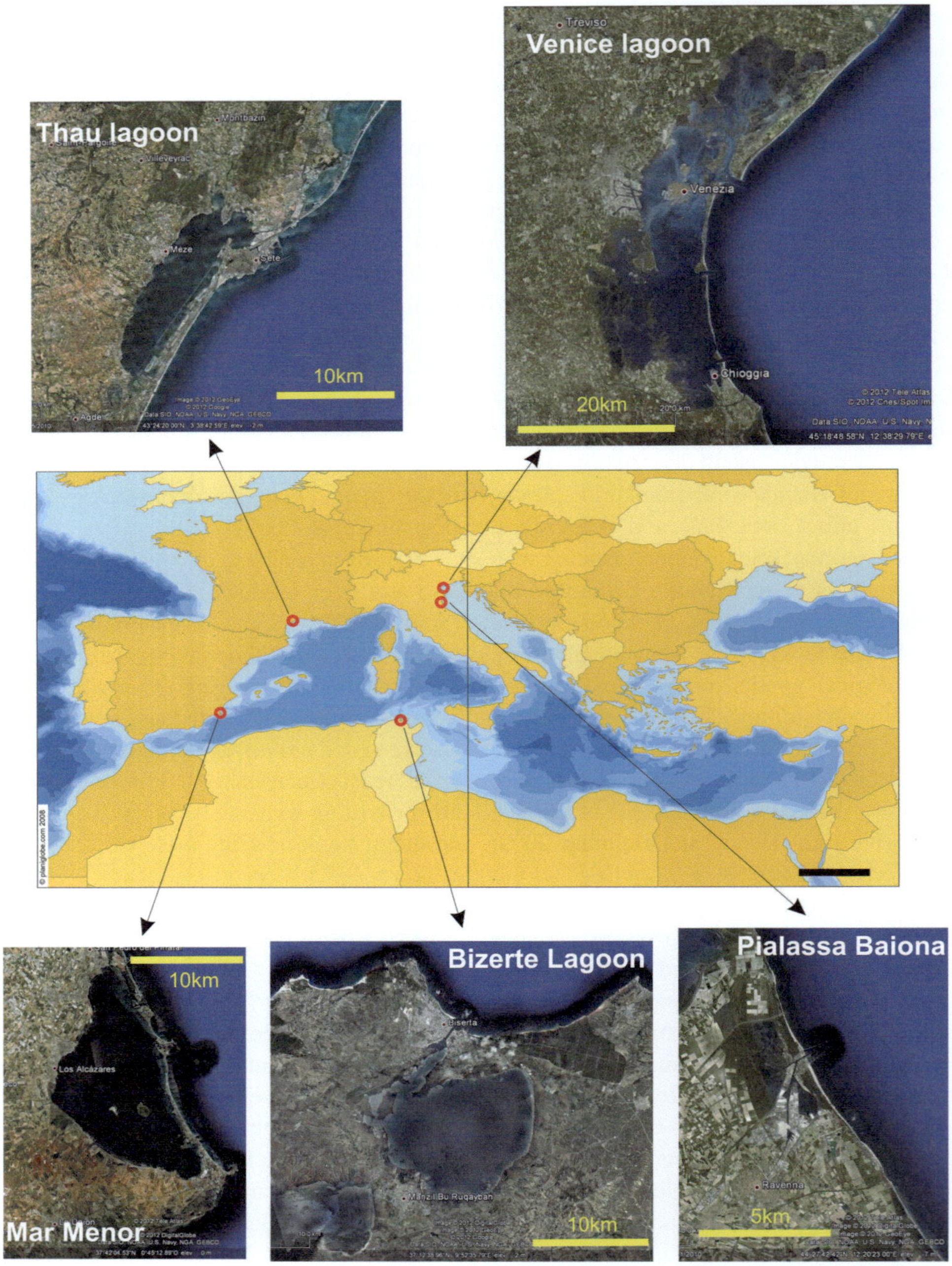

Fig. 23.2 The Mediterranean basin with indication of the location of the five coastal lagoons described in the text and detail of their features (Images from Google Earth)

widespread contamination of the sedimentary compartment due to heavy metals (e.g., Cd, Cr, Cu, Pb, Zn) occurring in surface and suspended sediments (i.e. Fabbri et al. 1998). Moreover, a north–south increasing gradient was observed for Hg and several organic pollutants (Trombini et al. 2003; Fabbri et al. 2006). The highest concentrations of contaminants were detected in the Magni channel, at the southernmost part of the lagoon. Moreover, during the month of September 2006 temperature fluctuation wider (10–24 °C) than the lagoon average (16–18 °C) were recorded in this channel, which was also subjected to variations of about 10 °C within the 24 h (Franzellitti et al. 2010).

Cause of these variations was mainly the cooling water discharge from two power plants located at the two channel extremes. Indeed, temperature stress might occur due to climatic events or discharges of heated waste water by power plants; although this latter is limited by law, episodes are still identified (Franzellitti et al. 2010). The high average water temperature in Magni channel is a constant feature confirmed by following recordings (Dinelli et al., in preparation) with summer temperatures of about 31 °C against an average of 25 °C throughout the lagoon. Moreover, pH values of about 7.5 against a lagoon average of 8.0 were registered during high tide; a significantly lower salinity was also measured in this channel, about 7 psu lower than the lagoon average (Dinelli et al. in preparation). Global change may entail increases in temperature and precipitation (thus hyposmotic stress), as well as decreases in pH in estuarine-like systems, therefore mussels exposed to the Magni channel are subjected to the typical climate-driven stressors and simultaneously to high levels of pollutants. In this view, mussels transplanted to different channels within the lagoon served as the reference organisms, as exposed "only" to different pollution levels.

Translocation procedures using filter-feeder organisms were carried out. This practice is suggested by the Water Directive Framework (EC 2010) as it may assess the potential for bioaccumulation of a water body being therefore recommended to improve evaluation of the potential risk of bioaccumulation of a specific substance. About ten biomarkers at the cellular and tissue levels were measured in mussels, *Mytilus galloprovincialis*, during the different campaigns following the procedures proposed by the UNEP/RAMOGE Manual (1999) and by more recent works (Viarengo et al. 2007). These included lysosome membrane stability, neutral lipid and lipofuscin accumulation, sensitive to general stress factors; malondhyaldeide accumulation and antiodixant enzyme activities, sensitive to oxidative stress factors; metallothionein levels, sensitive to heavy metals bioaccumulation; acethylcholinesterase activity, sensitive to organophosphate and carbamate pesticides exposure; DNA fragmentation, as a consequence of genotoxic compounds. Expressions of proteins and genes known to be target for environmental stress were also evaluated as very early signals of health risk (Franzellitti et al. 2010).

Physiological changes measured in mussels transplanted into the Magni channel for 2, 4, 7, 14 and 30 days showed that cellular biomarkers, including lysosome functionality, were severely altered already after 7 days of exposure and further worsen over time. Transcription of genes coding cytoprotective proteins against stress stimuli, including metallothionein and heat shock protein expressions, was dramatically raised after only 2 days of exposure and increased thereafter (Franzellitti et al. 2010). During the 2003 survey, temperatures raised well above the typical summer levels and mussels transplanted into the Magni channel were not able to maintain their homeostasis and died within 7–10 days (Donnini et al. 2007). In the rest of the lagoon all mussels survived, although they experienced similar temperature raises. The study within the Magni channel underlined that the combination of multiple stressors has a strong impact on wildlife; moreover, animals already exposed to a high pollution level are less capable to survive to a temperature stress. In a further survey during which temperatures were no higher than 26 °C (Franzellitti et al. 2010), mussels transplanted in the Magni channel survived the 30 day-exposure. These animals showed the increase of several cytoprotective responses at the molecular levels soon after 2 days of exposure. Moreover they showed biomarker alterations after 1 week. Alteration of temperature, as the main climate change parameter, acts therefore as a co-stressors with chemical pollutants affecting the ability of wildlife to maintain homeostasis through the development of cytoprotective responses. As previously reported, species living at the edge of their physiological tolerance become less able to cope with the dual stressors of climate change and contaminant exposure. This observation is well in agreement with several indications (Somero 2012).

Mussels represent sentinel organisms useful for biomonitoring of coastal lagoons and, in general, coastal environments exposed to climate change. When possible, it is necessary to establish more intensive monitoring of vulnerable water bodies where the levels of bioaccumulative compounds are relevant in the biota species detected. The detection of dangerous bioaccumulative compounds should preferably be performed on biota and sediments because the concentrations in the water column, especially in marine-coastal waters, are not representative of the contamination; to this aim, the appropriate selection of the biota species will be crucial to evaluate the bioaccumulation potential.

According to our experience, oysters and fish may be utilized as well, while clams seem to be less useful since not enough sensitive (*Tapes phylippinarum*, Caselli and Fabbri 2005) or viceversa too sensitive (*Chamelea gallina*, Caselli and Fabbri 2005) to pollution as well as to temperature changes, dissolved oxygen, etc. However, clams were fruitfully used by other groups, who analysed behavioural responses and/or biomarkers different from those we employed (Da Ros and Nesto 2005). Biomarkers are potent early warning signals addressing the environmental quality, therefore their use is encouraged in line with the conclusions reached by main EU-funded projects and the Marine Strategy (Directive 2008/56/EC). Our data indicate that they are able to point out progressive reductions also in a climate change scenario.

Mediterranean Coastal Lagoons, Similar But Rather Different Habitats

Despite the Mediterranean region has been identified as one of the most prominent hot spots in future climate change projections (Giorgi and Lionello 2008), the low number of lagoons studied at this regard contrasts with their natural and economic importance and, at the same time, their vulnerability and usefulness as sentinel ecosystems (Brito et al. 2012b).

Besides the Pialassa Baiona, other Mediterranean coastal lagoons are subjected to multiple stressors related to climate change including temperature, pH, and salinity and at the same time to inorganic and organic pollutant differently combined throughout the system and harmful to the living fauna. Unfortunately, reported studies are incomplete and not always useful for the present discussion.

For example, more investigations are advisable within the Bizerta lagoon (Fig. 23.2) to evaluate potential consequences of climate change. The Bizerta lagoon is situated in northern Tunisia and occupies an area of about 15 km^2 with a mean depth of 7 m. It represents an economically important water body that hosts a variety of fishing and aquaculture activities. As most Mediterranean lagoons, this environment is characterized by a great variability in its ambient conditions, mainly salinity, temperature, and dissolved oxygen levels, and is subjected to strong eutrophication phenomena during summer. This natural fragility is exacerbated by anthropogenic disturbances, including domestic, agricultural and industrial wastes discharges. Heavy metals (Zn, Cd, Ni, Pb and Cu), organo-chlorinated pesticides, halogenated aromatic compounds like polychlorobiphenyls (PCBs), organotins and polycyclic aromatic hydrocarbons (PAHs) were found in the Bizerta lagoon (Ben Ameur et al. 2012). These contaminants are likely the causes of oxidative stress, genotoxicity, and histopathological alterations highlighted through the analysis of biomarkers observed in molluscs and fish (Mahmoud et al. 2010; Tlili et al. 2010; Ben Ameur et al. 2012) and imposex phenomena in snails (Lahbib et al. 2011).

Also the Thau lagoon (Fig. 23.2) deserves more studies on potential effects of climate change, given its great importance for the economic market. The Thau lagoon is a microtidal coastal enclosed bay of about 75 km^2 surface area, with mean depth of 3.5 m and salinity between 31 and 39 psu, located to the south of France and connected to the Mediterranean sea. Twenty percent of its surface is occupied by shellfish farming, mainly oysters, *Crassostrea gigas*, and mussels, *Mytilus galloprovincialis*. At present the areas hosting farming activities experience strong depletion in dissolved oxygen, accumulation of organic matter and higher rates of sulphate reduction (Dedieu et al. 2007; Thouzeau et al. 2007; Point et al. 2007), intense early diagenetic reactions affect also the fluxes of some elements (e.g. Mn, Cd, Cu, Hg) that show different behaviour within the sediment and at the sediment water interface (Point et al. 2007; Muresan et al. 2007; Monperrus et al. 2007; Metzger et al. 2007).

Mar Menor (SE Spain) is one of the largest coastal lagoons along the coast of the Mediterranean sea (Fig. 23.2). It covers an area of about 135 km^2, reaches a maximum depth of 7 m and a mean depth of 4.5 m. The present-day precipitation regime makes it a sheltered hypersaline lagoon which is increasingly affected by several impacts represented by agriculture, tourism, fishery and effluents from former mining activities (Conesa and Jiménez-Cárceles 2007). Critical for the state of the lagoon was the expansion of agricultural activity along the shores leading to increase nutrient supply and changing the ecological regime from oligotrophic to eutrophic (Pérez-Ruzafa et al. 2002; Lloret et al. 2005). Effects of climate change were evaluated by De Pascalis et al. (2012) using numerical tools that suggested an increase in temperature and a decrease in salinity under the A2 IPCC scenario for the 2100. A temperature increase would affect also the nutrient cycling actually mediated by the nutrient uptake provided by the widespread primary producer *Caulerpa prolifera* Forsskal (Lamououx) whose photosynthetic capacity would be greatly reduced leading to increased eutrophication (Lloret et al. 2008).

The Venice Lagoon is the largest lagoon system in Italy (about 550 km^2), and one of the largest in the Mediterranean Sea (Fig. 23.2). It surrounds the city of Venice, and reaches Chioggia and Mestre comprising one of the most important industrial areas of Italy. Climate change is considered a major threat to the survival of Venice Lagoon mainly because of the projected increase in sea level to which adds local problems due to subsidence (Lambeck et al. 2011). The lagoon ecosystem is affected by several stressors that include land-based activities sources of nutrients, heavy metals, and other pollutants, aquaculture activities, fishing and tourism, groundwater extraction and subsidence that affect physical and morphological features (Solidoro et al. 2010; Pastres and Solidoro 2012). Nutrients derived from agricultural and industrial activities and from sewage discharges, as already mentioned, can promote eutrophication leading to hypoxia and mortality episodes. Sediments are polluted by heavy metals and various types of organic chemicals (Dalla Valle et al. 2005; Frignani et al. 2005) derived from industrial waste and other sources (i.e. drainage basins, urban areas, etc.) through direct discharge or atmospheric fallout (Guerzoni et al. 2007). Atmospheric temperature and sea level rise related to climate change are predicted to increase significantly over the next 50–100 years, leading to greater frequency of flooding. According to Gambolati et al. (2002) and Lambeck et al. (2011), extensive areas of the western part of the northern Adriatic Sea coastline could be permanently lost by 2100. In addition, changes in the pattern of rainfall over the Venice drainage basin (Salon et al. 2008), with drier spring and summer and wetter autumn and winter, will change the pattern of nutrient contribution to the coastal area and modifying primary and secondary production (Cossarini et al. 2008).

Given the complicated interactions between organisms and variable physical environments, the physiological mechanisms by which environmental factors drive organism behaviour, fitness and survival, and the indirect effects of climate change related impacts on species interactions, can we find common traits and impacts on animal life in these Mediterranean lagoons?

The main changes observed are in all cases related to a rise in water temperature. These may depend from sea temperature and/or from air temperature, according to the morphology of the specific lagoon as well described for Mar Menor (De Pascalis et al. 2012). We are aware that fluctuations in body temperature are not uncommon in intertidal environments (Finke et al. 2009; Marshall et al. 2010) and may be rather large. This means that organismal performance changes throughout the course of the day or the season in parallel with changes in body temperature, although mobile organisms can potentially ameliorate much of that variability through microhabitat selection. A time series of the mean monthly values of gonado-somatic index, coupled with thermal profiles of the Venice lagoon water temperatures over the last 14 years (Zucchetta et al. 2012) showed that the reproductive investment of the goby fish *Zosterisessor ophiocephalus* was positively affected by water temperature changes, and in particular the reproductive peak tended to occur earlier during warmer than during colder years. The analysis of 30 year simulations in the Venice lagoon underlined that changes in the seasonal dynamics of temperature, biogeochemical properties, and plankton productivity affect habitat suitability for clam growth and aquaculture, and highlight the need for management policies to mitigate the adverse effects of climate change on rearing activities (Canu et al. 2010).

Furthermore, temperature changes are added to concomitant impacts of climate change that are generally observed in these lagoons, such as a rise in sea-level, changes in water salinity and pH, modifications of the hydrodynamics of water masses, and increase in frequency of extreme events. Again, lagoon features influence the consequences of these factors; for example salinity is influenced by temperature and/or by increased exchanges (inward-outward) between the lagoon and the open sea (De Pascalis et al. 2012). Due to their restricted exchange with the adjacent ocean, at a smaller or greater extent, all Mediterranean lagoons are vulnerable to eutrophication. Indeed, an increased nutrient input into these areas is expected due to the increasing population in coastal zones; furthermore, the use of fertilizers for agriculture in the surrounding areas contribute to this effect. Eutrophication processes affect benthic primary producers inhabiting coastal zones through increases in water column light attenuation caused by increased phytoplankton concentration or even more severe phenomena, such as oxygen depletion at the bottom. A further common impact originate from changes in bioavailability of pollutants. Regions subject to decreases in precipitation may experience enhanced volatilization of POPs and pesticides to the atmosphere. Differently, increases in the intensity and frequency of storm events linked to climate change could lead to more severe episodes of chemical contamination of water bodies and surrounding watersheds.

Conclusions

This review underlined that physiological studies have the potential to support researchers in understanding the basic mechanisms at the basis of ecosystem alterations, and provide mechanistic knowledge for predicting future changes and animal adaptation phenomena.

The overall picture that emerges from the present analysis is that climate changes influence a multitude of physical, chemical, and biological processes in coastal areas, that deeply affect the species populations in Mediterranean coastal lagoons. We may expect the most sensitive species to decrease, due to some main factors (increase of temperature, reduced salinity, and acidification) and their consequences, and the tolerant species to increase. Several investigations suggest that climate-induced increases in nutrient loading will likely cause increasing phytoplankton production, which combined with higher temperatures and decreased oxygen solubility will lead to more intense and frequent episodes of hypoxia impacting animal life.

We observe that the direction of many trends is fairly certain, however quantitative relationships are few, particularly for temperature impacts. As an overall, we have very few whole-system views of the response of lagoon biogeochemistry to temperature. This is unfortunate because estuarine biogeochemical processes are complex and highly non-linear, and efforts to untangle the biogeochemical impacts of multiple climate forcings (temperature, CO_2, sea level) from each other and from other agents (i.e. land-use and land cover change) will require deep investigations and dedicated resources. In this sense, baseline studies and further insights on existing information will be an essential step towards understanding the way Mediterranean species will cope with ongoing changes. We also highlighted that the combination of both organismal and suborganismal scale measurements with studies at the population and community levels may provide a much more comprehensive view of the drivers of ecological thresholds with respect to simple correlations between climate change and community responses.

To conclude, we must admit that many important aspects were not covered in this paper. We do hope the rich bibliographic section may provide further information of interest to the readers.

References

Andersson AJ, Kuffner IB, Mackenzie FT, Jokiel PL, Rodgers KS, Tan A (2009) Net Loss of $CaCO_3$ from a subtropical calcifying community due to seawater acidification: mesocosm-scale experimental evidence. Biogeosciences 6:1811–1823

Angilletta MJ (2009) Thermal adaptation: a theoretical and empirical synthesis. Oxford University Press, Oxford

Bambrick HJ, Kjellström TE (2004) Good for your heart but bad for your baby? Revised guidelines for fish consumption in pregnancy. Med J Aust 181:61–62

Ben Ameur WB, de Lapuente J, El Megdiche Y et al (2012) Oxidative stress, genotoxicity and histopathology biomarker responses in mullet (*Mugil cephalus*) and sea bass (*Dicentrarchus labrax*) liver from Bizerte Lagoon (Tunisia). Mar Pollut Bull 64:241–251

Berge JA, Bjerkeng B, Pettersen O et al (2006) Effects of increased sea water concentrations of CO_2 on growth of the bivalve *Mytilus edulis* L. Chemosphere 62:681–687

Blackford JC (2010) Predicting the impacts of ocean acidification: challenges from an ecosystem perspective. J Mar Syst 81:12–18

Booth S, Zeller D (2005) Mercury, food webs, and marine mammals: implications of diet and climate change for human health. Environ Health Perspect 113:521–526

Breitburg DL, Pihl L, Kolesar SE (2001) Effects of low dissolved oxygen on the behavior, ecology and harvest of fishes: a comparison of the Chesapeake Bay and Baltic-Kattegat systems. Coastal Estuar Stud 58:241–267

Brito AC, Newton A, Tett P, Fernandes TF (2012a) How will shallow coastal lagoons respond to climate change? A modelling investigation. Estuar Coast Shelf Sci 112:98–104

Brito AC, Newton A, Tett P et al (2012b) Changes in the yield of microphytobenthic chlorophyll from nutrients: considering denitrification. Ecol Indic 19:226–230

Bury NR, McGeer JC, Wood CM (1999) Effects of altering freshwater chemistry on physiological responses of rainbow trout to silver exposure. Environ Toxicol Chem 18:49–55

Bussell JA, Gidman EA, Causton DR, Gwynn-Jones D, Malham SK, Jones MLM et al (2008) Changes in the immune response and metabolic fingerprint of the mussel, *Mytilus edulis* (Linnaeus) in response to lowered salinity and physical stress. J Exp Mar Biol Ecol 358:78–85

Cajaraville MP, Olabarrieta I, Marigomez I (1996) In vitro activities in mussel hemocytes as biomarkers of environmental quality: a case study in the Abra Estuary (Biscay Bay). Ecotoxicol Environ Saf 35:253–260

Canu DM, Solidoro C, Cossarini G, Giorgi F (2010) Effect of global change on bivalve rearing activity and the need for adaptive management. Climate Res 42:13–26

Carere M, Miniero R, Cicero MR (2011) Potential effects of climate change on the chemical quality of aquatic biota. Trac-Trends Anal Chem 30:1214–1221

Caselli F, Fabbri E (2005) Use of *Mytilus gralloprovincialis* and *Tapes philippinarium* as sentinel organisms for the development of a biosurveillance program in the Pialassa Baiona coastal lagoon (Ravenna, Italy). Chem Ecol 21:465–477

Chapman PM (2012) Management of coastal lagoons under climate change. Estuar Coast Shelf Sci 110:32–35

Christen R, Schackmann RW, Shapiro BM (1986) Ionic regulation of sea-urchin sperm motility, metabolism and fertilizing-capacity. J Physiol 379:347–365

Clark D, Lamare M, Barker M (2009) Response of sea urchin pluteus larvae (Echinodermata: Echinoidea) to reduced seawater pH: a comparison among a tropical, temperate, and a polar species. Mar Biol 156:1125–1137

IPCC (Intergovernmental Panel on Climate Change) (2007a) Climate change 2007: the scientific basis, contribution of working group I to the Third assessment report. Cambridge University Press, Cambridge

IPCC (Intergovernmental Panel on Climate Change) (2007b) Summary for policymakers. In: Parry ML, Canziani OF, Palutikof JP et al (eds) Climate change 2007: impacts, adaptation and vulnerability. Contribution of working group II to the Fourth assessment report of the intergovernmental panel on climate change. Cambridge University Press, Cambridge

Conesa HM, Jiménez-Cárceles FJ (2007) The Mar Menor lagoon (SE Spain): a singular natural ecosystem threatened by human activities. Mar Pollut Bull 54:839–849

Cook T, Folli M, Klinck J et al (1998) The relationship between increasing sea-surface temperature and the northward spread of *Perkinsus marinus* (Dermo) disease epizootics in oysters. Estuar Coast Shelf Sci 46:587–597

Cossarini G, Libralato S, Salon S, Gao X, Giorgi F, Solidoro C (2008) Downscaling experiment for the Venice Lagoon. II. Effects of changes in precipitation on biogeochemical properties. Climate Res 38:43–59

Coutant CC (1985) Striped bass, temperature and dissolve-oxygen: a speculative hypothesis for environmental risk. Trans Am Fish Soc 114:31–61

Crain CM, Kroeker K, Halpern BS (2008) Interactive and cumulative effects of multiple human stressors in marine systems. Ecol Lett 11:1304–1315

Da Ros L, Nesto N (2005) Cellular alterations in *Mytilus galloprovincialis* (LMK) and *Tapes philippinarum* (Adams and Reeve, 1850) as biomarkers of environmental stress: field studies in the Lagoon of Venice (Italy). Environ Int 31:1078–1088

Dagnino A, Allen JI, Moore MN et al (2007) Development of an expert system for the integration of biomarker responses in mussels into an animal health index. Biomarkers 12:155–172

Dalla Valle M, Marcomini A, Jones K, Sweetman AJ (2005) Reconstruction of historical trends of PCDD/Fs and PCBs in the Venice Lagoon, Italy. Environ Int 31:1047–1052

Day J, Ibáñez C, Scarton F, Pont D, Hensel P, Day J, Lane R (2011) Sustainability of Mediterranean deltaic and Lagoon wetlands with sea-level rise: the importance of river input. Estuar Coasts 34:483–493

De Pascalis F, Pérez-Ruzafa A, Gilabert J, Marcos C, Umgiesser G (2012) Climate change response of the Mar Menor coastal lagoon (Spain) using a hydrodynamic finite element model. Estuar Coast Shelf Sci 114:118–129

Dedieu K, Rabouille C, Thouzeau G, Jean F et al (2007) Benthic O_2 distribution and dynamics in a Mediterranean lagoon (Thau, France): an in situ microelectrode study. Estuar Coast Shelf Sci 72:393–405

Denny M, Helmuth B (2009) Confronting the physiological bottleneck: a challenge from ecomechanics. Integr Comp Biol 49:197–201

Denny MW, Dowd WW, Bilir L et al (2011) Spreading the risk: small-scale body temperature variation among intertidal organisms and its implications for species persistence. J Exp Mar Biol Ecol 400:175–190

Directive 2008/56/EC of the European Parliament and of the Council of 17 June 2008 Official Journal of the European Union 25.6.2008

Doney SC, Ruckelshaus M, Emmett Duffy J, Barry JP et al (2012) Climate change impacts on marine ecosystems. Ann Rev Mar Sci 4:11–37

Donker MH, Abdel-Lateif HM, Khalil MA et al (1998) Temperature, physiological time, and zinc toxicity in the isopod *Porcellio scaber*. Environ Toxicol Chem 17:1558–1563

Donnini F, Dinelli E, Sangiorgi F, Fabbri E (2007) A biological and geochemical integrated approach to assess the environmental quality of a coastal lagoon (Ravenna, Italy). Environ Int 33:919–928

Durán D, Durán I, Costescu C (2009) The impact of regional development on the environment. Environ Engin Manag J 8:721–725

European Commission (EC) (2010) Guidance document 25, Guidance on chemical monitoring of sediment and biota under the Water Framework Directive, Technical Report 2010-041. (http://circa.europa.eu/Public/irc/env/wfd/library?l=/framework_directive/guidance_documents/guidance_monitoring/_EN_1.0_&a=d). Lastly accessed 13 Aug 2012

Fabbri D, Felisatti O, Lombardo M et al (1998) The lagoon of Ravenna (Italy): characterisation of mercury-contaminated sediments. Sci Total Environ 213:121–128

Fabbri D, Baravelli V, Giannotti K et al (2006) Bioaccumulation of cyclopenta[cd]pyrene and benzo[ghi]fluoranthene by mussels transplanted in a coastal lagoon. Chemosphere 64:1083–1092

Fabbri E, Valbonesi P, Franzellitti S (2008) HSP expression in bivalves. Inv Surv J 5:135–161

Finke GR, Bozinovic F, Navarrete SA (2009) A mechanistic model to study the thermal ecology of a Southeastern Pacific dominant intertidal mussel and implications for climate change. Physiol Biochem Zool 82:303–313

Fisher SW, Wadleigh RW (1985) Effects of temperature on the acute toxicity and uptake of lindane by *Chironomus riparius* (Meighen) (*Diptera, Chironomidae*). J Econ Entomol 78:1222–1226

Fitzgerald D, Fenster M, Argow B, Buynevich I (2008) Coastal impacts due to sea-level rise. Ann Rev Earth Planet Sci 36:601–647

Franzellitti S, Buratti S, Donnini F, Fabbri E (2010) Exposure of mussels to a polluted environment: insights into the stress syndrome development. Comp Biochem Physiol C Toxicol Pharmacol 152:24–33

Frignani M, Bellucci LG, Favotto M, Albertazzi S (2005) Pollution historical trends as recorded by sediments at selected sites of the Venice Lagoon. Environ Int 31:1011–1022

Gambolati G, Teatini P, Gonnella M (2002) GIS Simulations of the inundation risk in the coastal lowlands of the northern Adriatic Sea. Math Comp Model 35:963–972

Gatti JL, Christen R (1985) Regulation of internal pH of sea-urchin sperm: a role for the Na/K pump. J Biol Chem 260:7599–7602

Gattuso J-P, Hansson L (2011) Ocean acidification: background and history. In: Gattuso J-P, Hansson L (eds) Ocean acidification. Oxford University Press, Oxford

Gazeau F, Quiblier C, Jansen JM, Gattuso JP, Middelburg JJ, Heip CHR (2007) Impact of elevated CO_2 on shellfish calcification. Geophys Res Lett 34, L07603

Gazeau F, Gattuso J, Dawber C, Pronker AE et al (2010) Effect of ocean acidification on the early life stages of the blue mussel *Mytilus edulis*. Biogeosciences 7:2051–2060

Gazeau F, Martin S, Hansson L, Gattuso J-P (2011) Ocean acidification in the coastal ocean. LOICZ In print 2011/3,pp 5–14

Geyer HJ, Scheuners I, Rapp K et al (1993) The relevance of fat content in toxicity of lipophilic chemicals to terrestrial animals with special reference to dieldrin and 2,3,7,8-tetrachlorodibenzo-P-dioxin (TCDD). Ecotoxicol Environ Saf 26:45–60

Giorgi F, Lionello P (2008) Climate change projections for the Mediterranean region. Glob Plan Change 63:90–104

Gosling SN, Warren R, Arnell NW, Good P et al (2011) A review of recent developments in climate change science. Part II: the global-scale impacts of climate change. Prog Phys Geo 35:443–464

Guerzoni S, Rossini P, Sarretta A, Raccanelli S, Ferrari G, Molinaroli E (2007) POPs in the lagoon of Venice: budgets and pathways. Chemosphere 67:1776–1785

Hall LW Jr, Anderson RD (1995) The influence of salinity on the toxicity of various classes of chemicals to aquatic biota. Crit Rev Toxicol 25:281–346

Hamer B, Jaksic Z, Pavicic-Hamer D et al (2008) Effect of hypoosmotic stress by low salinity acclimation of Mediterranean mussels *Mytilus galloprovincialis* on biological parameters used for pollution assessment. Aquat Toxicol 89:137–151

Harley CDG, Hughes AR, Hultgren KM (2006) The impacts of climate change in coastal marine systems. Ecol Lett 9:228–241

Havenhand JN, Schlegel P (2009) Near-future levels of ocean acidification do not affect sperm motility and fertilization kinetics in the oyster *Crassostrea gigas*. Biogeosciences 6:3009–3015

Havenhand JN, Buttler FR, Thorndyke MC et al (2008) Near-future levels of ocean acidification reduce fertilization success in a sea urchin. Curr Biol 18:R651–R652

Helmuth B (2009) From cells to coastlines: how can we use physiology to forecast the impacts of climate change? J Exp Biol 212:753–760

Helmuth B, Kingsolver JG, Carrington E (2005) Biophysics, physiological ecology, and climate change: does mechanism matter? Ann Rev Physiol 67:177–201

Helmuth B, Broitman BR, Yamane L et al (2010) Organismal climatology: analyzing environmental variability at scales relevant to physiological stress. J Exp Biol 213:995–1003

Helmuth B, Yamane L, Lalwani S, Matzelle A et al (2011) Hidden signals of climate change in intertidal ecosystems: what (not) to expect when you are expecting. J Exp Mar Biol Ecol 400:191–199

Hernroth B, Baden S, Thorndyke M et al (2011) Immune suppression of the echinoderm *Asterias rubens* (L.) following long-term ocean acidification. Aquat Toxicol 103:222–224

Hoffmann AA, Sgrò CM (2011) Climate change and evolutionary adaptation. Nature 470:479–485

Hofmann GE, Barry JP, Edmunds PJ (2010) The effect of ocean acidification on calcifying organisms in marine ecosystems: an organism-to-ecosystem perspective. Ann Rev Ecol Evol Syst 41:127–147

Jacobson PJ, Neves RJ, Cherry DS et al (1997) Sensitivity of glochidial stages of freshwater mussels (Bivalvia: Unionidae) to copper. Environ Toxicol Chem 16:2384–2392

Karen DJ, Ownby DR, Forsythe BL et al (1999) Influence of water quality on silver toxicity to rainbow trout (*Oncorhynchus mykiss*), fathead minnows (*Pimephales promelas*), and water fleas (*Daphnia magna*). Environ Toxicol Chem 18:63–70

Kearney M (2006) Habitat, environment and niche: what are we modelling? Oikos 115:186–191

Kurihara H, Asai T, Kato S et al (2009) Effects of elevated pCO_2 on early development in the mussel *Mytilus galloprovincialis*. Aquat Toxicol 4:225–233

Lahbib Y, Abidli S, Rodriguez Gonzalez P (2011) Monitoring of organotin pollution in Bizerta channel (Northern Tunisia): temporal trend from 2002 to 2010. Bull Environ Contam Toxicol 86:531–534

Lambeck K, Antonioli F, Anzidei M, Ferranti L, Leoni G, Scicchitano G, Silenzi S (2011) Sea level change along the Italian coast during the Holocene and projections for the future. Quat Int 232:250–257

Lee B-G, Wallace WG, Luoma SN (1998) Uptake and loss kinetics of Cd, Cr and Zn in the bivalves *Poamocorbula amurenis* and *Macoma balthica*: effects of size and salinity. Mar Ecol Prog Ser 175:177–189

Lenihan JM, Drapek R, Bachelet D et al (2003) Climate change effects on vegetation distribution, carbon, and fire in California. Ecol Appl 13:1667–1681

Lloret J, Marín A, Marín-Guirao L, Velasco J (2005) Changes in macrophytes distribution in a hypersaline coastal lagoon associated with the development of intensively irrigated agriculture. Ocean Coast Manag 48:828–842

Lloret J, Marín A, Marín-Guirao L (2008) Is coastal lagoon eutrophication likely to be aggravated by global climate change? Estuar Coast Shelf Sci 78:403–412

Lotze HK, Lenihan HS, Bourque BJ et al (2006) Depletion, degradation, and recovery potential of estuaries and coastal seas. Science 312:1806–1809

Mahmoud N, Dellali M, El Bour M (2010) The use of *Fulvia fragilis* (Mollusca: Cardiidae) in the biomonitoring of Bizerta lagoon: a multimarkers approach. Ecol Indic 10:696–702

Marshall DJ, McQuaid CD, Williams GA (2010) Non-climatic thermal adaptation: implications for species' responses to climate warming. Biol Lett 6:669–673

Metzger E, Simonucci C, Viollier E, Sarazin G, Prévot F, Elbaz-Poulichet F, Seidel J-L, Jézéquel D (2007) Influence of diagenetic processes in Thau lagoon on cadmium behaviour and benthic fluxes. Estuar Coast Shelf Sci 72:497–510

Michaelidis B, Ouzounis C, Paleras A (2005) Effects of long-term moderate hypercapnia on acid-base balance and growth rate in marine mussels *Mytilus galloprovincialis*. Mar Ecol Prog Ser 293:109–118

Miles H, Widdicombe S, Spicer JI et al (2007) Effects of anthropogenic seawater acidification on acid-base balance in the sea urchin *Psammechinus miliaris*. Mar Pollut Bull 54:89–96

Miller AW, Reynolds AC, Sobrino C, Riedel GF (2009) Shellfish face uncertain future in high CO_2 world: influence of acidification on oyster larvae calcification and growth in estuaries. PLoS One 4:e5661

Monaco CJ, Helmuth B (2011) Tipping points, thresholds and the keystone role of physiology in marine climate change research. Adv Mar Biol 60:123–160

Monperrus M, Tessier E, Point D, Vidimova K et al (2007) The biogeochemistry of mercury at the sediment–water interface in the Thau Lagoon. 2. Evaluation of mercury methylation potential in both surface sediment and the water column. Estuar Coast Shelf Sci 72:485–496

Morita M, Awata S, Takahashi T et al (2010) Sperm motility adaptation to ion-differing aquatic environments in the Tanganyikan Cichlid, *Astatotilapia burtoni*. J Exp Zool A Ecol Genet Physiol 313A:169–177

Muresan B, Cossa D, Jézéquel D, Prévot F, Kerbellec S (2007) The biogeochemistry of mercury at the sediment–water interface in the Thau lagoon. 1. Partition and speciation. Estuar Coast Shelf Sci 72:472–484

Newton A, Icely JD, Falcao M et al (2003) Evaluation of eutrophication in the Ria Formosa coastal lagoon. Portugal Cont Shelf Res 23:1945–1961

Nicholls RJ (2002) Rising sea level: potential impacts and responses. Glob Environ Change 17:83–107

Nicholls RJ (2004) Coastal flooding and wetland loss in the 21st century: changes under the SRES climate and socio-economic scenarios. Glob Environ Change Hum Policy Dimens 14:69–86

Nicholls RJ (2007) The impacts of sea level rise. Ocean Challenge 15:13–17

Niklitschek EJ, Secor DH (2005) Modeling spatial and temporal variation of suitable nursery habitats for Atlantic sturgeon in the Chesapeake Bay. Estuar Coast Shelf Sci 64:135–148

Noyes PD, McElwee MK, Miller HD, Clark BW et al (2009) The toxicology of climate change: environmental contaminants in a warming world. Environ Int 35:971–986

O'Donnell MJ, Todgham AE, Sewell MA et al (2010) Ocean acidification alters skeletogenesis and gene expression in larval sea urchins. Mar Ecol Prog Ser 398:157–171

Orr JC (2011) Recent and future changes in ocean carbonate chemistry. In: Gattuso J-P, Hansson L (eds) Ocean acidification. Oxford University Press, Oxford

Pastres R, Solidoro C (2012) Monitoring and modeling for investigating driver/pressure-state/impact relationships in coastal ecosystems: examples from the Lagoon of Venice. Estuar Coast Shelf Sci 96:22–30

Peck LS (2011) Organisms and responses to environmental change. Mar Genomics 4:237–243

Pérez-Ruzafa A, Gilabert J, Gutiérrez JM, Fernández AI, Marcos C, Sabah S (2002) Evidence of a planktonic food web response to changes in nutrient input dynamics in the Mar Menor coastal lagoon, Spain. Hydrobiologia 475/476:359–369

Persoone G, Vandevel A, Vansteertegem M et al (1989) Predictive value of laboratory tests with aquatic invertebrates: influence of experimental conditions. Aquat Toxicol 14:149–167

Philippart CJM, Anadón R, Danovaro R, Dippner JW, Drinkwater KF et al (2011) Impacts of climate change on European marine ecosystems: observations, expectations and indicators. J Exp Mar Biol Ecol 400:52–69

Piano A, Asirelli C, Caselli F, Fabbri E (2002) Hsp70 expression in thermally stressed *Ostrea edulis*, a commercially important oyster in Europe. Cell Stress Chap 7:250–257

Piano A, Valbonesi P, Fabbri E (2004) Expression of cytoprotective proteins, heat shock protein 70 and metallothioneins, in tissues of *Ostrea edulis* exposed to heat and heavy metals. Cell Stress Chap 9:134–142

Point D, Monperrus M, Tessier E, Amouroux D, Chauvaud L et al (2007) Biological control of trace metal and organometal benthic fluxes in a eutrophic lagoon (Thau Lagoon, Mediterranean Sea, France). Estuar Coast Shelf Sci 72:457–471

Pörtner HO, Farrell AP (2008) Ecology physiology and climate change. Science 322:690–692

Pörtner HO, Knust R (2007) Climate change affects marine fishes through the oxygen limitation of thermal tolerance. Science 315:95–97

Ramachandran S, Sweezey MJ, Hodson PV, Boudreau M et al (2006) Influence of salinity and fish species on PAH uptake from dispersed crude oil. Mar Pollut Bull 52:1182–1189

Range P, Chícharo MA, Ben-Hamadou R, Piló D, Matias D et al (2011) Calcification, growth and mortality of juvenile clams *Ruditapes decussatus* under increased pCO_2 and reduced pH: variable responses to ocean acidification at local scales? J Exp Mar Biol Ecol 396:177–184

Riba Lopez I, Kalman J, Vale C et al (2010) Influence of sediment acidification on the bioaccumulation of metals in *Ruditapes philippinarum*. Environ Sci Pollut Res 17:1519–1528

Ries JB, Cohen AL, McCorkle DC (2009) Marine calcifiers exhibit mixed responses to CO_2-induced ocean acidification. Geology 37:1131–1134

Ripley JL, Foran CM (2007) Influence of estuarine hypoxia on feeding and sound production by two sympatric pipefish species (*Syngnathidae*). Mar Environ Res 63:350–367

Roberts DA (2012) Causes and ecological effects of resuspended contaminated sediments (RCS) in marine environments. Environ Int 40:230–243

Rodríguez-Rodríguez M (2007) Hydrogeology of ponds, pools, and playa-lakes of southern Spain. Wetlands 27:819–830

Rombough PJ, Moroz BM (1997) The scaling and potential importance of cutaneous and branchial surfaces in respiratory gas exchange in larval and juvenile walleye *Stizostedion vitreum*. J Exp Biol 200:2459–2468

Rosas C, Ramirez P (1993) Effect of chromium and cadmium on the thermal tolerance of the prawn *Macrobrachium rosenbergii* exposed to hard and soft water. Bull Environ Contam Toxicol 51:568–574

Sadok S, Uglow RF, Haswell SJ (1997) Haemolymph and mantle fluid ammonia and ninhydrin positive substances variations in salinity-challenged mussels (*Mytilus edulis* L). J Exp Mar Biol Ecol 211:195–212

Salon S, Cossarini G, Libralato S, Gao X, Solidoro C, Giorgi F (2008) Downscaling experiment for the Venice lagoon. I. Validation of the present-day precipitation climatology. Climate Res 38:31–41

Schiedek D, Sundelin B, Readman JW, Macdonald RW (2007) Interactions between climate change and contaminants. Mar Pollut Bull 54:1845–1856

Secor DH, Gunderson TE (1998) Effects of hypoxia and temperature on survival, growth, and respiration of juvenile Atlantic sturgeon, *Acipenser oxyrinchus*. Fish Bull 96:603–613

Shimps EL, Rice JA, Osborne JA (2005) Hypoxia tolerance in two juvenile estuary-dependent fishes. J Exp Mar Biol Ecol 325:146–162

Smit CE, Van Gestel CAM (1997) Influence of temperature on the regulation and toxicity of zinc in *Folsomia candida* (Collembola). Ecotoxicol Environ Saf 37:213–222

Solidoro C, Bandelj V, Aubry Bernardi F, Camatti E, Ciavatta S et al (2010) Response of the Venice Lagoon ecosystem to natural and anthropogenic pressures over the last 50 years. In: Kennish MJ, Paerl HW (eds) Coastal lagoons: critical habitats of environmental change. CRC Press, Boca Raton

Somero GN (2010) The physiology of climate change: how potentials for acclimatization and genetic adaptation will determine 'winners' and 'losers'. J Exp Biol 213:912–920

Somero GN (2012) The physiology of global change: linking patterns to mechanisms. Ann Rev Mar Sci 4:39–61

Sorte CJB, Jones SJ, Miller LP (2011) Geographic variation in temperature tolerance as an indicator of potential population responses to climate change. J Exp Mar Biol Ecol 400:209–217

Stachowicz JJ, Terwin JR, Whitlatch RB et al (2002) Linking climate change and biological invasions: ocean warming facilitates nonindigenous species invasions. Proc Natl Acad Sci USA 99:15497–15500

Talmage S, Gobler CJ (2009) The effects of elevated carbon dioxide concentrations on the metamorphosis, size, and survival of larval hard clams (*Mercenaria mercenaria*), bay scallops (*Argopecten irradians*), and Eastern oysters (*Crassostrea virginica*). Limnol Oceanogr 54:2072–2080

The MerMex Group: Durrieu de Madron X, Guieu C, Sempéré R, Conan P, Cossa D, et al (2011) Marine ecosystems' responses to climatic and anthropogenic forcings in the Mediterranean. Prog Oceanogr 91:97–166

Thouzeau G, Grall J, Clavier J, Chauvaud L et al (2007) Spatial and temporal variability of benthic biogeochemical fluxes associated with macrophytic and macrofaunal distributions in the Thau lagoon (France). Estuar Coast Shelf Sci 72:432–446

Tlili S, Jebali J, Banni M et al (2010) Multimarker approach analysis in common carp *Cyprinus carpio* sampled from three freshwater sites. Environ Monit Assess 168:285–298

Todgham AE, Hofmann GE (2009) Transcriptomic response of sea urchin larvae *Strongylocentrotus purpuratus* to CO_2-driven seawater acidification. J Exp Biol 212:2579–2594

Trombini C, Fabbri D, Lombardo M et al (2003) Mercury and methylmercury contamination in surficial sediments and clams of a coastal lagoon (Pialassa Baiona, Ravenna, Italy). Cont Shelf Res 23:1821–1831

Tyler RM, Targett TE (2007) Juvenile weakfish *Cynoscion regalis* distribution in relation to diel-cycling dissolved oxygen in an estuarine tributary. Mar Ecol Prog Ser 333:257–269

UNEP/RAMOGE Manual (1999) Manual of the biomarkers recommended for the MEDPOL biomonitoring programme. UNEP/RAMOGE, Athens

Van Wezel AP, Jonker MTO (1998) Use of the lethal body burden in the risk quantification of field sediments; influence of temperature and salinity. Aquat Toxicol 42:287–300

Vargas-Yáñez M, Moya F, García-Martínez MC, Zunino P et al (2010) Climate change in the western Mediterranean Sea 1900–2008. J Mar Syst 82:171–176

Verreault J, Dietz R, Sonne C et al (2008) Comparative fate of organohalogen contaminants in two top carnivores in Greenland: captive sledge dogs and wild polar bears. Comp Biochem Physiol C Toxicol Pharmacol 147:306–315

Viarengo A, Lowe D, Bolognesi C, Fabbri E, Koehler A (2007) The use of biomarkers in biomonitoring: a 2-tier approach assessing the level of pollutant-induced stress syndrome in sentinel organisms. Comp Biochem Physiol C Toxicol Pharmacol 146:281–300

Viaroli P, Lasserre P, Campostini P (2007) Lagoons and coastal wetlands in the global change context: impacts and Management issues: preface. Hydrobiologia 577:1–3

Wernberg T, Russell BD, Moore PJ, Ling SD, Smale DA et al (2011) Impacts of climate change in a global hotspot for temperate marine biodiversity and ocean warming. J Exp Mar Biol Ecol 400:7–16

Wood HL, Spicer JI, Widdicombe S (2008) Ocean acidification may increase calcification rates, but at a cost. Proc R Soc B Biol Sci 275:1767–1773

Wright DA (1995) Trace metals and major ion interactions in aquatic animals. Mar Pollut Bull 31:8–18

Zanchettin D, Traverse P, Tomasino M (2007) Observations on future sea level changes in the Venice lagoon. Hydrobiologia 577:41–53

Zucchetta M, Cipolato G, Pranovia F, Antonetti P, Torricelli P et al (2012) The relationships between temperature changes and reproductive investment in a Mediterranean goby: insights for the assessment of -climate change effects. Estuar Coast Shelf Sci 101:15–23

Part III

Global Climate Change and Ecosystem Effects

A Comparative Analysis of Trophic Structure and Functioning in Large-Scale Mediterranean Marine Ecosystems

24

Alberto Barausse and Luca Palmeri

Abstract

The Mediterranean is a sea rich with many kinds of diversity. It is a hotspot of marine biodiversity which covers many habitats and environmental conditions, and is surrounded by three continents characterized by different cultures and degrees of socio-economic development, whose coastal human activities exert multiple pressures on the marine environment. Yet, surprisingly, the diversity in the structure and functioning of Mediterranean marine ecosystems has not been analyzed rigorously, especially on large spatial scales. Such information are critical to implement an Ecosystem Approach to the management of the Mediterranean Sea. To fill this gap, a comparative analysis of the South Catalan, the Northern-Central Adriatic, the Northern Adriatic and the North Aegean Seas was performed. Trophic network models of the marine pelagic environment in each system were assembled with Ecopath software, based on published datasets. To facilitate the comparison, models were constructed with the same number and kind of trophic groups. Multiple indicators from ecological network analysis were calculated and consistently highlighted similarities and differences among Mediterranean pelagic food webs. Shared traits included the key role of intermediate-trophic level species, the low overall impact exerted by large predators, and inefficiencies in the exploitation of phytoplankton and detritus production giving rise to high export flows fuelling the benthic compartment. Primary productivity markedly influenced food web properties, but additional differences in the global structure of trophic flows emerged, highlighting a great ecosystem diversity. The systems could be ranked in a clear order of development and maturity (from high to low): South Catalan, Northern-Central Adriatic, North Aegean, Northern Adriatic Sea.

Keywords

Mediterranean marine systems • Food webs • Ecological network analysis • Ecopath • Ecosystem structure and functioning • Ecosystem maturity

A. Barausse • L. Palmeri (✉)
Laboratorio di Analisi dei Sistemi Ambientali, Università di Padova,
via Marzolo 9, 35131 Padova, Italy
e-mail: lpalmeri@unipd.it

S. Goffredo and Z. Dubinsky (eds.), *The Mediterranean Sea: Its history and present challenges*,
DOI 10.1007/978-94-007-6704-1_24, © Springer Science+Business Media Dordrecht 2014

Abbreviation List

TST	Total system throughput
PPR	Primary production required (to sustain catches)
FCI	Finn's cycling index
PCI	Predatory cycling index
FMPL	Finn's mean path length
FSTPL	Finn's straight-through path length

Introduction

A Sea Rich with Diversity

The Mediterranean is a semi-enclosed sea with a relatively small area compared to other large marine ecosystems worldwide, yet its waters and coasts are characterized by a disproportionately huge diversity from the environmental and socio-economical points of view. Three continents (Europe, Asia, Africa) and 21 countries surround this basin, a number which alone suggests the great variety in human culture and socio-economic development found in Mediterranean coastal zones, which have been inhabited for millennia. Marine environmental and ecological conditions and habitats greatly vary: a wide range of depths is covered, and primary productivity displays a clear west-east and north-south decrease, similarly to fishery landings, a spatial pattern opposite to that of temperature and salinity (Caddy 1993; Caddy et al. 1995; Coll et al. 2010; Fig. 24.1). The Mediterranean Sea is also a hotspot of biodiversity: more than 17,000 marine species occur there, and about a fifth are endemic (Lejeusne et al. 2009; Coll et al. 2010).

This highly-diverse environment interacts with the numerous pressures generated by the heterogeneous coastal human activities, as well as with climate variability, in a complex, intriguing manner, to the point that the Mediterranean has been defined as a "miniature ocean" or a "giant mesocosm" where the effects of climate and anthropic pressures on the world's oceans could be studied (Lejeusne et al. 2009). Climate change should impact this region strongly, acting together and, possibly, synergistically with multiple anthropogenic pressures which are already high in the basin, such as fishing, habitat degradation and loss, pollution and nutrient enrichment.

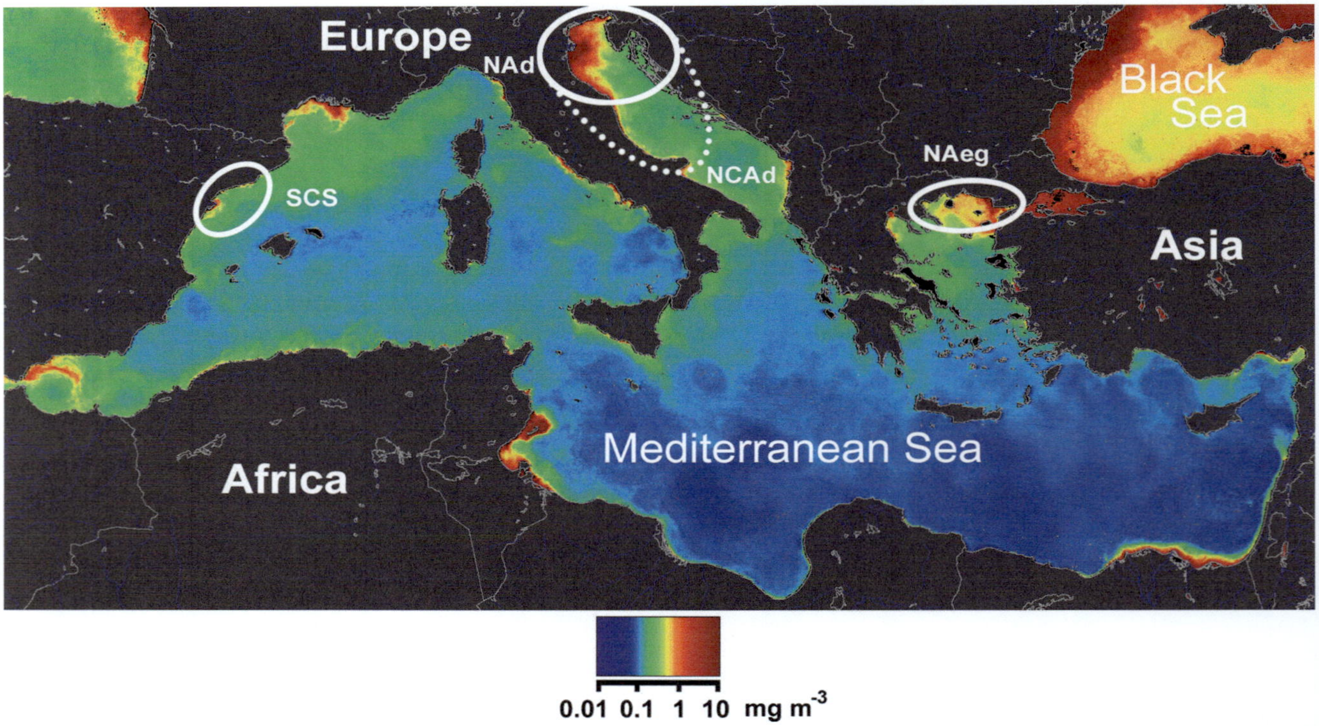

Fig. 24.1 The location of the four ecosystems analyzed, as indicated by the ovals. *SCS* indicates the South Catalan Sea; *NAd* is the Northern Adriatic Sea; *NCAd* is the Northern and Central Adriatic Sea (*dotted oval*, also including the *continuous oval* of the Northern Adriatic); *NAeg* is the Northern Aegean Sea. The *color map* represents satellite-based estimates of chlorophyll-a concentration in Mediterranean marine waters (mg m^{-3}, see scale bar; monthly composite of SeaWiFS data for June 1998; image downloaded from the OceanColour portal, Institute for Environment and Sustainability, European Commission – Joint Research Centre, http://oceancolour.jrc.ec.europa.eu/). The map is not supposed to mirror exactly the spatial patterns of primary productivity over one or multiple years, nevertheless marked differences in algal biomass can be seen across the basin

Invasions of alien species represent another key impact, possibly driven by both climate and human factors (Halpern et al. 2008; Lejeusne et al. 2009; Coll et al. 2010). The Mediterranean appears to be one of the most human-impacted seas in the world (Halpern et al. 2008; Lejeusne et al. 2009).

The Need for an Ecosystem Approach in the Mediterranean Sea

Diversity, or variety, seems to be the keyword for the Mediterranean Sea. Thus, a sustainable management of human activities impacting the marine environment should be adapted the peculiarities of it local sub-basins and, also, integrated. The latter word means that management should adopt a network perspective, explicitly considering that the ecological impacts of different socio-economic activities can combine, that ecosystem components, both biotic and abiotic, are interconnected and this influences their response to human impacts, and that feedbacks exist between socio-economic and ecological systems: human activities impact marine ecosystems, but the subsequent ecological change can impact human welfare and activities. Such view of an integrated management strategy for human activities impacting the marine environment and its resources is known as the Ecosystem Approach (e.g. CBD 2000; Garcia et al. 2003). This approach has greatly influenced fishery science, which is currently switching from the classical management focus on single species towards multispecies and food web approaches taking predatory interactions into account (Garcia et al. 2003; Worm et al. 2009). The Ecosystem Approach is becoming more and more popular and is being integrated into marine environmental policies worldwide, an outstanding example being the European Marine Strategy Framework Directive (EU 2008).

The Need for Ecological Network Analysis

As an Ecosystem Approach can only be applied based on a good knowledge of the integrated functioning of marine ecosystems and their response to human pressures and environmental forcing, several ecological questions appear to be of practical importance. For example: how strongly does one species influence the others? Are there species playing particularly important roles in the marine ecosystem? What are the main pathways through which energy flows from primary producers to other species? Is the ecosystem configuration resilient and resistant to external disturbances? Are whole-system properties such as productivity, efficiency and stability affected by human pressures? Is the ecosystem found in an early stage of ecological succession, e.g. is it overexploited and stressed, or is it mature and developed (Odum 1969, 1985; Ulanowicz 1997)?

To answer these questions, ecologists have studied the structure and function of food webs, i.e. the networks of trophic interactions linking populations in the ecosystem, by quantifying energy and material flows within them.[1] This approach to the study of ecosystems is called Ecological Network Analysis and has partly been inspired by economics, a scientific field in which data on commodity flows are more frequently available than explanations concerning what caused such flows[2] (Ulanowicz 1986, 1997, 2004; Baird et al. 1991).

The analysis of marine trophic networks is important because it helps to refine our conceptual models of how ecosystems function as a whole. A conceptual model is the first, key step to construct a quantitative model of an ecological system to be used for management purposes (Jørgensen and Bendoricchio 2001). It is well-known that the Mediterranean Sea encompasses greatly-diverse ecosystems, reflecting the abovementioned diversity in species, habitats, environmental conditions and pressures, but, surprisingly, the diversity in ecosystem structure and functioning has not been analyzed rigorously. Relatively few food web models and ecological network analysis applications have been published for Mediterranean marine systems, mostly during the last decade and covering limited areas of the basin. According to the review of Coll and Libralato (2011), only one publication dealt with the Southern Mediterranean, analyzing the Gulf of Tunis food web. Moreover, few comparisons of trophic networks from different Mediterranean sub-basins exist (Coll et al. 2008; Libralato et al. 2010; Tsagarakis et al. 2010; Coll and Libralato 2011). Comparative studies are vital to highlight differences in the functioning of Mediterranean ecosystems, which would result in different management plans depending on the local ecological peculiarities.

The goal of this work is to perform a comparative analysis of four trophic network models of large-scale marine ecosystems in the Mediterranean Sea during recent years, focusing on differences and similarities in their structure, functioning and development *sensu* Odum (1969). They are the South Catalan Sea (Coll et al. 2006), the Northern and Central Adriatic Sea (Coll et al. 2007), the Northern Adriatic Sea (Barausse et al. 2009) and the North Aegean Sea (Tsagarakis et al. 2010). These four models were carefully selected among published ones (Coll and Libralato 2011), based both on their quality and on common characteristics (e.g. system surface) needed to exclude artificial sources of bias from the

[1] Such flow reconstruction results in a trophic network "model", see section "Materials and Methods".

[2] Although marine ecosystems are influenced by many factors besides predation, the assumption made when focusing on food webs is that, in phenomenological fashion, the status of a system is mirrored by the structure of its trophic flows. No direct causal relationship between system status and flows is claimed, much like temperature reflects, but is not the cause of the health or sickness of a person.

comparison (see Materials and Methods). All the original trophic networks were standardized into a common structure, i.e. models were aggregated into the same number of network nodes, each node representing similar species or functional groups. Model standardization is fundamental because ecosystem indicators and properties from ecological network analysis can strongly depend on, and be biased by, the number of nodes in a network (e.g. Abarca-Arenas and Ulanowicz 2002). To our knowledge, very few papers have performed comparative analyses of standardized Mediterranean food webs. Coll et al. (2008) analyzed only the South Catalan and the North-Central Adriatic Sea, while Libralato et al. (2010) compared the North-Central Adriatic with a small marine reserve (1.2 km^2) in the Northern Adriatic Sea. Tsagarakis et al. (2010) compared three marine systems (South Catalan, North-Central Adriatic and North Aegean Sea), but the models showed some differences in the degree of aggregation. In this work, to ensure the best aggregation possible, only pelagic organisms (living in the water column) were included in the standardized model structure, while benthic and demersal ones (living on or close to the bottom) were excluded. Why only pelagic ecosystems? On large spatial scales, they seem to be better studied and understood with respect to benthic ones, possibly because data are more widely available and food webs are simpler. Thus, pelagic data were probably of better quality and it was easier to find a standardized model structure suitable for all systems. A forced and therefore partly-inadequate standardization of the benthic compartment would have been a potential source of bias in the comparative analyses.

Materials and Methods

Models of Trophic Networks and the Ecopath Methodology

In a trophic network model, a food web is represented as a set of nodes (or groups) each representing some organisms, e.g. a population, taxon or functional group, and as the trophic flows leaving and entering such nodes. Flows include consumption and predation, i.e. flows from one node to another, imports and exports from the system (e.g. migratory flows, catches), respiration, and flows to detritus (e.g. excretion, egestion, dead organisms due to illness or old age). Each node is characterized by the total biomass of its organisms. When constructing a model, usually, information on flows and biomasses in that ecosystem is obtained from literature, but it can be difficult to collect the huge amount of data needed, especially with good precision. Approaches to tackle this issue and reconstruct the values of trophic flows and biomasses include inverse modeling (Vézina and Platt 1988) and the Ecopath methodology (Christensen et al. 2005), which, originally, was used to build the four models and, then, to standardize and analyze them.

Ecopath is based on the principle of mass conservation and the typical assumption that the food web is in an approximate steady state over the modeled period. Based on these considerations, two mass budgets can be written for each model group: (1) its production is equal to the sum of different mortality sources (predation, fishing, other mortality sources such as illness and old age) plus net exports and emigration from the system; (2) flows into the node (consumption) are equal to those leaving it, i.e. production plus respiration and unassimilated food. When written for all N groups in the network, these budgets result in a linear system:

$$\begin{aligned} B_i(P/B)_i = &\sum_{j=1}^{N} B_j(Q/B)_j DC_{ij} + Y_i \\ &+ B_i(P/B)_i \cdot (1 - EE_i) + E_i + Ex_i \end{aligned} \tag{24.1}$$

$$B_i(Q/B)_i = B_i(P/B)_i + B_i(R/B)_i + GS_i B_i(Q/B)_i \tag{24.2}$$

B_i is the biomass of group i, $(P/B)_i$ is its production rate, $(Q/B)_i$ is the consumption rate, DC_{ij} is a matrix whose elements represent the fraction that group i represents in the diet of j, Y_i are fishery yields, EE_i is the ecotrophic efficiency and represents the fraction of production "used" in the system, so that $(1-EE_i)$ represents the fraction of production flowing into detritus due to mortality sources such as illness and old age, E_i is net emigration from the system, Ex_i represents net exports, $(R/B)_i$ is the respiration rate and GS_i the fraction of unassimilated food. The modeller has to specify values for all such data but two unknowns for each group, generally EE and R/B, which are estimated by solving the linear system. The model is considered acceptable if it is mass-balanced, i.e. $EE<1$ and $R/B>0$ for all groups, otherwise input data are mutually inconsistent and have to be varied within uncertainty ranges. Thus, the mass-balance approach reduces the uncertainty about input values, ensuring that the network is, at least, a physically-feasible representation of reality where mass is conserved and flows are non-negative.

The Four Trophic Network Models and Their Standardization

Only published and high-quality Ecopath models, whose construction was well-documented, were chosen for standardization and subsequent analysis. Also, models had to describe relatively-large marine systems (surface at least 1,000 km^2) during recent years (1990s or 2000s). These criteria restricted the number of systems for which models were available in the literature, but allowed us to select only the ones which were fully comparable, as ecosystem properties can be strongly influenced by surface and depth (Libralato

Table 24.1 The common structure of the standardized models of Mediterranean trophic networks

Standardized network model	North Aegean Sea	Northern Adriatic Sea	Northern and Central Adriatic Sea	South Catalan Sea
1. Detritus	Detritus	Detritus	Detritus	Detritus
2. Phytoplankton	Phytoplankton	Phytoplankton	Phytoplankton	Phytoplankton
3. Zooplankton	Mesozooplankton	Zooplankton	Micro and mesozooplankton	Micro and mesozooplankton
4. Jellyfish	Jellyfish and *M. leidyi*	Jellyfish	Jellyfish	Jellyfish
5. Pelagic – benthopelagic cephalopods	Squids	Squids	Benthopelagic cephalopods	Benthopelagic cephalopods
6. Benthopelagic fish	Hake; Benthopelagic fishes; Picarels and bogue	European hake; Nectobenthic zooplanktivorous fish	Hake (1); Hake (2); Benthopelagic fish	Poor cod; Juvenile hake; Adult hake; Blue whiting; Benthopelagic fishes
7. Anchovy	Juvenile anchovy; Adult anchovy	Anchovies	European anchovy	European anchovy
8. Sardine	Juvenile sardine; Adult sardine	Sardines	European pilchard	European pilchard
9. Other small pelagic fish	Other small pelagic fishes	Other small pelagics	Other small pelagic fish	Small pelagic fishes
10. Medium-sized pelagic fish	Horse mackerels; Mackerels; Medium pelagic fishes	Mackerel; Horse mackerel	Horse mackerel; Mackerel; Atlantic bonito	Horse mackerel; Mackerel; Atlantic bonito
11. Large pelagic fish	Large pelagic fishes	–	Large pelagic fish	Large pelagic fishes
12. Seamammals	Dolphins	Dolphins	Dolphins	Dolphins; Fin whale
13. Seabirds	Seabirds	Seabirds	Seabirds	Audouins gull; Other sea birds

For each of the four ecosystems analyzed, the table reports which groups of the original model were used and, in some cases, lumped together to form the corresponding group of the standardized network

et al. 2010) and by the intensity of anthropogenic pressures. For example Piroddi et al. (2010) described the Northeastern Ionian ecosystem in 1964, before the decline of most fishing resources due to overexploitation and changes in primary productivity. Such model was not included in this study because, most probably, it would have been exceedingly different from the others due to the relatively "pristine" state described.

The selected ecosystems (Fig. 24.1) are now briefly described. More details are found in the cited publications describing the original models. All systems represented continental shelves except the South Catalan Sea, which also included the upper slope. The model of the Northern Adriatic food web during 1996–1998 originally comprised 34 groups (Barausse et al. 2009). The basin has a surface of 32,000 km^2, depth range about 0–50 m and is very productive and eutrophic due to the nutrient loads discharged by Italian rivers. For this reason, the system has been strongly exploited for decades by fisheries targeting mostly small pelagic fish and invertebrates, and is one of the most fished Mediterranean basins (Barausse et al. 2009, 2011). Coll et al. (2007) constructed a 40-group model of the Northern and Central Adriatic Sea during the 1990s, covering a surface of 55,500 km^2 (depth range from 10 to 200 m) which partially overlaps with the area of the Northern Adriatic model. This second model was included to represent a deeper, more oligotrophic ecosystem where more large pelagic predators (e.g. tunas) were found with respect to the more productive Northern basin. Similarly to the Northern Adriatic, fishing pressure is intense in this system (Coll et al. 2007). The North Aegean Sea food web in the mid-2000s (Tsagarakis et al. 2010) had 40 groups and covered a surface of 8,374 km^2, depth range from 20 to 300 m. The ecosystem is oligotrophic but pelagic productivity is high compared to other sub-basins of the Eastern Mediterranean due to the influence of nutrient-rich Black Sea waters, river discharges and the extended continental shelf. A major fraction of trophic flows is originated in the pelagic compartment. Anchovy and sardine dominate fishery landings and the system is probably unsustainably fished (Tsagarakis et al. 2010). The 40-group network of the South Catalan Sea (Coll et al. 2006) represents the system during 1994; its surface is 4,500 km^2 and depth ranges from 50 to 400 m. The ecosystem is oligotrophic and has been highly exploited for a long time. Again, the pelagic domain dominates trophic flows and biomasses (Coll et al. 2006).

In all networks, biomasses were expressed as wet weight on a per area basis (t km^{-2}) and biomass flows as yearly averages (t km^{-2} y^{-1}). Food web models were digitized based on the original publications, imported into Ecopath software version 5.1.218 (www.ecopath.org) and standardized into a common structure of 13 groups (Table 24.1). Such structure comprised one detritus group, representing non-living bio-available matter, and three plankton compartments: primary producers (phytoplankton), zooplankton and jellyfish. Fish were sorted according to size: three small pelagic fish groups, representing two abundant species in the systems (sardine and anchovy) and other small pelagics; medium-sized pelagics, such as mackerel or horse mackerel; large pelagics, representing species such as tuna, swordfish or sharks. A bentho-pelagic fish group was also included. Other groups were cephalopods (bentho-pelagic and pelagic, e.g. squid), sea-mammals and sea-birds. Model standardization involved the lumping of some groups in the original models, e.g. mackerel, horse mackerel and medium pelagic fishes in Tsagarakis et al. (2010) were aggregated into the medium pelagics group. Each aggregated group had biomass and flows (production, consumption, landings, etc.) equal to the sum of those of the groups aggregated to compose it. Unlike other models, the Northern Adriatic standardized network only included 12 and not 13 nodes, since large pelagics were not incorporated in the original model due to their low biomass. Such ecologically-meaningful difference among models did not introduce artificial biases into the results: even if a large pelagics node had been added to the Northern Adriatic model, its low biomass and flows would not have influenced the network analysis indicators computed for the inter-system comparison.

Exchanges of matter between living groups in the pelagic domain and those found in the original models but not included in the standardized version were represented as import or export flows through the system boundary. Such external groups included benthos and demersal fish, and groups which were found only in some of the analyzed food webs and thus, to obtain a common network structure, could not be included in the present analysis, e.g. ciliates for the North Aegean model, or sea turtles. No imports or exports for detritus were include in the standardized models. In the original networks there were groups representing fishery discards and by-catches, but here they were excluded since, generally, a large fraction of discarded catches was benthos or demersal fish. Consequently groups feeding on discard were assumed to feed on imports.

Indicators from Ecological Network Analysis

To compare the four models, indicators based on ecological theory, thermodynamics and information theory were computed for each network based on its flows and biomasses. The focus was on properties related to whole-system functioning or the global role of groups in the food web. Indicators based on trophic levels (Christensen et al. 2005) were not calculated, because the not-negligible imports from the benthic compartment in the diets of some groups could have led to biased assessments of trophic levels, misrepresenting the actual trophic positions.

Indicators Based on Flows, Biomasses and Their Ratios

Several types of flows were assessed in each network: total primary production, the sum of all consumption, of all respiratory flows, of all flows into detritus, of all production, of all catches, of all exports. The sum of all network flows, or Total System Throughput (*TST*), was computed to gauge the system activity and energetic size (Ulanowicz 1986; Christensen et al. 2005). The total biomass in the network was computed by summing over all living groups, i.e. excluding detritus.

Based on such values, other indicators were computed such as the net system production (primary production minus respiration flows) which, according to the theory of ecological succession, should approach zero in more mature ecosystems (*sensu* Odum 1969), where primary production is more efficiently exploited and more biomass is stored, as reflected by higher respiration (Odum 1969; Christensen et al. 2005). Consequently, the primary production-total respiration ratio (dimensionless) should be close to one in mature systems. The primary production-total biomass ratio (y^{-1}) should be higher in ecosystems in their early developmental stages, when populations are characterized by high turnover rates (i.e. production-biomass ratios), and should decrease as systems mature and biomass is stored (Odum 1969; Christensen et al. 2005). For similar reasons, the total biomass-*TST* ratio (y), i.e. the biomass sustained by a unit of flow in the system, should be higher in developed ecosystems, where biomass is accumulated, K-selected organisms with lower metabolic rates dominate, and energy transfers in the food web are more efficient and organized (Odum 1969; Peters 1983; Ulanowicz 1997; Christensen et al. 2005).

Fishing

Fishery gross efficiency (dimensionless) was calculated as the ratio of catch (including discards) to primary production. High values can indicate that fishing pressure is high or, alternatively, that catches mainly consist of abundant low-trophic level species (Christensen et al. 2005). The equivalent units of primary production required (PPR) to sustain catches represented the sum of primary and detritus production consumed at the bottom of all the trophic chains leading to fished groups and, ultimately, associated catches, similarly to Odum's emergy (Christensen et al. 2005). Unlike fishery gross efficiency, PPR weights catches of high-trophic level predatory species more than an equal amount of caught herbivorous species. PPR was expressed as percentage of the overall detrital and phytoplankton production in the ecosystem.

Food Web Structure and Cycling

Connectance (dimensionless) is the number of existing predator-prey links in the food web divided by the number of potential links, i.e. the fraction of realized links (Christensen et al. 2005), and it gauges the connectedness of the network. The System Omnivory index (dimensionless; Christensen et al. 2005) provides similar information, but it takes the magnitude of trophic flows into account: it is the average Omnivory index of consumer groups, weighted by the logarithm of their food intake. The Omnivory index of a group is the variance of the trophic levels of its preys. The System Omnivory index should increase with system maturity, as the flow structure changes from linear to web-like (Odum 1969).

The number of cycles in the network was computed, as well as their average length (the mean number of trophic interactions in a cycle, i.e. without taking the magnitude of flows involved in different cycles into account). The amount of cycled flows, i.e. passing more than once through the same compartment, was calculated, as well as its ratio with *TST*, named Finn's Cycling index (FCI). Such dimensionless fraction should be higher as systems mature, becoming less dependent on exogenous inputs, and trophic flows become more diverse (Odum 1969; Finn 1980; Ulanowicz 1986, 2004; Christensen et al. 2005). The Predatory Cycling Index (PCI; Christensen et al. 2005) is the ratio of cycled to total system throughput, but calculated after cycles involving detritus have been removed.

Finn's mean path length (FMPL) was computed as *TST* divided by the sum of exports and respiratory flows (Finn 1980; Christensen et al. 2005). FMPL represents the average number of compartments that a unit of inflow goes through before leaving the food web, and should grow with ecosystem maturity in response to increased flow and species diversity, connectedness and cycling (Odum 1969; Christensen et al. 2005). Finn's straight-through path length (FSTPL) has a similar meaning: it is the average path length in the network if cycles are neglected, i.e. the ratio of the throughput going straight through the system (i.e. *TST* minus cycled flows) to the sum of exports and respiration (Finn 1980).

Information Theory and Ascendency

Matter/energy flows in ecological networks can be translated into flows of information (Ulanowicz 1986). In general, information is carried by changes in observable physical quantities (e.g. matter or energy) through which it can be stored and propagated. In analogy with Shannon's definition of entropy, the diversity of flows in a trophic network is:

$$H = -K \cdot \sum_{i,j} P_{ij} \log_2 P_{ij} \tag{24.3}$$

$P_{ij} = J_{ij} / TST$ is the unconditional probability that flow J_{ij} from i to j occurs, and the sum is over all i, j combinations. H is related to the structure of the network of flows, but not to system size, which is measured by *TST*. Conversely, *TST* does not say anything about network organization. Since the scalar K in (24.3) defines the units of information (the scale) and *TST* is a feature of the network scale, Ulanowicz (1986) proposed $K = TST$ and termed (24.3) the network capacity C, which can be factorized as $C = A + \Phi$, where:

$$A = TST \bullet \sum_{i,j} P_{ij} \log_2 \frac{P_{i|j}}{P_j^*} \tag{24.4}$$

$$\Phi = -TST \bullet \sum_{i,j} P_{ij} \log_2 \frac{P_{i|j} P_{ij}}{P_j^*} \tag{24.5}$$

A is called ascendency and Φ is the overhead. $P_{i|k} = J_{ik} / \sum_q J_{iq}$ is the conditional probability that a unit of medium enters k knowing *a priori* that it originated in i, and $P_k^* = \sum_q J_{qk} / TST$ is the *a priori* probability that one unit of medium flowing in the network enters k. C measures the maximum exchangeable information in all potential network configurations attainable with the available machinery (connections and flow intensities), while A accounts for the information actually realized in the current configuration. Φ measures the degrees of freedom available to the network, i.e. its capability to rearrange in response to external disturbances.

Ulanowicz (1986, 1997, 2004) proposed that ecosystems show a propensity to increase in A, because it measures both the growth (or activity, an extensive property represented by *TST*) and development (or organization, an intensive property given by the summation in (24.4), called average mutual information or *AMI*) of the system. An increase in *TST* means that the system has grown in the usual economic sense. An increase in *AMI* means that the system has developed new constraints to channel the flows into more specific and efficient pathways. Thus increasing A indicates that the system is experiencing growth and/or development (Ulanowicz 1997). Processes that increase A include a greater number of species, a higher retention of resources and the specialization of the food chain, which are among the primary signs that the ecosystem is maturing (Odum 1983). Such propensity to achieve an efficient, streamlined network configuration is accompanied by a trade-off: high A can mean low Φ, i.e. the system becomes vulnerable to perturbations, if their sum C is fixed (Ulanowicz 1986, 1997, 2004).

C, A and Φ can all be expressed as the sum of four contributions due to different flow types: flows between groups (internal flows), respiratory flows, imports and exports. In particular, the ratio of internal ascendency to internal capacity (A_i/C_i) is the realized fraction of the potential organization of prey-predator flows and measures system maturity, resistance and efficiency. The internal overhead-capacity ratio (Φ_i/C) originates from the redundancy of predatory connections and measures system resilience (Baird et al. 1991; Ulanowicz 1997, 2004).

Input-Output Analysis and Keystone Species

The global importance of different groups in the food web was evaluated using the mixed trophic impact matrix (Ulanowicz and Puccia 1990; Libralato et al. 2006), whose elements m_{ij} represent the total effect that an infinitesimal change in the biomass of one group *i* has on another group *j*. m_{ij} is the sum of all direct and indirect (i.e. mediated through trophic interactions) effects, and can be positive or negative. The overall impact of a group *i* on the ecosystem can be estimated as:

$$\varepsilon_i = \sqrt{\sum_{j \neq i} m_{ij}^2} \qquad (24.6)$$

Where impacts are summed over all living groups but *i*. To identify keystone species, i.e. groups having a disproportionately high total impact on the system with respect to the fraction p_i of the total living biomass that they represent, the keystoneness *KS* (Libralato et al. 2006) was calculated for each group *i*:

$$KS_i = \log_{10}\left[\varepsilon_i \left(1 - p_i\right)\right] \qquad (24.7)$$

Results

Flows and Biomasses

Table 24.2 reports indicators from ecological network analysis computed for the four standardized pelagic models. Absolute biomass flows showed that the Adriatic Sea was clearly the most vigorous and dynamic system, particularly the Northern basin, where *TST* was 1.6 times as large as that of the model including the Central basin, and nearly 4 times as those of the North Aegean and South Catalan networks, showing similar total flows. Primary productivity (Table 24.2) partly explained these differences: the latter two systems are oligotrophic while the Adriatic is quite productive, and eutrophic in its Northern area (Barausse et al. 2009).

However, differences in total primary production and in net system production among systems were much larger than those in *TST*: in the Northern Adriatic, phytoplankton productivity was twice as large as that of the Northern and Central Adriatic, 4.3 times as that of the North Aegean and 6 times as that of the South Catalan Sea, while net system production was 2.2 times as large as in the Northern and Central Adriatic, 4.5 times as large as in the North Aegean and 14.8 times as large as in the South Catalan Sea. These results suggested that not only primary production, but also the flow structure differed among food webs, e.g. the South Catalan network was probably more complex and longer while the Northern Adriatic food web was shorter. Indeed, *TST* partitioning among flow typologies greatly varied across systems: absolute consumption flows were largest in the Northern and Central Adriatic Sea, where they represented about 27 % of *TST*, similarly to the North Aegean model. Consumption dominated total flows in the South Catalan Sea (47 % of *TST*), while in the Northern Adriatic Sea it corresponded only to 17 % of *TST*. In this last network, *TST* was dominated by exports (43 %), mainly unconsumed detrital production and predatory flows towards the benthic compartment (catches were negligible), and by flows to detritus (33 % of *TST*), suggesting that a large fraction of production in the system was not immediately consumed or not used at all. Similarly, flows to detritus and exports dominated total flows in the Northern and Central Adriatic (respectively, 35 and 28 % of *TST*) and in the North Aegean Sea (respectively, 24 and 42 % of *TST*). The sum of flows to detritus and exports was less than half of the total flows (respectively, 21 and 12 % of *TST*) only in the South Catalan model.

In line with these results, the ecotrophic efficiency of detritus, i.e. the fraction of pelagic flows into detritus which was consumed in the pelagic ecosystem, was high only in the South Catalan Sea (0.55), being 0.22 in the Northern and Central Adriatic, 0.17 in the North Aegean and only 0.08 in the Northern Adriatic Sea. Such system ranking was exactly opposite to that based on the fraction of *TST* represented by exports and detrital flows. The ratio of all production to primary production in the system suggested, similarly, that basal resources were used more intensively, or more efficiently, in the South Catalan Sea (ratio = 1.57) than in the Northern-Central Adriatic (1.25), the Northern Aegean (1.18) and, lastly, the Northern Adriatic Sea (1.11).

Total biomasses seemed to be only partially related to primary productivity or other flows in the systems (e.g. *TST*, production, consumption), suggesting differences in the development of the flow networks. Living biomass was highest in the Northern and Central Adriatic network, while the more productive Northern basin showed a similar value to that of the South Catalan Sea. The North Aegean Sea showed the lowest biomass, despite a primary productivity 1.4 times as large as that of the South Catalan ecosystem. The primary production-total biomass ratio, representing the algal production needed to sustain a unit of biomass in the food web, suggested that the systems could be ranked in this order of decreasing specific productivity and of increasing efficiency and maturity: Northern Adriatic, North Aegean, Northern and Central Adriatic, South Catalan Sea. The primary production-respiration ratio provided the same ranking, suggesting that the South Catalan Sea was the most mature system since, there, the largest fraction of energy fixed by phytoplankton was respired, i.e. used to sustain the living biomass stored in the ecosystem. Indeed, in the South Catalan Sea the total biomass-*TST* ratio, i.e. the biomass sustained by a unit of flow in the food web, was twice as large as in the

Table 24.2 Indicators from ecological network analysis, computed for the four standardized trophic network models of Mediterranean marine ecosystems

Indicator	Currency	North Aegean	Northern Adriatic	North-Central Adriatic	South Catalan
Flows, biomasses and their ratios					
Consumption flows	t km^{-2} y^{-1}	316.2	801.4	846.0	606.6
Export flows	t km^{-2} y^{-1}	506.9	2068.3	876.5	148.7
Respiration flows	t km^{-2} y^{-1}	100.1	359.6	282.8	255.3
Flows into detritus	t km^{-2} y^{-1}	293.6	1612.8	1078.8	268.2
Total system throughput (*TST*)	t km^{-2} y^{-1}	1217.0	4842.0	3084.0	1279.0
Sum of all production	t km^{-2} y^{-1}	635.0	2553.0	1442.0	609.0
Total primary production	t km^{-2} y^{-1}	536.1	2310.0	1150.0	386.7
Tot. primary prod./tot. respiration	–	5.35	6.42	4.07	1.51
Net system production (p.p. – resp.)	t km^{-2} y^{-1}	435.9	1950.4	867.3	131.4
Total primary prod./total biomass	y^{-1}	35.44	71.89	30.63	11.88
Total biomass/*TST*	y	0.0124	0.0066	0.0100	0.0254
Total biomass (excluding detritus)	t km^{-2}	15.13	32.13	37.54	32.55
Fishing					
Total catches	t km^{-2} y^{-1}	1.45	2.95	1.17	4.45
Fishery gross efficiency (catch/p.p.)	–	0.0027	0.0013	0.0010	0.0115
PPR to sustain catches	%	4.39 %	3.31 %	2.92 %	18.03 %
Food web structure and cycling					
Connectance	–	0.35	0.40	0.30	0.30
System Omnivory Index	–	0.26	0.18	0.33	0.37
Finn's mean path length (FMPL)	–	2.00	1.99	2.66	3.17
Finn's straight-through path length (FSTPL)	–	1.91	1.96	2.48	2.68
Number of cycles in the food web	–	322	318	114	104
Mean length of cycles	–	5.30	5.82	4.91	4.82
Throughput cycled	t km^{-2} y^{-1}	59.9	78.2	206.0	196.3
Finn's cycling index (FCI)	% of *TST*	4.92 %	1.61 %	6.68 %	15.35 %
Throughput cycled (excl. detritus)	t km^{-2} y^{-1}	13.4	14.5	36.3	25.7
Predatory cycling index (PCI)	% of *TST* w/o detr.	2.13 %	0.89 %	3.92 %	3.46 %
Information theory					
Ascendency (*A*)	t bits km^{-2} y^{-1}	839.8	4025.7	3014.4	1149.9
Overhead (Φ)	t bits km^{-2} y^{-1}	3513.9	10267.2	6909.7	3584.4
Capacity (*C*)	t bits km^{-2} y^{-1}	4353.7	14292.9	9924.1	4734.3
A/*C*	%	19.3 %	28.2 %	30.4 %	24.3 %
Internal *A*/Internal *C*	%	11.8 %	22.1 %	18.7 %	19.7 %
Internal Φ/*C*	%	44.7 %	39.7 %	53.6 %	50.3 %

North Aegean Sea, 2.5 times as large as in the Northern and Central Adriatic, and 3.8 times as large as in the Northern Adriatic Sea.

Fishing

Catches per unit of system surface greatly varied: by far, the highest values were observed in the South Catalan Sea, 3.8 times as large as the lowest catches of the Northern and Central Adriatic Sea. Such difference was even larger (11.5 times) when rescaling catches on primary production. PPR to sustain pelagic fisheries was markedly larger in the South Catalan Sea with respect to other systems, suggesting quite a strong fishing pressure: nearly a fifth of the whole primary production indirectly sustained exploitation. Again, the Northern and Central Adriatic Sea scored lowest in PPR among the four systems.

Pathways and Cycles

The Northern Adriatic and North Aegean networks were the most connected according to connectance, but the complexity of the food webs was ranked differently if the magnitude of flows along different pathways was weighted through the System Omnivory index. The South Catalan Sea had the most web-like network, as well as the longest average flow pathway (FMPL>3 only in this system), and this was true even if cycles were neglected (FSTPL). The System Omnivory index, FMPL and FSTPL consistently showed that the Northern and Central Adriatic had the second most complex network, while the Northern Adriatic and Aegean Sea had a short (both path length measures<2) and linear flow structure.

The number of cycles and their mean length was markedly higher in the Northern Adriatic and North Aegean Seas, about 320, but, similarly to connectance, these indicators were not much informative because they did not weight the magnitude of flows along trophic pathways. FCI showed that cycling was actually most intense in the South Catalan network, followed by far by the Northern and Central Adriatic Sea, while it was lowest in the Northern Adriatic Sea. This rank was partly confirmed when detritus and the cycles involving it were not considered, but the top positions were exchanged: PCI was slightly higher in the Northern and Central Adriatic than in the South Catalan Sea, for which FCI >>PCI. Therefore, in the latter system, a large fraction of cycled flows involved the detrital pool. In general, cycling involved a small fraction of total flows, especially when neglecting detritus (mean FCI=7.14 %, mean PCI=2.60 %).

Indicators from Information Theory

System ranking according to ascendency, overhead and capacity mirrored exactly that based on *TST*, confirming that these indices weight disproportionately the magnitude of the scaling factor *TST* in Eqs. 24.3, 24.4 and 24.5 (Fath et al. 2001). When rescaling internal overhead on capacity (Φ_i/C), system ranking became the same as the one based on cycling (PCI), confirming that fewer redundant pathways were present in the North Aegean and Northern Adriatic pelagic food webs, while the other two networks were more complex. The other results were apparently not consistent with most of the other findings previously shown in this work: the *A*/*C* ratio, i.e. the fraction of the potential system organization that is realized, identified the two Adriatic networks as the most developed ones. The North Aegean was ranked last among the four networks based on this index, as well as according to A_i/C_i. Surprisingly, A_i/C_i identified the Northern Adriatic network as the most efficient and mature among the analyzed models. However, *A*/*C* and A_i/C_i ratios did not generally differ much among systems.

Keystones and Overall Impacts

Indices of keystoneness and overall trophic impacts (Libralato et al. 2006) are reported on Fig. 24.2. Groups belonging to intermediate trophic levels such as zooplankton and medium-sized pelagics exerted strong impacts on the pelagic food webs, moreover disproportionately high impacts with respect to their abundance in the system (*KS* index). To a lesser extent, other groups belonging to medium-low trophic levels appeared to influence the food webs noticeably, such as phytoplankton, cephalopods and anchovy or, alternatively, sardine (these two species have similar trophic roles and, typically, opposite abundance trends in marine ecosystems, Chavez et al. 2003). Other small pelagic fish had weak effects in all systems, indicating that anchovy and sardine were the dominating small pelagic fish species in these Mediterranean basins. These results were generally in line with previous findings for the original food web models including the benthic compartment (Coll et al. 2006, 2007; Barausse et al. 2009; Tsagarakis et al. 2010). Jellyfish, sea-birds, mammals and large pelagic fish also had generally-negligible ecosystem impacts. The only exceptions were dolphins and fin whales in the South Catalan Sea and large pelagics in the Northern and Central Adriatic Sea, representing keystone species with medium-high overall impacts on the pelagic system.

Discussion

To our knowledge, this is the first comparative analysis of several standardized trophic network models of large-scale Mediterranean marine ecosystems. To ensure a proper comparison, published models were standardized to represent the pelagic compartment of the food web through a common structure. Does this choice of focusing exclusively on the pelagic ecosystem provide a distorted picture of its functioning? In all systems, the pelagic compartment dominated the demersal one in terms of flows (Coll et al. 2006, 2007; Barausse et al. 2009; Tsagarakis et al. 2010), and phytoplankton was always, by far, the most important energy source: from a quantitative point of view the choice of focusing on the pelagic compartment would be justified. This consideration does not exclude, however, that the benthic compartment can influence the pelagic one through key trophic interactions, and a good benthic-pelagic coupling appears to be a characteristic of several large-scale Mediterranean marine systems (e.g. Coll et al. 2007; Tsagarakis et al. 2010). Therefore the present study only represents a first step to understand the functioning of Mediterranean ecosystems through comparative analysis, and future research should extend network standardization to include the demersal compartment.

The standardized networks were based on models which had previously been built according to different criteria and for different purposes, all factors that complicated the process of aggregation. Some standardization choices, such as the common model structure and lumped or excluded groups, were partly forced by data availability and could have been

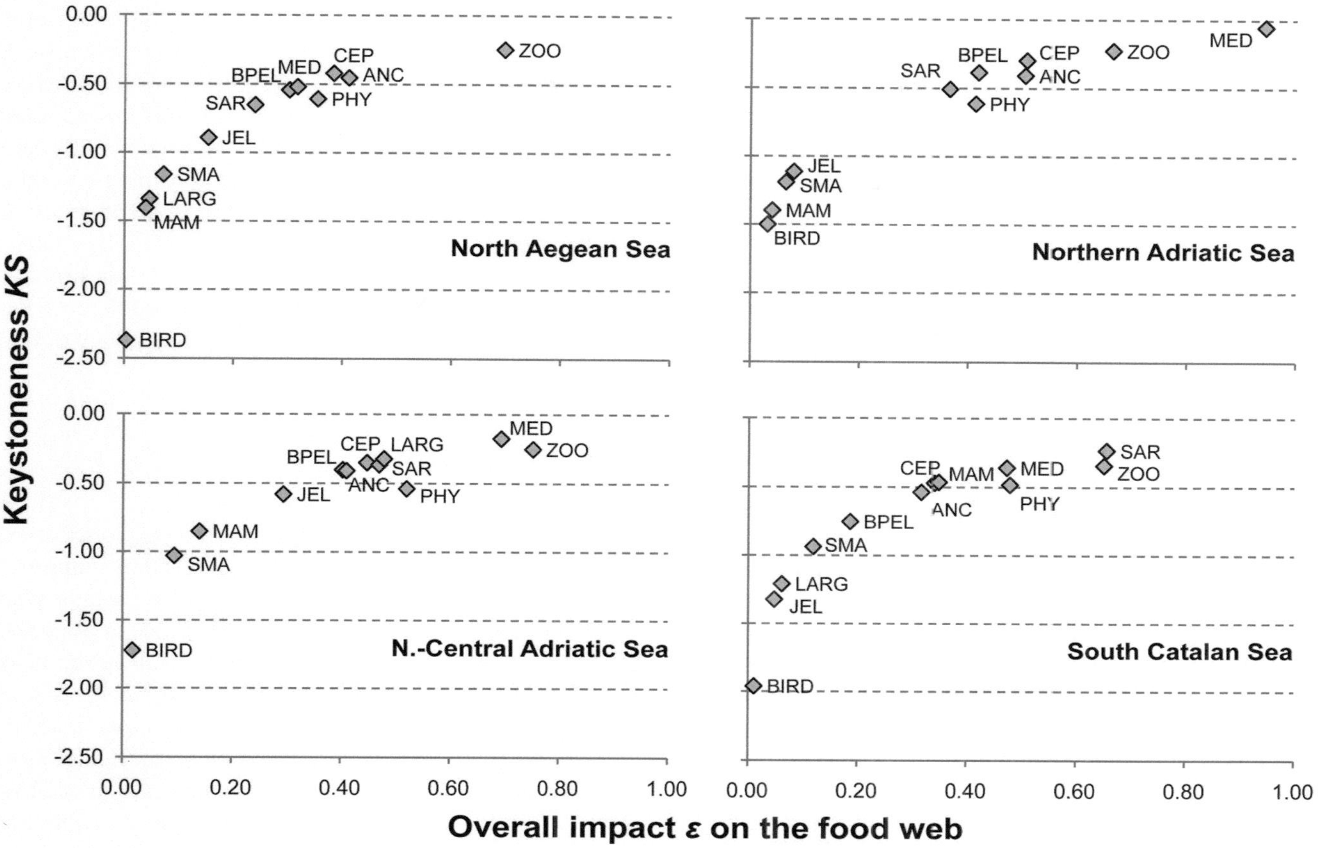

Fig. 24.2 Keystoneness *sensu* Libralato et al. (2006) plotted against the overall impact on the ecosystem for the living groups of the four trophic network models. Labels indicate phytoplankton (*PHY*), zooplankton (*ZOO*), jellyfish (*JEL*), cephalopods (*CEP*), bentho-pelagic fish (*BPEL*), anchovy (*ANC*), sardine (*SAR*), medium pelagics (*MED*), other small pelagics (*SMA*), large pelagics (*LARG*), sea-birds (*BIRD*) and sea-mammals (*MAM*). Generally, intermediate-trophic level organisms such as zooplankton, medium pelagics, anchovy and sardine had great impacts, both on an absolute scale and with respect to their relative biomass in the system, while sea-birds and sea-mammals ranked low

different if the networks had been built for the only purpose of comparing the functioning of their pelagic compartments. The effect of such bias on results was difficult to evaluate, but it was probably not big, since lumping did not involve many groups in the models (Table 24.1).

Structure, Functioning and Stages of Development of Mediterranean Ecosystems

The comparative network analysis highlighted clear differences in the trophic structure and functioning of Mediterranean marine systems. The South Catalan Sea emerged as the most developed and mature ecosystem *sensu* Odum (1969) according to several indicators: its food web sustained the largest amount of living biomass per unit of primary production or of flow in the network, its primary production-respiration ratio was the closest to one, it had the highest fraction of cycled flows, the longest pathways and the most web-like structure. Conversely, the Northern Adriatic Sea appeared to be in the earliest developmental stage, with high specific productivity and low sustained biomass per unit of flow, and a short, linear trophic network with low cycling. The North Aegean and the Northern and Central Adriatic Seas showed an intermediate degree of maturity with respect to these two extremes. According to most indicators (e.g. the primary production-respiration ratio, path lengths, System Omnivory index, cycling indices), the latter appeared to be more developed than the former, but this difference was not always clear, e.g. the total biomass-*TST* ratio was highest in the North Aegean Sea.

Primary production emerged as a key process, clearly visible in ecosystem properties such as the amount of total flows, but it did not explain every difference among systems. In particular, it did not mirror the maturity ranking, since the two Adriatic systems were much more productive than the others. Interestingly, the simple inter-system comparison of total primary production or net system production with other, equally-uncomplicated indicators such as *TST* or total production was able to anticipate some of the differences in development highlighted by more sophisticated analyses. Higher primary productivity corresponded to a comparatively-weaker increase in *TST* suggesting that, for example, the flow structure of the eutrophic Northern Adriatic had to be shorter and simpler compared to other systems, particularly the South Catalan Sea. In general, most indicators led to consistent interpretations, in line with the indicator redundancy expected from ecological theory and observed in past studies (Odum 1969; Christensen 1995).

The composition of *TST* in terms of exports and flows to detritus *versus* consumption confirmed the maturity ranking of the systems, and provided additional insights. The most mature ecosystem was again the South Catalan Sea, where consumption dominated, thus suggesting that resources were intensely and efficiently exploited in the pelagic compartment, followed in order of decreasing development by the Northern and Central Adriatic, the North Aegean and the Northern Adriatic Sea, where exports and flows to detritus were higher, symptomatic of low organization and inefficient resource exploitation. A large amount of unused or exported primary production could be a signal of systemic stress (Odum 1985) and of disruption of the predator-prey pathways transferring energy to higher trophic levels. Such high inefficiencies suggest that the demersal compartment in Mediterranean food webs could be strongly dependent on the energy provided by the pelagic compartment. As previously discussed, the pelagic compartment dominates the demersal one in several Mediterranean sub-basins, and bentho-pelagic coupling based on detritus flows and an inefficient exploitation of production in the pelagic domain has previously been identified as a typical feature of Mediterranean marine systems (Coll et al. 2008; Barausse et al. 2009; Tsagarakis et al. 2010).

The analysis of cycles and connectivity highlighted the importance of weighting the magnitude of trophic flows when studying food webs. Most probably, the differences in connectance, number of cycles and mean cycle lengths among systems simply highlighted different approaches to the construction of the original models: more but weaker trophic interactions were included in the North Aegean and Northern Adriatic models. Only such interpretation can explain the nearly-threefold difference in the number of cycles between the two Adriatic systems.

Two information theory indicators, A/C and A_i/C_i, did not agree with the maturity ranking identified by other analyses. A justification is that differences among systems in the values of A/C and A_i/C_i are too small to be meaningful, and should not be interpreted. A different, speculative explanation is that, in the case of the Adriatic, the high value of these two ratios actually portrayed the long-term adaptation of the ecosystem to continuously-stressful conditions, rather than system maturity (Baird et al. 1991). The Northern Adriatic could have scored high among the systems in both ratios because its flow configuration has adapted to a long history of intense fishing and nutrient enrichment (Fortibuoni et al. 2010; Barausse et al. 2011), leading to a depressed state (Barausse et al. 2009) which could have become the usual system condition. Within such existing constraints, trophic connections have become efficient, organized and resistant (see Baird et al. 1991). Instead, attributes of maturity characterized by shorter-term dynamics, such as a low primary production-total biomass ratio, are not observed because the ecosystem is, indeed, continuously perturbed (Barausse et al. 2009).

Exploitation

The Adriatic systems appeared to be the least exploited ones. Indeed their high primary productivity, fuelled by anthropogenic nutrient loads especially in the Northern basin, is able to sustain large fisheries (Table 24.2; Barausse et al. 2009, 2011). The Northern and Central Adriatic food web appears less exploited than the Northern one, possibly because pelagic fisheries concentrate in the more productive Northern area. A bit surprisingly, the highest pelagic catches were observed in the oligotrophic South Catalan Sea, as well as the most intense exploitation (PPR). This was probably not in disagreement with its greatest development among the four Mediterranean networks: only a system where energy is efficiently channeled towards fishery resources can sustain a fishing pressure indirectly consuming 18 % of its primary production.

The analysis of overall impacts and keystones highlighted some similarities among Mediterranean pelagic food webs and, possibly, the impact of fishing: r-selected species with medium-low trophic levels appeared to play key roles in the functioning of all systems while, generally, large predators had negligible impacts, even when taking their low biomass into account through the keystoneness index, with partial exceptions in the more mature South Catalan and Northern and Central Adriatic Seas. Large predators are the first species to succumb to fishing pressure because they are less abundant, grow more slowly, mature later and are easier to catch than smaller species (Pauly et al. 1998). Indeed, several Mediterranean large predatory species have strongly declined because of overexploitation (e.g. Ferretti et al. 2008; Coll et al. 2008; Barausse et al. 2011), and the weak impacts of large predators could suggest that the analyzed ecosystems are all strongly impacted by fishing. The key role of zooplankton and small-medium pelagic fish may be related to their trophic position, as they represent the main channel through which energy from phytoplankton reaches pelagic food webs, and suggests that these Mediterranean ecosystems could be wasp-waist controlled (Cury et al. 2001).

Conclusions

This comparative study highlighted that the Mediterranean is a sea rich with diversity also in the structure and functioning of its large-scale marine pelagic ecosystems. Multiple indicators from ecological network analysis provided consistent and, thus, robust indications concerning similarities and differences among Mediterranean pelagic trophic networks. Shared traits included the important role of intermediate-trophic level species, the low impact on the system exerted by large predators, and inefficiencies in the exploitation of phytoplankton and detritus production, which gave rise to high export flows fuelling the benthic compartment. Primary production was a fundamental process influencing the properties of the food webs, but other marked differences in the global structure of trophic flows emerged, unrelated to phytoplankton productivity. In particular, the ecosystems could be clearly ranked according to their maturity and development (from high to low): South Catalan, Northern and Central Adriatic, North Aegean and, lastly, Northern Adriatic Sea. Despite its higher maturity, the South Catalan ecosystem sustained a strong fishing impact: it was speculated that the higher efficiency of its trophic network made this possible. The Northern Adriatic appeared to be the system in the earliest developmental stage. This basin is heavily exploited and other anthropogenic pressures, such as nutrient enrichment, and environmental factors, such as climate variations, probably act synergistically with fishing and exert a strong stress on the ecosystem (Barausse et al. 2009, 2011).

Additional research is needed to relate anthropogenic pressures to ecosystem structure and functioning in the Mediterranean Sea on large spatial scales.[3] In particular, it remains to be seen whether the varying degrees of development in Mediterranean ecosystems highlighted by this work are directly related to the intensity of human pressures or, instead, mirror different natural conditions.[4] This is a complex research question deserving more analyses. Future work will also, hopefully, focus upon the Eastern and Southern Mediterranean areas, where few or no efforts appear to have been made to construct food web models and study the structure and functioning of ecological networks. These information are critical in order to implement an effective Ecosystem Approach to the management of the Mediterranean Sea and its sub-basins.

Acknowledgments The research leading to these results has received funding from the European Community's Seventh Framework Programme (FP7/2007-2013) under grant agreement n° 226675, KnowSeas project.

[3]For a review of what has been done so far in the Mediterranean, see Coll and Libralato (2011).

[4]See Barausse et al. (2009) for the Northern Adriatic Sea.

References

Abarca-Arenas LG, Ulanowicz RE (2002) The effects of taxonomic aggregation on network analysis. Ecol Model 149:285–296

Baird D, McGlade JM, Ulanowicz RE (1991) The comparative ecology of six marine ecosystems. Philos Trans R Soc Lond Ser B 333:15–29

Barausse A, Duci A, Mazzoldi C, Artioli Y, Palmeri L (2009) Trophic network model of the Northern Adriatic Sea: analysis of an exploited and eutrophic ecosystem. Estuar Coast Shelf Sci 83:577–590

Barausse A, Michieli A, Riginella E, Palmeri L, Mazzoldi C (2011) Long-term changes in community composition and life-history traits in a highly exploited basin (northern Adriatic Sea): the role of environment and anthropogenic pressures. J Fish Biol 79:1453–1486

Caddy JF (1993) Contrast between recent fishery trends and evidence for nutrient enrichment in two large marine ecosystems: the Mediterranean and the Black Seas. In: Sherman K, Alexander LM, Gold BD (eds) Large marine ecosystems. Stress, mitigation and sustainability. American Association for the Advancement of Science, Washington, DC

Caddy JF, Refk R, Do-Chi T (1995) Productivity estimates for the Mediterranean: evidence of accelerating ecological change. In: Effects of riverine inputs on coastal ecosystems and fisheries resources. FAO Fisheries Technical Paper 349. FAO, Rome

CBD (2000) Conference of the parties to the convention on biological diversity. Report of the fifth meeting of the conference of the parties to the convention on biological diversity. Nairobi, 15–26 May 2000. UNEP/CBD/COP/5/23, Decision V/6

Chavez FP, Ryan J, Lluch-Cota SE, Niquen CM (2003) From anchovies to sardines and back: multidecadal change in the Pacific Ocean. Science 299:217–221

Christensen V (1995) Ecosystem maturity – towards quantification. Ecol Model 77:3–32

Christensen V, Walters CJ, Pauly D (2005) Ecopath with Ecosim: a user's guide. Nov 2005 edn. Fisheries Centre, University of British Columbia, Vancouver

Coll M, Libralato S (2011) Contributions of food web modelling to the ecosystem approach to marine resource management in the Mediterranean Sea. Fish Fish 13:60–88

Coll M, Palomera I, Tudela S, Sardà F (2006) Trophic flows, ecosystem structure and fishing impacts in the South Catalan Sea, Northwestern Mediterranean. J Mar Syst 59:63–96

Coll M, Santojanni A, Palomera I, Tudela S, Arneri E (2007) An ecological model of the Northern and Central Adriatic Sea: analysis of ecosystem structure and fishing impacts. J Mar Syst 67:119–154

Coll M, Lotze HK, Romanuk TN (2008) Structural degradation in Mediterranean Sea food webs: testing ecological hypotheses using stochastic and mass-balance modelling. Ecosystems 11:939–960

Coll M, Piroddi C, Steenbeek J et al (2010) The biodiversity of the Mediterranean Sea: estimates, patterns, and threats. PLoS One 5(8):e11842. doi:10.1371/journal.pone.0011842

Cury P, Shannon L, Shin YJ (2001) The functioning of marine ecosystem. In: Reykjavik conference on responsible fisheries in the marine ecosystem, Reykjavik, 2001

EU (2008) Directive 2008/56/EC of the European Parliament and of the Council of 17 June 2008, establishing a framework for community action in the field of marine environmental policy (Marine Strategy Framework Directive). Official Journal of the European Union, L 164:19–40

Fath BD, Patten BC, Choi JS (2001) Complementarity of ecological goal functions. J Theor Biol 208:493–506

Ferretti F, Myers RA, Serena F, Lotze HK (2008) Loss of large predatory sharks from the Mediterranean Sea. Conserv Biol 22:952–964

Finn JT (1980) Flow analysis of models of the Hubbard Brook ecosystem. Ecology 61:562–571
Fortibuoni T, Libralato S, Raicevich S, Giovanardi O, Solidoro C (2010) Coding early naturalists' accounts into long-term fish community changes in the Adriatic Sea (1800–2000). PLoS One 5(11):e15502. doi:10.1371/journal.pone.0015502
Garcia SM, Zerbi A, Aliaume C, Do Chi T, Lasserre G (2003) The ecosystem approach to fisheries. Issues, terminology, principles, institutional foundations, implementation and outlook. FAO Fisheries Technical Paper 443. FAO, Rome
Halpern BS, Walbridge S, Selkoe KA et al (2008) A global map of human impact on marine ecosystems. Science 319:948–952
Jørgensen SE, Bendoricchio G (2001) Fundamentals of ecological modelling, 3rd edn. Elsevier, Amsterdam
Lejeusne C, Chevaldonné P, Pergent-Martini C, Boudouresque CF, Pérez T (2009) Climate change effects on a miniature ocean: the highly diverse, highly impacted Mediterranean Sea. Trends Ecol Evol 25:250–260
Libralato S, Christensen V, Pauly D (2006) A method for identifying keystone species in food web models. Ecol Model 195:153–171
Libralato S, Coll M, Tempesta M et al (2010) Food-web traits of protected and exploited areas of the Adriatic Sea. Biol Conserv 143:2182–2194
Odum EP (1969) The strategy of ecosystem development. Science 164:262–270
Odum HT (1983) Systems ecology. Wiley, New York
Odum EP (1985) Trends expected in stressed ecosystems. Bioscience 35:419–422
Pauly D, Christensen V, Dalsgaard J, Froese R, Torres F Jr (1998) Fishing down marine food webs. Science 279:860–863
Peters RH (1983) The ecological implications of body size. Cambridge University Press, Cambridge
Piroddi C, Bearzi G, Christensen V (2010) Effects of local fisheries and ocean productivity on the northeastern Ionian Sea ecosystem. Ecol Model 221:1526–1544
Tsagarakis K, Coll M, Giannoulaki M et al (2010) Food-web traits of the North Aegean Sea ecosystem (Eastern Mediterranean) and comparison with other Mediterranean ecosystems. Estuar Coast Shelf Sci 88:233–248
Ulanowicz RE (1986) Growth and development: ecosystems phenomenology. Springer, New York
Ulanowicz RE (1997) Ecology, the ascendent perspective. Columbia University Press, New York
Ulanowicz RE (2004) Quantitative methods for ecological network analysis. Comput Biol Chem 28:321–339
Ulanowicz RE, Puccia CJ (1990) Mixed trophic impacts in ecosystems. Coenoses 5:7–16
Vézina AF, Platt T (1988) Food web dynamics in the ocean. I. Best-estimates of flow networks using inverse methods. Mar Ecol Prog Ser 42:269–287
Worm B, Hilborn R, Baum JK et al (2009) Rebuilding global fisheries. Science 325:578–585

Bioconstructions in the Mediterranean: Present and Future

25

Petar Kružić

Abstract

In the Mediterranean Sea, most important habitat formers are bioconstructors. Bioconstructors provide habitats for a large variety of organisms and these organisms rely on bioconstructors as a source of food and shelter. Marine bioconstructors in temperate seas have been recognized to have a structural and functional role of marine biodiversity (as a habitat formers and ecosystem engineers), the same as coral reefs in tropical regions. Bioconstructors are ranging from coralligenous formations (formed usually by coralline algae, sponges, cnidarians, and bryozoans) to vermetid reefs, deep-sea white corals and oyster banks. Some habitats like coral banks formed by shallow-water coral *Cladocora caespitosa* od deep-water coral *Lophelia pertusa*, together with coralligenous buildups and maerl beds are of special interest for scientists and people involving with nature protection. Habitat degradation, destruction, fragmentation and loss are the most dramatic consequences of anthropogenic pressures on natural ecosystems and marine bioconstructors as a part of that. Under the present climate warming trend, together with marine acidification, new mass mortality events may occur in the near future, possibly driving a major biodiversity crisis in the Mediterranean Sea, especially in Mediterranean bioconstructors.

Keywords:

Anthropogenic pressures • Bioconstructors • Biodiversity crisis • *Cladocora caespitosa* • Climate warming trend • Coralligenous buildups • Coralligenous formations • Deep-sea white corals • Ecosystem engineers • Habitat formers • *Lophelia pertusa* • Maerl beds • Marine acidification • Mass mortality events • Mediterranean bioconstructors • Mediterranean Sea • Temperate seas • Vermetid reefs

Bioconstructors and Mediterranean Biodiversity

Diversity Pattern

Marine bioconstructors are, in simplified sense, marine calcifying organisms, both pelagic and benthic. The term applies to organisms with hard skeletons that remain in place after death, becoming a secondary substrate for other organisms of the same species or non-bioconstructors. They are best represented by tropical coral reefs, but are also present in temperate seas. Their impact in temperate seas is often neglected, and knowledge of the carbonate production of

P. Kružić (✉)
Laboratory for Marine Biology, Department of Zoology, Faculty of Science, University of Zagreb, Rooseveltov trg 6, HR-10000 Zagreb, Croatia
e-mail: pkruzic@zg.biol.pmf.hr

S. Goffredo and Z. Dubinsky (eds.), *The Mediterranean Sea: Its history and present challenges*, DOI 10.1007/978-94-007-6704-1_25, © Springer Science+Business Media Dordrecht 2014

Fig. 25.1 The *Cladocora caespitosa* bank in the Mljet National Park (Southern Adriatic Sea)

non-tropical species is scarce. In the Mediterranean Sea, most important habitat formers are bioconstructors and they should have an adequate consideration in conservations policies (Bianchi and Morri 2000). Major carbonate marine frameworks are formed by long-living organisms such as encrusting red algae, corals, polychaetes, and bryozoans (Laborel 1987; Bianchi 1997, 2002; Bianchi and Morri 2004; Cocito and Ferdeghini 2001). Their growth is mainly temperature-dependent, and when abundant, these organisms may grow to a fairly large size. The largest biologically constructed features are coral reefs in tropical regions, although similar features (*Cladocora caespitosa* banks) are also known from the Mediterranean Sea during early Pleistocene (Peirano et al. 1998, 2004). Marine benthic bioconstructors have great significance in generating and maintaining marine biodiversity. A wide range of marine organisms acts as bioconstructors: coralline algae, sponges, corals (both symbiotic and non-symbiotic), vermetids, mussels, oysters, polychaetes (serpulids and sabellerids), barnacles and bryozoans. There is a great difference in bioconstructors diversity among tropical seas and temperate seas. Coral reefs in tropical seas are formed by large number of coral species, while in temperate seas, only a few species participate in bioconstructions. Similar examples could be found also in bryozoan species (Cocito 2004). Bryozoans and corals are known to form biohermal mounds in tropical, but also in temperate seas, like Mediterranean Sea. Bryozoan *Pentapora fascialis*, serpulid polychaete *Ficopomatus enigmaticus* and scleractinian coral *Cladocora caespitosa* are the most famous of all biconstructors in the Mediterranean Sea. The scleractinian coral *Cladocora caespitosa* belongs to the family Faviidae, a species known as important coral reef builders. Carbonate production of this coral can reach up to 10 kg $CaCO_3$ per m^2 $year^{-1}$, comparable to data from constructional, tropical corals (Peirano et al. 2001). The coral *C. caespitosa* deserves a special place in this group of Mediterranean bioconstructors. It is a bushy coral (dendroid form), with parallel growing calices. Living at depths of 4–30 m, this coral builds formations that are possibly related to the true coral reefs of tropical seas. *C. caespitosa* is an obligatorily zooxanthellate, characteristic of hermatypic tropical corals. It is an endemic Mediterranean coral and is found in beds of hemispherical colonies or in a form of coral bank produced by the fusion of adjacent colonies, covering several square meters (Morri et al. 1994; Peirano et al. 1998, 2001; Kružić and Požar-Domac 2003). Large banks of *C. caespitosa*, both living and fossil, have been found at several sites in the Mediterranean Sea: near the Tunisian coast and in the Aegean and Ligurian seas (Laborel 1961; Zibrowius 1980; Peirano et al. 1998). In the Adriatic Sea, Abel (1959) also described extensive colonies of *C. caespitosa* in the Lim channel near Rovinj, while Pax and Müller (1962) mention banks of *C. caespitosa* near the island of Mljet. The *C. caespitosa* bank in the Mljet National Park (Southern Croatia, the Adriatic Sea) ranges from depths of 6 to 18 m and covers an area of 650 m^2, forming one of the largest banks of *C. caespitosa* found in the Mediterranean Sea (Fig. 25.1). The strong currents, which occur as a result of tidal exchange in the channel, appear to favour the growth of the bank (Kružić 2002; Kružić and Požar-Domac 2003; Kružić and Benković 2008). These larger formations are very rare nowadays, compared to the Pleistocene when *C. caespitosa* formed true "reefs" during the warmer climatic stages (Bernasconi et al. 1997). The reason for this impoverishment is not known and could possibly be connected with climatic changes in the Mediterranean Sea and recent sea pollution. *C. caespitosa* tolerates relatively low winter temperatures in the Mediterranean Sea (as low as 6 °C measured in the northern Adriatic Sea) (Kružić and Benković 2008). Peirano et al. (2004) consider this coral a true biological recorder of recent and past climates. Recent studies of stable

Fig. 25.2 Coralligenous community in the Mljet National Park (Southern Adriatic Sea)

isotopes and trace elements in *C. caespitosa* have confirmed that this coral species can be used as an ideal candidate for reconstructing detailed climate and environmental changes (Montagna et al. 2007, 2008; Rodolfo-Metalpa et al. 2008). These authors provided evidence that this non-tropical coral represents a new, potentially important, high-resolution archive for climate variability in the Mediterranean Sea. Deep sea corals (most often stony corals) are also important for the Mediterranean biodiversity. Most of what we know about deep sea corals has come from exploration and research within the past few decades. With the development of underwater imaging technologies and complex vehicles like ROV (remotely operated vehicles), scientists have been able to begin the study of corals within their natural environment. Deep-water corals are widely distributed within the earth's oceans, with large reefs in the far North and far South Atlantic, as well as in the tropics in places such as the Florida coast. In the Mediterranean Sea, the principal coral species that contribute to reef formation are *Lophelia pertusa*, *Madrepora oculata, Desmophyllum dianthus*, *Dendrophyllia ramea* and *D. cornigera*. The last two are known as 'yellow corals' and live in the circalittoral zone, especially in the south-western areas of the Mediterranean, therefore showing a distribution typical of warm-water species (Bianchi 2007). *Madrepora oculata*, *Desmophyllum dianthus* and *Lophelia pertusa* are known as 'white corals', live in the bathyal zone and have a strong affinity for cold waters. Information on distribution of Mediterranean deep-water corals is still too scarce.

Coralligenous Community and Maërl Beds

Coralligenous buildups (hard bottom of biogenic origin always develop in almost vertical walls, deep channels, or overhangs) seem to be common all around the Mediterranean coasts, but one of the major gaps concerning the current state of knowledge of the coralligenous habitat is the absence of cartographical data. Coralligenous concretions are the result of the building activities of algal and animal builders and the physical as well as biological eroding processes (Ballesteros 2006). The final result is a very complex structure composed of several microhabitats (Fig. 25.2). Light is the most important environmental factor regarding the distribution of benthic organisms along the rocky bottoms of the continental shelf (Ballesteros 1992). Habitats situated in open waters (horizontal and vertical surfaces) can be easily distinguished from those situated in overhangs and cavities. Holes and cavities within the coralligenous structure always bear a complex community dominated by suspension feeders (anthozoans, bryozoans, serpulids and molluscs). Overhangs and big cavities of the coralligenous have a different species composition from that found in open waters. Algae are usually completely absent because light is very reduced.

Algae dominate in horizontal to sub-horizontal surfaces and their abundance decreases with depth and decreasing irradiance. Coralline algae are the main coralligenous builders (Laborel 1961). *Mesophyllum lichenoides* and *Mesophyllum* (*Pseudolithophyllum*) *expansum* has probably been identified by most authors as being the most common coralline algae in the coralligenous (Fig. 25.3).

Animal assemblages can greatly differ according to light levels reaching the coralligenous outcrop but also according to current intensity, sedimentation rates and geographical areas (Ballesteros 2006). In the richest, relatively more eutrophic zones, with rather constant and low water temperature, gorgonians usually dominate the community, but they are completely absent or rare in the more oligotrophic or low current areas with rather high or seasonally variable temperature, being replaced by poriferans, bryozoans or ascidians (Ballesteros 2006). Coralligenous animal builders are studied by Hong (1980) from the region of Marseilles. He found

Fig. 25.3 Coralline algae *Mesophyllum lichenoides* from the coralligenous community in the Mljet National Park (Southern Adriatic Sea)

a total number of 124 species contributing to the buildups. He found that the most abundant animal group are the bryozoans, accounting for 62 % of the species, followed by the serpulid polychaetes, with the 23.4 % of the species. Minor contributors are the cnidarians (4 %), molluscs (4 %), sponges (4 %), crustaceans (1.6 %) and foraminifers (0.8 %). Hong (1980) distinguished three different types of animal builders: (a) those directly contributing to the buildups, which are relatively big (bryozoans from the genus *Schizomavella*, *Celleporina*, *Myriapora*, serpulids *Serpula vermicularis* and *Serpula concharum*, the molluscs *Vermetus* sp. and *Thylacodes* (*Serpulorbis*) *arenarius*, and the scleractinian corals *Hoplangia durotrix*, *Leptopsammia pruvoti*, *Caryophyllia inornata* and *Caryophyllia smithii*); (b) those with a reduced buildup activity because of their small sizes (small bryozoans like *Crassimarginatella maderensis* and *Mollia patellaria*, serpulids like *Hydroides* spp., *Filogranula* spp., and *Spirorbis* spp., the cirripeds *Verruca strömia* and *Balanus perforatus*, and the foraminifer *Miniacina miniacea*) and (c) those which agglomerate carbonate particles (sponges like *Geodia* spp., *Spongia virgultosa* and *Faciospongia cavernosa*, bryozoans *Beania* spp. and the alcyonarian *Epizoanthus arenaceus*). Gili and Ballesteros (1991) described the species composition and abundance of the cnidarian populations in coralligenous concretions dominated by the gorgonian *Paramuricea clavata* from the Medes islands. Kružić (2002, 2007) described the cnidarian species composition in the Mljet National Park and Nature Park "Telašćica". In these papers species contributing the most to the total biomass of the taxocenosis were anthozoans *Leptopsammia pruvoti*, *Caryophyllia inornata*, *Caryophyllia smithii*, *Paramuricea clavata*, *Eunicella cavolini*, *Corallium rubrum*, *Alcyonium acaule* and *Parerythropodium coralloides*.

The assemblage of red coral in the Mediterranean is dominated by cnidarians *Corallium rubrum*, *Caryophyllia inornata*, *Caryophyllia smithii*, *Hoplangia durotrix*, *Phyllangia mouchezi*, and *Leptopsammia pruvoti*, sponges *Petrosia ficiformis* and *Aplysina cavernicola* and bryozoans *Celleporina caminata*, *Myriapora truncata*, *Smittina cervicornis* and *Reteporella beaniana*. Among the anthozoans, red coral is exploited commercially in almost all Mediterranean countries, and its stocks have strongly declined in most areas, particularly in shallow waters (Weinberg 1991).

Canals and Ballesteros (1997) estimated the carbonate production of the phytobenthos in the coralligenous and other communities from the continental shelf of the Balearic Islands. The coralligenous community thriving in relatively shallow waters (with *Mesophyllum alternans* and *Halimeda tuna* as dominant algae) was the one with the highest production (around 465 g $CaCO_3$ m^{-2} $year^{-1}$). Production of deep water coralligenous concretions dominated by *Lithophyllum cabiochae* was much lower (around 170 g $CaCO_3$ m^{-2} $year^{-1}$). However, estimates of the animal carbonate production in deep water (25–50 m depth) rocky bottoms from the Alboran Sea (southwestern Mediterranean), mainly dominated by big suspension feeders with calcareous skeletons (predominantly the coral *Dendrophyllia ramea*, the bryozoans *Pentapora fascialis*, *Smittina cervicornis* and *Myriapora truncata*, and polychaetes such as *Salmacina dysteri*, *Protula* sp. and other serpulids), are very high (around 660 g $CaCO_3$ m^{-2} $year^{-1}$) (Cebrian et al. 2000).

Maërl beds are also very diverse as coralligenous buildups. Coralline red algae (Corallinaceae) are usually the main constituents (most common are genus *Lithothamnion*, *Phymatolithon*, *Peyssonnelia* and *Litophyllum*) and live unattached. These species can form extensive beds, mostly in coarse clean sediments of gravels and clean sands or muddy mixed sediments, which occur either on the open coast, in tide-swept channels or in sheltered areas of marine inlets with weak current. Maërl beds form a complex three-dimensional habitat and have been found to harbour greater diversity of life in comparison to surrounding habitats and have high productivity (Martin et al. 2005; Grall et al. 2006). The distribution of maërl is dependent on water movement, light and occurs in the photic zone, up to 120 m deep in the Mediterranean Sea. Maërl beds can act as nursery areas for the juvenile stages of commercial species of bivalves, fish, crabs and scallops.

Threats to Mediterranean Bioconstructors

Major Threats

Trawling and Dredging

Trawling is probably the most destructive impact currently affecting marine bioconstructors (like impact on coralligenous communities). As this fishing gear drags and rolls along and digs into the sea floor, the result is often habitat destruction. Trawling is also completely destructive in maërl beds, being the main cause of maërl disappearance in large Mediterranean areas. The action of trawling gear over coralligenous and maërl assemblages leads to the death of most engineering, dominant and builder species, completely changing the environmental conditions of the coralligenous microhabitats and from the maërl environment. Before the trawling the habitat was described as a clean maërl gravel with low silt content supporting abundant suspension feeding bivalves, but after the trawling the habitat is of muddy sand dominated by deposit feeders and omnivores. As most of these bioconstructional species are particularly long-lived, have low recruitment and complex demographic patterns, destruction of the coralligenous and maërl structure is critical as their recovery will probably take several decades or even centuries (Ballesteros 2006). Trawling close to coralligenous outcrops or maërl beds affects negatively to algal growth and suspension-feeding due to an increase in turbidity and sedimentation. Maërl has been dredged from both living beds and fossilised deposits for use as an agricultural soil conditioner as well as use in animal food additives and water filtration systems. Although quantities were initially small, by the 1970s a peak of around 600,000 tonnes were extracted per year in France (Briand 1991). Due to the very slow rate of growth, maërl is considered to be a non-renewable resource and, even if the proportion of living maërl in commercially collected material is low, extraction has major effects on the wide range of species present in both live and dead maërl deposits (Bárbara et al. 2003).

Dynamite Fishing

This illegal practice can be extremely destructive to the surrounding ecosystem, as the explosion often destroys the underlying habitat (such as coralligenous community). Dynamite fishing is set off under water to kill fish for easy collection or to attract fish from deeper bottom. The explosives used destroy the marine environment, leaving it devoid of life. Large blasted areas are very slow to recover because coralligenous communities have difficulty establishing on loose substrate. Dynamite fishing is one of the most pervasive fishing techniques and it is banned by most countries, but is, unfortunately, still frequent in Mediterranean.

Anchoring

Anchoring has a very severe impact in coralligenous concretions, as most of the engineering organisms are very fragile and are easily detached or broken by anchors and chains. Anchors can crush and destroy this valuable bottom through sediment disturbance and fragmentation. Coralligenous concretions of frequently visited sites by recreational fishing or diving activities are degraded by the destructive potential of anchors (Ballesteros 2006). While it is possible for sea-bottom habitats, such as coralligenous community, to recover from anchor damage, this is a very slow process that can usually take many decades. In areas of intense anchor damage, it is unlikely that this community will ever make a full recovery. In some areas in the Mediterranean, the anchoring is a huge problem even in protected areas, because of lack of monitoring control and management.

Invasive Species

There is a lack of knowledge on the effects that lessepsian species have on coralligenous and maërl communities in the Mediterranean. Currently, at least three algal species are threatening coralligenous and maërl communities in the Mediterranean: *Womersleyella setacea*, *Caulerpa racemosa v. cylindracea* and *Caulerpa taxifolia*. All of them are only invasive in relatively shallow water coralligenous outcrops and maërl beds (until 60 m), where irradiance levels are sufficient to permit their growth (Ballesteros 2006). However, they are especially dangerous, because they completely cover the basal stratum of encrusting corallines and increase sedimentation rates which lead to a total shut down of coralligenous growth or the survival of rhodolits. Probably the most dangerous alien species for the coralligenous community is the red alga *Womersleyella setacea*, which is currently distributed along most of the Mediterranean basin (Athanasiadis 1997). The species richness found in sites invaded by *W. setacea* is lower than that observed in noncolonised sites (Piazzi et al. 2002). The other two algae, *Caulerpa taxifolia* and *C. racemosa var. cylindracea* are species that can threaten the coralligenous community and in some areas they have totally invaded the coralligenous community, especially the second one (Meinesz and Hesse 1991; Piazzi et al. 2005). In the summer of 2004, *C. racemosa* var. *cylindracea* was found in the National Park Mljet in Veliko jezero between 8 and 14 m depth, affecting around 150 m of coastline and the

Fig. 25.4 Invasive alga *C. racemosa* var. *cylindracea* on the coral *Cladocora caespitosa*

marginal part of the coral bank of *Cladocora caespitosa* (Kružić et al. 2008) (Fig. 25.4). The alga was observed smothering live coral colonies, while the coral produced excessive mucus preceded by the complete retraction of the polyps, leaving the calyx rim deprived of tissue coverage.

Waste Water Discharges and Aquaculture

Waste waters profoundly affect the structure of coralligenous communities by inhibiting coralline algal growth, increasing bioerosion rates, decreasing species richness and densities of the largest individuals of the epifauna, eliminating some taxonomical groups and increasing the abundance of highly tolerant species (Hong 1980, 1982; Ballesteros 2006). Although there is no information available on the impact of eutrophication in Mediterranean maërl beds, the effects must be similar to those reported for coralligenous concretions. The number of coralligene species decreased with increased pollution, both in terms of number and density of individuals. It is also known that orthophosphate ions inhibit calcification (Simkiss 1964). Water turbidity seems to be the main factor causing degradation and homogenisation of the phytobenthos and some coral species. Because of algal blooms due high eutrophication, algae tend to grow very quickly under high nutrient availability and the result is a high concentration of dead organic matter which starts to decay. The decay process consumes dissolved oxygen in the water, resulting in hypoxic conditions. Without sufficient dissolved oxygen in the water, coralligene species may die off in large numbers. Marine mucilage on benthic species is also common in the area of eutrophication.

Although there are no studies on the impact of aquaculture facilities situated over or at the proximity of coralligenous outcrops, nor maërl beds, their effects should match those produced by waste water dumping. Coastal construction and increases in agricultural and sewage discharges may also have some impact if they increase sediment loads or result in the excessive growth of ephemeral species of macroalgae around maërl beds (Birkett et al. 1998).

Recreational Diving

The coralligenous community is one of the most popular sites for recreational diving in the Mediterranean due to its great variety of life and great visual appeal (Harmelin 1993). Uncontrolled or over-frequentation of divers in coralligenous communities has been described to produce an important effect over certain large or fragile suspension feeders inhabiting coralligenous communities (Sala et al. 1996; Garrabou et al. 1998; Coma et al. 2004; Linares et al. 2005; Ballesteros 2006). Impacts caused by recreational scuba activity vary widely among different divers with differing underwater behaviour. The proportion of damage varied significantly with the frequency of SCUBA diving and sometimes resulting in damage of bioconstructors like red algae and bryozoan and coral colonies.

Bioconstructors and Acidification

Anthropogenic carbon dioxide (CO_2) has already caused ocean pH to decrease by an average of 0.1 units since pre-industrial time (Raven et al. 2005). By 2100, it is projected to fall by 0.3–0.5 pH units (Orr et al. 2005). Short-term experiments have revealed that this degree of seawater acidification can alter calcification rates in certain planktonic and benthic organisms, although the effects recorded may be shock responses and the long-term ecological effects are unknown.

The level of acidification in the Mediterranean Sea is still poorly known, but certain characteristics of this semi-enclosed ocean makes it especially sensitive to increasing atmospheric CO_2 (Calvo et al. 2011). The relationship between the rising concentration of atmospheric CO_2 and ocean acidification, together with the possible effect of the decline in seawater pH has been widely studied, especially over the last years. A first estimate indicates a pH decrease of up to 0.14 units since the pre-industrial era affecting the entire water column, especially in the western Mediterranean basin (Touratier and Goyet 2011). Only a few studies have tested the effects of ocean acidification in corals from the Mediterranean Sea. Faster growing colonies were the most affected by a decreased pH. Corals like *Balanophyllia europaea* and *Cladocora caespitosa* will also be detrimentally affected by the near future progressive ocean acidification. Rodolfo-Metalpa et al. (2010) combined short (1 month) and long (1 year) exposure of *C. caespitosa* to low pH (7.8 units) in experimental aquaria and found no detrimental effects in calcification rate of the colonies. In the second study, the transplantation of *C. caespitosa* corals to a gradient of naturally acidified areas close to CO_2 vents (pH level range between 8.1 and 7.5 units) showed evidences of dissolution (Rodolfo-Metalpa et al. 2011). The mussel *Mytilus galloprovincialis* and the limpet *Patella caerulea* showed similar trends in their ability to accrete shell under high CO_2 conditions. This suggested that the effects of acidification on these species may show up below a certain threshold of pH decrease. This study did not find evidence of dissolution on *Balanophyllia europaea* corals exposed to the same gradient of natural acidification. The authors also make the potentially important observation that the positive relationship between CO_2 and gross calcification for the coral *B. europaea* and the mussel *M. galloprovincialis* in early to mid-summer becomes negative following late summer warming of the Mediterranean Sea. They conclude that the combination of warming and acidification was simply too much for these calcifiers to overcome. Cigliano et al. (2010) reported that six main taxonomic groups (foraminiferans, nematodes, polychaetes, molluscs, crustaceans and chaetognaths) showed highly significant reductions in recruitment caused by acidified conditions near the vents (pH 7.08–7.79) at Ischia (Gulf of Naples, Tyrrhenian Sea), Italy. At the same site Rodolfo-Metalpa et al. (2010) reported that organic tissues enveloping the skeleton of the Mediterranean bryozoan *Myriapora truncata* play a protective role from the effects of ocean acidification on the species calcareous skeleton. Skeletons of live colonies in acidic pH site after 45 days of exposure were less corroded than those of dead bryozoan colonies.

Coralline algae are of significant importance, especially in the Mediterranean Sea where they are the major contributors to biogenic carbonate production. The study of Martin et al. (2008) shows a significant reduction in epiphytic coralline algal cover with increasing acidification of seawater due to natural CO_2 vents. Although a range of factors may be responsible for this observed shift in seagrass epiphytism, lowered pH and reduced calcite saturation levels are the most likely factors affecting coralline algal cover. Coralline algae were absent where the pH periodically fell below 7 and their calcimass was greatly affected where the pH ranged from 7.7 to 8.2. Jokiel et al. (2008) reported skeleton dissolution rather than growth for the species *Lithophyllum*, *Hydrolithon* and *Porolithon* sp. at a pH of approximately 7.9, relative to a normal pH of 8.2, while Kuffner et al. (2008) reported a drop in recruitment rate and per cent cover of 78 and 92 %, respectively, at a pH of 7.9.

The wide variety of experimental studies on the effects of ocean acidification on calcifying marine organisms like algae, corals and other carbonate dominated systems showed reduced calcification rates during increased acidification.

Bioconstructors and Mass Mortalities

Several episodes of mass mortalities have been detected in the northwestern Mediterranean (Rivoire 1991; Bavestrello et al. 1994; Cerrano et al. 2000; Garrabou 1999; Garrabou et al. 2001, 2009). In most cases, sessile marine organisms were victims of these large-scale mortality events that affected shallow water assemblages (5–40 m depth). The characteristic summer conditions of reduced resources, high water column stability and high temperatures (normally during July and August) lasted much longer than usual and this coincided with a mass mortality of benthic organisms, affecting also a coralligenous community, situated at a depth of <40 m. Red coral populations thriving above a depth of 30 m were also affected (Garrabou et al. 2001). This large-scale mortality, together with other small-scale mass mortalities recorded during the past decade in the Mediterranean, may be related to seawater temperature increase and global warming (Cerrano et al. 2000). Some suspension feeders might be able to withstand the normal duration of adverse summer conditions but not an anomalous prolongation of these conditions, resulting from an energy shortage of suspension feeders related to low food availability in summer (Coma et al. 2000; Coma and Ribes 2003). A mass mortality event of the red coral *Corallium rubrum* occurred in the NW Mediterranean region in summer 1999 (Garrabou et al. 2001). During late summer 1999, the NW Mediterranean area affected by the mass mortality event experienced high temperatures and hydrographic stability over a period of several weeks. This temperature anomaly could have caused physiological stress and triggered the development of pathogenic agents that otherwise would have remained non-virulent. There were no signs of mass mortality below 40 m depth in red coral populations. In late summer 2003, extensive mass mortality of at least 25 rocky benthic macroinvertebrate species (mainly gorgonians and sponges) was observed in the entire NW Mediterranean region (Garrabou et al. 2009).

Fig. 25.5 Bleached polyps of the coral *Cladocora caespitosa*

The heat wave of 2003 in Europe caused an anomalous warming of seawater, which reached the highest temperatures ever recorded in the studied regions, between 1 and 3 °C above the climatic values. It seems likely that the anomalous temperature played a key role in the observed mortality event. The western areas (Catalan coast and Balearic Islands) were the least affected, the central areas (Provence coast and Corsica-Sardinia) showed a moderate impact and the northernmost and eastern areas (Gulf of Genoa and Gulf of Naples) displayed the highest impact. The Adriatic Sea has been also affected by several severe mortality events of benthic invertebrate species during the last few decades (Kružić et al. 2012). The first unprecedented large-scale mass mortality event was recorded during summer 1997, affecting about 30 benthic species. Anomalous high-temperature conditions during summer 1997 were identified as the main triggering factor of these mortalities. Mortality events of the zooxanthellate corals *Cladocora caespitosa*, *Madracis pharensis* and *Balanophyllia europaea* and non zooxanthellate *Caryophyllia inornata* and *Leptopsammia pruvoti* were recorded in several sites in the Adriatic Sea during the summers of 1997, 1999, 2003, 2006 and 2009. At the end of the summer of 2003, about 30 % of the colonies of the coral *C. caespitosa* were damaged and more than 10 % of the damaged colonies were completely dead, all caused by bleaching events (Fig. 25.5). In most cases bleaching has been attributed to elevated temperature, but other instances involving high solar irradiance, and sometimes disease, have been also documented. These high temperature events involved coral colonies living in shallow water (6–25 m depth). Mortality was preceded by the complete retraction of the polyp, leaving the calyx rim deprived of tissue coverage. However, we cannot ascertain whether temperature had a direct effect on organisms or acted in synergy with a latent or waterborne agent (microbiological or chemical). Taking into account the global warming context in the Mediterranean Sea, monitoring programs of physical-chemical parameters and vulnerable coral populations should rapidly be set up.

Under the present climate warming trend, new mass mortality events may occur in the near future, possibly driving a major biodiversity crisis in the Mediterranean Sea, especially in Mediterranean bioconstructors.

Climate Change and Mediterranean Bioconstructors

How Climate Change Influences Bioconstructors

Current climate warming is superimposed over natural phenomena, some of which are cyclic. During the Last Glacial Maximum, ~23,000 to 19,000 years ago, the Mediterranean Sea surface temperature was 2 °C lower than today in the eastern basin, and up to 8 °C lower in the northwestern basin (Kuhlemann et al. 2008). The thermohaline circulation of the Mediterranean is determined by the flux of incoming Atlantic water by the Gibraltar Straits and by the sinking of waters formed at the three coldest areas of the sea: the Gulf of Lions, the northern Adriatic and the North Aegean Sea (Bianchi 2007). In the last few decades this general pattern has been disrupted in several ways. The first definite signs of changes in sea temperature came from Mediterranean deep waters. A 30-year time series (1959–1989) of deep-water temperatures acquired in the northwestern Mediterranean (NWM) demonstrated a general warming trend of ~0.12 °C (~0.004 °C per year), a possible result of human-induced global warming (Béthoux et al. 1990; Vargas-Yáñez et al. 2002). Some unexpected effects on thermohaline circulation have been documented from 1987 through the 1990s; these changes have been termed the Eastern Mediterranean Transient (EMT). Strong anomalies in temperature and rainfall are believed to be involved in triggering the EMT, which has dramatically changed the hydrology of the deep eastern Mediterranean (Bianchi 2007; Roether et al. 2007). Temperature, salinity, stratification and circulation of water masses have been affected. A 30-year data set (1974–2005) from the Spanish Catalan coast first demonstrated the warming of the littoral zone at four different depths from the surface to 80 m and showed a clear trend at all depths, with an impressive warming of 1.4 °C at 20 m (Coma et al. 2009). Most of the introduced species in the Mediterranean originates from the Red Sea, as Lessepsian migrants (67 %), with an additional set of species (7 %) from other tropical areas (Bianchi 2007). In the colder Northwestern Mediterranean, tens of significant range expansions of species of warmwater affinity have been recorded. One of the examples is the warm stenotherm orange coral *Astroides calycularis*, very common in the eastern and southern shores and recently shifts its distribution northwestwards (Francour et al. 1994; Kružić 2002).

Climate change combines with Atlantic influx, lessepsian migration and the introduction of exotic species by humans to the establishment of tropical marine biota in the Mediterranean Sea (Bianchi 2007). Present-day warming ultimately favours the spread of warm-water species through direct and indirect effects, and especially by changing water circulation. Coral *Cladocora caespitosa* will certainly play the role of constructional coral in a more and more 'tropical' Mediterranean Sea (Bianchi 2007). In coincidence with positive anomalies of sea surface temperature recorded in these last few summers, this coral species, together with *Balanophyllia europaea* and *Madracis pharensis*, underwent mass-mortality events, recalling those observed in the tropics (Rodolfo-Metalpa et al. 2000; Garrabou et al. 2001).

If mass mortalities are indeed related to the global warming trend, such events might occur again and become more frequent, which would cause profound changes in the specific composition and structure of marine communities (Ballesteros 2006). It is not known whether all parts of the Mediterranean will be equally affected by global change. Different climatic conditions are involved in the Mediterranean basin, so answers might come from current efforts to ensure that several eastern and southern Mediterranean stations begin continuous records of temperature and of some indicator species (sessile invertebrates like corals and bryozoans). This will help us to understand the climatic changes in the Mediterranean and how will species cope with ongoing changes.

Temperature Anomalies in the Mediterranean

Anomalous high water temperatures seem to trigger large scale mortalities of several suspension feeders growing in coralligenous assemblages (Cerrano et al. 2000; Pérez et al. 2000). Most of the organisms living in coralligenous communities are able to support the normal seasonal temperature range characteristic of Mediterranean waters which are in the range of 10–20 °C. Pascual and Flos (1984) found temperatures ranging between 12 and 20 °C at the shallowest limit of the coralligenous communities of the Medes Islands (20 m depth), although temperatures ranged from 12 to 16 °C close to their deepest limit (60 m depth). Ballesteros (1992) reported more or less the same temperatures for the coralligenous assemblages developing at depths of 20 and 40 m at Tossa de Mar between the end of November and the end of June (13–16 °C), but differences of up to 9 °C in summer, when the thermocline is situated at a depth of around 35 m; however, peak temperatures of 22 °C were detected at the end of August at a depth of 40 m. In the Balearic Islands, where coralligenous communities are restricted to waters >40 m deep, water temperature ranges from 14.5 to 17 °C for most of the year, although occasional peaks of 22 °C are detected at the end of October, when the thermocline is at its deepest (Ballesteros and Zabala 1993). In the Adriatic Sea water temperature of the coralligenous communities In National Parks "Mljet" (southern Adriatic) and "Kornati" (central Adriatic) peak temperatures of 24 °C were detected at the in August and September at a depth of 40 m (Kružić unpublished data). Some organisms living in coralligenous assemblages from deep waters seem to be highly stenothermal, as they are never found in shallow waters. Recent large-scale mortality events of benthic suspension feeders thriving in coralligenous communities have been attributed to unusually long-lasting periods of high temperatures during summer (like in the Adriatic Sea).

Sea temperature anomalies have also an impact on calcification in bioconstructors. The results of Bessat and Buiges (2001) and Crabbe (2008) indicated that a 1 °C rise in temperature would lead to an increase in the density rate of about 10.5 % and an increase in the calcification rate of about 4.5 %. The growth rates of the coral *Cladocora caespitosa* were studied in the Ligurian and Adriatic Sea (Peirano et al. 1999, 2004, 2005; Morri et al. 2001; Peirano and Kružic 2004; Kružić and Benković 2008) and near the Columbretes Islands (NW Mediterranean, Spain) (Kersting and Linares 2012). The results showed that the growth rate seemed to correlate with climate fluctuations. Coral calcification rates and extension rates have been highly correlated with sea surface temperatures (SSTs), and to a lesser extent with incoming solar radiation (Lough and Barnes 2000). Although there are numerous studies of the relationships between environmental parameters and coral growth in the tropics, such studies are scarce for temperate zones. Unlike most studies on other tropical and temperate corals, in which calcification was positively correlated with solar radiation and SST, studies in the Mediterranean showed that calcification was not correlated with solar radiation, whereas it was negatively correlated with SST (Goffredo et al. 2009). Kushmaro et al. (1998)

reported that high temperatures (about 29 °C) caused rapid bleaching and dying of the coral *Oculina patagonica* from the Mediterranean Sea as a result of a bacterial infection. It is recognized that heat stress triggers diseases in marine waters, and the frequency of this event clearly increased over the last two decades. Mortality of the coral *C. caespitosa* caused by elevated water temperatures has also been described by Rodolfo-Metalpa et al. (2000, 2005, 2006) in the Ligurian Sea, as well as by Cerrano et al. (2000), Ferrier-Pagès et al. (2009), and Garrabou et al. (2001, 2009) in other sessile marine invertebrates. Laboratory observations on the calcification rates of the coral *C. caespitosa* and *O. patagonica* indicate that prolonged periods of high temperatures lead to a decrease in calcification (Rodolfo-Metalpa et al. 2006). Recent extreme temperature events in the Mediterranean Sea are likely to have disturbed the normal functioning of biological systems. Exposure to anomalous high temperatures can act directly to cause metabolic dysfunctions (Coma and Ribes 2003; Coma et al. 2009) or indirectly by favoring the development of pathogens (Bally and Garrabou 2007).

Future of Bioconstructors in the Mediterranean

Projections available for the Mediterranean basin region for the twenty-first century indicate warmer and drier conditions with a dramatic increase in the frequency and persistence of extreme events such as heat waves (Stott et al. 2004; Diffenbaugh et al. 2007). In the Black Sea and Mediterranean Sea there is no danger of surface waters becoming corrosive to calcium carbonate before 2100, but they will suffer sharp reductions in carbonate ion concentrations (Mediterranean Sea – 37 %; Black Sea – 45 %) (McNeil and Matear 2008; Steinacher et al. 2009). These rapid chemical changes are an added pressure on marine bioconstructors and ecosystems of the Mediterranean that are already heavily suffering from other anthropogenic influences. Consequences of these events could severely expose populations of affected species to local ecological extinction processes, threatening the conservation of the rich Mediterranean biodiversity (Bianchi and Morri 2000; Ballesteros 2006; Bianchi 2007). This climate change is the latest form of human disturbance affecting the Mediterranean coastal communities at the regional scale through mortality outbreaks and shifts in species distributions (Francour et al. 1994; Bianchi and Morri 2000). The expansion and intensification of climate change effects together with the effects of other strong disturbances (pollution, mechanical habitat destruction, invasive species, and overfishing) may act together, driving the Mediterranean to a major biodiversity crisis during the twenty-first century (Bianchi and Morri 2000; Sala 2004).

Although most experimental studies on ocean acidification to date have focussed on tropical assemblages, recent work has shown that effects of ocean acidification on temperate marine systems may be equally complex and deleterious (Hall-Spencer et al. 2008; Martin and Gattuso 2009; Russell et al. 2009). Atmospheric CO_2 partial pressure (pCO_2) is expected to increase to 700 μatm or more by the end of the present century. Anthropogenic CO_2 is absorbed by the oceans, leading to decreases in pH and the $CaCO_3$ saturation state (Ω) of the seawater. Elevated pCO_2 was shown to drastically decrease calcification rates in tropical zooxanthellate corals. Therefore, the conventional belief that calcification rates will be affected by ocean acidification may not be widespread in temperate corals. Rodolfo-Metalpa et al. (2010) showed, using the Mediterranean zooxanthellate coral *Cladocora caespitosa*, that an increase in pCO_2, in the range predicted for 2100, does not reduce its calcification rate. Seasonal change in temperature is the predominant factor controlling photosynthesis, respiration, calcification and symbiont density. An increase in pCO_2, alone or in combination with elevated temperature, had no significant effect on photosynthesis, photosynthetic efficiency and calcification. The lack of sensitivity *C. caespitosa* to elevated pCO_2 might be due to its slow growth rates, which seem to be more dependent on temperature than on the saturation state of calcium carbonate in the range projected for the end of the century.

Global ocean models predict surface pH reductions of 0.3–0.5 units by the year 2100 due to the uptake of anthropogenic CO_2 (Caldeira and Wickett 2005). This is likely to have strong negative impacts on coralline algae (Hall-Spencer et al. 2008). Ocean warming could also have particularly severe effects on some maërl species due to their fragmented ranges and poor dispersal but only limited knowledge of temperature tolerances and optima are available for some species of maërl (Blake and Maggs 2003).

In some areas of the Mediterranean, there are recurrent downwellings (>40 m) during summer, which carries increased temperature further down in the water column, resulting in warm conditions over longer periods. This phenomenon could be enhanced with the influence of Mistral (NWN) and Bora (NEN) winds. The lower frequency and shorter duration of these winds prevent the upwelling of deep, cold waters and the subsequent breakage of the thermocline, thus allowing the temperature to increase. Continued sea warming is likely to affect the successful recruitment, growth and productivity of prominent coastal organisms like marine bioconstructors.

There is currently no evidence of changes attributable to non-temperature related climate impacts, potentially due to a lack of long-term observational data, experimental evidence suggests that ocean acidification will result in negative effects on calcifying algae and animals (Wernberg et al. 2011). More importantly, recent experiments suggest the combined effects of climate change and non-climate stressors (overharvesting, reduced water quality) will lower the resilience of temperate marine communities to perturbations (storms, diseases, and invasive species), many of which are also predicted to increase in frequency and severity.

Since most of the predictions about future of the bioconstructors were made upon experimental laboratory work, the future of the Mediterranean climatic change could show us different picture than what scientists predicted. Coral *Cladocora caespitosa* have huge range of temperature tolerance (from 6 °C in winter period to 29 °C in summer period) without bleaching episodes. Bleaching of *C. caespitosa* took place with prolonged periods of low or high sea temperatures or when these extreme temperatures appear suddenly causing stress phenomena. Corals or other constructional organisms are not getting more abundant. On the contrary, the native constructional coral *Cladocora caespitosa* and other large invertebrate or coraline bioconstructors that 'shape' the submarine seascape of coastal Mediterranean ecosystems are perhaps going to face more frequent mass mortality events (Bianchi 2007).

The novel environmental conditions generated by climate change (increases in temperature and CO_2 concentrations) are likely to drive greater change in localities that have lower resilience (e.g. nutrient pollution on oligotrophic coasts) to combinations of global and local stressors (Wernberg et al. 2011). Future patterns of species diversity and their individual functions will be a product of the temporal modification of spatial opportunities and constraints for colonisation and persistence.

References

Abel EF (1959) Zur Kenntnis der marinen Höhlenfauna unter besonderer Berücksichtigung der Anthozoen. Pubbl Staz Zool Napoli 30:1–94

Athanasiadis A (1997) North Aegean Marine Algae IV. *Womersleyella setacea* (Hollenberg) R.E. Norris (Rhodophyta, Ceramiales). Bot Mar 40:473–476

Ballesteros E (1992) Els vegetals i la zonació litoral: espècies, comunitats i factors que influeixen en la seva distribució. Institut d'Estudis Catalans, Barcelona

Ballesteros E (2006) Mediterranean coralligenous assemblages: a synthesis of present knowledge. Oceanogr Mar Biol Ann Rev 44:123–195

Ballesteros E, Zabala M (1993) El bentos: El marc físic. In: Història Natural de l'Arxipèlag de Cabrera, JA Alcover et al (eds) Monografies de la Societat d'Història Natural de Balears 2. Palma de Mallorca, CSIC-Ed. Moll, pp 663–685

Bally M, Garrabou J (2007) Thermodependent bacterial pathogens and mass mortalities in temperate benthic communities: a new case of emerging disease linked to climate change. Glob Change Biol 13:2078–2088

Bárbara C, Bordehore C, Borg JA et al (2003) Conservation and mangaement of northeast Atlantic and Mediterranean maërl beds. Aquat Conserv Mar Freshw Ecosyst 13:S65–S76

Bavestrello G, Bertone S, Cattaneo-Vietti R, Cerrano C, Gaino E, Zanzi D (1994) Mass mortality of *Paramuricea clavata* (Anthozoa: Cnidaria) on Portofino Promontory cliffs (Ligurian Sea). Mar Life 4:15–19

Bernasconi MP, Corselli C, Carobene L (1997) A bank of the scleractinian coral *Cladocora caespitosa* in the Pleistocene of the Crati valley (Calabria, Southern Italy): growth versus environmental conditions. Boll Soc Paleont Ital 36:53–61

Bessat F, Buiges D (2001) Two centuries of variation in coral growth in a massive *Porites* colony from Moorea (French Polynesia): a response of ocean-atmosphere variability from south central Pacific. Palaeogeogr Palaeoclimatol Palaeoecol 175:381–392

Béthoux J-P, Gentili B, Raunet J, Tailliez D (1990) Warming trend in the western Mediterranean deep water. Nature 347:660–662

Bianchi CN (1997) Climate change and biological response in the marine benthos. Proc Ital Assoc Ocean Limnol 12:3–20

Bianchi CN (2002) Bioconstruction in marine ecosystems and Italian marine biology. Biol Mar Mediterr 8:112–130

Bianchi CN (2007) Biodiversity issues for the forthcoming tropical Mediterranean Sea. Hydrobiologia 580:7–21

Bianchi CN, Morri C (2000) Marine biodiversity of the Mediterranean Sea: situation, problems and prospects for future research. Mar Pollut Bull 40:367–376

Bianchi CN, Morri C (2004) Climate change and biological response in Mediterranean Sea ecosystems – a need for broad-scale and long-term research. Ocean Chall 13:32–36

Birkett DA, Maggs CA, Dring MJ (1998) Maërl. An overview of dynamic and sensitivity characteristics for conservation management of marine SACs. Scottish Association for Marine Science, Oban

Blake C, Maggs CA (2003) Comparative growth rates and internal banding periodicity of maërl species (Corallinales, Rhodophyta) from northern Europe. Phycologia 42:606–612

Briand X (1991) Seaweed harvesting in Europe. In: Guiry MD, Blunden G (eds) Seaweed resources in Europe: uses and potential. Wiley, Chichester

Caldeira K, Wickett ME (2005) Ocean model predictions of chemistry changes from carbon dioxide emissions to the atmosphere and ocean. J Geophys Res 110:C09S04

Calvo E, Pelejero C, Pena LD, Cacho I, Logan GA (2011) Eastern equatorial pacific productivity and related-CO_2 changes since the last glacial period. Proc Natl Acad Sci USA 108:5537–5541

Canals M, Ballesteros E (1997) Production of carbonate sediments by phytobenthic communities in the Mallorca-Minorca Shelf, Northwestern Mediterranean Sea. Deep Sea Res Pt II 44:611–629

Cebrian E, Ballesteros E, Canals M (2000) Shallow rocky bottom benthic assemblages as calcium carbonate producers in the Alboran Sea (Southwestern Mediterranean). Oceanol Acta 23:311–322

Cerrano C, Bavestrello G, Bianchi CN et al (2000) A catastrophic mass-mortality episode of gorgonians and other organisms in the Ligurian Sea (northwestern Mediterranean), summer 1999. Ecol Lett 3:284–293

Cigliano M, Gambi MC, Rodolfo-Metalpa R, Patti FP, Hall-Spencer J (2010) Effects of ocean acidification on invertebrate settlement at volcanic CO_2 vents. Mar Biol 157:2489–2502

Cocito S (2004) Bioconstruction and biodiversity: their mutual influence. Sci Mar 68:137–144

Cocito S, Ferdeghini F (2001) Carbonate standing stock and carbonate production of the bryozoan *Pentapora fascialis* in the north-western Mediterranean. Facies 45:25–30

Coma R, Ribes M (2003) Seasonal energetic constraints in Mediterranean benthic suspension feeders: effects at different levels of ecological organization. Oikos 101:205–215

Coma R, Ribes M, Gili JM, Zabala M (2000) Seasonality in coastal benthic ecosystems. Trends Ecol Evol 15:448–453

Coma R, Polà E, Ribes M, Zabala M (2004) Long-term assessment of temperate octocoral mortality patterns, protected vs. unprotected areas. Ecol Appl 14:1466–1478

Coma R, Ribes M, Serrano E et al (2009) Global warming enhanced stratification and mass mortality events in the Mediterranean. Proc Natl Acad Sci USA 106:6176–6181

Crabbe MJC (2008) Climate change, global warming and coral reefs: modelling the effects of temperature. Comput Biol Chem 32:311–314

Diffenbaugh NS, Pal JS, Giorgi F, Gao XJ (2007) Heat stress intensification in the Mediterranean climate change hotspot. Geophys Res Lett 34:1–6

Ferrier-Pagès C, Tambutté E, Zamoum T, Segonds N, Merle P-L, Bensoussan N, Allemand D, Garrabou J, Tambutté S (2009) Physiological response of the symbiotic gorgonian *Eunicella singularis* to a long-term temperature increase. J Exp Biol 212:3007–3015

Francour P, Boudouresque CF, Harmelin JG, Harmelin-Vivien M, Quignard JP (1994) Are the Mediterranean waters becoming warmer? Information from biological indicators. Mar Poll Bull 28:523–526

Garrabou J (1999) Life history traits of *Alcyonium acaule* and *Parazoanthus axinellae* (Cnidaria, Anthozoa), with emphasis on growth. Mar Ecol Prog Ser 178:193–204

Garrabou J, Sala E, Arcas A, Zabala M (1998) The impact of diving on rocky sublittoral communities: a case study of a bryozoan population. Conserv Biol 12:302–312

Garrabou J, Perez T, Sartoretto S, Harmelin JG (2001) Mass mortality event in red coral (*Corallium rubrum*, Cnidaria, Anthozoa, Octocorallia) population in the provence region (France, NW Mediterranean). Mar Ecol Prog Ser 217:263–272

Garrabou J, Coma R, Bensoussan N et al (2009) Mass mortality in Northwestern Mediterranean rocky benthic communities: effects of the 2003 heat wave. Glob Change Biol 15:1090–1103

Gili JM, Ballesteros E (1991) Structure of cnidarian populations in Mediterranean sublitoral benthic communities as a result of adaptation to different environmental conditions. Oecol Aquat 10:243–254

Goffredo S, Caroselli E, Mattioli G, Pignotti E, Dubinsky Z, Zaccanti F (2009) Inferred level of calcification decreases along an increasing temperature gradient in a Mediterranean endemic coral. Limnol Oceanogr 54:930–937

Grall J, Le Loc'h F, Guyonnet B, Riera P (2006) Community structure and food web based on stable isotopes ($\delta^{15}N$ and $\delta^{13}C$) analyses of a North Eastern Atlantic maërl bed. J Exp Mar Biol Ecol 338:1–15

Hall-Spencer JM, Rodolfo-Metalpa R, Martin S, Ransome E, Fine M, Turner SM, Rowley SJ, Tedesco D, Buia M-C (2008) Volcanic carbon dioxide vents show ecosystem effects of ocean acidification. Nature 454:96–99

Harmelin JG (1993) Invitation sous l'écume. Cah Parc Nat Port-Cros 10:1–83

Hong JS (1980) Étude faunistique d'un fond de concrétionnement de type coralligène soumis à un gradient de pollution en Méditerranée nord-occidentale (Golfe de Fos). Disertation, Université d'Aix-Marseille II, Aix-en-Provence

Hong JS (1982) Contribution à l'étude des peuplements d'un fond coralligène dans la région marseillaise en Méditerranée Nord-Occidentale. Bull KORDI 4:27–51

Jokiel PL, Rodgers KS, Kuffner IB, Andersson AJ, Cox EF, Mackenzie FT (2008) Ocean acidification and calcifying reef organisms: a mesocosm investigation. Coral Reefs 27:473–483

Kersting DK, Linares C (2012) *Cladocora caespitosa* bioconstructions in the Columbretes Islands Marine Reserve (Spain, NW Mediterranean): distribution, size structure and growth. Mar Ecol 33:425–436

Kružić P (2002) Marine fauna of the Mljet National Park (Adriatic Sea, Croatia). 1. Anthozoa. Nat Croat 11:265–292

Kružić P (2007) Polyp expulsion of the coral *Cladocora caespitosa* (Anthozoa, Scleractinia) in extreme sea temperature conditions. Nat Croat 16:211–214

Kružić P, Benković L (2008) Bioconstructional features of the coral *Cladocora caespitosa* (Anthozoa, Scleractinia) in the Adriatic Sea (Croatia). Mar Ecol-Evol Persp 29:125–139

Kružić P, Požar-Domac A (2003) Banks of the coral *Cladocora caespitosa* (Anthozoa, Scleractinia) in the Adriatic Sea. Coral Reefs 22:536

Kružić P, Zuljević A, Nikolić V (2008) The highly invasive alga Caulerpa racemosa var. Cylindracea poses a new threat to the banks of the coral Cladocora caespitosa in the Adriatic Sea. Coral Reefs 27:441

Kružić P, Sršen P, Benković L (2012) The impact of seawater temperature on coral growth parameters of the colonial coral *Cladocora caespitosa* (Anthozoa, Scleractinia) in the eastern Adriatic Sea. Facies 58:477–491

Kuffner IB, Andersson AJ, Jokiel PL, Rodgers KS, Mackenzie FT (2008) Decreased abundance of crustose coralline algae due to ocean acidification. Nat Geosci 1:114–117

Kuhlemann J, Rohling EJ, Krumrei I, Kubik P, Ivy-Ochs S, Kucera M (2008) Regional synthesis of Mediterranean atmospheric circulation during the last glacial maximum. Science 321:1338–1340

Kushmaro A, Rosenberg E, Fine M, Ben Haim Y, Loya Y (1998) Effect of temperature on bleaching of coral *Oculina patagonica* by *Vibrio* AK-1. Mar Ecol Prog Ser 171:131–137

Laborel J (1961) Le concretionnement algal "coralligène" et son importance géomorphologique en Méditerranée. Rev Trav Stat Mar Endoume 23:37–60

Laborel J (1987) Marine biogenic constructions in the Mediterranean. Sci Rep Port-Cros Natl Park 13:97–126

Linares C, Coma R, Diaz D, Zabala M, Hereu B, Dantart L (2005) Immediate and delayed effects of mass mortality event on gorgonian population dynamics and benthic community structure in the NW Mediterranean. Mar Ecol Prog Ser 305:127–137

Lough JM, Barnes DJ (2000) Environmental controls on growth of the massive coral *Porites*. J Exp Mar Biol Ecol 245:225–243

Martin S, Gattuso JP (2009) Response of Mediterranean coralline algae to ocean acidification and elevated temperature. Glob Change Biol 15:2089–2100

Martin S, Clavier J, Guarini J-M, Chauvaud L, Hily C, Grall J, Thouzeau G, Jean F, Richard J (2005) Comparison of *Zostera marina* and maërl community metabolism. Aquat Bot 83:161–174

Martin S, Rodolfo-Metalpa R, Ransome E, Rowley S, Buia MC, Gattuso JP, Hall-Spencer J (2008) Effects of naturally acidified seawater on seagrass calcareous epibionts. Biol Lett 4:689–692

McNeil BI, Matear RJ (2008) Southern ocean acidification: a tipping point at 450-ppm atmospheric CO_2. Proc Natl Acad Sci USA 105:18860–18864

Meinesz A, Hesse B (1991) Introduction et invasion de l'algue tropicale *Caulerpa taxifolia* en Méditerranée nord-occidentale. Oceanol Acta 14:415–426

Montagna P, McCulloch M, Mazzoli C, Silenzi S, Odorico R (2007) The non-tropical coral *Cladocora caespitosa* as the new climate archive for the Mediterranean: high-resolution (weekly) trace element systematics. Quat Sci Rev 26:441–462

Montagna P, Silenzi S, Devoti S, Mazzoli C, McCulloch M, Scicchitano G, Taviani M (2008) Climate reconstructions and monitoring in the Mediterranean Sea: a review on some recently discovered high-resolution marine archives. Rend Fis Acc Lincei 19:121–140

Morri C, Peirano A, Bianchi CN, Sassarini M (1994) Present-day bioconstructions of the hard coral, *Cladocora caespitosa* (L.) (Anthozoa, Scleractinia), in the eastern Ligurian Sea (NW Mediterranean). Biol Mar Mediterr 1:371–372

Morri C, Peirano A, Bianchi CN (2001) Is the Mediterranean coral *Cladocora caespitosa* an indicator of climatic change? Arch Oceanogr Limnol 22:139–144

Orr JC, Fabry VJ, Aumont O et al (2005) Anthropogenic ocean acidification over the twenty-first century and its impact on calcifying organisms. Nature 437:681–686

Pascual J, Flos J (1984) Meteorología y Oceanografía. In: Ros JD, Olivella I, Gili JM (eds) Els sistemes naturals de les Illes Medes. Institut d'Estudis Catalans, Barcelona

Pax F, Müller I (1962) Die Anthozoenfauna der Adria. Institut für Ozeanographie und Fischerei, Split

Peirano A, Kružic P (2004) Growth comparison between Ligurian and Adriatic samples of the coral *Cladocora caespitosa*: first results. Biol Mar Mediterr 11:166–168

Peirano A, Morri C, Mastronuzzi G, Bianchi CN (1998) The coral *Cladocora caespitosa* (Anthozoa, Scleractinia) as a bioherm builder in the Mediterranean Sea. Mem Descr Carta Geol d'Italia 52:59–74

Peirano A, Morri C, Bianchi CN (1999) Skeleton growth and density pattern of the temperate, zooxanthellate scleractinian *Cladocora caespitosa* from the Ligurian Sea (NW Mediterranean). Mar Ecol Prog Ser 185:195–201

Peirano A, Morri C, Bianchi CN, Rodolfo Metalpa R (2001) Biomass, carbonate standing stock and production of the Mediterranean coral *Cladocora caespitosa* (L.). Facies 44:75–80

Peirano A, Morri C, Bianchi CN, Aguirre J, Antonioli F, Calzetta G, Carobene L, Mastronuzzi G, Orrù P (2004) The Mediterranean coral *Cladocora caespitosa*: a proxy for past climate fluctuations? Glob Planet Change 40:195–200

Peirano A, Abbate M, Cerrati G, Difesca V, Peroni C, Rodolfo-Metalpa R (2005) Monthly variations in calix growth, polyp tissue and density banding of the Mediterranean scleractinian *Cladocora caespitosa* (L.). Coral Reefs 24:404–409

Pérez T, Garrabou J, Sartoretto S, Harmelin JG, Francour P, Vacelet J (2000) Mortalité massive d'invertébrés marins: un événement sans précédent en Méditerranée nordoccidentale. CR Acad Sci Paris III 323:853–865

Piazzi L, Pardi G, Balata D, Cecchi E, Cinelli F (2002) Seasonal dynamics of a subtidal north-western Mediterranean macroalgal community in relation to depth and substrate inclination. Bot Mar 45:243–252

Piazzi L, Ceccherelli G, Meinesz A et al (2005) Invasion of *Caulerpa racemosa* (Caulerpales, Chlorophyta) in the Mediterranean Sea: the balance of thirteen years of spread. Cryptogam Algol 26:189–202

Raven J, Caldeira K, Elderfield H, Hoegh-Guldberg O, Liss P, Riebesell U, Shepherd J, Turley C, Watson A (2005) Ocean acidification due to increasing atmospheric carbon dioxide. Royal Society, London

Rivoire G (1991) Mortalité du corail et des gorgones en profondeur au large des côtes provençales. In: Boudouresque CF, Avon M, Gravez V (eds) Les espèces marines à protéger en Méditérranée. Groupement d'Intérêt Scientifique pour la Posidonie, Marseille

Rodolfo-Metalpa R, Bianchi CN, Peirano A, Morri C (2000) Coral Mortality in NW Mediterranean. Coral Reefs 19:24

Rodolfo-Metalpa R, Bianchi CN, Peirano A, Morri C (2005) Tissue necrosis and mortality of the temperate coral *Cladocora caespitosa*. Ital J Zool 72:271–276

Rodolfo-Metalpa R, Richard C, Allemand D, Ferrier-Pagès C (2006) Growth and photosynthesis of two Mediterranean corals, *Cladocora caespitosa* and *Oculina patagonica*, under normal and elevated temperatures. J Exp Biol 209:4546–4556

Rodolfo-Metalpa R, Peirano A, Houlbrèque F, Abbate M, Ferrier-Pagès C (2008) Effects of temperature, light and heterotrophy on the growth rate and budding of the temperate coral *Cladocora caespitosa*. Coral Reefs 27:17–25

Rodolfo-Metalpa R, Martin S, Ferrier-Pagès C, Gattuso J-P (2010) Response of the temperate coral *Cladocora caespitosa* to mid- and long-term exposure to pCO_2 and temperature levels projected for the year 2100. Biogeosciences 7:289–300

Rodolfo-Metalpa R, Houlbrèque F, Tambutté E et al (2011) Coral and mollusc resistance to ocean acidification adversely affected by warming. Nat Clim Change 1:308–312

Roether W, Klein B, Manca BB, Theocharis A, Kioroglou S (2007) Transient Eastern Mediterranean deep waters in response to the massive dense-water output of the Aegean Sea in the 1990s. Prog Oceanogr 74:540–571

Russell BD, Thompson J-AI, Falkenberg LJ, Connell SD (2009) Synergistic effects of climate change and local stressors: CO_2 and nutrient-driven change in subtidal rocky habitats. Glob Change Biol 15:2153–2162

Sala E (2004) The past and present topology and structure of Mediterranean subtidal rocky-shore food webs. Ecosystems 7:333–340

Sala E, Garrabou J, Zabala M (1996) Effects of diver frequentation on Mediterranean sublittoral populations of the bryozoan *Pentapora fascialis*. Mar Biol 126:451–459

Simkiss K (1964) Phosphates as crystalpoisons of calcification. Biol Rev 39:487–505

Steinacher M, Joos F, Frölicher TL, Plattner G-K, Doney SC (2009) Imminent ocean acidification in the Arctic projected with the NCAR global coupled carbon cycle-climate model. Biogeosciences 6:515–533

Stott PA, Stone DA, Allen R (2004) Human contribution to the European heatwave of 2003. Nature 432:610–614

Touratier F, Goyet C (2011) Impact of the eastern Mediterranean Transient on the distribution of anthropogenic CO_2 and first estimate of acidification for the Mediterranean Sea. Deep Sea Res Pt I 58:1–15

Vargas-Yáñez M, Salat J, de Puelles MLF, López-Jurado J, Pascual J, Ramírez T, Corteś D, Franco I (2002) About the seasonal variability of the Alboran Sea circulation. J Mar Syst 35:229–248

Weinberg S (1991) Faut-il protéger les gorgones de Méditerranée? In: Boudouresque CF, Avon M, Gravez V (eds) Les espèces marines à protéger en Méditérranée. Groupement d'Intérêt Scientifique pour la Posidonie, Marseille

Wernberg T, Russell BD, Moore PJ, Ling SD, Smale DA, Campbell A, Coleman MA, Steinberg PD, Kendrick GA, Connell SD (2011) Impacts of climate change in a global hotspot for temperate marine biodiversity and ocean warming. J Exp Mar Biol Ecol 400:7–16

Zibrowius H (1980) Les Scléractiniaires de la Méditerranée et de l'Atlantique nord-oriental. Mem Inst Oceanogr (Monaco) 11:1–284

Marine Bioerosion

26

Christine H.L. Schönberg and Max Wisshak

Abstract

Research conducted in the Mediterranean significantly contributed to our understanding of bioerosion, providing faunistic records and key information about the succession that occurs when fresh substrate is colonized by eroding biota. Bioeroders that have a substantial role in the Mediterranean are microendoliths, sponges, boring mollusks and various grazers. A multitude of environmental factors controls their abundances, diversities and eroding capacities. With ongoing climate change, several of these factors are likely to magnify the effects of bioerosion in the Mediterranean and worldwide. We regard eutrophication as the most important in the Mediterranean, but climate change, especially ocean acidification, will also have an important effect. Should bioerosion levels change, characteristic limestone coasts will be impacted, as will be community and sediment compositions, enigmatic cold-water coral ecosystems, mollusk aquaculture and man-made materials that are submerged. Understudied topics in Mediterranean bioerosion include rates, interactions at community level, as well as direct effects of climate change.

Keywords

Bioeroder densities • Bioeroders • Bioeroding organisms • Bioeroding taxa • Bioerosion rates of microbes • Bioerosion rates • Bioerosion • Chemical bioerosion • Coastal damage • Degradation of hard substrates • Eutrophication and pollution • Global warming • Lithic bioerosion • Macrobioerosion • Mediterranean bioerosion • Microbioerosion • Microendolithic bioerosion • Microendoliths • Ocean acidification • Sponge bioerosion • Wood borers

C.H.L. Schönberg (✉)
Australian Institute of Marine Science (AIMS),
Oceans Institute at the University of Western Australia,
Crawley, WA 9006, Australia
e-mail: C.Schoenberg@aims.gov.au

M. Wisshak
Marine Research Department, Senckenberg am Meer,
26382 Wilhelmshaven, Germany

S. Goffredo and Z. Dubinsky (eds.), *The Mediterranean Sea: Its history and present challenges*,
DOI 10.1007/978-94-007-6704-1_26,

Bioerosion in the Mediterranean

The Past of Mediterranean Bioerosion

Endolithic microbes belong to the oldest known life forms on earth with a confirmed age of at least 3.35 Ga (Fliegel et al. 2010). Endolithic cyanobacterian body fossils reach back to the Proterozoic (Golubic and Seong-Joo 1999), other microendoliths appeared in the Cambrian (Glaub and Vogel 2004). 'Deep borers' such as mollusks and worms were first reported from the Silurian (Beuck et al. 2008), and mobile, grazing epiliths from the Triassic (Wilson 2006). According to Bertling (1999, 2002), significant levels of bioerosion in prehistoric Europe co-occurred with the arrival of reefal substrate, and fluctuated with environmental conditions such as sedimentation levels and oxygenation, which favored or suppressed different bioeroders. Sea-level change resulted in changing successions of bioeroder communities and created a high diversity (Santos et al. 2008 and references therein). Bertling (1999, 2002) related increasing evidence of bioerosion to higher turbidity and nutrient levels, which is a recurrent theme in the past of bioerosion (e.g. Highsmith 1980a; Hallock and Schlager 1986; Hallock 1988). Being very responsive to changes in nutrients and having bathymetric distribution ranges, bioeroding sponges and their traces have probably received most attention in the Mediterranean paleontological record (Hartman 1957; Bromley and D'Alessandro 1984, 1990; Wisshak 2008). But due to their narrow ecological niche separation, microendoliths are even better paleoindicators for nutrients, temperature, light, moisture and salinity, particularly from the Late Ordovician onwards (e.g. Perry and Macdonald 2002; Vogel and Brett 2009; Wisshak 2012).

State of Art of Marine Bioerosion Research – What Is Contributed from the Mediterranean?

Since Neumann's (1966) definition of 'bioerosion' as the **degradation of hard substrates by living organisms**, related research has steadily increased (Schönberg and Tapanila 2006, their Fig. 1). While bioerosion research in general is mainly concentrating on tropical coral reefs (8.5 % of the publications are from the Great Barrier Reef alone), the Mediterranean is nevertheless well represented with almost 11 % (Fig. 26.1a). In numbers of publications on bioerosion listed in Thomson Reuter's Web of Knowledge (2011) for the last decade, study locations in southern France and Italy are on place 3 and 4 worldwide (Fig. 26.1b, however, 'France' data include publications on studies conducted in French Polynesia and at Reunion Island). Sites along the Mediterranean coast of Africa were not represented, but European countries contributed about 78 % of studies conducted in the Mediterranean Sea, and Asian countries 22 % (Fig. 26.1b).

As in other areas of research, studies on bioerosion evolved from more descriptive to more applied topics, and the overall knowledge is somewhat patchy and sometimes even contradictive (Schönberg 2008). Presently, faunistic studies on Mediterranean bioeroders are well represented, especially for the Aegean Sea, but functional studies are comparatively rare. Some of the few results available indicate that grazers control settlement success in bioeroders (e.g. Bromley et al. 1990), while other workers claim that by reducing algal cover, grazers are enhancing bioeroder growth (e.g. Cebrian and Uriz 2006). Results on bioerosion rates are scarce from the Mediterranean, with one publication on sponge bioerosion (Calcinai et al. 2011) and a thesis with results on accretion and bioerosion in the Ionian Sea in relation to light and temperature with water depth (Pyko 2009). Long-term studies described the succession of bioeroders in marble blocks over time (e.g. Bromley et al. 1990), with one taxon group preparing the substrate for the next, although it has never been explained what is involved in such a preparation (see Hutchings 2008). In the Mediterranean, but also elsewhere, we do not yet fully understand why endolithic bioeroders evolved to live within hard substrates (e.g. Golubic et al. 1975; Tribollet 2008a). Proposed reasons for this life style include: protection against excessive light or water turbulence, avoidance of predators, competition, parasites and disease, saving maintenance costs, and obtaining nutrients from the substrate (Yonge 1963; Bromley 1970; Ward and Risk 1977; Vénec-Peyré 1996; Glynn 1997; Golubic et al. 2005; Tribollet 2008a; Schönberg and Wisshak 2012). The *Lithophaga* etching agent is known (a calcium-binding mucoprotein, Jaccarini et al. 1968), and microendolithic bioerosion has lately received critical attention (Garcia-Pichel 2006; Garcia-Pichel et al. 2010). However, for most bioeroder taxa using chemical bioerosion we do not know how they proceed, especially at cell level (Vénec-Peyré 1996; Martin and Britayev 1998; Golubic et al. 2005; Hutchings 2008; McLoughlin et al. 2008; Schönberg 2008). And lastly: Because of their endolithic life style, bioeroders are usually ignored in general studies and biodiversity assessments. In combination with their often difficult (ichno) taxonomy and with many species insufficiently or not yet described (e.g. Hutchings 2008; Xavier et al. 2010), we have a long way to go to understand their effects on our seas and oceans.

Agents of Mediterranean Bioerosion

Bioeroders function as a guild (Tapanila 2008) and are linked by numerous interactions, which occasionally make it

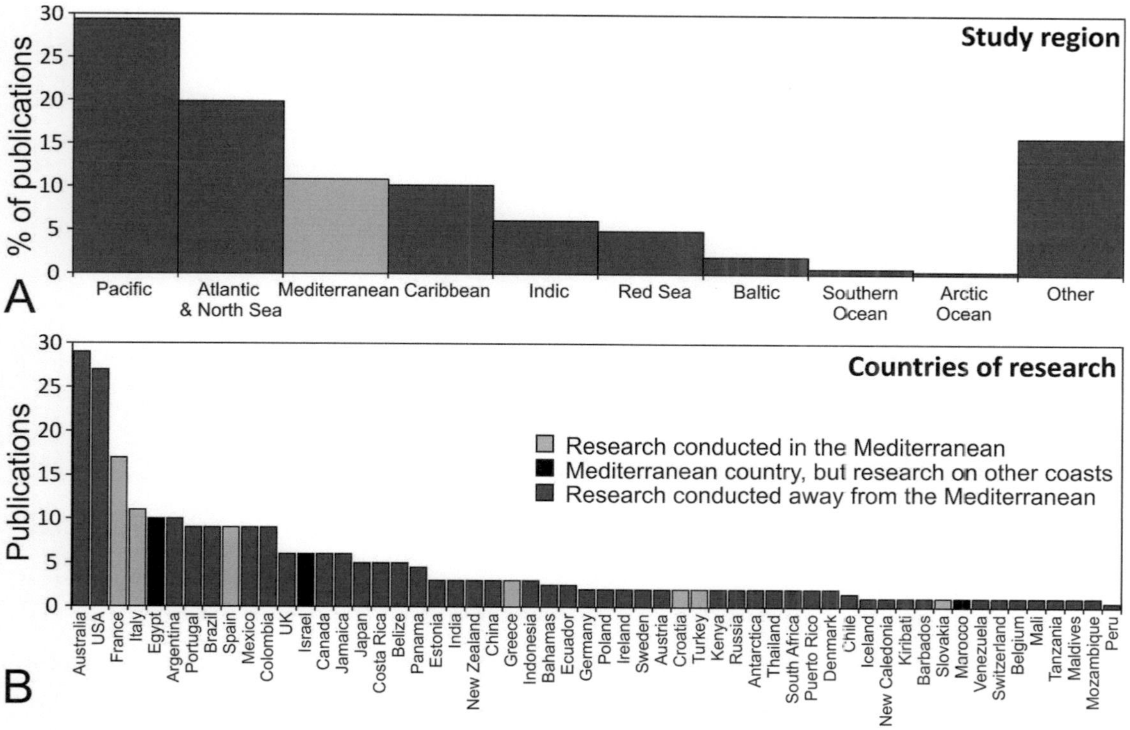

Fig. 26.1 Study locations for bioerosion research over the last 10 years (according to Thomson Reuters Web of Knowledge 2011). (**a**) By ocean or sea. 'Other' is comprised to 1/3 of terrestrial research, to 2/3 of research at unspecified locations. (**b**) By country. Denmark incl. Greenland, USA incl. Hawaii, and France incl. French Polynesia and Reunion Island

difficult to pry apart the roles of separate agents, especially in the warm-temperate biogeographic realm which appears to have a higher biodiversity of bioeroders compared to polar high or tropical low latitudes (e.g. Wisshak et al. 2011). Overall, we regard the following bioeroder groups as the most important in the Mediterranean: microendoliths, sponges, mollusks, and sea urchins. In contrast to tropical settings, fish do not play a significant role in Mediterranean bioerosion (Ballesteros 2006). Wood borers can also be key contributors, but were beyond the scope of the present publication with focus on lithic substrates. Additional details can be obtained e.g. from Yonge (1963), Bromley (1970) and Risk and MacGeachy (1978).

Microendoliths encompass the group of microbial bioeroders such as bacteria, cyanobacteria, microalgae and fungi (e.g. Golubic et al. 1975, 2005; Risk and MacGeachy 1978; Tribollet 2008a). Cyanobacteria and chlorophyte algae are key agents (Fig. 26.2) that were investigated in detail in the Mediterranean near Marseille by Le Campion-Alsumard (1975, 1979), building upon the historic studies carried out by Bornet and Flahault (1888, 1889) and Ercegović (1927, 1929, 1932). Wherever fresh substrate becomes available it is colonized and attacked within days to weeks by pioneering microbes, which then go through a succession of different taxa, reaching an equilibrium after months to years (e.g. Hong 1980; Le Campion-Alsumard et al. 1995; Sartoretto 1998; Naylor and Viles 2002; Beuck et al. 2010). In shallow water, substrate surfaces are grazed by mollusks and echinoids as soon as microendoliths have taken hold, intensifying the effect imparted by the microbes and shaping limestone coasts (e.g. Schneider 1976, 1977; Bromley et al. 1990). Bathymetric patterns and penetration depths within substrates can be related to the compensation depth of photosynthetic microendoliths, with only fungal and bacterial microbioerosion in aphotic depths (e.g. Golubic et al. 1975; Sartoretto 1998; Tribollet 2008b). On the other hand, where light suddenly increases it can cause blooms in the microendolithic communities, possibly producing excess nutrients passed on to neighboring or host organisms (Fine and Loya 2002; Fine et al. 2004). Pace and diversity in development and linked bioerosion rates of microbes are thus highly dependent on bathymetry and the extent of photic zones, as well as the orientation of the substrate (Wisshak et al. 2011).

Sponges are among the best studied bioeroders in the Mediterranean, especially from Spain, France and Italy (e.g. Coll et al. 2010). Excavating sponges establish in newly available substrates after about 2 years and often thrive better on the undersides of substrates or ledges, where competition with coralline algae is lower (Bromley et al. 1990; Cerrano et al. 2001). They etch fine, cup-shaped grooves and mechanically remove lentil-shaped chips that are expelled with the exhalant water to form fine-grained sediment

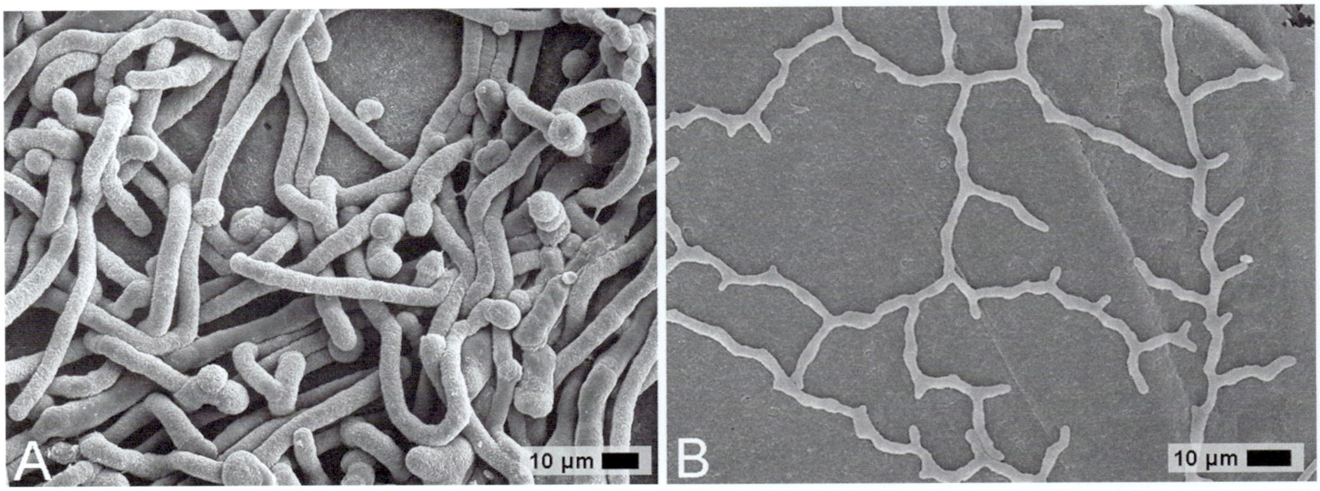

Fig. 26.2 SEM images of epoxy resin casts of microborings produced by the cyanobacterium *Mastigocoleus testarum* (**a**) and the chlorophyte alga *Ostreobium queketti* (**b**) in 15 and 50 m water depth, respectively, recorded on a settlement experiment off the southern Peloponnes, Ionian Sea (Images courtesy of I. Pyko)

Fig. 26.3 The most common and important bioeroding sponges of the Mediterranean: *Cliona viridis* (**a**, *grayish tissue*) and *Cliona celata* (**b**, *yellow tissue*) (Photographs courtesy of T. Perez)

(Fütterer 1974; Rützler and Rieger 1973). The proportion of chemical etching compared to what is removed as chip has been estimated between 2:98 and 70:30 %, a discrepancy that may well be related to environmental conditions (Rützler and Rieger 1973; Zundelevich et al. 2007; Nava and Carballo 2008). In the Mediterranean *Cliona viridis* is the most abundant and destructive sponge, partly rivaled by species of the *Cliona celata* species complex (Fig. 26.3; e.g. Rosell et al. 1999; Calcinai et al. 2011). Bioeroding sponges have been recognized to be good pollution indicators and biomonitors for environmental conditions (e.g. Hong 1980; Carballo et al. 1994, 1996).

Many mollusks such as chitons and limpets produce 'home scars' or resting scars at places where they most often sit (Fig. 26.4a), and generate grazing traces by removing substrate layer by layer either by radular scraping or by use of acid (e.g. Hutchings 1986). Their intertidal activity is strongly governed by access to water, and areas closer to the water line and spray zone are usually more strongly eroded, which can lead to coastal notch-formation (e.g. Palmer et al. 2003; Neumann 1966). Locally more important than these surface marks are the large hollows mostly produced by bivalves such as pholads, gastrochaenids and mytilids. These mollusks erode by mechanical or chemical means (e.g. Jaccarini et al. 1968; Ansell and Balakrishna Nair 1969; Appukuttan 1969). For reefal areas pholad densities between 50 and 500 individuals per m^2 can occur (Warme 1975; Highsmith 1980b; Loya 1982), but we have no comparable data for the Mediterranean. Here, the slow-growing mytilid lithophags play a considerable role in

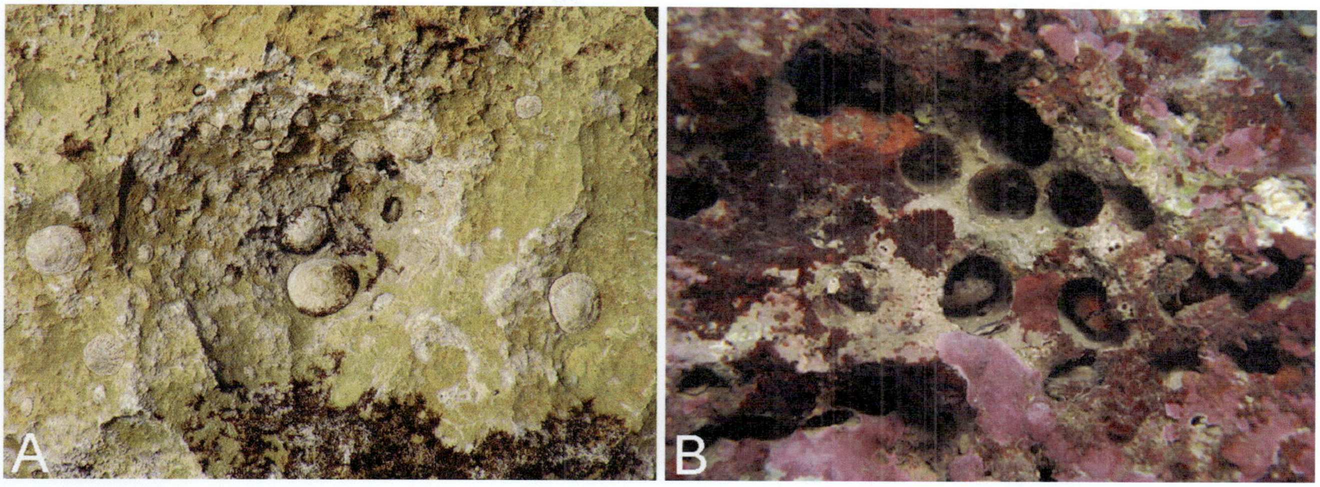

Fig. 26.4 (**a**) *Patella caerulea* and *Chiton olivaceus* with *Patella* resting scars at the Costa Daurada, Spain. (**b**) Borehole battery made by *Lithophaga lithophaga* in the French Mediterranean (**a** – courtesy of J. Martinell, **b** – of A. Antonioli)

Fig. 26.5 Bioerosion caused by regular sea urchins. (**a**) *Star-shaped* scrape marks made by *Paracentrotus lividus* on an *Anomia ephippium* valve from the Costa Brava, Spain. (**b**) Resting scars produced by *Paracentrotus lividus* at the Costa Daurada, Spain (Photographs courtesy of J. Martinell)

weakening coastal structures (Fig. 26.4b; Kleemann 1973; Devescovi 2009), a process that is magnified by lithophags being harvested as a local delicacy (Fanelli et al. 1994; Russo and Cicogna 1991; Hrs-Brenko et al. 1991; Fraschetti et al. 2001; Guidetti et al. 2003; Devescovi et al. 2005). Therefore, despite being abundant, since 2005 *Lithophaga lithophaga* is a CITES-protected species (Boudouresque et al. 1991; CITES 2011), a restriction which is not necessarily respected by the people (Bianchi and Morri 2000).

Sea urchins appear to play a comparatively small role in Mediterranean bioerosion (Fig. 26.5; Laubier 1966; Sartoretto and Francour 1997), which may in part be related to the fact that they are locally overfished for their roe (Guidetti et al. 2004; Micael et al. 2009). But where urchins have large diameters or their densities reach high levels of at least 20 individuals per 25 m^2 (Sartoretto and Francour 1997) their continuous grazing can maintain 'barren grounds' of mostly coralline algae (Guidetti et al. 2003; Privitera et al. 2005), they can create cup-shaped borings or scrape deep channels into the substrate surface (e.g. Asgaard and Bromley 2008; Fig. 26.5b). Where they graze, sea urchins remove live tissue of other invertebrates or infaunal bioeroders (e.g. Tribollet and Golubic 2005) and can dislodge smaller sessile fauna by bulldozing (Schönberg, pers. obs.). By their rasping activity they produce substrate debris that is defecated as sediments and will accumulate in the area, a process that has been studied in the neighboring Red Sea (Mokady et al. 1996).

Bioeroding worms should be mentioned as well, as they cause the mud blister syndrome in mollusk shells (Fig. 26.6a; e.g. Martin and Britayev 1998). Worms establish themselves in fresh substrate at a rate similar to sponges and appear in a succession of taxa after 2–3 years (e.g. Hutchings 2008).

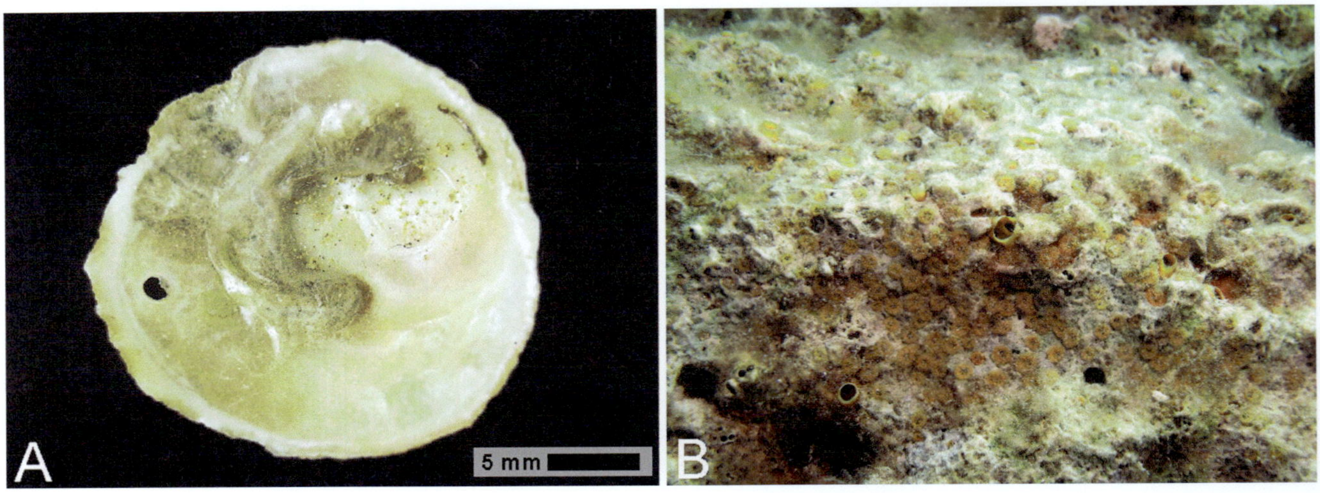

Fig. 26.6 (**a**) Mud blister in an *Anomia ephippium* valve from the Costa Brava, Spain. (**b**) This benthic community of bioeroders in the French Mediterranean is a result of a complicated interplay of e.g. nutrient availability to the filter feeders (*Cliona celata* and endolithic molluscs), competition with coralline algae and other benthic organisms and predation (**a** – courtesy of J. Martinell, **b** – of T. Perez)

Their occurrences vary with environmental conditions, and they can be quite useful as pollution indicators (Hong 1980; Dean 2008). New species can become introduced pests and invasive species (Simboura and Zenetos 2005; Çinar et al. 2011; Zenetos et al. 2011). Polychaetes are a comparatively prominent worm group in the Mediterranean (Feldmann 1937; Hong 1980), while only three species of local endolithic sipunculids are presently identified (Laubier 1966; Sartoretto and Francour 1997). Ironically, Mediterranean worm reefs built by other polychaetes can become subject to bioerosion (Porras et al. 1996; Fornós et al. 1998).

Environmental Factors Acting on Mediterranean Bioerosion and Possible Trends

Many bioeroding organisms such as microendoliths, sponges, worms and endolithic mollusks are sessile and depend on nutrients and/or particulate organic matter being carried to them in the water column. But even mobile grazers can benefit from fertilization if it induces increased algal growth (e.g. Le Bris et al. 1998; Carreiro-Silva et al. 2005, 2009). We therefore assume that in the Mediterranean the strongest local drivers of future trends of bioerosion will be eutrophication and pollution (e.g. Coll et al. 2010). Both factors are thought to aggravate the effects of climate change (Le Grand and Fabricius 2011), which will be of importance in a region that is highly industrialized, densely populated, and has comparatively little exchange with surrounding water masses (e.g. Bianchi and Morri 2000). In particular urban development and river runoff bring about anthropogenic stress known to affect shifts in community structures and to increase bioeroder densities and bioerosion rates (Le Grand and Fabricius 2011 and their table 1). However, the magnitude of nutrient effects are probably the most difficult to determine at community level, because a multitude of cascades and feed-back loops exist that may either enhance the overall effect or confuse the results (e.g.: pollution → increased bacteria concentrations in water column → increased abundances of filter feeding bioeroders → intensive recycling of nutrients → fertilization of microendoliths → increased porosity of substrate → increased efficiency of grazer bioerosion → creation of fresh surfaces aiding settlement of bioeroders; e.g. Corredor et al. 1988; Rose and Risk 1985; Carreiro-Silva et al. 2009; Fig. 26.6b). Further complications arise from the complex interplay of dissolved inorganic nutrients, particulate organic matter, siliciclastic material or pollutants such as heavy metals in the water, with different effects on the various types of bioeroding taxa (fertilization, shading, poisoning, etc.). It is thus hardly a surprise that studies on the influence of the trophic regime on bioerosion yielded partly contradictive results (e.g. Chazottes et al. 2002; Szmant 2002; Carreiro-Silva et al. 2005; Holmes et al. 2009).

To date not many data are available on how climate change may affect bioerosion and whether global warming or ocean acidification may bring about a pertinent change. Within physiological tolerance levels chemical bioerosion itself may be enhanced by increased temperatures due to the simple fact that the reaction rate is accelerated. Overall, global warming is expected to have less severe effects on bioeroding organisms than on calcifiers, because grazers will move into shade and endoliths are extremely well sheltered and insulated by living within porous, water-retaining substrates that provide shade, and many endoliths can produce cooling currents passing through or over their bodies. Some very common species of bioeroding sponges live in symbiosis with dinoflagellates similar to those in corals, yet they have been observed to be very bleaching resistant during periods of abnormal heating (Vicente 1990; Schönberg and Wilkinson 2001), but the symbiosis may be quite susceptible to light stress (Hill and Wilcox 1998; Schönberg and Suwa 2007; Schönberg et al. 2008). If other bioeroders are as heat resistant as these sponges, the entire guild may indirectly benefit from thermal damage to the calcifiers, which in response to heat stress may be weakened or dead and thus an easy target. In the Mediterranean, warming events are expected to increase, and eastern parts of the basin are more likely to be affected than central and western parts (e.g. Bianchi and Morri 2000; Garrabou et al. 2001, 2009; Coll et al. 2010).

Ocean acidification may have a larger impact on bioerosion than global warming. Most macrobioeroders are either filter feeders or grazers, groups that were found to not be negatively affected or even more common near Mediterranean CO_2 vents (Kroeker et al. 2011). Moreover, many bioeroders at least in part use chemical dissolution to remove substrate, a process which is often thought to be pH dependent or relying on acid production (e.g. Pomponi 1980). Where this is the case, bioerosion is likely to increase in acidified waters (Tribollet et al. 2006). Experimental evidence on how bioerosion might change as a direct reaction to climate change and ocean acidification is scarce. With higher pCO_2 Tribollet et al. (2006, 2009) observed increased penetration depth and bioerosion by microendoliths with a stable community composition. Studies by the present authors investigate the impacts of climate change on macrobioerosion, and results suggest that temperature has little effect on sponge bioerosion (Wisshak et al. unpubl. data), but it will be significantly accelerated by ocean acidification (Wisshak et al. 2012). However, if acidification ever reaches catastrophic levels, reducing the abundance of suitable substrate in the Mediterranean (e.g. Fine and Tchernov 2007; Kroeker et al. 2011), lithic bioerosion will likewise come under pressure. Volcanic activity and hydrothermal seepage in the Mediterranean provides an ideal environment to study consequences of ocean acidification on bioerosion in natural settings (Kroeker et al. 2011; Rodolfo-Metalpa et al. 2011).

Which Assets Will Be Most Likely Affected by Mediterranean Bioerosion?

Mediterranean bioeroder communities are constantly modifying and structuring biogenic as well as abiogenic hardground of coastal areas. This includes the characteristic coralline alga assemblages that are mostly shaped by grazers such as sea urchins (Fig. 26.7; Ballesteros 2006). In extreme cases the urchins can create and maintain 'barrens' devoid of high structural complexity (Privitera et al. 2005). Cover with red algae will in turn reduce settlement success of the larvae of many bioeroders and withstands bioerosion better than other substrates (Hong 1980; Smyth 1989; Chazottes et al. 2002, Ballesteros 2006), although bioeroding sponges frequently penetrate it (Cotte 1914; Cerrano et al. 2001).

Many bioeroders rework and produce considerable amounts of sediment. For example, sponge-generated sediments in the North Adriatic Sea make up about 2–3 % (Fütterer 1974). Higher bioeroder densities or shifts in their community compositions would mean that both their influence on the hard-bottom morphology and the sediment quality in their habitat will change (Hutchings 1986).

Diverse bioeroder communities inhabit slow-growing solitary, (pseudo)colonial and cold-water Mediterranean corals (Fig. 26.8a; Bromley 2005; Wisshak 2008; Mastrototaro et al. 2010), which they can infest to a high degree (20–75 %, Bavestrello et al. 1997; Corriero et al. 1997; Cerrano et al. 2001), deform them (Beuck et al. 2007), cause a higher risk of fragmentation by removing up to 70 % of the skeleton (Beuck et al. 2010), overwhelm early life stages (Calcinai et al. 2000) or cause wide-spread mortality (Corriero et al. 1997). Bioeroding sponges in particular are a devastating pest in Mediterranean corals and among others attack the precious red coral *Corallium rubrum* (Fig. 26.8b; e.g. Melone 1965; Maldonado 1992; Bavestrello et al. 1996; Calcinai et al. 2000, 2002, 2008), the overexploited Mediterranean 'red gold'. *Corallium* has been used as a gemstone since ancient times (Pronzato 2000), historically supported trade between Mediterranean countries and India, and is part of the Greek mythology (Medusa's head, Poseidon's Palace; De Simone 2010 and other contributions in the same volume). Bioeroding sponges are very common in Mediterranean *Corallium* colonies (Barletta and Vighi 1968). Corriero et al. (1997) stated that coral survival is strongly correlated with severity of infestation with these borers that are able to kill colonies that reach 4 years of age, while they only become reproductive at 2 years or later (Santangelo et al. 2003). Should the infestation rate by sponges increase, the traditional Mediterranean culture and trade based on the already threatened red precious coral may well become a thing of the past (Weinberg 1991; Garrabou et al. 2001, 2009; Costantini et al. 2007). As many cold-water reefs support a very diverse associated fauna, a reduced abundance of such corals would affect other taxa as well (e.g. Mastrototaro et al. 2010).

Bioeroders furthermore invade mollusk shells and can cause considerable damage on local oysters and mussels. Aquaculture hatcheries generate considerable profit for at least six Mediterranean countries (Fisheries and Aquaculture

Fig. 26.7 The 'coralligène' (**a**) and some of the main players involved in shaping this habitat (**b**): the sea urchin *Paracentrotus lividus*, the sponge *Cliona viridis* (*brown papillae* in the *lower left*) and coralline algae (in the background) (**a** – courtesy of T. Perez)

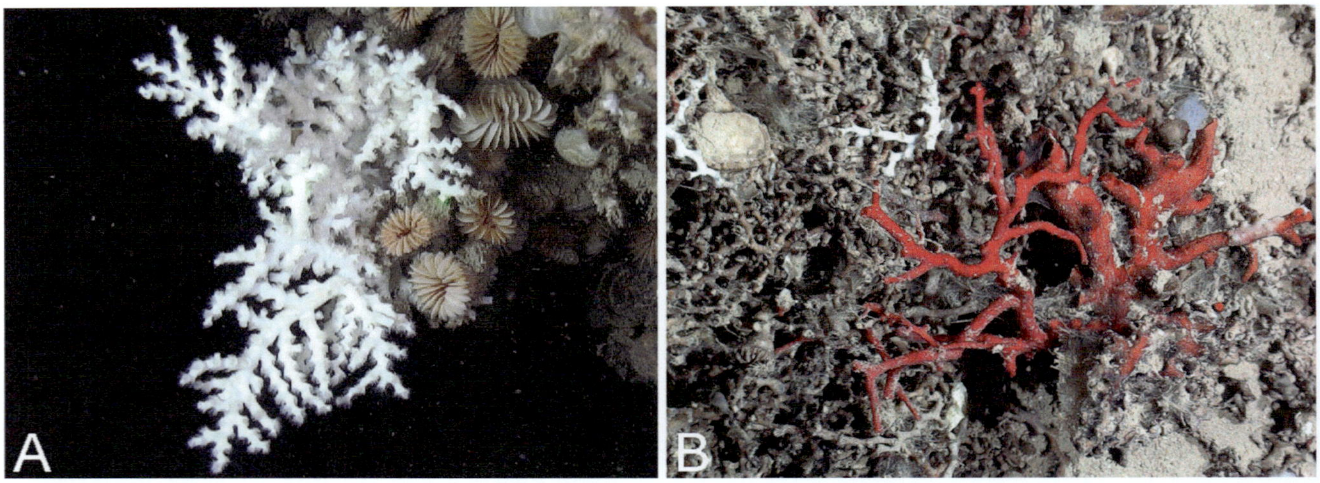

Fig. 26.8 (**a**) The cold-water corals *Madropora oculata* and the solitary *Desmophyllum dianthus* from about 400 m water depth in the Lacaze-Duthiers Canyon near Banyuls-sur-Mer, France. (**b**) Colony of the red coral *Corallium rubrum* from the Linosa Trough in the Pelagian Island group showing clear evidence of bioerosion and subsequent deformation (**a** – courtesy of Senckenberg/MARUM, **b** – courtesy of L. Beuck and A. Freiwald)

Department 2011). France is a large, traditional oyster producer, ranking among the top producer and consumer countries of the world, with an annual production of 9500 t in the Thau Lagoon (Héral 1989; Buestel et al. 2009). Losses related to bioerosion have been historically observed (Carazzi 1895) and may significantly increase in the future. While oysters are readily attacked by various bioeroders (Labura and Hrs-Brenko 1990: up to 89 % oysters by *Polydora hoplura*, Rosell et al. 1999: 100 % of oysters by two species of sponges), mussels are to a much lesser extent and degree, but often with more devastating effects to their shell stability and thus to survival from predation (Lauckner 1983).

Bioerosion in combination with changing weather patterns and increased frequencies of severe storms may also result in more coastal damage, because bioerosion reduces the security of attachments and breaking strength of the invaded organisms and materials (Tunnicliffe 1979; Highsmith 1980b; Mitchell-Tapping 1983; Clark and Morton 1999; Scott and Risk 1988). Microendoliths and urchins can attack volcanic materials such as dolerites, balsanites and volcanic glass (Allouc et al. 1996; McLoughlin et al. 2008), which would play a role in those parts of the Mediterranean that have a geological history of volcanism (e.g. Aiello et al. 2001). And as final reminder of cause-and-response to our actions: Man-made environmental changes influencing bioerosion will also damage man-made or man-shaped substrates. A number of bioeroding organisms such as sponges and pholad bivalves are able to riddle and whittle away artificial materials or

Fig. 26.9 Greek statues from the Athens National Museum of Archaeology retrieved from a shipwreck in the Aegean. They were exposed to marine bioerosion for about 2100 years, the degree of damage depending on the sediment cover. (**a**) The statue's eyes were replaced by a brachiopod and a pholad boring. (**b**) Parts of the statue that were not covered by sand were strongly affected by bioerosion (Photographs courtesy of R. G. Bromley and U. Asgaard)

structures (Scott 1991; Brusco et al. 2005). Other mollusks are destabilizing breakwater boulders in the Adriatic (Devescovi and Iveša 2008) and can easily penetrate concrete containing calcium, as can some sponges and polychaetes (Yonge 1963; Scott et al. 1988). Echinoids have been suspected to attack even steel pilings (Emery 1960) and are known to pry grains out of granite at a faster rate than they would erode limestone (Bromley 1970). And in conclusion, bioerosion also affects submerged works of art, e.g. Greek marble statues or submerged mosaics in the Aegean (Fig. 26.9; Bromley and Asgaard 2004; Ricci et al. 2008).

Acknowledgements The authors thank Z. Dubinsky and S. Goffredo for inviting us to contribute a chapter on bioerosion to this book, and L. Walker for unearthing some of the literature on Mediterranean mud blister worms. We are indebted to the following colleagues who gave us permission to use some of their photographs: A. Antonioli, L. Beuck and A. Freiwald (IPAL 2006 and ROV Quest, MARUM 2006), T. Perez, and I. Pyko, J. Martinell, and R. Bromley and U. Asgaard.

References

Aiello G, Budillon V, Cristofalo G, D'Argenio B, de Alteriis G, De Lauro M, Ferraro L, Marsella E, Pelosi N, Sacchi M, Tomielli R (2001) Marine geology and morphobathymetry in the Bay of Naples (south-eastern Tyrrhenian Sea, Italy). In: Faranda FM, Guglielmo L, Spezie G (eds) Mediterranean ecosystems. Structures and processes. Springer, Milano, pp 1–8

Allouc J, Le Campion Alsumard T, Tack DL (1996) Bioerosion of magmatic rock in a coastal environment: the example of the Cap Vert Peninsula (western Senegal). Geobios 29:485–502

Ansell AD, Balakrishna Nair N (1969) A comparative study of bivalves which bore mainly by mechanical means. Am Zool 9:857–868

Appukuttan KK (1969) Coral-boring bivalves of Gulf of Mannar and Palk Bay. In: Proceedings of the symposium corals and coral reefs India, Marine Biological Association of India, Mandapam Camp, India, pp 379–398

Asgaard U, Bromley RG (2008) Echinometrid sea urchins, their trophic styles and corresponding bioerosion. In: Wisshak M, Tapanila L (eds) Current developments in bioerosion. Springer, Heidelberg, pp 279–303

Ballesteros E (2006) Mediterranean coralligenous assemblages: a synthesis of present knowledge. Oceanogr Mar Biol Ann Rev 44:123–195

Barletta G, Vighi M (1968) Richerche sul corallo rosso: V. Poriferi perforanti lo sclerasse di *Corallium rubrum* Lamarck. Rend Ist Lombardo Sci Lett Milano B 102:145–159

Bavestrello G, Calcinai B, Sarà M (1996) *Delectona ciconiae* sp. nov. (Porifera, Demospongiae) boring in the scleraxis of *Corallium rubrum*. J Mar Biol Assoc UK 76:867–873

Bavestrello G, Calcinai B, Cerrano C, Sarà M (1997) *Delectona madreporica* n. sp. (Porifera, Demospongiae) boring the corallites of some scleractinians from the ligurian Sea. Ital J Zool 64:273–277

Bertling M (1999) Late Jurassic reef bioerosion – the drawing of a new era. Bull Geol Soc Den 45:173–176

Bertling M (2002) Coral reef bioerosion in times of crises – the Late Triassic/Early Jurassic example. In: Proceedings of the 9th international coral reef symposium, Bali, vol 1, pp 23–27

Beuck L, Vertino A, Stepina E, Karolczak M, Pfannkuche O (2007) Skeletal response of *Lophelia pertusa* (Scleractinia) to bioeroding sponge infestation visualised with micro-computed tomography. Facies 53:157–176

Beuck L, Wisshak M, Munnecke A, Freiwald A (2008) A giant boring in a Silurian stromatoporoid analysed by computer tomography. Acta Palaeontol Pol 53:147–158

Beuck L, Freiwald A, Taviani M (2010) Spatiotemporal bioerosion patterns in deep-water scleractinians from off Santa Maria die Leuca (Apulia, Ionian Sea). Deep-Sea Res II 57:458–470

Bianchi CN, Morri C (2000) Marine biodiversity of the Mediterranean Sea: situation, problems and prospects for future research. Mar Poll Bull 40:367–376

Bornet E, Flahault C (1888) Note sur deux nouveaux genres d'algues perforantes. J Bot Morot 2:161–165

Bornet E, Flahault C (1889) Sur quelques plantes vivant dans le test calcaire des mollusques. Bull Soc Bot France 36:147–179

Boudouresque CF, Avon M, Gravez V (1991) Les espèces marines à protéger en Méditerrannée. GIS Posidonie, Marseille, pp 1–447

Bromley RG (1970) Borings as trace fossils and *Entobia cretacea* Portlock, as an example. In: Crimes TP, Harper JC (eds) Trace fossils. Geol J special issue 3:49–90

Bromley RG (2005) Preliminary study of bioerosion in the deep-water coral *Lophelia*, Pleistocene, Rhodes, Greece. In: Freiwald A, Roberts JM (eds) Cold-water corals and ecosystems. Springer, Heidelberg, pp 898–914

Bromley RG, Asgaard, U (2004) Archaeological disaster a bioerosion goldmine: test-blocks retrieved 2100 years after deposition, Antikythira, Greece. In: Mikuláš R (ed) Abstracts for the 4th bioerosion workshop, Czech Republic, pp 5–6

Bromley RG, D'Alessandro A (1984) The ichnogenus *Entobia* from the Miocene, Pliocene and Pleistocene of southern Italy. Riv It Paleont Stratigr 90:227–296

Bromley RG, D'Alessandro A (1990) Comparative analysis of bioerosion in deep and shallow water, Pliocene to recent, Mediterranean Sea. Ichnos 1:43–49

Bromley RG, Hanken N-M, Asgaard U (1990) Shallow marine bioerosion: preliminary results of an experimental study. Bull Geol Soc Den 38:85–99

Brusco F, Calcinai B, Bavestrello G, Cerrano C (2005) Attivita di erosione di *Cliona celata* Grant, 1826 e *Pione vastifica* (Hancock, 1849) (Porifera, Demospongiae) perforanti substrati artificiali. Biol Mar Medit 12:240–243

Buestel D, Ropert M, Prou J, Goulletquer P (2009) History, status, and future of oyster culture in France. J Shellfish Res 28:813–820

Calcinai B, Cerrano C, Bavestrello G, Milanese M, Sarà M (2000) Il popolamento di spugne perforatrici di *Corallium rubrum* e di alcuni madreporari del Promontorio di Portofino. Boll Mus Inst Biol Univ Genova 64–65:53–59

Calcinai B, Cerrano C, Bavestrello G (2002) A new species of *Scantiletta* (Demospongiae, Clionaidae) from the Mediterranean precious coral with some remarks on the genus. Bull Mar Sci 70:919–926

Calcinai B, Cerrano C, Iwasaki N, Bavestrello G (2008) Sponges boring into precious corals: an overview with description of a new species of *Alectona* (Demospongiae, Alectonidae) and a worldwide identification key for the genus. Mar Ecol 29:273–279

Calcinai B, Bavestrello G, Cuttone G, Cerrano C (2011) Excavating sponges from the Adriatic Sea: description of *Cliona adriatica* sp. nov. (Demospongiae: Clionaidae) and estimation of its boring activity. J Mar Biol Assoc UK 91:339–346

Carazzi D (1895) Revisione del genre *Polydora* Bosc e cenni su due specie che vivono sulle ostriche. Mitt zool Stat Neapel 11:4–45

Carballo JL, Sanchez-Moyano JE, García-Gómez JC (1994) Taxonomic and ecological remarks on boring sponges (Clionidae) from the Straits of Gibraltar (southern Spain): tentative bioindicators? Zool J Linn Soc Lond 112:407–424

Carballo JL, Naranjo SA, García-Gómez JC (1996) Use of marine sponges as stress indicators in marine ecosystems at Algeciras Bay (southern Iberian Peninsula). Mar Ecol Prog Ser 135:109–122

Carreiro-Silva M, McClanahan TR, Kiene WE (2005) The role of inorganic nutrients and herbivory in controlling microbioerosion of carbonate substratum. Coral Reefs 24:214–221

Carreiro-Silva M, McClanahan TR, Kiene WE (2009) Effects of inorganic nutrients and organic matter on microbial euendolithic community composition and microbioerosion rates. Mar Ecol Prog Ser 392:1–15

Cebrian E, Uriz M-J (2006) A facilitation example involving sponges: grazing on fleshy seaweeds by sea urchins enhances *Cliona viridis* abundance. Mar Ecol Prog Ser 323:83–89

Cerrano C, Bavestrello G, Bianchi CN, Calcinai B, Cattaneo-Vietti R, Morri C, Sarà M (2001) The role of sponge bioerosion in Mediterranean coralligeneous accretion. In: Faranda FM, Guglielmo L, Spezie G (eds) Mediterranean ecosystems. Structures and processes. Springer, Milano, pp 235–240

Chazottes V, Le Campion-Alsumard T, Peyrot-Clausade M, Cuet P (2002) The effects of eutrophication-related alterations to coral reef communities on agents and rates of bioerosion (Reunion Island, Indian Ocean). Coral Reefs 21:375–390

Çinar ME, Bilecenoğlu M, Öztürk B, Katağan T, Yokeş MB, Aysel V, Dağli E, Açik S, Özcan T, Erdoğan H (2011) An updated review of alien species on the coasts of Turkey. Medit Mar Sci 12:257–315

CITES (2011) Consideration of proposals for amendment of Appendices I and II, and Appendices I, II and II valid from 27 April 2011. http://www.cites.org/eng/cop/13/prop/e13-p35.pdf; and http://www.cites.org/eng/app/E-Apr27.pdf. Accessed 14 Sept 2011

Clark T, Morton B (1999) Relative role of bioerosion and typhoon-induced disturbance on the dynamics of a high latitude scleractinian coral community. J Mar Biol Assoc UK 79:83–820

Coll M, Piroddi C, Steenbeek J, Kaschner K, Ben Rais Lasram F, Aguzzi J, Ballesteros E, Bianchi CN, Corbera J, Dailianis T, Danovaro R, Estrada M, Froglia C, Galil BS, Gasol JM, Gertwagen R, Gil J, Guilhaumon F, Kesner-Reyes K, Kitsos M-S, Koukouras A, Lampadariou N, Laxamana E, de la Cuadra L-F, Lotze HK, Martin D, Mouillot D, Oro D, Raicevich S, Ruis-Barile J, Saiz-Salinas JI, San Vicente C, Somot S, Templado J, Turon X, Vafidis D, Villanueva R, Voultsiadou E (2010) The biodiversity of the Mediterranean Sea: estimates, patterns, and threats. PLoS One 5(e11842):1–36

Corredor JE, Wilkinson CR, Vicente VP, Morell JM, Otero E (1988) Nitrate release by Caribbean reef sponges. Limnol Oceanogr 33:114–120

Corriero G, Abbiati M, Santangelo G (1997) Sponges inhabiting a Mediterranean red coral population. PSZN Mar Ecol 18:147–155

Costantini F, Fauvelot C, Abbiati M (2007) Genetic structuring of the temperate gorgonian coral (*Corallium rubrum*) across the Western Mediterranean Sea revealed by microsatellites and nuclear sequences. Mol Ecol 16:5168–5182

Cotte J (1914) L'association de *Cliona viridis* (Schmidt) (spong.) et de *Lithophyllum expansum* (Philippi) (algues). Comptes Rendus des Séances de la Société Biologie Paris 76:739–740

De Simone AE (2010) 'The Coral Routes': From the Mediterranean Sea to the world-wide jewelry. In: Bussoletti E, Cottingham D, Bruckner A, Roberts G, Sandulli R (eds) Proceedings of the international workshop on red coral science, management and trade. Lessons from the Mediterranean Naples 2009, NOAA Technical Memorandum CRCP-13, Silver Spring, MD, USA, pp 102–105

Dean HK (2008) The use of polychaetes (Annelida) as indicator species of marine pollution: a review. Rev Biol Trop 56:11–38

Devescovi M (2009) Biometric differences between date mussels *Lithophaga lithophaga* colonizing artificial and natural structures. Acta Adriatica 50:129–138

Devescovi M, Iveša L (2008) Colonization patterns of the date mussel *Lithophaga lithophaga* (L., 1758) on limestone breakwater boulders of a marina. Periodicum Biol 110:339–345

Devescovi M, Ozretić B, Iveša L (2005) Impact of date mussel harvesting on the rocky bottom structural complexity along the Istrian coast (northern Adriatic, Croatia). J Exp Mar Biol Ecol 325:134–145

Emery KO (1960) The sea off southern California, a modern habitat of petroleum. Wiley, New York, p 147

Ercegović A (1927) Tri nova roda litofitskih cijanoficeja sa jadranske obale. Acta Bot Zagreb Univ 2:78–84

Ercegović A (1929) Sur quelques nouveaux types des Cyanophycées lithophytes de la côte Adriatique. Arch Protistenkunde 66:164–174

Ercegović A (1932) Ekološke i sociološke studije o litofitskim cijanoficejama sa jugoslavenske obale Jadrana. Rad Jugoslav akad znan umjet. Mat-prirod razred Knj 75:129–220

Fanelli G, Piraino S, Belmonte G, Geraci S, Boero F (1994) Human predation along Apulian rocky coast: desertification caused by *Lithophaga lithophaga* (Mollusca) fisheries. Mar Ecol Prog Ser 110:1–8

Feldmann J (1937) Recherches sur la végétation marine de la Méditerranée: la côte des Albères. Wolf, Rouen, p 339

Fine M, Loya Y (2002) Endolithic algae: an alternative source of photoassimilates during coral bleaching. Proc R Soc Lond 269:1205–1210

Fine M, Tchernov D (2007) Scleractinian coral species survive and recover from decalcification. Science 315:1811

Fine M, Steindler L, Loya L (2004) Endolithic algae photoacclimate to increased irradiance during bleaching. Mar Freshw Res 55:115–121

Fisheries and Aquaculture Department (2011) Mediterranean fisheries. http://www.fao.org/docrep/006/N7865E/N7865E01.htm. Accessed 13 July 2011

Fliegel D, Kosler J, McLoughlin N, Simonetti A, de Wit MJ, Wirth R, Furnes H (2010) In-situ dating of the earth's oldest trace fossil at 3.34 Ga. Earth Planet Sci Lett 299:290–298

Fornós JJ, Forteza V, Martinez-Taberner A (1998) Modern polychaete reefs in Western Mediterranean lagoons: *Ficipotamus enigmatus* (Fauvel) in the Albufera of Menorca, Balearic Islands. Palaeogeogr Palaeoclimatol Palaeoecol 128:175–186

Fraschetti S, Bianchi CN, Terlizzi A, Fanelli G, Morri C, Boero F (2001) Spatial variability and human disturbance in shallow subtidal hard substrate assemblages: a regional approach. Mar Ecol Prog Ser 212:1–12

Fütterer DK (1974) Significance of the boring sponge *Cliona* for the origin of fine grained material of carbonate sediments. J Sed Petrol 48:1149–1157

Garcia-Pichel F (2006) Plausible mechanisms for the boring in carbonates by microbial phototrophs. Sed Geol 185:205–213

Garcia-Pichel F, Ramirez-Reinat E, Gao Q (2010) Microbial excavation of solid carbonates powered by P-type ATPase-mediated transcellular Ca^{2+} transport. PNAS 107:21749–21754

Garrabou J, Pérez T, Sartoretto S, Harmelin JG (2001) Mass mortality event in red coral *Corallium rubrum* populations in the Provence region (France, Mediterranean). Mar Ecol Prog Ser 217:263–272

Garrabou J, Coma R, Bensoussan N, Bally M, Chevaldonné P, Cigliano M, Diaz D, Harmelin JG, Gambi MC, Kersting DK, Ledoux JB, LeJeusne C, Linares C, Marschall C, Pérez T, Ribes M, Romano JC, Serrano E, Teixido N, Torrents O, Zabala M, Zuberer F, Cerrano C (2009) Mass mortality in northwestern Mediterranean rocky benthic communities: effects of the 2003 heat wave. Glob Chang Biol 15:1090–1103

Glaub I, Vogel K (2004) The stratigraphic record of microborings. Fossils Strata 51:126–135

Glynn PW (1997) Bioerosion and coral reef growth: a dynamic balance. In: Birkeland C (ed) Life and death of coral reefs. Chapman and Hall, New York, pp 69–98

Golubic S, Seong-Joo L (1999) Early cyanobacterial fossil record: preservation, palaeoenvironments and identification. Eur Jr Phycol 34:339–348

Golubic S, Perkins RD, Lukas KJ (1975) Boring microorganisms and microborings in carbonate substrates. In: Frey EW (ed) The study of trace fossils. Springer, New York, pp 229–259

Golubic S, Radtke G, Le Campion-Alsumard T (2005) Endolithic fungi in marine ecosystems. Trends Microbiol 13:229–235

Guidetti P, Fraschetti S, Terlizzi A, Boero F (2003) Distribution patterns of sea urchins and barrens in shallow Mediterranean rocky reefs impacted by the illegal fishery of the rock-boring mollusc *Lithophaga lithophaga*. Mar Biol 143:1135–1142

Guidetti P, Terlizzi A, Boero F (2004) Effects of the edible sea urchin, *Paracentrotus lividus*, fishery along the Apulian rocky coast (SE Italy, Mediterranean Sea). Fish Res 66:287–297

Hallock P (1988) The role of nutrient availability in bioerosion: consequences to carbonate buildups. Palaeogeogr Palaeoclimatol Palaeoecol 63:275–291

Hallock P, Schlager W (1986) Nutrient excess and the demise of coral reefs and carbonate platforms. Palaios 1:389–398

Hartman W (1957) Ecological niche differentiation in the boring sponges (Clionidae). Evolution 11:294–297

Héral M (1989) Traditional oyster culture in France. In: Barnabé G, Solbé JF, Laird L (eds) Aquaculture, vol 1. Ellis Horwood, London, pp 342–387

Highsmith RC (1980a) Geographic patterns of coral bioerosion: a productivity hypothesis. J Exp Mar Biol Ecol 46:177–196

Highsmith RC (1980b) Passive colonization and asexual colony multiplication in the massive coral *Porites lutea* Milne Edwards & Haime. J Exp Mar Biol Ecol 47:55–67

Hill M, Wilcox T (1998) Unusual mode of symbiont repopulation after bleaching in *Anthosigmella varians*: acquisition of different zooxanthellae strains. Symbiosis 25:279–289

Holmes G, Ortiz J-C, Schönberg CHL (2009) Bioerosion rates of the sponge *Cliona orientalis* Thiele, 1900: spatial variation over short distances. Facies 55:203–211

Hong JS (1980) Étude faunistique d'un fond de concrétionnement de type coralligène soumis à un gradient de pollution en Méditerranée nord-occidentale (Golfe de Fos). Dissertation, University Aix-Marseille II, 134 pp

Hrs-Brenko M, Zavodnik D, Zahtila E (1991) The date shell *Lithophaga lithophaga* Linnaeus, and its habitat calls for protection in the Adriatic Sea. In: Boudouresque CF, Avon M, Gravez V (eds) Les espèces marines à protéger en Méditerrannée. GIS Posidonie, Marseille, pp 151–158

Hutchings PA (1986) Biological destruction of coral reefs. Coral Reefs 4:239–252

Hutchings PA (2008) Role of polychaetes in bioerosion of coral substrates. In: Wisshak M, Tapanila L (eds) Current developments in bioerosion. Springer, Heidelberg, pp 249–264

Jaccarini V, Bannister WH, Miccalef H (1968) The pallial glands and rock boring in *Lithophaga lithophaga* (Lamellibranchia, Mytilidae). J Zool Lond 154:397

Kleemann KH (1973) Der Gesteinsabbau durch Ätzmuscheln an Kalkküsten. Oecologia 13:377–395

Kroeker KJ, Micheli F, Gambi MC, Martz TR (2011) Divergent ecosystem responses within a benthic marine community to ocean acidification. PNAS 108:14515–14520

Labura Ž, Hrs-Brenko M (1990) Infestation of European flat oyster (*Ostrea edulis*) by polychaete (*Polydora hoplura*) in the northern Adriatic Sea. Acta Adriatica 31:173–181

Laubier L (1966) Le coralligène des Albères: monographie biocénotique. Ann Inst Océanogr Monaco 72:1–112

Lauckner G (1983) Diseases caused by metazoans. Agents: Porifera. In: Kinne O (ed) Diseases of Mollusca, vol II, Introduction, Bivalvia to Scaphopoda. Biologische Anstalt Helgoland, Hamburg, pp 617–626

Le Bris S, Le Campion-Alsumard T, Romano J-C (1998) Caractéristiques du feutrage algal des récifs coralliens de Polynésie française soumis à différentes intensités de bioérosion. Oceanol Acta 21:695–708

Le Campion-Alsumard T (1975) Etude expérimentale de la colonisation d'éclats de calcite par les Cyanophycées endolithes marines. Cah Biol Mar 16:177–185

Le Campion-Alsumard T (1979) Les Cyanophycées endolithes marines. Systématique, ultrastructure, écologie et biodestruction. Oceanol Acta 2:143–156

Le Campion-Alsumard T, Golubic S, Hutchings PA (1995) Microbial endoliths in skeletons of live and dead corals: *Porites lobata* (Moorea, French Polynesia). Mar Ecol Prog Ser 117:149–157

Le Grand HM, Fabricius KE (2011) Relationship of internal macrobioeroder densities in living massive *Porites* to turbidity and chlorophyll on the Australian Great Barrier Reef. Coral Reefs 30:97–107

Loya Y (1982) Life history strategies of boring bivalves in corals. In: Proceedings of the 4th international coral reef symposium, Marine Science Centre, University of the Philippines, Manila, Philippines, vol 2, pp 756

Maldonado M (1992) Demosponges of the red coral bottoms from the Alboran Sea. J Nat Hist 26:1131–1161

Martin D, Britayev TA (1998) Symbiotic polychaetes: review of known species. In: Ansell AD, Gibson RN, Barnes M (eds) Oceanogr Mar Biol Ann Rev 36:217–340

Mastrototaro F, D'Onghia G, Corriero G, Matarrese A, Maiorano P, Panetta P, Gherardi M, Longo C, Rosso A, Scutio F, Sanfilippo R, Gravili C, Boero F, Taviani M, Tursi A (2010) Biodiversity of the white coral bank of Cape Santa Maria di Leuca (Mediterranean Sea): an update. Deep-Sea Res II 57:412–430

McLoughlin N, Furnes H, Banerjee NR, Staudigel H, Muehlenbachs K, de Wit M, Van Kranendonk MJ (2008) Micro-bioerosion in volcanic glass: extending the ichnofossil record to Archaean basaltic crust. In: Wisshak M, Tapanila L (eds) Current developments in bioerosion. Springer, Heidelberg, pp 371–396

Melone N (1965) I poriferi associate a *Corallium rubrum* (L.) della Sardegna. Ann Mus Civ Storia Nat 'Giacomo Doria' Genova 75:343–358

Micael J, Alves MJ, Costa AC, Jones MB (2009) Exploitation and conservation of echinoderms. Oceanogr Mar Biol Ann Rev 47:191–208

Mitchell-Tapping HJ (1983) Experimental fracturing of various corals in the reef environment. Caribb J Sci 19:49–52

Mokady O, Lazar B, Loya Y (1996) Echinoid bioerosion as a major structuring force of Red Sea coral Reefs. Biol Bull 190:367–372

Nava H, Carballo JL (2008) Chemical and mechanical bioerosion of boring sponges from Mexican Pacific coral reefs. J Exp Biol 211:2827–2831

Naylor LA, Viles HA (2002) A new technique for evaluating short-term rates of coastal bioerosion and bioprotection. Geomorphology 47:31–44

Neumann AC (1966) Observations on coastal erosion in Bermuda and measurements of the boring rate of the sponge, *Cliona lampa*. Limnol Oceanogr 11:92–108

Palmer M, Fornós JJ, Balaguer P, Gómez-Pujol LI, Pons GX, Villanueva G (2003) Spatial and seasonal variability of the macro-invertebrate community of a rocky coast in Mallorca (Balearic Islands): implications for bioerosion. Hydrobiologia 501:13–21

Perry CT, Macdonald IA (2002) Impacts of light penetration on the bathymetry of reef microboring communities: implications for the development of microendolithic trace assemblages. Palaeogeogr Palaeoclimatol Palaeoecol 186:101–113

Pomponi SA (1980) Cytological mechanisms of calcium carbonate excavation by boring sponges. Int Rev Cytol 65:301–319

Porras R, Bataller JV, Murgui E, Torregrosa MT (1996) Trophic structure and community composition of polychaetes inhabiting some *Sabellaria alveolata* (L.) reefs along the Valencia Gulf Coast, Western Mediterranean. Mar Ecol 17:583–602

Privitera D, Chiantore M, Mangialajo L, Glavic N, Kozul W, Cattaneo-Vietti R, Giancarlo A (2005) Spatial scale of variability of sea urchins and algal communities in the Dubrovnik area. In: Proceedings of the 15th meeting of the Italian Society of Ecology, pp 1–6

Pronzato R (2000) Il corallo. L'oro rosso del Mediterraneo. Boll Mus Inst Biol Univ Genova 64–65:1–94

Pyko IA (2009) Ichnodiversity and bioerosion patterns in the Eastern Mediterranean – an experimental study. Masters, Friedrich-Alexander-Universität Erlangen-Nürnberg, 68 pp

Ricci S, Priori GF, Bartolini M (2008) Bioerosione di pavimentazioni musive sommerse ad opera della spugna endolitica *Cliona celata*. Boll Inst Sup Conserv Restauro 15:7–18

Risk MJ, MacGeachy JK (1978) Aspects of bioerosion of modern Caribbean reefs. Rev Biol Trop suppl 26:85–105

Rodolfo-Metalpa R, Houlbrèque F, Tambutté É, Boisson F, Baggini C, Patti FP, Jeffree R, Fine M, Foggo A, Gattuso J-P, Hall-Spencer JM (2011) Coral and mollusc resistance to ocean acidification adversely affected by warming. Nat Clim Chang 1:308–312

Rose CS, Risk MJ (1985) Increase in *Cliona delitrix* infestation of *Montastrea cavernosa* heads on an organically pollutes portion of the Grand Cayman fringing reef. PSZNI Mar Ecol 6:345–363

Rosell D, Uriz M-J, Martin D (1999) Infestation by excavating sponges on the oyster (*Ostrea edulis*) populations of the Blanes littoral zone (north-western Mediterranean). J Mar Biol Assoc UK 79:409–413

Russo GF, Cicogna F (1991) The date mussel (*Lithophaga lithophaga*), a 'case' in the Gulf of Naples. In: Boudouresque CF, Avon M, Gravez V (eds) Les espèces marines à protéger en Méditerrannée. GIS Posidonie, Marseille, pp 141–150

Rützler K, Rieger G (1973) Sponge burrowing: fine structure of *Cliona lampa* penetrating calcareous substrate. Mar Biol 21:144–162

Santangelo G, Carletti E, Maggi E, Bramanti L (2003) Reproduction and population sexual structure of the overexploited Mediterranean red coral *Corallium rubrum*. Mar Ecol Prog Ser 248:99–108

Santos A, Mayoral E, da Silva CM, Cachao M, Domènech R, Martinell J (2008) Trace fossil assemblages on Miocene rocky shores of southern Iberia. In: Wisshak M, Tapanila L (eds) Current developments in bioerosion. Springer, Heidelberg, pp 431–450

Sartoretto S (1998) Bioérosion des concrétions coralligènes de Méditerranée par les organisms perforants: essai de quantification des processus. CR Acad Sci Paris Sci Terre Planèt 327:839–844

Sartoretto S, Francour P (1997) Quantification of bioerosion by *Sphaerechinus granularis* on 'coralligene' concretions of the western Mediterranean. J Mar Biol Assoc UK 77:565–568

Schneider J (1976) Biological and inorganic factors in the destruction of limestone coasts. Contrib Sed Geol 6:1–123

Schneider J (1977) Carbonate construction and decomposition by epilithic and endolithic micro-organisms in salt- and freshwater. In: Flügel E (ed) Fossil algae. Springer, Berlin, pp 248–260

Schönberg CHL (2008) A history of sponge erosion: from past myths and hypotheses to recent approaches. In: Wisshak M, Tapanila L (eds) Current developments in bioerosion. Springer, Heidelberg, pp 165–211

Schönberg CHL, Suwa R (2007) Why bioeroding sponges may be better hosts for symbiotic dinoflagellates than many corals. In: Custódio MR, Lôbo-Hajdu G, Hajdu E, Muricy G (eds) Porifera research: biodiversity, innovation and sustainability. Museu Nacional, Rio de Janeiro, Brazil Série Livros, vol 28, 569–580

Schönberg CHL, Tapanila L (2006) Bioerosion research before and after 1996 – a discussion of what has changed since the first international bioerosion workshop. Ichnos 13:99–102

Schönberg CHL, Wilkinson CR (2001) Induced colonization of corals by a clionid bioeroding sponge. Coral Reefs 20:69–76

Schönberg CHL, Wisshak M (2012) The perks of being endolithic. Aquatic Biol 17:1-5

Schönberg CHL, Suwa R, Hidaka M (2008) Sponge and coral zooxanthellae in heat and light: preliminary results of photochemical efficiency monitored with pulse amplitude modulated fluorometry. Mar Ecol 29:247–258

Scott PJB (1991) Rapid destruction of PVC piping by boring bivalves. Int Biodeterior 27:87–92

Scott PJB, Risk MJ (1988) The effect of *Lithophaga* (Bivalvia: Mytilidae) bore holes on the strength of the coral *Porites lobata*. Coral Reefs 7:145–151

Scott PJB, Moser KA, Risk MJ (1988) Bioerosion of concrete and limestone by marine organisms: a 13 year experiment from Jamaica. Mar Poll Bull 19:219–222

Simboura N, Zenetos A (2005) Increasing polychaete diversity as a consequence of increasing research effort in Greek waters: new records and exotic species. Mediterr Mar Sci 6:75–88

Smyth MJ (1989) Bioerosion of gastropod shells: with emphasis of effects of coralline algal cover and shell microstructure. Coral Reefs 8:119–125

Szmant AM (2002) Nutrient enrichment on coral reefs: is it a major cause of coral reef decline? Estuaries 25:743–766

Tapanila L (2008) The endolithic guild: an ecological framework for residential cavities in hard substrates. In: Wisshak M, Tapanila L (eds) Current developments in bioerosion. Springer, Heidelberg, pp 3–20

Thomson Reuters (2011) Web of knowledge. http://apps.webofknowledge.com/UA_GeneralSearch_input.do?product=UA&search_mode=GeneralSearch&SID=Q18Il06BF4i@pOobL6e&preferencesSaved=. Accessed 14 Sept 2011

Tribollet A (2008a) The boring microflora in modern coral reef ecosystems: a review of its roles. In: Wisshak M, Tapanila L (eds) Current developments in bioerosion. Springer, Heidelberg, pp 67–94

Tribollet A (2008b) Dissolution of dead corals by euendolithic organisms across the northern Great Barrier Reef (Australia). Mircobiol Ecol 55:569–580

Tribollet A, Golubic S (2005) Cross-shelf differences in the pattern and pace of bioerosion of experimental carbonate substrates exposed for 3 years on the northern Great Barrier Reef, Australia. Coral Reefs 24:422–434

Tribollet A, Atkinson MJ, Langdon C (2006) Effects of elevated pCO_2 on epilithic and endolithic metabolism of reef carbonates. Glob Chang Biol 12:220–2208

Tribollet A, Godinot C, Atkinson M, Langdon C (2009) Effects of elevated pCO_2 on dissolution of coral carbonates by microbial euendoliths. Global Biogeochem Cycles 23:1–7

Tunnicliffe V (1979) The role of boring sponges in coral fracture. In: Lévi C, Boury-Esnault N (eds) Biologie des spongiares, vol 291, Colloques internationaux du centre national de la recherche scientifique. CNRS, Paris, pp 309–315

Vénec-Peyré M-T (1996) Bioeroding foraminifera: a review. Mar Micropaleontol 28:19–30

Vicente VP (1990) Response of sponges with autotrophic endosymbionts during the coral-bleaching episode in Puerto Rico. Coral Reefs 8:199–202

Vogel J, Brett CE (2009) Record of microendoliths in different facies of the Upper Ordovician in the Cincinnati Arch region USA: the early history of light-related microendolithic zonation. Palaeogeogr Palaeoclimatol Palaeoecol 281:1–24

Ward P, Risk MJ (1977) Boring pattern of the sponge *Cliona vermifera* in the coral *Montastrea annularis*. J Paleontol 51:520–526

Warme JE (1975) Borings as trace fossils, and the process of marine bioerosion. In: Frey RW (ed) The study of trace fossils. Springer, Berlin, pp 181–229

Weinberg S (1991) Faut-il protéger les gorgones de Méditerranée? In: Boudouresque CF, Avon M, Gravez V (eds) Les espèces marines à protéger en Méditerrannée. GIS Posidonie, Marseille, pp 47–52

Wilson MA (2006) Macroborings and the evolution of marine bioerosion. In: Miller W III (ed) Trace fossils: concepts, problems, prospects. Elsevier, Amsterdam, pp 356–367

Wisshak M (2008) Two new dwarf *Entobia* ichnospecies in a diverse aphotic ichnocoenosis (Pleistocene/Rhodes, Greece). In: Wisshak M, Tapanila L (eds) Current developments in bioerosion. Springer, Heidelberg, pp 213–233

Wisshak M (2012) Microbioerosion. In: Knaust D, Bromley RG (eds) Trace fossils as indicators of sedimentary environments. Developments in sedimentology 64. Elsevier, Amsterdam, pp 213–243

Wisshak M, Tribollet A, Golubic S, Jakobsen J, Freiwald A (2011) Temperate bioerosion: Ichno- and biodiversity from intertidal to bathyal depths (Azores). Geobiology 9:492–520

Wisshak M, Schönberg CHL, Form A, Freiwald A (2012) Ocean acidification accelerates reef bioerosion. PLoS ONE 7(9):e45124

Xavier JR, Rachello-Dolmen PG, Parra-Velandia F, Schönberg CHL, Breeuwer JAJ, van Soest RWM (2010) Molecular evidence of cryptic speciation in the 'cosmopolitan' excavating sponge *Cliona celata* (Porifera, Clionaidae). Mol Phylogenet Evol 56:13–20

Yonge CM (1963) Rock-boring organisms. In: Soggnaes RF (ed) Mechanisms of hard tissue destruction. Publ Am Association Advancem Sci 75: 1–24

Zenetos A, Katsanevakis S, Poursanidis D, Crocetta F, Damalas D, Apostolopoulos C (2011) Marine alien species in Greek Seas: additions and amendments by 2010. Medit Mar Sci 12:95–120

Zundelevich A, Lazar B, Ilan M (2007) Chemical versus mechanical bioerosion of coral reefs by boring sponges – lessons from *Pione* cf. *vastifica*. J Exp Biol 210:91–96

Metamorphoses: Bioinvasions in the Mediterranean Sea

27

B.S. Galil and Menachem Goren

Abstract

Six hundred and eighty alien marine multicellular species have been recorded in the Mediterranean Sea, with many establishing viable populations and dispersing along its coastline. A brief history of bioinvasions research in the Mediterranean Sea is presented. Particular attention is paid to gelatinous invasive species: the temporal and spatial spread of four alien scyphozoans and two alien ctenophores is outlined. We highlight few of the discernible, and sometimes dramatic, physical alterations to habitats associated with invasive aliens in the Mediterranean littoral, as well as food web interactions of alien and native fish. The propagule pressure driving the Erythraean invasion is powerful in the establishment and spread of alien species in the eastern and central Mediterranean. The implications of the enlargement of Suez Canal, reflecting patterns in global trade and economy, are briefly discussed.

Keywords

Alien • Vectors • Trends • Propagule pressure • Trophic levels • Jellyfish • Mediterranean Sea

Brief History of Bioinvasion Research in the Mediterranean Sea

The eminent European marine naturalists of the sixteenth century – Belon, Rondelet, Salviani, Gesner and Aldrovandi – recorded solely species native to the Mediterranean Sea, though mercantile horizons have already expanded with geographical discoveries and shipping was no longer confined to European seas.

Recognition that marine species may be introduced into the Mediterranean Sea from other parts of the world came suddenly with the much publicized plans of the Saint-Simonians for a “Canal de jonction des deux mers” at the Isthmus of Suez. Even before the Suez Canal was fully excavated, the French zoologist Léon Vaillant (1865) argued that the breaching of the isthmus will bring about species migration and mixing of faunas, and advocated what would be considered nowadays a ‘baseline study’. Paul Fischer (1865), who studied molluscs from Port Said, Suez and the Bitter Lakes, was certain that the Canal will modify the faunas at its Mediterranean and Erythraean ends. Yet for the next 50 years the documentation of the changes in the biota in the adjacent marine environments was left to “Forschungsreisender” – a now extinct breed of learned amateurs and natural scientists who styled themselves ‘scientific travelers’. Keller (1883), a professor of Zoology in Zurich, traveled to Egypt in 1882 and 1886 and identified 10 Mediterranean species and 17 Red Sea species from the Canal itself, 5 Mediterranean species in the Red Sea, and 4 Red Sea species in the Mediterranean. Doubts had been raised concerning some of his identifications, as well as his sources,

B.S. Galil (✉)
National Institute of Oceanography, Israel Oceanographic and Limnological Research, 8030, Haifa 31080, Israel
e-mail: bella@ocean.org.il

M. Goren
Department of Zoology, George S. Wise Faculty of Life Sciences, Tel Aviv University, Tel Aviv 69978, Israel
e-mail: GorenM@tauex.tau.ac.il

S. Goffredo and Z. Dubinsky (eds.), *The Mediterranean Sea: Its history and present challenges*,
DOI 10.1007/978-94-007-6704-1_27, © Springer Science+Business Media Dordrecht 2014

however, the importance of his work lies in confirming the role of the canal as a 'corridor' for the movement of species and in the transformation of the adjacent marine biotas.

At the same time ship fouling too was recognized as a vector for introduction of marine alien species: in 1873 the tri-masted *Karikal* arrived at the port of Marseille from India carrying on its hull a small forest of "living beings" including three grapsoid crabs species numbering in the hundreds (Catta 1876). Catta, with farsighted acumen, warned of faunal changes stemming from ship-transported organisms. In fact, fouling serpulid polychaetes were already identified as the first alien species in the Mediterranean: *Hydroides dianthus* (Verrill, 1873) was documented in the port of Izmir as early as 1865, and *H. dirampha* Mörch, 1863 and *H. elegans* Haswell, 1883 were recorded in the harbour of Naples in 1870 and 1888, respectively (Carus 1889; Zibrowius 1973).

Fifty-five years after the opening of the Canal, 'The Cambridge Expedition to the Suez Canal' embarked on an investigation "…why some forms of life can get through the Canal and others cannot, and … throw light on marine migrations in general" (Gardiner 1924:521). Indeed, anxieties whether "… such an immigration caused important changes in the fauna of the eastern basin of the Mediterranean?" had already been raised (Steinitz 1919). The investigations of the biota of the southern Levantine continental shelf in the 1940s highlighted the extent the Erythraean taxa have been established along the coast (Ben-Tuvia 1953a, b; Gottlieb 1953, 1960; Holthuis and Gottlieb 1958; Pérès 1958; Tebble 1959). By the 1970s 140 alien species were known to have crossed the Suez Canal into the Mediterranean (Steinitz 1970) and it was perceived that the littoral biota of the Levantine sea has been undergoing a rapid and profound change (Holthuis 1961; George and Athanassiou 1967; Geldiay and Kocatas 1968; Por 1978; Galil and Lewinshon 1981). But as this phenomenon seemed limited to the Levant Por (1990), the scientific community by and large ignored it.

Records of species introduced by shipping and mariculture into the Mediterranean appeared sporadically in the scientific literature (Rioja 1924; Hamel 1926; Berner 1944; Mendez-Domingo 1957; Raimbault 1964; Bascheri 1965), but as their number and impact were considered negligible they "… have not been the subject of inventories as representative as those of lessepsian migrants" (Zibrowius 1992, 1994). This state of affairs changed once the impacts of a pair of invasive chlorophytes have been noted off the Mediterranean coast of France. As Samuel Johnson famously observed "nothing so concentrates the mind as the sight of the gallows", or "killer algae" (*Caulerpa taxifolia* (Vahl) C. Agardh and *Caulerpa racemosa* var. *cylindracea* (Sonder) Verlaque, Huisman and Boudouresque). Soon, inventories of alien macrophytes were compiled (Verlaque 1994, 1996, 2001; Ribera and Boudouresque 1995; Ribera Siguan 2002). The European Commission Environmental Programme and the Mediterranean Science Commission (CIESM) organized a workshop on "Introduced species in European coastal waters" (Boudouresque et al. 1994), followed by workshops entitled "Ship-transported alien species in the Mediterranean and Black Sea" and "Impact of mariculture on Mediterranean coastal ecosystems" (www.ciesm.org), and a widely acclaimed illustrated atlas provided information on alien fish, decapod crustaceans, mollusks and macrophytes in the Mediterranean (www.ciesm.org/atlas). Still, research remains Eurocentric and much focused on the *Caulerpa* pair, with a tally of 361 and 150 research articles on *C. taxifolia* and *racemosa* respectively (Klein and Verlaque 2008). Intensive efforts and much funding have been invested in assembling national, regional and EU-wide inventories of marine alien species. More recently, molecular genetic analyses have increasingly been utilized in studying bioinvasions in the Mediterranean: establishing species identity, differentiating among cryptic species, and identifying cryptic invasions, and the history of introduction (Estoup and Guillemaud 2008).

The EU's ambitious Marine Strategy Framework Directive (MSFD) (2008/56/EC), aiming to achieve good environmental status by 2020 (http://ec.europa.eu/environment/water/marine/directive_en.htm), places great emphasis on the "trends in abundance, temporal occurrence and spatial distribution in the wild of non-indigenous species … in relation to the main vectors and pathways" (http://eur-lex.europa.eu/lexUriServ). Expanding the "state of the art" tools to invasion biology will doubtlessly improve documenting the full extent of the bioinvasion of the Mediterranean Sea and provide information for the development of effective policies and competent management.

'Alien' and 'Invasive' – Which and Where

Over the past three decades the study of marine bioinvasions has thrived and spawned a profusion of specialized terms, liable to misapplication and error. A computerized search of scientific databases disclosed a surfeit of synonyms, partial synonyms and neologisms that hinder communication (Occhipinti-Ambrogi and Galil 2004). The terminology varies among taxonomic kingdoms, scientific disciplines, and linguistic and national borders. Certain frequently used terms such as adventive, alien, allochthonous, exotic, introduced, invasive, naturalized, non-indigenous and non-native have several commonly established interpretations, leaving us with vague and occasionally overlapping definitions.

We follow the definitions set out by the International Council for the Exploration of the Sea (ICES) (as "Introduced species") and the International Union for Conservation of Nature (IUCN) (as "Alien species"), though recognizing that 'population' rather than 'species' is "transported intentionally or accidentally by a human-mediated vector into aquatic habitats outside its native range" and "... could not occupy without direct or indirect introduction or care by humans" respectively (http://www.ices.dk/reports/general/2003/codemarineintroductions2003.pdf; http://iucn.org/about/union/secretariat/offices/iucnmed/iucn_med_programme/species/invasive_species/). We use the term 'invasive' to denote an alien population which maintains itself without human assistance and spreads beyond its point of introduction, without assumption as to adverse environmental or economic impact.

Populations with naturally shifting distributions (e.g. due to climate change) are not to be considered alien, but a founder alien population spreading beyond its initial range in the recipient region, even if unmediated by humans, is still considered an alien population.

Six hundred eighty multicellular species are recognized as aliens in the Mediterranean Sea, and a great many are invasive – 118 alien species were recorded in 5 or more countries, and some 19 hyper-invasive species were recorded in 10 or more countries. As noted in previous analyses (Galil 2012), molluscs (31 %), crustaceans (17 %) and bony fish (16 %) make the largest contributions to the number of documented alien species in the Mediterranean and comprise 63 % of total alien species richness in the sea. Macrophytes (14 %), annelid polychaetes (9 %), bryozoans and cnidarians (each 3 %), and all other taxa (7 %) complete the number. The great majority of these are littoral benthic or demersal species. The data is presumably most accurate for large conspicuous species, yet on closer examination some dissolved into "species complexes" (Bucciarelli et al. 2002; Holland et al. 2004; Tsoi et al. 2007; Terranova et al. 2007; Golani and Bernardi 2012). The record grossly underestimates the total number of alien species. The magnitude of the gap is difficult to assess and differs among taxa, locations and habitats as research efforts vary greatly along the coasts of the Mediterranean. Data is entirely absent for many of the small-sized invertebrate phyla because of limited search effort and erosion of taxonomic expertise. The alien unicellular biota is nearly entirely invisible though it is well established that anthropic dispersal and redistribution of propagules in ballast water and sediments and shellfish transplantation facilitate range expansions not only of harmful algal blooms (HABs) causuatives and other microalgae, but of other microbial loop components as well, including viruses, algicidic bacteria, and microbial loop grazers such as protoperidinians, ciliates and other protistans (Hallegraeff and Bolch 1992; Pierce et al. 1997; Galil and Hülsmann 1997; Smayda 2007).

There are considerable differences among the peri Mediterranean countries in the number of alien species recorded (Galil 2012). Far more alien species have been documented in the Levantine Basin than the entire western Mediterranean: 346 multicellular alien species were reported along the 180 km long coast of Israel, whereas only 56 alien species were reported off the 1,660 km long Mediterranean coast of peninsular Spain. And whereas most of the alien species in the eastern Mediterranean have presumably entered through the Suez Canal, shipping and massive shellfish farming are the drivers in the western Mediterranean (Fig. 27.1).

Food Web Interactions of Alien and Native Fish

An analysis of the trophic level of alien species is essential for estimating their impact on native species through competition and predation. To date, little information is available regarding the diet of most native and alien species in the Mediterranean, nor do we have solid information on their biomass. This gap is especially significant in the Levant, where aliens have been gradually replacing native species. Fish are the only taxa for which sufficient and reliable information on the trophic level is available. A comparison of alien and native species in the Mediterranean based on data extracted from FISHBASE (Froese and Pauly 2012a, b) revealed that the average trophic levels of native and alien fish species are similar (3.509 ± 0.496 (S.D.) and 3.456 ± 0.531 respectively), though the distribution of the trophic levels differs (Fig. 27.2, Table 27.1). Herbivorous species comprise 5 % of aliens but only 2 % of native fish. Moreover, the native herbivores are either small sized (e.g. *Scartella cristata* (Linnaeus, 1758), *Parablennius sanguinolentus* (Pallas, 1811) and *Parablennius gattorugine* (Brünnich, 1768)) of negligible biomass (Goren and Galil 2001) or relatively rare (e.g. *Sarpa salpa* (Linnaeus 1758)), while the two alien herbivores, *Siganus rivulatus* Forsskål, 1775 and *S. luridus* (Rüppell, 1829) comprise about one third of the fish biomass in shallow rocky habitats, impacting the local macrophytes and food web (Goren and Galil 2001; Lundberg et al. 2004). Sala et al. (2011) who studied the impact of the two siganids on the Mediterranean coast of Turkey, where they comprise 46–57 % of the total fish biomass in rocky habitats, found they outcompeted native sea urchins, denuded rocky ledges from their algal cover and formed extensive "barrens". Also among the carnivorous fish (trophic levels 3.4–4.5; Fig. 27.2) in rocky habitats the proportion of aliens is slightly higher than natives (64 % and 59 % respectively). However, on soft-bottom habitats the trophic status of the alien fish differs greatly. The biomass of herbivorous fish, both alien and native is negligible, and the distribution of the native and alien species among the higher trophic guilds differs. A study of fish collected off the southern Israeli coast revealed that the weighted trophic level of the alien species is much higher (and statistically significant) than that of the

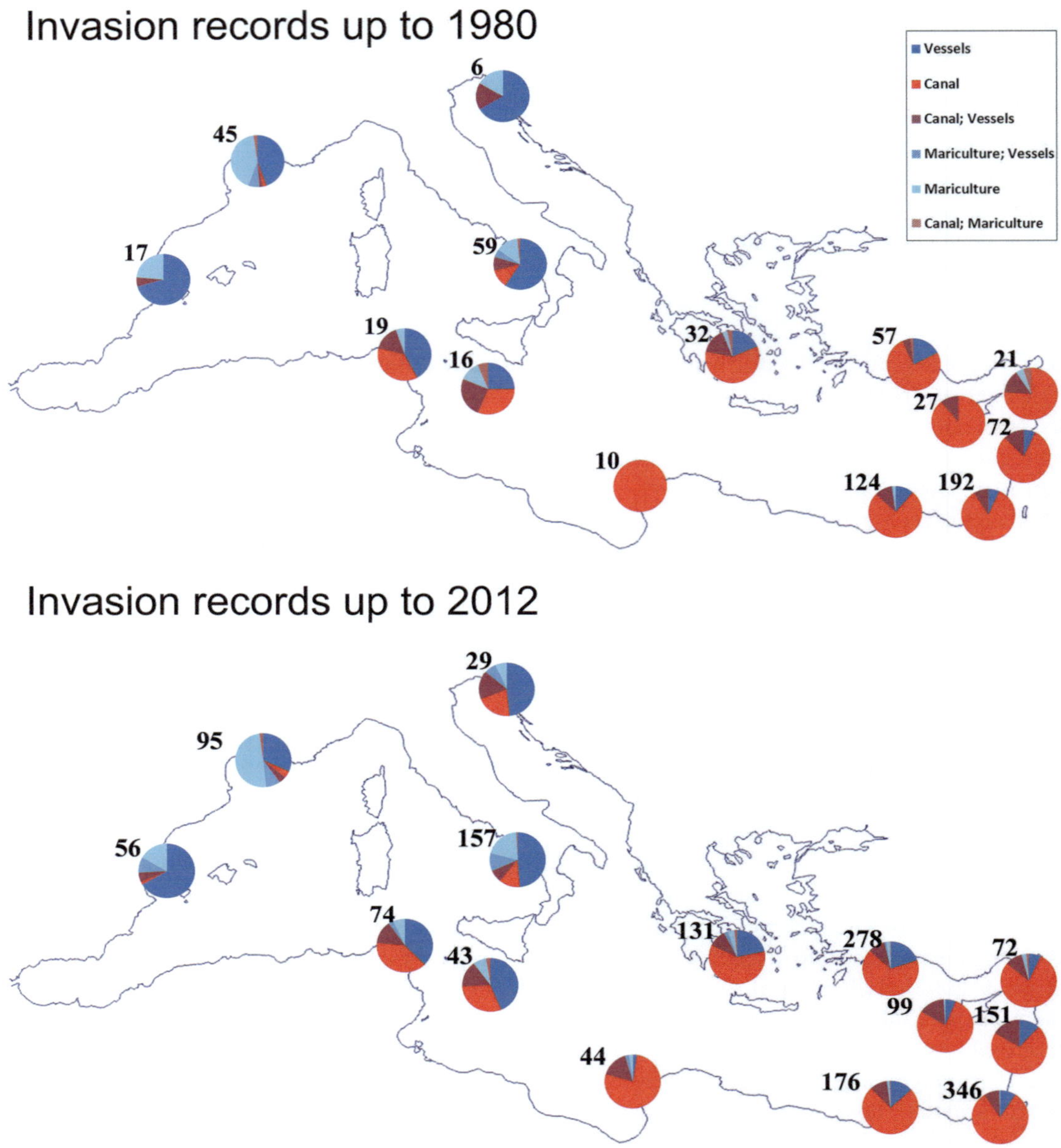

Fig. 27.1 Number of multicellular marine alien species in peri-Mediterranean countries, and their means of introduction (After Galil 2012)

natives (Levit 2012). At trophic level of 3.7 and higher, aliens comprise 38.5 % of the total catch while natives comprise 15.7 % (data extracted and calculated from Levit 2012). Trophic interactions affect ecosystem processes by altering fluxes of energy and material, or by impacting the abundance of populations that control those fluxes. The sharp increase in the weighted trophic level due to the expansion of the alien fish populations may explain in part the decline in local fishery catch (Shapiro 2007). The alien species in this trophic guild (abundance > 1 % of total catch) are of high commercial value: *Scomberomorus commerson* (Lacepède, 1800), *Decapterus russelli* (Rüppell, 1830), *Saurida macrolepis* Tanaka, 1917, *Nemipterus randalli* Russell, 198 and *Upeneus moluccensis* (Bleeker, 1855), whereas the native species at this trophic level, *Raja clavata* Linnaeus, 1758 and *R. miraletus* Linnaeus, 1758, lack commercial value.

Alien Jellyfish and Comb Jellies in the Mediterranean

Periodic outbreaks of indigenous scyphozoan jellyfish have long been noted in the Mediterranean (UNEP 1991; CIESM 2001). Various anthropogenic perturbations including eutrophication, overfishing, global warming and the increase of littoral man-made hard substrates have been

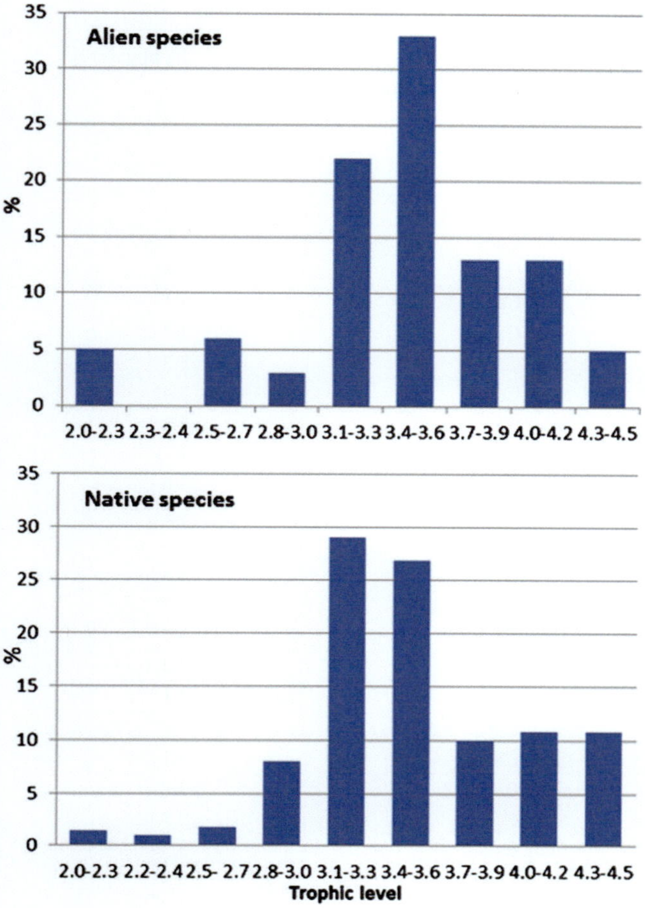

Fig. 27.2 Comparison of the trophic levels of native and alien fish species in the Mediterranean Sea (Data on trophic levels extracted from Froese and Pauly (2012a))

suggested as contributing to the proliferation of jellyfish populations in recent decades (CIESM 2001; Daskalov et al. 2007; Purcell 2007, 2012; Richardson et al. 2009; Riisgard et al. 2012). Boero et al. (2008) proposed that the removal of top predators and the formation of oligotrophic temperature-stable water masses may cause the suppression of the high energy fish and mammal-dominated food web and the re-emergence of medusozoan-dominated food web. But whereas most jellyfish outbreaks in the western and central Mediterranean consist of native species, alien species have taken over in the eastern part of the sea. The SE Levant is unique in hosting four alien scyphozoan jellyfish concurrently, in addition to two alien ctenophores (Fig. 27.3).

Already in 1886 *Cassiopea andromeda* (Forskål, 1775) occurred in the Suez Canal (Keller 1888), and a short time later was recorded off Cyprus (Maas 1903). It established populations along the Levantine coastline (Spanier 1989; Bilecenoğlu 2002), but was recently reported from further afield – Malta (Schembri et al. 2010). *Phyllorhiza punctata* Lendenfeld, 1884 was first sighted in the Mediterranean off the Israeli coast in 1965; where, since 2005, it established a regularly recorded resident population (Galil et al. 1990, 2009). Also off Lefkada Island, Greece, where it was recorded in 2005 and 2006, the population has apparently occurred for a number of years (Abed-Navandi and Kikinger 2007). Recently, a single specimen was collected in Iskenderun Bay, Turkey, and another photographed off Tavolaro Island, Sardinia, Italy (Boero et al. 2009; Çevik et al. 2011). The recently described *Marivagia stellata* (Galil et al. 2010), established a small population off Israel and the adjacent Lebanese coast. It is unclear whence *M. stellata* has arrived; but as the native range of nine out of ten alien species recorded off the SE Levant is the Indo-Pacific Ocean, the Indian Ocean or the Red Sea (Galil 2007), and the Indo-Pacific is a hot-spot for cepheid jellyfish, it may have originated there. Though found occasionally in small aggregations, the Mediterranean populations of these alien scyphzoans have remained small. In contrast, the Erythraean *Rhopilema nomadica* Galil, 1990, first recorded in the Mediterranean in the early 1970s, is notorious for the swarms it has formed each summer since the early 1980s along the SE Levantine coast (Galil et al. 1990). *Rhopilema* swarms adversely affect tourism, fisheries and coastal installations. The annual swarming results each year in envenomation victims suffering adverse effects that may last weeks and even months after the event (Benmeir et al. 1990; Silfen et al. 2003; Yoffe and Baruchin 2004; Sendovski et al. 2005). Coastal trawling and purse-seine fishing are disrupted for the duration of the swarming due to net clogging and inability to sort yield. Jellyfish-blocked water intake pipes pose a threat to desalination plants and seawater cooling systems of coastal power plants: in the summer of 2011 Israel Electric removed tons of jellyfish from its seawater intake pipes at its largest power plants (www.bbc.co.uk/news/world-middle-east-14038729) (Fig. 27.4).

The comb jelly *Mnemiopsis leidyi* A. Agassiz, 1865, indigenous in western Atlantic coastal waters (40°N to 46°S), has spread in the past three decades to the Black, Caspian, Baltic and North seas (Mianzan 1999; Shiganova et al. 2001a, b; Javidpour et al. 2006; Faasse et al. 2006). The introduction of the zooplanktivorous *M. leidyi* to the Black Sea in the 1980s set in motion a dramatic chain of events that culminated in a crash of the sea's major fishery and massive economic losses that garnered the species a slot on the International Union for Conservation of Nature (IUCN) list of 100 'World's Worst' invaders (http://www.issg.org/worst100_species.html). Given the severe ecological and economical harm elsewhere, the introduction of *M. leidyi* into the Mediterranean is of major concern. It was first recorded in the Mediterranean Sea in 1990 in the western Aegean Sea presumably swept with the outflow of the Black Sea water masses (Shiganova et al. 2001b), shortly thereafter

Table 27.1 List of alien fish species in the Mediterranean Sea, October 2012

Species	Introduction	Trophic level
Abudefduf vaigiensis (Quoy & Gaimard, 1825)	Suez Canal; Aquaculture	2.6
Alepes djedaba (Forsskål, 1775)	Suez Canal	3.3
Aphanius dispar (Rüppell, 1829)	Suez Canal	2.2
Apogonichthyoides pharaonis Bellotti, 1874	Suez Canal	3.5
Apogon queketti Gilchrist, 1903	Suez Canal	3.5
Apogon smithi (Kotthaus, 1970)	Suez Canal	3.5
Atherinomorus forskalii (Rüppell, 1838)	Suez Canal	3.3
Bregmaceros atlanticus Goode & Bean, 1886	Suez Canal; Vessels	3.1
Callionymus filamentosus Valenciennes, 1837	Suez Canal	3.3
Chaetodon austriacus Rüppell, 1836	Suez Canal	3.3
Chaetodon larvatus Cuvier, 1831	Suez Canal	3.4
Champsodon nudivittis (Ogilby, 1895)	Suez Canal	4.2
Chanos chanos (Forsskål, 1775)	Aquaculture	2
Cheilodipterus novemstriatus Rüppell, 1836	Suez Canal	3.6
Coryogalops ochetica (Norman, 1927)	Suez Canal	3.2
Crenidens crenidens (Forsskål, 1775)	Suez Canal	2.8
Cyclichthys spilostylus (Leis & Randall, 1982)	Suez Canal	3.5
Cynoglossus sinusarabici (Chabanaud, 1913)	Suez Canal	3.3
Decapterus russelli (Rüppell, 1830)	Suez Canal	3.7
Dussumieria elopsoides Bleeker, 1849	Suez Canal	3.4
Elates ransonetti (Steindachner, 1876)	Vessels	4.1
Epinephelus coioides (Hamilton, 1822)	Suez Canal	3.9
Epinephelus fasciatus (Forsskål, 1775)	Suez Canal	3.7
Epinephelus malabaricus (Bloch & Schneider, 1804)	Suez Canal	3.8
Equulites elongatus (Günther, 1874)	Suez Canal	3.5
Equulites klunzingeri (Steindachner, 1898)	Suez Canal	3.5
Etrumeus golanii DiBattista, Randall & Bowen, 2012	Suez Canal	3.5
Fistularia commersonii (Rüppell, 1835)	Suez Canal	4.3
Glaucostegus halavi (Forsskål, 1775)	Suez Canal	4.1
Hemiramphus far (Forsskål, 1775)	Suez Canal	2.9
Heniochus intermedius Steindachner, 1893	Suez Canal	3.5
Herklotsichthys punctatus (Rüppell, 1837)	Suez Canal	3.2
Hippocampus fuscus Rüppell, 1838	Suez Canal	3.4
Hyporhamphus affinis (Günther, 1866)	Suez Canal	3.5
Iniistius pavo Valenciennes, 1840	Suez Canal	3.5
Lagocephalus sceleratus (Gmelin, 1788)	Suez Canal	3.6
Lagocephalus spadiceus (Richardson, 1844)	Suez Canal	3.5
Lagocephalus suezensis Clark & Gohar, 1953	Suez Canal	3.5
Liza carinata (Valenciennes, 1836)	Suez Canal	2.6
Liza haematocheilus (Temminck & Schlegel, 1845)	Aquaculture	2.5
Lutjanus argentimaculatus Forsskål, 1775)	Suez Canal	3.6
Monotaxis grandoculis (Forsskål, 1775)	Suez Canal	3.2
Muraenesox cinereus (Forsskål, 1775)	Suez Canal	4.1
Mycteroperca fusca (Lowe, 1838)	Aquaculture	4.4
Nemipterus randalli Russell, 1986	Suez Canal	3.7
Omobranchus punctatus (Valenciennes, 1836)	Suez Canal; Vessels	2.5
Oplegnathus fasciatus (Temminck & Schlegel, 1944)	Vessels	3.6
Ostorhinchus fasciatus (White, 1790)	Suez Canal	3.5
Ostracion cubicus Linnaeus, 1758	Suez Canal	3.4
Oxyurichthys petersi (Klunzinger, 1871)	Suez Canal	3.5
Pagrus major (Temminck & Schlegel, 1843)	Aquaculture	3.7
Pampus argenteus (Euphrasen, 1788)	Vessels	3.1
Papilloculiceps longiceps (Cuvier, 1829)	Suez Canal	4.1

(continued)

Table 27.1 (continued)

Species	Introduction	Trophic level
Papillogobius melanobranchus (Fowler, 1934)	Suez Canal	3.3
Parexocoetus mento (Valenciennes, 1846)	Suez Canal	3.3
Parupeneus forsskali (Fourmanoir & Guézé, 1976)	Suez Canal	3.3
Pelates quadrilineatus (Bloch, 1790)	Suez Canal	3.5
Pempheris rhomboidea Kossmann & Räuber 1877	Suez Canal	3.4
Petroscirtes ancylodon Rüppell, 1838	Suez Canal	2.1
Pinguipes brasilianus Cuvier, 1829	Vessels	3.8
Platax teira (Forsskål, 1775)	Suez Canal	4
Platycephalus indicus (Linnaeus, 1758)	Suez Canal	3.6
Plotosus lineatus (Thunberg, 1787)	Suez Canal	3.5
Pomacanthus imperator (Bloch, 1787)	Suez Canal	2.7
Pomacanthus maculosus (Forsskål, 1775)	Suez Canal	2.7
Pomadasys stridens (Forsskål, 1775)	Suez Canal	4
Priacanthus sagittarius Starnes, 1988	Suez Canal	4.1
Pteragogus pelycus Randall, 1981	Suez Canal	3.5
Pterois miles (Bennet, 1803)	Suez Canal	3.7
Rachycentron canadum (Linnaeus, 1766)	Suez Canal	4
Rastrelliger kanagurta (Cuvier, 1816)	Suez Canal	3.2
Rhabdosargus haffara (Forsskål, 1775)	Suez Canal	3.5
Rhynchoconger trewavasae Ben-Tuvia, 1993	Suez Canal	3.9
Sargocentron rubrum (Forsskål, 1775)	Suez Canal	3.5
Saurida macrolepis Shindo & Yamada, 1972	Suez Canal	4.2
Scarus ghobban Forsskål, 1775	Suez Canal	2
Sciaenops ocellatus (Linnaeus, 1766)	Aquaculture	4.1
Scomberomorus commerson Lacepède, 1800	Suez Canal	4.5
Siganus luridus (Rüppell, 1829)	Suez Canal	2
Siganus rivulatus Forsskål, 1775	Suez Canal	2
Silhouetta aegyptia (Chabanaud, 1933)	Suez Canal	3.3
Sillago sihama (Forsskål, 1775)	Suez Canal	3.4
Sorsogona prionota (Sauvage, 1873)	Suez Canal	3.7
Sphyraena chrysotaenia Klunzinger, 1884	Suez Canal	4
Sphyraena flavicauda Rüppell, 1838	Suez Canal	3.8
Spratelloides delicatulus (Bennett, 1831)	Suez Canal	3.1
Stephanolepis diaspros Fraser-Brunner, 1940	Suez Canal	2.8
Stolephorus insularis Hardenberg, 1933	Suez Canal	3.2
Synagrops japonicus (Doderlein, 1884)	Vessels	3.5
Synanceia verrucosa Bloch & Schneider, 1801	Suez Canal	4.2
Terapon jarbua (Forsskål, 1775)	Suez Canal	3.9
Terapon puta (Cuvier, 1929)	Suez Canal	3.1
Terapon theraps Cuvier, 1829	Suez Canal	3.5
Tetrosomus gibbosus (Linnaeus, 1758)	Suez Canal	3.5
Torquigener flavimaculosus Hardy & Randall, 1983	Suez Canal	3.2
Trachurus indicus Necrasov, 1966	Suez Canal	3.8
Tridentiger trigonocephalus (Gill, 1859)	Vessels	3.3
Trypauchen vagina (Bloch & Scheider, 1801)	Suez Canal	3.5
Tylerius spinosissimus (Regan, 1908)	Suez Canal	3.2
Tylosurus choram (Rüppell, 1837)	Suez Canal	4.4
Tylosurus crocodilus (Péron & Lesueur, 1821)	Suez Canal	4.5
Upeneus moluccensis (Bleeker, 1855)	Suez Canal	3.6
Upeneus pori Ben-Tuvia & Golani, 1989	Suez Canal	3.3
Vanderhorstia mertensi Klausewitz, 1974	Suez Canal	3.3
Himantura uarnak (Gmelin, 1789)	Suez Canal	3.6
Torpedo sinuspersici Olfers, 1831	Suez Canal	4.5

Data on trophic levels extracted from Froese and Pauly (2012a)

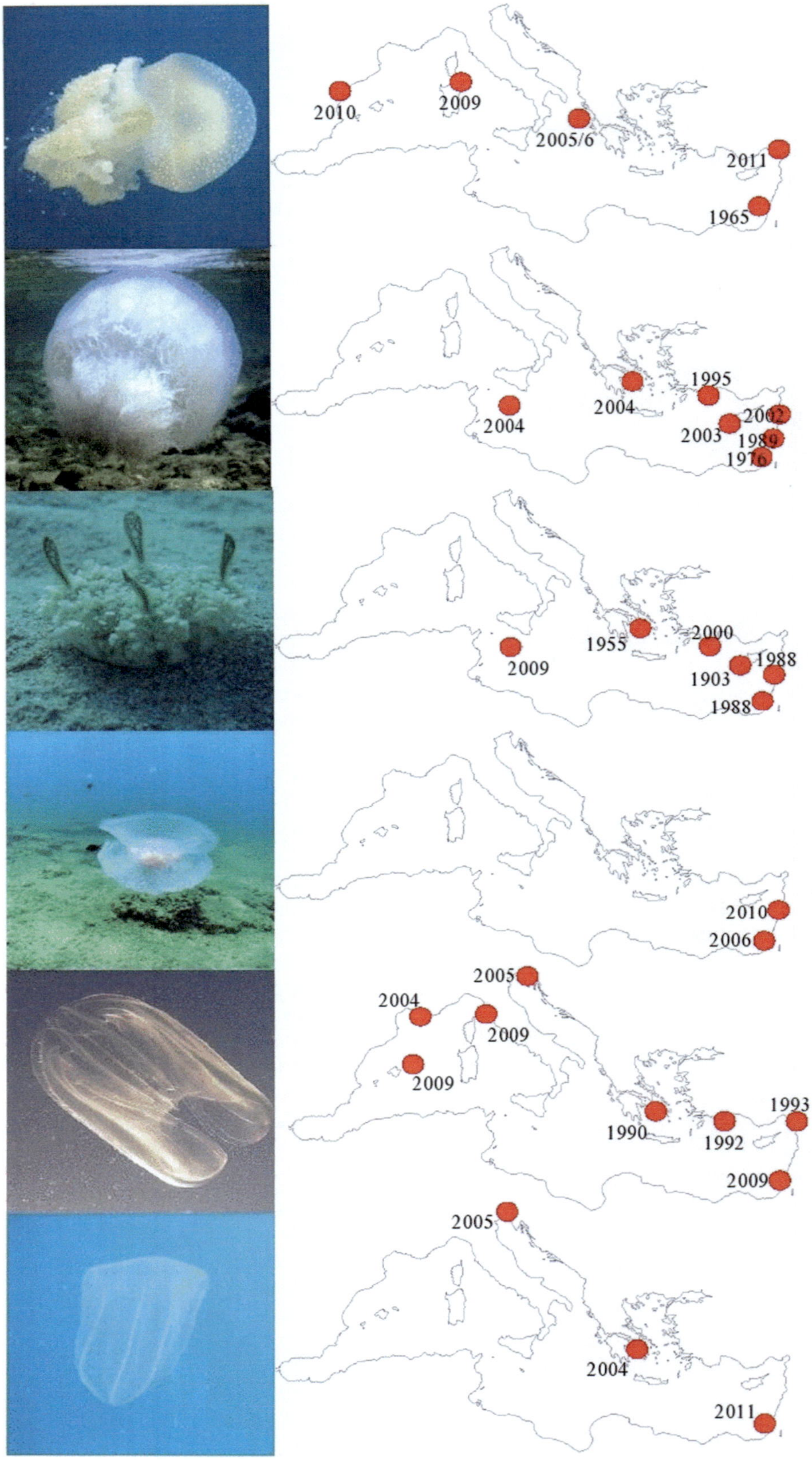

Fig. 27.3 Gelatinous alien species in the Mediterranean Sea, with distribution maps and date of first record per country. From *top*: *Phyllorhiza punctata* Lendenfeld, 1884, *Rhopilema nomadica* Galil, 1990, *Cassiopea andromeda* (Forsskål, 1775), *Marivagia stellata* Galil & Gershwin, 2010, *Mnemiopsis leidyi* A. Agassiz, 1865, *Beroe ovata* Bruguière, 1789 (Photos by IOLR, R. Gevili, G. Paz, G. Rilov, S. Shafir) (After Galil 2012)

Fig. 27.4 *Rhopilema nomadica* Galil, 1990. Jellyfish removed from blocked seawater cooling systems of a coastal power plant, Israel, summer 2011

off Turkey's Aegean and Mediterranean coasts (Uysal and Mutlu 1993; Kideys and Niermann 1994) and off the Syrian coast (Shiganova 1997). As all those locales are in the vicinity of ports, and no population persisted long, it was suggested that it had been introduced with ballast water (Shiganova et al. 2001b). Its presence in the Bay of Piran, northern Adriatic Sea (Shiganova and Malej 2009) too was presumably caused by "… ballast water originating from the Black Sea, as there is direct connection between the Port of Koper and various Black Sea ports" (Shiganova and Malej 2009: 64). Suddenly, in 2009, large swarms appeared along the Ligurian, Tyrrhenian, and Ionian shores of Italy, the Mediterranean coast of Spain including the Balearic Islands and the SE Levant (Boero et al. 2009; Fuentes et al. 2009; Galil et al. 2009). Since January 2009 swarms of *M. leidyi* have been intermittently observed along the Israeli coast, where summer SST tops 30 °C. *Mnemiopsis*-fouled fishing gear poses a hardship to local artisanial fisheries. *Mnemiopsis* blocked water intake pipes threaten desalination plants and force plant engineers to modify their operation by increasing the frequency of backwash cycles in the pretreatment stage and consequently raising the discharge of coagulants such as ferric sulfate into the sea, and ultimately reducing output. The ctenophore *Beroe ovata* Bruguière, 1789, a specialized predator of *M. leidyi,* was first noted in the Mediterranean in Greece, in 2004 (Shiganova et al. 2007), a year later in the northern Adriatic Sea (Shiganova and Malej 2009), and recently (2011–2013), it was recorded near the Port of Ashdod, Israel (Galil et al. 2011).

Invasive alien scyphozoan and ctenophorans may impact the ecosystem in ways we neither expect nor understand, and which are more significant than their obvious impacts in economic and human health terms. As gelatinous plankton plays a pivotal role in marine food webs and elemental fluxes, invasive alien populations may affect production cycles in plankton and benthos. Exploiting their high feeding and growth rate potentials, their massive consumption of plankton, including eggs and larvae of fish and invertebrates, may directly impact recruitment, to the point it may be impossible for the pre-invasion ecosystem to recover.

Metamorphoses: The Changing of the Mediterranean Littoral

Alien macrophytes, invertebrates and fish – all of 680 species that have been recorded in the Mediterranean Sea (Galil 2012) – are found in most littoral habitats, forming some prominent micro-communities and biological facies (*sensu* Last et al. 2010). With few exceptions, the role invasive alien species play *vis a vis* the native biota is poorly known, though it is believed that engineering and keystone invasive aliens cause major changes in community composition and structure and alter ecosystem processes over spatial and temporal scales and a wide range of impacts (Ehrenfeld 2011). A large minority (as much as 30 %) of established invasive aliens are highly disruptive and affect ecosystem functions substiantially (Strayer 2010; Simberloff 2011).

Physical alteration is the most frequently recorded pathways of ecosystem change, either through substantial change in species population size and density, or through structural conversion of the physical character of the habitat (Crooks 2002; Simberloff 2011; Strayer and Hillebrand 2012). Such transformations are often accompanied by changes in local species composition and relative abundances, inducing further changes in ecosystem processes and pathways.

We highlight a few of the discernible and sometimes dramatic physical alterations to habitats associated with invasive aliens in the Mediterranean Sea (Fig. 27.5).

An invasive Australian strain of the tropical alga *C. taxifolia*, unintentionally introduced into the Mediterranean in 1984 with aquaria outflow (Jousson et al. 1998), has since spread to Spain, France, Italy, Croatia and Tunisia. Its propensity to form dense meadows (up to 14,000 blades per m^2) on various infralittoral bottom types, especially in areas of elevated nutrient loads, led to formation of homogenized microhabitats and replacement of native *Cymodocea nodosa* and *Posidonia oceanica*. The alga's dense clumps of rhizomes and stolons and the defensive secondary metabolites alter total species richness, density and biomass of resident biota, from epiphytic Foraminifera and invertebrates to fish assemblages. It is considered "a real threat for the balance of the marine coastal biodiversity". (Boudouresque et al. 1994; Verlaque and Fritayre 1994; Francour et al. 1995; Villele and Verlaque 1995; Ceccherelli and Cinelli 1999; Harmelin-Vivien et al. 1999; Levi and Francour 2004; Longpierre et al. 2005; Francour et al. 2009; Guillem et al. 2010). The extremely invasive *C. racemosa* var. *cylindracea*, was

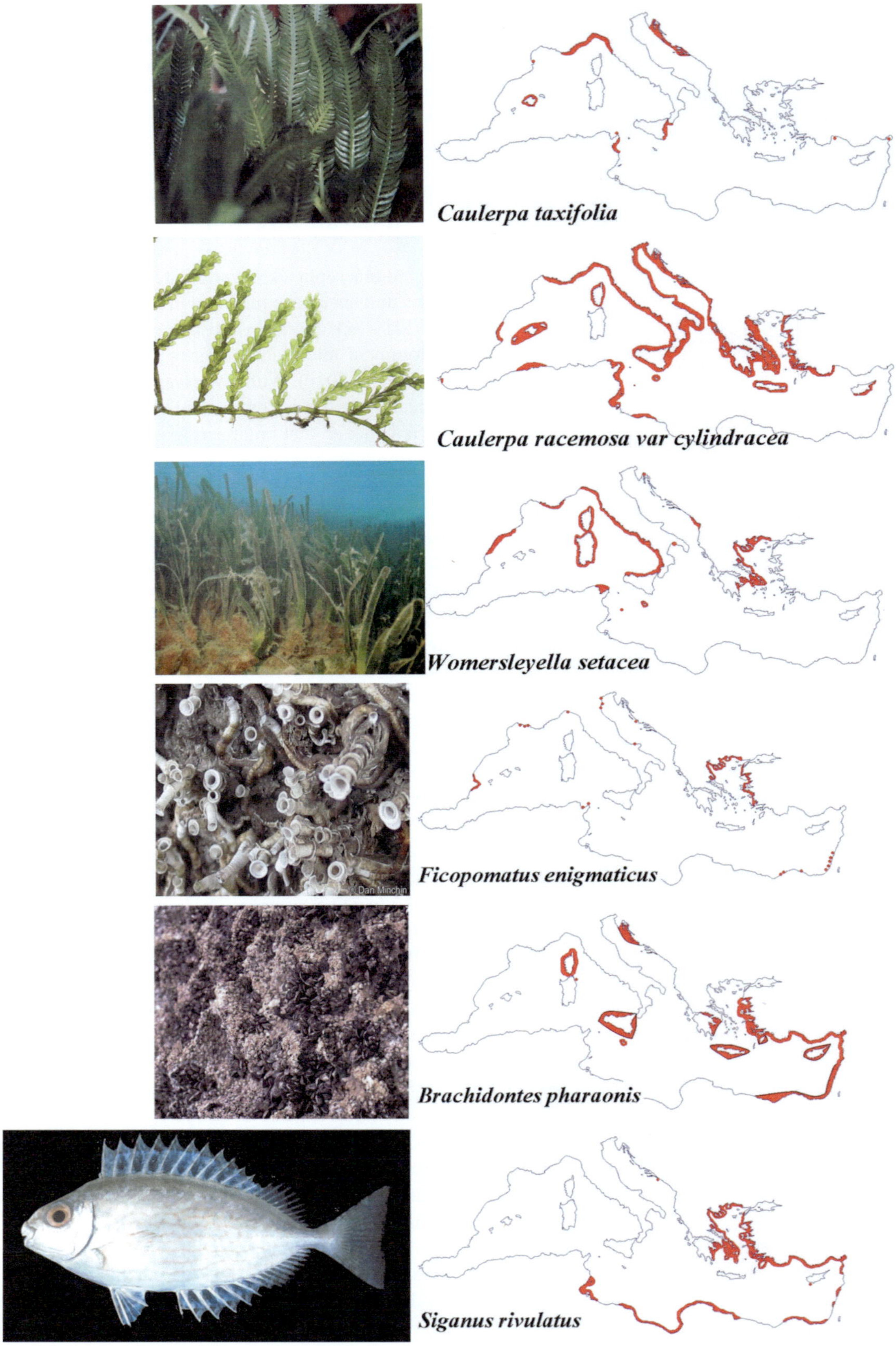

Fig. 27.5 Invasive alien species in the Mediterranean Sea, with distribution maps. From top: *Caulerpa taxifolia* (Vahl) C. Agardh, *Caulerpa racemosa* var. *cylindracea* (Sonder) Verlaque, Huisman and Boudouresque, *Womersleyella setacea* (Hollenberg) R.E.Norris 1992, *Ficopomatus enigmaticus* (Fauvel, 1923), *Brachidontes pharaonis* (Fischer P., 1870), *Siganus rivulatus* Forsskål, 1775

discovered in the Mediterranean in 1990, and has since spread from Cyprus to Spain. Its fast growing stolons allow it to overgrow other macroalgae, mainly turf and encrusting species, and to curtail species number, percent cover and diversity of the macroalgal community, even in highly diverse, native macroalagal assemblages with dense coverage. The drastic change in the composition of the phytobenthos brought about a modification of the macrobenthos and consequent trophic guild, niche breadth shift and ecosystem functioning. A study of the meiofauna revealed an increase in average density but a significant decrease of diversity, and profound change in the crustacean assemblage, where harpacticoid copepods proliferate at the expense of ostracods, cumaceans, isoppds, amphipods and tanaidaceans (Argyrou et al. 1999; Carriglio et al. 2003; Verlaque et al. 2003; Piazzi et al. 2001, 2003; Klein and Verlaque 2008; Piazzi and Balata 2008; Baldacconi and Corriero 2009; Vazquez-Luis et al. 2009; Box et al. 2010; Deudero et al. 2011; Lorenti et al. 2011; Zuljevic et al. 2011). The invasive rhodophyte *Womersleyella setacea* was discovered in Italy in 1986, presumably introduced with vessel fouling. It forms thick, persistent carpets in deep sublittoral rocky substrata, substantial negative effects on native communities, modifies benthic assemblages and outcompetes gorgonians. Lower values of diversity and significant differences in species composition and abundance. Changes in assemblage structure and reduced species richness were noted in coralligenous communities, reducing diversity and abundance of other turf algae; avoided by herbivorous fishes (Benedetti-Cecchi and Cinelli 1989; Verlaque 1994; Piazzi and Cinelli 2000; Piazzi et al. 2007; Antoniadou and Chintiroglou 2007; Tomas et al. 2011; Cebrian and Rodriguez-Prieto 2012). The reef building serpulid worm *Ficopomatus enigmaticus* is found in brackish to hypersaline sheltered coasts, estuaries and lagoons. An engineering species, it forms extensive reefs - large, complex structures where no hard substrate had existed before. The structural conversion from flat soft-bottom to dimensionally complex dense tube colonies provide refuge for invertebrates (Bianchi and Morri 1996, 2001; Fornos et al. 1997; Ben Eliahu and Ten Hove 2011).

Macrophytes and reef-building invertebrates such as corals, molluscs or polychaetes, are obvious examples of species able to modify their physical environment and create habitats for other species. Planktonic ecosystem engineering seems at first oxymoronic, yet jellyfish clearly alter their physical environment by trophic interactions that control abundances of other species (predator-mediated landscape structure *sensu* Breitburg et al. 2010). Swarms of the Erythraean scyphozoan *R. nomadica*, some over 100 km long. must play havoc with resources in the oligotrophic Levant. It had replaced the native *Rhizostoma pulmo* (Galil 2007).

Changes in food web architecture provide another mechanism by which ecosystem processes are altered. The gregarious intertidal Erythraean mytilid bivalve *Brachidontes pharaonis* (Fischer P., 1870) is one of the earliest invasive aliens recorded in the Mediterranean. Where suitable conditions are present it forms extremely dense aggregations (1,000/m^2) on natural and man-made hard substrates. It locally displaced the smaller-sized and thinner shelled native mytilid *Mytilaster minimus*, possibly affecting trophic guilds. (Fuchs 1878; Safriel et al. 1980; Safriel and Sasson-Frostig 1988; Rilov et al. 2002; Mienis 2003; Sara et al. 2006; Crocetta et al. 2009; Cilia and Deidun 2012). Similarly, the schooling herbivorous Erythraean rabbitfish *Siganus rivulatus*, first recorded in the Mediterranean in 1924, replaced native herbivorous fish along the Levant. It formed "barrens" in the rocky infralittoral- dramatically reducing habitat complexity, altered the community structure and the native food web along the Levantine rocky infralittoral, accelerated transfer of energy from the producer to the consumer levels and has served as major prey item for larger predators (Steinitz 1927; Goren and Galil 2001, 2005; Aronov 2002; Sala et al. 2011).

The biology and ecology of the great majority of invasive alien species in the Mediterranean have never been studied, so it is impossible to say to what degree they impacted the ecosystem.

Invasiveness, Invasibility and Propagule Pressure

It was a truth universally acknowledged that species and ecosystems traits, "invasiveness" and "invasibility" respectively, determined in large part invasion success (Ehrlich 1989; Catford et al. 2012). Researchers were determined to elucidate the biological characteristics of a "successful invader" and thus identify potential invasive species into the Mediterranean Sea, or characterize an habitat vulnerable to invasion (Safriel et al. 1980; Safriel and Ritte 1980). They puzzled over the "universal correlates of colonizing ability" of Erythraean aliens that "distinguishes these species from the many hundreds of Red Sea species that have not (yet?) used the opportunity to colonize the Mediterranean" (Safriel and Ritte 1983, p. 231). Safriel and Ritte (1986) set out to examine these correlates in a pair of closely related mollusks, *Cerithium caeruleum* G.B. Sowerby II, 1855 and *C. scabridum* Philippi, 1848. Though both species occur in the Suez Canal and the life history traits of the former are such that it "should have migrated through the Suez Canal with much greater ease than the colonizer [*C. scabridum,* BG]" (there, p. 572), only the latter invaded the Mediterranean. Following studies of a small Erythraean mytilid mussel, *B. pharaonis*, that in the early 1970s was "c. 250 times rarer" than the native mytilid *Mytilaster minimus* (Poli, 1795), it was predicted the Erythraean alien would neither establish dense populations nor displace the native from the intertidal rocky ledges along the southern Levant (Safriel et al. 1980, p. 39). How disappointing then that their sincere and ardent efforts to relate species traits to "invasiveness" have failed to alight on the "universal correlates": these very ledges are todate covered with dense clumps of the alien mytilid ".. while *M. minimus* is only rarely encountered." (Mienis 2003, p. 15).

Since the Levant Sea was recognized early as pregnable (Steinitz 1927), it was *ipso facto* an "undersaturated niche" (Diamant et al. 1986), "biologically underexploited by marine life … a sort of ecological vaccum where many ecological niches are available" (Oliverio and Taviani 2003, p. 314), where "The Lessepsian migrants may be considered, in a figurative sense "welcome guests" in the impoverished, subtropical cul-de-sac" (Por 1978, p. 123), even, to have "biologically enriched" the sea (Tortonese 1973, p. 327). Lately however, such heavily invaded "Godot basins" (Oliverio and Taviani 2003) as the Levant and coastal lagoons, are considered impacted by 'propagule pressure' (Verlaque 2001; Galil 2007, 2009; Boudouresque and Verlaque 2012). A composite measure of the number of viable alien individuals, genotypes and taxa, the number of discrete introduction events, their frequency and duration, 'propagule pressure' is recognized as the primary determinant of invasion success even to overwhelming disturbance and biodiversity (Wonham et al. 2001; Verling et al. 2005; Von Holle and Simberloff 2005; Simberloff 2009). Indeed, a meta-analysis of invasiveness and invasibility studies found it was a significant predictor of both (Colautti et al. 2006). It stands to reason that, when controlling for donor and recipient regions, large inocula are more likely to lead to establishment and that the establishment chance increases with introduction frequency. Large and recurring introductions enable nascent alien populations to overcome limitations associated with small population sizes, including deleterious genetic effects, and weather adverse environmental or demographic events. Recurring introductions of individuals from the same donor location into the same recipient location sustain nascent populations even if initial transfers were of insufficient size or badly timed (Lockwood et al. 2005; Roman and Darling 2007). Indeed, the genetic diversity of the Mediterranean populations of Erythraean aliens proved to be high, and no evidence of founder effect or bottleneck was found. Instead, high levels of gene flow were found between the Red Sea and the Mediterranean populations (Safriel and Ritte 1986; Golani and Ritte 1999; Bucciarelli et al. 2002; Bonhomme et al. 2003; Hassan et al. 2003; Hassan and Bonhomme 2005; Azzurro et al. 2006; Terranova et al. 2006; Golani et al. 2007; Bariche and Bernardi 2009; Sanna et al. 2011; Rius et al. 2012; Rius and Shenkar 2012). A likely explanation is that the propagule pressure driving the Erythraean invasion is large enough to overcome the loss of genetic diversity. Of the three high propagule conveyances – shipping, mariculture, Suez Canal – the latter supplies the largest number of successfully established aliens in the eastern and central Mediterranean Sea by virtue of the magnitude, frequency, and duration of the trans-isthmian corridor invasion and the common evolutionary history of its components. Such high propagule pressure of co-evolved invaders greatly increases the risk of further invasions (Ricciardi et al. 2011).

The sea level lockless Suez Canal that opened in 1869 was 8 m deep, its cross section 304 m^2. Repeatedly deepened and widened, the present canal is 24 m deep and its cross section 5,200 m^2. Already the Suez Canal Authority (SCA) is evaluating the feasibility of further increasing the Canal's depth and doubling its width to increase its capacity and reduce transit time (www.suezcanal.gov.eg). The implications of a deeper Canal combined with higher through-current velocities on propagule pressure of Erythraean aliens are all too clear- increasing the delivery of multiple invaders, increasing the supply of deeper living taxa, thus accounting for the observed long term changes in the Erythraean invasion. The influx of thermophilic Erythraean aliens has not abated, quite the opposite (Galil 2012), and recently they have been recorded in deeper waters (Goren et al. 2011; Kalogirou and Corsini-Foka 2012), increasing the spatial extent of the recipient ecosystem in the Mediterranean Sea and establishing an ever larger invasion bridgehead in the Mediterranean Sea.

Acknowledgements This research was partly supported by the European Community's Seventh Framework Programme (FP7/2007-2013) for the projects *Vectors of Change in Oceans and Seas Marine Life, Impact on Economic Sectors* (VECTORS) and *Options for Delivering Ecosystem-Based Marine Management* (ODEMM).

References

Abed-Navandi D, Kikinger R (2007) First record of the tropical scyphomedusa *Phyllorhiza punctata* von Ledenfeld, 1884 (Cnidaria: Rhizostomeae) in the Central Mediterranean Sea. Aquat Invasion 2:391–394

Antoniadou C, Chintiroglou C (2007) Zoobenthos associated with the invasive red alga Womersleyella setacea (*Rhodomelacea*) in the northern Aegean Sea. J Mar Biol Assoc UK 87:629–641

Argyrou M, Demetropoulous M, Hadjichristophorou M (1999) Proceedings of the workshop on invasive *Caulerpa* species in the Mediterranean, Heraklion, Crete, Greece. MAP technical report series 125, pp 139–158

Aronov A (2002) Comparative study of the ecology of three groupers (Epinephelidae, Serranidae) at the shallow rocky habitats of the Israeli Mediterranean coast. Dissertation, Tel Aviv University

Azzurro E, Golani D, Bucciarelli G (2006) Genetics of the early stages of invasion of the Lessepsian rabbitfish *Siganus luridus*. J Exp Mar Biol Ecol 333:190–201

Baldacconi R, Corriero G (2009) Effects of the spread of the alga *Caulerpa racemosa* var. *cylindracea* on the sponge assemblage from coralligenous concretions of the Apulian coast (Ionian Sea, Italy). Mar Ecol 30:337–345

Bariche M, Bernardi G (2009) Lack of a genetic bottleneck in a recent Lessepsian bioinvader, the blue-barred parrotfish, *Scarus ghobban*. Mol Phyl Evol 53:592–595

Bascheri MC (1965) Essai d'acclimatation du clam *Venus mercenaria* en milieu lagunaire mediterraneen (note preliminaire). Rapp p-v réunion Comm int explor sci Mer Méditerr Monaco 18, p 70

Benedetti-Cecchi L, Cinelli F (1989) Note on a *Polysiphonia* sp. (Rhodohyta, Ceramiales) collected at Rosignano Solvay (Western Mediterranean). Giorn Bot Ital 123:49–54

Ben-Eliahu MN, Ten Hove HA (2011) Serpulidae (Annelida: Polychaeta) from the Suez Canal- from a Lessepsian migration perspective. Zootaxa 2848:1–147

Benmeir P, Rosenberg L, Sagi A, Vardi D, Eldad A (1990) Jellyfish envenomation: a summer epidemic. Burns 16:471–472

Ben-Tuvia A (1953a) Mediterranean fishes of Israel. Bull Sea Fish Res Stat Haifa 8:1–40

Ben-Tuvia A (1953b) New Erythrean fishes from the Mediterranean coast of Israel. Nature 172:464–465

Berner L (1944) Le peuplement des coques de bateaux a Marseille. Bull Inst oceanogr Monaco 858:1–44

Bianchi CN, Morri C (1996) *Ficopomatus* 'Reefs' in the Po River delta (Northern Adriatic): their constructional dynamics, biology, and influences on the brackish-water biota. Mar Ecol 17:51–66

Bianchi CN, Morri C (2001) The battle is not to the strong: serpulid reefs in the lagoon of orbetello (Tuscany, Italy). Estuar Coast Shelf Sci 53:215–220

Bilecenoğlu M (2002) Türkiye kiyilalarinda ilk kez rastlanan bir denizanasi tütü: *Cassiopea andromeda* (Forsskål, 1775) (Cassiopeidae, Scyphozoa). Sualti Düny Derg 72:42–43

Boero F, Bouillon J, Gravili C, Miglietta MP, Parsons T, Piraino S (2008) Gelatinous plankton: irregularities rule the world (sometimes). Mar Ecol Prog Ser 356:299–310

Boero F, Putti M, Trainito E, Prontera E, Piraino S, Shiganova TA (2009) First records of *Mnemiopsis leidyi* (Ctenophora) from the Ligurian, Thyrrhenian and Ionian Seas (Western Mediterranean) and first record of *Phyllorhiza punctata* (Cnidaria) from the Western Mediterranean. Aquat Invasion 4:675–680

Bonhomme F, Baranes A, Golani D, Harmelin-Vivien M (2003) Lack of mitochondrial differentiation between Red Sea and Mediterranean populations of the Lessepsian rabbitfish, *Siganus rivulatus* (Perciformes: Siganidae). Sci Mar (Barc) 67:215–217

Boudouresque CF, Verlaque M (2012) An overview of species introduction and invasion processes in marine and coastal lagoon habitats. Cah Biol Mar 53:309–317

Boudouresque CF, Briand F, Nolan C (1994) Introduced species in European Coastal waters. Ecosystems Research Reports No 8. European Commission, Luxembourg

Box A, Martin D, Deudero S (2010) Changes in seagrass polychaete assemblages after invasion by *Caulerpa racemosa* var. *cylindracea* (Chlorophyta: Caulerpales): community structure, trophic guilds and taxonomic distinctness. Sci Mar 74:317–329

Breitburg DL, Crump BC, Dabiri JO, Gallegos CL (2010) Ecosystem engineers in the pelagic realm: alteration of habitat by species ranging from microbes to jellyfish. Integr Comp Biol 50:188–200

Bucciarelli G, Golani D, Bernardi G (2002) Genetic cryptic species as biological invaders: the case of a Lessepsian fish migrant, the hardyhead silverside *Atherinomorus lacunosus*. J Exp Mar Biol Ecol 273:143–149

Carriglio D, Sandulli R, Deastis S, Gall D'Addabbo M, Grimaldi de Zio S (2003) *Caulerpa racemosa* spread effects on the meiofauna of the Gulf of Taranto. Biol Mar Medit 10:509–511

Carus JV (1889) Prodromus Faunae mediterraneae sive descriptio animalium maris Mediterranei incolarum, 2: Brachiostomata. Mollusca. Tunicata. Vertebrata. Schweizerbart Science Publishers, Stuttgart

Catford JA, Vesk PA, Richardson DM, Pysek P (2012) Quantifying levels of biological invasion: towards the objective classification of invaded and invasible ecosystems. Glob Change Biol 18:44–62

Catta JD (1876) Note sur quelques crustacés erratiques. Ann Sci Nat Zool Paleo Biol Anim 6:1–33

Cebrian E, Rodríguez-Prieto C (2012) Marine invasion in the Mediterranean Sea: the role of abiotic factors when there is no biological resistance. PLoS One 7:e31135

Ceccherelli G, Cinelli F (1999) Effects of *Posidonia oceanica* canopy on *Caulerpa taxifolia* size in a north-western Mediterranean bay. J Exp Mar Biol Ecol 240:19–36

Çevik C, Derici OB, Çevik F, Cavas L (2011) First record of *Phyllorhiza punctata* von Lendendfeld, 1884 (Scyphozoa: Rhizostomeae: Mastigiidae) from Turkey. Aquat Invasion 6:S27–S28

CIESM (2001) Gelatinous zooplankton outbreaks: theory and practice. CIESM workshop series Monaco 14, pp 1–112

Cilia D, Deidun A (2012) Branching out: mapping the spatial expansion of the Lessepsian invader mytilid *Brachidontes pharaonis* (Fischer, 1870) around the Maltese Islands. Mar Biodiv Rec 5:e28

Colautti RI, Grigorovich IA, MacIsaac HJ (2006) Propagule pressure: a null model for biological invasions. Biol Invasion 8:1023–1037

Crocetta F, Renda W, Vazzana A (2009) Alien Mollusca along the Calabrian shores of the Messina Strait area and a review of their distribution in the Italian seas. Boll Mal 45:15–30

Crooks JA (2002) Characterizing ecosystem-level consequences of biological invasions: the role of ecosystem engineers. Oikos 97(2):153–166

Daskalov GM, Grishin AN, Rodionov S, Mihneva V (2007) Trophic cascades triggered by overfishing reveal possible mechanisms of ecosystem regime shifts. Proc Natl Acad Sci U S A 104: 10518–10523

de Villèle X, Verlaque M (1995) Changes and degradation in a *Posidonia oceanica* bed invaded by the introduced tropical alga *Caulerpa taxifolia* in the north western Mediterranean. Bot Mar 38:79–87

Deudero S, Box A, Alós J, Arroyo NL, Marbà N (2011) Functional changes due to invasive species: food web shifts al shallow *Posidonia oceanica* seagrass beds colonized by the alien macroalga *Caulerpa racemosa*. Estuar Coast Shelf Sci 93:106–116

Diamant A, Ben Tuvia A, Baranes A, Golani D (1986) An analysis of rocky coastal eastern Mediterranean fish assemblages and a comparison with an adjacent small artificial reef. J Exp Mar Biol Ecol 97:269–285

Ehrenfeld JG (2011) Ecosystem consequences of biological invasions. Annu Rev Ecol Evol Syst 41:59–80

Ehrlich PR (1989) Attributes of invaders and the invading process: vertebrates. In: Drake JA, Mooney HA, di Castri F, Groves RH, Kruger FJ, Rejmánek M, Williamson M (eds) Biological invasions: a global perspective. Wiley, New York

Estoup A, Guillemaud T (2010) Reconstructing routes of invasion using genetic data: why, how and so what? Mol Ecol 19:4113–4130

Faasse MA, Ligthart AHM, Bayha KM (2006) The first record of the ctenophore *Mnemiopsis leidyi* in Dutch estuaries: an unrecognized invasion? Aquat Invasion 1:270–277

Fischer P (1865) Note sur les faunes conchyliologiques des deux rivages de l'isthme de Suez. J Conchyl 13:241–248

Fornós JJ, Forteza V, Martínez-Taberner A (1997) Modern polychaete reefs in Western Mediterranean lagoons: *Ficopomatus enigmaticus* (Fauvel) in the Albufera of Menorca, Balearic Islands. Palaeogeogr Palaeoclimateol Palaeoecol 128:175–186

Francour P, Harmelin-Vivien M, Harmelin JG, Duclerc J (1995) Impact of *Caulerpa taxifolia* colonization on the littoral ichthyfauna of north-western Mediterranean: preliminary results. Hydrobiologia 345–353

Francour P, Pellissier V, Mangialajo L, Buisson E, Stadelmann B, Veillard N, Meinesz A, Thibaut T (2009) Changes in invertebrate assemblages of *Posidonia oceanica* beds following *Caulerpa taxifolia* invasion. Vie Milieu 59:31–38

Froese R, Pauly D (2012a) FishBase. World Wide Web electronic publication. http://www.fishbase.org. Accessed 6 Oct 2012

Froese R, Pauly D (2012b) Food webs and trophic levels. FishBase. http://www.fishbase.org/FishOnLine/English/FOL_FishAsPart.htm#Food_webs_and_trophic_levels. Accessed 6 Oct 2012

Fuchs T (1878) Die geologische Beschaffenheit der Landenge von Suez. Denk Akad Wiss Wien Math Naturwiss 38:25
Fuentes VL, Atienza D, Gili JM, Purcell JE (2009) First records of *Mnemiopsis leidyi* A. Agassiz 1865 off the NW Mediterranean coast of Spain. Aquat Invasion 4:671–674
Galil B, Hülsmann N (1997) Protist transport via ballast water – biological classification of ballast tanks by food web interactions. Eur J Protistol 34:244–253
Galil BS (2007) Loss or gain? Invasive aliens and biodiversity in the Mediterranean Sea. Mar Pollut Bull 55:314–322
Galil BS (2009) Taking stock: inventory of alien species in the Mediterranean Sea. Biol Invasion 11:359–372
Galil BS (2012) Truth and consequences: the bioinvasion of the Mediterranean Sea. Integr Zool 7:299–311
Galil BS, Lewinsohn C (1981) Macrobenthic communities of the eastern Mediterranean continental shelf. Mar Ecol 2:343–352
Galil BS, Spanier E, Ferguson WW (1990) The Scyphomedusae of the Mediterranean coast of Israel, including two Lessepsian migrants new to the Mediterranean. Zool Meded 64:95–105
Galil BS, Shoval L, Goren M (2009) *Phyllorhiza punctata* (Scyphozoa: Rhizostomeae: Mastigiidae) reappeared off the Mediterranean coast of Israel. Aquat Invasion 4:481–483
Galil BS, Gershwin LA, Douek J, Rinkevich B (2010) *Marivagia stellata* gen. et sp. nov. (Scyphozoa: Rhizostomeae: Cepheidae), another alien jellyfish from the Mediterranean coast of Israel. Aquat Invasion 5:331–340
Galil BS, Gevili R, Shiganova T (2011) Not far behind: first record of *Beroe ovata* Mayer 1912 (Ctenophora; Beroida; Beroidae) off the Mediterranean coast of Israel. Aquat Invasion 6:S89–S90
Gardiner JS (1924) The biology of the Suez Canal. Nature 114:520–521
Geldiay R, Kocataş A (1968) The species of Penaeidae family from the Mediterranean Coast of Turkey. Science reports Faculty of Sciences, Ege University, Izmir 58, pp 1–15
George CJ, Athanassiou V (1967) A two year study of the fishes appearing in the seine fishery of St. George Bay, Lebanon. Ann Mus Civ Stor Nat Genova 76:237–294
Golani D, Bernardi G (2012) Differential invading potential among cryptic species of a Lessepsian bioinvader, the blotchfin dragonet *Callionymus filamentosus*. Mar Ecol Prog Ser 450:159–166
Golani D, Ritte U (1999) Genetic relationships in goatfishes (Mullidae: Perciformes) of the Red Sea and the Mediterranean, with remarks on Suez Canal migrants. Sci Mar 63:129–135
Golani D, Azzurro E, Corsini-Foka M (2007) Genetic bottlenecks and successful biological invasions: the case of a recent lessepsian migrant. Biol Lett 3:541–545
Goren M, Galil BS (2001) Fish biodiversity and dynamics in the vermetid reef of Shiqmona (Israel). Mar Ecol 22:369–378
Goren M, Galil BS (2005) A review of changes in the fish assemblages of Levantine Inland and marine ecosystems following the introduction of non-native fishes. J Appl Ichthyol 21:364–370
Goren M, Stern N, Galil BS, Diamant A (2011) On the occurrence of the Indo-Pacific *Champsodon nudivittis* (Ogilby, 1985) (Perciformes, Champsodontidae) from the Mediterranean coast of Israel, and the presence of the species in the Red Sea. Aquat Invasion 6:115–117
Gottlieb E (1953) Decapod in the collection of the Sea Fisheries Research Station, Caesarea, Israel. Bull Res Counc Isr 2:440–441
Gottlieb E (1960) Benthonic Amphipoda of the Mediterranean coast of Israel. I. Notes on the geographical distribution. Bull Res Counc Isr 9B:157–166
Guillem MV, Box A, Deudero S, Rodriguez B (2010) Comparative analysis of epiphytic foraminifera in sediments colonized by seagrass *Posidonia oceanica* and invasive macroalgae *Caulerpa* spp. J Foraminifer Res 40:134–147
Hallegraeff GM, Bolch CJ (1992) Transport of diatom and dinoflagellate resting spores in ships' ballast waters: implications for phytoplankton biogeography and aquaculture. J Plankton Res 14:1067–1084
Hamel G (1926) Quelques Algues rares ou nouvelles pour la flore Mediterraneenne. Bull Mus Nat Hist Nat 32:420
Harmelin-Vivien M, Francour P, Harmelin JG (1999) Impact of *Caulerpa taxifolia* on Mediterranean fish assemblages: a six year study. In: Proceedings of the workshop on invasive Caulerpa in the Mediterranean. UNEP, Athens. Mediterranean Action Plan technical report series 125, pp 127–138
Hassan M, Bonhomme F (2005) No reduction in neutral variability of mitochondrial and nuclear genes for a Lessepsian migrant, *Upeneus moluccensis*. J Fish Biol 66:865–870
Hassan M, Harmelin-Vivien M, Bonhomme F (2003) Lessepsian invasion without bottleneck: Example of two rabbitfish species (*Siganus rivulatus* and *Biology*). J Exp Mar Biol Ecol 291:219–232
Holland BS, Dawson MN, Crow GL, Hofmann DK (2004) Global phylogeography of *Cassiopea* (Scyphozoa: Rhizostomae): molecular evidence for cryptic species and multiple Hawaiian invasions. Mar Biol 145:1119–1128
Holthuis LB (1961) Report on a collection of Crustacea Decapoda and Stomatopoda from Turkey and the Balkans. Zool Verhand Leiden 47:1–67
Holthuis LB, Gottlieb E (1958) An annotated list of the decapod Crustacea of the Mediterranean coast of Israel, with an appendix listing the Decapoda of the Eastern Mediterranean. Bull Res Counc Isr Zool 7B:1–126
Javidpour J, Sommer U, Shiganova T (2006) First record of *Mnemiopsis leidyi* in the Baltic Sea. Aquat Invasion 1:299–302
Jousson O, Pawlowski J, Zaninetti L, Meinesz A, Boudouresque CF (1998) Molecular evidence for the aquarium origin of the green alga Caulerpa taxifolia introduced to the Mediterranean Sea. Mar Ecol Prog Ser 172:275–280
Kalogirou S, Corsini-Foka M (2012) First record of the Indo-Pacific *Champsodon nudivittis* (Ogilby, 1895) (Perciformes, Champsodontidae) in the Aegean waters (eastern Mediterranean Sea). BioInvasion Rec 1:229–233
Keller C (1883) Die Fauna im Suez-Kanal und die Diffusion der Mediterranean und Erythräischen Tierwelt. Neue Denkschr Allg schweiz Ges ges Naturw 28:1–39
Keller C (1888) Die Wanderung der marinen Thierwelt im Suezcanal. Zool Anz 11(359–364):389–395
Kideys AE, Niermann U (1994) Occurrence of *Mnemiopsis* along the Turkish coast. ICES J Mar Sci 51:423–427
Klein J, Verlaque M (2008) The *Caulerpa racemosa* invasion: a critical review. Mar Pollut 56:205–225
Last PR, Lyne VD, Williams A, Davies CR, Butler AJ, Yearsley GK (2010) A hierarchical framework for classifying seabed biodiversity with application to planning and managing Australia's marine biological resources. Biol Conserv 143:1675–1686
Levi F, Francour P (2004) Behavioural response of *Mullus surmuletus* to habitat modification by the invasive macroalga *Caulerpa taxifolia*. J Fish Biol 64:55–64
Levitt Y (2012) The impact of depth gradient on the status of alien species along the Mediterranean Sea coast of Israel. Dissertation, Tel Aviv University
Lockwood JL, Cassey P, Blackburn T (2005) The role of propagule pressure in explaining species invasions. Trends Ecol Evol 20:223–228
Longpierre S, Robert A, Levi F, Francour P (2005) How an invasive alga species (*Caulerpa taxifolia*) induces changes in foraging strategies of the benthivorous fish *Mullus surmuletus* in coastal Mediterranean ecosystems. Biodivers Conserv 14:365–376
Lorenti M, Gambi MC, Guglielmo R, Patti FP, Scipione MB, Zupo V, Buia MC (2011) Soft-bottom macrofaunal assemblages in the Gulf of Salerno, Tyrrhenian Sea, Italy, an area affected by the invasion of the seaweed *Caulerpa racemosa* var. *cylindracea*. Mar Ecol 32:320–334
Lundberg B, Ogorek R, Galil BS, Goren M (2004) Dietary choices of siganid fish at Shiqmona reef, Israel. Isr J Zool 50:39–53

Maas O (1903) Die Scyphomedusen der Siboga Expedition 1901. Siboga Expeditie 11:1–91

Mendez-Domingo C (1957) Sur l'existence du *Colpomenia peregrina* (Sauv.) Hamel dans la Méditerranée. Vie Milieu 8:92–98

Mianzan HW (1999) Ctenophora. In: Boltovskoy D (ed) South Atlantic zooplankton. Backhuys Publishers, Leiden

Mienis HK (2003) Native marine molluscs replaced by Lessepsian migrants. Tentacle 1:15–16

Occhipinti-Ambrogi A, Galil BS (2004) A uniform terminology on bioinvasions: a chimera or an operative tool? Mar Pollut Bull 49:688–694

Oliverio M, Taviani M (2003) The Eastern Mediterranean Sea: tropical invasions and niche opportunities in a "Godot Basin". Biogeographia 24:313–327

Pérès JM (1958) Ascidies recoltées sur les côtes Méditerranéennes d'Israel. Bull Res Counc Isr 7B:143–150

Piazzi L, Balata D (2008) The spread of *Caulerpa racemosa* var. *cylindracea* in the Mediterranean Sea: an example of how biological invasions can influence beta diversity. Mar Environ Res 65:50–61

Piazzi L, Cinelli F (2000) Effets de l'expansion des Rhodophyceae introduites *Acrothamnion preisii* et *Womersleyella setacea* sur les communautés algales des rhizomes de *Posidonia oceanica* de Méditerranée occidentale. Cryptogam Algol 21:291–300

Piazzi L, Ceccherelli G, Cinelli F (2001) Threat to macroalgal diversity: effects of the introduced alga *Caulerpa racemosa* in the Mediterranean. Mar Ecol Prog Ser 210:149–159

Piazzi L, Balata D, Cecchi E, Cinelli F (2003) Co-occurrence of *Caulerpa taxifolia* and C *racemosa* in the Mediterranean Sea: interspecific interactions and influence on native macroalgal assemblages. Cryptogam Algol 24:233–243

Piazzi L, Balata D, Cinelli F (2007) Invasions of alien macroalgae in Mediterranean coralligenous assemblages. Cryptogam Algol 28:289–301

Pierce RW, Carlton JT, Carlton DA, Geller JB (1997) Ballast water as a vector for tintinnid transport. Mar Ecol Prog Ser 149:295–297

Por FD (1978) Lessepsian migration – the influx of Red Sea Biota into the Mediterranean by way of the Suez Canal, vol 23, Ecological studies. Springer, Berlin

Por FD (1990) Lessepsian migration. An appraisal and new data. Bull Inst Océanogr (Monaco) 7:1–10

Purcell JE (2012) Jellyfish and ctenophore blooms coincide with human proliferations and environmental perturbations. Annu Rev Mar Sci 4:209–235

Purcell JE, Uye S, Lo WT (2007) Anthropogenic causes of jellyfish blooms and their direct consequences for humans: a review. Mar Ecol Prog Ser 50:153–174

Raimbault R (1964) Croissance des huitres atlantiques elevees dans les eaux mediterraneennes francaises. Sci Peche 126:1–10

Ribera Siguan M (2002) Review of non native marine plants in the Mediterranean Sea. In: Leppakoski E et al (eds) Invasive aquatic species in Europe. Distribution, impacts and management. Kluwer Academic, Dordrecht, pp 291–310

Ribera MA, Boudouresque CF (1995) Introduced marine plants with special reference to macroalgae: mechanisms and impacts. Progr Phycol Res 11:187–268

Ricciardi A, Jones LA, Kestrup ÅM, Ward JM (2011) Expanding the propagule pressure concept to understand the impact of biological invasions. In: Richardson DM (ed) Fifty years of invasion ecology: the legacy of Charles Elton. Blackwell, Oxford

Richardson AJ, Bakun A, Hays GC, Gibbons MJ (2009) The jellyfish joyride: causes, consequences and management responses to a more gelatinous future. Trends Ecol Evol 24:312–322

Riisgard HU, Andersen P, Hoffmann E (2012) From fish to jellyfish in the eutrophicated Limfjorden (Denmark). Estuar Coast 35: 701–713

Rilov G, Gasith A, Benayahu Y (2002) Effect of an exotic prey on the feeding pattern of a predatory snail. Mar Environ Res 54:85–98

Rioja E (1924) La Mercierella *enigmatica* Fauvel, serpulido de agua salobre, en Espana. Bol Soc Esp Hist Nat 24:160–169

Rius M, Shenkar N (2012) Ascidian introductions through the Suez Canal: the case study of an Indo-Pacific species. Mar Pollut Bull 64:2060–2068

Rius M, Turon X, Ordonez V, Pascual M (2012) Tracking invasion histories in the sea: facing complex scenarios using multilocus data. PLoS One 7:e35815

Roman J, Darling JA (2007) Paradox lost: genetic diversity and the success of aquatic invasions. Trends Ecol Evol 22:454–464

Safriel UN, Ritte U (1980) Criteria for the identification of potential colonizers. Biol J Linn Soc 13:287–297

Safriel UN, Ritte U (1983) Universal correlates of colonizing ability. In: Swingland IR, Greenwood PJ (eds) The ecology of animal movement. Clarendon, Oxford

Safriel UN, Ritte U (1986) Population biology of Suez Canal migration – which way, what kind of species and why. In: Karlin S, Nevo E (eds) Evolutionary processes and theory. Academic, Orlando

Safriel UN, Sasson-Frostig Z (1988) Can colonizing mussel outcompete indigenous mussel? J Exp Mar Biol Ecol 117:211–226

Safriel UN, Felsenburg T, Gilboa A (1980) Distribution of rocky intertidal mussels in the Red Sea coasts of Sinai, the Suez Canal and the Mediterranean coasst of Israel, with special reference to recent colonizers. J Biogeogr 7:39–62

Sala E, Kizilkaya Z, Yildirim D, Ballesteros E (2011) Alien marine fishes deplete algal biomass in the Eastern Mediterranean. PLoS One 6:e17356

Sanna D, Merella P, Lai T et al (2011) Combined analysis of four mitochondrial regions allowed the detection of several matrilineal lineages of the lessepsian fish *Fistularia commersonii* in the Mediterranean Sea. J Mar Biol Assoc UK 91:1289–1293

Sarà G, Romano C, Mazzola A (2006) A new Lessepsian species in the western Mediterranean (Brachidontes pharaonis Bivalvia: Mytilidae): density, resource allocation and biomass. Mar Biodiv Rec 1:e8

Schembri PJ, Deidun A, Vella PJ (2010) First record of *Cassiopea andromeda* (Scyphozoa: Rhizostomeae: Cassiopeidae) from the central Mediterranean Sea. Mar Biodiv Rec 3:e6

Sendovski U, Goffman M, Goldshlak L (2005) Severe delayed cutaneous reaction due to Mediterranean jellyfish (*Rhopilema nomadica*) envenomation. Contact Dermat 52:282–283

Shapiro J (2007) The fisheries and aquaculture of Israel 2007. http://www.fishery.moag.gov.il/NR/rdonlyres/5EB9733E-D5BB-4F81-810D-C51805EA8142/0/Shnaton 2007.pdf. Accessed 6 Oct 2012

Shiganova TA (1997) *Mnemiopsis leidyi* abundance in the Black Sea and its impact on the pelagic community. In: Ozsoy E, Mikaelyan A (eds) Sensitivity of the North Baltic Sea and Black Sea to anthropogenic and climatic changes. Kluwer Academic Publishers, Dordrecht/Boston/London, pp 117–130

Shiganova TA, Christou ED, Siokou-Frangou I (2007) First finding of alien species *Beroe ovata* Mayer 1912 in the Aegean Sea. Med Mar Sci 8:5–14

Shiganova TA, Kamakin A, Zhukova DP, Ushivtzhev VB, Dulimov AB, Musaeva EI (2001a) New invader in the Caspian Sea – ctenophore *Mnemiopsis* and its initial effect on the pelagic ecosystem. Oceanology 41:542–549

Shiganova TA, Malej A (2009) Native and non-native ctenophores in the Gulf of Trieste, northern Adriatic Sea. J Plank Res 31(1): 61–71

Shiganova TA, Mirzoyan ZA, Studenikina EA, Volovik SP, Siokou-Frangou I, Zervoudaki S, Christou ED, Skirta AY, Dumont HJ (2001b) Population development of the invader ctenophore *Mnemiopsis leidyi* in the Black Sea and other seas of the Mediterranean basin. Mar Biol 139:431–445

Silfen R, Vilan A, Wohl I, Leviav A (2003) Mediterranean jellyfish (*Rhopilema nomadica*) sting. Burns 29:868–870

Simberloff D (2009) The role of propagule pressure in biological invasions. Annu Rev Ecol Evol Syst 40:81–102

Simberloff D (2011) How common are invasion-induced ecosystem impacts? Biol Invasion 13:1255–1268

Smayda TJ (2007) Reflections on the ballast water dispersal-harmful algal bloom paradigm. Harmful Algae 6:601–622

Spanier E (1989) Swarming of jellyfishes along the Mediterranean coast of Israel. Isr J Zool 36:55–56

Steinitz W (1919) Memorial on the founding of a sea-station on the coast of Palestine for zoological investigations. Mimeographed publication. 16 pp

Steinitz W (1927) Beiträge zur Kenntnis der Küstenfauna Palästinas. I. Pubbl Staz Zool Napoli 8:331–353

Steinitz H (1970) A critical List of Immigrants via the Suez Canal. Biota of the Red Sea and Eastern Mediterranean 1970/1971. Mimeographed publication pp. 64–75

Strayer DL (2010) *Alien species* in fresh waters: ecological effects, interactions with other stressors, and prospects for the future. Freshw Biol 55:S152–S174

Strayer DL, Hillebrand H (2012) Eight questions about invasions and ecosystem functioning. Ecol Lett 15:1199–1210

Tebble N (1959) On a collection of Polychaetes from the Mediterranean coast of Israel. Bull Res Counc Isr 8B:9–30

Terranova MS, Lo Brutto S, Arculeo M, Mitton JB (2006) Population structure of *Brachidontes pharaonis* (P. Fischer, 1870) (Bivalvia, Mytilidae) in the Mediterranean Sea, and evolution of a novel mtDNA polymorphism. Mar Biol 150:89–101

Terranova MS, Lo Brutto S, Arculeo M, Mitton JB (2007) A mitochondrial phylogeography of *Brachidontes variabilis* (Bivalvia: Mytilidae) reveals three cryptic species. J Zool Syst Evol Res 45:289–298

Tomas F, Cebrian E, Ballesteros E (2011) Differential herbivory of invasive algae by native fish in the Mediterranean Sea. Estuar Coast Shelf Sci 92:27–34

Tortonese E (1973) Facts and perspectives related to the spreading of Red Sea organisms into the eastern Mediterranean. Ann Mus Civ Stor Nat Giacomo Doria 79:322–329

Tsoi KH, Chan TY, Chu KH (2007) Population structure of the kuruma shrimp *Penaeus japonicus* species complex in western Pacific. Mar Biol 150:1345–1364

UNEP (1991) Jellyfish blooms in the Mediterranean. Marine Action Plan technical reports series 47, pp 1–320

Uysal Z, Mutlu E (1993) Preliminary note on the occurrence and biometry of ctenophore *Mnemiopsis leidyi* finally invaded Mersin Bay. Turk J Zool 17:229–239

Vaillant L (1865) Recherches sur la faune malacologique de la baie de Suez. J Conchyl 13:97–127

Vázquez-Luis M, Sanchez-Jerez P, Bayle-Sempere JT (2009) Comparison between amphipod assemblages associated with *Caulerpa racemosa* var. *cylindracea* and those of other Mediterranean habitats on soft substrate. Estuar Coast Shelf Sci 84:161–170

Verlaque M (1994) Inventaire des plantes introduites en Mediterranee: origines et repercussions sur l'environnement et les activites humaines. Oceanol Acta 17:1–23

Verlaque M (1996) L'étang de Thau (France), un site majeur d'introduction d'espèces en Méditerranée - Relations avec l'ostréiculture. In: Ribera A, Ballesteros E, Boudouresque CF, Gomez A, Gravez V (eds) Second international workshop on *Caulerpa taxifolia*. Publicacions Universitat Barcelona, Barcelona

Verlaque M (2001) Checklist of the macroalgae of Thau Lagoon (Herault, France), a hot spot of marine species introduction in Europe. Oceanol Acta 24:29–49

Verlaque M, Fritayre P (1994) Modification des communautés algales méditerranéennes en presence de l'algue envahissante *Caulerpla taxifolia* (Vahl) C. Agardh. Oceanol Acta 17:659–672

Verlaque M, Durand C, Huisman JM, Boudouresque CF, Le Parco Y (2003) On the identity and origin of the Mediterranean invasive *Caulerpla racemosa* (Caulerpales, Chlorophyta). Eur J Phycol 38:325–339

Verling E, Ruiz GM, Smith LD, Galil B, Miller AW, Murphy KR (2005) Supply-side invasion ecology: characterizing propagule pressure in coastal ecosystems. Proc R Soc London, Ser B 272:1249–1257

Von Holle B, Simberloff D (2005) Ecological resistance to biological invasion overwhelmed by propagule pressure. Ecology 86: 3212–3218

Wonham MJ, Walton WC, Ruiz GM, Frese AM, Galil BS (2001) Going to the source: role of the invasion pathway in determining potential invaders. Mar Ecol Prog Ser 215:1–12

Yoffe B, Baruchin AM (2004) Mediterranean jellyfish (*Rhopilema nomadica*) sting. Burns 30:503–504

Zibrowius H (1973) Remarques sur trois espèces de Serpulidae acclimatées en Méditerranée: *Hydroides dianthus* (Verrill, 1873), *Hydroides dirampha* Mörch, 1863, et *Hydroides elegans* (Haswell, 1883). Rapp p-v réunion Comm int explor sci Mer Méditerr 21:683–686

Zibrowius H (1992) Ongoing modifications of the Mediterranean marine fauna and flora by the establishment of exotic species. Mésogée 51:83–107

Zibrowius H (1994) Introduced invertebrates: examples of success and nuisance in the European Atlantic and in the Mediterranean. In: Boudouresque CF et al (eds) Introduced species in European coastal waters. European Commission, Luxembourg

Zuljevic A, Thibaut T, Despalatovic M, Cottalorda JM, Nikolic V, Cvitkovic I, Antolic B (2011) Invasive alga *Caulerpa racemosa* var. *cylindracea* makes a strong impact on the Mediterranean sponge *Sarcotragus spinosulus*. Biol Invasion 13:2303–2308

Future Trends of Mediterranean Biodiversity

28

José Templado

Abstract

This chapter focuses on analysing the current biodiversity of the Mediterranean Sea and the changes that are taking place on a human time scale (decades). Some of the changes observed may be sometimes interpreted as natural changes (cyclical, episodic or catastrophic), but most of them are of human origin. Each of the main anthropogenic impacts (habitat fragmentation and loss, overfishing and exploitation of living resources, pollution, species introductions and others) are analysed separately, although it is noted that predictions of how all the impacts interact synergistically are necessary. Furthermore, the effects of the so called "global change" (including both global warming and ocean acidification) on Mediterranean biodiversity are highlighted. It also deals about some episodic events (mass mortalities, jellyfish blooms, noxious algal blooms, proliferation of mucilages) caused by a combination of different impacts. Finally, some predictions are done about the near future of marine biodiversity in the Mediterranean Sea and some suggestions to address the problem are given.

Keywords

Mediterranean Sea • Biodiversity • Natural changes • Anthropogenic impacts • Global change • Episodic events

Changing Mediterranean Marine Biodiversity

Change is the rule in biodiversity. Biodiversity, at all its levels, is a changing entity at very different time-scales, from evolutionary to seasonal, or even daily. Nevertheless, global biodiversity is changing nowadays at an unprecedented rate as a complex response to several human-induced changes in global environment. This chapter focuses on analysing the current biodiversity of the Mediterranean Sea and the changes that are taking place on a human time scale (decades). The changes caused by each of the identified impacts (both natural and anthropogenic) are analysed separately, but cause and effect may not be so simple, and often all the links in the chain of ecological consequences can be traced only by painstaking detective work. Ecology continuously teaches us that ecosystems rarely operate in a simple, linear fashion with causes leading to obvious effects. Threats do not operate alone, therefore predictions of how all the different impacts interact synergistically are necessary. Furthermore, the wide range of processes that are affected by climate change make it difficult to predict the magnitude and rate of its effects. Only an extensive knowledge of how the ecosystem operates will allow us to understand, interpret and predict the consequences of the whole set of impacts that are affecting the marine biodiversity of the Mediterranean Sea. Nevertheless, some predictions about the near future (focusing in decadal scale) of marine biodiversity in the Mediterranean Sea are advanced and some possible ways to address the problem are commented on.

J. Templado (✉)
Department of Biodiversity and Evolutionary Biology, Museo Nacional de Ciencias Naturales, Consejo Superior de Investigaciones Científicas, Madrid, Spain
e-mail: templado@mncn.csic.es

S. Goffredo and Z. Dubinsky (eds.), *The Mediterranean Sea: Its history and present challenges*,
DOI 10.1007/978-94-007-6704-1_28,

To place recent rapid changes into context, it is necessary to know both past (long-term) and present (short-term) changes to elucidate the dominant scales of variation and what current biodiversity trends are different from historical trends. We have to retrieve the past and, from the past, we can estimate the future. To make inferences about the future of biodiversty in the coming decades, we basically need to know what has happened over the past decades, but a historical perspective is necessary to compare rates of change across evolutionary and ecological time-scales in the absence of human disturbance.

The Mediterranean Fauna and Flora have evolved over millions of years into a unique mixture of temperate and subtropical elements, with a large proportion of endemic species. From its complex history, the high species richness of the Mediterranean is largely due to both its long evolutionary history through the Tertiary and the post Pliocene diversity pump from the Atlantic (Bianchi 2007). The Mediterranean Sea has been subjected to extensive changes in configuration and climate since the Miocene. A historical review (Taviani 2002) shows that tropical biota survived in the Mediterranean till the beginning of the Pleistocene. Shift from the middle Miocene to Quaternary temperate-glacial conditions brought about radical biological changes in the Mediterranean basin. The combination of events such as the opening/closure of the Strait of Gibraltar, ice cycles and changes in temperature, salinity and current patterns, has apparently made this area a notable generator of diversity. The alternation of the ice ages with the warm interglacials resulted in different immigration waves of Atlantic fauna of boreal or subtropical origin, respectively. Because of reduced opportunities for north–south migration in response to changing sea temperatures, the species present within the Mediterranean were subjected to higher evolutionary pressure.

The Mediterranean Sea is a remnant of the once extensive Tethys Ocean, an open equatorial water body that bit into Pangea during the Triasic. In the Cretaceous, after the opening of the Atlantic Ocean, Tethys connected the newly born ocean to the older Indo-Pacific Ocean through this equatorial water-body. The circum-tropical Tethys ceased to exist when the Mediterranean basin lost contact with the nascent Indian Ocean about 16 mya (Vermeij 2012), but its biota continued to be tropical and highly diverse, roughly comparable to that to be found today in the tropical Indo-Pacific. At the end of the Miocene, during the Messinian (about 6 mya), the connection with the Atlantic also closed, and the Mediterranean become an isolated sea. The progressive desiccation of this basin due to its negative water balance during the so-called "Messinian Salinity Crisis" (MSC) caused a mass extinction of the Tethyan biota. At the dawn of the Miocene (5.3 mya), the Strait of Gibraltar re-opened and the Mediterranean was repopulated by species of Atlantic Origin, that were prevalently of (sub)tropical affinity (Bianchi 2007).

The first phases of the Pliocene (the Zancladian and the Piazenzian) were warm, especially during the so-called Pliocene Climate Optimum (between 3.6 and 2.6 mya), with a temperature 5 °C warmer than today and high sea levels of +20 to +35 m (Por 2009). Therefore, the Mediterranean biota preserved its tropical character. At the beginning of the Pleistocene (the Gelasian) the glacial cycles started, with a sudden cooling about 2.6 mya (the Artic Glaciation) causing the end of the tropical biota of the Mediterranean, because most of Early Pliocene marine species became extinct.

During the Pliocene the shallow-water marine fauna of southern Europe and West Africa, including the Mediterranean basin, constituted a more or less unified biogeographical entity extending from northwestern France to southern Angola (Vermeij 2012). Nevertheless, due to the low sea water temperatures of glacial periods, the Mediterranean was invaded by cold water species from the northeastern Atlantic. On the other hand, the tropical post-Pliocene biota that lived along the West African coast and the Cape Verde islands was isolated from the Gibraltar portal by the cold Saharan upwelling and the Canaries current, and still is separated to some extent, constituting the so-called Senegalese biota. Furthermore, the west–east gradient of increasing temperatures within the Mediterranean was steeper during the glacial period than today. While the winter temperature fell as low as 7 °C (Thunell 1979) in the western basin, in the Levant basin the winter temperature was never lower than 16 °C. Today the gradient between west and east is only of about 13–18 °C of minimum surface temperature. So, the Levant basin functioned as a "cul de sac" of warm water, which was out of reach for the cold-water species entering the Mediterranean.

During the last interglacial (the Eemian Interglacial), between 125,000 and 110,000 ya, with global temperatures 2–3 °C higher than today, the West African fauna succeeded in overcoming the Saharan upwelling and the temperature gradient in the Mediterranean to reach the Levant. Thus, the input of the Senegalese biota through the Gibraltar portal has been the only possible tropical input during the Pleistocene. After this warm period, and especially during the Last Glacial Maximum (20,000 ya), the Mediterranean was invaded by cold-adapted North Atlantic species. In recent decades, the ongoing global increase in temperature is noticeable all over the Mediterranean.

Recent Mediterranean Biodiversity

According to a recent review of Mediterranean biodiversity (Coll et al. 2010), the number of macroscopic living species at present inventoried in this sea can be estimated at about 13.200 species (nearly 11,600 metazoans and 1,600 macrophytes). This is a very high figure, if we consider that the Mediterranean only represents less than 0.8 % of the overall world ocean area, while the number of species that inhabit it represents about 5 % of total known recent marine species.

The present-day extraordinarily rich biota of the Mediterranean Sea is due to the outcome of the above dramatic events, which subjected the biota to high evolutionary pressures and may have been the cause of frequent phenomena of extinction and speciation. On the other hand, the complex topography of its coastline, with a number of islands and archipelagos of different sizes and fairly isolated sub-basins subject to a great variety of climatic and hydrological conditions, also promotes a high species diversity. Lejeusne et al. (2010) defined the Mediterranean as a factory designed to produce endemics; in fact, about a quarter of Mediterranean marine species are endemic. These endemics live together with species deriving from several waves of colonization of either temperate or tropical organisms.

Although the Mediterranean Sea is recognized as one of the most diverse regions on Earth, its depths are quite species-poor. Less than 10 % of Mediterranean animal species are present below a depth of 1,000 m, and less than 3 % below 2,000 m (Boudouresque 2004). The Mediterranean deep waters are homothermic at about 13 °C and this limits the establishment of truly abyssal groups, which are typically adapted to colder waters (Emig and Geistdoerfer 2004). The low diversity of the Mediterranean deep fauna may also be due to the Gibraltar sill (less than 350 m depth) acting as a physical barrier to the colonization from the richer Atlantic deep fauna (Bouchet and Taviani 1992).

Most of the present-day Mediterranean species are of Atlantic origin that entered after the MSC, and therefore the Mediterranean biota presents a strong similarity to that of the northeastern Atlantic (European and African coasts). This caused the biogeographic physiognomy of the Mediterranean Sea to become that of an Atlantic province (Briggs 1974).

Recently, the man-made contact through the Suez Canal, opened in 1869, started to give access to hundreds of Indo-Pacific species to the Mediterranean. On the other hand, an increasing settlement by tropical Atlantic newcomers occurs entering the Mediterranean through the strait of Gibraltar. According to Por (2009), the congruence of these two events, the warming of the sea and the influx of the Indo-Pacific and tropical West African biota, led to the present re-establishment of the Tethyan biota in the Mediterranean. It should be noted here that thousands of alien species have been introduced by man in recent times, most of them lessepsian immigrants.

The main characteristics of the present-day Mediterranean Sea and its biota can be summarized as follow:

- Quasi-enclosed sea characterized by rich and complex physical dynamics with distinctive traits, especially in regard to the thermohaline circulation.
- Warm, salty, nutrient-poor water (widespread P deficit).
- Irradiance about 20 % greater than the mean irradiance incident at similar latitudes in the Atlantic Ocean.
- Seasonal thermocline from May–June to September–October.
- Homogeneous deep-water layers down 250–300 m that do not get colder than 12–13 °C.
- Late winter phytoplankton boom.
- High diversity of habitats.
- High species richness and low abundance. High number of rare species.
- Mainly species of Atlantic origin with less abundant populations and generally smaller individual sizes.
- High rate of endemism (more than one-quarter of the species).
- Shallow-water rocky bottoms dominated by frondose algae, mainly fucales of the genus *Cystoseira*.
- Lush meadows of the seagrass *Posidonia oceanica*, which cover large areas between 0 and 40 m deep and are a key ecosystem.
- Proliferation of long-lived organisms within scyaphilous benthic communities.
- Low primary production, low fish production, and poor development of higher levels of the food chain (low pressures from top predators).
- Overall, the open waters are largely dominated by small autotrophs, microheterotrophs and egg-carrying copepod species.
- Low levels of herbivory and low levels of chemical defences.

In short, the Mediterranean is an oligotrophic sea (rich in oxygen and poor in nutrients). Nutrient concentration decreases along both the west–east and the north–south directions, resulting in variations in the structure of the pelagic food web. In the same way, the general pattern of species richness decreases from west to east. The Siculo-Tunisian sill (400 m deep) separates two distinct basins, the western and eastern basins, and has been traditionally considered as a geographical and hydrological barrier.

The Mediterranean also contains a high diversity of habitats. A description of the great variety of marine Mediterranean habitats and communities was first given by Pérès and Picard (1958), and later amended by Augier (1982), Bellan-Santini et al. (1994), (RAC/SPA 2006), among others. The basic scheme of classification, coming from these authors and widely adopted, is based on substrate and sediment type, depth, irradiance, hydrodynamic, and plant distribution. Recently Fraschetti et al. (2008) reviewed existing marine habitat classification systems in Europe and proposed a reduced habitat classification for the Mediterranean. This list contains 94 habitats that follow a hierarchical structure and refers to the level on the shore, the primary geological substrate, and common biological assemblages and foundation taxa.

Compared with the Atlantic, the Mediterranean marine communities generally have more species, with smaller individuals having a shorter life cycle (Bellan-Santini et al. 1994).

Rocky reefs, *Posidonia oceanica* meadows and coralligenous assemblages are among the most productive habitats, sustaining high biodiversity due to their heterogeneity and three-dimensional complexity. They supply food resources, nurseries and shelters for a variety of organisms.

Loss of Mediterranean Marine Biodiversity in Recent Times

Present day biodiversity is undergoing rapid alteration under the combined pressure of both global change and human impact. There is a growing evidence that human activities are directly or indirectly resulting in the extensive loss of biodiversity and in an impoverishment of the Mediterranean marine biota. Human-dominated Mediterranean ecosystems are experiencing accelerating loss of populations and species, and significant changes in food webs due to anthropogenic disturbances and global change are evident.

Many biologists or divers who have observed marine communities over a period of time have noted local disappearance of some species or significant declines of their populations, but regrettably they do not have accurate data to confirm or refute these personal impressions. No species are known to have totally disappeared from the Mediterranean in recent years, but many species may have disappeared unnoticed. A possible example may be the hydroid *Tricyclusa singularis*, that has never been recorded after its original description (Boero and Bonsdorff 2007), and several species of elasmobranch have virtually disappeared from some areas of the Mediterranean (Ferretti et al. 2008) In any case, it is very difficult to confirm the disappearance of a species in the marine environment. But, of course, extinction of an entire species in merely the end point of an incremental process of extinction of individual populations. As a species loses its component populations it also slowly loses its ability to adapt, its role in the ecosystem and, ultimately, its ability to survive. Local depletion of some species has been pointed out in some places. As an example, in the case of the Venice lagoons, where 141 algal species were found in 1938, only 104 were found in 1962, and 95 in 1987 (Zenetos et al. 2002). In the same way, 9 of 14 species of canopy-forming algae have disappeared since the 1930s from rocky shores of the northwestern Mediterranean, most of the genera *Cystoseira* and *Sargassum* (Thibaut et al. 2005).

Nevertheless, global or regional loss of species are only the last steps of marine biodiversity decrease (Sala and Knowlton 2006), but ecological (or functional) extinction occurs long before species completely disappear. Population declines precede ecological extinction, that occurs when a species is so rare that it no longer fulfills its natural ecosystem function and, hence, becomes ecologically irrelevant. Furthermore, size/density population declines involve genetic diversity loss, that will likely affect all levels of biodiversity, since genetic diversity is the first level of biodiversity. Low genetic diversity may enhance lower resistance to disturbance.

Natural Changes (Cyclical, Episodic or Catastrophic)

Nature is subject to continuous changes, which not always happen gradually over the long-term. Some of them are cyclical or respond to episodic disturbances. Causal events may be unusual in a human scale, but common in an evolutionary time-scale. Fluctuations or pulses in populations and ecosystems are part of the natural dynamics of ecological systems. In the Mediterranean marine climate is characterized by oscillatory variation with a periodicity of ca. 22-year (Duarte et al. 1999). Changes in biota may also occur in cycles of variable duration (Southward 1995).

Many species show episodic recruitment pulses. After long periods of steady growth of the populations of some species, they may experience mass mortality events. These high mortality events can lead to episodes of high recruitment that result in the recovery of populations and ultimately regulate their dynamics and structure (Navarro et al. 2011). Cyclical or episodic mortality events may serve to renew populations, to maintain diversity preventing aggressive species from outcompeting other species, and to allow the persistence of a mosaic of successional stages (Cerrano et al. 2000). The study of catastrophic episodic events may contribute to the understanding of life history patterns of many long-lived organisms (Boero 1996; Navarro et al. 2011). Otherwise, episodic disturbances may act synergistically with punctual small-scale anthropogenic disturbances to put at the brink of disappearance highly complex and species-rich benthic communities.

Natural disturbances, such as extreme storms, diseases or herbivore outbreaks, are well known and there are numerous examples in the literature. Recurrent diseases have been noted in species of some groups, such as bivalves (*Spondylus gaederopus* and some arcids), sea urchins (*Paracentrotus lividus, Arbacia lixula, Sphaerechinus granulatus*), or large corneous demosponges (Cerrano and Bavestrello 2009). By contrast, outbreaks of common sea urchins can completely remove erect algae and other organisms, producing barren grounds. On the other hand, severe storm events cause massive mortalities of benthic organisms by direct wave action, or as a consequence of sediment and boulder displacement (García Rubiés et al. 2009). Such extreme severe storms normally occur every 40–50 years, but their frequency and/or intensity may be more sensitive to global warming. Aumann et al. (2008) pointed out that frequency of exceptional storms increases globally at the rate of 6 % per decade, whereas Jiménez et al. (2012) noted that storm-induced damaged has increased at a rate of about 40 % per decade in the NW Mediterranean since the 1950s.

Tu sum up, when analyzing the changes caused by human activities it should be taken into account that some of the observed changes could be just natural episodic or cyclical changes.

Main Anthropogenic Impacts

Human activities are now the strongest driver of change in marine biodiversity at all levels of organization; hence, future trends will depend largely on human-related threats (Sala and Knowlton 2006). Most human activities have local impacts, but some and, above all, the overall total acting synergistically may have global impact through cumulative processes. The building of the Aswan High Dam in 1968 may not only have deleteriously influenced the productivity, biochemistry and food web structure in the Nile delta and Eastern Mediterranean, but also the hydrological functioning and structure of the whole of the Mediterranean, which itself will influence the chemical and biological characteristics in a feedback loop (Turley 1999). Likewise, the man made Suez Canal (in 1869) has allowed the entry of hundreds of exotic species into the eastern Mediterranean, many of which have already spread over the entire basin.

The most widespread human activities causing impacts on biodiversity are coastal development, overfishing and exploitation of living resources, pollution and introduction of exotic species, which are commented below.

Coastal Development

Habitat modification, fragmentation, degradation and loss are widely considered some of the most serious threats to all components of diversity, structure and functioning of marine coastal ecosystems and to the goods and services they provide (Claudet and Fraschetti 2010). Fragmentation of natural habitats is itself a major cause of local species extinction, even if the surviving patches of habitat still look healthy. Coastal development, along with pollution and sediment loading have notably reduced the extension of important habitats for marine diversity, such as seagrass meadows. The loss of habitat structure generally leads to a decline in species richness and to reductions in overall abundance and biomass. This raises the question of a significant loss of biodiversity, the impoverishment of many species and the fragmentation and decline of their populations. Ultimately, it results in a large-scale degradation, simplification and homogenization of marine communities. In the transition from complex native habitats to simplified habitats, many resident and characteristic species are replaced by fewer opportunistic species with generalistic and/or invader traits (Airoldi et al. 2008).

Due to loss and fragmentation of habitats, connectivity between populations of the same species (genetic flux) is interrupted or reduced. Connectivity is increasingly recognized as a key conservation objective because of its importance for species replenishment (Saenz-Agudelo et al. 2011). Moreover, when the population density of many species decreases, their reproductive success is also considerably reduced, especially in the case of broadcasting invertebrates ("Allee effect") (Courchamp et al. 2008). This has important impacts on population stability and community dynamics. Populations become more isolated and decrease in size and abundance due to fragmentation of habitats, and there is a strong relationship between population size/density and fitness. Changes in fertilization success, larval supply and recruitment play a major role in future population dynamics, and in long-term viability of species.

Recruitment patterns govern the establishment of new populations and hinder the extinction of local populations. If there are no suitable habitats for recruitment, larvae of many species will die without having found an appropriate place to settle. The decrease in available habitats for a species leads to a reduction in the number of both donors and potential recipient populations. Furthermore, the local extinction of species without dispersal ability prevents their recovery. The role of marine protected areas (MPAs) as exporters of larvae has been highlighted often, but this is not much use if the larvae that are exported do not find suitable places to settle outside of MPAs. At the same time, populations within MPAs will suffer a progressive genetic impoverishment if they only act as donors but do not import larvae from surrounding areas.

Overfishing and Exploitation of Living Resources

The most important negative consequence of fishing activities is the degradation of marine ecosystems by removal of target or non-target species and by physical disturbance inflicted by some fishing gear. Most of the major impacts of fishing on the ecosystems recorded around the world severely affect the Mediterranean. They vary from local effects on the sea bottom caused by trawler gear to large-scale impacts on cetacean populations driven by driftnet bycatch. The huge diversity of fishing gear and practices, the very high intensity of fishing, and the presence of a vast array of large vulnerable species (that include emblematic sharks, turtles, whales and seals) make the Mediterranean an area especially sensitive to the effects of fishing (Tudela 2004). Fisheries in the Mediterranean are diverse, catches are usually low and sometimes show marked seasonal differences.

In short, fishing has resulted in the overexploitation of several fish stocks and has notably increased in the past decades, but the depletion of species is also evident on historical time scales. As for particular fishing gear, bottom trawling, longlining and driftnets appear as those with most impact on marine ecosystems over the whole Mediterranean region. Bottom trawling causes continuous alteration and destruction of continental shelf benthic communities and produces large amounts of discards of benthic invertebrates, accounting for up to 80 % of the catch. Additionally, it can be responsible for the alteration of the benthic biogeochemical cycles (Pusceddu et al. 2005).

Evidence shows that the effects of fishing in the Mediterranean go far beyond the isolated impacts on overfished target species, vulnerable non-commercial groups or sensitive habitats. The effects of fishing are also conspicuous at the systemic level, and constitute the major factor causing changes in the food webs in the Mediterranean (Sala 2004; Coll et al. 2008). A holistic approach should therefore be adopted if the overall changes to the structure and the functioning of marine ecosystems caused by fishing are to be remedied (Tudela 2004), but multispecies catch hinders the implementation of appropriate management actions.

On the other hand, overexploitation by intensive and long time harvesting has led to a serious decline in some benthic invertebrates, such as the red coral *Corallium rubrum* (Santangelo et al. 1993), date-mussel *Lithophaga lithophaga* (Fanelli et al. 1994), large demosponges (Pronzato and Marconi 2008) or many shellfish (Volulsiadou et al. 2009). Some aggressive practices affect rocky bottoms, such as dynamite fishing for coral and date mussels.

Pollution

Eutrophication and pollution arise from agriculture, industrial activity, tourism and human population growth that have become serious problems in most of the densely populated and industrialized regions of the world. Hydrocarbon spills, heavy metal contamination, and their biological effects cause increasing concern in the Mediterranean Sea. Furthermore, special attention is now being paid to the "new pollution" processes, that is the introduction of novel substances with biological activity that might have synergic effects with "classical pollutants" (Danovaro 2003). The immunosuppressive effects of contaminants arising from agriculture, industrial activity and human population growth may have contributed to the severity of mass mortalities among marine mammals, and the additional chronic effects of organochlorines could hinder, or even prevent, recovery of individuals from pathogenic disease. All this leads to an impoverishment of species diversity, an increase in bioproductivity, and the intensification of potentially toxic cyanobacterial blooms. Furthermore, the presence of macro- and micropollutants may have a strong impact on microbial-loop functioning and cause the increase of viral infection (Danovaro et al. 2003).

On the other hand, water transparency shows a declining trend of -0.1 m year^{-1} in the NW Mediterranean (Marbà and Duarte 1997), that provides further evidence for deterioration of water quality along the Mediterranean littoral. The increase of water turbidity in the last decades can be the result of an increase in suspended sediment in the water, an increase in suspended particulate organic matter, or both. The decreasing water transparency has important consequences for the ecosystems, mainly in their vegetal components. Many algal species, but also *Posidonia oceanica*, are sensitive to a reduction in water transparency due to pollution and turbidity. The decline or loss of some deep-water species, mainly fucales, such as *Cystoseira spinosa* and *C. zosteroides*, has been detected in recent decades (Thibaut et al. 2005) accompanied by a tendency to shift their deeper distribution limits to an upper level.

Species Introduction

Current changes and human activities are favouring the increasing introduction of non-indigenous species ("alien species"). In the last decades, enhanced mainly by the opening of the Suez Canal, aquaculture and ship transport, hundreds of alien species reached and established themselves in the Mediterranean. Most of the species introduced to the basin are thermophilic originating in the tropical Indo-Pacific (Lessepsian migration). Arrival of alien species can be taken as natural when considering the Atlantic newcomers from the newly active Gibraltar Portal, but anthropogenic when considering Lessepsian migration from the Red Sea through the artificial Suez Canal, or the many species actively or passively introduced by humans. The list of exotic species that have invaded the Mediterranean Sea is continuously increasing, particularly in the eastern basin. A growing bulk of literature on the subject has been published in recent years, with subsequent revisions. Today more than 600 alien species have been reported in the Mediterranean (Galil 2008), or nearly 1,000 if we consider newcomers of Atlantic origin (Zenetos et al. 2010). Their number has nearly doubled every 20 years since the beginning of the twentieth century (Galil 2008, 2009). Most alien species are lessepsian migrants coming from the Red Sea, with an additional set of species from other tropical areas. Therefore, the bulk of the introduced species are of tropical origin and they have been confined for long to the easternmost Levantine shores, but the general warming of the Mediterranean favours their spread (Occhipinti-Ambrogi 2007).

The lack of evidence of species extinction coupled with the establishment of alien species is apparently leading to an increased species richness of the Mediterranean. Therefore, a biodiversity increase might be perceived as a positive consequence of alien arrival and establishment, especially in the eastern basin, where some thermophilic species have now attained commercial relevance (Galil 2007). Nevertheless, the spread of these species may lead to biotic homogenization, increasing risk of local extinction of native species, reduction of genetic diversity, loss of ecosystem functions, and alteration of both habitat structure and ecosystem processes (Boero et al. 2008). According to Zenetos et al. (2010) at least 134 alien species are or may be invasive in the Mediterranean, i.e. with a negative effect on native species and communities.

Other Minor or Local Impacts

A number of other minor or local impacts (beach replenishment programmes, marine farms, diving, anchoring, hypersaline brine discharges from desalination plants, …) do not cause significant damage on their own, but the sum of them all acting synergistically prevents the recovery of the entire marine ecosystem. Cumulative human impacts and synergistic interactions between multiple stressors can exacerbate nonlinear responses of ecosystems to human impacts and limit their adaptive capacity. To this should be added a number of other sporadic or accidental anthropogenic impacts, such as oil spills, accidental polluting, or occasional effluents.

Global Change

Impacts of Global Warming ("Tropicalization"?)

Global warming in now recognised as the predominant threat to biodiversity in many parts of the world. The global warming observed since the end of the nineteenth century has been caused not only by natural climate changes on the decades-centuries scale but also by the impact of human activities on the Earth's climate system. The enclosed Mediterranean basin is a miniature ocean where the effects of climate change are likely to be more apparent earlier than in other more open seas and oceans (Coll et al. 2010). In the Mediterranean, a positive trend in temperatures is clearly seen after the mid 1970s (0.040 °C/year for surface water and 0.025 °C/year at 80 m depth since 1975), and the number of days that the temperature was >18 °C shows an increasing trend of 1.25 days/year since the mid 1970s (Coma et al. 2009). According to Marbá and Duarte (2010), the average annual maximum temperature for the first decade of the present century was 1 °C above temperatures recorded for the last decade of the last century. In addition, over this general trend, cyclical oscillations are overlayed, and thermal anomaly events are increasingly frequent in recent years (Coma et al. 2009; Garrabou et al. 2009; Crisci et al. 2011). At least three well-documented heat waves impacted the NW Mediterranean in the summers of 1999, 2003 and 2006. The occurrence of heat waves, rather than a gradual, smooth elevation of the temperature also reduces the opportunities for adaptive changes. Furthermore, the frecuency of other extreme events (severe storms, droughts, floods) has increased in recent decades and is likely to further increase in the future (Calvo et al. 2011).

Rapid warming of the Mediterranean and heat waves threaten marine biodiversity and, particularly, marine ecosystems already stressed by other impacts, such as *Posidonia oceanica* meadows. A clear relationship between increasing seagrass shoot mortality rates and increased warming has been highlighted by Marbà and Duarte (2010), even in pristine meadows, which provides evidence that warning alone has the potential to cause abrupt mortality in *Posidonia* meadows. According to these authors, a pronounced *P. oceanica* shoot mortality occurs when seawater exceeds 28 °C associated with the occurrence of heat waves. Thus, they observed that those heat waves recorded in 2003 and 2006 lead to a mortality of 13 % of the *Posidonia* shoots population the subsequent year at Cabrera Island (a protected marine area in the Balearic Islands).

As a result of warming, one expectation could be a homogenisation of the Mediterranean biota, disrupting present biogeographical entities (Lejeusne et al. 2010). Climate change is causing shifts in the geographical distribution of species to keep within their thermal regime, associated with local extinction and contraction/expansion of their distribution ranges. Therefore, the most evident phenomenon correlated with global warming is the successful geographical spread of species of warm water affinity (Boero et al. 2008). These distributional changes are clearly evident for highly mobile species (mainly fish), but also for some benthic invertebrates (Fig. 28.1). Coinciding with the recent change of temperature, many native thermophilic species appeared or became common in northern sectors, where they were formerly absent or rare (Bianchi and Morri 1993; Francour et al. 1994; Despalatović et al. 2008; Puce et al. 2009). More than 30 warm-water indigenous fish species have now been reported north of their last geographical distribution (Boero et al. 2008). However, northward displacement in the Mediterranean Sea is limited due to the physical presence of the European continent. Concurrently, cold water species have practically disappeared from many areas of the Mediterranean. Besides, the warming of the Mediterranean waters could also modify the migration pattern of some species, such as bluefin tunna *Thunnus thynnus* and the amberjack *Seriola dumerili*, which have lengthened their stay in northern and central Mediterranean waters before migrating towards their winter areas (Bombace 2001).

Fig. 28.1 Many species of warm water affinity, such as the sea star *Ophidiaster ophidianus*, are spreading to the northern sectors of the Mediterranean Sea (Author: J.C. Calvín)

In the Mediterranean Sea climate change combines with Atlantic influx, lessepsian migration and the introduction of exotic species by humans lead to the establishment of tropical marine biota (Bianchi 2007). The Suez Canal serves as a gateway for natural migration but also facilitates the expansion of ship-borne fouling biota. There are thousands of species that settle in the Mediterranean coming from the Red Sea in a stepwise ("Lessepsian migration") or as one-jump noxious immigrants ("Erythrean alien") (Galil 2006). There is also an increasing settlement by tropical Atlantic newcomers entering the Mediterranean through the Strait of Gibraltar. In this sense, Ben Rais Lasram and Mouillot (2009) pointed out that the currently warmer Mediterranean is acting increasingly as a "catchment basin" for southern species. They recorded 127 thermophilic fish species that supplemented in the last decades the Mediterranean fauna, 65 lessepsian species and 62 coming from the Atlantic. Initial northward and westward expansion of the range of Atlantic and Levantine species, respectively, has also been detected (Bianchi and Morri 1993; Guidetti and Boero 2001; Azzurro 2008; Relini 2009). Furthermore, several tropical invaders have already reached the northernmost sectors of the Mediterranean Sea (e.g. Dulčić et al. 2008; Francour and Mouine 2008; Kružić 2008; Daniel et al. 2009; Katsanevakis et al. 2011).

A warming Mediterranean is becoming more receptive also to species arriving accidentally with ship ballast and other artificial ways, adding to the number of successful establishment cases. This influx of thousands of tropical species into the Mediterranean is one of the most remarkable biogeographic phenomena of today. Observed changes due to climate change involve both indigenous species subjected to a process named "meridionalization" and non-indigenous species, subjected to a process defined as "tropicalization" (Boero et al. 2008). Such changes are causing a progressive homogeneization of Mediterranean marine biotas (Philippart et al. 2011). Some authors consider the use of the term "tropicalization" to be exaggerated (Lejeusne et al. 2010) and suggest that "meridionalization" is a more realistic term to refer to the increase of the proportion of thermophilic species.

On the other hand, if the Mediterranean Sea is becoming 'tropical' those species of boreo-Atlantic origin that entered the Mediterranean during glacial periods and established themselves in the northern colder areas of the basin cannot migrate further northward due to the geographic configuration of the Mediterranean. Therefore there is the risk of extinction of these cold-water species from the Mediterranean. Rarefaction or even disappearance of cold-water species from their refuges in the Gulf of Lion-Ligurian Sea and the northern Adriatic has already been recorded (Chevaldonné and Lejeusne 2003; Boero and Bonsdorff 2007; Ben Rais Lasram et al. 2010). On the other hand, the colder and less saline western basin of the Mediterranean Sea, will become more similar to the eastern basin, allowing exotic and native warm-water species to spread and thrive in the northern areas. As a result, sub-regional peculiarities in biodiversity might eventually disappear, leading to a taxonomic, genetic and functional homogenization (Boero et al. 2008).

Climate warming can also affect competitive interactions between native species of different thermal affinity (Chevaldonné and Lejeusne 2003). For example, the increase of *Sardinella aurita* in the western Mediterranean might have contributed to the decrease of both the anchovy *Engraulis encrasicolus* and the sardine *Sardina pilchardus* (Sabatés et al. 2006). An illustration of cascading effects caused by global warming is exemplified by the competitive interactions between the common native sea urchins *Paracentrotus lividus* and *Arbacia lixula* and their effects on the invasive coral *Oculina patagonica*. In the Mediterranean Sea both urchins coexist in rocky shallow waters and are considered the main controlling factors of the dynamics, structure and composition of infralittoral macroalgal assemblages. *Arbacia lixula*, a thermophilic species, seems to exert a major role in barren areas formation and maintenance. The two species have a clearly different thermal optimum at the larval stage: *A. lixula* showed higher larval survival at higher temperature than *P. lividus* (Privitera et al. 2011). This allows us to predict a potential decline of *P. lividus* populations, while *A. lixula* should benefit from higher temperatures, as indicated by higher recruitment after particularly hot summers. In turn, this leads us to expect an increase of the barren areas. Furthermore, Coma et al. (2011) found evidence that barren areas facilitate the recruitment and proliferation of the alien coral *Oculina patagonica*, which shows the synergistic effects of global warming on the dynamics of invasion of exotic species.

Even populations in protected areas are susceptible to global changes, because of the pervasive effects of temperature shifts. Therefore, habitat protection alone may not be enough to ensure persistence of a population and so the conservation and harvesting management of species requires the development of objective criteria to assess species' vulnerability to climate change (Lo Brutto et al. 2011).

Impacts of Acidification

Around half of all extra CO_2 produced so far by human activities has dissolved in the oceans. The anthropogenic increase in the concentration of atmospheric carbon dioxide is reducing the pH of seawater on a global scale (a process termed "ocean acidification"), changing the chemistry and biogeochemical cycling of carbon and carbonate. A growing number of studies have demonstrated adverse impacts of acidification on marine organisms. The combination of elevated temperature and acidification has been shown to be detrimental to the calcification process, hence marine organisms with calcareous skeletons, shells or plates will experience problems. These include major bioconstructing organisms, such as scleractinian corals, bryozoans or red coralline algae, but also molluscs, crustaceans, echinoderms, foraminifera and some calcifying phytoplankton (mainly coccolithophorides) as well as zooplancton (Pteropods and larvae of many groups). Ocean acidification reduces the abundance of carbonate ions, which are an essential component of the mineral calcium carbonate that these organisms use to build their protective shells and skeletons. An update of the impacts of acidification on biological, chemical and physical systems in the Mediterranean and Black sea was published by CIESM (2008). Recently an estimation of a pH decrease of up to 0.14 units has been identified in the western Mediterranean Sea since preindustrial times, which is of higher magnitude than the global surface ocean decrease of about 0.1 units over this time period (Calvo et al. 2011).

Ocean acidification is also predicted to reduce the absorption of low frequency sound, leading to a noisier environments for marine mammals (Hester et al. 2008). On the other hand, recent studies suggested that jellyfish increase in abundance as pH declines, probably due to the negative impact of more acidic conditions on calcifying plankton, opening up an ecological space for gelatinous plankton (Attrill et al. 2007).

Episodic Events

Mass Mortalities

Temperature anomalies and higher sea surface temperatures have severely impacted entire shallow coastal ecosystems, causing the elimination of sensitive species as well as mass mortalities (Boero et al. 2008). Increasing frequency, severity and expansion of mass mortalities related to abnormal summer heat waves and to seasonal stratification were observed in different parts of the Mediterranean. Two major well-documented multispecies mass mortality events impacted the NW Mediterranean after the summer heat waves of 1999 and 2003 (Cerrano et al. 2000; Perez et al. 2000; Garrabou et al. 2001, 2009; Coma et al. 2009; Crisci et al. 2011). They affected many long-lived filter-feeders, structural invertebrates (Fig. 28.2) over several hundred kilometres of coastline in this area. Structural species provide biogenic structure, so other species not directly affected by the mass mortalities could also indirectly suffer the impact through modification of habitat conditions (Garrabou et al. 2009). The bottlenecks induced by mass mortality events reduce the genetic diversity of these key structural species and thus influence their next evolutionary trajectory. This strongly suggests that temperature anomalies, events of short duration, can, directly or indirectly, dramatically change the structural complexity of Mediterranean biodiversity, because of their essential role in maintaining the structure and species composition of the communities where they live. The lush coralligenous community (Fig. 28.3) has been one of the more affected in recent times by mass mortality events of some of its main structural species. Furthermore, it must be taken into account that although species known to have been affected by mass mortalities are structural or conspicuous species, small, cryptic, or poorly known species may be affected also by such events.

In addition to these two multispecies mass mortality major events, other mass mortality events that affected fewer species or were geographically restricted have been recorded in the Mediterranean Sea. For instance, the three common littoral Mediterranean sea urchins (*Paracentrotus lividus, Arbacia lixula* and *Sphaerechinus granulatus*) have been sometimes

Fig. 28.2 The scleractinian coral *Cladocora caespitosa* is one of the long-lived filter-feeders structural invertebrates affected by mass mortalities after heat waves (Author: D.K. Kersting)

Fig. 28.3 The lush and colorful coralligenous community has been affected in recent times by mass mortality events of some of its main structural species (Author: J.C. Calvín)

involved in local mass mortalities (Miller 1985, obs, pers.). Sometimes such mortalities are caused by disease, possibly linked to overcrowding (Boero 1996). Episodic disease outbreaks of large Mediterranean sponges (mostly Dictyoceratida) have been documented also (Pronzato 1999; Maldonado et al. 2010; Cebrian et al. 2011). The bivalve *Spondylus gaederopus* (Fig. 28.4) suffered widespread mortality in the 1981–1983 summers (Meinesz and Mercier 1983). Because there were no temperature anomalies during these summers Meinesz and Mercier (1983) hypothesised a viral, bacterial or fungal infection. Local mortalities of this bivalve have been reported later (Kersting et al. 2006), along with that of other bivalves, such as the arcids *Arca noae* and *Barbatia barbata*.

Thermal stress may directly affect the physiology of organisms or reduce their resistance, resulting in numerous diseases affecting natural populations (Cerrano and Bavestrello 2009). Mass mortalities may be due to the exposure to lethal temperature and/or to pathogenic microorganisms, more probably to physiological stress that makes them more susceptible to opportunistic, residential, and/or pathogens, including bacterians, some fungi and protozoans (Cerrano et al. 2000; Coma et al. 2009). Many pathogens have the capacity to cause disease only in a host under stressful conditions (such as increased seawater temperature) that can trigger mortality by inducing a microbe to be more virulent or the host to become more vulnerable.

Fig. 28.4 The bivalve *Spondylus gaederopus* suffered widespread mortality in the 1981–1983 summers and local mortalities have been reported recently (Author: D.K. Kersting)

In recent years mortality of the coral *Oculina patagonica* has been caused by the pathogenic bacterium *Vibrio shiloi* (Rosenberg and Loya 1999). After the 2003 mass mortality, a dominant strain affecting the Mediterranean gorgonian *Paramuricea clavata* was identified as *Vibrio coralliilyticus*, previously identified as a thermodependent pathogen of a tropical coral species (Balli and Garrabou 2007).

Predicted global warming leading to long-lasting hot summer periods together with stratification resulting in energetic constraints, represent a major threat to the survival of benthic invertebrates in the temperate NW Mediterranean Sea due to potential disease outbreaks associated with *Vibrio* pathogens (Vezzulli et al. 2010). Bacteria belonging to the genus *Vibrio* are of particular concern as they constitute a considerable part of marine halophilic bacterial populations, are strongly thermodependent and are often associated with human and marine animals' diseases, including diseases of several benthic organisms, such as corals and bivalves. With predicted further warming of sea surface temperatures over the coming century, the severity of disease in invertebrate populations in the NW Mediterranean Sea is likely to increase, placing further pressure on the health of this marine ecosystem.

Jellyfish Blooms

Although cyclical or episodic dense jellyfish aggregations are a natural feature of pelagic ecosystems (Piraino et al. 2002), a picture is now emerging of more severe and frequent outbreaks in many areas. Furthermore, mounting evidence suggests that the structure of pelagic ecosystems can change rapidly from one that is dominated by fish to a less desirable gelatinous state, with lasting ecological, economic and social consequences (Richardson et al. 2009). Available evidence suggests a suite of human activities might act separately or synergistically to result in outbreaks of some jellyfish species. Of these human activities, overfishing, eutrophication and climate change seem to play a role in increasing jellyfish blooms, particularly in coastal areas. The results are that the energy that previously went into production of fish in the pelagic ecosystem may be switched over to the production of Cnidaria and Ctenophora.

Jellyfish and fish interact both as predators and competitors of each other. Many fish species compete for the same zooplankton prey as jellyfish, some fish are predators of jellyfish, and jellyfish also predate on fish larvae and eggs. Modern commercial fishing removes predators and competitors of jellyfish with increasing efficiency, enabling many gelatinous zooplancton species to proliferate (Sabatés et al. 2010). On the other hand, nutrients added to the coastal zone are rich in nitrogen and phosphorous, but poor in silica. Under such conditions non-siliceous phytoplankton, such as flagellates (that include some harmful red-tide species), proliferate and replace diatoms, resulting in a reduction in the size of primary and secondary producers. This modifies and simplifies the pelagic food webs. Diatoms are the main food of copepods, which are the main prey of planktivorous fish (anchovies, sardines, herrings), which are, in turn, eaten by larger pelagic fishes. Yet, jellyfish have a wide range of prey, including flagellates. Therefore, a future pelagic food web has been hypothesized supporting fewer fish, marine mammals, and seabirds, because of the replacement of diatoms and copepods by flagellates and gelatinous plankton, with a smaller average of food size (Pearson and Lalli 2002). Furthermore, warming of the sea surface can enhance water column stratification, leading to nutrient-poor surface water where flagellates can also outcompete diatoms. Such flagellate-dominated food webs are more favourable for jellyfish than for fish. Warmer waters also accelerate medusae growth and ephyrae production in many species.

Besides, jellyfish have a suite of successful attributes that enable them to survive in disturbed marine ecosystems, such as broad diet, fast growth rates, the ability to shrink when starved, the capacity to fragment and regenerate, and the ability to tolerate hypoxia (Richardson et al. 2009). These are characteristic of opportunistic species that give jellyfish advantages over fish in environments stressed by global change, eutrophication and overfishing.

In the last decades there has been a marked increase in the frequency and extent of jellyfish outbreaks in the whole Mediterranean Sea. The scyphomedusa *Pelagia noctiluca* have been especially noteworthy because the medusae sting and the summer blooms are highly harmful to summer bathers. *Pelagia noctiluca* is an opportunistic predator that consumes a wide variety of prey, but more than 10 % of which are fish larvae (Sabatés et al. 2010). Regular population fluctuations of this species are well known, with population peaks occurring in the past on average every 10–12 years (Goy et al. 1989; Mills 2001), mainly in spring and summer months. Nevertheless, nowadays plagues of this and other jellyfish species are becoming more frequent and happen at any time of year.

Noxious Algal Blooms

Harmful algae blooms are another emerging phenomenon causing health and economic concern, especially in tourist areas. The term 'harmful algal blooms' (HABs) covers a heterogeneous set of events that share two characteristics: they are caused by microalgae (mainly Dinoflagellates) and they have a negative impact on human activities. The concept of explosive high abundance is implicit in the term "bloom". Despite these common features, HABs are very diverse in terms of causative organisms, dynamics of blooms and type of impact (Zingone and Enevoldsen 2000), and have usually been associated with nutrients derived from anthropogenic activities. These blooms are often natural phenomena, but sometimes may be dangerous for human health and deleterious for the commercial exploitation of coastal areas. Problems related to HABs are not only associated with food safety and commercial shellfish activities. Environmental damage, health problems (not associated to shellfish consuming), such as allergic reactions, recreational shellfish harvesting or even aesthetic issues have important social implications and can prevent the use of coastal waters for recreational purposes, causing evident economic damage. Furthermore, a number of HABs affect organisms that may have no commercial value but nonetheless are functional components of the marine ecosystem. The degradation of high biomass blooms can exhaust oxygen supplies, thus killing not only commercially important species, but also other plants and animals that are unable to leave the anoxic area. In short, degradation of coastal water quality by HABs occurrence has direct economical impact on some Mediterranean tourism locations.

Records of toxic benthic dinoflagellates have dramatically increased along the Mediterranean coast over the last decades and the list of harmful species is growing (Garcés et al. 2000). In particular, *Ostreopsis ovata* has bloomed in the Mediterranean region in recent years, with increasing frequency, intensity, and distribution (both in western and eastern coasts of the Mediterranean), causing mortality of benthic organisms and human health problems. In turn, the genus *Alexandrium* is the group of dinoflagellates which causes most HABs, and *A. taylori* is one of the noxious species. It is known that blooms of this species have occurred in the northern Mediterranean beaches since the 1980s (Basterretxea et al. 2005). This dinoflagellate is a non toxin-producing organism but a high-biomass bloom-former that provokes greenish-brown discoloration of the water during the summer months in protected pocket beaches causing an evident water deterioration.

Furthermore, the geographical distribution of some specific harmful benthic dinoflagellates, such as the ciguatera-causing genus *Gambierdiscus*, mainly restricted to circumtropical areas until recently, have spread to temperate regions in recent times, including the Mediterranean sea (Aligizaki et al. 2009; Faimali et al. 2012). This poses a potential risk for a future occurrence of ciguatera in the area.

Proliferation of Mucilages

The production of mucilaginous aggregates is a well-known phenomenon in seawater occurring during intensive nuisance blooms of certain phytoplanktonic taxa, mainly diatoms, blue-green algae and Prymnesiophyceae (Metaxatos et al. 2003). Mucilage is made of expolymeric compounds with highly colloidal properties that are released by marine organisms through different processes and through death and decomposition of cell-wall debris. The dominant components of the algal mucilaginous material are exudates composed by polysaccharides, making the bulk of mucous biomass, but proteins, lipids and unidentified molecules are also components. Mucilages heavily affect marine ecosystems, mainly when the massive development of aggregates covers large areas of rocky substrate, suffocating benthic biocoenoses. Besides, marine mucilage can represent a new, though ephemeral, substrate for microbial colonization, including pathogenic forms (Danovaro et al. 2009), because it is a hot spot of viruses and bacterial diversity. Mucilage, in turn, can induce hypoxic phenomena and even promote extensive anoxia resulting in a decreased production of "ecosystem goods and services".

The proliferation of mucilages is a sporadic and recurrent phenomenon well documented in the Mediterranean, but it has increased almost exponentially in the last decades, spreading to most regions of the Mediterranean basin. Many environmental stresses seem to cause the hyperproduction of exudates (high irradiance, temperature anomalies, water column stability, nutrient availability). According to Innamoranti et al. (2001) nutrients are one of the main causes, particularly when the N/P ratio shifts toward a phosphorous deficiency. Nevertheless, Danovaro et al. (2009) note that mucilage is not necessarily associated with eutrophic conditions, as several mucilage outbreaks have been recently observed in oligotrophic areas. According to these authors, mucilage represents a symptomatic response of the Mediterranean ecosystem to direct and indirect anthropogenic impacts and additionally, a potentially expanding carrier of viruses and bacteria, including pathogenic forms that are harmful to the health of human and marine organisms.

Global Ongoing Changes and Predicted Trends

Humans are greatly altering Mediterranean marine biodiversity in many ways, including climatic change (Bianchi and Morri 2000), which is no longer to be considered as natural because of the increased anthropogenic emission of carbon dioxide (CO_2) and other 'greenhouse' gases into the atmosphere. Climatic models predict that the Mediterranean basin will be one of the regions most affected by the ongoing warming trend and by an increase in extreme events (Lejeusne et al. 2010). Therefore, climate and humans are combining their effects on Mediterranean Sea biodiversity. Since climate change interacts synergically with many other disturbances, there are reasons to believe that the Mediterranean is one of the most impacted seas in the world. An overview of ongoing changes in the Mediterranean Sea is summarized below:

Physico-chemical trends

- Temperature increase and higher frequency of short-term extreme events. Increase of heat waves, storm frequency and changes in wind speed and direction.
- Decline of water transparency (increase in water turbidity).
- Longer and deeper summer thermocline.
- Strengthening of the stratification and slowing down of thermohaline circulation.
- Changes in circulation patterns.
- Shift of the downward transfer of carbon (dioxide) by means of the "biological pump".
- Appearance of areas of hypoxia and stratification processes associated. Emergence and spread of anoxic zones.
- Increase of nutrient concentrations.
- Progressive acidification and decrease of the carbonate ion concentration.
- Decrease in the capacity to absorb atmospheric CO_2.
- A noisier environment for marine mammals.
- Reduction of sediment delivery due to widespread construction of dams.
- Towards a permanent stratification of the basin?

Biological trends

- Homogenization and impoverishment of the marine biota.
- Shift from endemic to "common" species.
- Extinction of some rare species.
- Decline of species abundance and increase of microbial communities.
- Loss of genetic variability.
- Decline of *Posidonia oceanica* meadows, fucoid algal beds and many other key species.
- The decline of *P. oceanica* meadows involves a reduction in natural carbon sink capacity.
- Replacement of canopy erect algae on shallow rocky shores by turf forming or filamentous algae cover and sea urchin barrens.
- Changes in the life cycles of the species, in their reproductive effort and in their demography. Progressive reduction of the size of species.
- Progressive decline of "cold" stenothermal species from the surface layer, which move to deeper layers, and increasing abundance and range extension of thermotolerant species.
- "Tropicalization" or "meridionalization" of the biota. Northward extension and increase in the abundance of native thermophilic species (meridionalization). Increase in arrival, establishment and range extension of tropical exotic species (tropicalization).
- Subtle adaptive responses (such as physiological adjustments and microevolutionary processes).
- Increase in primary production and shift in food web structure.
- Shift towards non-siliceous phytoplankton (from diatoms to coccolithophorides and flagellates).
- Shift from fish to jellyfish (decline of small pelagic fish and increase of gelatinous plankton).
- Modification of both bottom-up and bottom-down interactions.
- Spread of mucilages.
- Increasing episodes of mass mortalities.
- Increase in blooms of potentially toxic blue-green algae (cyanobacteria) and toxic dinobionts or harmful algae in coastal areas.
- Complex interactions and synergies between the various disturbance factors.

Human activities can lead to homogenization and impoverishment of ecosystems due to reductions in diversity within functional groups, food-web complexity, distribution range, biogenic habitat structure, and size of organisms (Claudet and Fraschetti 2010). Entire ecosystems may cease to function in their current form, potentially leading to a loss of the goods and services derived from them. Lost canopies of frondose algae (mainly fucales) tend to be replaced by species with lower structural complexity, such as turf-forming, filamentous or other ephemeral seaweeds, mussels, or sea urchin "barrens" (Perkol-Finkel and Airoldi 2010), and the extensive *Posidonia oceanica* meadows are being progressively decimated (Boudouresque et al. 2009). These meadows are considered to be among the most important Mediterranean marine ecosystems, with regard to both ecology and biodiversity, and for the services they provide.

In turn, the Mediterranean Sea is undergoing a rapid and dramatic transformation from a warm-temperate region to a warmer sea in which thousands of tropical species are becoming established owing largely to the human-constructed Suez Canal (Vermeij 2012). This so-called "tropicalization" (Bianchi 2007) will restore a biogeographical link between the Mediterranean and Indo-West Pacific (IWP) that last existed during the Middle Miocene (until 16 mya). This new tropical signature will, however, differ substantially from that prevailing during the warm Early Pliocene (5–3 mya), when most tropical elements in southern Europe had West African affinities. The future biota of a particular region depends on the species pools from which spreading taxa are drawn as well as on the nutritional and evolutionary regimes in both the donor and recipient regions (Vermeij 2012). So, if further warming should reduce the barrier of cold Canary Current and Saharan upwelling, as occurred during the productive and warm Early Pliocene, the tropical Atlantic elements will again be able to enter the Mediterranean as they did during its reflooding after the MSC. Moreover, taxa entering through the Suez Canal may spread toward West Africa. An increasingly oligotrophic Mediterranean will probably be more receptive to IWP immigrants from the Red Sea than to eastern Atlantic newcomers from the productive waters of West Africa, but a future Mediterranean biota with both African and IWP elements can be expected. Thus, the Mediterranean could become a fascinating meeting place (Vermeij 2012) for two tropical biota of contrasting source origins (tropical West Africa and Indo-Pacific), which have had separate histories for at least 16 million years.

The different driving forces of change act on very different time scales, but all have apparently accelerated in the last decades. According to Bianchi (2007), the time scale of the Atlantic influx is of the order of 10^4 years (since the beginning of the last interglacial), the introduction of exotic species by humans has acted on a time scale of 10^3 years, the time scale of lessepsian migration is of the order of 10^2 years (it started soon after the opening of the Canal in 1869), and finally sea warming has a time scale of 10^1 years, despite large cyclic fluctuation.

Concurrent expansion of the range of warm-water species and northward contraction of that of cold-water species are disrupting the present biogeographic patterns within the basin, and some authors have drawn attention to the homogenisation of the Mediterranean biota (Bianchi 2007; Boero et al. 2008; Lejeusne et al. 2010; Philippart et al. 2011). The idea of fading the commonly accepted biogeographic boundary between the western and eastern Mediterranean basins began to be raised in favour of a series of gradients in a south–north direction.

Uncertainties

Whereas the effects of each of the different impacts are now well documented, major unknowns remains as to how they will ultimately affect the functioning of ecosystems through cascade effects. The interaction of all these impacts acting synergistically may be greater than the sum of individual impacts. It is therefore necessary to know how all of them interact.

According to Por (2009), the Mediterranean is possibly and partly reverting to its original tropical warm-water biological condition, which was only relatively recently interrupted by the start of the glaciation cycles about 2.6 million years ago. Under a geological perspective, what is now happening is considered by this author to be a return to normal conditions, possibly a normalization event. The Present Climatic Optimum represents a return to the Pliocene Climatic Optimum and thus, can be seen as a repetition, a cyclic event and not as an artificial disruption (Por 2009). However, it should be noted that marine communities change in a progressive way, and do not fluctuate around a hypothetical "optimal" or "typical" configuration (Bianchi and Morri 2000). The fact is that the observed ongoing changes lead to a progressive degradation and impoverishment, which seem to veer away from the idyllic rich tropical sea of the past.

It is impossible to foresee to what extent the exuberance of warm-water species will affect the composition of marine communities, the trophic webs and the functioning of the whole Mediterranean ecosystem. While Mediterranean communities are being modified in their species composition pattern, they do not seem to be acquiring a more marked tropical physiognomy (Bianchi 2007). Mediterranean coastal marine ecosystems are still dominated by frondose algae (even if the species that are gaining ascendency are of tropical origin) and not by corals, as is the rule in tropical seas. Instead of coral reef, different Mediterranean organisms build significant monospecific or oligospecific bioconstructions, mainly coralline algae, plus a few vermetid gastropods, briozoans, serpulids or corals. But corals or other constructional organisms are not becoming more abundant; on the contrary they are perhaps going to face more frequent mass mortalities or weakening, such as the scleractinian coral *Cladocora caespitosa* (Bianchi 2007). So, the Mediterranean biota might loose in the near future what have been called their peculiarities and acquire a different and unprecedented configuration structure.

What Is the Role of Rare Species?

Rarity is a common state in the life of most species (commonness of rarity). A significant proportion of biodiversity at species level is composed of small and inconspicuous rare species. Most communities are composed of a few dominant species and a high number of rare species. Nevertheless, rare species are the neglected component of biodiversity and they are often considered as 'noise' in ecological studies and ignored (Boero 1994). Likewise, biodiversity conservation focuses mainly on a few charismatic species (generally either vertebrates or large and conspicuous invertebrates) and disregards what represents the bulk of biodiversity: a host of small rare species (Piraino et al. 2002). On the contrary, the importance of rare species could be crucial as a reservoir of potential diversity and provide the information for the possible future composition of a community after changes in environmental conditions. Boero (1994) argued that very small populations might represent the last representatives of a declining species or the first stages of an emerging one, or finally the normal abundance in a given time window. Rare species are the insurance for the continuation of biological diversity and, at least some of them, will take the place of species that are common now. Therefore, they are particularly important from the point of view of conservation, ecology and evolutionary biology (Lim et al. 2012). A flexible species composition may allow ecosystems to maintain their functioning unfalteringly (Bianchi and Morri 2000) and a high number of species serve as a buffer against the effects of environmental impacts and enhance the recovery potential of communities. Ecosystems with more species perform better and have higher resilience (the ability to recover after adverse impacts). Focusing conservation on a few charismatic species is a very biased and incomplete view of the problem. To understand the magnitude of changes that biodiversity is undergoing today, it is necessary to know what is happening to rare species. Unfortunately, we only know what is happening to a few species of vertebrates and conspicuous invertebrates.

Is It Possible for Marine Populations and Ecosystems to Recover?

Marine populations and ecosystems have a high recovery potential on timescales of a few years to a few decades after major disturbances. Despite long periods of intense human impact, most marine species persist and some populations show signs of recovery. This could provide a promising outlook on the future of Mediterranean ecosystems and biodiversity if those measures necessary for this to happen are urgently taken. Recovery often depends on intrinsic factors, such as life-history characteristics, meta-population structure and genetic diversity, but also on extrinsic factors, such as the type and magnitude of disturbance, and the conservation and management measures applied to reduce human impacts (Lotze et al. 2011). Small short lived organisms, such as most invertebrates and algae, have a much higher recovery potential than large vertebrates or long-lived, structural invertebrates.

The most obvious drivers of recovery are the reduction of those human impacts that caused the depletion or degradation, especially habitat loss, pollution and overexploitation, combined with recovery favorable environmental and life-history conditions. Awareness, legal protection and enforcement plans are also important. Once the ability of natural ecosystems to absorb and recover from multiple stressors of natural or human origin is eroded, the mere restoration of environmental conditions preceding the loss, if at all possible, may be a necessary but perhaps not sufficient condition for recovery (Perkol-Finkel and Airoldi 2010). In most cases a full recovery of the initial state after disturbance is not possible, and most often only a partial recovery to a reduced, altered or alternative stable state is achieved. The species with reduced fertility and/or low dispersal ability recover with more difficulty than those of high fertility and high dispersal capacity. New strategies for restoration and conservation are necessary, taking into account the resilience of the system in light of cumulative impacts, incorporating threshold models and feedback mechanisms (Perkol-Finkel and Airoldi 2010). Recovery is always the result of interactions between species and extrinsic factors, and there are circumstances where such interactions can feedback negatively, limiting recovery for decades.

Marine Protected Areas (MPAs) are created mainly for the recovery and conservation of marine populations and ecosystems. In fact, marine reserves lead to increases in total abundance, biomass and size of fish within their boundaries and adjacent areas, especially for species targeted by fisheries (Guidetti and Sala 2007). However, their role is less clear for the other components of the marine ecosystem, especially small invertebrates. Protection and recovery of fish assemblages may affect benthic ecosystems through cascading effects. Much has been written about MPAs as exporters of fish biomass and larvae to surrounding areas, but little has been said about them as recipients of larvae. On the one hand, if the exported larvae do not find appropriate places to settle outside the MPAs, most larvae exported are wasted. On the other hand, if MPAs do not import larvae from outside, then there will be a genetic impoverishment of their populations.

Therefore, although the establishment of MPAs is positive and necessary, it is not enough, and an integrated conservation of the entire marine ecosystem as a whole is required.

Suggestions

The Mediterranean is an ideal laboratory in which to investigate the kinds of biogeographical alterations that will become much more common as global warming intensifies (Vermeij 2012) and to understand how a variety of species respond to it.

To mitigate environmental degradation and biodiversity loss, first we need to mitigate the effect of those impacts we are more able to control (overfishing, habitat destruction, pollution). The effects of climate change would be attenuated if we were able to reduce the effect of the other disturbing factors (Calvo et al. 2011). Furthermore, a better knowledge of the basic components of biodiversity is necessary to detect the changes that are occurring. Hence research work in those scientific areas currently unfashionable by funding agencies (taxonomy, systematics or biogeography) is also necessary. Regrettably, while biodiversity problems are growing, we are losing expertise in biodiversity (Boero 2001; Giangrande 2003, among others), namely those experts able to identify, describe and classify species (taxonomists).

The Convention on Biological Diversity is calling for more systematic inventories. Promote systematic inventories and taxonomy may provide a permanent scientific record for documenting patterns of diversity, endemism and alien species across zones, ecosystems and habitats, baselines for monitoring programmes, it could identify indicator species of environmental changes and it is essential for identifying and establishing conservation priorities (Mikkelsen and Cracraft 2001). Small cryptic invertebrates, i.e. polychaetes, molluscs, crustaceans, etc. are numerically dominant in most marine ecosystems, but the inventory of these small invertebrates requires the participation of trained taxonomists.

Another issue related to the above is the need to make existing information available (especially historical records) on the distribution of both warm- and cold-water species at the basin scale, and to identify historical reference points to detect and understand recent changes that may occur in marine ecosystems and in species populations.

Inventories of selected taxa should be ideally coupled with long-term monitoring programmes. Large scale monitoring programs, and internationally co-ordinated networks of monitoring hydrological parameters and networks of MPAs are necessary. These programs would be pivotal to quantify trends of changes in habitats and species distribution and unequivocally attribute the causes.

Most studies have investigated the individual effects of a number of natural and human disturbances, but processes driving observed changes or those that may influence recovery of marine benthic communities remain untested (Sala et al. 2012). Currently, the methods used to evaluate the effects of climate warming on biodiversity are largely ecological, but it is necessary to look into physiological optima and thermal tolerance limits of species that, combined with genetic capacity of adaptation (genetic capacity variability), influence ecological and adaptive potentials. How potential for acclimatization and genetic adaptation will determine "winners" and "losers" (Somero 2010). Search on thermal tolerance limits and acclimatization capacities of the species will provide insight into potential rates of adaptive responses (adaptive capabilities vs rate of change; external forcing vs internal dynamics).

Genetic studies may allow inferences on past and contemporary temporal variations in effective sizes of the populations and genetic variability. These parameters, associated with ecological studies, are of primary importance to detect when a population is endangered, and to predict the influence of environmental changes on individual species (Lo Brutto et al. 2011). Over long times scales, phylogeography converges with biogeography, and these disciplines together shed light on past dispersal vicariance events and, possibly, may allow predictions of future population expansions or contractions prompted by climate change.

Both range expansion and local adaptation may be responses to climate warming and can often be detected with genetic analysis, which can also be used to test competing hypotheses. Furthermore, addressing the causes and consequences of adaptive genetic differentiation between populations promises to advance community ecology, climate change research, and the effective management of marine ecosystems (Sandford and Kelly 2011).

On the other hand, long-lived organisms can be useful witnesses of long-term changes in Mediterranean ecosystems. Long-lived species have records of short-scale climatic changes within their structures and could provide an insight into the ecosystem responses to changes (Duarte et al. 1999), and hence bridge the gap resulting from the paucity of direct records. Adequate techniques to reconstruct these responses are available, such as the use of growth marks in seagrass shoots, and sclerochronologic and stable isotope records in corals or molluscs. As an example, *Pinna nobilis* have the potential to provide records of changes in the isotopic composition of dissolved organic carbon in the Mediterranean surface waters (García-March et al. 2011). Likewise, vermetid reefs or the scleractinian coral *Cladocora caespitose* have been used as natural archives of past sea-level and surface temperature variations (Silenzi et al. 2004; Montagna et al. 2007; Chemello and Silenzi 2011).

To sum up, a number of research needs and suggestions that might be useful in order to deal as best as possible with the loss of biodiversity in the Mediterranean and its negative effects associated are provided bellow:

- Promote systematic inventories and taxonomy.
- Make existing information available (especially historic records) on the distribution of both warm- and cold-water species at the basin scale.
- Extend marine monitoring efforts: establish long-term, large scale monitoring programs, and an internationally co-ordinated network of monitoring the main hydrological, chemical and biological parameters and a network of MPAs.
- Standardize and simplify methodologies for monitoring.
- Find out the role of rare species in the maintenance of biodiversity and in the functionality and resilience of ecosystems.
- Study thermotolerance features of model species of different groups, and adaptive and acclimatization processes. Extend our knowledge on sensitivities and adaptation capabilities of marine key species to climate change and identify key species as suitable descriptors.
- Obtain more data on the population structure and genetic diversity of selected species (the "gene-climate approach").
- Study relationship between genetic diversity and resistance to warming.
- Examine the resilience (as a measure of resistance and recovery) of long-lived marine structural invertebrate species facing global change, inquiring into the interplays between processes underlying this resilience, such as recruitment, connectivity, genetic diversity and resistance to disturbances.
- Use long-lived organisms as witnesses of long-term changes in Mediterranean ecosystems.
- Not only document declines, but also examine the processes of resistance and recovery.
- Develop multidisciplinary approach, linking functional ecology with invasion biology, population genetics, macrophysiology and hydrology.
- Develop modelling tools and new methodological approaches to answer new and relevant questions in the fields of thermotolerances, supply side ecology (larval supply and source and sink populations), population genetics, and conservation biology.
- Extend our knowledge on the regional factors that determine the vulnerability and resilience of marine communities to climate change.
- Assess the synergies between multiple pressures on marine ecosystems and biodiversity.

Conclusion

We know something about the major impacts that nowadays affect the marine biodiversity of the Mediterranean. We also know something about the effects that individually produce each of these impacts, but we know very little about how the sum of all these impacts act together. However, when asked how will the Mediterranean be affected in the near future, the answer is clear: we address towards a poorer and most vulnerable Mediterranean Sea.

References

Airoldi L, Balata D, Beck MW (2008) The gray zone: relationships between habitats loss and marine diversity and their applications in conservation. J Exp Mar Biol Ecol 366:8–15

Aligizaki K, Katikou P, Nikolaidis G (2009) Toxic benthic dinoflagellates and potential risk in the mediterranean Sea. International Conference on Molluscan Shellfish Safety, Nantes

Attrill MJ, Wright J, Edwards M (2007) Climate-related increases in jellyfish frequency suggest a more gelatinous future for the North Sea. Limnol Oceanogr 52:480–485

Augier H (1982) Inventaire et classification des biocènoses marines benthiques de la Méditerranée. Conseil de l'Europe, Strasbourg

Aumann HH, Ruzmaikon A, Teixeira J (2008) Frequency of severe storms and global warming. Geophys Res Lett 35, L19805

Azzurro E (2008) The advance of thermophilic fishes in the Mediterranean Sea: overview and methodological questions. CIESM workshop monographs 35, Monaco

Balli M, Garrabou J (2007) Thermodependant bacterial pathogens and mass mortalities in temperate benthic communities: a new case of emerging disease linked to climate change. Glob Change Biol 13:2078–2088

Basterretxea G, Garcés E, Jordi A, Masó M, Tintoré J (2005) Breeze conditions as a favoring mechanism of *Alexandrium taylori* blooms at a Mediterranean beach. Estuar Coast Shelf Sci 62:1–12

Bellan-Santini D, Lacaze J-C, Poizat C (1994) Les biocénoses marines et littorales de Méditerranée, synthèse, menaces et perspectives. Muséum National d'Histoire Naturelle, Collection Patrimoines Naturels 19, Paris

Ben Rais Lasram F, Mouillot D (2009) Increasing southern invasion enhances congruence between endemic and exotic the Mediterranean fish fauna. Biol Invasions 11:697–711

Ben Rais Lasram F, Guilhaumon F, Albouy C et al (2010) The Mediterranean Sea as a "cul-de-sac" for endemic fishes facing climate change. Glob Change Biol 16:3233–3245

Bianchi CN (2007) Biodiversity issues for the forthcoming tropical Mediterranean Sea. Hydrobiologia 580:7–21

Bianchi CN, Morri C (1993) Range extension of warm-water species in the northern Mediterranean: evidence for climatic fluctuations? Porcup Newsl 5:156–159

Bianchi CN, Morri C (2000) Marine biodiversity of the Mediterranean Sea: situation, problems and prospects for future research. Mar Pollut Bull 40:367–376

Boero F (1994) Fluctuations and variations in coastal marine environments. Mar Ecol 15:3–25

Boero F (1996) Episodic events: their relevance to ecology and evolution. Mar Ecol 17:237–250

Boero F (2001) Light after the dark: the partnership for enhancing expertise in taxonomy. Trends Ecol Evol 16:266

Boero F, Bonsdorff E (2007) A conceptual framework for marine biodiversity and ecosystem functioning. Mar Ecol 28(supp. 1): 134–145

Boero F, Féral JP, Azzurro E et al (2008) Climate warming and related changes in the Mediterranean marine biota (Executive summary of CIESM workshop 35). CIESM workshop monographs 35, Monaco

Bombace G (2001) Influence of climate changes on stocks, fish species and marine ecosystems in the Mediterranean Sea. Arch Oceanogr Limnol 22:67–72

Bouchet P, Taviani M (1992) The Mediterranean deep-sea fauna: pseudopopulations of Atlantic species? Deep-Sea Res 39:169–184

Boudouresque CF (2004) Marine biodiversity in the Mediterranean: status of species, populations and communities. Sci Rep Port-Cros Natl Park 20:97–146

Boudouresque CF, Bernard G, Pergent G, Shili A, Verlaque M (2009) Regression of Mediterranean seagrasses caused by natural processes and anthropogenic disturbances and stress: a critical review. Bot Mar 52(39):395–418

Briggs JC (1974) Marine zoogeography. McGraw-Hill, New York

Calvo E, Simó R, Coma R et al (2011) Effects of climate change on Mediterranean marine ecosystems: the case of the Catalan Sea. Clim Res 50:1–29

Cebrian E, Uriz MJ, Garrabou J, Ballesteros E (2011) Sponge mass mortalities in a warming Mediterranean Sea: are cyanobacteria-harboring species worse off? PLoS One 6:e20211

Cerrano C, Bavestrello G (2009) Mass mortalities and extinctions. In: Wahl M (ed) Marine hard bottom communities, Ecological studies 206. Springer, Berlin

Cerrano C, Bavestrello G, Bianchi CN et al (2000) A catastrophic mass mortality episode of gorgonians and other organisms in the Ligurian Sea (north-western Mediterranean), summer 1999. Ecol Lett 3:284–293

Chemello R, Silenzi S (2011) Vermetid reefs in the Mediterranean Sea as archives of sea-level and surface temperature changes. Chem Ecol 27:121–127

Chevaldonné P, Lejeusne C (2003) Regional warming-induced species shift in NW Mediterranean marine caves. Ecol Lett 6:371–379

CIESM (2008) Impacts of acidification on biological, chemical and physical systems in the Mediterranean and Black Sea (F Briand edit). CIESM workshop monographs 36, Monaco

Claudet J, Fraschetti S (2010) Human-driven impacts on marine habitats: a regional meta-analysis in the Mediterranean Sea. Biol Conserv 143:2195–2206

Coll M, Lotze HK, Romanuk TN (2008) Structural degradation in Mediterranean Sea food webs: testing ecological hypotheses using stochastic and mass-balance modelling. Ecosystems 11:939–960

Coll M, Piroddi C, Steenbeek J et al (2010) The biodiversity of the Mediterranean Sea: estimates, patterns, and threats. PLoS One 5:e11842

Coma R, Ribes M, Serrano E et al (2009) Global warming-enhanced stratification and mass mortality events in the Mediterranean. Proc Natl Acad Sci USA 106:6176–6181

Coma R, Serrano E, Linares C et al (2011) Sea urchins predation facilitates coral invasion in a marine reserve. PLoS One 6:e22017

Courchamp F, Berec L, Gascoigne J (2008) Allee effects in ecology and conservation. Oxford University Press, Oxford

Crisci C, Bensoussan N, Romano JC, Garrabou J (2011) Temperature anomalies and mortality events in marine communities: insights on factors behind differential mortality impacts in the NW Mediterranean. PLoS One 6:e23814

Daniel B, Piro S, Charbonnel E, Francour P, Letourneur Y (2009) Lessepsian rabbitfish *Siganus luridus* reached the French Mediterranean coasts. Cybium 33:163–164

Danovaro R (2003) Pollution threats in the Mediterranean Sea: an overview. Chem Ecol 19:15–32

Danovaro R, Armeni M, Corinaldesi C, Mei ML (2003) Viruses and marine pollution. Mar Pollut Bull 46:301–304

Danovaro R, Fonda Unami S, Pusceddu A (2009) Climate change and the potential spreading of marine mucilage and microbial pathogens in the Mediterranean Sea. PLoS One 4:e7006

Despalatović M, Grubelić I, Nikolić V et al (2008) Allochthonous warm water species in the benthic communities and ichthyofauna of the eastern part of the Adriatic Sea. CIESM workshop monographs 35, Monaco

Duarte CM, Agustí S, Kennedy H, Vaqué D (1999) The Mediterranean climate as a template for Mediterranean marine ecosystems: the example of the northeast Spain littoral. Prog Oceanogr 44:245–270

Dulčić J, Scordella G, Guidetti P (2008) On the record of the Lessepsian migrant *Fistularia commersonii* (Rüppell, 1835) from the Adriatic Sea. J Appl Ichthyol 24:101–102

Emig C, Geistdoerfer P (2004) The Mediterranean deep-sea fauna: historical evolution, bathymetric variations and geographical changes. Carnets Géol 1:1–10

Faimali M, Giussani V, Piazza V et al (2012) Toxic effects of harmful benthic dinoflegellate *Ostreopsis ovata* on invertebrate and vertebrate marine organisms. Mar Environ Res 76:97–107

Fanelli G, Piraino S, Belmonte G et al (1994) Human predation along Apulian rocky coasts (SE Italy): desertification caused by *Lithophaga lithophaga* (Mollusca) fisheries. Mar Ecol Prog Ser 110:1–8

Ferretti F, Myers RA, Serena F, Lotze HK (2008) Loss of large predatory sharks from the Mediterranean Sea. Conserv Biol 22:952–964

Francour P, Mouine N (2008) First record of *Kyphosus sectator* (Kyphosidae) along the French Mediterranean coast. Cybium 32:275–276

Francour P, Boudouresque CF, Harmelin JG et al (1994) Are the Mediterranean waters becoming warmer? Information from biological indicators. Mar Pollut Bull 28:523–526

Fraschetti S, Terlizzi A, Boero F (2008) How many habitats are there in the sea (and where)? J Exp Mar Biol Ecol 366:309–332

Galil BS (2006) The marine caravan: the Suez Canal and the Erythrean invasion. In: Golash S, Galil BS, Cohen AN (eds) Bridging divides: maritime canals as invasion corridors. Springer, Heidelberg

Galil BS (2007) Loss or gain? Invasive aliens and biodiversity of the Mediterranean Sea. Mar Pollut Bull 55:314–322

Galil BS (2008) Alien species in the Mediterranean Sea: which, when, where, why? Hydrobiologia 606:105–116

Galil BS (2009) Taking stock: inventory of alien species in the Mediterranean Sea. Biol Invasions 11:359–372

Garcés E, Masó M, Vila M, Camp J (2000) HABs events in the Mediterranean Sea: are they increasing? A case study the last decade in the NW Mediterranean and the genus *Alexandrium*. Harmful Algal News 20:1–11

García-March JR, Surge D, Lees JM, Kersting DK (2011) Ecological and water mass properties in the Mediterranean recorded by stable isotope ratios in *Pinna nobilis*. J Geophys Res 116, G02009

García-Rubiés A, Mateo MA, Hereu B et al (2009) Preliminary assessment of the impact of an extreme storm on Catalan Mediterranean benthic communities. 11th Plinius Conference on Mediterranean storms, Barcelona

Garrabou J, Perez T, Sartoretto S, Harmelin JG (2001) Mass mortality event in red coral (*Corallium rubrum*, Cnidaria, Anthozoa, Octocorallia) population in the Provence region (France NW Mediterranean). Mar Ecol Prog Ser 217:263–272

Garrabou J, Coma R, Bensoussan N et al (2009) Mass mortality event in northwestern Mediterranean rocky benthic communities: effects of the 2003 heat wave. Glob Change Biol 15:1090–1103

Giangrande A (2003) Biodiversity, conservation and the 'taxonomic impediment'. Aquat Conserv 13:451–459

Goy J, Morand P, Etienne M (1989) Long-term fluctuations of *Pelagia noctiluca* (Cnidaria, Scyphomedusa) in the western Mediterranean Sea. Prediction by climatic variables. Deep-Sea Res 36:269–279

Guidetti P, Boero F (2001) Occurrence of the Mediterranean parrfish *Sparisoma cretense* (Perciformes: SCaridae) in south-eastern Apulia (south-east Italy). J Mar Biol Assoc UK 82:717–719

Guidetti P, Sala E (2007) Community-wide effects of marine reserves in the Mediterranean Sea. Mar Ecol Prog Ser 335:43–56

Hester KC, Peltzer ET, Kirkwood WJ, Brewer PG (2008) Unanticipated consequences of ocean acidification: a noisier ocean at lower pH. Geophys Res Lett 35, L19601

Innamoranti M, Nuccio C, Massi L et al (2001) Mucilages and climate changes in the Tyrrhenian Sea. Aquat Conserv 11:289–298

Jiménez JA, Sancho-García A, Bosom A et al (2012) Storm-induced damages along the Catalan coast (NW Mediterranean) during the period 1958–2008. Geomorphology 143–144:24–33

Katsanevakis S, Poursanidis D, Yokes MB et al (2011) Twelve years after the first report of the crab *Percnon gibbesi* (H. Milne Edwards, 1853) in the Mediterranean: current distribution and invasion rates. J Biol Res 16:224–236

Kersting DK, García-March JR, Templado J (2006) Evaluation of *Spondylus gaederopus* Linné, 1758, mass mortality event in the Columbretes Islands Marine Reserve (western Mediterranean, Spain). Org Divers Evol 6, Electr Suppl 16, part 1:45

Kružić P (2008) First records of *Cladopsammia rolandi* (Cnidaria: Antozoa) in the Adriatic Sea. Nat Croat 17:9–14

Lejeusne C, Chevaldonné P, Pergent-Martini C et al (2010) Climate change effects on a miniature ocean: the highly diverse, highly impacted Mediterranean Sea. Trends Ecol Evol 25:250–260

Lim GS, Balke M, Meier R (2012) Determining species boundaries in a world full of rarity: singletons, species delimitation methods. Syst Biol 61:165–169

Lo Brutto S, Arculeo M, Grant WS (2011) Climate change and population genetic structure of marine species. Chem Ecol 27:107–119

Lotze HK, Coll M, Magera AM, Ward-Paige C, Airoldi L (2011) Recovery of marine animal populations and ecosystems. Trends Ecol Biol 26:595–605

Maldonado M, Sánchez-Tocino L, Navarro C (2010) Recurrent disease outbreaks in corneous demosponges of the genus *Ircinia*: epidemic incidence and defense mechanisms. Mar Biol 157:1577–1590

Marbà N, Duarte CM (1997) Interannual changes in seagrass (*Posidonia oceanica*) growth and environmental change in the Spanish Mediterranean littoral. Limnol Oceanogr 42:800–810

Marbà N, Duarte CM (2010) Mediterranean warming triggers seagrass (*Posidonia oceanica*) shoot mortality. Glob Change Biol 16:2366–2375

Meinesz A, Mercier D (1983) Sur les fortes mortalités de Spondyles (Spondylus gaederopus Linné) observées sur les cotes de Méditerranée. Trav Sci Parc nation Port-Cros 9:89–95

Metaxatos A, Panagiotopoulos C, Ignatiades L (2003) Monosaccharide and aminoacid composition of mucilage material produced from a mixture of four phytoplanktonic taxa. J Exp Mar Biol Ecol 294:203–217

Mikkelsen PM, Cracraft J (2001) Marine biodiversity and the need for systematic inventories. Bull Mar Sci 69:525–534

Miller RJ (1985) Succession in sea urchin and seaweed abundance in Nova Scotia, Canada. Mar Biol 84:275–286

Mills CE (2001) Jellyfish blooms: are populations increasing globally in response to changing ocean conditions. Hydrobiologia 451:55–68

Montagna P, McCulloch M, Mazzoli C et al (2007) The non-tropical coral *Cladocora caespitosa* as the new climate archive for the Mediterranean: high-resolution (similar to weekly) trace elements systematics. Quat Sci Rev 26:441–462

Navarro L, Ballesteros E, Linares C, Hereu B (2011) Spatial and temporal variability of deep-water algal assemblages in the northwestern Mediterranean: the effects of an exceptional storm. Estuar Coast Shelf Sci 95:52–58

Occhipinti-Ambrogi A (2007) Global change and marine communities: alien species and climate change. Mar Pollut Bull 55:342–352

Pearson TR, Lalli CM (2002) Jellyfish population explosions: revisiting a hypothesis of possible causes. Mer (Paris) 40:111–121

Pérès JM, Picard J (1958) Manuel de bionomie benthique de la mer Méditerranée. Rec Trav Stat Mar Endoume 31:1–137

Perez T, Garrabou J, Sartoretto S et al (2000) Mortalité massive d'invertébrés marins: un événement sans precedent en Méditerranée nord-occidental. C R Acad Sc Paris sésrie III 323:853–865

Perkol-Finkel S, Airoldi L (2010) Loss and recovery potential of marine habitats: an experimental study of factors maintaining resilience in subtidal algal forests at the Adriatic Sea. PLoS One 5:e10791

Philippart CJM, Anadón R, Danovaro R et al (2011) Impact of climate change on European marine ecosystems: observations, expectations and indicators. J Exp Mar Biol Ecol 400:52–69

Piraino S, Fanelli G, Boero F (2002) Variability of species' roles in marine communities: change of paradigms for conservation priorities. Mar Biol 140:1067–1074

Por FD (2009) Tethys returns to the Mediterranean: success and limits of tropical recolonization. Bio Risk Assess 3:5–19

Privitera D, Noli M, Galugi C, Chiantore M (2011) Benthic assemblages and temperature effects on *Paracentrotus lividus* and *Arbacia lixula* larvae and settlement. J Exp Mar Biol Ecol 407:6–11

Pronzato R (1999) Sponge-fishing, disease and farming in the Mediterranean *Sea*. Aquat Conserv 9:485–493

Pronzato R, Marconi R (2008) Mediterranean commercial sponges: over 500 years of natural history and cultural heritage. Mar Ecol-Evol Persp 29:146–166

Puce S, Bavestrello G, Di Camillo CG, Boero F (2009) Long-term changes in hydroid (Cnidaria, Hydrozoa) assemblages: effect of Mediterranean warming? Mar Ecol 30:313–326

Pusceddu A, Fiordelmondo C, Polymenakou P et al (2005) Effects of bottom trawling on the quantity and biochemical composition of organic matter in coastal marine sediments (Thermaikos Gulf, northwestern Aegean Sea). Cont Shelf Res 425:2491–2505

RAC/SPA (2006) Classification of benthic marine habitat types for the Mediterranean region. RAC/SPA, Tunisia

Relini G (2009) *Ocypode cursor* in Sicilia. Not Soc Ital Biol Mar 56:49

Richardson AJ, Bakun A, Hays GC, Gibbons MJ (2009) The jellyfish joyride: causes, consequences and management responses to a more gelatinous future. Trends Ecol Evol 24:312–322

Rosenberg E, Loya Y (1999) *Vibrio shiloi* is the etiological (causative) agent of *Oculina patagonica* bleaching: general implications. Reef Encount 25:8–10

Sabatés A, Martín P, Lloret J, Raya V (2006) Sea warming and fish distribution: the case of the small pelagic fish, *Sardinella aurita*, in the western Mediterranean. Glob Change Biol 12:2209–2219

Sabatés A, Pagés F, Atienza D et al (2010) Planktonic cnidarian distribution and feeding of *Pelagia noctiluca* in the NW Mediterranean Sea. Hydrobiologia 645:153–165

Saenz-Agudelo P, Jones GP, Thorrold SR, Planes S (2011) Connectivity dominates larval replenishment in a coastal reef fish metapopulation. Proc R Soc B 278:2954–2961

Sala E (2004) The past and present topology and structure of Mediterranean subtidal rocky-shore food webs. Ecosystems 7:333–340

Sala E, Knowlton N (2006) Global marine biodiversity trends. Annu Rev Environ Resour 31:93–122

Sala E, Ballesteros E, Dendrinos P et al (2012) The structure of Mediterranean rocky reef ecosystems across environmental and human gradients, and conservation implications. PLoS One 7:e32742

Sandford E, Kelly MW (2011) Local adaptation in marine invertebrates. Annu Rev Mar Sci 3:509–535

Santangelo G, Abbiati M, Giannini F, Cicogna F (1993) Red coral fishing trends in the western Mediterranean Sea. Sci Mar 57:139–143

Silenzi S, Antonioli F, Chemello R (2004) A new marker for sea surface temperature trend during the last centuries in the temperate areas: Vermetid reef. Glob Planet Change 40:105–114

Somero GN (2010) The physiology of climate change: how potential for acclimatation and genetic adaptation will determine "winners" and "losers". J Exp Biol 213:912–920

Southward AJ (1995) The importance of long time-series in understanding the variability of natural systems. Helgol Wissenshafttliche Meeressunters 49:329–333

Taviani M (2002) The Mediterranean benthos from late Miocene up to present: ten million years of dramatic climatic and geologic vicissitudes. Biol Mar Mediterr 9:445–463

Thibaut T, Pinedo S, Torras X, Ballesteros E (2005) Long-term decline of the population of fucales (*Cystoseira* spp. and *Sargassum* spp.) in the Alberes coast (France, north-western Mediterranean). Mar Pollut Bull 50:1472–1489

Thunell RC (1979) Climatic evolution of the Mediterranean Sea during the last 5.0 million years. Sediment Geol 23:67–79

Tudela S (2004) Ecosystem effects of fishing in the Mediterranean: an analysis of the major threats of fishing gear and practices to biodiversity and marine habitats. General Fisheries Commission for the Mediterranean 74, FAO, Rome

Turley CM (1999) The changing Mediterranean Sea: a sensitive ecosystem? Prog Oceanogr 44:387–400

Vermeij GJ (2012) The tropical history and future of the Mediterranean biota and the west African enigma. J Biogeogr 39:31–41

Vezzulli L, Previati M, Pruzzo C et al (2010) *Vibrio* infections triggering mass mortality events in warming Mediterranean Sea. Environ Microbiol 12:2007–2019

Volulsiadou E, Koutsoubas D, Achparaki M (2009) Bivalve mollusc exploitation in Mediterranean coastal communities: an historical approach. J Biol Res – Thessloniki 12:35–45

Zenetos A, Siokou-frangou I, Gotsis-Skretas O (2002) The Mediterranean Sea – blue oxygen-rich, nutrient-poor waters. European Environment Agency, Copenhagen

Zenetos A, Gofas S, Verlaque M et al (2010) Alien species in the Mediterranean Sea by 2010. A contribution to the application of European Union's Marine Strategy Framework Directive (MSFD). Part I. Spatial distribution. Mediterr Mar Sci 11:381–493

Zingone A, Enevoldsen HO (2000) The diversity of harmful algal blooms: a challenge for science and management. Ocean Coast Manag 43:725–748

Regime Shifts in the Marine Environment: How Do They Affect Ecosystem Services?

29

Alexandra C. Kraberg and Karen H. Wiltshire

Abstract

Marine ecosystems are facing unprecedented pressures for instance rising water temperatures, changing current patterns and ocean acidification. Coastal systems in particular are also challenged with additional anthropogenic pressures caused by accelerating rates of human settlement near the coast. This trend places increasing strains on the delivery of ecosystem services associated with recreation but also coastal fisheries. Here we introduce the concepts related to ecosystem stability and ecosystem services and review the evidence for regime shifts in the world's ecosystems with particular reference to the Mediterranean, before reviewing mechanisms for ecosystem valuations, ending with recommendations for increasing the practical relevance of future ecosystem evaluations. We stress the need for regional approaches, taking into account the views and needs of local populations, which might differ greatly geographically even for the same issue being considered.

Keywords

Ecosystem services • Ecosystem stability • Regime shifts • Regional management

Introduction

Defining coastal zones as those areas less than 10 m above sea level implies that 10 per percent of the worlds' population live in this coastal zone. In Europe alone this amounts to roughly 50 million people (McGranahan et al. 2007; Nicholls et al. 2007). With rapidly growing human populations living in coastal areas, coastal zones are subject to increasingly conflicting needs and uses. The human population increasingly relies on coastal seas for food, and coastal areas are also important for recreation, shipping and power generation (Remoundou et al. 2009). These uses are not necessarily mutually compatible and put increasing pressures on coastal ecosystem diversity and function. These pressures will increase even further as climate change forces upon us the greater need for coastal defences against sea level rise, storms and flooding with subsequent impacts on coastal ecosystems e.g. via habitat loss, changing hydrodynamics in estuaries etc. For a sustainable ecosystem and resource management some capacity for forward planning is absolutely crucial (Folke et al. 2004). However, the still accelerating anthropogenic climate change is making this increasingly difficult.

Biological Regime Shifts: Some Definitions

One class of abrupt changes in the environment is termed regime shift. This refers to non-linear responses of a system to a driver even if that driver is showing only gradual change. Rather than responding gradually, the system will initially show little response to the driver, but then, reaching a threshold 'x' (also called tipping point) move to a different system state (an alternative stable state) (Petraitis and Dudgeon 2004). Importantly a small shift of the driver back in the direction of the original state will not be sufficient to move the system

A.C. Kraberg (✉) • K.H. Wiltshire
Biologische Anstalt Helgoland, Alfred-Wegener Institute for Polar and Marine Research, Kurpromenade 201, 27498 Helgoland, Germany
e-mail: Alexandra.Kraberg@awi.de; Karen.Wiltshire@awi.de

S. Goffredo and Z. Dubinsky (eds.), *The Mediterranean Sea: Its history and present challenges*, DOI 10.1007/978-94-007-6704-1_29, © Springer Science+Business Media Dordrecht 2014

back to this original state a process known as hysteresis (Kinzig et al. 2006). The set of conditions in which the two system states will be stable is known as basins of attraction. Originally such shifts were defined as description of atmospheric processes on multi-decadal timescales and their consequences for the physical environment (Rahmstorf 1999). In recent years however usage of the term 'regime shift' has extended to shifts observed in freshwater (Scheffer et al. 2001b) as well as in marine and estuarine waters (Polovina 2005; Petersen et al. 2008) and includes not only shifts in the physical or chemical environment but also ecological shifts in response to these changing drivers (de Young et al. 2004, 2008) (reviewed in Kraberg et al. (2011)). These 'ecosystem consequences' are in the main related to changes in biological species composition. But even changes in the trophic organization within an ecosystem could constitute a regime shift. Importantly biological components of the system can also be drivers of the regime shift e.g. as a result of shifts in predator abundance.

In the context of these abrupt shifts a further important concept is the 'resilience' of a system. This resilience refers to the ability of a system to absorb or adapt to disturbances without losing its overall structure and function. With respect to regime shifts this can be regarded as the ease with which the system can be forced to move between the alternative states and this is to some extent a function of the biodiversity and structure of the ecosystem. The species in a biological community are intricately linked through interactions such as competition and predation and alterations in these interactions can influence the ability of a system to recover from a given stress (Scheffer et al. 2001a; Kinzig et al. 2006).

Human intervention can play a considerable role in these processes for instance through excessive nutrient inputs or the introduction of neobiota, such as the comb jelly *Mnemiopsis leidyi* in the Black and Caspian Seas (Oguz et al. 2008).

However, the consequences of climate change, such as rising air and water temperatures, sea level rise etc. are emerging as the dominant threats to ecosystem stability (Tol 2009). All of the above can lead to a simplification of the foodweb, if they cause not only a species loss *per se* but a loss of species functional types, which play a particular ecological role in the ecosystem (e.g. certain feeding types such as detritivores, planktivores etc.) and therefore a reduction in ecosystem resilience (Elmquist et al. 2003). Regime shifts' considerable effects on ecosystem services will be discussed in the following sections, after a brief review of evidence for regime shifts globally and particularly in the Mediterranean.

Global Evidence for Regime Shifts

Regime shifts have been detected in many regions of the world using a range of statistical methods. In the North Sea for instance analyses of the SAHFOS long-term data set have shown a shift from diatom to dinoflagellate dominated phytoplankton communities with a change point around 1988 (see also Weijerman et al. (2005)). Similar changes were also observed by Wiltshire et al. (2008) in the Helgoland Roads plankton datasets. In the northern North Sea, this change in the plankton was also reflected in increased catches due to a northward shift of horse mackerel (Reid et al. 2001). Simultaneous regime shifts were also detected in the Baltic (Alheit et al. 2005; Möllmann et al. 2009) and Pacific (Benson and Trites 2002; Wooster and Zhang 2004; Chiba et al. 2008). An earlier regime shift has also been detected in the Pacific in 1977 reflected in abrupt abundance changes of many commercial fish stocks, including several salmon species (Hare and Mantua 2000). However, not all observed even abrupt changes e.g. shifts between anchovy and sardine dominated systems and *vice versa* necessarily constitute regime shifts, but can be the result of cyclical shifts driven by climatic signals (Alheit and Bakun 2010).

Regime Shifts in the Mediterranean

A number of recent studies have also described regime shifts in the Mediterranean Sea. Molinero et al. (2008) for instance described a shift in 1987/1988 from crustacean zooplankton to an assemblage dominated by gelatinous zooplankton for the Ligurian Sea. Their conclusions were based on a study of selected scyphozoan and ctenophore species. Their findings were corroborated by Conversi et al. (2010) analyzing the same time series. In an earlier study Conversi et al. (2009) also showed an abrupt increase in abundance of selected zooplankton taxa (mostly smaller species) in the Gulf of Trieste (see also Matić et al. (2011)). This shift also seemed to affect other trophic levels with harmful dinoflagellate blooms decreasing and mucilage events increasing around the same time. Examining broad species groups rather than distinct species however, Garcia-Comas et al. (2011) did not detect the same regime shift-like changes but instead oscillations on a 8–9 year time scale, which would not constitute regime shifts (see also Katara et al. (2011)). It is particularly important to understand such ecological phenomena including regime shifts in the Mediterranean as it is known as a biodiversity hotspot and holds an estimated 4–18 % of global marine biodiversity (Bianchi and Morri 2000).

Regime Shifts and Ecosystem Services

Ecosystem Services

Ecosystem services can be broadly defined as the sum of all benefits for mankind that are derived directly or indirectly from ecosystems and facilitate a healthy and safe life for

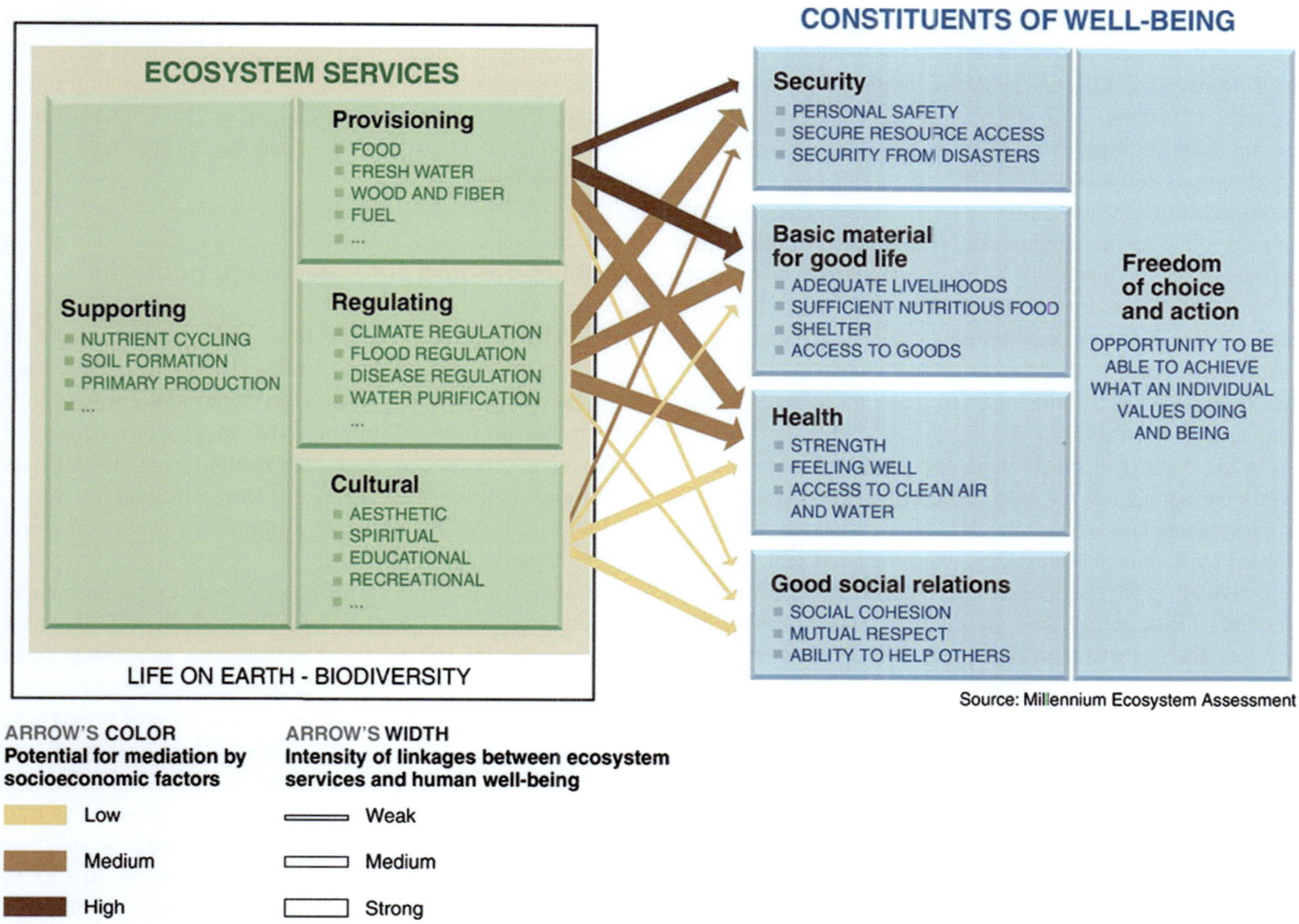

Fig. 29.1 Types and components of ecosystem services as defined by the Millenium Ecosystem Assessment (synthesis report), graphic downloaded, January 5th, 2012 (http://millenniumassessment.org/documents/document.354.aspx.pdf)

local human populations (Fisher et al. 2008). This definition includes a diverse array of different services, which the Millennium Assessment report has classed into four groups: Provisioning (e.g. agriculture and fisheries), regulating (e.g. flood control), cultural and supporting ecosystem services (Millennium Ecosystem Assessment 2005a, b; Fig. 29.1). This subdivision already indicates the complexity of the different services, but it has been criticized by Fisher et al. (2008) as not lending itself easily to formal economic valuations (see also Fisher and Turner (2008)).

Ecosystem valuations are often carried out as 'Marginal valuations' where marginality refers to the assumption that the value of a particular service is a function of small changes in the flow of that service. This means that a marginal valuation considers, not the total cost of an existing service (e.g. a forest), but the cost of delivering an additional unit of that service. However, this method is not suitable for the study of the costs of regime shifts, which are characterized by sudden alternations between very different ecosystem states including associated services. Here an additional 'unit cost' would have to be applied to an entirely different system (e.g. an anchovy as opposed to a sardine fishery or a system dominated by jellyfish rather than crustacean zooplankton as is increasingly seen in the Mediterranean (see case studies in CIESM 2008)). In such cases a risk analysis of the probability of attaining the new state might be more appropriate, at least in theory. The situation is further complicated by the fact that remedial and mitigation measures by humans, even action undertaken to protect key species in an ecosystem, can in themselves cause further ecosystem change (feedback loops) (Estes et al. 1998; Horan et al. 2011). All of these complexities and inter-linkages between different services or components thereof have to be considered and valued without introducing 'double-counting' of individual costs.

Biodiversity, Regime Shifts and Ecosystem Services

As mentioned above biodiversity plays a central role in the functioning and stability of any ecosystem. 'Manipulation' of one part of this system can have dramatic and unforeseen, non-linear consequences in another. This includes the occurrence of biological regime shifts and these have been shown in a number of different systems ranging from coral reef systems (Elmhirst et al. 2009), via the Canadian otter-sea urchin system (Estes et al. 1998) to the growing importance of gelatinous zooplankton in different coastal areas, including in the Mediterranean (Molinero et al. 2008). The economic impact of gelatinous plankton 'outbreaks' in the Mediterranean was valuated by a range of methods, a summary of which is given in CIESM (2008) see also Remoundou et al. (2009). Valuation methods here included market analyses of projected losses of commercial fish stocks as well as the use of non-monetary indicators to value the impact on general ecosystem stability.

In the context of biodiversity and ecosystem services it is worth noting that it is not always biodiversity on the whole that is the subject of a valuation but individual species that might be valued by humans because they are associated with a certain region or are seen as iconic or representative of biodiversity in general. These might also significantly enhance ecosystem services such as tourism (examples in the Mediterranean are: sharks, tuna, marlins, dolphins, monk seals) irrespective of their true ecological or economic significance.

Management decisions with respect to marine and coastal areas are therefore, not surprisingly, primarily geared towards maintaining the economic potential and safety of the human population of those areas (Fisher et al. 2008). This includes the maintenance of food resources and is therefore concerned, directly and indirectly, with the ecological consequences of regime shifts particularly the impacts of biodiversity changes on ecosystem resilience and stability (Millenium Ecosystem Assessment 2005a, b). Meanwhile the importance of the link between biodiversity and ecosystem services has also now been recognized by the World Economic Forum (see World Economic Forum (2010)).

Conclusions and Final Recommendations

These examples demonstrate the complexities involved in the valuation of ecosystems. While ecosystem services such as fisheries and tourism (which are of particular importance in the Mediterranean) might have an obvious economic dimension, others such as nutrient cycling or photosynthesis are more indirect and therefore more difficult to value, although they might constitute a significant service function within a particular ecosystem.

However, the value of different ecosystem components, monetary or otherwise, also depends on the needs, demands and perceptions of the local populations benefiting from them. Only if these are taken into account and assessed accurately, while also reconciling the management of potentially conflicting services such as commercial species with the protection of individual species and habitats will it become possible to convert the valuation of ecosystem services into efficient and coherent policy action and to address existing environmental legislation (Fig. 29.2) (Birol et al. 2006; Seppelt et al. 2011). This can be regarded as particularly

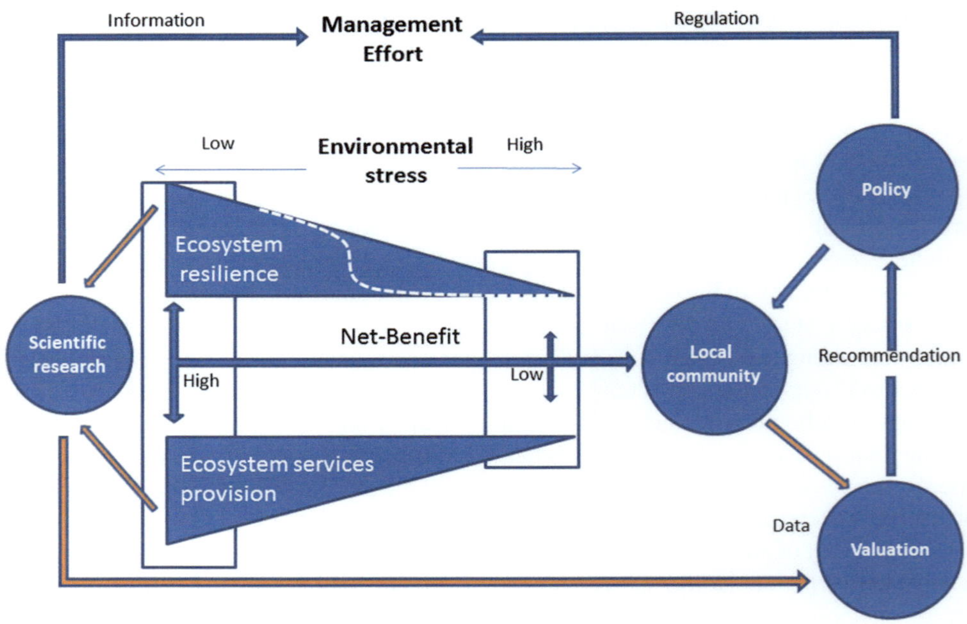

Fig. 29.2 Summary of the different players involved in the study and management of ecosystem services in the face of environmental stress including regime shifts. The *triangles* indicate the level of ecosystem resilience and resulting ecosystem services provision. The *stippled line* refers indicates a regime shift i.e. a sudden rather than gradual change in resilience (or ecosystem provision)

urgent in the Mediterranean, which has faced many pressures in the past e.g. nutrient enrichment/pollution from untreated sewage, oil discharges and overfishing while a large number of invasive species have led considerable modifications in marine biodiversity (Remoundou et al. 2009).

Clearly the overall goal is the sustainable management of coastal, and pelagic, resources and this is likely only to be achieved via an increase in resilience of the system under investigation. However, the most difficult step in this process will probably be the transition from theory to practical assessment and choosing the right metric (i.e. what do we need to measure?), spatial and temporal scales for both the valuation and underlying investigations of ecosystem dynamics. This leads to two recommendations:

1. Strengthening of regional approaches to the management of regime shifts and affected ecosystem services: The same ecosystem service might be valued very differently in different regions, both objectively in terms of monetary value, and in the public perception (Österblom et al. 2010). This is particularly true for non-market ecosystem services where benefits for the local population might be social or cultural in nature. The marginal value of these might vary considerably between ecosystems, but without concrete monetary figures that can easily be associated with these services it is difficult to motivate managers, policy makers and industry that action needs to be taken in a particular case (Fisher et al. 2008). The number of valuations combining the market and non-market values of ecosystem services are still rare particularly in the Mediterranean (Remoundou et al. 2009).
2. Not surprisingly, the second recommendation is the continued maintenance but also the better co-ordination of data collection and large-scale joint data analyses as well as better co-operation at different spatial scales and different types of agencies – essentially an interdisciplinary approach. Demonstrating the occurrence of a regime shift requires, by definition, the availability of detailed long-term biodiversity and environmental data as well as economic data for a given region. However, equally important is the development of robust (and standardized) tools for the analysis of these data collections. As we are dealing with dynamical systems these tools will likely include the application of techniques rarely used by marine biologists e.g. bifurcation theory and Bayesian techniques, for which close co-operation with mathematicians and statisticians will also be essential. Lastly, for such studies to be of relevant for the valuation of ecosystem services they should also address the information needs of economists/social scientists involved in ecosystem valuations. Collaborative ties between economists and ecologists do not exactly have a long tradition but these have to be encouraged and fostered (Turner 2000; Beaumont et al. 2007; Remoundou et al. 2009).

At any rate future solutions to the management of regime shifts will require a great deal of adaptability not only by resource managers but also coastal populations who might be faced with new food sources and different coastal landscapes.

References

Alheit J, Bakun A (2010) Population synchronies within and between ocean basins: apparent teleconnections and implications as to physical-biological linkage. J Mar Syst 70:267–285

Alheit J, Moellmann C, Dutz J, Kornilovs G, Loewe P, Mohrholz V, Wasmund G (2005) Synchronous ecological regime shifts in the central Baltic basin and the North Sea in the late 1980s. ICES J Mar Sci 62:1205–1215

Beaumont NJ, Austen MC, Atkins JP, Burdon D, Degraer S, Dentinho TP, Derous S, Holm P, Horton T, van Ierland E, Marboe AH, Starkey DJ, Townsend M, Zarzycki T (2007) Identification, definition and quantification of goods and services provided by marine biodiversity: implications for the ecosystem approach. Mar Pollut Bull 54:253–265

Benson AJ, Trites AW (2002) Ecological effects of regime shifts in the Bering Sea and eastern North Pacific Ocean. Fish Fish 3:95–103

Bianchi CN, Morri C (2000) Marine biodiversity of the Mediterranean Sea: situation, problems and prospects for future research. Mar Pollut Bull 40:367–376

Birol E, Karousakis K, Koundouri P (2006) Using economic methods and tools to inform water management policies: a survey and critical appraisal of available methods and an application. Sci Total Environ 12:1040–1051

Chiba S, Aita MN, Tadokoro K, Saino T, Sugisaki H, Nakata K (2008) From climate regime shifts to lower-trophic level phenology: synthesis of recent progress in retrospective studies of the western North Pacific. Prog Oceanogr 77:112–126

CIESM (2008) Economic valuation of natural coastal and marine ecosystems. No 37. In: Briand F (ed) CIESM workshop monographs. CIESM, Monaco

Conversi A, Peluso T, Fonda Umani S (2009) Gulf of Trieste: a changing ecosystem. J Geophys Res. doi:10.1029/2008JC004763

Conversi A, Umani SF, Peluso T, Molinero JC, Santojanni A, Edwards M (2010) The Mediterranean Sea regime shift at the end of the 1980s, and intriguing parallelisms with other European basins. PLoS One 5:e10633

de Young B, Harris R, Alheit J, Beaugrand G, Mantua N, Shannon LJ (2004) Detecting regime shifts in the ocean: data considerations. Prog Oceanogr 60:143–164

de Young B, Barange M, Beaugrand G, Harris R, Perry RI, Scheffer M, Werner F (2008) Regime shifts in marine ecosystems: detection, prediction and management. Trends Ecol Evol 23:402–409

Elmhirst T, Connolly SR, Hughes TP (2009) Connectivity, regime shifts and the resilience of coral reefs. Coral Reef 28:949–957

Elmquist T, Folke C, Nyström M, Peterson G, Bengtsson J, Walker B, Norberg J (2003) Response diversity, ecosystem change and resilience. Front Ecol Environ 1:488–494

Estes JA, Tinker MT, Williams TM, Doak DF (1998) Killer whale predation on sea otters linking oceanic and nearshore ecosystems. Science 282:473–476

Fisher B, Turner RK (2008) Ecosystem services: classification for valuation. Biol Conserv 141:1167–1169

Fisher B, Turner K, Zylstra M, Brouwer R, De Groot R, Farber S, Ferraro P, Green R, Hadley D, Harlow J, Jefferiss P, Kirkby C, Morling P, Mowatt S, Naidoo R, Paavola J, Strassburg B, Yu D,

Balmford A (2008) Ecosystem services and economic theory: integration for policy-relevant research. Ecol Appl 18:2050–2067

Folke C, Carpenter S, Walker B, Scheffer M, Elmqvist T, Gunderson L, Holling CS (2004) Regime shifts, resilience and biodiversity in ecosystem management. Annu Rev Ecol Syst 35:557–581

Garcia-Comas C, Semmann L, Ibanez F, Berline L, Mazzocchi MG, Gasparini S, Picheral M, Gorsky G (2011) Zooplankton long-term changes in the NW Mediterranean Sea: decadal periodicity forced by winter hydrographic conditions related to large-scale atmospheric changes? J Mar Syst 87:216–226

Hare SR, Mantua NJ (2000) Empirical evidence for North Pacific (climatic) regime shifts in 1977 and 1989. Prog Oceanogr 47: 103–145

Horan D, Fenichel EP, Drury KLS, Lodge DM (2011) Managing ecological thresholds in coupled environmental-human systems. Proc Natl Acad Sci U S A 108:7333–7338

Katara I, Pierce GJ, Illian J, Scott BE (2011) Environmental drivers of the anchovy/sardine complex in the Eastern Mediterranean. Hydrobiologia 670:49–65

Kinzig AP, Etienne RM, Allison H, Walker BH (2006) Resilience and regime shifts: assessing cascading effects. Ecol Soc 11:20

Kraberg AC, Wasmund N, Vanaverbeke J, Schiedek D, Wiltshire KH, Mieszkowska N (2011) Regime shifts in the marine environment: the scientific basis and political context. Mar Pollut Bull 62:7–20

Matić F, Grbec B, Morović M (2011) Indications of climate regime shifts in the Middle Adriatic Sea. Acta Adriat 52:235–246

McGranahan G, Balk D, Anderson B (2007) The rising tide: assessing the risks of climate change and human settlements in low elevation coastal zones. Environ Urban 19:17–37

Millenium Ecosystem Assessment (2005a) Ecosystems and human well being: biodiversity synthesis. World Resource Institute, Washington, DC

Millenium Ecosystem Assessment (2005b) Living beyond our means: statement from the board. World Resource Institute, Washington, DC

Molinero JC, Ibanez F, Souissi S, Buecher E, Dallot S, Nival P (2008) Climate control on the long-term anomalous changes of zooplankton communities in the Northwestern Mediterranean. Global Change Biol 14:11–26

Möllmann C, Diekmann R, Müller-Karulis B, Kornilovs G, Plikshs M, Axe P (2009) The reorganization of a large marine ecosystem due to atmospheric and anthropogenic pressure – a discontinuous regime shift in the central Baltic Sea. Global Change Biol 15:1377–1393

Nicholls RJ, Wong PP, Burkett JO, Codignotto JO, Hay JE, McLean RF, Ragoonaden S, Woodroffe CD (2007) Coastal systems and low-lying areas. In: Parry ML, Canziani OF, Palutikof JP, van der Linden PJ, Hanson CE (eds) Climate change 2007: impacts, adaptation and vulnerability. Contribution of working group II to the fourth assessment report of the intergovernmental panel on climate change. Cambridge University Press, Cambridge

Oguz T, Fach B, Salihoglu B (2008) Invasion dynamics of the alien ctenophore *Mnemiopsis leidyi* and its impact on anchovy collapse in the Black Sea. J Plankton Res 30:1385–1397

Österblom H, Gardmark A, Bergstrom L, Müller-Karulis B, Folke C, Lindegren M, Casini M, Olsson P, Diekmann R, Blenckner T, Humborg C, Möllmann C (2010) Making the ecosystem approach operational: can regime shifts in ecological and governance systems facilitate the transition? Mar Policy 34:1290–1299

Petersen JK, Hansen JW, Jaursen MB, Clausen P, Carstensen J, Conley DJ (2008) Regime shift in a coastal marine ecosystem. Ecol Appl 18:497–510

Petraitis PS, Dudgeon SR (2004) Detection of alternative stable states in marine communities. J Exp Mar Biol Ecol 300:343–371

Polovina JJ (2005) Climate variation, regime shifts and implications for sustainable fisheries. Bull Mar Sci 76:233–244

Rahmstorf S (1999) Shifting seas in the greenhouse? Nature 378: 145–149

Reid PC, de Fatima Borges M, Svendsen E (2001) A regime shift in the North Sea circa 1988 linked to changes in the North Sea horse mackerel fishery. Fish Res 50:163–171

Remoundou K, Koundouri P, Kontogianni A, Nunes PALD, Skourtos M (2009) Valuation of natural marine ecosystems: an economic perspective. Environ Sci Policy 12:1040–1051

Scheffer M, Carpenter S, Foley J, Folke C, Walker B (2001a) Catastrophic regime shifts in ecosystems. Nature 413:591–596

Scheffer M, Straile D, Van Nes EH, Hosper H (2001b) Climate warming causes regime shifts in lake food webs. Limnol Oceanogr 46:1780–1783

Seppelt R, Dormann CF, Eppink FV, Lautenbach S, Schmidt S (2011) A quantitative review of ecosystem service studies: approaches, shortcomings and the road ahead. J Appl Ecol 48:630–636

Tol RSJ (2009) The economic effects of climate change. J Econ Perspect 23:29–51

Turner RK (2000) Integrating natural and socio-economic science in coastal management. J Mar Syst 25:447–460

Weijerman M, Lindeboom H, Zuur AF (2005) Regime shifts in marine ecosystems of the North Sea and Wadden Sea. Mar Ecol Prog Ser 298:21–39

Wiltshire KH, Malzahn AM, Wirtz K, Greve W, Janisch S, Mangelsdorf P, Manly BFJ, Boersma M (2008) Resilience of North Sea phytoplankton spring bloom dynamics: an analysis of long-term data at Helgoland Roads. Limnol Oceanogr 53:1294–1302

Wooster WS, Zhang CI (2004) Regime shifts in the North Pacific: early indications of the 1976–1977 event. Prog Oceanogr 60:183–200

World Economic Forum (2010) Biodiversity and business risk: a global network briefing. World Economic Forum, Geneva

Part IV

Mediterranean Man and Sea: Myths, Origins, Challenges and Opportunities

Mediterranean Culture and Climatic Change: Past Patterns and Future Trends

30

Anna Maria Mercuri and Laura Sadori

Abstract

This chapter illustrates the main trends of interlaced forces acting on the development of the Mediterranean landscape during the Holocene. The mosaic of habitats distributed in the Mediterranean basin has been continuously transformed by climatic changes occurring at a global scale during the early, mid and late Holocene. In the meantime, the environment has been exploited and the landscape shaped by different civilizations. Climate changes and human activities are observed through the lens of pollen found in terrestrial and marine sediment cores and in archaeological layers. Joint actions of increasing dryness, climate oscillations, and human impact are hard to disentangle, and this becomes particularly true after the mid-Holocene onset of Bronze Age cultures. Regional differences and similarities are reported for eastern, central and western Mediterranean, and for northern Africa and Sahara. The mixing of cultures accelerated the exchanges of ideas, technologies, raw materials and people along the coasts of this 'great lake', making the different civilizations linked between them as one network of regions belonging to the 'Mediterranean culture'.

Keywords

Great lake • Mediterranean culture • Africa • Archaeological layers • Bronze Age cultures • Civilizations • Climate oscillations • Climatic changes • Development of the Mediterranean landscape • Dryness • Holocene • Human activities • Human impact • Mediterranean • Mixing of cultures • Pollen • Sahara • Sediment cores

Introduction

The Mediterranean basin has always featured, and still has, extremely rich environmental and cultural biodiversity. The mosaic of habitats distributed around the Mediterranean basin was primarily transformed by climatic changes occurring at a global scale. In the meantime, the environment has been continuously exploited and the landscape shaped. Mediterranean is in fact a key region that is worldwide as the house for many of the most ancient civilizations.

Since the passage from hunting and gathering to stockbreeding and cultivation, that would have meant a more stable occupation of lands and intense action on a limited area, the different scenarios for resource exploitation have been natural sets for the advance of different cultures (Table 30.1). This led to the onset and expansion of the relevant different cultural landscapes (Bottema et al. 1990; Butzer 2005; Mercuri et al. 2011; Mikhail 2012; Pons and Quézel 1985). The presence of large human groups and well-developed civilizations had remarkable effects on the environment. Changes to biotic and abiotic systems have been so dramatic that human ecologists focused their attention on the impact of human activities on the biotic world,

A.M. Mercuri (✉)
Dipartimento di Scienze della Vita, Università di Modena e Reggio Emilia, Viale Caduti in Guerra 127, 41123 Modena, Italy
e-mail: annamaria.mercuri@unimore.it

L. Sadori
Dipartimento di Biologia Ambientale, Università "La Sapienza", Piazzale Aldo Moro 5, 00185 Rome, Italy
e-mail: laura.sadori@uniroma1.it

S. Goffredo and Z. Dubinsky (eds.), *The Mediterranean Sea: Its history and present challenges*,
DOI 10.1007/978-94-007-6704-1_30, © Springer Science+Business Media Dordrecht 2014

Table. 30.1 Schematic description of relationships between human societies and ecosystems (Diamond 2002; Richerson et al. 1996)

Cultures	Chronology	Resource exploitation		Knowledge of nature	Organization
Advanced agrarian					
From Bronze to Medieval ages, and then to modern times	Mid- and late Holocene – from about 5.0 ka BP to present	Human-manipulated agricultural ecosystems	Cultivations of cereals and other crops; wood management and few plant collection in the wild; animal breeding and herding	Good knowledge and manipulation of ecosystems, increase of control on nature; permanent, large-scale, plowed farming style with more technologically advanced and more efficient techniques for planting and harvesting; large-scale irrigation systems; writing and arts developed and accelerated cultural evolution	Dense human populations and large cities because many people do not have to work to procure food; the increased specialization sometimes results in highly stratified social structures
Early agrarian					
Neolithic; Eneolithic	Early and mid Holocene – from about 11.5 ka BP to about 5.0 ka BP	Mixed from natural ecosystems and agricultural activities	Cultivations of cereals and other crops; animal breeding; integrated activities with hunting and foraging	Good knowledge and first control of natural ecosystems; presence of domesticated species; small-scale slash and burn agriculture, and small-scale irrigation systems	Higher population density than hunter-gatherers; permanent villages and cities; greater division of labor because not everyone needs to work to procure food
Hunter-gatherers					
Late Palaeolithic; Epipalaeolithic; Mesolithic	Pleistocene; early Holocene – before about 11.5–7.5 ka BP	Directly from "natural" ecosystems	Plant harvesting/collection in the wild; hunting wild animals	High knowledge of plant and animal species; horticulture practices	Low population density; mobility rather than living in settlements; little division of labor because everyone works to procure food

and geologists proposed a new epoch the Anthropocene (Ruddiman et al. 2011; Zalasiewicz et al. 2011).

A large number of research works carried out recently shows how many environmental and cultural changes have taken place in the last 20,000 years. After the last Pleniglacial (ca. 20 ka BP), sudden and drastic changes as well as minor oscillations have occurred in global climate. The general climatic instability largely affected vegetation as well as human populations that, at that time, were represented by Late Palaeolithic or Epipalaeolithic hunter-gatherers acting as an inseparable part of natural ecosystems (Flannery 1994; Kuneš 2008; Mercuri and Garcea 2007; Richerson et al. 1996).

An afforestation occurring after the last Pleniglacial is evident in many central and southern European pollen diagrams (e.g. Combourieu Nebout et al. 2009; Lowe et al. 1996; Pons and Reille 1988; Reille and Andrieu 1995). This suggests that there were impressive modifications of habitats and of ecological niches requiring redistribution, new species entries, shift or extinction of flora and fauna. At sites with sufficient moisture availability, a pre-temperate phase of open woodland (with *Juniperus* and *Pinus*) was followed by a temperate phase characterized by early expansion of deciduous *Quercus*, *Carpinus*, *Corylus*, *Ostrya*, *Tilia* and *Ulmus*, and by Mediterranean sclerophylls (Roberts et al. 2011a; Sadori 2013; Tzedakis 2007).

The increase of forest cover, for example, has been stressed among the factors determining the transition from the Upper Palaeolithic (named Magdalenian after the type site of the rock shelter La Madeleine, in France, dated to about 18–11 ka BP) to the Late Palaeolithic in Central and Western Europe. As part of the hypothesis of a 'prehistoric overkill' (Martin 1967), some authors stressed that humans probably contributed to the final extinction of the megafauna. This was constituted by large herbivores such as mammoth, rhinoceros or European bison (Floss 1994; Kuneš 2008). Burney and Flannery (2005) observed that, in the Americas

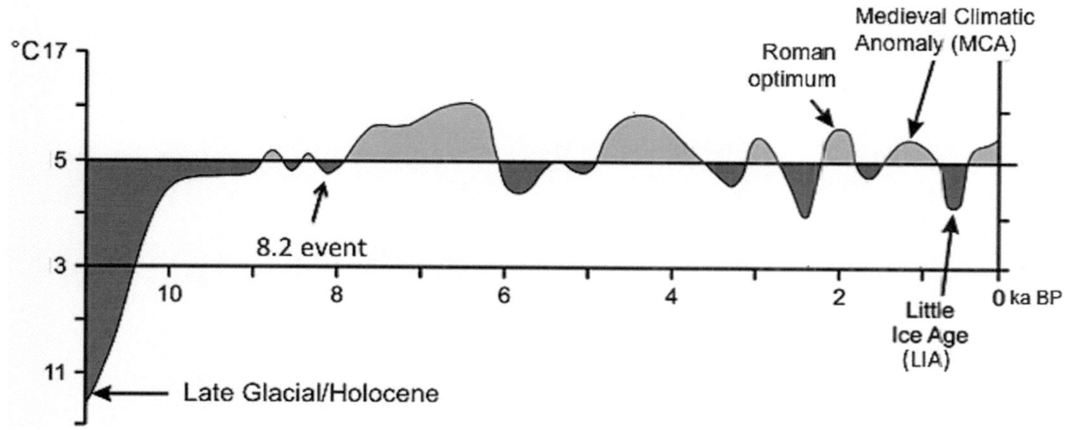

Fig. 30.1 Simplified trend of climate oscillations (temperature estimates) in Europe during the Holocene (Adapted from Schönwiese 1995)

and Australia, the megafaunal collapses point to a variety of disturbance of the equilibrium, including rapid overharvesting, biological invasions, habitat transformations and diseases. On one hand, the large herds need open grasslands, and their survival was threatened by forests. On the other hand, the reduction of herbivores played a key role in vegetation development, and therefore their disappearance has naturally caused a significant reduction of the grazing that had contributed to maintain grasslands. As a matter of fact, the gradual disappearance of large herbivores demanded new hunting strategies adapted to forest game, and changes in mobility of populations (Linstädter 2008).

In this chapter, all the dates are expressed in ka (1,000 years) BP (before 1950 AD).

Early Holocene

With the onset of the Holocene (ca. 11.5 ka BP), while a general trend from wetter to drier conditions occurred, the nature-dominated environment more or less gradually but inexorably moved to a human-dominated one (Berger and Guilaine 2009). This process was profoundly rooted in the dynamic nature-culture system that has established under the Holocene global climate oscillations.

The early Holocene is broadly considered a phase of cool and wet climate, abruptly interrupted by the 8.2 ka event, but it was characterised by increasing temperature with respect to Late Glacial conditions, and several oscillations (Fig. 30.1). In the Mediterranean basin, the early Holocene relative warming favoured environmental change from open steppe-like or grassland vegetation to broadleaf (or evergreen) forest dominated landscapes (Sadori 2013). The maximum extension of deciduous *Quercus* forest marked the so-called Holocene climatic optimum centred between 9.0 and 7.0 ka BP (Combourieu Nebout et al. 2009; Lowe et al. 1996). This maximum of afforestation was more or less evident throughout the Mediterranean. Many records from central-western regions show early Holocene forests (Padul: Pons and Reille 1988; Lagaccione: Magri 1999; Lago di Vico: Magri and Sadori 1999; Lago Grande di Monticchio: Allen et al. 2002) while in south-eastern Iberia, coastal Sicily and south-eastern Mediterranean regions showed a delay in postglacial afforestation (Roberts et al. 2011b; Sadori et al. 2011).

Forest expansion must have continued to influence land use by prehistoric people who became even less mobile and affected more intensively the environments in which they lived. Pollen records register a delay in oak forest spreading into Turkey, suggesting that prehistoric landscape management could have been among the causes of this vegetation behaviour (Roberts 2002). Humans could have intentionally or unintentionally supported the survival, migration and maintenance of useful plants, including food, in the changing environments (Mercuri 2008b).

A rapid extension of the Neolithic populations in eastern Mediterranean regions, from Turkey and Cyprus as far as Bulgaria and Corfu, occurred between 8.5 and 8.2 ka BP. Farming technology spread with them probably at a faster rate than the 1 km per year proposed for 'demic flow' by Ammerman and Cavalli-Sforza (1971). Deforestation due to natural fires caused by climatic conditions may have offered wide-open spaces to first Neolithic colonists favouring the expansion of agriculture and pastoralism. The extension may have been favoured by the earliest agro-pastoralists escaping the arid zones of the Near East and avoiding settling the wet and thickly forested regions of central Europe. According to the modern socio-environmental perspective, Neolithisation was therefore influenced by broad-scale climatic changes transforming both the geomorphology and ecology of the Mediterranean basin and central Europe (Berger and Guilaine 2009).

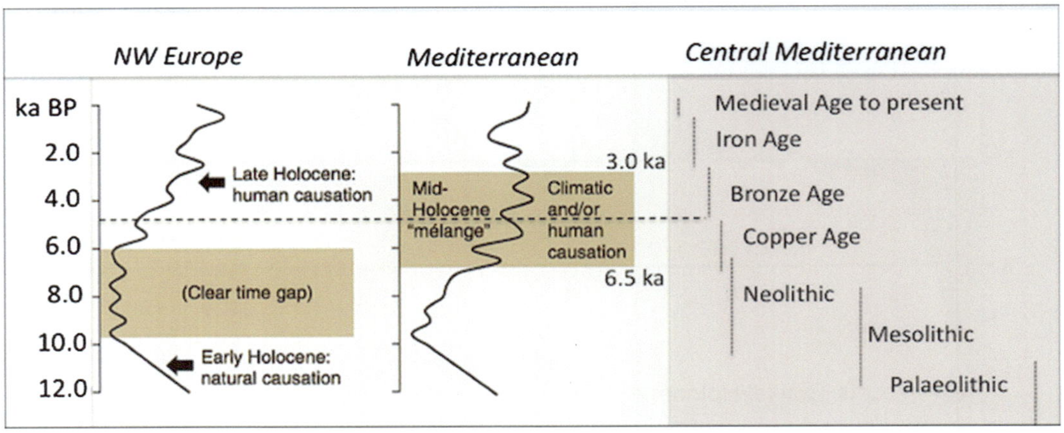

Fig. 30.2 Holocene changes in landscape disturbance by Roberts et al. (2011b), and corresponding cultures in Central Mediterranean regions

Mid-Holocene

The mid-Holocene (around 8.0–3.0 ka BP) was a time of great instability due to both increasing climate variability and cultural changes in many regions of the world. Though climatic and cultural trends were neither global nor synchronous, climate was warmer and less variable for several millennia before 5.8 ka BP than in the immediately following period. Humans had the ability to exhibit adaptive behaviours, not all of which led in the same direction or were equally successful (Sandweiss et al. 1999).

Sadori et al. (2011) selected pollen records from central and central-eastern Mediterranean to detect climate changes during the mid-Holocene. The role of changing seasonality (Magny et al. 2012; Peyron et al. 2011) appears to be crucial for this region. Precipitation seasonality increased during the early to mid Holocene with winter precipitation attaining a maximum, and summer precipitation a minimum. Moreover, at least one rapid dry climate event with changes in plant biomass is centred on ca. 6.0 ka BP in central-eastern Mediterranean and North African regions.

Climate became progressively more arid. Many southern European pollen records show substantial differences between early and late Holocene vegetation, suggesting a general evolution from wetter to drier environmental conditions, in agreement with arid phases recognised by other, independent palaeoclimatic methods. Nevertheless pollen records show different trends and climatic impact varies with latitude (Roberts et al. 2011b). Human groups everywhere adapted by economic switching, or responded by social stress, to environmental changes. These latter included rapid climatic changes such as decadal to multi-century droughts (Weninger et al. 2006). The short-term rapid changes probably impact more on survival than long-term trends (Berglund 2003).

The emergence of many cultures in the biodiversity-rich Mediterranean regions led to the 'mid-Holocene mélange' (Roberts et al. 2011a; Fig. 30.2). This means that climate and human forces have continuously and fairly simultaneously operated in landscape disturbance.

The transition from the Neolithic to the Chalcolithic period started and ended at different times from place to place. The first evidence occurred just after 8.2 ka BP (Çatalhöyük, Turkey), and then from 7.4 to 5.3 ka BP in the eastern Mediterranean, while in the western and central Mediterranean it began about one to two millennia later (Mercuri et al. 2011).

The onset and development of Bronze age complex urban and state-level societies started in eastern regions, showing a east–west time-shift, and may have been a response to increasing aridity (Brooks 2006; Finné et al. 2011; Roberts et al. 2004; Fig. 30.3). Then, from ca. 4.5 ka BP human impact overlapped with a further climate change, probably bipartite, towards dryness (Magny et al. 2009).

The varied responses of the diverse Mediterranean vegetation ecosystems to the gradual mid-Holocene climatic change probably accounts for the time-transgressive nature of changes in regional pollen spectra. Many authors, however, attributed this change to human impact which led to the evergreen sclerophyll dominance in the Mediterranean over several millennia. The high degree of present-day xerophytization is not a unicausal process, but the result of a combination of anthropogenic disturbance and resource exhaustion under increasing aridity.

Late Holocene

During the late Holocene human impact strengthened climate forcing, leading to the well-developed Mediterranean vegetation. Pollen records covering the last millennia are not so numerous, due both the difficulty of recovering recent sediments from extant lakes and agricultural practices that destroyed the top sediments in drained lakes.

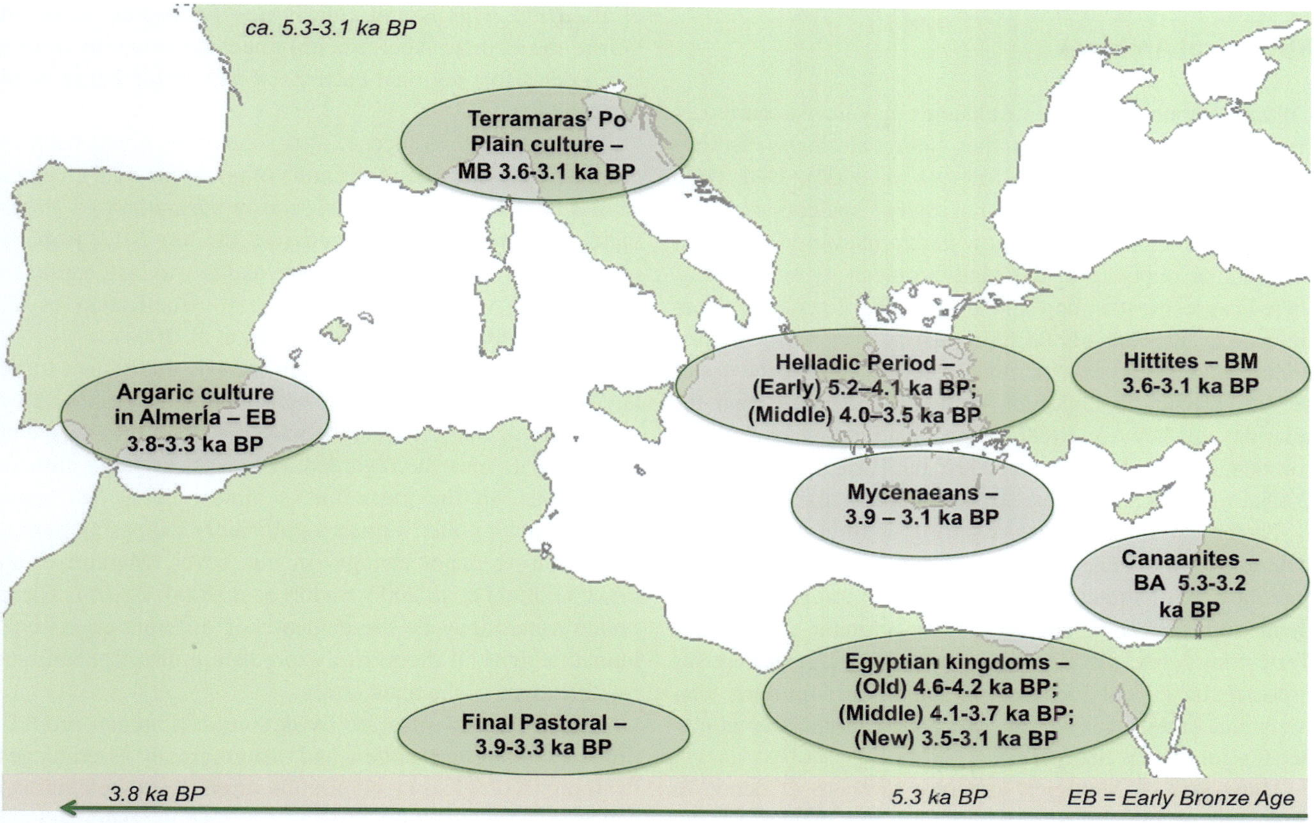

Fig. 30.3 Examples of Bronze Age cultures that developed in the Mediterranean basin during mid-Holocene. Bronze came into use at different times in different regions (*EA* Early Bronze, *MB* Middle Bronze; all dates are to be considered as approximate and general)

Particularly for the last 2,000 years of environmental history it is essential to understand the response of the terrestrial ecosystems to climatic changes under scenarios similar to the present, to better evaluate the interplays and feedbacks of climate variability in the context of human landscapes and activities. Prolonged climatic phases such as the Little Ice Age (LIA) and the Medieval Climatic Anomaly (MCA) stand out as significant periods in history (Fig. 30.1). Both have been related to variations in solar activity, but this linkage remains controversial (Bard and Frank 2006).

Distinguishing between climate changes and human impact on the plant landscape is quite hard for the mid-Holocene, and impossible for the late Holocene. To understand the main forcing causes leading to the present situation we must use climatic archives not influenced by humans. We should therefore rely on speleothemes (Drysdale et al. 2006), and, when possible, on tree rings (Büntgen et al. 2011) and partly on lacustrine level changes. Lakes have, in fact, been often artificially drained in the last 1,000 years (Giraudi et al. 2011). Historical climatic archives may often provide useful records, especially for the last centuries (Camuffo and Enzi 1995).

A recent paper (Büntgen et al. 2011) reconstructed the climate of central Europe for the past 2,500 years using tree rings. It is clear that recent warming is unprecedented, but modern hydroclimatic variations may have at times been exceeded in magnitude and duration. Wet and warm summers occurred during periods of Roman and Medieval prosperity, while enhanced climate variability from ~250 to 600 AD coincided with the end of the western Roman Empire. Such an environmental proxy record, independent from human pressure, is not available yet for the Mediterranean.

Roman times are well represented in pollen samples from Italian harbour basins, of riverine (Pisa port: Mariotti Lippi et al. 2006), deltaic (Rome port: Mazzini et al. 2011; Pepe et al. 2013; Sadori et al. 2010b) and marine (Naples port: Allevato et al. 2010, and Lago d'Averno: Grüger and Thulin 1998) origin. Looking at these records, the general impression is that a complex mosaic of natural vegetation stands and human landscapes coexisted at the time. It is once again clear that only a multidisciplinary approach can help to distinguish between human agency and climate changes in recent times.

Biological Archives

Cultural responses to climate changes may have occurred so frequently and so rapidly that they cannot be clearly discriminated in palaeorecords (Roberts et al. 2004). Data from palynology, hydrology, oceanography, speleothems and marine sedimentology show how the events were sometimes different or opposite in the Mediterranean regions. Today, there is agreement among scholars that the climate or human induced events can only be addressed by integrating detailed archaeological and palaeoclimatic records, and by the use of various proxies to highlight, correlate and understand the relationships between Mediterranean and global palaeoenvironmental events (deMenocal 2001; Jalut et al. 2009; Mercuri 2008a; Mercuri and Sadori 2012; Roberts et al. 2011a).

Biological variables are favouring or limiting factors in human life, including subsistence and activities, and therefore they are of strategic importance in the study of development and spread of cultures under climate oscillations. Terrestrial cores, marine cores and archaeological deposits preserve biological archives, including foraminifera, diatoms and other algae, fungi, and pollen grains mirror palaeoenvironmental changes.

Huge sets of biological archives preserve evidence of flora and vegetation changes (pollen and plant macroremains) in layers from off-site and on-site, archaeological, deposits in the different Mediterranean regions. Obviously the human land transformation and managing are more evident and unambiguous in archaeological sites. In this chapter, we mainly focus attention on data obtainable from pollen, which is among the most abundant and ubiquitous biological remain, preserved in different but complementary contexts. When compared with off-site records, therefore, pollen from archaeological sites helps to evaluate the extent of human influence in different regions (Mercuri et al. 2010b).

Pollen Records Mirroring Cultural Adaptations to Climate Changes

Disentangling climatic and human impact is an everlasting dilemma, and its solution is not easy. Potentially the best way to detect the cause of landscape change comes from combining continuous records from lake cores obtained as close as possible to archaeological sites, thus providing a direct link to cultural history.

The high value of pollen analysis in palaeoenvironmental studies, and its use in reconstructing the climate of the past and evaluating human impact are well known. However, human impact is hard to detect in pollen diagrams from the Mediterranean region (Roberts et al. 2011a, b; Sadori et al. 2011). This is particularly true for sediments dated before the Bronze Age, probably because previous to this it did not affect regional patterns of vegetation being local in scale.

Since the mid-Holocene, also sometimes going back to the Late Glacial, pollen indicators of environmental changes cannot be everywhere unambiguously attributed to a single factor as climate or human activities (Sadori 2013; Roberts et al. 2011a). They both have been, and are, in fact, causes in determining vegetation changes (e.g., Hoelzmann et al. 2001; Oldfield et al. 2003; Zolitschka et al. 2003).

Continental and marine pollen records are known to be key tools for palaeoenvironmental reconstructions as they show fairly synchronous signals from different regions, and these signals must be regarded as indexes of large climate oscillations. In the meantime, sediment cores can show anthropogenic pollen signals significantly increasing in diagrams during major changes in tree cover (Mercuri et al. 2002; Oldfield et al. 2003; Sadori et al. 2004, 2010b). These pollen trends may be the evidence of different degrees of human activity in the territory, producing direct or indirect modifications in the plant cover.

Archaeological sites are by definition a human production, and therefore pollen and macroremain assemblages from archaeological layers are the direct results of 'human action' (Mercuri 2008b). Human presence in a site, and the relevant influence on the environment, can be observed by the presence of houses, storage structures, land-use managing and burials. Most of the plant remains found in archaeological sites are evidence of voluntary or involuntary transport by humans in houses, settlements and burials, while only a minor part comes by air or water. Macroscopic and microscopic plants, stratigraphically accumulated in layers, represent the biological archives which preserve key evidence of plant cover, territory management and resource exploitation made by past populations in the area surrounding the site (Birks et al. 1988; Harvey and Fuller 2005; Mercuri et al. 2006; Pearsall 2000; van Zeist et al. 1991). As a consequence, the archaeobotanical record gives insight into ethnobotanical aspects. Which plants have been harvested or cultivated in the different chronological phases? Were the selected species the preferred ones or simply the most available in the territory? How many changes in human behaviour have occurred when manipulating plants? (Diamond 2002; Peña-Chocarro 1999; Valamoti 2004; Weiss et al. 2006).

Much information is classically obtained from plant macroremains that have a high taxonomic accuracy. Despite its lower specific determination, however, pollen is the best tool for studying the diachronical dimension of events, permitting sequential changes in vegetation and ecology as well as in human impact to be followed (Faegri et al. 1989).

Pollen of trees and herbs highlights high resolution climatic inferences while pollen linked to human activities reflects land use and economy changes. Human actions, such as clearing and settling-in, cultivation and breeding/pasture practices, affect spectra determining variations in quality and abundance of pollen. Thus, making fires destroys pollen (samples from hearths), and accumulation of plants concentrated it (samples from dung and organic sand) (Mercuri 2008b). High pollen concentrations, or percentages, are generically indicative of human behaviour such as food storage in rooms, flower offer in rituals, preparation of drinks, deposition of faeces, forest or wood clearance, changes in subsistence strategies (Behre 1986; Behre and Jacomet 1991; Birks et al. 1988; Jahns 2003; Mercuri 1999; Mercuri et al. 2005, 2006; Rösch 2005; Sadori et al. 2010c; Tipping 1994).

From Natural to Human Environments

Natural and human environments are intimately interlaced as the latter inevitably roots in the former. The relation between natural and human landscapes, in the past as in the present, must be investigated to understand to what extent environmental changes depend on different biological variables.

Humans adapted to the natural environment but in the meantime tried to model the landscape to be more suitable for their activities. They were able to make different degrees of changes in the environment, depending on their culture and economy. Different types of land use are recognisable through plant records (Faegri et al. 1989): (a) Exploitation: the direct use of food and economic plant resources, such as harvesting, selective collection, forest clearance or wood coppicing; (b) Cultivation: the planting and care of crops, by which some native plants can be spread, or exotics-aliens introduced; (c) Stockbreeding: indirect use, by increase of pastureland and selection of unpalatable plants by animal browsing; (d) Settlement construction: spreading of ruderals, nitrophilous plants, or plants preferably living in trampled areas.

Pollen and NPPs – Non Pollen Palynomorphs (e.g., Florenzano et al. 2012; van Geel et al. 2003) are especially useful to discriminate these types of actions. In archaeological sites, the useful or economic plants are represented in decidedly higher amounts than the normal component of the background pollen rain (Kelso et al. 2006). Obviously, the diverse typologies of land use can coexist, and their 'botanical traces', both micro- and macroremains, are recognisable in the same layers.

Pollen data converge showing that the joint influence of dryness and human activities must have increased the extension of xeric environments with shrubs and open areas including grasslands (Jalut et al. 2009; Mercuri et al. 2010a; Fig. 30.4). Landscape modifications moved the natural environments towards human environments deeply impressed by settlements and activities, a transformed land that everywhere has been shaped into the Mediterranean cultural landscapes (Fig. 30.5).

The understanding of the genesis and evolution of the cultural landscape, though not explicit, is at the centre of the debate on environmental transformations in the Mediterranean basin. How much did vegetation changes depend upon climatic changes? Was the response of vegetation synchronous with the forces of change? How to evaluate the effects of human activities on natural vegetation in the past? To resolve the latter question we must observe what happened before a settlement was established, and at its onset. What happened after the abandonment of a settlement: how does the socio-ecosystem react? What kind of resilience mechanism is acting? We must, therefore, use the same approach to study present day cultural landscapes, and to compare their past and present natural history.

Past Patterns

The genesis of human influence on the Mediterranean plant landscape has roots that possibly precede the Holocene. The early Holocene was characterised by wet and cool conditions, and the climate was considered as Mediterranean, close to that of the present meso-Mediterranean or supra-Mediterranean sub-humid or humid stages (Jalut et al. 2009; Quézel and Barbero 1982). During the last decades, pollen analyses from continental deposits and archaeological sites have shown that during the early Holocene, deciduous broad-leaf trees, and especially mixed oak woods, were frequently dominant while sclerophyllous and evergreen trees were present, but rare. This was the time of early Epipalaeolithic/Mesolithic transitions in Mediterranean and North African regions.

The Neolithic dispersal, progressing from east to west Mediterranean Europe, occurred between 8.5 and 7.2 ka BP (Berger and Guilaine 2009). An abrupt cool and dry event is recognised at around 8.2 ka BP in several records, and noticeable lake-level lowering, which correlates with a dry phase, was observed at ca. 8.6–7.9 ka BP in central Italy (Lago dell'Accesa, Tuscany: Drescher-Schneider et al. 2007; Magny et al. 2007). The expansion of drought-tolerant taxa all over the Mediterranean basin occurred principally after the 8.2 ka BP event, associated with a re-organisation of regional climate (Sadori et al. 2011).

In the mid-Holocene, these regions became progressively more arid, especially after 6.5–6.0 ka BP, and then around 5.0–4.0 ka BP. Sclerophyllous and evergreen forests developed in western-central Mediterranean regions. According to different interpretations, their spread was

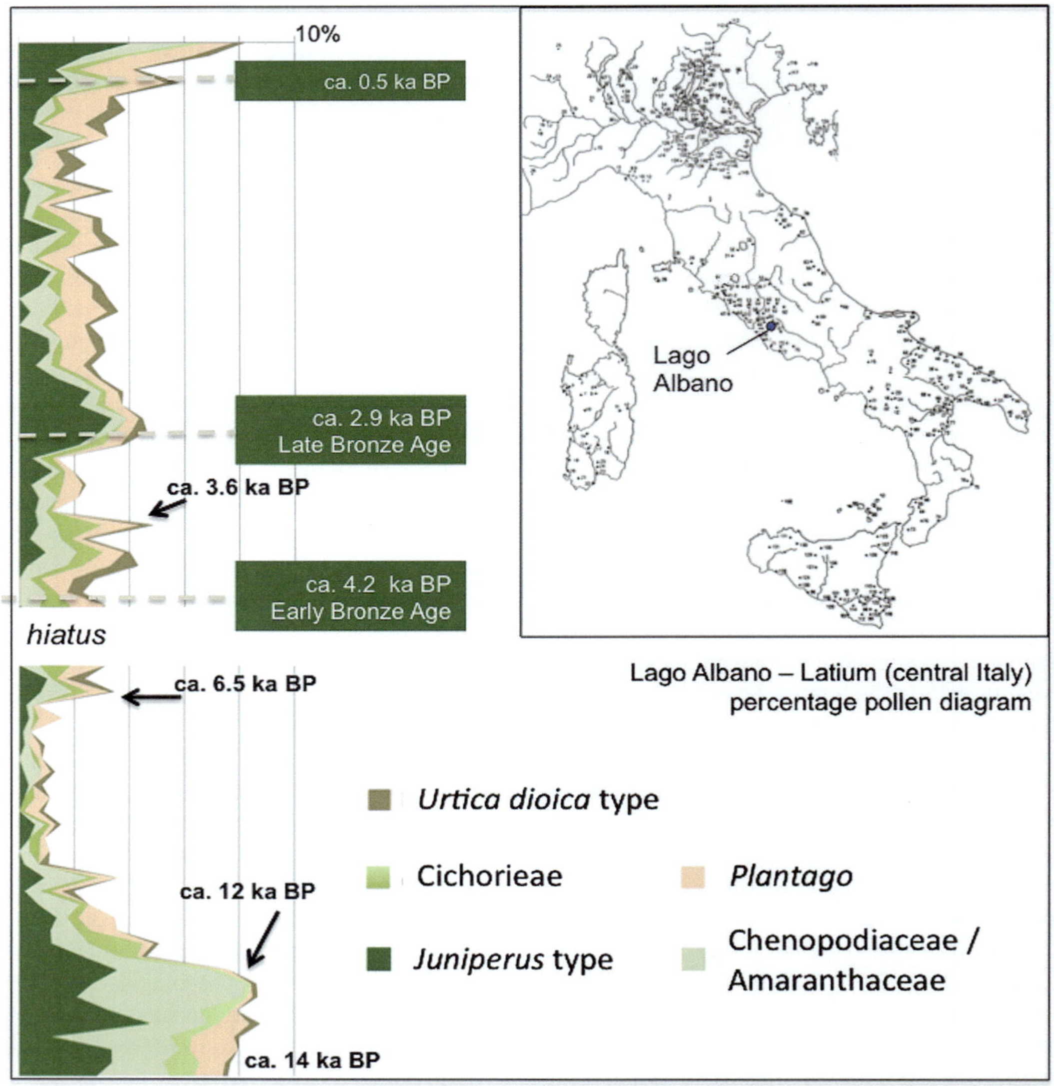

Fig. 30.4 Xeric environments were widespread in the Late Glacial. They became less common, and their mid-Holocene rise, especially from around 6.5 ka BP onwards, coincided sometimes with human disturbance. At around 3.6 ka B.P., the increase of xeric vegetation types marked the presence of Middle Bronze Age settlements and pastoral activities near the shores of the lakes (Lago di Albano and Lago di Mezzano) in Central Italy. On the *right*, the map reports the dense number of archaeological sites dated to metal ages discovered in Italy (From Guidi and Piperno 1992, redrawn)

considered either an effect of human impact or as the natural development of Mediterranean vegetation favoured by the drying trend of climate oscillations (Carriòn et al. 2010a, b; Mercuri et al. 2012; Pons and Quézel 1998; Roberts et al. 2011a; Sadori et al. 2004, 2011). Two other lake-level lowerings were observed at ca. 4.6–4.3 and 3.7–2.8 ka BP in central Italy (Lago dell'Accesa). Central and northern Italy and the interior parts of the Balkans were less affected by mid-Holocene 'Mediterraneanization' than other regions such as Spain, Sicily, Croatia or southern Greece. Even today, central and northern Italy is one of the wettest and greenest regions around the Mediterranean Sea (Sadori et al. 2011).

Eastern Mediterranean

The eastern Mediterranean, cradle of some of the oldest and most important civilisations of the world, was affected by very significant changes in climate during the Holocene. Therefore, joint effects of climate and increasing human impact are particularly hard to disentangle. Moreover there is no clear division separating an early Holocene period of nature-dominated environment from a human-dominated late Holocene one (Roberts et al. 2011a). Palaeoclimate records seem to indicate that the eastern Mediterranean experienced an increase in winter-season precipitation during the early Holocene, followed by an oscillatory decline at

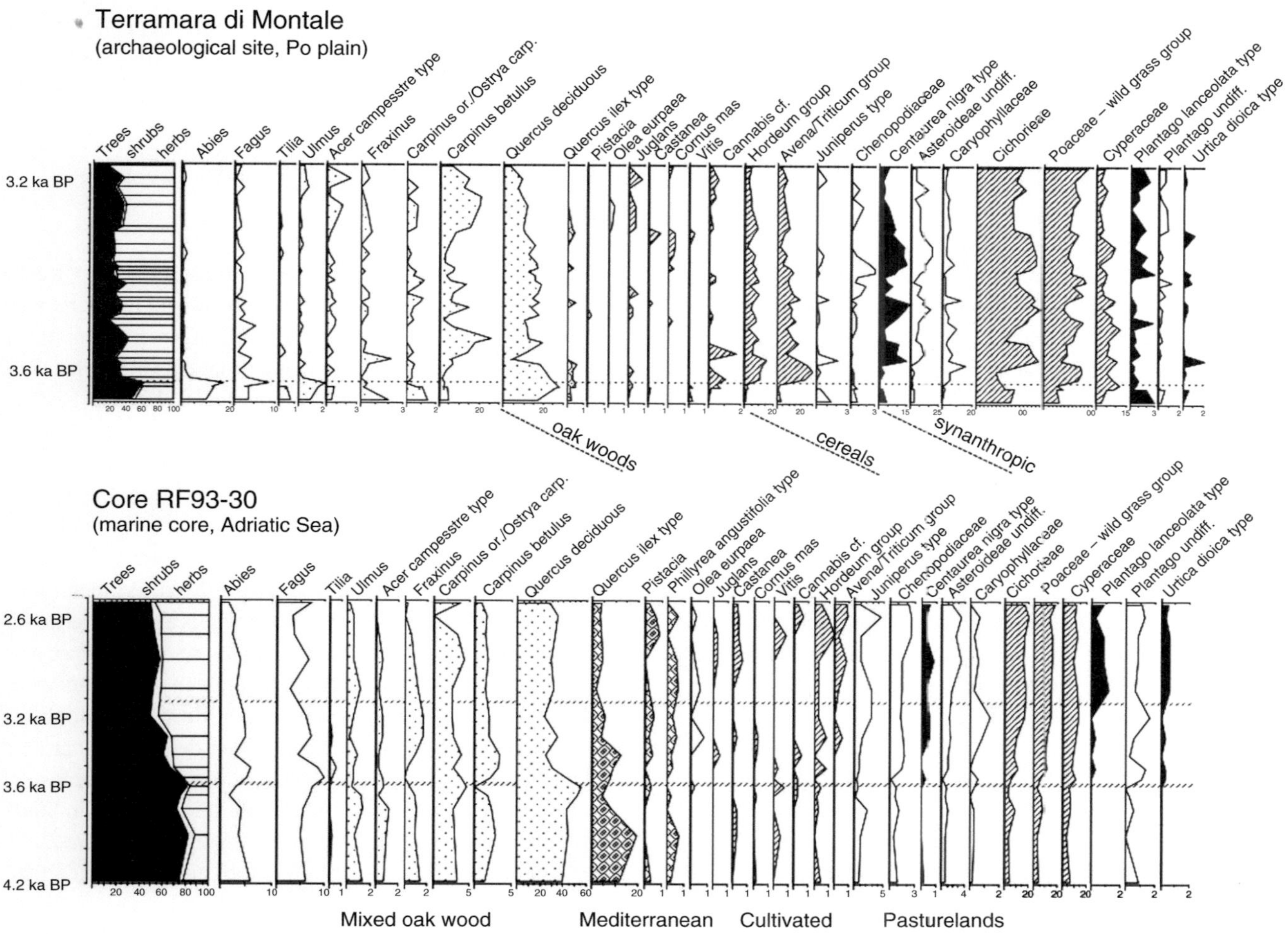

Fig. 30.5 The cultural landscape has clearly developed in the Italian peninsula since the Bronze Age. Pollen curves representing main changes in vegetation cover under human pressure: decrease of deciduous oaks, increase of cereals and wild synanthropic plants are taken as the evidence of human impact. The 2 % pollen diagrams refer to two very different contexts: *top* – Terramara di Montale, archaeological site of Middle Bronze Age, Po plain; *bottom* – RF93-30, marine core from the central Adriatic Sea (From Mercuri et al. 2012, redrawn)

ca 6.0 ka BP. Since the review by Willis (1994) for the Balkans and the data gathered by Bottema and colleagues (e.g. Bottema 1987, 1995; Bottema et al. 1990, 2001) for the Near East many new data from the region have become available. New pollen records have been published from Crete (Bottema and Sarpaki 2003), the Dalmatian islands of Mljet (Jahns and van den Bogaard 1998) and Cres (Schmidt et al. 2000) and from western Greece (Jahns 2005). A particular increase in information about eastern Mediterranean palynology regards Albania, where big lacustrine basins are present. Two of them are wide and trans-boundary lakes, Lake Ohrid and Lake Shkodra. There are current (Lake Shkodra, Sadori et al. 2012) and published (ancient Lake Maliq: Bordon et al. 2009, Denèfle et al. 2000, and Lake Ohrid: Wagner et al. 2009) research studies about the Postglacial vegetation and climate history of the region.

Holocene forest expansion was mainly by deciduous oak and accompanying taxa, but it varied widely across the Balkan Peninsula, suggesting different possible locations of single taxa refugia. The spread of forest was not synchronous. A dense canopy forest was established in north-western Greece by 12.5 ka BP, while in northern Greece and in southern Bulgaria, woodlands remained fairly open until ca. 10.2 ka BP. Around 11.9 ka BP an intermediate forest cover was established on the island of Cres (Croatia). In Anatolia region there was a significant delay (some millennia) in deciduous oaks reafforestation, probably hindered by human action (Roberts 2002).

A slight but noteworthy expansion of *Pistacia* (probably *P. terebinthus* in most sites) occurred during the early Holocene throughout much of Greece (Lawson et al. 2004). Its peaks between 10.2 and 8.9–8.3 ka BP suggest an aridity phase. This spread was also recorded in Crete, with maximum values at 8.9 ka BP.

A change was found in most upland sites from Greece and Albania (Denèfle et al. 2000) between 8.3 and 6.8 Kyear

BP. This was marked by decreasing levels of deciduous oak pollen, and by increasing levels of one or more of the following taxa: hazel, hornbeam (*Carpinus orientalis*/*Ostrya*), elm, linden, beech, fir and pine. As striking similar vegetation dynamics were found during previous interglacials, many possible explanations have been advanced, ruling out anthropogenic intervention. The expansion of hornbeam and beech found at Lake Maliq may point to conditions of lower available moisture and warmer winters, as these trees are tolerant of dryness, but intolerant of winter cold (Denèfle et al. 2000).

An important change was found in a quite different landscape, on the Adriatic island of Mljet, around 8.1 ka BP. There, a partial replacement of deciduous oak by juniper and *Phillyrea*, resembles the start of the slow spread of many sclerophyllous trees found in Sicily (Sadori and Narcisi 2001) and in central Italy (Magri and Sadori 1999). This last change seems to be synchronous with the aridity phase in the southwestern Mediterranean from 8.3 to 7.8 ka BP. The chronology of each sediment record has to be improved to carry on unequivocal correlations of vegetation and climate changes and a good help in this sense is coming from tephrochronology (Zanchetta et al. 2011).

Even if it is ascertained that human societies developed particularly in the eastern parts of the Mediterranean basin since the Neolithic, unequivocal human intervention is detected only very late in the pollen record (Sadori 2007). The first sign of disturbance consists in a reduction of both forest density (increasing pasture and secondary anthropogenic indicators) and/or diversity (reduction of deciduous oaks and appearance/increase of fruit trees and/or Ericaceae). Such a sign is found in most pollen diagrams only after 6.0 ka BP. A "Mediterraneization" of the climate could likewise have produced such a vegetation change. We should consider that an important change towards more arid conditions is found in speleothemes from Soreq cave, in Israel (Bar-Matthews et al. 1997; Bar-Matthews and Ayalon 2011).

Cereal pollen type is always recorded in pollen diagrams of Near East, as it is the area of origin and domestication of most edible plants. The huge amount of cereal pollen grains found in the sediment record of Çatalhöyük (Eastwood et al. 2007) suggests that the best way to detect clear signs of human presence is to sample near human settlements, as at the beginning the human impact was very local in scale.

Pollen of fruit trees appeared very early, although in low amounts. Most of them were present in the region since early Holocene. In some sites, walnut pollen grains appeared or increased at ca. 3.5 ka BP (together with *Ceratonia* in Crete, Bottema and Sarpaki 2003); some centuries later the pollen of these cultivars was spread throughout Greece (Jahns 2003) and Dalmatia. The onset of olive cultivation is difficult to define, as it is not clear if the expansion found at least since the Bronze Age could be ascribed to a natural spread of this native plant. From Greek times onwards, human impact produced strong landscape changes culminating in the present conditions.

Central Mediterranean

During the Holocene, a moist period from ca. 9.5 to 7.8 ka BP, interrupted by the 8.2 event, was followed by increasing dryness from ca. 7.8 to 5.0 ka BP. Changes in seasonality might have been responsible for climate oscillations. The earlier phase, in fact, seems to have been characterised by wet winters and dry summers inducing a seasonal hydrological contrast stronger than today, while the later phase had a reduced seasonal contrast and lower overall precipitation in central Italy (Peyron et al. 2011).

After these changes, over time the cultural landscape characterises more and more the central Mediterranean with different degrees of evidence mainly observable in the trends of arboreal pollen grains. Their changes seem to respond to different forces. In Italy, a decline of evergreen oak occurs when the frequency of fires increased at the Mesolithic/Neolithic transition in Tuscany, the entire tree cover decreased at around 7.7 ka BP in Latium, and the biomass drop and an increase of olive pollen were evident from the Neolithic onwards in Sicily (Colombaroli et al. 2008; Magri 1999; Magri and Parra 2002; Pérez-Obiol and Sadori 2007; Sadori and Giardini 2007; Sadori and Narcisi 2001; Sadori et al. 2008).

The onset of the Bronze age, archaeologically dated from about 5.0 to 3.5 ka BP in the Mediterranean basin is among the most problematic examples of a cultural change probably driven by climatic changes. The mid-Holocene was in fact a period of important climate instability. In Italy, the Bronze age started at ca. 4.3–4.2 ka BP (Cardarelli 2009). A mid-Holocene forest opening as en effect of a dry phase has become evident in many central Italian lakes since these times, approximately at around 4.3–3.6 ka BP (Lago di Mezzano: Sadori et al. 2004; Sadori and Giardini 2008; Lago Albano and Lago di Nemi: Mercuri et al. 2002; Rolph et al. 2004; Lago di Vico: Magri and Sadori 1999; Lago dell'Accesa: Drescher-Schneider et al. 2007). This dry phase was mostly contemporary with the hydrological changes in Lago Fucino and other lakes in central Italy (at 4.2–4.1 ka BP; Giraudi 1998, 2004; Giraudi et al. 2011) and with a decrease in pollen concentration recorded at Lagaccione (Magri 1999). Forest clearance might have happened twice in the range of few centuries (Magny et al. 2009). However, a sure correlation of these events should be assessed using the position of tephra layers as it was proved that radiocarbon chronology might be some centuries older than tephrochronology. In fact, the Avellino tephra layer contained in sediments radiocarbon dated at ca. 4.3–4.2 ka BP was

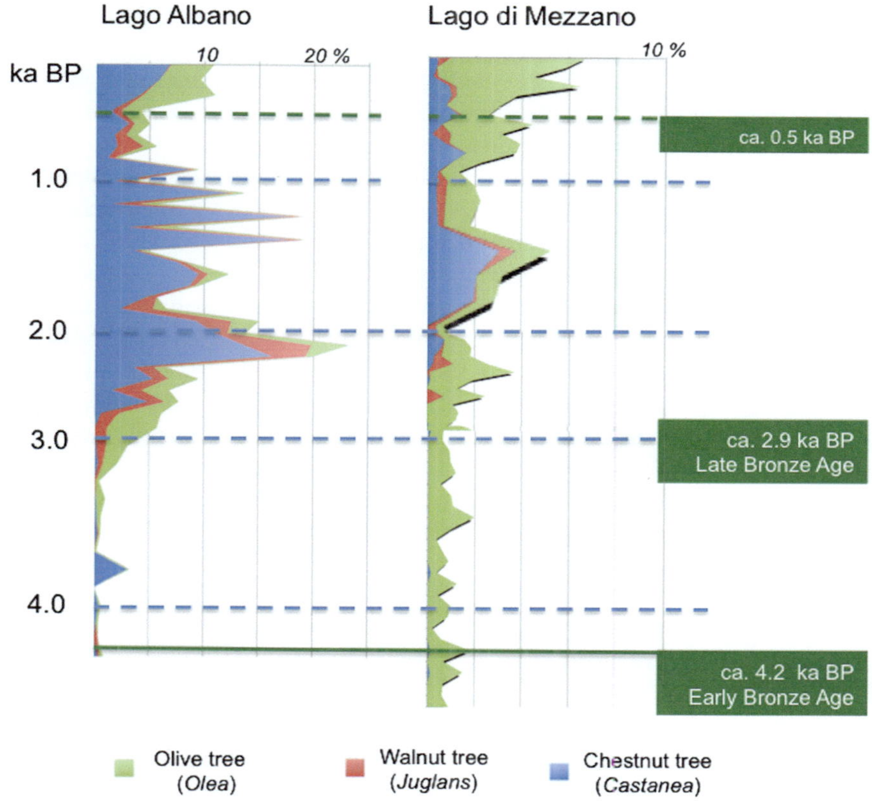

Fig. 30.6 Human impact and the evolution of the cultural landscape in Italy can mainly be traced through the OJC group curve that includes pollen of *Olea*, *Juglans* and *Castanea* (Examples from Lago Albano (Mercuri et al. 2002) and Lago di Mezzano (Sadori et al. 2004; Sadori and Giardini 2008)

recently dated at ca. 3.8–3.9 ka BP (Sevink et al. 2011; Sulpizio et al. 2008; Zanchetta et al. 2011).

Few centuries after the occurrence of the Avellino tephra, a sudden and strong forest thinning signed the advance of drier conditions at Lago di Mezzano, Lago Albano and Lago dell'Accesa (Sadori et al. 2011). An important change in tree concentrations, that did not match an increase in fires and that therefore may be considered as a natural opening of the forest, preceded the establishment of Middle Bronze age populations near the shores of Lago di Mezzano (Sadori and Giardini 2008; Sadori et al. 2004). In the meantime, the terramara civilization developed in the Po plain, and people moved near rivers. The socio-economic system consisted of villages, sometimes settled on natural hillocks surrounded by embankment and ditch (Mercuri et al. 2006, 2013; Ravazzi et al. 2004).

The movements of populations were probably favoured by the decrease of water availability due to an increasing dryness of climate. Therefore, xeric environments spread near the lakes of central Italy (Fig. 30.4) and lave level lowering probably provided fertile and flat beaches available for cultivation. A human-modified landscape became visible in many contexts by 3.6 ka BP, with clear evidence of the local performance of agrarian activities such as cereal and legume cultivations, forest clearance of selected species, fire, and grazing activities. In the meantime, the increase in anthropogenic indicators such as olive and cereal pollen became evident in Lago Albano, Lago di Nemi, Lago di Mezzano and Lago di Vico (Magri and Sadori 1999; Mercuri et al. 2002; Sadori et al. 2004).

Human impact became unquestionably evident soon after the Bronze age in central Italy, since by 3.0 ka BP cultivated trees including olive-*Olea*, walnut-*Juglans* and chestnut-Castanea trees (OJC) spread simultaneously, and cereals were grown in fields (Fig. 30.6). This suggests that the pre-Roman Iron age populations had an advanced agricultural economy (Sadori et al. 2010a). A well-established agriculture was evident in the strong expansion of chestnut in pollen diagrams dated at around 2.9–2.7 ka BP (Valle di Castiglione, Lago di Vico, Lago Albano and Lago di Nemi), and some later around 2.4 ka BP at Lago di Mezzano. The cultivation of trees matched an enhanced forest cover after the forest clearance of Bronze age, e.g. at Lago di Vico. A '*Cannabis-phase*', i.e. a hemp pollen peak which followed the peaks of the cultivated trees, marked Roman times while the 'chestnut landscape' definitively spread at around 0.8–0.7 ka BP (Mercuri et al. 2002, 2012).

In addition to central Italy lakes, often very close to archaeological sites, research mainly concentrated in coastal areas such as lagoons, marshy zones, coastal plains and ancient port areas. From 9.0 to 6.0 ka BP the records from the Ligurian and Tuscany coasts were characterized by high percentages of fir, probably growing close to the sea (Bellini et al. 2009). At that time in Sicily Mediterranean vegetation was already established both at Gorgo Basso and Biviere di Gela (Tinner et al. 2009), while the development of open meso-thermophilous woods and maquis on the coastal plains of southern Liguria and Tuscany dates back at 7.0 ka BP. Just north of the Tiber delta, at Stagno di Maccarese, a quite similar situation is found, with riparian fresh water trees playing an important role (Di Rita et al. 2010). From the Adriatic side of the peninsula, in Apulia the available data come from Lago Alimini Piccolo (Di Rita and Magri 2009) and the former Lago Battaglia (Caroli and Caldara 2007), with evergreen Mediterranean vegetation dominant from the beginning of the records, respectively at ca. 5.5 and at ca. 6.0 ka BP. The last couple of millennia are well represented in Roman harbours (Allevato et al. 2010; Grüger and Thulin 1998; Mariotti Lippi et al. 2006; Sadori et al. 2010b).

Western Mediterranean

Within the wet early Holocene, two dry intervals at 10.9–9.7 ka BP and 8.4–7.6 ka BP were recognised on the basis of the pollen ratio between deciduous broadleaf and evergreen sclerophyllous trees. The upper limit of the second dry interval corresponded to a major oceanographic circulation change recorded in the western Mediterranean at 7.7 ka BP. The Holocene vegetation of this area appeared rather diversified. In fact, clear differences of vegetation development and composition between the regions with continental and Mediterranean climates were found (Beaulieau et al. 2005; Jimenez-Espejo et al. 2007).

In Spain the scientific debate about the role of climatic change has been mostly restricted to vegetation dominated by oak and pine. Many sites show high values of pine persisting since the early Holocene, suggesting that mountain pine forests were partly spread in the inner and Atlantic regions of the Iberian Peninsula. However, during periods with open woodlands, the effects of over-representation of pine pollen in percentage diagrams should be considered. As suggested by Carrión et al. (2010a) different causes such as continentality, altitude and aridity may have favoured pines in opposition to oaks. Thus, the existence of *Pinus* woodlands during the Lateglacial and Early Holocene can be considered a consequence of the orographic influence in several of the Iberian mountain pollen records.

The recent study of the Planells de Perafita fen and of other sites in Andorra (Ejarque et al. 2011; Miras et al. 2007) led Miras et al. (2010) to hypothesize a quite early human presence in the Pyrenees, probably of seasonal nature, up to the sub-alpine belt, during the Mesolithic-early Neolithic transition dating back to ca. 8.4–8.0 ka BP.

Then during the whole mid-Holocene, in some Iberian continental and mountain zones pine forests seem to have persisted where conifers were not replaced by broadleaf trees, as recorded in other areas of south-western Europe.

The close examination of several pollen sequences of south-western Spain carried out by Pérez-Obiol et al. (2011) revealed climatic transformations in the flora and vegetation between 7.0 and 4.0 ka BP. Carrión and colleagues (2010b) carried out a recent analysis of the aridification trends and processes occurred in Murcian-Almerian bioprovince, south-eastern Spain, during the Holocene.

The mid- Holocene sudden and sharp decrease of box (*Buxus*), which led to its extinction at Minorca in about one millennium, interpreted as a sign of increased xeric conditions (Yll et al. 1997), was recorded in all the Balearic Islands.

The ratio deciduous broadleaf and evergreen sclerophyllous trees was useful to identify a number of dry phases during the mid and late Holocene intervals 5.3–4.2, 4.3–3.4, 2.8–1.7 and 1.3–0.7 ka BP (Jalut 2005; Jalut et al. 1997, 2000). Some of them were correlated with phases of decreasing fluvial activity described by Magny (1999) and Magny et al. (2002) in the western Mediterranean around 4.0, 3.0, 2.0 and 0.8 ka BP (besides those at around 11.5, 10.5, 9.0, 7.0 ka BP).

The human impact, highly variable in the investigated region, complicated the possibility to distinguish between natural causes and human inputs. Such an impact on the environment was highly variable over south-eastern Spain, starting earlier in low-elevation areas and river basins, after 5.0 ka BP. Forest degradation in the mountains started even later (4.3–3.6 ka BP) and reached its maximum during the Roman occupation, ca. 2.0 ka BP.

The hypothesis of a climatic link between southern Spain and North-Western Africa is strengthened by data concerning the transport of pollen of *Cedrus* from North Africa to the northern shore of the Mediterranean (Magri and Parra 2002). This would explain the existence of a Mediterranean vegetation cover since the beginning of the Holocene in southern Spain (Jalut et al. 1997, 2000).

Northern Africa and Sahara

During the Late Pleistocene and early to mid-Holocene, the Maghreb and Sahara had more luxuriant vegetation and water environments than today. These regions were cooler, less arid, more heavily forested at higher elevations and along the coast, and more densely populated by large herbivores. All these conditions, even considering the local differences, would have made them attractive to hunter-gatherer populations (Cremaschi and Zerboni 2009; Lézine 2009; Lubell 1984, 2001).

The early Holocene was characterised by relatively favourable environmental conditions, with trend of increasing temperature and precipitation in North Africa. New cultural patterns developed under these conditions (Linstädter 2008).

Between about 9.0 and 6.3 ka BP, stronger monsoons produced wetter conditions than earlier in the Holocene, and in the region west of the Nile Valley, the villages which were seasonally occupied by cattle herders expanded. Then, the region became arid again as a result of changes in the monsoon regime, and it was eventually abandoned (Sandweiss et al. 1999).

The wet phases, which in northern Africa are generally attributed to stronger monsoon activity, were marked by the formation of lakes in the Sahara desert (Cremaschi 1998, 2002; Cremaschi and Zerboni 2010). In the Sahara, there was a progressive desiccation taking the form of short phases of increasing aridity (Gasse 2002). The arid phases, attributed to weak monsoon activity, can be dated around 12.4, 8.2, 6.6 and ca. 4.0 ka BP (Gasse 2000; Jousse 2006; Mercuri 2008a). These dates are consistent with wind-borne sediments matching the drying up oscillations of the Gobero Lake in Niger (Giraudi and Mercuri 2013).

In Northwest Africa, during the early Holocene, the Capsian and the Mediterranean Epipalaeolithic cultures (ca. 11.5–7.0 ka BP) developed (Linstädter 2008). They came from the split of the Ibero-maurusian population, which occurred at the Late Glacial (ca. 14.7–11.5 ka BP), into two major units. Though there has been considerable destruction of Capsian sites by erosion (Lubell et al. 2009), it seems that there was continuity between the late Pleistocene Iberomaurusian and early Holocene Caspian industries, in terms of lithic assemblages, cranial morphology, site location, and chronology (Lubell 1984). Capsian populations colonized the previously semi-arid areas of the Northern Sahara as the increased precipitation of the African Humid Period allowed occupation of the former hyper arid Sahara (Cremaschi 2002).

Later, under the influence of the neighbouring Neolithic cultures – the Mediterranean in the northwest and the Saharan in the south – the Capsian economy may have been transformed into the so-called Neolithic of Capsian Tradition at around 7.0 ka cal BP. During the mid-Holocene, in the eastern Maghreb, Capsian sites first occupied thousands of years previously continued to be used by later groups thus giving a further indication of continuity during the succeeding periods in the Maghreb (Lubell et al. 2009).

The Mediterranean Epipalaeolithic remained in the traditional territory and adapted to environment changes such as afforestation. This transition is not clear, sites are scarce and data are incomplete. New subsistence strategies, including changes in mobility and hunting adapted to forest game, may include collectors who move away from long term residential camps for hunting and collecting (residential mobility; archaeologically well detectable), and foragers who displace their residential camps when the resources of the local exploitation were exhausted (logistical mobility; archaeologically more difficult to detect) (Binford 1983).

Then, the Early Neolithic and the Epipalaeolithic with pottery (7.8–6.9 ka BP) developed. The early Neolithic, with scattered sites along the coastline, displayed a subsistence including heavy exploitation of marine resources, domesticated animals and domesticated plants.

In Tunisia, the economy of the coastal site of Hergla (SHM-1) was epipalaeolithic linked to hunting, fishing and gathering activities, and the material culture was characterized by a Capsian substratum (Mulazzani et al. 2010). In eastern Morocco and western Algeria, the local hunter-gatherer Epipalaeolithic populations did not seem to have either domestic animals or plants but adopt Neolithic innovations (Ballouche and Marinval 2004; Linstädter 2008). This process started with the production of pottery, with both local styles of decoration and overlapping influences, and continues later with the introduction of domesticated animals.

When the retreating monsoonal rains caused the onset of desiccation of the Eastern Sahara at about 6.3–6.2 ka BP, sites were abandoned (Nabta Playa in Egypt: Wendorf and Schild 1998) and people moved to the Nile valley or ecological refuges, or into the Sudanese Sahara where rainfall and surface water were still sufficient (Kuper and Kröpelin 2006). In the Nile Valley, however, people began to worship cattle and create monumental architecture (Brookfield 2010; Sandweiss et al. 1999). In the Central Sahara, pastoral cultures passed from cattle to sheep/goat herding, better adapted to a dry climate, and wild cereal exploitation changed towards more drought-resistant species (Cremaschi and di Lernia 1999; di Lernia 2002; Mercuri et al. 1998; Fig. 30.7).

The aridity associated with the southward retreat of the monsoon system was definitively established at around 5.0–4.0 ka BP (Gasse and van Campo 1994; Kuper and Kröpelin 2006). The timing of the local environmental response to the cessation of rainfall around this time depended on geographical location, topography, hydrogeology, and regional climatic systems, and culminated in the present-day desert.

In the Wadi Teshuinat, Libyan Sahara, the action of herbivores on plant cover favoured the increase of shrubby vegetation (Mercuri 2008a). In stable climatic conditions, this would have led to a more stable, denser vegetation, and moderate grazing would also have resulted in grass cover. However, in Saharan areas, the pressure of grazing, in addition to the drying climate, contributed to Poaceae decrease and Chenopodiaceae increase, bringing the environment into overgrazing conditions (Mercuri et al. 1998). Grazing and browsing animals, both domestic and wild, play an important role in African grasslands, savannas and woodlands. Besides denudation of the soil, effects on vegetation include bush encroachment and increase in unpalatable species.

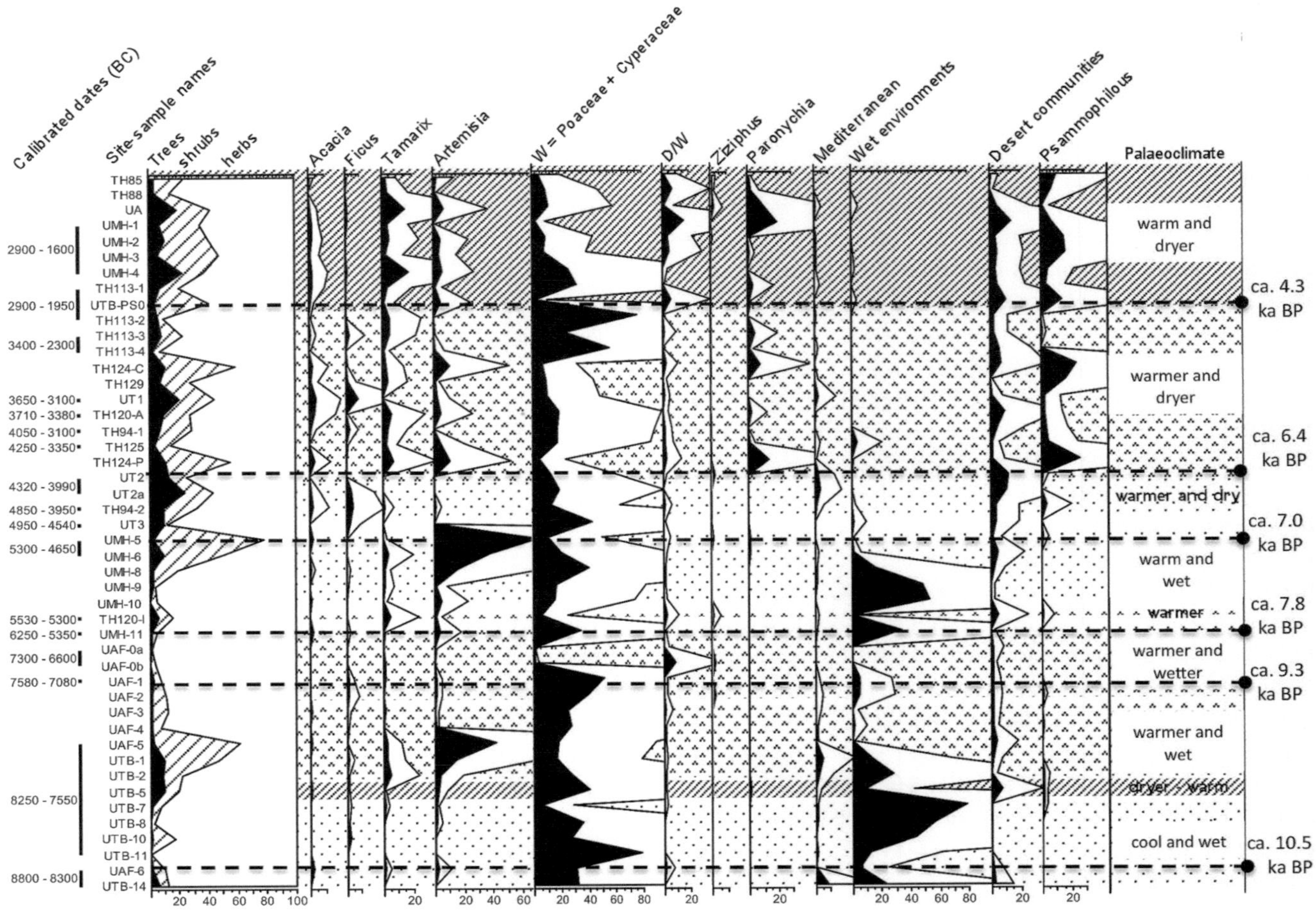

Fig. 30.7 Percentage pollen diagram of the archaeological sites of the Wadi Teshuinat region, in central Sahara, south-western Libya (From Mercuri 2008a, redrawn). According to the original paper, names of the samples and chronology in calibrated ages BC are reported at the *left* of the diagram. Samples are ordered per chronology and stratigraphy. Though spectra are largely marked by anthropogenic pollen (grazing plants in pastoral sites, food and medicinal plants in hunter-gatherer sites), main climatic changes can be inferred. Main early and mid-Holocene oscillations are reported in the last column. (W = sum of Poaceae + Cyperaceae, indicative of relatively wetter conditions; D = sum of Chenopodiaceae + Asteraceae, indicative of relatively dryer conditions)

The simultaneous increase of aridity and ovicaprine grazing pressure favoured the spread of xerophilous vegetation and therefore the economy enhanced the effects of dryness on vegetation (Mercuri 2008a).

The late Holocene was characterised by Garamantian, Roman and then Medieval cultures, whose rural and trade economies developed in the Saharan oases (Liverani 2005; Mercuri et al. 2005). This was documented as long ago as the early explorers, and was well known in historical times.

Future Trends

The deep sense of the palaeoenvironmental research is to understand what can be the lesson of the past to estimate future trends of world history. Oldfield (2005) and Dearing et al. (2006) stress that "our current predicament focuses attention on the question of how best to anticipate human–environment interactions in the future. To what extent will we be able to anticipate, project, forecast or predict future conditions in human-dominated landscapes and highly valued ecosystems? What is realistic with respect to understanding climate–human activity interactions and defining the needs for sustainable management?"

The current environment developed in response to climatic and human factors, with plant cover transformations and human influence on the landscape closely interconnected. Thus, while, subsistence strategies were adjusted to suit climatic and environmental changes, at the same time, humans slowly and continuously shaped the plant landscape (Mercuri 1999).

Humans act as part of ecosystems, and have always been actors, in the past as in the present, of environmental changes. In the meantime, they have been influenced and have been dependent upon the natural geomorphological set of territories and availability of resources. It is obvious from this that what was indisputable for the past must be true also for the future.

The archaeobotanical data suggest that human exploitation of wild resources has been capable of changing the landscape system at least since the early Holocene, and then in a more evident way from the Bronze age onwards. Therefore, today, we live in a world that it is objectively less 'natural' than we think.

Palaeolithic, Epipalaeolithic or Mesolithic hunter-gatherers were relatively mobile, relied on materials available in the wild, so far as we know, usually without storing surplus food and moving on as food resources become exhausted.

Then, Neolithic agrarian cultures increased their capability of control on nature but the new economy obliged people to become more sedentary. Diamond (2002) argues that a chain reaction of new behaviours and social organisations came after food production. After the agriculture subsistence strategy was established, "an explosion of technology was possible because sedentary living permitted the accumulation of heavy technology that nomadic hunter–gatherers could not carry, and because the storable food surpluses resulting from agriculture could be used to feed full-time craftspeople and inventors. By also feeding full-time kings, bureaucrats, nobles and soldiers, those food surpluses led to social stratification, political centralization and standing armies".

Complex societies were so characteristic of mid-Holocene Mediterranean history, had relations with environments more complex than those of the previous civilizations, and in general tended to shape the landscape *via* their culture rather than adapt it to the environment. The more we study the more we discover how much the Mediterranean geomorphological set and biodiversity have had a critical role in orienting or conditioning the development of cultures, and eventually resulted in the modern social organisation.

What remains of the past processes in modern times? What will remain in the next years? Probably, only modern interdisciplinary investigation will give unforeseen insights.

Most studies on environmental changes are led by scientists who pay no attention to the complexity and capacity of cultural dynamics to act on the environment, while most studies on humans ability in adapting to their environment are led by archaeologists or social scientists who are often indifferent to the physical and biological dynamics of ecosystems (Berglund 2003; Stoffle et al. 2003). Only the interdisciplinary approach, involving geo-biological and anthropological sciences, permits the dual nature environment/humans to emerge. Integrated palaeoenvironmental-humanistic research may contribute to the development of templates for sustainable management of landscapes and ecosystems in modern times. As the models used by present day climatologists and ecologists are mainly based on inferences from short-term observations, they are not likely to be fully used for developing future trajectories on long timescales as well as data from palaeo-research (Dearing et al. 2006).

Concluding Remarks

There is an inescapable liaison between humans and environmental sustainability and therefore cultural trajectories have shown, and still show, trends that in many cases seem to be fairly coincident with climatic changes (Mercuri and Sadori 2012). Water and food, limiting subsistence factors, were the primary environmental features that assured human survival and development of quality in human living.

After the wet early Holocene phase, Mediterranean regions became more and more arid, especially after the end of the 7th millennium BP, and then again during the 5th millennium BP. Several adaptation processes should have had rapid shifts following a climate-environment-humans linear direction, and soon after their onset they should have entered complex interactions between climate and environment and between environment and humans.

Pronounced increases in social complexity seem to have coincided with climatic deterioration, particularly with increased aridity in mid-Holocene times. Three dry events of key relevance for climatic-cultural changes seem to have occurred at ca. 8.2 ka BP, ca. 6.0 ka BP, and ca. 4.2 ka BP.

Earlier cultural adaptations to environmental and climatic changes are evident in archaeological sites, and fairly evident in many pollen records while cultural adaptations to environmental changes are more and more evident in many pollen records from the late mid-Holocene, or from the Bronze Age period onwards.

Archaeobotanical records clearly show that when human groups occupied a site or region, they exploited the territory and selected plants for food, building and firewood. They used and selected what the territory offered. When climatic/environmental changes occurred, humans changed their subsistence strategies also moving towards what was available on territory. Overexploitation of thinned plant resources, including overgrazing, accelerated trend towards aridity during drying climatic phases. As humans enforced the aridity crisis, the climate signal in vegetation, and consequently in palaeoclimatic records, was enhanced. When dry climate depleted water and plant resources below a sustainable level, humans necessarily moved to new places.

The onset of the metal age civilizations, in a period of high climatic instability, is a good example of the close interaction between environment and humans. Patterns of landscape evolution suggest that climate was the first agent of the shift towards aridity visible in pollen diagrams at around 4.0 ka BP. This water shortage probably obliged Bronze Age populations to settle along the lakeshores in pile dwellings or close to rivers in protected villages. Human populations left clear clues of their presence in the lake sediments: their role did not consist in producing devastating effects on the environment, but emphasised a climatic process, causing strong modifications in an already deforested open landscape.

We make our own the words by Wroe and colleagues (2006) who stressed that humans undoubtedly must take full responsibility for the current biodiversity, but "by prematurely minimizing the role of climatic influences [from the Upper Palaeolithic to the more recent times, …] we risk overlooking salient lessons from the past in a world now facing significant climate change".

Besides climate changes, after the agriculture discovery, many factors were involved in the development of our social model, and the Mediterranean basin offered extraordinary favourable conditions of water availability, biodiversity, places for herding and cultivation, defensive strategy, migration and trade. The mixing of cultures accelerated the exchanges of ideas, technologies, raw materials and people along the coasts of this 'great lake' making the different civilizations found there as part of one network of regions belonging to the well recognized 'Mediterranean culture'. As far as we know, humans were not able, and we think that they are still not able, to change definitively the natural trends. Although increasing levels of instability, such as war or economic equilibrium, population growth, societal responses to internal and environmental problems are very important causes of crisis in our society (Diamond 2005), the environmental damages, such as deforestation, soil erosion, salinization, overexploitation, competition between introduced and native species, were and will continue to be among the main risks for our future.

Acknowledgements The authors wish to thank Stefano Goffredo and Zvy Dubinsky for the invitation to write this chapter. A special acknowledgement to David Lubell and Yannick Miras who greatly improved the manuscript with helpful comments and suggestions. This chapter was conceived in the frame of the PRIN (2008FJCEF4) project PICAR (Cultural landscape and human impact in circum-Mediterranean countries).

References

Allen JRM, Watts WA, McGee E, Huntley B (2002) Holocene environmental variability – the record from Lago Grande di Monticchio, Italy. Quat Int 88:69–80. doi:10.1016/S1040-6182(01)00074-X

Allevato E, Russo Ermolli E, Boetto G, Di Pasquale G (2010) Pollen-wood analysis at the Neapolis harbour site (1st-3rd century AD, southern Italy) and its archaeobotanical implications. J Archaeol Sci 37:2365–2375. doi:10.1016/j.jas.2010.04.010

Ammerman AJ, Cavalli-Sforza LL (1971) Measuring the rate of spread of early farming in Europe. Man (New Series) 6:674–688

Ballouche A, Marinval P (2004) At the origin of agriculture in the Maghreb. Palynological and carpological data on the early Neolithic of Northern Morocco. In: XIth congress of Panafrican Association, Soro Print Color, Bamako

Bard E, Frank M (2006) Climate change and solar variability: what's new under the sun? Earth Planet Sci Lett 248:480–493. doi:10.1016/j.epsl.2006.06.016

Bar-Matthews M, Ayalon A (2011) Mid-Holocene climate variations revealed by high-resolution speleothem records from Soreq Cave, Israel and their correlation with cultural changes. Holocene 21:163–171. doi:10.1177/0959683610384165

Bar-Matthews M, Ayalon A, Kaufman A (1997) Late Quaternary paleoclimate in the Eastern Mediterranean region from stable isotope analysis of speleothems at Soreq Cave, Israel. Quat Res 47:155–168. doi:10.1006/qres.1997.1883

Behre K-E (1986) Anthropogenic indicators in pollen diagrams. Balkema, Rotterdam

Behre K-E, Jacomet S (1991) The ecological interpretation of archaeobotanical data. In: van Zeist W, Wasylikowa K, Behre K-E (eds) Progress in old world palaeoethnobotany. Balkema, Rotterdam

Bellini C, Mariotti-Lippi M, Montanari C (2009) The Holocene landscape history of the NW Italian coasts. Holocene 19:1161–1172. doi:10.1177/0959683609345077

Berger JC, Guilaine J (2009) The 8200 cal BP abrupt environmental change and the Neolithic transition: a Mediterranean perspective. Quat Int 200:31–49. doi:10.1016/j.quaint.2008.05.013

Berglund BE (2003) Human impact and climate changes – synchronous events and a causal link? Quat Int 105:7–12. doi:10.1016/S1040-6182(02)00144-1

Binford L (1983) In pursuit of the past: decoding the archaeological record. Thames & Hudson, London

Birks HH, Birks JB, Kaland PE, Moe D (1988) The cultural landscape. Cambridge University Press, Cambridge

Bordon A, Peyron O, Lézine AM, Brewer S, Fouache E (2009) Pollen-inferred Late-Glacial and Holocene climate in southern Balkans (lake Maliq). Quat Int 200:19–30. doi:10.1016/j.quaint.2008.05.014

Bottema S (1987) Chronology and climatic phases in the Near East from 16,000 to 10,000 years BP. In: Aurenche O, Evin J, Hous F (eds) Chronologies du Proche Orient/Chronologies in the Near East. Relative chronologies and absolute chronology 16,000—4,000 B.P., British archaeological reports international series 379. Archaeopress, Oxford

Bottema S (1995) Holocene vegetation of the Van Area: palynological and chronological evidence from Söğütlü, Turkey. Veg Hist Archaeobot 4:187–193. doi:10.1007/BF00203937

Bottema S, Sarpaki A (2003) Environmental change in Crete: a 9000-year record of Holocene vegetation history and the effect of the Santorini eruption. Holocene 13:733–749. doi:10.1191/0959683603hl659rp

Bottema S, Entjes-Nieborg G, van Zeist W (1990) Man's role in the shaping of the eastern Mediterranean landscape. Balkema, Rotterdam

Bottema S, Woldring H, Kayan I (2001) The late Quaternary vegetation history of western Turkey. In: Roodenberg JJ, Thissen LC (eds) The Ilipinar excavations II. Nederlands Instituut voor het Nabije Oosten, Leiden

Brookfield M (2010) The desertification of the Egyptian Sahara during the Holocene (the last 10,000 years) and its influence on the rise of Egyptian Civilization. In: Martini IP, Chesworth W (eds) Landscapes and societies. Springer, Dordrecht

Brooks N (2006) Cultural responses to aridity in the Middle Holocene and increased social complexity. Quat Int 151:29–49. doi:10.1016/j.quaint.2006.01.013

Büntgen U, Tegel W, Nicolussi K, McCormick M, Frank D et al (2011) 2500 years of European climate variability and human susceptibility. Science 331:578–582. doi:10.1126/science.1197175

Burney DA, Flannery TF (2005) Fifty millennia of catastrophic extinctions after human contact. Trends Ecol Evol 20:395–401. doi:10.1016/j.tree.2005.04.022

Butzer KW (2005) Environmental history in the Mediterranean world: cross-disciplinary investigation of cause-and-effect for degradation and soil erosion. J Archaeol Sci 32:1773–1800. doi:10.1016/j.jas.2005.06.001

Camuffo D, Enzi S (1995) The analysis of two bi-millennial series: Tiber and Po river floods. In: Jones PD, Bradley RS, Jouzel J (eds) Climatic variations and forcing mechanisms of the last 2000 years. Springer, Berlin

Cardarelli A (2009) Insediamenti dell'Età del Bronzo fra Secchia e Reno. Formazione, affermazione e collasso delle Terramare. In: Cardarelli A, Malnati L (eds) Atlante dei Beni Archeologici della Provincia di Modena – vol III: Collina e Alta Pianura. All'Insegna del Giglio, Firenze

Caroli I, Caldara M (2007) Vegetation history of Lago Battaglia (Eastern Gargano coast, Apulia Italy) during the middle–late Holocene. Veg Hist Archaeobot 16:317–327. doi:10.1007/s00334-006-0045-y

Carrión JS, Fernández S, González-Sampériz P, Gil-Romera G, Badal E, Carrión-Marco Y, López-Merino L, López-Sáez JA, Fierro E, Burjachs F (2010a) Expected trends and surprises in the Lateglacial and Holocene vegetation history of the Iberian Peninsula and Balearic Islands. Rev Palaeobot Palynol 162:458–475. doi:10.1016/j.revpalbo.2009.12.007

Carrión JS, Fernández S, Jiménez-Moreno G, Fauquette S, Gil-Romera G, González-Sampériz P, Finlayson C (2010b) The historical origins of aridity and vegetation degradation in southeastern Spain. J Arid Environ 74:731–736. doi:10.1016/j.jaridenv.2008.11.014

Colombaroli D, Vannière B, Emmanuel C, Magny M, Tinner W (2008) Fire–vegetation interactions during the Mesolithic–Neolithic transition at Lago dell'Accesa, Tuscany, Italy. Holocene 18:679–692. doi:10.1177/0959683608091779

Combourieu Nebout N, Peyron O, Dormoy I, Desprat S, Beaudouin C, Kotthoff U, Marret F (2009) Rapid climatic variability in the west Mediterranean during the last 25 000 years from high resolution pollen data. Clim Past 5:503–521. www.clim-past.net/5/503/2009/

Cremaschi M (1998) Late Quaternary geological evidence for environmental changes in south-western Fezzan (Libyan Sahara). In: Cremaschi M, di Lernia S (eds) Wadi Teshuinat: palaeoenvironment and prehistory in south-western Fezzan (Libyan Sahara). Ed. All'Insegna del Giglio, Firenze

Cremaschi M (2002) Late Pleistocene and Holocene climatic changes in the Central Sahara. The case of study of the south-western Fezzan, Libya. In: Hassan FA (ed) Droughts, food and culture. Kluwer Academic/Plenum Publishers, New York

Cremaschi M, di Lernia S (1999) Holocene climatic changes and cultural dynamics in the Libyan Sahara. Afr Archaeol Rev 16:211–238. doi:10.1023/A:1021609623737

Cremaschi M, Zerboni A (2009) Early to Middle Holocene landscape exploitation in a drying environment: two case studies compared from the central Sahara (SW Fezzan, Libya). CR Geosci 341:689–702. doi:10.1016/j.crte.2009.05.001

Cremaschi M, Zerboni A (2010) Human communities in a drying landscape: Holocene climate change and cultural response in the Central Sahara. In: Martini IP, Chesworth W (eds) Landscapes and societies. Springer, Dordrecht

de Beaulieu J-L, Miras Y, Andrieu V, Guiter F (2005) Vegetation dynamics in the north-west Mediterranean regions: instability of the Mediterranean bioclimate. Plant Biosyst 139:114–126. doi:10.1080/11263500500197858

Dearing JA, Battarbee RW, Dikau R, Laroque I, Oldfield F (2006) Human-environment interactions: towards synthesis and simulation. Reg Environ Change 6:115–123. doi:10.1007/s10113-005-0011-8

deMenocal PB (2001) Cultural responses to climate change during the late Holocene. Science 292:667–672. doi:10.1126/science.1059287

Denèfle M, Lezine AM, Fouache E, Dufaure JJ (2000) A 12,000-year pollen record from Lake Maliq, Albania. Quat Res 54:423–432. doi:10.1006/qres.2000.2179

Diamond J (2002) Evolution, consequences and future of plant and animal domestication. Nature 418:700–707. doi:10.1038/nature01019

Diamond J (2005) Collapse: how societies choose to fail or succeed. Viking, New York

di Lernia S (2002) Dry climatic events and cultural trajectories: adjusting middle Holocene pastoral economy of the Libyan Sahara. In: Hassan F (ed) Droughts, food and culture: ecological change and food security in Africa's later prehistory. Kluwer Academic/Plenum Publishers, New York

Di Rita F, Magri D (2009) Holocene drought, deforestation and evergreen vegetation development in the central Mediterranean: a 5500 year record from Lago Alimini Piccolo, Apulia, southeast Italy. Holocene 19:295–306. doi:10.1177/0959683608100574

Di Rita F, Celant A, Magri D (2010) Holocene environmental instability in the wetland north of the Tiber delta (Rome, Italy): sea-lake-man interaction. J Paleolimnol 44:51–67. doi:10.1007/s10933-009-9385-9

Drescher-Schneider R, de Beaulieu J-L, Magny M, Walter-Simonnet A-V, Bossuet G, Millet L, Brugiapaglia E, Drescher A (2007) Vegetation history, climate and human impact over the last 15,000 years at Lago dell'Accesa (Tuscany, Central Italy). Veg Hist Archaeobot 16:279–299. doi:10.1007/s00334-006-0089-z

Drysdale R, Zanchetta G, Hellstrom J, Maas R, Fallick A, Pickett M, Cartwright I, Piccini L (2006) Late Holocene drought responsible for the collapse of Old World civilizations is recorded in an Italian cave flowstone. Geology 34:101–104. doi:10.1130/G22103.1

Eastwood WJ, Roberts N, Boyer P (2007) Pollen analysis at Çatalhöyük. In: Hodder I (ed) Excavations at Çatalhöyük: the 1995–1999 seasons. McDonald Institute and the British Institute of Archaeology (BIAA), Ankara

Ejarque A, Miras Y, Riera S (2011) Pollen and non-pollen palynomorph indicators of vegetation and highland grazing activities obtained from modern surface and dung datasets in the eastern Pyrenees. Rev Palaeobot Palynol 167:123–139. doi:10.1016/j.revpalbo.2011.08.001

Faegri K, Kaland PE, Krzywinski K (1989) Textbook of pollen analysis, 4th edn. Wiley, London

Finné M, Holmgren K, Sundqvist HS, Weiberg E, Lindblom M (2011) Climate in the eastern Mediterranean, and adjacent regions, during the past 6000 years – a review. J Archaeol Sci 38:3153–3173. doi:10.1016/j.jas.2011.05.007

Flannery T (1994) The future eaters: an ecological history of Australasia lands and peoples. Reed New Holland, Sydney

Florenzano A, Mercuri AM, Pederzoli A, Torri P, Bosi G et al (2012) The significance of intestinal parasite remains in pollen samples from Mediaeval pits in the Piazza Garibaldi of Parma, Emilia Romagna, Northern Italy. Geoarchaeology 27:34–47. doi:10.1002/gea.21390

Floss H (1994) Rohmaterialversorgung im Paläolithikum des Mittelrheingebietes, Monographien des Römisch-Germanischen Zentralmuseums 21. Dr. Rudolf Habelt, Bonn

Gasse F (2000) Hydrological changes in the African tropics since the Last Glacial Maximum. Quat Sci Rev 19:189–211. doi:10.1016/S0277-3791(99)00061-X

Gasse F (2002) Diatom-inferred salinity and carbonate oxygen isotopes in Holocene waterbodies of the western Sahara and Sahel (Africa). Quat Sci Rev 21:737–767. doi:10.1016/S0277-3791(01)00125-1

Gasse F, van Campo E (1994) Abrupt postglacial climate events in West Asia and North Africa monsoon domains. Earth Planet Sci Lett 126:435–456. doi:10.1016/0012-821X(94)90123-6

Giraudi C (1998) Late Pleistocene and Holocene lake-level variations in Fucino Lake (Abruzzo, Central Italy) inferred from geological, archaeological and historical data. In: ESF workshop 'Palaeohydrology as reflected in lake-level changes as climatic evidence for Holocene times' Paläoklimaforschung 25. G Fischer Verlag, Stuttgart

Giraudi C (2004) Le oscillazioni di livello del lago di Mezzano (Valentano-VT): variazioni climatiche e interventi antropici. Quaternario 17:221–230

Giraudi C, Mercuri AM (in press) Early to middle Holocene environmental variations in the Gobero basin. In: Garcea EAA (ed) Gobero: the no-return frontier archaeology and landscape at the Saharo-Sahelian Borderland

Giraudi C, Magny M, Zanchetta G, Drysdale RN (2011) The Holocene climatic evolution of Mediterranean Italy: a review of the continental geological data. Holocene 21:105–115. doi:10.1177/0959683610377529

Grüger E, Thulin B (1998) First results of biostratigraphical investigations of Lago d'Averno near Naples relating to the period 800 BC-800 AD. Quat Int 47(48):35–40. doi:10.1016/S1040-6182(97)00068-2

Guidi A, Piperno M (1992) Italia preistorica. Editori Laterza, Bari

Harvey EL, Fuller DQ (2005) Investigating crop processing using phytolith analysis: the example of rice and millets. J Archaeol Sci 32:739–752. doi:10.1016/j.jas.2004.12.010

Hoelzmann P, Keding B, Berke H, Kröpelin S, Kruse H-J (2001) Environmental change and archaeology: lake evolution and human occupation in the eastern Sahara during the Holocene. Palaeogeogr Palaeoclimtol Palaeoecol 169:193–217. doi:10.1016/S0031-0182(01)00211-5

Jahns S (2003) A late Holocene pollen diagram from the Megaris, Greece, giving possible evidence for cultivation of *Ceratonia siliqua* L. during the last 2000 years. Veg Hist Archaeobot 12: 127–130. doi:10.1007/s00334-003-0013-8

Jahns S (2005) The Holocene history of vegetation and settlement at the coastal site of Lake Voulkaria in Acarnania, western Greece. Veg Hist Archaeobot 14:55–66. doi:10.1007/s00334-004-0053-8

Jahns S, van den Bogaard C (1998) New palynological and tephrostratigraphical investigations of two salt lagoons on the island of Mljet, south Dalmatia, Croatia. Veg Hist Archaeobot 7:219–234. doi:10.1007/BF01146195

Jalut G (2005) Le climat, la végétation et l'Homme en Méditerranée à l'Holocène. In: Guilaine J (ed) Populations néolithiques et environnments. Séminaires du Collège de France. Errance, Paris

Jalut G, Esteban Amat A, Riera i Mora S, Fontugne M, Mook R, Bonnet L, Gauquelin T (1997) Holocene climatic changes in the western Mediterranean: installation of the Mediterranean climate. CR Acad Sci Paris Earth Planet Sci 325:327–334. doi:10.1016/S1251-8050(97)81380-8

Jalut G, Esteban Amat A, Bonnet L, Gauquelin T, Fontugne M (2000) Holocene climatic changes in the Western Mediterranean, from south-east France to south-east Spain. Palaeogeogr Palaeoclimnol Palaeoecol 160:255–290. doi:10.1016/S0031-0182(00)00075-4

Jalut G, Dedoubat JJ, Fontugne M, Otto T (2009) Holocene circum-Mediterranean vegetation changes: climate forcing and human impact. Quat Int 200:4–18. doi:10.1016/j.quaint.2008.03.012

Jimenez-Espejo F, Martinez-Ruiz F, Sakamoto T, Iijima K, Gallego-Torres D, Harada N (2007) Paleoenvironmental changes in the western Mediterranean since the last glacial maximum: high resolution multiproxy record from the Algero-Balearic basin. Palaeogeogr Palaeoclimnol Palaeoecol 246:292–306. doi:10.1016/j.palaeo.2006.10.005

Jousse H (2006) What is the impact of Holocene climatic changes on human societies? Analysis of West African Neolithic populations dietary customs. Quat Int 151:63–73. doi:10.1016/j.quaint.2006.01.015

Kelso GK, Dimmick FR, Dimmick DH, Largy TB (2006) An ethnopalynological test of task-specific area analysis: Bay View Stable, Cataumet, Massachusetts. J Archaeol Sci 33:953–960. doi:0.1016/j.jas.2005.11.002

Kuneš P (2008) Human-driven and natural vegetation changes during the last glacial and early Holocene. Dissertation, Charles University

Kuper R, Kröpelin S (2006) Climate-controlled Holocene occupation in the Sahara: motor of Africa's evolution. Science 313:803–807. doi:10.1126/science.1130989

Lawson I, Frogley M, Bryant C, Preece R, Tzedakis P (2004) The Lateglacial and Holocene environmental history of the Ioannina basin, northwest Greece. Quat Sci Rev 23:1599–1625. doi:10.1016/j.quascirev.2004.02.003

Lézine AM (2009) Climatic history of the African and Arabian deserts. CR Geosci 341:569–574

Linstädter J (2008) The Epipalaeolithic-Neolithic-Transition in the Mediterranean region of Northwest Africa. Quartär 55:41–62. doi:10.1016/j.crte.2009.09.002

Liverani M (ed) (2005) Aghram Nadharif. The Barkat Oasis (Sha'abiya of Ghat, Libyan Sahara) in Garamantian time, Arid zone archaeology monographs 5. All'Insegna del Giglio, Firenze

Lowe JJ, Accorsi CA, Bandini Mazzanti M, Bishop A, Forlani L et al (1996) Pollen stratigraphy of sediment sequences from crater lakes (Lago Albano and Lago di Nemi) and the Central Adriatic spanning the interval from Oxygen isotope Stage 2 to the present day. Mem Ist Ital Idrobiol 55:71–98

Lubell D (1984) Palaeoenvironments and Epipalaeolithic economies in the Maghreb (ca. 20 000 to 5 000 BP). In: Clark JD, Brandt SA (eds) From hunters to farmers: considerations of the causes and consequences of food production in Africa. University of California Press, Berkley

Lubell D (2001) Late Pleistocene-Early Holocene Maghreb. In: Peregrine PN, Ember M (eds) Encyclopedia of prehistory 1: Africa. Kluwer Academic/Plenum Publ, New York

Lubell D, Feathers J, Schwenninger J-L (2009) Post-Capsian occupation in the eastern Maghreb: implications of a revised chronological assessment for the adult burial at Aïn Misteheyia. J Afr Archaeol 7:175–189. doi:10.3213/1612-1651-101

Magny M (1999) Lake-level fluctuations in the Jura and French subalpine ranges associated with ice-rafting events in the north Atlantic and variations in the polar atmospheric circulation. Quaternaire 10:61–64

Magny M, Miramont C, Sivan O (2002) Assessment of the impact of climate and anthropogenic factors on Holocene Mediterranean vegetation in Europe on the basis of palaeohydrological records. Palaeogeogr Palaeoclimnol Palaeoecol 186:47–59. doi:10.1016/S0031-0182(02)00442-X

Magny M, De Beaulieu J-L, Drescher-Schneider R, Vannière B, Walter-Simonnet A-V, Miras Y et al (2007) Holocene climate changes in the central Mediterranean as recorded by lake-level fluctuations at Lake Accesa (Tuscany, Italy). Quat Sci Rev 26:1736–1758. doi:10.1016/j.quascirev.2007.04.014

Magny M, Vannière B, Zanchetta G, Fouache E, Touchais G, Petrika L, Coussot C, Walter-Simonnet A-V, Arnaud F (2009) Possible complexity of the climatic event around 4300–3800 cal. BP in the central and western Mediterranean. Holocene 19:1–11. doi:10.1177/0959683609337360

Magny M, Peyron O, Sadori L, Ortu E, Zanchetta G, Vannière B, Tinner W (2012) Contrasting patterns of precipitation seasonality during the Holocene in the south- and north-central Mediterranean. J Quat Sci 27:290–296. doi:10.1002/jqs.1543

Magri D (1999) Late Quaternary vegetation history at Lagaccione near Lago di Bolsena (central Italy). Rev Palaeobot Palynol 106:171–208. doi:10.1016/S0034-6667(99)00006-8

Magri D, Parra I (2002) Late Quaternary western Mediterranean pollen records and African winds. Earth Planet Sci Lett 200:401–408. doi:10.1016/S0012-821X(02)00619-2

Magri D, Sadori L (1999) Late Pleistocene and Holocene pollen stratigraphy at Lago di Vico, central Italy. Veg Hist Archaeobot 8:247–260. doi:10.1007/BF01291777

Mariotti Lippi M, Bellini C, Trinci C, Benvenuti M, Pallecchi P, Sagri M (2006) Pollen analysis of the ship site of Pisa San Rossore,

Tuscany, Italy: the implications for catastrophic hydrological events and climatic change during the late Holocene. Veg Hist Archaeobot 16:453–465. doi:10.1007/s00334-006-0070-x

Martin PS (1967) Prehistoric overkill. In: Martin PS, Wright HE (eds) Pleistocene extinctions: the search for a cause. Yale University Press, New Haven

Mazzini I, Faranda S, Giardini M, Giraudi C, Sadori L (2011) Late Holocene palaeoenvironmetal evolution of the Roman harbour of *Portus*, Italy. J Paleolimnol 46:243–256. doi:10.1007/s10933-011-9536-7

Mercuri AM (1999) Palynological analysis of the early Holocene sequence. In: di Lernia S (ed) The Uan Afuda cave Hunter-gatherer Societes of Central Sahara, Arid zone archaeology, monographs 1. All'Insegna del Giglio, Firenze

Mercuri AM (2008a) Human influence, plant landscape, evolution and climate inferences from the archaeobotanical records of the Wadi Teshuinat area (Libyan Sahara). J Arid Environ 72:1950–1967. doi:10.1016/j.jaridenv.2008.04.008

Mercuri AM (2008b) Plant exploitation and ethnopalynological evidence from the Wadi Teshuinat area (Tadrart Acacus, Libyan Sahara). J Archaeol Sci 35:1619–1642. doi:10.1016/j.jas.2007.11.003

Mercuri AM, Garcea EAA (2007) The impact of hunter/gatherers in the Central Sahara during the Early Holocene. In: Cappers R (ed) Fields of changes. Progress in African archaeobotany, Groningen archaeological studies 5. Groningen University Library, Groningen

Mercuri AM, Sadori L (2012) Climate changes and human settlements since the Bronze Age period in central Italy. Rendiconti Online della Società Geologica Italiana 18:26–28. doi:10.3301/ROL.2011.63

Mercuri AM, Trevisan Grandi G, Mariotti Lippi M, Cremaschi M (1998) New pollen data from the Uan Muhuggiag rockshelter (Libyan Sahara, VII–IV millennia BP). In: Cremaschi M, di Lernia S (eds) Wadi Teshuinat: palaeoenvironment and prehistory in South-western Fezzan (Libyan Sahara). Quaderni Geodinamica Alpina e Quaternaria, C.N.R, Firenze

Mercuri AM, Accorsi CA, Bandini Mazzanti M (2002) The long history of *Cannabis* and its cultivation by the Romans in central Italy, shown by pollen records from Lago Albano and Lago di Nemi. Veg Hist Archaeobot 11:263–276. doi:10.1007/s003340200039

Mercuri AM, Trevisan Grandi G, Bosi G, Forlani L, Buldrini F (2005) The archaeobotanical remains (pollen, seeds/fruits and charcoal). In: Liverani M (ed) Aghram Nadharif. The Barkat Oasis (Sha'abiya of Ghat, Libyan Sahara) in Garamantian time, Arid zone archaeology, monographs 5. All'Insegna del Giglio, Firenze

Mercuri AM, Accorsi CA, Bandini Mazzanti M, Bosi G, Cardarelli A, Labate D et al (2006) Economy and environment of Bronze Age settlements – Terramaras – in the Po Plain (Northern Italy): first results of the archaeobotanical research at the Terramara di Montale. Veg Hist Archaeobot 16:43–60. doi:10.1007/s00334-006-0034-1

Mercuri AM, Florenzano A, Massamba N'siala I, Olmi L, Roubis D, Sogliani F (2010a) Pollen from archaeological layers and cultural landscape reconstruction: case studies from the Bradano valley (Basilicata, southern Italy). Plant Biosyst 144:888–901. doi:10.1080/11263504.2010.491979

Mercuri AM, Sadori L, Blasi C (2010b) Editorial: Archaeobotany for cultural landscape and human impact reconstructions. Plant Biosyst 144:860–864. doi:10.1080/11263504.2010.514137

Mercuri AM, Sadori L, Uzquiano Ollero P (2011) Mediterranean and north-African cultural adaptations to mid-Holocene environmental and climatic change. Holocene 21:189–206. doi:10.1177/0959683610377532

Mercuri AM, Bandini Mazzanti M, Torri P, Vigliotti L, Bosi G, Florenzano A, Olmi L, Massamba N'siala I (2012) A marine/terrestrial integration for mid-late Holocene vegetation history and the development of the cultural landscape in the Po Valley as a result of human impact and climate change. Veg Hist Archaeobot 21:351–372. doi:10.1007/s00334-012-0352-4

Mercuri AM, Bandini Mazzanti M, Florenzano A, Montecchi MC, Rattighieri E (2013) *Olea*, *Juglans* and *Castanea*: the OJC group as pollen evidence of the development of human-induced environments in the Italian peninsula. Quat Int 300 (in press). doi:10.1016/j.quaint.2013.01.005

Mikhail A (2012) Global implications of the Middle Eastern Environment. Hist Compass 12:252–270. doi:10.1111/j.1478-0542.2011.00811.x

Miras Y, Ejarque A, Riera S, Palet JM, Orengo H, Euba I (2007) Dynamique holocene de la vegetation et occupation des Pyrenees andorranes depuis le Neolithique ancien, d'apres l'analyse pollinique de la tourbière de Bosc dels Estanyons (2180 m, Vall del Madrid, Andorra). CR Palevol 6:291–300. doi:10.1016/j.crpv.2007.02.005

Miras Y, Ejarque A, Orengo H, Riera S, Palet JM, Poiraud A (2010) Prehistoric impact on landscape and vegetation at high altitudes: an integrated palaeoecological and archaeological approach in the eastern Pyrenees (Perafita valley, Andorra). Plant Biosyst 144:924–939. doi:10.1080/11263504.2010.491980

Mulazzani S, Cavulli F, Aouadi-Abdeljaouad N, Belhouchet L, Boussoffara R et al (2010) Il sito olocenico costiero SHM-1 (Hergla, Tunisia), tra sostrato capsiano e le prime attestazioni della più antica navigazione. Rivista di Scienze Preistoriche LX:49–61

Oldfield F (2005) Environmental change: key issues and alternative approaches. University Press Cambridge, Cambridge

Oldfield F, Asioli A, Accorsi CA, Mercuri AM, Juggins S et al (2003) A high resolution late Holocene palaeoenvironmental record from the central Adriatic Sea. Quat Sci Rev 22:319–342. doi:10.1016/S0277-3791(02)00088-4

Pearsall DM (2000) Palaeoethnobotany: a handbook of procedures, 2nd edn. Academic, London

Peña-Chocarro L (1999) Prehistoric agriculture in Spain. The application of ethnographic models, British archaeobotany reports international series 818. Archaeopress, Oxford

Pepe C, Giardini M, Giraudi C, Masi A, Mazzini I, Sadori L (2013) Plant landscape and environmental changes recorded in marginal marine environments: the ancient Roman harbour of Portus (Rome, Italy). Quat Int 300 (in press). doi: 10.1016/j.quaint.2012.11.008

Pérez-Obiol R, Sadori L (2007) Similarities and dissimilarities, synchronisms and diachronisms in the Holocene vegetation history of the Balearic Islands and Sicily. Veg Hist Archaeobot 16:259–265. doi:10.1007/s00334-006-0038-x

Pérez-Obiol R, Jalut G, Julià R, Pèlachs A, Iriarte MJ, Otto T, Hernández-Beloqui B (2011) Mid-Holocene vegetation and climatic history of the Iberian Peninsula. Holocene 21:75–93. doi:10.1177/0959683610384161

Peyron O, Goring S, Dormoy I, Kotthoff U, Pross J, de Beaulieu J-L, Drescher-Schneider R, Magny M (2011) Holocene seasonality changes in the central Mediterranean region reconstructed from the pollen sequence of Lake Accesa (Italy) and Tenaghi Philippon (Greece). Holocene 21:131–146. doi:10.1177/0959683610384162

Pons A, Quézel P (1985) The history of the flora and vegetation and past and present human disturbance in the Mediterranean region. In: Gomez-Campo C (ed) Plant conservation in the Mediterranean area. Geobotany 7. W Junk, Dordrecht

Pons A, Quézel P (1998) A propos de la mise en place du climat méditerranéen. CR Acad Sci Paris 327:755–760. doi:10.1016/S1251-8050(99)80047-0

Pons A, Reille M (1988) The Holocene- and upper Pleistocene pollen record from Padul (Granada, Spain): a new study. Palaeogeogr Palaeoclimnol Palaeoecol 66:243–263. doi:10.1016/0031-0182(88)90202-7

Quézel P, Barbero M (1982) Definition and characterization of Mediterranean-type ecosystems. Ecologia Mediterranea 8:15–29

Ravazzi C, Cremaschi M, Forlani L (2004) Studio archeobotanico della Terramara di S.Rosa di Poviglio (RE). Nuovi dati e analisi floristica e sintassonomica della vegetazione nell'età del Bronzo. In: Bernabò Brea M, Cremaschi M (eds) Gli scavi nell'abitato piccolo della Terramara Santa Rosa di Poviglio (Reggio nell'Emilia). All'Insegna del Giglio, Firenze

Reille M, Andrieu V (1995) The late Pleistocene and Holocene in the Lourdes Basin, Western Pirénéès, France: new pollen analytical and chronological data. Veg Hist Archaeobot 4:1–21. doi:10.1007/BF00198611

Richerson PJ, Borgerhoff Mulder M, Vila BJ (1996) Principles of human ecology. Simon & Schuster Custom Pub, New York

Roberts N (2002) Did prehistoric landscape management retard the post-glacial spread of woodland in Southwest Asia? Antiquity 76:1002–1010

Roberts N, Stevenson T, Davis B, Cheddadi R, Brewster S, Rosen A (2004) Holocene climate, environment and cultural change in the circum-Mediterranean region, Chapter 17. In: Battarbee RW, Gasse F, Stickley CE (eds) Past climate variability through Europe and Africa, Developments in paleoenvironmental research 6. Springer, Dordrecht

Roberts N, Brayshaw D, Kuzucuoğlu C, Perez R, Sadori L (2011a) The mid-Holocene climatic transition in the Mediterranean: causes and consequences. Holocene 21:3–13. doi:10.1177/0959683610388058

Roberts N, Eastwood WJ, Kuzucuoğlu C, Fiorentino G, Caracuta V (2011b) Climatic, vegetation and cultural change in the eastern Mediterranean during the mid-Holocene environmental transition. Holocene 21:147–162. doi:10.1177/0959683610386819

Rolph TC, Vigliotti L, Oldfield F (2004) Mineral magnetism and geomagnetic secular variation of marine and lacustrine sediments from central Italy: timing and nature of local and regional Holocene environmental change. Quat Sci Rev 23:1699–1722. doi:10.1016/j.quascirev.2004.02.005

Rösch M (2005) Pollen analysis of the contents of excavated vessels – direct archaeobotanical evidence of beverages. Veg Hist Archaeobot 14:179–188. doi:10.1007/s00334-005-0015-9

Ruddiman WF, Crucifix MC, Oldfield FA (2011) Introduction to the early-Anthropocene special issue. Holocene 21:713. doi:10.1177/0959683611398053

Sadori L, Giardini M (2007) Charcoal analysis, a method to study vegetation and climate of the Holocene: the case of Lago di Pergusa, Sicily (Italy). Geobios 40:173–180. doi:10.1016/j.geobios.2006.04.002

Sadori L, Giardini M (2008) Environmental history in the Mediterranean basin: microcharcoal as a tool to disentangle human impact and climate change. In: Fiorentino G, Magri D (eds) Charcoals from the past: cultural and palaeoenvironmental implications, British archaeological reports international series 1807. Archaeopress, Oxford

Sadori L, Narcisi B (2001) The postglacial record of environmental history from Lago di Pergusa (Sicily). Holocene 11:655–671. doi:10.1191/09596830195681

Sadori L, Giraudi C, Petitti P, Ramrath A (2004) Human impact at Lago di Mezzano (central Italy) during the Bronze Age: a multidisciplinary approach. Quat Int 113:5–17. doi:10.1016/S1040-6182(03)00077-6

Sadori L, Zanchetta G, Giardini M (2008) Last glacial to Holocene palaeoenvironmental evolution at Lago di Pergusa (Sicily, Southern Italy) as inferred by pollen, microcharcoal, and stable isotopes. Quat Int 181:4–14. doi:10.1016/j.quaint.2007.02.024

Sadori L, Giardini M, Giraudi C, Mazzini I (2010a) The plant landscape of the imperial harbour of Rome. J Archaeol Sci 37:3294–3305. doi:10.1016/j.jas.2010.07.032

Sadori L, Mercuri AM, Mariotti Lippi M (2010b) Reconstructing past cultural landscape and human impact using pollen and plant macroremains. Plant Biosyst 144:940–951. doi:10.1080/11263504.2010.491982

Sadori L, Giardini M, Susanna F (2010c) The plant landscape as inferred from a basket of the Roman town of Privernum (Latium, central Italy). Plant Biosyst 144:874–887. doi:10.1080/11263504.2010.491981

Sadori L, Jahns S, Peyron O (2011) Mid-Holocene vegetation history of the central Mediterranean. Holocene 21:117–129. doi:10.1177/0959683610377530

Sadori L, Zanchetta G, van Welden A, Baneschi I, Drysdale R, Giardini M, Gliozzi E, Mazzini I, Roberts N (2012) Climate changes at Lake Shkodra (Albania): the last 4500 years. ROL Società Geologica Italiana 18:35–38. doi:10.1016/j.quaint.2007.02.024

Sadori L (2013) Postglacial pollen records of Southern Europe. In: Elias SA (ed) Encyclopedia of quaternary science, vol 4. Amsterdam, Elsevier, pp 179–188

Sandweiss DH, Maasch KA, Anderson DG (1999) Transitions in the Mid-Holocene. Science 283:499–500. doi:10.1126/science.283.5401.499

Schmidt R, Müller J, Drescher-Schneider R, Krisai R, Szeroczyńska K, Barić A (2000) Changes in lake level and trophy at Lake Vrana, a large karstic lake on the Island of Cres (Croatia), with respect to palaeoclimate and anthropogenic impacts during the last approx. 16,000 years. J Limnol 59:113–130. doi:10.4081/jlimnol.2000.113

Schönwiese C-D (1995) Klimaänderungen 224 S. Springer, Heidelberg/Berlin

Sevink J, van Bergen M, van der Plicht J, Feiken H, Anastasia C, Huizinga A (2011) Robust date for the Bronze Age Avellino eruption (Somma-Vesuvius): 3945 ± 10 cal BP (1995 ± 10 cal BC). Quat Sci Rev 30:1035–1046. doi:10.1016/j.quascirev.2011.02.001

Stoffle RW, Toupal RS, Zedeño MN (2003) Landscape, nature, and culture: a diachronic model of human-nature adaptations. In: Selin H (ed) Nature across cultures: views of nature and the environment in non-Western cultures. Kluwer Academic Publishers, Dordrecht

Sulpizio R, Bonasia R, Dellino P, Di Vito MA, La Volpe L, Mele D, Zanchetta G, Sadori L (2008) Discriminating the long distance dispersal of fine ash from sustained columns or near ground ash clouds: the example of the Pomici di Avellino eruption (Somma-Vesuvius, Italy). J Volcanol Geotherm Res 17:263–276. doi:10.1016/j.jvolgeores.2007.11.012

Tinner W, van Leeuwen JFN, Colombaroli D, Vescovi E, van der Knaap WO et al. (2009) Holocene environmental changes at Gorgo Basso, a Mediterranean coastal lake in southern Sicily, Italy. Quat Sci Rev:1–13. doi:10.1016/j.quascirev.2009.02.001

Tipping R (1994) Ritual floral tributes in the Scottish Bronze Age, palynological evidence. J Archaeol Sci 21:133–139. doi:10.1006/jasc.1994.1014

Tzedakis PC (2007) Seven ambiguities in the Mediterranean palaeoenvironmental narrative. Quat Sci Rev 26:2042–2066. doi:10.1016/j.quascirev.2007.03.014

Valamoti SM (2004) Plants and people in late Neolithic and early Bronze Age Northern Greece: an archaeobotanical investigation, British archaeological reports international series 1258. Archaeopress, Oxford

van Geel B, Buurman J, Brinkkemper O, Schelvis J, Aptroot A, van Reenen G et al (2003) Environmental reconstruction of a Roman Period settlement site in Uitgeest (The Netherlands), with special reference to coprophilous fungi. J Archaeol Sci 30:873–883. doi:10.1016/S0305-4403(02)00265-0

van Zeist W, Wasylikowa K, Behre K-E (1991) Progress in old world palaeoethnobotany. Balkema, Rotterdam

Wagner B, Lotter AF, Nowaczyk N, Reed JM, Schwalb A et al (2009) A 40,000-year record of environmental change from ancient Lake Ohrid (Albania and Macedonia. J Paleolimnol 41:407–430. doi:10.1007/s10933-008-9234-2

Weiss E, Kislev ME, Hartmann A (2006) Autonomous cultivation before domestication. Science 312:1608–1610. doi:10.1126/science.1127235

Wendorf F, Schild R (1998) Holocene settlement of the Egyptian Sahara 1: the archaeology of Nabta Playa. Kluwer, New York

Weninger B, Alram-Stern E, Bauer E, Clare L, Danzeglocke U, Jöris O et al (2006) Climate forcing due to the 8200 cal yr BP event observed at Early Neolithic sites in the eastern Mediterranean. Quat Res 66:401–420. doi:10.1016/j.yqres.2006.06.009

Willis KJ (1994) The vegetational history of the Balkans. Quat Sci Rev 13:769–788. doi:10.1016/0277-3791(94)90104-X

Wroe S, Field J, Grayson DK (2006) Megafaunal extinction: climate, humans and assumptions. Trends Ecol Evol 21:61–62. doi:10.1016/j.tree.2005.11.010

Yll EI, Pérez-Obiol R, Pantaléon-Cano J, Roure JM (1997) Palynological evidence for climatic change and human activity during the Holocene on Minorca (Balearic Islands). Quat Res 48:339–347. doi:doi.org/10.1006/qres.1997.1925

Zalasiewicz J, Williams M, Haywood A, Ellis M (2011) The Anthropocene: a new epoch of geological time? Philos Trans A Math Phys Eng Sci 369:835–841. doi:10.1098/rsta.2010.0339

Zanchetta G, Sulpizio R, Roberts N, Cioni R, Eastwood WJ, Siani G, Caron B, Paterne M, Santacroce R (2011) Tephrostratigraphy, chronology and climatic events of the Mediterranean basin during the Holocene: an overview. Holocene 21:33–52. doi:10.1177/0959683610377531

Zolitschka B, Behre K-E, Schneider J (2003) Human and climatic impact on the environment as derived from colluvial, fluvial and lacustrine archives e examples from the Bronze Age to the migration period, Germany. Quat Sci Rev 22:81–100. doi:10.1016/S0277-3791(02)00182-8

The Mediterranean Human Population: An Anthropological Genetics Perspective

31

Marco Sazzini, Stefania Sarno, and Donata Luiselli

Abstract

The Mediterranean Sea has long been one of the most important and crowded natural hubs for the expansion of human genes and cultures, representing a tri-continental crossroads for human migrations since the first dispersals of anatomically modern humans out of Africa. Both its ancient and modern history, with its amazing chronicle of biological and cultural transitions, has substantially influenced the current patchwork of anthropological types existing within this area. For a deep dissection of this patchwork, Anthropological Genetics combines information related to the population dynamics able to shape the genetic structure of human populations (i.e. geographical constraints, language, cultural, social and political barriers) to those provided by the powerful tools of molecular biology and population genetics. This comprehensive approach allows to trace genetic profiles of Mediterranean populations into the past to discover and reconstruct their origins and demographic histories, as well as their evolutionary relationships. Nevertheless, the genetic landscape of Mediterranean populations is far from being exhaustively drawn. Several Anthropological Genetics projects, basing on an even deeper genetic characterization of dense and accurately selected geographic samples, have been just launched and promise to shed new light on the pivotal role of the Mediterranean basin as a genetic barrier and/or a bridge between human groups characterized by different African, Near Eastern or European cultural backgrounds.

Keywords

Anthropological Genetics • Mediterranean human populations • Polymorphic genetic markers • Paleolithic and Neolithic genetic signatures

Introduction

The Mediterranean Sea has long been one of the most important and crowded natural hubs for the expansion of human genes and cultures, occupying a pivotal geographical position within a broader passageway that has represented a tri-continental crossroads for human migrations since the first dispersals of anatomically modern humans out of Africa (Calderon 2000).

Its name means "*the Sea in the Middle of the lands*" and, accordingly, its history is a patchwork of migration and interaction events which have occurred among the several cultures and peoples inhabiting the lands surrounding its basin.

To understand the complex processes which have shaped this history from an Anthropological Genetics perspective, the range of population dynamics that can exercise substantial influence on the genetic structure of human populations has to be taken into account. For instance, geographical constraints are generally present in all biological populations, representing effective limits to reproductive exchanges, but

M. Sazzini • S. Sarno • D. Luiselli (✉)
Department of Biological, Geological, and Environmental Sciences, University of Bologna, Bologna, Italy
e-mail: donata.luiselli@unibo.it

S. Goffredo and Z. Dubinsky (eds.), *The Mediterranean Sea: Its history and present challenges*,
DOI 10.1007/978-94-007-6704-1_31,

without blocking them completely. In fact, attempts to assess the effects of migrant exchanges among human populations according to their geographical distances and using mathematical models based on isolation by distance have shown that the adjustment is not too strong. This may be due to the fact that a certain number of immigrants per generation are received in all human groups. That being so, it can be hypothesized that these recurrent movements, occurring both at group and at individual level, may have played a pivotal role in shaping the currently observed patterns of genetic diversity. According to this view, although many studies have detected a spatial pattern of genetic distances, especially on large geographical scale (i.e. at the continental or even world-wide level) (Lao et al. 2008), a high correlation should not be expected between geography and genetics in human populations which are settled in less extensive areas.

Language has also been a crucial component of evolutionary processes in *H. sapiens*, as well as in the differentiation of human populations, often acting as a genetic barrier for different ethnic groups. Recently, several studies have focused on an attempt to measure correlation between linguistic frontiers and genetic variation (Barbujani and Colonna 2010; Colonna et al. 2010). Considering Europe as a whole, significant relationships have been found between linguistics and genetics. However, areas of relatively sharp genetic change are irregularly spread and do not necessarily surround large population clusters, often overlapping with geographic and linguistic barriers. Therefore, we can presume that language only partially explains genetic differences among populations.

Finally, cultural, social and political barriers have also influenced human mating patterns through preferences or avoidances in the choice of a mate, a fact that still today represents an obstacle to gene flow and thus an "evolutionary" force able to generate significant structure within and among human populations.

For a deep dissection of the current patchwork of anthropological types existing within the Mediterranean basin, Anthropological Genetics combines information related to the above mentioned factors to those provided by the powerful tools and methods of molecular biology and population genetics. This comprehensive approach allows description of patterns of genetic variation in current Mediterranean populations, to help understand how these patterns arose. It also allows the tracing of genetic profiles into the past to discover and reconstruct origins and demographic histories of Mediterranean populations, as well as the development of inferences concerning their evolutionary relationships. This is done by assigning an ancestral place of origin to each population and reconstructing how and when subsequent dispersal division or migration took place, thus acquiring the genetic peculiarities which differentiate them from other contemporary populations.

The Mediterranean Region, a Tri-Continental Crossroads for Human Migrations

According to its pivotal role in the origin and diffusion of several *H. sapiens* populations and cultures, the Mediterranean region has become an ideal subject of cultural and biological anthropological studies, which have focused on modern human groups characterized by considerably tangled features from both geographical and chronological points of view (González-Pérez et al. 2010). These studies present an extremely precious opportunity to explore the genetic basis of the evolution of the entire human species. Most of them have investigated a time interval that broadly spans between the Upper Paleolithic and Mesolithic cultures, dating respectively to around 40,000 and 8,000 YBP, to the more recent Neolithic ones. Moreover, in addition to the prehistoric settlements which provide clear evidence for the ancient peopling of the Mediterranean area, a number of independent and/or intersecting population movements have also historically contributed to complicate the present patterns of distribution of the Mediterranean human groups (Cunliffe 2008).

Paleolithic Occupation of the Mediterranean Area by *H. sapiens* and Post Glacial Expansions from Refuges

Both paleoanthropological studies based on skeletal remains and Anthropological Genetics studies focused on the genetic diversity of present-day human populations agree in identifying the place of origin of the *H. sapiens* species in Africa, at about 150,000 YBP (Mellars 2006). From here, anatomically modern humans moved to colonize the remaining continents via the so-called Out-of-Africa migration (Stoneking 2008). Although during these movements, *H. sapiens* populations experienced severe demographic bottlenecks, they finally succeeded in the complete replacement of pre-existing human species in both Europe (*H. neanderthalensis*) and Asia (*H. erectus*) (Hublin 2011).

The Mediterranean area has thus represented a key point in early human colonization processes, especially for what concerns the entry and subsequent spread of *H. sapiens* in the European continent. Until the arrival of anatomically modern humans, most of the European and Western Asian territories, including those surrounding the Mediterranean basin, were inhabited by an archaic human form, the Neanderthals, which locally evolved in the last 350,000 years (Klein 2003). Although there is debate regarding the question of whether Neanderthals actually hybridized with *H. sapiens* (Krause et al. 2010; Prat et al. 2011; Currat and Excoffier 2011), it is certain that the appearance of its groups in Europe perfectly coincided with the beginning of the Neanderthal decline (Stringer and Davies 2001).

The currently accepted hypothesis stating that the fast spread of early groups of modern humans into Europe was strictly related to considerable improvements of climate during glacial intervals, assumes migration events that followed a southeast-northwest direction.

The earliest reliable evidence of the presence of modern humans in the European continent is dated to less than 48,000 YBP and has been proved by Bohunician sites of South-Central and Eastern Europe (Kozlowski 2007), having been linked to a migration wave in the Balkans during the Greenland Interstadial 12 warm climate interval. A second potential movement of *H. sapiens* groups through Mediterranean Europe is proven by Proto-Aurignacian artifact assemblages (Hoffecker et al. 2008) dating to the short Greenland Interstadials 11 and 9, which occurred approximately 40,000 YBP. Later, artifact assemblages assigned to the classic Aurignacian culture were associated with several modern humans' remains in Central-Western and Eastern Europe (Hoffecker et al. 2008).

The Levant is the most probable place of origin of both Bohunician and Proto-Aurignacian Upper Palaeolithic cultures, so that related archeological and fossil remains from the northern Mediterranean coastline (Northern Italy, Southeastern France, Northeastern Spain) likely reflect modern human population movements from the Near East to Europe via the Balkans (Prat et al. 2011), although Central Asia probably represented another early source of modern human populations for Eastern Europe (Otte 2007). This is in agreement with the fact that Iberian and Caucasian Neanderthal populations survived until more recent times compared to Southern European ones (Adler et al. 2008), demonstrating that early contacts between the two species did not occur in the extreme western and eastern boundaries of the Neanderthal distribution (Hoffecker et al. 2008).

Interestingly, recent re-examination of findings from two archeological sites has pointed to a reassessment of the timing of the earliest modern human colonization of Europe (Mellars 2011). Based on the most recent calibration of the radiocarbon timescale, Higham et al. (2011) document the dispersion of modern humans in northwestern Europe (Kent's Cavern, UK) at 42,000–43,000 YBP. By contrast, re-analysis of teeth discovered in the Italian Grotta del Cavallo dates the presence of *H. sapiens* in the extreme southern heel of Italy back to 43,000–45,000 YBP, thus supporting its dispersal in Southern Europe before the Aurignacian culture spread along the Danube river (Benazzi et al. 2011). Furthermore, Italian remains were associated with the Uluzzian technology, geographically confined only to the Italian Peninsula and southern Greece, which appeared to be a mixture of both Middle Paleolithic and Upper Paleolithic elements. This suggests that *H. sapiens* co-existed with Neanderthals at least in the Mediterranean part of Europe for several thousand years (Mellars 2011) (Fig. 31.1).

Approximately 25,000 YBP, dramatic worldwide climatic changes occurred, rapidly leading to the Last Glacial Maximum (LGM). This was characterized by the expansion of large ice sheets which covered a considerable part of the northern hemisphere, thus rendering much of the European continent almost uninhabitable (Mithen 2006). Accordingly, gaps in archaeological records suggest a complete abandonment of Northern and Central Europe, with a forced contraction of human populations towards southern European regions (Pala et al. 2009). In fact, the northern human groups which did not become extinct rapidly retreated to the Mediterranean coastlines, occupying refuge areas such as Northern Iberia and Southwest France (the so-called Franco-Cantabrian refuge), as well as the Italian and Balkan peninsulas. Here, they came into contact with pre-existing southern populations and remained until the glaciers began to recede approximately 16,000–13,000 YBP. The re-peopling process from these refuge regions, and thus the wider geographical re-distribution of new admixed and/or extremely reduced populations, was relatively slow, but increasing in rate with gradual climatic warming (Achilli et al. 2004). That being so the LGM undoubtedly exerted a great influence on the genetic landscape of present-day Mediterranean populations, having contributed heavily to the re-shaping of early Paleolithic patterns of variation (Rootsi 2006) (Fig. 31.2).

Main Migration Patterns Associated with Mesolithic Processes and the Neolithic Transition

The diffusion of agricultural techniques from their Near Eastern places of origin, and thus the extent of the Neolithic transition impact on Mediterranean human populations, has been widely debated and still represents a subject of discussion among archaeologists, linguists, anthropologists and geneticists (Ammerman and Cavalli-Sforza 1984; Simoni et al. 2000; Chikhi et al. 2002; Richards et al. 2002).

Undoubtedly, domestication of plants and animals, as well as a consequent increasingly sedentary lifestyle, were the main features that changed the human communities that adopted Neolithic practices (Haak et al. 2010).

Although a broad tendency to explain the diffusion of these changes as a result of a single homogeneous process exists, several sources of evidence take a contrasting view. The Neolithic impact on the European and Mediterranean gene pool is still highly debated in terms of both gene flow and dispersal routes (Lacan et al. 2011).

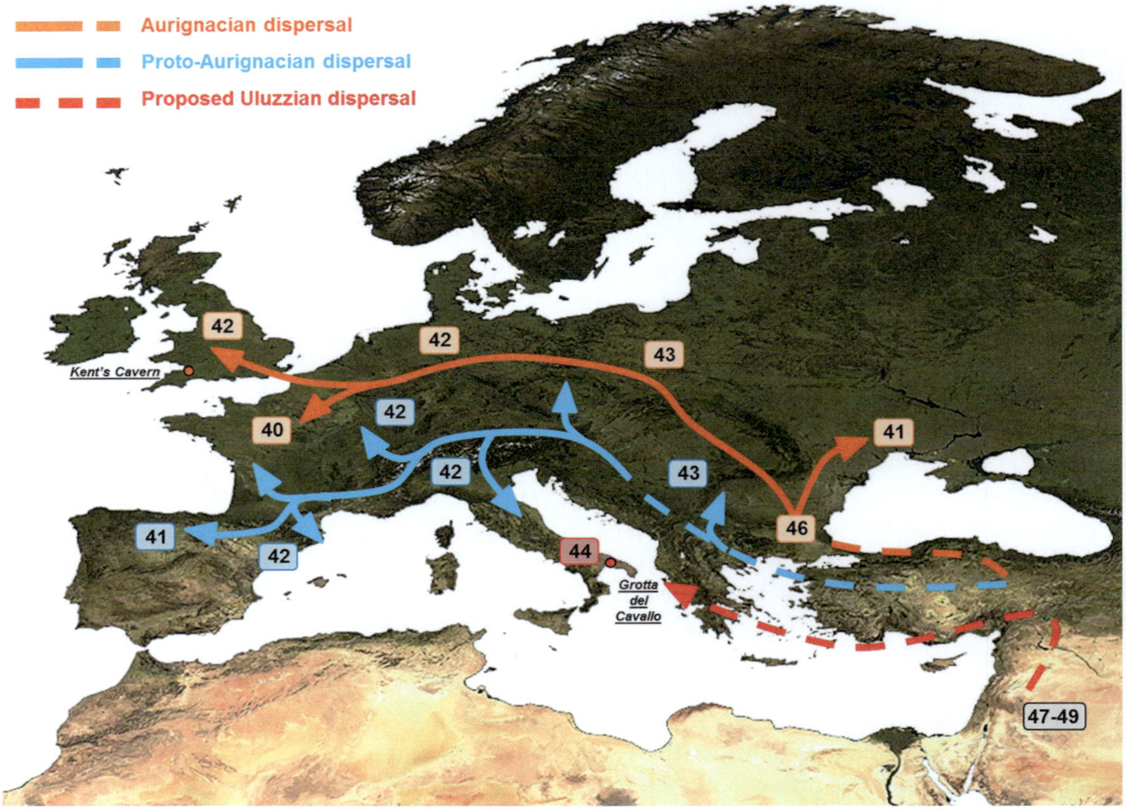

Fig. 31.1 Paleolithic dispersal routes of modern humans across the Mediterranean area. *Numbers* indicate kilo years before the present (KYBP) (Modified from Mellars 2011)

Two main hypotheses have been formulated to account for the demographic and genetic effects that this cultural revolution entailed: demic diffusion and cultural diffusion.

The first model, based on both archaeological and genetic data, implies population movements and thus supposes that an appreciable Near Eastern genetic signature is detectable in the Mediterranean human groups involved in the passage from hunter-gatherer to farmer societies (Ammerman and Cavalli-Sforza 1984). In particular, it hypothesizes that Near Eastern colonists, driven by considerable population growths due to the completely new resources offered by agriculture, progressively pushed aside local hunter–gatherer communities along a westward route of farming diffusion.

The second model, on the other hand, assumes that transition to agriculture, especially in the western part of the Mediterranean basin, was a predominantly cultural process (Whittle 1996), according to which the selective adoption of Neolithic innovations by local pre-existing Mediterranean populations was a consequence of trade and technology transfer alone. It thus assumes that there was no significant gene flow between hunter–gatherers and colonizing eastern farming groups, so preventing any appreciable influence of Neolithic transition on the gene pools of most Mediterranean populations (Zeder 2008).

However, as new archaeological data were collected and completely new methods have been applied to the analysis of even already acquired data, a more complex picture of such phenomena can be drawn, pointing out both migrations and local admixture events (Colledge et al. 2004; Zvelebil 2004), as well as outstanding regional variations in the process of agriculture spread, with differential and intersecting levels of demic and cultural diffusion (Pinhasi et al. 2005; Battaglia et al. 2008). It is currently accepted that none of the above mentioned models alone adequately succeed in describing the processes of Neolithic emergence in the Mediterranean Basin.

Archaeological data has identified the earliest Mediterranean agricultural settlements, dating to 6,800–7,000 YBP, on the western coast of the Aegean Sea and on Crete (Roewer et al. 2005; Tresset and Vigne 2011). From there, agriculture rapidly spread to the Balkan Peninsula, to Epirus and Corfu, and westwards via two different routes, the Vardar-Danube-Rhine corridor and the coasts of the Adriatic Sea, probably according to divergent necessities of adaptation of such practices to different micro-climates and environments. A geographical stratification in the local adoption of agricultural techniques has been thus observed in agreement with the two main cultures that co-existed in Early Neolithic:

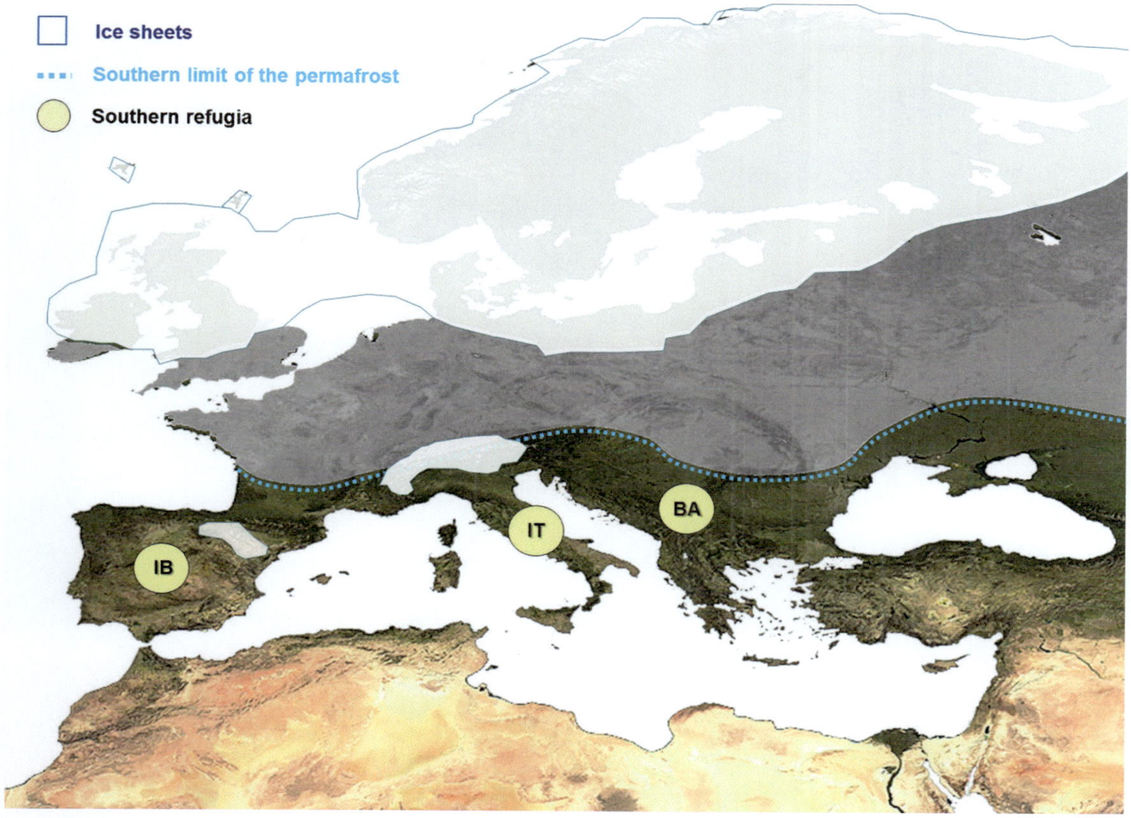

Fig. 31.2 Locations of the three main refugia during the last glacial maximum (LGM). *IB* stands for Iberian Peninsula, *IT* stands for Italy, *BA* stands for Balcans

the Linienbandkeramik culture, which spread along the Danubian river, and the Impressed Ware culture, which diffused along the Mediterranean basin. In particular, such latter geographical area probably was less suitable for the hunter-gatherer subsistence with respect to Central Europe and the Atlantic, leading to less demographic numbers in local Palaeolithic groups and thus favoring Neolithic diffusion (Sampietro et al. 2007).

A huge variation in population dispersal rates has been also inferred, since approximately 2,000 years seem to be required to Neolithic farmers to move from Cyprus to the Aegean, 500 years were then necessary for them to reach the Italian Peninsula, and, surprisingly, only another 600–800 years were needed for these people to move from Italy to the Atlantic (Zeder 2008; Tresset and Vigne 2011; Leonardi et al. 2012).

Environmental archaeology, showing that climatic factors may have strongly influenced such agriculture diffusion, allows the investigation of the Neolithic transition in a completely new socio-environmental perspective and may provide plausible explanations for its unexpected speed. Increased deforestation due to natural fires probably offered a good opportunity for the earliest Neolithic colonists, favoring the expansion of agriculture and herding groups and plausibly explaining their rapid expansion from Turkey and Cyprus as far as Bulgaria and Corfu. In particular, this Meso-Mediterranean region may have played a pivotal role in favoring the earliest agro-pastoralists which were fleeing arid Near Eastern areas, as well as the wet and thickly forested Central European regions (Berger and Guilaine 2009).

A detailed re-evaluation of archaeological evidence has shown greater discontinuities between Mesolithic and Neolithic cultures compared to what was previously thought (Skeates 2003).

The Late Mesolithic probably represented a period of particular population decline and relocation for many Mediterranean regions, independently from the westward movements of Near Eastern farmers. In the Iberian Peninsula, Neolithic settlements appeared on the eastern and southern coasts around 5,500 YBP (Tresset and Vigne 2011; Leonardi et al. 2012) on limestone-based territories which had already been abandoned by Mesolithic groups, whereas similar communities arrived in analogous places of the Atlantic coast of Portugal a few hundred years later. Mesolithic groups which based their subsistence on the intensive exploitation of estuary resources actually survived for several hundred years after the arrival of the described farming groups on the coast. The same was true for northern Italy, in which the introduction of agriculture arrived 300–500 years later than the

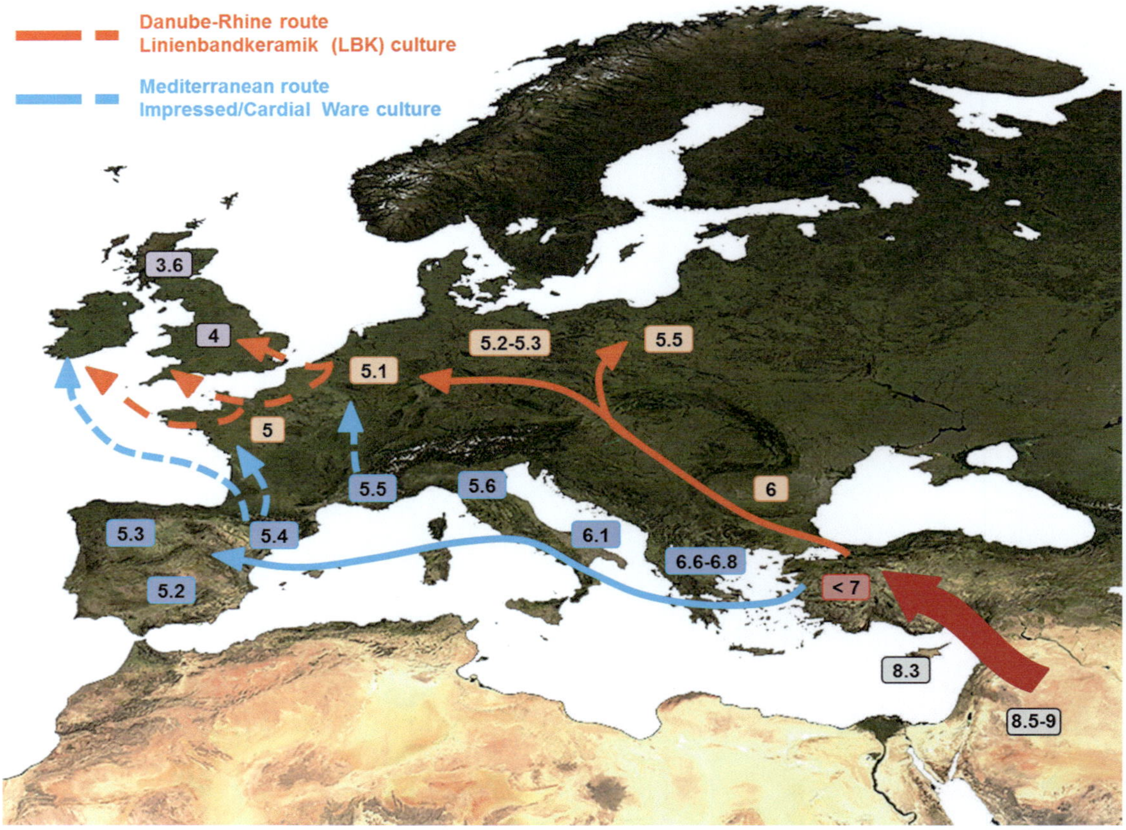

Fig. 31.3 Main routes of Neolithic spread across the Mediterranean area. *Number* indicate kilo years before the present (KYBP) (Modified from Tresset and Vigne 2011)

southeastern region (Biagi 2003), southern France (Tresset and Vigne 2007), in which sharp geographic, ecological, and cultural mismatches between inland Mesolithic and coastal Neolithic can be observed (Guilaine et al. 2008). This is true for the whole Aegean area (Pérles 2001), where the subsequent spread of agriculture into inland areas came via a combined process of colonist expansion, selective adoption of Neolithic technologies by local populations and their integration with farmer peoples, and a progressive blending of Neolithic and Mesolithic traditions after an initial process of colonization (Zeder 2008).

Another climatic crisis struck the Mediterranean regions around 6,200 YBP, leading to hyper-arid conditions in the northern Mediterranean coastlines and the appearance of new farming communities in North West Anatolia, Thrace, Macedonia, Thessaly and Bulgaria (Fig. 31.3).

Population Dynamics in Historical Times

In addition to the complex interplay of Paleolithic and Neolithic population movements, a number of historical events have also contributed to complicate the patterns of human genetic variation in the Mediterranean area. Population movements from the Eastern Mediterranean, including both the already described ancestral Paleolithic peopling by *H. sapiens* and subsequent Neolithic farmer migrations, have broadly the potential to shape genetic patterns deriving from the Levant with evident clines which decrease toward western regions. However, the dissection of independent contributions of each historical event on the gene pools of several Mediterranean populations is difficult to establish.

The collapse of Eastern Mediterranean Late-Bronze-Age societies around 3,000 YBP led to a cascade of demographic events peaking with the expansion of the Phoenicians from the coastal Levant (Cunliffe 2001) and Greek populations from the Aegean Sea (Murray 1993). Both of these societies established several colonies along European and North African Mediterranean coastlines, engaging in extensive trade.

The apogee of Phoenician colonization took place approximately 2,850 YBP and was characterized by the founding of new cities, including Carthage, mainly on the Northern African and Southern Iberian coastlines.

The Greek expansion ended approximately 2,600 YBP having mainly involved Southern Italian and Sicilian coastlines, the so called *Magna Graecia*, as well as the Iberian Peninsula and coastal France, and having exerted a noteworthy demographic and socio-cultural impact in the central and western parts of the Mediterranean basin (King et al. 2011).

The colonization processes of the Romans were even more complex, exploiting different typologies of colonies due to the highly unstable political climate, and covering a longer time interval spanning from 2,300 YBP to 1,800 YBP. In the early phase of this colonization, colonies were established only along the Italian Peninsula, whereas after the republican phase a major colonization impulse progressively led to the creation of the Roman Empire and the complete domination of the entire Mediterranean coastline with the definition of the Mediterranean Sea as the *Mare nostrum*. Since the Roman Empire was actually centered on the Mediterranean Sea, this whole area further increased its importance, definitively assuming a pivotal role for wide commercial routes and contacts among populations until the Empire fell around 1,500 YBP (Scarre 1995).

After the collapse of the Roman Empire, the Arab expansion across the Mediterranean basin represented one of the most important historical events. Approximately 1,400 YBP, the Arabs spread along the Southern Mediterranean coastlines and succeeded in occupying North Africa, colonizing native African Berber groups (Hitti 1990). One hundred years later, they occupied a great part of the Iberian Peninsula, as well as all Sicily, ruling over these regions up to the fifteenth and thirteenth centuries, respectively (Norman 1975) and thus contributing to a further increase in the ethno-linguistic heterogeneity of the southern and western areas of the Mediterranean basin.

According to this brief overview, multiple waves of population expansion occurred in the Mediterranean area during antiquity, each one being characterized by significantly different backgrounds and conditions.

Finally, the very recent migratory events of the last two centuries (Norwich 2007; Cunliffe 2008), as well as the present-day processes of extreme worldwide globalization, represent additional factors that have shaped, and are still currently shaping, patterns of genetic variation in the Mediterranean populations. These demographic and socio-cultural events further increase rates of gene flow across different ethnic groups and thus potentially favor the progressive reduction of genetic differences that can be observed across the region, with an unavoidable loss of more ancient and faint genetic signatures (Capelli et al. 2005).

That being so, the modern history of the Mediterranean area also involves several population movements across the lands that surround the basin, which have the potential to actually challenge the investigation of the extent to which human migrations have influenced the genetic profiles of present Mediterranean ethnic groups (Athanasiadis et al. 2010).

Mediterranean Populations: A paradigm for the Recent Evolutionary History of Humans

If we consider what has been mentioned in the previous paragraphs, it is clear that both the ancient and modern history of the Mediterranean basin, with its amazing chronicle of biological and cultural transitions, human migrations and of being a melting pot of anthropological types and cultures, represents a challenge for Anthropological Genetics.

A huge number of genetic studies have been carried out on the Mediterranean region, both on a wide and micro-geographical perspective, exploiting information from very different sets of genetic markers and aiming at dissecting very different population histories. Nevertheless, no definitive consensus has been reached on the whole genetic landscape of Mediterranean human groups, once again demonstrating the outstanding complexity of its population dynamics (Tomas et al. 2008).

Anthropological Genetics and the Use of Molecular Markers

The study of the evolution of *H. sapiens*, the differentiation processes of its populations, as well as the historical and phylogenetic relationships among different human groups and their patterns of migration and admixture, has long required a multidisciplinary approach. Accordingly, historical, linguistic, anthropological, as well as, more recently, bio-statistical and molecular biology disciplines have joined their efforts in the attempt to improve the knowledge of human history and evolution (Cavalli-Sforza and Feldman 2003).

In particular, Anthropological Genetics has succeeded in conjugating a detailed anthropological perspective able to rigorously describe the human groups under investigation, with the more recent advances in molecular genetics. This has resulted in a careful interpretation of genetic data that contextualizes them not only in the landscape of human biological variability, but also in that of human population histories and socio-cultural habits, making it possible to reliably infer population origins and subsequent demographic and/or evolutionary events which have created and shaped the genetic diversity of present-day human groups (Flores et al. 2003; Schaffner 2004).

Data on several classes of DNA variation (i.e. polymorphisms) in human populations have been thus accumulated, reinforcing the idea that the investigation of these patterns of genetic diversity represents an extremely precious chance not only to understand the evolutionary history of our species (Lander 2011), but also to explore the genetic basis of human adaptation and its crucial medical implications (Benjeddou et al. 2006).

The availability of different typologies of genetic markers, each characterized by specific inheritance patterns and evolving with completely different rates, makes it possible to distinguish populations separated since long or short evolutionary times, and to disentangle the evolutionary forces which have then shaped their differentiation during time.

Genetic similarities and dissimilarities among human groups are actually due to a relatively small number of evolutionary factors (i.e. mutation, recombination, genetic drift, migration and selection) which have differentially acted on the gene pool of human populations during their dispersal out of Africa. For example, the vast majority of newly acquired mutations are lost by chance because of genetic drift, especially in groups made up of a little number of reproductive individuals, whereas few of these new variants raise in frequency and succeed in spreading worldwide by migration processes. Moreover, different selective pressures may have also led human groups, living in different geographic areas, to develop divergent genetic profiles to adequately adapt to very specific environments.

Markers of mitochondrial DNA (mtDNA) and Y-chromosome genetic systems have long played a pivotal role in Anthropological Genetics, being able to provide reliable reconstructions of phylogenetic and phylogeographic relationships among populations. Thanks to their peculiar inheritance, being mtDNA maternally transmitted (only from the mother to offspring) and Y-chromosome representing its paternally inherited counterpart (transmitted only from the father to sons), these markers are indeed able to account for patrilineal and matrilineal mobility patterns (Jobling et al. 2004).

Therefore, being transmitted without the reshuffling of maternal and paternal genomes in each generation (i.e. recombination), they can change over generations only by accumulating mutations over time. The subsequent accumulation of these mutations during evolutionary times marks paternal and maternal phylogenetic lineages (also named *haplogroups*) which can be used to infer, going back in time, human history and population movements. Current geographic distribution of these lineages (i.e. phylogeographic patterns) records and reflects past population migrations, mergers and divisions that account for the current spectrum of human variability (Underhill et al. 2001).

It can be also assumed that polymorphisms occurring on mtDNA and Y-chromosome are selectively neutral (i.e. they do not confer a selective advantage in terms of survival and/or reproductive success to the individual carrying the mutation), thus offering a reliable tool for inferring the impact of genetic drift, migration and cultural practices on population genetic histories.

Moreover, the genetic variation described by these uniparentally inherited non-recombining systems is more sensitive to the effects of genetic drift (especially founder effect), leading to increased and faster geographic clustering and/or differentiation between groups of lineages observed in different populations, and is ideal for assessing the origins of contemporary population diversity (Pericić et al. 2005).

In very recent years, a considerable amount of genetic data has started to become available for autosomal and X-chromosome loci (Tomas et al. 2008). Individually, each of them provides much less information compared to both mtDNA and Y-chromosome data, as a consequence of recombination that strongly limits the possibility of inferring a reliable genealogical history for a given autosomal or X-chromosome segment. However, the combination of these data has the potential to provide much greater statistical power for testing specific hypotheses about past demographic events (Schaffner 2004; Wilkins 2006).

Recent advances in single nucleotide polymorphism (SNP) genotyping technologies have led Anthropological Genetics to an unprecedented level of resolution, shifting the research focus from a loci-based to a genome-wide perspective. It is thus possible to contemporaneously investigate hundreds of thousands of autosomal SNPs, so succeeding in shedding new light on the genetic variability of our species, as well as on different origins, demographic histories and also disease susceptibilities of its populations (Stoneking and Krause 2011). In addition to this, technological improvements of high-throughput sequencing platforms have recently enabled the complete sequencing of several human genomes, bringing the study of our species' genetic variation to a completely new level of complexity. The "1000 Genome Project" has just reported information on approximately 15 million SNPs (1000 Genomes Project Consortium 2010). According to this genome wide autosomal approach, and to the new theoretical tools developed for the analysis of population structure and admixture (Falush et al. 2003; Alexander et al. 2009), a finer description of the variable level of geographical structuring of human groups, and thus of relationships and/or dissimilarities even among closely-related populations, was achieved (Rasmussen et al. 2010; Reich et al. 2010). This has provided a step forward in the attempt to disentangle to what extent human populations within continental regions represent discrete genetic clusters or halfway points of a genetic continuum.

Demographic Events and Their Legacy on the Current Genetic Landscape

As already mentioned in the previous paragraphs, the history of *H. sapiens* in the Mediterranean area can be divided into clearly-distinct episodes of population migration having occurred both in pre-historical and historical times. While there is still an intense debate on the actual dynamics that have characterized these events, each of them has left an impressive footprint in the gene pool of modern Mediterranean populations (Soares et al. 2010).

Unfortunately, inferring the exact genetic contribution of each migration process is very difficult, since most migrations followed the same direction of dispersion within the Mediterranean Basin, being also separated by short evolutionary time-frames.

That being so although many studies have focused on this issue, a definitive conclusion has been not obtained as a consequence of often-contradictory results (Sampietro et al. 2007).

The Pre-Neolithic Genetic Inheritance of Mediterranean populations

According to the mtDNA perspective, between 40 and 80 % of the mitochondrial lineages that can be observed in most Mediterranean populations are related with Palaeolithic demographic processes (Richards et al. 2000; Semino et al. 2000), especially in the late Upper Palaeolithic and the Late Glacial or postglacial expansion times. Palaeolithic population movements from the Near East seem therefore to be those with the most appreciable impact on the present Mediterranean mtDNA pool, with Central Asian lineages (I, X, W) accounting for only 7 % of current mitochondrial variation (Semino et al. 2000; Richards et al. 2000; Soares et al. 2010).

Two different dispersal routes from the Near-East have been proposed, marked by different sub-lineages of mtDNA haplogroup U, U5 into Europe and U6 into North-Africa, respectively (Soares et al. 2010). It is widely accepted that the most ancient mtDNA lineages observed in the European coastline of the Mediterranean Basin belonged to the U5 haplogroup, which arose in Europe with the first settlers in the Upper Paleolithic and is supposed to be locally-originated around 37,000 YBP (Soares et al. 2010; Malyarchuk et al. 2010). The age of U5 may be an underestimate, as a consequence of significant postglacial diffusion of its major sub-clades (U5a and U5b). Its real age is potentially between the current estimate and the 56,000 YBP time of origin of its ancestral lineage (U). Regardless, mtDNA evidence broadly supports a modern human colonization of the European territories surrounding the Mediterranean Basin roughly occurred between 55,000 and 30,000 YBP, in accordance with the main paleoanthropological data (Soares et al. 2010).

Y-chromosome studies have drawn a similar picture, once again suggesting that a considerable fraction of the Mediterranean gene pool has an Upper Paleolithic origin, although it was subsequently reshaped and relocated after the LGM expansions (Semino et al. 2000).

In particular, Y-chromosome data support the diffusion of the Paleolithic Aurignacian and Gravettian cultures, showing that almost 80 % of European and Mediterranean Y-chromosomes are mainly originated from two lineages (R1-M173 and I-M170).

The R1-M173 lineage, whose derived clades (R1a and R1b) are just highly spread in Europe, has been proposed as an ancient Eurasian marker which arrived on the European coastlines of the Mediterranean Basin around 35,000 YBP, perfectly in concordance with the westwards diffusion of the Aurignacian culture found in several archeological remains, such as crafted tools of antler, bone, and ivory, as well as rock-art paintings. The I-M170 lineage was instead related to a second wave of migration into the European continent occurred approximately 22,000 YBP, coming from the Near East together with the Gravettian culture (Gibbons 2000).

The Y-chromosome I lineage, being mainly restricted to the European continent, in the same way of its mtDNA counterpart U5, can be therefore considered a good marker of early Upper Palaeolithic population movements (Soares et al. 2010).

It has been proposed that its ancestral lineage (IJ-M429*) spread within the European continent via the Balkan route, then evolving into lineage J in the Near East and lineage I in Europe (Battaglia et al. 2008).

However, since the age of the I lineage is far from being reliably estimated, no Y-chromosome markers perfectly corresponding to mtDNA haplogroup U5 are found to have arrived from the Near East in the same period, highlighting that some discrepancies actually exist between the maternal and paternal perspectives provided by the two described genetic systems.

These differences may be due to the more recent molecular age proposed for the Y-chromosome with respect to other loci, implying a faster replacement of previous Y-chromosomes, and/or to different migration patterns in Paleolithic males and females (Semino et al. 2000).

During the LGM (25,000–19,000 YBP) the climate severely worsened and human populations became concentrated in refuge areas scattered from Southwestern Europe, along the Mediterranean coastline, to the Balkans and the Levant, as well as to Eastern Europe. In particular, peoples belonging to the Aurignacian culture, carrying the Y-chromosome R-M173 marker, mainly concentrated in Iberian and Ukrainian refuges, whereas groups characterized by the I-M170 Gravettian marker congregated in the Balkans. Nevertheless, no clear conclusions have been drawn about the modalities of such population concentration, because it is difficult to accurately distinguish local extinction events and migration ones. Very rapid and extensive population expansions subsequently followed the contraction of glaciers, thus explaining why the genetic markers mentioned above are predominant in present-day European and Mediterranean populations.

Different evidence has been used in indicating that the major signal in the modern mtDNA European gene pool can be traced back to the re-expansion and resettlement from southern refuges during the warming phase after LGM (Soares et al. 2010).

An important contribution to the Mediterranean and, more in general, to the European mitochondrial landscape was offered by the H lineage, which exhibits significantly high frequencies (45 % in modern Europeans) across a wide geographic distribution. It has been proposed that this lineage probably has a Near Eastern origin (around 30,000 YBP) and that it spread into the European continent during the diffusion of the Gravettian culture (25,000–20,000 YBP), being finally involved in postglacial expansions (Torroni et al. 1998; Richards et al. 2000).

Similarly, mtDNA lineage V, which is also supposed to have re-expanded from south-west after LGM, has been proposed to have arrived earlier in Europe. Geographical distribution in the Mediterranean and the estimated age support the scenario that preV lineages originated in Europe before the LGM (like its larger sister lineage H), perhaps in eastern Europe, whence both spread along an east–west axis with Gravettian contacts (Torroni et al. 2001; Soares et al. 2010).

Interestingly, the analysis of several Western Eurasian and North African populations revealed that H1 and H3 subclades show frequency peaks in the Iberian Peninsula and in its surrounding areas, with declining frequencies toward the northeast and southeast. Both this distribution pattern, and H1, H3 ages of origin are perfectly comparable to those already described for lineage V (Torroni et al. 1998, 2001), suggesting that the Franco-Cantabrian refuge played a major role in postglacial expansions towards Central and Northern Europe, in accordance to early results obtained with classical genetic markers (Cavalli-Sforza et al. 1994).

Most of the present observable mtDNA lineages, such as V, H1, H3 (11,000–11,500 YBP) and H5 (13,900 YBP), as well as U5b1 (13,000 YBP), originated during or immediately after the LGM mainly in Southwestern Europe and experienced a postglacial spread along different routes into Western, Central and Northern Europe, together with the pre-existing lineages of Near Eastern origin dating back to the Paleolithic colonization of the European continent.

With the exception of the rare U5b3 lineage that probably expanded from an Italian refuge (Pala et al. 2009), most mtDNA evidence suggests a postglacial Western and Central Europe resettlement from Franco-Cantabrian refuge areas (Soares et al. 2010). An Eastern Europe re-colonization has been also suggested, especially by lineage U4, from Ukrainian refuges (Malyarchuk et al. 2010).

This scenario is also supported by archaeological and Y-chromosome data (Semino et al. 2000; Cinnioğlu et al. 2004; Rootsi et al. 2004; Achilli et al. 2004). Y-chromosome analysis revealed that the majority of lineages making up the modern European paternal gene pool were implicated in Late Glacial or postglacial expansions.

The Y-chromosome I lineage, which originated from a pool of European Upper Paleolithic Y-chromosomes (Semino et al. 2000) and which subsequently diffused together with the Gravettian culture, encompasses several sub-clades with geographically-specific patterns of distribution within Europe.

Several of these sub-lineages (I1, I2a and I2a1, according to the Karafet et al. 2008 nomenclature) have been proposed to have expanded from specific southern European refuges, so providing reliable records of Mediterranean and European-specific paternal heritage involved in contraction, isolation, and subsequent postglacial diffusion processes (Soares et al. 2010).

In particular, the I1-M253 clade is mostly found in northern Europe, reaching its highest frequencies in Scandinavian populations (88–100 %). Since the Scandinavian Peninsula was completely depopulated during the LGM, a post-glacial origin from southern Mediterranean Europe has been suggested for Scandinavian I1 chromosomes (Rootsi et al. 2004). Interesting, this sub-clade shows a distribution pattern similar to those observed in the second PC of the synthetic maps based on classical genetic markers (Cavalli-Sforza et al. 1994), as well as showing a significant positive correlation with mtDNA V and U5b lineages already proposed as markers of postglacial population expansions from the Iberian Peninsula (Torroni et al. 2001; Rootsi et al. 2004).

With respect to the I2a-P37.2 lineage (Karafet et al. 2008), it is the predominant Y-chromosome lineage in Eastern Europe and the Balkans (31–40 %), whereas its sub-clade I2a1-M26 (Karafet et al. 2008) is found in Western Europe at a very low frequency with the exception of Sardinia (41 %) (Rootsi et al. 2004).

It has been suggested that the I2a and I2a1 lineages were separated before the LGM and that the I2a1-M26 mutation arose in Iberia or Southern France, whereas I2a has probably expanded from a Balkan refuge area mainly towards Eastern Europe (Rootsi et al. 2004).

A different history can be inferred for the Y-chromosome R1 lineage, which seems to have originated in South Asia during the early dispersal of modern humans along the so-called southern route (Kivisild et al. 2003; Soares et al. 2010). Its R1b sub-lineage arrived in the European continent immediately after the LGM (paralleling mtDNA lineages H and V), and then further originated the R1b1b2-M269 subclade that subsequently expanded into Western and Central Europe. R-M269 has split into geographically localized sub-lineages, appearing to have spread from different areas of highly localized frequencies within Europe (Busby et al. 2011; Myres et al. 2011; Cruciani et al. 2010; Balaresque et al. 2010; Morelli et al. 2010).

The R1a sub-lineage may instead mark an expansion from an Ukrainian refuge, but it might also have been the result of more recent dispersals (Underhill et al. 2010; Soares et al. 2010).

Impact of Early Agriculture on the Mediterranean Gene Pool

The debate on the interplay of demic dispersal and cultural diffusion processes underlying the Neolithic transition has been a central topic for Anthropological Genetics for at least two decades, being heavily complicated by challenges in modeling complex demographic events, made up of multiple migrations, and in assigning specific genetic signatures to their actual geographical areas of origin.

The demic diffusion model has been supported by several investigations on the diversity patterns of nuclear proteins (Cavalli-Sforza et al. 1994; Barbujani et al. 1994), uniparentally inherited genetic markers and autosomal data (Chikhi et al. 2002; Currat and Excoffier 2005; Balaresque et al. 2010). These studies suggest the expansion of Near Eastern agricultural groups throughout the European continent, their early genetic isolation with respect to indigenous hunter-gatherers, as well as final large-scale population replacements (Ammerman and Cavalli-Sforza 1984), according to broad European and Mediterranean gradients of variation (Dupanloup et al. 2004).

Recently, ancient mtDNA analysis of skeletons from pre-Neolithic hunter-gatherers, as well as early Neolithic farmers, have shown mtDNA lineage frequency distributions of in modern Europeans to be more similar to early farmers, with pre-Neolithic populations having quite different genetic compositions. This points to a replacement of hunter-gatherers by in-coming farming groups during Neolithic colonization. In particular, two main mtDNA lineages (namely H and U) have been proposed as typical for early farmers and hunter-gatherers respectively.

Population expansion between 15,000 and 10,000 YBP has been proposed for hunter-gatherers mtDNA U-types, in accordance with the end of the LGM in Europe and the subsequent northwards post-glacial expansion. Population increase for early farmers, carrying the H-type, dated instead approximately to 9,000 YBP. After 4,000 YBP both U-type and H-type mtDNAs showed similar patterns of population growth, thus probably suggesting a fusion between the two populations (assimilation of resident hunter-gatherers), after the primarily different histories (Fu et al. 2012).

Other studies of mtDNA and Y-chromosome variation in present and past populations have drawn opposite conclusions (Richards et al. 2000; Semino et al. 2000; Richards 2003; Achilli et al. 2004; Battaglia et al. 2008; Soares et al. 2010), highlighting the association of European and Mediterranean patterns of diversity to Upper Paleolithic and postglacial events and identifying genetic profiles for early Neolithic farmers that are quite rare in present-day populations, thus supporting a mainly cultural diffusion of agriculture.

According to these studies, only approximately 20 % of the mtDNA and Y-chromosome landscape of Mediterranean populations show Neolithic ancestry (Richards et al. 2000; Semino et al. 2000).

The cultural diffusion model is especially supported by some mtDNA investigations (Richards et al. 2000; Torroni et al. 1998) which assume that small groups of Near Eastern farmers moved into the European continent and that a progressive wide-scale adoption of agricultural techniques subsequently occurred in indigenous Mesolithic populations.

In the same way, investigations of Y-chromosome variation confirmed what has already been described for mtDNA markers, identifying the Neolithic paternal contribution for current Mediterranean populations in approximately 20 % of them (Semino et al. 2000). However, reinterpretation of the previous data through application of an admixture model performed by Chikhi et al. (2002) suggests that the contribution of genes from the Near East to Europe was 50 and 65 % when using Basques and Sardinians respectively, as proxy Paleolithic parental populations.

Further resolution of the Y-chromosomal lineages, has provided the ability to distinguish European Mesolithic Y lineages (E1b1b1a2-V13 and I2a2-M423) from subsequent Neolithic range expansions from the Near East (J2b2-M241), instead supporting the cultural diffusion model in Southeast Europe (Balkans). In particular, they provide further insights on the impact of Neolithic transition in the Balkans and in their role as a longstanding gateway to the European continent from the Near East.

Several J sub-clades originating from Anatolia have been linked to the demic expansion of farmers in the Mediterranean basin due to their diffusion times (7,000–8,000 YBP) (Di Giacomo et al. 2004; King et al. 2008), which are perfectly comparable with archaeological estimates for the early appearance of Neolithic pottery in the Adriatic region. Moreover, the J-M410 and G-P15 lineages have been hypothesized to record demic farming colonization processes in Crete, the Aegean Sea and the Italian Peninsula (King et al. 2008; Martinez et al. 2007; Di Giacomo et al. 2003, 2004).

On the contrary, the age of the expansion of the V13 lineage into Western Europe from Southern Balkans, which is consistent with late Mesolithic (King et al. 2008), as well as the I-M423 Southeastern European autochthonous clade, suggested that also cultural adoption of agricultural techniques by Mesolithic hunter-gatherers took place in the Balkans.

Once again genetic data depict a more complex pattern of interactions between Neolithic farmers and pre-existing foragers rather than a simple replacement of the latters. Within the traditional dichotomy between demic diffusion process and acculturation phenomenon, recent models have in fact proposed a more complex scenario, suggesting a varying intensity of the genetic impact from Neolithic farming communities through Europe.

In this respect, ancient DNA analyses, exploiting the potential to directly investigate genetic patterns of early

Neolithic and late hunter-gatherer groups, have been also performed in attempts to clarify the diffusion processes of agricultural techniques.

Up to now, studies conducted on ancient DNA samples from different parts of Europe have suggested two different routes to explain the models of Neolithic dispersion in the European continent (Hervella et al. 2012), supporting an acculturation model through North-Central Europe (Haak et al. 2005, 2010) and a demic diffusion model through the Mediterranean route (Sampietro et al. 2007; Lacan et al. 2011), respectively.

A study based on Central European individuals belonging to the Linienbandkeramik culture observed a high frequency (25 %) of the currently rare (0.2 %) mtDNA lineage N1a, which is instead found to be absent in neighboring Mesolithic groups. This suggests that N1a can be considered a candidate signature for early Neolithic farming populations (Haak et al. 2010) and that they have had relatively small impact on the genetic background of present-day European and Mediterranean groups. Pre-existing hunter-gatherer communities that adopted Neolithic technological improvements then outnumbered the early colonizer, diluting the N1a frequency to present values. This represents additional support for the cultural diffusion model.

Alternatively, another study based on Neolithic remains from the Iberian Mediterranean region and belonging to the Impressed Ware culture, did not find mtDNA differences from present-day populations. The absence of N1a lineages and the presence of lineages typical of present-day populations in that area point to a demic diffusion model of Neolithic culture in the Mediterranean part of Europe (Sampietro et al. 2007). A Mediterranean origin for the Neolithic contribution was also suggested for the French Mediterranean area, based on the absence of the mtDNA lineage N1a and the R1b Y-chromosome lineage, both considered Neolithic dispersion markers through Central Europe (Lacan et al. 2011).

Similarly, the more recent mtDNA analysis of Upper Paleolithic and Early Neolithic individuals from the Northern Spain region (Hervella et al. 2012), reaffirm the aforementioned different patterns between Central-European and Western-Mediterranean Neolithic groups, suggesting that the genetic influence of Neolithic colonization in Central Europe was different to that in the Mediterranean area. These genetic data may thus provide support for a random dispersion model, in which Neolithic farmers influenced various geographic regions of Europe differently, contradicting the more simplistic models of total acculturation or replacement previously proposed to explain Neolithisation (Hervella et al. 2012).

The results described above highlight that Neolithic spread was neither a genetically nor geographically perfectly homogeneous across the Mediterranean and European areas, but that it differentially involved demographic and cultural diffusion in relation to different regions of Europe (Sampietro et al. 2007).

Signatures of Recent Historical Events in the Mediterranean Genetic Landscape

Among the historical events listed in section "The Mediterranean Region, a Tri-Continental Crossroads for Human Migrations", the following ones have been proved to have left detectable footprints in the genetic landscape of Mediterranean populations according to several Anthropological Genetics studies. However the dissection of independent contributions of each historical factor is often difficult to disentangle.

Several studies have identified Y-chromosome lineages which show clines of variation from the Near East to Western Mediterranean regions, some of which have been specifically related to the Phoenician expansion. The comparison of genetic profiles from historically well-documented areas of Phoenician colonization, and from neighboring locations without such historical features but with plausibly shared genetic backgrounds, indeed allows the distinction of genetic patterns representing a Phoenician signature and overall geographical genetic gradients. According to this, the overall east-to-west background variation, mainly reflecting Paleolithic and Neolithic migrations, as well as the influence of the more widespread Greek colonization, can be discarded.

The Y-chromosome J2 lineage, as well as six haplotypes associated with it and with E3b and J* lineages, contribute to more than 6 % of the current Y-chromosome variation in Phoenician-influenced populations of the Iberian Peninsula and of coastal Tunisia, suggesting that the more recent destruction of Carthage by Romans did not coincide with the complete replacement of the Phoenician gene pool in this geographical area (Zalloua et al. 2008).

After Phoenician colonization, the Greek expansion from both Greece and Anatolia undoubtedly played a major role in the cultural development of Western Mediterranean populations during the Iron Age, by virtue of the foundation of several colonies in Southern Italy, France, and the Iberian Peninsula.

Genetic signatures of such a colonization process have been detected in the Sicilian island, in which approximately 37 % of Y-chromosome variation, represented by the Balkan/Greek E-V13 marker (Cruciani et al. 2007; Battaglia et al. 2008), can be traced back to Greek immigration (Di Gaetano et al. 2009), in contrast to the lower contribution hypothesized by previous demographic studies. Moreover, the E-V13 lineage also shows low frequency in Anatolia (Cruciani et al. 2007). This is useful for identifying Greek colonization from early Neolithic westwards expansion.

Y-chromosome data from individuals belonging to Provence, Smyrna and Asia Minor Phokaia, the source point of Greek colonization for Marseilles and Aleria (Corsica), show that approximately 12–19 % of the observed lineages in Smyrna and Asia Minor Phokaia are represented by E-V13 haplogroup. This frequency decreases to 4 and 4.6 % in Provence and Eastern Corsica, respectively, and to 1.6 % in Western Corsica. Nevertheless, overall 17 % of Provencal Y-chromosome variation can be attributed to Greek colonization, suggesting that additional and potentially confounding demographic events, such as the subsequent Roman dominance, actually had a minor influence on the genetic background of this geographical area. This is also confirmed by the fact that in England, Southern Spain, Morocco and Sardinia, as well as in other Roman colonies, the E-V13 frequency reaches a maximum value of 1 % (Cruciani et al. 2007). These findings thus support the hypothesis of a male-mediated Greek asymmetric gene flow into the Provencal autochthonous Celto-Ligurian groups, as a consequence of differential mating practices, elite dominance and/or enslavement processes (King et al. 2011).

The Arab expansion seems to have had a substantial influence on Eastern and Southern Mediterranean coastlines, also involving sex-specific gene flow (Capelli et al. 2005).

The North African E-M81 Y-chromosome lineage has been long considered the most important marker associated with medieval Arab contributions to the Southern European genetic landscape. However, according to historical evidence indicating a mainly Berber origin for the Arabs invading Southern Europe (Hitti 1990; Norman 1975), lineages such as E-M81 derived, E-M78 derived showing the rare DYS439 allele 10, and a subset of J1 (M267 derived), identified as being NW-Africa specific and accounting together for 58–90 % of Northwestern African Y-chromosome variation and for 13 % of the European one (Arredi et al. 2004; Cruciani et al. 2004; Semino et al. 2004; Francalacci and Sanna 2008), have been identified as good candidates for an Arab genetic signature. Particularly high frequencies of such lineages are found in the Iberian Peninsula, especially in the Pas Valley (Cantabria), and in the Italian Peninsula, mainly in Sicily, in accordance with the long-term Arab rule in these areas (Capelli et al. 2009).

The Middle Sea: A Bridge or a Barrier Between Northern and Southern Mediterranean Populations?

The Mediterranean pivotal role in the origin, dispersion and admixture processes of human groups is mainly due to the fact that it represents a natural geographic boundary between the African and European continents. Populations characterized by considerably different cultural backgrounds are thus distributed along its northern and southern coastlines and, to date, Anthropological Genetics studies have achieved contrasting results about the extent of gene flow and genetic affinities between them (Currat et al. 2010).

In particular, great attention has been long addressed to the Western Mediterranean area, often considered apart from the overall African and European coasts, which have instead acted as termini of population movements (Plaza et al. 2003).

North Africa: The Southern Mediterranean Shore

The peopling of North Africa has been characterized by completely different patterns compared to those observed for sub-Saharan African regions, representing an anthropological island within the African continent and being heavily influenced by the geographical barriers represented by the Sahara Desert and the Mediterranean Sea. Its geographical location thus constrained migrations mainly into an east–west direction and through the Suez Isthmus and the Strait of Gibraltar, the two most probable passageways for intercontinental human dispersals.

Pre-historical and Historical Backgrounds

According to fossil records and archaeological remains belonging to the Aterian culture and dating back to around 40,000 YBP, North Africa has been peopled by anatomically modern humans since the Upper Palaeolithic (Rando et al. 1998; Ennafaa et al. 2009). The Capsian Mesolithic culture instead seems to have given way to the Neolithic transition in that area, spreading from the Near East to Egypt (Ennafaa et al. 2009; Fadhlaoui-Zid et al. 2004) and, around 9,000 YBP, to part of the Sahara, exploiting the period of its maximum wetness (Rando et al. 1998). Increase in Sahara desertification then forced people to retreat northwards to mountainous regions of Maghreb and southwards to the Sahel, whereas Neolithic European Mediterranean migrations to Northwest Africa finally occurred from around 6,000 YBP.

Despite the presence of the geographic barriers mentioned above, trade routes were established across the desert, as well as across the Mediterranean basin, especially since the Phoenician colonization around 2,000 YBP. Subsequently, Romans, Vandals and Byzantines also occupied North African coasts, but probably slightly contributing to the local genetic background (Rando et al. 1998). In fact, diffusion of Pharaonic and classical Mediterranean cultures in that area imposed strong cultural influences, but only a negligible demic impact (Ennafaa et al. 2009).

Finally, Arab Berbers from Northern Morocco and Algeria expanded up to the Iberian Peninsula for almost eight centuries (Hitti 1990; Rando et al. 1998; Comas et al. 2000). Between the sixteenth and twentieth centuries Tunisia was also ruled by Turks, Spaniards and French, as well as being involved in Sub-Saharan African slave trade from Sudan (Fadhlaoui-Zid et al. 2004).

Despite the described gene-flow, North Africa has also experienced long periods of isolation which led local groups to increase the frequency of specific markers, resulting in an overall genetic landscape that represents a patchwork of autochthonous, European, Sub-Saharan African and even West Asian components.

Thus, Anthropological Genetics studies continue to debate the relative contributions of demographic events to the current North African genetic landscape.

North African Genetic Landscape

Early mtDNA studies show that most of North African maternal genetic background is phylogenetically related to the Eurasian one, with an appreciable Sub-Saharan African component (Ennafaa et al. 2009). It has been shown that Western Mediterranean populations from both Europe and North Africa have similar mtDNA pools made up of European and Near Eastern lineages, (such as H, J, T, U, I, W, X), autochthonous lineages (U6 and V), and distinct L and M sub-clades introduced by gene flow from the south and east, respectively. In particular, sub-Saharan African L lineages mainly differentiate North African and Southwestern European groups, accounting for approximately 26 % of the mtDNA variation in the former (Rando et al. 1999), whereas only 8 % of their Y-chromosome variation shows an analogous sub-Saharan African origin (E1 and E3a lineages according to the nomenclature of the Y Chromosome Consortium 2002; Bosch et al. 2001). Increased frequencies of the East African M1 lineage in Algeria and Tunisia instead suggest a more consistent gene flow from the east for these specific regions (Quintana-Murci et al. 1999).

Of particular interest is the U6 lineage that has been firstly hypothesized to be originated in Northwest Africa, in a time interval spanning from 63,000 to 31,000 YBP, then remaining geographically restricted to this area, although with an overall moderate frequency of 10 %, to the Canaries and the Iberian Peninsula, as well as, with extremely low frequencies to Southern Italy, Near East and the Sahel. This peculiar distribution pattern has been proposed to be the consequence of a very limited number of population movements from North Africa, with the almost sole exception of the Arab expansion, and of ancient locus-specific drift and/or male-biased bottleneck, since U6 paternal counterpart, the autochthonous Y-chromosome lineage E-M81, instead reaches an outstanding higher frequency (64 %) in the same North African territories (Bosch et al. 2001; Plaza et al. 2003). Subsequent studies have also considered the U6 lineage as a sister clade of U5, which spread westwards from the Near East along the northern Mediterranean coast, hypothesizing that it similarly originated in the Near East and then diffused along the southern Mediterranean coast into North Africa during modern human Paleolithic settlement (around 30,000 YBP) (Maca-Meyer et al. 2003; Olivieri et al. 2006; González et al. 2007; Ennafaa et al. 2009). According to this view, a common Near Eastern source population settling both the northern and southern Mediterranean coastlines can be invoked (Cherni et al. 2005).

This scenario is once again supported by recent studies that show an overall genetic similarity for North African Berbers and Southern Europeans, as well as a non random distribution of some mtDNA markers in different Berber-speaking groups that allows them to be distinguished from the Maghreb and Egyptian groups. The former seem to be more related to Europeans, according to their H1 and V frequencies, whereas the latter show more genetic affinities with groups from East Africa and the Nile Valley, according to high M1 frequencies and to the presence of specific L sub-lineages (L0a1, L3i, L4 and L4b2). Such internal differentiation may be explained by different gene flow between local and migrating populations in North West and North East Africa. The former is located at the end of a long migration route, whereas the latter represents a crossroads between Near East, East Africa, sub-Saharan Africa and North Africa, so that influences from the Near East and East Africa are still marked in Egypt, whereas those from Southwestern Europe are greater in the Maghreb (Coudray et al. 2009).

North African maternal genetic background is well represented by lineage H, the most frequent Western Eurasian one, the frequency of which decreases from Northwest Africa (37 % in Morocco, 32 % in Algeria, 26 % in Tunisia) (Rando et al. 1998; Plaza et al. 2003; Fadhlaoui-Zid et al. 2004), to the south (24 % in Saharans, 23 % in Mauritanians) (Rando et al. 1998; González et al. 2006) and to the east (21 and 14 % in Egyptians) (Krings et al. 1999; Stevanovitch et al. 2004). In particular, its H1 and H3 sub-clades seem to have had similar Palaeolithic expansion times in North Africa and Europe. This ancient link between the northern and southern Mediterranean coasts is in accordance with estimates obtained for classical and Y-chromosome markers. Sub-clades of E-M78 and E-M81 Y-chromosome lineages were initially considered H1 and H3 paternal counterparts (Bosch et al. 2001), even if they were subsequently associated with Neolithic or historic events and a Near East or Northeast African origin (Arredi et al. 2004; Cruciani et al. 2004, 2007; Semino et al. 2004). Regardless, patterns of H distribution confirm the significant genetic differentiation between Northwestern and Eastern North African populations, in agreement with several studies focused on classical (Bosch et al. 1997), Y-chromosome (Arredi et al. 2004; Cruciani et al. 2004, 2007; Semino et al. 2004) and X-chromosome markers (Tomas et al. 2008). They also suggest that the main contribution to the North African genetic background has Near Eastern Paleolithic and Neolithic origins, as supported by autosomal (Bosch et al. 2000; Comas et al. 2000; Ennafaa et al. 2006), Y-chromosome (Bosch et al. 2001; Flores et al. 2001) and other mtDNA polymorphisms (Plaza et al. 2003; Fadhlaoui-Zid et al. 2004; Cherni et al. 2005). Historical invasions, such as the Arab conquest, were indeed mainly cultural processes, rather than replacements of local populations, as pointed out by the absence of differentiation between North African Arabs and Berbers (Ennafaa et al. 2009; Fadhlaoui-Zid et al. 2004), showing instead a sharp cultural differentiation, according to classical (Bosch et al. 1997) autosomal (Bosch et al. 2000; Comas et al. 2000) and Y-chromosome markers (Bosch et al. 2001).

Y-chromosome analyses have long pointed out an evident pattern of genetic dissimilarity between North African and Southern Europe. Early hypotheses have assumed that it was mainly due to independent Upper Paleolithic population movements, with subsequent differential Neolithic gene flow from the Near East, and even less significant historical bidirectional genetic exchanges across the Mediterranean basin (Bosch et al. 2001; Fadhlaoui-Zid et al. 2004).

The most prevalent Y-chromosome lineage in North Africa, E-M81 (now E1b1b1b according to Karafet et al. 2008) differs from those characterizing Southern Europe and sub-Saharan Africa. Lineage R1b indeed reaches an overall Southern European frequency of 55 % (Underhill et al. 2000), with a peak of 77 % in the Iberian Peninsula (Bosch et al. 2001), whereas lineage E-M2 (now E1b1a according to Karafet et al. 2008) predominates in sub-Saharan Africa, with an overall frequency of 64 % (Underhill et al. 2000; Cruciani et al. 2002). These discrepancies suggest that, despite lineages shared at low frequency, North African groups experienced a different paternal genetic history compared to both Southern European and sub-Saharan African ones, and a limited gene flow occurred over the timescales needed for Y-chromosome differentiation (Arredi et al. 2004).

A strong geographical structure for Y-chromosome variation was further observed within North Africa overall, providing a description of changing diversity patterns from Morocco to Egypt that is consistent with the hypothesis of a demic expansion from the Near East. Moreover, although Moroccan Y-chromosomes were initially supposed to have a predominant Upper Paleolithic origin from East Africa (Bosch et al. 2001), subsequent age estimates for most of them of approximately 7,000 YBP, demonstrated that this Paleolithic contribution was not substantial. Early Neolithic sites are indeed documented in the eastern part of North Africa and later ones in the west, in agreement with westwards farmers' migrations. This represents further support for the hypothesis that Neolithic diffusion originated in Near East and branched into two flows separated by the geographical barrier of the Mediterranean Sea (Arredi et al. 2004).

Gene Flow Across the Two Mediterranean Shores: Genetic Barriers Matching the Geographical Barrier

Early studies based on classical (Bosch et al. 1997; Simoni et al. 1999), autosomal (Bosch et al. 2000), mtDNA (Rando et al. 1998) and Y-chromosome (Bosch et al. 1999) genetic markers pointed out an east–west pattern of genetic variation in North Africa and an appreciable differentiation between populations from northern and southern Mediterranean coastlines, which reflect the presence of considerable geographical barriers, such as the Sahara Desert and the Mediterranean Sea. This suggests that independent human colonization processes characterized the different Mediterranean shores and that limited gene flow occurred between North Africa and Southern Europe, in spite of recent migration events (Plaza et al. 2003). It thus has been proposed that the Gibraltar Straits was not crossed during early settlement of anatomically modern humans either from Africa to Europe or from Europe to Africa.

Irrespective of the fact that the Mediterranean Sea draws water from the Atlantic Ocean through the Gibraltar Straits, thus producing strong maritime currents that might have made navigation difficult in very ancient times, genetic differentiation between North Africa and Iberia probably increased during the Neolithic. Agricultural diffusion may have run in parallel along the different Mediterranean coastlines (Bosch et al. 1997; Simoni et al. 1999), generating substantial cultural differences due to the adoption of Indo-European languages on the northern shore and of Afro-Asiatic languages on the southern one (Renfrew 1991; Barbujani et al. 1994). These cultural factors may have further enhanced genetic separation of pre-existing populations.

Nevertheless, although Northwest Africa has had a deep cultural influence on the Iberian Peninsula, its demic impact is supposed to have been limited, since recent demographic events, spanning from the Neolithic to the historical Mediterranean trade routes and the Arab expansion, seem to have had a small genetic influence on migrations along the south–north direction (Comas et al. 2000).

Even if a certain degree of genetic exchange occurred between Northwest Africa and the Southern part of the Iberian Peninsula, sharp changes in the frequency of genetic markers have been described in this area (Tomas et al. 2008). However, there is no general agreement on the role of the Gibraltar Strait as a barrier to gene flow, and the extent of gene flow detected between North African and South European populations strongly depends on the type of genetic marker used.

For example, the overall Northwest African contribution to the Iberian Y-chromosome variation has been estimated at around 7 %, with the highest values (14 %) being found in Andalusians from the Southern part of the peninsula.

Conversely, the European contribution to the Northwestern African paternal genetic background was estimated at about 4 % (Bosch et al. 2001). This demonstrates a strong barrier played by the Gibraltar Strait (Currat et al. 2010), in accordance also to what has been observed with classical (Bosch et al. 1997), autosomal (Comas et al. 2000; Flores et al. 2000a) and X-chromosome (Tomas et al. 2008) markers.

An overall homogeneity of X-chromosome genetic variation has also been observed in several Mediterranean populations, with only Moroccans showing significant genetic distances from the other groups, once again suggesting the existence of a north–south barrier to gene flow in the western part of the Mediterranean basin that can be identified in the Strait of Gibraltar (Comas et al. 2000; Bosch et al. 2001). The observed X-chromosome homogeneity supports the influence of a number of historical migration events (Semino et al. 2004) on the genetic background of the Mediterranean area, for which the sea has long represented a crucial connection between distant territories and populations (Malaspina et al. 2000; Tomas et al. 2008; Currat et al. 2010).

More recently, a study based on polymorphisms located on the F7 and F12 genes, which encode for coagulation factors VII and XII and are associated with susceptibility to cardiovascular disease in groups from the Mediterranean area, showed that Mediterranean populations broadly cluster into South Europeans and North Africans, despite the low genetic differentiation between these two groups. The evident distinction between these groups is thus in agreement with the Mediterranean Sea acting as a genetic barrier, with the sole exception being represented by people from Tunisia, which appear to be closer to South European groups (Athanasiadis et al. 2007, 2010).

Alternatively, a greater female permeability of the Gibraltar Straits has been inferred by several mtDNA studies (González et al. 2003; Plaza et al. 2003; Reidla et al. 2003), as well as by some other autosomal and GM polymorphisms (Coudray et al. 2006), which have highlighted substantial maternal Iberian input on North Africa (Rando et al. 1998; Plaza et al. 2003) and a contribution of North-West Africa to Iberia of 18 % (Plaza et al. 2003), thus suggesting that the Strait of Gibraltar does not constitute such a strong barrier as proposed by Y-chromosome data (Currat et al. 2010). In fact, mtDNA polymorphisms defining the H and V clades are shared between several Berber and South European groups. Their age estimates, together with those for H1 (14,000 YBP), H3 (10,000 YBP) and U5b sub-clades (Torroni et al. 2001; Achilli et al. 2004, 2005), suggest that they were probably exchanged by gene flow through the Straits of Gibraltar during the southward expansion of Paleolithic hunter-gatherers from the Franco-Cantabrian refuge after the LGM. Accordingly, H1 and H3 lineages are indeed detected mainly in Northwestern African populations and decrease in frequency eastwards (Coudray et al. 2009).

The presence of the U6 mtDNA lineage in Iberia is also considered as evidence for female gene flow from North Africa that is considerably greater compared to the male one (Bosch et al. 2001; Plaza et al. 2003; Cruciani et al. 2004, 2007). Sub-Saharan L lineages were also observed in both the Iberian and Italian peninsulas at frequencies of 3 and 2.4 %, respectively, whereas no Y-chromosome counterparts were detected in the same regions (Bosch et al. 2001; Rosser et al. 2000).

The differences in levels of gene flow detected throughout the Strait of Gibraltar, according to Y-chromosome and mtDNA markers, have been attributed to sex-specific migratory patterns, with females showing more permeability than males as a consequence of social practices such as patrilocality and polygyny (Flores et al. 2000a, b; Plaza et al. 2003), and to genetic drift differently affecting both sexes (Cherni et al. 2005). In detail, a greater genetic flow mediated by females than males was indeed observed in Mediterranean pre-agricultural and post-agricultural societies, with a generally low impact of the Neolithic expansion and of more recent events on North African populations.

However, historically documented population movements from Northwest Africa to the Iberian Peninsula were all associated with the Islamic military conquest, and thus to a gene flow that was mainly male-driven. According to this, if genetically important, this expansion would have homogenized the paternal gene pool of Iberia and North Africa to a greater extent than the female one, in contrast to results from the Anthropological Genetics studies described here.

Early modern human migrations repeatedly crossing the Gibraltar Strait to establish patrilocality seem quite improbable, so that probably a higher mtDNA heterogeneity between North Africa and Southern Europe than proposed in previous studies has to be invoked, with female levels of gene flow which would match approximately those of males, identifying the Strait of Gibraltar as a barrier similarly affecting male and female exchanges (Ennafaa et al. 2009).

Nevertheless, the exact role of the Mediterranean Sea as a vector or barrier to migration may have been extremely variable in time and in different regions. As already mentioned, it may have promoted east–west migration along its coasts, but its actual influence on north–south migration is still uncertain.

Moreover, demographic and population genetics models of constant gene flow are probably too simplistic to capture the impact of maritime movements over the Mediterranean Sea (Currat et al. 2010).

Although there is no general consensus on the role of the Gibraltar Strait as a genetic barrier between Northwest Africa and the Iberian Peninsula, due also to the fact that different genetic markers have revealed different patterns of gene flow, the Mediterranean Sea as a whole has played, at least partially, the role of a barrier to gene flow between its northern and southern shores.

According to this view, more data, especially from less studied, but very important geographic regions, such as the Italian Peninsula, the Adriatic Sea and the Northeast Africa, would be necessary in order to reach a definite conclusion (Athanasiadis et al. 2010).

Future Perspectives: Shedding Light on the Role of the Italian Peninsula in the Mediterranean Genetic Landscape

According to the overview described in the previous chapters, the genetic landscape of Mediterranean populations is far from being exhaustively drawn by Anthropological Genetics, since some crucial geographical regions still lack sufficient data. This limits the reliability with which the Mediterranean basin can be considered a genetically homogeneous area, as well as with which such Sea is thought to have played the role of genetic barrier and/or bridge between human groups characterized by different African, Near Eastern or European cultural backgrounds.

Moreover, as broadly observed at the worldwide level, also in the Mediterranean area human genetic variation is geographically distributed in an unequal way among populations, with the addition of several geographic or linguistic isolates, such as the Basques from Iberia, people from Corsica (Latini et al. 2004), Sardinia (Angius et al. 2002; Calò et al. 2008), Balearic Islands (Picornell et al. 2005) and Morocco (Tomas et al. 2008), which appear to be genetic isolates as well (Nelis et al. 2009; Athanasiadis et al. 2010). Therefore, to better understand origins, migration and admixture processes, as well as evolutionary forces, which have shaped the genetic diversity of Mediterranean populations as a whole, further studies focused on very specific geographic areas and/or ethnic groups are needed.

The reconstruction of worldwide human migrations through the analysis of uniparentally-inherited genetic markers (Y-chromosome and mtDNA) is just the purpose of one of the most ambitious Anthropological Genetics study, the Genographic Project (http://genographic.nationalgeographic.com), launched in April 2005 by National Geographic and IBM, together with the Waitt Family Foundation. In the context of this international project, which aims to sample at least 100,000 DNA samples from worldwide populations, a huge amount of genetic data will be collected also for the Mediterranean area, as well as, for its surrounding African, Near Eastern and European regions (Comas 2010). Y-chromosome, as well as mtDNA genetic variation, will be investigated for each sampled individual that will be selected according to specific bio-demographical criterion (at least three generations of paternal and maternal ancestry in their country). This exceptional sampling and genotyping effort will achieve an unprecedented level of resolution for the analysis of movements and admixture events occurred in the melting pot of Mediterranean populations, promising to substantially improve the answer to crucial questions that have been long debated, but not totally clarified, by Anthropological Genetics. It will allow more accurate disentanglement of the impact of both Paleolithic and Neolithic migrations, as well as of postglacial re-colonization processes from refuge areas and of historical population movements, on the present-day Mediterranean genetic landscape, but also to define whether the Neolithic diffusion was an homogeneous process across the Mediterranean basin or was made up by different migratory waves (Comas 2010).

In particular, accurate and extensive collection of data from the Italian Peninsula, as well as from the surrounding Adriatic and Tyrrhenian regions, will fill a significant gap in the genetic picture of the Mediterranean basin, offering an extremely important opportunity to further investigate the demographic and historical dynamics that have characterized its peopling. The Italian Peninsula, together with its neighboring coastlines, was indeed involved in the main colonization phases of the Mediterranean due to its pivotal geographical position located in the heart of that area, but, at the same time, tightly connected with several European continental territories.

Nevertheless, despite this strategic importance, only few studies have specifically focused on the genetic history of the Italian population as a whole, since the pioneering study carried out by Cavalli-Sforza and collogues (1994) exploiting classical genetic markers. This study described the presence of clines of genetic variation along the peninsula, which were subsequently confirmed also by mtDNA analyses (Barbujani et al. 1995; Turchi et al. 2008; Ottoni et al. 2009). Investigation of Y-chromosome diversity instead identified a single North–South major cline, suggesting local drift and founder effects as the main causes for the observed pattern of variation (Di Giacomo et al. 2003). More recently, a study based on a much larger set of genetic markers, as well as on a more specific sampling scheme, has revealed a further degree of genetic differentiation within Italians, with northern groups which turned out to be closer to western European populations and southern ones which showed more genetic affinities with Southeastern and South-Central Europeans (Capelli et al. 2007).

The contemporaneous investigation of several hundreds of thousands of autosomal SNPs, made possible by the recent advances in high-throughput genotyping technologies, has also refined the description of the overall European genetic variation. Clusters of genetically related populations that strongly reflect the geographic map of the European continent have been identified (Lao et al. 2008) and the southeast-northwest gradient of population genetic structure already observed with classical (Cavalli-Sforza et al. 1994; Sokal et al. 1989) and Y-chromosome markers (Chikhi et al. 2002;

Roewer et al. 2005), but in contrast with mtDNA analyses (Simoni et al. 2000; Torroni et al. 2006), has been confirmed. In that extremely detailed picture, whereas a general statistically significant positive correlation between genetic and geographic population distances was observed, potential genetic barriers between Italians and other European populations are invoked (Lao et al. 2008), together with a further internal differentiation between northern and southern Italian groups (Nelis et al. 2009).

According to these findings, a more complex and heterogeneous genetic landscape can be hypothesized for the Italian Peninsula in comparison to what was previously supposed.

Within the worldwide framework of the Genographic Project, the Laboratory of Molecular Anthropology of the University of Bologna, in collaboration with the Human Genome Diversity Laboratory of the Pompeu Fabra University of Barcelona, is analyzing a panel of more than 1,000 individuals, representative of most of the Italian subpopulations distributed along the peninsula. This study will represent a substantial improvement to the current knowledge of the Italian genetic variation due to both the specificity of the sampling strategy and the resolution of the exploited uniparentally-inherited genetic markers.

In addition to this, the shift from Population Genetics to Population Genomics, launched by genome-wide genotyping studies and sped up by the amazing advances in the massive parallel sequencing technologies, which are racing towards a more and more cost-effective whole genome sequencing (1000 Genomes Project Consortium 2010), will help to dissect with an unprecedented detail the multiple contributions to the complex genetic history of the Italian Peninsula.

The Genographic Project, together with even larger genome-wide studies, basing on dense and accurately selected geographic samples, thus promises to shed new light on the pivotal role of the Italian Peninsula as a crossroads within and between the European and Mediterranean genetic contexts, in accordance to the belief that every increase in the knowledge of its genetic structure will substantially contribute to clarifying the genetic history of the whole Mediterranean basin.

Acknowledgements We are very grateful to Prof. Stanley Ulijaszek for his insightful remarks on the manuscript and for the language revision of it.

References

1000 Genomes Project Consortium (2010) A map of human genome variation from population-scale sequencing. Nature 467:1061–1073. doi:10.1038/nature09534

Achilli A, Rengo C, Magri C, Battaglia V, Olivieri A, Scozzari R, Cruciani F, Zeviani M, Briem E, Carelli V, Moral P, Dugoujon JM, Roostalu U, Loogväli EL, Kivisild T, Bandelt HJ, Richards M, Villems R, Santachiara-Benerecetti AS, Semino O, Torroni A (2004) The molecular dissection of mtDNA haplogroup H confirms that the Franco-Cantabrian glacial refuge was a major source for the European gene pool. Am J Hum Genet 75:910–918. doi:10.1086/425590

Achilli A, Rengo C, Battaglia V, Pala M, Olivieri A, Fornarino S, Magri C, Scozzari R, Babudri N, Santachiara-Benerecetti AS, Bandelt HJ, Semino O, Torroni A (2005) Saami and Berbers – an unexpected mitochondrial DNA link. Am J Hum Genet 76:883–886. doi:10.1086/430073

Adler DS, Bar-Yosef O, Belfer-Cohen A, Tushabramishvili N, Boaretto E, Mercier N, Valladas H, Rink WJ (2008) Dating the demise: Neanderthal extinction and the establishment of modern humans in the southern Caucasus. J Hum Evol 55:817–833. doi:10.1016/j.jhevol.2008.08.010

Alexander DH, Novembre J, Lange K (2009) Fast model-based estimation of ancestry in unrelated individuals. Genome Res 19:1655–1664. doi:10.1101/gr.094052.109

Ammerman AJ, Cavalli-Sforza LL (1984) The Neolithic transition and the genetics of populations in Europe. Princeton University Press, Princeton

Angius A, Bebbere D, Petretto E, Falchi M, Forabosco P, Maestrale B, Casu G, Persico I, Melis PM, Pirastu M (2002) Not all isolates are equal: linkage disequilibrium analysis on Xq13.3 reveals different patterns in Sardinian sub-populations. Hum Genet 111:9–15. doi:10.1007/s00439-002-0753-z

Arredi B, Poloni ES, Paracchini S, Zerjal T, Fathallah DM, Makrelouf M, Pascali VL, Novelletto A, Tyler-Smith C (2004) A predominantly Neolithic origin for Y-chromosomal DNA variation in North Africa. Am J Hum Genet 75:338–345. doi:10.1086/423147

Athanasiadis G, Esteban E, Via M, Dugoujon JM, Moschonas N, Chaabani H, Moral P (2007) The X chromosome Alu insertions as a tool for human population genetics: data from European and African human groups. Eur J Hum Genet 15:578–583. doi:10.1038/sj.ejhg.5201797

Athanasiadis G, González-Pérez E, Esteban E, Dugoujon JM, Stoneking M, Moral P (2010) The Mediterranean Sea as a barrier to gene flow: evidence from variation in and around the F7 and F12 genomic regions. BMC Evol Biol 10:84. doi:10.1186/1471-2148-10-84

Balaresque P, Bowden GR, Adams SM, Leung HY, King TE, Rosser ZH, Goodwin J, Moisan JP, Richard C, Millward A, Demaine AG, Barbujani G, Previderè C, Wilson IJ, Tyler-Smith C, Jobling MA (2010) A predominantly Neolithic origin for European paternal lineages. PLoS Biol 8:e1000285. doi:10.1371/journal.pbio.1000285

Barbujani G, Colonna V (2010) Human genome diversity: frequently asked questions. Trends Genet 26:285–295. doi:10.1016/j.tig.2010.04.002

Barbujani G, Pilastro A, De Domenico S, Renfrew C (1994) Genetic variation in North Africa and Eurasia: Neolithic demic diffusion vs. Paleolithic colonisation. Am J Phys Anthropol 95:137–154. doi:10.1002/ajpa.1330950203

Barbujani G, Bertorelle G, Capitani G, Scozzari R (1995) Geographical structuring in the mtDNA of Italians. Proc Natl Acad Sci U S A 92:9171–9175

Battaglia V, Fornarino S, Al-Zahery N, Olivieri A, Pala M, Myres NM, King RJ, Rootsi S, Marjanovic D, Primorac D, Hadziselimovic R, Vidovic S, Drobnic K, Durmishi N, Torroni A, Santachiara-Benerecetti AS, Underhill PA, Semino O (2008) Y-chromosomal evidence of the cultural diffusion of agriculture in Southeast Europe. Eur J Hum Genet 17:820–830. doi:10.1038/ejhg.2008.249

Benazzi S, Douka K, Fornai C, Bauer CC, Kullmer O, Svoboda J, Pap I, Mallegni F, Bayle P, Coquerelle M, Condemi S, Ronchitelli A, Harvati K, Weber GW (2011) Early dispersal of modern humans in Europe and implications for Neanderthal behaviour. Nature 479:525–528. doi:10.1038/nature10617

Benjeddou M, Leat N, Davison S (2006) Use of molecular genetics and historical records to reconstruct the history of local communities. Afr J Biotechnol 5:2516–2519

Berger JF, Guilaine J (2009) The 8200 cal BP abrupt environmental change and the Neolithic transition: a Mediterranean perspective. Quat Int 200:31–49. doi:10.1016/j.quaint.2008.05.013

Biagi P (2003) A review of the late Mesolithic in Italy and its implication for the Neolithic transition. In: Ammerman A, Biagi P (eds) The widening harvest. Archaeological Institute of America, Boston

Bosch E, Calafell F, Pérez-Lezaun A, Comas D, Mateu E, Bertranpetit J (1997) Population history of North Africa: evidence from classical genetic markers. Hum Biol 69:295–311

Bosch E, Calafell F, Santos FR, Pérez-Lezaun A, Comas D, Benchemsi N, Tyler-Smith C, Bertranpetit J (1999) Variation in short tandem repeats is deeply structured by genetic background on the human Y chromosome. Am J Hum Genet 65:1623–1638. doi:10.1086/302676

Bosch E, Calafell F, Pérez-Lezaun A, Clarimón J, Comas D, Mateu E, Martínez-Arias R, Morera B, Brakez Z, Akhayat O, Sefiani A, Hariti G, Cambon-Thomsen A, Bertranpetit J (2000) Genetic structure of north-west Africa revealed by STR analysis. Eur J Hum Genet 8:360–366. doi:10.1038/sj.ejhg.5200464

Bosch E, Calafell F, Comas D, Oefner PJ, Underhill PA, Bertranpetit J (2001) High-resolution analysis of human Y-chromosome variation shows a sharp discontinuity and limited gene flow between northwestern Africa and the Iberian Peninsula. Am J Hum Genet 68:1019–1029. doi:10.1086/319521

Busby GB, Brisighelli F, Sánchez-Diz P, Ramos-Luis E, Martinez-Cadenas C, Thomas MG, Bradley DG, Gusmão L, Winney B, Bodmer W, Vennemann M, Coia V, Scarnicci F, Tofanelli S, Vona G, Ploski R, Vecchiotti C, Zemunik T, Rudan I, Karachanak S, Toncheva D, Anagnostou P, Ferri G, Rapone C, Hervig T, Moen T, Wilson JF, Capelli C (2011) The peopling of Europe and the cautionary tale of Y chromosome lineage R-M269. Proc Biol Sci 279:884–892. doi:10.1098/rspb.2011.1044

Calderon R (2000) Population and peopling in the Mediterranean world. Int J Anthropol 15:271–278. doi:10.1007/BF02445138

Calò CM, Melis A, Vona G, Piras IS (2008) Sardinian population (Italy): a genetic review. Int J Mod Anthropol 1:39–64

Capelli C, Redhead N, Romano V, Calì F, Lefranc G, Delague V, Megarbane A, Felice AE, Pascali VL, Neophytou PI, Poulli Z, Novelletto A, Malaspina P, Terrenato L, Berebbi A, Fellous M, Thomas MG, Goldstein DB (2005) Population structure in the Mediterranean basin: a Y chromosome perspective. Ann Hum Genet 70:207–225. doi:10.1111/j.1529-8817.2005.00224.x

Capelli C, Brisighelli F, Scarnicci F, Arredi B, Caglia' A, Vetrugno G, Tofanelli S, Onofri V, Tagliabracci A, Paoli G, Pascali VL (2007) Y chromosome genetic variation in the Italian peninsula is clinal and supports an admixture model for the Mesolithic-Neolithic encounter. Mol Phylogenet Evol 44:228–239. doi:10.1016/j.ympev.2006.11.030

Capelli C, Onofri V, Brisighelli F, Boschi I, Scarnicci F, Masullo M, Ferri G, Tofanelli S, Tagliabracci A, Gusmao L, Amorim A, Gatto F, Kirin M, Merlitti D, Brion M, Verea AB, Romano V, Cali F, Pascali V (2009) Moors and Saracens in Europe: estimating the medieval North African male legacy in southern Europe. Eur J Hum Genet 17:848–852. doi:10.1038/ejhg.2008.258

Cavalli-Sforza LL, Feldman MW (2003) The application of molecular genetic approaches to the study of human evolution. Nat Genet 33:266–275. doi:10.1038/ng1113

Cavalli-Sforza LL, Menozzi P, Piazza A (1994) History and geography of human genes. Princeton University Press, Princeton

Cherni L, Loueslati BY, Pereira L, Ennafaâ H, Amorim A, El Gaaied AB (2005) Female gene pools of Berber and Arab neighboring communities in central Tunisia: microstructure of mtDNA variation in North Africa. Hum Biol 77:61–70. doi:10.1353/hub.2005.0028

Chikhi L, Nichols RA, Barbujani G, Beaumont MA (2002) Y genetic data support the Neolithic demic diffusion model. Proc Natl Acad Sci U S A 99:11008–11013. doi:10.1073/pnas.162158799

Cinnioğlu C, King R, Kivisild T, Kalfoğlu E, Atasoy S, Cavalleri GL, Lillie AS, Roseman CC, Lin AA, Prince K, Oefner PJ, Shen P, Semino O, Cavalli-Sforza LL, Underhill PA (2004) Excavating Y-chromosome haplotype strata in Anatolia. Hum Genet 114:127–148. doi:10.1007/s00439-003-1031-4

Colledge S, Conolly J, Shennan S (2004) Archaeobotanical evidence for the spread of farming in the Eastern Mediterranean. Curr Anthropol 45:35–45. doi:10.1086/422086

Colonna V, Boattini A, Guardiano C, Dall'ara I, Pettener D, Longobardi G, Barbujani G (2010) Long-range comparison between genes and languages based on syntactic distances. Hum Hered 70:245–254. doi:10.1159/000317374

Comas D (2010) The Genographic Project: insights into Western/Central European variation. J Anthropol Sci 88:243–244

Comas D, Calafell F, Benchemsi N, Helal A, Lefranc G, Stoneking M, Batzer MA, Bertranpetit J, Sajantila A (2000) Alu insertion polymorphisms in NW Africa and the Iberian Peninsula: evidence for a strong genetic boundary through the Gibraltar Straits. Hum Genet 107:312–319. doi:10.1007/s004390000370

Coudray C, Guitard E, Kandil M, Harich N, Melhaoui M, Baali A, Sevin A, Moral P, Dugoujon JM (2006) Study of GM immunoglobulin allotypic system in Berbers and Arabs from Morocco. Am J Hum Biol 18:23–34. doi:10.1002/ajhb.20465

Coudray C, Olivieri A, Achilli A, Pala M, Melhaoui M, Cherkaoui M, El-Chennawi F, Kossmann M, Torroni A, Dugoujon JM (2009) The complex and diversified mitochondrial gene pool of Berber populations. Ann Hum Genet 73:196–214. doi:10.1111/j.1469-1809.2008.00493.x

Cruciani F, Santolamazza P, Shen P, Macaulay V, Moral P, Olckers A, Modiano D, Holmes S, Destro-Bisol G, Coia V, Wallace DC, Oefner PJ, Torroni A, Cavalli-Sforza LL, Scozzari R, Underhill PA (2002) A back migration from Asia to sub-Saharan Africa is supported by high-resolution analysis of human Y-chromosome haplotypes. Am J Hum Genet 70:1197–1214. doi:10.1086/340257

Cruciani F, La Fratta R, Santolamazza P, Sellitto D, Pascone R, Moral P, Watson E, Guida V, Colomb EB, Zaharova B, Lavinha J, Vona G, Aman R, Cali F, Akar N, Richards M, Torroni A, Novelletto A, Scozzari R (2004) Phylogeographic analysis of haplogroup E3b (E-M215) Y chromosomes reveals multiple migratory events within and out of Africa. Am J Hum Genet 74:1014–1022. doi:10.1086/386294

Cruciani F, La Fratta R, Trombetta B, Santolamazza P, Sellitto D, Colomb EB, Dugoujon JM, Crivellaro F, Benincasa T, Pascone R, Moral P, Watson E, Melegh B, Barbujani G, Fuselli S, Vona G, Zagradisnik B, Assum G, Brdicka R, Kozlov AI, Efremov GD, Coppa A, Novelletto A, Scozzari R (2007) Tracing past human male movements in northern/eastern Africa and western Eurasia: new clues from Y-chromosomal haplogroups E-M78 and J-M12. Mol Biol Evol 24:1300–1311. doi:10.1093/molbev/msm049

Cruciani F, Trombetta B, Antonelli C, Pascone R, Valesini G, Scalzi V, Vona G, Melegh B, Zagradisnik B, Assum G, Efremov GD, Sellitto D, Scozzari R (2010) Strong intra- and inter-continental differentiation revealed by Y chromosome SNPs M269, U106 and U152. Forensic Sci Int Genet 5:49–52. doi:10.1016/j.fsigen.2010.07.006

Cunliffe B (2001) The Oxford illustrated history of prehistoric Europe. Oxford University Press, Oxford

Cunliffe B (2008) Europe between the oceans: 9000 BC–AD 1000. Yale University Press, New Haven

Currat M, Excoffier L (2005) The effect of the Neolithic expansion on European molecular diversity. Proc Biol Sci 272:679–688. doi:10.1098/rspb.2004.2999

Currat M, Excoffier L (2011) Strong reproductive isolation between humans and Neanderthals inferred from observed patterns of introgression. Proc Natl Acad Sci U S A 108:15129–15134. doi:10.1073/pnas.1107450108

Currat M, Poloni ES, Sanchez-Mazas A (2010) Human genetic differentiation across the Strait of Gibraltar. BMC Evol Biol 10:237. doi:10.1186/1471-2148-10-237

Di Gaetano C, Cerutti N, Crobu F, Robino C, Inturri S, Gino S, Guarrera S, Underhill PA, King RJ, Romano V, Cali F, Gasparini M, Matullo G, Salerno A, Torre C, Piazza A (2009) Differential Greek and northern African migrations to Sicily are supported by genetic evidence from the Y chromosome. Eur J Hum Genet 17:91–99. doi:10.1038/ejhg.2008.120

Di Giacomo F, Luca F, Anagnou N, Ciavarella G, Corbo RM, Cresta M, Cucci F, Di Stasi L, Agostiano V, Giparaki M, Loutradis A, Mammi' C, Michalodimitrakis EN, Papola F, Pedicini G, Plata E, Terrenato L, Tofanelli S, Malaspina P, Novelletto A (2003) Clinal patterns of human Y chromosomal diversity in continental Italy and Greece are dominated by drift and founder effects. Mol Phylogenet Evol 28:387–395. doi:10.1016/S1055-7903(03)00016-2

Di Giacomo F, Luca F, Popa LO, Akar N, Anagnou N, Banyko J, Brdicka R, Barbujani G, Papola F, Ciavarella G, Cucci F, Di Stasi L, Gavrila L, Kerimova MG, Kovatchev D, Kozlov AI, Loutradis A, Mandarino V, Mammi' C, Michalodimitrakis EN, Paoli G, Pappa KI, Pedicini G, Terrenato L, Tofanelli S, Malaspina P, Novelletto A (2004) Y chromosomal haplogroup J as a signature of the post-Neolithic colonization of Europe. Hum Genet 115:357–371. doi:10.1007/s00439-004-1168-9

Dupanloup I, Bertorelle G, Chikhi L, Barbujani G (2004) Estimating the impact of prehistoric admixture on the genome of Europeans. Mol Biol Evol 21:1361–1372. doi:10.1093/molbev/msh135

Ennafaa H, Amor MB, Yacoubi-Loueslati B, Khodjet el-khil H, Gonzalez-Perez E, Moral P, Maca-Meyer N, Elgaaied A (2006) Alu polymorphisms in Jerba Island population (Tunisia): comparative study in Arab and Berber groups. Ann Hum Biol 33:634–640

Ennafaa H, Cabrera VM, Abu-Amero KK, González AM, Amor MB, Bouhaha R, Dzimiri N, Elgaaïed AB, Larruga JM (2009) Mitochondrial DNA haplogroup H structure in North Africa. BMC Genet 10:8. doi:10.1186/1471-2156-10-8

Fadhlaoui-Zid K, Plaza S, Calafell F, Ben Amor M, Comas D, Bennamar El gaaied A (2004) Mitochondrial DNA heterogeneity in Tunisian Berbers. Ann Hum Genet 68:222–233. doi:10.1046/j.1529-8817.2004.00096.x

Falush D, Stephens M, Pritchard JK (2003) Inference of population structure using multilocus genotype data: linked loci and correlated allele frequencies. Genetics 164:1567–1587

Flores C, Maca-Meyer N, González AM, Cabrera VM (2000a) Northwest African distribution of the CD4/Alu microsatellite haplotypes. Ann Hum Genet 64:321–327. doi:10.1046/j.1469-1809.2000.6440321.x

Flores C, Hernández M, González AM, Cabrera VM (2000b) Genetic affinities among human populations inhabiting the Subsaharan area, Northwest Africa, and the Iberian Peninsula. In: Arnaiz-Villena A (ed) Proceedings of prehistoric Iberia: genetics, anthropology, and linguistics. Kluwer Academic/Plenum Press, New York

Flores C, Maca-Meyer N, Pérez JA, Hernández M, Cabrera VM (2001) Y-chromosome differentiation in Northwest Africa. Hum Biol 73:513–524

Flores C, Maca-Meyer N, Pérez JA, González AM, Larruga JM, Cabrera VM (2003) A predominant European ancestry of paternal lineages from Canary Islanders. Ann Hum Genet 67:138–152. doi:10.1046/j.1469-1809.2003.00015.x

Francalacci P, Sanna D (2008) History and geography of human Y-chromosome in Europe: a SNP perspective. J Anthropol Sci 86:59–89

Fu Q, Rudan P, Pääbo S, Krause J (2012) Complete mitochondrial genomes reveal Neolithic expansion into Europe. PLoS One 7:e32473. doi:10.1371/journal.pone.0032473

Gibbons A (2000) Evolutionary genetics. Europeans trace ancestry to Paleolithic people. Science 290:1080–1081. doi:10.1126/science.290.5494.1080

González AM, Brehm A, Pérez JA, Maca-Meyer N, Flores C, Cabrera VM (2003) Mitochondrial DNA affinities at the Atlantic fringe of Europe. Am J Phys Anthropol 120:391–404. doi:10.1002/ajpa.10168

González AM, Cabrera VM, Larruga JM, Tounkara A, Noumsi G, Thomas BN, Moulds JM (2006) Mitochondrial DNA variation in Mauritania and Mali and their genetic relationship to other Western Africa populations. Ann Hum Genet 70:631–657. doi:10.1111/j.1469-1809.2006.00259.x

González AM, Larruga JM, Abu-Amero KK, Shi Y, Pestano J, Cabrera VM (2007) Mitochondrial lineage M1 traces an early human backflow to Africa. BMC Genomic 8:223. doi:10.1186/1471-2164-8-223

González-Pérez E, Esteban E, Via M, Gayà-Vidal M, Athanasiadis G, Dugoujon JM, Luna F, Mesa MS, Fuster V, Kandil M, Harich N, Bissar-Tadmouri N, Saetta A, Moral P (2010) Population relationships in the Mediterranean revealed by autosomal genetic data (Alu and Alu/STR compound systems). Am J Phys Anthropol 141:430–439. doi:10.1002/ajpa.21161

Guilaine J, Manen C, Vigne JD (2008) New insights into the Neolithization of Mediterranean France. Centre d'Anthropologie, Toulouse

Haak W, Forster P, Bramanti B, Matsumura S, Brandt G, Tänzer M, Villems R, Renfrew C, Gronenborn D, Alt KW, Burger J (2005) Ancient DNA from the first European farmers in 7500-year-old Neolithic sites. Science 310:1016–1018. doi:10.1126/science.1118725

Haak W, Balanovsky O, Sanchez JJ, Koshel S, Zaporozhchenko V, Adler CJ, Der Sarkissian CS, Brandt G, Schwarz C, Nicklisch N, Dresely V, Fritsch B, Balanovska E, Villems R, Meller H, Alt KW, Cooper A, Genographic Consortium (2010) Ancient DNA from European early Neolithic farmers reveals their near eastern affinities. PLoS Biol 8:e1000536. doi:10.1371/journal.pbio.1000536

Hervella M, Izagirre N, Alonso S, Fregel R, Alonso A, Cabrera VM, de la Rúa C (2012) Ancient DNA from hunter-gatherer and farmer groups from northern Spain supports a random dispersion model for the Neolithic expansion into Europe. PLoS One 7:e34417. doi:10.1371/journal.pone.0034417

Higham T, Compton T, Stringer C, Jacobi R, Shapiro B, Trinkaus E, Chandler B, Gröning F, Collins C, Hillson S, O'Higgins P, FitzGerald C, Fagan M (2011) The earliest evidence for anatomically modern humans in northwestern Europe. Nature 479:521–524. doi:10.1038/nature10484

Hitti P (1990) The Arabs: a short history. Gateway, Washington, DC

Hoffecker JF, Holliday VT, Anikovich MV, Sinitsyn AA, Popov VV, Lisitsyn SN, Levkovskaya GM, Pospelova GA, Forman SL, Giaccio B (2008) From the Bay of Naples to the River Don: the Campanian Ignimbrite eruption and the Middle to Upper Paleolithic transition in Eastern Europe. J Hum Evol 55:858–870. doi:10.1016/j.jhevol.2008.08.018

Hublin JJ (2011) Palaeoanthropology: African origins. Nature 476:395. doi:10.1038/476395a

Jobling MA, Hurles ME, Tyler-Smith C (2004) Human evolutionary genetics: origins, peoples and disease. Garland Science, New York

Karafet TM, Mendez FL, Meilerman MB, Underhill PA, Zegura SL, Hammer MF (2008) New binary polymorphisms reshape and increase resolution of the human Y chromosomal haplogroup tree. Genome Res 18:830–838. doi:10.1101/gr.7172008

King RJ, Ozcan SS, Carter T, Kalfoğlu E, Atasoy S, Triantaphyllidis C, Kouvatsi A, Lin AA, Chow CE, Zhivotovsky LA, Michalodimitrakis M, Underhill PA (2008) Differential Y-chromosome Anatolian influences on the Greek and Cretan Neolithic. Ann Hum Genet 72:205–214. doi:10.1111/j.1469-1809.2007.00414.x

King RJ, Di Cristofaro J, Kouvatsi A, Triantaphyllidis C, Scheidel W, Myres NM, Lin AA, Eissautier A, Mitchell M, Binder D, Semino O, Novelletto A, Underhill PA, Chiaroni J (2011) The coming of the Greeks to Provence and Corsica: Y-chromosome models of archaic Greek colonization of the western Mediterranean. BMC Evol Biol 11:69. doi:10.1186/1471-2148-11-69

Kivisild T, Rootsi S, Metspalu M, Mastana S, Kaldma K, Parik J, Metspalu E, Adojaan M, Tolk HV, Stepanov V, Gölge M, Usanga E,

Papiha SS, Cinnioğlu C, King R, Cavalli-Sforza L, Underhill PA, Villems R (2003) The genetic heritage of the earliest settlers persists both in Indian tribal and caste populations. Am J Hum Genet 72:313–332. doi:10.1086/346068

Klein RG (2003) Paleoanthropology. Whither the Neanderthals? Science 299:1525–1527. doi:10.1126/science.1082025

Kozlowski JK (2007) In: Mellars P, Boyle K, Bar-Yosef O, Stringer C (eds) Rethinking the human revolution. McDonald Institute, Cambridge

Krause J, Fu Q, Good JM, Viola B, Shunkov MV, Derevianko AP, Pääbo S (2010) The complete mitochondrial DNA genome of an unknown hominin from southern Siberia. Nature 464:894–897. doi:10.1038/nature08976

Krings M, Salem AE, Bauer K, Geisert H, Malek AK, Chaix L, Simon C, Welsby D, Di Rienzo A, Utermann G, Sajantila A, Pääbo S, Stoneking M (1999) mtDNA analysis of Nile River Valley populations: a genetic corridor or a barrier to migration? Am J Hum Genet 64:1166–1176

Lacan M, Keyser C, Ricaut FX, Brucato N, Duranthon F, Guilaine J, Crubézy E, Ludes B (2011) Ancient DNA reveals male diffusion through the Neolithic Mediterranean route. Proc Natl Acad Sci U S A 108:9788–9791. doi:10.1073/pnas.1100723108

Lander ES (2011) Initial impact of the sequencing of the human genome. Nature 470:187–197. doi:10.1038/nature09792

Lao O, Lu TT, Nothnagel M, Junge O, Freitag-Wolf S, Caliebe A, Balascakova M, Bertranpetit J, Bindoff LA, Comas D, Holmlund G, Kouvatsi A, Macek M, Mollet I, Parson W, Palo J, Ploski R, Sajantila A, Tagliabracci A, Gether U, Werge T, Rivadeneira F, Hofman A, Uitterlinden AG, Gieger C, Wichmann HE, Rüther A, Schreiber S, Becker C, Nürnberg P, Nelson MR, Krawczak M, Kayser M (2008) Correlation between genetic and geographic structure in Europe. Curr Biol 18:1241–1248. doi:10.1016/j.cub.2008.07.049

Latini V, Sole G, Doratiotto S, Poddie D, Memmi M, Varesi L, Vona G, Cao A, Ristaldi MS (2004) Genetic isolates in Corsica (France): linkage disequilibrium extension analysis on the Xq13 region. Eur J Hum Genet 12:613–619. doi:10.1038/sj.ejhg.5201205

Leonardi M, Gerbault P, Thomas MG, Burger J (2012) The evolution of lactase persistence in Europe. A synthesis of archaeological and genetic evidence. Int Dairy J 22:88–97

Maca-Meyer N, González AM, Pestano J, Flores C, Larruga JM, Cabrera VM (2003) Mitochondrial DNA transit between West Asia and North Africa inferred from U6 phylogeography. BMC Genet 4:15. doi:10.1186/1471-2156-4-15

Malaspina P, Cruciani F, Santolamazza P, Torroni A, Pangrazio A, Akar N, Bakalli V, Brdicka R, Jaruzelska J, Kozlov A, Malyarchuk B, Mehdi SQ, Michalodimitrakis E, Varesi L, Memmi MM, Vona G, Villems R, Parik J, Romano V, Stefan M, Stenico M, Terrenato L, Novelletto A, Scozzari R (2000) Patterns of male-specific inter-population divergence in Europe, West Asia and North Africa. Ann Hum Genet 64:395–412. doi:10.1046/j.1469-1809.2000.6450395.x

Malyarchuk B, Derenko M, Grzybowski T, Perkova M, Rogalla U, Vanecek T, Tsybovsky I (2010) The peopling of Europe from the mitochondrial haplogroup U5 perspective. PLoS One 5:e10285. doi:10.1371/journal.pone.0010285

Martinez L, Underhill PA, Zhivotovsky LA, Gayden T, Moschonas NK, Chow CE, Conti S, Mamolini E, Cavalli-Sforza LL, Herrera RJ (2007) Paleolithic Y-haplogroup heritage predominates in a Cretan highland plateau. Eur J Hum Genet 15:485–493. doi:10.1038/sj.ejhg.5201769

Mellars P (2006) Why did modern human populations disperse from Africa ca. 60,000 years ago? A new model. Proc Natl Acad Sci U S A 103:9381–9386. doi:10.1073/pnas.0510792103

Mellars P (2011) Palaeoanthropology: the earliest modern humans in Europe. Nature 479:483–485. doi:10.1038/479483a

Mithen S (2006) After the ice: a global human history, 20.000 – 5.000 BC. Harvard University Press, Cambridge

Morelli L, Contu D, Santoni F, Whalen MB, Francalacci P, Cucca F (2010) A comparison of Y-chromosome variation in Sardinia and Anatolia is more consistent with cultural rather than demic diffusion of agriculture. PLoS One 5:e10419. doi:10.1371/journal.pone.0010419

Murray O (1993) Early Greece. Harvard University Press, Cambridge

Myres NM, Rootsi S, Lin AA, Järve M, King RJ, Kutuev I, Cabrera VM, Khusnutdinova EK, Pshenichnov A, Yunusbayev B, Balanovsky O, Balanovska E, Rudan P, Baldovic M, Herrera RJ, Chiaroni J, Di Cristofaro J, Villems R, Kivisild T, Underhill PA (2011) A major Y-chromosome haplogroup R1b Holocene era founder effect in Central and Western Europe. Eur J Hum Genet 19:95–101. doi:10.1038/ejhg.2010.146

Nelis M, Esko T, Mägi R, Zimprich F, Zimprich A, Toncheva D, Karachanak S, Piskácková T, Balascák I, Peltonen L, Jakkula E, Rehnström K, Lathrop M, Heath S, Galan P, Schreiber S, Meitinger T, Pfeufer A, Wichmann HE, Melegh B, Polgár N, Toniolo D, Gasparini P, D'Adamo P, Klovins J, Nikitina-Zake L, Kucinskas V, Kasnauskiene J, Lubinski J, Debniak T, Limborska S, Khrunin A, Estivill X, Rabionet R, Marsal S, Julià A, Antonarakis SE, Deutsch S, Borel C, Attar H, Gagnebin M, Macek M, Krawczak M, Remm M, Metspalu A (2009) Genetic structure of Europeans: a view from the North-East. PLoS One 4:e5472. doi:10.1371/journal.pone.0005472

Norman D (1975) The Arabs and medieval Europe. Longman Group Limited, London

Norwich JJ (2007) The Middle Sea: a history of the Mediterranean. Secker and Warburg, London

Olivieri A, Achilli A, Pala M, Battaglia V, Fornarino S, Al-Zahery N, Scozzari R, Cruciani F, Behar DM, Dugoujon JM, Coudray C, Santachiara-Benerecetti AS, Semino O, Bandelt HJ, Torroni A (2006) The mtDNA legacy of the Levantine early Upper Palaeolithic in Africa. Science 314:1767–1770. doi:10.1126/science.1135566

Otte M (2007) In: Mellars P, Boyle K, Bar-Yosef O, Stringer C (eds) Rethinking the human revolution. McDonald Institute, Cambridge

Ottoni C, Martinez-Labarga C, Vitelli L, Scano G, Fabrini E, Contini I, Biondi G, Rickards O (2009) Human mitochondrial DNA variation in Southern Italy. Ann Hum Biol 36:785–811. doi:10.3109/03014460903198509

Pala M, Achilli A, Olivieri A, Kashani BH, Perego UA, Sanna D, Metspalu E, Tambets K, Tamm E, Accetturo M, Carossa V, Lancioni H, Panara F, Zimmermann B, Huber G, Al-Zahery N, Brisighelli F, Woodward SR, Francalacci P, Parson W, Salas A, Behar DM, Villems R, Semino O, Bandelt HJ, Torroni A (2009) Mitochondrial haplogroup U5b3: a distant echo of the epipaleolithic in Italy and the legacy of the early Sardinians. Am J Hum Genet 84:814–821. doi:10.1016/j.ajhg.2009.05.004

Peričić M, Barać Lauc L, Martinović Klarić I, Janićijević B, Rudan P (2005) Review of Croatian genetic heritage as revealed by mitochondrial DNA and Y chromosomal lineages. Croat Med J 46:502–513

Pérles C (2001) The early Neolithic in Greece. Cambridge University Press, Cambridge

Picornell A, Gómez-Barbeito L, Tomàs C, Castro JA, Ramon MM (2005) Mitochondrial DNA HVRI variation in Balearic populations. Am J Phys Anthropol 128:119–130. doi:10.1002/ajpa.10423

Pinhasi R, Fort J, Ammerman AJ (2005) Tracing the origin and spread of agriculture in Europe. PLoS Biol 3:e410. doi:10.1371/journal.pbio.0030410

Plaza S, Calafell F, Helal A, Bouzerna N, Lefranc G, Bertranpetit J, Comas D (2003) Joining the pillars of Hercules: mtDNA sequences show multidirectional gene flow in the western Mediterranean. Ann Hum Genet 67:312–328. doi:10.1046/j.1469-1809.2003.00039.x

Prat S, Péan SC, Crépin L, Drucker DG, Puaud SJ, Valladas H, Lázničková-Galetová M, van der Plicht J, Yanevich A (2011) The oldest anatomically modern humans from far southeast Europe: direct dating, culture and behavior. PLoS One 6:e20834. doi:10.1371/journal.pone.0020834

Quintana-Murci L, Semino O, Bandelt HJ, Passarino G, McElreavey K, Santachiara-Benerecetti AS (1999) Genetic evidence of an early exit of Homo sapiens sapiens from Africa through eastern Africa. Nat Genet 23:437–441. doi:10.1038/70550

Rando JC, Pinto F, González AM, Hernández M, Larruga JM, Cabrera VM, Bandelt HJ (1998) Mitochondrial DNA analysis of northwest African populations reveals genetic exchanges with European, Near-Eastern, and sub-Saharan populations. Ann Hum Genet 62:531–550. doi:10.1046/j.1469-1809.1998.6260531.x

Rando JC, Cabrera VM, Larruga JM, Hernández M, González AM, Pinto F, Bandelt HJ (1999) Phylogeographic patterns of mtDNA reflecting the colonization of the Canary Islands. Ann Hum Genet 63:413–428. doi:10.1046/j.1469-1809.1999.6350413.x

Rasmussen M, Li Y, Lindgreen S, Pedersen JS, Albrechtsen A, Moltke I, Metspalu M, Metspalu E, Kivisild T, Gupta R, Bertalan M, Nielsen K, Gilbert MT, Wang Y, Raghavan M, Campos PF, Kamp HM, Wilson AS, Gledhill A, Tridico S, Bunce M, Lorenzen ED, Binladen J, Guo X, Zhao J, Zhang X, Zhang H, Li Z, Chen M, Orlando L, Kristiansen K, Bak M, Tommerup N, Bendixen C, Pierre TL, Grønnow B, Meldgaard M, Andreasen C, Fedorova SA, Osipova LP, Higham TF, Ramsey CB, Hansen TV, Nielsen FC, Crawford MH, Brunak S, Sicheritz-Pontén T, Villems R, Nielsen R, Krogh A, Wang J, Willerslev E (2010) Ancient human genome sequence of an extinct Palaeo-Eskimo. Nature 463:757–762. doi:10.1038/nature08835

Reich D, Green RE, Kircher M, Krause J, Patterson N, Durand EY, Viola B, Briggs AW, Stenzel U, Johnson PL, Maricic T, Good JM, Marques-Bonet T, Alkan C, Fu Q, Mallick S, Li H, Meyer M, Eichler EE, Stoneking M, Richards M, Talamo S, Shunkov MV, Derevianko AP, Hublin JJ, Kelso J, Slatkin M, Pääbo S (2010) Genetic history of an archaic hominin group from Denisova Cave in Siberia. Nature 468:1053–1060. doi:10.1038/nature09710

Reidla M, Kivisild T, Metspalu E, Kaldma K, Tambets K, Tolk HV, Parik J, Loogväli EL, Derenko M, Malyarchuk B, Bermisheva M, Zhadanov S, Pennarun E, Gubina M, Golubenko M, Damba L, Fedorova S, Gusar V, Grechanina E, Mikerezi I, Moisan JP, Chaventré A, Khusnutdinova E, Osipova L, Stepanov V, Voevoda M, Achilli A, Rengo C, Rickards O, De Stefano GF, Papiha S, Beckman L, Janicijevic B, Rudan P, Anagnou N, Michalodimitrakis E, Koziel S, Usanga E, Geberhiwot T, Herrnstadt C, Howell N, Torroni A, Villems R (2003) Origin and diffusion of mtDNA haplogroup X. Am J Hum Genet 73:1178–1190. doi:10.1086/379380

Renfrew C (1991) Before Babel: speculations on the origins of linguistic diversity. Camb Archaeol J 1:3–23. doi:10.1017/S0959774300000238

Richards M (2003) The Neolithic invasion of Europe. Annu Rev Anthropol 32:135–162. doi:10.1146/annurev.anthro.32.061002.093207

Richards M, Macaulay V, Hickey E, Vega E, Sykes B, Guida V, Rengo C, Sellitto D, Cruciani F, Kivisild T, Villems R, Thomas M, Rychkov S, Rychkov O, Rychkov Y, Gölge M, Dimitrov D, Hill E, Bradley D, Romano V, Calì F, Vona G, Demaine A, Papiha S, Triantaphyllidis C, Stefanescu G, Hatina J, Belledi M, Di Rienzo A, Novelletto A, Oppenheim A, Nørby S, Al-Zaheri N, Santachiara-Benerecetti S, Scozari R, Torroni A, Bandelt HJ (2000) Tracing European founder lineages in the Near Eastern mtDNA pool. Am J Hum Genet 67:1251–1276. doi:10.1016/S0002-9297(07)62954-1

Richards M, Macaulay V, Torroni A, Bandelt HJ (2002) In search of geographical patterns in European mitochondrial DNA. Am J Hum Genet 71:1168–1174. doi:10.1086/342930

Roewer L, Croucher PJ, Willuweit S, Lu TT, Kayser M, Lessig R, de Knijff P, Jobling MA, Tyler-Smith C, Krawczak M (2005) Signature of recent historical events in the European Y-chromosomal STR haplotype distribution. Hum Genet 116:279–291. doi:10.1007/s00439-004-1201-z

Rootsi S (2006) Y Chromosome haplogroup I prehistoric gene flow in Europe. Documenta Praehistorica 33:17–20

Rootsi S, Magri C, Kivisild T, Benuzzi G, Help H, Bermisheva M, Kutuev I, Barać L, Pericić M, Balanovsky O, Pshenichnov A, Dion D, Grobei M, Zhivotovsky LA, Battaglia V, Achilli A, Al-Zahery N, Parik J, King R, Cinnioğlu C, Khusnutdinova E, Rudan P, Balanovska E, Scheffrahn W, Simonescu M, Brehm A, Goncalves R, Rosa A, Moisan JP, Chaventre A, Ferak V, Füredi S, Oefner PJ, Shen P, Beckman L, Mikerezi I, Terzić R, Primorac D, Cambon-Thomsen A, Krumina A, Torroni A, Underhill PA, Santachiara-Benerecetti AS, Villems R, Semino O (2004) Phylogeography of Y-chromosome haplogroup I reveals distinct domains of prehistoric gene flow in Europe. Am J Hum Genet 75:128–137. doi:10.1086/422196

Rosser ZH, Zerjal T, Hurles ME, Adojaan M, Alavantic D, Amorim A, Amos W, Armenteros M, Arroyo E, Barbujani G, Beckman G, Beckman L, Bertranpetit J, Bosch E, Bradley DG, Brede G, Cooper G, Côrte-Real HB, de Knijff P, Decorte R, Dubrova YE, Evgrafov O, Gilissen A, Glisic S, Gölge M, Hill EW, Jeziorowska A, Kalaydjieva L, Kayser M, Kivisild T, Kravchenko SA, Krumina A, Kucinskas V, Lavinha J, Livshits LA, Malaspina P, Maria S, McElreavey K, Meitinger TA, Mikelsaar AV, Mitchell RJ, Nafa K, Nicholson J, Nørby S, Pandya A, Parik J, Patsalis PC, Pereira L, Peterlin B, Pielberg G, Prata MJ, Previderé C, Roewer L, Rootsi S, Rubinsztein DC, Saillard J, Santos FR, Stefanescu G, Sykes BC, Tolun A, Villems R, Tyler-Smith C, Jobling MA (2000) Y-chromosomal diversity in Europe is clinal and influenced primarily by geography, rather than by language. Am J Hum Genet 67:1526–1543. doi:10.1086/316890

Sampietro ML, Lao O, Caramelli D, Lari M, Pou R, Martí M, Bertranpetit J, Lalueza-Fox C (2007) Palaeogenetic evidence supports a dual model of Neolithic spreading into Europe. Proc Biol Sci 274:2161–2167. doi:10.1098/rspb.2007.0465

Scarre C (1995) The Penguin historical atlas of Ancient Rome. The Bath Press, Bath

Schaffner SF (2004) The X chromosome in population genetics. Nat Rev Genet 5:43–51. doi:10.1038/nrg1247

Semino O, Passarino G, Oefner PJ, Lin AA, Arbuzova S, Beckman LE, De Benedictis G, Francalacci P, Kouvatsi A, Limborska S, Marcikiae M, Mika A, Mika B, Primorac D, Santachiara-Benerecetti AS, Cavalli-Sforza LL, Underhill PA (2000) The genetic legacy of Paleolithic Homo sapiens sapiens in extant Europeans: a Y chromosome perspective. Science 290:1155–1159. doi:10.1126/science.290.5494.1155

Semino O, Magri C, Benuzzi G, Lin AA, Al-Zahery N, Battaglia V, Maccioni L, Triantaphyllidis C, Shen P, Oefner PJ, Zhivotovsky LA, King R, Torroni A, Cavalli-Sforza LL, Underhill PA, Santachiara-Benerecetti AS (2004) Origin, diffusion, and differentiation of Y-chromosome haplogroups E and J: inferences on the Neolithization of Europe and later migratory events in the Mediterranean area. Am J Hum Genet 74:1023–1034. doi:10.1086/386295

Simoni L, Gueresi P, Pettener D, Barbujani G (1999) Patterns of gene flow inferred from genetic distances in the Mediterranean region. Hum Biol 71:399–415

Simoni L, Calafell F, Pettener D, Bertranpetit J, Barbujani G (2000) Geographic patterns of mtDNA diversity in Europe. Am J Hum Genet 66:262–278. doi:10.1086/302706

Skeates R (2003) Radiocarbon dating and interpretations of the Mesolithic-Neolithic transition in Italy. In: Ammerman A, Biagi P (eds) The widening harvest. Archaeological Institute of America, Boston

Soares P, Achilli A, Semino O, Davies W, Macaulay V, Bandelt HJ, Torroni A, Richards MB (2010) The archaeogenetics of Europe. Curr Biol 20:174–183. doi:10.1016/j.cub.2009.11.054

Sokal RR, Harding RM, Oden NL (1989) Spatial patterns of human gene frequencies in Europe. Am J Phys Anthropol 80:267–294

Stevanovitch A, Gilles A, Bouzaid E, Kefi R, Paris F, Gayraud RP, Spadoni JL, El-Chenawi F, Béraud-Colomb E (2004) Mitochondrial DNA sequence diversity in a sedentary population from Egypt. Ann Hum Genet 68:23–39. doi:10.1046/j.1529-8817.2003.00057.x

Stoneking M (2008) Human origins. The molecular perspective. EMBO Rep 9:46–50

Stoneking M, Krause J (2011) Learning about human population history from ancient and modern genomes. Nat Rev Genet 12:603–614. doi:10.1038/nrg3029

Stringer C, Davies W (2001) Archaeology. Those elusive Neanderthals. Nature 413:791–792. doi:10.1038/35101688

Tomas C, Sanchez JJ, Barbaro A, Brandt-Casadevall C, Hernandez A, Ben Dhiab M, Ramon M, Morling N (2008) X-chromosome SNP analyses in 11 human Mediterranean populations show a high overall genetic homogeneity except in North-west Africans (Moroccans). BMC Evol Biol 8:75. doi:10.1186/1471-2148-8-75

Torroni A, Bandelt HJ, D'Urbano L, Lahermo P, Moral P, Sellitto D, Rengo C, Forster P, Savontaus ML, Bonné-Tamir B, Scozzari R (1998) MtDNA analysis reveals a major late Paleolithic population expansion from southwestern to northeastern Europe. Am J Hum Genet 62:1137–1152. doi:10.1086/301822

Torroni A, Bandelt HJ, Macaulay V, Richards M, Cruciani F, Rengo C, Martinez-Cabrera V, Villems R, Kivisild T, Metspalu E, Parik J, Tolk HV, Tambets K, Forster P, Karger B, Francalacci P, Rudan P, Janicijevic B, Rickards O, Savontaus ML, Huoponen K, Laitinen V, Koivumäki S, Sykes B, Hickey E, Novelletto A, Moral P, Sellitto D, Coppa A, Al-Zaheri N, Santachiara-Benerecetti AS, Semino O, Scozzari R (2001) A signal, from human mtDNA, of postglacial recolonization in Europe. Am J Hum Genet 69:844–852. doi:10.1086/323485

Torroni A, Achilli A, Macaulay V, Richards M, Bandelt HJ (2006) Harvesting the fruit of the human mtDNA tree. Trends Genet 22:339–345. doi:10.1016/j.tig.2006.04.001

Tresset A, Vigne JD (2007) Substitution of species, techniques and symbols at the Mesolithic-Neolithic transition in western Europe. Proc Br Acad 144:189–210

Tresset A, Vigne JD (2011) Last hunter-gatherers and first farmers of Europe. C R Biol 334:182–189

Turchi C, Buscemi L, Previderè C, Grignani P, Brandstätter A, Achilli A, Parson W, Tagliabracci A, Ge.F.I. Group (2008) Italian mitochondrial DNA database: results of a collaborative exercise and proficiency testing. Int J Legal Med 122:199–204. doi:10.1007/s00414-007-0207-1

Underhill PA, Shen P, Lin AA, Jin L, Passarino G, Yang WH, Kauffman E, Bonné-Tamir B, Bertranpetit J, Francalacci P, Ibrahim M, Jenkins T, Kidd JR, Mehdi SQ, Seielstad MT, Wells RS, Piazza A, Davis RW, Feldman MW, Cavalli-Sforza LL, Oefner PJ (2000) Y chromosome sequence variation and the history of human populations. Nat Genet 26:358–361. doi:10.1038/81685

Underhill PA, Passarino G, Lin AA, Shen P, Mirazón Lahr M, Foley RA, Oefner PJ, Cavalli-Sforza LL (2001) The phylogeography of Y chromosome binary haplotypes and the origins of modern human populations. Ann Hum Genet 65:43–62. doi:10.1046/j.1469-1809.2001.6510043.x

Underhill PA, Myres NM, Rootsi S, Metspalu M, Zhivotovsky LA, King RJ, Lin AA, Chow CE, Semino O, Battaglia V, Kutuev I, Järve M, Chaubey G, Ayub Q, Mohyuddin A, Mehdi SQ, Sengupta S, Rogaev EI, Khusnutdinova EK, Pshenichnov A, Balanovsky O, Balanovska E, Jeran N, Augustin DH, Baldovic M, Herrera RJ, Thangaraj K, Singh V, Singh L, Majumder P, Rudan P, Primorac D, Villems R, Kivisild T (2010) Separating the post-Glacial coancestry of European and Asian Y chromosomes within haplogroup R1a. Eur J Hum Genet 18:479–484. doi:10.1038/ejhg.2009.194

Whittle A (1996) Europe in the Neolithic. The creation of new worlds. Cambridge University Press, Cambridge

Wilkins JF (2006) Unraveling male and female histories from human genetic data. Curr Opin Genet Dev 16:611–617. doi:10.1016/j.gde.2006.10.004

Y Chromosome Consortium (2002) A nomenclature system for the tree of human Y-chromosomal binary haplogroups. Genome Res 12:339–348. doi:10.1101/gr.217602

Zalloua PA, Platt DE, El Sibai M, Khalife J, Makhoul N, Haber M, Xue Y, Izaabel H, Bosch E, Adams SM, Arroyo E, López-Parra AM, Aler M, Picornell A, Ramon M, Jobling MA, Comas D, Bertranpetit J, Wells RS, Tyler-Smith C, Genographic Consortium (2008) Identifying genetic traces of historical expansions: Phoenician footprints in the Mediterranean. J Hum Genet 83:633–642. doi:10.1016/j.ajhg.2008.10.012

Zeder MA (2008) Domestication and early agriculture in the Mediterranean basin: origins, diffusion, and impact. Proc Natl Acad Sci U S A 105:11597–11604. doi:10.1073/pnas.0801317105

Zvelebil M (2004) Who were we 6,000 years ago? In search of prehistoric identities. In: Jones M (ed) Traces of ancestry: studies in honour of Colin Renfrew. McDonald Institute Monographs, Cambridge

The Mediterranean and the Voices Transported by Time

32

Maria Helena Trindade Lopes

Abstract

The Mediterranean was the great sea of Antiquity. And its waters, symbol of a changing world, allowed many men – Egyptians, Phoenicians, Greeks, etc. – to find and to recognize themselves in their particularities. Later, Rome, which in just three centuries redesigned the world, ensured that the messages of the early times – of Egyptian, Phoenician and Greek spaces – penetrated this world. It added others – law and religion – and thus it was shaped Western civilization, which is Roman and Greek, but it is also African and Oriental, via a river whose waters "drew" the history of the world.

Keywords

Mediterranean Sea • Egyptians • Phoenicians • Greeks • Romans • Western Civilization

Barcino, The Mediterranean Sea (2011)

"Nature is a petrified magic city"

Novalis

The reading of spaces and their intervention in the course of history is simultaneously a primordial and passionate task. There are spaces that seem devoid of meaning, as if time had passed by them without leaving any impressions. Some deserts – not all of course – fall into this category. And then there are other spaces that appear to carry the history of the world on their shoulders as if they were a metaphor of the world encapsulated in an eternity of time. The Mediterranean, the great sea as it was called by the ancients, is the most expressive and effulgent example of this second type.

In terms of imagery, the waters of the sea have always been described as the waters of the creation. Almost all the ancient cosmogonies confirm this. But in history the Mediterranean, through its privileged position in the course of so many peoples and cultures, quickly embodied the role of "true sea", the "sea between the lands" which Hebrews spoke of as well as some of the classical authors such as Lucretius, Pliny and Seneca. Its waters, the symbol of a changing world, represented and evoked the dangers and difficulties of the passage from a known world to a world yet to be known. And by conquering it, human beings individually

M.H.T. Lopes (✉)
Department of History, F.C.S.H. – Universidade Nova de Lisboa,
Av. de Berna, nº 26 C, 1069-061 Lisbon, Portugal
e-mail: helenatrindadelopes@hotmail.com

S. Goffredo and Z. Dubinsky (eds.), *The Mediterranean Sea: Its history and present challenges*,
DOI 10.1007/978-94-007-6704-1_32, © Springer Science+Business Media Dordrecht 2014

and collectively, enjoyed the possibility of becoming a hero, changing their identity and their life. In doing so they also transfigured the history of their people and their land by means of the contacts established with the world.

If we go back to a time where everything seems to have been drawn, where history began to write the first accounts of its existence, the ancient world of pre-classical and, afterwards, classical civilizations, we find that the areas which contributed the most to the development of this particular relationship between the "great sea" and the land, "which is always mother," were Egypt, Cyprus, Crete, the Cyclades, Anatolia and mainland Greece. The nature and extent of relations established from the end of prehistory to the late Bronze Age (1070 BC), led to the formulation of the so-called Western civilization that, contrary to what was thought for a long time, is not only Greek but also African and Oriental.

Before the 2nd millennium, contacts between these people had been very sparse and we can only detect them through Egyptian materials and items spread mainly throughout the islands and in western Anatolia. After that, contacts intensified in both directions and archaeological and textual sources from areas such as Amarna, Thebes Avaris, Mari, Ugarit and Bogazkoy show the presence of a very active and diversified international trade. Thus, the different features of a world that was being drawn established the first bridges between these features by business needs. And the Mediterranean, in resembling a road, allowed the coexistence and the proximity between peoples and cultures. By doing so it was rewriting a page of history.

Simultaneously, the "great sea" conditioned the development and political organization of each of the territories involved in this adventure. Not all organized similarly nor progressed in the same way. Let us look, for example, at the civilization that was born and raised in the Nile Valley. Egypt's great originality against all its neighbours, in the 4th millennium, lay in its evolution, in a few hundred years, from a system of independent principalities to the formation of a centralized state, while the entire Middle East stayed for a long time with the system of city-states. But when we look closely at the map of ancient Egypt, we can easily understand how its physical geography conditioned and determined its history and its administrative organization.

Egyptian civilization was born, grew and developed in a privileged space. A protected area delimited by natural borders and blessed by the fruitful waters of a river that became the real ruler of all of Egypt. Borders open and close doors … expedite and protect… To the north, Egypt opened out to the world through the Mediterranean Sea, that sea which grew in our imagination in proportion to the dimensions of its history. The sea enabled contact with other worlds, other cultures … whetted its curiosity, and unfastened it, slowly and subtly, from its ancestral roots. To the south, Elephantine, marking the first cataract, defended the territory and its humans, enclosing them in their world and in their oldest traditions. To the west and east, two deserts, the Libyan and the Arab, guaranteed the protection of the people, working as magic doors, whose passage would inevitably lead to Non-Existence. These arid and austere regions, beyond the oasis and the high mountains, hid the wealth of their mines and covered up access routes which represented fundamental strategic assets. The image conveyed by the apparent death of the silence of the desert was overcome by the force of the overflowing waters that crossed the country from south to north, and became lost in the immensity of the sea. The country was thus was thus constituted in the poetical game of contrasts.

Enclosed and protected in its world, Egypt slowly built "bridges" to the outside. In doing so it created certain routes, paths … "roads" … pathways. Its motivations were always objective and well defined. Trade and the exploitation of mines and quarries determined its steps… The Mediterranean allowed the Egyptian world to enter into contact with the Aegean world. Thus, from the Middle Kingdom (c. 2022–1650 BC) onwards we can witness the establishment of trade relations with the Mediterranean islands. The famous Cyprus copper "invaded" Egypt, as well as other luxury goods also famous for its perfection and beauty: the trappings of gold, faience beads and the magnificent marble vases. Relating to the coast of the eastern Mediterranean, the first contacts date back to the Old Kingdom (c. 2635–2140 BC) and illustrate the first expeditions to Byblos, the large emporium of Semitic trade, famous for its cedar and pine wood, which the Egyptians imported for the manufacture of their boats. These trade relations would be intensified during the Middle Kingdom and spread to Syria in the New Kingdom (c. 1539–1069 BC), giving rise to new imports: boats, cars, weapons, musical instruments (lyres and flutes), horses, wine, beer, etc. And finally, in Sinai, in the Eastern Desert and in Nubia activity was concentrated on the exploitation of mines and quarries. With Anatolia Egypt developed another type of relationship based mainly on armed conflict and the dispute over spheres of influence. And when this conflict did not seem to be heading towards any solution, diplomacy emerged as an extraordinary resource to resolve all the clashes between the empires.

So these people, enabled by the Mediterranean Sea to learn about the world, shaped a civilization founded on two types of integrating elements: those which presided over the construction of their own identity and the others… the ones which ran the world, bringing to different men the message of a unique time and civilization. Among the first we can highlight essentially four: the pharaoh, the guarantor of divinity and the perpetuity of the monarchical institution; Maat, the guide and regulator of the cosmic, political, social and individual principle; the sacred dimension of the cosmos in all its manifestations; finally, language, the unifying element par excellence of any civilization. Regarding the second type of integrating elements, these belong to the set of archetypes and messages that the Egyptians bequeathed to the world at the beginning of time. For example, one of the major contributions of Egypt to Western civilization refers to the model of creation through the word. Ptah, the creator god of Memphis, conceived the cosmos in its different manifestations in his heart and realized it through the creative and operative force of the word. The doctrine of the creator verb, usually recognized from the biblical text (Gen. 1) and situated in a particular historical, geographical and temporal context, actually dates back to a time and a place which was very different, the Nile Valley. But the same phenomenon occurs with respect to the explanation of man's creation: it was modelled after the "dust of the earth" by Khnum, the potter god, recalling, again, the Jewish tradition recognized in Gen. 2.7. The similarity between the Hymn to Aten, the programmatic text of the "monotheistic" reform of Amenhotep IV, and Psalm 104 is another example of these constant "coincidences" that ancient history consecrated, demonstrating that the human beings of the past, just as today's human beings do, travelled, and in doing so they took and brought with them ancestral values, derived from their dwellings, or singularities that they incorporated in their own traditions. The road, which was the Mediterranean, connected Egypt to the world and by doing so carried traces of this civilization to other spaces, other peoples, other cultures.

Unlike the Egyptians, the Phoenicians were not protected "by the gods" in terms of geography, but they were also challenged. The challenge of the Mediterranean pushed them to a destination from which they could not escape.

The Phoenicians built their story caught between two worlds: the Hittite and Amorite world to the north and the Egyptian world to the south. In addition, its beautiful coastline, with its harbours, became the target of many and varied raids over time. Its boundaries were never barriers, but simply routes to be crossed: the northern border was located at Tell Sukas, south of Ugarit, while the southern border, Acco, was a short distance from Mount Carmel.

In this geographic strip, where we can distinguish three distinct zones (Northern Syria, Lebanon and Palestine), the landscape is dominated by limited coastal plains and the hinterland by mountain ranges, which make lines of communication further East almost impossible.

To the west, Phoenicia was faced with the "great sea" and turned it into its "privileged dwelling" where it moved at ease. And so the Phoenicians turned the Mediterranean into a second territory, into a path that allowed them to set off in conquest of markets and raw materials. Across the Mediterranean they traced the routes from Tyre to Gadir, the passage of the Strait of Gibraltar, the route from Gadir to Tyre and the Atlantic route.

All of this traversed landscape is not a geographical unit nor is it uniform from the physical point of view... the plains, mountains, rivers, and... the sea. This diversity created the wealth necessary for the establishment of the first populations. This challenge can also be a gift.

Thus, the people settled in the coastline building up the height of their cities, which they walled, and which never had a very large area – Byblos, for example. In the 3rd millennium, an Egyptian community was established in Byblos, carrying with them their customs and traditions, which blended with the customs and traditions of the local populations. Byblos was a city gifted for its agriculture, for the exploitation of forests, ports and metallurgy, but despite this great diversity of resources its "prime" destiny was trade. However, the westward expansion of the Phoenicians marked their decline. Its attitude of neutrality in the face of different invaders did not cause a good impression nor create good neighbours, even though the Phoenician never had had a strong sense of nationhood. This attitude can be explained by the fact that the "Phoenician political space" was a much wider territory than Phoenicia itself. Rather than feeling a political unity, that had never existed, they were part of a cultural community that had fulfilled a determinant historical mission: the unity of the Mediterranean world, the union between the East and the distant West. In terms of integrating elements, this was the great legacy that Phoenicia left to the world.

The purely commercial attitude of the Phoenicians not only alerted the peoples to the possibilities of navigation, but also between 1200 BC and 332 BC put them in contact with all the great civilizations of antiquity.

With this attitude, Phoenicia fulfilled a crucial cultural mission, by establishing contacts, exchanges, intimacies between the East and the far distant West. Moreover, it took its unifying role even further, by leaving us one of the most remarkable inventions of the cultural history of mankind: the alphabet. This resulted from a long historical process of maturation, which manifested for the first time in the passage from the 2nd to the 1st millennium. This alphabet would spread eastwards and westwards. The Hebrews and the Aramaeans used it to transcribe their own language and the Greeks, around 800 BC, adapted it to the needs of a non-Semitic language and vocalized it. From that moment on, the "genetic" relationship between the Phoenician alphabet and the Greek alphabet was completed. The former went its own way and the latter, via the Etruscan and Roman world, gave origin to our modern alphabet.

Differently in the Greek case, we find that the most striking feature of ancient Greece was its breakdown into a multitude of cities (poleis) that then formed an almost corresponding number of states. And, if on the one hand this division manifested itself at times as the main reason for all its greatness and glory, at others it was the real cause of all its weaknesses.

The history of Greece, its definition of space, its organization into cities that constituted themselves as sovereign states, each with a physiognomy, personality and life of its own, was strongly influenced by its geographic environment, by the subtle and symbolic dialogue established between the mountains, which cover almost the entire territory, and the sea.

Indeed, looking at the "Greek world" we encounter small communities who settled in the south of the Balkan Peninsula, in the narrow depressions along the coast, access to which was only possible along the coast. The difficulty of communications made humans a prisoner of a particular model of space occupation. Thus, the mountains dictated the "division" of Greek space into small towns independent of each other. But to this geographical condition other intrinsic factors were added: the dry climate and poor soil did not allow the development of an agricultural activity that supported "the men of every mountain" and thus, these human beings turned to the sea and made the Mediterranean "the great highway of the Greek world." In this way, the poverty of the soil and the need to develop trade to increase wealth determined the beginning of colonization.

The first stage of colonization (775–675 BC) led the Greeks to Italy and Sicily. And the colonists marvelled at their extent and especially at the fertility of the plains. In a second stage (675–550 BC), they turned to the Black Sea, reached Egypt and even the distant shores of southern France and the Spanish Levant. The Western Mediterranean appeared as an area particularly rich in metals. In the meantime, Marseille becomes the major commercial hub, from which Greek products would be spread throughout the South of France and the Iberian Levant. In Africa, Cyrene also became a major centre of commerce – its port of Apollonia reached caravans coming from Egypt, Sudan and Carthage. But in Egypt, everything was different. For over a century, the Greeks of Asia had offered their services to the successors of the ancient pharaohs. Naukratis became a flourishing city, foreshadowing Hellenistic Alexandria. In the north, they crossed the Hellespont and the Bosphorus, resuming the links between the Mediterranean and the Black Sea. They reached Pontus. And they felt they knew the world… The Mediterranean became a lake of innumerable and wonderful messages.

Thus, when the borders of Greece widened (from the eighth century BC until the sixth century AD) and spread to the south of Italy, to the Black Sea, to the coasts of France and the Iberian Peninsula, to Africa, where the geographical conditions are in some cases quite different. Interestingly, the Greeks reproduced and recreated the same model of space organization, the polis. And why? Because the polis definitely marks the way "of being" of the Greeks. So when the Athenians, after the reforms of Cleisthenes (sixth century BC), presented a new constitution, democracy, to the world, everyone recognized it as the natural consequence of the form of government represented by the "city state": the participation of citizens, free individuals, in meetings where everything was discussed and decided. However, when the Athenian Confederation, formed from within the Pan-Hellenic League in about 478 BC, intended to show the Greek world what the political unity of a large number of cities united under the supremacy of one of them would be, what was achieved was the progressive suppression of the autonomy of the different cities, "educating them" in the democratic regime, in its weights and measures, as well as in the use of its currency. But it did not win, because the spirit of autonomy of the Greek cities eventually rebelled, only to limp without any possibility of escape, to the Hellenes' Koinon, created by Philip of Macedonia in 337 BC. This definitely marked the end of the Greek conception of political space, the end of classical times and the freedom to be isolated.

Thus, the polis was the model invented by the Greeks in response to the geographical condition, which determined the history of its time: "dividing the land, uniting the people". A great slogan!

But simultaneously these individuals, who received the ancient world in their hearts, gave us the basis of our "conceptual geography." It was they who laid the roots of science we use today, and they knew how to "invent" History and Philosophy, essential foundations to construct our memory and thinking beings and the world. Then they created the Theatre, the Olympic games and "inaugurated" democracy. Who in Europe does not recognize the value of these archetypes?

These three geographical zones – Egypt, Lebanon and Greece, with the Aegean islands and Crete, where the first polis were born – have not only been influenced in their history and their formulation as political spaces by the Mediterranean, as well as then allowing that the same sea to transport their "messages", their civilization archetypes to all the known world.

Hellenism undertook the task of transmitting all knowledge from the East to the Western world. Then Rome, in just three centuries, redesigned the world by conquering all the lands surrounding the Mediterranean, from the Italian peninsula to Carthage, its great rival in economic terms, from Carthage to the western Mediterranean (Iberian Peninsula, Gaul) and to the eastern Mediterranean (Macedonia, Greece, Asia Minor), turning the great sea of the ancient world into the "Mare Nostrum" of the Romans. Rome, which we now eruditely think of as a Mediterranean civilization, ensured that the messages of the early times – from Egyptian,

Phoenician and Greek spaces – penetrated this world by creating an empire nearly the size of the world. However, these came to us already reinforced by elements provided by the Roman Empire itself, the law and the religion, Christianity, fundamental pillars of our Western identity. The "Roman law" configured the western world and was the support for the rational spirit of the systems of the occidental nations. Furthermore, Christianity flourished as the most important religion of the western world. But all this was possible because the Romans, with their practicality, built roads that covered the known world. Then the men travelled on those roads, carrying with them the Roman law, the Christian religion, the cultures and customs of the mixed world and the magnificence and the genius of Rome.

Later, the medieval crusaders and the discoverers of the "new world" perfected only what time and history had been encharged with forging.

Thus, when someone in Europe today is moved when listening to Mozart or reading a page of Byron, this is because this person recognizes in these different melodies, ancient and recalcitrant voices, founding voices, voices that summarize the African, Oriental and classical past of Western civilization. This has been the great legacy of the Mediterranean.

> Those who live by the sea can hardly form a single thought of which the sea would not be part.
>
> Hermann Broch

Bibliography

AAVV (2000–2003) Enciclopedia del Mediterraneo. Jaca Book, Milano
Abulafia D (2011) The Great Sea: a human history of the Mediterranean. Allen Lane, London
Arnason J, Eisenstadt SN, Wittrock B (2004) Axial civilizations and world history. Brill, Leiden
Aubet ME (1993) The phoenicians and the west. Politics, colonies and trade. Cambridge University Press, Cambridge
Baurain C, Bonnet C (1992) Les Phéniciens, marins des trois continents. Armand Colin, Paris
Braudel F (1977–1978) La Méditerranée. I, L'éspace et l'histoire; II, Les hommes et l'héritage. Arts et Métiers Graphiques, Paris
Braudel F (1985) La Méditerranée: histoire et espace. Flammarion, Paris
Braudel F (1995) The Mediterranean and the Mediterranean world in the age of Philip II, 2 vols. University of California Press, Los Angeles
Braudel F (1998) Les mémoires de la Méditerranée. De Fallois, Paris
Carpentier J, Lebrun F (2001) Histoire de la Méditerranée. Seuil, Paris
Casson L (1991) The ancient mariners: Seafarers and sea fighters of the Mediterranean in ancient times, 2nd edn. Princeton University Press, Princeton
Constable OR (2003) Housing the stranger in the Mediterranean world: lodging, trade, and travel in late antiquity and the Middle Ages. Cambridge University Press, Cambridge
Demand NH (2011) The Mediterranean context of early Greek history. Wiley/Blackwell, Oxford
Freeman C (1999) Egypt, Greece and Rome – civilizations of the ancient Mediterranean. Oxford University Press, Oxford
Harris WV (2005) Rethinking the Mediterranean. Oxford University Press, Oxford
H-Mediterranean, a site for history discussion. http://www2.h-net.msu.edu/~mediter. Accessed 8 Nov 2011
Horden P, Purcell N (2000) The corrupting sea: a study of Mediterranean history. Blackwell, Oxford
Huskinson J (2000) Experiencing Rome – culture, identity and power in the Roman Empire. Routledge, London
Malkin I (2005) Mediterranean paradigms and classical antiquity. Routledge, New York
Mediterranean Historical Review (1986). http://www.tau.ac.il/humanities/cmc/mhr/mhr.html. Accessed 21 Nov 2011
Mediterranean Studies: The Journal of the Mediterranean Studies Association http://www.mediterraneanstudies.org/ms/medstud.html. Accessed 24 Nov 2011
Mota G (2010) The Mediterranean between myth and history. Mediterr J Soc Sci 1:7–25
Munford GD (2001) Mediterranean area. In: Redford DB (ed) The Oxford encyclopaedia of ancient Egypt, vol 2. Oxford University Press, Oxford
Randsborg K (1989) The demise of antiquity: Europe and the Mediterranean in the first millennium AD. Annu Rev Anthropol 18:227–244
Scripta Mediterranea: Bulletin of the Society for Mediterranean Studies. http://www.utoronto.ca/cims/files/jourcurr.htm. Accessed 26 Nov 2011

Mediterranean Royal Purple: Biology Through Ritual

33

David Iluz

Abstract

The eastern Mediterranean coast was the center of the dyeing industry in the ancient world. Royal purple, the most famous indigoid dye of antiquity, was derived from hypobranchial-gland extracts of various marine gastropod mollusks. Extensive marine biological surveys have revealed that the only snails in the Mediterranean that produce stable dyes are those of the murex family. The dye obtained from the common snail *Murex trunculus* is very stable and steadfast. The extracts were processed in a lengthy, elaborate procedure, according to the Roman writer of the first century AD, Pliny the Elder; other, more cursory texts are known from the Classical Greek period up until Byzantine times. Purple dyeing was also discovered by cultures on the western and eastern shores of the Pacific Ocean in China and Peru, respectively, apparently with no connection to the Canaanites or Phoenicians. We describe here the history of this dye from its earliest traces, through its role in Jewish rituals, its rise to Roman regal attire, and its subsequent decline and final replacement by synthetic products.

Keywords

Royal purple • *Murex trunculus* • *tekhelet* • Indigo

Introduction

Royal purple, also known as Tyrian purple, is produced from marine snails. It originated from the port of Tyre in Phoenicia (now southern Lebanon). The Phoenicians made their wealth trading in dyestuffs, and dye houses were ubiquitous in the region. Because of its lucrative nature, purple and blue dyeing eventually came under imperial control. The Romans issued edicts that only royalty could wear garments colored with these dyes, and only imperial dye houses were permitted to manufacture it. Nero and the Byzantine emperor, Theodosius, issued edicts prohibiting (upon threat of capital punishment) the wearing of some types of purple cloth by anyone other than themselves. Suetonius, in his biography of the Roman emperor, Caligula (*De Vita Cae-sarum-Liber 4, Caligula, 35*), records how, in 40 AD, Caligula had Ptolemy (king of Mauretania and son of Judah II) murdered because the fine purple cloak worn by Ptolemy attracted universal admiration during his entrance into the amphitheatre during a gladiatorial show (Born 1937; Reinhold 1970). This ruling apparently drove the Jewish *tekhelet* ((תְּכֵלֶת industry underground. The dye was used in garments worn for ritual purposes, such as the *Talit*, the shawl worn under the clothing, where a single purple thread was required, as well as in the high priest's ritual dress, where a weave of purple yarn was included. The Bible referred to that same valuable *tekhelet* dye commonly used by royalty throughout the ancient world. The Torah consistently included *tekhelet* along with gold, silver, and other familiar, precious materials, recognized by all for their worth. Later, with the Arab conquest of the Middle East, including today's Israel (683 CE), the secret of *tekhelet* was essentially lost, and the dyeing process using it was forgotten.

D. Iluz (✉)
The Mina and Everard Goodman Faculty of Life Sciences, Bar-Ilan University, Ramat Gan 52900, Israel
e-mail: iluzda@gmail.com

S. Goffredo and Z. Dubinsky (eds.), *The Mediterranean Sea: Its history and present challenges*, DOI 10.1007/978-94-007-6704-1_33, © Springer Science+Business Media Dordrecht 2014

Fig. 33.1 The murex are widespread in the Mediterranean, and are used for purple dye. From *left*: *Bolinus brandaris, Murex trunculus*, and *Stramonita haemastoma*

Royal purple is the most renowned of ancient dyes (Brunello 1973). Even before Nero issued a decree in the first century AD that gave the emperor the exclusive right to wear royal purple garments (Born 1937; Reinhold 1970), the association of this dye with royalty and high ecclesiastics was well established, as it was the color used for the togas of the Roman emperors. Some 1,500 years ago it was a capital offense for anyone other than the Emperor to wear it. Catholic cardinals wore purple, but nowadays they wear scarlet. The process for making Tyrian purple fell into obscurity after the fall of Constantinople (1453) and only recently scientists have rediscovered how it was processed (Edmonds 2000; Boesken Kanold 2005; Koren 2005).

Royal purple is a natural purple-red dye extracted from marine snails. The taxonomy of marine mollusks changed considerably during the second half of the last century. Today, all marine mollusks used as sources of purple in all parts of the world – including the historical purple-producing species of the Mediterranean, are classified as part of the *Muricide* family. Today they are classified in three subfamilies: the murexes are classified as Muricinae; the Ocenebrinae include two purple-giving species, one of which was previously considered a murex and the other a rockshell; and most rockshells, on the other hand, are classified as Rapaninae (Houart 2001). There are about 1,300 species of Muricidae in the entire world, living mainly in warm or temperate seas. Of these, some 370 are species of Muricinae, 130 of Ocenebrinae, and 140 of Rapaninae. Very few of them were ever used as a source of purple dye and it is not known if they are all capable of producing it (Radwin and d'Attilio 1976).

The dye substance consists of a mucous secretion from the hypobranchial gland of one of several common, medium-sized carnivorous sea snails found in the eastern Mediterranean. These are the marine gastropods *Bolinus brandaris*, the spiny dye-murex (originally known as *Murex brandaris* (Linnaeus 1758)), the banded dye-murex *Hexaplex trunculus*, and the rock-shell *Stramonita haemastoma* (Radwin and D'Attilio 1986; Ziderman 1986b). The synonyms of these murex are: *Bolinus brandaris Hexaplex* for *murex brandaris*; *Phyllonotus trunculus*, *Trunculariopsis trunculus*, *Murex turbinate*, *Murex goldi*, *Muricanthus trunculus*, and *Hexaplex trunculus* for *murex trunculus*; and *Stramonita haemastoma*, *Thais haemastoma Buccinum haemastoma, and Purpura haemastoma* for *murex haemastoma* (Fig. 33.1). The dye is an organic compound of bromine (i.e., an organobromine compound), a class of compounds often found in algae and other sea life but rarely found in land animals.

Murex snails are active carnivores. They use a rasp-like muscular tongue called the radula to bore into the shells of clams or other bivalves and eat the animal inside. Murex snails produce fluids from a gland that aid in the boring action and may also anesthetize the prey.

The History of Mediterranean Royal Purple

Royal purple was especially associated in history with the Phoenicians (Jensen 1963; Bruin 1970). Their preeminent role in this Mediterranean industry has been substantiated by the chemical identification of dibromoindigotin (DBI), the main component of the dye, on the interior of "Canaanite" jars used for storage and shipment of goods, found at a thirteenth century BC dyeing factory at Sarepta, a site on the coast of Lebanon (Pritchard 1978; McGovern and Michel 1985). However, the history of Tyrian purple does not, in fact, begin in Phoenicia, but in Crete and the small

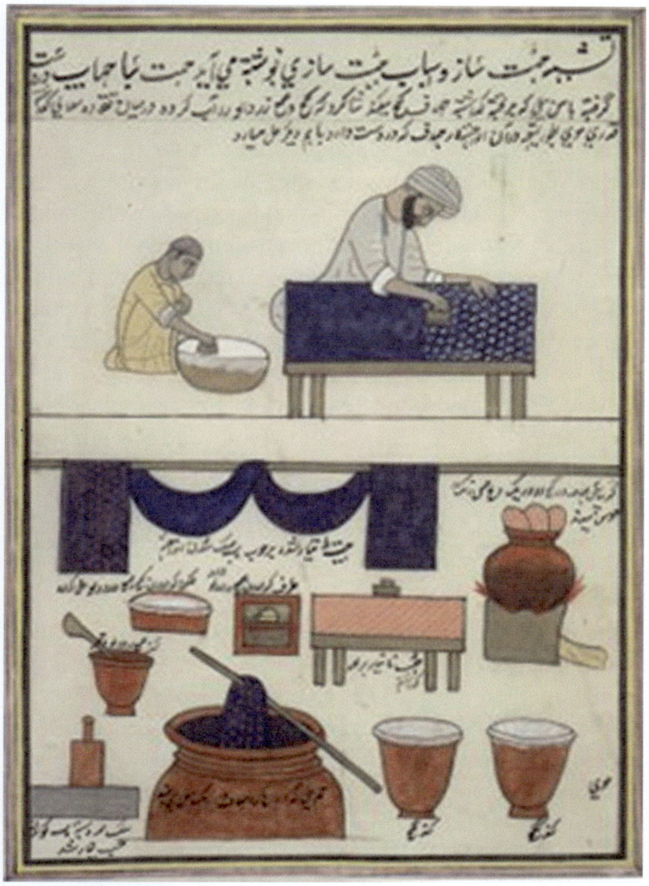

Fig. 33.2 Dying and printing indigo cloth Partner, 1850–1860. Artist unknown. British Library, Image reference: BL.Add.Or.1714. Origin: Kashmir, India

surrounding islands where the oldest piles of purple-yielding mollusk shells (consisting mainly of *Hexaplex trunculus*) were discovered. Dating from between 1800 and 1600 BC, they were, therefore, at least a century older than those found on the shores of the Levant. The expression 'royal purple' appears for the first time on a thirteenth century BC Linear B tablet discovered at Knossos in Crete (Reese 1987).

The use of mollusks to produce purple or violet dyes is not a cultural phenomenon unique to the Mediterranean world. Purple dyeing was practiced on the Atlantic coasts of Europe and, since pre-Colombian times, in South and Central America, where it is still used today to decorate the traditional *posahuancos* (wrap skirts) worn by many village women in the province of Oaxaca in Mexico. Historic texts and fabrics provide evidence that purple was also used in several ancient civilizations in Asia (Fig. 33.2), notably in Japan, where recent analysis has identified purple from the mollusk called akanishi (*Rapana venosa*) on a fragment of silk dating from the mid-first century BC, discovered in a funerary urn from the Yayoi civilization found on the island of Kyushu (Shimoyama and Noda 1993). It is also now known that purple-producing mollusks can be found in all the hot and temperate seas (Jackson 1917; Baker 1974; Lammens and Verhecken 1984; Verhecken 1989; Cooksey 1994; Haubrichs 2004). The Phoenicians established an ancillary production facility on the Iles Purpuraires at Mogador, in Morocco. The gastropod harvested at this western Moroccan dye-production facility was *Hexaplex trunculus*. This second species of dye murex is found today on the Mediterranean and Atlantic coasts of Europe and Africa (Spain, Portugal, Morocco, and the Canary Islands) http://en.wikipedia.org/wiki/Tyrian_purple-cite_note-Radwin.2C_G._E_1986._p93-3#cite_note-Radwin.2C_G._E_1986._p93-3 (Radwin and D'Attilio 1986).

The color-fast (non-fading) dye was an item of luxury trade prized by the Romans, who used it in ceremonial robes. It is believed that the intensity of the purple hue improved rather than faded as the dyed cloth aged. A few classical authors have written about the techniques of dyeing with purple in the Mediterranean area. The most helpful are Aristotle (384–322 BC), Vitruvius (born 80–70 BC, died after c. 15 BC) and, particularly, Pliny the Elder (77–79 AD) (Doumet 1980; Elsner 1992; Edmonds 2000; Koren 2005). Vitruvius mentioned the production of Tyrian purple from shellfish (*De Architectura* VII:13). Aristotle described the shellfish from which Tyrian purple was obtained and the process of extracting the tissue that produced the dye (Aristotle 2004). Pliny the Elder described the production of Tyrian purple in his Natural History (Pliny the Elder 1855).

The Roman mythographer Julius Pollux (1846), who wrote in the second century BC, asserted that purple dye was first discovered by Heracles, or rather by his dog, whose mouth was stained purple from chewing on snails along the shores of the Levant. As mentioned above, the archaeological discovery of substantial numbers of murex shells in Crete suggests that the Minoans may have pioneered the extraction of imperial purple centuries before the Tyrians. Dating from collocated Cretan pottery suggests that the dye may have already been produced as early as during the Middle Minoan period in the twentieth to eighteenth centuries BC (Reese 1987; Stieglitz 1994). The accumulation of crushed murex shells from a hut at the Coppa Nevigata site in southern Italy may, likewise, indicate the production of purple dye there from at least the eighteenth century BC (Cazzella and Moscoloni 1998) (Fig. 33.3).

Despite the importance of purple dyeing to the Phoenicians, they rarely mentioned it in their writings, perhaps in order to protect closely guarded secrets (Oppenheim et al. 1970) or because only a small textual Phoenician corpus was recovered from excavations. The first lengthy, probably eyewitness, account of the industry occurred over 1,000 years after it began. The Roman scholar of the mid-first century AD recounted in great detail how the carnivorous mollusks were captured using baited wicker baskets.

Archaeological discoveries appear to confirm Pliny's description of the procedure that was used to dye fabrics purple (Fig. 33.4). Piles of shells were found at a number of sites with large open pits nearby that could have been used for the initial maceration of purple mollusk glands in salt. At Kerkouane, a Punic (Phoenician) site at Cape Bon in Tunisia, pits or basins of this type were hollowed out on the rocky shore (Fantar 1986). In Greece, two large granite vats, complete with holes for decanting or emptying out liquid, were found at Delos. Three granite blocks found close by might have been used as anvils to break the murex shells (Bruneau 1969). A pot found in an archaeological layer at a Phoenician site at Sarepta dating from the thirteenth century BC seems to have had a similar function. Large numbers of murex shells were found at the same site, the great majority being *H. trunculus*. Another find at the same site was a fragment of an amphora on which traces of purple sediment identified as 6, 6′-dibromoindigotin survived (McGovern and Michel 1984). A group of cylindrical stone vats used for purple dyeing were found at Tel Beit Mirsim in Israel (c. 700 BC). Some pots and basin shards were also found in Israel at Tel Keisan, Tel Siqmona, and Tel Kabri, in archaeological layers dating from the eleventh, ninth or eighth, and seventh centuries BC, respectively (Briend and Humbert 1980; Karmon and Spanier 1988; Koren 1995).

Fig. 33.3 Detail of a Roman bath mosaic depicting a murex (ca. late second to early first century BC)

Archaeological data from Tyre indicate that the snails were collected in large vats and left to decompose. This produced a hideous stench that was mentioned by several ancient authors (Aristotle, Vitruvius, and Pliny the Elder). Not much is known about the subsequent dyeing steps, and the actual ancient method for mass-producing the two murex dyes has not yet been successfully reconstructed. The special "blackish clotted blood" color, which was prized above all others, is believed to have been achieved by double-dipping the cloth, once in the indigo dye of *H. trunculus* and once in the purple-red dye of *Bolinus brandaris*. *H. trunculus* was

Fig. 33.4 Map of archaeology sites discovered as *purple*-dying centers in the ancient Mediterranean world

the dominant or only species in the earliest huge piles of shells that were found in the eastern Mediterranean and Carthage (Reese 1987, 2000; Wilson 2004).

Only short, often cryptic, references to molluscan purple were found in other Greek and Roman writings. Plutarch (1959) noted in his biography of Alexander the Great that the well-preserved textiles recovered by the conqueror at the Persian court of Susa had been dyed 190 years earlier at Hermione in Greece by using either honey or oil. The alchemical texts written in Egypt in the third century AD, in particular, Papyri Leidensis (Berthelot 1887) and Graecus Holmiensis, (Lagercrantz 1913), with probable roots in much earlier traditions (Reinking 1925), also mention specific materials and procedures that entered into the processing of purple dyes. But since a goal of the dyeing was to produce less-expensive substitutes for royal purple, these papyri likely describe plant- (Fig. 33.5) or insect-derived dyes rather than authentic molluscan royal purple.

Talmudic and later period descriptions of purple processing refer to marine animal dyes, probably including royal purple, that were processed by heating the "blood" of the animal (presumably, the hypobranchial glandular secretions). These imprecise recipes, including mixtures of urine, alumina, and fenugreek, or fermented barley-flour dough, were attempted unsuccessfully in modern studies (Herzog 1987).

On the basis of available scientific, historical, and textual knowledge, scholars have attempted to reconstruct the ancient Mediterranean industry from Pliny's account and the other ancient references with varying degrees of success. To assess current interpretations and direct future research, this account reviews the field from the perspective of modern chemistry (McGovern and Michel 1990).

The production of murex purple for the Byzantine court came to an abrupt end with the sack of Constantinople in 1453, this being the most critical episode of the Fourth Crusade. David Jacoby concludes that "neither a Byzantine emperor, nor any Latin ruler in former Byzantine territories could muster the financial resources required for the pursuit of murex purple production. On the other hand, murex fishing and dyeing with genuine purple are attested for Egypt in the tenth to thirteenth centuries" (Jacoby 2004). However, Jacoby finds that there is no mention of purple fishing, dyeing, or trade in the colorant in any Western source, even in the Frankish Levant. At about that time, the European West turned instead to vermilion, provided by the insect *Kermes vermilio* (Fig. 33.6) (Donkin 1977; Ziderman 1986a; Sandberg 1997; Amar et al. 2005).

Fig. 33.5 *Isatis tinctoria* – one of the plant dyes, including indigo, used instead of authentic mollusk royal purple

Fig. 33.6 *Left*: The female coccoid scale *Kermes echinatus* ("*tolaat shani*") with eggs, grown in Israel and used for dye production. *Right*: Dry Armenian insects also used for red dye

Dyeing Use

Since mollusks were a common food source, the unusual and intense coloration of their secretions and their suitability for dyeing was readily discovered and valued. Ancient peoples from around the globe, including the Phoenicians, Chinese, and Peruvians (Jackson 1917), discovered dyeing processes based on molluscan secretions.

Many other species within the Muricidae family, for example *Plicopurpura pansa* from the tropical eastern Pacific, and *Plicopurpura patula* (Linnaeus 1758) from the Caribbean zone of the western Atlantic, also produce a similar substance and this ability has sometimes been historically exploited by local inhabitants in the areas where these snails live. (Some other predatory gastropods, such as some wentletraps from the Epitoniidae family, also seem to produce a similar substance, although this has not been studied or exploited commercially.) The dog whelk *Nucella lapillus* from the North Atlantic can also be used to produce red-purple and violet dyes (Biggam 2006).

The sociopolitical and religious significance of royal purple was closely tied to its economic value. In some periods, it was worth as much as 10–20 times its weight in gold (Born 1937). This circumstance can be traced to the fact that the precursors of 6,6′-dibromoindigotin, which is converted to the dye in air and light, are found in nature only in the hypobranchial secretions of certain marine mollusks (Baker 1974). Paul Friedländer, the first man to elucidate the chemical structure of Tyrian purple (1909), required 12,000 *Murex brandaris* snails in order to obtain 1.4 g of the substance (Friedlaender 1909; Bruin 1970). Koren (2005) wrote that it should be possible to produce uniform purple dyeing with a ratio of 1 g wool: 7 medium size snails: 70 ml alkaline solution (Fig. 33.7).

Fig. 33.7 Wool dyed with *Murex tranculum* for *tzizit* in Israel ("Ptil Tekhelet" non-profit organization)

The Chemistry of Royal Purple

It has been suggested that, in nature, snails use their secretions as part of their predatory behavior and for their survival throughout all life stages. Until recently, the biosynthesis of indigoid colorants and their function in the animals were a matter of conjecture. A group of Australian biologists present strong evidence in support of the hypothesis of the *de novo* biosynthesis of brominated precursors from the dietary amino acid tryptophan by biosynthetic enzymes present in the hypobranchial gland of the purple-producing muricids. The production of indole derivatives in Muricidae should be seen as a naturally selected, multifaceted function of reproduction and defense: at an earlier stage they may play a role in reproduction and contribute to the antimicrobial defense and chemical ripening of the egg masses; at a later stage, they may also function as an antimicrobial defense mechanism in the immune system of adults. During feeding, the mercaptans released during photo-oxidative reactions may serve as an olfactory cue, directly deterring predator competition or indirectly masking prey alarm signals (Westley et al. 2006; Benkendorff 2008). The snail also secretes this substance when it is poked or physically attacked by humans. Therefore, the dye can be collected either by "milking" the snails, which is more labor-intensive but it is a renewable resource, or by collecting and then crushing the snails completely, which, of course, is destructive (Jacoby 2004).

Inside the hypobranchial gland of the snail, the precursors of the dye exist as clear watery liquid. When these are exposed to air and sunlight in the presence of the enzyme purpurase, which is also found in the gland, they turn into the dye. Purpurase quickly decomposes, so in order for this reaction to take place, the gland must be crushed soon after being taken from the live snail, in accordance with the Talmudic passage that the *tekhelet* is taken from the snail (=*Chilazon)* while still alive. The liquid from *M. trunculus* produces a mixture of dibromoindigo (purple) and indigo (Wolk and Frimer 2010). These molecules must be put into solution in order for them to bind tightly to textile. In this state, if dibromoindigo is exposed to ultraviolet light, it will transform to indigo, turning the trunculus mixture from purplish-blue to pure blue.

The main chemical constituent of the Tyrian dye was discovered by Paul Friedländer (1909) to be 6,6′-dibromoindigo (DBI) (Fig. 33.8) and The synthetic scheme of 6,6′-dibromoindirubin was investigated in details (Tanoue et al. 2007; Szalda et al. 2012). DBI was the dye obtained from one of the Mediterranean molluscan species, *Murex brandaris.* Later research by the same author (Friedländer 1922) showed that this compound was also a major component of the dyes from other Mediterranean species *(Murex trunculus*

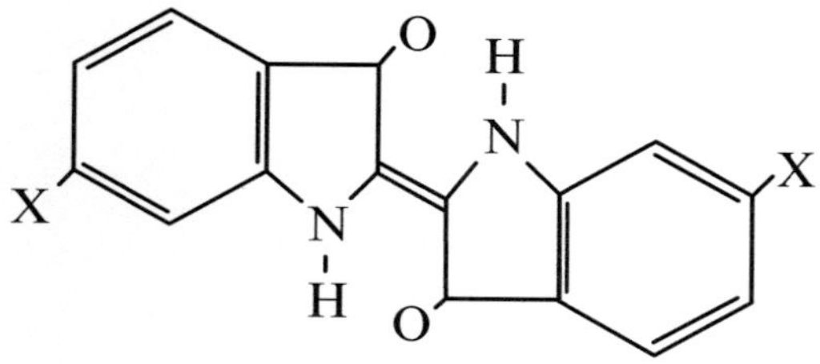

Fig. 33.8 The formula of organobromine compound of purple dye: dibromoindigotin and indigo

and *Purpura haemastoma),* as well as mollusk species occurring in other parts of the world (e.g., *Purpura patula* and *Nucella (Purpura) lapillus* from the Gulf of Mexico and the Atlantic Ocean, respectively). The dye was thus shown to be an organobromine compound (Friedlaender 1909). However, it has never been synthesized commercially (Cooksey 2001).

This dye is made up of mixed bromine-substituted indigo and indirubin isomers. *Hexaplus trunculus* (60 kg) was processed to yield indirubin (3.5 mg), 6′-bromoindirubin (5.5 mg), 6-bromoindirubin (2.8 mg), and 6,6′-dibromoindirubin (3 mg) (Meijer et al. 2003).

In 1998, through a lengthy trial-and-error process, an English engineer named John Edmonds rediscovered the secret of how to dye cloth with Tyrian purple (Edmonds 2000). He researched the recipes and observations of dyers from the fifteenth to the eighteenth centuries. He explored the biotechnology process behind the production of the blue "dyer's woad" from the annual Brassicaceae plant *Isatis tinctoria* by fermentation. After collaborating with a chemist, Edmonds hypothesized that an alkaline fermentation vat was necessary. He studied the above-mentioned ancient recipe for Tyrian purple as recorded by Pliny the Elder. By altering the percentage of sea salt in the dye vat and adding potash, he was able to successfully dye wool a deep purple color (Chenciner 2000).

The Dyeing Process

A synthetic purple pigment, if one can afford to make it, appears 'sterile' compared to the color produced from fresh murex. The snail contains in its body a secretion of a most-complex nature, carrying the precursor of a color. This color needs to be created and revealed, i.e., made visible with the help of a person, a dyer, an artist, someone who knows the method. For an artist, this is exactly the moment or the chance to discover a different approach, the possibility of new expression, to explore the unknown.

In order to bring to life the so-called royal purple, the snail has to die. The hypobranchial gland, which contains the precursor of the color, is found diagonally opposite the shell's natural opening. When the shell is broken and the gland exposed to air and light, the originally transparent mucus enters a metamorphosis from pus-like yellow to green to indigo blue and, finally, to one of the many violet nuances. It is shown that more dye is produced by larger snails; females produce more dye than males of the same size; individuals larger than 5 cm recover from milking in 7 days and smaller ones in 15–21 days (Rios-Jara et al. 1994).

If each animal is treated separately, the different purple nuances become clearly visible. This observation inspired a technique for working with murex in pictorial art. A snail was opened with the help of a small hammer, and then placed on a support. Old bed linen from past centuries with their worn, soft fibers was extremely suitable to catch the seawater-drenched mucus. The "juice" soaked the surface in a spectacular way: large patches, with strong blue or violet in the center where the gland touched, appeared on the textile. Similar to the batik technique, melted wax was applied beforehand to control the running of the watery substance. This provided a means for keeping all purple stains together, as wanted for the composition. On one occasion, hundreds of cracked-open murexes were placed along both sides of the seam of particularly worn, heavy linen. The entire watercolor tended to strive for the outer edges away from the central seam, thus forming an oval symmetrical pattern.

The artist is the 'chef d'orchestre', directing the snails, supervising the time they need to leave their mark on cloth or paper. Even coffee filter paper has proven to be a good means of catching every single colored molecule and, once opened, is reminiscent of a beautiful fan.

Economy

The economic importance of purple-producing mollusks during all of antiquity was such that it was inevitable that governments would seek to derive some financial benefit from it. At the end of the fourth century BC, the tax on murex fishing imposed by the hyerops (priest administrators) of the sanctuary on the island of Delos brought in as much as 650 drachmas in a good year (Koren 2005). In the time of the late Roman Empire, the harvesting of purple-producing mollusks became a state monopoly. The murex fishermen had to belong to guilds governed by very strict rules. They were obliged to supply a fixed quantity of mollusks, and failure to do so resulted in the forfeiture of their property. A fisherman was only allowed to leave his trade if he was prepared to renounce any right to his goods and could find someone to replace him. Anyone taking over ownership of the property of a *murilegulus* having difficulties was obliged to join the guild and provide the state with any outstanding quantity of mollusks owned by the seller (Spanier 1987).

The mollusks with the purple-dye precursors were probably also a source of food in each region. The Mediterranean species, for example, are still a great delicacy in France and Italy, and the Chinese are renowned for exploiting every food source in their environment.

The artisanal muricid fishery is an important local fishing industry because the banded murex (*Hexaplex trunculus*) and the purple-dye murex (*Bolinus brandaris*) are greatly popular seafoods, with high commercial value in the Portuguese seafood market.

Gastropods represent around 2 % of the marine mollusks fished worldwide, but some species play significant social roles in small-scale artisanal fisheries.

In Europe, France, the United Kingdom and Ireland account for over 90 % of the European gastropod catches, mainly composed of the common periwinkle (*Littorina littorea*) and the whelk (*Buccinum undatum*) (Leiva and Castilla 2002).

The banded murex, *Hexaplex trunculus* (Linnaeus 1758), is mainly distributed in the Mediterranean Sea, but can also be found in the adjacent Atlantic Ocean, from the Portuguese coast southward to Morocco and to the Madeira and Canary Archipelagos (Poppe and Goto 1991; Houart 2001). The purple-dye murex *Bolinus brandaris* (Linnaeus 1758) is very common throughout the Mediterranean Sea, whereas its distributional range in the Atlantic Sea is limited to Portugal and Morocco (Poppe and Goto 1991; Houart 2001).

Historical sources such as Aristotle and Pliny described a method of catching muricid gastropods (namely *H. trunculus* and *B. brandaris*) using baited wicker baskets for the subsequent extraction of the purple dye (Spanier and Karmon 1987). Presently, both *H. trunculus* and *B. brandaris* are commercially valuable species in several Mediterranean countries (Poppe and Goto 1991; Houart 2001), and are traditionally fished for human consumption using diverse types of artisanal fishing gears (e.g., pots, basket traps, dredges, scoop nets, trawl nets, gill nets, and trammel nets) (Gaillard 1987; Martin et al. 1995) (Fig. 33.9). The banded murex is regularly or occasionally fished/harvested in Italy, Cyprus, Turkey (Gaillard 1987), Croatia (Peharda and Morton 2006), and Tunisia (Gharsallah et al. 2004), while the purple-dye murex is fished for seafood in France (Bartolome 1985), Italy (Ramon and Amor 2001), and, occasionally, Tunisia and Turkey (Ramon and Flos 2001). In Iberian waters, *H. trunculus* and *B. brandaris* are commercially exploited by artisanal fisheries both in Spain, mostly along the Mediterranean and Atlantic coasts of Andalucia (Martin et al. 1995; Anon. 2001; Ramon and Amor 2001; Tirado et al. 2002), and in Portugal, along the Algarve coast, mainly in the Ria Formosa lagoon (Muzavor and Morenito 1999; Carneiro et al. 2006).

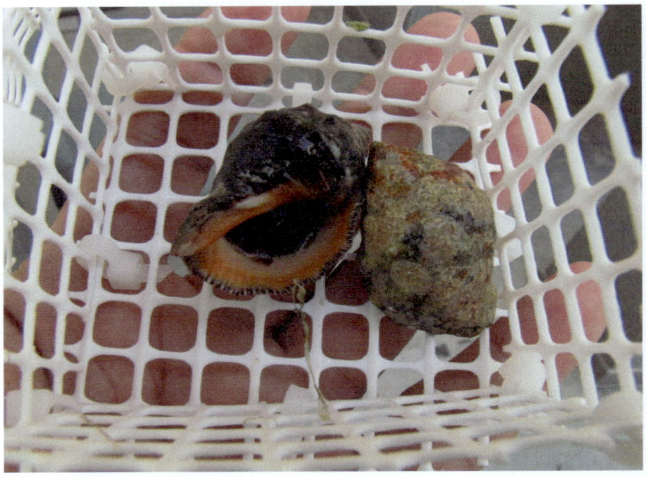

Fig. 33.9 Method of catching murex using baited wicker baskets for the subsequent extraction of the purple dye

In the Ria Formosa, the artisanal fishery targeting these muricid gastropods is an important local industry because the banded murex and purple-dye murex are greatly popular seafoods with high commercial value on the market (reaching values of around 10–15 €/kg and 20 €/kg for first sale, respectively).

In addition to hand-harvesting in intertidal areas, the snails are traditionally caught with artisanal fishing gear, locally known as the "wallet-line", considered illegal fishing gear according to the fishery regulations for lagoons (DR 1990). Due to their easy capture and high fishing yield, the massive aggregations of *H. trunculus* and/or *B. brandaris* females during collective spawning are also subjected to hand-harvesting by recreational users and professional fishermen during low tide (Vasconcelos et al. 2008), hand-harvesting by scuba divers operating illegally in the lagoon, and fishing by professional fishermen using prohibited entangling nets (Muzavor and Morenito 1999), amounting to a considerable total catch of these species.

Probably due to overfishing, detrimental harvesting practices, and a disregard for the minimum landing size stipulated for these species, the abundance of *H. trunculus* and *B. brandaris* in the Ria Formosa has been decreasing for more than a decade, possibly indicating overexploitation of fishing resources.

Other Uses

Roméo et al. (2006) and Gharbi-Bouraoui et al. (2008) studied the biomarker responses of *Murex trunculus* exposed to selected pollutant metals (cadmium and copper) and organic compounds (carbofuran and lindane) under laboratory conditions. They suggested that *M. trunculus* could be an interesting sentinel species for monitoring pollution. This species is found in both relatively clean and polluted areas, and is easy to find and sample. They are predatory animals that feed on bivalves (including clams and mussels).

Tyrian purple precursors in the egg masses of the Australian muricid *Dicathais orbita* contain a high proportion of tyrindoleninone, which reacts to form tyriverdin and, subsequently, Tyrian purple and 6-bromoisatin as the eggs develop and the larvae hatch. Tyrindoleninone and especially tyriverdin exhibit antimicrobial activity (Benkendorff et al. 2000).

Glover (1995) illustrates in his article the feasibility and/or desirability of using natural dyes, including Tyrian purple and cochineal, as compared to the alternative use of synthetic products.

Purple dye was found on powdered material at Akrotiri. It was probably used for cosmetics (Aloupi et al. 1990).

Violet–purple residues collected from a Gallo–Roman burial dated back to the second half of the third century AD and excavated in Naintré, France, were chemically investigated and dibromoindigo was found. These results are the earliest chemical evidence of purple colorant used during funeral rituals (not as textile dye) and enabled us to highlight new aspects of funeral practices in Roman times (Devièse et al. 2011).

In Biblical Hebrew, the dye extracted from the *Bolinus brandaris*, violet-red purple, was known as *argaman* (ארגמן). Another dye extracted from a related sea snail, *Hexaplex trunculus*, produced an indigo color (blue-purple) called *tekhelet* (תְּכֵלֶת), which was used in garments worn for ritual purposes (Elsner 1992). Biblical texts (Exod. 26:1, 31; 2844; 391, 28–29. 1 Kings 5:L-12; 7:13–14; 9112–14, 26–28; 1011, 22. 2 Chron. 2:7, 14; 314. Ezek. 27:7, 16, 24) incorporating late-Bronze and Iron-Age traditions, prescribed that the tabernacle curtains and the high priest's ritual vestments were to be dyed with royal purple.

These two dyes are invariably mentioned together, along with the kermes (*tola'at shani*), as the three dyes prescribed for hangings and other textiles to be used in the Temple. Most importantly, the hanging separating the 'Holy' from the 'Holy of Holies', as well as the priests' vestments, were to be dyed with all three colors. In contrast, the high priest's cloak, the bands that attached the pectoral to the *ephod* (ceremonial belt), and the diadem, were to be dyed solely with blue-purple (Exodus 39: 21, 22, 31).

God tells Moses "to speak to the children of Israel and tell them that they make for themselves a tassel (*tzitzit*) on the corners of their garments for generations, and they shall put on a corner tassel a twisted thread of *tekhelet*" (Fig. 33.10) (Numbers XV; 13).

Jews are commanded to place a thread of blue on their *tzitzit* as a constant and conspicuous reminder of their stature, i.e., that they are noble sons of the King of the Universe, always pursuing His *mitzvot*.

Fig. 33.10 A tassel (*tzitzit*) dyed with *Murex tranculum* on corners of garment

Fig. 33.11 A squid with hypobranchial dye glands

The Biblical commandment to wear *tzitzit* is still observed today by Orthodox Jews, but the prominent blue thread has all but been forgotten. What did remain are passages in the Talmud describing the source of the blue dye – a snail known as the *Chilazon*, defined in modern Hebrew as snail. This marine creature had a shell, could be found along the northern coast of Israel, and its body was "similar to the sea." The dye's color was "similar to the sky and sea," it was steadfast, extracted from the snail while still alive, and was indistinguishable from a dye of vegetable origin called *kala ilan* (indigo) (Fig. 33.5), which was considered inferior and unfit for ritual purposes (McGovern 1990).

There were a few steps in the search for blue:

Rabbi Leiner from Radzyn discovered the blue in Italy. He believed that squid was the dye source for *tekhelet*. He wrote three books about this, and his Hassidim produced their *tzitzit* from this dye and wore it (Fig. 33.11).

In 1913, the Chief Rabbi of Ireland, Rabbi Isaac Herzog (later Chief Rabbi of Israel), wrote a doctoral dissertation on the subject of *tekhelet* in Israel. He sent samples of the Radzyn *tekhelet* to chemists and dye experts for analysis. The dye was found to be inorganic – a synthetically manufactured color known as Prussian blue (Spanier 1990).

At first Rabbi Herzog thought that the common purple sea snail *Jantina jantina* was the appropriate snail for *tekhelet* (Fig. 33.12). Then he realized that all evidence pointed to *Murex trunculus* as the most likely candidate to be the source of *tekhelet*. Two problems, however, prevented Rabbi Herzog from positively identifying that snail as the one associated with the Talmudic *chilazon*. First, the dye obtained from the *trunculus* was purplish-blue, not pure blue, as tradition maintained. Second, this snail has an off-white shell with brown stripes, hardly fitting the Talmudic description of the *Chilazon* as appearing similar to the sea.

Fig. 33.12 The common purple sea snail *Jantina jantina*, thought to be the appropriate snail for *tekhelet*

Current research (Ziderman 1986b) has supplied solutions to these objections. The shell appears off-white with brown stripes when it is out of the water, cleaned, and polished. However, in its natural element, trunculus is covered with a coat of sea fouling that is the color of the ocean (Fig. 33.13). Everything in its vicinity is covered with the same fouling, so that it is almost impossible to distinguish the snail from the sea bed on which it is found. The Talmud's description is of the *Chilazon* in its natural habitat!

The riddle of producing a pure blue color from the snail was serendipitously solved. While researching the methods used by the ancient dyers, it was found that at a certain stage of the dyeing process, exposure to sunlight will alter the dye, changing its color from purple to blue. To the dye masters of old, working in the bright Mediterranean sunlight, this was certainly no secret (Elsner 1992).

The evidence for identifying the *Murex trunculus* as the source of *tekhelet* now seems decisive, and goes beyond merely fitting the general descriptions of the *Chilazon* as found in the Talmud (Ziderman 2008):

1. The Jerusalem Talmud (as quoted by the Raavyah) translates *tekhelet* as *porphiron* (the Latin and Greek name for *trunculus*-like shells). Pliny and Aristotle described these shells as the source of the ancient dyes.
2. The Talmud indicates that true *tekhelet* is indistinguishable from the blue dye of vegetable origin – *kala ilan* (indigo). The dye ultimately derived from *trunculus* is molecularly equivalent to indigo.
3. Extensive marine biological surveys have revealed that the only snails in the Mediterranean that produce stable dyes are those of the murex family. The dye obtained from *trunculus* is very stable and steadfast, which accords with the Rabbinical description of *tekhelet*.

Fig. 33.13 *Murex haemastoma*: cleaned and polished (*left*) against murex is covered with a coat of sea fouling the color of the ocean

4. Archaeologists in Tyre and elsewhere uncovered mounds of murex shells dating from the Biblical period that were broken in the exact spot necessary to obtain the dyestuff. Chemical analyses of blue stains on vats from 1200 BCE revealed patterns consistent with those of modern-day *trunculus*.
5. When listing the precious commodities used in building the *Mishkan* (tabernacle), the Torah consistently included *tekhelet al*ong with gold, silver, and other familiar materials, recognized by all for their worth. The prophet Yechezkel spoke of the *tekhelet* from Tyre and the "Isles of Elisha", and the *Megillah* tells us that in Persia, Mordechai wore royal clothes made of *tekhelet*. Surely, the Torah is referring to that same valuable dye commonly used by royalty throughout the rest of the ancient world.

It is noteworthy that the purple dye was even the basis of an ancient proverb, still in use: טלית שכולה תכלת (*Talit shekula tekhelet*), literally meaning an entirely purple *Talit*. Since the purple was so extremely precious, one thread was considered sufficient for the ritual vest, the *Talit*. Hence, the proverb is used to describe a person who considers himself more virtuous than his peers.

References

Aloupi E, Maniatis Y, Paradellis T, Karali-Yannacopoulou L (1990) Analysis of a purple material found at Akrotiri. In: Hardy DA (ed) Thera and the Aegean World III, vol I: archaeology. Thera Foundation, London

Amar Z, Gottlieb H, Varshavsky L, Iluz D (2005) The scarlet dye of the Holy Land. BioScience 55:1080–1083

Anon (2001) Especies de interés pesquero en el litoral de Andalucía. Vol II. Invertebrados. Junta de Andalucía. Consejería de Agricultura y Pesca, Sevilla

Aristotle (2004) History of animals, Book V (trans: D'Arcy Wentworth Thompson). Kessinger Publishing, Whitefish

Baker JT (1974) Tyrian purple: ancient dye, a modern problem. Endeavor 33:11–17

Bartolome C (1985) Contribution à l'étude du gastéropode *Murex brandaris* (Linnaeus, 1758) dans le Golfe du Lion. Dissertation, Université des Sciences et Techniques du Languedoc

Benkendorff K (2008) Bioactive molluscan resources and their conservation: biological and chemical studies on the egg masses of marine molluscs. http://web.archive.org/web/20070830143907/http://www.library.uow.edu.au/adt-NWU/public/adt-NWU20011204.154039/index.html. Accessed 10 Nov 2012

Benkendorff K, Bremner JB, Davis AR (2000) Tyrian purple precursors in the egg masses of the Australian muricid, *Dicathais orbita*: a possible defensive role. J Chem Ecol 26:1037–1050

Berthelot MPE (1887) Collection des anciens alchimistes grecs. G. Steinheil, Paris

Biggam CP (2006) Whelks and purple dye in Anglo-Saxon England. In: Ridout Sharpe J (ed) The Archaeo+Malacology Group newsletter. Archaeomalacology, Oxfordshire

Boesken Kanold I (2005) The purple fermentation vat: dyeing or painting parchment with Murex trunculus. In: Cooksey C, van Bommel MR (eds) Dyes in history and archaeology. Archetype Books, London

Born W (1937) Purple. Ciba Rev 1:106–111

Briend J, Humbert J-B (1980) Tell Keisan (1971–1976): Une cité phénicienne en Galilée. Éditions Universitaires Friebourg Suisse, Paris

Bruin F (1970) Royal purple and the dye industries of the Mycenaeans and Phoenicians. In: Mollat M (ed) Sociétés et compagnes de commerce en Orient et dans L'Océan Indien. Bibliothèque Générale de l'École Pratique, Paris

Bruneau P (1969) Documents sur l'industrie delienne de la pourpre. Bull Corresp Hellénique 93:759–791

Brunello F (1973) The art of dyeing in the history of mankind. Neri Pozza Editore, Venice

Carneiro M, Martins R, Rebordão FR (2006) Contribuição para o conhecimento das artes de pesca utilizadas no Algarve, 13. Publicações Avulsas do IPIMAR, Lisboa

Cazzella A, Moscoloni M (1998) Coppa Nevigata: un insediamento fortificato dell'età del Bronzo. In: Drago Troccoli L (ed) Scavi e ricerche archeologiche dell'Università di Roma "La Sapienza". "L'Erma" di Bretscneider, Roma

Chenciner R (2000) Madder red: a history of luxury and trade: plant dyes and pigments in world commerce and art. Curzon Press, Richmond

Cooksey C (1994) Bibliography of Tyrian (shellfish) purple. Dyes Hist Archaeol 12:57–65

Cooksey CJ (2001) Tyrian purple: 6,6'-dibromoindigo and related compounds. Molecules 6:736–769

D R (1990) Regulamento da pesca na Ria Formosa. Portaria No. 560/90 de 19 de Julho, Dário da República, Iª Série, No. 165

Devièse T, Ribechini E, Baraldi P, Farago-Szekeres B, Duday H, Regert M, Colombini MP (2011) First chemical evidence of royal purple as a material used for funeral treatment discovered in a Gallo–Roman burial (Naintré, France, third century AD). Anal Bioanal Chem 401:1739–1748

Donkin RA (1977) The insect dyes of western and west-central Asia. Anthropos 72:847–880

Doumet J (1980) A study on ancient purple. Imprimerie Catholique, Beirut

Edmonds J (2000) Tyrian or imperial purple dye: the mystery of imperial purple dye. John Edmonds, High Wycombe

Elsner O (1992) Solution of the enigmas of dyeing Tyrian purple and the Biblical *tekhelet*. Dyes Hist Archaeol 10:11–16

Fantar MH (1986) Kerkouane, cite punique du Cap Bon (Tunisie), vol 3. Institut National d'Archeologie et d'Art, Tunis

Friedlaender P (1909) Zur kenntnis des farbstoffes des antiken purpurs aus *Murex brandaris*. Monatsh Chem 30:247–253

Friedländer P (1922) Über die farbstoffe aus purpura aperta und purpura lapillus. Ber Dtsch Chem Ges 55:1655–1658

Gaillard JM (1987) Gastéropodes. In: Fischer W, Bauchot ML, Schneider M (eds) Fiches FAO d'identification des espèces pour les besoins de la pêche. Vol I – Méditerranée et Mer Noire. FAO, Rome

Gharbi-Bouraoui S, Gnassia-Barelli M, Roméo M, Dellali M, Aïssa P (2008) Field study of metal concentrations and biomarker responses in the neogastropod, *Murex trunculus*, from Bizerta Lagoon (Tunisia). Aquat Living Resour 21:213–220

Gharsallah IH, Zamouri-Langar N, Missaoui H, El Abed A (2004) Étude de la croissance relative et de la biomasse d'*Hexaplex trunculus* dans la lagune de Bizerte. Bull Soc Zool Fr 129:427–436

Glover B (1995) Are natural colorants good for your health – are synthetic ones better. Text Chem Color 27:17–20

Haubrichs R (2004) L'etude de la pourpre: Histoire d'une couleur, chimie et experimentations. Preist Alp 20:133–160

Herzog I (1987) The royal purple and the Biblical blue *srgaman* and *tekhelet*. In: Spanier E (ed) The study of Chief Rabbi Dr. Isaac Herzog on the dye industries of ancient Israel and recent scientific contributions. Keter Publishing House, Jerusalem

Houart R (2001) A review of the recent Mediterranean and Northeastern Atlantic species of Muricidae. Edizioni Evolver, Rome

Jackson JW (1917) Shells as evidence of the migrations of early culture. Manchester University Press, London

Jacoby D (2004) Silk economics and cross-cultural artistic interaction: Byzantium, the Muslim world, and the Christian west. DOP 58:197–240

Jensen LB (1963) Royal purple of Tyre. J Near East Stud 22:104–118

Karmon N, Spanier E (1988) Remains of a purple dye industry found at Tel Shiqmona. Isr Explor J 38:184–186

Koren ZC (1995) High-performance liquid chromatographic analysis of an ancient Tyrian purple dyeing vat from Israel. Isr J Chem 35:117–124

Koren ZC (2005) The first optimal all-Murex all-natural purple dyeing in the Eastern Mediterranean in a millennium and a half. Dyes Hist Archaeol 20:136–149

Lagercrantz O (1913) Papyrus graecus holmiensis (P. Holm.). Recepte für silber, steine und purpur. Almquist and Wiksells, Uppsala

Lammens C, Verhecken A (1984) 'Het Molluskenpurple' and 'Lets over de Chimie van het Molluskenpurple'. Gloria Maris 23:147–175

Leiva GE, Castilla JC (2002) A review of the world marine gastropod fishery: evolution of the catches, management and the Chilean experience. Rev Fish Biol Fish 11:283–300

Linnaeus C (1758) Systema naturae per regna tria naturae, secundum classes, ordines, genera, species, cum characteribus, differentiis, synonymis, locis. Tomus I. Editio decima, reformata. Laurentii Salvii, Holmiae

Martin P, Sanchez P, Ramon M (1995) Population structure and exploitation of *Bolinus brandaris* (Mollusca: Gastropoda) off the Catalan coast (northwestern Mediterranean). Fish Res 23:319–331

McGovern PE (1990) The royal purple and the Biblical blue (*argaman* and *tekhelet*) – the study of Chief Rabbi Dr. Isaac Herzog on the dye industries in ancient Israel and recent scientific contributions. ISIS 81:563–565

McGovern PE, Michel RH (1984) Royal purple and the pre-Phoenician dye industry of Lebanon. MASCA J 3:67–70

McGovern PE, Michel RH (1985) Royal purple dye: tracing the chemical origins of the industry. Anal Chem 57:1514A–1522A

McGovern PE, Michel RH (1990) Royal purple dye: the chemical reconstruction of the ancient Mediterranean industry. Acc Chem Res 23:152–158

Meijer L, Skaltsounis A-K, Magiatis P, Polychronopoulos P, Knockaert M, Leost M, Ryan XP, Vonica CA, Brivanlou A, Dajani R, Crovace C, Tarricone C, Musacchio A, Roe SM, Pearl L, Greengard P (2003) GSK-3-selective inhibitors derived from Tyrian purple indirubins. Chem Biol 10:1255–1266

Muzavor S, Morenito PM (1999) Roteiro ecologico da Ria Formosa, vol IV – Moluscos gastropodos. University of Algarve Press, Faro

Oppenheim AL, Brill RH, Barag D, von Saldern A (1970) Glass and glassmaking in ancient Mesopotamia. Corning Museum of Glass, Corning

Peharda M, Morton B (2006) Experimental prey species preferences of *Hexaplex trunculus* (Gastropoda: Muricidae) and predator–prey interactions with the black mussel *Mytilus galloprovincialis* (Bivalvia: Mytilidae). Mar Biol 148:1011–1019

Pliny the Elder (1855) The natural history. Taylor & Francis, London

Plutarch (1959) Lives of the noble Greeks. Dell, New York

Pollux J (1846) Onomasticon, Book 1. Friedrich Nicolai, Berlin

Poppe GT, Goto Y (1991) European seashells, vol 1 Polyplacophora, Claudofoveata, Solenogastra, Gastropoda. Verlag Christa Hemmen, Wiesbaden

Pritchard JB (1978) Recovering Sarepta, A Phoenician city. Princeton University, Princeton

Radwin G, D'Attilio A (1976) Murex shells of the world. Stanford University Press, Stanford

Radwin GE, D'Attilio A (1986) Murex shells of the world. An illustrated guide to the Muricidae. Stanford University Press, Stanford

Ramon M, Amor MJ (2001) Increasing imposex in populations of *Bolinus brandaris* (Gastropoda: Muricidae) in the north-western Mediterranean. Mar Environ Res 52:463–475

Ramon M, Flos R (2001) First trials to cultivate the muricid gastropod *Bolinus brandaris* (Linnaeus). Eur Aquac Soc Spec Publ 29:219–220

Reese DS (1987) Palaikastro shells and Bronze Age purple-dye production in the Mediterranean Basin. Annu Br Sch Archaeol Athens 82:201–206

Reese DS (2000) Iron Age shell purple-dye production in the Aegean. In: Shaw JW, Shaw MC (eds) Kommos: an excavation on the south coast of Crete, vol IV: The Greek sanctuary, Part 1. Princeton University Press, Princeton

Reinhold M (1970) History of purple as a status symbol in antiquity. Latomus collection, vol 116. Latomus, Brussels

Reinking K (1925) Ueber die älteste beschreibung der küpenfärberei im papyrus graecus holmiensis. Mell Textilber 5:349–351

Rios-Jara E, León HG, Lizárraga-Chávez L, Michel-Morfín JE (1994) Produccíon y tiempo de recuperacíon del tinte de Plicopurpura patula pansa (Neogastropoda: Muricidae) en Jalisco, México. Rev Biol Trop 42:537–545

Roméo M, Gharbi-Bouraoui S, Gnassia-Barelli M, Dellali M, Aïssa P (2006) Responses of *Hexaplex* (*Murex*) *trunculus* to selected pollutants. Sci Total Environ 359:135–144

Sandberg G (1997) The red dyes: cochineal, madder and murex purple. Lark Books Press, New York

Shimoyama S, Noda Y (1993) Non-destructive determination of natural dyestuffs used for ancient coloured cloths using a three-dimensional fluorescence spectrum technique. Dyes Hist Archaeol 12:45–56

Spanier E (ed) (1987) The royal purple and the Biblical blue. *Argaman* and *tekhelet*. Keter Publishing House, Jerusalem

Spanier E (1990) Rediscovering royal purple and Biblical blue. Oceanus 33:75

Spanier E, Karmon N (1987) Muricid snails and the ancient dye industries. In: Spanier E (ed) The royal purple and the Biblical blue: *Argaman* and *tekhelet*. Keter Publishing House, Jerusalem

Stieglitz RR (1994) The Minoan origin of Tyrian purple. Biblic Archaeol 57:46–54

Szalda DJ, Ramig K, Lavinda O, Koren ZC, Massa L (2012) 6-bromoindigo dye. Acta Crystallogr C68:160–163

Tanoue Y, Hara A, Kai N, Sakata K, Hashimoto M, Nagai T (2007) A revised synthetic scheme of 6,6'-dibromoindirubin. J Heterocycl Chem 39:1135–1138

Tirado C, Rodríguez de la Rúa A, Bruzón MA, López JI, Salas C, Márquez I (2002) La reproducción de bivalvos y gasterópodos de interés pesquero en Andalucía. Consejería de Agricultura y Pesca, Huelva

Vasconcelos P, Carvalho S, Margarida Castro M, Gaspar MB (2008) The artisanal fishery for muricid gastropods (banded murex and purple dye murex) in the Ria Formosa lagoon (Algarve coast, southern Portugal). Sci Mar 72:287–298

Verhecken A (1989) Dyeing with kermes is still alive. J Soc Dye Colour 105:389–391

Westley CB, Vine KL, Benkendorff K (2006) A proposed functional role for indole derivatives in reproduction and defense of the Muricidae (Neogastropoda: Mollusca). In: Meijer L, Guyard N, Skaltsounis AL, Eisenbrand G (eds) Indirubin, the red shade of indigo. Life in Progress, Roscoff

Wilson AI (2004) Archaeological evidence for textile production and dyeing in Roman North Africa. In: Alfaro C, Wild JP, Costa B (eds) Purpureae vestes. Textiles y tintes del Mediterráneo en época romana. Publicaciones de la Universidad de Valencia, Valencia

Wolk JL, Frimer AA (2010) Preparation of Tyrian purple (6,6'-dibromoindigo): past and present. Molecules 15:5473–5508

Ziderman II (1986a) Biblical dyes of animal origin. Chem Britain 22:419–421

Ziderman II (1986b) Purple dye made from shellfish in antiquity. Rev Prog Coloration 16:46–52

Ziderman II (2008) The biblical dye *tekhelet* and its use in Jewish textiles. Dyes Hist Archaeol 21:36–44

Socioeconomic Aspects: Human Migrations, Tourism and Fisheries

34

Dafna M. DiSegni and Mordechai Shechter

Abstract

In the present chapter we qualitatively survey the impacts of changes in Mediterranean Ecosystem on human wellbeing via three economic indicators: the fishery sector, tourism industry and human migration. The indicators selected cover impacts on different dimensions of human being needs: labor and welfare, living, and leisure.

Keywords

Economics of climate change • Commercial fisheries • Climate migration • Climate and tourism

Abbreviations

GDP Gross Domestic Production
DC Direct Cost

Since the beginning of the 1990s, increasing number of studies have focused on worldwide phenomena of climate change likely to change characteristics of oceans and seas and the way these are likely to impact economies lying at their shores. For many parts of the world there is considerable uncertainty about the value of marine, land and capital endowments threatened by climate change. This in turn generates uncertainty about the level of protection, adaptation and overall mitigation enacted by countries and their economies. We surveyed these studies in order to qualitatively characterize the impact of climate change on human wellbeing via three economic indicators: the fishery sector, tourism industry and human migration.[1] The indicators selected cover impacts on different dimensions of human being needs: labor and welfare, living, and leisure.

The indicators selected also fulfill two important macroeconomic features: First, they generate significant direct impacts on the Gross Domestic Production (GDP) in economies within the Mediterranean region, either in specific sectors (fishery and tourism), or across sectors (migration); second, they are considered highly sensitive to climate in general, and the impacts of climate change in particular. The sources of the impacts of climate change on economies are mostly due to (a) sea level rise and storms that lead to higher risk of inundation and flooding, resulting in shoreline erosion and realignment, and leading to loss of amenity or damage to assets (natural and man-made); (b) warmer ocean temperatures, with potential impact on biodiversity, through

[1] We do not provide a quantitative survey of the impacts. Direct-cost estimates found in the literature commonly used to measure the economic damages of specific phenomenon (for example, sea level rise). We believe such estimates are important for the purpose of a site specific analysis. However, such estimates suffer from three key limitations that bound our understanding as of the impact on human being at large: (i) values of threatened endowments are not well-known; (ii) loss of endowments often does not include the effect on consumer prices, and (iii) in studying mitigation at the international level international trade is often overlooked.

D.M. DiSegni
Berglas School of Economics, Tel Aviv University, Tel Aviv, Israel
e-mail: deshel@post.tau.ac.il

M. Shechter (✉)
Natural Resource and Environmental Research Center, University of Haifa, and School of Sustainability, Interdisciplinary Center (IDC), Herzlyia, Israel
e-mail: Mordechai@idc.ac.il

S. Goffredo and Z. Dubinsky (eds.), *The Mediterranean Sea: Its history and present challenges*,
DOI 10.1007/978-94-007-6704-1_34, © Springer Science+Business Media Dordrecht 2014

effects on the distribution and reproductive patterns of marine organisms and, consequently, food; (c) ocean acidification: increased CO_2 concentration in sea water is altering ocean chemistry, making it more difficult for calcite organisms, such as coccolithophores, corals and mollusks, to grow and function; (d) decreased rainfall and drought: warmer temperatures will cause greater evaporation, increase the severity of drought for a given decrease in rainfall; and, finally, (e) ocean stability and currents: changes to ocean currents may affect dispersal and distribution patterns of marine organisms (Hobday and Matear 2005; Hobday et al. 2006; Hennessy et al. 2007).

In this chapter we do not attempt to survey all results in the field related to marine economics and climate change, but offer a *qualitative* and speculative assessment of their relative importance. In the sections that follows we summarize key litterateurs related to the impacts on the fishery sector, tourism and migrations, accordingly. Mostly in the Mediterranean basin, due to climate change.

Coastal Commercial Fisheries and Aquaculture

Costal fisheries and aquaculture are highly important to many of the countries surrounding the Mediterranean Sea, and strongly sensitive to changes in the marine environment. Fishery makes a major contribution to the human food supply and development. During the last three decades, fish catch rose to 30–40 % of worldwide total international trade, and to 20 % of international trade among Mediterranean countries. Total production of fish in the Mediterranean marine and inland aquaculture catch reached 100,000 million metric tons in 2000. It was lower than that for several years thereafter, prompting speculations about the collapse of the fisheries due to overall changes in the marine environment, principally due to over-fishing and climate change.

Climate Change is considered to have adverse effects on Mediterranean mariculture, which mainly includes production of sea bass (*Dicentrarchus labrax*), sea bream (*Sparus aurata*), mussels (*Mytilus galloprovincialis*) and oysters (*Ostrea edulis*). The impact of climate change on the fishery sector is evaluated on the basis of volume of consumption of fish and shellfish at the regional markets, and on changes in the market price of fishing harvests. Sometimes it is measured in terms of the impact on fishing fleets, employment in the sector, and changes in development policies.

The impacts affecting aquatic systems and fisheries in general are mainly the rising of sea levels, increasing acidity of marine waters, increasing global temperature. These impacts have been analyzed comprehensively in two recent reviews commissioned by the Australian Greenhouse Office (Hobday et al. 2006; Hobday and Matear 2005) the study by Roessig et. al. 2004, and Cooley and Doney study on the impact of ocean acidification on US commercial fishery (Cooley and Doney 2009). According to these studies the impact of species such as the freshwater fish carp may diminish because of reduced spawning opportunities in floodplains. For other exotics, such as the marine alga caulerpa, the distribution may increase, due to more suitable conditions in estuaries. Oyster farmers might be affected as the nutrient supply shifts away from estuaries to near-shore coastal waters, as a side-effect of climate change. Ocean acidification and declining carbonate concentration could directly damage organisms, specifically corals and mollusks, by decreasing calcification rates. Reduced calcification is observed in response to rising CO_2 and declining carbonate concentration even in waters that are thermodynamically supersaturated for calcium carbonate. Many organisms, some commercially valuable, also exhibit a range of negative consequences for metabolism, reproduction, development, intracellular chemistry, and immunity (Fabry et al. 2008; Burgents et al. 2005). Ocean acidification's impact is not yet known for every commercially and recreationally valuable species, but emerging data suggest that the number or quality of many high-value, aragonite-forming mollusks could decrease, and declining economic revenues in that fishery sector may follow. This possibility is supported by findings such as decreased mollusk populations in acidified ecosystems (Wootton et al. 2008; Hall-Spencer et al. 2008).

If economic viability is reduced as predicted for impacts resulting from climate change, fisheries resource harvest strategies will affect social wellbeing, often negatively. Harvest strategies change because of changes in the recruitment patterns for fish, crustaceans and mollusks, which occur due to changes in physical habitats and ecological processes in estuaries and the coastal environment. In such cases, structural adjustment and adaptive management of fisheries industries may be necessary.

Xiaozi and Heino (2010) have offered a model to simulate how climate warming induced stock displacement affects international fishery management based on annually updated harvest policies. According to this study, with the moderation of winter water temperature in the past two decades, invaders native to warmer waters are increasingly able to establish and persist. Thermopiles species have been arriving for much of the twentieth century yet few established outside the Levant until the 1980s and 1990s, when the current trend began in the Mediterranean. Galil and Zenetos (2002) has proposed that the rising sea temperature has favoured the thermophilic aliens' reproduction, growth, and survival, and provided them with a distinct advantage over native temperate Mediterranean taxa.

The influx of the invasive medusa is already teetering fisheries through proliferation of noxious and poisonous species, displacement of commercially-important native species, and the alteration of the food web. Nakar et al. (2011) have estimated the impact of invasive medusa (*Rhopilema nomadic Ctenophore, Mnemiopsis Leidyi, and A. Agassiz*) on the Israeli fishery sector. There, the authors have estimated an annual loss of profits of approximately 3 % for trawl fishing and 40 % for trammel and gill net fishing. The difference in the impact of jellyfish on the two distinct fishing techniques is explained by the significantly higher impact on days of work cancellation in trammel and gill net fishing relative to cancellations in trawl fishing as well as a greater negative direct impact on the efficiency of net-fishing efforts in the presence of jellyfish.

Tourism

Tourism is an industry of primary importance for the world economy. For some Mediterranean shoreline countries, tourism is the first source of income. Spain and France are now among the most visited countries in the world, and in these countries international tourism receipts account for just over 2 % of gross domestic product. In Greece 10 % of total employment is in the tourist industry. There is large scale capital transfer from the tourist demand areas of NW Europe to the generally poorer Mediterranean countries. Climate constitutes an important part of the environmental context in which recreation and tourism takes place; because tourism is a voluntary and discretionary activity, participation will often depend on favorable conditions. Climate and weather are among the most important factors affecting participation in leisure and tourism activities, since for many activities there are critical threshold levels beyond which participation and enjoyment levels fall and safety may be endangered.

We should recall, however, that tourism is a continuously adapting industry, responding to changing demographic and economic conditions as well as to new demands and technologies. Climate change will present new challenges, but also lead to opportunities for tourist investment to capitalize on new environmental conditions. Work has only just begun on "translating" the suggested future climate scenarios into their impacts on tourism but already some interesting adaptations are emerging. Perry (2000) has focused on the adaptive capacities to climate change, considering the following effects:

(a) Higher air and sea temperatures are likely to encourage a longer tourist season. If the summer becomes widely perceived as too hot, the season could become "doughnut shaped", with peaks in spring and autumn months and a "hole" in high summer. Resorts need to discourage a closing down attitude at the end of summer Higher temperatures will allow a prolongation of the season and – if possible – added cultural and sporting attractions such as arts festivals, regattas, food and drink events, and local fiestas can help this process. Breaking the traditional seasonal patterns has as much to do with changing consumer attitudes as with developing new attractions and more targeted advertising could help in this respect. A longer tourist season would allow quicker returns on investment with more intensive utilization of facilities over a longer period.
(b) Tourists will increasingly expect holiday accommodation to be air conditioned. Such accommodation will attract a premium price, whilst poorer quality self catering apartments and rooms without air conditioning will be much less attractive in the summer. At present only a fifth of rooms in hotels in Mediterranean countries are in the 4 and 5 star categories. Increased demands will be made on electricity supply from the demand for additional cooling systems.
(c) There will be a higher risk of epidemics of cholera and typhoid as well as infectious diseases such as malaria, dengue fever, etc. Adverse publicity would follow such public health scares and frighten tourists away as happened at Salou, in Spain, a few years ago. More care will be needed on hygiene matters, particularly in hotel and restaurant food preparation if the incidence of dysentery and e-coli is not to dramatically increase. Adverse publicity for tourism could also accompany an increase in illegal refugees from North Africa, fleeing environmental degradation in their home countries.
(d) The beaches that are the principal assets of coastal tourism will be at considerable threat from erosion as sea levels rise (Nicholls and Hoozemans 1996). This will also apply in many cases where coastlines are flat, with tourist accommodation and other infra-structure located immediately behind the beaches. New tourist resorts should locate all long-life infra-structure inland from the coast and only place easily removed structures near the beach line. Expensive beach nourishment schemes may be appropriate in some places .In some places there is a shortage of sandy beaches and their erosion or removal will severely disadvantage tourism. Additional costs would be incurred by escalating costs of coastal defense protection and repairs. Nautical tourism has stimulated the building of large numbers of marinas and berths which could be affected by sea level rise. Management efforts need to be expanded to consider the implications of climate change and to conduct vulnerability surveys.

Hamilton et al. (2005) offer qualitative analysis of a large database, on the impact of climate change on demand through changes in population, per capita income and climate change. According to this study, climate change would lead to a gradual shift of tourism destinations towards higher latitudes and altitudes. Climate change would also imply that the current dominant group of international tourists - sun and beach lovers from Western Europe would stay home, implying a

Table 34.1 Impact of SLR on area and population of Mediterranean developing countries

	1 m	2 m	3 m	4 m	5 m
Area (total-10,050,556 sq. km.)					
Impacted area	24,654	33,864	43,727	53,615	63,120
% of total area	0.25	0.34	0.44	0.53	0.63
Population (total = 259,396,000)					
Impacted population	8,307,472	10,912,744	13,684,993	16,454,655	19,439,678
% of total population	3.2	4.21	5.28	6.34	7.49
GDP (total = 1,404,470 million USD)					
Impacted GDP	20,870	30,365	39,037	46,209	54,853
% of total DGP	1.49	2.16	2.78	3.29	3.91
Urban extent (total – 354,294 sq. km.)					
Impacted area	3,679	5,037	6,529	7,951	9,384
% of total area	1.94	2.65	3.44	4.18	4.94
Agricultural extent (total = 354,294 sq. km.)					
Impacted area	4,086	6,031	8,007	9,819	11,451
% of total area	1.15	1.70	2.26	2.77	3.23
Wetlands area (total = 342,185 sq. km.)					
Impacted area	11,361	14,758	18,224	21,417	24,277
% of total area	3.32	4.31	5.33	6.26	7.09

Source: Dasgupta et al. (2007, Table 5)

relatively small fall of total international tourist numbers and total distance travelled. Changes induced by climate change are generally much smaller than those resulting from population and economics growth are. Hall and Higbam (2005) have emphasized that tourism is impacted by other factors aside from the influence of global and local climate change, such as processes of political, economic and cultural globalization, technology, mega-urbanization and related environmental changes. Such processes may themselves interact with global climate change in ways that we are only starting to identify and understand.

Migrations

Migration refers to a permanent departure from the country of origin that is often induced by economic and/or environmental changes. Climate change may increase rates of migration, and requires communities to adapt accordingly. Migration is an ever-growing global phenomenon that has important development implications for both the sending and the receiving countries. For the country of origin migration and the resulting potential remittances lead to increased incomes and poverty reduction, and improved health and educational outcomes, and promote economic development. Yet these gains might come at substantial social costs to the migrants and their families. If developed countries are the recipients of international migrants, they face challenges of integration of immigrants, job competition between migrant and native workers, and fiscal costs associated with the provision of social services to the migrants. Climate refugees are defined as people who have to leave their habitats because of sudden or gradual alterations in their natural environment related to one of three impacts of climate change: sea-level rise, extreme weather events, and drought and water scarcity. That is, we take into account only direct impact of climate change on human migration and neglect the indirect effect of migration due to changing economic environment.

Dasgupta et al. (2009) have analyzed the impacts of sea level rise (SLR) on six socioeconomic indicators, including land, population, gross domestic production (GDP), urban extent, agricultural extent and wetlands. These impacts are calculated for SLR scenarios ranging from 1 to 5 m. Table 34.1 summarizes the findings of SLR on area and population on Mediterranean developing countries. This study reveals a potentially high impact of climate change on the population in East Mediterranean and North Africa. Climate change impact is expected to affect 3.2 up to 7.49 % of the population in developing countries in the Middle East and North Africa. The impacts are not uniformly distributed across the regions and countries of the developing world (considering 84 coastal developing countries in Latin America and Caribbean, Middle East and North Africa, Sub Saharan Africa, East Asia, and South Asia), and are expected to be severe only in a limited number of countries. According to the above study, the Egyptian population would be the most severely impacted by SLR. With a 1 m SLR, approximately 10 % of the population would be impacted. Most of this impact would take place in the Nile delta; it reaches 20 % with a 5 m SLR. Migration is expected to be wide-spread, resulting from the combined impact on the indicators studied above, including direct impact on agricultural land, wetland, GDP, population, urban extend and flooded area. It has been shown that the Middle-East would experience the largest percentage impacts from SLR. Egypt and Tunisia are among the ten most affected nations worldwide, with an estimated impact on GDP of about 6 and 3 %, respectively.

Nevertheless, affirmative trends are hard to identify as a result of a number of shortcomings. First, data are rough and often incomplete Welfare estimates are often confined to direct cost (DC), as for example, the value of land and/or capital lost plus investments is coastal protection. Such estimates suffer from three limitations, of which research should be aware of: First, the value of land and capital located in coastal areas threatened by sea level rise is not well known. Because of their influence on the optimal level of coastal protection, different assumptions about land and capital values have an indirect as well as an immediate effect on DC estimates. Second, DC is only a first-order approximation of welfare losses. By assuming constant prices, it neglects second order effects. Third, DC is generally estimated for specific regions in isolation; because of international trade, however, the economic impacts of environmental changes (sea level rise, acidification and so on) are likely to spill across regional and national boundaries and affect areas with little or no immediate damages. Thus, Darwin and Tol (2001) have evaluated the extent to which these limitations may distort estimates of the economic losses. They estimate those discrepancies at about 36 % in general, when estimating the impact of sea level rise alone.

References

Burgents JE, Burnett KG, Burnett LE (2005) Effects of hypoxia and hypercapnic hypoxia on the localization and the elimination of *Vibrio campbelli* in *Litopenaeus vannamei*, the Pacific white shrimp. Biol Bull 208:159–168

Cooley R, Doney SC (2009) Anticipating ocean acidification's economic consequences for commercial fisheries. Environ Res Lett 4:1–8

Darwin RF, Tol RSJ (2001) Estimates of the economic effects of sea level rise. Environ Res Econ 19:113–129

Dasgupta S, Laplante B, Meisner C, Wheeler D, Yan J (2009) The impact of sea level rise on developing countries: a comparative analysis. Clim Change 93:379–388

Fabry VJ, Seibel BA, Feely RA, Orr JC (2008) Impacts of ocean acidification on Marine Fauna and ecosystem processes. ICES J Mar Sci 65:414–432

Galil BS, Zenetos A (2002) A sea change: exotics in the eastern Mediterranean. In: Leppäkoski E, Gollasch S, Olenin S (eds) Invasive aquatic species of Europe: distribution, impacts and management. Kluwer Academic Publishers, Dordrecht

Hall CM, Higbam J (2005) Tourism recreation and climate change. Channel View Publications, Clevedon

Hall-Spencer JM, Rodolfo-Metalpa R, Martin S, Ransome E, Fine M, Turner SM, Rowley SJ, Tedesco D, Buia MC (2008) Volcanic carbon dioxide vents show ecosystem effects of ocean acidification. Nature 454:96–99

Hamilton JM, Maddison DJ, Tol RSJ (2005) Climate change and international tourism: a simulation study. Glob Environ Chang 15:253–266

Hennessy KB, Fitzharris B, Bates BC, Harvey N, Howden M, Hughes L, Salinger J, Warrick R (2007) Australia and New Zealand: climate change 2007: impacts, adaptation and vulnerability. In: Parry ML, Canziani OF, Palutikof JP, van der Linden PJ, Hanson CE (eds) Fourth assessment report of the intergovernmental panel on climate change. Cambridge University Press, Cambridge

Hobday AJ, Matear R (2005) Review of climate impacts on Australian fisheries and aquaculture: implications for the effects of climate change. Report to the Australian Greenhouse, Canberra

Hobday AJ, Okey TA, Poloczanska ES, Kuns TJ, Richardson AJ (2006) Impacts of climate change on Australian marine life: part C literature review. Report to the Australian Greenhouse Office, Canberra

Nakar N, DiSegni DM, Angel D (2011) Economic valuation of jellyfish bloom on the fishery sector. Selected paper, Biodiversity and Economics for Conservation Organization (BIOECON). Annual conference, Venice

Nicholls RJ, Hoozemans FMJ (1996) The Mediterranean: vulnerability to coastal implications of climate change. Ocean Coast Manag 31:105–132

Perry A (2000) Impacts of climate change on tourism in the Mediterranean: adaptive responses. Working paper 35/2000. Fondazione Eni Enrico Mattei, Venice

Roessig JM, Woodley CM, Cech JJ, Hansen LJ (2004) Effect of Global climate change on marine and estuarine fishes and fisheries. Rev Fish Biol Fish 14:251–275

Wootton JT, Pfister CA, Forester JD (2008) Dynamic patterns and ecological impacts of declining ocean pH in a high-resolution multi-year dataset. Proc Natl Acad Sci U S A 105:18848–18853

Xiaozi Liu, Mikko Heino (2010) Global warming and international fishery management: does anticipation of the temperature change matter? Discussion paper, Norwegian School of Economics and Business Administration, Helleveien

Effect of Climatic and Anthropogenic Disturbances on Sponge Fisheries

35

Thierry Pérez and Jean Vacelet

Abstract

The commercial sponge industry is a fascinating cultural heritage of several Mediterranean countries, where it continues to represent an important economic activity. Mediterranean bath sponges are of the highest quality and the commercial demand for them is still significant, however, sponges are suffering from environmental disturbances that seem to be occurring more frequently in recent decades. Here we present some general data about commercial sponges of the Mediterranean Sea, and examine probable consequences of both overfishing, which has been occurring for many centuries on most sponge beds, and the effects of climatic change, which appears to be responsible for increased disease outbreaks and mass mortality events. Together these disturbances may alter the species distribution. We also examine the potential future of sponge cultivation under these conditions.

Keywords

Porifera • Bath sponges • Biology • Ecology • Population • Global change

Introduction

Sponge fishery has a long history in the Mediterranean Sea, where it represents a fascinating cultural heritage (Pronzato and Manconi 2008). Commercial sponges have been used for domestic purposes since high Antiquity, especially in Greece (Arndt 1937; Voultsiadou et al. 2011). Traces of sponge use have been found in the Minoan civilization in Crete dating back to approximately the nineteenth century BC. Sponge harvesting was undertaken for a long time by free divers (Castritsi-Catharios 1998; Warn 2000; Voultsiadou et al. 2011). It became more intensive with the development of diving techniques, and improvements in dredging technology in the late nineteenth century. At the same time, demand has increased worldwide, which has been satisfied by an extension of exploitation to other seas such as the Caribbean Sea and the Pacific Ocean. Simultaneously there has been severe depletion of Mediterranean sponge grounds. Overfishing was already occurring in the first half of the twentieth century, causing the emigration of many Greek sponge fishermen, especially to Florida. Comparisons of ancient and recent sponge populations in Cyprus, Crete and Sardinia show dramatic falls of sponge abundance in recent decades (Pronzato et al. 1999). Furthermore, several instances of diseases and mass mortality events have severely affected sponge fisheries in the twentieth century. Nowadays, although commercial demand is declining due to the use of cheaper artificial sponges, bath sponge stocks face serious decline in the Mediterranean due to the combined effects of overfishing, diseases and climate change. In this chapter, we aim to present the current status of Mediterranean commercial sponges and how it will be affected by environmental disturbances.

T. Pérez (✉) • J. Vacelet
CNRS UMR 7263 Institut Méditerranéen de Biodiversité et d'Ecologie marine et continentale (IMBE), Aix-Marseille Université, Station Marine d'Endoume, rue Batterie des Lions, 13007 Marseille, France
e-mail: thierry.perez@imbe.fr; jean.vacelet@imbe.fr

S. Goffredo and Z. Dubinsky (eds.), *The Mediterranean Sea: Its history and present challenges*, DOI 10.1007/978-94-007-6704-1_35, © Springer Science+Business Media Dordrecht 2014

Fig. 35.1 Underwater photographs of (**a**) *Hippospongia communis*, (**b**) *Spongia officinalis*, (**c**) *Spongia lamella* (Photos T. Pérez/CNRS)

Fig. 35.2 Treatment of sponges in Tunisia (Photo Nicole Boury-Esnault)

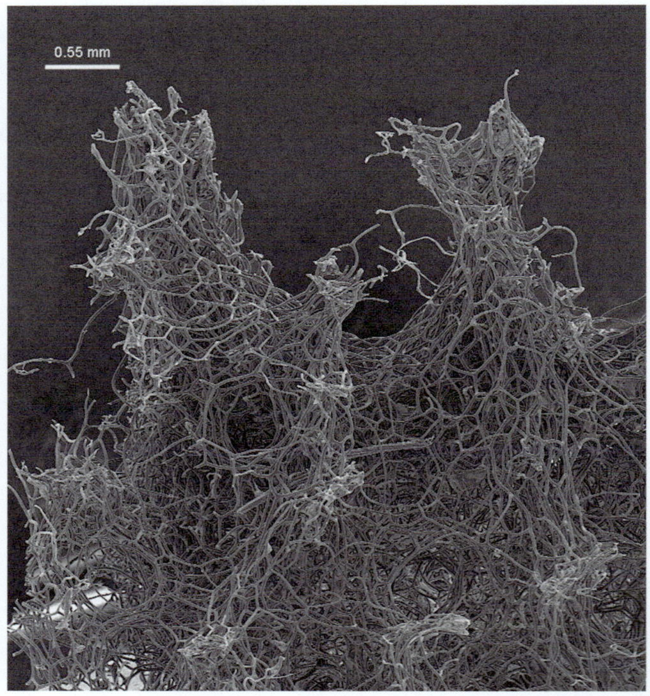

Fig. 35.3 Scanning electron microscope view of the skeleton of *Spongia* (*Spongia*) *officinalis*

The Mediterranean Commercial Sponges

Among the numerous species of sponges currently reported from the Mediterranean Sea, only five species of the genera *Spongia* and *Hippospongia* (Fig. 35.1) are presently used for domestic purposes. These commercial sponges are classified in the Class Demospongiae, Order Dictyoceratida in which the skeleton consists only of a reticulation of fibres, without the siliceous spicules found in most other demosponges.

The commercial sponge is the cleaned skeleton of the species of the Spongiidae family (Fig. 35.2). The skeleton is a dense network of two types of fibres (Fig. 35.3). The primary fibres, 40–100 μm in diameter, are rather scarce. They generally contain a small amount of foreign bodies, sand grains or foreign spicules. They ascend towards the surface, where they generally protrude in small conules. The secondary fibres, 6–35 μm in diameter, build a very dense reticulation (Fig. 35.3). These fibres consist of a collagen named 'spongin'. The reticulation made by these fibres is remarkable by its softness, elasticity, resistance and high water retention capacity, which are all of top quality and clearly superior to that of artificial or plant 'sponges'.

Five species are commonly harvested in the Mediterranean (Pronzato and Manconi 2008): *Spongia (Spongia) officinalis* Linnaeus, 1759; *Spongia (Spongia) mollissima* Schmidt, 1862; *Spongia (Spongia) lamella* (Schulze, 1879); *Spongia (Spongia) zimocca* Schmidt, 1862; *Hippospongia communis* (Lamarck, 1814). The distinction between *S. mollissima* and *S. officinalis* as distinct species (Pronzato and Manconi 2008) is still under discussion, and Voultsiadou et al. (2011) and Dailianis et al. (2011) still prefer to maintain its classification as a subspecies of *S. officinalis* together with the subspecies *adriatica*. They are known by different commercial names by the sponge traders, sometimes depending on local varieties and local traditions. Their respective abundance is variable in the diverse areas of the Mediterranean, as well as the depth to which they may occur in sufficient abundance to permit profitable exploitation. Most of the productive sponge banks are located in the eastern basin of the Mediterranean Sea (Tunisia, Libya, Greece, Egypt), and there is also a significant production in the Adriatic Sea (Croatia). In other countries such as Spain, France or Italy, the sponge fishery is trivial, with small amounts of sponges being sold mainly on local markets.

All Mediterranean commercial sponges are included as protected species in the Appendix III of the Bern Convention on the Conservation of European Wildlife and Natural Habitat.

Spongia (Spongia) officinalis, 'Greek bath sponge' (English), 'Fine Grecque' (French), 'Matapas' (Greek), is massive, spherical or lobate. The primary fibres contain foreign bodies, the secondary fibres, 20–35 μm in diameter, are free of inclusions. It is common throughout the whole Mediterranean, from 0.5 m (under overhangs) to 40 m depth, but may occur at greater depths, especially in the Aegean Sea where it has been recorded as deep as 100 m (Kefalas et al. 2003). This species is of excellent commercial quality and is highly regarded, especially for bath use. It constitutes a large part of the Mediterranean production, for instance 37.3 % of the Kalymnian sponge fishing fleet production in 1996 (Castritsi-Catharios 1996; Castritsi-Catharios et al. 2007).

Spongia (Spongia) mollissima, 'Turkey cup' or 'Levantine' (English), 'Melati' (Greek), 'Fine Syrie' (French), is roughly cylindrical or in the shape of an inverse cone with an enlarged flat or depressed summit, or cup-shaped with thick walls. The fibres are similar to those of *S. officinalis*, with slightly thinner secondary fibres (10–25 μm in diameter). The species is recorded only from the eastern Mediterranean, generally 10–30 m in depth, although there are records deeper than 120 m in the Aegean Sea (Kefalas et al. 2003). It is considered as the best bath sponge in the world, but represents only a small part of the Mediterranean production; for instance 15.3 % of the Kalymnian sponge fishing fleet production in 1996 (Castritsi-Catharios 1996; Castritsi-Catharios et al. 2007).

Spongia (Spongia) lamella, 'Elephant Ear' (English), 'Oreille d'éléphant' (French), 'Lagophyto' or 'Psathouri' (Greek) has been called *Spongia agaricina* Pallas, 1766 for a long time due to a confusion with an Indian Ocean species (Pronzato and Manconi 2008). This sponge is lamellar, most often forming a more or less regular cup up to 1 m in diameter, exceptionally up to 3 m in diameter (Castritsi-Catharios et al. 2011a). The primary fibres are cored by foreign bodies, the secondary fibres are 20–40 μm in diameter. The species is found in the whole Mediterranean area. Its depth distribution, however, differs in the two Mediterranean basins. In the western basin, it is present from a few metres to 110 m in depth, whereas in the eastern one it is generally found in depths from 50 to 110 m. Its commercial quality is considered to be very good, with a higher resistance to tearing than other commercial sponges (Castritsi-Catharios et al. 2011b). Due to its lamellar shape, it is mainly used for cosmetic and for polishing glasses or pottery, and there is still a significant demand according to a French sponge trader (C. Cypreos, personal communication, www.sponges.fr). Its production, however, is not large, for instance 3.1 % of the Kalymnian sponge fishing fleet production in 1996 (Castritsi-Catharios 1996; Castritsi-Catharios et al. 2007).

Spongia (Spongia) zimocca, 'Tsimoucha' (Greek), 'Chimousse' (French), 'Leather sponge' (English), 'Zimoukha' (Arabic/Tunisian), is a massive sponge, generally irregularly lobate but rather variable in shape, up to 25–30 cm in diameter (Castritsi-Catharios et al. 2011b). The primary fibres are uncored; the secondary fibres are generally of two size classes, the thinner being 5–18 μm in diameter. The skeleton is rather rough and harsh when compared to the other Mediterranean commercial sponges, with the lowest elasticity and resistance to tearing (Castritsi-Catharios et al. 2011b). Accordingly, its commercial value is relatively low. The sponge is recorded at depths of 25–85 m, and is harvested only in the eastern basin. Its presence in the southwest of the western basin is possible, but not confirmed. This species constitutes a small part of the sponge production in the Mediterranean, for instance 4.4 % of the Kalymnian sponge fishing fleet production in 1996 (Castritsi-Catharios 1996; Castritsi-Catharios et al. 2007).

Hippospongia communis, 'Honey Comb' or 'Horse sponge' (English), 'Eponge commune' or 'Eponge cheval' (French), 'Kapadiko' (Greek), is massive, most often subspherical, with few conules. It is characterized by very large internal cavities. The primary fibres are reduced and contain foreign bodies. The secondary fibres are 20–30 μm in diameter. The commercial value is good, but its skeleton is less soft than that of *S. officinalis* and *S. lamella*, which are preferred for bath use. The species is widely distributed throughout the Mediterranean, often occurring in seagrass beds, sometimes very shallow especially in Tunisia, but also on rocky substrates up to 40 m deep and possibly more. It constitutes the main part of the Mediterranean sponge production, for instance 40 % of the Kalymnian sponge fishing fleet production in 1996 (Castritsi-Catharios 1996; Castritsi-Catharios et al. 2007).

There are two other species of *Spongia* in the Mediterranean, which are not, or rarely, collected for domestic use. *Spongia (Spongia) virgultosa* (Schmidt, 1868) is a small, encrusting sponge (Pronzato et al. 1998) with a rather harsh skeleton. *Spongia (Spongia) nitens* (Schmidt, 1862), with a massive lobate shape is generally smaller than the five commercial species. It has a very soft skeleton and a high water retention capacity, but is very fragile.

Biology and Reproduction of the Commercial Species

Nutrition and Filtering Activity

Commercial sponges are regarded as among the most efficient benthic suspension feeders of the Mediterranean Sea as they are capable of intense pumping activity. Overall, sponges have already been shown to feed on ultraplankton (<10 μm) and particularly on picoplankton (<2 μm) (Pile et al. 1996, 1997). Savarese et al. (1997) evaluated that the grazing ability of large sponge beds could significantly deplete picoplankton near the bottom. New technologies such as flow cytometry were used recently to determine more accurately which group of planktonic cells constitutes the main source of carbon for a number of sponge species. Most of these studies involved sampling both the inhaled and exhaled waters of sponges living under various environmental conditions in order to estimate the natural filter feeding capacity (see for instance Yahel et al. 2003, 2005, 2007; Pile and Young 2006; Trussel et al. 2006). Only one recent study has aimed to investigate the feeding capacity of a Mediterranean commercial sponge. *Spongia officinalis* was studied under natural conditions in order to assess its efficiency at retaining very small particles (<20 μm), and determine the origin of assimilated carbon by coupling flow cytometry and stable isotope analysis (Topçu et al. 2010). This study showed that *S. officinalis* efficiently retained the main picoplanktonic groups, and it also indicated a size selectivity of particles with significantly lower retention efficiencies for nanoplankton than picoplankton. Whereas picoplankton, such as cyanobacteria and picoeucaryotes, constituted the major food source of *S. officinalis* in terms of particle abundance, most of the carbon retained in term of biomass originated from nanoeukaryotes. Thus, this study demonstrated that even though sponges were known to retain smaller particles than other filter-feeding organisms, which is an adaptive advantage in oligotrophic environments, they are also able to efficiently assimilate carbon from larger organisms such as nanoeukaryotes. This feeding strategy permits sponges to optimize their energy intake. This study also showed a significant increase in retention efficiency in Spring compared to Winter, suggesting an increase in energy needs related to reproduction or growth. Finally, this study indicated that *S. officinalis* excretes faecal pellets as observed in previous studies of other sponge species (*e.g.* Wolfrath and Barthel 1989).

This first novel combination of techniques (Topçu et al. 2010) demonstrated that comprehensive studies of sponge feeding strategies would help to better understand the importance of sponges and their potential role in organic matter recycling. Moreover, by transposing this approach to a selection of other suspension feeders growing on Mediterranean hard substrates, it will be possible to better understand the functional role of sponges in benthic-pelagic coupling. Commercial sponges appear to be highly sensitive to environmental stress related to global climate change (Lejeusne et al. 2010), thus this approach using these species as models, may be useful to assess the potential impact of environmental disturbances, such as the effect of climate change on the function of benthic systems.

Reproduction and Genetics

Knowledge of bath sponge reproductive biology is well advanced. Allemand-Martin (1906) first observed the sexual reproduction of *H. communis* from Sfax and Kerkennah. A second significant study by Tuzet and Pavans de Ceccatty (1958) effectively described its gametogenesis and embryonic development, which was confirmed later by ultrastructural investigations of spermatogenesis (Gaino et al. 1984) and embryogenesis (Scalera-Liaci et al. 1971). Finally, some aspects of the reproductive cycle of the species from the Spongiidae family were investigated in more recent studies of: *Spongia* spp. and *Hippospongia* spp. from the Gulf of Mexico and the Caribbean Sea (Kaye 1991; Kaye and Reiswig 1991a, b), and *Coscinoderma matthewsi* from the Pacific (Abdul Wahab et al. 2012). Some observations have been obtained recently for other Mediterranean species *S. officinalis* (Baldacconi et al. 2007), and *H. communis* (Zarrouk et al. in press).

The general reproductive pattern of bath sponges seems to be similar to those reported for other Dictyoceratida (Scalera-Liaci et al. 1971; Hoppe 1988; Kaye and Reiswig 1991a, b; Baldacconi et al. 2007; Whalan et al. 2007; Chung et al. 2010; Ereskovsky 2010). Sperm cells originate from choanocytes, with choanocyte chambers evolving into widespread spermatic cysts throughout the choanosome, whereas oocytes are derived from archaeocytes. *H. communis* and *S. officinalis* are gonochoric and ovoviviparous. Successive hermaphroditism, with alternate production of oocytes and spermatic cysts in the same reproductive season, has been rarely observed (Baldacconi et al. 2007). Oocytes and embryos are produced all year round, whereas

Fig. 35.4 Sponge fishery in Syria (from 'L'Univers Illustré', 13 September 1865). Syria was the main producer of the Levantine sponge, or 'fine Syrie' in the nineteenth century. Its production is now completely stopped

spermatogenesis generally occurs in a shorter period (Autumn or early Winter depending on the Mediterranean species). In females, a given individual is able to develop young oocytes and develop larvae at the same time. Whatever the Mediterranean species considered, the larvae are generally released between May and July (Baldacconi et al. 2007; Zarrouk et al. in press, and unpublished data). However, the most recent study showed a difference in the timing of release of larvae between two locations experiencing very distinct thermal regimes. In *H. communis*, the warmest temperatures tend to increase the reproductive effort, accelerate the embryogenesis and trigger larval release (sooner in the warmest region) (Zarrouk et al. in press). This work indicated that temperature is a key factor controlling the reproductive strategy of bath sponges, as already shown in other demosponges (Fell 1976; Witte and Barthel 1994; Witte et al. 1994; Meroz-Fine et al. 2005; Ereskovsky 2010).

Thus, in the context of global warming of the Mediterranean Sea, some effects on sponge reproduction patterns can be expected, but we do not know yet if these changes would constitute true adaptive responses. Moreover, further investigations of larval biology and the recruitment of bath sponge juveniles are needed to obtain a comprehensive understanding of the population dynamics and establish a conservation plan for sponge beds.

The first studies on the genetic diversity of commercial sponges (Noyer et al. 2009; Dailianis et al. 2011) have shown they have a high level of genetic variability. This suggests that after mass mortalities due to global warming, re-colonization could occur from more tolerant populations. Such studies will also provide a basis for effective management of the species (Dailianis et al. 2011).

Sponge Fishery in the Mediterranean Sea

Most of the Mediterranean sponge production is harvested by divers, using SCUBA diving, or hookah or narghile diving. Free-diving has been the main method for centuries, but is now rarely used (Fig. 35.4). Similarly, hard-hat diving equipment is only used for tourist exhibitions nowadays. Sponge divers rarely work deeper than 40 or 50 m. Decompression tables were ignored or badly respected during most of the twentieth century, and are still sometimes inappropriately followed, resulting in decompression diseases (*e.g.* the so-called 'bends'). There are many paralysed sponge fishermen within these communities (*e.g.* Flégel 1910; Warn 2000). A special dredge named 'gangava', which allows the exploitation of sponges from deeper depths, is still used in the eastern Mediterranean (Castritsi-Catharios et al. 2011b), although it is highly destructive of other benthic fauna. In shallow waters, especially on seagrass beds or shallow rocky substrates up to 15 m depth, *H. communis* and sometimes also *S. officinalis* may be collected by surface visual methods with the fisherman looking from the surface through a cylindrical tube fitted with glass at one end and using a harpoon ('kamaki') to collect the sponge.

The production of commercial sponges has been highly variable, partly due to the occurrence of sponge diseases, which have altered the balance between the Mediterranean and the Central West Atlantic production. The maximum production was in the 1930s. According to Arndt (1937), the world production of sponges averaged 1346 metric ton/year between 1927 and 1936, the Caribbean and Gulf of Mexico being the main producers (998.5 t/year compared to 347.5 t/year from the Mediterranean). The figures were considerably lower in the 1980s, with a global production recorded by

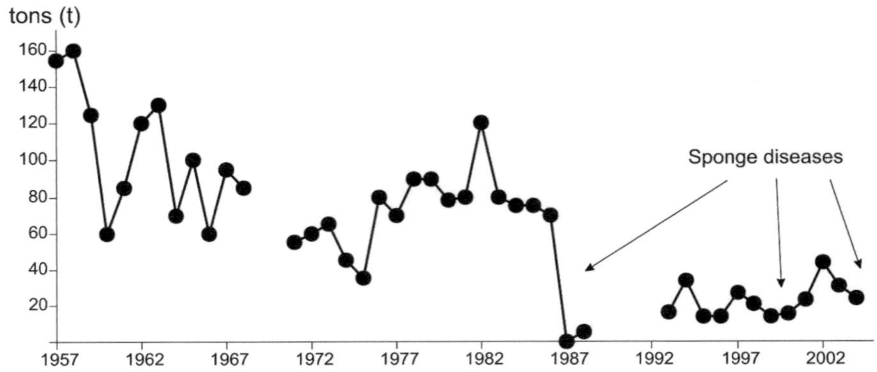

Fig. 35.5 Survey of the Tunisian Sponge fishery between 1957 and 2004. *Arrows* indicate putative effects on catches of successive sponge diseases (Data from the 'Commissariat général aux pêches' (1957–1988) and the 'Direction Générale de la Pêche et de l'Aquaculture' (1993–2004) of Tunisia)

FAO of only 130 t (Wells 1983). This decrease was mainly due to a disease which swept through the sponge beds in the West Atlantic in 1938 and in the following years; there was practically no sponge production for many years. During that time the Mediterranean was the main sponge producer, resulting in severe depletion of the sponge beds due to overfishing. The sponge industry has now somewhat recovered in the West Atlantic, although it has still not fully revived (Storr 1964; Stevely and Sweat 1985), due to the disappearance of local fishermen, most of them being of Greek origin, and possibly also to a decrease in sponge abundance linked to recent climatic change (West et al. 2011).

The Mediterranean sponge fisheries are again facing serious threats of overexploitation, and have also suffered from several episodes of mortality. In Tunisia, which is the main producer of the honey comb (*Hippospongia communis*), production from 1957 to 2002 was highly variable, with severe depletions due to diseases and mass mortality events (Ben Mustapha and Vacelet 1991; and Fig. 35.5). It appears from accounts of old fishermen that the commercial sponges were incredibly more abundant during the 1930s, with more than 200–300 specimens/100 m^2 whereas more recently, the mean density on unexploited sponge banks is often less than 50 specimens/100 m^2 (Pronzato 1999; Pronzato et al. 1999). According to Voultsiadou et al. 2011, bath sponge populations are considerably reduced compared to previous time periods. Sponge beds from the easternmost part of the Mediterranean Sea (Syria, Lebanon, Cyprus), which were famous for the production of the high quality 'turkey cup', now contain very low populations of commercial sponges; a depletion which may be partly due to the influence of the Aswan dam on the Nile River reducing the input of seasonal nutrients.

Sponge Culture

Over-exploitation of sponge beds has long been a problem and is becoming worse with the occurrence of diseases, which appear to be increasing in frequency, possibly due to global warming. It is appealing to combat overexploitation by culturing sponges, which appears at first sight to be easy due to the great regeneration power of sponges. A 'mother sponge' can be cut into several pieces, which when secured on ropes or various substrata in suitable environments, can grow to commercial size in a few years. This method, called culture by fragmentation, has been tried often in the Mediterranean as early as 1862 (Marenzeller 1879), in the Caribbean and in the Pacific (review in Vacelet 1985), and experiments and pilot exploitation are currently being undertaken. Other methods, such as the reorganization of small sponges from dissociated cells or collection of larvae on substrate collectors, were never successful but may be reassessed in the future (Abdul Wahab et al. 2012).

Sponge culture by fragmentation, although simple and experimentally successful, has never been a commercial success in the Mediterranean (Vacelet 1985; Verdenal and Vacelet 1990; Pronzato et al. 1999). The main problem is the rather slow growth rate of the cuttings, which increases risks of destruction of the 'sponge farm' by storms or epizootic diseases, and the variable percentage of mortality. Rapid growth, with explants doubling or tripling in size in a year, has been reported, but rarely occurs on a regular basis in the Mediterranean, especially in areas where the temperatures are rather low (Celik et al. 2011). Some 'sponge farms' in the Pacific (Adams et al. 1995; Croft 1995; Duckworth 2009) have provided the market with productive output for several years. However, their profitability has never been particularly rewarding, although they may be interesting for coastal indigenous communities (Duckworth 2009). Several large-scale experiments, with up to 700,000 cuttings in the Caribbean in 1938, were destroyed by an epizootic disease (Smith 1941), possibly exacerbated by the high concentration of sponges in the farm. However, the profitability and interest in sponge farming has to be reconsidered with the present increase in prices due to the depletion of natural production and an increased interest in natural products (Verdenal and Verdenal 1987). Furthermore, sponge farming, even though not fully profitable by itself, may be interesting in an integrated mariculture system, in which intensive culture of filter-feeders such as sponges has been proposed in order to control increased eutrophication under fish culture cages (Pronzato 1999; Abdul Wahab et al. 2012).

Pollution Effects, Diseases and Mass Mortality Events

It is conceivable that sponges may be highly sensitive to pollution, considering their natural filter feeding activities. Actually, the most productive sponge beds are generally located in areas that are not subjected to heavy pollution. However, Mediterranean commercial sponges are able to withstand some level of urban pollution, provided they are protected from heavy sedimentation. For instance, a large population of *S. officinalis* is present under overhangs some 300 m from the sewage outlet of the city of Marseille. Under these conditions, however, the sponge skeleton is heavily encrusted with particles of iron oxide, decreasing its resistance and commercial value, and culture experiments suffered from high mortality and low growth rates (Vacelet et al. 1988; Verdenal and Vacelet 1990). *Spongia officinalis* appears to have some of the fundamental characteristics of the ideal 'biomonitor' of trace pollutants. It may live in diverse pollution conditions, has a long life span, and filters a considerable amount of water with a rather well known particle retention ability (Topçu et al. 2010). Several studies conducted on this species showed its ability to strongly accumulate metals in higher concentrations than those found in other biomonitor species, to reflect accurately the level of contamination at a given site (Verdenal et al. 1990; Perez et al. 2005), and even to degrade some complex organic compounds (Perez et al. 2002, 2003).

Bath sponges could be affected by the introduction of invasive species, which already play a key role in changes underway in the Mediterranean Sea (Lejeusne et al. 2010). An example is the macrophyte *Caulerpa racemosa*, which has already resulted in a significant decrease in the percentage cover of sponge assemblages (Baldacconi and Corriero 2009). Harmful effects of invasive *Caulerpa* spp. on commercial sponges have not been reported, but have been shown for massive keratose sponges such as *Sarcotragus spinosulus* (Zuljevic et al. 2011).

Commercial sponge mass mortalities occurred in the Mediterranean at the end of the nineteenth and in the early twentieth centuries (Arndt 1937), but as far as we know they were never as devastating as the disease which completely stopped the sponge industry for years in the West Atlantic after 1938 (Galtsoff 1942). It appears, however, that episodes of mass mortality since the 1980s are more frequent and more severe among commercial sponges as well as among other diverse invertebrates; these events are likely linked to thermal anomalies and global warming (Lejeusne et al. 2010). This appears to be a general observation which is not only restricted to the Mediterranean (Webster 2007). Heat stress is often recognized to trigger the disease outbreak, and the frequency of these events has clearly increased over the last three decades because of climate change (Lejeusne et al. 2010). The first signs of sponge disease are often associated with a change in colour or bleaching (similar to 'coral bleaching' in tropical coral reefs), the development of a bacterial white veil on the epidermis, followed by a rather rapid decomposition of the skeleton (e.g. Pérez et al. 2000). Among the best documented events was the severe mortality on the Mediterranean sponge beds in 1986 which dramatically decreased sponge production, especially in Tunisia (Fig. 35.5). This disease outbreak motivated an inquiry mandated by the FAO to assess the incidence and virulence of the phenomenon, and to identify its possible causes (Vacelet 1994). Several studies showed that the disease resulted in a bacterial attack on the spongin skeleton (Gaino and Pronzato 1989; Vacelet et al. 1994). The authors argued that the virulence of this pathogen could have been triggered by high seawater temperatures, an explanation that was also proposed in a case of coral bleaching in the eastern Mediterranean (see for instance Kushmaro et al. 1996). The possible agent of subsequent sponge mass mortality events in 1999 and 2003 remains unknown. After the 1999 Mediterranean multispecies mass mortality, the potential role of pathogenic microorganisms has often been proposed (Cerrano et al. 2000; Pérez et al. 2000; Martin et al. 2002), but the infectious origin of the disease outbreak has been demonstrated only for gorgonians (Bally and Garrabou 2007). In that case, the authors also proposed that the introduction and spread of a pathogenic *Vibrio* might have been promoted by climate warming (Bally and Garrabou 2007).

The context of warming reported around the Mediterranean is probably behind these mass mortality events which have affected the Mediterranean sponge beds. The 1980s corresponded to a succession of positive North Atlantic Oscillations. Both in 1999 and 2003, the analysis of meteorological data and seawater temperature records highlighted exceptional temperature patterns in the NW Mediterranean. Unfortunately, there is a lack of data outside this region, although the 1999 mortalities have been related to similar temperature anomalies in Tunisia and the Aegean Sea (Pérez et al. 2000; Lejeusne et al. 2010). In the NW Mediterranean, seawater temperatures in the summer of 2003 were the warmest in 30 years (Coma et al. 2009; Garrabou et al. 2009). In both years 1999 and 2003, not only were temperatures high, but also the warm and stable conditions lasted for an unusually long time; a positive correlation was observed between mortality rates and exposure to heat stress (Garrabou et al. 2009). The severity of the 1999 event was very high, with for instance about 50 % of *S. officinalis* individuals

Fig. 35.6 Survey of a labeled individual of the *Spongia officinalis* population in the Port-Cros National Park. During the last 10 years, this individual suffered several necroses which affected its shape and made its size highly fluctuate between 3,000 and 7,000 cm³. Here it is illustrated by underwater photographs of this individual taken in 2007 with no necrosis, in 2009 with a high necrosis rate, and in 2010 with the remaining healed part of the individual (Photos T. Pérez/CNRS)

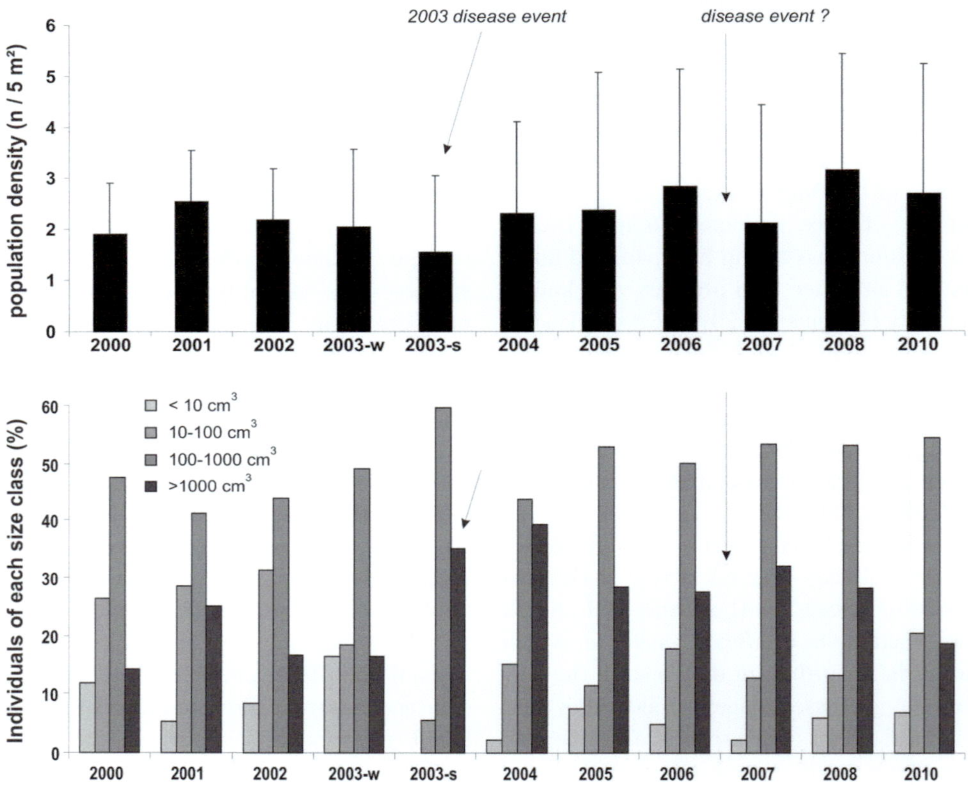

Fig. 35.7 *Spongia officinalis* populations in the Port-Cros National Park were monitored in 1 × 5 m quadrats deployed on a permanent transect. Each year, a minimum of 20 quadrats were studied during winter, with an additional survey in summer of 2003 after the mass mortality event. No surveys were performed in 2009. Sponge volumes were assessed by measuring their three largest dimensions. The resulting indicator of size expressed in cm³ corresponds to the smallest box where the sponge can be inserted without any squeezing. Size classes: <10 cm³ correspond to newly recruited individuals; 10–100 cm³ to individuals of 1–3 previous year classes; 100–1,000 cm³ to individuals of average size; and >1,000 cm³ to large individuals

affected in the Port-Cros National Park, and mortality rates of about 90 % in some other locations. The other commercial species that were known in Port-Cros, *S. lamella* and *H. communis*, totally disappeared, and only scattered individuals of *S. officinalis* have survived in this part of the NW Mediterranean. Since then, population monitoring has been implemented in locations where this species is most abundant (Figs. 35.6 and 35.7). The surveys conducted over

10 years showed a population density fluctuating slightly between 2 and 3 individuals per 5 m^2 (on average). The effects of the 2003 event has been detected both in overall incidence (about 25 % of sponges affected, and a decrease in population density to 1.5 individuals per 5 m^2), as well as a change in population size structure (Fig. 35.7). Whereas small sponges of less than 100 cm^3 in size represented about 35 % of the surveyed population before the mortality (38, 34 39, 35 % in respectively 2000, 2001, 2002 and winter 2003), they only represented approximately 5 % after the summer of 2003. The youngest specimens thus appeared to be more sensitive to heat stress; although the mortality was followed by a strong recruitment of new sponges, this category remained less represented in the population than previously (17, 19 and 23 % in respectively 2004, 2005 and 2006). A new drop in population density observed in 2007 (from 2.8 individuals per m^2 in winter 2006 to 2.1 individuals in winter 2007), together with a decrease in percentage of young sponges (from 23 to 15 %), might be related to a new but moderate mortality event that occurred in summer 2006 (Fig. 35.7). Thus, since 2003, some localized mortality episodes have been observed almost every year in different regions of the NW Mediterranean Sea, but the extent of these episodes has been very limited.

Conclusions

The sponge fishery is still an important economic activity in the Mediterranean Sea, although the sponge trade has declined since its peak at the end of the nineteenth century and the beginning of the twentieth. It also represents a valuable cultural heritage, especially in Greece, Tunisia and other countries of the eastern Mediterranean Sea. This activity has been threatened several times by overexploitation and by disease outbreaks. Recent scientific investigations have provided more precise information about the biology of commercial sponges to explain some of the population crashes, which may help to implement more sustainable exploitation with better control and regulation, including harvesting by cutting rather than tearing the sponge loose (Stevely and Sweat 1985), and possibly result in a profitable spongiculture industry. However, it also appears that the global change in the Mediterranean, especially unusual warming episodes, the extension of invasive species, and the global level of pollution, may increase the threats to the traditional sponge beds and the culture of sponge fishing.

Acknowledgements We gratefully acknowledge the financial and logistical support of the National Park of Port-Cros. Some data on the Tunisian fisheries were provided by Hechmi Missaoui from Institut National d'Agronomie de Tunis. We thank Catherine Cypreos (Cypreos France Eponges, www.sponges.fr) for her information on the sponge trade. Figures 35.2 and 35.4 were kindly provided by respectively Nicole Boury-Esnault and Daniel Faget (Aix-Marseille Université). The English text was thoroughly revised by Clive Wilkinson.

References

Abdul Wahab MA, De Nys R, Whalan S (2012) Closing the lifecycle for the sustainable aquaculture of the bath sponge *Coscinoderma matthewsi*. Aquaculture 324–325:281–289

Adams C, Stevely JM, Sweat D (1995) Economic feasibility of small-scale sponge farming in Pohnpei, Federated States of Micronesia. J World Aquac Soc 26:132–142

Allemand-Martin A (1906) Etude de physiologie appliquée à la spongiculture sur les côtes de Tunisie. Dissertation, University of Lyon

Arndt W (1937) Schwämme. In: Pax F, Arndt W (eds) Die Rohstoffe des Tierreichs, vol 1. Verlag von Gebrüder Borntraeger, Berlin

Baldacconi R, Corriero G (2009) Effects of the spread of the alga *Caulerpa racemosa* var. *cylindracea* on the sponge assemblage from coralligenous concretions of the Apulian coast (Ionian Sea, Italy). Mar Ecol Evol Perspect 30:337–345

Baldacconi R, Nonnis-Marszano C, Gaino E, Corriero G (2007) Sexual reproduction, larval development and release in *Spongia officinalis* L. (Porifera, Demospongiae) from the Apulian coast. Mar Biol 152:969–979

Bally M, Garrabou J (2007) Thermodependent bacterial pathogens and mass mortalities in temperate benthic communities: a new case of emerging disease linked to climate change. Glob Change Biol 13:2078–2088

Ben Mustapha K, Vacelet J (1991) Etat actuel des fonds spongifères de Tunisie. In: Boudouresque CF et al (eds) Les espèces marines à protéger en Méditerranée. GIS Posidonie Publ, Marseille

Castritsi-Catharios J (1996) Sponges species in the Mediterranean. EEC DG XIV-C-1, Research Contract MED/92/024, final report, Athens

Castritsi-Catharios J (1998) Kalymnos and the secrets of the sea. Contribution to the sponge fisheries. Commission of the European Communities, Athens

Castritsi-Catharios J, Magli M, Vacelet J (2007) Evaluation of the quality of two commercial sponges by tensile strength measurement. J Mar Biol Assoc UK 87:1765–1771

Castritsi-Catharios J, Miliou H, Kapiris K, Kefalas E (2011a) Recovery of the commercial sponges in the central and southeastern Aegean Sea (NE Mediterranean) after an outbreak of sponge disease. Mediterr Mar Sci 12:5–20

Castritsi-Catharios J, van Soest RM, Kefalas E, Vacelet J (2011b) Revised description of a poorly known Mediterranean Dictyoceratid bath sponge, *Spongia (Spongia) zimocca* (Schmidt, 1862) (Porifera: Demospongiae: Dictyoceratida). Zootaxa 2812:41–62

Celik I, Cirik S, Altmagc U, Ayaz A, Celik P, Tekesoglu H, Yilmaz H, Oztekin A (2011) Growth performance of bath sponge (*Spongia officinalis* Linnaeus, 1759) farmed on suspended ropes in the Dardanelles (Turkey). Aquac Res 42:1807–1815

Cerrano C, Bavestrello G, Bianchi CN, Cattaneo-Vietti R, Bava S, Morganti C, Morri C, Picco P, Sara G, Schiaparelli S, Siccardi A, Sponga F (2000) A catastrophic mass-mortality episode of gorgonians and other organisms in the Ligurian Sea (North-western Mediterranean), Summer 1999. Ecol Lett 3:284–293

Chung IF, Huang YM, Lee TH, Liu LL (2010) Reproduction of the bath sponge *Spongia ceylonensis* (Dictyoceratida: Spongiidae) from Penghu, Taiwan. Zool Stud 49:601–607

Coma R, Ribes M, Serrano E, Jimenez E, Salat J, Pascual J (2009) Global warming-enhanced stratification and mass mortality events in the Mediterranean. Proc Natl Acad Sci U S A 106: 6176–6181

Croft RA (1995) Commercial sponge survey in Kiribati and sponge farming development in the South Pacific. FAO South Pacific Aquaculture Development Project, Suva

Dailianis T, Tsigenopoulos CS, Dounas C, Voultsiadou E (2011) Genetic diversity of the imperilled bath sponge *Spongia officinalis* Linnaeus, 1759 across the Mediterranean Sea: patterns of population differentiation and implications for taxonomy and conservation. Mol Ecol 20:3757–3772

Duckworth A (2009) Farming sponges to supply bioactive metabolites and bath sponges: a review. Mar Biotechnol 11:669–679

Ereskovsky AV (2010) The comparative embryology of sponges. Springer, Saint-Petersburg

Fell PE (1976) The reproduction of *Haliclona loosanoffi* and its apparent relationship to water temperature. Biol Bull 150:200–210

Flégel CH (1910) The abuse of the scaphander in the sponge fisheries. Bull Bur Fish 28:513–543

Gaino E, Pronzato R (1989) Ultrastructural evidence of bacterial damage to *Spongia officinalis* fibres (Porifera, Demospongiae). Dis Aquat Org 6:67–74

Gaino E, Burlando B, Zunino L, Pansini M, Buffa P (1984) Origin of male gametes from choanocytes in *Spongia officinalis* (Porifera, Demospongiae). Int J Invertebr Reprod Dev 7:83–93

Galtsoff PS (1942) Wasting disease causing mortality of sponges in the West Indies and Gulf of Mexico. Proc 8th Am Sci Congr 3:411–421

Garrabou J, Coma R, Bally M, Bensoussan N, Chevaldonné P, Cigliano M, Diaz D, Harmelin J-G, Gambi MC, Pérez T, Ribes M, Romano J-C, Serrano E, Teixido N, Torrents O, Zabala M, Zuberer F, Cerrano C (2009) Mass mortality in northwestern Mediterranean rocky benthic communities: effects of the 2003 heat wave. Glob Change Biol 15:1090–1103

Hoppe WF (1988) Reproductive patterns in three species of large coral reef sponges. Coral Reef 7:45–50

Kaye HR (1991) Sexual reproduction in four Caribbean commercial sponges. 2. Oogenesis and transfer of bacterial symbionts. Invertebr Reprod Dev 19(1):13–24

Kaye HR, Reiswig HM (1991a) Sexual reproduction in four Caribbean commercial sponges. 1. Reproductive cycles and spermatogenesis. Invertebr Reprod Dev 19(1):1–11

Kaye HR, Reiswig HM (1991b) Sexual reproduction in four Caribbean commercial sponges. 3. Larval behaviour, settlement and metamorphosis. Invertebr Reprod Dev 19(1):25–35

Kefalas E, Tsirtsis G, Castritsi-Catharios J (2003) Distribution and ecology of Demospongiae from the circalittoral of the islands of the Aegean Sea (Eastern Mediterranean). Hydrobiologia 499:125–134

Kushmaro A, Loya Y, Fine M, Rosenberg E (1996) Bacterial infection and coral bleaching. Nature 380:396

Lejeusne C, Chevaldonné P, Pergent-Martini C, Boudouresque CF, Pérez T (2010) Climate change effects on a miniature ocean: the highly diverse, highly impacted Mediterranean Sea. TREE 25:250–260

Marenzeller E (1879) Die Aufsucht des Badeschwämmes aus Theilstücken. Verhandl zool Bot Ges Wien 28:687–694

Martin Y, Bonnefont JL, Chancerelle L (2002) Gorgonians mass mortality during the 1999 late summer in French Mediterranean coastal waters: the bacterial hypothesis. Water Res 36:779–782

Meroz-Fine E, Shefer S, Ilan M (2005) Changes in morphology and physiology of an East Mediterranean sponge in different habitats. Mar Biol 147:243–250

Noyer C, Agell G, Pascual M, Becerro MA (2009) Isolation and characterization of microsatellite loci from the endangered Mediterranean sponge *Spongia agaricina* (Demospongiae: Dictyoceratida). Conserv Genet 10:1895–1898

Pérez T, Garrabou J, Sartoretto S, Harmelin J-G, Francour P, Vacelet J (2000) Mortalité massive d'invertébrés marins : un événement sans précédent en Méditerranée nord-occidentale. C R Acad Sci Paris 323:853–865

Pérez T, Sarrazin L, Rebouillon P, Vacelet J (2002) First evidences of surfactant biodegradation by marine sponges (Porifera): an experimental study with a linear alkylbenzenesulfonate. Hydrobiologia 489:225–233

Pérez T, Wafo E, Fourt M, Vacelet J (2003) Marine sponges as biomonitor of polychlorobiphenyl contamination: concentration and fate of 24 congeners. Environ Sci Technol 37:2152–2158

Pérez T, Longet D, Schembri T, Rebouillon P, Vacelet J (2005) Effects of 12 years' operation of a sewage treatment plant on trace metal occurrence within a Mediterranean commercial sponge (*Spongia officinalis*, Demospongiae). Mar Pollut Bull 50:301–309

Pile AJ, Young CM (2006) The natural diet of a hexactinellid sponge: Benthic-pelagic coupling in a deep-sea microbial food web. Deep-Sea Res I 53:1148–1156

Pile AJ, Patterson MR, Witman JD (1996) *In situ* grazing on plancton <10 μm by the boreal sponge *Mycale lingua*. Mar Ecol Prog Ser 141:95–102

Pile AJ, Patterson MR, Savarese M, Chernykh VI, Fialkov VA (1997) Trophic effects of sponge feeding within Lake Baikal's littoral zone. 2. Sponge abundance, diet, feeding efficiency, and carbon flux. Limnol Oceanogr 42:178–184

Pronzato R (1999) Sponge-fishing, disease and farming in the Mediterranean Sea. Aquat Conserv 9:485–493, Aquatic Conservation: Marine and Freshwater Ecosystems

Pronzato R, Manconi R (2008) Mediterranean commercial sponges: over 5000 years of natural history and cultural heritage. Mar Ecol 29:146–166

Pronzato R, Bavestrello G, Cerrano C (1998) Morphofunctional adaptations of three species of *Spongia* (Porifera, Demospongiae) from a Mediterranean vertical cliff. Bull Mar Sci 63:317–328

Pronzato R, Bavestrello G, Cerrano C, Magnino G, Manconi R, Pantelis J, Sarà A, Sidri M (1999) Sponge farming in the Mediterranean sea: new perspectives. Meml Qld Mus 44:485–491

Savarese M, Patterson MR, Chernykh VI, Fialkov VA (1997) Trophic effects of sponge feeding within Lake Baikal's littoral zone. 1. *In situ* pumping rates. Limnol Oceanogr 42:171–178

Scalera-Liaci L, Sciscioli M, Matarrese A, Giove C (1971) Osservazioni sui cicli sesuali di alcune Keratosa (Porifera) e loro interesse negli studi filogenetici. Atti Soc Pelorit Sci Fisic Matem Nat 17:33–52

Smith FGW (1941) Sponge disease in British Honduras, and its transmission by water currents. Ecology 22:415–421

Stevely J, Sweat D (1985) Survival and growth of cut versus hooked commercial sponges in the Florida keys. Technical paper no 38. Florida Sea Grant Program, University of Florida, Gainesville, 12 pp

Storr JF (1964) Ecology of the Gulf of Mexico commercial sponges and its relation to the fishery. Spec Sci Rep Fish 466:1–73

Topçu EN, Pérez T, Gregori G, Harmelin-Vivien M (2010) In situ investigation of *Spongia officinalis* (Demospongiae) particle feeding: coupling flow cytometry and stable isotope analysis. J Exp Mar Biol Ecol 389:61–69

Trussel GC, Lesser MP, Patterson MP, Genovese SJ (2006) Depth-specific differences in growth of the reef sponge *Callyspongia vaginalis*: role of bottom-up effects. Mar Ecol Prog Ser 323: 149–158

Tuzet O, Pavans de Ceccatty M (1958) La spermatogenèse, l'ovogenèse, la fécondation et les premiers stades du développement d'*Hippospongia communis* LMK (= *H. equina* O.S.). Bull Biol Fr Belg 92:1–18

Vacelet J (1985) Bases historiques et biologiques d'une éventuelle spongiculture. Oceanis 11:551–584

Vacelet J (1994) The struggle against the epidemic which is decimating Mediterranean sponges. FAO technical report, Rome

Vacelet J, Verdenal B, Perinet G (1988) The iron mineralization of *Spongia officinalis* L. (Porifera, Dictyoceratida) and its relationships with the collagen skeleton. Biol Cell 62:189–198

Vacelet J, Gaino E, Gallissian M-F, Vacelet E (1994) Bacterial attack of spongin skeleton during the 1986–1990 Mediterranean sponge disease. In: van Soest RWM, van Kempen TMG, Braekman JC (eds) Sponges in time and space; biology, chemistry, paleontology. Balkema, Rotterdam

Verdenal B, Vacelet J (1990) Sponge culture on vertical ropes in the northwestern Mediterranean Sea. In: Rützler K (ed) New perspectives in sponge biology. Smithsonian Institution Press, Washington, DC

Verdenal B, Verdenal M (1987) Evaluation de l'intérêt économique de la culture d'éponges commerciales sur les côtes méditerranéennes françaises. Aquaculture 64:9–29

Verdenal B, Diana C, Arnoux A, Vacelet J (1990) Pollutant levels in Mediterranean commercial sponges. In: Rützler K (ed) New perspectives in sponge biology. Smithsonian Institution Press, Washington, DC

Voultsiadou E, Dailanis T, Antoniadou C, Vafidis D, Dounas C, Chintigoglou CC (2011) Aegean bath sponges: historical data and current status. Rev Fish Sci 19:34–51

Warn F (2000) Bitter Sea, the real story of Greek sponge diving. Guardian Angel Press, South Woodham Ferrers

Webster NS (2007) Sponge disease: a global threat? Environ Microbiol 9:1363–1375

Wells SM (1983) Porifera. In: Wells SM, Pyle RM, Collins NM (eds) The IUCN invertebrate red data book. IUCN, Gland

West R, McKinney FK, Fagestrom JA, Vacelet J (2011) Biological interactions among extant and fossil clonal organisms. Facies 57:351–374

Whalan S, Battershill C, Nys RD (2007) Sexual reproduction of the brooding sponge *Rhopaloeides odorabile*. Coral Reef 26:655–663

Witte U, Barthel D (1994) Reproductive cycle and oogenesis of *Halichondria panicea* (Pallas) in Kiel Bight. In: van Soest RWM, van Kempen TMG, Braekman JC (eds) Sponges in time and space; biology, chemistry, paleontology. Balkema, Rotterdam

Witte U, Barthel D, Tendal O (1994) The reproductive cycle of the sponge *Halichondria panicea* Pallas (1766) and its relationship to temperature and salinity. J Exp Mar Biol Ecol 183:41–52

Wolfrath B, Barthel D (1989) Production of faecal pellets by the marine sponge *Halichondria panicea* Pallas (1766). J Exp Mar Biol Ecol 129:81–94

Yahel G, Sharp JH, Marie D, Häse C, Genin A (2003) *In situ* feeding and element removal in the symbiont-bearing sponge *Theonella swinhoei*: bulk DOC is the major source for carbon. Limnol Oceanogr 48:141–149

Yahel G, Marie D, Genin A (2005) InEx – a direct in situ method to measure filtration rates, nutrition, and metabolism of active suspension feeders. Limnol Oceanogr 3:46–58

Yahel G, Whitney F, Reiswig HM, Eerkes-Medrano DI, Leys SP (2007) In situ feeding and metabolism of glass sponges (Hexactinellida, Porifera) studied in a deep temperate fjord with a remotely operated submersible. Limnol Oceanogr 52:428–440

Zarrouk S, Ereskovsky AV, Ben Mustapha K, El Abed A, Pérez T (in press) Sexual Reproduction of *Hippospongia communis* (Lamarck, 1814) (Dictyoceratida, Demospongiae): comparison of two populations living in contrasted environmental conditions. Mar Ecol Evol Perspect

Zuljevic A, Thibaut T, Despalatovic M, Cottalorda J-M, Nikolic V, Cvitkovic I, Antolic B (2011) Invasive alga *Caulerpa racemosa* var. *cylindracea* makes a strong impact on the Mediterranean sponge *Sarcotragus spinosulus*. Biol Invasion 13:2303–2308

Seafood Safety and Human Health Implications

36

António Marques, Rui Rosa, and Maria Leonor Nunes

Abstract

Current projections for the twenty-first century show that global warming will accelerate, with stronger storms, extreme precipitation, dry spells and rising sea levels as the primary symptoms. Such changes will have implications in seafood production, security, and safety, as well as in human health, due to the increase in the frequency of harmful algal blooms, levels and bioaccumulation of several chemical contaminants, prevalence and virulence of common foodborne pathogenic microorganisms. The Mediterranean will be particularly affected by climate change due to water scarcity in the region. Assuring seafood safety in such scenario requires the active involvement of all stakeholders to elaborate and implement adaptation and mitigation plans. In this context, the current chapter aims to provide an overview of the potential effects of climate change in the Mediterranean seafood safety and human health, taking into account chemical and biological contaminants, and to discuss potential adaptation and mitigation measures.

Keywords:

Climate change • Seafood safety • Benefit and risk assessment • Adaptation • Mitigation

Introduction

Seafood is a balanced, nutritious, readily digestible and healthy food for human consumption, able to prevent coronary heart diseases, hypertension, diabetes and cancer, since contains low cholesterol levels, high quality proteins with all essential amino acids, polyunsaturated n-3 fatty acids, liposoluble vitamins and essential elements like calcium, iodine and selenium (Simopolpoulos 1997). Compared with land animals, seafood has a far higher percentage of edible flesh and little waste, with few exceptions like shellfish. Nonetheless, some seafood is affected by the accumulation of microbiological and chemical contaminants, particularly toxic elements (e.g. Hg, Cd, Pb and As), toxins from harmful algal blooms (HABs) and pathogenic microorganisms, which can be extremely dangerous for human health.

The Mediterranean is an oligothrophic, semi-enclosed and water deficitary sea with 2.5 million km^2 surface, incorporating water from the Atlantic Ocean and Black Sea, and contributing with 0.8 % to the total world marine surface and 0.3 % volume (Lleonart and Maynou 2003). The Mediterranean Sea is located in the temperate zone of the Northern hemisphere with a marked seasonal cycle, i.e. absence of precipitation and stratified water masses in summer with constant temperature 13 °C below 400 m and surface temperatures ranging between 14 and 26 °C (Lleonart and Maynou 2003; EEA 2006). Its salinity is high (mean around 39 g L^{-1}), higher in the Eastern basin and lower in the western. Oxygen levels are almost saturated in the surface

A. Marques (✉) • M.L. Nunes
Divisão de valorização e Aquacultura, Instituto Português do Mar e da Atmosfera (IPMA, I.P.), Avenida de Brasília, 1449-006 Lisbon, Portugal
e-mail: amarques@ipma.pt; mlnunes@ipma.pt

R. Rosa
Laboratório Marítimo da Guia, Centro de Oceanografia, Faculdade de Ciências da, Universidade de Lisboa, Av. Nossa Senhora do Cabo, 939, 2750-374 Cascais, Portugal
e-mail: rrosa@fc.ul.pt

S. Goffredo and Z. Dubinsky (eds.), *The Mediterranean Sea: Its history and present challenges*,
DOI 10.1007/978-94-007-6704-1_36, © Springer Science+Business Media Dordrecht 2014

layer (6 mg L^{-1} in winter and 4.8 mg L^{-1} in summer), and slightly lower in deep waters (4.5 mg L^{-1} in the western and 4.2 mg L^{-1} in the eastern) (EEA 2006). The Mediterranean counts 22 border countries, only possess territorial waters within 12 nautical miles, and the continental shelf is mostly a narrow coastal fringe (Ronzitti 1999). Consequently, the international management structures have not been enforced sufficiently and regular assessment by international working groups have only recently started (Lleonart and Maynou 2003). From fisheries and aquaculture point of view, the Mediterranean has a rich marine fauna and flora, representing 8–9 % of world seas species richness (4–18 % according to the group of species considered e.g. mollusc, echinoderms, crustaceans, etc.), including 18 % of the flora, of which 28 % are endemic (Caddy 1998; EEA 2006). Mediterranean seafood production in 2009 accounted 3.5 % of the world production (Eurostat 2011).

The Mediterranean has long been identified as a "hot spot" for substantial impact of climate change in the future because of water scarcity in the region, a rapidly increasing population. Mediterranean climate modelling project increased risk of changes in temperature, precipitation, moisture, extreme events (droughts and floods) and sea level rise (EEA 1999).

Importance of Seafood in Mediterranean Countries

Seafood is widely consumed in the Mediterranean (average 16.5 kg/capita/year in the 22 countries, ranging from 5.2 in Algeria and 40.0 in Spain; data from 2007; FAO 2011a), being the consumption of several seafood species strongly linked to religious traditions and social (e.g. cod in Christmas). Interestingly, Mediterranean fisheries provide only 7.2 kg of the total consumption, with the rest being imported. The growth of seafood demand in the Mediterranean is expected to increase in the future, especially in southern countries (Cochrane and de Young 2007).

One fourth of the Mediterranean seafood supply comes from aquaculture activities, whereas the remaining is from fisheries. Aquaculture production in the Mediterranean reached 1.3 million tonnes in 2009, representing approximately 1.8 % of the world aquaculture production (72 million tonnes) (FAO 2011a). Although Mediterranean aquaculture was mostly focused on mollusc production in the past (62 % in 1992), the share of fish production is constantly increasing (from 37 % in 1992 to 84 % in 2009) (FAO 2011a). Fish aquaculture production is mainly represented by tilapia (*Oreochromis niloticus*), carp (*Cyprinus carpio*), trout (*Oncorhynchus mykiss*), seabream (*Sparus aurata*), seabass (*Dicentrarchus labrax*), mullet (*Mugil cephalus*), European eel (*Anguilla anguilla*) and turbot (*Psetta maxima*) (Basurco and Lovatelli 2003). In contrast, molluscan shellfish aquaculture production is mainly represented by mussels (*Mytilus edulis* and *Mytilus galloprovincialis*), oysters (*Crassostrea gigas*) and clams (*Ruditapes philippinarum*) (Basurco and Lovatelli 2003).

Fisheries production in the Mediterranean region attained four million tonnes in 2009, approximately 4.5 % of the world fisheries production (90 million tonnes) (FAO 2011a). Mediterranean fisheries are mostly dependent on small scale artisanal fisheries (80 % vessels are lower than 12 m length) with bottom otter trawls, purse seines and coastal gears (e.g. gillnets, trammel nets, long-lines, hand-lines with hooks, traps and pots) that mostly capture fish (89 %), molluscs (8 %) and crustaceans (3 %) (data from 2009, FAO 2011a). The most important species captured in Mediterranean are European anchovy (*Engraulis encrasicolus*), sardine (*Sardina pilchardus*), sardinellas (*Sardinella* sp.), horse mackerel (*Trachurus* spp.), sprat (*Sprattus sprattus*), bonito (*Sarda* sp.), bogue (*Boops boops*), hake (*Merluccius* sp.), bluefin tuna (*Thunnus thynnus*), blue whiting (*Micromesistius poutassou*), swordfish (*Xiphias gladius*), octopus (*Octopus* sp.) and striped venus clam (*Chamelea gallina*) (Eurostat 2008). Currently, several Mediterranean seafood stocks are overexploited: (a) demersal fish are almost fully exploited, if not over-exploited, with a general trend towards smaller individual sizes; (b) small pelagic fish stocks are highly variable in abundance (depending on environmental conditions) and not fully exploited, except for anchovy; (c) large pelagic fish (tuna and swordfish) are overexploited, especially red tuna for which the Mediterranean is an important spawning area; and (d) habitats of high biological significance, such as the *Posidonia oceanica* meadows, are frequently destroyed by trawl-nets operating close to the shore (EEA 1999).

Current Problems Affecting the Mediterranean Sea

Socio-economical

The concentration of resident and non-resident populations and human activities around the Mediterranean Sea represent considerable threats to coastal ecosystems and seafood resources that are expected to increase in the future. The resident population of the Mediterranean was 246 million in 1960, is currently 450 million, and is expected to rise to 520–570 million in 2030 and 600 million in 2050 (EEA 1999). Population density is greater in coastal regions, especially near big cities. Additionally, the Mediterranean is the world's leading tourist destination, accounting 30 % of international tourism. Coastal tourism is strongly seasonal and has been steadily increasing annually (135 million in 1990 and is expected to reach 350 million in 2025, EEA 1999).

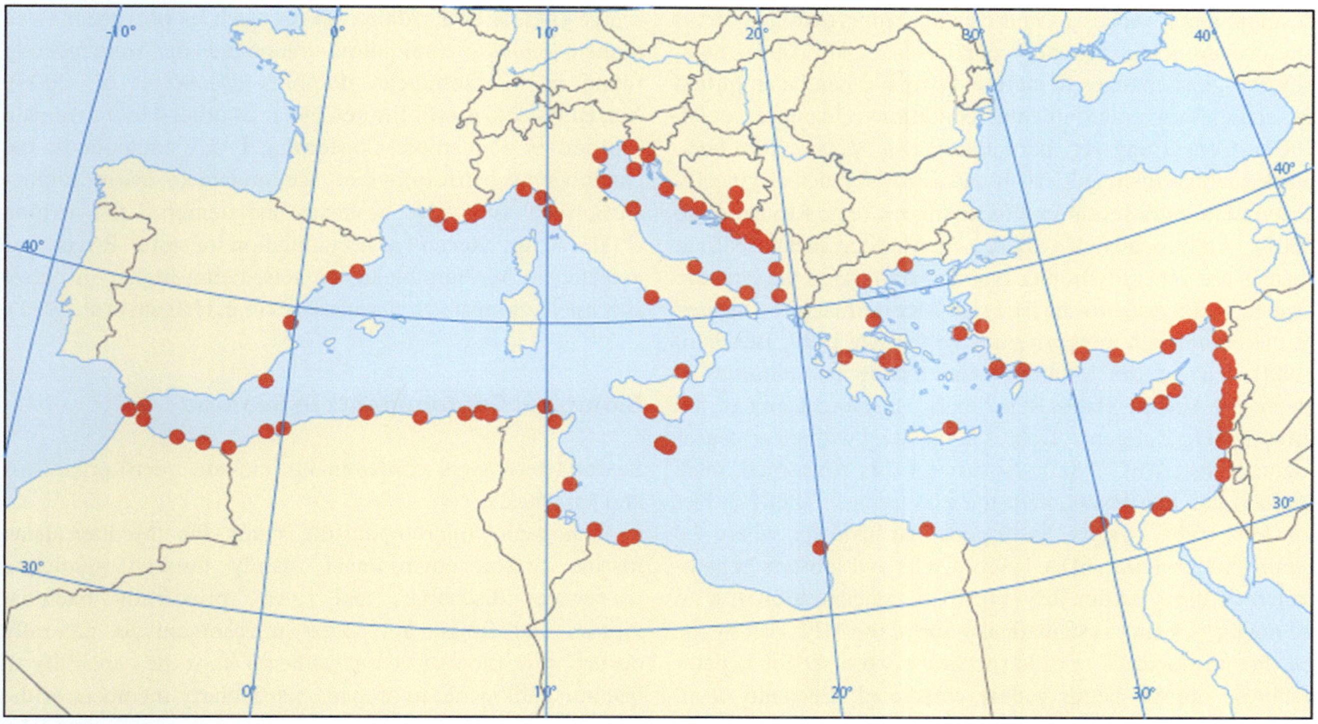

Fig. 36.1 Pollution hot spots (*red dots*) along the Mediterranean coast (Adapted from EEA 2006)

The specific morphology of the Mediterranean basin allows intense agricultural activity in the limited coastal plains. The main pressures from agriculture are soil erosion and nutrient surplus when excessive fertilisers are applied (EEA 1999).

There is a wide range of industrial activities scattered around the Mediterranean basin, with 161 identified hot-spots of heavy industrial complexes (e.g. chemical/petrochemical and metallurgy) and big commercial harbours concentrated mainly in the north-west (EEA 1999, 2006). Other industrial sectors include treatment of wastes and solvent regeneration, surface treatment of metals, production of paper, paints and plastics, and tanneries (EEA 1999).

Concerning maritime traffic, the Mediterranean is one of the most important routes in the world. It is estimated that about 220,000 vessels of more than 100 tonnes cross the Mediterranean each year, representing 30 % of the total merchant shipping in the world and 20 % of oil shipping (EEA 1999). This represents a huge concern with the introduction of exotic species and pollution.

Chemical Contaminants in Seafood

Mediterranean health authorities and consumers are deeply concerned with the exposure to toxic chemical contaminants in seafood. In recent years, the contamination of the marine environment by chemical contaminants has risen due to the global increase of population and industrial development (e.g. non controlled discharges from chemical industries, sewage and agriculture, EEA 1999, see pollution hot-spots in Fig. 36.1). In order to ensure seafood stakeholders confidence, EU authorities have established Maximum Permissible Levels (MPLs) for the presence of several contaminants in seafood (e.g. Commission regulation 1881/2006 and later amendments for Hg, Cd and Pb).

Toxic elements like Hg, Cd, As and Pb are the chemical contaminants of major concern in aquatic environments. These contaminants bioaccumulate along the food chain, though only part is used for metabolic purposes, while the remaining is excreted (through faeces, eggs or moulting) or detoxified (binding to proteins, such as metallothioneins, or to insoluble metaliferous granules) (Rainbow 2002). Bioaccumulation is a serious problem due to the toxicity for fish and humans when reaching a substantially high level (Islam and Tanaka 2004; Francesconi 2007).

The main toxicity mechanisms of these contaminants are related to the osmotic disturbances and alterations of enzyme synthesis and activity (Jezierska et al. 2009). Additionally, toxic metals in seafood can affect various physiological processes, including tissue damages, inability to regenerate damaged tissues, growth inhibition, damages to genetic material and changes in breeding and development (Jezierska et al. 2009). The embryonic development, soon after fertilization, and the hatching period are the most sensitive periods to toxic elements intoxication, resulting in disturbances of developmental processes and causing embryonic and larval malformation and mortality (Jezierska et al. 2009). Such physiological changes in marine organisms can be amplified by changes in environmental conditions (Jezierska et al. 2009). Concerning Hg, particularly fish species, like tuna, black scabbardfish, anglerfish and elasmobranchs, typically accumulate high levels in muscle tissues (e.g. Afonso et al. 2007). Additionally, Hg values in Mediterranean seafood species are generally higher than those found in the Atlantic Ocean, mainly due to the fact that Mediterranean is located in the Himalayan mercuriferous belt (EEA 1999; Bernhard 1988). Part of the Mediterranean fishing communities is above the United States Benchmark Hg Dose Limit of ten times the Hg Reference Dose, i.e. the level with clear neurological effects (EC 2005). As far as Cd is concerned, most seafood has low levels, with the exception of shellfish like lobsters, crabs, oysters, gastropods and bivalves, where Cd binds with proteins (FDA 1993). Crabs and lobsters hepatopancreas have particularly high Cd concentration (up to 40 mg kg^{-1}), being systematically above the MPLs set by the EU for crustacean's muscle (no limits were set for hepatopancreas despite being widely consumed; Barrento et al. 2008, 2009; Marques et al. 2010a). The release of Cd from sediments has been reported in several Mediterranean regions, including the Gulf of Trieste, in the northern Adriatic Sea off the mouth of the river Po (Zago et al. 2000). Arsenic and lead are commonly found in seafood, particularly Pb in shrimps and bivalves, and As in algae, fish and crustaceans (e.g. Anacleto et al. 2009).

Persistent organic pollutants (POPs) are also of concern in Mediterranean seafood, since they have the ability to accumulate in the biota, being mostly released from anthropogenic sources (Islam and Tanaka 2004). These pollutants include certain prohibited pesticides and industrial chemicals like polycyclic aromatic hydrocarbons (PAHs), dicloro-difenil-tricloroetano (DDT), aldrin, dieldrin, endrin, chlordane, hexachlorocyclohexanes (HCHs, like lindane), heptachlor, hexachlorobenzene, mirex, toxaphene, polychlorinated biphenyls (PCBs), dioxins and furans. Recently, particular attention is being paid to the presence of emerging contaminants in seafood like pharmaceutical and personal care products, new endocrine disruptors, perfluorinated compounds (PFCs), brominated flame retardants (BFRs) and marine litter released into the environment. The use of organochlorine pesticides (OCPs), like DDT and PCBs, was banned more than 30 years ago in Mediterranean countries, being reflected in the low values registered in Mediterranean seafood (average PCBs below 30 and 20 ng g^{-1} for DDT, EEA 1999). Nonetheless, marine mammals, sardines and swordfish are still affected by high levels of OCPs that tend to accumulate in seafood body fat. In contrast, PFCs tend to accumulate in blood proteins and liver, being recently found in dolphins, swordfish and tuna, despite data is still scarce (Alessi et al. 2006). BFRs, such as polybrominated diphenylethers, are ubiquitous compounds that were recently found in Mediterranean dolphins (Alessi et al. 2006). Yet, BFRs data is still limited as far as other Mediterranean seafood is concerned. Concerning PAHs compounds, the carcinogenic benzo(a)pyrene accumulate in marine organisms, such as bivalve molluscs and demersal fish (Moon et al. 2010). Recently, accumulation of total PAHs was detected in Mediterranean mussels, being higher in native specimens compared to farmed ones (e.g. Galgani et al. 2011).

Biological Contaminants in Seafood

Seafood biological contaminants include micro-organisms and parasites.

Pathogenic microorganisms enter the Mediterranean marine coastal environment mainly through municipal wastewater discharges and rivers, principally bacteria, viruses and fungi. Yet, other microorganisms naturally occurring in the marine environment have the capability to become pathogenic to humans, particularly members of the *Vibrio* genus. These microorganisms are accumulated by seafood, particularly filter-feeding organisms like bivalves. *Vibrio cholerae* is still the leading cause of Vibrio-associated illnesses worldwide, being usually transmitted to humans through contaminated water and not generally considered to be a threat to human health through seafood consumption (Faruque et al. 1998). Yet, *V. vulnificus* and *V. parahaemolyticus* also cause a significant number of clinical infections, usually through the ingestion of raw or incompletely cooked fish or shellfish (Bonner et al. 1983).

Apart from pathogenic microorganisms discharged into the marine environment, another group of naturally occurring marine microorganisms can pose a similar threat to human health when present in large numbers, i.e. marine algae that produce toxins, also known as Harmful Algal Blooms (HABs), mainly composed by dinoflagellates (Table 36.1), to which man is exposed mainly through the consumption of contaminated shellfish. HABs usually occur in areas with excessive organic material, i.e. eutrophication areas (Fig. 36.2), being responsible for the consumption or even depletion of oxygen, and causing a series of secondary problems, including mortality of marine organisms, formation of corrosive and other undesirable substances (e.g. CH_4, H_2S, and NH_3), taste and odour-producing substances, organic acids, mucilage and toxins (EEA 1999, 2006). Marine toxins originating from HABs are generally tasteless and odorless, and heat- and acid-stable, and can cause gastrointestinal, cardiological and neurological problems or induce mortalities after the consumption of contaminated seafood (EEA 2006). Therefore, legislation is available in the EU to protect consumers from ASP (Amnesic Shellfish Poisoning), PSP

Table 36.1 Pathogenic bacteria, viruses and microalgae responsible for algal blooms and toxin production, reported in the Mediterranean Sea

Microalgae		Bacteria	Virus
Diatoms	*Gonyaulax* sp. (PSP/DSP)	*Salmonella* spp.	Enteroviruses
Cerataulina bergonii	*Gymnodinium* sp. (PSP/NSP)	*Shigella* spp.	Poliovirus
Chaetoceros sp.	*Katodinium rotundatum*	*Vibrio cholerae*	Echovirus
Cyclotella sp.	*Peridinium* sp.	*V. algynolyticus*	Coxsackie virus A/B
Leptocylindrus spp.	*Prorocentrum* sp. (DSP)	*V. parahaemolyticus*	Hepatitis A virus
Nitzschia closterium	*Ptrotogonyaulax tamarensis*	*Staphylococcus aureus*	Other viruses
Pseudo-Nitzschia sp. (ASP)	*Scrippsiella trochoidea*	*Pseudomonas aeruginosa*	Adenovirus
Rhizosolenia sp.	Coccolithophores	*Clostridium perfringens*	Rotavirus
Skeletonema costatum	*Coccolithus pelagicus*	*Campylobacter* spp.	
Thalassiosira sp.	*Emiliania huxleyi*	*Aeromonas hydrophila*	
Dinoflagellates	Other flagellates		
Alexandrium sp. (PSP)	*Chlamydomonadaceae*		
Amphidinium curvatum	*Cryptomonas* sp.		
Cachonina sp.	*Cyanobacteria*		
Chattonella subsalsa	*Microflagellates*		
Dinophysis spp. (DSP)	*Noctiluca* sp.		
Gambierdiscus sp. (CFP)	*Pyramimonas* sp.		
Glenodinium sp.	*Spirulina jenneri*		

Adapted from EEA (1999, 2006), Spatharis et al. (2007), Caillaud et al. (2010)
Abbreviations: *ASP* amnesic shellfish poisoning, *DSP* diarrheic shellfish poisoning, *PSP* paralytic shellfish poisoning, *NSP* neurotoxic shellfish poisoning, *CFP* ciguatera fish poisoning

(Paralytic Shellfish Poisoning), and some lipophilic toxins, including okadaic acid, dinophysistoxins, pectenotoxins and azaspiracids. Yet, many other natural toxins produced by HAB organisms can affect human health due to the ingestion of contaminated seafood (primarily shellfish), such as diarrheic shellfish poisoning (DSP), neurotoxic shellfish poisoning (NSP) and ciguatera fish poisoning. Recent detection of the tropical genus *Gambierdiscus* responsible for the production of ciguatera fish poisoning (CFP) ciguatoxins occurred in the Mediterranean Sea, as well as in north-eastern Atlantic Ocean, Canary Islands and Madeira (Caillaud et al. 2010). CFP is responsible for severe human disorders due to the consumption contaminated fish. Palytoxin is one of the most potent toxins known in nature, reported in Mediterranean seafood, including sardines, but was also reported to occur in other fish, molluscs, crustaceans and equinoderms (Yasumoto and Satake 1998; Onuma et al. 1999). Gymnodimine toxin (GYM) was also recently identified in the digestive gland of clams *Ruditapes decussatus* from Tunisia, as well as in greenshell mussel, blue mussel, scallop, cockle, surf clam, oyster and abalone (Stirling 2001; Biré et al. 2002; MacKenzie et al. 2002). Despite the GYM chronic toxicity remains unclear, its role in the development of neurodegenerative illnesses like Alzheimer or Parkinson's diseases has been debated (Alonso et al. 2011).

About 35 parasitic diseases have been reported in the Mediterranean Sea (Athanassopoulous et al. 2009). The main parasitic diseases in freshwater seafood are trichodiniasis, costiasis, white spot diseases, *Dactylogyrus* and gyrodactilosis. In contrast, the main parasitic diseases in marine seafood are trichodiniasis, costiasis, *Enteromyxum leei*, *Ceratomyxa*, amyloodiniosis, mycrocotylosis and sea lice disease. There is an increasing concern of parasitic diseases in seafood from the Mediterranean area, particularly farmed seafood, since some parasites triggers mortality episodes, such as *Amyloodinium* (Dinoflagellates), Scuticociliatida (Ciliates), *Enteromyxum* spp. (Myxosporea) or Mycrocotylidae (Monogenea). The geographical spread of parasites (and their hosts) to areas where previously they were inhibited by lower temperature is a reality despite being moderated by several ecological and environmental factors (Harvell et al. 1999, 2002).

Climate Change in the Mediterranean and Seafood Safety

Projected Environmental Fluctuations

Several reports from the Intergovernmental Panel on Climate Change (IPCC) indicate that variations in world climate will be reflected in the Mediterranean region, with a discernible trend of increased salinity and warmer temperature in key water masses registered over the last 50 years (IPCC 2007). According to IPCC, potential impacts in the Mediterranean associated to climate change include drought, decline of water quality, floods, changes in soil erosion and desertification, storms, coastal erosion, changes in seawater temperature, salinity and pH, sea level rise and biodiversity reduction

Fig. 36.2 Mediterranean areas where eutrophication phenomena (**a**), harmful algal blooms (HAB); and (**b**) seafood toxin blooms (STB) were reported (Adapted from EEA 1999, 2006)

(EEA 1999). Sea level rise, potentially accelerated by anthropogenic activities, is one of the most important impacts of climate change in coastal zones. New models project an increase in sea level of 3–61 cm until 2100 due to thermal warming, melting glaciers/ice sheets, and aerosol concentrations (Marcos and Tsimplis 2008). The IPCC (2007) estimates that global mean seawater surface temperature will increase 1.1–6.4 °C until 2100. Recent studies indicate that a global temperature rise of 2 °C is likely to lead to a corresponding warming of 1–3 °C in the Mediterranean region (Tin et al. 2005). A general warming trend has been observed in deep waters of the western Mediterranean, where temperatures have increased by 0.12 °C in the past 30 years as a possible result of global warming (Bethoux et al. 1990). Climate change is also expected to trigger seawater pH decrease by 0.14–0.35 units (IPCC 2007). Concerning salinity, the Mediterranean is expected to experience different increasing rates until 2100: (a) the global, Eastern, Western, and Levantine basins, with an average increase of 0.23 g L^{-1}; and (b) the Aegean and Adriatic Seas, with 0.61 and 0.70 g L^{-1} increases, respectively (Sevault et al. 2004). The Earth has not experienced variations of this magnitude in such short timescale and the consequences to future generations are largely unknown (IPCC 2007).

Direct and Indirect Impacts on Seafood

Climate change is expected to have economic effects on fisheries and aquaculture worldwide, such as losses in revenues

Fig. 36.3 Schematic diagram indicating the biophysical and socio-economic impacts of climate change at different levels of organizations, from seafood specimens to the society (Adapted from Sumaila et al. 2011)

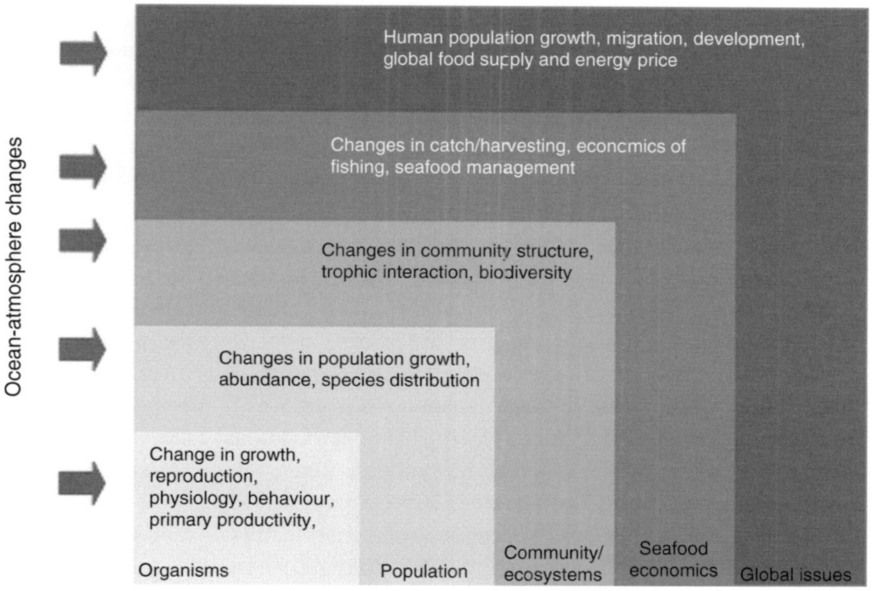

and reduced availability of seafood to consumers, among other impacts (Fig. 36.3). Although the evolution of climate change is uncertain, high diversity coastal ecosystems like the Mediterranean Sea are more vulnerable to environmental perturbations than low diversity places (May 1973). Consequently, a reduction of biodiversity of marine species is expected in Mediterranean with climate change. Among the first species to disappear under heavy stress conditions are benthic organisms with large body size (EEA 1999). Climate change may lead to large-scale redistribution of catch potential, with an average of 5–15 % drop in the Mediterranean region (Cheung et al. 2010).

Climate fluctuations play a predominant role in marine ecosystems and seafood by: (a) directly affecting the organisms through changes in survival, reproductive success and dispersal pattern; (b) promoting modifications in biotic interactions; and (c) affecting ocean currents indirectly (EEA 1999; Gambaiani et al. 2009).

As seawater temperatures warm at a large spatial and temporal scale, the timing of ecological events or phenology may also change (Nye 2010). Many organisms time their migrations and spawning to changes in temperature and photoperiod. As temperature, salinity, and hydrography changes, organisms will likely shift the timing of spawning and migration. If species do not change in unison, the reproductive success for many organisms may be dramatically reduced. In the last decades, the northward expansion of the geographical range of warm-water marine species has been observed in Mediterranean areas due to increase in seawater temperature (e.g. Ligurian and Adriatic Seas, Astraldi et al. 1995; Bello et al. 2004). New species can have an impact on indigenous species through inter-specific competition, predation, and possible genetic degradation of indigenous stock (EEA 1999). In addition to northward migration, bathymetric displacements may occur among populations of invasive and endemic species (Galil and Zenetos 2002). This is the case of indigenous red mullet (*Mullus barbatus*), hake (*Merluccius merluccius*) and spottail mantis shrimp (*Squilla mantis*) that were reported to move into deeper and cooler waters to avoid warm-water competitors (Oren 1957; Galil and Zenetos 2002). In contrast, cold-water species tend to disappear or move to more favourable habitats. Recently, anomalous increase of summer temperatures (2–3 °C) and the deepening of the thermocline in western Mediterranean coastal areas have resulted in massive mortalities of the benthic fauna (e.g. sponges and gorgonians) inhabiting hard substrates (Romano et al. 2000). Mortality was equally attributed to the surface water warming and stability of high sea temperatures over long periods (i.e. several months). Additionally, small temperature shift (0.05–0.10 °C) in Eastern Mediterranean deep sea is sufficient to considerable change species biodiversity (Danovaro et al. 2004).

The tropicalization of the Mediterranean has been recently confirmed in the Adriatic Sea, where among other the jellyfish *Pelagia noctiluca* and three toxic dinoflagellate tropical species occurred in the last years (*Ostreopsis lenticularis, Coolia monotis* and *Prorocentrum mexicanum*) (Bello et al. 2004; Licandro et al. 2010). The expansion of toxic microorganisms combined with the expected higher Mediterranean coastal eutrophication due to the increase of temperature and incidence of extreme events (e.g. floods), is likely to promote the frequency of HABs. HABs and microbial pollution in euthrophized areas generate the depletion of oxygen in the water, with detrimental impacts on productivity, nursery grounds, biodiversity, habitat and human health risks related to the ingestion of contaminated seafood (EEA 2006).

Recently, a sudden input of high nutrient water in Easter Mediterranean lead to an increase in phytoplankton biomass, with the dominance of the toxic HAB *Pseudo-nitzschia calliantha* (Spatharis et al. 2007). Climate change has the potential of increasing the incidence and spread of several foodborne pathogens and parasites, either through the emergence of new ones or through the selection of existing pathogenic strains that differ in survival, persistence, habitat range and ability to be transmitted or infect humans (Gamble 2008). Evidence of the impact of climate change on the transmission of seafood and waterborne diseases comes from a number of sources, e.g. the seasonality of foodborne and diarrhoeal diseases, changes in disease patterns due to temperature (e.g. *Vibrio*), and associations between the incidence of seafood and waterborne illness and severe weather events (Cook et al. 2002; De Paola et al. 2003; FAO 2008a). Changes in other environmental factors, such as salinity and pH, may also result in changes in the distribution and virulence of pathogens (Elena and Lenski 2003; Sokurenko et al. 2006). In aquaculture, the expected change in the incidence of diseases in seafood due to climate change may exacerbate the use of veterinary drugs, leading to higher and unacceptable levels of such drugs in seafood (FAO 2008b).

Ocean acidification (pH decrease) is expected to threat particularly marine organisms with calcified shells, since they may not be able to make the hard calcified shells and their growth is affected (Nye 2010). Some species like the blue crab and American lobster may respond favourably to ocean acidification, whereas most organisms will respond unfavourably (e.g. bivalve species that constitute important commercial fisheries) (Green et al. 2009).

The expected extreme weather events induced by climate change can result in escapes of farmed seafood stock and contribute to reductions in genetic diversity of wild stocks, thus affecting biodiversity (FAO 2008b).

The availability and toxicity of chemical contaminants is expected to vary due to the effect of climate change (Marques et al. 2010b). In regions where intense rainfall is expected to increase, pesticides, fertilisers, organic matter, heavy metals, among other, will be increasingly washed from soils to water bodies (FAO 2008b). Concerning salinity, metals like Cd, Cr, Cu, Hg, Ni and Zn are taken up more slowly by phytoplankton/fungi, annelids, bacteria, molluscs and crustaceans at higher salinities despite their toxicity increases (Hall and Anderson 1995; Modassir 2000). In contrast, no consistent trend has been detected for the toxicity of most POPs with salinity, except organophosphate insecticides (e.g., parathion, mevinphos, terbufos, trichlorfon) that increase at higher salinities (Hall and Anderson 1995). As far as temperature is concerned, high temperatures promote the uptake, bioaccumulation and toxicity of toxic elements (e.g. Cu, Zn, Cd, Pb) and POPs in several marine organisms, including crustaceans, echinoderms and molluscs (Sullivan 1977; Hutchins et al. 1996; Wang et al. 2005; Khan et al. 2006). Warmer seawater temperatures facilitate Hg methylation, and the subsequent uptake of methyl Hg by fish and mammals by 3–5 % for each °C rise in seawater temperature (Booth and Zeller 2005). Mubiana and Blust (2007) revealed a positive correlation between temperature and Cd and Pb accumulation in mussels (*Mytilus edulis* maintained between 6 and 26 °C), while Co and Cu were independent and inversely related to temperature. Temperature increase can promote the inhibitory effects of toxic elements on respiration of several other marine organisms that utilize the Cu-based hemocyanins as respiratory pigments, such as the zebra mussels *Dreissena polymorpha* (Rao and Khan 2000). Monserrat and Bianchini (1995) detected tenfold increase in acute lethality of crabs (*Chasmagnathus granulata*) to methyl parathion with temperature (12–30 °C).

Human Health Implications

The main risks to human health induced by climate change in the Mediterranean Sea arise from intake of pathogenic microorganisms or toxins from infected sea water and beach sand (e.g. faecal streptococci and coliform bacteria), and consumption of seafood contaminated by pathogens (e.g. *Salmonella*, *Shigella*, hepatitis A, *Candida albicans*), toxins from HABs or chemical pollutants (EEA 1999). The extent of damage to the Mediterranean population health still has to be determined. If climate change projections are correct for the Mediterranean Sea, and the expected increase in the occurrence of contaminants indeed occurs, the health risks for human populations will be amplified when eating seafood items like bivalves and predator fish species. In contrast, in Mediterranean areas where contaminant load diminishes, the expected seafood consumption risks will decrease, prevailing the nutritional benefits of eating seafood.

Health Risks from Microbiologically and Toxins Contamination

Pathogenic microorganisms and toxins produced by HABs present in seawater, sediments, beaches and shellfish can be broadly divided into two categories: those that affect the gastrointestinal tract, and those that affect other parts of the body (EEA 1999). As far as the former category is concerned, diseases spread by the faecal/toxins-oral route can occur, including: (a) bacterial diseases such as salmonellosis (including typhoid and paratyphoid fevers), shigellosis (bacillary dysentery), cholera and gastro-enteritis caused by enteropathogenic *Escherichia coli* and *Yersinia enterocolitica*; (b) viral diseases (e.g. hepatitis A and E), illnesses caused by enteric viruses (polioviruses, coxsackie viruses A and B, echoviruses, reoviruses and adenoviruses) and gastroenteritis caused by human rotavirus (Norwalk virus, Adenovirus serotype, calicivirus, parvo-like viruses); (c) diseases caused by a variety of protozoan and metazoan parasites, such as amoebic dysentery, giardiasis, and ascariasis; and (d) diseases from HABs toxins (WHO 1992, 1996; Marques et al. 2010b).

There is ample evidence that the major source of illness in areas where the sea is polluted, resulted from consumption of sewage contaminated shellfish and/or bathing near sewage contaminated beaches, as well as with toxins or microorganisms that affect the gastrointestinal tract (EEA 1999). This is particularly relevant for farmed shellfish, due to the animal density in mariculture systems and site location, usually in coastal urbanised areas subjected to intense sewage contamination. Diseases affecting the human gastrointestinal tract are usually caused by the consumption of raw or partially cooked seafood. Salmonellosis is one of the most important food-borne diseases in Europe, currently accounting 70 % of all laboratory-confirmed outbreaks of (WHO 2001; Kovats et al. 2004; Britton et al. 2010). The outbreaks caused by this food-borne pathogen is directly affected by temperature, as it has been reported an increase of 5–10 % in the number of cases for each °C increase in weekly temperatures has been estimated above a threshold of approximately 5 °C with inappropriate food preparation and adequate storage preceding consumption being important determining factors (Kovats et al. 2004). Recently, Vibrionaceae bacteria have been responsible for several disease outbreaks due to the ingestion of contaminated shellfish. As example, an unprecedented outbreak of *V. parahaemolyticus* gastroenteritis occurred in Alaska with more than 400 confirmed cases, when cruise ship passengers ate raw oysters harvested from Prince William Sound (McLaughlin et al. 2005). Increased water temperature was considered to be a major factor in the emergence of *V. parahaemolyticus* in Alaska, as the summer of 2004 was exceptionally warm with water temperature remaining above 15 °C over a 2 month period. The number of epidemics and outbreaks of various diseases attributed to the consumption of contaminated shellfish is also increasing in the Mediterranean, with reports in new regions where such outbreaks were previously absent (EEA 1999; Marques et al. 2010b).

Apart from diseases affecting the gastrointestinal tract, disorders affecting eyes, ears, skin, upper respiratory tract and other parts of the body have been associated with bathing in contaminated seawater. This particular category of infective conditions is caused by microorganisms like *Staphylococcus aureus*, *Pseudomonas aeruginosa*, *Clostridium welchii*, *Candida albicans* and adenoviruses (EEA 1999). These microorganisms cause infection after entering into skin, ear or nose wounds, including those resulting from diving or bathing.

While such records provide evidence of occurrence and indications of magnitude, they are still scarce and the extent of damage caused to local and tourist Mediterranean population health by the expected higher occurrence of pathogenic microorganisms and toxins from HABs in seafood due to climate change still has to be determined.

Health Risks from Chemical Contamination

The most relevant chemicals found in Mediterranean seafood and marine environment likely to create health risks are from the presence of toxic or heavy metals, POPs and illicit dumping of contaminated waste. Following entry into the marine environment, these chemicals accumulate in algae and animals and bioaccumulate along the trophic chain, reaching their highest levels in filter-feeders (e.g. bivalve molluscs) and large predatory fish (e.g. tuna and swordfish). Effects on human health are mainly caused by the consumption of chemically contaminated seafood and are essentially long-term, depending on the chemicals themselves, and the rate and amount of intake (reviewed by Marques et al. 2010b). In general, the principal risk is restricted to those individuals consuming seafood more than two times a week, although the risk varies with the type of seafood, the concentration of pollutants and the circumstances of the consumer (EEA 1999).

The European Rapid Alert System for Food and Feed (RASFF) has been created to provide food and feed control authorities with an effective tool to exchange information about measures taken against serious risks detected in relation to food or feed. According to 2009 data, among all food items, the highest number of alerts in Europe occurred with seafood (716 alerts; Fig. 36.4) representing almost 25 % of all food and feed alerts in 2009 (3,272 alerts) (RASFF 2010). Environmental contaminants represented more than 25 % of all European alerts in 2009, but the proportion of biotoxins and priority contaminants was much lower. Food products imported from Asia or produced in Europe had the highest amounts of alerts compared to food imported from other continents. However, the number of people contaminated with chemicals following seafood ingestion is still limited. In many cases mild effects are not noticed, or have not been correctly associated with contaminated seafood because the symptoms affecting the nervous system are not specific, and the condition can easily be attributed to other causes (e.g. Hg, WHO 1995).

Benefit and Risk Assessment Tools

The risks associated with fish products are toxic metals, PCBs and other organic pollutants, toxins, pathogenic microorganisms and parasites. Assessing potential risks from chemical and biological contaminants in seafood is

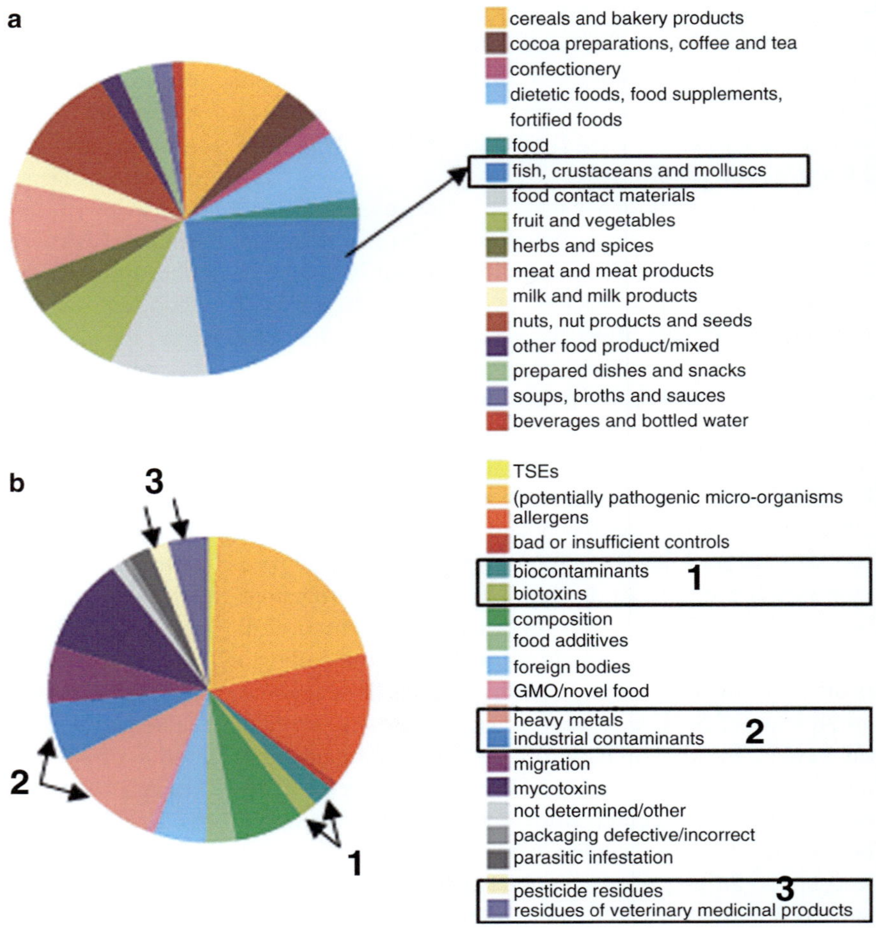

Fig. 36.4 (**a**) RASFF alert notification by hazard category in 2009 (seafood is *highlighted*); (**b**) RASFF alert notification by product category in 2009 (environmental chemical contaminants are *highlighted*) (Adapted from RASFF 2010)

difficult, since the amount of seafood ingested is only a fraction of the total food intake (WHO 1992). In addition, seafood consumption patterns are greatly influenced by food preference, price and availability. Presently, there is a lack of seafood consumption studies carried out in Mediterranean countries. In general, seafood is more available in coastal areas and to specific population sectors (e.g. fishermen, fish vendors and their families, and people on diet).

Concerning chemical contaminants, many studies have been conducted on risk-benefit assessment of seafood consumption in Europe (e.g. Domingo et al. 2007; Sioen et al. 2008; Cardoso et al. 2010; FAO 2011b). This body of evidence shows the pertinence and global interest in this theme. An accurate estimation of risk requires accounting for all the variability of the data, i.e. a full probabilistic approach, taking into account uncertainties in analytical data and interindividual differences in consumers' consumption pattern (e.g. neuroprobabilistic hazard index; Nadal et al. 2008) (Sioen et al. 2008). This procedure ensures the estimation of the probability that the individual exposure to a specific component of seafood surpasses a threshold or a reference value for that component. In a large population, an accurate estimation of this quantity is crucial since even a difference of 1 % involves many individuals. However, the estimation of this probability depends on the tail behaviour of the distribution, since reference values are usually higher than most individual intakes.

The application of some statistical estimators has led to inaccurate estimation of risks. Recently, promising statistical techniques have been developed, such as the extreme value theory (EVT) coupled with recent statistical innovations like bias correction techniques for the Hill estimator (Cardoso et al. 2010). However, further research and monitoring is still needed in this area, since the marine environment is continuously exposed to chemicals present in the environment for a long time, but also to emergent and priority chemicals. Such information is crucial to help health authorities to accurately measure the risks associated to seafood consumption in a changing environment in order to set rigorous adaptation and mitigation strategies.

Microbiological risk assessment and predictive microbiology are emerging tools for the evaluation of the safety of seafood supplies, which involve quantitative exposure

Table 36.2 Adaptation and mitigation measures for the effect of climate change in seafood safety from fisheries

Impact of climate change	Adaptation mitigation measures
Catch reduction	Increase fishing effort
	Shift targeted species
	Protect fishing stocks
	Shift to farmed seafood
Increase in catch variability	Promote catch and consumption of new species
	Shift targeted species
Change in distribution of fisheries	Shift fishing effort and strategies
	Shift target species
Less seafood available	Promote farmed seafood consumption
	Shift to farmed seafood
Increase virulence and expansion of contaminants and diseases	Increase biosecurity measures
	More monitoring and early warning systems
	Implement genetic improvements for higher resistance
	Adopt solutions to reduce contaminant load, e.g. processing seafood, cooking, phycoremediation, etc.
	Develop guidelines and predictive modelling tools for stakeholders
Calcareous shell formation/deposition	Adapt production and handling techniques
	Move to other production zones

assessment in a series of stages (tiered approach). A rough estimate is first made of the order of magnitude that individual factors or parameters may contribute to exposure or risk. This could be considered as analogous to preparing a risk profile. For those that contribute most significantly, a more detailed assessment is performed, or more data are gathered and combined in, for instance, a deterministic approach. Where relevant, an even higher level of detail can be achieved using, for instance, a stochastic approach (see more details in FAO 2008b). Several mathematical modelling approaches can be used (e.g., Event trees, Fault trees, Dynamic Flow Trees, PRM, MPRM, etc.) according to the step of the trade chain (e.g. primary production, processing and post processing). Several variables are crucial in the assessment, such as temperature, product formulation, time, cross-contamination and consumption data.

Adaptation and Mitigation Strategies

Adaptation is a mechanism for management and prevention of climate change impacts in seafood safety, whereas mitigation consists of limiting the process of climate change in seafood safety. A wide range of adaptations and mitigations can be implemented to minimize the effect of climate change in seafood safety, from products originated from the fisheries sector (Table 36.2). A comprehensive analysis for the aquaculture sector is given in Rosa et al., in this volume. These measures must be carefully selected, as excessive protective measures can have negative social and economic impacts. As far as Mediterranean seafood safety is concerned, it is necessary to establish strengthened communication and cooperation among professionals of the seafood sector, including public health, veterinary health, environmental health, and food safety services. Such cooperation must be combined with the thorough understanding of the fate of contaminants in the environment and of the links between seafood, contaminants and environment. Additionally, it is implement accurate monitoring programmes and predictive modelling tools for more effective risk management and prediction of seafood safety for consumers. The implementation of adaptive holistic, integrated and participatory approaches to fisheries and aquaculture management and practices is required in an ecosystem-based perspective. Such approaches should be the best and most immediate forms of adaptation, providing a sound basis for seafood production able to accommodate climate change impacts.

The expected catch reduction and increased species variability may promote a shift to aquaculture from fisheries coastal communities. Integrating aquaculture with agro/multitrophic aquaculture and culture-based fisheries, offers the possibility to recycle nutrients and efficiently use energy and water (FAO 2008c).

The application of waste water treatment with cost-effective and environmental-friendly techniques can also limit the impact of climate change in seafood safety. Short-cycle aquaculture may also be valuable, using new species/strains and new technologies/management practices to fit into seasonal opportunities (FAO 2008c). Aquaculture can be a useful adaptation option for other activity sectors, such as coastal agriculture under salinization threats and biofuel production (e.g. algal biomass or discards and by-products of fish processing), though the current aquaculture technologies will need to be improved to cope climate change impacts. For feed-based aquaculture, the dependence on fish meal and oil from fisheries, and growing competition for terrestrial raw materials is of concern as climate change can limit seafood meal supply. Feeding materials and formulation strategies will be particularly important in maintaining and expanding aquaculture production, while containing costs and energy inputs, and improving resilience to climate change (FAO 2008c). Adaptations also include changing to less carnivorous species, genetic improvements, feed source diversification, better formulation, quality control and management. Genetic knowledge and management in aquaculture are still insufficient, and will be a major challenge and opportunity in the future. Examples include genetic improvement for more efficient feeding and diet specificity, and for species resistance to higher temperature, lower oxygen and pathogens/contaminants (FAO 2008c). Since aquatic pathogen risks may be exacerbated with climate change, biosecurity and prevention measures may need to change accordingly, including early identification and detection mechanisms, suitable treatment strategies and developed products. The creation of certification systems, including sustainability, organic, fair-trade and other criteria will need to be addressed more carefully in the context of climate change (FAO 2008c).

The primary mitigation route for the seafood safety lies in fisheries and aquaculture energy consumption, through fuel, raw material use and production. The knowledge of contaminant levels in seafood feeds and possibilities to mitigate contaminants in those feeds e.g. physical adsorption on activated carbon is used during refining fish oil to remove organic contaminants such as dioxins/furans and dioxin-like PCBs (Maes et al. 2005; Oterhals et al. 2007). The use of vegetable ingredients in seafood feed may also reduce contaminant levels, though the proportion of vegetable constituents needs to be limited to avoid reducing the beneficial constituents in seafood, mainly omega-3 fatty acids (Berntssen et al. 2010). Phycoremediation, i.e. the use of micro or macroalgae to sequester contaminants in aquatic environments has been extensively reported to be extremely useful with toxic elements (Rajamani et al. 2007). Their ability to adsorb and metabolize trace metals is associated with their large surface-volume ratios (up to 10 % of their biomass), the presence of high-affinity, metal-binding groups on their cell surfaces, and efficient metal uptake and storage systems. Other mitigation strategies may involve the elaboration of stakeholders modelling tools and guidelines/recommendations with information for stakeholders like the type, size, season and origin of seafood that should be avoided, pre-treatment/culinary procedures that decreases contaminants levels (e.g. discard water from seafood boiled with water soluble biotoxins), etc. The careful selection of sites where aquaculture sites could be located, with precise definition of their environmental carrying capacity, will contribute to minimise nutrient load to ecosystems. At last, new technologies should be developed involving nanosciences and nanotechnologies able to remove chemical and microbiological contaminants from water in a simple and inexpensive way. Such tools can greatly facilitate improvements in food hygiene and safety management.

Conclusions

Assuring seafood safety is a complex task. Climate change-induced seafood safety hazards can arise at any stage of the trade chain from primary production to consumption. A better understanding of changes that might arise is an essential step to implement mitigation measures by stakeholders. Seafood from fisheries and aquaculture and its safety can be affected by climate change in several ways: (a) the spatial distribution of seafood stocks may change due to migration from one region to another in search of suitable conditions; (b) surface winds can alter the delivery of nutrient into the photic zone and the strength and distribution of ocean currents; (c) high CO_2 levels will alter ocean acidity and affect calcified marine organisms; (d) changes in sea levels and salinity will affect marine organisms; (e) productivity of aquaculture systems will be affected; (f) increase vulnerability of cultured fish to diseases and contaminants; (g) extreme weather events can result in escape of farmed stock and contribute to the reduction of wild stock genetic diversity; (h) eutrophication due to nutrient loading will cause phytoplankton growth and increased frequencies of HABs, including toxin-producing species; and (i) increase in water temperatures will promote the growth of pathogenic and foodborne microorganisms and facilitate methylation of Hg, availability and bioaccumulation of chemical contaminants in seafood. It is therefore necessary for governments and international authorities to be prepared for those changes. The complexity of seafood safety requires interdisciplinary approaches due to the inter-relationships between environment, seafood and contaminants. The current principles of good hygiene, aquaculture and fishery practices need to be adapted to address climate change challenges. Integrated monitoring and surveillance of environment and seafood is critical for the early identification of emerging problems. Accurate predictive modelling and risk assessment will depend on the quality and quantity of available data. Enhanced early warning systems and stakeholders' education are essential elements to reduce the exposure to contaminants in seafood due to climate change. At last, the creation of new technological tools able to remove contaminants from environment and seafood is also of utmost importance to ensure safe seafood for Mediterranean consumers.

Acknowledgments The Portuguese Foundation for Science and Technology (FCT) supported this study through Senior Research Positions Ciência 2007 and 2008 to R.R. and A.M., respectively.

References

Afonso C, Lourenço HM, Dias A, Nunes ML, Castro M (2007) Contaminant metals in black scabbard fish (*Aphanopus carbo*) caught off Madeira and the Azores. Food Chem 101:120–125

Alessi E, Tognon G, Sinesi M, Guerranti C, Perra G, Focardi S (2006) Chemical contamination in the Mediterranean: the case of swordfish. World Wide Fund for Nature. http://wwf.fi/mediabank/1092.pdf. Accessed 22 Dec 2011

Alonso E, Vale C, Vieytes MR, Laferla FM, Giménez-Llort L, Botana LM (2011) The cholinergic antagonist Gymnodimine improves Aβ and Tau neuropathology in an *in vitro* model of Alzheimer disease. Cell Physiol Biochem 27:783–794

Anacleto P, Lourenço HM, Ferraria V, Afonso C, Carvalho ML, Martins MF, Nunes ML (2009) Total arsenic content in seafood consumed in Portugal. J Aquat Food Prod Technol 18:1–14

Astraldi MF, Bianchi CN, Gasparini GP, Morri C (1995) Climatic fluctuations, current variability and marine species distribution: a case study in the Ligurian Sea (north-western Mediterranean). Oceanol Acta 18:139–149

Athanassopoulous F, Pappas IS, Bitchava K (2009) An overview of the treatments of parasitic disease in Mediterranean aquaculture. Option Méditerranéennes A 86:65–83

Barrento S, Marques M, Teixeira B, Vaz-Pires P, Carvalho ML, Nunes ML (2008) Composition of essential elements and contaminants in edible tissues of European and American lobsters. Food Chem 111:862–867

Barrento S, Marques A, Teixeira B, Carvalho ML, Vaz-Pires P, Nunes ML (2009) Contaminants in two populations of edible crab Cancer pagurus: environmental and human health implications. Food Chem Toxicol 47:150–156

Basurco B, Lovatelli A (2003) The aquaculture situation in the Mediterranean sea predictions for the future. Ocean docs. http://hdl.handle.net/1834/543. Accessed in 22 Dec 2011

Bello G, Casavola N, Rizzi E (2004) Aliens and visitors in the Southern Adriatic Sea: effects of tropicalisation. Rapport du 37ème Congrès de la Commission Internationale pour l' Exploration Scientifique de la Mer Méditerranée. CIESM congress proceedings, Mónaco

Bernhard M (1988) Mercury in the Mediterranean. Regional seas reports and studies no. 98. UNEP. http://www.unep.org/regionalseas/publications/reports/RSRS/pdfs/rsrs098.pdf. Accessed 22 Dec 2011

Berntssen MHG, Olsvik PA, Torstensen BE, Julshamn K, Midtun T, Goksøyr A, Johansen J, Sigholt T, Joerum N, Jakobsen JV, Lundebye AK, Lock EJ (2010) Reducing persistent organic pollutants while maintaining long chain omega-3 fatty acid in farmed Atlantic salmon using decontaminated fish oils for an entire production cycle. Chemosphere 81:242–252

Bethoux JP, Gentili B, Raunet J, Tailliez D (1990) Warming trend in the Western Mediterranean deep water. Nature 347:660–662

Biré R, Krys S, Frémy JM, Dragacci S, Stirling D, Kharrat R (2002) First evidence on occurrence of gymnodimine in clams from Tunisia. J Nat Toxins 11:269–275

Bonner JR, Coker AS, Berryman CR, Pollock HM (1983) Spectrum of *Vibrio* infections in a Gulf Coast community. Ann Intern Med 99:464

Booth S, Zeller D (2005) Mercury, food webs, and marine mammals: implications of diet and climate change for human health. Environ Health Perspect 113:521–526

Britton E, Hales S, Venugopal K, Baker MG (2010) Positive association between ambient temperature and salmonellosis notifications in New Zealand, 1965–2006. Aust N Z J Public Health 34:126

Caddy JF (1998) GFCM and its future relationship to marine science. Workshop on gaps in fishery science. CIESM workshop series 5, Dubrovnik, pp 7–10

Caillaud A, de la Iglesia P, Darius HT, Pauillac S, Aligizaki K, Fraga S, Chinain M, Diogène J (2010) Update on methodologies available for ciguatoxin determination: perspectives to confront the onset of ciguatera fish poisoning in Europe. Mar Drugs 8:1838–1907

Cardoso C, Bandarra N, Lourenço H, Afonso C, Nunes M (2010) Methylmercury risks and EPA + DHA benefits associated with seafood consumption in Europe. Risk Anal 30:827–840

Cheung WWL, Lam VWY, Sarmiento JL, Kearney K, Watson R, Zeller D, Pauly D (2010) Large-scale redistribution of maximum fisheries catch potential in the global ocean under climate change. Glob Change Biol 16:24–35

Cochrane K, de Young C (2007) Ecosystem approach to fisheries management in the Mediterranean. Cah Option Mediterraneenes B 62:71–85

Cook DW, Bowers JD, DePaola A (2002) Density of total and pathogenic (tdh+) *Vibrio parahaemolyticus* in Atlantic and Gulf Coast molluscan shellfish at harvest. J Food Prot 65:1873–1880

Danovaro R, Dell'Anno A, DePaola AP (2004) Biodiversity response to climate change in a warm deep sea. Ecol Lett 7:821–828

De Paola A, Nordstrom JL, Bowers JC, Wells JG, Cook DW (2003) Seasonal abundance of total and pathogenic *Vibrio parahaemolyticus* in Alabama oysters. Appl Environ Microbiol 69:1521–1526

Domingo JL, Bocio A, Falcó G, Llobet JM (2007) Benefits and risks of fish consumption: part I. A quantitative analysis of the intake of omega-3 fatty acids and chemical contaminants. Toxicology 230:219–226

EC (2005) Commission staff working paper. Annex to the communication from the commission to the council and the European Parliament on Community Strategy Concerning Mercury. Extended Impact Assessment European Commisision. COM(2005)20 final. http://europa.eu.int/comm/environment/chem-icals/mercury/pdf/extended_impact_assessment.pdf. Accessed 22 Dec 2011

EEA (1999) State and pressures of the marine and coastal Mediterranean environment. European Environmental Agency, Copenhagen. http://www.eea.europa.eu/publications/medsea/at_download/file. Accessed 22 Dec 2011

EEA (2006) Priority issues in the Mediterranean environment. Report European Environmental Agency, Copenhagen. http://www.eea.europa.eu/publications/eea_report_2006_4/at_download/file. Accessed 22 Dec 2011

Elena SF, Lenski RE (2003) Evolution experiments with microorganisms: the dynamics and genetic bases of adaptation. Nat Rev Genet 4:457–469

Eurostat (2008) Half of Mediterranean fish catches are by Mediterranean Partner Countries. Agriculture and fisheries Statistics in focus. EUROSTAT 88/2008, 7p

Eurostat (2011) Fisheries statistics. European Commission, Brussels. http://epp.eurostat.ec.europa.eu/portal/page/portal/fisheries/introduction. Accessed 22 Dec 2011

FAO (2008a) Climate change for fisheries and aquaculture. Technical background document from the expert consultation held on 7–9 April 2008. Food and Agriculture Organization of the United Nations, Rome. ftp://ftp.fao.org/docrep/fao/meeting/013/ai787e.pdf. Accessed 22 Dec 2011

FAO (2008b) Exposure assessment of microbiological hazards in food – guidelines. World Health Organization. Food and Agriculture Organization of the United Nations, Rome. http://www.fao.org/docrep/010/a0251e/a0251e00.htm. Accessed 22 Dec 2011

FAO (2008c) Climate change: implications for food safety. Food and Agriculture Organization of the United Nations, Rome. http://www.fao.org/docrep/010/i0195e/i0195e00.htm. Accessed 22 Dec 2011

FAO (2011a) Fisheries stat. Food and Agriculture Organization of the United Nations, Rome. http://www.fao.org/fishery/statistics/global-consumption/en. Accessed 22 Dec 2011

FAO (2011b) Report of the joint FAO/WHO expert consultation on the risks and benefits of seafood consumption. Food and Agriculture Organization of the United Nations, Rome. World Health Organization, Geneva. http://www.fao.org/docrep/014/ba0136e/ba0136e00.pdf. Accessed 22 Dec 2011

Faruque SM, Albert MJ, Mekalanos JJ (1998) Epidemiology, genetics, and ecology of toxigenic *Vibrio cholerae*. Microbiol Mol Biol Rev 62:1301–1314

FDA (1993) Guidance document for cadmium in shellfish. Food and Drug Administration, Washington, DC. http://dino.wiz.uni-kassel.de/dain/ddb/x270.html. Accessed 22 Dec 2011

Francesconi KA (2007) Toxic metal species and food regulations-making a healthy choice. Analyst 132:17–20

Galgani F, Martínez-Gómez C, Giovanardi F, Romanelli G, Caixach J, Cento A, Scarpato A, BenBrahim S, Messaoudi S, Deudero S, Boulahdid M, Benedicto J, Andral B (2011) Assessment of polycyclic aromatic hydrocarbon concentrations in mussels (*Mytilus galloprovincialis*) from the Western basin of the Mediterranean Sea. Environ Monit Assess 172:301–317

Galil BS, Zenetos A (2002) A sea change—exotics in the eastern Mediterranean. In: Leppakoski E, Gollasch S, Olenin S (eds) Invasive aquatic species of Europe. Distribution, impacts and management. Kluwer Academic Publishers, Dordrecht

Gambaiani DD, Mayol P, Isaac SJ, Simmonds MP (2009) Potential impacts of climate change and greenhouse gas emissions on Mediterranean marine ecosystems and cetaceans. J Mar Biol Assoc UK 89:179–201

Gamble JL (2008) Analyses of the effects of global change on human health and welfare and human systems: final report, synthesis and assessment product. Report by the U.S. climate change science program and the Subcommittee on Global Change Research, Washington, DC. http://www.climatescience.gov/Library/sap/sap4-6/default.php. Accessed 22 Dec 2011

Green MA, Waldbusser GG, Reilly SL, Emerson K, O'Donnell S (2009) Death by dissolution: sediment saturation state as a mortality factor for juvenile bivalves. Limnol Oceanogr 54:1037–1047

Hall LW, Anderson RD (1995) The influence of salinity on the toxicity of various classes of chemicals to aquatic biota. Crit Rev Toxicol 25:281–346

Harvell CD, Kim K, Burkholder JM et al (1999) Emerging marine diseases-climate links and anthropogenic factors. Science 285:1505–1510

Harvell CD, Mitchell CE, Ward JR, Altizer S, Dobson AP, Ostfeld RS, Samuel MD (2002) Climate warming and disease risks for terrestrial and marine biota. Science 296:2158–2162

Hutchins DA, Teyssii J-L, Boisson F, Fowler SW, Fisher NS (1996) Temperature effects on uptake and retention of contaminant radionuclides and trace metals by the brittle star *Ophiothrix fragilis*. Mar Environ Res 41:363–378

IPCC (2007) In: Solomon S, Qin D, Manning M, Chen Z, Marquis M, Averyt KB, Tignor M, Miller HL (eds) Contribution of working group I to the fourth assessment report of the Intergovernmental Panel on Climate Change, 2007. United Nations Intergovernmental Panel on Climate Change. Cambridge University Press, Cambridge. http://www.ipcc.ch/publications_and_data/ar4/wg1/en/contents.html. Accessed 22 Dec 2011

Islam MD, Tanaka M (2004) Impacts of pollution on coastal and marine ecosystems including coastal and marine fisheries and approach for management: a review and synthesis. Mar Pollut Bull 48:624–649

Jezierska B, Lugowska K, Witeska M (2009) The effects of heavy metals on embryonic development of fish. Fish Physiol Biochem 35:625–640

Khan MAQ, Ahmed SA, Catalin B, Khodadoust A, Ajayi O, Vaughn M (2006) Effect of temperature on heavy metal toxicity to juvenile crayfish, *Orconectes immunis* (Hagen). Environ Toxicol 21:513–520

Kovats RS, Edwards S, Hajat S, Armstrong B, Ebi KL, Menne B (2004) The effect of temperature on food poisoning: time series analysis in 10 European countries. Epidemiol Infect 132:443–453

Licandro P, Conway DVP, Daly Yahia MN, Fernandez de Puelles ML, Gasparini S, Hecq JH, Tranter P, Kirby RR (2010) A blooming jellyfish in the northeast Atlantic and Mediterranean. Biol Lett 6:688

Lleonart J, Maynou F (2003) Fish stock assessments in the Mediterranean: state of the art. Sci Mar 67:37–49

MacKenzie L, Holland P, McNabb P, Beuzenberg V, Selwood A, Suzuki T (2002) Complex toxin profiles in phytoplankton and Greenshell mussels (*Perna canaliculus*), revealed by LC-MS/MS analysis. Toxicon 40:1321–1330

Maes J, de Meulenaer B, Van Heerswynghels P, De Greyt W, Eppe G, De Pauw E, Huyghebaert A (2005) Removal of dioxins and PCB from fish oil by activated carbon and its influence on the nutritional quality of the oil. J Am Oil Chem Soc 82:593–597

Marcos M, Tsimplis MN (2008) Comparison of results of AOGCMs in the Mediterranean Sea during the 21st century. J Geophys Res 113:C12028

Marques A, Teixeira B, Barrento S, Anacleto P, Bandarra N, Mendes R, Carvalho ML, Nunes ML (2010a) Compositional characteristics of spider crab *Maja brachydactyla*: human health implications. J Food Compos Anal 23:230–237

Marques A, Nunes ML, Moore S, Strom M (2010b) Climate change and seafood safety: human health implications. Food Res Int 43:1766–1779

May RM (1973) Stability and complexity in model ecosystems. Princeton University Press, Princeton

McLaughlin JB, DePaola A, Bopp CA, Martinek KA, Napolilli NP, Allison CG, Murray SL, Thompson EC, Bird MM, Middaugh JP (2005) Outbreak of *Vibrio parahaemolyticus* gastroenteritis associated with Alaskan oysters. N Engl J Med 353:1463–1470

Modassir Y (2000) Effect of salinity on the toxicity of Mercury in mangrove clam, *Polymesoda erosa* (Lightfoot 1786). Asian Fish Sci 13:335–341

Monserrat J, Bianchini A (1995) Effects of temperature and salinity on the toxicity of a commercial formulation of methyl parathion to *Chasmagnathus granulata* (Decapoda, Grapsidae). Braz J Med Biol Res 28:74–78

Moon H-B, Kim H-S, Choi M, Choi H-G (2010) Intake and potential health risk of polycyclic aromatic hydrocarbons associated with seafood consumption in Korea from 2005 to 2007. Arch Environ Contam Toxicol 58:214–221

Mubiana VK, Blust R (2007) Effects of temperature on scope for growth and accumulation of Cd, Co, Cu and Pb by the marine bivalve *Mytilus edulis*. Mar Environ Res 63:219–235

Nadal M, Kumar V, Schuhmacher M, Domingo JL (2008) Applicability of a neuroprobabilistic integral risk index for the environmental management of polluted areas: a case study. Risk Anal 28:271–286

Nye J (2010) Climate change and its effects on ecosystems, habitats and biota: state of the gulf of Maine report. Gulf of Maine Council of the Marine Environment. http://www.gulfofmaine.org/state-of-the-gulf/docs/climate-change-and-its-effects-on-ecosystems-habitats-and-biota.pdf. Accessed 22 Dec 2011

Onuma Y, Satake M, Ukena T, Roux J, Chanteau S, Rasolofonirina N, Ratsimaloto M, Naoki H, Yasumoto T (1999) Identification of putative palytoxin as the cause of clupeotoxism. Toxicon 37:55–65

Oren OH (1957) Changes in the temperature of the Eastern Mediterranean Sea in relation to the catch of the Israel trawl fishery during the years 1954/55 and 1955/56. Bull Inst Océánographique Monaco 1102:1–12

Oterhals A, Solvang M, Nortvedt R, Berntssen MHG (2007) Optimization of activated carbon-based decontamination of fish oil by response surface methodology. Eur J Lipid Sci Technol 109:691–705

Rainbow PS (2002) Trace metal concentrations in aquatic invertebrates: why and so what? Environ Pollut 120:497–507

Rajamani S, Siripornadulsil S, Falcao V, Torres M, Colepicolo P, Sayre R (2007) Phycoremediation of heavy metals using transgenic microalgae. Adv Exp Med Biol 616:99–109

Rao DGVP, Khan MAQ (2000) Zebra mussels: temperature potentiation of copper toxicity. Water Environ Res 72:175–178

RASFF (2010) The Rapid Alert System for Food and Feed (RASFF) annual report 2009. European Commission. http://ec.europa.eu/food/food/rapidalert/docs/report2009_en.pdf. Accessed 22 Dec 2011

Romano J-C, Bensoussan N, Younes WA, Arlhac D (2000) Thermal anomaly in waters of the Gulf of Marseilles during the summer of 1999. A partial explanation of the mortality of certain fixed invertebrates? CR Acad Sci 323:415–427

Ronzitti N (1999) Le zone di pesca nel Mediterraneo e la tutela degli interessi italiani. Riv Maritima 6:96

Sevault F, Somot S, Déqué M (2004) Climate change scenario for the Mediterranean sea. Geophys Res Abstr 6:1–10

Simopolpoulos AP (1997) Seafood from producer to consumer, natural aspects of the fish. In: Lutten JB, Borrensen T, Oehlenschläger J (eds) Seafood from producer to consumer, integrated approach to quality, vol 38. Elsevier, Amsterdam

Sioen I, Leblanc JC, Volatier JL, De Henauw S, Van Camp J (2008) Evaluation of the exposure methodology for risk-benefit assessment of seafood consumption. Chemosphere 73:1582–1588

Sokurenko EV, Gomulkiewicz R, Dykhuizen DE (2006) Source-sink dynamics of virulence evolution. Nat Rev Microbiol 4:548–555

Spatharis S, Tsirtsis G, Danielidis DB, Chi TD, Mouillot D (2007) Effects of pulsed nutrient inputs on phytoplankton assemblage structure and blooms in an enclosed coastal area. Estuar Coast Shelf Sci 73:807–815

Stirling DJ (2001) Survey of historical New Zealand shellfish samples for accumulation of gymnodimine. N Z J Mar Freshw Res 35:851–857

Sullivan JK (1977) Effects of salinity and temperature on the acute toxicity of cadmium to the estuarine crab *Paragrapsus gaimardii* (Milne Edwards). Aust J Mar Freshw Res 28:739–743

Sumaila UR, Cheung WWL, Lam VWY, Pauly D, Herrick S (2011) Climate change impacts on the biophysics and economics of world fisheries. Nat Clim Change 1:449–456. doi:10.1038/NCLIMATE1301

Tin T, Giannakopoulos G, Bindi M (2005) Climate change impacts in the Mediterranean resulting from a 2 °C global temperature rise. Report of WWF. http://www.panda.org/downloads/climate_change/medreportfinal8july05.pdf. Accessed 22 Dec 2011

Wang J, Chuang C-Y, Wang W-X (2005) Metal and oxygen uptake in the green mussel *Perna viridis* under different metabolic conditions. Environ Toxicol Chem 24:2657–2664

WHO (1992) Health risks from bathing in marine waters. Report of a joint WHO/UNEP meeting, Athens, 15–18 May 1991. Document EUR/ICP/CEH 103. World Health Organization Regional Office for Europe, Copenhagen. http://www.who.int/water_sanitation_health/bathing/recreadis.pdf. Accessed 22 Dec 2011

WHO (1995) Health risks from marine pollution in the Mediterranean, Part II, review of hazards and health risks, Document EUR/ICP/EHAZ94 01/MT01(2). World Health Organization Regional Office for Europe, Copenhagen. http://bases.bireme.br/cgi-bin/wxislind.exe/iah/online/?IsisScript=iah/iah.xis& src=google&base=WHOLIS&lang=p&nextAction=lnk&exprSearch=54757&indexSearch=ID. Accessed 22 Dec 2011

WHO (1996) Assessment of the state of microbiological pollution of the Mediterranean Sea. MAP technical reports series no.108, Athens. http://www.unepmap.org/index.php?module=library&mode=pub&action=results&_stype=3&s_category=&s_descriptors=Microbiological%20pollution. Accessed 22 Dec 2011

WHO (2001) WHO surveillance programme for control of food-borne infections and intoxications in Europe. In: Schmidt K, Tirado C (eds) Seventh report 1993–1998. http://www.bfr.bund.de/internet/7threport/7threp_ctryreps_fr.htm. Accessed 22 Dec 2011

Yasumoto T, Satake M (1998) New toxins and their toxicological evaluations. In: Reguera B, Blanco J, Fernandez ML, Wyatt T (eds) Harmful algae. Xunta de Galicia and Intergovernmental Oceanographic Commission of UNESCO, Paris

Zago C, Capodaglio G, Ceradini S, Ciceri G, Abelmoschi L, Soggia F, Cescon P, Scarponi G (2000) Benthic fluxes of cadmium, lead, copper and nitrogen species in the northern Adriatic Sea in front of the river Po outflow, Italy. Sci Total Environ 246:121–137

Mediterranean Aquaculture in a Changing Climate

37

Rui Rosa, António Marques, and Maria Leonor Nunes

Abstract

The semi-enclosed Mediterranean Sea is the biggest marginal sea of the Earth, and it is at the centre of the life of several millions of people. Seafood is widely consumed in this region, with an average of 16.5 Kg/capita/year, and one fourth of the seafood supply comes from aquaculture activities. The Mediterranean aquaculture sector has expanded over the last decades. It increased 77 % over the last decade reaching *circa* 1.3 million tons in 2009. The value ranged around 3.700 millions US dollars, representing 3.4 % of the global aquaculture value. The growth of seafood demand in the Mediterranean is expected to increase in the future, especially in southern countries. Yet, during the 21st century, the Mediterranean basin is expected to observe: (i) an increase in air temperature between 2.2 °C to 5.1 °C, (ii) a significant decrease in rainfall (ranging between −4 and −27), (iii) an increase in drought periods related to high frequency of days during which the temperature would exceed 30 °C, (iv) an increase of the sea level (around 35 cm) and saline intrusion. Moreover, extreme events, such as heat waves, droughts or floods, are also likely to be more frequent and violent. Here, we review the present status of the Mediterranean aquaculture (e.g. production trends, main farmed species, production systems, major producing countries), the most relevant impacts of climate change to this sector (e.g. temperature, eutrophication, harmful algae blooms, water stress, sea level rise, acidification and diseases), and provide a wide range of adaptation and mitigation strategies that can be implemented to minimize such climatic effects.

Keywords

Aquaculture • Climate change • Mediterranean Sea

Introduction

The semi-enclosed Mediterranean Sea is the biggest marginal sea of the Earth, representing 0.69 % of the global ocean-surface, 0.27 % of the global ocean-volume. It has a surface area of 2.5 million square kilometres and a total water body of about 4 million cubic kilometres (Goudie 2001). The Mediterranean Sea is one of the most oligotrophic seas in the world (UNEP 1989), with some of the largest rivers of Europe and Africa draining the nutrient- and sediment-rich waters directly or indirectly into it. The basin is directly linked to the Atlantic ocean through the Strait of Gibraltar, so that its full-body water is entirely renewed every ~100 years. Located between the mid-latitude storm rainband and the Sahara Desert, it experiences a profound seasonal cycle (Peixoto et al. 1982). The hydrological cycle is especially sensitive to the timing and the location of the winter storms as they move into the region. Interannual climate variability is closely related to the variability in the

R. Rosa (✉)
Laboratório Marítimo da Guia, Centro de Oceanografia, Faculdade de Ciências da, Universidade de Lisboa, Av. Nossa Senhora do Cabo, 939, 2750-374 Cascais, Portugal
e-mail: rrosa@fc.ul.pt

A. Marques • M.L. Nunes
Unidade de Investigação em Valorização dos Produtos da Pesca e Aquacultura, INRB, I.P./L-IPIMAR, Avenida de Brasília, 1449-006 Lisboa, Portugal

S. Goffredo and Z. Dubinsky (eds.), *The Mediterranean Sea: Its history and present challenges*,
DOI 10.1007/978-94-007-6704-1_37,

Atlantic sector such as the North Atlantic Oscillation (NAO) (Hurrell 1995, 1996; Rodo et al. 1997; Eshel and Farrel 2000).

The Mediterranean basin displays a great variety of climatic, physical, ecological, social, economic and cultural traits. Nevertheless, an in spite of apparent diversity, the Mediterranean region has long been recognized as a single functional climatic, ecological, economic and social system. The basin is characterized by a general mild climate, winter-dominated rainfall, dry summers, and a profusion of microclimates, ranging from the hot dry conditions of the Nile delta to the cooler and somewhat wetter conditions of the Ebro region. Despite the differences in microclimate, all areas are characterized by warm and hot dry summers and wetter cooler winters, and the influence of the sea, which reduces daily and seasonal temperature extremes and results in diurnal land-ocean breezes (Jeftic et al. 1996). Yet, the prospect of major climate change is a source of growing concern, raising serious questions over the sustainability of the region.

Aquaculture in the Mediterranean Sea

Production Trends and Current Status

In 2009, the total world seafood production reached 162 million tones, which represented a 19 % increase in production over the last decade (Fig. 37.1). This was entirely due to the 84 % increase in aquaculture production, since capture fisheries declined in the same period by 3 % (from 1999 to 2009). The technology applied has evolved rapidly as a result of the modifications of existing farming facilities and the development of new farming concepts, e.g. off-shore submerged netcage technology. As a consequence, presently, a wide-range of production activities of marine species co-exist in different Mediterranean environments using a variety of technologies, from extensive mollusc or fish production to highly intensive raceways or netcage fish farming (Barazi-Yeroulanos 2010).

The Mediterranean aquaculture production increased 77 % over the last decade, especially in the brackish environment (Fig. 37.2), reaching *circa* 1.3 million tons in 2009. Yet, it only represented 1.8 % of the world's total production. The value of Mediterranean aquaculture ranged around 3.700 millions US dollars in 2009 (Fig. 37.3), representing 3.4 % of the global aquaculture value. The Mediterranean aquaculture have grown steadily over the last two decades (Fig. 37.2), but with lower annual growth rates for marine and brackish fish production in the last decade (1999–2009, Fig. 37.4). The fish species that have contributed the most for these high growth rates were the nile tilapia (*Oreochromis niloticus*) and the flathead grey mullet (*Mugil cephalus*) in brackish systems (Fig. 37.5), and the gilthead seabream (*Sparus aurata*) and European seabass (*Dicentrarchus labrax*) in marine systems (Fig. 37.5). The rapid increase in the production of these marine carnivorous fish is quite striking. The rapid output growth for these two species has resulted from

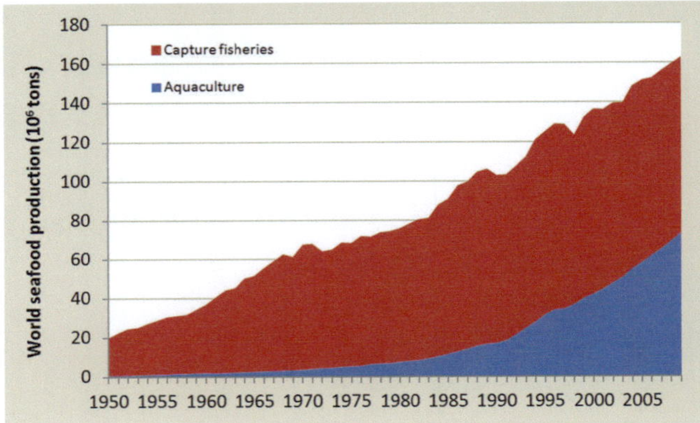

Fig. 37.1 Evolution of world capture fisheries and aquaculture production (million tons) between 1950 and 2009 (Source: FAO Fishstat)

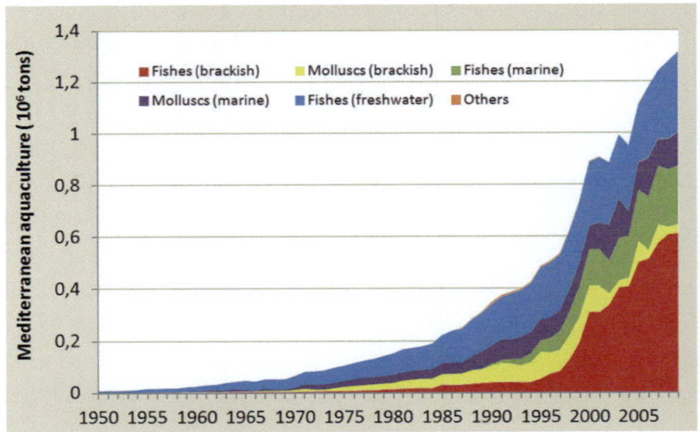

Fig. 37.2 Evolution of Mediterranean aquaculture production (million tons), in brackish, marine and freshwater systems, between 1950 and 2009 (Source: FAO Fishstat)

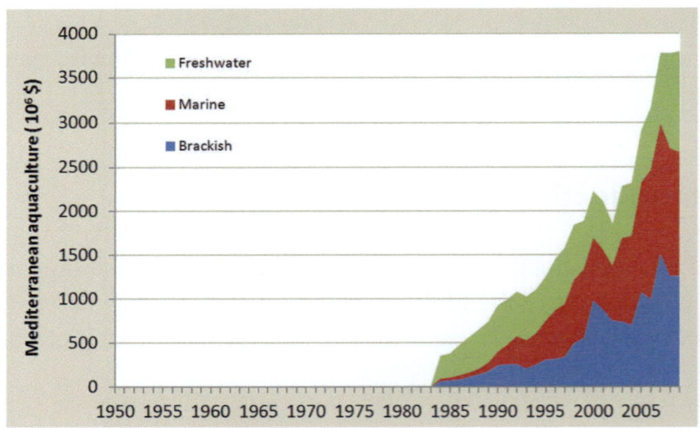

Fig. 37.3 Evolution of total Mediterranean aquaculture production (in US dollars) between 1950 and 2009 (Source: FAO Fishstat)

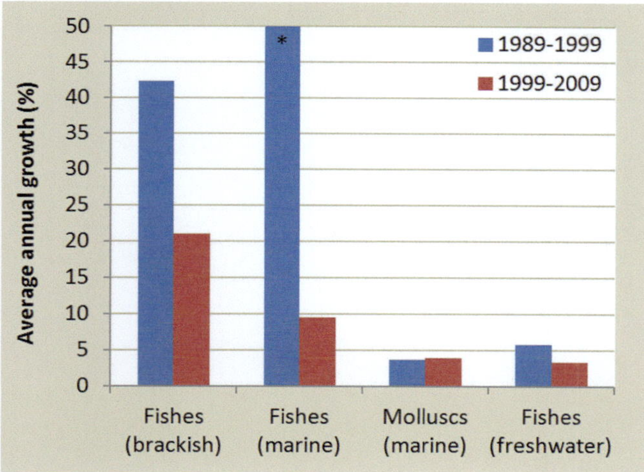

Fig. 37.4 Average annual growth (% of increase) of the major FAOStat groups, between 1989 and 1999, and 1999 and 2009, in the Mediterranean Sea (Source: FAO Fishstat). * out of scale, representing a 367 % increase

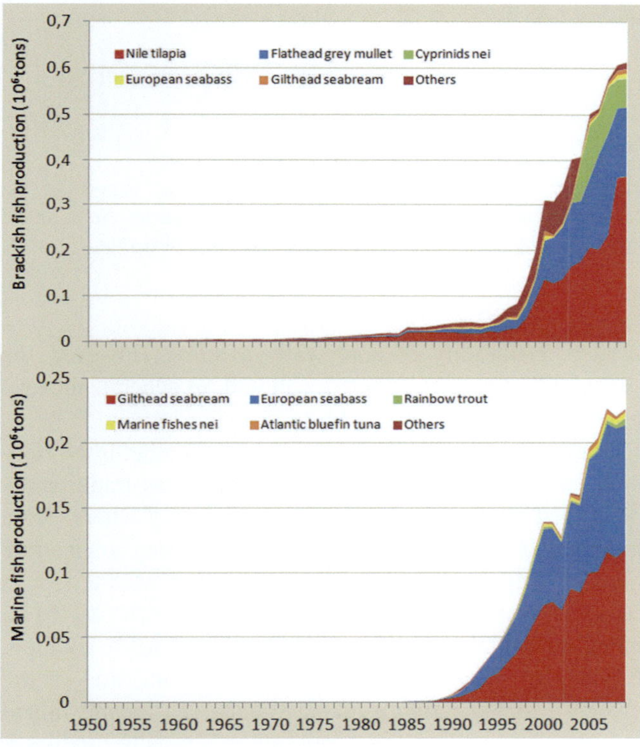

Fig. 37.5 Evolution of total Mediterranean aquaculture fish production (million tons), in brackish and marine systems, between 1950 and 2009 (Source: FAO Fishstat)

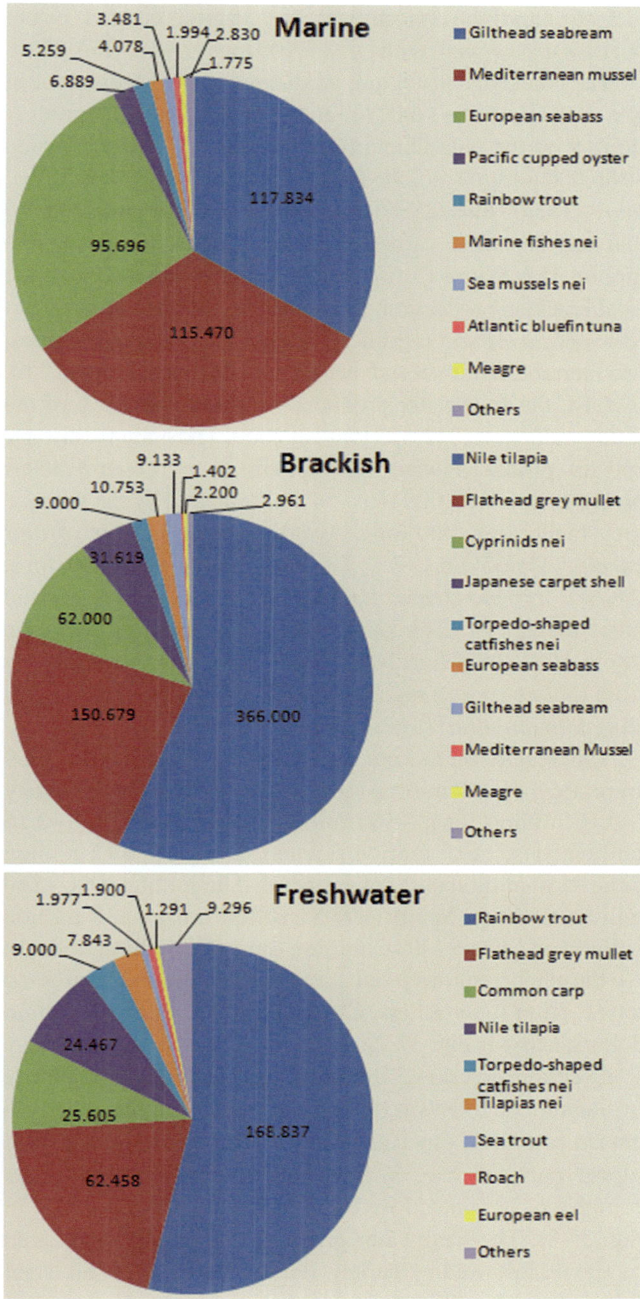

Fig. 37.6 Main species produced (tons) in the marine, brackish and freshwater aquaculture systems of the Mediterranean Sea in 2009 (Source: FAO Fishstat)

the development of reliable seed production techniques, the formulation of specialized feeds and the application of intensive production systems, particularly cages. Furthermore, the support from the EU and strong markets from the late 1980s and early 1990s have also played a significant role.

In 2009, marine aquaculture production attained 355,306 tons, representing 27 % of the total Mediterranean production. The gilthead seabream (*Sparus aurata*) was the most important resource reaching over 118,000 tons (i.e. 33 %), the European seabass (*Dicentrarchus labrax*) over 95,000 tons (27 %), followed by the Atlantic bluefin tuna (*Thunnus thynnus*) with only 1,994 tons (~0.6 %) and meager (*Argyrosomus regius*) with 1,775 tons (0.5 %) (upper panel in Fig. 37.6). Nonetheless, the second most important marine resource (in quantity) was the Mediterranean mussel (*Mytilus galloprovincialis*) attaining over 115,000 tons

(32 %), and the fourth was the Pacific cupped oyster (*Crassostrea gigas*), with *circa* 6,900 tons (1.9 %) (Fig. 37.6).

The brackish aquaculture production in the Mediterranean Sea in 2009 reached 645,747 tons (49 % of the total production), with the Nile tilapia (*Oreochromis niloticus*) reaching ~366,000 tons (57 %), the flathead grey mullet (*Mugil cephalus)* ~150,000 tons (23 %), the cyprinids ~62,000 tons (9.6 %) and the Japanese carpet shell (*Ruditapes philippinarum*) around 31,600 tons (5 %). Representing less than 2 % of the total brackish production, there were the torpedo-shaped catfishes (*Clarias* spp.), gilthead seabream, European seabass, Mediterranean mussel and meagre (middle panel in Fig. 37.6).

Last, the freshwater production represented 24 % of the total production (312,673 tons) in 2009. The main freshwater species produced were the rainbow trout (*Oncorhynchus mykiss*) with ~169,000 tons (54 %), followed by the flathead grey mullet (~62,460 tons, 20 %), common carp (*Cyprinus carpio*, 25,605 tons, 8.2 %), Nile tilapia (~24,460 tons, 7.8 %), torpedo-shaped catfishes (9,000 tons, 7.8 %) and other tilapias (~7,843 tons, 2.5 %). Representing less than 0.6 % of the total freshwater production, there were the sea trout (*Salmo trutta*), roach (*Rutilus rutilus*) and European eel (*Anguilla anguilla*) (lower panel in Fig. 37.6).

The major Mediterranean producing countries in 2009 were: Egypt (~0.7 million tons; ~1,240 million dollars), Italy (~0.16 million tons; ~660 million dollars), Turkey (~0.16 million tons; ~615 million dollars), Greece (~0.12 million tons; ~550 million dollars), France (~0.06 million tons; ~260 million dollars), Spain (~0.05 million tons; ~230 million dollars) and Israel (~0.02 million tons; ~70 million dollars). All the other countries production varied between 13 thousand tons (~55 million dollars, Croatia) and 240 tons (~1 million dollars, Libya) (Fig. 37.7).

Excluding Turkey, almost all the marine aquaculture production in 2009 was undertaken by the northern Mediterranean countries, especially Greece, Italy, Spain and France (upper panel in Fig. 37.8). The brackish production was completely dominated by Egypt and Italy (middle panel in Fig. 37.8). The freshwater production was also dominated by Egypt, followed by Turkey, France, Italy, Spain and Israel (lower panel Fig. 37.8).

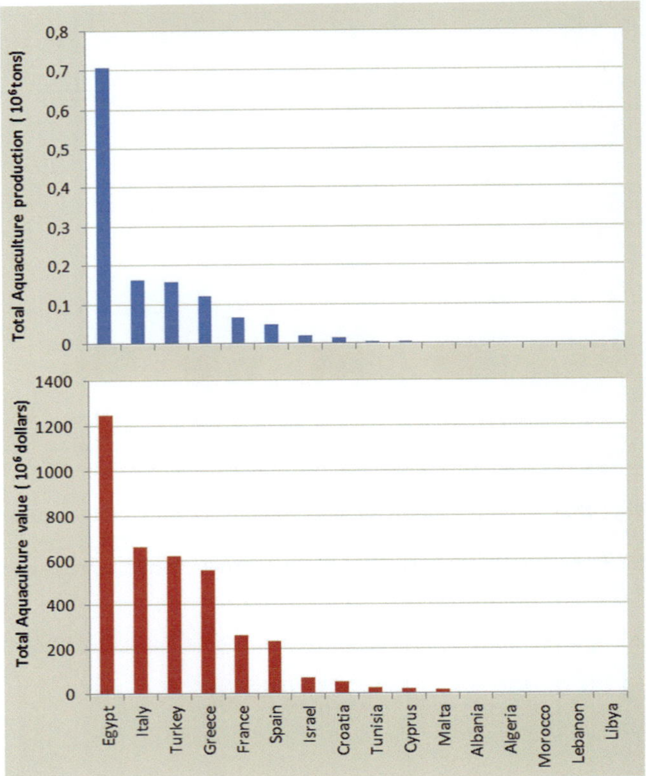

Fig. 37.7 Aquaculture production (quantity and value) of the Mediterranean countries in 2009 (Source: FAO Fishstat). *Note*: Serbia and Montenegro have not shown any production in that year

Production Systems

Intensive and Semi-intensive Inland Aquaculture (Including Integrated Aquaculture)

This sub-sector is mainly dominated by trout, carp and tilapia farming. Trout farming (mainly *Oncorhynchus mykiss* and *Salmo trutta*) is concentrated in the EU Mediterranean countries and Turkey, where it represents the most widespread form of intensive inland aquaculture. In the other Mediterranean countries, trout culture is rare because of the limited availability of cold water sources. However, the sector still attracts new investment and, in general, shows a gradual expansion (FAO 2011).

Carp aquaculture, which is based on the rearing of common carp and introduced Chinese carps, was traditionally developed in the Eastern European countries where, for a long period of time, it was the most widespread aquaculture practice. Egypt and Israel are the only other Mediterranean countries which base their production mainly (70–80 %) on inland aquaculture, with carp and tilapia farming. In addition, Egypt also produces mullets, sea bass and, due to the presence of low salinity inland waters, sea bream. Inland semi-intensive and intensive aquaculture is usually combined, in ponds and canals, with water use for irrigation. Both countries reported a progressive intensification of production (from semi-intensive to intensive pond management) to limit water consumption. This process of intensification appears to be more urgent in Israel, because of the limited availability of freshwater, but more dramatic in Egypt, where it involves the elimination or restoration of the most primitive production units which represents about 70 % of the farmed area. Intensive or hyper-intensive eel culture was reported by Italy and Greece. However, this activity is significantly

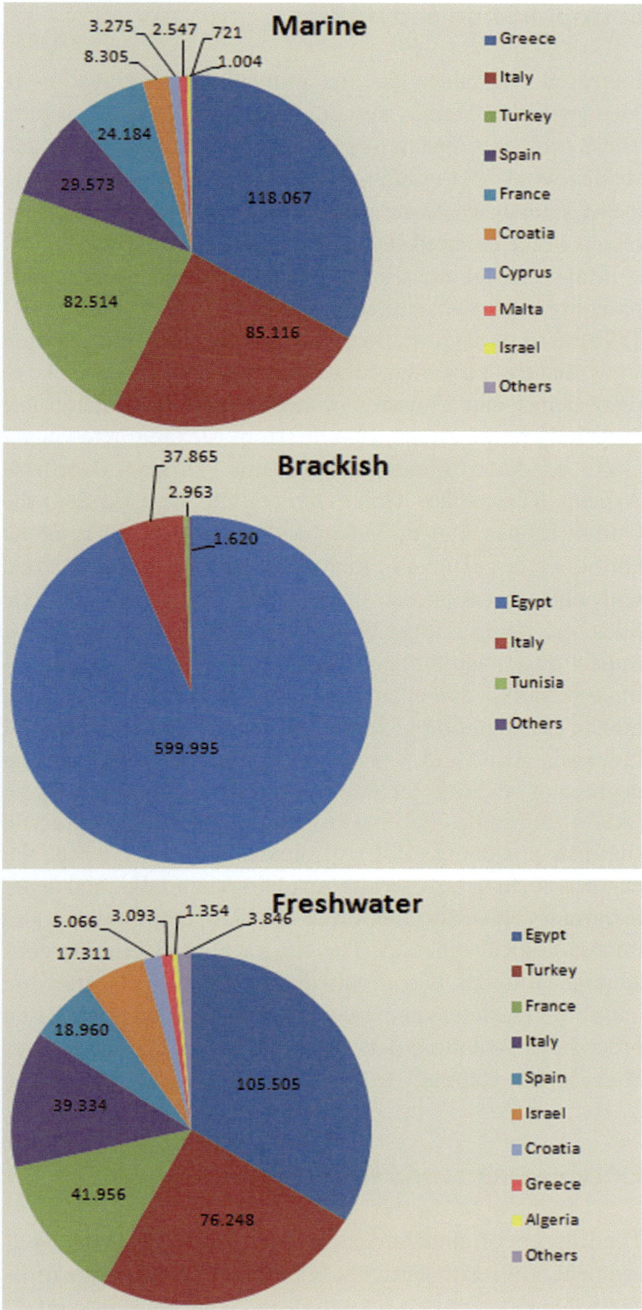

Fig. 37.8 Relative contribution of the main producing Mediterranean countries (values in tons) for the marine, brackish and freshwater aquaculture production in 2009 (Source: FAO Fishstat)

constrained by the limited availability of wild seed (glass eels). Egypt is the only Mediterranean country which reported the development of integrated aquaculture (rice-fish cultivation) stating, however, that its existence depends highly on governmental subsidies (often in the form of free fingerlings supply). However, production has gone down significantly now that fingerlings are no longer distributed free.

In-Shore Aquaculture

It represents the bulk of coastal aquaculture production, having been developed in protected areas such as bays, gulfs, canals and coastal lagoons, mostly based on mollusc farming and finfish cage culture. Despite the fact that it is subject to the same numerous constraints to on-land aquaculture (e.g. deterioration in coastal water quality), which limit the availability of adequate sites, a few countries still plan a significant expansion of in-shore aquaculture, as well as its gradual transfer into open waters (see next section). Mollusc culture, which has been developed since the beginning of the twentieth century, involves the use of various techniques (fixed and floating structures) basically for the production of mussels (*Mytilus galloprovincialis*) and oysters (*Crassostrea gigas* and *Ostrea edulis*). Other species, such as scallops, clams and abalone, have also been farmed in order to diversify the production, but none attained the same commercial scale as mussel and oyster rearing. In particular, the promising culture of the imported Japanese carpet shell was strongly constrained by its rapid adaptation to the natural environment (the Po river delta in Italy and the Morbihan Gulf in France) which gave rise to a competitive fisheries production. Mollusc culture is highly developed in the EU Mediterranean countries (Spain, France, Italy and Greece, with a total of several hundred thousands of tons per year) which can also rely on a traditional domestic consumption of these species groups. The sector has reached a high level of technology and mechanization, and farms are operated both by private and corporate producers.

The cage farming of marine finfish started in the early 1980s. This in-shore activity provided the opportunity to avoid expensive land-based facilities but was initially confined to protected coastal areas until the more recent development of reliable off-shore cages. This farming technique therefore appears to be highly constrained by the competitive use of the littoral zone, and it is not expected to show further significant expansion. Greece, Turkey, Italy, France, Spain and Croatia (Fig. 37.8) are the most important producers of marine fish in the Mediterranean, using in-shore cage farming. Marine fish farming is mostly developed by private enterprises and less so by corporate groups.

Off-Shore Aquaculture

This form of aquaculture has the potential to solve most of the difficulties which hamper the expansion of coastal aquaculture, and therefore attracts great interest in most of the Mediterranean countries. Conflicts with the tourism industry or scarcity of appropriate sites forced the producers to move far from the coast. As a result, most production of intensive marine finfish aquaculture in the Mediterranean takes place today in (semi) offshore floating cages (open sea system). This system type has been preferred to land based facilities, but requires high capital investments and has high running costs. Yet, in the right site, open sea culture offers a better environment for the fish welfare, and for intensive marine aquaculture is believed to have fewer negative impacts on the environment. Most cages are situated in semi-offshore sites with depths between 15 and 30 m. In Spain, gilthead seabream and European seabass are mainly produced in off-shore cage systems along the Mediterranean coast and in the Canary Islands. The cages are floating, gravity, circular structures of the same type used throughout the Mediterranean. In Greece, close to 80 % of the aquaculture fish is cultured in cages, with the remaining 20 % produced in land-based race-ways. Nonetheless, many countries still have no local suppliers and must import cages and equipment from abroad. The increased management costs related with daily routine operations concerning cages located far from the coast is also limiting off-shore aquaculture development. Off-shore aquaculture has also been used, but at a lesser extent, to mussel production (in long-lines). High investment and management costs are also required, namely strong mooring systems and long-line structures as well as large workboats.

Climate Change and Mediterranean Aquaculture

Temperature

Since 1970, south-western Europe has reported a temperature rise of about 2 °C. Yet, by the end of the twenty-first century, it is expected fan increase in air temperature between 2.2 and 5.1 °C in Mediterranean region (IPCC 2007, scenario A1B). This warming will have diverse impacts on aquaculture, depending on the production system, farmed species and country/region. The intensive and semi-intensive inland aquaculture in northern (EU) Mediterranean countries is dominated by trout farming. These fish species have a very narrow optimal range of temperature and a relatively low upper thermal limit, and consequently, warming may significantly enhance trout mortality and affect productivity (Ficke et al. 2007). On the other hand, southern countries, like Egypt, base their inland production in carp and tilapia farming, a group of fish with wider optimal range of temperature and higher thermal limits. Water temperature rise will result in increased metabolism, growth rates and hence in overall production. Yet, the increased air temperature may also enhance vaporization and cloud cover, and, consequently, reduce solar radiation (IPCC 2007). As a result, this may not induce any predicted enhancement in pond-based production. It is worth noting that increased temperatures may also affect pond evaporation rates, especially in Mediterranean North Africa, and the resultant increases in pond salinity could adversely affect less salt-tolerant species.

Eutrophication and HABs

Increased eutrophication and pronounced stratification in Mediterranean lenthic systems is expected in the future. These processes lead to oxygen depletion in dawn hours and sudden wind- and rainfall-driven upwelling processes bring the hypoxic waters to the surface, with consequent deleterious effects to the cultured stocks (Cochrane et al. 2009).

Many coastal areas in the Mediterranean Sea are eutrophic, especially in semi enclosed areas (UNEP/FAO/WHO 1996). Use of satellite imagery on chlorophyll distribution show that the very high concentrations are located close to river deltas and estuaries or near urban agglomerations, especially in the estuary of Nile (from Alexandria to Gaza), Gulfs of Antalya and Alexandretta (Turkey), Northern Aegean, Thermaikos Gulf (Greece), the Adriatic Sea, the Gulf of Lions (France), Valencia-Barcelona (Spain) and the Gulf of Gabes (Tunisia). Increased temperatures associated with eutrophication can enhance the occurrence of toxic tides, i.e., harmful algae blooms (HABs), and have negative impact on aquaculture production (specially the farming of filter feeders), and increase human health risks. In fact, human consumption of seafood contaminated with harmful biotoxins results in a variety of illnesses with varying degrees of severity, including amnesic shellfish poisoning (ASP), diarrhetic shellfish poisoning (DSP), and paralytic shellfish poisoning (PSP). Biotoxins primarily target the nervous system, but can also result in potentially fatal acute respiratory distress and other chronic neurological and immunological illnesses. The danger of inadvertently consuming biotoxins is compounded by the fact that they are odour- and tasteless and are unaffected by food preparation procedures (see more details of seafood safety in Marques et al., in this volume).

Extreme Events and Water Stress

The impacts of climate change on the Mediterranean environment will relate particularly to water, via a change of its cycle due to a rise in evaporation and a decrease in rainfall. Extreme events, such as heat waves, droughts or floods, are likely to be more frequent and violent. A significant decrease in rainfall, ranging between −4 and −27 % for the countries of Southern Europe and the Mediterranean region is expected (while the countries of Northern Europe will report a rise between 0 and 16 %) (IPCC 2007, scenario A1B). Concomitantly, an increase in drought periods related to high frequency of days during which the temperature would exceed 30 °C is also predicted (Giannakopoulos et al. 2005; Tourre et al. 2008).

This water problem will be of crucial importance with regard to the issue of sustainable development in the region.

Floods (resulting from altered rainfall patterns) will affect nutrient loads in the coastal aquaculture areas. High inorganic sediment loads can reduce or arrest the filtration rates of bivalves. Elevated nutrient levels can also stimulate HABs. For coastal and offshore aquaculture, more frequent and intense storms result in increased physical damage and stock losses, both of which are costly to operations. Many coastal processes, such as sediment transport, happen mostly during high-energy events (storms). An increase in storm activity may therefore initiate erosion. Any severe flooding event could result in mass mortalities of animals in aquaculture ponds, open-water rafts, and lines or cages in coastal and offshore areas. Regarding drought, over much of the Mediterranean basin the general tendency is towards decreasing rainfall. The predicted water stress is thought to result in decreasing water availability in the major Mediterranean freshwater systems, areas where there are important aquaculture activities (see production in Fig. 37.6).

Sea Level Rise

Rising global temperatures are very likely to raise the sea level by expanding ocean water and melting mountain ice caps and glaciers (IPCC 2007). Based on the existing models available for assessment, the central values for projections of sea level rise by 2100 in the Mediterranean Sea range from about 30 to 40 cm (Marcos et al. 2009), and about 60 % of this increase would be due to the thermal expansion of sea water (Tourre et al. 2008). This rise is expected to destroy areas where sand belts are essential for the protection of lagoons and other low-lying areas. A significant part of Mediterranean aquaculture activities occur in the deltaic areas (e.g. Nile) and estuaries, at the middle to upper levels of the tidal ranges. Sea level rise, coupled with increased saline water intrusion, will change the water quality and affect most freshwater and brackish production. Higher temperatures and evaporation would also cause rises in the salinity of lakes and reservoirs. In Malta, a 1 m rise in sea level could reduce water from the main reservoir by 40 % (Attard et al. 1996), while in France, the Vaccares and lower lakes of the Camargue are anticipated to become hyper-saline (Corre 1992). Problems of saline intrusion would be further exacerbated by reductions in runoff and by increased withdrawals in response to higher demand. Excessive demand already contributes to saline intrusion problems in many coastal areas of Italy, Spain, Greece and North Africa (Aru 1996). Changing production to more salinity tolerant strains and/or to farming saline tolerant species, are possible adaptive measures to these problems. Yet, such shifts are going to be costly and will also impact on the socio-economic status of the Mediterranean communities involved.

Acidification

In the year 2000, seventy-two per cent (72 %) of the Mediterranean greenhouse gases emissions were due to CO_2 connected with energy use (77 % in the northern Mediterranean countries and 64 % in the southern ones). In 2025, the CO_2 emissions due to energy use will be twice as high as they were in 1990. The share of the southern countries in the total emissions generated in the Mediterranean would be of about 50 % in 2025. In 2006, the northern Mediterranean countries accounted for about 2/3 (two thirds) of the CO_2 emissions due to energy use of the whole Mediterranean basin (Tourre et al. 2008). However, the growth of CO_2 emissions seems to be far more rapid in the south than in the north. Indeed, while the northern part reported an increase by 18 % between 1990 and 2004, the emissions of the southern rim increased by 58 % over the same period. This growth rate is 20 points higher than the world rate.

Although oceanic uptake of anthropogenic CO_2 will lessen the extent of global warming, the direct effect of CO_2 on ocean chemistry may affect marine biota profoundly (Kleypas et al. 2006; Fabry et al. 2008; Doney et al. 2009). In fact, the future changes in the chemistry of Mediterranean Sea may pose particular problems for marine organisms with $CaCO_3$ shells and skeletons. The degree to which these organisms are affected depends largely upon the $CaCO_3$ saturation state. With increasing seawater pCO_2 both pH and carbonate availability will decrease in the Mediterranean Sea, which may result in reduced calcification rates and shell dissolution (Michaelidis et al. 2005a, b; Gazeau et al. 2007). In order to accurately assess socio-economic and ecological impacts of ocean acidification, it is urgent to investigate the adaptive response of calcareous organisms to long-term CO2 enrichment and the interactive effect of the concomitant increase in seawater temperature. Although some studies highlighted the negative effect of seawater acidification on calcification by benthic mollusks under projected atmospheric CO2 rising scenarios by the end of this century, longer-term studies still need to be achieved in order to take into account a potential adaptation of the organisms to low pH environments.

Diseases

Mediterranean aquaculture is threatened by pathogen range expansions induced by a changing climate as increased temperatures stimulate the growth, transmission and survival of aquatic parasites. Rising sea surface temperatures will afford aquatic parasites the conditions necessary to proliferate in areas where previously they were inhibited by lower temperature thresholds (Ward and Lafferty 2004; Handisyde et al. 2006). Additionally, aquaculture practice unnaturally increases host densities thereby making species even more vulnerable to infectious diseases (Snieszko 1974). Yet, the geographical spread of parasites (and their hosts) will inevitably be moderated by a variety of ecological and environmental factors, rising temperatures notwithstanding (Harvell et al. 1999, 2002; De Silva and Soto 2009). Aquatic diseases in the Mediterranean basin will adversely affect human health through consumption of contaminated seafood and ingestion of water-borne pathogens. Although the transmission of these diseases is influenced by a variety of social, economic and ecological conditions and human immunity, climatic conditions will play an increasingly important role.

Bacterial and Viral Fish Diseases

It is expected that most future disease reports in the Mediterranean basin will be related to finfish intensive production, especially seabass and seabream in marine systems and trout in freshwater ones. It is worth noting that scarce information has been obtained about diseases occurring in the extensive and semi-intensive systems of the southern margin of the Mediterranean basis (i.e. in tilapia, carp and mullet).

The main diseases reported in the Mediterranean region are vibriosis, pasteurellosis, enteric red mouth (ERM) disease, furunculosis and marine flexibacteriosis. Within the genus *Vibrio*, the species causing the most economically serious diseases in marine culture are *Vibrio anguillarum*, *V. vulnificus*, *V. ordalii*, and *V. salmonicida*, being the first two the most significant diseases for the Mediterranean region (Arias et al. 1995, 1997). ERM is the most reported disease in trout farming in Mediterranean countries. The typical furuncolosis is described in both fresh water fish (mainly cultured salmonids) and marine fish (mainly seabass, seabream and turbot). Other significant diseases, with lower reports are rainbow trout fry syndrome (RTFS), columnaris disease, motile *Aeromonas septicemia*, pseudomoniasis, streptococcocis, mycobacteriosis, epitheliocystis, and rainbow trout gastrointestinal syndrome (RTGS) (Rodgers and Furones 1998; Le Breton 1999; Ariel and Olesen 2002; Toranzo 2004)

Regarding viral diseases, about nine viruses, namely lymphocystis, nodavirus, infectious pancreatic necrosis (IPN), infectious hematopoietic necrosis (IHN), viral hemorrhagic septicemia (VHS), spring viraemia of carp (SVC), catfish iridovirus, catfish herpervirus and eel herpervirus have been reported for the Mediterranean area, although only six represent some threat (see Barja 2004).

Table 37.1 Main reported fish parasites/diseases in the Mediterranean Sea

Major groups	Parasite/disease	Most frequent fish host
Amoebozoan	Amebiasis	1, 2, 3, 4
Dinoflagellates	Amyloodiniosis	1, 2
Flagellates		
Ectoparasitic	Costiasis, cryptobiasis	Freshwater and marine nei
Endoparasitic	Trypanosoma sp., Hexamitiasis	3, 5
Ciliates	Chilodonellosis	Freshwater nei
	Cryptocaryosis	2, 6
	Trichodiniasis	1, 2, 5, 7
	White spot disease	3
	Scuticociliatida	1, 2, 4
Protozoans	Coccidiosis	1, 2, 4, 6
	Microsporidiosis	1, 4, 6, 8
Myxosporea	Whirling disease	3
	Proliferative kidney disease (PKD)	3
	Spaherospora renicola	5
	Ceratomyxa spp.	1, 2
	Enteromyxum leei	1, 6
	Enteromyxum scophthalmi	4
	Sphaerospora dicentrarchi	1
	Sphaerospora testicularis	2
Monogenea	Gyrodactylus spp.	1, 2, 3, 5
	Dactylogyrus spp.	3, 5, 7
	Diplectanum spp.	1, 2
	Furnestinia spp.	1, 2
	Microcotylosis	1
Trematoda digenea	Diplostomum spp.	3, 5
	Sanguinicolosis	1, 5
Cestodes	Pseudophyllidea	Freshwater nei
	Protecephalidea	Freshwater nei
Crustaceans	Copepoda	1, 2, 3, 5
	Isopoda	2, 6

Adapted from Alvárez-Pellitero (2004), Athanassopoulous et al. (2009)
1 Seabream, 2 seabass, 3 salmonids, 4 turbot, 5 cyprinids, 6 other sparids, 7 eel, 8mugilids

Fish Parasitic Diseases

In the Mediterranean Sea, there are about 35 different parasitic diseases (Table 37.1), which is significantly higher than those reported for fish bacterial and viral diseases (see previous section).

The main parasitic diseases in freshwater fish (salmonids, cyprinids and eels) are trichodiniasis, costiasis, white spot diseases, *Dactylogyrus* and gyrodactilosis. The main parasitic diseases in marine fish (seabass, seabream, turbot, other sparids, etc.) are trichodiniasis, costiasis, *Enteromyxum leei*, *Ceratomyxa*, amyloodiniosis, mycrocotylosis and sea lice disease. There is an increasing concern of parasitic diseases in intensively cultured finfish in the Mediterranean area, since some parasites can be considered a serious threat for mariculture, such as *Amyloodinium* (Dinoflagellates), Scuticociliatida (Ciliates), *Enteromyxum* spp. (Myxosporea) or Mycrocotylidae (Monogenea). Other parasites are rarely reported in mortality episodes. However, their pathological concern should not be neglected, considering their increasing presence in the cultures and their direct or indirect effects, even when they are not the direct cause of mortality (Alvárez-Pellitero et al. 1993, 1995; Rodgers and Furones 1998; Le Breton 1999; Ariel and Olesen 2002; Alvárez-Pellitero 2004; Athanassopoulous et al. 2009).

Mollusc Diseases

As discussed in previous sections, the main mollusc species produced in the Mediterranean region are the Mediterranean mussel, the blue mussel, the Pacific cupped oyster, the Japanese carpet shell clam and the European flat oyster. The main diseases that have been reported for these species are bonamiosis, marteiliosis, perkinsosis, haplosporidiosis, mytilicolosis, brown ring disease, larval/juvenile vibriosis and herpes-like virus infection (see Berthe 2004).

Sustainable Development and Adaptive Measures

Domestication and Introduced Species

Domestication can contribute to a future sustainable Mediterranean aquaculture since it avoids the need to capture wild stocks. It is noteworthy that the potential impact on the wild ecosystem of fish escapes may be minimized since cultured organisms can be selected to be unable to survive in wild conditions, dying in a short period of time and with a high percentage of organisms unable to reproduce (sterile organisms). Yet, some negative effects of domestication are related to the emergence of genetic drift and inbreeding problems (Agnese et al. 1995), since only a small population of parents is maintained.

An alternative is the introduction of new species, but the risks are quite relevant. The consequences of the escape of those species may dictate major impacts on Mediterranean biodiversity and ecosystems (see McNeely and Schutyser 2003), such as alterations in the genetic pattern of wild populations. Such organisms may compete with native species for food and space, and might also transfer diseases and parasites. In fact, there is already a high number of introduced marine species which are increasing, mainly in ports and lagoons. Over 600 marine exotic species have been recorded in the Mediterranean Sea (UNEP/MAP 2004) and the mode of introduction is different within the basin. In the eastern Mediterranean, penetration via the Suez Canal is the main mode of introduction, while in the western Mediterranean, shipping and aquaculture (at a lesser extent) are responsible for the great majority of exotic species (EEA 2006). Introduction of species should be carried out only in special cases and taking all required precautions. The recommendations and suggestions mentioned in ICES Code of Practices (2005) as well as in Hewitt et al. (2006) should be followed. Regional and international collaboration in the Mediterranean basin should be supported to address transboundary biodiversity impacts of introduced species (see UNEP/MAP 2005).

Capture of Wild Stocks and Aquafeeds

One of the most important issues for the future and sustainable development of Mediterranean aquaculture is the sustainability of the source of cultivated fish or shellfish. Aquaculture should help to relieve this pressure on wild stocks and promote the maintenance of biodiversity, whilst satisfying the growing market demand for aquatic products (IUCN 2007). Therefore, as a principle, the stocking of aquaculture farms should not affect the natural status or viability of wild populations and their ecosystems or biodiversity in general. An example is the expansion of the Atlantic bluefin tuna farming activities in the Mediterranean that has generated a growing demand for wild fish specimens. Hence, one of the main concerns about this demand is the current and potential pressure to increase fishing. In general, it is preferable that organisms that are to be raised in aquaculture farms should have been produced in hatcheries.

Regarding aquafeeds, the Mediterranean market is focused on the production of formulated feeds for intensively produced carnivorous fishes, namely seabass, seabream and trout. Other important fish species, albeit with a much lower production, are turbot, eel, and salmon (Martín 1999). Due to the growth in production of marine fish species, the Mediterranean aquafeed market has significantly grown from an estimated production of 315,000 tons in 1995 to 500,000 tons in 2001, a 58 % increase (Montero et al. 2005). It is highlighted that parallel to these increases in production of aquafeed for marine fish there has been a substantial improvement in feed efficiency. Practically the whole of the aquafeed market is concentrated in only five countries: Spain, Greece, France, Italy and Turkey. The great development of intensive fish farming in recent years is noteworthy in countries such as Greece and Turkey, meaning an important change in the distribution of the market. The future development of Mediterranean aquaculture is strongly linked to the possibility of providing sustainable aquafeed ingredients. The current marked increase in aquaculture production has to take into account that fish meal and fish oil are worldwide limited resources (Tacon 2004). If the Mediterranean aquaculture of carnivorous species wishes to continue further growth, improvements must be achieved in the feeding of these animals, and alternative raw ingredients for aquafeeds must be found. Partial replacement of fish meal and fish oil in the diets of farmed fish by vegetable proteins

Table 37.2 Summary of the effects of climate change in Mediterranean aquaculture, and potential adaptive measures

Impact of climate change	Adaptation/mitigation measures
Temperature rise (above optimal range of tolerance)	Use better feeds
	Use selective breeding and genetic improvements
	Use short-cycle aquaculture
Enhanced growth/production	Increase feed input
Sea level rise/salt water intrusion	Changing farmed species
	Moving operations away from the shore
Increase in euthrophication	Improve monitoring and early warning systems
	Implement waste water treatment with cost-effective and environmental-friendly techniques
Limitations on seafood meal and fish oil supplies	Shift to non-carnivorous, bivalve and seaweed species
	Genetic improvement for alternative feeds
Less wild seed stocks	Use of hatchery seed
	Protect nursery habitats
	Improve seed quality and production efficiency
	Close the species life cycle
Water quality	Implement waste water treatment with cost-effective and environmental-friendly techniques
	Improve efficiency of water use
	Avoid using contaminated water
	Shift to faster growing seafood species
	Use agro or multitrophic aquaculture
Loss of stocks	Improve site and design to prevent losses and escapes
	Encourage use of indigenous or non-reproducing species
Increase resistance to diseases (increase use of veterinary drugs)	Replace veterinary synthetic drugs by natural control of diseases (e.g. probiotics, green water technique, natural immunostimulants, vaccines)
	Implement genetic improvements for higher resistance
	Shift to resistant farmed species
	Improve management practices
Increase virulence and expansion of contaminants and diseases	Increase biosecurity measures
	More monitoring and early warning systems
	Implement genetic improvements for higher resistance
	Adopt solutions to reduce contaminant load, e.g. processing seafood, cooking, phycoremediation, etc.
	Develop guidelines and predictive modelling tools for stakeholders
Acidification and calcareous shell formation/deposition	Adapt production and handling techniques
	Move to other production zones

and oils is taking place without compromising the fish quality, as well as the health benefits of PUFA ω3 diet (Rosa et al. 2010). But, the complete substitution or replacement of fish meal by more sustainable and renewable protein sources, like oilseeds or vegetable meals, has brought up several issues, partially because of an inappropriate aminoacid balance and poor protein digestibility (Sargent and Tacon 1999).

Policy and Planning Measures

Planning systems are diverse and complex, but they are critical to the sustainable development of Mediterranean aquaculture. For instance, aquaculture site selection and site management can be facilitated through Integrated Coastal Zone Management (ICZM), which is an adaptive process based on clear and transparent governance and thorough knowledge to support decision making (IUCN 2009). ICZM is a tool for decision-makers in the planning, zoning and licensing of all human activities and all stakeholder rights in relation to a defined coastal area. In the Mediterranean, the major ICZM issues include the uncontrolled urban growth near the coast, tourism, and loss of coastal biodiversity (IUCN 2009). Future planning should look at suitable zones for aquaculture development as well as potential zones for the development of aquaculture. In fact, locating and identifying areas of interest or areas that are suitable for aquaculture is a key factor in ensuring the sustainable development of this sector in the Mediterranean. The process facilitates administrative procedures and allows for better management and forecasting of growth (IUCN 2009).

Another priority area for the development of this sector is the ecosystem approach to aquaculture (EAA), which aims to integrate aquaculture within the wider ecosystem in such a way that it promotes sustainability of interlinked socialeco-logical systems. It is a tool for the integrated management of human activities, based on the protection of land, water and living resources; a strategy that promotes conservation and sustainable use of the ecosystem in an equitable way. EAA emphasizes the need to integrate aquaculture with other sectors (e.g. fisheries, agriculture, and urban development) that share and affect common resources (see Shepherd 2004; Soto et al. 2008; De Silva and Soto 2009; IUCN 2009).

Conclusions

The Mediterranean region, also known as the cradle of western civilization, has been subject to human intervention for millennia. Due to religious and social traditions, seafood is widely consumed in this region, with an average of 16.5 kg/capita/year, and ranging from 5.2 in Algeria and 40.0 in Spain (FAO 2011). The growth of seafood demand in the Mediterranean is expected to increase in the future, especially in southern countries (Cochrane and de Young 2007). One fourth of the Mediterranean seafood supply comes from aquaculture activities, whereas the remaining is from fisheries.

The Mediterranean aquaculture sector has expanded over the last decades. It increased 77 % over the last decade reaching *circa* 1.3 million tons in 2009. Along with predicting vulnerable areas, the selection of suitable sites in relation to specific culture methods and species will be valuable in maximizing profit and food production in the face of a changing climate. The impacts of climate change will relate particularly to water availability and/or quality, due to the expected rise in evaporation, decrease in rainfall, and more frequent and violent extreme events (heat waves, droughts or floods) in this region. For adaptation these climate trends, and consequent environmental changes, the key focus on Mediterranean aquaculture will be on selecting suitable species and culture methods. A summary of the most important predicted impacts of climate change in the Mediter ranean aquaculture is given in Table 37.2, with a wide range of adaptation and mitigation strategies that can be implemented to minimize such effects.

Implementing adaptation and mitigation pathways for Mediterranean communities dependent on aquatic ecosystems will need special attention from policy-makers and planners. The solution to climate change is not straight forward, and many extrinsic factors to the aquaculture sector have to be integrated collectively, with an ecosystem perspective.

References

Agnese J-F, Oteme ZJ, Gilles S (1995) Effects of domestication on genetic variability, fertility, survival and growth rate in a tropical siluriform: *Heterobranchus longifilis* Valenciennes 1840. Aquaculture 131:197–204

Alvárez-Pellitero P (2004) Report about fish parasitic diseases. Opt Méd B Et Rech 49:103–130

Alvárez-Pellitero P, Sitjà-Bobadilla A, Franco-Sierra A (1993) Protozoan parasites of wild and cultured sea bass, *Dicentrarchus labrax*, from the Mediterranean area. Aquac Fish Manag 24:101–108

Alvárez-Pellitero P, Sitjà-Bobadilla A, Franco-Sierra A, Palenzuela O (1995) Protozoan parasites of gilthead sea bream (*Sparus aurata*) from different culture systems in Spain. J Fish Dis 18:105–115

Arias CR, Garay E, Aznar R (1995) Nested PCR method for rapid and sensitive detection of *Vibrio vulnificus* in fish, sediments, and water. Appl Environ Microbiol 61:3476–3478

Arias CR, Verdonck L, Swings J, Garay E, Aznar R (1997) Intraspecific differentiation of *Vibrio vulnificus* biotypes by amplified fragment length polymorphism and ribotyping. Appl Environ Microbiol 63:2600–2666

Ariel E, Olesen NJ (2002) Finfish in aquaculture and their diseases – a retrospective view on the European Community. Bull Eur Assos Fish Pathol 22:72–85

Aru A (1996) The Rio Santa Lucia site: an integrated study of desertification. In: Brandt CJ, Thornes JB (eds) Mediterranean desertification and land use. Wiley, Chichester

Athanassopoulous F, Pappas IS, Bitchava K (2009) An overview of the treatments of parasitic disease in Mediterranean aquaculture. Option Méditerr A 86:65–83

Attard DJ, Axiak V, Borg S et al (1996) Implications of expected climatic changes for Malta. In: Jeftic L, Keckes S, Pernetta JC (eds) Climate change and the Mediterranean, vol 2. Arnold Press, London

Barazi-Yeroulanos L (2010) Synthesis of Mediterranean marine finfish aquaculture: a marketing and promotion strategy. FAO, Rome

Barja JL (2004) Report about fish viral diseases. Opt Méd B Et Rech 49:91–101

Berthe F (2004) Report about mollusk diseases. Opt Méd B Et Rech 49:33–48

Cochrane K, de Young C (2007) Ecosystem approach to fisheries management in the Mediterranean. Cah Option Mediterr B 62:71–85

Cochrane K, De Young C, Soto D, Bahri T (2009) Climate change implications for fisheries and aquaculture: overview of current scientific knowledge. FAO, Rome

Corre JJ (1992) Implications des changements climatiques etude de cas: Le Golfe du Lion (France). In: Jeftic L, Milliman JD, Sestini G (eds) Climatic change and the Mediterranean. Arnold Press, London

De Silva SS, Soto D (2009) Climate change and aquaculture: potential impacts, adaptation and mitigation. In: Cochrane K, Young C, De Soto D, Bahri T (eds) Climate change implications for fisheries and aquaculture overview of current scientific knowledge, FAO fisheries and aquaculture technical paper: no 530. FAO, Rome

Doney SC, Fabry VJ, Feely RA, Kleypas JA (2009) Ocean acidification: the other CO_2 problem. Ann Rev Mar Sci 1:169–192

EEA (2006) Priority issues in the Mediterranean environment, European Environmental Agency report 4/2006. European Environmental Agency, Copenhagen

Eshel G, Farrel BF (2000) Mechanisms of Eastern Mediterranean rainfall variability. J Atmos Sci 57:3219–3232

Fabry VJ, Seibel BA, Feely RA, Orr JC (2008) Impacts of ocean acidification on marine fauna and ecosystem processes. ICES J Mar Sci 65:414–432

FAO (2011) The state of world fisheries and aquaculture 2010. Food and Agriculture Organization of the United Nations, Rome

Ficke AD, Myrick CA, Hansen LJ (2007) Potential impacts of global climate change on fresh water fisheries. Rev Fish Biol Fish 17: 581–613

Gazeau F, Quiblier C, Jansen JM, Gattuso JP, Middelburg JJ, Heip CHR (2007) Impact of elevated CO_2 on shellfish calcification. Geophys Res Lett 34, L07603

Giannakopoulos C, Bindi M, Moriondo M, LeSager P, Tin T (2005) Climate change impacts in the Mediterranean resulting from a 2°C global temperature rise. WWF Report for a Living Planet. The Global Conservation Organization, Gland

Goudie A (2001) Encyclopedia of global change: environmental change and human society, vol 1. Oxford University Press, Oxford

Handisyde NT, Ross LG, Badjeck M-C, Allison EH (2006) The effects of climate change on world aquaculture: a global perspective. Final technical report, DFID/Aquaculture and Fish Genetics Research Programme/Stirling Institute of Aquaculture, Stirling

Harvell CD, Kim K, Burkholder JM et al (1999) Emerging marine diseases-climate links and anthropogenic factors. Science 285: 1505–1510

Harvell CD, Mitchell CE, Ward JR, Altizer S, Dobson AP, Ostfeld RS, Samuel MD (2002) Climate warming and disease risks for terrestrial and marine biota. Science 296:2158–2162

Hewitt CL, Campbell ML, Gollasch S (2006) Alien species in aquaculture. Considerations for responsible use. IUCN, Gland

Hurrell JW (1995) Decadal trends in the North Atlantic Oscillation: regional temperatures and precipitation. Science 269:676–679

Hurrell JW (1996) Influence of variations in extratropical wintertime teleconnections on northern hemisphere temperature. Geophys Res Lett 23:665–668

ICES (2005) ICES Code of practice on the introductions and transfers of marine organisms. International Council for the Exploration of the Sea, Copenhagen, 30 pp

IPCC (2007) Climate change 2007: the physical science basis: contribution of working group I to the fourth assessment report of the intergovernmental panel on climate change. Cambridge University Press, New York

IUCN (2007) Guide for the sustainable development of Mediterranean aquaculture. Interaction between aquaculture and the environment. IUCN, Gland

IUCN (2009) Guide for the sustainable development of Mediterranean aquaculture 2. Aquaculture site selection and site management. IUCN, Malaga

Jeftic L, Keckes S, Pernetta JC (1996) Climate change and the Mediterranean, vol 2. United Nations Environment Programme (UNEP)/Arnold Press, London

Kleypas JA, Feely RA, Fabry VJ, Langdon C, Sabine CL, Robbins LL (2006) Impacts of ocean acidification on coral reefs and other marine calcifiers. A guide for future research. NSF/NOAA/The U.S. Geological Survey, St Petersburg

Le Breton AD (1999) Mediterranean finfish pathologies: present status and new developments in prophylactic methods. Bull Eur Assos Fish Pathol 19:250–253

Marcos M, Tsimplis MN, Shaw AGP (2009) Sea level extremes in Southern Europe. J Geophys Res 114(C1), C01007

Martín A (1999) Aquaculture feed manufacturing practice in Mediterranean countries. Option Méditerr 37:123–128

McNeely JA, Schutyser F (2003) Invasive species: a global concern bubbling to the surface. In: International conference on the impact of global environmental problems on continental and coastal marine waters, Geneva

Michaelidis B, Haas D, Grieshaber MK (2005a) Extracellular and intracellular acid–base status with regard to the energy metabolism in the oyster *Crassostrea gigas* during exposure to air. Physiol Biochem Zool 78:373–383

Michaelidis B, Ouzounis C, Paleras A, Pörtner HO (2005b) Effects of long-term moderate hypercapnia on acid–base balance and growth rate in marine mussels *Mytilus galloprovincialis*. Mar Ecol Prog Ser 293:109–118

Montero D, Basurco B, Nengas I, Alexis M, Izquierdo M (2005) Mediterranean fish nutrition. Cah Option Méditerr 63:1–158

Peixoto JP, De Almeida M, Rosen RD, Salstein DA (1982) Atmospheric moisture transport and the water balance of the Mediterranean Sea. Water Resour Res 18:83–90

Rodgers CJ, Furones MD (1998) Disease problems in cultured marine fish in the Mediterranean. Fish Pathol 33:157–164

Rodo X, Baert E, Comin FA (1997) Variations in seasonal rainfall in southern Europe during present century: relationships with the North Atlantic Oscillation and the El Nino Southern Oscillation. Clim Dyn 13:275–284

Rosa R, Andrade AM, Bandarra NM, Nunes ML (2010) Physiological and biochemical effects of conjugated linoleic acid and its use in aquaculture. Rev Aquac 2:59–72

Sargent JR, Tacon AG (1999) Development of farmed fish: a nutritionally necessary alternative to meat. Proc Nutr Soc 58:377–383

Shepherd G (2004) The ecosystem approach: five steps to implementation. The World Conservation Union (IUCN), Gland

Snieszko SF (1974) The effects of environmental stress on outbreaks of infectious diseases of fishes. J Fish Biol 6:197–208

Soto D, Aguilar-Manjarrez J, Brugère C et al (2008) Applying an ecosystem-based approach to aquaculture: principles, scales and some management measures, FAO fisheries proceedings no 14. FAO, Rome

Tacon AGJ (2004) Use of fish meal and fish oil in aquaculture: a global perspective. Aquat Resour Cult Dev 1:3–14

Toranzo AE (2004) Report about fish bacterial diseases. Opt Méd B Etud Recherch 49:49–89

Tourre Y, Van Grunderbeeck P, Allal H et al (2008) Climate change and energy in the Mediterranean. Plan Bleu Regional Activity Center, Sophia Antipolis

UNEP (1989) State of the Mediterranean marine environment, MAP technical series nº 28. UNEP, Athens

UNEP/FAO/WHO (1996) Assessment of the state of eutrophication in the Mediterranean Sea, MAP technical series nº 106. UNEP, Athens

UNEP/MAP (2004) Marine pollution indicators fact. United Nations Environment Programme/Mediterranean Action Plan, Athens

UNEP/MAP (2005) Action plan concerning species introductions and invasive species in the Mediterranean Sea. United Nations Environment Programme/Mediterranean Action Plan, Tunis

Ward JR, Lafferty KD (2004) The elusive baseline of marine disease: are diseases in ocean ecosystems increasing? PLoS Biol 2:542–547

Climate Change, Extreme Weather Events and Health Effects

38

Francesca de' Donato and Paola Michelozzi

Abstract

Since the 1970s an increase in temperatures has been observed, and future climate predictions suggest an increase in the frequency and intensity of extreme weather events. Conversely to other environmental risks, climate change is a global phenomenon which will have an impact on the population's health at the global scale. In recent years, many studies have documented the impact of extreme events and climate variability on health as well as estimating the potential future risks under different climate change scenarios. An overview of the potential effects of climate change on health in the Mediterranean region is presented focussing on the increase in health burden associated with heat waves and extreme events, floods and droughts, air pollution episodes and, to the rise in the indirect health effects of infectious disease.

Keywords

Climate change • Extreme weather events • Health effects • Health impact

Climate Change and the Impact on Health in the Mediterranean

A rapid increase in global temperatures has been observed from the mid-1970s and has been substantially attributable to anthropogenic forcings; in particular to the increased production of greenhouse gases. In Europe, a progressive warming has been observed (+0.90 °C from 1901 to 2005), while precipitation trends have shown an increase in northern Europe (between 10 and 40 %) and a decrease of up to 20 % in southern Europe (IPCC 2007a, b; Alcamo et al. 2007). Climate prediction models forecast that the warming will be associated with more frequent, more intense and longer lasting heat waves especially in Southern and Eastern Europe but also in areas not currently susceptible to these events (Räisänen et al. 2004; IPCC 2007a; Christensen and Christensen 2007; Kjellström et al. 2007). An increase in the frequency of hot days, tropical nights and heat waves is expected for summer months, while cold spells and frost nights will decline in winter (Meehl and Tebaldi 2004; Giorgi et al. 2004; Alcamo et al. 2007; Della-Marta et al. 2007). In regards to precipitation, climate change prediction models indicate that precipitation will become more variable, with greatest decline in mean precipitation in the Mediterranean (Alcamo et al. 2007). Throughout Europe, the higher intensity of rainfall events will increase the likelihood of floods, in particular flash floods, while the occurrence of heat waves and droughts will increase the risk of desertification especially around the Mediterranean basin (Alcamo et al. 2007). Furthermore, an increase in the risk of wildfires is expected especially in areas with high forest cover in Mediterranean countries (IPCC 2007a, b). Ice sheets and mountain glaciers are expected to melt and sea level to rise progressively as a result of temperature rise. Furthermore, sea surface temperatures are warming and oceans are becoming more acidic (IPCC 2007a).

In recent years, many studies have given evidence of the impact of extreme events and climate variability on health in Europe as well as estimating the potential future risks

F. de' Donato (✉) • P. Michelozzi
Department of Epidemiology, Lazio Regional Health Service, Via di Santa Costanza 53, 00198 Rome, Italy
e-mail: f.dedonato@deplazio.it; p.michelozzi@deplazio.it

S. Goffredo and Z. Dubinsky (eds.), *The Mediterranean Sea: Its history and present challenges*,
DOI 10.1007/978-94-007-6704-1_38,

(Kovats et al. 2003; McMichael et al. 2003, 2006; Haines et al. 2006; Menne and Ebi 2006; APAT and WHO 2007; Michelozzi et al. 2011). However, recent events such as the 2003 heat wave, Russian heat wave and forest fires of 2010, floods occurred in 2010 and in Italy and France in 2011, and more recently cold spell in February 2012 in Europe have demonstrated that populations and health systems are unable to cope with these extreme events throughout Europe. During these episodes the resilience of communities is reduced, vulnerable areas are affected and the coping capacities of societies are besieged. Policies to reduce the impact of climate change on health in both the short and long term are to date limited on a global scale.

Changing patterns of disease will occur in response to changing environmental conditions (McMichael et al. 2006). The strongest adverse effects on natural and human systems are expected as a result of altered frequencies and intensities of extreme weather, together with sea level rise. Figure 38.1 and Table 38.1 show the potential effects of climate change on the environment and their impacts on health. Among the climate-related hazards (heatwaves, floods, storms, fires) to which populations will be exposed, the direct health effects, in particular morbidity, mortality and various infectious diseases have been studied and are easier to assess. On the contrary, indirect effects such as variations in regional food yields, disruption of fisheries, loss of livelihoods, and population displacement (because of sea-level rise, water shortages) are more complex to study (Kovats and Ebi 2006).

How populations respond and adapt to these extreme events depends on the severity of the event and on local socio-demographic, behavioural, cultural characteristics and technological and public health resources available. Conversely to other environmental risks, climate change is a global phenomenon which will have an impact on the population's health at the global scale. In particular, the differential in inequalities between rich and poor countries will increase, social and political conflicts will arise as a consequence of mass migration putting health and social systems at risk. Although the rich developed countries are the major greenhouse gas emitters, the less develop countries, with a limited adaptive capacity, are the ones which will suffer the most from the consequences of climate change, thus increasing the health inequalities (Kovats and Ebi 2006; McMichael et al. 2006; Bargagli and Michelozzi 2011).

This chapter will give an overview of the potential effects of climate change on health in the Mediterranean region. With particular focus on the increase in health burden associated with heat waves and extreme events, floods and droughts, air pollution episodes and, to the rise in the indirect health effects of infectious disease.

Extreme Temperatures

Recent multi-city studies conducted in Europe have shown a greater impact of heat and heat waves in the Mediterranean areas (Baccini et al. 2008; Michelozzi et al. 2009; D'Ippoliti et al. 2010). Considering climate change scenarios the burden on health is expected to increase (Baccini et al. 2011; McMichael et al. 2006).

The health effects of exposure to heat and cold have been studied in several populations worldwide (Basu and Samet 2002; Basu 2009; Gosling et al. 2007, 2009; Åström et al. 2011; Ye et al. 2012). Most epidemiological studies on extreme temperatures have been carried out in Europe and North America (Curriero et al. 2002; Analitis et al. 2008; Baccini et al. 2008; Hajat et al. 2002) while more recently in other areas of the world (McMichael et al. 2003; Hajat et al. 2005; Kim et al. 2006; Wang et al. 2012; Ye et al. 2012).

The first studies quantified the impact of individual heatwave events on mortality (Albertoni et al. 1984; Katsouyanni et al. 1988; Semenza et al. 1996; Rooney et al. 1998; Naughton et al. 2002) while more recently, the relation between temperature and mortality is explored using the time series approach. This method allows to investigate the dose–response relationship between daily temperatures and mortality in each population; the association is typically described by a non-linear U-, or J- shaped function, with the lowest mortality rates recorded at moderate temperatures, rising progressively as temperatures increase or decrease (Huynen et al. 2001; Curriero et al. 2002). These studies have shown that the temperature-mortality relationship varies geographically due to climate and local population characteristics which determine population susceptibility and adaptive capacity (Basu and Samet 2002; Basu 2009; Gosling et al. 2009; Baccini et al. 2008) (Fig. 38.2). In the multi-city European project PHEWE the minimum mortality (threshold) during summer ranged from 23.3 °C in North-Continental cities to 29.4 °C in Mediterranean cities (Baccini et al. 2008) (Fig. 38.2). Among the European cities included, total mortality associated with a 1 °C increase in maximum apparent temperature above the threshold is of +3.12 % (95 % CrI: 0.60–5.72 %) in the Mediterranean region and 1.84 % (95 % CrI: 0.06–3.64 %) in the North-Continental region (Baccini et al. 2008). Similarly, Analitis et al. found that the cold effect was also greater in the warmer areas with a +1.62 (CI95 %: 1.38, 1.85) increase in mortality for every 1 °C decrease in minimum apparent temperature in the Mediterranean cities, compared to +1.15 % (CI 95 %: 0.96, 1.34) in Northern-Continental cities (Analitis et al. 2008).

Several studies have shown that not all the population is affected in the same way, but there are subgroups more

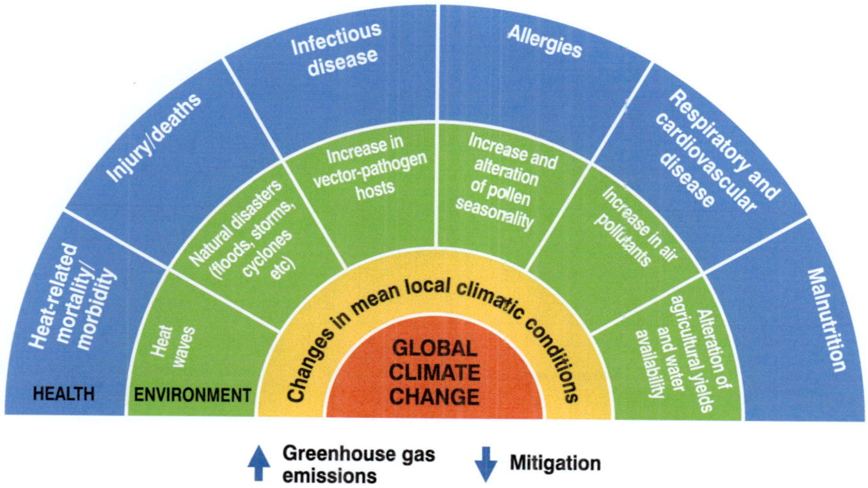

Fig. 38.1 Environmental hazards and the health impacts due to climate change (Adapted from: Bargagli and Michelozzi 2011)

Table 38.1 Summary of the impact of climate change on health

Health outcomes associated to climate change	Mechanisms
Heat and cold-related mortality and morbidity	Increase heat stroke deaths during heat waves
	Worsening of health conditions during heat waves in patients with pre-existing chronic disease
	Decline in cold-related deaths
Deaths and injury	Increase in injuries and deaths related to extreme events (floods, cyclones, wildfires)
Allergic rhinitis	Increase in asthma and rhinitis due to change in the distribution, seasonality and production of aeroallergens under climate change scenarios
Cardiovascular and respiratory disease	Increase in high temperatures and air pollutants concentrations increase cardio-respiratory disease
Infectious disease	Floods provide breeding ground for mosquito vectors and lead to disease outbreaks
Mental disorders	Floods may increase post-traumatic stress disorders
Mosquito, tick-borne and rodent-borne diseases (malaria, dengue, tick-borne encephalitis and Lyme diseases)	Higher temperatures shorten the development time of pathogens in vectors and increase the potential transmission to humans
	Climate conditions determine the necessary conditions to maintain transmission of each vector specie
Waterborne and food-borne disease	Climate conditions affect water availability and quality
	Temperature influences the survival-rate of disease-causing organisms
	Extreme rainfall episodes affect the transport of specific organisms causing disease into the water supply
Malnutrition	Climate change will decrease food supplies and access to food

Adapted from McMichael et al. (2006) and Kovats et al. (2003)

susceptible to the effect of extreme temperatures. Subjects with pre-existing chronic illnesses such as cardiovascular disease, respiratory disease, disease of the central nervous system, psychological illnesses and persons with metabolic/endocrine gland disorders and diabetes have been described to have an increased risk of dying during heat waves (Michelozzi et al. 2005; Stafoggia et al. 2006; Vandentorren et al. 2004; Bouchama et al. 2007; Schifano et al. 2009; Basu 2009; Gosling et al. 2009). Studies conducted on hospital admissions and exposure to high temperatures have shown and significant impact on respiratory admissions especially among Mediterranean cities (Michelozzi et al. 2009). Furthermore, the elderly (Basu 2009), persons with low socio-economic status (Michelozzi et al. 2005; Filleul et al. 2006; Stafoggia et al. 2006), persons living alone (Semenza et al. 1996), and those having no/limited access to air conditioning (Vandentorren et al. 2004; Bouchama et al. 2007) have been shown to be more at risk during heat wave

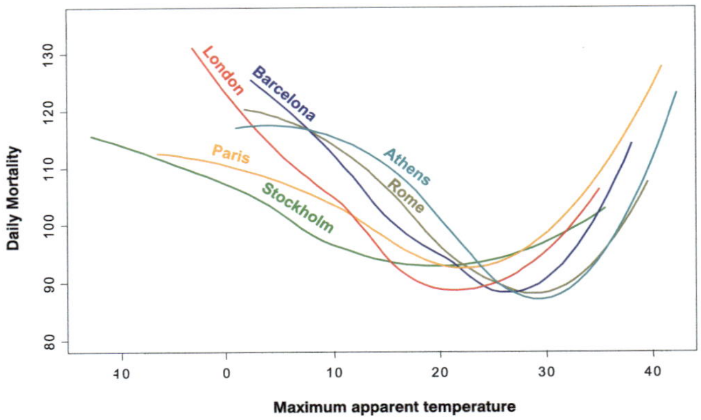

Fig. 38.2 Temperature-mortality relationship in European cities (Source: Baccini et al. 2008)

episodes. People living in urban environments are at greater risk due to the higher thermal exposure under the urban heat island effect (Smargiassi et al. 2009; de'Donato et al. 2008), the greater exposure to traffic-derived air pollutants (Samoli et al. 2005) and poor quality housing and socio-economic degradation of some areas within the city (Semenza et al. 1996; Naughton et al. 2002).

The predicted increase in heat waves in the future, will increase the attributed burden of deaths and hospitalizations related to heat, even if a shift in the threshold towards higher temperatures due to some adaptation is expected (McMichael et al. 2006). Conversely, the decline in extreme cold episodes will reduce the number of cold-related deaths, however, this reduction will only partly compensate the predicted increase in heat-related impact on health which is expected to increase worldwide. Heat waves will interest a more extensive area of Europe, including regions which to date are not typically interested by this phenomenon, making a greater proportion of the European population at risk. Furthermore, most European countries are experiencing an aging of their populations which will signify a greater burden of deaths related to extreme events in these at-risk population subgroups, especially in urban areas.

Floods and Droughts

Floods result from the interaction of rainfall, surface run-off, evaporation, wind, sea level, and local topography. In recent years an increase in floods has been observed and this trend could increase with climate change affecting human health. Floods are rare events which have a high-impact on both the environmental as well as human populations and can overwhelm infrastructures, human resilience and social organisations (McMichael et al. 2006). The risk associated to these events can be altered by water management practices, urbanisation, intensified land use and forestry. In recent years flooding episodes in Europe and other areas worldwide, have increased and events are been becoming more intense (Alcamo et al. 2007). Populations residing in settlements located along the coast, on river deltas or on small islands are at great risk when we consider climate change predictions of sea level rise.

Deaths related to storms and floods are from drowning and severe injuries. Populations with poor sanitation infrastructure and high burdens of infectious disease often experience increased rates of diarrhoeal diseases after flood events. Increases in cholera, cryptosporidiosis and typhoid fever have been reported in low- and middle-income countries. In high income countries, the risk of infectious disease following flooding is generally low, although increases in respiratory and diarrhoeal diseases have been reported after floods (McMichael et al. 2006; Haines et al. 2006). In Europe, the greatest impact of flooding is in terms of economic losses by the population and local infrastructures, while the direct health effects (death and drowning) are to some extent limited. Flooding may also lead to contamination of waters with dangerous chemicals, heavy metals or other hazardous substances (Haines et al. 2006). Furthermore, little is known about the health impacts of floods derived from both large and smaller flooding events both in the short and long-term. These health effects are difficult to quantify and directly attributable to flooding and mostly include an increases in psychological distress, in particular as anxiety and depression.

According to the IPCC 2007 report, drought-affected areas around the world are likely to increase in extent in future years (IPCC 2007b). In response to the future decline in precipitation and increase in temperatures in the Mediterranean region, warmer and drier conditions and consequently to more frequent and prolonged droughts are expected. Furthermore, drought and high temperatures will increase the areas at risk of fires in the Mediterranean (Alcamo et al. 2007). The effects of drought on health include deaths, malnutrition, infectious diseases and respiratory diseases (Menne and Bertollini 2005; Haines et al. 2006). More than two billion people live in dry areas of the world and suffer from malnutrition, infant mortality and diseases related to contaminated or insufficient water (WHO 2005). Drought and the consequent loss of livelihoods is also a major trigger for population movements, particularly rural to urban migration. Population displacement can lead to increases in communicable diseases and poor nutritional status resulting from overcrowding and a lack of safe water, food and shelter. Climate change is also likely to affect water quality and quantity in Europe. The microbial loads in freshwater may increase in response to extreme rainfall and droughts thus having severe implications for disease outbreaks and water quality monitoring. Higher temperatures have implications for food safety, as transmission of salmonellosis is temperature sensitive (Kovats et al. 2003).

Air Pollution

Climate change can affect summer concentrations of pollutants and the levels of tropospheric ozone especially as stable conditions and high temperatures are expected to increase and persist. Air pollution is a well known risk factor for health and the negative effects on health are exacerbated during heat waves (Ren et al. 2009). Air-pollution concentrations are the result of interactions between variations in the physical and dynamic properties of the atmosphere on time-scales from hours to days, atmospheric circulation features, wind, topography and energy use. Air pollution episodes are often associated with stationary or slowly migrating anti-cyclonic or high pressure systems, which reduce pollution dispersion and diffusion. Concentrations of air pollutants in general, and fine particulate matter (PM) and ozone concentrations in particular, may change in response to climate change because their formation depends, in part, on temperature and humidity.

Morbidity and mortality, in particular for cardiovascular and respiratory causes, increase during high air pollution episodes (Samoli et al. 2005; Stafoggia et al. 2006, 2008; Forastiere et al. 2008; Rückerl et al. 2011). Weather variables, and in particular high temperatures, and air pollution have been shown to interact producing negative health effects. Furthermore, wildfires and dust storms episodes may increase pollution levels in a specific region well above the background levels with potential consequences on health. In Europe, the predicted increase in heatwaves and wild fires, under climate change scenarios, will further worsen the negative impact of air pollution on health outcomes, in particular cardiopulmonary and allergic disease (Kovats et al. 2003; Confalonieri et al. 2007). Finally, air pollutants may enhance the allergenic potential of pollens, and the variation in pollen distribution and concentrations expected under climate change represent a further risk for health.

Infectious Disease

Climate change is expected to increase the risk of several infectious disease and so a limited extent also in Europe (McMichael et al. 2006; IPCC 2007d; Semenza and Menne 2009). Many factors determine the transmission of infectious diseases including socio-economic and ecological conditions, access to health care, and intrinsic human immunity. Many infectious agents, vector organisms, non-human reservoir species, and rate of pathogen replication are sensitive to climatic conditions. In recent years, different hypotheses have been proposed to explain the complex association between climate change and infectious diseases. For example the higher proliferation and reproduction rates at higher temperatures, the extension of the transmission season, changes in ecological balances, and climate-related migration of vectors, reservoir hosts, or human populations. Vector-borne diseases are infections transmitted by the bite of infected arthropod species, such as mosquitoes, ticks, triatomine bugs, sandflies, and blackflies. The main risks of infectious disease due to climate change in Europe have been presented in a recent review (Semenza and Menne 2009). Although the literature is not conclusive in demonstrating that recent changes in climate have increased the incidence of vector-borne disease, the review suggests that there are some indications that in the past decades ticks have progressively spread into higher latitudes and altitudes of Europe, have become more prevalent and intensified the transmission season (Semenza and Menne 2009). The risk of reintroduction of malaria into parts of Europe is very low and not determined by climate change. While the introduction of dengue, West Nile fever, and chikungunya into new regions of Europe is a more immediate consequence of virus importation into competent vector habitats, and climate change is one of many factors that influence vector habitat. Conversely, in drought and flood-ridden locations the risk for Lyme borreliosis is projected to be reduced.

Vulnerability and Adaptation

Vulnerability of human health to climate change depends on exposure to extreme weather or environmental-related events, sensitivity to these changes on the basis of population susceptibility (socio-economic, demographic characteristics, health conditions and lifestyle) and their adaptation capacity (depends on how actions are modified to reduce risks and minimize the impact on health) at both the individual and population level (Kovats et al. 2003; Ebi et al. 2006). The interaction of these three elements defines an individual or populations' health risk and vulnerability (Fig. 38.3). The vulnerability of a population to a health risk depends on the local environment, as well as on the effectiveness of, and access to, socio-political infrastructures, in particular the effectiveness of governance and civil institutions and the quality of the public health system. Climate change will have a greater impact on those populations who are more susceptible to extreme weather events and variations in climate and who adopt limited and inadequate adaptation measures.

Adaptation to climate change includes the strategies, policies and measures undertaken to reduce the potential adverse health effects related to climate under present and future conditions (Kovats et al. 2003). Adaptation capacity of individuals, populations and institutions is their ability to adjust to potential damages, to take advantage of opportunities and to cope with the consequences. Understanding a population's capacity to adapt to the changes in extreme weather events is

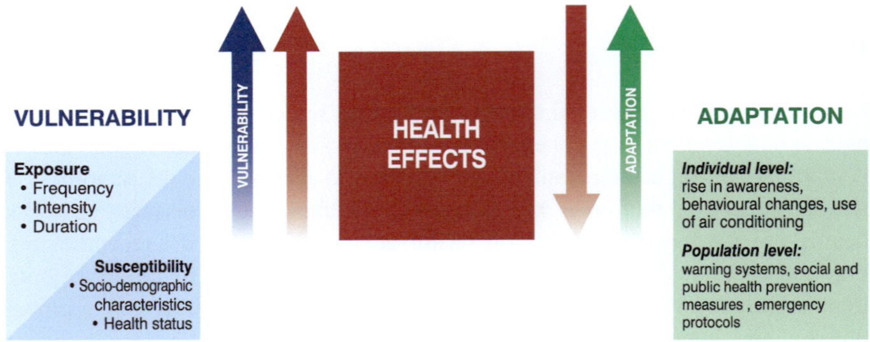

Fig. 38.3 Vulnerability and adaptation to climate change (Adapted from: Bargagli and Michelozzi 2011)

crucial to assess the potential health effects of climate change. The estimation of populations adaptive capacity is an element of uncertainty in climate change scenarios and the estimation of future health impacts. Adaptation measures can be both in the short term to deal with current climate change effects and in the long term which include measures promoted and introduced to reduce future vulnerability of populations and the health impact of climate change. Figure 38.2 shows some examples of vulnerability and adaptation of biophysical systems. To date, adaptation measures related to climate change put in place are still limited. Most diseases associated with environmental exposure have many, interrelated causal factors. These multiple causal factors, need to be addressed and understood when investigating complex associations between disease and exposure in order to adopt the most cost-efficient measures to reduce the impact. Adaptation measures have to be taken not only by the health sector but also by decision-makers and social institutions.

An example of adaptation measure which has been widely adopted in Europe, especially after the 2003 heat wave is the introduction of heat warning systems and heat prevention plans (Pascal et al. 2006; Matthies et al. 2008; Michelozzi et al. 2010; Lowe et al. 2011). Heat prevention plans are both national and local, especially in terms of social and health care prevention activities put in place. Heat prevention measures include informative campaigns, dedicated helplines, such as informative leaflets for the general population, guidelines for health and social work professionals and definition of susceptible subgroups to whom active surveillance by GPs, health and social care workers should be targeted to. Recent studies have suggested that the introduction of warning systems may have contributed in reducing the impact of heat waves on mortality (Michelozzi et al. 2005; Morabito et al. 2012; Filleul et al. 2006). Other measures are aimed at reducing the effect of 'heat islands' through urban planning, shifts in work patterns.

Another example is linked to water management strategies in response to climate change and the increasing water stress mainly in southeastern Europe, and increasing risk of floods throughout most of the continent. Principal strategies to lessen the risks of flooding include warning systems, evacuation from lowlands and adequate structural measures to protect against floods (Alcamo et al. 2007). To date, limited work has been done to formally evaluate warning systems and prevention plans as a whole and specific measures. This represents an essential aspect for the better management of the available resources and an important tool for policy makers when implementing adaptation strategies to climate change.

Conclusions

Climate change is a global phenomenon which will affect areas and populations around the globe in different ways, often exacerbating socio-economic inequalities. According to the IPCC and WHO the most significant threats to human health include: extreme weather events, air pollution, infectious diseases, and disruptions of food and water supplies, along with effects of socio-political changes (Kovats et al. 2003, 2005; Alcamo et al. 2007; IPCC 2007c; Ebi and McGregor 2008). The Mediterranean region will be particular vulnerable to the consequences of climate change especially due to the expected increase in heat waves and reduction in precipitation (Alcamo et al. 2007).

Adaptation measures should be both in the short and long-term as well as at different scales, from single individuals to local communities, and from regional to national governments and institutions. Public health prevention measures are an important component of adaptation to climate change. However, to date the level of implementation and the resources are heterogeneous among countries. Research efforts should be focused on understanding and identifying which subgroups are most at risk to each extreme weather event in order to optimize the available resources and effectively reduce the impact on health. Adaptation should be paralleled with mitigation strategies aimed at reducing greenhouse gases emissions now and in the future. The role of local governments is crucial for the effective implementation of both mitigation and adaption measures. This can be achieved by the introduction of policies, financial incentives as well as education and awareness campaigns. At the global level, this requires a reduction in fossil fuel dependency and a major investment in renewable energy resources (Smith et al. 2009; Haines et al. 2009). Mitigation technologies and strategies should involve different sectors from energy supply to transport, agriculture, industry and waste management (IPCC 2007d). For example more fuel efficient and clearer vehicles, shift to rail and public transport systems and non-motorized individual transport (cycling, walking, car sharing, etc.). At the local level, individuals and communities need to adopt a more sustainable lifestyle in particular by adopting more efficient lighting and heating systems in homes and offices, promoting recycling and waste minimization and implementation of non-CO_2 emission transport (Smith et al. 2009; Haines et al. 2009).

References

Albertoni F, Arca M, Borgia P et al (1984) Heat-related mortality-latium region, Italy, summer. Morb Mortal Wkly Rep 33:518–521

Alcamo J, Moreno JM, Nováky B et al (2007) Europe. In: Parry ML, Canziani OF, Palutikof JP, van der Linden PJ, Hanson CE (eds) Climate change 2007: impacts, adaptation and vulnerability. Contribution of working group II to the fourth assessment report of the intergovernmental panel on climate change. Cambridge University Press, Cambridge

Analitis A, Katsouyanni K, Biggeri A et al (2008) Effects of cold weather on mortality: results from 15 European cities within the PHEWE project. Am J Epidemiol 168:1397–1408

APAT and WHO Europe (2007) Cambiamenti climatici ed eventi estremi: rischi per la salute in Italia. APAT – Agenzia per la Protezione dell'Ambiente e per i servizi Tecnici, Roma

Åström DO, Forsberg B, Rocklöv J (2011) Heat wave impact on morbidity and mortality in the elderly population: a review of recent studies. Maturitas 69:99–105

Baccini M, Biggeri A, Accetta G et al (2008) Effects of apparent temperature on summer mortality in 15 European cities: results of the PHEWE project. Epidemiology 19:1–9

Baccini M, Kosatsky T, Analitis A et al (2011) Impact of heat on mortality in 15 European cities: attributable deaths under different weather scenarios. J Epidemiol Community Health 65:64–70

Bargagli AM, Michelozzi P (2011) Clima e salute. Come contrastare i rischi immediati e a lungo termine delle ondate di calore. Pensiero Scientifico, Roma

Basu R (2009) High ambient temperature and mortality: a review of epidemiologic studies from 2001 to 2008. Environ Health 16:8–40

Basu R, Samet J (2002) Relation between elevated ambient temperature and mortality: a review of the epidemiologic evidence. Epidemiol Rev 24:190–202

Bouchama A, Dehbi M, Mohamed G et al (2007) Prognostic factors in heat wave related deaths: a meta-analysis. Arch Intern Med 167:2170–2176

Christensen JH, Christensen OB (2007) A summary of the PRUDENCE model projections of changes in European climate during this century. Clim Chang 81:S7–S30

Confalonieri U, Menne B, Akhtar R et al (2007) Human health. In: Parry ML, Canziani OF, Palutikof JP et al (eds) Climate change 2007: impacts, adaptation and vulnerability. Contribution of working group II to the fourth assessment report of the intergovernmental panel on climate change. Cambridge University Press, Cambridge

Curriero F, Heiner K, Samet J et al (2002) Temperature and mortality in 11 cities of the eastern United States. Am J Epidemiol 155:80–87

D'Ippoliti D, Michelozzi P, Marino C et al (2010) The impact of heat waves on mortality in 9 European cities: results from the EuroHEAT project. Environ Health 9:37

de'Donato FK, Stafoggia M, Rognoni M et al (2008) Airport and city-centre temperatures in the evaluation of the association between heat and mortality. Int J Biometeorol 52:301–310, Epub 2007 Nov 21

Della-Marta PM, Haylock MR, Luterbacher J et al (2007) Doubled length of western European summer heat waves since 1880. J Geophys Res 112, D15103. doi:10.1029/2007JD008510

Ebi KL, McGregor G (2008) Climate change, tropospheric ozone and particulate matter, and health impacts. Environ Health Perspect 116:1449–1455

Ebi KL, Kovats SR, Menne B (2006) An approach for assessing human health vulnerability and public health interventions to adapt to climate change. Environ Health Perspect 114:1930–1934

Filleul L, Cassadou S, Médina S et al (2006) The relation between temperature, ozone, and mortality in nine French cities during the heat wave of 2003. Environ Health Perspect 114:1344–1347

Forastiere F, Stafoggia M, Berti G et al (2008) Particulate matter and daily mortality: a case-crossover analysis of individual effect modifiers. Epidemiology 19:571–580

Giorgi F, Bi X, Pal J (2004) Mean, interannual variability and trends in a regional climate change experiment over Europe. II: climate change scenarios (2071–2100). Clim Dyn 23:839–858

Gosling SN, McGregor GR, Paldy A (2007) Climate change and heat-related mortality in six cities part 1: model construction and validation. Int J Biometeorol 51:525–540

Gosling S, McGregor GR, Lowe J (2009) Climate change and heat-related mortality in six cities part 2: climate model evaluation and projected impacts from changes in the mean and variability of temperature with climate change. Int J Biometeorol 53:31–51

Haines A, Kovats RS, Campbell-Lendrum D et al (2006) Climate change and human health: impacts, vulnerability, and mitigation. Lancet 367:2101–2109

Haines A, McMichael AJ, Smith KR et al (2009) Public health benefits of strategies to reduce greenhouse-gas emissions: overview and implications for policy makers. Lancet 374:2104–2114

Hajat S, Kovats R, Atkinson R et al (2002) Impact of hot temperatures on death in London: a time series approach. J Epidemiol Community Health 56:367–372

Hajat S, Armstrong BG, Gouveia N et al (2005) Mortality displacement of heat-related deaths – a comparison of Delhi, São Paulo, and London. Epidemiology 16:613–620

Huynen M, Martens P, Schram D et al (2001) The impact of heat waves and cold spells on mortality rates in the Dutch population. Environ Health Perspect 109:463–470

IPCC (2007a) Climate change 2007: the physical science basis. In: Solomon S, Qin D, Manning M et al (eds) Contribution of working group I to the fourth assessment report of the intergovernmental panel on climate change. Cambridge University Press, Cambridge/New York

IPCC (2007b) Summary for policymakers. In: Solomon S, Qin D, Manning M, Chen Z, Marquis M et al (eds) Climate change 2007: the physical science basis. Contribution of working group I to the fourth assessment report of the intergovernmental panel on climate change. Cambridge University Press, Cambridge/New York

IPCC (2007c) Summary for policymakers. In: Parry ML, Canziani OF, Palutikof JP et al (eds) Climate change 2007: impacts, adaptation and vulnerability. Contribution of working group II to the fourth assessment report of the intergovernmental panel on climate change. Cambridge University Press, Cambridge

IPCC (2007d) Summary for policymakers. In: Metz B, Davidson OR, Bosch PR et al (eds) Climate change 2007: mitigation. Contribution of working group III to the fourth assessment report of the intergovernmental panel on climate change. Cambridge University Press, Cambridge/New York

Katsouyanni K, Trichopoulos D, Zavitsanos X et al (1988) The 1987 Athens heatwave. Lancet S3:573

Kim H, Ha JS, Park J (2006) High temperature, heat index, and mortality in 6 major cities in South Korea. Arch Environ Occup Health 61:265–270

Kjellström E, Bärring L, Jacob D et al (2007) Modelling daily temperature extremes: recent climate and future changes over Europe. Clim Chang 81:S249–S265

Kovats RS, Ebi KL (2006) Heatwaves and public health in Europe. Eur J Public Health 16:592–599

Kovats S, Ebi KL, Menne B et al (2003) Methods of assessing human health vulnerability and public health adaptation to climate change. WHO/Health Canada, Copenhagen

Kovats RS, Campbell-Lendrum D, Matthies F (2005) Climate change and human health: estimating avoidable deaths and disease. Risk Anal 25:1409–1418

Lowe D, Ebi KL, Forsberg B (2011) Heatwave early warning systems and adaptation advice to reduce human health consequences of heatwaves. Int J Environ Res Public Health 8:4623–4648

Matthies F, Bickler G, Marin N et al (2008) Heat health action plans guidance. WHO, Geneva

McMichael A, Campbell-Lendrum D, Corvalán C et al (2003) Climate change and human health: risks and responses. In: WHO (ed) World Health Organization (WHO), Geneva. http://whqlibdoc.who.int/publications/2003/9241590815.pdf

McMichael AJ, Woodruff RE, Hales S (2006) Climate change and human health: present and future risks. Lancet 367:859–869

Meehl GA, Tebaldi C (2004) More intense, more frequent, and longer lasting heatwaves in the 21st century. Science 305:994–997

Menne B, Bertollini R (2005) Health and climate change: a call for action. BMJ 331:1283–1284

Menne B, Ebi KL (2006) Climate change and adaptation strategies for human health. Steinkopff Verlag, Darmstadt

Michelozzi P, de'Donato F, Bisanti L et al (2005) The impact of the summer 2003 heat waves on mortality in four Italian cities. Euro Surveill 10:161–165

Michelozzi P, Accetta G, De Sario M et al (2009) High temperature and hospitalizations for cardiovascular and respiratory causes in 12 European cities. Am J Respir Crit Care Med 179:383–389

Michelozzi P, de' Donato FK, Bargagli AM et al (2010) Surveillance of summer mortality and preparedness to reduce the health impact of heat waves in Italy. Int J Environ Res Public Health 7:2256–2273. doi:10.3390/ijerph7052256

Michelozzi P, Bargagli AM, de'Donato F et al (2011) Geographical heterogeneity of short term effects of heat in Italian cities. Epidemiol Prev 35:118–121

Morabito M, Profili F, Crisci A et al (2012) Heat-related mortality in the Florentine area (Italy) before and after the exceptional 2003 heat wave in Europe: an improved public health response? Int J Biometeorol 56:801–810. doi:10.1007/s00484-011-0481-y

Naughton MP, Henderson A, Mirabelli MC, Kaiser R, Wilhelm JL, Kieszak SM, Rubin CH, McGeehin MA (2002) Heat-related mortality during a 1999 heat wave in Chicago. Am J Prev Med 22:221–227

Pascal M, Laaidi K, Ledrans M et al (2006) France's heat health watch warning system. Int J Biometeorol 50:144–153

Räisänen J, Hansson U, Ullerstig A et al (2004) European climate in the late 21st century: regional simulations with two driving global models and two forcing scenarios. Clim Dyn 22:13–31

Ren C, Williams GM, Mengersen K et al (2009) Temperature enhanced effects of ozone on cardiovascular mortality in 95 large US communities, 1987–2000: assessment using the NMMAPS data. Arch Environ Occup Health 64:177–184

Rooney C, McMichael AJ, Kovats RS et al (1998) Excess mortality in England and Wales. And in Greater London, during the 1995 heatwave. J Epidemiol Community Health 52:482–486

Rückerl R, Schneider A, Breitner S et al (2011) Health effects of particulate air pollution: a review of epidemiological evidence. Inhal Toxicol 23:555–592

Samoli E, Analitis A, Touloumi G et al (2005) Estimating the exposure-response relationships between particulate matter and mortality within the APHEA multicity project. Environ Health Perspect 113:88–95

Schifano P, Cappai G, De Sario M et al (2009) Susceptibility to heat wave-related mortality: a follow-up study of a cohort of elderly in Rome. Environ Health 8:8–50

Semenza JC, Menne B (2009) Climate change and infectious diseases in Europe. Lancet Infect Dis 96:365–375

Semenza J, Rubin C, Falter K et al (1996) Heat-related deaths during the July 1995 heat wave in Chicago. N Engl J Med 335:84–90

Smargiassi A, Goldberg MS, Plante C et al (2009) Variation of daily warm season mortality as a function of micro-urban heat islands. J Epidemiol Community Health 63:659–664

Smith KR, Jerrett M, Anderson HR et al (2009) Public health benefits of strategies to reduce greenhouse-gas emissions: health implications of short-lived greenhouse pollutants. Lancet 374:2091–2103

Stafoggia M, Forastiere F, Agostini D et al (2006) Vulnerability to heat-related mortality: a multicity, population-based, case-crossover analysis. Epidemiology 17:315–323

Stafoggia M, Forastiere F, Agostini D et al (2008) Factors affecting in-hospital heat-related mortality: a multi-city case-crossover analysis. J Epidemiol Community Health 62:209–215

Vandentorren S, Suzan F, Medina S et al (2004) Mortality in 13 French cities during the August 2003 heat wave. Am J Public Health 94:1518–1519

Wang XY, Barnett AG, Yu W et al (2012) The impact of heatwaves on mortality and emergency hospital admissions from non-external causes in Brisbane, Australia. Occup Environ Med 69:163–169

WHO (2005) Ecosystems and human well-being: health synthesis, A report of the millennium ecosystem assessment. World Health Organization, Geneva

Ye X, Wolff R, Yu W et al (2012) Ambient temperature and morbidity: a review of epidemiological evidence. Environ Health Perspect 120:19–28

Conflicts and Security Risks of Climate Change in the Mediterranean Region

39

Jürgen Scheffran and Hans Günter Brauch

Abstract

Global warming poses a significant challenge for the Mediterranean region (Southern Europe, North Africa and Middle East). The assessment starts from a description of the physical effects of climate change, including variations in temperature, precipitation, weather extremes and sea-level, that affect soil erosion, desertification, river flows, coastal zones, rural and urban areas. The vulnerability of the region is analysed in various dimensions, including water, food, migration and energy. Climate hotspots are discussed which are particularly vulnerable, including North Africa and the Nile River basin. The potential consequences of these developments for population, economy and societal stability as well as human security and violent conflict in the region are considered. Finally, we investigate policy responses and institutional frameworks for climate adaptation, conflict resolution and cooperation, with a particular focus on renewable energy collaboration across the Mediterranean region.

Keywords

Adaptation • Climate change • Environmental conflict • Cooperation • Energy security • Food security • Human security • Migration • Mediterranean • Vulnerability • Water security

Introduction

Climate change poses a significant challenge for the Mediterranean region (Southern Europe, North Africa, and Middle East), affecting both environmental and anthropogenic systems and their interaction.[1] According to the fourth Assessment Report (AR4) of the Intergovernmental Panel on Climate Change (IPCC 2007a) the physical effects of climate change include rising temperature and declining precipitation, exacerbating existing pressures on limited water resources. Heat waves and forest fires compromise vegetation cover and add to existing environmental problems. Ecosystem change affects soil quality and moisture, the carbon cycle and local climate. These factors add to existing problems of desertification, water scarcity, and food production (Giannakopoulos et al. 2009; Stern et al. 2006; IPCC 2007b; Brauch et al. 2003), having adverse impacts on agricultural and forestry yields. Population pressure and water-intensive activities such as irrigation already impose stress on water supplies, posing dangers to human health, ecosystems and national economies of countries. Projected sea-level rise affects low lying and densely populated coastal regions. Declining reserves of oil and gas will place additional stress on the economy.

With various impacts interconnected and contributing to vulnerability, in the long run, the whole Mediterranean

[1] This analysis focuses on the EU countries in Southern Europe and on the five North African countries. Brauch (2012) offers detailed analysis also on the Balkans and the countries of the Eastern Mediterranean.

J. Scheffran (✉)
Institute of Geography, Research Group Climate Change and Security (CLISEC), KlimaCampus Excellence Initiative, University of Hamburg, Hamburg, Germany
e-mail: juergen.scheffran@zmaw.de

H.G. Brauch
Faculty of Political and Social Sciences, Free University of Berlin, Berlin, Germany

Peace Research and European Security Studies (AFES-PRESS), Mosbach, Germany
e-mail: hg.brauch@onlinehome.de

S. Goffredo and Z. Dubinsky (eds.), *The Mediterranean Sea: Its history and present challenges*,
DOI 10.1007/978-94-007-6704-1_39, © Springer Science+Business Media Dordrecht 2014

region could become a hotspot of global warming. Particular severe impacts are projected for the Middle East and North Africa (MENA), where the combination of climate change, environmental degradation, water scarcity, food insecurity, population growth, economic, and societal instability could contribute to security risks and conflicts in the region. The complex interactions between global climate change and regional impacts do not respect the natural boundaries in the Mediterranean and require an integrated and interdisciplinary perspective of natural and social phenomena.

Starting from a description of the physical effects of climate change on the Mediterranean, this article offers an overview of the vulnerability to climatic changes and looks at various dimensions relevant to climate change in the Mediterranean region, including water, food, migration and energy. This analysis allows us to identify hotspots which are particularly vulnerable to climate change and discuss security concerns of climate change. Finally, we investigate policy responses and institutional frameworks for adaptation, conflict resolution and cooperation.

Projections and Impacts of Future Climate Change in the Mediterranean

The Mediterranean climate of southern Europe and the coastal areas of North Africa are affected by hot, dry summers and cooler, wetter weather in winter. The MENA area away from the coast is characterized by a hot, dry desert climate. The AR4 (IPCC 2007a, b) offered no specific assessment on the Mediterranean region but provided results for Southern Europe, North Africa and West Asia that are relevant for the Mediterranean. The projections for temperature change in the region differ widely, depending on the assumptions of the global circulation models assessed by the IPCC.

Temperature Change

According to model calculations based on the A1B scenario, a temperature rise of 2.2–5.1 °C (mean 3.4 °C) in the Mediterranean area is likely by 2100 compared to the baseline period of 1980–1999 (IPCC 2007a). The greatest warming and temperature fluctuations are expected to occur in summer while temperature variability in winter will be smaller (Giorgi and Bi 2005). The IPCC Synthesis report (2007a) concluded that "in southern Europe, climate change is projected to worsen conditions (high temperatures and drought) in a region already vulnerable to climate variability". This report noted that Africa will be most severely affected "because of low adaptive capacity and projected climate change impacts".

More specific observations and projections of climate change in the Mediterranean are given in other studies. The UNEP-MAP report of 2009 noted that "over the twentieth century and with a clear acceleration since 1970, South-Western Europe (the Iberian Peninsula, South of France) recorded an increase in temperature of almost 2 °C. The same increase can also be noticed for the North of Africa … The rise in temperature is more marked in winter than in summer and for the minimum rather than the maximum figures." Regarding future trends, the UNEP-MAP report claimed (p.22):

"By the end of the century, the average annual temperature increase is likely to be between 2.2 and 5.1 °C for 2080–2099, compared with 1980–1999. The probability of temperatures rising by between 3 and 4 °C is estimated at 50 %. The expected rise in surface temperature varies from one region to the next – in the Sub-Saharan regions it could well be as much as 4 °C in summer. On the other hand, on the Northern shores, the rise is likely to be more marked in winter, at around 3 °C. … The greatest rises in temperature … are likely to be recorded in the Mashreq (Palestinian Territories, Jordan, Lebanon, Syria and Iraq). Higher temperatures should thus produce summers with an increasing number of very hot days."

Based on IPCC (2007a) findings, the AFED Report (Abou and Ayman 2009) noted that by the end of the twenty-first century, the Arab region will face an increase of 2–5.5 °C in the surface temperature. This increase will be coupled with a projected decrease in precipitation up to 20 %. These changes will lead to shorter winters and dryer summers, hotter summers, more frequent heat wave occurrence, and more variability and extreme weather events occurrence.[2]

For all North African countries the trend of increasing annual mean temperatures that has been observed for the second half of the twentieth century in North Africa is likely to continue and to cause warmer and drier conditions (for an overview of climatic conditions and trends in Northern Africa, see Schilling et al. 2012). Future projections point to an increasing temperature, between 2 and 3 °C by 2050 under A1B scenario conditions (Paeth et al. 2009). The temperature assessment indicates increases of mean temperature in all months of the year, with largest warming rates in summer (June/July) and autumn (Oct./Nov.) of partly more than 4 °C until the end of the twenty-first century (Hertig and Jacobeit 2008a, b, 2010). Worst case scenarios assume a very strong warming of about 6 °C over north-western Africa in the twenty-first century compared to the twentieth century (Patricola and Cook (2010). Future projections of climate change for Africa show considerable uncertainties (including

[2]This section is based on Brauch (2012) where many additional references can be found, especially on the debate in Arab countries.

the possibility of the greening of the Sahara, Claussen et al. 2003). The IPCC concludes that it is necessary to improve the assessments for the African regions, including specific feedback mechanisms such as the oceans and of land use change.

Changes in Precipitation

The IPCC Synthesis Report (2007a) observed that from 1900 to 2005 "precipitation declined in the Sahel, the Mediterranean, southern Africa and parts of southern Asia". Winter precipitation has declined in the Mediterranean region since the middle of last century (WBGU 2008), a trend that has been ascribed to the North Atlantic Oscillation (Xoplaki et al. 2004). For the future the greatest decline in precipitation is anticipated in summer. Precipitation will likely fall by 4–27 % in southern Europe and by 20 % in North Africa on average by end of the century (WBGU 2008; Brauch 2007). Changes in precipitation and temperature lead to changes in runoff and water availability that is projected to "decrease by 10–30 % over some dry regions at mid-latitudes and dry tropics, due to decreases in rainfall and higher rates of evapotranspiration. There is also high confidence that many semi-arid areas (e.g. the Mediterranean Basin, western United States, southern Africa and north-eastern Brazil) will suffer a decrease in water resources due to climate change" (IPCC 2007a).

Climate change will affect "Mediterranean-type ecosystems because of reduction in rainfall; and tropical rainforests where precipitation declines." The UNEP-MAP (2009) report noted that "rainfall … has … diminished in Southern Europe. In the Mediterranean, some regions to the south have recorded a 20 % drop in rainfall. The trend is less even as far as North Africa is concerned." According to a precipitation trend analysis by the *Potsdam Institute for Climate Impact Research* (PIK) for 1975–2004, precipitation dropped between 0–20 % in South France and in the mountainous regions of Central Italy they rose by 10 %. But in North Italy precipitation dropped by 20–40 %, along the Ligurian coast by 20–30 % and in South Spain up to 20 %.

According to the scenarios of the Acacia project the precipitation in South Europe will decline until 2020 in summer in parts of Central Spain by 13–14 % and in the Northern part of Greece by 10 %. Up to 2050 the projected precipitation in summer will drop in Greece by 14–20 %, in South France by 13–15 % and along the Spanish Mediterranean coast by 21–23 %. By the year 2080 the decline in summer precipitation in Greece will be between 18–27 %, in South France between 17–20 % and in Spain between 27–42 %. The A-2 scenario of the Acacia project projected a decline of summer precipitation for Spain amounting to 51–73 % and to Greece by 41–52 %. For parts of the Atlas mountain region in North Africa precipitation may decline in summer by 20–25 % until 2020, by 31–49 % until 2050 and by 47–75 % until 2090. The precipitation trend maps of PIK for 1975–2004 indicated an increase between 0–30 % for the Moroccan Atlantic coast and in the Northwest of Egypt by 10–20 %. However, according to these data, precipitation declined along the Moroccan and Algerian coast by 0–30 % and in Tunisia and Libya by 20–40 %.[3]

Only in two countries a reduction in precipitation was already observed while the other three experienced no change. Climate change studies indicate a decline in precipitation in all of North Africa during the twenty-first century (Schilling et al. 2012), and precipitation in North Africa is likely to decrease between 10 and 20 % until the year 2050 under SRES A1B scenario conditions (Paeth et al. 2009). A more detailed analysis shows decreased rainfall and drier conditions in spring (April/May), precipitation reductions at the beginning of the rainy season in October/November, substantial increases in high winter (December/January) and more intense precipitation events (Christoph et al. 2010).

In the north-western parts of northern Africa observed changes point towards warmer and drier conditions in the last decades, with significant reduction in precipitation. For north-eastern Morocco and north-western Algeria, several studies indicate below average annual rainfall rates since about the mid-1970s (Hertig and Jacobeit 2008a, b). Moroccan rainfall might be reduced in the period 2011–2050 between 5 % (mountainous areas) and 30 % in the southern regions for the SRES A1B scenario and by 5 and 20 % for the B1 scenario (Christoph et al. 2010). No pronounced precipitation trends have been observed for the eastern regions such as north-eastern Algeria, Mediterranean Tunisia, central Tunisia, and the Mediterranean parts of Libya and Egypt during the last decades of the twentieth century (Schilling et al. 2012).

Extreme Weather Events and Hazards

The projected increase in the number and intensity of natural hazards (IPCC 2007a, 2012) due to anthropogenic climate change often coincides with an increase in environmental stress due to the scarcity and degradation of water and soil resulting from rapid population growth and the overuse of these environmental resources. Due to the difference in social vulnerability between Southern Europe and MENA,

[3] The data by the Potsdam Institute for Climate Change Impact Research (PIK) were made available H.G. Brauch in summer 2006 as background information for the preparation of the expert study for the WBGU; see Brauch (2007, 2010).

the number of persons affected has been much higher in North Africa and in the Eastern Mediterranean (Guha-Sapir and Vos 2011).

Extreme weather events have severely affected the Mediterranean region that has been confronted with drought throughout history. When the precipitation comes in autumn, it is often connected to intensive flash floods that cause a high number of casualties and affect people and economic damages in many MENA countries due to the higher degree of social vulnerability (Brauch 2003). Both on the global and regional Mediterranean level an increase of the number and intensity of natural hazards has been projected that may affect "many cities and industrial regions in coastal zones". An indicator of potential impacts has been the heat wave of August 2003 when between 35,000 and up to more than 70,000 people died in Central and Southern Europe.

According to the IPCC (2007a) "Negative impacts will include increased risk of inland flash floods and more frequent coastal flooding and increased erosion (due to storminess and sea level rise)." By the 2080s, many millions more people than today are projected to experience floods every year due to sea level rise. The numbers affected will be largest in the densely populated and low-lying megadeltas of Asia and Africa while small islands are especially vulnerable (*very high confidence*).

In Northern Africa, the risk of extreme events, in particular droughts, is likely to increase (Schilling et al. 2012). In Europe and western North Africa average precipitation reduction is associated with a reduced number of precipitation days (Räisänen et al. 2004). For the future a significant prolongation of very long dry spells is expected by end of the century (Voss et al. 2002) and considerable drying over western Mediterranean and North Africa (Beniston et al. 2007).

Projected Sea-Level Rise

The projections on the average sea-level rise have been disputed in the preparation of the IPCC's AR4 which could only agree on an average increase during the twenty-first century of 18–59 cm. Rahmstorf (2007, 2009) referred to recent studies of an observed average increase of the sea level by 1.6 mm/year between 1961 and 2003 and 2.5 mm/year between 2003 and 2008. According to the A1B scenario in the IPCC Special Report on Emission Scenarios (SRES) by the mid-2090s, global sea level reaches 0.22 to 0.44 m above 1990 levels, and is rising at about 4 mm/yr (IPCC 2007a). Towards the end of the twenty-first century, projected sea level rise will affect low-lying coastal areas with large populations. The cost of adaptation could amount to at least 5–10 % of GDP (IPCC 2007a).

Impact of Climate Change on Water Supply and Water-Related Conflicts

Water Scarcity and Conflict in the Mediterranean

Water is the resource most directly and strongly affected by climatic change, and the MENA region is already one of the regions with high water scarcity and severe droughts. Higher soil water evaporation exacerbates summer water shortages, and river water levels are higher in winter and lower in summer when need is greatest. Higher temperatures and less (reliable) precipitations will likely decrease overall water availability. Increasing water scarcity has negative impact on agricultural and forestry yields and on hydropower. Increased risk of fires in droughts will compromise vegetation cover (from forest to bush and patchy cover). The North African countries are either termed water stressed or water scarce, with Algeria and Tunisia facing the highest level of water scarcity while in Egypt and Morocco water is less scarce (Schilling et al. 2012). The dependency on water and its importance for the economy is particularly high in Morocco.

Water has traditionally been a strategic issue in the Middle East, closely interconnected with the region's deeply rooted conflicts. The arid climate, the imbalance between water demand and supply, and the on-going confrontation between key political actors exacerbate the water crisis. Since much of the water resources are trans-boundary, water disputes often coincide with land disputes. Competition over utilizing shared resources has been observed for the rivers Nile, Euphrates, Tigris and Jordan. Most intense conflicts occurred among the co-riparians of the Jordan River basin where the available water resources are not shared in an equitable manner (Dombrowsky 2003; Selby 2009; Salem 2011).

Although water scarcity sometimes contributed to tensions, statements on 'water wars' remained exaggerated. So far there have been no interstate wars over water but rather water cooperation and numerous agreements (for a survey, see Biswas 1994; Wolf 1998; Amery and Wolf 2000; Shuval and Dweik 2007). Conflicts are largely determined by deep political differences (in particular, between Israel and the Palestinians), where hydrological matters represent an additional dimension that contributes to social instability, ethnic clashes or border disputes (Messerschmidt 2012). Hydrological issues have been treated in all major agreements, including the bilateral Peace Treaty of October 1994 which lists a series of common water projects. The conditions for cooperative solutions may decline due to persistent population growth, over-exploitation and pollution.

Global warming is projected to increase the likelihood and intensity of droughts in the region. Water shortages in summer are already widespread in many countries of southern Europe and North Africa. With higher temperatures the evaporation rate of water in the soil will increase and in rivers the seasonal differences in water flows are becoming bigger, adding to the water problems in summer, when demand for water is greatest. In Southern Europe, a temperature rise of 2 °C could decrease summer water availability by 20–30 %; a rise of 4 °C by 40–50 % (Stern et al. 2006). Decreasing water availability has multiple effects, including negative impacts on agricultural and forestry yields and on the generation of hydro electricity (IPCC 2007b).

Due to projected population growth and precipitation decline, the access to safe drinking water and to green water for agriculture will further worsen. "This dynamic", the WBGU (2008) argues "triggers distributional conflicts and poses major challenges to water management systems in the countries concerned". In the MENA region "the countries which will suffer the greatest water stress are generally those which already lack the political and institutional framework necessary for the adaptation of water and crisis management systems. This could overstretch existing conflict resolution mechanisms, ultimately leading to destabilization and violence".

The Case of the Nile River Basin

Reduced water supply over an extended period bears a conflict potential among the countries in the Nile River basin (Peichert 2003; Mason 2004; Mason et al. 2009; Piontek 2010; Link et al. 2012). With nearly 6,700 km the Nile is the world's longest river and flows through ten countries, feeding its water into the Eastern Mediterranean Sea. The drainage basin comprises approximately one tenth the size of Africa and more than one fifth of the African population lives within the river basin.

The growing population in the riparian countries increases the pressure on the Nile River water resources. Water supply into the tributaries of the Blue and While Nile shows large fluctuations which poses a problem for the production of hydroelectric power. Considerable evaporation in the swamps of southern Sudan decreases the available water further downstream. The future development of the overall water supply in the Nile River basin is highly uncertain, and its water resources are sensitive to changes in climatic conditions (Beyene et al. 2010; Conway et al. 2007). There has been only little climate modelling on the Nile Basin, especially on the precipitation in the upstream countries (Buontempo et al. 2011). Large-scale simulations for a doubling of atmospheric CO_2 show large variations in future river flows which could range from a 30 % increase to a 77 % decrease (Strzepek et al. 1996).

Large variability and reduced water supply over an extended period bear a conflict potential among the countries in the Nile basin which have great discrepancies in economic power, individual welfare and societal stability (Piontek 2010; Link et al. 2012). While the water supply is greatest in the upstream countries which mainly consist of grassland and shrubland, especially in Ethiopia, the demand for water is largest in the downstream countries Egypt and Sudan where most of the arable land is in the direct vicinity of the river. Since many of the upstream countries are among the countries most sensitive to hunger, they have growing needs for water to overcome this situation. Egypt depends on the Nile for 95 % of its drinking and industrial water and thus is particularly vulnerable to the impacts of climate change. If Egypt feels threatened by increasing water demands of upstream countries, it could be tempted to use its hegemonic power, aggravating the chances for political crisis and violent clashes (Piontek 2010; Link et al. 2012). Lack of usable land and water resources may add to impoverishment and migration from rural areas to cities. Cairo's infrastructure is already under pressure due to its rapid growth which endangers water availability, hygiene, waste disposal, and housing. Climate change is likely to worsen existing problems.

Due to temperature rise, the agriculturally productive Nile river delta is at risk from sea-level rise and salinization in the Mediterranean Sea which may affect between two and four million Egyptians (FoEME 2007) many of whom could seek refuge in Cairo's suburbs. In the Nile delta, "an increase of 50 cm would imply that the salty water would intrude about 9 km into coastal aquifers" (Sherif and Singh 1999; WBGU 2006). Without protective countermeasures a sea-level rise of 50 cm would affect in the administrative districts of Alexandria and Port Said about "1.5 million people" (El Raey 1991, 2011; Brauch 2002a).

If disputes can be resolved, joint management schemes can strengthen cooperation as countries increasingly work together to allocate resources to the greatest possible benefit of all. One way is to establish formal agreements to manage the water resources. Egypt currently uses most of the water according to prevailing agreements (Nile Waters Agreement of 1929, agreement between Egypt and Sudan of 1959) which are not recognized by the upstream riparians since they were not involved. A possible way out of the water crisis is to increase the efficiency of water use, reallocate water to other sectors than agriculture and share waste resources in a fair way (Mason 2004). The Nile Basin Initiative can provide a framework to resolve crises cooperatively.

Consequences for Food Security

Even without climate change the food situation is already critical in the Mediterranean region. The overuse and decline of water and soil resources, with the additional pressure from climate change, has detrimental effects on food security. A major problem is salinization which may result from drought and an increase in irrigation. The degradation of ecosystems further affects soil quality and moisture, carbon cycle and local climate. Self-reinforcing consequences are lower air humidity and less precipitation. If this trend continues it can lead to desertification, with land then being lost for agricultural use. These developments are a growing problem in parts of Southern Europe, for instance, soils in large areas of Spain and some regions of Italy and Greece are already salinized. Particularly vulnerable are countries with a resource-intensive agriculture, such as Greece where 7–8 % of GDP depend on agriculture.

The situation is much more critical in North Africa due to a combination of factors (see the overview in Schilling et al. 2012): The agricultural land is limited by the Sahara, in Egypt to 2.6 %, in Libya to 1.2 %, in Tunisia to 10.0 %, in Algeria to 3 % and in Morocco to 20 % of land area (Galil 2011). Morocco has about 8.4 million ha agricultural land, Algeria more than 7.5 million ha and Tunisia more than 4.6 million ha. The areas suitable for agriculture are already largely exploited. Water scarcity strongly affects the agricultural sector which in all North African countries is by far the largest consumer of water (mostly above 80 %). Except from Egypt's agricultural sector which depends on the Nile, all other countries in North Africa rely almost entirely on precipitation for agriculture (Schilling et al. 2012). It is estimated that wheat and maize production in Egypt could significantly drop by middle of the twenty-first century.

Rapid population growth (see section "Population and migration in the mediterranean") and human activities, such as deforestation, land clearance, overgrazing and non-sustainable irrigation, put the soils under serious stress. The risk of widespread soil salinization is increasing, leading to the loss of pasture and arable land. The overuse of the land as a consequence of the intensification of agricultural methods has eroded the soils (Chourou 2011; Galil 2011). The rising evapotranspiration will contribute in most cases to crop yield declines. For instance, in Morocco between 1987 and 1991, the cropland increased from 4.4 to 5.4 million ha primarily by extending to marginal rangeland which contributed to overgrazing in poorer land and soil erosion. The importance of the agricultural sector for the economy varies strongly among the considered states. In all countries the contribution of agriculture to GDP is less than 20 %, reaching the highest level in Morocco (17 %). A large population is directly employed in agriculture, reaching more than 30 % of the labour force in Egypt (in 2001) and more than 40 % in Morocco (in 2006) (CIA 2010).

Since agricultural productivity is very sensitive to changes in temperature and precipitation, for a global temperature rise of 2–4 °C a drop in agricultural productivity is expected. Growing aridity and reduced agricultural yields will likely affect food security in the Mediterranean region, particularly in North Africa where climate change adds pressure to current non-sustainable land uses. Surface run-off will fall significantly by 2050, and the risk of desertification will increase in all countries, in particular in the northern Sahara. The IPCC (2007b) estimates that in North Africa due to climate change agricultural production will decline by 0.4–1.3 % of GDP by 2100. For Algeria and Morocco a decrease in productivity of more than one third is projected without carbon fertilization (Schilling et al. 2012). Altogether climate change will diminish availability, accessibility, stability and utilization of food. This was assessed in the most recent National Communications (NCs) to the UNFCCC.[4]

Despite agricultural policies the goal for self-sufficiency in food remained unachievable due to a continued high population growth and environmental problems. Self-sufficiency has been rather declining in the past five decades and will continue to decline by mid-century, increasing the need in food imports (Alexandratos 1995, 2003). In consequence, food prices may rise further as will be financial resources needed to pay for these imports. All MENA countries must spend more for importing cereals and making potable water available to their people, to tourists and for irrigation. Due to declining oil revenues there will be less income to invest into food imports. These factors together "may trigger regional food crises and further undermine the economic performance of weak and unstable states, thereby encouraging or exacerbating destabilization, the collapse of social systems, and violent conflicts" (Bruinsma 2003).

While the North may adapt to declining precipitation more easily with a slightly growing population, for the countries on the southern and eastern shore declining precipitation will require major efforts that affect the water available for irrigation and food production (Hayek and Al Hmoud 2011; Haddad 2011; Issar 2011). Self-sufficiency in cereals will drop in MENA countries and the need for food imports or 'virtual water' (Allan 2003) will rise. In reaction to a region-wide economic crisis in the mid-1980s Algeria, Morocco and Tunisia launched agricultural reforms to reduce food imports and subsidies that contributed to foreign debt.

Population and Migration in the Mediterranean

Population and Demographic Trends in North Africa

Demographic trends aggravate the physical effects of climate change in the MENA region. The strong population growth and urbanization is increasing the demand for water and food on the national and regional scale.

- From 1950 to 2010, the population of 11 MENA countries grew from 72,512,000 to 280,411,000 (UNDP 2011)

[4]See the NCs of Non-Annex 1 Mediterranean countries at: http://unfccc.int/national_reports/non-annex_i_natcom/items/2979.php

and is projected to grow by an additional 111,169,000 people from 2010 to 2050, reaching about 370,352,000 people by 2100, based on the medium projection of the UN Population Division of May 2011. Population growth is especially significant in North African countries, where the population has increased from 43.8 million in 1950 to 165.4 million in 2010 and it is projected to reach 230.6 million by 2050, with a drop to 215.2 million by 2100 due to the process of demographic transition. The population is increasingly concentrated in large cities and along the coast.

- The urbanization trends between Southern Europe and North Africa have differed between 1950 and 2010. While in Southern Europe the relative urban population is projected to grow and the rural population will decline until 2050, in North Africa the rural population will stabilize while nearly all projected population growth will be in cities that are projected to grow from 107.3 million in 2010 to 223.3 million in 2050 (UNDP 2010; Brauch 2012). This rapid and often chaotic urbanization has resulted in mega cities (Brauch 2011) where the people are highly socially vulnerable to natural hazards. These population growth and urbanization trends pose different demands for water and food.
- Population in coastal cities of the Mediterranean region has grown between 1950 and 2000 from 25 to 70 million and it has been projected to rise up to 90 million by 2025. In southern Europe the coastal population doubled between 1950 and 2000 from 20 to 40 million, and it will probably stabilize at this level by 2030, while it is projected to rise in the MENA region from 40 to 50 million between the year 2000 and 2030. This will most seriously affect the deltas of the major rivers, most particularly the Nile Delta (Brauch 2002a). The UNEP-MAP report (2009) noted based on satellite monitoring "between January 1993 and June 2006 … an obvious east–west differentiation, with a clear trend towards a sea level rise in the Eastern Mediterranean".
- The recent Human Development Index (HDI) in 2011 was lowest for Morocco (place 130), for Egypt (113), Algeria (96), Tunisia (94) and highest for Libya (64). On the Southern shores of the Mediterranean income per capita is far less than on the Northern shores, e.g. the income between Spain and Morocco in 2009 GDP/cap was 32,150 (PPP $) for Spain and 4,494 (PPP $) for Morocco (UNDP 2011).

Impact of Climate Change on Migration

The discrepancies in income and development between Southern Europe and the MENA countries are a major driver for people to migrate across the Mediterranean Sea. With a rising warming, it may be increasingly difficult to sustain the living standards and provide development opportunities for a growing population in the Mediterranean region. Lack of usable land and water resources adds to impoverishment and forces people to move from rural areas to cities, or to go abroad. North Africa is both a destination and transit region for migrants. Population pressures are amplified by migration from the Sahel region.

Water scarcity and lower agricultural productivity may add to migration pressure from the rural areas to urban centres like Cairo and contribute to degradation of sanitary conditions and increasing social unrest as well. Changes in environmental conditions have an influence on water and land availability, which in turn affects economic production. Since human welfare and societal stability depend on wealth, a deterioration of the economy has negative implications on society as well. Together with the continuing demographic pressure, this may intensify competition over remaining arable land and increase the pressure to migrate. Most of the transboundary environmental migration prevails as south-south migration.

Since the fourth assessment report of the IPCC (2007a, b), the link between climate change, security and migration has raised increasing attention in public statements and the research literature which identifies environmental migration as a potential security problem (Gemenne et al. 2011). According to WBGU (2008), "the increase in drought, soil degradation and growing water scarcity in combination with high population growth, unstable institutions, poverty or a high level of dependency on agriculture means that there is a particularly significant risk of environmental migration occurring and increasing in scale". In this view, migration of a large number of people could potentially increase the likelihood of conflict in transit and target regions because environmental migrants compete with the resident population for scarce resources such as farmland, housing, water, employment, and basic social services; in certain cases they may upset the regional "ethnic balance" (Reuveny 2007).

Estimating the number of climate-induced migrants is a controversial subject. For some researchers, the number of "climate refugees" will dramatically rise, up to several hundred millions globally (Myers 2002). These numbers have been questioned as speculative and exaggerated, lacking justification and empirical evidence (Jakobeit and Methmann 2012; Oels 2012). Globally, regionally and in host countries there are no reliable international statistics on environmentally and climate-induced migration as environmental factors do not entitle the refugee status. A comprehensive study (Black et al. 2011) shows that migration has to be considered as a multicausal and complex process that precludes isolating environmental factors from other migration drivers. Going beyond the threat-victim dichotomy, migrants are seen as active social agents that shape and create their livelihood under changing environmental conditions and find collective responses to the climate challenge that maintain community resilience. Remittances and other resources from migrants can contribute to climate adaptation in their home countries (Scheffran et al. 2012a).

In European policy debates migration has been described as an increasing security challenge. In this view, both push and pull factors increase the migration pressure from some MENA countries to Europe, but also to the Arab Gulf, North America and even to Australia. Accordingly, migration from the MENA region to the EU has become a major issue of domestic or internal security and of the intergovernmental policy coordination on justice and home affairs. Migration has already been a major driver for European initiatives, both in the framework of the European Mediterranean Partnership (EMP) of the Barcelona Process (BP) and of the new Union for the Mediterranean (UfM). With the Frontex organization, the EU is handling migrants as an issue of border security.

Human Security, Environmental Conflict and Climate Adaptation

Environmental Conflict and Climate Security Risks

Global climate change can connect several regional challenges (pollution, desertification, water scarcity, food insecurity, population growth, unemployment, economic recession, urbanization, and political instability) and make them more severe. Whether and to which degree these developments will possibly trigger large-scale migration and violent conflict, not only depends on climate stress factors and human or socio-economic vulnerability, but also on adaptive problem-solving capabilities and policy decisions.

The linkages between natural resource scarcity, environmental degradation, social stress and violent conflicts were analysed in the 1990s by the Toronto research group (Homer-Dixon 1991, 1994) and by the ENCOP project in Zürich (Bächler and Spillmann 1996; Bächler et al. 1996). Homer-Dixon's environmental scarcity theory claims that environmental scarcity may lead to violent conflict when it is combined with population growth and unequal resource distribution. The concept of environmental stress was applied to land degradation and drought, but not systematically to climate change.

After a decade of research a consensus emerged that environmental stress is rarely considered to be the sole or dominant factor in precipitating conflict within and between nations. A major effect of several types of environmental stress is economic decline that will affect the poor more than rich groups and countries. Environmental stress and rapid population growth contribute to soil degradation, internal displacement, socially precarious slum formation and larger-scale migration in life threatening conditions. A second pathway from environmental stress to conflict is through forced migration that may be caused by floods, droughts, loss of natural soil fertility, locusts or famine linked to deteriorated land. Additional related push factors are water and air pollution and deforestation due to lacking fuel wood. It is subject to research how these phenomena contribute to urban and rural violence, e.g. between nomads and peasants in the Sudano-Sahelian region.

While the environmental conflict literature remained inconclusive, it received a new push with the discourse on the 'securitization' of climate change that emerged since the publication of the 2007 IPCC fourth assessment report. The main question in this debate is whether climate change could overwhelm the adaptive capacities of societies and contribute to their destabilization, possibly leading to security risks and violent conflicts (for an overview see Brauch 2009; Brzoska 2009; Buhaug et al. 2008; Maas and Tänzler 2009; Webersik 2010; Scheffran and Battaglini 2011; Scheffran et al. 2012b). There is an ongoing scientific debate on the climate-conflict link. Several quantitative studies and data-bases provide empirical material together with statistical analysis to test hypotheses about relations between climatic variables (temperature, precipitation) and conflict-related variables (number of armed conflicts or casualties) (Barnett and Adger 2007; Nordås and Gleditsch 2007; Raleigh and Urdal 2007; Scheffran et al. 2012c). Empirical studies do not provide a clear picture yet, which may be due to lack of data or the fact that the impact of climate has been weak in the past compared to what is expected for the future. The assessment of potential conflict implications of climate change is complicated since it is still not possible to systematically and directly measure resource distribution or small scale violent events (such as riots). Increasing attention is being paid to data bases of low-level instability events (Nardulli and Leetaru 2012).

Climate Vulnerability and Impacts in the Mediterranean

The generic results have to be differentiated for the Mediterranean region that is affected by major interstate wars and low-level instability events (Pfetsch 2003), as well as by environmental degradation, resource scarcity, population growth and low development that together could contribute to conflict. The impact of climate change on the Mediterranean region depends on its vulnerability which according to the IPCC (2007b) "is a function of the character, magnitude, and rate of climate change and variation to which a system is exposed, its sensitivity, and its adaptive capacity". Exposure, sensitivity and adaptive capacity vary significantly in the region.

There is so far no systematic and regionally integrated scientific assessment of the physical climate change impacts for the whole Mediterranean and especially for the MENA region. The report of the regional seminar of the Blue Plan on climate change in the Mediterranean (2008) addressed the impact on the environment and human health, most particularly the vulnerability of Mediterranean coasts that are to take the more drastic sea level rise into account, on the migration of ecological niches and shift of invasive species in the Mediterranean, on the risks of spreading of infectious diseases and on the health impacts of heat waves. Another focus was on the discussion of the impact on natural resources, natural risks and anthropogenic activities (fisheries, tourism, agriculture, energy).

The UNEP-MAP (2009) analysis listed among the major climate change impacts "water availability, biodiversity and the economic activities on which they depend" and the implications of these projected changes on fisheries, forest fires, tourism and on human health. But both recent reports as well as the previous extensive studies of the Blue Plan (Grenon and Batisse 1989; Benoit and Comeau 2005) did not discuss the implications of climate change on crop yields and food production as well as on the declining self-sufficiency rates in cereals for the MENA region countries (Brauch 2010).

A report on the environment and security in the southern Mediterranean region briefly discussed the "pressures on water, agriculture, desertification" but it did not meet the systematic requirements of an authoritative IPCC assessment (MEDSEC 2009). Regarding the physical impacts projected until 2020, 2050 and 2100, the IPCC Synthesis Report concluded that "in southern Europe, climate change is projected to worsen conditions (high temperatures and drought) in a region already vulnerable to climate variability, and to reduce water availability, hydropower potential, summer tourism and, in general, crop productivity. Climate change is also projected to increase the health risks due to heat waves and the frequency of wildfires" (IPCC 2007b). The same report also noted that Africa will be most severely affected "because of low adaptive capacity and projected climate change impacts". According to the IPCC assessment by 2020, "between 75 and 250 million of people are projected to be exposed to increased water stress due to climate change". In some countries, "yields from rain-fed agriculture could be reduced by up to 50 %".

The physical and societal effects of climate change on the Mediterranean region may be analysed from different security perspectives: the national security of riparian Mediterranean countries, international and regional Mediterranean security, as well as human security for the affected human beings. Last not least, the above physical effects and impacts on societies produce also various environmental security impacts for the region, the states and the peoples.

In its implementation paper of the European Security Strategy (EU 2003) the 27 countries of the European Council concluded that "natural disasters, environmental degradation and competition for resources exacerbate conflict, especially in situations of poverty and population growth, with humanitarian, health, political and security consequences, including greater migration" (EU 2008b). This decision was based on a report of the High Representative and Commission presented in March 2008 "which described climate change as a 'threat multiplier'" (EU 2008a). The European Council noted that "climate change can also lead to disputes over trade routes, maritime zones and resources previously inaccessible" (EU 2008b).

Climate Vulnerability and Conflict in North Africa

A literature synopsis on regional security implications of climate change (Adelphi 2009) reviewed reports on six target regions in the Commission Report, of which the MENA region is of special interest: "Climate change is likely to worsen divisions, particularly those running along social divisions and unequal wealth distribution and resource access. Crisis events such as sudden food shortages may trigger violent riots and – if left unaddressed – could destabilise states and increase public support for extremist groups offering viable alternatives. Given the geopolitical currents in MENA, this could have global repercussions." This synopsis argued that "the two major sources of wealth, income and employment in the MENA region, oil and agriculture, will diminish," and that "climate change will likely decrease agricultural output due to heat stress and reduced available water". The review sees a future security challenge "when the combination of converging climate and socio-economic trends will reach critical thresholds, catalysing processes that lead to state fragility". Particular conflict-prone are unresolved water disputes (Israel, Palestine; Nile river) and the high sensitivity to food price hikes which fuelled riots in MENA countries.

Due to increasing demand for food and water (driven by population growth) and the projected decline in agricultural productivity (driven by climate change), concerns have been raised about food and water security and its implications for conflict in North Africa. Regarding the IPCC indicators for vulnerability, trends suggest an increasing climate change exposure, a significant sensitivity and limited adaptation capacities which make the region particularly vulnerable to climate change (Schilling et al. 2010, 2012). Due to its high dependency on rain-fed agriculture (except for Egypt) the region is very sensitive to reduction in precipitation. Morocco is seen as the most vulnerable state in North Africa, due to the combination of high climate exposure, low generic adaptive capacity and high sensitivity (Yohe et al. 2006; Sullivan and Huntingford 2009; Iglesias et al. 2009). Social stability has already been affected by past droughts, and future droughts additionally have the potential to increase social inequality, and threaten social stability by severe shocks in food prices. Climate change is likely to cause a decrease in primary agricultural production in Morocco which already has low per capita income, an indicator for conflict sensitivity (Collier 2007).

The high vulnerability to climate change could contribute to social instability and violent conflict, in addition to factors such as ethnic diversity and political marginalization. Algeria, Egypt and Libya have experienced several conflicts in the recent two decades. Some authors see a "potential for more pronounced water conflicts of neighbouring countries" in North Africa (Iglesias et al. 2010), and a high risk of political instability or armed conflict as a knock-on consequence of climate change for Egypt, Libya, Algeria and Morocco (Smith and Vivekananda 2007).

Adaptive Capacities in the Mediterranean

In the Mediterranean region there are significant differences in adaptive capacities (WBGU 2008; Brauch 2007). Southern Europe is characterized by relatively high economic and social capabilities. Countries are consolidated democracies with largely functioning governance mechanisms, market economy structures, social security networks and supranational equalization systems (e.g. European Structural Funds). There is rather developed know-how, resources and technical capabilities to handle drought or flooding. Individual risks can be diminished through private insurance and further supported through European subsidies, e.g. the European Disaster Relief Fund. Adaptive capacities can be further backed up by support from the EU to mitigate the impact, strengthen long-term adaptation and resolve conflict. Despite expected environmental changes, outbreaks of violence and conflict are less likely in the foreseeable future. The economic crisis in some of the South European countries challenges the stability of the region as well as the management capabilities of the EU.

In contrast, the MENA countries are highly vulnerable, have low adaptive capacities and significant conflicts (Pfetsch 2003; Brauch 2007). In North Africa climate change interacts with the region's other problems that involve high population growth and unemployment, large social discrepancies, dependence on agriculture, unsustainable agricultural practices, weak governance and institutional structures (Schilling et al. 2010). The significant challenges posed by climate change increase the importance of major political reforms and socio-economic adaptation measures which have to address the specific elements of exposure and sensitivity to efficiently reduce vulnerability. The adaptive capacity of a society can be divided into generic indicators such as education, income and health, and impact specific indicators that relate to institutions, knowledge and technology (IPCC 2007b).

The adaptive capacities of the North African states are lower than in Europe but higher compared to the average of the African continent. Low per capita income and its unequal distribution (most unequal in Morocco and Tunisia) limit the generic adaptive capacity while the high level of corruption is a general weakness of the impact specific adaptive capacity (Schilling et al. 2012). The percentage of undernourished population has not been critical over the past 20 years (FAO 2010a, b), although low-income groups are sensitive to food price increases. Morocco performs poorest in economic resources, human development, health and education, while economic and development indicators are strongest in Egypt and Libya (UNDP 2011). Regarding the impact-specific adaptive capacity widespread corruption limits the efficient use of economic assets to cope with the effects of climate change (Transparency International 2011). The level of knowledge, technology and foreign investment in North Africa is significantly higher than in the rest of Africa (Schilling et al. 2012).

Adequate policies are needed to diminish the risk that climate change contributes to social inequality and instability in North Africa. A continued mismanagement of soil quality, rangeland vegetation and groundwater depletion amplifies the impacts of climate change. Examples are the cropping of marginal lands, irrigation from fossil groundwater and the growing of wheat instead of the traditionally used barley. There is a variety of options for adaptation to climate change available in agriculture that focus on conservation of productive assets instead of depleting them to maximize present day output (Schilling et al. 2012). In order to take advantage of the full spectrum of innovations for sustainable agriculture it is necessary to have strong monitoring mechanisms that include local and specific knowledge. Examples are the shifting of planting patterns and adjusting of planted crops in rain fed agriculture, a better monitoring of irrigation practices and soil conditions, area based rainfall insurances, as well as improvement of seasonal weather predictions. Further options include commercial pastoralism and replacement of firewood by solar power which creates new job opportunities and mitigates climate change (Schilling et al. 2012; Freier et al. 2011, 2012). Some of these issues are addressed in new agricultural strategies, for instance Morocco's "Plan Maroc Vert" (Maroc 2011). To strengthen adaptive capacities of poorer parts of the population, further measures are required to balance social inequalities.

Energy Security as Field of Conflict and Cooperation

Conflict Dimensions of Fossil and Nuclear Energy

Like other parts of the world, the Mediterranean region is still heavily dependent on fossil fuels, in particular oil and natural gas. MENA countries, especially Libya, Algeria, Syria, Iraq, Saudi Arabia, Iran and the Gulf Cooperation Council countries, are rich in these resources and much of its wealth is built on fossil fuel exports, at the cost of a strong dependence of the world on fossil fuels from MENA. In the past, this dependence has contributed to a number of violent conflicts that absorbed significant amounts of the region's resources for military purposes and caused environmental damage. The oil and gas revenues were partly used to ensure access to water (e.g. through desalinization plants) and to food imports ('virtual water'). By mid-century the reserves of oil and gas are projected to decline or to be exhausted in some countries (Mason and Kumetat 2011).

So far, nuclear power was limited to a few countries in Southern Europe (France, Spain, Slovenia) and discussions in the MENA region (in Algeria, Egypt, Jordan, Morocco, Turkey and in Israel). In this conflict-prone region the dual potential of nuclear energy for military purposes has increased mutual threat perceptions and has become an additional driver of conflict, as in the case of Iraq and more recently Iran (Kubbig and Fikenscher 2012). A number of countries in MENA aim for access to nuclear power as a low-carbon energy alternative to fossil fuels, which could further increase the threat of nuclear proliferation.

Renewable Energy Cooperation in the Mediterranean: A Contribution to Energy Security and Climate Protection

Due to its high solar input, the Mediterranean region has significant renewable energy resources. While in Southern Europe the full range of renewables is accessible (including hydropower, bioenergy, wind, solar and geothermal), the most promising resource in the MENA region is solar power, with some wind power potential in mountain and coastal areas (see the overview in Mason and Kumetat 2011). A strong cooperation between Europe, North Africa and the Middle East on energy security and climate security could highly benefit the entire region, increase adaptive capacity, substantially contribute to emission reduction, especially in the power sector, and improve the conditions for long term stability. An opportunity for strengthened cooperation between Europe and MENA is the vision of connecting renewable energy systems by an electric power grid for high-voltage direct-current transmission (Nitsch and Staiß 1997; Brauch 1997, 2012; Czisch and Giebel 2007; Trieb et al. 2009; IPCC 2011). The DESERTEC concept offers a policy vision for a sustainable supply of electricity for both the MENA region and Europe.

Producing a large quantity of electricity from renewable sources in the Mediterranean region offers a number of benefits: to prevent climate change and meet the emission reduction and renewable energy obligations of the whole region, decrease energy dependency from fossil fuels and engage the whole EU-MENA region in close cooperation on renewable energy. North African countries see the opportunity to meet the increasing local energy demand, to attract substantial foreign investments, generate export benefits and open the way to technology sharing, income and employment opportunities. Before such a project can be realized, a number of hurdles and criteria need to be addressed and sustainability criteria are to be developed (Klawitter 2010; Abaza et al. 2011).

Two strategies have been launched since 2008: (a) the Mediterranean Solar Plan (MSP) within the Union for the Mediterranean (UfM) that was launched in Paris in 14 July 2008, and (b) the Desertec Industrial Initiative (Dii) that was announced in Munich a year later on 14 July 2009. The cooperative renewable energy concept for the MENA has been promoted by the Club of Rome and developed further by the DESERTEC Foundation in Hamburg. The DESERTEC White Book "describes a scenario of electricity demand and supply opportunities by renewable energy in the integrated EUMENA region up to the middle of the century, and confirms the importance of international cooperation to achieve economic and environmental sustainability."[5] The DESERTEC Foundation is to contribute to the realization of this concept and works "for creating a global alliance to ensure security of energy supplies, to promote economic development, and to stabilize the world's climate".

The Dii pursues the goal "to analyse and develop the technical, economic, political, social and ecological framework for carbon-free power generation in the deserts of North Africa." The aim is to produce sufficient power to meet around 15 % of Europe's electricity requirements and a substantial portion of the power needs of the producer countries. The Dii plans to cooperate closely with the Mediterranean Solar Plan (MSP) which could serve as a policy framework for the development of DESERTEC.[6] On 3 November 2010 the DESERTEC Foundation together with

[5] See summary of the DESERTEC project; at: http://www.desertec.org/en/concept/summary/

[6] See 30 October 2009, Press release, "Joint venture DII established and ready to take up work"; at: http://www.desertec.org/en/press/press-releases/091030-01-formation-dii-gmbh/

18 universities and research facilities launched a DESERTEC university network. Both initiatives (MSP and Dii) could serve as conceptual components of a far more ambitious *EU-MENA Survival Pact* that links together two essential commodities: food or 'virtual water' and sustainable solar energy or 'virtual sun' (Brauch 2002b, 2010, 2012).

Besides business opportunities for the companies, there are other economic, ecological and social potentials: (a) greater energy security in the EU and MENA countries; (b) growth and development opportunities for the MENA region as a result of substantial private investment; (c) safeguarding the future water supply in the MENA countries by utilizing excess energy in seawater desalinization plants; and (d) reducing CO_2 emissions and thus making a significant contribution to achieving the climate change targets of the EU up to 2020 and 2050 (Energy Roadmap EU 2011).[7]

Given the policy implications of the emerging outcomes of the Arab Spring, a major initiative needs to be launched by the respective countries with support of the EU, to create in the medium term new and additional jobs and national income. This requires major advances to improve the scientific, technological and administrative knowledge base to overcome the deficits the Arab Human Development reports have stressed. Thus, the Desertec concept is no panacea to address the short-term dissatisfaction of the Arab youth, but it can become a major component of a longer-term Trans-Mediterranean sustainable co-development starting with the energy sector, and materializing the green growth potentials the Arab Forum on Environment and Development (Abaza et al. 2011) has addressed.

Political and Economic Frameworks for Cooperation in the Mediterranean

To prevent the security risks in the Mediterranean, it is necessary to address the new challenges in a multilateral and cooperative way to match the asymmetric distribution of risks and adaptive capacities. Yet the region suffers from a lack of cooperation, and attempts for a stronger integration and institutionalization remained with minor success. A number of Mediterranean dialogues coexist with little interaction, at the Euromed, NATO and OSCE level (Brauch 2010). It should therefore be a priority for the European institutions to promote deeper cooperation in assessing climate change and its impacts, as well as in the field of renewable energies. A *Mediterranean and Human Security Initiative* (MEH-SEC) would allow a balanced economic co-development across the Mediterranean, especially in the agricultural and energy sectors (Brauch 2010, 2012). To overcome these concerns, institutional frameworks for cooperation are needed to mitigate the impacts and strengthen long-term adaptation that reduce the likelihood of violent conflict.

Concerns about the security risks of climate change have not yet been mainstreamed into the eight priority projects of the Union for the Mediterranean (UfM). During the third Euro-Mediterranean ministerial conference on the environment on 20 November 2006, its Cairo Declaration only once referred to climate change welcoming an EU initiative for a "Communication establishing an environment strategy for the Mediterranean that outlines the approach of the European Commission regarding its environmental cooperation in the region for the coming years across a range of sectors such as climate change, desertification and biodiversity."

EU's environmental policy for the Mediterranean in the framework of activities of Horizon 2020 has among its four key priorities: (a) projects to reduce the most significant pollution sources; (b) capacity-building measures to help neighbouring countries create national environmental administrations; (c) using the Commission's research budget to develop and share knowledge of environmental issues; and (d) developing indicators to monitor the success of Horizon 2020. Climate change has not been specifically emphasized. The timetable of Horizon 2020 for the first phase (2007–2013) does not even once refer to climate change.

The "Communication from the Commission to the Council and the European Parliament – Establishing an Environment Strategy for the Mediterranean" (SEC 2006) referred to climate change in the framework of ongoing and completed research under the 5th and 6th Research Framework Programmes "addressing water issues under the EU Water Initiative's Mediterranean component and those covering accidental marine pollution, marine and coastal research and climate change impacts." Furthermore, "global environmental threats such as climate change and biodiversity loss" are identified in this Communication within a thematic programme for sustainable management of natural resources.

While Mediterranean climate change impacts have been so far no major concern of EU environment policy, the discussion on security implications of climate change in the Euro-Mediterranean region have increasingly become a concern for the DG on External Relations that launched the EU roadmap process on the security impacts of climate change.

In the security realm, several security–related dialogues have emerged: (a) the Euro-Arab dialogue of EC and Arab League states since 1973; (b) the CSCE/OSCE Mediterranean partners for cooperation since 1975; (c) the French initiative for a 5+5 dialogue of 1990; (d) the Egyptian proposal of 1994

[7]Press release, Munich, 13 Jul 2009: "12 companies plan establishment of a Desertec Industrial Initiative"; at http://www.desertec.org/fileadmin/downloads/press/09-07-13_PM_DII_english.pdf

for a Mediterranean Forum for Dialogue and Cooperation, (e) the WEU Mediterranean Subgroup that has ended, and (f) the NATO Mediterranean Initiative of 1994 (de Santis 2003) and since 2004 the Istanbul Cooperation Initiative (Bin 2008). In early 2012, still three of these Mediterranean security dialogues have co-existed without coordination: (a) the OSCE with its *Mediterranean partners for cooperation* (MPCs); (b) NATO's Mediterranean dialogue with the Mediterranean Cooperation Group; (c) and the *Barcelona process* aiming at a *Euro-Mediterranean partnership* (EMP) launched in November 1995 and reinvigorated as the UfM in 2008. After the transfer of the WEU's operational activities to the EU, the WEU Assembly has acted as the interim European Security and Defence Assembly[8] providing a forum for political discussion and reflection on the *European Security and Defence Policy* (ESDP).

The *Barcelona Declaration* of 1995 establishing a *Euro-Mediterranean Partnership* (EMP) and its successor the *Union for the Mediterranean* (UfM) are the only multi-issue regimes in the making covering most Mediterranean riparians and MENA countries. It is the only politically relevant forum for an increasing pan-Mediterranean functional cooperation on the three baskets aiming at (a) a security partnership, (b) an economic and financial partnership, and (c) a partnership in social, cultural and human affairs.

While the 'space' of the *Barcelona Convention* (1976) has focused on riparian nations to protect the Mediterranean Sea (*environmental space*), the *Barcelona Declaration* (1995) relies on a larger Euro-Mediterranean space as common area of peace and stability, of shared prosperity and of understanding between cultures and exchanges between civil societies (*political space*). While this environmental space remained unchanged, the political space has significantly changed due to EU enlargement and the emergence of the UfM. The 'Mediterranean space' used by international organizations, regimes and dialogues depends on their goals given the lack of commonly accepted geographic and political criteria.

Only the European Union has the resources to deal with the long-term environmental challenges the southern and eastern shores have experienced since 1950 and will be confronted with in the twenty-first century. To cope with these non-military challenges in the Mediterranean, close political and economic cooperation becomes crucial. The EMP partnership and the UfM offer a political institution where competence, legitimacy and resources are present. Modest funding for environmental projects in the Mediterranean has been granted by UN agencies and programmes (UNESCO, ECE, ECA, ESCWA, UNEP, UNDP), and by international financial institutions. Since 1995, major funding for joint environmental activities has been provided by the European Union within the EMP of the Barcelona Process and by the European Neighbourhood Policy (ENP).

Between the environmental regimes and the security dialogues there was little political cooperation. With the *Athens Declaration* a closer cooperation between SMAP and MAP and synergies with other organizations, programmes (Mediterranean Action Programme MAP) and donors (Mediterranean Technical Assistance Programme METAP) were encouraged. The Blue Plan secretariat has also analysed many environmental challenges of the twenty-first century for the Mediterranean Commission on Sustainable Development (MCSD). At the *intergovernmental level* and in the *EuroMesco* network[9] there has been no conceptual debate on environmental security issues in the Mediterranean until MEDSEC was proposed during 2009.

Due to the unresolved Middle East conflict a major breakthrough in the Barcelona process and in the UfM remained stalled. In 2010 the UfM experienced severe setbacks with the postponement of the second summit of the heads of state. But the functional cooperation continued with the establishment of a UfM secretariat in Barcelona in 2011. The UfM builds on the achievements of the Barcelona Process whose goals and cooperation areas (political dialogue, economic cooperation and free trade, and human, social and cultural dialogue) remain valid. The UfM enhances the Barcelona Process by (a) upgrading the political level of the relationship; (b) reinforcing co-ownership with a system of co-presidencies with a Secretariat and a Joint Permanent Committee; (c) making these relations more concrete and visible. At the Paris Summit among the six adopted priority projects two are of direct relevance for achieving sustainable development solutions focusing on (a) a de-pollution of the Mediterranean; and (b) on alternative energies within the MSP.

Acknowledgment Research for this study was funded in part by the German Science Foundation (DFG) through the Cluster of Excellence CliSAP (EXC177).

References

Abaza H, Saab N, Zeitoon B (2011) Sustainable transition in a changing Arab world. Arab Forum for Environment and Development, Beirut
Abou H, Ayman F (2009) Food production. In: Tolba MK, Mostafa K, Saab NW (eds) Impact of climate change on Arab countries. Arab Forum on Environment and Developments, Beirut
Adelphi (2009) Regional security implications of climate change – a synopsis. Adelphi, Berlin
Alexandratos N (1995) World agriculture: towards 2010. FAO, Paris
Alexandratos N (2003) Mediterranean countries and world markets: basic foods and Mediterranean products. In: Brauch HG et al (eds) Security and environment in the Mediterranean. Springer, Berlin/Heidelberg/New York

[8]See for details at: http://www.assembly-weu.org/en/index.php

[9]See: http://www.euromesco.net/

Allan JA (2003) Water security in the Mediterranean and the Middle East. In: Brauch HG et al (eds) Security and environment in the Mediterranean. Springer, Berlin/Heidelberg/New York

Amery HA, Wolf AT (2000) Water in the Middle East: a geography of peace. University of Texas Press, Austin

Bächler G, Spillmann KR (1996) Environmental degradation as a cause of war: regional and country studies of research fellows. Rüegger, Chur/Zürich

Bächler G, Böge V, Libiszewski S, Spillmann KR (1996) Kriegsursache Umweltzerstörung. Rüegger, Chur/Zürich

Barnett J, Adger WN (2007) Climate change, human security and violent conflict. Polit Geogr 26:639–655. doi:10.1016/j.polgeo.2007.03.003

Beniston M et al (2007) Future extreme events in European climate: an exploration of regional climate model projections. Clim Chang 81:1–95. doi:10.1007/s10584-006-9226-z

Benoit G, Comeau A (2005) A sustainable future for the Mediterranean. The blue plan's environment & development outlook. Earthscan, London

Beyene TD, Lettenmaier DP, Kabat P (2010) Hydrologic impacts of climate change on the Nile River Basin: implications of the 2007 IPCC scenarios. Clim Chang 100:433–461

Bin A (2008) NATO's role in the Mediterranean and broader middle east region. In: Brauch HG et al (eds) Globalization and environmental challenges: reconceptualizing security in the 21st century. Springer, Berlin/Heidelberg/New York

Biswas AK (1994) International waters of the Middle East: from Euphrates-Tigris to Nile. Oxford University Press, Oxford

Black R et al (2011) Foresight: migration and global environmental change. Final project report. The Government Office for Science, London

Brauch HG (1997) Energy policy in North Africa (1950–2050): from hydrocarbons to renewables. UNISCI, Madrid

Brauch HG (2002a) Climate change, environmental stress and conflict. Federal Ministry for the Environment, Nature Conservation and Nuclear Safety, Berlin. http://www.afes-press.de/pdf/Brauch_ClimateChange_BMU.pdf. Accessed 4 Feb 2012

Brauch HG (2002b) A Survival pact for the Mediterranean: linking 'virtual water' and 'virtual sun'. In: Pachauri RK, Vasudeva G (eds) Johannesburg and beyond. Towards concrete action. Proceedings of the colloquium held on 24 March 2002 in New York. Teri, New Delhi

Brauch HG (2003) Natural disasters in the Mediterranean (1900–2001): from disaster response to disaster preparedness. In: Brauch HG et al (eds) Security and environment in the Mediterranean. Springer, Berlin/Heidelberg/New York

Brauch H (2007) Environment and security in the Middle east: conceptualizing environmental, human, water, food, health and gender security. In: Lipchin C, Pallant E, Saranga D, Amster A (eds) Integrated water resources management and security in the Middle East. Springer, Berlin/Heidelberg/New York

Brauch HG (2009) Securitizing global environmental change. In: Brauch HG et al (eds) Facing global environmental change. Springer, Berlin/Heidelberg/New York

Brauch HG (2010) Climate change and Mediterranean security. European Institute of the Mediterranean, Barcelona

Brauch HG (2011) Global climate change impacts for the Mediterranean in the 21st century: challenges for human and environmental security. In: Brauch HG et al (eds) Coping with global environmental change, disasters and security. Springer, Berlin/Heidelberg/New York

Brauch HG (2012) Policy responses to climate change in the Mediterranean and MENA region during the anthropocene. In: Scheffran J et al (eds) Climate change, human security and violent conflict. Springer, Berlin/Heidelberg/New York

Brauch HG et al (2003) Security and environment in the Mediterranean: conceptualising security and environmental conflicts. Springer, Berlin/Heidelberg/New York

Bruinsma J (2003) World agriculture: towards 2015/2030. FAO perspective. Earthscan, London

Brzoska M (2009) The securitization of climate change and the power of conceptions of security. Sicherheit und Frieden 27:137–145

Buhaug H, Gleditsch NP, Theisen OM (2008) Implications of climate change for armed conflict. 'Social dimensions of climate change'. Washington, DC

Buontempo C, Lørup JK, Sanderson M et al (2011) Dealing with uncertainties in climate change impacts assessments: a case study on the Nile Basin. In: Brauch HG et al (eds) Coping with global environmental change, disasters and security. Springer, Berlin/Heidelberg/New York

Chourou B (2011) Environmental challenges and risks in North Africa. In: Brauch HG et al (eds) Coping with global environmental change, disasters and security. Springer, Berlin/Heidelberg/New York

Christoph M, Fink AH, Paeth H, Born K, Kerschgens M, Piecha K (2010) Climate scenarios. In: Speth P, Christoph M, Diekkrüger B (eds) Impacts of global change on the hydrological cycle in West and Northwest Africa. Springer, Berlin/Heidelberg/New York

CIA (2010) CIA world factbook. https://www.cia.gov/library/publications/the-world-factbook. Accessed 4 Feb 2012

Claussen M, Brovkin V, Ganopolski A, Kubatzki C, Petoukhov V (2003) Climate change in Northern Africa: the past is not the future. Clim Change 57:99–118. doi:10.1023/a:1022115604225

Collier P (2007) The bottom billion: why the poorest countries are failing and what can be done about it. Oxford University Press, New York

Conway D, Hanson CE, Doherty R, Persechino A (2007) GCM simulations of the Indian Ocean dipole influence on East African rainfall: present and future. Geophys Res Lett 34, L03705

Czisch G, Giebel G (2007) Realisable scenarios for a future electricity supply based 100 % on renewable energies. In: Proceedings of the Risø international energy conference – energy solutions for sustainable development. http://www.risoe.dk/rispubl/reports/ris-r-1608_186-195.pdf. Accessed 18 Sept 2012

de Santis N (2003) NATO's agenda and the Mediterranean dialogue. In: Brauch HG et al (eds) Security and environment in the Mediterranean. Conceptualising Security and Environmental Conflict. Springer, Berlin/Heidelberg/New York

Dombrowsky I (2003) Water accords in the Middle East peace process: moving towards cooperation. In: Brauch HG et al (eds) Security and environment in the Mediterranean. Conceptualising security and environmental conflicts. Springer, Berlin/Heidelberg/New York

El Raey M (1991) Responses to the impacts of greenhouse-induced sea-level rise on Egypt. In: Titus JG (ed) Changing climate and the coast, vol 2. UNEP/USEPA, Washington, DC

El Raey M (2011) Mapping areas affected by sea-level rise due to climate change in the Nile delta until 2100. In: Brauch HG et al (eds) Coping with global environmental change, disasters and security. Springer, Berlin/Heidelberg/New York

EU (2003) European union security strategy. European Council, Brussels. http://www.eu2003.gr/en/articles2003/6/20/31/21. Accessed 4 Feb 2012

EU (2008a) Climate change and international security. Paper from the high representative and the European Commission to the European Council. European Commission, Brussels

EU (2008b) Report on the implementation of the European security strategy – providing security in a changing world. European Commission, Brussels

EU (2011) Energy roadmap 2050. Communication from the European Commission, COM(2011)885/2 European Commission, Brussels

FAO (2010a) AQUASTAT main country database. http://www.fao.org/nr/water/aquastat/main/index.stm. Accessed 4 Feb 2012

FAO (2010b) FAOSTAT online statistical service. faostat.fao.org. Accessed 4 Feb 2012

FoEME (2007) Climate change: a new threat to Middle East security. Friends of the earth Middle East, Amman/Bethlehem/Tel Aviv.

http://www.boell.org.il/downloads/Climate_A_New_Threat_to_Middle_East_Security_December_2007_Final.pdf. Accessed 4 Feb 2012

Freier KP, Schneider U, Finckh M (2011) Dynamic interactions between vegetation and land-use in arid environments: using a Markov process for modeling Moroccan rangelands under climate change. Agric Ecosyst Environ 140:462–472

Freier KP, Brüggemann R, Scheffran J, Finckh M, Schneider U (2012) Assessing the predictability of future livelihood strategies of pastoralists in semi-arid Morocco under climate change. Technol Forecast Soc Change 79:371–382

Galil I (2011) Desertification process in Egypt. In: Brauch HG et al (eds) Coping with global environmental change, disasters and security. Springer, Berlin/Heidelberg/New York

Gemenne F, Brücker P, Glasser J (2011) The state of environmental migration 2010. Iddri/IOM. http://www.iddri.org/Publications/Collections/Analyses/The-State-of-Environmental-Migration-%28SEM%29-2010. Accessed 3 Feb 2012

Giannakopoulos C, Le Sager P, Bindi M, Moriondo M, Kostopoulou E, Goodess CM (2009) Climatic changes and associated impacts in the Mediterranean resulting from a 2 degrees C global warming. Glob Planet Change 68:209–224

Giorgi F, Bi X (2005) Regional changes in surface climate interannual variability for the 21st century from ensembles of global model simulations. Geophys Res Lett 32, L13701

Grenon M, Batisse M (1989) Futures for the Mediterranean basin. The Blue Plan. Oxford University Press, Oxford

Guha-Sapir D, Vos F (2011) Quantifying global environmental change impacts: methods, criteria and definitions for compiling data on hydro-meteorological disasters. In: Brauch HG et al (eds) Coping with global environmental change, disasters and security. Springer, Berlin/Heidelberg/New York

Haddad M (2011) Water scarcity and degradation in Palestine as challenges, vulnerabilities, and risks for environmental security. In: Brauch HG et al (eds) Coping with global environmental change, disasters and security. Springer, Berlin/Heidelberg/New York

Hayek BO, Al Hmoud ND (2011) Water degradation as a human security challenge in Jordan. In: Brauch HG et al (eds) Coping with global environmental change, disasters and security. Springer, Berlin/Heidelberg/New York

Hertig E, Jacobeit J (2008a) Assessments of Mediterranean precipitation changes for the 21st century using statistical downscaling techniques. Int J Climatol 28:1025–1045

Hertig E, Jacobeit J (2008b) Downscaling future climate change: temperature scenarios for the Mediterranean area. Glob Planet Change 63:127–131

Hertig E, Jacobeit J (2010) Predictability of Mediterranean climate variables from oceanic variability. Part II: statistical models for monthly precipitation and temperature in the Mediterranean area. Clim Dyn 36:825–843

Homer-Dixon TF (1991) On the threshold: environmental changes as causes of acute conflict. Int Secur 16:76–116

Homer-Dixon T (1994) Environmental scarcities and violent conflict – evidence from cases. Int Secur 19:5–40

Iglesias A, Moneo M, Quiroga S (2009) Methods for evaluating social vulnerability to drought. In: Iglesias A et al (eds) Coping with drought risk in agriculture and water supply systems. Springer, Berlin/Heidelberg/New York

Iglesias A, Mougou R, Moneo M, Quiroga S (2010) Towards adaptation of agriculture to climate change in the Mediterranean. Reg Environ Change 11:S159–S166. doi:10.1007/s10113-010-0187-4

IPCC (2007a) Climate change 2007. Synthesis report. IPCC http://www.ipcc.ch/pdf/assessment-report/ar4/syr/ar4_syr.pdf. Accessed 4 Feb 2012

IPCC (2007b) Climate change 2007. Climate change impacts, adaptation and vulnerability. Working group II to fourth assessment report of the intergovernmental panel on climate change. Cambridge University Press, Geneva

IPCC (2011) Renewable energy sources and climate change mitigation (SRREN). IPCC, Geneva

IPCC (2012) Managing the risks of extreme events and disasters to advance climate change adaptation (SREX). Synthesis report. IPCC, Geneva

Issar AS (2011) Progressive development of the water resources of Israel and Palestine to mitigate the negative impact of global w arming. In: Brauch HG et al (eds) Coping with global environmental change, disasters and security. Springer, Berlin/Heidelberg/New York

Jakobeit C, Methmann C (2012) 'Climate refugees' as dawning catastrophe? A critique of the dominant quest for numbers. In: Scheffran J et al (eds) Climate change, human security and violent conflict. Springer, Berlin/Heidelberg/New York

Klawitter J (2010) Towards a sustainability framework for the desertec concept. Master's Thesis University of Applied Sciences Eberswalde, Eberswalde. http://clisec.zmaw.de/fileadmin/user_upload/fks/publications/theses/Thesis_Klawitter_2010.pdf. Accessed 4 Feb 2012

Kubbig BW, Fikenscher SE (2012) Arms control and missile proliferation in the Middle East. Routledge, London/New York

Link PM, Piontek F, Scheffran J, Schilling J (2012) On foes and flows: vulnerabilities, adaptive capacities and transboundary relations in the Nile river basin in times of climate change. L'Europe en Formation 365:99–138

Maas A, Tänzler D (2009) Regional security implications of climate change. A synopsis. Adelphi Consult, Berlin

Maroc (2011) Projet d'intégration du changement climatique dans la mise en oeuvre du plan Maroc vert. Ministère de l'agriculture et de la pêche maritime and Agence pour le développement agricole. http://www.ada.gov.ma/uplds/pars/ECIES_PICCPMV.pdf. Accessed 4 Feb 2012

Mason SA (2004) From conflict to cooperation in the Nile Basin. Interaction between water availability, water management in Egypt and Sudan, and international relations in the Eastern Nile Basin. Swiss Federal Institute of Technologie, Zurich

Mason M, Kumetat D (2011) At the crossroads: energy futures for North Africa. Energy Policy 39:4407–4410

Mason SA, Hagmann T, Bichsel C, Arsano Y (2009) Linkages between sub-national and international water conflicts: the eastern Nile Basin. In: Brauch HG et al (eds) Facing global environmental change. Springer, Berlin/Heidelberg/New York

MEDSEC (2009) Environment and security issues in the Southern Mediterranean region. Zoi Environment Network, Geneva. http://www.medsecnet.org/en/publications.html. Accessed 4 Feb 2012

Messerschmidt C (2012) Nothing new in the Middle East – reality and discourses of climate change in the Israeli-Palestinian conflict. In: Scheffran J et al (eds) Climate change, human security and violent conflict. Springer, Berlin/Heidelberg/New York

Myers N (2002) Environmental refugees: a growing phenomenon of the 21st century. Philos Trans R Soc Lond Ser B 357:609–613

Nardulli P, Leetaru KH (2012) Climate change, societal stability and the SID project. In: Scheffran J et al (eds) Climate change, human security and violent conflict. Springer, Berlin/Heidelberg/New York

Nitsch J, Staiß F (1997) Perspektiven eines solaren Energieverbundes für Europa und den Mittelmeerraum. In: Brauch HG (ed) Energiepolitik. Springer, Berlin/Heidelberg/New York

Nordås R, Gleditsch NP (2007) Climate change and conflict. Polit Geogr 26:627–638

Oels A (2012) From 'securitization' of climate change to 'climatization' of the security field: comparing three theoretical perspectives. In: Scheffran J et al (eds) Climate change, human security and violent conflict. Springer, Berlin/Heidelberg/New York

Paeth H, Born K, Girmes R, Podzun R, Jacob D (2009) Regional climate change in tropical and Northern Africa due to greenhouse forcing and land use changes. J Climate 22:114–132. doi:10.1175/2008JCLI2390.1

Patricola C, Cook K (2010) Northern African climate at the end of the twenty-first century: an integrated application of regional and global climate models. Clim Dyn 35:193–212. doi:10.1007/s00382-009-0623-7

Peichert H (2003) The Nile basin initiative: a catalyst for cooperation. In: Brauch HG et al (eds) Security and environment in the Mediterranean. Conceptualising security and environmental conflicts. Springer, Berlin/Heidelberg/New York

Pfetsch FR (2003) Conflicts in and among Mediterranean countries (1945–2001). In: Brauch HG et al (eds) Security and environment in the Mediterranean. Springer, Berlin/Heidelberg/New York

Piontek F (2010) The impact of climate change on conflict and cooperation in the Nile Basin. Master's Thesis, University of Hamburg, Hamburg. http://clisec.zmaw.de/fileadmin/user_upload/fks/staff/piontek/Thesis_Piontek_2010.pdf. Accessed 4 Feb 2012

Rahmstorf S (2007) A semi-empirical approach to projecting sea-level rise. Science 315:368–370

Rahmstorf S (2009) Sea level rise. Plenary speech at Copenhagen conference on climate change. http://climatecongress.ku.dk/speakers/stefanrahmstorf-plenaryspeaker-10march2009.pdf. Accessed 4 Feb 2012

Räisänen J et al (2004) European climate in the late twenty-first century: regional simulations with two driving global models and two forcing scenarios. Clim Dyn 22:13–31. doi:10.1007/s00382-003-0365-x

Raleigh C, Urdal H (2007) Climate change, environmental degradation and armed conflict. Polit Geogr 26:674–694

Reuveny R (2007) Climate change-induced migration and violent conflict. Polit Geogr 26:656–673

Salem HS (2011) Social, environmental and security impacts of climate change on the Eastern Mediterranean. In: Brauch HG et al (eds) Coping with global environmental change, disasters and security. Springer, Berlin/Heidelberg/New York

Scheffran J, Battaglini A (2011) Climate and conflicts: the security risks of global warming. Reg Environ Change 11:S27–S39

Scheffran J, Marmer E, Sow P (2012a) Migration as a contribution to resilience and innovation in climate adaptation: social networks and co-development in Northwest Africa. Appl Geogr 33:119–127

Scheffran J, Brzoska M, Brauch HG, Link PM, Schilling J (2012b) Climate change, human security and violent conflict: challenges for societal stability. Springer, Berlin/Heidelberg/New York

Scheffran J, Brzoska M, Kominek J, Link M, Schilling J (2012c) Climate change and violent conflict. Science 336:869–871

Schilling J, Scheffran J, Link PM (2010) Climate change and land use conflicts in Northern Africa. In: Endlicher W, Gerstengarbe FW (eds) Continents under climate change. Nova Acta Leopoldina, Halle

Schilling J, Freier K, Hertig E, Scheffran J (2012) Climate change, vulnerability and adaptation in North Africa with focus on Morocco. Agric Ecosyst Environ 156:12–26

SEC (2006) Communication from the commission to the council and the European parliament – establishing an environment strategy for the Mediterranean (SEC(2006) 1082/*COM/2006/0475 final). http://eur-lex.europa.eu/LexUriServ/LexUriServ.do?uri=COM:2006:0475:FIN:EN:HTML. Accessed 4 Feb 2012

Selby J (2009) New security thinking' in Israeli-Palestinian water relations. In: Brauch HG et al (eds) Facing global environmental change: environmental, human, energy, food, health and water security concepts. Springer, Berlin/Heidelberg/New York

Sherif MM, Singh VP (1999) Effect of climate change on sea water intrusion in coastal aquifers. Hydrol Process 13:1277–1287

Shuval H, Dweik H (2007) Water resources in the Middle East: Israel-Palestinian water issues – from conflict to cooperation. Springer, Berlin/Heidelberg/New York

Smith D, Vivekananda J (2007) A climate of conflict – the links between climate change, peace and war. International Alert, London

Stern N et al (2006) The economics of climate change. Treasury, London

Strzepek KM, Yates DN, El Quosy DED (1996) Vulnerability assessment of water resources in Egypt to climatic change in the Nile basin. Clim Res 6:89–95

Sullivan CA, Huntingford C (2009) Water resources, climate change and human vulnerability. In: 18th world IMACS/MODSIM congress. http://www.mssanz.org.au/modsim09/I13/sullivan_ca.pdf. Accessed 18 Sept 2012

Transparency International (2011) Global corruption report: climate change. Earthscan, London

Trieb F, Krewitt W, May N (2009) Solar energy as a key for power and water in the Middle East and North Africa. In: Brauch HG (ed) Facing global environmental change: environmental, human, energy, food, health and water security concepts. Springer, Berlin/Heidelberg/New York

UNDP (2010) World urbanization prospects: the 1999 revision UNPD [UN Population Division]. United Nations, New York. http://esa.un.org/unpd/wup/CD-ROM_2009/WUP2009-F02-Proportion_Urban.xls

UNDP (2011) Human development report 2011 – sustainability and equity: a better future for all. Palgrave Macmillan, New York

UNEP/MAP (2009) State of the environment and development in the Mediterranean. UNEP/MAP, Athens

Voss R, May W, Roeckner E (2002) Enhanced resolution modelling study on anthropogenic climate change: changes in extremes of the hydrological cycle, vol 22. Wiley, Hoboken

WBGU (2006) Die Zukunft der Meere – zu warm, zu hoch, zu sauer. German Advisory Council Global Environmental Change, Berlin

WBGU (2008) World in transition – climate change as a security risk. Earthscan. http://www.wbgu.de/wbgu_jg2007_engl.html. Accessed 12 Dec 2010

Webersik C (2010) Climate change and security – a gathering storm of global challenges. Praeger, Denver

Wolf AT (1998) Conflict and cooperation along international waterways. Water Policy 1:251–265

Xoplaki E, Gonzàlez-Rouco JF, Luterbacher J, Wanner H (2004) Wet season Mediterranean precipitation variability: influence of large-scale dynamics and trends. Clim Dyn 23:63–78

Yohe G, Malone E, Brenkert A, Schlesinger M, Meij H, Xing X (2006) Global distributions of vulnerability to climate change. Integr Assess 6:35–44

40 Developing Resource Management and Marine Policy for the Eastern Mediterranean Sea

Michelle E. Portman

Abstract

Man's ability to exploit marine resources has improved in recent decades and as expected, efforts of exploitation have intensified. To avoid conflicts and support ecosystem-based principles, determination of pro-active marine policy based on an understanding of current and future uses of the marine environment is essential. This chapter highlights major uses of the marine and near shore environment of Israel. I relate trends in marine uses in this part of the Mediterranean Sea to corresponding past policy developments in coastal and near-shore management. Current policies fail to address new and increasing uses, especially those that are exclusively marine and/or beyond the country's territorial waters. The lack of marine policy in Israel is particularly worrisome when one considers the value of the country's offshore resources and the pressures on its terrestrial resources. Approaches such as marine ecosystem based management, integrated coastal zone management and mechanisms for the pro-active mediation of develop and conservation interests, such as marine spatial planning, are discussed in terms of their potential contribution for the management of evolving and emerging uses of the sea.

Keywords

Coastal zone management • Conflicting uses • Ecosystem-based management • Exclusive economic zone • Marine policy • Marine resources • Territorial waters

Introduction

In the past few decades, man's ability to exploit resources of the sea has improved and as expected, efforts to make use of this improved ability have intensified. Today, for example, in contrast to 50 years ago, electronic devices aid fishermen in locating and tracking schools of wild fish; as another example, newly developed synthetic materials enable the operation of offshore wind turbines at depths and at distances from shore that were impossible just a decade ago. The Mediterranean, like other seas, bears the brunt of much intense development, not only from maritime activities but also from increased development on land. Studies have indicated that most of the sea's pollutants are land-based (see Talitman et al. 2003). As an enclosed sea with a long and complex history of human exploitation, impact and intervention, the Mediterranean provides an important case study in the development of marine policy and resource management.

The marine environment engenders many valuable uses including those for energy production, food, water supply, mineral mining, transport of goods and people, the provision of raw materials for construction (i.e., sand) and substances for the pharmaceuticals industry. The value of the world seaborne trade (measured in transport tariffs[1]) was estimated to

M.E. Portman (✉)
Faculty of Architecture and Town Planning,
Technion–Israel Institute of Technology, Haifa, Israel
e-mail: michellep@cc.technion.ac.il

[1] This amount is approximately 5 % of total global trade. It is difficult to quantify the volume of marine trade in monetary terms because traditionally, maritime trade is reported in tons or ton-miles of goods transported.

S. Goffredo and Z. Dubinsky (eds.), *The Mediterranean Sea: Its history and present challenges*,
DOI 10.1007/978-94-007-6704-1_40,

be approximately 380 billion dollars in 2009. The international trade in fish products from the sea was estimated at around 90 billion dollars in 2008.[2] Despite the declining state of the region's fisheries, fish products exported from the Mediterranean Sea's riparian countries were estimated to be worth $ 2.1 billion in 2008.[3] Mediterranean coasts in particular are important sites for recreation and tourism, attracting visitors from all corners of the globe.

Beyond the value of their goods and other direct uses, the Mediterranean Sea and coasts also have great value as providers of myriad ecosystems services (Costanza et al. 1997). They provide climate regulation through the absorption of CO_2 and maintain sea grass meadows that serve as fishery breeding grounds. Some coasts contain natural barrier islands that protect developed areas from floods and hazardous storms. Man is invariably dependent on the well-being and continuous functioning of all elements of marine ecosystems in order to maintain the very resources to be exploited; for example, clean beaches, or fish and or marine water for desalination.

Not all marine uses are detrimental to ecosystem health and well-being, however, based on experience with environmental issues related to land policy, we know that it is essential to understand use trends, identify and mitigate conflicts before they become unmanageable, and work proactively to develop appropriate management regimes for marine and coastal areas. To some extent, biogeography dictates the uses of the sea and country contexts determine regulation and juridical frameworks. Therefore, to inform about resource management and marine policy we need to differentiate between sub-regions within the Mediterranean basin. We also need to be cognizant of regional and national planning and regulatory frameworks that pertain to the sea that are already in existence in these areas or that have a realistic potential of being adopted in the future.

Spalding et al. (2007) provide a prominent classification for the world's coastal and shelf areas that helps distinguish between particular biogeographic characteristics. They developed a topology, called the Marine Ecoregions of the World, based on physical conditions and representation of ongoing ecological processes. The topology uses a nested system of 12 realms, 62 provinces and 232 eco-regions worldwide (Spalding et al. 2007). Due to its relatively high spatial resolution compared to earlier global classifications and practical utility, the topology has been used in many previous studies including some involving spatial distribution of human impacts (e.g., Abdulla et al. 2009; Halpern et al. 2008). The Mediterranean Sea is part of the Temperate Northern Atlantic Realm; it is characterized as a province with seven ecoregions identified within: the Adriatic Sea, Aegean Sea, Levantine Sea, Tunisia Plateau/Gulf of Sidra, Ionian Sea, and the Western Mediterranean and Alboran Sea (Spalding et al. 2007).

The riparian countries of the Levantine Sea are Egypt, Lebanon, Israel, Syria, Turkey, and a relatively small portion of Libya (Fig. 40.1). The independent Gaza Strip also has a small part of the most eastern part of the coast. Today and historically this is an area of high geopolitical tension. In addition to the isolation of the central country, Israel, whose borders with three out of five of its neighbors have resulted from cease fire agreement rather than peace accords, internal political events in the early years of the new millennium in Lybia, Egypt, Syria and to a less extent in Lebanon, coined "the Arab Spring", will undoubtedly leave their mark on governance regimes in the region.

This chapter focuses mostly on the marine resource management and policy of Israel, at the extreme eastern side of the Levantine Sea (see Fig. 40.1). As a country with a relatively stable and growing economy, yet with significant geopolitical challenges related to the environment, use of resources and security concerns, Israel's management of the sea may epitomize the promise and perils of the region. In any case, there are signs that its relationship with the sea, both from public and policymaker's perspectives, is likely to change rapidly within the coming decades, making the country an interesting and important case study.

Marine Uses Past, Present and Future

Beyond Israel's shoreline, within the territorial sea[4] which extends to 12 nautical miles (22.2 km) seaward and also beyond it, many important natural resources can be exploited for public benefit: fish, water for desalination and cooling power plants, and sources of energy including large recently-discovered natural gas reserves. The size of Israel's Exclusive Economic Zone (EEZ),[5] although not officially declared through United Nations Convention on the Law of the Sea (UNCLOS), at 27,317 km^2 in size is much greater than its terrestrial area and over 6 times greater than its territorial sea; the territorial sea measures about 4,200 km^2. As opportunities arise, it is clear that business interests as well as the

[2] ftp://ftp.fao.org/fi/stat/summary/default.htm

[3] FAO Yearbook of Fishery and Aquaculture Statistics 2008 at ftp://ftp.fao.org/FI/CDrom/CD_yearbook_2008/navigation/index_content_commodities_e.htm

[4] Territorial waters, or territorial sea, is a boundary limitation stipulated by the United Nations Convention on the Law of the Sea (UNCLOS) indicating a belt of coastal water area regarded as sovereign territory of the adjacent coastal nation. It extends, in most cases, to 12 nm from the baseline, usually the mean low-water mark.

[5] The EEZ is a seazone which according to UNCLOS stretches from the outer territorial sea boundary to 200 nm from the base shoreline of the adjacent coastal nation. The coastal state/nation has special rights in this area over the exploration and use of marine resources, including for energy production.

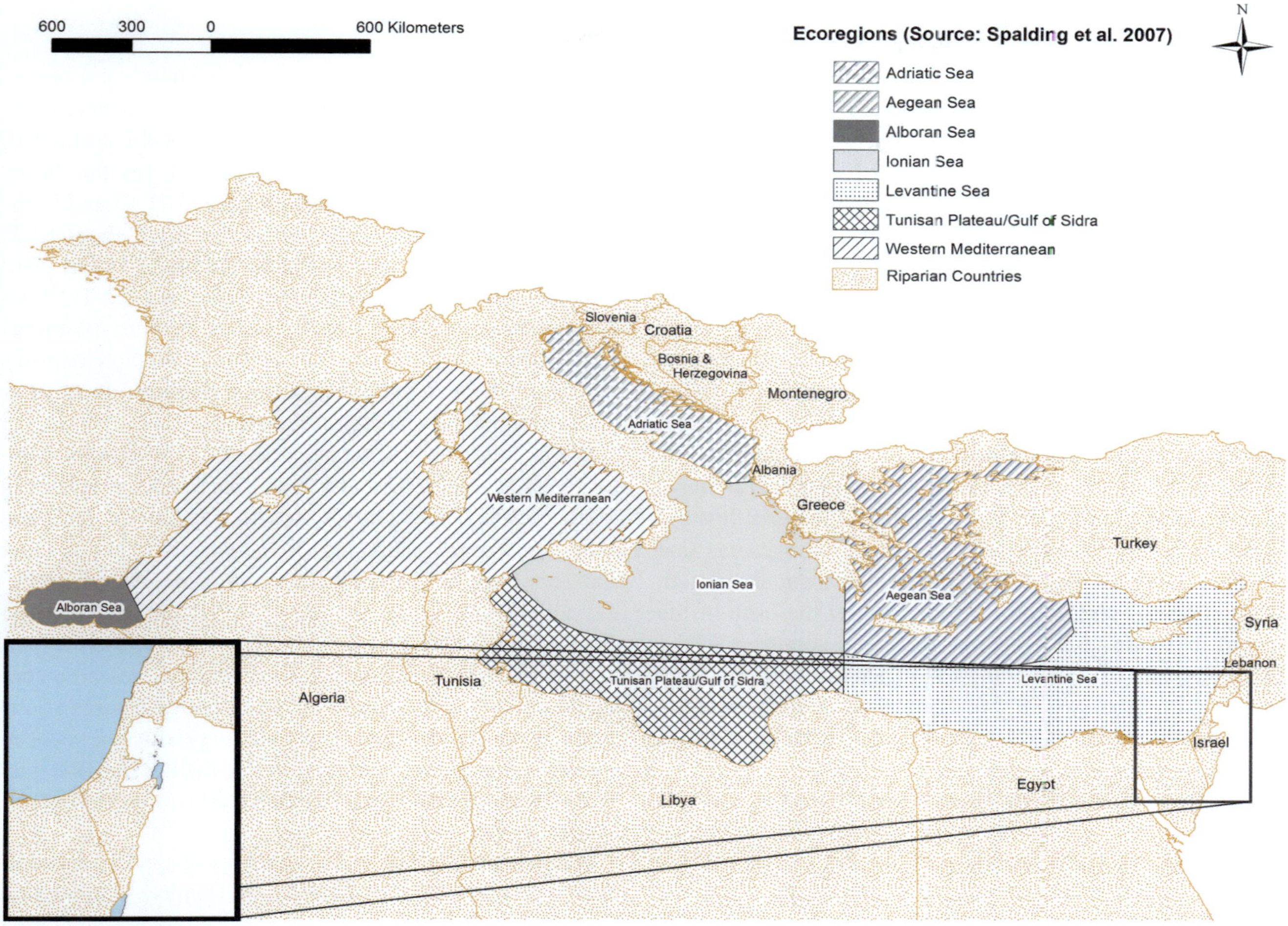

Fig. 40.1 Ecoregions of the Mediterranean Sea (According to Spalding et al. 2007). *Inset*: Israel on the eastern edge of the Levantine Sea

public-at-large will wish to further exploit adjacent marine resources, both in the territorial sea and in the EEZ.

Ocean space is likely to prove the most valuable of all marine resources for this relatively crowded country with limited and highly exploited terrestrial resources. Population density in Israel is already one of the highest in the world, at 332 inhabitants per square kilometer in 2010. Its population is growing at a rate of approximately 1.6 % annually (Israel Central Bureau of Statistics 2011). Waters off the coast of Israel, including those past the 12 nm territorial sea boundary will no doubt prompt great interest on the part of planners, policy makers and society as competition for ocean space increases and conflicts arise.

Marine space off the coast of Israel is already used for vast systems of underground communication cables, shipping lanes, military purposes (training and operational), dumping of dredge materials and extraction of materials including natural gas and sand. Marine sites are being proposed for locating relatively new uses and activities considered undesirable on land, e.g., offshore liquid natural gas (LNG) terminals and locations for the discharge of concentrated waste from desalination. Dumping of wastes from sewage water treatment out at sea has been going on for years.

Simultaneously, as knowledge of the marine environment improves, policy makers and planners realize that marine waters and the seabed are made up of fragile ecosystems full of rich life and diversity that must be preserved so that the continued existence and well-being of resources that humans currently exploit and will exploit in the future are guaranteed. Ecosystems off the shores of Israel include hundreds of fish species, invertebrates, underwater vegetation, reptiles (endangered sea turtles) and marine mammals (Yahel 2011). As for most marine ecosystems throughout the Mediterranean basin, a history of intensive use by Israel and its neighbors have put ecosystems of the Levantine Sea in jeopardy. There are numerous political tensions between neighbors throughout the Levant that perpetuate a lack of cooperation between countries and therefore the inability to approach marine planning and management on a regional basis. Another aggravation is the relatively low public awareness in these countries (and others in the Mediterranean basin) about environmental protection and sustainable development principles (Laubier 2005). Capacity to deal with environmental problems is also low. For example, due to the lack of proper sewage collections infrastructure, 60 % of the sewage from coastal cities along the coast with population over 100,000 inhabitants flows directly into the sea.

In contrast to other Mediterranean peoples such as the Greeks or Portuguese, Israelis, and the Jewish people in general, lack a maritime history. During biblical times, Jews settled and governed the mountain regions while others, such as the Philistines, controlled coastal areas (Firestone and Han 1998). Other than for limited uses, mostly local fishing operations and the military, the marine area and its resources has not been part of the national public dialogue. The sea does play a role, however, in the national historic narrative in that hundreds of thousands of immigrants came to the country by sea during the British mandate period, many illegally, and in the early years after the birth of the nation in 1948. To this day, as for many countries of the Mediterranean, Israel is highly dependent on seaborne transport of goods with approximately 99 % of its international trade passing through its seaports in Haifa, Ashdod and Eilat (on the Red Sea) (Israel Ports 2008).

Despite the marine environment's low profile, the country's approach to planning and management of marine and coastal areas has changed significantly over the past 15 years. Shifts in policy were first articulated in an important document, the Coastal Waters Policy Paper (CWPP) published in 1999 (Coastal Waters Committee 1999; Alfasi 2009). The publication of this policy was followed by the passage of groundbreaking legislation in 2004, the Law for the Protection of the Coastal Environment (LPCE). But these policy and regulatory changes transpired as a response to increasing development occurring along the terrestrial coast; they involved on-shore uses, tangentially related to the marine environment. By and large, the general public has been unconcerned and unaware of what goes on at sea. It is considered far away, unknown and irrelevant to daily life. This perspective on ocean resources appears to be quite common (e.g., Steel et al. 2005).

Although perhaps not as prominent in day-to-day political discourse as terrestrial issues, the uses of marine areas away from shore are important to Israel with all its geopolitical and environmental challenges. Israel's most important marine uses are energy production, fishing, shipping and recreational boating. Emerging uses are very place-based (stationary as opposed to transient): the potential for fill land expansion, protection from coastal hazards (e.g., the construction of break waters), the construction of offshore islands and terminals, infrastructure needs (including for desalination and outfalls), and marine protection. The following sections describe important uses in further detail with their relevance for the country.

Energy Production

A major existing and expanding use of Israel's sea area is natural gas production. Private exploration companies confirmed the existence of more than 37 trillion cubic feet (TCF) off the coast of Israel in 2010. This includes the Tamar Reserve (8.5 TCF), Leviathan Reserve (16 TCF) and in the Noah Reserve located in the Tethys Sea area (about 12.25 TCF). Seismological surveys conducted at the end of 2010 estimated a 90 % probability of an additional 6.2 TCF at a distance of about 40–70 km from shore in three additional tracks: Shimshon, Mira and Sara. All these offshore deposits are beyond Israel's territorial sea and therefore Israeli law and environmental regulations do not apply.

While these new natural gas discoveries bring opportunities for the country, they also have sparked deep debates in Israeli society. The country's archaic Gas Law of 1952 and other legislation regulating the extraction of offshore resources, supports exploration activities more than extraction activities. Fiscally, the laws favored the gas company conglomerates, leaving a very small cut of profits to the public from extraction. According to comparative research conducted in 2009 at the behest of Israeli policy makers on the public income from offshore gas and oil production between western countries, Israel ranked among the last. By law, only 24 % of revenues went to benefit the public as taxes or royalties paid to the government.[6]

Following public outcry, the government appointed a commission in early 2010, named the Shashinksy Commission after its leading member, to review fiscal policy (integrating taxes, commission and fees) related to gas production. Public advocacy groups took great interest in the commission and rallied to influence outcomes to the greatest extent possible. The Shashinsky Commission found that the existing fiscal policies were indeed relics of earlier times, reflecting outdated geopolitical conditions, some dating back as far as the time of the British Mandate when the price of oil was all of $2 a barrel. In 2011, recommendations of the Shashinsky Commission were largely adopted by the government. The outcome was a new law, the Fees on Gas Revenues Law of 2011. Even though the call of activists for a split of 20 % to the corporations and 80 % to public coffers failed, the government cut went up to 52–62 % of gas revenues (see Shashinsky Commission 2010).

Extraction of conventional fuels from new sources offshore of Israel, and other countries in the Levantine Sea (e.g., Cyprus), contrasts with what is transpiring in other western countries, mostly in Europe, with regards to development of

[6] For the sake of comparison, in 2009 gas exploration companies offshore of Australia paid between 53–56 %, those subject to Norwegian fiscal policy between 75–84 % and companies drilling in areas neighboring to Israel – offshore of Egypt – paid royalties (including taxes) of around 79–82 % of their production revenues.

marine renewable energy. Siting of facilities for the generation of renewable energy offshore is in fact driving pro-active marine planning in territorial waters. In some cases, such as in Germany and the UK, these uses are the motivations for marine planning farther offshore in the EEZ (Portman et al. 2009). A recent report found that marine spatial planning is driven in US states by entrepreneurs' interest in having policy makers proactively designate areas suitable for offshore wind farms and thus ensure greater security of investment (Eastern Research Group 2010). Although most of these energy infrastructure projects are wind farms, at least on the east coast of the US, marine renewable energy makes use of various technologies: hydrokinetic exploitation of wave, tidal and current power. Even ocean thermal energy conversion is being explored (Portman 2010).

European interest in reducing dependence on foreign, unstable sources of fossil fuels and commitments to cut emissions of green house gases has motivated new emphases on renewable energies including those offshore. European countries are committed to generating 20 % of the electricity consumed from renewable sources by 2020. In the next 10 years, 30,000 new wind turbines are expected to be constructed in the North Sea, offshore of countries such as Denmark and Germany. Offshore wind is the favored source of marine renewable production in Europe, but there are other projects as well, such as hydrokinetic energy, currently offshore of Ireland and the UK. Whether any of these types of energy production technologies are suitable for conditions in the Mediterranean Sea and particularly offshore of Israel, is as yet unknown (Laster 2011). In any case, it is clear that energy production, conventional, renewable and alternative, is one of the uses that will play a major role in the future in determining marine policy in Israel and the region.

Artificial Islands and Offshore Infrastructure

The idea of constructing artificial islands off the coast of Israel was first proposed in the 1970s. Feasibility and planning studies were conducted in the 1990s. In 2004, the National Planning and Building Commission (NPBC) ordered the preparation of a policy document on the subject (by the Interior Ministry) followed by a physical planning survey (by the Ministry of the Infrastructure). The policy document, which the NPBC published in 2007 based on surveys, determined that the islands would be used for infrastructure alone and that a number of different technologies for construction the islands would have to be tested. The first stage of development would consist of the creation of a small "pilot" island of several hundred square kilometers that would serve various infrastructure needs, particularly those unwanted on land, e.g., airports, storage for hazardous chemicals or fuels, power plants, military testing, recycling of construction and demolitions materials, etc. (Aviv Ltd. 2007).

Unfortunately, this approach consisting of siting undesirable land uses out at sea perpetuates rather dated approaches that view the oceans as the backyards of terrestrial areas. The locating of a conglomeration of 'threshold' uses that relieve the NIMBY pressures[7] on land neglect to consider widely accepted approaches to coastal and marine management that call for integration among sectors at a regional scale (see section "Ways forward for Israel" of this chapter). In some places where artificial islands have been constructed, they are seen as mistakes that severely altered the natural environment and would not be permitted today, or they are in areas where costs and environmental trade-offs are not fully understood or considered.

Aquaculture and Fishing

Compared to other marine eco-regions of the Mediterranean, the Levantine is poor in nutrients and as such, it makes for relatively unproductive fishing grounds. More commercial fishing occurs in the Western part of the Mediterranean, although landings have decreased in recent years throughout the entire sea. Fishing in international waters is the exception; small coastal fisheries are by far the majority throughout the Mediterranean (Laubier 2005). In Israel in 2009, the landings of wild fish were only around 3,000 tons. This contrasts with 19,400 tons of fish raised in mostly inland fish farms (FAO 2009). Today there are three mariculture enterprises in the Mediterranean Sea; two close to shore and one located in deep waters (Policy Research Corporation 2011).

There are about 30 trawl fishing boats operating offshore of Israel. These vessels are responsible for about 40 % of the landings from Israeli waters. Otherwise, the Israel Department of the Agriculture issues about 400 fishing permits to smaller fishing vessels of various types and there is some 'pirate' (unlicensed) fishing for commercial purposes occurring as well.

Aquaculture is rapidly developing in Israel as a solution to the rise in demand for fish products and the decreasing wild fish stock however, most of it occurs on land. Fish production is well suited to the country's arid climate especially since fish ponds don't require fresh water (Rosen 1999). However, inland fish ponds do impact the sea as the most commonly-used technology involves the disposal of waste water, some of it going into rivers and the sea. A relatively large enterprise existed in the Northern Gulf of Aqaba, in the Red Sea, until public outcry about the impact of the fish cages on the sensitive coral reefs resulted in a national government decision in 2005 to remove them within the course of 3 years (Portman 2007).

[7]NIMBY: Not In My Back Yard.

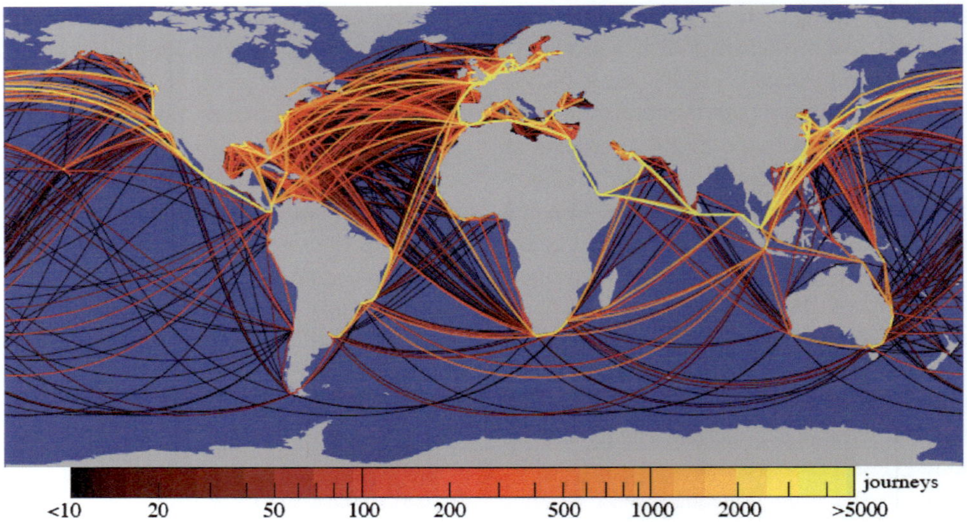

Fig. 40.2 The global marine transport network. The *bottom ruler* illustrates the number of trips by color for the year 2007 (Kaluza et al. 2010)

Despite its potential, in the near to medium time horizon of 20 years, aquaculture is unlikely to greatly influence activities in Israel's Mediterranean waters far from shore. Due to the high cost of operation, maintenance and transport of fish beyond the territorial waters and the competition for space closer in to shore, aquaculture's contribution to Israel economy will likely remain marginal (Rosen 1999).

Unless trends change, wild fishing for commercial purposes will steadily decrease in coming years. Not only are the numbers of fish caught falling from year to year, but other negative trends are observed such as smaller and smaller individuals are caught by trawlers during the summer months (Spanier and Sunin 2008). The fishing sector is regulated by anachronistic laws dating back to the time of the British Mandate. Although updated from time to time, fishermen claim that the amendments to the laws are insufficient and fail to reflect the changing realities in an accurate and timely manner.[8] Compliance and enforcement is almost non-existent and therefore, it is likely that commercial fishing off the Israeli coast will not be an important marine use in future unless these trends are reversed or halted.

Marine Transport

Transportation networks play an essential role in human travel, commerce and trade and also in the transmission of invasive species. Ninety percent of world trade is transported by sea and cargo ship routes are among the most important channels of transit the world over (Fig. 40.2). Most cargo vessels are one of three types: bulk dry carriers, container ships and oil tankers. These three categories do not only differ in the ships' physical characteristics, but also in their mobility patterns. Container ships follow regularly repeating paths whereas bulk dry carriers and oil tankers move less predictably between ports.

As mentioned ninety-nine percent of Israel's international commerce is transported by sea. Foreign trade has the potential to grow at about 5–8 % annually, an increase that will double the container traffic in Israel's ports in 10 year (Israel Ports 2008). The country has three main seaports and seven marinas. Official navigational right-of-ways are indicated on nautical maps and updated from time to time with the extent of these lanes and the traffic they handle being a function of container volume and trade. The Israel Shipping and Port Authority estimates a rise in work capacity from 900 to 3,000 TEUs (20 ft equivalent units) per meter dock space by 2055, based on the expected improved stacking efficiency (i.e., improved structural container technology) and therefore anticipated shorter storage time on wharfs and piers. This represents about 2.5 % annual increase in capacity.

Passenger travel by sea to and from Israel is limited. About 500 tourists entered the country by sea in 2010 (<0.2 % of total arrivals). However, close to 170,000 cruise ship passengers visited the country. This constitutes about 2.5 times the amount in 2009 (Israel Ministry of Tourism 2010). The Ministry of the Transportation licenses about 16,000 boats registered as small fishing vessels or sports and recreation boats and the number is growing from year to year. This increase in seafaring vessels for myriad private uses indicates a need for marina expansion. Similarly, the expected increase in trade and cruise ship travelers makes port expansion an important coastal planning issue in the

[8]See http://www.israelfishingforum.co.il/

country. However, expected impacts are mostly to developed urban and industrial areas adjacent to existing port facilities and close in to shore.

Military Uses

With Israel's security concerns, the history of conflict in the region, and the current geopolitical volatility of the region, military activities are among the most important offshore uses for the nation. At times this use is uncompromising and inflexible. The territorial sea boundary of the country is its western border and relative to other borders of Israel, there is constant traffic coming and going across it. Like on land, military activities in the sea are varied in that some occur underwater, some over water (airspace) and some on the surface. Like with other uses, as Israel continues to be isolated, and the geopolitical situation remains tense, military uses of Israel marine areas will likely increase and intensify, reinforcing the need to work out compatibilty or at least coordination between uses.

According to international law, military forces of riparian countries have mobility rights and operational rights in the adjacent sea. Only movement rights, which include all rights associated with the mobility of seaborne forces, were codified in the UNCLOS. This was largely a result of the reluctance to address operational precedent among countries.

Despite the potential conflict of uses that can arise between military and civilian uses, one advantage could be for nature protection. One submerged area along the central coast of Israel to which entrance is prohibited for military reasons (Atlit) has served as a site for comparative research on species of fish targeted for consumption. Spanier and Sunin (2008) found clear evidence of greater fish biomass (i.e., larger individuals) within the military 'protected' zone. This suggests that the nature protection authorities, marine conservationists and the military could perhaps all benefit from further partnerships.

Desalination

Although desalination plants are located on the coast and do not take up marine space, they do extract a major resource from the sea: water. Israel is making desalination the mainstay of its solution to the shortage of fresh water, a long-standing problem now further exacerbated by population growth and the effects of climate change. At the extent planned, production of approximately 720 million cubic meters of water by 2020 (Feitelson and Rosenthal 2012), desalination plants that require coastal locations will undoubtedly have numerous impacts on the marine environment. Although the scope of the likely impacts of desalination on the marine environment is unclear, main influences will be from physical construction jutting into submerged areas, the discharge of waste streams contaminated with spent cleaning solution and consisting of a high salt concentrate (Khan et al. 2009).

Today there are three central desalination plants operating on the coast (in Hadera, Palmachim and Ashkelon) and three more in advanced planning stages (Ashdod, Soreq and for the Palestinian Authority in Hadera). These plants are designed to use reverse osmosis technology. The impact of desalination effluents will be different from place to place depending on bathymetry, tidal flows, sediment flows, currents and flushing. However, in general, as an enclosed sea, the flushing and recharge rates in the Mediterranean are comparatively low (Hinrichsen 1998) and Israel is not the only Mediterranean coastal country interested in desalination. Therefore its future influence on the Levantine Sea may be great.

Nature Conservation

As for much of the global ocean, the Mediterranean marine ecosystems are severely threatened by over-exploitation of natural resources, pollution and climate change. The limited marine protected areas (MPAs) provide the proverbial 'drop in the sea' for conservation and are very likely to be insufficient and unable to reverse trends. A recent analysis of protected areas within which human activities are limited, regulated and/or prohibited indicates that only 3.8 % of the sea has some level of protection. Most of the protected areas (51 %) are in marine parks or reserves located less than 2 km from continental or large island shorelines; seventy percent of these are smaller than 100 km^2 and almost half (48 %) include terrestrial shore lands within them (Portman et al. 2012b). Some experts contend that the main reasons for such a dismal level of effective protection (which may be as low as 2 % of the Mediterranean Sea) are the fragmented nature of maritime governance, the poor enforcement of existing regulations and the difficult harmonization between the European Union (EU) and non-EU countries in the Mediterranean basin (CIESM 2001).

As directed by the Barcelona Convention, MPAs should safeguard natural ecosystems in danger of disappearance, including areas most vital to habitat and species survival, by ensuring that endangered species, endemic flora and fauna, and sites with scientific, ecological and cultural value are protected. Principles for the establishment of MPAs recommend that they be large enough to protect marine life throughout their range of existence. There should also be a significant measure of connectivity between reserves with a priority for protecting large portions of rare habitats over the protection of plentiful ones. The difficulty in implementing these ideal principles is observed throughout the Mediterranean; Israel is no exception.

There are six reserves established along the coast of Israel that include some portions of submerged lands. Their total area is less than a quarter percent of the territorial waters of the country and they extend at most only several hundred meters out to sea from the shore. To improve the situation, the Israel Nature and Parks Authority (NPA) is working to enlarge the protected area along the coast of Israel by expanding existing reserves and proposing new ones. This work is justified based on the Israel Coastal Waters Policy Paper (1999) stipulating that between 10–20 % of the territorial waters' area should be protected and as much as possible protection should implement the principles of the Barcelona Convention mentioned in the previous paragraph (Yahel 2011).

Expanded large MPAs are planned that will extend from the shoreline to the seaward limit of the territorial waters and will include representative deep habitats (from 1 km depths in the north to 150 m opposite Ashkelon beach in the south). New reserves – (from north to south) at Bustan HaGalil, the Carmel headland, the Sharon, and the Nitzanim area – will protect unique seascapes with valued benthos including those with unique geological formations and habitat for rare or endangered bottom-dweller species. Additionally, NPA plans to gain approval for marine national parks that will protect antiquities such as the submerged remains of ancient coastal cities, ports and underwater heritage sites.

Although significant resources have been spent on data collection (surveying, inventorying, evaluating, etc.) to indentify the most suitable areas for protection, it is unclear what the legal framework will be for their management. Recognizing the limited space available for all activities in the territorial sea, the NPA is committed to allowing as much human activity as possible within the MPAs (Yahel 2011). The farther away from shore the MPAs are, the more difficult it may be to ensure compliance and enforcement of restrictions on activities within them. National authorities in the country, including the NPA, have jurisdiction only to the limits of the territorial waters. Activities occurring in the adjacent EEZ, past the territorial limit beyond the oversight of the NPA and other regulators may impact efforts, leaving managers without recourse unless new types of protection policies are pursued.

Conflicts, Competition and Compatibility as Policy Drivers

There are basically four types of conflicts expected in the marine environment related to human exploitation (Miles 1991). First there are conflicts expected from competition over ocean space. The second type consists of conflicts resulting from negative externalities of one use on another, (for example, mining for materials from the seabed will impact ground fish habitat). Thirdly, there are conflicts that arise from the disruption of natural processes that support and ensure continued health and functioning of ecosystems (e.g., ecosystem services). The fourth type of conflict includes those resulting from activities conducted far from the shore that engender competition for space on shore (such as increased recreational boat traffic requiring enlarged marinas or a rise in international seaborne trade that requires port expansion in intensely developed urban areas where competition with other water-dependent industries is rife).

By and large, conflicts or the potential for conflict drives the formulation of policy. A good policy process will include assessment, evaluation, proactive planning, public participation and adaptive management such that some conflicts are mitigated or avoided altogether. Some uses will be compatible, reconcilable or only temporarily at odds. Clear, well thought-out marine policy will aid all those involved in the long term. In the short-term much time and effort needs to be invested to "get [policy] right".

Vallega (1995) was one of the first marine policy analysts to use a matrix to describe 29 different marine activities occurring in the Mediterranean. Douvere and Ehler (2009) suggest using a similar, albeit simplified, matrix for sorting out existing and planned uses as an initial step in a marine spatial planning process (see section "Ways forward for Israel"). Vallega's matrix grouped the 29 uses into 8 overarching categories: navigation and communication, mineral resources, biological resources, waste disposal and pollution, defense, research, recreation, and protection. Using the matrix, he showed that all of the 29 "use" pairs can be categorized as reciprocally beneficial, conflicting, reciprocally hazardous, or singularly beneficial or hazardous to one use or the other. This type of understanding of the interaction between resource uses in the Mediterranean basin within their economic contexts could lead to the establishment of a continental shelf-based maritime jurisdictional framework that would replace or complement the Barcelona Convention (Vallega 1995).

Such a framework has not yet become a reality. However, interestingly enough, over the last 30 years the Mediterranean Sea is considered to be the marine area whose protection has brought about extraordinary international co-operation (Talitman et al. 2003; Laubier 2005; European Commission 2010). Great progress has been achieved, but it is generally admitted that these efforts have not yet achieved their potential. Conflicts in the region, between Israel, the Palestinian Authority and its Arab neighbors, as well as between other peoples, including in the West Balkans, Turkey, Greece and Syria, has intensified in recent years. This has slowed adoption and implementation of some regionally promoted policies impacting marine use.

International Efforts

Before reviewing approaches to marine policy that should be considered by policy makers in Israel, it is worthwhile to provide a brief history of international and regional efforts that the country has been party to. These efforts have influenced Israel's activities, as it has that of other countries in the region, particularly regarding marine pollution abatement. Despite, Israel's lack of a coherent articulated national marine policy and the effects of other countries on its adjacent waters, significant strides have been made in the country to improve marine water quality. These improvements have occurred mostly through land-based infrastructure improvements and effective monitoring programs.

In June 1973, the UN Environmental Programme (UNEP) expaned the monitoring of marine pollution and its impact on marine ecosystems, paying particular attention to problematic water bodies, such as semi-enclosed seas. The coastal countries of the Mediterranean, recognizing the grave degradation of the neighboring marine environment, called for global action at both national and regional levels and initiated the Mediterranean Action Plan (MAP) within the framework of the UNEP regional seas program in 1975. The plan includes three different parts: an initiative focused on socio-economic aspects of development (known as Blue Plan), a scientific directive (known as MED POL) and a legal initiative (leading to the Barcelona Convention, see below). MAP has contributed to Integrated Coastal Zone Management through its Coastal Area Management Program (CAMP), which has been oriented towards practical coastal management projects in selected countries, including Israel. The Barcelona Convention for the Protection of the Mediterranean Sea Against Pollution was signed a year after MAP began, in 1976, by the same 16 countries that adopted the MAP.

In regards to pollution abatement, two protocols of action were initially adopted: the first deals with the prevention of pollution by dumping operations from ships and aircraft (Dumping Protocol) and the second one with pollution by hydrocarbons and other noxious chemicals (Protocol Concerning Cooperation in Combating Pollution of the Mediterranean Sea by Oil and other Harmful Substances in Cases of Emergency). Subsequently the countries adopted protocols against land-based pollution in 1980 (Protocol for the Protection of the Mediterranean Sea against Pollution from Land-Based Sources), on Specially Protected Areas of Mediterranean Importance (SPAMI) in 1982, against pollution of the sea floor and subsoil resulting from continental shelf exploration and exploitation in 1994 and on the Prevention of Pollution of the Mediterranean Sea by Transboundary Movements of Hazardous Wastes and their Disposal in 1996 (Talitman et al. 2003).

Around the same time, the UN Environment Conference in Rio de Janeiro, held in 1992, adopted several global concepts such as sustainable development and the precautionary principle. To act on these new concepts, the EU launched an important program for development of Mediterranean third-world countries, called MEDA in 1995. This was followed by the European Union member states and the European Commission joining the Barcelona Convention, which was revised accordingly and renewed. A Euro-Mediterranean partnership was established with the aim of guaranteeing peace, stability, and prosperity in the region through enhanced dialog, free trade, and cooperation. A new Mediterranean Action Plan, MAP II, was also adopted. A special advisory body to the contracting parties of the Barcelona Convention – the Mediterranean Commission for Sustainable Development (MCSD) – was established. The MCSD produced recommendations and proposals for actions on water demand management, coastal zone management, tourism, industry, urban development and more (Laubier 2005).

Since 2007 the European Neighborhood and Partnership Instrument, known as ENPI, has replaced MEDA that was supporting the Euro-Med Partnership. It is a cooperation instrument managed by DG EuropAid through which political decisions taken at all levels of government are turned into actions on the ground. One of its main programs, the Horizon2020 Initiative related to the Mediterranean, was an outcome of the 10th anniversary of the Barcelona Convention. The project was also endorsed under the Union for the Mediterannean (UfM) in 2008.

Horizon2020 aims to build on existing institutions and results, filling gaps and operating within the framework of existing and developing policy instruments. A major component is aimed at reducing pollution. It also has a capacity building project within it that is coupled with its Mediterranean Environmental Program (Horizon2020 CB/MEP) that works to develop programs and initiatives that integrate policy, science, and education in many fields. In general, the project supports the implementation of the commitments undertaken within the framework of the Barcelona Convention, particularly those launched by the MCSD and MAP's Strategic Action Programme (SAP). The project also supports other European initiatives that will impact southern Mediterranean countries, such as the EU Marine Strategy Framework Direction and the EU Communication on Integrated Coastal Zone Management described in the next section of this chapter.

Ways Forward for Israel

In 2002, during her tenure as head of Israel's Planning Authority within its national Ministry of the Interior, Architect Dina Rachevsky wrote: "Attention of the authorities is no longer turned only toward land…Planning

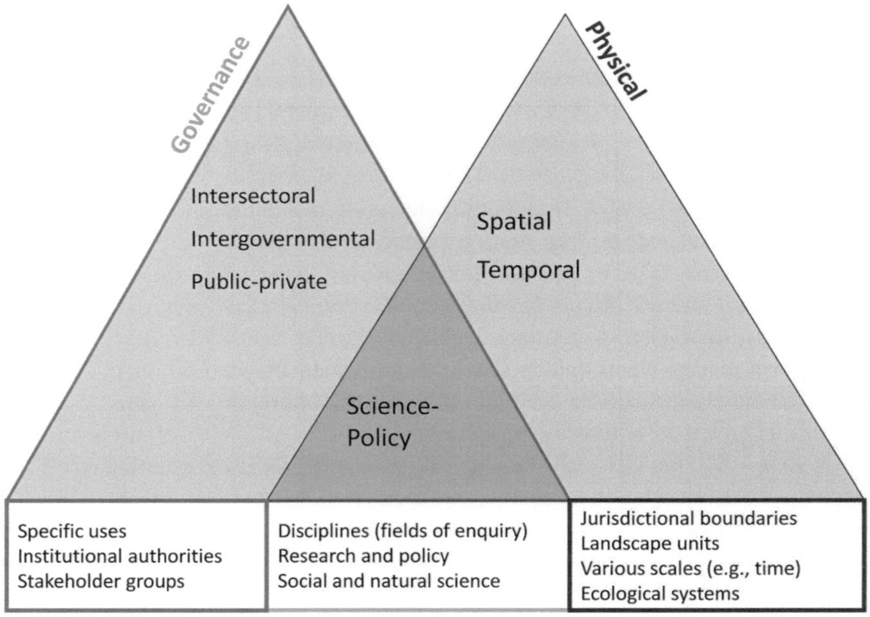

Fig. 40.3 Common dimensions of integration. The *bottom boxes* indicate what could be integrated within each dimension in more detail (Source: Portman 2011)

authorities understand that there is a need to preserve open spaces at sea, areas for which there was no fear of losing due to aggressive development in the past" (Rachevsky 2002). Rachevsky was referring to the new protection status bestowed on open (terrestrial) landscapes throughout the country by its newest National Outline Scheme (No. 35) suggesting that such consideration be given to seascapes as well. In the decade since those words were written, Israeli policymakers have done little to address either protection or development of the marine environment. Part of the problem is the lack of a clearly defined and articulated national marine policy.

Israel should begin developing a national marine policy that coincides with international frameworks, addresses uses occurring or expected to become relevant in the next few decades, and builds on past efforts within the country to promote wise use of coastal resources. Policy makers would be wise to learn from management approaches adopted and implemented in other parts of the world. Many of these incorporate principles of integration, ecosystem-based management and use new tools for marine spatial planning. Some have simultaneously developed new institutions dedicated exclusively to marine management.

Integrated Coastal Zone Management

One of the earliest articulations of the importance of integrated environmental and resource management is in the report of the Brundtland Commission, published in 1987. Not long after in 1992, the text of Agenda 21 adopted at the UN's environmental summit in Rio de Janeiro established integration as a sought-after principle of sustainable resource management. The failure of traditional approaches for management of the fragile coastal and marine environment, particularly those applied to manage commercial fisheries and for coastal conservation, led to the adoption of integrated coastal zone management (ICZM) by many countries the world over. The fundamental concept of this management approach revolves around principle of integration, a widely-practiced (or at least desired practice) for resource management fields such as energy, water and general environmental policy (Portman and Fishhendler 2011).

Integrated management is distinctly different from sectoral management that addresses one species or one use sector or one specific geographical unit or activity. Integration in the coastal zone specifically addresses management challenges related to the interface of land and sea (Cicin-Sain and Knecht 1998), among other dimensions of integration (Fig. 40.3). Successful ICZM will also involve the bringing together of myriad disciplines (science-policy integration), the consideration of future generations (temporal integration), stakeholders from many sectors of government (intergovernmental) and more.

European Union countries are obligated to adopt and implement ICZM. In 2002 the EU ratified the "Recommendations concerning the implementation of ICZM in Europe (2002/413/EC)" (European Parliament 2002). Among the most important contributions of the recommendations is the establishment of eight principles to guide member states in ICZM projects and the requirement that countries report at set intervals on the progress in ICZM. In the US, ICZM has been implemented for some time through the national Coastal Zone Management Act of 1972. Many past studies and a wealth of literature are available on the benefits and challenges of various ICZM plans and programs.

In Israel ICZM has been an accepted approach to planning and development along the coast since the 1980s. It is realized through most of the country's regulatory guidance pertaining to the coast: the National Outline Scheme 13 (NOS 13–1983), the Coastal Waters Policy Paper (1999), and the Law for the Protection of the Coastal Environment (2004). A recent study conducted as part of a large EU-funded eight-country study on coastal development, explored the different mechanisms used for ICZM including those used in Israel. The study identified and evaluated the effectiveness of the following mechanisms: statutory regulatory commissions, setback lines, environmental impact assessment and the planning hierarchy, as these are required by various laws and plans, including but not limited to those mentioned above (Portman et al. 2012a).

Despite the many achievements of these and other mechanisms for ICZM in Israel, as one moves farther out to sea the tendency is still towards sectoral planning and management. For example, the siting of LNG terminals offshore is being debated within the framework of the National Outline Scheme 37 for Natural Gas as opposed to being considered as part of a comprehensive ocean plan. The difficulty in implementing ICZM corroborates with findings of progress reports on EU member country implementation. Reviewers identified three salient impediments to implementation: difficulties addressing land and sea resources together simultaneously; translating ICZM into practical action, and integrating policy and science on the ground (Rupprecht Consult 2006).

Israel has yet to adopt the ICZM protocol of the Barcelona Convention proposed in 2009, perhaps indicating its reluctance to commit to the actions specified within. More importantly, while Israel's planners and spatial managers are focused on coastal development and near-shore submerged areas, its seascapes are largely neglected. Development in the marine environment is addressed largely on a case-by-case basis.

Ecosystem-Based Management

This approach to development has evolved over the years from the knowledge that the well-being of the marine environment is necessary to guarantee the well-being of nearby human populations and their continued use of marine resources. Central to ecosystem-based management (EBM) is the need to develop a common vision of healthy and resilient seas that considers the ecosystem, society, and the economy. To be EBM-focused, the regional vision should strive to maintain the full suite of ecosystem services upon which humans rely, both in the short and long term (Mengerink et al. 2009).

The Millennium Ecosystem Assessment conducted by a large interdisciplinary cadre of experts, defined the following four types of ecosystem services: product services, regulatory resource services, those with cultural purposes and those providing support services (Millennium Ecosystem Assessment 2005). Product services from the marine environment are those that provide consumer goods. They include food, minerals, clean water and generally all extractable goods. Regulatory resource services are those derived from natural processes that benefit human populations. These could be flood control, or even atmospheric processes that maintain the earth's environment. Cultural services are non-material resources such as aesthetic values or heritage sites that represent and preserve historic knowledge or experience. Supporting services consist of those processes that guarantee the continued functioning of all the other services. Today, it is clear that many of these services provided by ocean ecosystems are disrupted or endangered due to human intervention. It is hoped that ecosystem-based management applied to the marine environment can help remedy or improve the situation while maintaining some level of development and exploitation.

Marine policy experts are aware of the need to incorporate principles of marine EBM (MEBM) into development approaches pertaining to the Mediterranean Sea. An advantage in the Levantine Sea is that it is relatively well-studied (e.g., Abdulla et al. 2009; Coll et al. 2010) which makes it possible to identify services that must be protected. The hope is that a MEBM approach will balance conservation measures with development (Ehler and Douvere 2007). These can best be achieved through a comprehensive marine spatial planning process guided by an MEBM perspective. Marine spatial planning efforts in other countries have already implemented MEBM, such as the Massachusetts Ocean Management Plan and Netherlands' Integrated Management Plan for the North Sea 2015[9] (Douvere and Ehler 2009), thus acquiring experience that can be replicated elsewhere.

Marine Spatial Planning

The intense and growing competition among uses of the marine environment in Israel is similar to what is happening in other countries that have already embarked on planning processes. These processes, commonly referred to as marine

[9]In 2005, the EU Marine Thematic Strategy (Marine Strategy), which is the environmental pillar of the EU's Maritime Policy, introduced the principle of ecosystem-based marine spatial planning and provided a supportive framework for national initiatives toward spatial planning designed for improving the health of marine environments.

spatial planning (MSP), will determine the spatial distribution of uses and activities in the marine environment. Numerous countries, many in Europe and North America, are initiating MSP due to the lack of clear marine regulation and public sector planning, as well as problems of fragmented governance. MSP attempts to gain consensus, identify potential conflicts and possible solutions in a proactive manner. It has the potential, depending on goals and objectives, to balance both protection of the environment and development. In many respects there is a mutual, respective relationship between MSP and MEBM. Ideally, MEBM should guide management decisions made about uses of the sea within a MSP process which constitutes a tool for implementation.

Marine spatial planning is "a public process of analyzing and allocating the spatial and temporal distribution of human activities in marine areas to achieve ecological, economic, and social objectives that are usually specified through a political process" (Ehler and Douvere 2009). Some scholars and practitioners have related MSP to terrestrial planning and zoning that have been used for decades on land, calling it "analogous to land use planning" (e.g., Turnipseed et al. 2009). Although there are some similarities, most aspects of the marine environment are structurally and functionally distinct from those of the terrestrial environment. These different attributes pose both challenges and opportunities for policy makers and planners working on the marine environment and for those attempting to transfer well-established planning tools, institutional constructs and conservation approaches from terrestrial to marine applications.

By and large, the resources of the sea are public. Resources at sea are more dynamic, fluid, and transient than those on land. Agardy (2000) provides a comprehensive and useful description of the differences between marine and terrestrial systems: nebulous versus clear boundaries, large versus small spatial scales, fine versus coarse temporal scales, three-dimensional versus two-dimensional living space, unstructured versus structured food webs, and non-linear versus linear system dynamics. Additionally, marine resources are less understood and, by and large, they hold open access and/or common property goods. The ocean is three-dimensional in that the depth of ocean resources determines the characteristics and interactions among organisms and elements of the environment. Use technologies are frequently limited by depth factors. Human constructs, such as property rights, will be different when applied to the oceans. Resource use rights are well established on land, but less so in the sea.

Recognizing grave problems in the health and well-being of ocean systems and resources, such as overfishing and pollution, marine policy experts in the US have called for a comprehensive marine policy, ocean zoning and integrated marine management at the national level for decades (Underdal 1980; Stokstad 2009; Lubchenco and Sutley 2010). Following the catastrophic "Deep Horizon" oil spill in the Gulf of Mexico that occurred in April of 2010, President Obama signed the Executive Order for Stewardship of the Ocean, Coasts, and the Great Lakes. This Order, signed in July 2010, created the National Ocean Council to coordinate the work of the multiple federal agencies already involved in marine conservation and planning. It also established advisory committees for the development of regional coastal and marine spatial plans (CMS Plans). Efforts are now ongoing to prepare these plans on a regional basis.

Following a series of communications about the marine environment (such as the Thematic Strategy on the Protection and Conservation of the Marine Environment and an Impact Assessment on the proposed Marine Strategy Directive in 2005), the European Commission published its guidelines for integrated marine policy in June 2008 (European Commission 2008a). The Roadmap for Maritime Spatial Planning: Achieving Common Principles in the European Union followed in November 2008 (European Commission 2008b). EU member countries immediately began implementing the roadmap, but over varying schedules. The Directorate-General for Maritime Affairs and Fisheries has sponsored a study on the prospects for MSP in the Mediterranean. The country reports (see Policy Research Corporation 2011), covering all littoral countries of the Mediterranean Sea including Israel, indicate EU interest in capacity building for non-EU member countries to promote the use of this tool.

It is very likely that an MSP initiative will be undertaken in the near future in Israel. A first step in such a process would be the development of a clear marine policy along the lines of the CWPP, but that explicitly addresses marine uses and characteristics of the marine environment. An ocean plan could be a non-statutory master plan. As such it would have the advantage of being longer-term and more suitable to incorporate adaptive management (i.e., an iterative process that involves frequent adjustment as time goes on). In any case, the public nature of resources of the sea, anticipated effects of climate change on the marine environment and the growing national importance of maintaining ecosystems of the sea, particularly as terrestrial resources become scarce, will likely drive a MSP process in Israel. Policy makers and planners would be wise to take advantage of lessons learned by other countries.

Conclusions

While the marine environment of the Mediterranean clearly engenders increasing economic and social value, it is only recently that the public and policy makers have acknowledged that the health and well-being of marine ecosystems are essential for continued exploitation. Using this knowledge for the development of appropriate policies thus laying the foundation for significant institutional changes is the next challenge for the littoral countries of the Mediterranean Sea.

Policy makers are faced with challenges in the realm of marine policy the world over as use patterns change and intensify. However, the lack of marine policy in Israel is particularly worrisome when one considers the value of its offshore resources and the limitation of terrestrial resources in the country, including land (space). The Mediterranean Sea is Israel's western border, a fact that impacts national goals of social, environmental, economic, security and geo-political importance.

The Israeli ocean advocacy association, Zalul (in Hebrew: clear water), in their 2011 Annual Report on the Status of the Sea, call for the creation of a national ocean management agency. They based this call on an inventory of threats to the country's fragile marine environment, listing offshore drilling for gas, port expansion, land-based effluents and solid waste, coastal development and seawater desalination (Gidron et al. 2011). It is unlikely that such an agency would completely curb impacts to the marine environment resulting from development, but the fact that such an advocacy group and others, are pressing for new institutions and the extension of existing regulations to areas beyond territorial waters is a sign of public concern and interest in action destined to make their mark.

In the past, public controversy over proposed mega developments slated for construction along the coast threatening the remaining open and publically accessible areas of the shore (including at most about 13 km of bathing beaches from a total coastline of almost 200 km (Papay 2007)) proved the impetus for significant regulatory changes, such as the LPCE of 2004. Increasingly conflicting uses of the submerged areas of sea have yet to be addressed, but they will be. When they become high-profile concerns in the public's mind, likely only a matter of time, Israel would do well to be a regional player, to continue to act according to agreed-upon principles for marine and coastal development such as ICZM and MEBM, and to incorporate lessons learned from successful experiences in marine spatial planning.

References

Abdulla A, Gomei M, Hyrenbach D, Notarbartolo-di-Sciara G, Agardy T (2009) Challenges facing a network of representative marine protected areas in the Mediterranean: prioritizing the protection of underrepresented habitats. ICES J Mar Sci 66:22–28

Agardy T (2000) Information needs for marine protected areas: scientific and societal. Bull Mar Sci 66:875–888

Alfasi N (2009) Can planning protect the public interest? The challenge of coastal planning in Israel. Geogr Res Forum 29:83–102

Aviv Ltd. (2007) Policy document for artificial islands for infrastructure. Aviv Management Engineering and Information Systems Ltd. prepared for the Ministry of the Interior, Tel Aviv Israel

Cicin-Sain B, Knecht RW (1998) Integrated coastal and ocean management: concepts and practices. Island Press, Washington, DC

Coastal Waters Committee (1999) Coastal water policy paper. Government report, Jerusalem

Coll M, Piroddi CJS, Steenbeek J, Kaschner K, Ben Rais Lasram F et al (2010) The biodiversity of the Mediterranean Sea: estimates, patterns, and threats. PLoS One 5:e11842

Costanza R, d'Arge R, de Groot R, Farber S, Grasso M et al (1997) The value of the world's ecosystem services and natural capital. Nature 387:253–260

Douvere F, Ehler CN (2009) New perspectives on sea use management: initial findings from European experience with marine spatial planning. J Environ Manage 90:77–88

Eastern Research Group Inc (2010) Marine spatial planning stakeholder analysis. Consultant report, Lexington

Ehler C, Douvere F (2007) Visions for a sea change. Report of the first international workshop on marine spatial planning. UNESCO, Paris

Ehler C, Douvere F (2009) Marine spatial planning: a step-by-step approach toward ecosystem-based management. UNESCO, Paris

European Parliament (2002) Recommendation of the European Parliament and Council of 30 May 2002 concerning the implementation of ICZM in Europe (2002/413/EC). OJ L 148/24 6.6.2002. Publications Office of the European Union, Brussels

European Commission (2008a) Communication from the Commission to the Council, the European Parliament, the European Economic and Social Committee and the Committee of the Regions of 26 on "Guidelines for an Integrated Approach to Maritime Policy: Towards best practice in integrated maritime governance and stakeholder consultation" COM(2008) 395 final

European Commission (2008b) Communication from the Commission: roadmap for maritime spatial planning: achieving common principles in the EU. Brussels. 25.11.2008, COM(2008) 791 final

European Commission (2010) Towards an integrated maritime policy for better governance in the Mediterranean, Luxembourg. Publications Office of the European Union, Brussels. doi:10.2771/22665

FAO (2009) Yearbook of fishery statistics. Food and Agricultural Organization of the United Nations, Rome, Italy. ftp://ftp.fao.org/fi/stat/summary/default.htm. Accessed 24 Sept 2011

Feitelson E, Rosenthal G (2012) Desalination, space and power: the ramifications of Israel's changing water geography. Geoforum 43:272–284

Firestone S, Han I (1998) Integrated policy for planning and management of coastal areas: an international perspective. Society for the Protection of Nature in Israel, Tel Aviv

Gidron T, Barak B, Tal D (2011) State of the sea 2011: an inventory of threats to the Mediterranean Sea and a call for the establishment marine and coastal management authority. Zalul, Tel Aviv. http://www.zalul.co.il/artical756.asp. Accessed 7 Feb 2012

Halpern BS, Walbridge S, Selkoe KA, Kappel CV, Micheli F, D'Agrosa C, Bruno JF, Casey KS, Ebert C, Fox HE, Fujita R, Heinemann D, Lenihan HS, Madin EMP, Perry MT, Selig ER, Spalding M, Steneck R, Watson R (2008) A global map of human impact on marine ecosystems. Science 319:948–952

Hinrichsen D (1998) Coastal waters of the world: trends, threats, and strategies. Island Press, Washington, DC

Israel Ports (2008) Strategic master plan for the development of Mediterranean Sea port of Israel (2006–2055): from vision to reality. Company report, Haifa. http://www.israports.co.il/PortsDevelop/masterplan/Pages/StrategicMasterPlan.aspx. Accessed 15 Dec 2011

Israel Central Bureau of Statistics (2011) Statistical abstract of Israel. Jerusalem. Central Bureau of Statistics, State of Israel, Jerusalem

Israel Ministry of Tourism (2010). Annual statistical report for 2010. Government report, Jerusalem. http://www.tourism.gov.il/. Accessed 6 Dec 2011

Kaluza P, Kölzsch A, Gastner MT, Blasius B (2010) The complex network of global cargo ship movement. J R Soc Interface 7:1093–1103
Khan SJ, Murchland D, Rhodes M, Waite D (2009) Management of concentrated waste streams from high-pressure membrane water treatment systems. Crit Rev Environ Sci Technol 39:367–415
Laster Y (2011) Integrated environmental policy for the energy sector in Israel. Integrated environmental governance: opportunities and constraints – workshop organized by the Free University Berlin, Environmental Policy Research Centre and the Hebrew University Jerusalem, Department of Geography. Austrian Hospice, Jerusalem
Laubier L (2005) Mediterranean Sea and humans: improving a conflictual partnership. In: Saliot A (ed) The handbook of environmental chemistry. Springer, Berlin
Lubchenco J, Sutley N (2010) Proposed U.S. policy for ocean, coast, and great lakes stewardship. Science 328:1485–1486
Mediterranean Science Commission (CIESM) (2001) Marine Peace Parks in the Mediterranean – a CIESM proposal. CIESM workshop monographs, no 41. http://www.ciesm.org/online/monographs/Siracusa11.pdf. Accessed 24 Aug 2012
Mengerink K, Schempp A, Austin J (2009) Ocean and coastal ecosystem based management: implementation handbook. Environmental Law Institute, Washington, DC
Miles EL (1991) Future challenges in ocean management: towards integrated national ocean policy. In: Fabbri P (ed) Ocean management in global change. Elsevier Applied Science, New York
Millennium Ecosystem Assessment (2005) Ecosystems and human well-being: synthesis. Island Press, Washington, DC
Papay N (2007) Follow up report on the situation of the Israeli coast. Society for the Protection of Nature in Israel, Tel Aviv
Policy Research Corporation (2011) Exploring the potential of marine spatial planning in the Mediterranean Sea. European Commission, Brussels. http://ec.europa.eu/maritimeaffairs/documentation/studies/study_msp_med_en.htm. Accessed 17 Dec 2011
Portman ME (2007) Zoning design for cross-border marine protected areas: the Red Sea marine peace park case study. Ocean Coast Manag 50:499–522
Portman ME (2010) Marine renewable energy policy: some U.S. and international perspectives compared. Oceanography 23(2):98–105
Portman ME (2011) Marine spatial planning: achieving and evaluating integration. ICES J Mar Sci 68:2191–2200
Portman ME, Fishhendler I (2011) Towards integrated coastal zone management: a toolkit for practitioners. Hebrew University of Jerusalem. http://www.projectsecoa.eu. Accessed 2 Dec 2011
Portman ME, Duff JA, Köppel J, Reisert J, Higgins ME (2009) Offshore wind energy development in the exclusive economic zone: legal and policy supports and impediments in Germany and the U.S. Energy Policy 37:3596–3607
Portman ME, Esteves LS, Quynh LX, Mahsud AZK (2012a) Progress in ICZM: a handbook for scholars and practitioners. Hebrew University of Jerusalem. http://www.projectsecoa.eu/. Accessed 4 Feb 2012
Portman ME, Nathan D, Levin N (2012b) From the Levant to Gibraltar: a regional perspective for marine conservation in the Mediterranean Sea. Ambio– Human Environ J 41(7):670–681
Rachevsky D (2002) Institutional approaches to planning and open landscapes as a mirror of statutory planning. Eyunim Nat Resour Manag Environ 2:93–96
Rosen D (1999) Assessment of the best location for siting of offshore aquaculture in the Mediterranean coastal waters of Israel, Government report, Haifa
Rupprecht Consult (2006) Evaluation of integrated coastal zone management in Europe. Government report, Cologne
Shashinsky Commission (2010). Shashinsky report – executive summary. Finance Ministry of the State of Israel, Jerusalem. http://mof.gov.il/BudgetSite/Reform/Lists/List11/Attachments/7/managerBrief.pdf. Accessed 12 Dec 2011
Spalding MD, Fox HE, Allen GR, Davidson N, Ferdaña ZA et al (2007) Marine ecoregions of the world: a bioregionalization of coastal and shelf areas. Bioscience 57:573–583
Spanier E, Sunin O (2008) Research on marine protected areas – atlit. University of Haifa, Israel
Steel BS, Smith C, Opsommer L, Curie S, Warner-Steel R (2005) Public ocean literacy in the United States. Ocean Coast Manag 48:97–114
Stokstad E (2009) U.S. poised to adopt national ocean policy. Science 326:1618
Talitman D, Tal A, Brenner S (2003) The devil is in the details: increasing international law's influence on domestic environmental performance – the case of Israel and the Mediterranean Sea. NYU Environ Law J 11:414–478
Turnipseed M, Roady S, Sagarin R, Crowder LB (2009) The silver anniversary of the United States' exclusive economic zone: twenty-five years of ocean use and abuse, and the possibility of a blue water public trust doctrine. Ecol Law Q 36:1–70
Underdal A (1980) Integrated marine policy: what? why? how? Mar Policy 4:159–169
Vallega A (1995) Towards the sustainable management of the Mediterranean Sea. Mar Policy 19:47–64
Yahel R (2011) Marine conservation policy of the Mediterranean Sea: marine nature reserves as a tool for protection of the environment and diversity. Government report, Jerusalem

Index

S. Goffredo and Z. Dubinsky (eds.), *The Mediterranean Sea: Its history and present challenges*,
DOI 10.1007/978-94-007-6704-1, © Springer Science+Business Media Dordrecht 2014

D

E

F

H

I

O

MIX
Papier aus verantwortungsvollen Quellen
Paper from responsible sources
FSC® C105338

If you have any concerns about our products,
you can contact us on
ProductSafety@springernature.com

In case Publisher is established outside the EU,
the EU authorized representative is:
Springer Nature Customer Service Center GmbH
Europaplatz 3, 69115 Heidelberg, Germany

Printed by Libri Plureos GmbH
in Hamburg, Germany